Fundamental Constants

Quantity	Symbol	Approximate Value	Current Best Value[†]
Speed of light in vacuum	c	3.00×10^8 m/s	2.99792458×10^8 m/s
Gravitational constant	G	6.67×10^{-11} N·m²/kg²	$6.67384(80) \times 10^{-11}$ N·m²/kg²
Avogadro's number	N_A	6.02×10^{23} mol⁻¹	$6.02214129(27) \times 10^{23}$ mol⁻¹
Gas constant	R	8.314 J/mol·K = 1.99 cal/mol·K = 0.0821 L·atm/mol·K	8.3144621(75) J/mol·K
Boltzmann's constant	k	1.38×10^{-23} J/K	$1.3806488(13) \times 10^{-23}$ J/K
Charge on electron	e	1.60×10^{-19} C	$1.602176565(35) \times 10^{-19}$ C
Stefan-Boltzmann constant	σ	5.67×10^{-8} W/m²·K⁴	$5.670373(21) \times 10^{-8}$ W/m²·K⁴
Permittivity of free space	$\epsilon_0 = (1/c^2\mu_0)$	8.85×10^{-12} C²/N·m²	$8.854187817\ldots \times 10^{-12}$ C²/N·m²
Permeability of free space	μ_0	$4\pi \times 10^{-7}$ T·m/A	$1.2566370614\ldots \times 10^{-6}$ T·m/A
Planck's constant	h	6.63×10^{-34} J·s	$6.62606957(29) \times 10^{-34}$ J·s
Electron mass	m_e	9.11×10^{-31} kg = 0.000549 u = 0.511 MeV/c^2	$9.10938291(40) \times 10^{-31}$ kg = $5.48579909(25) \times 10^{-4}$ u
Proton mass	m_p	1.6726×10^{-27} kg = 1.00728 u = 938.27 MeV/c^2	$1.672621777(74) \times 10^{-27}$ kg = 1.007276467(45) u
Neutron mass	m_n	1.6749×10^{-27} kg = 1.008665 u = 939.57 MeV/c^2	$1.674927351(74) \times 10^{-27}$ kg = 1.008664916(45) u
Atomic mass unit (1 u)		1.6605×10^{-27} kg = 931.49 MeV/c^2	$1.660538921(73) \times 10^{-27}$ kg = 931.494061(21) MeV/c^2

[†] Numbers in parentheses indicate one-standard-deviation experimental uncertainties in final digits.
Values without parentheses are exact (i.e., defined quantities).

Other Useful Data

Joule equivalent (1 cal)	4.186 J
Absolute zero (0 K)	$-273.15°$C
Acceleration due to gravity at Earth's surface (avg.)	9.80 m/s² (= g)
Speed of sound in air (20°C)	343 m/s
Density of air (dry)	1.29 kg/m³
Earth: Mass	5.98×10^{24} kg
Radius (mean)	6.38×10^3 km
Moon: Mass	7.35×10^{22} kg
Radius (mean)	1.74×10^3 km
Sun: Mass	1.99×10^{30} kg
Radius (mean)	6.96×10^5 km
Earth–Sun distance (mean)	149.60×10^6 km
Earth–Moon distance (mean)	384×10^3 km

The Greek Alphabet

Alpha	A	α	Nu	N	ν	
Beta	B	β	Xi	Ξ	ξ	
Gamma	Γ	γ	Omicron	O	o	
Delta	Δ	δ	Pi	Π	π	
Epsilon	E	ϵ, ε	Rho	P	ρ	
Zeta	Z	ζ	Sigma	Σ	σ	
Eta	H	η	Tau	T	τ	
Theta	Θ	θ	Upsilon	Υ	υ	
Iota	I	ι	Phi	Φ	ϕ, φ	
Kappa	K	κ	Chi	X	χ	
Lambda	Λ	λ	Psi	Ψ	ψ	
Mu	M	μ	Omega	Ω	ω	

Values of Some Numbers

$\pi = 3.1415927$	$\sqrt{2} = 1.4142136$	$\ln 2 = 0.6931472$	$\log_{10} e = 0.4342945$
$e = 2.7182818$	$\sqrt{3} = 1.7320508$	$\ln 10 = 2.3025851$	1 rad = $57.2957795°$

Mathematical Signs and Symbols

$\propto$	is proportional to	$\leq$	is less than or equal to
$=$	is equal to	$\geq$	is greater than or equal to
$\approx$	is approximately equal to	Σ	sum of
$\neq$	is not equal to	$\bar{x}$	average value of x
$>$	is greater than	Δx	change in x
$\gg$	is much greater than	$\Delta x \to 0$	Δx approaches zero
$<$	is less than	$n!$	$n(n-1)(n-2)\ldots(1)$
$\ll$	is much less than		

Properties of Water

Density (4°C)	1.000×10^3 kg/m³
Heat of fusion (0°C)	334 kJ/kg (79.8 kcal/kg)
Heat of vaporization (100°C)	2260 kJ/kg (539.9 kcal/kg)
Specific heat (15°C)	4186 J/kg·C° (1.00 kcal/kg·C°)
Index of refraction	1.33

Unit Conversions (Equivalents)

Length

1 in. = 2.54 cm (defined)
1 cm = 0.3937 in.
1 ft = 30.48 cm
1 m = 39.37 in. = 3.281 ft
1 mi = 5280 ft = 1.609 km
1 km = 0.6214 mi
1 nautical mile (U.S.) = 1.151 mi = 6076 ft = 1.852 km
1 fermi = 1 femtometer (fm) = 10^{-15} m
1 angstrom (Å) = 10^{-10} m = 0.1 nm
1 light-year (ly) = 9.461×10^{15} m
1 parsec = 3.26 ly = 3.09×10^{16} m

Volume

1 liter (L) = 1000 mL = 1000 cm^3 = 1.0×10^{-3} m^3 = 1.057 qt (U.S.) = 61.02 in.^{3}
1 gal (U.S.) = 4 qt (U.S.) = 231 in.^{3} = 3.785 L = 0.8327 gal (British)
1 quart (U.S.) = 2 pints (U.S.) = 946 mL
1 pint (British) = 1.20 pints (U.S.) = 568 mL
1 m^3 = 35.31 ft^3

Speed

1 mi/h = 1.4667 ft/s = 1.6093 km/h = 0.4470 m/s
1 km/h = 0.2778 m/s = 0.6214 mi/h
1 ft/s = 0.3048 m/s = 0.6818 mi/h = 1.0973 km/h
1 m/s = 3.281 ft/s = 3.600 km/h = 2.237 mi/h
1 knot = 1.151 mi/h = 0.5144 m/s

Angle

1 radian (rad) = 57.30° = 57°18′
1° = 0.01745 rad
1 rev/min (rpm) = 0.1047 rad/s

Time

1 day = 8.640×10^4 s
1 year = 365.242 days = 3.156×10^7 s

Mass

1 atomic mass unit (u) = 1.6605×10^{-27} kg
1 kg = 0.06852 slug
[1 kg has a weight of 2.20 lb where $g = 9.80$ m/s^2.]

Force

1 lb = 4.44822 N
1 N = 10^5 dyne = 0.2248 lb

Energy and Work

1 J = 10^7 ergs = 0.7376 ft·lb
1 ft·lb = 1.356 J = 1.29×10^{-3} Btu = 3.24×10^{-4} kcal
1 kcal = 4.19×10^3 J = 3.97 Btu
1 eV = 1.6022×10^{-19} J
1 kWh = 3.600×10^6 J = 860 kcal
1 Btu = 1.056×10^3 J

Power

1 W = 1 J/s = 0.7376 ft·lb/s = 3.41 Btu/h
1 hp = 550 ft·lb/s = 746 W

Pressure

1 atm = 1.01325 bar = 1.01325×10^5 N/m^2
= 14.7 lb/in.^{2} = 760 torr
1 lb/in.^{2} = 6.895×10^3 N/m^2
1 Pa = 1 N/m^2 = 1.450×10^{-4} lb/in.^{2}

SI Derived Units and Their Abbreviations

Quantity	Unit	Abbreviation	In Terms of Base Units[†]
Force	newton	N	kg·m/s^2
Energy and work	joule	J	kg·m^2/s^2
Power	watt	W	kg·m^2/s^3
Pressure	pascal	Pa	kg/(m·s^2)
Frequency	hertz	Hz	s^{-1}
Electric charge	coulomb	C	A·s
Electric potential	volt	V	kg·m^2/(A·s^3)
Electric resistance	ohm	Ω	kg·m^2/(A^2·s^3)
Capacitance	farad	F	A^2·s^4/(kg·m^2)
Magnetic field	tesla	T	kg/(A·s^2)
Magnetic flux	weber	Wb	kg·m^2/(A·s^2)
Inductance	henry	H	kg·m^2/(A^2·s^2)

[†] kg = kilogram (mass), m = meter (length), s = second (time), A = ampere (electric current).

Metric (SI) Multipliers

Prefix	Abbreviation	Value
yotta	Y	10^{24}
zeta	Z	10^{21}
exa	E	10^{18}
peta	P	10^{15}
tera	T	10^{12}
giga	G	10^{9}
mega	M	10^{6}
kilo	k	10^{3}
hecto	h	10^{2}
deka	da	10^{1}
deci	d	10^{-1}
centi	c	10^{-2}
milli	m	10^{-3}
micro	μ	10^{-6}
nano	n	10^{-9}
pico	p	10^{-12}
femto	f	10^{-15}
atto	a	10^{-18}
zepto	z	10^{-21}
yocto	y	10^{-24}

SEVENTH EDITION

VOLUME II

PHYSICS

PRINCIPLES WITH APPLICATIONS

DOUGLAS C. GIANCOLI

PEARSON

Boston   Columbus   Indianapolis   New York   San Francisco   Upper Saddle River
Amsterdam   Cape Town   Dubai   London   Madrid   Milan   Munich   Paris   Montréal   Toronto
Delhi   Mexico City   São Paulo   Sydney   Hong Kong   Seoul   Singapore   Taipei   Tokyo

President, Science, Business and Technology: Paul Corey
Publisher: Jim Smith
Executive Development Editor: Karen Karlin
Production Project Manager: Elisa Mandelbaum / Laura Ross
Marketing Manager: Will Moore
Senior Managing Editor: Corinne Benson
Managing Development Editor: Cathy Murphy
Copyeditor: Joanna Dinsmore
Proofreaders: Susan Fisher, Donna Young
Interior Designer: Mark Ong
Cover Designer: Derek Bacchus
Photo Permissions Management: Maya Melenchuk
Photo Research Manager: Eric Schrader
Photo Researcher: Mary Teresa Giancoli
Senior Administrative Assistant: Cathy Glenn
Senior Administrative Coordinator: Trisha Tarricone
Text Permissions Project Manager: Joseph Croscup
Editorial Media Producer: Kelly Reed
Manufacturing Buyer: Jeffrey Sargent
Indexer: Carol Reitz
Compositor: Preparé, Inc.
Illustrations: Precision Graphics

Cover Photo Credit: North Peak, California (D. Giancoli); Insets: left, analog to digital (page 488); right, electron microscope image—retina of human eye with cones artificially colored green, rods beige (page 785).
Back Cover Photo Credit: D. Giancoli

Credits and acknowledgments for materials borrowed from other sources and reproduced, with permission, in this textbook appear on page A-61.

Library of Congress Cataloging-in-Publication Data on file

ISBN-10: 0-321-73362-2
ISBN-13: 978-0-321-73362-7

2 3 4 5 6 7 8 9 10—**CRK**—17 16 15 14

www.pearsonhighered.com

Contents, Volume 1

1 INTRODUCTION, MEASUREMENT, ESTIMATING   1

2 DESCRIBING MOTION: KINEMATICS IN ONE DIMENSION   21

3 KINEMATICS IN TWO DIMENSIONS; VECTORS   49

4 DYNAMICS: NEWTON'S LAWS OF MOTION   75

5 CIRCULAR MOTION; GRAVITATION   109

Force   Displacement

Contents, Volume 2

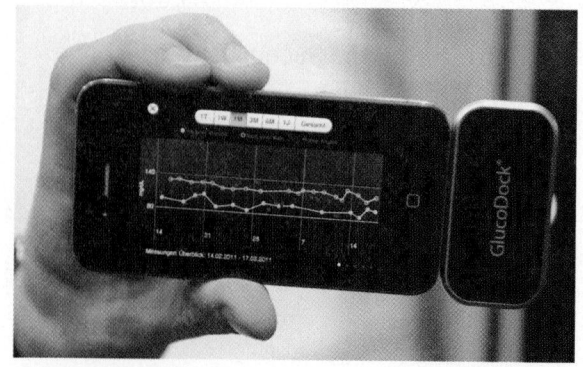

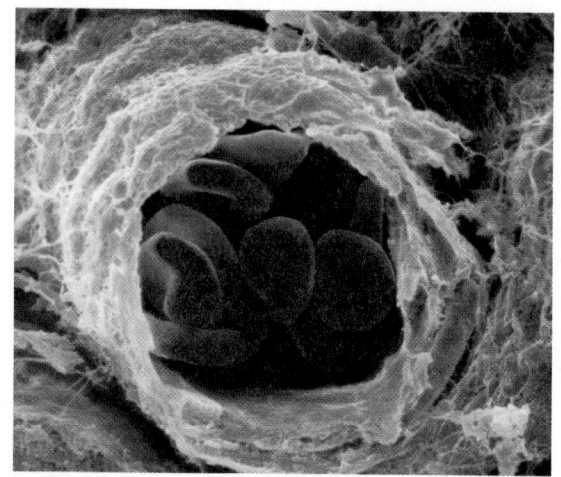

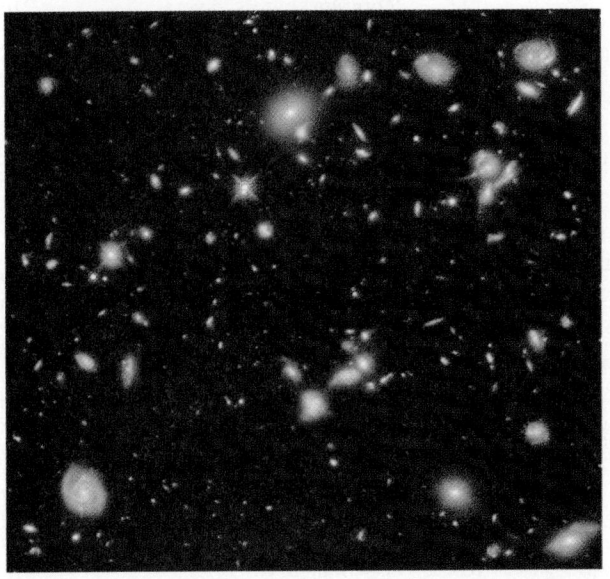

Applications to Biology and Medicine (Selected)

Applications to Other Fields and Everyday Life (Selected)

Student Supplements

- **MasteringPhysics**™ (www.masteringphysics.com) is a homework, tutorial, and assessment system based on years of research into how students work physics problems and precisely where they need help. Studies show that students who use MasteringPhysics significantly increase their final scores compared to hand-written homework. Mastering-Physics achieves this improvement by providing students with instantaneous feedback specific to their wrong answers, simpler sub-problems upon request when they get stuck, and partial credit for their method(s) used. This individualized, 24/7 Socratic tutoring is recommended by nine out of ten students to their peers as the most effective and time-efficient way to study.

- The **Student Study Guide with Selected Solutions**, **Volume I** (Chapters 1–15, ISBN 978-0-321-76240-5) and **Volume II** (Chapters 16–33, ISBN 978-0-321-76808-7), written by Joseph Boyle (Miami-Dade Community College), contains overviews, key terms and phrases, key equations, self-study exams, problems for review, problem solving skills, and answers and solutions to selected end-of-chapter questions and problems for each chapter of this textbook.

- **Pearson eText** is available through MasteringPhysics, either automatically when MasteringPhysics is packaged with new books, or available as a purchased upgrade online. Allowing students access to the text wherever they have access to the Internet, Pearson eText comprises the full text, including figures that can be enlarged for better viewing. Within eText, students are also able to pop up definitions and terms to help with vocabulary and the reading of the material. Students can also take notes in eText using the annotation feature at the top of each page.

- **Pearson Tutor Services** (www.pearsontutorservices.com): Each student's subscription to MasteringPhysics also contains complimentary access to Pearson Tutor Services, powered by Smarthinking, Inc. By logging in with their MasteringPhysics ID and password, they will be connected to highly qualified e-instructors™ who provide additional, interactive online tutoring on the major concepts of physics.

- **ActivPhysics OnLine**™ (accessed through the Self Study area within www.masteringphysics.com) provides students with a group of highly regarded applet-based tutorials (see above). The following workbooks help students work though complex concepts and understand them more clearly.

- **ActivPhysics OnLine Workbook Volume 1: Mechanics • Thermal Physics • Oscillations & Waves** (ISBN 978-0-805-39060-5)

- **ActivPhysics OnLine Workbook Volume 2: Electricity & Magnetism • Optics • Modern Physics** (ISBN 978-0-805-39061-2)

Preface

What's New?

Lots! Much is new and unseen before. Here are the big four:

1. Multiple-choice Questions added to the end of each Chapter. They are not the usual type. These are called **MisConceptual Questions** because the responses (*a*, *b*, *c*, *d*, etc.) are intended to include common student misconceptions. Thus they are as much, or more, a learning experience than simply a testing experience.

2. **Search and Learn Problems** at the very end of each Chapter, after the other Problems. Some are pretty hard, others are fairly easy. They are intended to encourage students to go back and reread some part or parts of the text, and in this search for an answer they will hopefully learn more—if only because they have to read some material again.

3. **Chapter-Opening Questions** (COQ) that start each Chapter, a sort of "stimulant." Each is multiple choice, with responses including common misconceptions—to get preconceived notions out on the table right at the start. Where the relevant material is covered in the text, students find an Exercise asking them to return to the COQ to rethink and answer again.

4. **Digital.** Biggest of all. Crucial new applications. Today we are surrounded by digital electronics. How does it work? If you try to find out, say on the Internet, you won't find much physics: you may find shallow hand-waving with no real content, or some heavy jargon whose basis might take months or years to understand. So, for the first time, I have tried to explain

    *   The basis of digital in bits and bytes, how analog gets transformed into digital, sampling rate, bit depth, quantization error, compression, noise (Section 17–10).

    *   How digital TV works, including how each pixel is addressed for each frame, data stream, refresh rate (Section 17–11).

    *   Semiconductor computer memory, DRAM, and flash (Section 21–8).

    *   Digital cameras and sensors—revised and expanded Section 25–1.

    *   New semiconductor physics, some of which is used in digital devices, including LED and OLED—how they work and what their uses are—plus more on transistors (MOSFET), chips, and technology generation as in 22-nm technology (Sections 29–9, 10, 11).

Besides those above, this new seventh edition includes

5. *New topics, new applications, principal revisions.*

    *   *You* can measure the Earth's radius (Section 1–7).

    *   Improved graphical analysis of linear motion (Section 2–8).

    *   Planets (how first seen), heliocentric, geocentric (Section 5–8).

    *   The Moon's orbit around the Earth: its phases and periods with diagram (Section 5–9).

    *   Explanation of lake level change when large rock thrown from boat (Example 10–11).

- Biology and medicine, including:
  - Blood measurements (flow, sugar)—Chapters 10, 12, 14, 19, 20, 21;
  - Trees help offset CO_2 buildup—Chapter 15;
  - Pulse oximeter—Chapter 29;
  - Proton therapy—Chapter 31;
  - Radon exposure calculation—Chapter 31;
  - Cell phone use and brain—Chapter 31.
- Colors as seen underwater (Section 24–4).
- Soap film sequence of colors explained (Section 24–8).
- Solar sails (Section 22–6).
- Lots on sports.
- Symmetry—more emphasis and using italics or boldface to make visible.
- Flat screens (Sections 17–11, 24–11).
- Free-electron theory of metals, Fermi gas, Fermi level. New Section 29–6.
- Semiconductor devices—new details on diodes, LEDs, OLEDs, solar cells, compound semiconductors, diode lasers, MOSFET transistors, chips, 22-nm technology (Sections 29–9, 10, 11).
- Cross section (Chapter 31).
- Length of an object is a script ℓ rather than normal l, which looks like 1 or I (moment of inertia, current), as in $F = I\ell B$. Capital L is for angular momentum, latent heat, inductance, dimensions of length $[L]$.

6. **New photographs** taken by students and instructors (we asked).

7. **Page layout**: More than in previous editions, serious attention to how each page is formatted. Important derivations and Examples are on facing pages: no turning a page back in the middle of a derivation or Example. Throughout, readers see, on two facing pages, an important slice of physics.

8. **Greater clarity**: No topic, no paragraph in this book was overlooked in the search to improve the clarity and conciseness of the presentation. Phrases and sentences that may slow down the principal argument have been eliminated: keep to the essentials at first, give the elaborations later.

9. Much use has been made of physics education research. See the new powerful pedagogic features listed first.

10. **Examples modified**: More math steps are spelled out, and many new Examples added. About 10% of all Examples are Estimation Examples.

11. **This Book is Shorter** than other complete full-service books at this level. Shorter explanations are easier to understand and more likely to be read.

12. **Cosmological Revolution**: With generous help from top experts in the field, readers have the latest results.

See the World through Eyes that Know Physics

I was motivated from the beginning to write a textbook different from the others which present physics as a sequence of facts, like a catalog: "Here are the facts and you better learn them." Instead of beginning formally and dogmatically, I have sought to begin each topic with concrete observations and experiences students can relate to: start with specifics, and after go to the great generalizations and the more formal aspects of a topic, showing *why* we believe what we believe. This approach reflects how science is actually practiced.

The ultimate aim is to give students a thorough understanding of the basic concepts of physics in all its aspects, from mechanics to modern physics. A second objective is to show students how useful physics is in their own everyday lives and in their future professions by means of interesting applications to biology, medicine, architecture, and more.

Also, much effort has gone into techniques and approaches for solving problems: worked-out Examples, Problem Solving sections (Sections 2–6, 3–6, 4–7, 4–8, 6–7, 6–9, 8–6, 9–2, 13–7, 14–4, and 16–6), and Problem Solving Strategies (pages 30, 57, 60, 88, 115, 141, 158, 184, 211, 234, 399, 436, 456, 534, 568, 594, 655, 666, and 697).

This textbook is especially suited for students taking a one-year introductory course in physics that uses algebra and trigonometry but not calculus.[†] Many of these students are majoring in biology or premed, as well as architecture, technology, and the earth and environmental sciences. Many applications to these fields are intended to answer that common student query: "Why must I study physics?" The answer is that physics is fundamental to a full understanding of these fields, and here they can see how. Physics is everywhere around us in the everyday world. It is the goal of this book to help students "see the world through eyes that know physics."

A major effort has been made to not throw too much material at students reading the first few chapters. The basics have to be learned first. Many aspects can come later, when students are less overloaded and more prepared. If we don't overwhelm students with too much detail, especially at the start, maybe they can find physics interesting, fun, and helpful—and those who were afraid may lose their fear.

Chapter 1 is *not* a throwaway. It is fundamental to physics to realize that every measurement has an *uncertainty*, and how significant figures are used. Converting units and being able to make rapid *estimates* are also basic.

Mathematics can be an obstacle to students. I have aimed at including all steps in a derivation. Important mathematical tools, such as addition of vectors and trigonometry, are incorporated in the text where first needed, so they come with a context rather than in a scary introductory Chapter. Appendices contain a review of algebra and geometry (plus a few advanced topics).

Color is used pedagogically to bring out the physics. Different types of vectors are given different colors (see the chart on page xix).

Sections marked with a star * are considered optional. These contain slightly more advanced physics material, or material not usually covered in typical courses and/or interesting applications; they contain no material needed in later Chapters (except perhaps in later optional Sections).

For a brief course, all optional material could be dropped as well as significant parts of Chapters 1, 10, 12, 22, 28, 29, 32, and selected parts of Chapters 7, 8, 9, 15, 21, 24, 25, 31. Topics not covered in class can be a valuable resource for later study by students. Indeed, this text can serve as a useful reference for years because of its wide range of coverage.

[†]It is fine to take a calculus course. But mixing calculus with physics for these students may often mean not learning the physics because of stumbling over the calculus.

Thanks

Many physics professors provided input or direct feedback on every aspect of this textbook. They are listed below, and I owe each a debt of gratitude.

Edward Adelson, The Ohio State University
Lorraine Allen, United States Coast Guard Academy
Zaven Altounian, McGill University
Leon Amstutz, Taylor University
David T. Bannon, Oregon State University
Bruce Barnett, Johns Hopkins University
Michael Barnett, Lawrence Berkeley Lab
Anand Batra, Howard University
Cornelius Bennhold, George Washington University
Bruce Birkett, University of California Berkeley
Steven Boggs, University of California Berkeley
Robert Boivin, Auburn University
Subir Bose, University of Central Florida
David Branning, Trinity College
Meade Brooks, Collin County Community College
Bruce Bunker, University of Notre Dame
Grant Bunker, Illinois Institute of Technology
Wayne Carr, Stevens Institute of Technology
Charles Chiu, University of Texas Austin
Roger N. Clark, U. S. Geological Survey
Russell Clark, University of Pittsburgh
Robert Coakley, University of Southern Maine
David Curott, University of North Alabama
Biman Das, SUNY Potsdam
Bob Davis, Taylor University
Kaushik De, University of Texas Arlington
Michael Dennin, University of California Irvine
Karim Diff, Santa Fe College
Kathy Dimiduk, Cornell University
John DiNardo, Drexel University
Scott Dudley, United States Air Force Academy
Paul Dyke
John Essick, Reed College
Kim Farah, Lasell College
Cassandra Fesen, Dartmouth College
Leonard Finegold, Drexel University
Alex Filippenko, University of California Berkeley
Richard Firestone, Lawrence Berkeley Lab
Allen Flora, Hood College
Mike Fortner, Northern Illinois University
Tom Furtak, Colorado School of Mines
Edward Gibson, California State University Sacramento
John Hardy, Texas A&M
Thomas Hemmick, State University of New York Stonybrook
J. Erik Hendrickson, University of Wisconsin Eau Claire
Laurent Hodges, Iowa State University
David Hogg, New York University
Mark Hollabaugh, Normandale Community College
Andy Hollerman, University of Louisiana at Lafayette
Russell Holmes, University of Minnesota Twin Cities
William Holzapfel, University of California Berkeley
Chenming Hu, University of California Berkeley
Bob Jacobsen, University of California Berkeley
Arthur W. John, Northeastern University
Teruki Kamon, Texas A&M
Daryao Khatri, University of the District of Columbia
Tsu-Jae King Liu, University of California Berkeley
Richard Kronenfeld, South Mountain Community College
Jay Kunze, Idaho State University
Jim LaBelle, Dartmouth College
Amer Lahamer, Berea College
David Lamp, Texas Tech University
Kevin Lear, SpatialGraphics.com
Ran Li, Kent State University
Andreí Linde, Stanford University
M.A.K. Lodhi, Texas Tech
Lisa Madewell, University of Wisconsin

Bruce Mason, University of Oklahoma
Mark Mattson, James Madison University
Dan Mazilu, Washington and Lee University
Linda McDonald, North Park College
Bill McNairy, Duke University
Jo Ann Merrell, Saddleback College
Raj Mohanty, Boston University
Giuseppe Molesini, Istituto Nazionale di Ottica Florence
Wouter Montfrooij, University of Missouri
Eric Moore, Frostburg State University
Lisa K. Morris, Washington State University
Richard Muller, University of California Berkeley
Blaine Norum, University of Virginia
Lauren Novatne, Reedley College
Alexandria Oakes, Eastern Michigan University
Ralph Oberly, Marshall University
Michael Ottinger, Missouri Western State University
Lyman Page, Princeton and WMAP
Laurence Palmer, University of Maryland
Bruce Partridge, Haverford College
R. Daryl Pedigo, University of Washington
Robert Pelcovitz, Brown University
Saul Perlmutter, University of California Berkeley
Vahe Peroomian, UCLA
Harvey Picker, Trinity College
Amy Pope, Clemson University
James Rabchuk, Western Illinois University
Michele Rallis, Ohio State University
Paul Richards, University of California Berkeley
Peter Riley, University of Texas Austin
Dennis Rioux, University of Wisconsin Oshkosh
John Rollino, Rutgers University
Larry Rowan, University of North Carolina Chapel Hill
Arthur Schmidt, Northwestern University
Cindy Schwarz-Rachmilowitz, Vassar College
Peter Sheldon, Randolph-Macon Woman's College
Natalia A. Sidorovskaia, University of Louisiana at Lafayette
James Siegrist, University of California Berkeley
Christopher Sirola, University of Southern Mississippi
Earl Skelton, Georgetown University
George Smoot, University of California Berkeley
David Snoke, University of Pittsburgh
Stanley Sobolewski, Indiana University of Pennsylvania
Mark Sprague, East Carolina University
Michael Strauss, University of Oklahoma
Laszlo Takac, University of Maryland Baltimore Co.
Leo Takahashi, Pennsylvania State University
Richard Taylor, University of Oregon
Oswald Tekyi-Mensah, Alabama State University
Franklin D. Trumpy, Des Moines Area Community College
Ray Turner, Clemson University
Som Tyagi, Drexel University
David Vakil, El Camino College
Trina VanAusdal, Salt Lake Community College
John Vasut, Baylor University
Robert Webb, Texas A&M
Robert Weidman, Michigan Technological University
Edward A. Whittaker, Stevens Institute of Technology
Lisa M. Will, San Diego City College
Suzanne Willis, Northern Illinois University
John Wolbeck, Orange County Community College
Stanley George Wojcicki, Stanford University
Mark Worthy, Mississippi State University
Edward Wright, UCLA and WMAP
Todd Young, Wayne State College
William Younger, College of the Albemarle
Hsiao-Ling Zhou, Georgia State University
Michael Ziegler, The Ohio State University
Ulrich Zurcher, Cleveland State University

New photographs were offered by Professors Vickie Frohne (Holy Cross Coll.), Guillermo Gonzales (Grove City Coll.), Martin Hackworth (Idaho State U.), Walter H. G. Lewin (MIT), Nicholas Murgo (NEIT), Melissa Vigil (Marquette U.), Brian Woodahl (Indiana U. at Indianapolis), and Gary Wysin (Kansas State U.). New photographs shot by students are from the AAPT photo contest: Matt Buck, (John Burroughs School), Matthew Claspill (Helias H. S.), Greg Gentile (West Forsyth H. S.), Shilpa Hampole (Notre Dame H. S.), Sarah Lampen (John Burroughs School), Mrinalini Modak (Fayetteville–Manlius H. S.), Joey Moro (Ithaca H. S.), and Anna Russell and Annacy Wilson (both Tamalpais H. S.).

I owe special thanks to Prof. Bob Davis for much valuable input, and especially for working out all the Problems and producing the Solutions Manual for all Problems, as well as for providing the answers to odd-numbered Problems at the back of the book. Many thanks also to J. Erik Hendrickson who collaborated with Bob Davis on the solutions, and to the team they managed (Profs. Karim Diff, Thomas Hemmick, Lauren Novatne, Michael Ottinger, and Trina VanAusdal).

I am grateful to Profs. Lorraine Allen, David Bannon, Robert Coakley, Kathy Dimiduk, John Essick, Dan Mazilu, John Rollino, Cindy Schwarz, Earl Skelton, Michael Strauss, Ray Turner, Suzanne Willis, and Todd Young, who helped with developing the new MisConceptual Questions and Search and Learn Problems, and offered other significant clarifications.

Crucial for rooting out errors, as well as providing excellent suggestions, were Profs. Lorraine Allen, Kathy Dimiduk, Michael Strauss, Ray Turner, and David Vakil. A huge thank you to them and to Prof. Giuseppe Molesini for his suggestions and his exceptional photographs for optics.

For Chapters 32 and 33 on Particle Physics and Cosmology and Astrophysics, I was fortunate to receive generous input from some of the top experts in the field, to whom I owe a debt of gratitude: Saul Perlmutter, George Smoot, Richard Muller, Steven Boggs, Alex Filippenko, Paul Richards, James Siegrist, and William Holzapfel (UC Berkeley), Andreí Linde (Stanford U.), Lyman Page (Princeton and WMAP), Edward Wright (UCLA and WMAP), Michael Strauss (University of Oklahoma), Michael Barnett (LBNL), and Bob Jacobsen (UC Berkeley; so helpful in many areas, including digital and pedagogy).

I also wish to thank Profs. Howard Shugart, Chair Frances Hellman, and many others at the University of California, Berkeley, Physics Department for helpful discussions, and for hospitality. Thanks also to Profs. Tito Arecchi, Giuseppe Molesini, and Riccardo Meucci at the Istituto Nazionale di Ottica, Florence, Italy.

Finally, I am grateful to the many people at Pearson Education with whom I worked on this project, especially Paul Corey and the ever-perspicacious Karen Karlin.

The final responsibility for all errors lies with me. I welcome comments, corrections, and suggestions as soon as possible to benefit students for the next reprint.

email: Jim.Smith@Pearson.com
Post: Jim Smith
        1301 Sansome Street
        San Francisco, CA 94111

D.C.G.

About the Author

Douglas C. Giancoli obtained his BA in physics (summa cum laude) from UC Berkeley, his MS in physics at MIT, and his PhD in elementary particle physics back at UC Berkeley. He spent 2 years as a post-doctoral fellow at UC Berkeley's Virus lab developing skills in molecular biology and biophysics. His mentors include Nobel winners Emilio Segrè and Donald Glaser.

He has taught a wide range of undergraduate courses, traditional as well as innovative ones, and continues to update his textbooks meticulously, seeking ways to better provide an understanding of physics for students.

Doug's favorite spare-time activity is the outdoors, especially climbing peaks. He says climbing peaks is like learning physics: it takes effort and the rewards are great.

To Students

HOW TO STUDY

1. Read the Chapter. Learn new vocabulary and notation. Try to respond to questions and exercises as they occur.

2. Attend all class meetings. Listen. Take notes, especially about aspects you do not remember seeing in the book. Ask questions (everyone wants to, but maybe you will have the courage). You will get more out of class if you read the Chapter first.

3. Read the Chapter again, paying attention to details. Follow derivations and worked-out Examples. Absorb their logic. Answer Exercises and as many of the end-of-Chapter Questions as you can, and all MisConceptual Questions.

4. Solve at least 10 to 20 end of Chapter Problems, especially those assigned. In doing Problems you find out what you learned and what you didn't. Discuss them with other students. Problem solving is one of the great learning tools. Don't just look for a formula—it might be the wrong one.

NOTES ON THE FORMAT AND PROBLEM SOLVING

1. Sections marked with a star (*) are considered **optional**. They can be omitted without interrupting the main flow of topics. No later material depends on them except possibly later starred Sections. They may be fun to read, though.

2. The customary **conventions** are used: symbols for quantities (such as m for mass) are italicized, whereas units (such as m for meter) are not italicized. Symbols for vectors are shown in boldface with a small arrow above: $\vec{\mathbf{F}}$.

3. Few equations are valid in all situations. Where practical, the **limitations** of important equations are stated in square brackets next to the equation. The equations that represent the great laws of physics are displayed with a tan background, as are a few other indispensable equations.

4. At the end of each Chapter is a set of **Questions** you should try to answer. Attempt all the multiple-choice **MisConceptual Questions**. Most important are **Problems** which are ranked as Level I, II, or III, according to estimated difficulty. Level I Problems are easiest, Level II are standard Problems, and Level III are "challenge problems." These ranked Problems are arranged by Section, but Problems for a given Section may depend on earlier material too. There follows a group of **General Problems**, not arranged by Section or ranked. Problems that relate to optional Sections are starred (*). Answers to odd-numbered Problems are given at the end of the book. **Search and Learn Problems** at the end are meant to encourage you to return to parts of the text to find needed detail, and at the same time help you to learn.

5. Being able to solve **Problems** is a crucial part of learning physics, and provides a powerful means for understanding the concepts and principles. This book contains many aids to problem solving: (a) worked-out **Examples**, including an Approach and Solution, which should be studied as an integral part of the text; (b) some of the worked-out Examples are **Estimation Examples**, which show how rough or approximate results can be obtained even if the given data are sparse (see Section 1–7); (c) **Problem Solving Strategies** placed throughout the text to suggest a step-by-step approach to problem solving for a particular topic—but remember that the basics remain the same; most of these "Strategies" are followed by an Example that is solved by explicitly following the suggested steps; (d) special problem-solving Sections; (e) "Problem Solving" marginal notes which refer to hints within the text for solving Problems; (f) **Exercises** within the text that you should work out immediately, and then check your response against the answer given at the bottom of the last page of that Chapter; (g) the Problems themselves at the end of each Chapter (point 4 above).

6. **Conceptual Examples** pose a question which hopefully starts you to think and come up with a response. Give yourself a little time to come up with your own response before reading the Response given.

7. **Math** review, plus additional topics, are found in Appendices. Useful **data, conversion factors**, and math **formulas** are found inside the front and back covers.

USE OF COLOR

Vectors

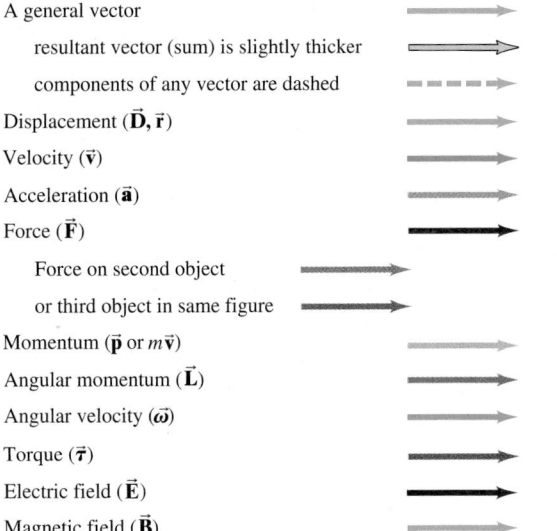

A general vector

   resultant vector (sum) is slightly thicker

   components of any vector are dashed

Displacement ($\vec{\mathbf{D}}, \vec{\mathbf{r}}$)

Velocity ($\vec{\mathbf{v}}$)

Acceleration ($\vec{\mathbf{a}}$)

Force ($\vec{\mathbf{F}}$)

   Force on second object

   or third object in same figure

Momentum ($\vec{\mathbf{p}}$ or $m\vec{\mathbf{v}}$)

Angular momentum ($\vec{\mathbf{L}}$)

Angular velocity ($\vec{\boldsymbol{\omega}}$)

Torque ($\vec{\boldsymbol{\tau}}$)

Electric field ($\vec{\mathbf{E}}$)

Magnetic field ($\vec{\mathbf{B}}$)

Electricity and magnetism

Electric field lines

Equipotential lines

Magnetic field lines

Electric charge (+)   +   or   ● +

Electric charge (−)   −   or   ● −

Electric circuit symbols

Wire, with switch S

Resistor

Capacitor

Inductor

Battery

Ground

Optics

Light rays

Object

Real image (dashed)

Virtual image (dashed and paler)

Other

Energy level (atom, etc.)

Measurement lines   ⊢—1.0 m—⊣

Path of a moving object

Direction of motion or current

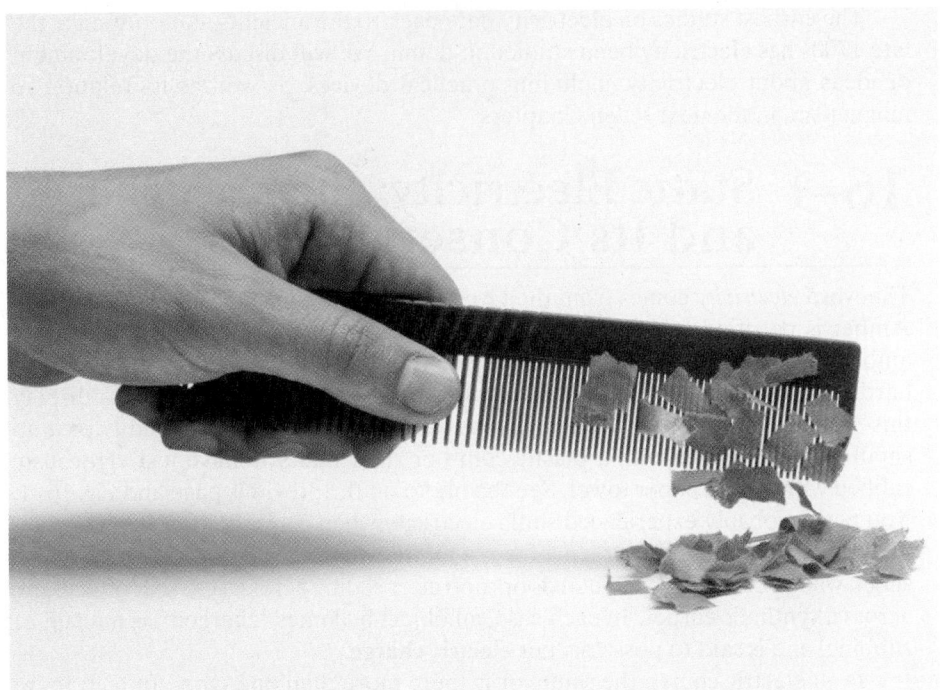

This comb has acquired a static electric charge, either from passing through hair, or being rubbed by a cloth or paper towel. The electrical charge on the comb induces a polarization (separation of charge) in scraps of paper, and thus attracts them.

Our introduction to electricity in this Chapter covers conductors and insulators, and Coulomb's law which relates the force between two point charges as a function of their distance apart. We also introduce the powerful concept of electric field.

Electric Charge and Electric Field

CHAPTER-OPENING QUESTION—Guess now!

Two identical tiny spheres have the same electric charge. If their separation is doubled, the force each exerts on the other will be
- **(a)** half.
- **(b)** double.
- **(c)** four times larger.
- **(d)** one-quarter as large.
- **(e)** unchanged.

CONTENTS

T he word "electricity" may evoke an image of complex modern technology: lights, motors, electronics, and computers. But the electric force plays an even deeper role in our lives. According to atomic theory, electric forces between atoms and molecules hold them together to form liquids and solids, and electric forces are also involved in the metabolic processes that occur within our bodies. Many of the forces we have dealt with so far, such as elastic forces, the normal force, and friction and other contact forces (pushes and pulls), are now considered to result from electric forces acting at the atomic level. Gravity, on the other hand, is a separate force.[†]

[†]As we discussed in Section 5–9, physicists in the twentieth century came to recognize four different fundamental forces in nature: (1) gravitational force, (2) electromagnetic force (we will see later that electric and magnetic forces are intimately related), (3) strong nuclear force, and (4) weak nuclear force. The last two forces operate at the level of the nucleus of an atom. Recent theory has combined the electromagnetic and weak nuclear forces so they are now considered to have a common origin known as the electroweak force. We discuss the other forces in later Chapters.

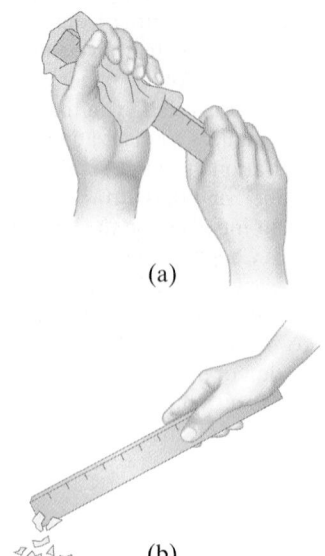

FIGURE 16–1 (a) Rub a plastic ruler with a cloth or paper towel, and (b) bring it close to some tiny pieces of paper.

FIGURE 16–2 Like charges repel one another; unlike charges attract. (Note color coding: we color positive charged objects pink or red, and negative charges blue-green. We use these colors especially for point charges, but not always for real objects.)

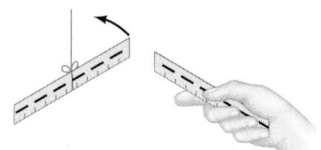

(a) Two charged plastic rulers repel

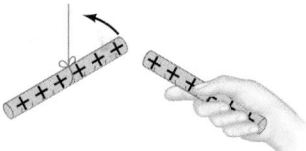

(b) Two charged glass rods repel

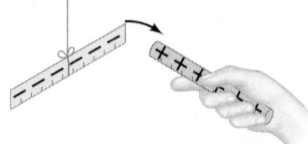

(c) Charged glass rod attracts charged plastic ruler

LAW OF CONSERVATION OF ELECTRIC CHARGE

The earliest studies on electricity date back to the ancients, but only since the late 1700s has electricity been studied in detail. We will discuss the development of ideas about electricity, including practical devices, as well as its relation to magnetism, in the next seven Chapters.

16–1 Static Electricity; Electric Charge and Its Conservation

The word *electricity* comes from the Greek word *elektron*, which means "amber." Amber is petrified tree resin, and the ancients knew that if you rub a piece of amber with a cloth, the amber attracts small pieces of leaves or dust. A piece of hard rubber, a glass rod, or a plastic ruler rubbed with a cloth will also display this "amber effect," or **static electricity** as we call it today. You can readily pick up small pieces of paper with a plastic comb or ruler that you have just vigorously rubbed with even a paper towel. See the photo on the previous page and Fig. 16–1. You have probably experienced static electricity when combing your hair or when taking a synthetic blouse or shirt from a clothes dryer. And you may have felt a shock when you touched a metal doorknob after sliding across a car seat or walking across a synthetic carpet. In each case, an object becomes "charged" as a result of rubbing, and is said to possess a net **electric charge**.

Is all electric charge the same, or is there more than one type? In fact, there are *two* types of electric charge, as the following simple experiments show. A plastic ruler suspended by a thread is vigorously rubbed with a cloth to charge it. When a second plastic ruler, which has been charged in the same way, is brought close to the first, it is found that one ruler *repels* the other. This is shown in Fig. 16–2a. Similarly, if a rubbed glass rod is brought close to a second charged glass rod, again a repulsive force is seen to act, Fig. 16–2b. However, if the charged glass rod is brought close to the charged plastic ruler, it is found that they *attract* each other, Fig. 16–2c. The charge on the glass must therefore be different from that on the plastic. Indeed, it is found experimentally that all charged objects fall into one of two categories. Either they are attracted to the plastic and repelled by the glass; or they are repelled by the plastic and attracted to the glass. Thus there seem to be two, and only two, types of electric charge. Each type of charge repels the same type but attracts the opposite type. That is: **unlike charges attract; like charges repel**.

The two types of electric charge were referred to as **positive** and **negative** by the American statesman, philosopher, and scientist Benjamin Franklin (1706–1790). The choice of which name went with which type of charge was arbitrary. Franklin's choice set the charge on the rubbed glass rod to be positive charge, so the charge on a rubbed plastic ruler (or amber) is called negative charge. We still follow this convention today.

Franklin argued that whenever a certain amount of charge is produced on one object, an equal amount of the opposite type of charge is produced on another object. The positive and negative are to be treated *algebraically*, so during any process, the net change in the amount of charge produced is zero. For example, when a plastic ruler is rubbed with a paper towel, the plastic acquires a negative charge and the towel acquires an equal amount of positive charge. The charges are separated, but the sum of the two is zero.

This is an example of a law that is now well established: the **law of conservation of electric charge**, which states that

the net amount of electric charge produced in any process is zero;

or, said another way,

no net electric charge can be created or destroyed.

If one object (or a region of space) acquires a positive charge, then an equal amount of negative charge will be found in neighboring areas or objects. No violations have ever been found, and the law of conservation of electric charge is as firmly established as those for energy and momentum.

16–2 Electric Charge in the Atom

Only within the past century has it become clear that an understanding of electricity originates inside the atom itself. In later Chapters we will discuss atomic structure and the ideas that led to our present view of the atom in more detail. But it will help our understanding of electricity if we discuss it briefly now.

A simplified model of an atom shows it as having a tiny but massive, positively charged nucleus surrounded by one or more negatively charged electrons (Fig. 16–3). The nucleus contains protons, which are positively charged, and neutrons, which have no net electric charge. All protons and all electrons have exactly the same magnitude of electric charge; but their signs are opposite. Hence neutral atoms, having no net charge, contain equal numbers of protons and electrons. Sometimes an atom may lose one or more of its electrons, or may gain extra electrons, in which case it will have a net positive or negative charge and is called an **ion**.

In solid materials the nuclei tend to remain close to fixed positions, whereas some of the electrons may move quite freely. When an object is *neutral*, it contains equal amounts of positive and negative charge. The charging of a solid object by rubbing can be explained by the transfer of electrons from one object to the other. When a plastic ruler becomes negatively charged by rubbing with a paper towel, electrons are transferred from the towel to the plastic, leaving the towel with a positive charge equal in magnitude to the negative charge acquired by the plastic. In liquids and gases, nuclei or ions can move as well as electrons.

Normally when objects are charged by rubbing, they hold their charge only for a limited time and eventually return to the neutral state. Where does the charge go? Usually the excess charge "leaks off" onto water molecules in the air. This is because water molecules are **polar**—that is, even though they are neutral, their charge is not distributed uniformly, Fig. 16–4. Thus the extra electrons on, say, a charged plastic ruler can "leak off" into the air because they are attracted to the positive end of water molecules. A positively charged object, on the other hand, can be neutralized by transfer of loosely held electrons from water molecules in the air. On dry days, static electricity is much more noticeable since the air contains fewer water molecules to allow leakage of charge. On humid or rainy days, it is difficult to make any object hold a net charge for long.

16–3 Insulators and Conductors

Suppose we have two metal spheres, one highly charged and the other electrically neutral (Fig. 16–5a). If we now place a metal object, such as a nail, so that it touches both spheres (Fig. 16–5b), the previously uncharged sphere quickly becomes charged. If, instead, we had connected the two spheres by a wooden rod or a piece of rubber (Fig. 16–5c), the uncharged ball would not become noticeably charged. Materials like the iron nail are said to be **conductors** of electricity, whereas wood and rubber are **nonconductors** or **insulators**.

Metals are generally good conductors, whereas most other materials are insulators (although even insulators conduct electricity very slightly). Nearly all natural materials fall into one or the other of these two distinct categories. However, a few materials (notably silicon and germanium) fall into an intermediate category known as **semiconductors**.

From the atomic point of view, the electrons in an insulating material are bound very tightly to the nuclei. In a good metal conductor, on the other hand, some of the electrons are bound very loosely and can move about freely within the metal (although they cannot *leave* the metal easily) and are often referred to as **free electrons** or **conduction electrons**. When a positively charged object is brought close to or touches a conductor, the free electrons in the conductor are attracted by this positively charged object and move quickly toward it. If a negatively charged object is brought close to the conductor, the free electrons in the conductor move swiftly away from it. In a semiconductor, there are many fewer free electrons, and in an insulator, almost none.

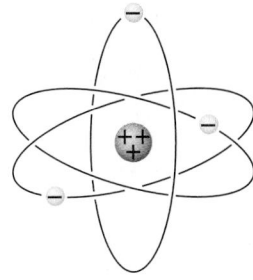

FIGURE 16–3 Simple model of the atom.

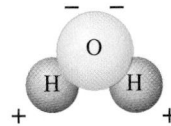

FIGURE 16–4 Diagram of a water molecule. Because it has opposite charges on different ends, it is called a "polar" molecule.

FIGURE 16–5 (a) A charged metal sphere and a neutral metal sphere. (b) The two spheres connected by a conductor (a metal nail), which conducts charge from one sphere to the other. (c) The original two spheres connected by an insulator (wood); almost no charge is conducted.

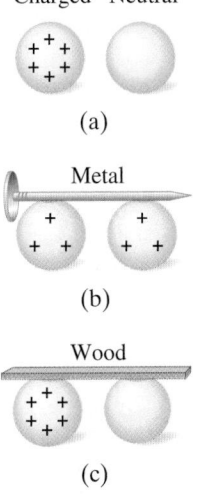

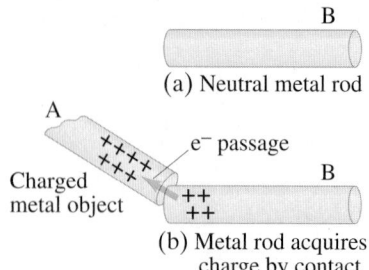

FIGURE 16–6 A neutral metal rod in (a) will acquire a positive charge if placed in contact (b) with a positively charged metal object. (Electrons move as shown by the green arrow.) This is called charging by conduction.

Suppose a positively charged metal object A is brought close to an uncharged metal object B. If the two touch, the free electrons in the neutral one are attracted to the positively charged object and some of those electrons will pass over to it, Fig. 16–6. Since object B, originally neutral, is now missing some of its negative electrons, it will have a net positive charge. This process is called **charging by conduction**, or "by contact," and the two objects end up with the same sign of charge.

Now suppose a positively charged object is brought close to a neutral metal rod, but does not touch it. Although the free electrons of the metal rod do not leave the rod, they still move within the metal toward the external positive charge, leaving a positive charge at the opposite end of the rod (Fig. 16–7b). A charge is said to have been *induced* at the two ends of the metal rod. No net charge has been created in the rod: charges have merely been *separated*. The net charge on the metal rod is still zero. However, if the metal is separated into two pieces, we would have two charged objects: one charged positively and one charged negatively. This is **charging by induction**.

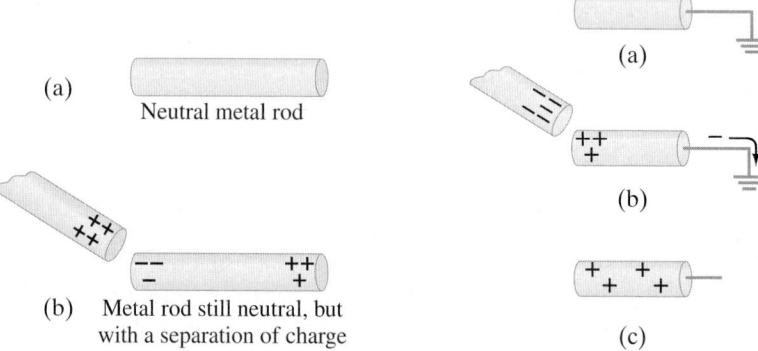

(a) Neutral metal rod

(b) Metal rod still neutral, but with a separation of charge

FIGURE 16–7 Charging by induction: if the rod in (b) is cut into two parts, each part will have a net charge.

(a)

(b)

(c)

FIGURE 16–8 Inducing a charge on an object connected to ground.

FIGURE 16–9 A charged object brought near a nonconductor causes a charge separation within the nonconductor's molecules.

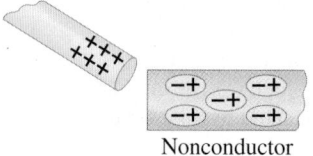

Nonconductor

Another way to induce a net charge on a metal object is to first connect it with a conducting wire to the ground (or a conducting pipe leading into the ground) as shown in Fig. 16–8a (the symbol ⏚ means connected to "ground"). The object is then said to be **grounded** or "earthed." The Earth, because it is so large and can conduct, easily accepts or gives up electrons; hence it acts like a reservoir for charge. If a charged object—say negative this time—is brought up close to the metal object, free electrons in the metal are repelled and many of them move down the wire into the Earth, Fig. 16–8b. This leaves the metal positively charged. If the wire is now cut, the metal object will have a positive induced charge on it (Fig. 16–8c). If the wire is cut *after* the negative object is moved away, the electrons would all have moved from the ground back into the metal object and it would be neutral again.

Charge separation can also be done in nonconductors. If you bring a positively charged object close to a neutral nonconductor as shown in Fig. 16–9, almost no electrons can move about freely within the nonconductor. But they can move slightly within their own atoms and molecules. Each oval in Fig. 16–9 represents a molecule (not to scale); the negatively charged electrons, attracted to the external positive charge, tend to move in its direction within their molecules. Because the negative charges in the nonconductor are nearer to the external positive charge, the nonconductor as a whole is attracted to the external positive charge (see the Chapter-Opening Photo, page 443).

An **electroscope** is a device that can be used for detecting charge. As shown in Fig. 16–10, inside a case are two movable metal leaves, often made of gold foil, connected to a metal knob on the outside. (Sometimes only one leaf is movable.)

FIGURE 16–10 Electroscope.

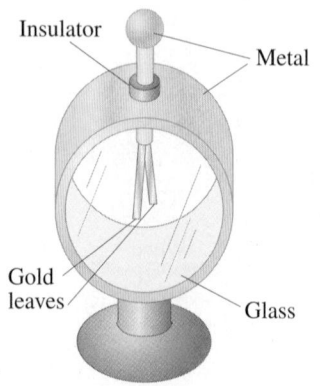

Insulator

Metal

Gold leaves

Glass

If a positively charged object is brought close to the knob, a separation of charge is induced: electrons are attracted up into the knob, and the leaves become positively charged, Fig. 16–11a. The two leaves repel each other as shown, because they are both positively charged. If, instead, the knob is charged by conduction (touching), the whole apparatus acquires a net charge as shown in Fig. 16–11b. In either case, the greater the amount of charge, the greater the separation of the leaves.

Note that you cannot tell the sign of the charge in this way, since negative charge will cause the leaves to separate just as much as an equal amount of positive charge; in either case, the two leaves repel each other. An electroscope can, however, be used to determine the sign of the charge if it is first charged by conduction: say, negatively, as in Fig. 16–12a. Now if a negative object is brought close, as in Fig. 16–12b, more electrons are induced to move down into the leaves and they separate further. If a positive charge is brought close instead, the electrons are induced to flow upward, so the leaves are less negative and their separation is reduced, Fig. 16–12c.

The electroscope was used in the early studies of electricity. The same principle, aided by some electronics, is used in much more sensitive modern **electrometers**.

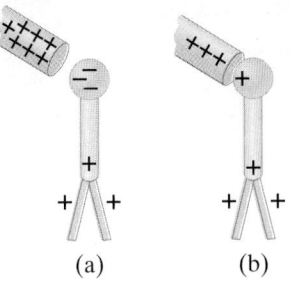

FIGURE 16–11 Electroscope charged (a) by induction, (b) by conduction.

FIGURE 16–12 A previously charged electroscope can be used to determine the sign of a charged object.

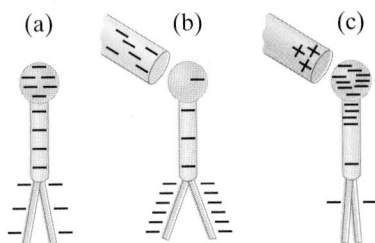

16–5 Coulomb's Law

We have seen that an electric charge exerts a force of attraction or repulsion on other electric charges. What factors affect the magnitude of this force? To find an answer, the French physicist Charles Coulomb (1736–1806) investigated electric forces in the 1780s using a torsion balance (Fig. 16–13) much like that used by Cavendish for his studies of the gravitational force (Chapter 5).

Precise instruments for the measurement of electric charge were not available in Coulomb's time. Nonetheless, Coulomb was able to prepare small spheres with different magnitudes of charge in which the *ratio* of the charges was known.[†] Although he had some difficulty with induced charges, Coulomb was able to argue that the electric force one tiny charged object exerts on a second tiny charged object is directly proportional to the charge on each of them. That is, if the charge on either one of the objects is doubled, the force is doubled; and if the charge on both of the objects is doubled, the force increases to four times the original value. This was the case when the distance between the two charges remained the same. If the distance between them was allowed to increase, he found that the force decreased with the *square of the distance* between them. That is, if the distance was doubled, the force fell to one-fourth of its original value. Thus, Coulomb concluded, the magnitude of the force F that one small charged object exerts on a second one is proportional to the product of the magnitude of the charge on one, Q_1, times the magnitude of the charge on the other, Q_2, and inversely proportional to the square of the distance r between them (Fig. 16–14). As an equation, we can write **Coulomb's law** as

$$F = k\frac{Q_1 Q_2}{r^2}, \qquad \text{[magnitudes]} \quad \textbf{(16–1)}$$

where k is a proportionality constant.[‡]

FIGURE 16–13 Coulomb's apparatus: when an external charged sphere is placed close to the charged one on the suspended bar, the bar rotates slightly. The suspending fiber resists the twisting motion, and the angle of twist is proportional to the force applied. With this apparatus, Coulomb investigated how the electric force varies as a function of the magnitude of the charges and of the distance between them.

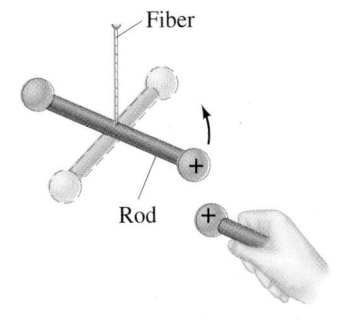

COULOMB'S LAW

FIGURE 16–14 Coulomb's law, Eq. 16–1, gives the force between two point charges, Q_1 and Q_2, a distance r apart.

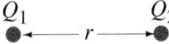

[†]Coulomb reasoned that if a charged conducting sphere is placed in contact with an identical uncharged sphere, the charge on the first would be shared equally by the two of them because of symmetry. He thus had a way to produce charges equal to $\frac{1}{2}$, $\frac{1}{4}$, and so on, of the original charge.

[‡]The validity of Coulomb's law today rests on precision measurements that are much more sophisticated than Coulomb's original experiment. The exponent 2 on r in Coulomb's law has been shown to be accurate to 1 part in 10^{16} [that is, $2 \pm (1 \times 10^{-16})$].

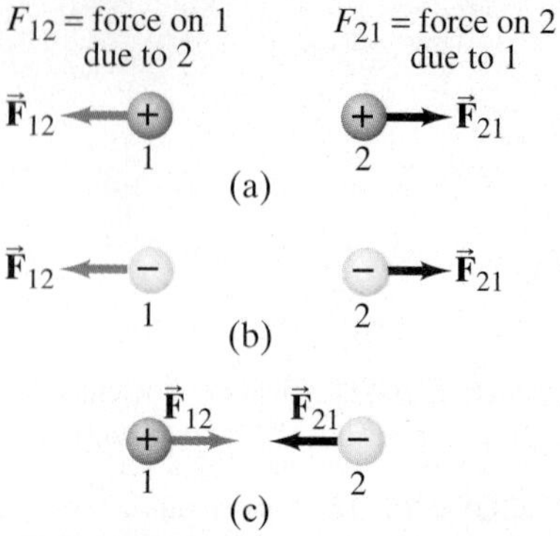

F_{12} = force on 1 due to 2

F_{21} = force on 2 due to 1

(a)

(b)

(c)

FIGURE 16–15 The direction of the static electric force one point charge exerts on another is always along the line joining the two charges, and depends on whether the charges have the same sign as in (a) and (b), or opposite signs (c).

As we just saw, Coulomb's law, Eq. 16–1,

$$F = k\frac{Q_1 Q_2}{r^2}, \qquad \text{[magnitudes]} \quad \textbf{(16–1)}$$

gives the *magnitude* of the electric force that either charge exerts on the other. The *direction* of the electric force *is always along the line joining the two charges*. If the two charges have the same sign, the force on either charge is directed away from the other (they repel each other). If the two charges have opposite signs, the force on one is directed toward the other (they attract). See Fig. 16–15. Notice that the force one charge exerts on the second is equal but opposite to that exerted by the second on the first, in accord with Newton's third law.

The SI unit of charge is the **coulomb** (C). The precise definition of the coulomb today is in terms of electric current and magnetic field, and will be discussed later (Section 20–6). In SI units, the constant k in Coulomb's law has the value

$$k = 8.988 \times 10^9 \, \text{N} \cdot \text{m}^2/\text{C}^2$$

or, when we only need two significant figures,

$$k \approx 9.0 \times 10^9 \, \text{N} \cdot \text{m}^2/\text{C}^2.$$

Thus, 1 C is that amount of charge which, if placed on each of two point objects that are 1.0 m apart, will result in each object exerting a force of $(9.0 \times 10^9 \, \text{N} \cdot \text{m}^2/\text{C}^2)(1.0 \, \text{C})(1.0 \, \text{C})/(1.0 \, \text{m})^2 = 9.0 \times 10^9 \, \text{N}$ on the other. This would be an enormous force, equal to the weight of almost a million tons. We rarely encounter charges as large as a coulomb.[†]

Charges produced by rubbing ordinary objects (such as a comb or plastic ruler) are typically around a microcoulomb $(1 \, \mu\text{C} = 10^{-6} \, \text{C})$ or less. Objects that carry a positive charge have a deficit of electrons, whereas negatively charged objects have an excess of electrons. The charge on one electron has been determined to have a magnitude of about $1.6022 \times 10^{-19} \, \text{C}$, and is negative. This is the smallest charge observed in nature,[‡] and because it is fundamental, it is given the symbol e and is often referred to as the **elementary charge**:

$$e = 1.6022 \times 10^{-19} \, \text{C} \approx 1.6 \times 10^{-19} \, \text{C}.$$

Note that e is defined as a positive number, so the charge on the electron is $-e$. (The charge on a proton, on the other hand, is $+e$.) Since an object cannot gain or lose a fraction of an electron, the net charge on any object must be an integral multiple of this charge. Electric charge is thus said to be **quantized** (existing only in discrete amounts: $1e$, $2e$, $3e$, etc.). Because e is so small, however, we normally do not notice this discreteness in macroscopic charges ($1 \, \mu\text{C}$ requires about 10^{13} electrons), which thus seem continuous.

Coulomb's law looks a lot like the *law of universal gravitation*, $F = Gm_1 m_2/r^2$, which expresses the magnitude of the gravitational force a mass m_1 exerts on a mass m_2 (Eq. 5–4). Both are **inverse square laws** $(F \propto 1/r^2)$. Both also have a proportionality to a property of each object—mass for gravity, electric charge for electricity. And both act over a distance (that is, there is no need for contact). A major difference between the two laws is that gravity is always an attractive force, whereas the electric force can be either attractive or repulsive. Electric charge comes in two types, positive and negative; gravitational mass is only positive.

The constant k in Eq. 16–1 is often written in terms of another constant, ϵ_0, called the **permittivity of free space**. It is related to k by $k = 1/4\pi\epsilon_0$. Coulomb's law can be written

$$F = \frac{1}{4\pi\epsilon_0} \frac{Q_1 Q_2}{r^2}, \qquad \textbf{(16–2)}$$

where

$$\epsilon_0 = \frac{1}{4\pi k} = 8.85 \times 10^{-12} \, \text{C}^2/\text{N} \cdot \text{m}^2.$$

[†]In the once common cgs system of units, k is set equal to 1, and the unit of electric charge is called the *electrostatic unit* (esu) or the statcoulomb. One esu is defined as that charge, on each of two point objects 1 cm apart, that gives rise to a force of 1 dyne.

[‡]According to the Standard Model of elementary particle physics, subnuclear particles called quarks have a smaller charge than the electron, equal to $\frac{1}{3}e$ or $\frac{2}{3}e$. Quarks have not been detected directly as isolated objects, and theory indicates that free quarks may not be detectable.

Equation 16–2 looks more complicated than Eq. 16–1, but other fundamental equations we haven't seen yet are simpler in terms of ϵ_0 rather than k. It doesn't matter which form we use since Eqs. 16–1 and 16–2 are equivalent. (The latest precise values of e and ϵ_0 are given inside the front cover.)[†]

Equations 16–1 and 16–2 apply to objects whose size is much smaller than the distance between them. Ideally, it is precise for **point charges** (spatial size negligible compared to other distances). For finite-sized objects, it is not always clear what value to use for r, particularly since the charge may not be distributed uniformly on the objects. If the two objects are spheres and the charge is known to be distributed uniformly on each, then r is the distance between their centers.

Coulomb's law describes the force between two charges when they are at rest. Additional forces come into play when charges are in motion, and will be discussed in later Chapters. In this Chapter we discuss only charges at rest, the study of which is called **electrostatics**, and Coulomb's law gives the **electrostatic force**.

When calculating with Coulomb's law, we usually use magnitudes, ignoring signs of the charges, and determine the direction of a force separately based on whether the force is attractive or repulsive.

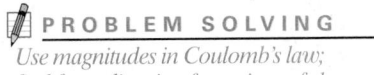

PROBLEM SOLVING
Use magnitudes in Coulomb's law; find force direction from signs of charges

EXERCISE A Return to the Chapter-Opening Question, page 443, and answer it again now. Try to explain why you may have answered differently the first time.

EXAMPLE 16–1 **Electric force on electron by proton.** Determine the magnitude and direction of the electric force on the electron of a hydrogen atom exerted by the single proton $(Q_2 = +e)$ that is the atom's nucleus. Assume the average distance between the revolving electron and the proton is $r = 0.53 \times 10^{-10}$ m, Fig. 16–16.

APPROACH To find the force magnitude we use Coulomb's law, $F = k Q_1 Q_2 / r^2$ (Eq. 16–1), with $r = 0.53 \times 10^{-10}$ m. The electron and proton have the same magnitude of charge, e, so $Q_1 = Q_2 = 1.6 \times 10^{-19}$ C.

SOLUTION The magnitude of the force is

$$F = k\frac{Q_1 Q_2}{r^2} = \frac{(9.0 \times 10^9 \ \text{N}\cdot\text{m}^2/\text{C}^2)(1.6 \times 10^{-19} \ \text{C})(1.6 \times 10^{-19} \ \text{C})}{(0.53 \times 10^{-10} \ \text{m})^2}$$

$$= 8.2 \times 10^{-8} \ \text{N}.$$

The direction of the force on the electron is toward the proton, because the charges have opposite signs so the force is attractive.

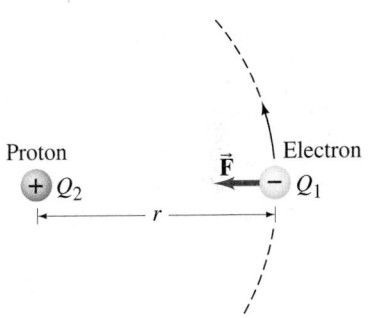

FIGURE 16–16 Example 16–1.

CONCEPTUAL EXAMPLE 16–2 **Which charge exerts the greater force?** Two positive point charges, $Q_1 = 50 \ \mu\text{C}$ and $Q_2 = 1 \ \mu\text{C}$, are separated by a distance ℓ, Fig. 16–17. Which is larger in magnitude, the force that Q_1 exerts on Q_2, or the force that Q_2 exerts on Q_1?

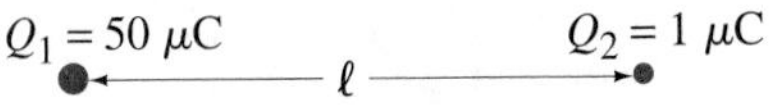

RESPONSE From Coulomb's law, the force on Q_1 exerted by Q_2 is

$$F_{12} = k\frac{Q_1 Q_2}{\ell^2}.$$

The force on Q_2 exerted by Q_1 is

$$F_{21} = k\frac{Q_2 Q_1}{\ell^2}$$

which is the same magnitude. The equation is symmetric with respect to the two charges, so $F_{21} = F_{12}$.

NOTE Newton's third law also tells us these two forces must have equal magnitude.

FIGURE 16–17 Example 16–2.
$Q_1 = 50 \ \mu\text{C}$ $Q_2 = 1 \ \mu\text{C}$

EXERCISE B In Example 16–2, how is the direction of F_{12} related to the direction of F_{21}?

[†]Our convention for units, such as $\text{C}^2/\text{N}\cdot\text{m}^2$ for ϵ_0, means m^2 is in the denominator. That is, $\text{C}^2/\text{N}\cdot\text{m}^2$ means $\text{C}^2/(\text{N}\cdot\text{m}^2)$ and does *not* mean $(\text{C}^2/\text{N})\cdot\text{m}^2 = \text{C}^2\cdot\text{m}^2/\text{N}$.

Keep in mind that Coulomb's law, Eq. 16–1 or 16–2, gives the force on a charge due to only *one* other charge. If several (or many) charges are present, the *net force on any one of them will be the vector sum of the forces due to each of the others*. This **principle of superposition** is based on experiment, and tells us that electric force vectors add like any other vector. For example, if you have a system of four charges, the net force on charge 1, say, is the sum of the forces exerted on charge 1 by charges 2, 3, and 4. The magnitudes of these three forces are determined from Coulomb's law, and then are added vectorially.

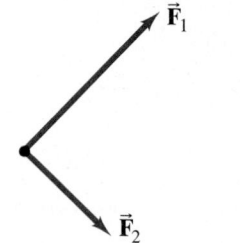

(a) Two forces acting on an object.

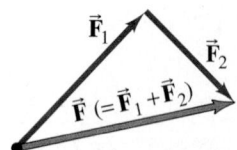

(b) The total, or net, force is $\vec{F} = \vec{F}_1 + \vec{F}_2$ by the tail-to-tip method of adding vectors.

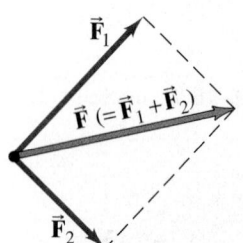

(c) $\vec{F} = \vec{F}_1 + \vec{F}_2$ by the parallelogram method.

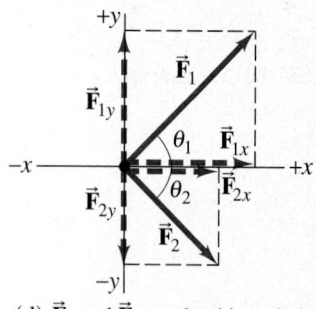

(d) $\vec{F}_1$ and $\vec{F}_2$ resolved into their *x* and *y* components.

FIGURE 16–18 Review of vector addition.

16–6 Solving Problems Involving Coulomb's Law and Vectors

The electric force between charged particles at rest (sometimes referred to as the **electrostatic force** or as the **Coulomb force**) is, like all forces, a vector: it has both magnitude and direction. When several forces act on an object (call them $\vec{F}_1$, $\vec{F}_2$, etc.), the net force $\vec{F}_{net}$ on the object is the vector sum of all the forces acting on it:

$$\vec{F}_{net} = \vec{F}_1 + \vec{F}_2 + \cdots.$$

This is the principle of superposition for forces. We studied how to add vectors in Chapter 3; then in Chapter 4 we used the rules for adding vectors to obtain the net force on an object by adding the different vector forces acting on it. It might be useful now to review Sections 3–2, 3–3, and 3–4. Here is a brief review of vectors.

Vector Addition Review

Suppose two vector forces, $\vec{F}_1$ and $\vec{F}_2$, act on an object (Fig. 16–18a). They can be added using the tail-to-tip method (Fig. 16–18b) or by the parallelogram method (Fig. 16–18c), as discussed in Section 3–2. These two methods are useful for *understanding* a given problem (for getting a picture in your mind of what is going on). But for *calculating* the direction and magnitude of the resultant sum, it is more precise to use the method of adding components. Figure 16–18d shows the forces $\vec{F}_1$ and $\vec{F}_2$ resolved into components along chosen *x* and *y* axes (for more details, see Section 3–4). From the definitions of the trigonometric functions (Figs. 3–11 and 3–12), we have

$$F_{1x} = F_1 \cos \theta_1 \qquad F_{2x} = F_2 \cos \theta_2$$
$$F_{1y} = F_1 \sin \theta_1 \qquad F_{2y} = -F_2 \sin \theta_2.$$

We add up the *x* and *y* components separately to obtain the components of the resultant force $\vec{F}$, which are

$$F_x = F_{1x} + F_{2x} = F_1 \cos \theta_1 + F_2 \cos \theta_2,$$
$$F_y = F_{1y} + F_{2y} = F_1 \sin \theta_1 - F_2 \sin \theta_2.$$

The magnitude of the resultant (or *net*) force $\vec{F}$ is

$$F = \sqrt{F_x^2 + F_y^2}.$$

The direction of $\vec{F}$ is specified by the angle θ that $\vec{F}$ makes with the *x* axis, which is given by

$$\tan \theta = \frac{F_y}{F_x}.$$

Adding Electric Forces; Principle of Superposition

When dealing with several charges, it is helpful to use double subscripts on each of the forces involved. The first subscript refers to the particle *on* which the force acts; the second refers to the particle that exerts the force. For example, if we have three charges, $\vec{F}_{31}$ means the force exerted *on* particle 3 *by* particle 1.

As in all problem solving, it is very important to draw a diagram, in particular a free-body diagram (Chapter 4) for each object, showing all the forces acting *on* that object. In applying Coulomb's law, we can deal with charge magnitudes only (leaving out minus signs) to get the magnitude of each force. Then determine separately the direction of the force physically (along the line joining the two particles: like charges repel, unlike charges attract), and show the force on the diagram. (You could determine direction first if you like.) Finally, add all the forces on one object together as vectors to obtain the net force on that object.

EXAMPLE 16–3 **Three charges in a line.** Three charged particles are arranged in a line, as shown in Fig. 16–19a. Calculate the net electrostatic force on particle 3 (the $-4.0 \, \mu C$ on the right) due to the other two charges.

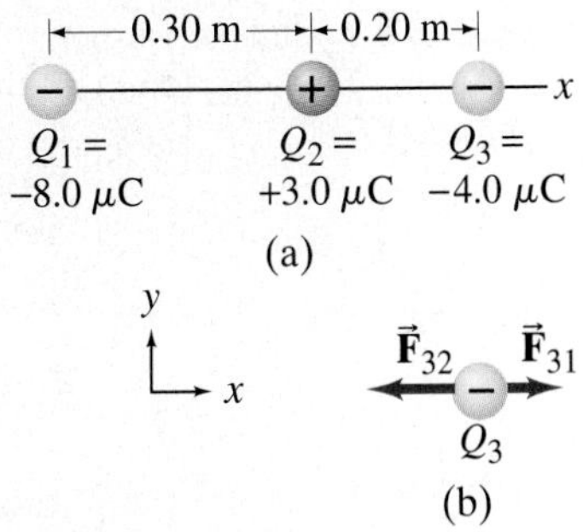

APPROACH The net force on particle 3 is the vector sum of the force $\vec{F}_{31}$ exerted on particle 3 by particle 1 and the force $\vec{F}_{32}$ exerted on 3 by particle 2:

$$\vec{F} = \vec{F}_{31} + \vec{F}_{32}.$$

SOLUTION The magnitudes of these two forces are obtained using Coulomb's law, Eq. 16–1:

$$F_{31} = k \frac{Q_3 Q_1}{r_{31}^2}$$

$$= \frac{(9.0 \times 10^9 \, \text{N} \cdot \text{m}^2/\text{C}^2)(4.0 \times 10^{-6} \, \text{C})(8.0 \times 10^{-6} \, \text{C})}{(0.50 \, \text{m})^2} = 1.2 \, \text{N},$$

where $r_{31} = 0.50 \, \text{m}$ is the distance from Q_3 to Q_1. Similarly,

$$F_{32} = k \frac{Q_3 Q_2}{r_{32}^2}$$

$$= \frac{(9.0 \times 10^9 \, \text{N} \cdot \text{m}^2/\text{C}^2)(4.0 \times 10^{-6} \, \text{C})(3.0 \times 10^{-6} \, \text{C})}{(0.20 \, \text{m})^2} = 2.7 \, \text{N}.$$

Since we were calculating the magnitudes of the forces, we omitted the signs of the charges. But we must be aware of them to get the direction of each force. Let the line joining the particles be the x axis, and we take it positive to the right. Then, because $\vec{F}_{31}$ is repulsive and $\vec{F}_{32}$ is attractive, the directions of the forces are as shown in Fig. 16–19b: F_{31} points in the positive x direction (away from Q_1) and F_{32} points in the negative x direction (toward Q_2). The net force on particle 3 is then

$$F = -F_{32} + F_{31}$$
$$= -2.7 \, \text{N} + 1.2 \, \text{N} = -1.5 \, \text{N}.$$

The magnitude of the net force is 1.5 N, and it points to the left.

NOTE Charge Q_1 acts on charge Q_3 just as if Q_2 were not there (this is the principle of superposition). That is, the charge in the middle, Q_2, in no way blocks the effect of charge Q_1 acting on Q_3. Naturally, Q_2 exerts its own force on Q_3.

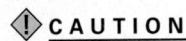

CAUTION

Each charge exerts its own force. No charge blocks the effect of the others

EXERCISE C Determine the magnitude and direction of the net force on charge Q_2 in Fig. 16–19a.

FIGURE 16–19 Example 16–3.

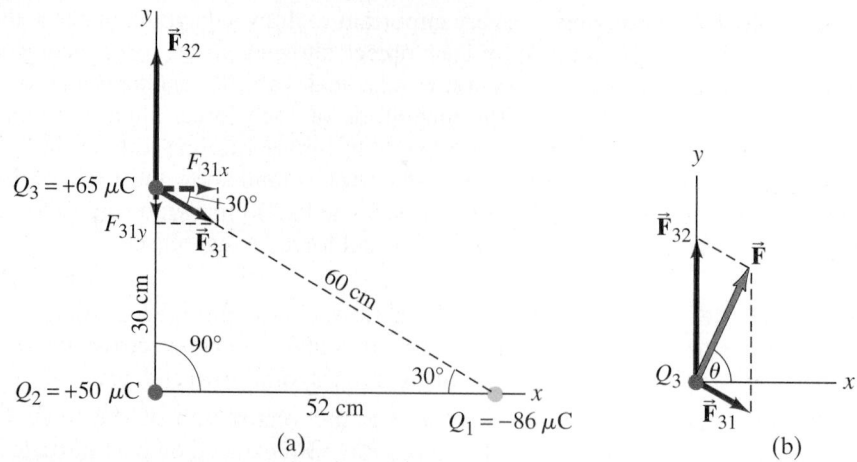

FIGURE 16–20 Determining the forces for Example 16–4. (a) The directions of the individual forces are as shown because $\vec{F}_{32}$ is repulsive (the force on Q_3 is in the direction away from Q_2 because Q_3 and Q_2 are both positive) whereas $\vec{F}_{31}$ is attractive (Q_3 and Q_1 have opposite signs), so $\vec{F}_{31}$ points toward Q_1. (b) Adding $\vec{F}_{32}$ to $\vec{F}_{31}$ to obtain the net force $\vec{F}$.

EXAMPLE 16–4 **Electric force using vector components.** Calculate the net electrostatic force on charge Q_3 shown in Fig. 16–20a due to the charges Q_1 and Q_2.

APPROACH We use Coulomb's law to find the magnitudes of the individual forces. The direction of each force will be along the line connecting Q_3 to Q_1 or Q_2. The forces $\vec{F}_{31}$ and $\vec{F}_{32}$ have the directions shown in Fig. 16–20a, since Q_1 exerts an attractive force on Q_3, and Q_2 exerts a repulsive force. The forces $\vec{F}_{31}$ and $\vec{F}_{32}$ are *not* along the same line, so to find the resultant force on Q_3 we resolve $\vec{F}_{31}$ and $\vec{F}_{32}$ into x and y components and perform the vector addition.

SOLUTION The magnitudes of $\vec{F}_{31}$ and $\vec{F}_{32}$ are (ignoring signs of the charges since we know the directions)

$$F_{31} = k\frac{Q_3 Q_1}{r_{31}^2} = \frac{(9.0 \times 10^9 \, \text{N} \cdot \text{m}^2/\text{C}^2)(6.5 \times 10^{-5} \, \text{C})(8.6 \times 10^{-5} \, \text{C})}{(0.60 \, \text{m})^2} = 140 \, \text{N},$$

$$F_{32} = k\frac{Q_3 Q_2}{r_{32}^2} = \frac{(9.0 \times 10^9 \, \text{N} \cdot \text{m}^2/\text{C}^2)(6.5 \times 10^{-5} \, \text{C})(5.0 \times 10^{-5} \, \text{C})}{(0.30 \, \text{m})^2} = 325 \, \text{N}.$$

(We keep 3 significant figures until the end, and then keep 2 because only 2 are given.) We resolve $\vec{F}_{31}$ into its components along the x and y axes, as shown in Fig. 16–20a:

$$F_{31x} = F_{31} \cos 30° = (140 \, \text{N}) \cos 30° = 120 \, \text{N},$$
$$F_{31y} = -F_{31} \sin 30° = -(140 \, \text{N}) \sin 30° = -70 \, \text{N}.$$

The force $\vec{F}_{32}$ has only a y component. So the net force $\vec{F}$ on Q_3 has components

$$F_x = F_{31x} = 120 \, \text{N},$$
$$F_y = F_{32} + F_{31y} = 325 \, \text{N} - 70 \, \text{N} = 255 \, \text{N}.$$

The magnitude of the net force is

$$F = \sqrt{F_x^2 + F_y^2} = \sqrt{(120 \, \text{N})^2 + (255 \, \text{N})^2} = 280 \, \text{N};$$

and it acts at an angle θ (see Fig. 16–20b) given by

$$\tan \theta = \frac{F_y}{F_x} = \frac{255 \, \text{N}}{120 \, \text{N}} = 2.13,$$

so $\theta = \tan^{-1}(2.13) = 65°$.

NOTE Because $\vec{F}_{31}$ and $\vec{F}_{32}$ are not along the same line, the magnitude of $\vec{F}_3$ is not equal to the sum (or difference as in Example 16–3) of the separate magnitudes. That is, F_3 is not equal to $F_{31} + F_{32}$; nor does it equal $F_{32} - F_{31}$. Instead we had to do vector addition.

CONCEPTUAL EXAMPLE 16–5 **Make the force on Q_3 zero.** In Fig. 16–20, where could you place a fourth charge, $Q_4 = -50\,\mu C$, so that the net force on Q_3 would be zero?

RESPONSE By the principle of superposition, we need a force in exactly the opposite direction to the resultant $\vec{F}$ due to Q_2 and Q_1 that we calculated in Example 16–4, Fig. 16–20b. Our force must have magnitude 290 N, and must point down and to the left of Q_3 in Fig. 16–20b, in the direction opposite to $\vec{F}$. So Q_4 must be along this line. See Fig. 16–21.

| **EXERCISE D** In Example 16–5, what distance r must Q_4 be from Q_3?

| **EXERCISE E** (*a*) Consider two point charges, $+Q$ and $-Q$, which are fixed a distance d apart. Can you find a location where a third positive charge Q could be placed so that the net electric force on this third charge is zero? (*b*) What if the first two charges were both $+Q$?

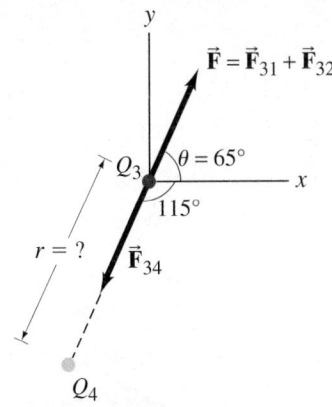

FIGURE 16–21 Example 16–5 and Exercise D: Q_4 exerts force ($\vec{F}_{34}$) that makes the net force on Q_3 zero.

16–7 The Electric Field

Many common forces might be referred to as "contact forces," such as your hands pushing or pulling a cart, or a tennis racket hitting a tennis ball.

In contrast, both the gravitational force and the electrical force act over a distance: there is a force between two objects even when the objects are not touching. The idea of a force *acting at a distance* was a difficult one for early thinkers. Newton himself felt uneasy with this idea when he published his law of universal gravitation. A helpful way to look at the situation uses the idea of the **field**, developed by the British scientist Michael Faraday (1791–1867). In the electrical case, according to Faraday, an *electric field* extends outward from every charge and permeates all of space (Fig. 16–22). If a second charge (call it Q_2) is placed near the first charge, it feels a force exerted by the electric field that is there (say, at point P in Fig. 16–22). The electric field at point P is considered to interact directly with charge Q_2 to produce the force on Q_2.

We can in principle investigate the electric field surrounding a charge or group of charges by measuring the force on a small positive **test charge** which is at rest. By a test charge we mean a charge so small that the force it exerts does not significantly affect the charges that create the field. If a tiny positive test charge q is placed at various locations in the vicinity of a single positive charge Q as shown in Fig. 16–23 (points A, B, C), the force exerted on q is as shown. The force at B is less than at A because B's distance from Q is greater (Coulomb's law); and the force at C is smaller still. In each case, the force on q is directed radially away from Q. The electric field is defined in terms of the force on such a positive test charge. In particular, the **electric field**, $\vec{E}$, at any point in space is defined as the force $\vec{F}$ exerted on a tiny positive test charge placed at that point divided by the magnitude of the test charge q:

$$\vec{E} = \frac{\vec{F}}{q}. \qquad (16\text{--}3)$$

More precisely, $\vec{E}$ is defined as the limit of $\vec{F}/q$ as q is taken smaller and smaller, approaching zero. That is, q is so tiny that it exerts essentially no force on the other charges which created the field. From this definition (Eq. 16–3), we see that the electric field at any point in space is a vector whose direction is the direction of the force on a tiny positive test charge at that point, and whose magnitude is the *force per unit charge*. Thus $\vec{E}$ has SI units of newtons per coulomb (N/C).

The reason for defining $\vec{E}$ as $\vec{F}/q$ (with $q \to 0$) is so that $\vec{E}$ does not depend on the magnitude of the test charge q. This means that $\vec{E}$ describes only the effect of the charges creating the electric field at that point.

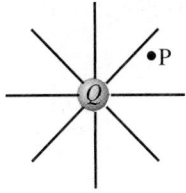

FIGURE 16–22 An electric field surrounds every charge. The red lines indicate the electric field extending out from charge Q, and P is an arbitrary point.

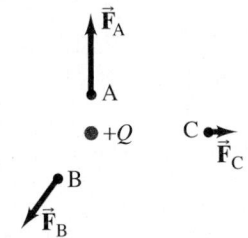

FIGURE 16–23 Force exerted by charge $+Q$ on a small test charge, q, placed at points A, B, and C.

The electric field at any point in space can be measured, based on the definition, Eq. 16–3. For simple situations with one or several point charges, we can calculate $\vec{E}$. For example, the electric field at a distance r from a single point charge Q would have magnitude

$$E = \frac{F}{q} = \frac{kqQ/r^2}{q}$$

$$E = k\frac{Q}{r^2};\qquad \text{[single point charge]}\quad \textbf{(16–4a)}$$

or, in terms of ϵ_0 as in Eq. 16–2 $\left(k = 1/4\pi\epsilon_0\right)$:

$$E = \frac{1}{4\pi\epsilon_0}\frac{Q}{r^2}.\qquad \text{[single point charge]}\quad \textbf{(16–4b)}$$

Notice that E is independent of the test charge q—that is, E depends only on the charge Q which produces the field, and not on the value of the test charge q. Equations 16–4 are referred to as the electric field form of Coulomb's law.

If we are given the electric field $\vec{E}$ at a given point in space, then we can calculate the force $\vec{F}$ on any charge q placed at that point by writing (see Eq. 16–3):

$$\vec{F} = q\vec{E}.\qquad\qquad\qquad \textbf{(16–5)}$$

This is valid even if q is not small as long as q does not cause the charges creating $\vec{E}$ to move. If q is positive, $\vec{F}$ and $\vec{E}$ point in the same direction. If q is negative, $\vec{F}$ and $\vec{E}$ point in opposite directions. See Fig. 16–24.

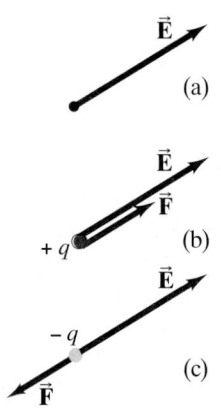

(a)

(b)

(c)

FIGURE 16–24 (a) Electric field at a given point in space. (b) Force on a positive charge at that point. (c) Force on a negative charge at that point.

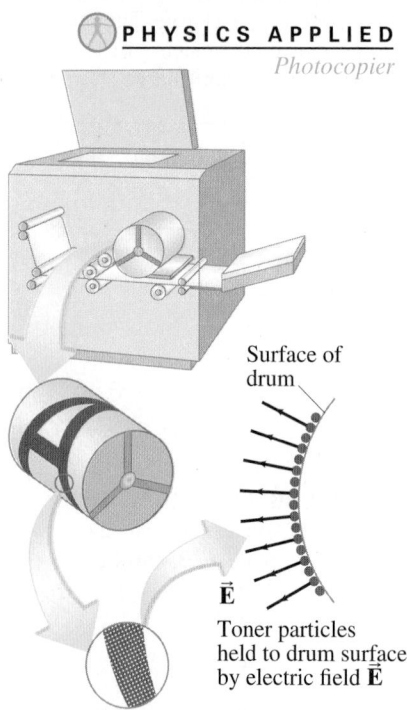

PHYSICS APPLIED

Photocopier

Surface of drum

$\vec{E}$

Toner particles held to drum surface by electric field $\vec{E}$

FIGURE 16–25 Example 16–6.

FIGURE 16–26 Example 16–7. Electric field at point P (a) due to a negative charge Q, and (b) due to a positive charge Q, each 30 cm from P.

——30 cm——

$Q = -3.0 \times 10^{-6}$ C $E = 3.0 \times 10^{5}$ N/C

P

(a)

$Q = +3.0 \times 10^{-6}$ C $E = 3.0 \times 10^{5}$ N/C

P

(b)

EXAMPLE 16–6 Photocopy machine. A photocopy machine works by arranging positive charges (in the pattern to be copied) on the surface of a drum, then gently sprinkling negatively charged dry toner (ink) particles onto the drum. The toner particles temporarily stick to the pattern on the drum (Fig. 16–25) and are later transferred to paper and "melted" to produce the copy. Suppose each toner particle has a mass of 9.0×10^{-16} kg and carries an average of 20 extra electrons to provide an electric charge. Assuming that the electric force on a toner particle must exceed twice its weight in order to ensure sufficient attraction, compute the required electric field strength near the surface of the drum.

APPROACH The electric force on a toner particle of charge $q = 20e$ is $F = qE$, where E is the needed electric field. This force needs to be at least as great as twice the weight (mg) of the particle.

SOLUTION The minimum value of electric field satisfies the relation

$$qE = 2mg$$

where $q = 20e$. Hence

$$E = \frac{2mg}{q} = \frac{2(9.0 \times 10^{-16}\,\text{kg})(9.8\,\text{m/s}^2)}{20(1.6 \times 10^{-19}\,\text{C})} = 5.5 \times 10^{3}\,\text{N/C}.$$

EXAMPLE 16–7 Electric field of a single point charge. Calculate the magnitude and direction of the electric field at a point P which is 30 cm to the right of a point charge $Q = -3.0 \times 10^{-6}$ C.

APPROACH The magnitude of the electric field due to a single point charge is given by Eq. 16–4. The direction is found using the sign of the charge Q.

SOLUTION The magnitude of the electric field is:

$$E = k\frac{Q}{r^2} = \frac{(9.0 \times 10^{9}\,\text{N}\cdot\text{m}^2/\text{C}^2)(3.0 \times 10^{-6}\,\text{C})}{(0.30\,\text{m})^2} = 3.0 \times 10^{5}\,\text{N/C}.$$

The direction of the electric field is *toward* the charge Q, to the left as shown in Fig. 16–26a, since we defined the direction as that of the force on a positive test charge which here would be attractive. If Q had been positive, the electric field would have pointed away, as in Fig. 16–26b.

NOTE There is no electric charge at point P. But there is an electric field there. The only real charge is Q.

This Example illustrates a general result: The electric field $\vec{E}$ due to a positive charge points away from the charge, whereas $\vec{E}$ due to a negative charge points toward that charge.

EXERCISE F Find the magnitude and direction of the electric field due to a $-2.5\,\mu C$ charge 50 cm below it.

If the electric field at a given point in space is due to more than one charge, the individual fields (call them $\vec{E}_1$, $\vec{E}_2$, etc.) due to each charge are added vectorially to get the total field at that point:

$$\vec{E} = \vec{E}_1 + \vec{E}_2 + \cdots.$$

The validity of this **superposition principle** for electric fields is fully confirmed by experiment.

EXAMPLE 16–8 $\vec{E}$ **at a point between two charges.** Two point charges are separated by a distance of 10.0 cm. One has a charge of $-25\,\mu C$ and the other $+50\,\mu C$. (a) Determine the direction and magnitude of the electric field at a point P between the two charges that is 2.0 cm from the negative charge (Fig. 16–27a). (b) If an electron (mass $= 9.11 \times 10^{-31}$ kg) is placed at rest at P and then released, what will be its initial acceleration (direction and magnitude)?

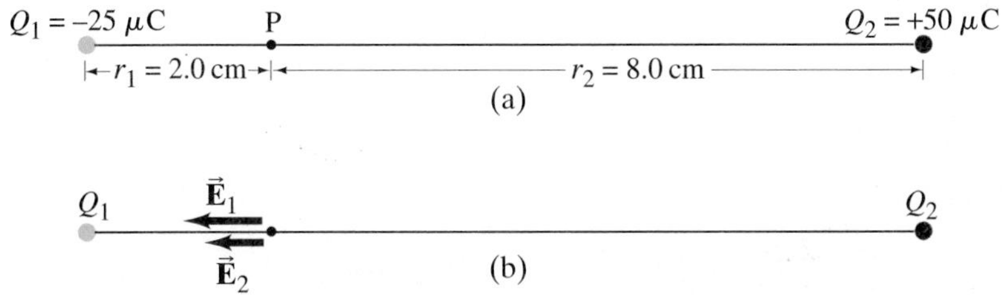

FIGURE 16–27 Example 16–8. In (b), we don't know the relative lengths of $\vec{E}_1$ and $\vec{E}_2$ until we do the calculation.

APPROACH The electric field at P will be the vector sum of the fields created separately by Q_1 and Q_2. The field due to the negative charge Q_1 points toward Q_1, and the field due to the positive charge Q_2 points away from Q_2. Thus both fields point to the left as shown in Fig. 16–27b, and we can add the magnitudes of the two fields together algebraically, ignoring the signs of the charges. In (b) we use Newton's second law ($\Sigma\vec{F} = m\vec{a}$) to find the acceleration, where $\Sigma\vec{F} = q\Sigma\vec{E}$.

SOLUTION (a) Each field is due to a point charge as given by Eq. 16–4, $E = kQ/r^2$. The total field points to the left and has magnitude

$$E = k\frac{Q_1}{r_1^2} + k\frac{Q_2}{r_2^2} = k\left(\frac{Q_1}{r_1^2} + \frac{Q_2}{r_2^2}\right)$$

$$= (9.0 \times 10^9\,\text{N}\cdot\text{m}^2/\text{C}^2)\left(\frac{25 \times 10^{-6}\,\text{C}}{(2.0 \times 10^{-2}\,\text{m})^2} + \frac{50 \times 10^{-6}\,\text{C}}{(8.0 \times 10^{-2}\,\text{m})^2}\right)$$

$$= 6.3 \times 10^8\,\text{N/C}.$$

(b) The electric field points to the left, so the electron will feel a force to the *right* since it is negatively charged. Therefore the acceleration $a = F/m$ (Newton's second law) will be to the right. The force on a charge q in an electric field E is $F = qE$ (Eq. 16–5). Hence the magnitude of the electron's initial acceleration is

$$a = \frac{F}{m} = \frac{qE}{m} = \frac{(1.60 \times 10^{-19}\,\text{C})(6.3 \times 10^8\,\text{N/C})}{9.11 \times 10^{-31}\,\text{kg}} = 1.1 \times 10^{20}\,\text{m/s}^2.$$

NOTE By considering the directions of *each* field ($\vec{E}_1$ and $\vec{E}_2$) before doing any calculations, we made sure our calculation could be done simply and correctly.

EXERCISE G Four charges of equal magnitude, but possibly different sign, are placed on the corners of a square. What arrangement of charges will produce an electric field with the greatest magnitude at the center of the square? (a) All four positive charges; (b) all four negative charges; (c) three positive and one negative; (d) two positive and two negative; (e) three negative and one positive.

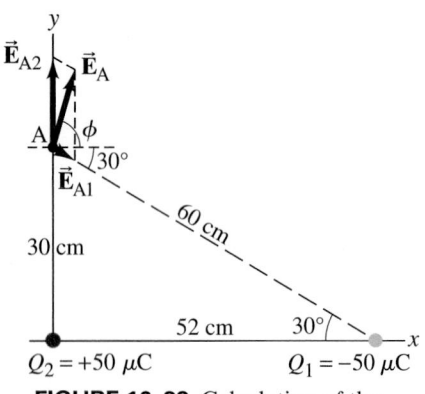

FIGURE 16–28 Calculation of the electric field at point A, Example 16–9.

EXAMPLE 16–9 $\vec{E}$ **above two point charges.** Calculate the total electric field at point A in Fig. 16–28 due to both charges, Q_1 and Q_2.

APPROACH The calculation is much like that of Example 16–4, except now we are dealing with electric fields instead of force. The electric field at point A is the vector sum of the fields $\vec{E}_{A1}$ due to Q_1, and $\vec{E}_{A2}$ due to Q_2. We find the magnitude of the field produced by each point charge, then we add their components to find the total field at point A.

SOLUTION The magnitude of the electric field produced at point A by each of the charges Q_1 and Q_2 is given by $E = kQ/r^2$, so

$$E_{A1} = \frac{(9.0 \times 10^9 \, \text{N·m}^2/\text{C}^2)(50 \times 10^{-6} \, \text{C})}{(0.60 \, \text{m})^2} = 1.25 \times 10^6 \, \text{N/C},$$

$$E_{A2} = \frac{(9.0 \times 10^9 \, \text{N·m}^2/\text{C}^2)(50 \times 10^{-6} \, \text{C})}{(0.30 \, \text{m})^2} = 5.0 \times 10^6 \, \text{N/C}.$$

The direction of E_{A1} points from A toward Q_1 (negative charge), whereas E_{A2} points from A away from Q_2, as shown; so the total electric field at A, $\vec{E}_A$, has components

$$E_{Ax} = E_{A1} \cos 30° = 1.1 \times 10^6 \, \text{N/C},$$
$$E_{Ay} = E_{A2} - E_{A1} \sin 30° = 4.4 \times 10^6 \, \text{N/C}.$$

Thus the magnitude of $\vec{E}_A$ is

$$E_A = \sqrt{(1.1)^2 + (4.4)^2} \times 10^6 \, \text{N/C} = 4.5 \times 10^6 \, \text{N/C},$$

and its direction is ϕ (Fig. 16–28) given by $\tan \phi = E_{Ay}/E_{Ax} = 4.4/1.1 = 4.0$, so $\phi = 76°$.

It is worthwhile summarizing here what we have learned about solving electrostatics problems.

PROBLEM SOLVING

Electrostatics: Electric Forces and Electric Fields

Whether you use electric field or electrostatic forces, the procedure for solving electrostatics problems is similar:

1. **Draw** a careful **diagram**—namely, a free-body diagram for each object, showing all the forces acting on that object, or showing the electric field at a point due to all significant charges present. Determine the **direction** of each force or electric field physically: like charges repel each other, unlike charges attract; fields point away from a + charge, and toward a − charge. Show and label each vector force or field on your diagram.

2. **Apply Coulomb's law** to calculate the magnitude of the force that each contributing charge exerts on a charged object, or the magnitude of the electric field each charge produces at a given point. Deal only with magnitudes of charges (leaving out minus signs), and obtain the magnitude of each force or electric field.

3. **Add vectorially** all the forces on an object, or the contributing fields at a point, to get the resultant. Use **symmetry** (say, in the geometry) whenever possible.

EXAMPLE 16–10 $\vec{E}$ **equidistant above two point charges.** Figure 16–29 (top of next page) is the same as Fig. 16–28 but includes point B, which is equidistant (40 cm) from Q_1 and Q_2. Calculate the total electric field at point B in Fig. 16–29 due to both charges, Q_1 and Q_2.

APPROACH We explicitly follow the steps of the Problem Solving Strategy above.

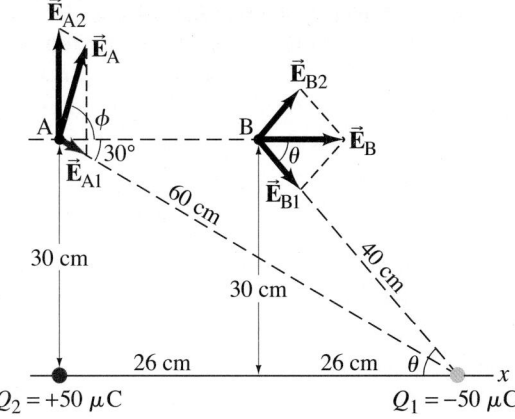

FIGURE 16–29 Same as Fig. 16–28 but with point B added. Calculation of the electric field at points A and B for Examples 16–9 and 16–10.

SOLUTION

1. **Draw** a careful **diagram**. The **directions** of the electric fields $\vec{\mathbf{E}}_{B1}$ and $\vec{\mathbf{E}}_{B2}$, as well as the net field $\vec{\mathbf{E}}_B$, are shown in Fig. 16–29. $\vec{\mathbf{E}}_{B2}$ points away from the positive charge Q_2; $\vec{\mathbf{E}}_{B1}$ points toward the negative charge Q_1.

2. **Apply Coulomb's law** to find the magnitudes of the contributing electric fields. Because B is equidistant from the two equal charges (40 cm by the Pythagorean theorem), the magnitudes of E_{B1} and E_{B2} are the same; that is,

$$E_{B1} = E_{B2} = \frac{kQ}{r^2} = \frac{(9.0 \times 10^9 \,\text{N} \cdot \text{m}^2/\text{C}^2)(50 \times 10^{-6} \,\text{C})}{(0.40 \,\text{m})^2} = 2.8 \times 10^6 \,\text{N/C}.$$

3. **Add vectorially**, and use **symmetry** when possible. The y components of $\vec{\mathbf{E}}_{B1}$ and $\vec{\mathbf{E}}_{B2}$ are equal and opposite. Because of this symmetry, the total field E_B is horizontal and equals $E_{B1}\cos\theta + E_{B2}\cos\theta = 2\,E_{B1}\cos\theta$. From Fig. 16–29, $\cos\theta = 26\,\text{cm}/40\,\text{cm} = 0.65$. Then

$$E_B = 2E_{B1}\cos\theta = 2(2.8 \times 10^6 \,\text{N/C})(0.65) = 3.6 \times 10^6 \,\text{N/C},$$

and the direction of $\vec{\mathbf{E}}_B$ is along the $+x$ direction.

16–8 Electric Field Lines

Since the electric field is a vector, it is sometimes referred to as a *vector field*. We could indicate the electric field with arrows at various points in a given situation, such as at A, B, and C in Fig. 16–30. The directions of $\vec{\mathbf{E}}_A$, $\vec{\mathbf{E}}_B$, and $\vec{\mathbf{E}}_C$ are the same as for the forces shown earlier in Fig. 16–23, but the magnitudes (arrow lengths) are different since we divide $\vec{\mathbf{F}}$ by q to get $\vec{\mathbf{E}}$. However, the relative lengths of $\vec{\mathbf{E}}_A$, $\vec{\mathbf{E}}_B$, and $\vec{\mathbf{E}}_C$ are the same as for the forces since we divide by the same q each time. To indicate the electric field in such a way at *many* points, however, would result in many arrows, which would quickly become cluttered and confusing. To avoid this, we use another technique, that of field lines.

To visualize the electric field, we draw a series of lines to indicate the direction of the electric field at various points in space. These **electric field lines** (or **lines of force**) are drawn to indicate the direction of the force due to the given field on a positive test charge. The lines of force due to a single isolated positive charge are shown in Fig. 16–31a, and for a single isolated negative charge in Fig. 16–31b. In part (a) the lines point radially outward from the charge, and in part (b) they point radially inward toward the charge because that is the direction the force would be on a positive test charge in each case (as in Fig. 16–26). Only a few representative lines are shown. We could draw lines in between those shown since the electric field exists there as well. We can draw the lines so that the *number of lines starting on a positive charge, or ending on a negative charge, is proportional to the magnitude of the charge.* Notice that nearer the charge, where the electric field is greater $(F \propto 1/r^2)$, the lines are closer together. This is a general property of electric field lines: *the closer together the lines are, the stronger the electric field in that region.* In fact, field lines can be drawn so that the number of lines crossing unit area perpendicular to $\vec{\mathbf{E}}$ is proportional to the magnitude of the electric field.

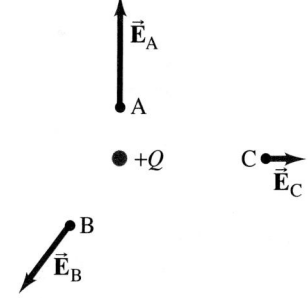

FIGURE 16–30 Electric field vector, shown at three points, due to a single point charge Q. (Compare to Fig. 16–23.)

FIGURE 16–31 Electric field lines (a) near a single positive point charge, (b) near a single negative point charge.

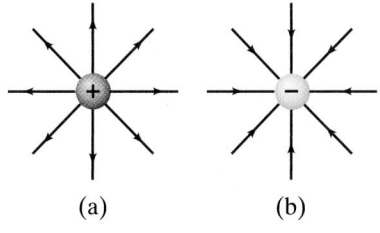

(a)　　　　(b)

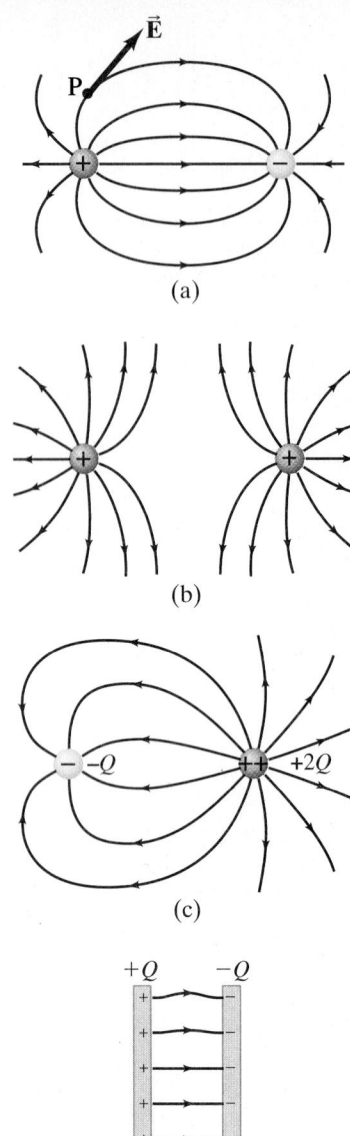

(a)

(b)

(c)

+Q   −Q

(d)

FIGURE 16–32 Electric field lines for four arrangements of charges.

FIGURE 16–33 The Earth's gravitational field, which at any point is directed toward the Earth's center (the force on any mass points toward the Earth's center).

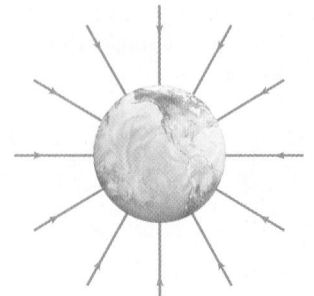

Figure 16–32a shows the electric field lines due to two equal charges of opposite sign, a combination known as an **electric dipole**. The electric field lines are curved in this case and are directed from the positive charge to the negative charge. The direction of the electric field at any point is tangent to the field line at that point as shown by the vector arrow $\vec{E}$ at point P. To satisfy yourself that this is the correct pattern for the electric field lines, you can make a few calculations such as those done in Examples 16–9 and 16–10 for just this case (see Fig. 16–29). Figure 16–32b shows the electric field lines for two equal positive charges, and Fig. 16–32c for unequal charges, $-Q$ and $+2Q$. Note that twice as many lines leave $+2Q$ as enter $-Q$ (number of lines is proportional to magnitude of Q). Finally, in Fig. 16–32d, we see in cross section the field lines between two flat parallel plates carrying equal but opposite charges. Notice that the electric field lines between the two plates start out perpendicular to the surface of the metal plates (we will see why this is true in the next Section) and go directly from one plate to the other, as we expect because a positive test charge placed between the plates would feel a strong repulsion from the positive plate and a strong attraction to the negative plate. The field lines between two close plates are parallel and equally spaced in the central region, but fringe outward near the edges. Thus, in the central region, the electric field has the same magnitude at all points, and we can write

$$E = \text{constant.} \quad \left[\begin{array}{l}\text{between two closely spaced, oppositely} \\ \text{charged, flat parallel plates}\end{array}\right] \quad \textbf{(16–6)}$$

The fringing of the field near the edges can often be ignored, particularly if the separation of the plates is small compared to their height and width.[†] We summarize the properties of field lines as follows:

1. Electric field lines indicate the direction of the electric field; the field points in the direction tangent to the field line at any point.
2. The lines are drawn so that the magnitude of the electric field, E, is proportional to the number of lines crossing unit area perpendicular to the lines. The closer together the lines, the stronger the field.
3. Electric field lines start on positive charges and end on negative charges; and the number starting or ending is proportional to the magnitude of the charge.

Also note that field lines never cross. Why not? Because it would not make sense for the electric field to have two directions at the same point.

Gravitational Field

The field concept can also be applied to the gravitational force (Chapter 5). Thus we can say that a **gravitational field** exists for every object that has mass. One object attracts another by means of the gravitational field. The Earth, for example, can be said to possess a gravitational field (Fig. 16–33) which is responsible for the gravitational force on objects. The *gravitational field* is defined as the *force per unit mass*. The magnitude of the Earth's gravitational field at any point above the Earth's surface is thus GM_E/r^2, where M_E is the mass of the Earth, r is the distance of the point from the Earth's center, and G is the gravitational constant (Chapter 5). At the Earth's surface, r is the radius of the Earth and the gravitational field is equal to g, the acceleration due to gravity. Beyond the Earth, the gravitational field can be calculated at any point as a sum of terms due to Earth, Sun, Moon, and other bodies that contribute significantly.

[†]The magnitude of the constant electric field between two parallel plates is given by $E = Q/\epsilon_0 A$, where Q is the magnitude of the charge on each plate and A is the area of one plate. We show this in the optional Section 16–12 on Gauss's law.

16–9 Electric Fields and Conductors

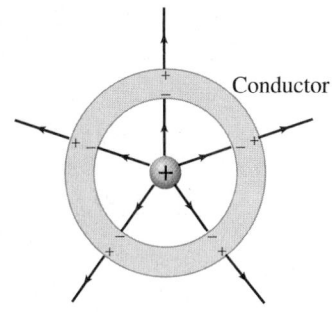

FIGURE 16–34 A charge inside a neutral spherical metal shell induces charge on its surfaces. The electric field exists even beyond the shell, but not within the conductor itself.

We now discuss some properties of conductors. First, *the electric field inside a conductor is zero in the static situation*—that is, when the charges are at rest. If there were an electric field within a conductor, there would be a force on the free electrons. The electrons would move until they reached positions where the electric field, and therefore the electric force on them, did become zero.

This reasoning has some interesting consequences. For one, *any net charge on a conductor distributes itself on the surface*. For a negatively charged conductor, you can imagine that the negative charges repel one another and race to the surface to get as far from one another as possible. Another consequence is the following. Suppose that a positive charge Q is surrounded by an isolated uncharged metal conductor whose shape is a spherical shell, Fig. 16–34. Because there can be no field within the metal, the lines leaving the central positive charge must end on negative charges on the inner surface of the metal. That is, the encircled charge $+Q$ induces an equal amount of negative charge, $-Q$, on the inner surface of the spherical shell. Since the shell is neutral, a positive charge of the same magnitude, $+Q$, must exist on the outer surface of the shell. Thus, although no field exists in the metal itself, an electric field exists outside of it, as shown in Fig. 16–34, as if the metal were not even there.

A related property of static electric fields and conductors is that *the electric field is always perpendicular to the surface outside of a conductor*. If there were a component of $\vec{E}$ parallel to the surface (Fig. 16–35), it would exert a force on free electrons at the surface, causing the electrons to move along the surface until they reached positions where no net force was exerted on them parallel to the surface—that is, until the electric field was perpendicular to the surface.

These properties apply only to conductors. Inside a nonconductor, which does not have free electrons, a static electric field can exist as we will see in the next Chapter. Also, the electric field outside a nonconductor does not necessarily make an angle of 90° to the surface.

FIGURE 16–35 If the electric field $\vec{E}$ at the surface of a conductor had a component parallel to the surface, $\vec{E}_\parallel$, the latter would accelerate electrons into motion. In the static case, $\vec{E}_\parallel$ must be zero, and the electric field must be perpendicular to the conductor's surface: $\vec{E} = \vec{E}_\perp$.

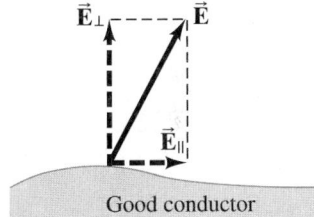

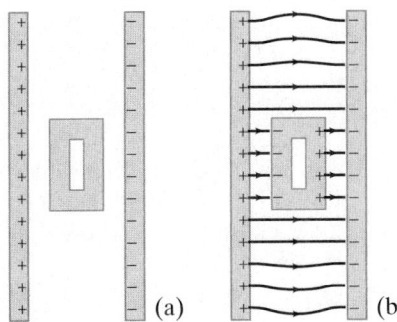

FIGURE 16–36 Example 16–11.

CONCEPTUAL EXAMPLE 16–11 **Shielding, and safety in a storm.** A neutral hollow metal box is placed between two parallel charged plates as shown in Fig. 16–36a. What is the field like inside the box?

RESPONSE If our metal box had been solid, and not hollow, free electrons in the box would have redistributed themselves along the surface until all their individual fields would have canceled each other inside the box. The net field inside the box would have been zero. For a hollow box, the external field is not changed since the electrons in the metal can move just as freely as before to the surface. Hence the field inside the hollow metal box is also zero, and the field lines are shown in Fig. 16–36b. A conducting box is an effective device for shielding delicate instruments and electronic circuits from unwanted external electric fields. We also can see that a relatively safe place to be during a lightning storm is inside a parked car, surrounded by metal. See also Fig. 16–37, where a person inside a porous "cage" is protected from a strong electric discharge. (It is not safe in a lightning storm to be near a tree which can conduct, or out in the open where you are taller than the surroundings.)

FIGURE 16–37 High-voltage "Van de Graaff" generators create strong electric fields in the vicinity of the "Faraday cage" below. The strong field accelerates stray electrons in the atmosphere to the KE needed to knock electrons out of air atoms, causing an avalanche of charge which flows to (or from) the metal cage. The metal cage protects the person inside it.

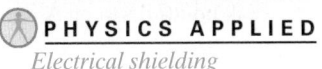

PHYSICS APPLIED
Electrical shielding

*16–10 Electric Forces in Molecular Biology: DNA Structure and Replication

PHYSICS APPLIED

Inside a cell: kinetic theory plus electrostatic force

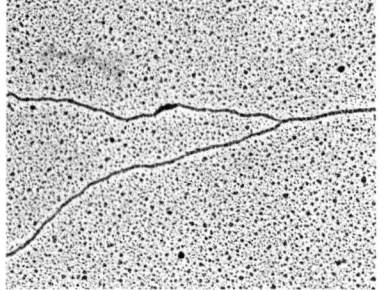

FIGURE 16–38 Image of DNA replicating, made by a transmission electron microscope.

PHYSICS APPLIED

DNA structure

The study of the structure and functioning of a living cell at the molecular level is known as molecular biology. It is an important area for application of physics. The interior of every biological cell is mainly water. We can imagine a cell as a thick soup of molecules continually in motion (kinetic theory, Chapter 13), colliding with one another with various amounts of kinetic energy. These molecules interact with one another because of the *electrostatic force* between molecules.

Indeed, cellular processes are now considered to be the result of *random ("thermal") molecular motion plus the ordering effect of the electrostatic force.* As an example, we look at DNA structure and replication. The picture we present is a model of what happens based on physical theories and experiment.

The genetic information that is passed on from generation to generation in all living cells is contained in the chromosomes, which are made up of genes. Each gene contains the information needed to produce a particular type of protein molecule, and that information is built into the principal molecule of a chromosome, DNA (deoxyribonucleic acid), Fig. 16–38. DNA molecules are made up of many small molecules known as nucleotide bases which are each *polar* (Section 16–2) due to unequal sharing of electrons. There are four types of nucleotide bases in DNA: adenine (A), cytosine (C), guanine (G), and thymine (T).

The DNA of a chromosome generally consists of two long DNA strands wrapped about one another in the shape of a "double helix." The genetic information is contained in the specific order of the four bases (A, C, G, T) along each strand. As shown in Fig. 16–39, the two strands are attracted by electrostatic forces—that is, by the attraction of positive charges to negative charges that exist on parts of the molecules. We see in Fig. 16–39a that an A (adenine) on one strand is always opposite a T on the other strand; similarly, a G is always opposite a C. This important ordering effect occurs because the shapes of A, T, C, and G are such that a T fits closely only into an A, and a G into a C. Only in the case of this close proximity of the charged portions is the electrostatic force great enough to hold them together even for a short time (Fig. 16–39b), forming what are referred to as "weak bonds."

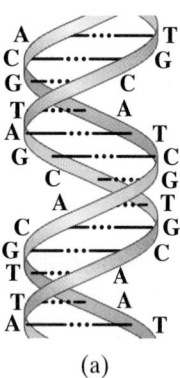

(a)

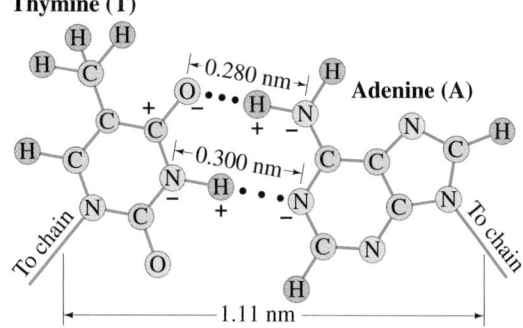

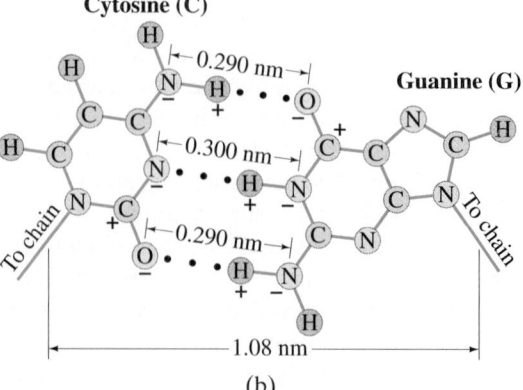

(b)

FIGURE 16–39 (a) Schematic diagram of a section of DNA double helix. (b) "Close-up" view of the helix, showing how A and T attract each other and how G and C attract each other through electrostatic forces. The + and − signs indicated on certain atoms represent net charges, usually a fraction of *e*, due to uneven sharing of electrons. The red dots indicate the electrostatic attraction (often called a "weak bond" or "hydrogen bond"—Section 29–3). Note that there are two weak bonds between A and T, and three between C and G.

The electrostatic force between A and T, and between C and G, exists because these molecules have charged parts. These charges are due to some electrons in each of these molecules spending more time orbiting one atom than another. For example, the electron normally on the H atom of adenine (upper part of Fig. 16–39b) spends some of its time orbiting the adjacent N atom (more on this in Chapter 29), so the N has a net negative charge and the H is left with a net positive charge. This H^+ atom of adenine is then attracted to the O^- atom of thymine. These net + and − charges usually have magnitudes of a fraction of e (charge on the electron) such as $0.2e$ or $0.4e$. (This is what we mean by "polar" molecules.)

[When H^+ is involved, the weak bond it can make with a nearby negative charge, such as O^-, is relatively strong (partly because H^+ is so small) and is referred to as a **hydrogen bond** (Section 29–3).]

When the DNA replicates (duplicates) itself just before cell division, the arrangement of A opposite T and G opposite C is crucial for ensuring that the genetic information is passed on accurately to the next generation, Fig. 16–40. The two strands of DNA separate (with the help of enzymes, which also operate via the electrostatic force), leaving the charged parts of the bases exposed.

PHYSICS APPLIED
DNA replication

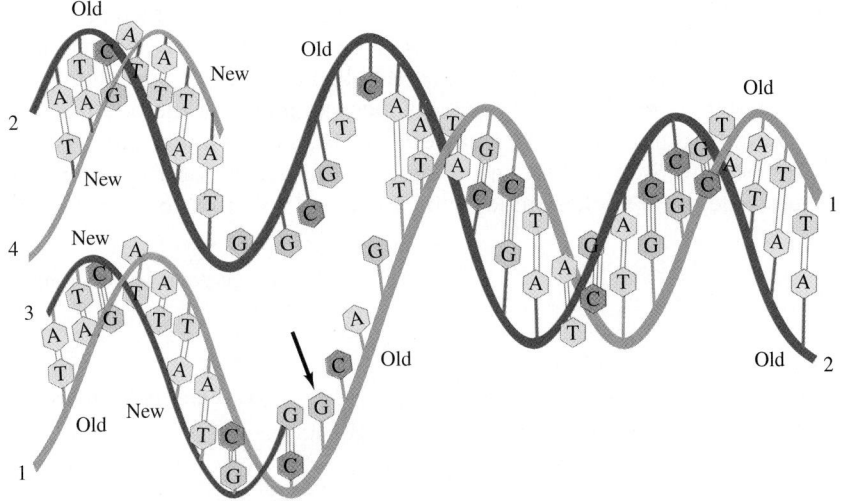

FIGURE 16–40 Replication of DNA.

Once replication starts, let us see how the correct order of bases occurs by looking at the G molecule indicated by the red arrow in Fig. 16–40. Many unattached nucleotide bases of all four kinds are bouncing around in the cellular fluid, and the only type that will experience attraction to our G, if it comes close to it, will be a C. The charges on the other three bases can not get close enough to those on the G to provide a significant attractive force—remember that the electrostatic (Coulomb) force decreases rapidly with distance $(\propto 1/r^2)$. Because the G does not attract an A, T, or G appreciably, an A, T, or G will be knocked away by collisions with other molecules before enzymes can attach it to the growing chain (number 3 in Fig. 16–40). But the electrostatic force will often hold a C opposite our G long enough so that an enzyme can attach the C to the growing end of the new chain. Thus electrostatic forces are responsible for selecting the bases in the proper order during replication. Note in Fig. 16–40 that the new number 4 strand has the same order of bases as the old number 1 strand; and the new number 3 strand is the same as the old number 2. So the two new double helixes, 1–3 and 2–4, are identical to the original 1–2 helix. Hence the genetic information is passed on accurately to the next generation.

This process of DNA replication is often presented as if it occurred in clockwork fashion—as if each molecule knew its role and went to its assigned place. But this is not the case. The forces of attraction are rather weak and become significant only when charged parts of the two molecules have "complementary shapes," meaning they can get close enough so that the electrostatic force $(\propto 1/r^2)$ is strong enough to form weak bonds. If the molecular shapes are not just right, there is almost no electrostatic attraction, which is why there are so few mistakes. Thus, out of the random motion of the molecules, the electrostatic force acts to bring order out of chaos.

The random (thermal) velocities of molecules in a cell affect *cloning*. When a bacterial cell divides, the two new bacteria have nearly identical DNA. Even if the DNA were perfectly identical, the two new bacteria would not end up behaving in exactly the same way. Long protein, DNA, and RNA molecules get bumped into different shapes, and even the expression[†] of genes can thus be different. Loosely held parts of large molecules such as a methyl group (CH_3) can also be knocked off by a strong collision with another molecule. Hence, cloned organisms are not identical, even if their DNA were identical. Indeed, there can not really be genetic determinism.

*16–11 Photocopy Machines and Computer Printers Use Electrostatics

Photocopy machines and laser printers use electrostatic attraction to print an image. They each use a different technique to project an image onto a special cylindrical drum (or rotating conveyor belt). The drum is typically made of aluminum, a good conductor; its surface is coated with a thin layer of selenium, which has the interesting property (called "photoconductivity") of being an electrical nonconductor in the dark, but a conductor when exposed to light.

PHYSICS APPLIED
Photocopy machines

In a **photocopier**, lenses and mirrors focus an image of the original sheet of paper onto the drum, much like a camera lens focuses an image on an electronic detector or film. Step 1, done in the dark, is the placing of a uniform positive charge on the drum's selenium layer by a charged roller or rod: see Fig. 16–41.

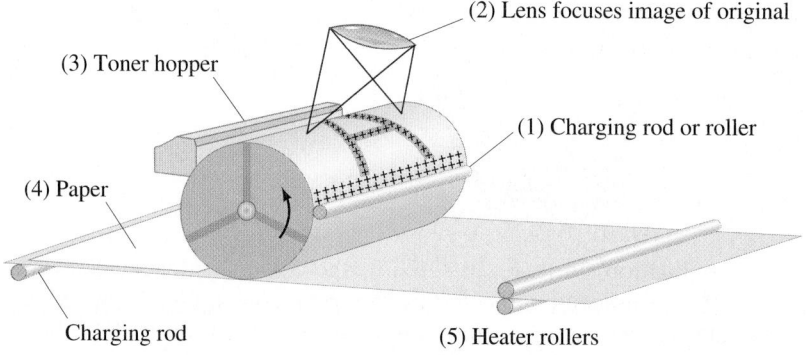
(3) Toner hopper
(2) Lens focuses image of original
(1) Charging rod or roller
(4) Paper
Charging rod
(5) Heater rollers

FIGURE 16–41 Inside a photocopy machine: (1) the selenium drum is given a + charge; (2) the lens focuses image on drum—only dark spots stay charged; (3) toner particles (negatively charged) are attracted to positive areas on drum; (4) the image is transferred to paper; (5) heat binds the image to the paper.

In step 2, the image to be copied is projected onto the drum. For simplicity, let us assume the image is a dark letter A on a white background (as on the page of a book) as shown in Fig. 16–41. The letter A on the drum is dark, but all around it is light. At all these light places, the selenium becomes conducting and electrons flow in from the aluminum beneath, neutralizing those positive areas. In the dark areas of the letter A, the selenium is nonconducting and so retains the positive charge already put on it, Fig. 16–41. In step 3, a fine dark powder known as *toner* is given a negative charge, and is brushed on the drum as it rotates. The negatively charged toner particles are attracted to the positive areas on the drum (the A in our case) and stick only there. In step 4, the rotating drum presses against a piece of paper which has been positively charged more strongly than the selenium, so the toner particles are transferred to the paper, forming the final image. Finally, step 5, the paper is heated to fix the toner particles firmly on the paper.

In a color copier (or printer), this process is repeated for each color—black, cyan (blue), magenta (red), and yellow. Combining these four colors in different proportions produces any desired color.

[†]The separate genes of a DNA double helix can be covered by protein molecules, keeping those genes from being "expressed"—that is, translated into the proteins they code for (see Section 29–3).

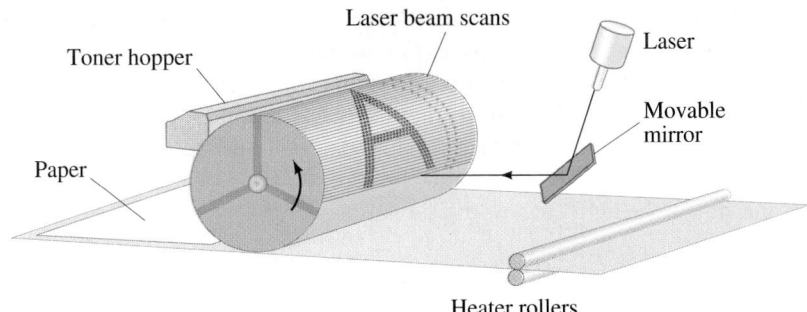

FIGURE 16–42 Inside a laser printer: a movable mirror sweeps the laser beam in horizontal lines across the drum.

A **laser printer** uses a computer output to program the intensity of a laser beam onto the selenium-coated drum of Fig. 16–42. The thin beam of light from the laser is scanned (by a movable mirror) from side to side across the drum in a series of horizontal lines, each line just below the previous line. As the beam sweeps across the drum, the intensity of the beam is varied by the computer output, being strong for a point that is meant to be white or bright, and weak or zero for points that are meant to come out dark. After each sweep, the drum rotates very slightly for additional sweeps, Fig. 16–42, until a complete image is formed on it. The light parts of the selenium become conducting and lose their (previously given) positive electric charge, and the toner sticks only to the dark, electrically charged areas. The drum then transfers the image to paper, as in a photocopier.

An **inkjet printer** does not use a drum. Instead nozzles spray tiny droplets of ink directly at the paper. The nozzles are swept across the paper, each sweep just above the previous one as the paper moves down. On each sweep, the ink makes dots on the paper, except for those points where no ink is desired, as directed by the computer. The image consists of a huge number of very tiny dots. The quality or resolution of a printer is usually specified in dots per inch (dpi) in each (linear) direction.

PHYSICS APPLIED
Laser printer

PHYSICS APPLIED
Inkjet printer

*16–12 Gauss's Law

An important relation in electricity is Gauss's law, developed by the great mathematician Karl Friedrich Gauss (1777–1855). It relates electric charge and electric field, and is a more general and elegant form of Coulomb's law.

Gauss's law involves the concept of **electric flux**, which refers to the electric field passing through a given area. For a uniform electric field $\vec{E}$ passing through an area A, as shown in Fig. 16–43a, the electric flux Φ_E is defined as

$$\Phi_E = EA\cos\theta,$$

where θ is the angle between the electric field direction and a line drawn perpendicular to the area. The flux can be written equivalently as

$$\Phi_E = E_\perp A = EA_\perp = EA\cos\theta, \qquad (16\text{–}7)$$

where $E_\perp = E\cos\theta$ is the component of $\vec{E}$ perpendicular to the area (Fig. 16–43b) and, similarly, $A_\perp = A\cos\theta$ is the projection of the area A perpendicular to the field $\vec{E}$ (Fig. 16–43c).

Electric flux can be interpreted in terms of field lines. We mentioned in Section 16–8 that field lines can always be drawn so that the number (N) passing through unit area perpendicular to the field $\left(A_\perp\right)$ is proportional to the magnitude of the field (E): that is, $E \propto N/A_\perp$. Hence,

$$N \propto EA_\perp = \Phi_E, \qquad (16\text{–}8)$$

so the flux through an area is proportional to the number of lines passing through that area.

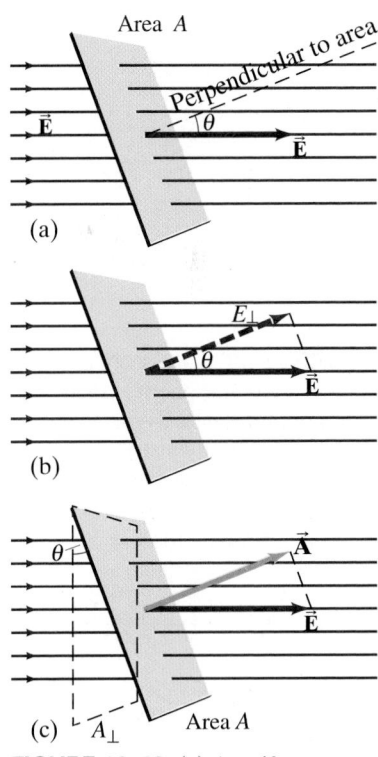

FIGURE 16–43 (a) A uniform electric field $\vec{E}$ passing through a flat square area A. (b) $E_\perp = E\cos\theta$ is the component of $\vec{E}$ perpendicular to the plane of area A. (c) $A_\perp = A\cos\theta$ is the projection (dashed) of the area A perpendicular to the field $\vec{E}$.

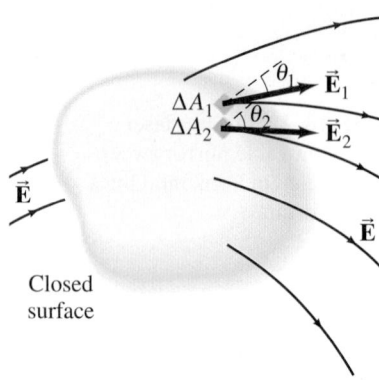

FIGURE 16–44 Electric field lines passing through a closed surface. The surface is divided up into many tiny areas, ΔA_1, ΔA_2, $\cdots$, and so on, of which only two are shown.

Gauss's law involves the *total* flux through a closed surface—a surface of any shape that encloses a volume of space. For any such surface, such as that shown in Fig. 16–44, we divide the surface up into many tiny areas, ΔA_1, ΔA_2, ΔA_3, $\cdots$, and so on. We make the division so that each ΔA is small enough that it can be considered flat and so that the electric field can be considered constant through each ΔA. Then the *total* flux through the entire surface is the sum over all the individual fluxes through each of the tiny areas:

$$\Phi_E = E_1 \, \Delta A_1 \cos \theta_1 + E_2 \, \Delta A_2 \cos \theta_2 + \cdots$$
$$= \sum E \, \Delta A \cos \theta = \sum E_\perp \, \Delta A,$$

where the symbol Σ means "sum of." We saw in Section 16–8 that the number of field lines starting on a positive charge or ending on a negative charge is proportional to the magnitude of the charge. Hence, the *net* number of lines N pointing out of any closed surface (number of lines pointing out minus the number pointing in) must be proportional to the net charge enclosed by the surface, Q_{encl}. But from Eq. 16–8, we have that the net number of lines N is proportional to the total flux Φ_E. Therefore,

$$\Phi_E = \sum_{\substack{closed \\ surface}} E_\perp \, \Delta A \propto Q_{encl}.$$

The constant of proportionality, to be consistent with Coulomb's law, is $1/\epsilon_0$, so we have

| GAUSS'S LAW |

$$\sum_{\substack{closed \\ surface}} E_\perp \, \Delta A = \frac{Q_{encl}}{\epsilon_0}, \tag{16–9}$$

where the sum (Σ) is over any closed surface, and Q_{encl} is the net charge enclosed within that surface. This is **Gauss's law**.

Coulomb's law and Gauss's law can be used to determine the electric field due to a given (static) charge distribution. Gauss's law is useful when the charge distribution is simple and symmetrical. However, we must choose the closed "gaussian" surface very carefully so we can determine $\vec{\mathbf{E}}$. We normally choose a surface that has just the **symmetry** needed so that E will be constant on all or on parts of its surface.

FIGURE 16–45 Cross-sectional drawing of a thin spherical shell (gray) of radius r_0, carrying a net charge Q uniformly distributed. The green circles A_1 and A_2 represent two gaussian surfaces we use to determine $\vec{\mathbf{E}}$. Example 16–12.

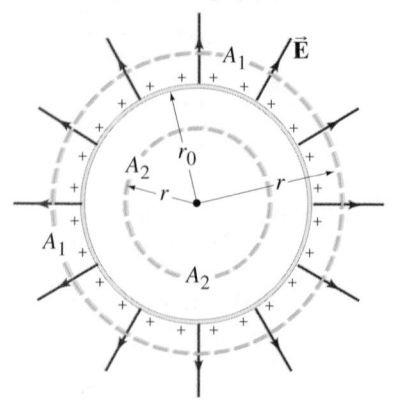

EXAMPLE 16–12 **Charged spherical conducting shell.** A thin spherical shell of radius r_0 possesses a total net charge Q that is uniformly distributed on it, Fig. 16–45. Determine the electric field at points (*a*) outside the shell, and (*b*) inside the shell.

APPROACH Because the charge is distributed symmetrically, the electric field must be *symmetric*. Thus the field outside the shell must be directed radially outward (inward if $Q < 0$) and must depend only on r.

SOLUTION (*a*) We choose any imaginary gaussian surface as a sphere of radius r ($r > r_0$) concentric with the shell, shown in Fig. 16–45 as the dashed circle A_1. Then, by symmetry, the electric field will have the same magnitude at all points on this gaussian surface. Because $\vec{\mathbf{E}}$ is perpendicular to this surface, Gauss's law gives (with $Q_{encl} = Q$ in Eq. 16–9)

$$\sum E_\perp \, \Delta A = E \sum \Delta A = E(4\pi r^2) = \frac{Q}{\epsilon_0},$$

where $4\pi r^2$ is the surface area of our sphere (gaussian surface) of radius r. Thus

$$E = \frac{1}{4\pi\epsilon_0} \frac{Q}{r^2}. \qquad [r > r_0]$$

We see that the field outside a uniformly charged spherical shell is the same as if all the charge were concentrated at the center as a point charge.

(*b*) Inside the shell, the electric field must also be symmetric. So E must again have the same value at all points on a spherical gaussian surface (A_2 in Fig. 16–45) concentric with the shell. Thus, E can be factored out of the sum and, with $Q_{encl} = 0$ because the charge inside surface A_2 is zero, we have

$$\sum E_\perp \, \Delta A \ = \ E \sum \Delta A$$

$$= \ E(4\pi r^2) \ = \ \frac{Q_{encl}}{\epsilon_0} \ = \ 0.$$

Hence

$$E \ = \ 0 \qquad\qquad\qquad\qquad [r < r_0]$$

inside a uniform spherical shell of charge (as claimed in Section 16–9).

The results of Example 16–12 also apply to a uniform *solid* spherical conductor that is charged, since all the charge would lie in a thin layer at the surface (Section 16–9). In particular

$$E \ = \ \frac{1}{4\pi\epsilon_0}\frac{Q}{r^2}$$

outside a spherical conductor. Thus, the electric field outside a spherically symmetric distribution of charge is the same as for a point charge of the same magnitude at the center of the sphere. This result applies also outside a uniformly charged nonconductor, because we can use the same gaussian surface A_1 (Fig. 16–45) and the same *symmetry* argument. We can also consider this a demonstration of our statement in Chapter 5 about the **gravitational force**, which is also a perfect $1/r^2$ force: The gravitational force exerted by a uniform sphere is the same as if all the mass were at the center, as stated on page 120.

| **EXAMPLE 16–13** | E **near any conducting surface.** Show that the magnitude of the electric field just outside the surface of a good conductor of any shape is given by

$$E \ = \ \frac{\sigma}{\epsilon_0},$$

where σ is defined as the surface charge density, Q/A, on the conductor's surface at that point.

APPROACH We choose as our gaussian surface a small cylindrical box, very small in height so that one of its circular ends is just above the conductor (Fig. 16–46). The other end is just below the conductor's surface, and the very short sides are perpendicular to it.

SOLUTION The electric field is zero inside a conductor and is perpendicular to the surface just outside it (Section 16–9), so electric flux passes only through the outside end of our cylindrical box; no flux passes through the very short sides or through the inside end of our gaussian box. We choose the area A (of the flat cylinder end above the conductor surface) small enough so that E is essentially uniform over it. Then Gauss's law gives

$$\sum E_\perp \, \Delta A \ = \ EA \ = \ \frac{Q_{encl}}{\epsilon_0} \ = \ \frac{\sigma A}{\epsilon_0},$$

and therefore

$$E \ = \ \frac{\sigma}{\epsilon_0}. \qquad\qquad [\text{at surface of conductor}]$$

This useful result applies for any shape conductor, including a large, uniformly charged flat sheet: the electric field will be constant and equal to σ/ϵ_0.

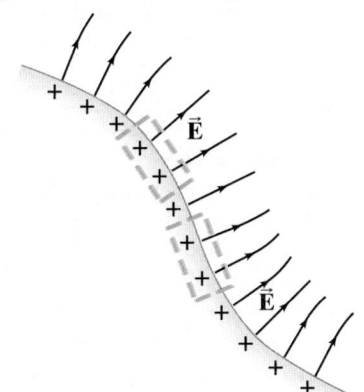

FIGURE 16–46 Electric field near the surface of a conductor. Two small cylindrical boxes are shown dashed. Either one can serve as our gaussian surface. Example 16–13.

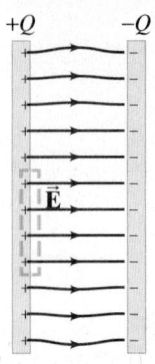

This last Example also gives us the field between the two parallel plates we discussed in Fig. 16–32d. If the plates are large compared to their separation, then the field lines are perpendicular to the plates and, except near the edges, they are parallel to each other. Therefore the electric field (see Fig. 16–47, which shows a similar very thin gaussian surface as Fig. 16–46) is also

$$E = \frac{\sigma}{\epsilon_0} = \frac{Q/A}{\epsilon_0} = \frac{Q}{\epsilon_0 A}, \quad \left[\begin{array}{l} \text{between two closely spaced,} \\ \text{oppositely charged, parallel plates} \end{array} \right] \quad \textbf{(16–10)}$$

where $Q = \sigma A$ is the charge on one of the plates.

FIGURE 16–47 The electric field between two closely spaced parallel plates is uniform and equal to $E = \sigma/\epsilon_0$.

Summary

There are two kinds of **electric charge**, positive and negative. These designations are to be taken algebraically—that is, any charge is plus or minus so many coulombs (C), in SI units.

Electric charge is **conserved**: if a certain amount of one type of charge is produced in a process, an equal amount of the opposite type is also produced; thus the *net* charge produced is zero.

According to atomic theory, electricity originates in the atom, which consists of a positively charged nucleus surrounded by negatively charged electrons. Each electron has a charge $-e = -1.60 \times 10^{-19}$ C.

Electric **conductors** are those materials in which many electrons are relatively free to move, whereas electric **insulators** or **nonconductors** are those in which very few electrons are free to move.

An object is negatively charged when it has an excess of electrons, and positively charged when it has less than its normal number of electrons. The net charge on any object is a whole number times $+e$ or $-e$. That is, charge is **quantized**.

An object can become charged by rubbing (in which electrons are transferred from one material to another), **by conduction** (which is transfer of charge from one charged object to another by touching), or **by induction** (the separation of charge within an object because of the close approach of another charged object but without touching).

Electric charges exert a force on each other. If two charges are of opposite types, one positive and one negative, they each exert an attractive force on the other. If the two charges are the same type, each repels the other.

The magnitude of the force one point charge exerts on another is proportional to the product of their charges, and inversely proportional to the square of the distance between them:

$$F = k\frac{Q_1 Q_2}{r^2} = \frac{1}{4\pi\epsilon_0}\frac{Q_1 Q_2}{r^2}; \quad \textbf{(16–1, 16–2)}$$

this is **Coulomb's law**.

We think of an **electric field** as existing in space around any charge or group of charges. The force on another charged object is then said to be due to the electric field present at its location.

The *electric field*, $\vec{\mathbf{E}}$, at any point in space due to one or more charges, is defined as the force per unit charge that would act on a tiny positive test charge q placed at that point:

$$\vec{\mathbf{E}} = \frac{\vec{\mathbf{F}}}{q}. \quad \textbf{(16–3)}$$

The magnitude of the electric field a distance r from a point charge Q is

$$E = k\frac{Q}{r^2}. \quad \textbf{(16–4a)}$$

The total electric field at a point in space is equal to the vector sum of the individual fields due to each contributing charge. This is the **principle of superposition**.

Electric fields are represented by **electric field lines** that start on positive charges and end on negative charges. Their direction indicates the direction the force would be on a tiny positive test charge placed at each point. The lines can be drawn so that the number per unit area is proportional to the magnitude of E.

The static electric field inside a conductor is zero, and the electric field lines just outside a charged conductor are perpendicular to its surface.

[*In the replication of DNA, the electrostatic force plays a crucial role in selecting the proper molecules so that the genetic information is passed on accurately from generation to generation.]

[*Photocopiers and computer printers use electric charge placed on toner particles and a drum to form an image.]

[*The **electric flux** passing through a small area A for a uniform electric field $\vec{\mathbf{E}}$ is

$$\Phi_E = E_\perp A, \quad \textbf{(16–7)}$$

where $E_\perp$ is the component of $\vec{\mathbf{E}}$ perpendicular to the surface. The flux through a surface is proportional to the number of field lines passing through it.]

[*Gauss's law states that the total flux summed over any closed surface (considered as made up of many small areas ΔA) is equal to the net charge Q_{encl} enclosed by the surface divided by ϵ_0:

$$\sum_{\substack{\text{closed} \\ \text{surface}}} E_\perp \, \Delta A = \frac{Q_{encl}}{\epsilon_0}. \quad \textbf{(16–9)}$$

Gauss's law can be used to determine the electric field due to given charge distributions, but its usefulness is mainly limited to cases where the charge distribution displays much symmetry. The real importance of Gauss's law is that it is a general and elegant statement of the relation between electric charge and electric field.]

Questions

1. If you charge a pocket comb by rubbing it with a silk scarf, how can you determine if the comb is positively or negatively charged?

2. Why does a shirt or blouse taken from a clothes dryer sometimes cling to your body?

3. Explain why fog or rain droplets tend to form around ions or electrons in the air.

4. Why does a plastic ruler that has been rubbed with a cloth have the ability to pick up small pieces of paper? Why is this difficult to do on a humid day?

5. A positively charged rod is brought close to a neutral piece of paper, which it attracts. Draw a diagram showing the separation of charge in the paper, and explain why attraction occurs.

6. Contrast the *net charge* on a conductor to the "free charges" in the conductor.

7. Figures 16–7 and 16–8 show how a charged rod placed near an uncharged metal object can attract (or repel) electrons. There are a great many electrons in the metal, yet only some of them move as shown. Why not all of them?

8. When an electroscope is charged, its two leaves repel each other and remain at an angle. What balances the electric force of repulsion so that the leaves don't separate further?

9. The balloon in Fig. 16–48 was rubbed on a student's hair. Explain why the water drip curves instead of falling vertically.

FIGURE 16–48 Question 9.

10. The form of Coulomb's law is very similar to that for Newton's law of universal gravitation. What are the differences between these two laws? Compare also gravitational mass and electric charge.

11. When a charged ruler attracts small pieces of paper, sometimes a piece jumps quickly away after touching the ruler. Explain.

12. We are not normally aware of the gravitational or electric force between two ordinary objects. What is the reason in each case? Give an example where we are aware of each one and why.

13. Explain why the test charges we use when measuring electric fields must be small.

14. When determining an electric field, must we use a *positive* test charge, or would a negative one do as well? Explain.

15. Draw the electric field lines surrounding two negative electric charges a distance ℓ apart.

16. Assume that the two opposite charges in Fig. 16–32a are 12.0 cm apart. Consider the magnitude of the electric field 2.5 cm from the positive charge. On which side of this charge—top, bottom, left, or right—is the electric field the strongest? The weakest? Explain.

17. Consider the electric field at the three points indicated by the letters A, B, and C in Fig. 16–49. First draw an arrow at each point indicating the direction of the net force that a positive test charge would experience if placed at that point, then list the letters in order of *decreasing* field strength (strongest first). Explain.

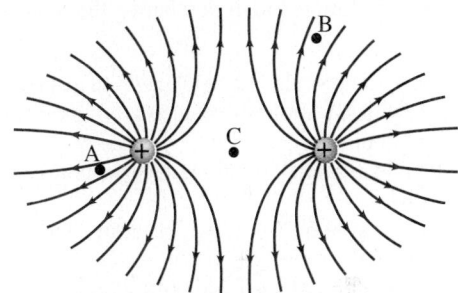

FIGURE 16–49 Question 17.

18. Why can electric field lines never cross?

19. Show, using the three rules for field lines given in Section 16–8, that the electric field lines starting or ending on a single point charge must be symmetrically spaced around the charge.

20. Given two point charges, Q and $2Q$, a distance ℓ apart, is there a point along the straight line that passes through them where $E = 0$ when their signs are (a) opposite, (b) the same? If yes, state roughly where this point will be.

21. Consider a small positive test charge located on an electric field line at some point, such as point P in Fig. 16–32a. Is the direction of the velocity and/or acceleration of the test charge along this line? Discuss.

*22. A point charge is surrounded by a spherical gaussian surface of radius r. If the sphere is replaced by a cube of side r, will Φ_E be larger, smaller, or the same? Explain.

MisConceptual Questions

1. $Q_1 = -0.10\,\mu C$ is located at the origin. $Q_2 = +0.10\,\mu C$ is located on the positive x axis at $x = 1.0\,m$. Which of the following is true of the force on Q_1 due to Q_2?
   - (a) It is attractive and directed in the $+x$ direction.
   - (b) It is attractive and directed in the $-x$ direction.
   - (c) It is repulsive and directed in the $+x$ direction.
   - (d) It is repulsive and directed in the $-x$ direction.

2. Swap the positions of Q_1 and Q_2 of MisConceptual Question 1. Which of the following is true of the force on Q_1 due to Q_2?
   - (a) It does not change.
   - (b) It changes from attractive to repulsive.
   - (c) It changes from repulsive to attractive.
   - (d) It changes from the $+x$ direction to the $-x$ direction.
   - (e) It changes from the $-x$ direction to the $+x$ direction.

3. Fred the lightning bug has a mass m and a charge $+q$. Jane, his lightning-bug wife, has a mass of $\frac{3}{4}m$ and a charge $-2q$. Because they have charges of opposite sign, they are attracted to each other. Which is attracted more to the other, and by how much?
   - (a) Fred, twice as much.
   - (b) Jane, twice as much.
   - (c) Fred, four times as much.
   - (d) Jane, four times as much.
   - (e) They are attracted to each other by the same amount.

4. Figure 16–50 shows electric field lines due to a point charge. What can you say about the field at point 1 compared with the field at point 2?
   - (a) The field at point 2 is larger, because point 2 is on a field line.
   - (b) The field at point 1 is larger, because point 1 is not on a field line.
   - (c) The field at point 1 is zero, because point 1 is not on a field line.
   - (d) The field at point 1 is larger, because the field lines are closer together in that region.

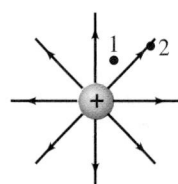

FIGURE 16–50
MisConceptual Question 4.

5. A negative point charge is in an electric field created by a positive point charge. Which of the following is true?
   - (a) The field points toward the positive charge, and the force on the negative charge is in the same direction as the field.
   - (b) The field points toward the positive charge, and the force on the negative charge is in the opposite direction to the field.
   - (c) The field points away from the positive charge, and the force on the negative charge is in the same direction as the field.
   - (d) The field points away from the positive charge, and the force on the negative charge is in the opposite direction to the field.

6. As an object acquires a positive charge, its mass usually
   - (a) decreases.
   - (b) increases.
   - (c) stays the same.
   - (d) becomes negative.

7. Refer to Fig. 16–32d. If the two charged plates were moved until they are half the distance shown without changing the charge on the plates, the electric field near the center of the plates would
   - (a) remain almost exactly the same.
   - (b) increase by a factor of 2.
   - (c) increase, but not by a factor of 2.
   - (d) decrease by a factor of 2.
   - (e) decrease, but not by a factor of 2.

8. We wish to determine the electric field at a point near a positively charged metal sphere (a good conductor). We do so by bringing a small positive test charge, q_0, to this point and measure the force F_0 on it. F_0/q_0 will be _____ the electric field $\vec{E}$ as it was at that point before the test charge was present.
   - (a) greater than
   - (b) less than
   - (c) equal to

9. We are usually not aware of the electric force acting between two everyday objects because
   - (a) the electric force is one of the weakest forces in nature.
   - (b) the electric force is due to microscopic-sized particles such as electrons and protons.
   - (c) the electric force is invisible.
   - (d) most everyday objects have as many plus charges as minus charges.

10. To be safe during a lightning storm, it is best to be
   - (a) in the middle of a grassy meadow.
   - (b) inside a metal car.
   - (c) next to a tall tree in a forest.
   - (d) inside a wooden building.
   - (e) on a metal observation tower.

11. Which are the worst places in MisConceptual Question 10?

12. Which vector best represents the direction of the electric field at the fourth corner of the square due to the three charges shown in Fig. 16–51?

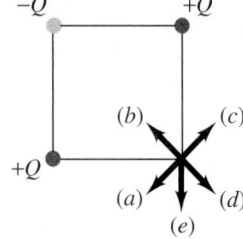

FIGURE 16–51
MisConceptual Question 12.

13. A small metal ball hangs from the ceiling by an insulating thread. The ball is attracted to a positively charged rod held near the ball. The charge of the ball must be
   - (a) positive.
   - (b) negative.
   - (c) neutral.
   - (d) positive or neutral.
   - (e) negative or neutral.

For assigned homework and other learning materials, go to the MasteringPhysics website.

Problems

16–5 and 16–6 Coulomb's Law

$\left[1\ \text{mC} = 10^{-3}\ \text{C}, \quad 1\ \mu\text{C} = 10^{-6}\ \text{C}, \quad 1\ \text{nC} = 10^{-9}\ \text{C}.\right]$

1. (I) What is the magnitude of the electric force of attraction between an iron nucleus $(q = +26e)$ and its innermost electron if the distance between them is $1.5 \times 10^{-12}\ \text{m}$?

2. (I) How many electrons make up a charge of $-48.0\ \mu\text{C}$?

3. (I) What is the magnitude of the force a $+25\ \mu\text{C}$ charge exerts on a $+2.5\ \text{mC}$ charge 16 cm away?

4. (I) What is the repulsive electrical force between two protons $4.0 \times 10^{-15}\ \text{m}$ apart from each other in an atomic nucleus?

5. (II) When an object such as a plastic comb is charged by rubbing it with a cloth, the net charge is typically a few microcoulombs. If that charge is $3.0\ \mu\text{C}$, by what percentage does the mass of a 9.0-g comb change during charging?

6. (II) Two charged dust particles exert a force of $4.2 \times 10^{-2}\ \text{N}$ on each other. What will be the force if they are moved so they are only one-eighth as far apart?

7. (II) Two small charged spheres are 6.52 cm apart. They are moved, and the force each exerts on the other is found to have tripled. How far apart are they now?

8. (II) A person scuffing her feet on a wool rug on a dry day accumulates a net charge of $-28\ \mu\text{C}$. How many excess electrons does she get, and by how much does her mass increase?

9. (II) What is the total charge of all the electrons in a 12-kg bar of gold? What is the net charge of the bar? (Gold has 79 electrons per atom and an atomic mass of 197 u.)

10. (II) Compare the electric force holding the electron in orbit $\left(r = 0.53 \times 10^{-10}\ \text{m}\right)$ around the proton nucleus of the hydrogen atom, with the gravitational force between the same electron and proton. What is the ratio of these two forces?

11. (II) Particles of charge $+65$, $+48$, and $-95\ \mu\text{C}$ are placed in a line (Fig. 16–52). The center one is 0.35 m from each of the others. Calculate the net force on each charge due to the other two.

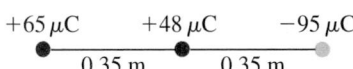

FIGURE 16–52 Problem 11.

12. (II) Three positive particles of equal charge, $+17.0\ \mu\text{C}$, are located at the corners of an equilateral triangle of side 15.0 cm (Fig. 16–53). Calculate the magnitude and direction of the net force on each particle due to the other two.

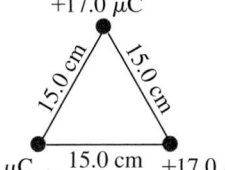

FIGURE 16–53 Problem 12.

13. (II) A charge Q is transferred from an initially uncharged plastic ball to an identical ball 24 cm away. The force of attraction is then 17 mN. How many electrons were transferred from one ball to the other?

14. (II) A charge of 6.15 mC is placed at each corner of a square 0.100 m on a side. Determine the magnitude and direction of the force on each charge.

15. (II) At each corner of a square of side ℓ there are point charges of magnitude Q, $2Q$, $3Q$, and $4Q$ (Fig. 16–54). Determine the magnitude and direction of the force on the charge $2Q$.

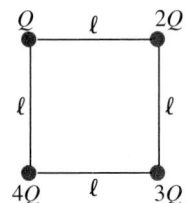

FIGURE 16–54 Problem 15.

16. (II) A large electroscope is made with "leaves" that are 78-cm-long wires with tiny 21-g spheres at the ends. When charged, nearly all the charge resides on the spheres. If the wires each make a $26°$ angle with the vertical (Fig. 16–55), what total charge Q must have been applied to the electroscope? Ignore the mass of the wires.

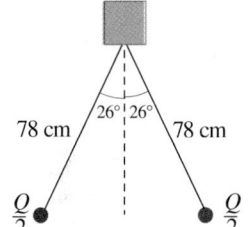

FIGURE 16–55 Problem 16.

17. (III) Two small nonconducting spheres have a total charge of $90.0\ \mu\text{C}$. (a) When placed 28.0 cm apart, the force each exerts on the other is 12.0 N and is repulsive. What is the charge on each? (b) What if the force were attractive?

18. (III) Two charges, $-Q$ and $-3Q$, are a distance ℓ apart. These two charges are free to move but do not because there is a third (fixed) charge nearby. What must be the magnitude of the third charge and its placement in order for the first two to be in equilibrium?

16–7 and 16–8 Electric Field, Field Lines

19. (I) Determine the magnitude and direction of the electric force on an electron in a uniform electric field of strength 2460 N/C that points due east.

20. (I) A proton is released in a uniform electric field, and it experiences an electric force of $1.86 \times 10^{-14}\ \text{N}$ toward the south. Find the magnitude and direction of the electric field.

21. (I) Determine the magnitude and direction of the electric field 21.7 cm directly above an isolated $33.0 \times 10^{-6}\ \text{C}$ charge.

22. (I) A downward electric force of 6.4 N is exerted on a $-7.3\ \mu\text{C}$ charge. Find the magnitude and direction of the electric field at the position of this charge.

23. (II) Determine the magnitude of the acceleration experienced by an electron in an electric field of 756 N/C. How does the direction of the acceleration depend on the direction of the field at that point?

24. (II) Determine the magnitude and direction of the electric field at a point midway between a $-8.0\ \mu\text{C}$ and a $+5.8\ \mu\text{C}$ charge 6.0 cm apart. Assume no other charges are nearby.

25. (II) Draw, approximately, the electric field lines about two point charges, $+Q$ and $-3Q$, which are a distance ℓ apart.

26. (II) What is the electric field strength at a point in space where a proton experiences an acceleration of 2.4 million "g's"?

27. (II) An electron is released from rest in a uniform electric field and accelerates to the north at a rate of $105 \, \text{m/s}^2$. Find the magnitude and direction of the electric field.

28. (II) The electric field midway between two equal but opposite point charges is $386 \, \text{N/C}$, and the distance between the charges is $16.0 \, \text{cm}$. What is the magnitude of the charge on each?

29. (II) Calculate the electric field at one corner of a square $1.22 \, \text{m}$ on a side if the other three corners are occupied by $3.25 \times 10^{-6} \, \text{C}$ charges.

30. (II) Calculate the electric field at the center of a square $42.5 \, \text{cm}$ on a side if one corner is occupied by a $-38.6 \, \mu\text{C}$ charge and the other three are occupied by $-27.0 \, \mu\text{C}$ charges.

31. (II) Determine the direction and magnitude of the electric field at the point P in Fig. 16–56. The charges are separated by a distance $2a$, and point P is a distance x from the midpoint between the two charges. Express your answer in terms of Q, x, a, and k.

FIGURE 16–56 Problem 31.

32. (II) Two point charges, $Q_1 = -32 \, \mu\text{C}$ and $Q_2 = +45 \, \mu\text{C}$, are separated by a distance of $12 \, \text{cm}$. The electric field at the point P (see Fig. 16–57) is zero. How far from Q_1 is P?

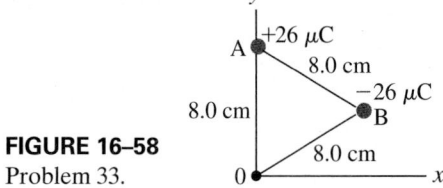

FIGURE 16–57 Problem 32.

33. (II) Determine the electric field $\vec{E}$ at the origin 0 in Fig. 16–58 due to the two charges at A and B.

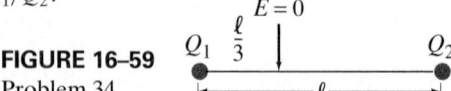

FIGURE 16–58 Problem 33.

34. (II) You are given two unknown point charges, Q_1 and Q_2. At a point on the line joining them, one-third of the way from Q_1 to Q_2, the electric field is zero (Fig. 16–59). What is the ratio Q_1/Q_2?

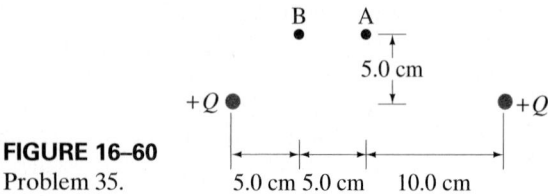

FIGURE 16–59 Problem 34.

35. (III) Use Coulomb's law to determine the magnitude and direction of the electric field at points A and B in Fig. 16–60 due to the two positive charges ($Q = 4.7 \, \mu\text{C}$) shown. Are your results consistent with Fig. 16–32b?

FIGURE 16–60 Problem 35.

36. (III) An electron (mass $m = 9.11 \times 10^{-31} \, \text{kg}$) is accelerated in the uniform field $\vec{E}$ ($E = 1.45 \times 10^4 \, \text{N/C}$) between two thin parallel charged plates. The separation of the plates is $1.60 \, \text{cm}$. The electron is accelerated from rest near the negative plate and passes through a tiny hole in the positive plate, Fig. 16–61. (a) With what speed does it leave the hole? (b) Show that the gravitational force can be ignored.

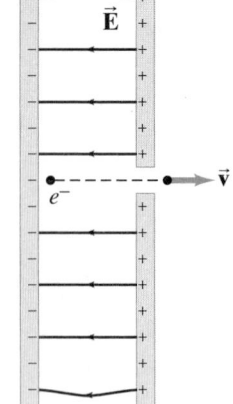

FIGURE 16–61
Problem 36.

16–10 DNA

*37. (III) The two strands of the helix-shaped DNA molecule are held together by electrostatic forces as shown in Fig. 16–39. Assume that the net average charge (due to electron sharing) indicated on H and N atoms has magnitude $0.2e$ and on the indicated C and O atoms is $0.4e$. Assume also that atoms on each molecule are separated by $1.0 \times 10^{-10} \, \text{m}$. Estimate the net force between (a) a thymine and an adenine; and (b) a cytosine and a guanine. For each bond (red dots) consider only the three atoms in a line (two atoms on one molecule, one atom on the other). (c) Estimate the total force for a DNA molecule containing 10^5 pairs of such molecules. Assume half are A–T pairs and half are C–G pairs.

16–12 Gauss's Law

*38. (I) The total electric flux from a cubical box of side $28.0 \, \text{cm}$ is $1.85 \times 10^3 \, \text{N} \cdot \text{m}^2/\text{C}$. What charge is enclosed by the box?

*39. (II) In Fig. 16–62, two objects, O_1 and O_2, have charges $+1.0 \, \mu\text{C}$ and $-2.0 \, \mu\text{C}$, respectively, and a third object, O_3, is electrically neutral. (a) What is the electric flux through the surface A_1 that encloses all three objects? (b) What is the electric flux through the surface A_2 that encloses the third object only?

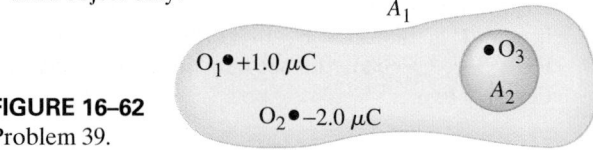

FIGURE 16–62 Problem 39.

*40. (II) A cube of side $8.50 \, \text{cm}$ is placed in a uniform field $E = 7.50 \times 10^3 \, \text{N/C}$ with edges parallel to the field lines. (a) What is the net flux through the cube? (b) What is the flux through each of its six faces?

*41. (II) The electric field between two parallel square metal plates is $130 \, \text{N/C}$. The plates are $0.85 \, \text{m}$ on a side and are separated by $3.0 \, \text{cm}$. What is the charge on each plate (assume equal and opposite)? Neglect edge effects.

*42. (II) The field just outside a 3.50-cm-radius metal ball is $3.75 \times 10^2 \, \text{N/C}$ and points toward the ball. What charge resides on the ball?

*43. (III) A point charge Q rests at the center of an uncharged thin spherical conducting shell. (See Fig. 16–34.) What is the electric field E as a function of r (a) for r less than the inner radius of the shell, (b) inside the shell, and (c) beyond the shell? (d) How does the shell affect the field due to Q alone? How does the charge Q affect the shell?

General Problems

44. How close must two electrons be if the magnitude of the electric force between them is equal to the weight of either at the Earth's surface?

45. Given that the human body is mostly made of water, estimate the total amount of positive charge in a 75-kg person.

46. A 3.0-g copper penny has a net positive charge of $32\,\mu C$. What fraction of its electrons has it lost?

47. Measurements indicate that there is an electric field surrounding the Earth. Its magnitude is about 150 N/C at the Earth's surface and points inward toward the Earth's center. What is the magnitude of the electric charge on the Earth? Is it positive or negative? [*Hint*: The electric field outside a uniformly charged sphere is the same as if all the charge were concentrated at its center.]

48. (a) The electric field near the Earth's surface has magnitude of about 150 N/C. What is the acceleration experienced by an electron near the surface of the Earth? (b) What about a proton? (c) Calculate the ratio of each acceleration to $g = 9.8\,\text{m/s}^2$.

49. A water droplet of radius 0.018 mm remains stationary in the air. If the downward-directed electric field of the Earth is 150 N/C, how many excess electron charges must the water droplet have?

50. Estimate the net force between the CO group and the HN group shown in Fig. 16–63. The C and O have charges $\pm 0.40e$, and the H and N have charges $\pm 0.20e$, where $e = 1.6 \times 10^{-19}\,C$. [*Hint*: Do not include the "internal" forces between C and O, or between H and N.]

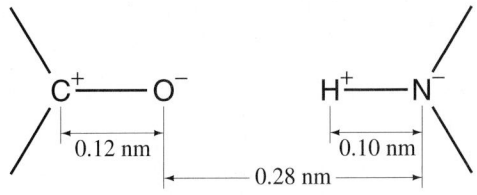

FIGURE 16–63 Problem 50.

51. In a simple model of the hydrogen atom, the electron revolves in a circular orbit around the proton with a speed of $2.2 \times 10^6\,\text{m/s}$. Determine the radius of the electron's orbit. [*Hint*: See Chapter 5 on circular motion.]

52. Two small charged spheres hang from cords of equal length ℓ as shown in Fig. 16–64 and make small angles θ_1 and θ_2 with the vertical. (a) If $Q_1 = Q$, $Q_2 = 2Q$, and $m_1 = m_2 = m$, determine the ratio θ_1/θ_2. (b) Estimate the distance between the spheres.

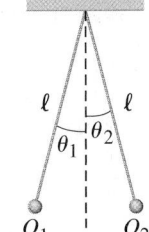

FIGURE 16–64
Problem 52.

53. A positive point charge $Q_1 = 2.5 \times 10^{-5}\,C$ is fixed at the origin of coordinates, and a negative point charge $Q_2 = -5.0 \times 10^{-6}\,C$ is fixed to the x axis at $x = +2.4\,\text{m}$. Find the location of the place(s) along the x axis where the electric field due to these two charges is zero.

54. Dry air will break down and generate a spark if the electric field exceeds about $3 \times 10^6\,\text{N/C}$. How much charge could be packed onto a green pea (diameter 0.75 cm) before the pea spontaneously discharges? [*Hint*: Eqs. 16–4 work outside a sphere if r is measured from its center.]

55. Two point charges, $Q_1 = -6.7\,\mu C$ and $Q_2 = 1.8\,\mu C$, are located between two oppositely charged parallel plates, as shown in Fig. 16–65. The two charges are separated by a distance of $x = 0.47\,\text{m}$. Assume that the electric field produced by the charged plates is uniform and equal to $E = 53{,}000\,\text{N/C}$. Calculate the net electrostatic force on Q_1 and give its direction.

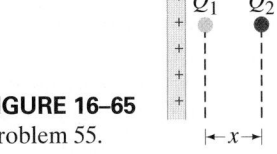

FIGURE 16–65
Problem 55.

56. Packing material made of pieces of foamed polystyrene can easily become charged and stick to each other. Given that the density of this material is about $35\,\text{kg/m}^3$, estimate how much charge might be on a 2.0-cm-diameter foamed polystyrene sphere, assuming the electric force between two spheres stuck together is equal to the weight of one sphere.

57. A point charge ($m = 1.0$ gram) at the end of an insulating cord of length 55 cm is observed to be in equilibrium in a uniform horizontal electric field of 9500 N/C, when the pendulum's position is as shown in Fig. 16–66, with the charge 12 cm above the lowest (vertical) position. If the field points to the right in Fig. 16–66, determine the magnitude and sign of the point charge.

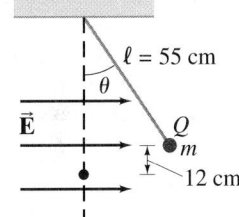

FIGURE 16–66
Problem 57.

58. Two small, identical conducting spheres A and B are a distance R apart; each carries the same charge Q. (a) What is the force sphere B exerts on sphere A? (b) An identical sphere with zero charge, sphere C, makes contact with sphere B and is then moved very far away. What is the net force now acting on sphere A? (c) Sphere C is brought back and now makes contact with sphere A and is then moved far away. What is the force on sphere A in this third case?

59. For an experiment, a colleague of yours says he smeared toner particles uniformly over the surface of a sphere 1.0 m in diameter and then measured an electric field of 5000 N/C near its surface. (a) How many toner particles (Example 16–6) would have to be on the surface to produce these results? (b) What is the total mass of the toner particles?

60. A proton ($m = 1.67 \times 10^{-27}\,\text{kg}$) is suspended at rest in a uniform electric field $\vec{E}$. Take into account gravity at the Earth's surface, and determine $\vec{E}$.

61. A point charge of mass 0.185 kg, and net charge +0.340 μC, hangs at rest at the end of an insulating cord above a large sheet of charge. The horizontal sheet of fixed uniform charge creates a uniform vertical electric field in the vicinity of the point charge. The tension in the cord is measured to be 5.18 N. Calculate the magnitude and direction of the electric field due to the sheet of charge (Fig. 16–67).

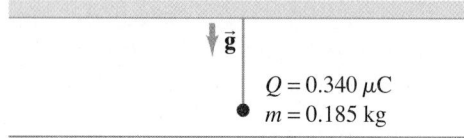

$Q = 0.340 \, \mu C$
$m = 0.185 \, kg$

Uniform sheet of charge

FIGURE 16–67 Problem 61.

62. An electron with speed $v_0 = 5.32 \times 10^6$ m/s is traveling parallel to an electric field of magnitude $E = 9.45 \times 10^3$ N/C. (a) How far will the electron travel before it stops? (b) How much time will elapse before it returns to its starting point?

63. Given the two charges shown in Fig. 16–68, at what position(s) x is the electric field zero?

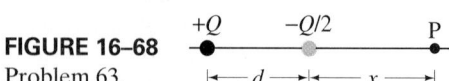

FIGURE 16–68 Problem 63.

64. What is the total charge of all the electrons in a 25-kg bar of aluminum? (Aluminum has 13 electrons per atom and an atomic mass of 27 u.)

65. Two point charges, $+Q$ and $-Q$ of mass m, are placed on the ends of a massless rod of length ℓ, which is fixed to a table by a pin through its center. If the apparatus is then subjected to a uniform electric field E parallel to the table and perpendicular to the rod, find the net torque on the system of rod plus charges.

66. Determine the direction and magnitude of the electric field at point P, Fig. 16–69. The two charges are separated by a distance of $2a$. Point P is on the perpendicular bisector of the line joining the charges, a distance x from the midpoint between them. Express your answers in terms of Q, x, a, and k.

FIGURE 16–69 Problem 66.

67. A mole of carbon contains 7.22×10^{24} electrons. Two electrically neutral carbon spheres, each containing 1 mole of carbon, are separated by 15.0 cm (center to center). What fraction of electrons would have to be transferred from one sphere to the other for the electric force and the gravitational force between the spheres to be equal?

Search and Learn

1. Referring to Section 16–4 and Figs. 16–11 and 16–12, what happens to the separation of the leaves of an electroscope when the charging object is removed from an electroscope (a) charged by induction and (b) charged by conduction? (c) Is it possible to tell whether the electroscope in Fig. 16–12a has been charged by induction or by conduction? If so, which way was it charged? (d) Draw electric field lines (Section 16–8) for the electroscopes in Figs. 16–11a and 16–11b, omitting the fields around the charging rod. How do the fields differ?

2. Four equal positive point charges, each of charge 6.4 μC, are at the corners of a square of side 9.2 cm. What charge should be placed at the center of the square so that all charges are at equilibrium? Is this a stable or an unstable equilibrium (Section 9–4) in the plane?

3. Suppose electrons enter a uniform electric field midway between two plates at an angle θ_0 to the horizontal, as shown in Fig. 16–70. The path is symmetrical, so they leave at the same angle θ_0 and just barely miss the top plate. What is θ_0? Ignore fringing of the field.

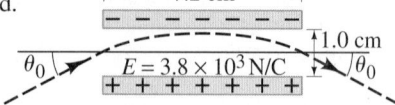

FIGURE 16–70 Search and Learn 3.

4. What experimental observations mentioned in the text rule out the possibility that the numerator in Coulomb's law contains the sum $(Q_1 + Q_2)$ rather than the product $Q_1 Q_2$?

5. Near the surface of the Earth, there is a downward electric field of 150 N/C and a downward gravitational field of 9.8 N/kg. A charged 1.0-kg mass is observed to fall with acceleration 8.0 m/s². What are the magnitude and sign of its charge?

6. Identical negative charges ($Q = -e$) are located at two of the three vertices of an equilateral triangle. The length of a side of the triangle is ℓ. What is the magnitude of the net electric field at the third vertex? If a third identical negative charge was located at the third vertex, then what would be the net electrostatic force on it due to the other two charges? Use symmetry and explain how you used it.

7. Suppose that electrical attraction, rather than gravity, were responsible for holding the Moon in orbit around the Earth. If equal and opposite charges Q were placed on the Earth and the Moon, what should be the value of Q to maintain the present orbit? Use data given on the inside front cover of this book. Treat the Earth and Moon as point particles.

ANSWERS TO EXERCISES

A: (d).
B: Opposite.
C: 0.3 N, to the right.
D: 0.32 m.

E: (a) No; (b) yes, midway between them.
F: 9.0×10^4 N/C, vertically upward.
G: (d), if the two + charges are not at opposite corners (use symmetry).

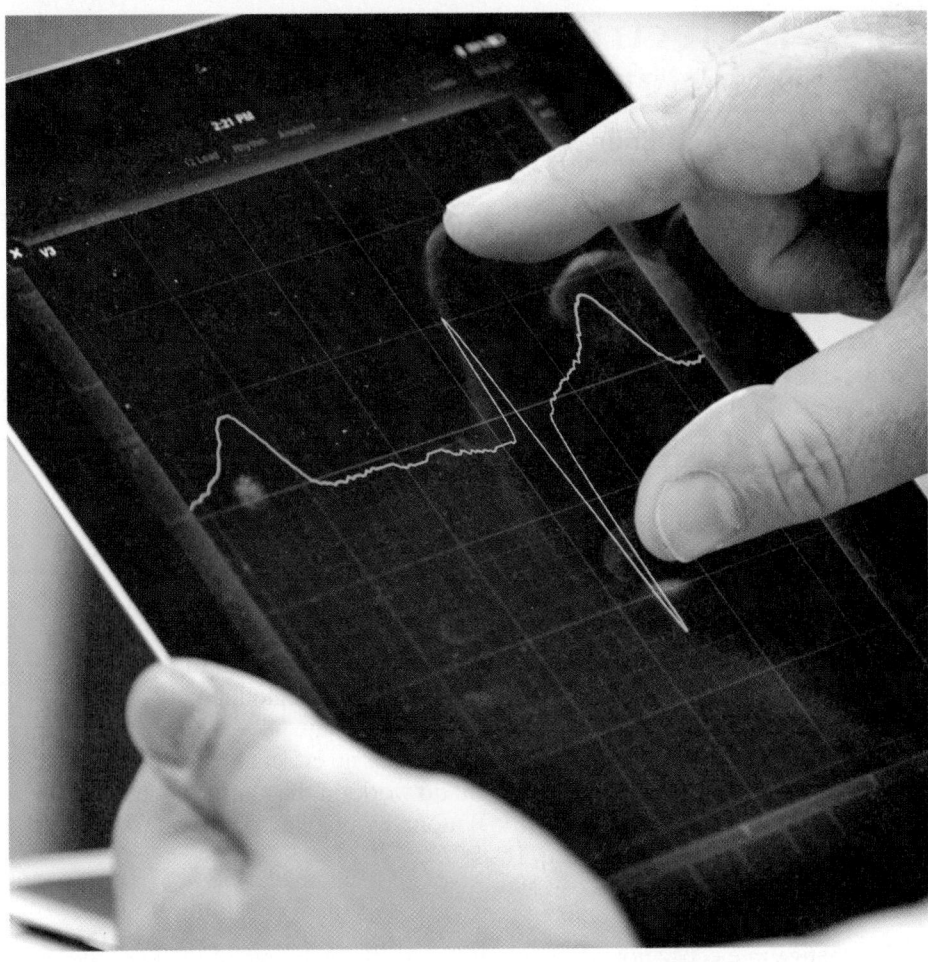

We are used to voltage in our lives—a 12-volt car battery, 110 V or 220 V at home, 1.5-volt flashlight batteries, and so on. Here we see displayed the voltage produced across a human heart, known as an electrocardiogram. Voltage is the same as electric potential difference between two points. Electric potential is defined as the potential energy per unit charge.

We discuss voltage and its relation to electric field, as well as electric energy storage, capacitors, and applications including the ECG shown here, binary numbers and digital electronics, TV and computer monitors, and digital TV.

Electric Potential

CHAPTER-OPENING QUESTION—Guess now!

When two positively charged small spheres are pushed toward each other, what happens to their potential energy?

(a) It remains unchanged.

(b) It decreases.

(c) It increases.

(d) There is no potential energy in this situation.

We saw in Chapter 6 that the concept of energy was extremely valuable in dealing with the subject of mechanics. For one thing, energy is a conserved quantity and is thus an important tool for understanding nature. Furthermore, we saw that many Problems could be solved using the energy concept even though a detailed knowledge of the forces involved was not possible, or when a calculation involving Newton's laws would have been too difficult.

The energy point of view can be used in electricity, and it is especially useful. It not only extends the law of conservation of energy, but it gives us another way to view electrical phenomena. The energy concept is also a tool in solving Problems more easily in many cases than by using forces and electric fields.

CONTENTS

473

17–1 Electric Potential Energy and Potential Difference

Electric Potential Energy

To apply conservation of energy, we need to define electric potential energy as we did for other types of potential energy. As we saw in Chapter 6, potential energy can be defined only for a conservative force. The work done by a conservative force in moving an object between any two positions is independent of the path taken. The electrostatic force between any two charges (Eq. 16–1, $F = kQ_1Q_2/r^2$) is conservative because the dependence on position is just like the gravitational force (Eq. 5–4), which is conservative. Hence we can define potential energy PE for the electrostatic force.

We saw in Chapter 6 that the change in potential energy between any two points, a and b, equals the negative of the work done by the conservative force on an object as it moves from point a to point b: $\Delta\text{PE} = -W$.

Thus we define the change in electric potential energy, $\text{PE}_b - \text{PE}_a$, when a point charge q moves from some point a to another point b, as the negative of the work done by the electric force on the charge as it moves from point a to point b. For example, consider the electric field between two equally but oppositely charged parallel plates; we assume their separation is small compared to their width and height, so the field $\vec{E}$ will be uniform over most of the region, Fig. 17–1. Now consider a tiny positive point charge q placed at the point "a" very near the positive plate as shown. This charge q is so small that it has no effect on $\vec{E}$. If this charge q at point a is released, the electric force will do work on the charge and accelerate it toward the negative plate. The work W done by the electric field E to move the charge a distance d is (using Eq. 16–5, $F = qE$)

$$W = Fd = qEd. \qquad \text{[uniform } \vec{E}]$$

The change in electric potential energy equals the negative of the work done by the electric force:

$$\text{PE}_b - \text{PE}_a = -qEd \qquad \text{[uniform } \vec{E}] \quad \textbf{(17–1)}$$

for this case of uniform electric field $\vec{E}$. In the case illustrated, the potential energy decreases (ΔPE is negative); and as the charged particle accelerates from point a to point b in Fig. 17–1, the particle's kinetic energy KE increases—by an equal amount. In accord with the conservation of energy, electric potential energy is transformed into kinetic energy, and the total energy is conserved. Note that the positive charge q has its greatest potential energy at point a, near the positive plate.[†] The reverse is true for a negative charge: its potential energy is greatest near the negative plate.

Electric Potential and Potential Difference

In Chapter 16, we found it useful to define the electric field as the force per unit charge. Similarly, it is useful to define the **electric potential** (or simply the **potential** when "electric" is understood) as the *electric potential energy per unit charge*. Electric potential is given the symbol V. If a positive test charge q in an electric field has electric potential energy PE_a at some point a (relative to some zero potential energy), the electric potential V_a at this point is

$$V_a = \frac{\text{PE}_a}{q}. \qquad \textbf{(17–2a)}$$

As we discussed in Chapter 6, only differences in potential energy are physically meaningful. Hence only the **difference in potential**, or the **potential difference**, between two points a and b (such as those shown in Fig. 17–1) is measurable.

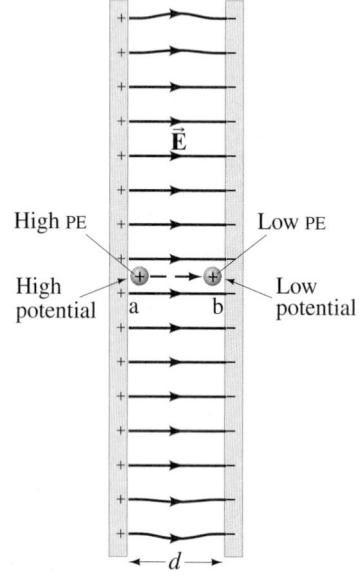

FIGURE 17–1 Work is done by the electric field $\vec{E}$ in moving the positive charge from position a to position b.

High PE   Low PE

High potential   a   b   Low potential

$\vec{E}$

$\leftarrow d \rightarrow$

[†]At point a, the positive charge q has its greatest ability to do work (on some other object or system).

When the electric force does positive work on a charge, the kinetic energy increases and the potential energy decreases. The difference in potential energy, $PE_b - PE_a$, is equal to the negative of the work, W_{ba}, done by the electric field to move the charge from a to b; so the potential difference V_{ba} is

$$V_{ba} = V_b - V_a = \frac{PE_b - PE_a}{q} = -\frac{W_{ba}}{q}. \qquad \textbf{(17-2b)}$$

Note that electric potential, like electric field, does not depend on our test charge q. V depends on the other charges that create the field, not on the test charge q; q acquires potential energy by being in the potential V due to the other charges.

We can see from our definition that the positive plate in Fig. 17–1 is at a higher potential than the negative plate. Thus a positively charged object moves naturally from a high potential to a low potential. A negative charge does the reverse.

The unit of electric potential, and of potential difference, is joules/coulomb and is given a special name, the **volt**, in honor of Alessandro Volta (1745–1827) who is best known for inventing the electric battery. The volt is abbreviated V, so 1 V = 1 J/C. Potential difference, since it is measured in volts, is often referred to as **voltage**. (Be careful not to confuse V for volts, with italic *V* for voltage.)

If we wish to speak of the potential V_a at some point a, we must be aware that V_a depends on where the potential is chosen to be zero. The zero for electric potential in a given situation can be chosen arbitrarily, just as for potential energy, because only differences in potential energy can be measured. Often the ground, or a conductor connected directly to the ground (the Earth), is taken as zero potential, and other potentials are given with respect to ground. (Thus, a point where the voltage is 50 V is one where the difference of potential between it and ground is 50 V.) In other cases, as we shall see, we may choose the potential to be zero at an infinite distance.

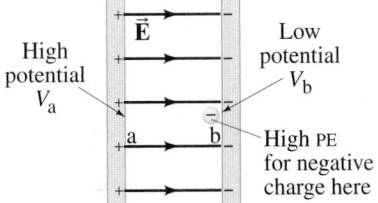

FIGURE 17–2 Central part of Fig. 17–1, showing a negative point charge near the negative plate. Example 17–1.

⚠ **CAUTION**

A negative charge has high PE when potential V is low

| CONCEPTUAL EXAMPLE 17–1 | **A negative charge.** Suppose a negative charge, such as an electron, is placed near the negative plate in Fig. 17–1, at point b, shown here in Fig. 17–2. If the electron is free to move, will its electric potential energy increase or decrease? How will the electric potential change?

RESPONSE An electron released at point b will be attracted to the positive plate. As the electron accelerates toward the positive plate, its kinetic energy increases, so its potential energy *decreases*: $PE_a < PE_b$ and $\Delta PE = PE_a - PE_b < 0$. But note that the electron moves from point b at low potential to point a at higher potential: $\Delta V = V_a - V_b > 0$. (Potentials V_a and V_b are due to the charges on the plates, not due to the electron.) The signs of ΔPE and ΔV are opposite because of the negative charge of the electron.

NOTE A positive charge placed next to the negative plate at b would stay there, with no acceleration. A positive charge tends to move from high potential to low.

Because the electric potential difference is defined as the potential energy difference per unit charge, then the change in potential energy of a charge q when it moves from point a to point b is

$$\Delta PE = PE_b - PE_a = q(V_b - V_a) = qV_{ba}. \qquad \textbf{(17-3)}$$

That is, if an object with charge q moves through a potential difference V_{ba}, its potential energy changes by an amount qV_{ba}. For example, if the potential difference between the two plates in Fig. 17–1 is 6 V, then a +1 C charge moved from point b to point a will gain (1 C)(6 V) = 6 J of electric potential energy. (And it will lose 6 J of electric potential energy if it moves from a to b.) Similarly, a +2 C charge will gain $\Delta PE = (2 C)(6 V) = 12$ J, and so on. Thus, electric potential difference is a measure of how much energy an electric charge can acquire in a given situation. And, since energy is the ability to do work, the electric potential difference is also a measure of how much *work* a given charge can do. The exact amount of energy or work depends both on the potential difference and on the charge.

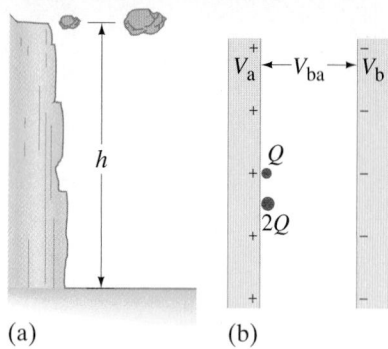

(a)   (b)

FIGURE 17–3 (a) Two rocks are at the same height. The larger rock has more potential energy. (b) Two positive charges have the same electric potential. The $2Q$ charge has more potential energy.

TABLE 17–1 Some Typical Potential Differences (Voltages)

Source	Voltage (approx.)
Thundercloud to ground	10^8 V
High-voltage power line	10^5–10^6 V
Automobile ignition	10^4 V
Household outlet	10^2 V
Automobile battery	12 V
Flashlight battery (AA, AAA, C, D)	1.5 V
Resting potential across nerve membrane	10^{-1} V
Potential changes on skin (ECG and EEG)	10^{-4} V

FIGURE 17–4 Electron accelerated, Example 17–2.

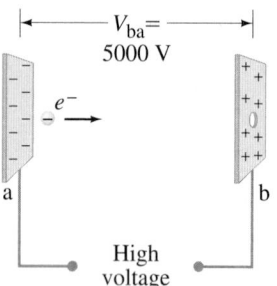

To better understand electric potential, let's make a comparison to the gravitational case when a rock falls from the top of a cliff. The greater the height, h, of a cliff, the more potential energy ($=mgh$) the rock has at the top of the cliff relative to the bottom, and the more kinetic energy it will have when it reaches the bottom. The actual amount of kinetic energy it will acquire, and the amount of work it can do, depends both on the height of the cliff and the mass m of the rock. A large rock and a small rock can be at the same height h (Fig. 17–3a) and thus have the same "gravitational potential," but the larger rock has the greater potential energy (it has more mass). The electrical case is similar (Fig. 17–3b): the potential energy change, or the work that can be done, depends both on the potential difference (corresponding to the height of the cliff) and on the charge (corresponding to mass), Eq. 17–3. But note a significant difference: electric charge comes in two types, $+$ and $-$, whereas gravitational mass is always $+$.

Sources of electrical energy such as batteries and electric generators are meant to maintain a potential difference. The actual amount of energy transformed by such a device depends on how much charge flows, as well as the potential difference (Eq. 17–3). For example, consider an automobile headlight connected to a 12.0-V battery. The amount of energy transformed (into light and thermal energy) is proportional to how much charge flows, which in turn depends on how long the light is on. If over a given period of time 5.0 C of charge flows through the light, the total energy transformed is (5.0 C)(12.0 V) = 60 J. If the headlight is left on twice as long, 10.0 C of charge will flow and the energy transformed is (10.0 C)(12.0 V) = 120 J. Table 17–1 presents some typical voltages.

EXAMPLE 17–2 Electron in TV tube. Suppose an electron is accelerated from rest through a potential difference $V_b - V_a = V_{ba} = +5000$ V (Fig. 17–4). (a) What is the change in electric potential energy of the electron? What is (b) the kinetic energy, and (c) the speed of the electron ($m = 9.1 \times 10^{-31}$ kg) as a result of this acceleration?

APPROACH The electron, accelerated toward the positive plate, will change in potential energy by an amount $\Delta PE = qV_{ba}$ (Eq. 17–3). The loss in potential energy will equal its gain in kinetic energy (energy conservation).

SOLUTION (a) The charge on an electron is $q = -e = -1.6 \times 10^{-19}$ C. Therefore its change in potential energy is

$$\Delta PE = qV_{ba} = (-1.6 \times 10^{-19}\,\text{C})(+5000\,\text{V}) = -8.0 \times 10^{-16}\,\text{J}.$$

The minus sign indicates that the potential energy decreases. The potential difference, V_{ba}, has a positive sign because the final potential V_b is higher than the initial potential V_a. Negative electrons are attracted toward a positive electrode (or plate) and repelled away from a negative electrode.

(b) The potential energy lost by the electron becomes kinetic energy KE. From conservation of energy (Eq. 6–11a), $\Delta KE + \Delta PE = 0$, so

$$\Delta KE = -\Delta PE$$
$$\tfrac{1}{2}mv^2 - 0 = -q(V_b - V_a) = -qV_{ba},$$

where the initial kinetic energy is zero since we are given that the electron started from rest. So the final KE $= -qV_{ba} = 8.0 \times 10^{-16}$ J.

(c) In the equation just above we solve for v:

$$v = \sqrt{-\frac{2qV_{ba}}{m}} = \sqrt{-\frac{2(-1.6 \times 10^{-19}\,\text{C})(5000\,\text{V})}{9.1 \times 10^{-31}\,\text{kg}}} = 4.2 \times 10^7\,\text{m/s}.$$

NOTE The electric potential energy does not depend on the mass, only on the charge and voltage. The speed *does* depend on m.

EXERCISE A Instead of the electron in Example 17–2, suppose a proton ($m = 1.67 \times 10^{-27}$ kg) was accelerated from rest by a potential difference $V_{ba} = -5000$ V. What would be the proton's (a) change in PE, and (b) final speed?

17–2 Relation between Electric Potential and Electric Field

The effects of any charge distribution can be described either in terms of electric field or in terms of electric potential. Electric potential is often easier to use because it is a scalar, whereas electric field is a vector. There is an intimate connection between the potential and the field. Let us consider the case of a uniform electric field, such as that between the parallel plates of Fig. 17–1 whose difference of potential is V_{ba}. The work done by the electric field to move a positive charge q from point a to point b is equal to the negative of the change in potential energy (Eq. 17–2b), so

$$W = -q(V_b - V_a) = -qV_{ba}.$$

We can also write the work done as the force times distance, where the force on q is $F = qE$, so

$$W = Fd = qEd,$$

where d is the distance (parallel to the field lines) between points a and b. We now set these two expressions for W equal and find $qV_{ba} = -qEd$, or

$$V_{ba} = -Ed. \qquad \text{[uniform } \vec{E}] \quad \textbf{(17–4a)}$$

If we solve for E, we find

$$E = -\frac{V_{ba}}{d}. \qquad \text{[uniform } \vec{E}] \quad \textbf{(17–4b)}$$

From Eq. 17–4b we see that the units for electric field can be written as volts per meter (V/m), as well as newtons per coulomb (N/C, from $E = F/q$). These are equivalent because $1 \, \text{N/C} = 1 \, \text{N·m/C·m} = 1 \, \text{J/C·m} = 1 \, \text{V/m}$. The minus sign in Eq. 17–4b tells us that $\vec{E}$ points in the direction of decreasing potential V.

EXAMPLE 17–3 | **Electric field obtained from voltage.** Two parallel plates are charged to produce a potential difference of 50 V. If the separation between the plates is 0.050 m, calculate the magnitude of the electric field in the space between the plates (Fig. 17–5).

APPROACH We apply Eq. 17–4b to obtain the magnitude of E, assumed uniform.

SOLUTION The magnitude of the electric field is

$$E = V_{ba}/d = (50 \, \text{V}/0.050 \, \text{m}) = 1000 \, \text{V/m}.$$

NOTE Equations 17–4 apply only for a uniform electric field. The general relationship between $\vec{E}$ and V is more complicated.

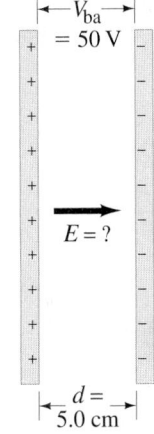

FIGURE 17–5 Example 17–3.

*General Relation between $\vec{E}$ and V

In a region where $\vec{E}$ is not uniform, the connection between $\vec{E}$ and V takes on a different form than Eqs. 17–4. In general, it is possible to show that the electric field in a given direction at any point in space is equal to the *rate at which the electric potential decreases over distance in that direction*. For example, the x component of the electric field is given by $E_x = -\Delta V/\Delta x$, where ΔV is the change in potential over a very short distance Δx.

Breakdown Voltage

When very high voltages are present, air can become ionized due to the high electric fields. Any odd free electron can be accelerated to sufficient kinetic energy to knock electrons out of O_2 and N_2 molecules of the air. This **breakdown** of air occurs when the electric field exceeds about 3×10^6 V/m. When electrons recombine with their molecules, light is emitted. Such breakdown of air is the source of lightning, the spark of a car's spark plug, and even short sparks between your fingers and a doorknob after you walk across a synthetic rug or slide across a car seat (which can result in a significant transfer of charge to you).

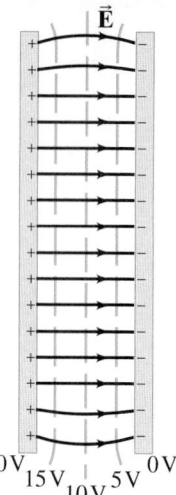

$\vec{E}$

20V 15V 10V 5V 0V

FIGURE 17–6 Equipotential lines (the green dashed lines) between two charged parallel plates are always perpendicular to the electric field (solid red lines).

FIGURE 17–7 Equipotential lines (green, dashed) are always perpendicular to the electric field lines (solid red), shown here for two equal but oppositely charged particles (an "electric dipole").

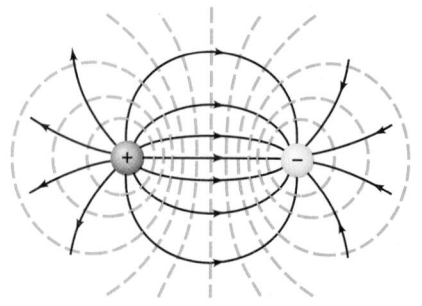

+   −

17–3 Equipotential Lines and Surfaces

The electric potential can be represented by drawing **equipotential lines** or, in three dimensions, **equipotential surfaces**. An equipotential surface is one on which all points are at the same potential. That is, the potential difference between any two points on the surface is zero, so no work is required to move a charge from one point on the surface to the other. An *equipotential surface must be perpendicular to the electric field* at any point. If this were not so—that is, if there were a component of $\vec{E}$ parallel to the surface—it would require work to move the charge along the surface against this component of $\vec{E}$; and this would contradict the idea that it is an *equi*potential surface.

The fact that the electric field lines and equipotential surfaces are mutually perpendicular helps us locate the equipotentials when the electric field lines are known. In a normal two-dimensional drawing, we show equipotential *lines*, which are the intersections of equipotential surfaces with the plane of the drawing. In Fig. 17–6, a few of the equipotential lines are drawn (dashed green lines) for the electric field (red lines) between two parallel plates at a potential difference of 20 V. The negative plate is arbitrarily chosen to be zero volts and the potential of each equipotential line is indicated. Note that $\vec{E}$ points toward lower values of V. The equipotential lines for the case of two equal but oppositely charged particles are shown in Fig. 17–7 as green dashed lines. (This combination of equal + and − charges is called an "electric dipole," as we saw in Section 16–8; see Fig. 16–32a.)

Unlike electric field lines, which start and end on electric charges, equipotential lines and surfaces are always continuous and never end, and so continue beyond the borders of Figs. 17–6 and 17–7. A useful analogy for equipotential lines is a topographic map: the contour lines are gravitational equipotential lines (Fig. 17–8).

We saw in Section 16–9 that there can be no electric field within a conductor in the static case, for otherwise the free electrons would feel a force and would move. Indeed the *entire volume of a conductor must be entirely at the same potential in the static case.* The surface of a conductor is thus an equipotential surface. (If it weren't, the free electrons at the surface would move, because whenever there is a potential difference between two points, free charges will move.) This is fully consistent with our result in Section 16–9 that the electric field at the surface of a conductor must be perpendicular to the surface.

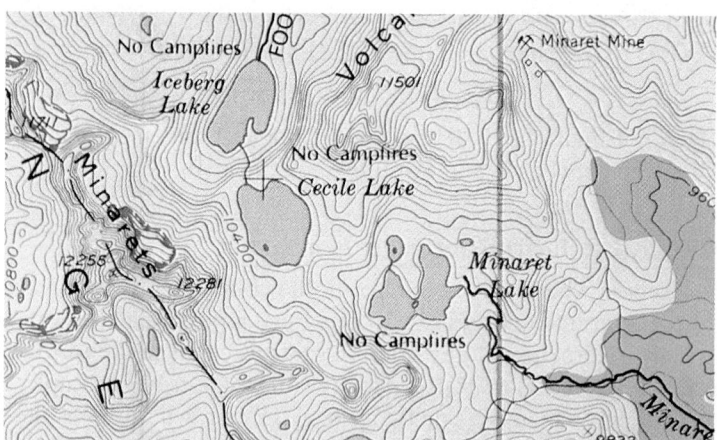

FIGURE 17–8 A topographic map (here, a portion of the Sierra Nevada in California) shows continuous contour lines, each of which is at a fixed height above sea level. Here they are at 80-ft (25-m) intervals. If you walk along one contour line, you neither climb nor descend. If you cross lines, and especially if you climb perpendicular to the lines, you will be changing your gravitational potential (rapidly, if the lines are close together).

17–4 The Electron Volt, a Unit of Energy

The joule is a very large unit for dealing with energies of electrons, atoms, or molecules. For this purpose, the unit **electron volt** (eV) is used. One electron volt is defined as the energy acquired by a particle carrying a charge whose magnitude equals that on the electron ($q = e$) as a result of moving through a potential difference of 1 V. The charge on an electron has magnitude 1.6022×10^{-19} C, and the change in potential energy equals qV. So 1 eV is equal to $(1.6022 \times 10^{-19}\,\text{C})(1.00\,\text{V}) = 1.6022 \times 10^{-19}$ J:

$$1\,\text{eV} = 1.6022 \times 10^{-19} \approx 1.60 \times 10^{-19}\,\text{J}.$$

Electron volt

An electron that accelerates through a potential difference of 1000 V will lose 1000 eV of potential energy and thus gain 1000 eV or 1 keV (kiloelectron volt) of kinetic energy.

On the other hand, if a particle with a charge equal to twice the magnitude of the charge on the electron $(= 2e = 3.2 \times 10^{-19}\,\text{C})$ moves through a potential difference of 1000 V, its kinetic energy will increase by 2000 eV = 2 keV.

Although the electron volt is handy for *stating* the energies of molecules and elementary particles, it is *not* a proper SI unit. For calculations, electron volts should be converted to joules using the conversion factor just given. In Example 17–2, for example, the electron acquired a kinetic energy of $8.0 \times 10^{-16}\,\text{J}$. We can quote this energy as 5000 eV $(= 8.0 \times 10^{-16}\,\text{J}/1.6 \times 10^{-19}\,\text{J/eV})$, but when determining the speed of a particle in SI units, we must use the KE in joules (J).

> **EXERCISE B** What is the kinetic energy of a He^{2+} ion released from rest and accelerated through a potential difference of 2.5 kV? (*a*) 2500 eV, (*b*) 500 eV, (*c*) 5000 eV, (*d*) 10,000 eV, (*e*) 250 eV.

17–5 Electric Potential Due to Point Charges

The electric potential at a distance r from a single point charge Q can be derived from the expression for its electric field (Eq. 16–4, $E = kQ/r^2$) using calculus. The potential in this case is usually taken to be zero at infinity $(= \infty$, which means extremely, indefinitely, far away); this is also where the electric field $(E = kQ/r^2)$ is zero. The result is

$$V = k\frac{Q}{r}$$
$$= \frac{1}{4\pi\epsilon_0}\frac{Q}{r},$$

$\left[\begin{array}{l}\text{single point charge}\\ V = 0 \text{ at } r = \infty\end{array}\right]$ **(17–5)**

where $k = 8.99 \times 10^9\,\text{N}\cdot\text{m}^2/\text{C}^2 \approx 9.0 \times 10^9\,\text{N}\cdot\text{m}^2/\text{C}^2$. We can think of V here as representing the absolute potential at a distance r from the charge Q, where $V = 0$ at $r = \infty$; or we can think of V as the potential difference between r and infinity. (The symbol ∞ means infinitely far away.) Notice that the potential V decreases with the first power of the distance, whereas the electric field (Eq. 16–4) decreases as the *square* of the distance. The potential near a positive charge is large and positive, and it decreases toward zero at very large distances, Fig. 17–9a. The potential near a negative charge is negative and increases toward zero at large distances, Fig. 17–9b. Equation 17–5 is sometimes called the **Coulomb potential** (it has its origin in Coulomb's law).

> **EXAMPLE 17–4** **Potential due to a positive or a negative charge.** Determine the potential at a point 0.50 m (*a*) from a $+20\,\mu\text{C}$ point charge, (*b*) from a $-20\,\mu\text{C}$ point charge.
>
> **APPROACH** The potential due to a point charge is given by Eq. 17–5, $V = kQ/r$.
>
> **SOLUTION** (*a*) At a distance of 0.50 m from a positive 20 μC charge, the potential is
>
> $$V = k\frac{Q}{r}$$
> $$= (9.0 \times 10^9\,\text{N}\cdot\text{m}^2/\text{C}^2)\left(\frac{20 \times 10^{-6}\,\text{C}}{0.50\,\text{m}}\right) = 3.6 \times 10^5\,\text{V}.$$
>
> (*b*) For the negative charge,
>
> $$V = (9.0 \times 10^9\,\text{N}\cdot\text{m}^2/\text{C}^2)\left(\frac{-20 \times 10^{-6}\,\text{C}}{0.50\,\text{m}}\right) = -3.6 \times 10^5\,\text{V}.$$
>
> **NOTE** Potential can be positive or negative, and we always include a charge's sign when we find electric potential.

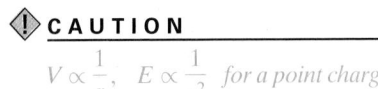

CAUTION

$V \propto \dfrac{1}{r}$, $E \propto \dfrac{1}{r^2}$ *for a point charge*

FIGURE 17–9 Potential V as a function of distance r from a single point charge Q when the charge is (a) positive, (b) negative.

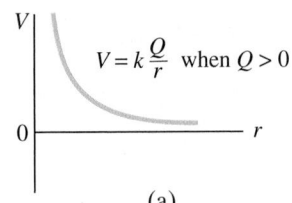

$V = k\dfrac{Q}{r}$ when $Q > 0$

(a)

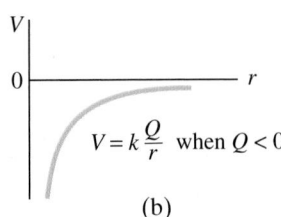

$V = k\dfrac{Q}{r}$ when $Q < 0$

(b)

PROBLEM SOLVING

Keep track of charge signs for electric potential

EXAMPLE 17–5 **Work required to bring two positive charges close together.** What minimum work must be done by an external force to bring a charge $q = 3.00\,\mu\text{C}$ from a great distance away (take $r = \infty$) to a point 0.500 m from a charge $Q = 20.0\,\mu\text{C}$?

APPROACH To find the work we cannot simply multiply the force times distance because the force is proportional to $1/r^2$ and so is not constant. Instead we can set the change in potential energy equal to the (positive of the) work required of an *external* force (Chapter 6, Eq. 6–7a), and Eq. 17–3: $W_{\text{ext}} = \Delta\text{PE} = q(V_{\text{b}} - V_{\text{a}})$. We get the potentials V_{b} and V_{a} using Eq. 17–5.

CAUTION

We cannot use $W = Fd$ if F is not constant

SOLUTION The external work required is equal to the change in potential energy:

$$W_{\text{ext}} = q(V_{\text{b}} - V_{\text{a}}) = q\left(\frac{kQ}{r_{\text{b}}} - \frac{kQ}{r_{\text{a}}}\right),$$

where $r_{\text{b}} = 0.500\,\text{m}$ and $r_{\text{a}} = \infty$. The right-hand term within the parentheses is zero ($1/\infty = 0$) so

$$W_{\text{ext}} = (3.00 \times 10^{-6}\,\text{C})\frac{(8.99 \times 10^9\,\text{N}\cdot\text{m}^2/\text{C}^2)(2.00 \times 10^{-5}\,\text{C})}{(0.500\,\text{m})} = 1.08\,\text{J}.$$

NOTE We could not use Eqs. 17–4 here because they apply *only* to uniform fields. But we did use Eq. 17–3 because it is always valid.

EXERCISE C What work is required to bring a charge $q = 3.00\,\mu\text{C}$ originally a distance of 1.50 m from a charge $Q = 20.0\,\mu\text{C}$ until it is 0.50 m away?

To determine the electric field at points near a collection of two or more point charges requires adding up the electric fields due to each charge. Since the electric field is a vector, this can be time consuming or complicated. To find the electric potential at a point due to a collection of point charges is far easier, because the electric potential is a scalar, and hence you only need to add numbers (with appropriate signs) without concern for direction.

EXAMPLE 17–6 **Potential above two charges.** Calculate the electric potential (*a*) at point A in Fig. 17–10 due to the two charges shown, and (*b*) at point B. [This is the same situation as Examples 16–9 and 16–10, Fig. 16–29, where we calculated the electric field at these points.]

APPROACH The total potential at point A (or at point B) is the algebraic sum of the potentials at that point due to each of the two charges Q_1 and Q_2. The potential due to each single charge is given by Eq. 17–5. We do not have to worry about directions because electric potential is a scalar quantity. But we do have to keep track of the signs of charges.

CAUTION

Potential is a scalar and has no components

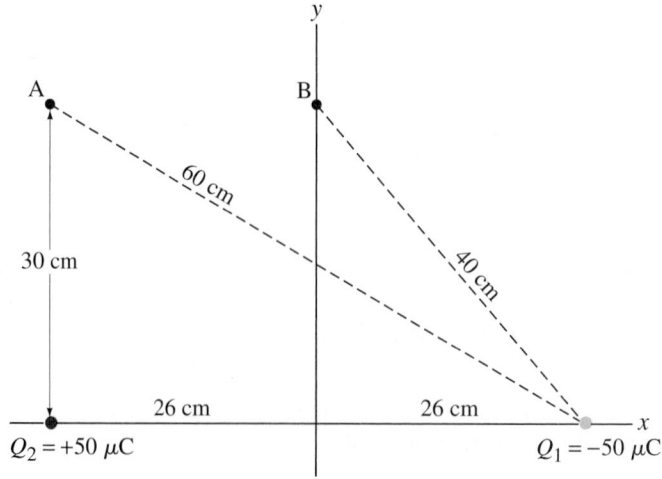

FIGURE 17–10 Example 17–6. (See also Examples 16–9 and 16–10, Fig. 16–29.)

SOLUTION (a) We add the potentials at point A due to each charge Q_1 and Q_2, and we use Eq. 17–5 for each:

$$V_A = V_{A2} + V_{A1}$$

$$= k\frac{Q_2}{r_{2A}} + k\frac{Q_1}{r_{1A}}$$

where $r_{1A} = 60\,\text{cm}$ and $r_{2A} = 30\,\text{cm}$. Then

$$V_A = \frac{(9.0 \times 10^9\,\text{N·m}^2/\text{C}^2)(5.0 \times 10^{-5}\,\text{C})}{0.30\,\text{m}} + \frac{(9.0 \times 10^9\,\text{N·m}^2/\text{C}^2)(-5.0 \times 10^{-5}\,\text{C})}{0.60\,\text{m}}$$

$$= 1.50 \times 10^6\,\text{V} - 0.75 \times 10^6\,\text{V}$$

$$= 7.5 \times 10^5\,\text{V}.$$

(b) At point B, $r_{1B} = r_{2B} = 0.40\,\text{m}$, so

$$V_B = V_{B2} + V_{B1}$$

$$= \frac{(9.0 \times 10^9\,\text{N·m}^2/\text{C}^2)(5.0 \times 10^{-5}\,\text{C})}{0.40\,\text{m}} + \frac{(9.0 \times 10^9\,\text{N·m}^2/\text{C}^2)(-5.0 \times 10^{-5}\,\text{C})}{0.40\,\text{m}}$$

$$= 0\,\text{V}.$$

NOTE The two terms in the sum in (b) cancel for any point equidistant from Q_1 and Q_2 ($r_{1B} = r_{2B}$). Thus the potential will be zero everywhere on the plane equidistant between the two opposite charges. This plane is an equipotential surface with $V = 0$.

Simple summations like these can be performed for any number of point charges.

CONCEPTUAL EXAMPLE 17–7 Potential energies. Consider the three pairs of charges shown in Fig. 17–11. Call them Q_1 and Q_2. (a) Which set has a positive potential energy? (b) Which set has the most negative potential energy? (c) Which set requires the most work to separate the charges to infinity? Assume the charges all have the same magnitude.

RESPONSE The potential energy equals the work required to bring the two charges near each other, starting at a great distance (∞). Assume the left (+) charge Q_1 is already there. To bring a second charge Q_2 close to the first from a great distance away (∞) requires external work

$$W_{\text{ext}} = Q_2 V = k\frac{Q_1 Q_2}{r}$$

where r is the final distance between them. Thus the potential energy of the two charges is

$$\text{PE} = k\frac{Q_1 Q_2}{r}.$$

(a) Set (iii) has a positive potential energy because the charges have the same sign. (b) Both (i) and (ii) have opposite signs of charge and negative PE. Because r is smaller in (i), the PE is most negative for (i). (c) Set (i) will require the most work for separation to infinity. The more negative the potential energy, the more work required to separate the charges and bring the PE up to zero ($r = \infty$), as in Fig. 17–9b.

EXERCISE D Return to the Chapter-Opening Question, page 473, and answer it again now. Try to explain why you may have answered differently the first time.

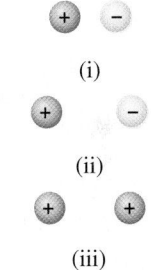

(i)

(ii)

(iii)

FIGURE 17–11 Example 17–7.

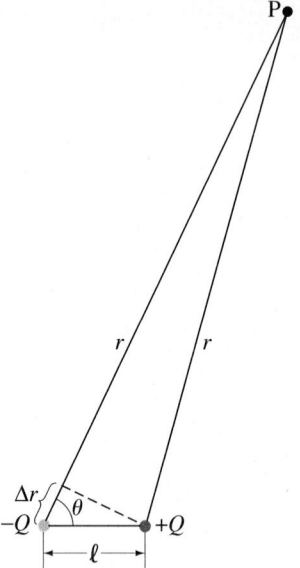

FIGURE 17-12 Electric dipole. Calculation of potential V at point P.

*17-6 Potential Due to Electric Dipole; Dipole Moment

Two equal point charges Q, of opposite sign, separated by a distance ℓ, are called an **electric dipole**. The electric field lines and equipotential surfaces for a dipole were shown in Fig. 17–7. Because electric dipoles occur often in physics, as well as in other disciplines such as molecular biology, it is useful to examine them more closely.

The electric potential at an arbitrary point P due to a dipole, Fig. 17–12, is the sum of the potentials due to each of the two charges:

$$V = \frac{kQ}{r} + \frac{k(-Q)}{r + \Delta r} = kQ\left(\frac{1}{r} - \frac{1}{r + \Delta r}\right) = kQ\frac{\Delta r}{r(r + \Delta r)},$$

where r is the distance from P to the positive charge and $r + \Delta r$ is the distance to the negative charge. This equation becomes simpler if we consider points P whose distance from the dipole is much larger than the separation of the two charges—that is, for $r \gg \ell$. From Fig. 17–12 we see that $\Delta r = \ell \cos\theta$; since $r \gg \Delta r = \ell \cos\theta$, we can neglect Δr in the denominator as compared to r. Then we obtain

$$V \approx \frac{kQ\ell \cos\theta}{r^2}. \qquad\qquad \text{[dipole; } r \gg \ell] \quad \textbf{(17–6a)}$$

We see that the potential decreases as the *square* of the distance from the dipole, whereas for a single point charge the potential decreases with the first power of the distance (Eq. 17–5). It is not surprising that the potential should fall off faster for a dipole: when you are far from a dipole, the two equal but opposite charges appear so close together as to tend to neutralize each other.

The product $Q\ell$ in Eq. 17–6a is referred to as the **dipole moment**, p, of the dipole. Equation 17–6a in terms of the dipole moment is

$$V \approx \frac{kp \cos\theta}{r^2}. \qquad\qquad \text{[dipole; } r \gg \ell] \quad \textbf{(17–6b)}$$

A dipole moment has units of coulomb-meters $(C\cdot m)$, although for molecules a smaller unit called a *debye* is sometimes used: 1 debye = $3.33 \times 10^{-30}\,C\cdot m$.

In many molecules, even though they are electrically neutral, the electrons spend more time in the vicinity of one atom than another, which results in a separation of charge. Such molecules have a dipole moment and are called **polar molecules**. We already saw that water (Fig. 16–4) is a polar molecule, and we have encountered others in our discussion of molecular biology (Section 16–10). Table 17–2 gives the dipole moments for several molecules. The + and − signs indicate on which atoms these charges lie. The last two entries are a part of many organic molecules and play an important role in molecular biology.

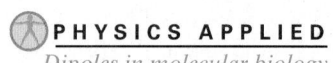

PHYSICS APPLIED
Dipoles in molecular biology

TABLE 17–2 Dipole Moments of Selected Molecules

Molecule	Dipole Moment $(C \cdot m)$
$H_2^{(+)}O^{(-)}$	6.1×10^{-30}
$H^{(+)}Cl^{(-)}$	3.4×10^{-30}
$N^{(-)}H_3^{(+)}$	5.0×10^{-30}
$>N^{(-)}-H^{(+)}$	$\approx 3.0 \times 10^{-30\ddagger}$
$>C^{(+)}=O^{(-)}$	$\approx 8.0 \times 10^{-30\ddagger}$

‡ These last two groups often appear on larger molecules; hence the value for the dipole moment will vary somewhat, depending on the rest of the molecule.

17-7 Capacitance

A **capacitor** is a device that can store electric charge, and normally consists of two conducting objects (usually plates or sheets) placed near each other but not touching. Capacitors are widely used in electronic circuits and sometimes are called **condensers**. Capacitors store charge for later use, such as in a camera flash, and as energy backup in devices like computers if the power fails. Capacitors also block surges of charge and energy to protect circuits. Very tiny capacitors serve as memory for the "ones" and "zeros" of the binary code in the random access memory (RAM) of computers and other electronic devices (as in Fig. 17–35). Capacitors serve many other applications as well, some of which we will discuss.

PHYSICS APPLIED
Uses of capacitors

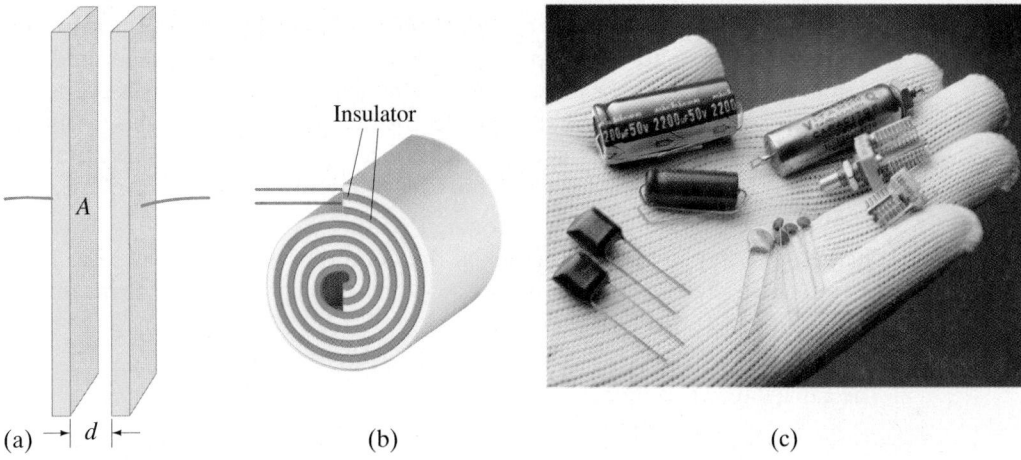

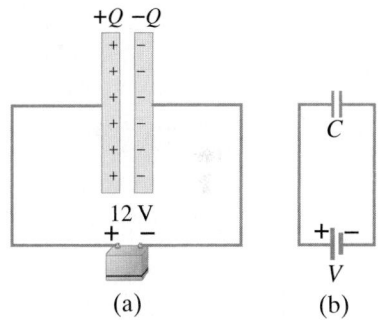

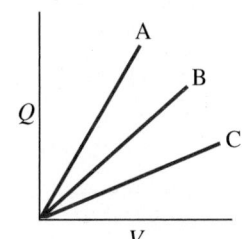

FIGURE 17–13
Capacitors: diagrams of
(a) parallel plate,
(b) cylindrical (rolled up
parallel plate). (c) Photo
of some real capacitors.

(a)   (b)   (c)

A simple capacitor consists of a pair of parallel plates of area A separated by a small distance d (Fig. 17–13a). Often the two plates are rolled into the form of a cylinder with paper or other insulator separating the plates, Fig. 17–13b; Fig. 17–13c is a photo of some actual capacitors used for various applications. In circuit diagrams, the symbol

⊣⊢   or   ⊣⊢   [capacitor symbol]

represents a capacitor. A battery, which is a source of voltage, is indicated by the symbol

⊣⊢   [battery symbol]
+ −

with unequal arms.

If a voltage is applied across a capacitor by connecting the capacitor to a battery with conducting wires as in Fig. 17–14, charge flows from the battery to each of the two plates: one plate acquires a negative charge, the other an equal amount of positive charge. Each battery terminal and the plate of the capacitor connected to it are at the same potential; hence the full battery voltage appears across the capacitor. For a given capacitor, it is found that the amount of charge Q acquired by each plate is proportional to the magnitude of the potential difference V between the plates:

$$Q = CV. \qquad \textbf{(17–7)}$$

The constant of proportionality, C, in Eq. 17–7 is called the **capacitance** of the capacitor. The unit of capacitance is coulombs per volt, and this unit is called a **farad** (F). Common capacitors have capacitance in the range of 1 pF $\left(\text{picofarad} = 10^{-12}\,\text{F}\right)$ to $10^3\,\mu\text{F}$ $\left(\text{microfarad} = 10^{-6}\,\text{F}\right)$. The relation, Eq. 17–7, was first suggested by Volta in the late eighteenth century.

In Eq. 17–7 and from now on, we will use simply V (in italics) to represent a potential difference, such as that produced by a battery, rather than V_{ba}, ΔV, or $V_b - V_a$, as previously.

Also, be sure not to confuse *italic* letters V and C which stand for voltage and capacitance, with non-italic V and C which stand for the units volts and coulombs.

The capacitance C does not in general depend on Q or V. Its value depends only on the size, shape, and relative position of the two conductors, and also on the material that separates them. For a parallel-plate capacitor whose plates have area A and are separated by a distance d of air (Fig. 17–13a), the capacitance is given by

$$C = \epsilon_0 \frac{A}{d}. \qquad \text{[parallel-plate capacitor]} \quad \textbf{(17–8)}$$

We see that C depends only on geometric factors, A and d, and not on Q or V. We derive this useful relation in the optional subsection at the end of this Section. The constant ϵ_0 is the *permittivity of free space*, which, as we saw in Chapter 16, has the value $8.85 \times 10^{-12}\,\text{C}^2/\text{N}\cdot\text{m}^2$.

| **EXERCISE E** Graphs for charge versus voltage are shown in Fig. 17–15 for three capacitors, A, B, and C. Which has the greatest capacitance?

FIGURE 17–14 (a) Parallel-plate capacitor connected to a battery. (b) Same circuit shown using symbols.

$+Q\ -Q$

12 V
+ −

(a)

C

+ ∥ −
V

(b)

⚠ **CAUTION**

V = potential difference from here on

FIGURE 17–15 Exercise E.

Q

A

B

C

V

EXAMPLE 17–8 | **Capacitor calculations.** (*a*) Calculate the capacitance of a parallel-plate capacitor whose plates are 20 cm × 3.0 cm and are separated by a 1.0-mm air gap. (*b*) What is the charge on each plate if a 12-V battery is connected across the two plates? (*c*) What is the electric field between the plates? (*d*) Estimate the area of the plates needed to achieve a capacitance of 1 F, assuming the air gap *d* is 100 times smaller, or 10 microns (1 **micron** = 1 μm = 10^{-6} m).

APPROACH The capacitance is found by using Eq. 17–8, $C = \epsilon_0 A/d$. The charge on each plate is obtained from the definition of capacitance, Eq. 17–7, $Q = CV$. The electric field is uniform, so we can use Eq. 17–4b for the magnitude $E = V/d$. In (*d*) we use Eq. 17–8 again.

SOLUTION (*a*) The area $A = (20 \times 10^{-2}\,\text{m})(3.0 \times 10^{-2}\,\text{m}) = 6.0 \times 10^{-3}\,\text{m}^2$. The capacitance C is then

$$C = \epsilon_0 \frac{A}{d} = (8.85 \times 10^{-12}\,\text{C}^2/\text{N}\cdot\text{m}^2)\frac{6.0 \times 10^{-3}\,\text{m}^2}{1.0 \times 10^{-3}\,\text{m}} = 53\,\text{pF}.$$

(*b*) The charge on each plate is

$$Q = CV = (53 \times 10^{-12}\,\text{F})(12\,\text{V}) = 6.4 \times 10^{-10}\,\text{C}.$$

(*c*) From Eq. 17–4b for a uniform electric field, the magnitude of E is

$$E = \frac{V}{d} = \frac{12\,\text{V}}{1.0 \times 10^{-3}\,\text{m}} = 1.2 \times 10^4\,\text{V/m}.$$

(*d*) We solve for A in Eq. 17–8 and substitute $C = 1.0\,\text{F}$ and $d = 1.0 \times 10^{-5}\,\text{m}$ to find that we need plates with an area

$$A = \frac{Cd}{\epsilon_0} \approx \frac{(1\,\text{F})(1.0 \times 10^{-5}\,\text{m})}{(9 \times 10^{-12}\,\text{C}^2/\text{N}\cdot\text{m}^2)} \approx 10^6\,\text{m}^2.$$

NOTE This is the area of a square 10^3 m or 1 km on a side. That is inconveniently large. Large-capacitance capacitors will not be simple parallel plates.

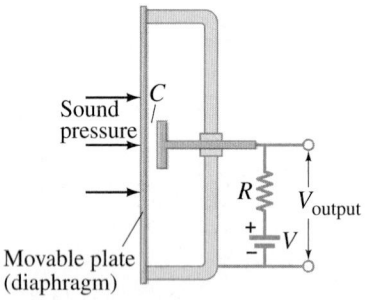
Capacitor as power backup; condenser microphone; computer keyboard

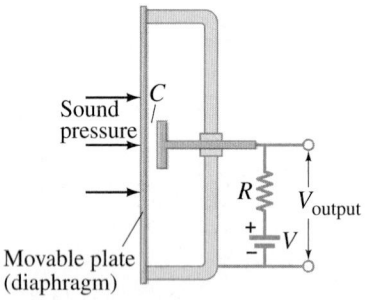

Sound pressure

Movable plate (diaphragm)

FIGURE 17–16 Diagram of a condenser microphone.

FIGURE 17–17 Key on a computer keyboard. Pressing the key reduces the plate spacing, increasing the capacitance.

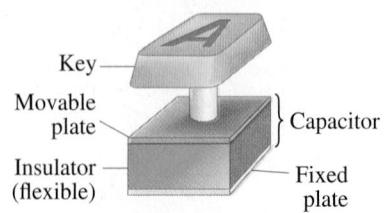

Key

Movable plate

Insulator (flexible)

Capacitor

Fixed plate

Not long ago, a capacitance greater than a few mF was unusual. Today capacitors are available that are 1 or 2 F, yet they are just a few cm on a side. Such capacitors are used as power backups, for example, in computer memory and electronics where the time and date can be maintained through tiny charge flow. [Capacitors are superior to rechargable batteries for this purpose because they can be recharged more than 10^5 times with no degradation.] Such high-capacitance capacitors can be made of *activated carbon* which has very high porosity, so that the surface area is very large; one-tenth of a gram of activated carbon can have a surface area of $100\,\text{m}^2$. Furthermore, the equal and opposite charges exist in an electric "double layer" about 10^{-9} m thick. Thus, the capacitance of 0.1 g of activated carbon, whose internal area can be $10^2\,\text{m}^2$, is equivalent to a parallel-plate capacitor with $C \approx \epsilon_0 A/d = (8.85 \times 10^{-12}\,\text{C}^2/\text{N}\cdot\text{m}^2)(10^2\,\text{m}^2)/(10^{-9}\,\text{m}) \approx 1\,\text{F}$.

The proportionality, $C \propto A/d$ in Eq. 17–8, is valid also for a parallel-plate capacitor that is rolled up into a spiral cylinder, as in Fig. 17–13b. However, the constant factor, ϵ_0, must be replaced if an insulator such as paper separates the plates, as is usual, as discussed in the next Section.

One type of microphone is a **condenser**, or capacitor, **microphone**, diagrammed in Fig. 17–16. The changing air pressure in a sound wave causes one plate of the capacitor C to move back and forth. The voltage across the capacitor changes at the same frequency as the sound wave.

Some computer keyboards operate by capacitance. As shown in Fig. 17–17, each key is connected to the upper plate of a capacitor. The upper plate moves down when the key is pressed, reducing the spacing between the capacitor plates, and increasing the capacitance (Eq. 17–8: smaller *d*, larger *C*). The *change* in capacitance results in an electric signal that is detected by an electronic circuit.

*Derivation of Capacitance for Parallel-Plate Capacitor

Equation 17–8 can be derived using the result from Section 16–12 on Gauss's law, namely that the electric field between two parallel plates is given by Eq. 16–10:

$$E = \frac{Q/A}{\epsilon_0}.$$

We combine this with Eq. 17–4a, using magnitudes, $V = Ed$, to obtain

$$V = \left(\frac{Q}{A\epsilon_0}\right)d.$$

Then, from Eq. 17–7, the definition of capacitance,

$$C = \frac{Q}{V} = \frac{Q}{(Q/A\epsilon_0)d} = \epsilon_0\frac{A}{d}$$

which is Eq. 17–8.

17–8 Dielectrics

In most capacitors there is an insulating sheet of material, such as paper or plastic, called a **dielectric** between the plates (Fig. 17–18). This serves several purposes. First, dielectrics break down (allowing electric charge to flow) less readily than air, so higher voltages can be applied without charge passing across the gap. Furthermore, a dielectric allows the plates to be placed closer together without touching, thus allowing an increased capacitance because d is smaller in Eq. 17–8. Thirdly, it is found experimentally that if the dielectric fills the space between the two conductors, it increases the capacitance by a factor K, known as the **dielectric constant**. Thus, for a parallel-plate capacitor,

$$C = K\epsilon_0\frac{A}{d}. \qquad (17\text{–}9)$$

This can be written

$$C = \epsilon\frac{A}{d},$$

where $\epsilon = K\epsilon_0$ is called the **permittivity** of the material.

The values of the dielectric constant for various materials are given in Table 17–3. Also shown in Table 17–3 is the **dielectric strength**, the maximum electric field before breakdown (charge flow) occurs.

FIGURE 17–18 A cylindrical capacitor, unrolled from its case to show the dielectric between the plates. See also Fig. 17–13b.

**TABLE 17–3
Dielectric Constants** (at 20°C)

Material	Dielectric constant K	Dielectric strength (V/m)
Vacuum	1.0000	
Air (1 atm)	1.0006	3×10^6
Paraffin	2.2	10×10^6
Polystyrene	2.6	24×10^6
Vinyl (plastic)	2–4	50×10^6
Paper	3.7	15×10^6
Quartz	4.3	8×10^6
Oil	4	12×10^6
Glass, Pyrex	5	14×10^6
Rubber, neoprene	6.7	12×10^6
Porcelain	6–8	5×10^6
Mica	7	150×10^6
Water (liquid)	80	
Strontium titanate	300	8×10^6

| **CONCEPTUAL EXAMPLE 17–9** | **Inserting a dielectric at constant V.** An air-filled capacitor consisting of two parallel plates separated by a distance d is connected to a battery of constant voltage V and acquires a charge Q. While it is still connected to the battery, a slab of dielectric material with $K = 3$ is inserted between the plates of the capacitor. Will Q increase, decrease, or stay the same?

RESPONSE Since the capacitor remains connected to the battery, the voltage stays constant and equal to the battery voltage V. The capacitance C increases when the dielectric material is inserted because K in Eq. 17–9 has increased. From the relation $Q = CV$, if V stays constant, but C increases, Q must increase as well. As the dielectric is inserted, more charge will be pulled from the battery and deposited onto the plates of the capacitor as its capacitance increases.

EXERCISE F If the dielectric in Example 17–9 fills the space between the plates, by what factor does (a) the capacitance change, (b) the charge on each plate change?

| **CONCEPTUAL EXAMPLE 17–10** | **Inserting a dielectric into an isolated capacitor.** Suppose the air-filled capacitor of Example 17–9 is charged (to Q) and then disconnected from the battery. Next a dielectric is inserted between the plates. Will Q, C, or V change?

RESPONSE The charge Q remains the same—the capacitor is isolated, so there is nowhere for the charge to go. The capacitance increases as a result of inserting the dielectric (Eq. 17–9). The voltage across the capacitor also changes—it *decreases* because, by Eq. 17–7, $Q = CV$, so $V = Q/C$; if Q stays constant and C increases (it is in the denominator), then V decreases.

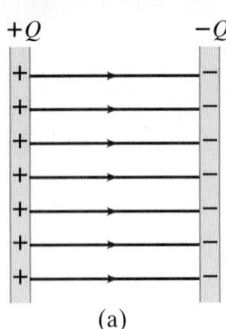

+Q   −Q

(a)

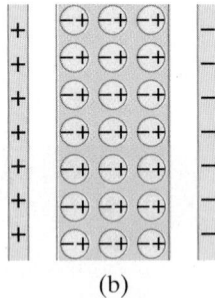

(b)

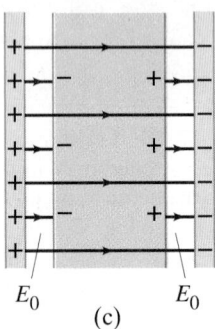

E_0 E_0

(c)

FIGURE 17–19 Molecular view of the effects of a dielectric.

*Molecular Description of Dielectrics

Let us examine, from the molecular point of view, why the capacitance of a capacitor should be larger when a dielectric is between the plates. A capacitor C_0 whose plates are separated by an air gap has a charge $+Q$ on one plate and $−Q$ on the other (Fig. 17–19a). Assume it is isolated (not connected to a battery) so charge cannot flow to or from the plates. The potential difference between the plates, V_0, is given by Eq. 17–7:

$$Q = C_0 V_0,$$

where the subscripts refer to air between the plates. Now we insert a dielectric between the plates (Fig. 17–19b). Because of the electric field between the capacitor plates, the dielectric molecules will tend to become oriented as shown in Fig. 17–19b. If the dielectric molecules are *polar*, the positive end is attracted to the negative plate and vice versa. Even if the dielectric molecules are not polar, electrons within them will tend to move slightly toward the positive capacitor plate, so the effect is the same. The net effect of the aligned dipoles is a net negative charge on the outer edge of the dielectric facing the positive plate, and a net positive charge on the opposite side, as shown in Fig. 17–19c.

Some of the electric field lines, then, do not pass through the dielectric but instead end on charges induced on the surface as shown in Fig. 17–19c. Hence the electric field within the dielectric is less than in air. That is, the electric field in the space between the capacitor plates, assumed filled by the dielectric, has been reduced by some factor K. The voltage across the capacitor is reduced by the same factor K because $V = Ed$ (Eq. 17–4) and hence, by Eq. 17–7, $Q = CV$, the capacitance C must increase by that same factor K to keep Q constant.

17–9 Storage of Electric Energy

A charged capacitor stores electric energy by separating $+$ and $−$ charges. The energy stored in a capacitor will be equal to the work done to charge it. The net effect of charging a capacitor is to remove charge from one plate and add it to the other plate. This is what a battery does when it is connected to a capacitor. A capacitor does not become charged instantly. It takes some time, often very little (Section 19–6). Initially, when the capacitor is uncharged, no work is required to move the first bit of charge over. As more charge is transferred, work is needed to move charge against the increasing voltage V. The work needed to add a small amount of charge Δq, when a potential difference V is across the plates, is $\Delta W = V \Delta q$. The total work needed to move total charge Q is equivalent to moving all the charge Q across a voltage equal to the *average* voltage during the process. (This is just like calculating the work done to compress a spring, Section 6–4, page 148.) The average voltage is $(V_f - 0)/2 = V_f/2$, where V_f is the final voltage; so the work to move the total charge Q from one plate to the other is

$$W = Q\frac{V_f}{2}.$$

Thus we can say that the electric potential energy, PE, stored in a capacitor is

$$\text{PE} = \text{energy} = \tfrac{1}{2}QV,$$

where V is the potential difference between the plates (we dropped the subscript), and Q is the charge on each plate. Since $Q = CV$, we can also write

$$\text{PE} = \tfrac{1}{2}QV = \tfrac{1}{2}CV^2 = \tfrac{1}{2}\frac{Q^2}{C}. \tag{17–10}$$

PHYSICS APPLIED

Camera flash

EXAMPLE 17–11 **Energy stored in a capacitor.** A camera flash unit (Fig. 17–20) stores energy in a 660-μF capacitor at 330 V. (a) How much electric energy can be stored? (b) What is the power output if nearly all this energy is released in 1.0 ms?

APPROACH We use Eq. 17–10 in the form PE $= \tfrac{1}{2}CV^2$ because we are given C and V.

SOLUTION (a) The energy stored is

$$\text{PE} = \tfrac{1}{2}CV^2 = \tfrac{1}{2}(660 \times 10^{-6}\,\text{F})(330\,\text{V})^2 = 36\,\text{J}.$$

(b) If this energy is released in $\frac{1}{1000}$ of a second $(= 1.0\,\text{ms} = 1.0 \times 10^{-3}\,\text{s})$, the power output is $P = \text{PE}/t = (36\,\text{J})/(1.0 \times 10^{-3}\,\text{s}) = 36{,}000\,\text{W}.$

| **EXERCISE G** A capacitor stores 0.50 J of energy at 9.0 V. What is its capacitance?

CONCEPTUAL EXAMPLE 17–12 | **Capacitor plate separation increased.**
A parallel-plate capacitor carries charge Q and is then disconnected from a battery. The two plates are initially separated by a distance d. Suppose the plates are pulled apart until the separation is $2d$. How has the energy stored in this capacitor changed?

RESPONSE If we increase the plate separation d, we decrease the capacitance according to Eq. 17–8, $C = \epsilon_0 A/d$, by a factor of 2. The charge Q hasn't changed. So according to Eq. 17–10, where we choose the form $\text{PE} = \tfrac{1}{2}Q^2/C$ because we know Q is the same and C has been halved, the reduced C means the PE stored increases by a factor of 2.

NOTE We can see why the energy stored increases from a physical point of view: the two plates are charged equal and opposite, so they attract each other. If we pull them apart, we must do work, so we raise the potential energy.

It is useful to think of the energy stored in a capacitor as being stored in the electric field between the plates. As an example let us calculate the energy stored in a parallel-plate capacitor in terms of the electric field.

We have seen that the electric field $\vec{\mathbf{E}}$ between two close parallel plates is nearly uniform and its magnitude is related to the potential difference by $V = Ed$ (Eq. 17–4), where d is the separation. Also, Eq. 17–8 tells us $C = \epsilon_0 A/d$ for a parallel-plate capacitor. Thus

$$\text{PE} = \tfrac{1}{2}CV^2 = \frac{1}{2}\left(\frac{\epsilon_0 A}{d}\right)(E^2 d^2)$$
$$= \tfrac{1}{2}\epsilon_0 E^2 A d.$$

The quantity Ad is the volume between the plates in which the electric field E exists. If we divide both sides of this equation by the volume, we obtain an expression for the energy per unit volume or **energy density**:

$$\text{energy density} = \frac{\text{PE}}{\text{volume}} = \tfrac{1}{2}\epsilon_0 E^2. \qquad \textbf{(17–11)}$$

The *electric energy stored per unit volume in any region of space is proportional to the square of the electric field* in that region. We derived Eq. 17–11 for the special case of a parallel-plate capacitor. But it can be shown to be true for any region of space where there is an electric field. Indeed, we will use this result when we discuss electromagnetic radiation (Chapter 22).

Health Effects

The energy stored in a large capacitance can give you a burn or a shock. One reason you are warned not to touch a circuit, or open an electronic device, is because capacitors may still be carrying charge even if the external power is turned off.

On the other hand, the basis of a heart *defibrillator* is a capacitor charged to a high voltage. A heart attack can be characterized by fast irregular beating of the heart, known as *ventricular* (or *cardiac*) *fibrillation*. The heart then does not pump blood to the rest of the body properly, and if the interruption lasts for long, death results. A sudden, brief jolt of charge through the heart from a defibrillator can cause complete heart stoppage, sometimes followed by a resumption of normal beating. The defibrillator capacitor is charged to a high voltage, typically a few thousand volts, and is allowed to discharge very rapidly through the heart via a pair of wide contacts known as "pads" or "paddles" that spread out the current over the chest (Fig. 17–21).

FIGURE 17–20 A camera flash unit. The 660-μF capacitor is the black cylinder.

PHYSICS APPLIED
Shocks, burns, defibrillators

FIGURE 17–21 Heart defibrillator.

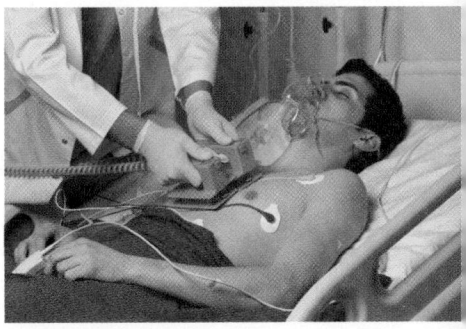

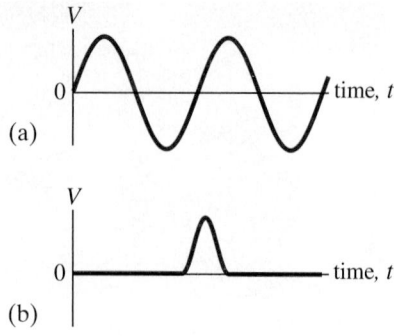

(a)

(b)

FIGURE 17–22 Two kinds of signal voltage: (a) sinusoidal, (b) a pulse, both analog. Many other shapes are possible.

TABLE 17–4
Binary to Decimal

Binary† number	Decimal number
00000000	0
00000001	1
00000010	2
00000011	3
00000100	4
00000111	7
00001000	8
00100101	37
11111111	255

†Note that we start counting from right to left: the 1's digit is on the far right, then the 2's, the 4's, the 8's, the 16's, the 32's, the 64's, and the 128's.

FIGURE 17–24 The red analog sine wave, which is at a 100-Hz frequency (1 wavelength is done in 0.010 s), has been converted to a 2-bit (4 level) digital signal (blue).

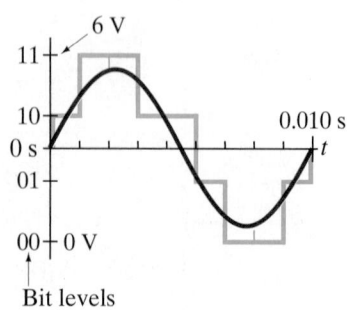

Bit levels

17–10 Digital; Binary Numbers; Signal Voltage

Batteries and a wall plug are meant to provide a constant **supply voltage** as power to operate a flashlight, an electric heater, and other electric and electronic devices.

A **signal voltage**, on the other hand, is a voltage intended to affect something else. A signal voltage varies in time and can also be very brief. For example, a sound such as a pure tone, which may be sinusoidal as we discussed in Chapters 11 and 12 (see Figs. 11–24 and 12–14), will produce an output voltage from a high quality microphone that is also sinusoidal. That signal voltage is amplified and reaches a loudspeaker, making it produce the sound we hear. Signal voltages (see Fig. 17–22) are sometimes a simple pulse (Fig. 17–22b; see also Figs. 11–23, 11–33), and often act to change some aspect of an electronic device.

Signal voltages are sent to cell phones ("I've got signal"), to computers from the Internet, or to TV sets with the information on the picture and sound. Not long ago, signal voltages were **analog**—the voltage varied continuously, as in Fig. 17–22a.

Today, television and computer signals are **digital** and use a binary number system to represent a numerical value. In a normal number, such as 609, there are *ten* choices for each digit—from 0 to 9—and normal numbers are called **decimal** (Latin for ten). In a **binary** number, each digit or **bit** has only *two* possibilities, 0 or 1 (sometimes referred to as "off" or "on"). In binary, 0001 means "one," 0010 means 2, 0011 means 3, and 1101 means $8 + 4 + 0 + 1 = 13$ in decimal. See Table 17–4, and note that counting starts from the right, just as in regular decimal (the "ones" digit is last, on the far right, then to the left is the "tens" and then "hundreds": for 609, the "ones" are 9, the "hundreds" are 6). Any value can be represented by a voltage pattern something like that shown in Fig. 17–23.

FIGURE 17–23 A traveling digital signal: voltage vs. position x or time t. If standing alone, this sequence would represent 10011001 or 153 ($= 128 + 0 + 0 + 16 + 8 + 0 + 0 + 1$).

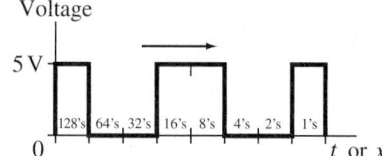

A "1" is a positive voltage such as +5 V, whereas a "0" is 0 V. The brightness signal, for example, that goes to each of the millions of tiny picture elements or "subpixels" of a TV or computer screen (Fig. 17–31, Section 17–11), is contained in a **byte**. One byte is 8 bits, which means

each byte of 8 bits allows $2^8 = 256$ possibilities

(that is, 0 to 255) or 256 shades for each of 3 colors: red, green, blue. The full color of each pixel (the three subpixel colors) has $(256)^3 = 17 \times 10^6$ possibilities. Digital television signals, which we discuss in the next Section, are transmitted at about 19 Mb/s = 19 Megabits per second. So 19×10^6 bits pass a given point per second, or one bit every 53 nanoseconds. We could write this in terms of bytes as 2.4 MB/s, where for bytes we use capital B.

When an analog signal, such as the pure sine wave of Fig. 17–22a, is converted to digital (**analog-to-digital converter**, ADC), the digital signal may look like the blue squared-off curve of Fig. 17–24. The digital signal has a limited number of discrete values. The difference between the original continuous analog signal and its digital approximation is called the **quantization error** or **quantization loss**. To minimize that loss, there are two important factors: (i) the **resolution** or **bit depth**, which is the number of bits or values for the voltage of each sample (= measurement); (ii) the **sampling rate**, which is the number of times per second the original analog voltage is measured ("sampled").

Consider a digital approximation for a 100-Hz sine wave: Figure 17–24 shows (i) a 0 to 6-V, 2-bit depth, measuring only 4 possible voltages (00, 01, 10, 11, or 0, 1, 2, 3 in decimal), and (ii) a sampling rate of (9 samples in one cycle or wavelength) × (100 cycles/s = 100 Hz) which is 900 samples/s or 900 Hz. This is very poor quality. For high quality reproduction, a greater bit depth and higher sampling rate are needed, which requires more memory, and more data to be transmitted.

For audio CDs, the sampling rate is 44.1 kHz (44,100 samplings every second) and 16-bit resolution, meaning each sampled voltage can have $2^8 \times 2^8 = 2^{16} \approx 65,000$ different voltage levels between, say, 0 and 5 volts. See Fig. 17–25 for details. Audio recording today typically uses 96 kHz and 24-bit ($2^{24} \approx 17 \times 10^6$ voltage levels) to give a better approximation of the original analog signal (on super-CDs or solid-state memory), but must be transferred down to 44.1 kHz and 16-bit to produce ordinary CDs. (DVDs can use 192 kHz sampling rate for sound.) But iPods and MP3 players have lower sampling rates and much less detail, which many listeners can notice.

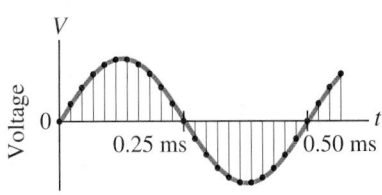

FIGURE 17–25 The sine wave shown could represent the analog electric signal from a microphone due to a pure 2000-Hz tone. (See Chapters 11 and 12.) The analog-to-digital electronics **samples** the signal—that is, measures and records the signal's voltage at intervals, many times per second. Each dot on the curve represents the voltage measured (sampled) at that point. The sampling rate in this diagram is 44,100 each second, or 44.1 kHz, like a CD. That is, a sample is taken every $(1\text{ s})/44,100 = 0.000023\text{ s} = 0.023$ ms. In 0.50 ms, as shown here, 22 samples (black dots) are taken. This is an alternate way to represent sampling compared to Fig. 17-24, and shows that we cannot see any changes that might happen between the samplings (dots).

Figure 17–25 gives some details about a pure 2000-Hz sound sampled at 44.1 kHz. Normal musical sounds are a complex summation of many such sine waves of different frequencies and amplitudes. A simple summation was shown in Fig. 12–14. Another example is shown in Fig. 17–26, where we can see that the fine details may be missed by a digital conversion. Look at Fig. 17–25: if that were 20,000 Hz (highest frequency of human hearing), it would be sampled only about two times per wavelength. Both those samples might be zero volts—obviously missing the entire waveform. Over many wavelengths, it might eventually reproduce the waveform somewhat well. But many sounds only last milliseconds, like the initial attack of a piano note or plucked guitar string. Many audiophiles hear the difference between an original vinyl record and its subsequent release as a CD at 44.1 kHz.

Digital audio signals must be converted back to analog (**digital-to-analog converter**, DAC) before being sent to a loudspeaker or headset. Even in a TV, the digital signals are converted to analog voltages before addressing the pixels (next Section), although the picture itself might be said to be digital since it is made up of separate pixels.

Digital photographs are made up of millions of "pixels" to produce a sharp image that is not "pixelated" or blurry. Also important (and complicated) are the number of bits provided for colors, plus the ability of the sensors (Chapter 25) to sustain a wide range of brightnesses under dim and bright light conditions.

Digital data has some real advantages: for one, it can be **compressed**, in the sense that repeated information can be reduced so that less memory space is required—fewer bits and bytes. For example, adjacent "pixels" on a photograph that includes a blue sky may be essentially identical. If 200 almost identical pixels can be coded as identical, that takes up less memory (or "size") than to specify all the 200 pixels individually. Compression schemes, like **jpeg** for photos, lose some information and may be noticeable. In audio, MP3 players use one-tenth the space that a CD does, but many listeners don't notice. Compression is one reason that more data or "information" can be transmitted digitally for a given **bandwidth**. [Bandwidth is the fixed range of frequencies allotted to each radio or TV station or Internet connection, and limits the number of bits transmitted per second.]

In audio, many listeners claim that digital does not match analog in full sound quality. And what about movies? Will digital ever match Technicolor?

*Noise

Digital information transmission has another advantage: any distortion or unwanted (external) electrical signal that intrudes from outside, broadly called **noise**, can badly corrupt an analog signal: Fig. 17–27a shows a time-varying analog signal, and Fig. 17–27b shows nasty outside noise interfering with it. But a digital signal is still readable unless the noise is very large, on the order of half the bit signal itself (Figs. 17–27c and d).

FIGURE 17–26 This type of complex signal is much more normal than the pure sine wave of Fig. 17–25. Sampling may not catch all the details, especially because the waveform is changing very fast in time.

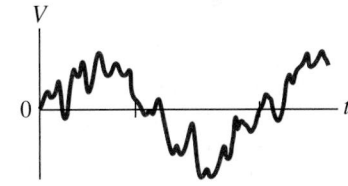

FIGURE 17–27 (a) Original analog signal and (b) the same signal dirtied up by outside signals (= noise). (c) A digital signal is still readable (d) without error if the noise is not too great.

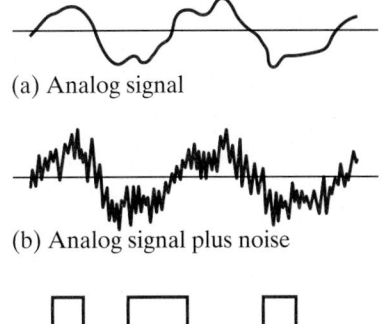

(a) Analog signal

(b) Analog signal plus noise

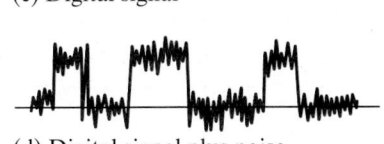

(c) Digital signal

(d) Digital signal plus noise

*17–11 TV and Computer Monitors: CRTs, Flat Screens

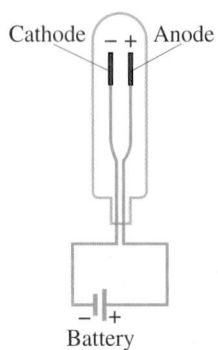

FIGURE 17–28 If the cathode inside the evacuated glass tube is heated to glowing (by an electric current, not shown), negatively charged "cathode rays" (= electrons) are "boiled off" and flow across to the anode (+), to which they are attracted.

PHYSICS APPLIED
CRT

The first television receivers used a **cathode ray tube** (**CRT**), and as recently as 2008 they accounted for half of all new TV sales. Two years later it was tough to find a new CRT set to buy. Even though new TV sets are flat screen plasma or **liquid crystal displays** (**LCD**), an understanding of how a CRT works is useful.

*CRT

The operation of a CRT depends on **thermionic emission**, discovered by Thomas Edison (1847–1931). Consider a voltage applied to two small electrodes inside an evacuated glass "tube" as shown in Fig. 17–28: the **cathode** is negative, and the **anode** is positive. If the cathode is heated (usually by an electric current) so that it becomes hot and glowing, it is found that negative charges leave the cathode and flow to the positive anode. These negative charges are now called electrons, but originally they were called **cathode rays** because they seemed to come from the cathode (more detail in Section 27–1 on the discovery of the electron).

Figure 17–29 is a simplified sketch of a CRT which is contained in an evacuated glass tube. A beam of electrons, emitted by the heated cathode, is accelerated by the high-voltage anode and passes through a small hole in that anode. The inside of the tube face on the right (the screen) is coated with a fluorescent material that glows at the spot where the electrons hit. Voltage applied across the horizontal and vertical deflection plates, Fig. 17–29, can be varied to deflect the electron beam to different spots on the screen.

FIGURE 17–29 A cathode-ray tube. Magnetic deflection coils are commonly used in place of the electric deflection plates shown here. The relative positions of the elements have been exaggerated for clarity.

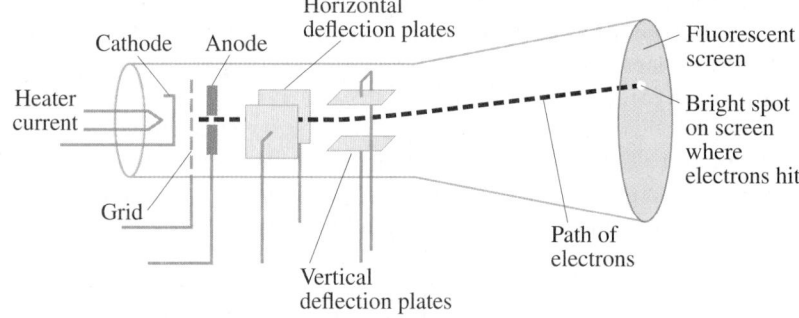

PHYSICS APPLIED
TV and computer monitors

FIGURE 17–30 Electron beam sweeps across a CRT television screen in a succession of horizontal lines, referred to as a **raster**. Each horizontal sweep is made by varying the voltage on the horizontal deflection plates (Fig. 17–29). Then the electron beam is moved down a short distance by a change in voltage on the vertical deflection plates, and the process is repeated.

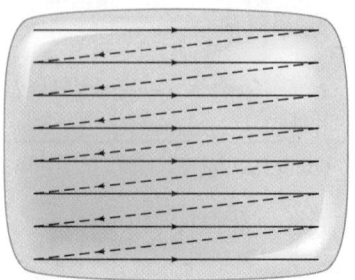

In TV and computer monitors, the CRT electron beam sweeps over the screen in the manner shown in Fig. 17–30 by carefully synchronized voltages applied to the deflection plates (more commonly by magnetic deflection coils—Chapter 20). During each horizontal sweep of the electron beam, the **grid** (Fig. 17–29) receives a signal voltage that limits the flow of electrons at each instant during the sweep; the more negative the grid voltage is, the more electrons are repelled and fewer pass through, producing a less bright spot on the screen. Thus the varying grid voltage is responsible for the brightness of each spot on the screen. At the end of each horizontal sweep of the electron beam, the horizontal deflection voltage changes dramatically to bring the beam back to the opposite side of the screen, and the vertical voltage changes slightly so the beam begins a new horizontal sweep slightly below the previous one. The difference in brightness of the spots on the screen forms the "picture." **Color screens** have red, green, and blue phosphors which glow when struck by the electron beam. The various brightnesses of adjacent red, green, and blue phosphors (so close together we don't distinguish them) produce almost any color. Analog TV for the U.S. provided 480 visible horizontal sweeps[†] to form a complete picture every $\frac{1}{30}$ s. With 30 new frames or pictures every second (25 in countries with 50-Hz line voltage), a "moving picture" is displayed on the TV screen. (Note: commercial movies on film are 24 frames per second.)

[†]525 lines in total, but only 480 form the picture; the other 45 lines contain other information such as synchronization. The sweep is **interlaced**: that is, every $\frac{1}{60}$ s every other line is traced, and in the next $\frac{1}{60}$ s, the lines in between are traced.

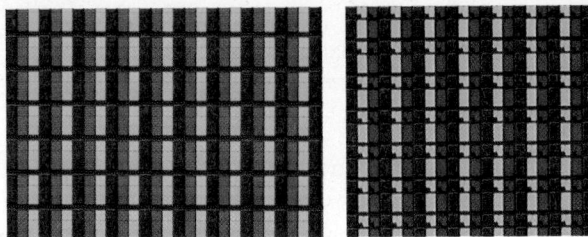

FIGURE 17–31 Close up of a tiny section of two typical LCD screens. You can even make out wires and transistors in the one on the right.

*Flat Screens and Addressing Pixels

Today's flat screens contain millions of tiny *picture elements*, or **pixels**. Each pixel consists of 3 **subpixels**, a red, a green, and a blue. A close up of a common arrangement of pixels is shown in Fig. 17–31 for an LCD screen. (How liquid crystals work in an LCD screen is described in Section 24–11.) Subpixels are so small that at normal viewing distances we don't distinguish them and the separate red (R), green (G), and blue (B) subpixels blend to produce almost any color, depending on the relative brightnesses of the three subpixels. Liquid crystals act as filters (R, G, and B) that filter the light from a white **backlight**, usually fluorescent lamps or *light-emitting diodes* (LED, Section 29–9).[†] The picture you see on the screen depends on the level of brightness of each subpixel, as suggested in Fig. 17–32 for a simple black and white picture.

High definition (HD) television screens have 1080 horizontal rows of pixels, each row consisting of 1920 pixels across the screen. That is, there are 1920 vertical columns, for a total of nearly 2 million pixels. Today, television in the U.S. is transmitted digitally at a rate of 60 Hz—that is, 60 frames or pictures per second (50 Hz in many countries) which makes the "moving picture." To form one frame, each subpixel must have the correct brightness. We now describe one way of doing this.

The brightness of each LCD subpixel (Section 24–11) depends on the voltage between its front and its back: if this voltage ΔV is zero, that subpixel is at maximum brightness; if ΔV is at its maximum (which might be $+5$ volts), that subpixel is dark.

Giving the correct voltage (to provide the correct brightness) is called **addressing** the subpixel. Typically the front of the subpixel is maintained at a positive voltage, such as $+5$ V. On the back of the display, the voltage at each subpixel is provided at the intersection of the 1080 horizontal wires (rows) and 1920×3 (colors) ≈ 6000 vertical wires (columns). See Fig. 17–33, which shows the array, or **matrix**, of wires. Each intersection of one vertical and one horizontal wire lies behind one subpixel. Because many frames are shown per second, the signal voltages applied are brief, like a pulse (see Fig. 17-22b or 11–23).

FIGURE 17–32 Example of an image made up of many small squares or *pixels* (picture elements). This one has rather low resolution.

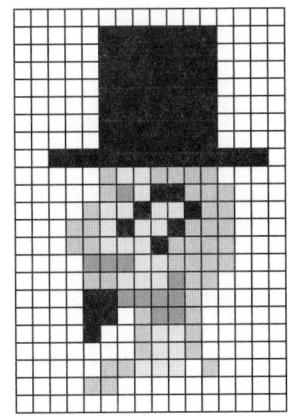

FIGURE 17–33 Array of wires (a matrix) behind all the pixels on an LCD screen. Each intersection of two wires is at a subpixel (red, green, or blue). One horizontal wire is activated at a time (the orange one at the moment shown) meaning it is at a positive voltage ($+20$ V) which allows that one row of pixels to be addressed at that moment; all other horizontal wires are at 0 V. At this moment, the data stream arrives to all the vertical wires, presenting the needed voltage (between 0 and 5 V) to produce the correct brightness for each of the nearly 6000 subpixels along the activated row.

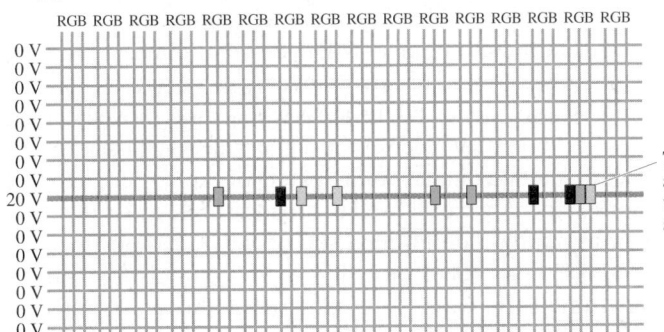

This pixel, with all 3 colors shining, would appear white at viewing distances

The video signal that arrives at the display **activates** only one horizontal wire at a time (the orange one in Fig. 17–33): that one horizontal wire has a voltage (let's say $+20$ V) whereas all the others are at 0 V. That 20 V is not applied directly to the pixels, but *allows* the vertical wires to apply briefly the proper "signal voltage" to each subpixel along that row (via a transistor, see below). These signal voltages, known as the **data stream**, are applied to all the vertical wires just as that one row is activated: they provide the correct brightness for each subpixel in that activated row. A few subpixels are highlighted in Fig. 17–33. Immediately afterward, the other rows are activated, one by one, until the entire frame has been completed (in $\frac{1}{60}$ s).

[†]LEDs are discussed in Section 29–9. Home TVs advertised as LED generally mean an LCD screen with an LED backlight. LED pixels small enough for home screens are difficult to make, but actual LED screens are found in very large displays such as at stadiums.

Then a new frame is started. The addressing of subpixels for each row of each frame serves the same purpose as the sweep of the electron beam in a CRT, Fig. 17–30.

*Active Matrix (advanced)

High-definition displays use an **active matrix**, meaning that a tiny **thin-film transistor** (**TFT**) is attached to a corner of the back of each subpixel. (Transistors are discussed in Section 29–10.) One electrode of each TFT, called the "source," is connected to the vertical wire which addresses that subpixel, Fig. 17–34, and the "drain" electrode is connected to the back of the subpixel. The horizontal wire that serves the subpixel is connected to the transistor's **gate** electrode. The gate's voltage, by attracting charge or not, functions as a switch to connect or disconnect the source voltage to the drain and to the back of the subpixel (its front is fixed at +5 V). The potential difference ΔV across a subpixel determines if that subpixel will be bright in color ($\Delta V = 0$), black ($\Delta V = $ maximum), or something in between. See Fig. 17–35. All the subpixel TFTs along the one activated horizontal wire (the orange one in Fig. 17–33) will have +20 V at the gate: the TFTs are turned "on," like a switch. That allows electric charge to flow, connecting the vertical wire signal voltage at each TFT source to its drain and to the back of the subpixel. Thus all subpixels along one row receive the brightness needed for that line of the frame.

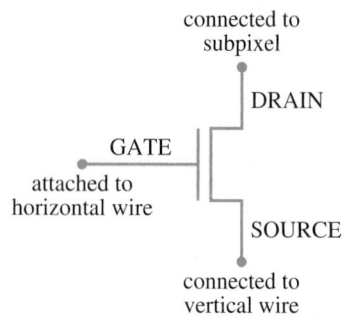

FIGURE 17–34 Thin-film transistor. One is attached to each screen subpixel.

FIGURE 17–35 Circuit diagram for one subpixel. The front of the subpixel is at +5 V. If the TFT gate is at 20 V (horizontal wire activated), the data stream voltage is applied to the back of the subpixel, and determines the brightness of that subpixel. If the gate is at 0 V (horizontal wire not activated), the TFT is "off": no charge passes through, and the capacitance helps maintain ΔV until the subpixel is updated $\frac{1}{60}$ s later.

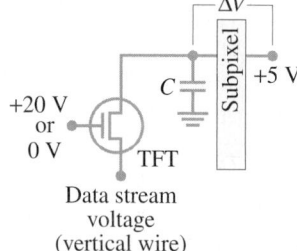

Within a subpixel's electronics is a capacitance that helps maintain the ΔV until that subpixel is **updated** with a new signal for the next frame, $\frac{1}{60}$ s later (≈ 17 ms). The row below the orange one shown in Fig. 17–33 is activated about 15 μs later $\left[= \left(\frac{1}{60}\,\text{s}\right)\left(\frac{1}{1080\,\text{lines}}\right)\right]$. The 6000 vertical wires (**data lines**) get their signal voltages (data stream) updated just before each row is activated in order to establish the brightness of each subpixel in that next row. All 1080 rows are activated, one-by-one, within $\frac{1}{60}$ s (≈ 17 ms) to complete that frame. Then a new frame is started.

New TV sets today can often refresh the screen at a higher rate. A **refresh rate** of 120 Hz (or 240 Hz) means that frames are interpolated between the normal ones, by averaging, which produces less blurring in fast action scenes.

Digital TV is transmitted at about 19 MB/s as mentioned in Section 17–10. (This rate is way too slow to do a full refresh every $\frac{1}{60}$ s—try the calculation and see—so a lot of compression is done and the areas where most movement occurs get refreshed.) The TV set or "box" that receives the digital video signal has to decode the signal in order to send analog voltages to the pixels of the screen, and at just the right time. TV stations in the U.S. are allowed to broadcast HD at 1080×1920 pixels or at 720×1280, or in standard definition (SD) of 480×704 pixels.

[When you read 1080p or 1080i for a TV, the "p" stands for "progressive," meaning an entire frame is made in $\frac{1}{60}$ s as described above. The "i" stands for "interlaced," meaning all the odd rows (half the picture) are done in $\frac{1}{60}$ s and then all the even rows are done in the next $\frac{1}{60}$ s, so a full picture is done at 30 per second or 30 Hz, thus reducing the data (or bit) rate. Analog TV (US) was 480i.]

*Oscilloscopes

An **oscilloscope** is a device for amplifying, measuring, and visually displaying an electrical signal as a function of time on an LCD or CRT monitor, or computer screen. The visible "trace" on the screen, which could be an electrocardiogram (Fig. 17–36), or a signal from an experiment on nerve conduction, is a plot of the signal voltage (vertically) versus time (horizontally). [In a CRT, the electron beam is swept horizontally at a uniform rate in time by the horizontal deflection plates, Figs. 17–29 and 17–30. The signal to be displayed is applied (after amplification) to the vertical deflection plates.]

FIGURE 17–36 An electrocardiogram (ECG) trace displayed on a CRT.

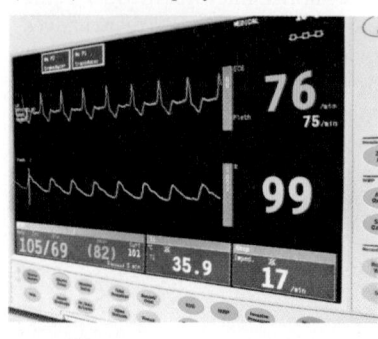

PHYSICS APPLIED

Oscilloscope

*17–12 Electrocardiogram (ECG or EKG)

Each time the heart beats, changes in electrical potential occur on its surface that can be detected using *electrodes* (metal contacts), which are attached to the skin. The changes in potential are small, on the order of millivolts (mV), and must be amplified. They are displayed with a chart recorder on paper, or on a monitor (CRT or LCD), Fig. 17–36. An **electrocardiogram** (ECG or EKG) is the record of the potential changes for a given person's heart. An example is shown in Fig. 17–37. We now look at the source of these potential changes and their relation to heart activity.

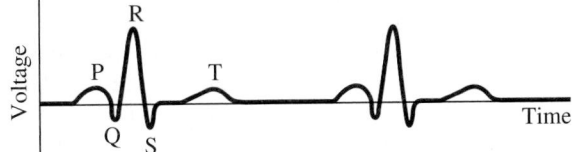

FIGURE 17–37 Typical ECG. Two heart beats are shown.

Both muscle and nerve cells have an electric dipole layer across the cell wall. That is, in the normal situation there is a net positive charge on the exterior surface and a net negative charge on the interior surface, Fig. 17–38a. The amount of charge depends on the size of the cell, but is approximately 10^{-3} C/m² of surface. For a cell whose surface area is 10^{-5} m², the total charge on either surface is thus $\approx 10^{-8}$ C. Just before the contraction of heart muscles, changes occur in the cell wall, so that positive ions on the exterior of the cell are able to pass through the wall and neutralize charge on the inside, or even make the inside surface slightly positive compared to the exterior. This "depolarization" starts at one end of the cell and progresses toward the opposite end, as indicated by the arrow in Fig. 17–38b, until the whole muscle is depolarized; the muscle then repolarizes to its original state (Fig. 17–38a), all in less than a second. Figure 17–38c shows rough graphs of the potential V as a function of time at the two points P and P′ (on either side of this cell) as the depolarization moves across the cell. The path of depolarization within the heart as a whole is more complicated, and produces the complex potential difference as a function of time, Fig. 17–37.

It is standard procedure to divide a typical electrocardiogram into regions corresponding to the various deflections (or "waves"), as shown in Fig. 17–37. Each of the deflections corresponds to the activity of a particular part of the heart beat (Fig. 10–42). The P wave corresponds to contraction of the atria. The QRS group corresponds to contraction of the ventricles as the depolarization follows a very complicated path. The T wave corresponds to recovery (repolarization) of the heart in preparation for the next cycle.

The ECG is a powerful tool in identifying heart defects. For example, the right side of the heart enlarges if the right ventricle must push against an abnormally large load (as when blood vessels become hardened or clogged). This problem is readily observed on an ECG, because the S wave becomes very large (negatively). *Infarcts*, which are dead regions of the heart muscle that result from heart attacks, are also detected on an ECG because they reflect the depolarization wave.

FIGURE 17–38 Heart muscle cell showing (a) charge dipole layer in resting state; (b) depolarization of cell progressing as muscle begins to contract; and (c) potential V at points P and P′ as a function of time.

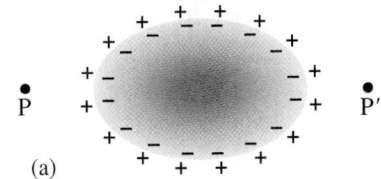

(a)

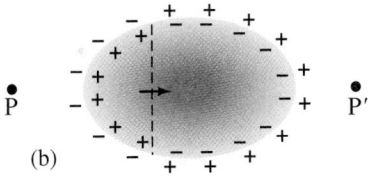

(b)

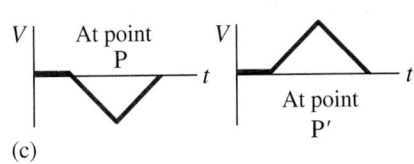

(c)

Summary

The **electric potential** V at any point in space is defined as the electric potential energy per unit charge:

$$V_{a} = \frac{PE_{a}}{q}. \qquad (17-2a)$$

The **electric potential difference** between any two points is defined as the work done to move a 1 C electric charge between the two points. Potential difference is measured in volts (1 V = 1 J/C) and is often referred to as **voltage**.

The change in potential energy when a charge q moves through a potential difference V_{ba} is

$$\Delta PE = qV_{ba}. \qquad (17-3)$$

The potential difference V_{ba} between two points a and b where a uniform electric field E exists is given by

$$V_{ba} = -Ed, \qquad (17-4a)$$

where d is the distance between the two points.

An **equipotential line** or **surface** is all at the same potential, and is perpendicular to the electric field at all points.

The electric potential at a position P due to a single point charge Q, relative to zero potential at infinity, is given by

$$V = \frac{kQ}{r}, \qquad (17-5)$$

where r is the distance from Q to position P and $k = 1/4\pi\epsilon_0$.

[*The potential due to an **electric dipole** drops off as $1/r^2$. The **dipole moment** is $p = Q\ell$, where ℓ is the distance between the two equal but opposite charges of magnitude Q.]

A **capacitor** is a device used to store charge (and electric energy), and consists of two nontouching conductors. The two conductors hold equal and opposite charges, of magnitude Q. The ratio of this charge Q to the potential difference V between the conductors is called the **capacitance**, C:

$$C = \frac{Q}{V}, \quad \text{or} \quad Q = CV. \qquad \textbf{(17–7)}$$

The capacitance of a parallel-plate capacitor is proportional to the area A of each plate and inversely proportional to their separation d:

$$C = \epsilon_0 \frac{A}{d}. \qquad \textbf{(17–8)}$$

The space between the two conductors of a capacitor contains a nonconducting material such as air, paper, or plastic. These materials are referred to as **dielectrics**, and the capacitance is proportional to a property of dielectrics called the **dielectric constant**, K (equal to 1 for air).

A charged capacitor stores an amount of electric energy given by

$$\text{PE} = \tfrac{1}{2}QV = \tfrac{1}{2}CV^2 = \tfrac{1}{2}\frac{Q^2}{C}. \qquad \textbf{(17–10)}$$

This energy can be thought of as stored in the electric field between the plates.

The energy stored in any electric field E has a density

$$\frac{\text{electric PE}}{\text{volume}} = \tfrac{1}{2}\epsilon_0 E^2. \qquad \textbf{(17–11)}$$

Digital electronics converts an analog **signal voltage** into an approximate digital voltage based on a **binary code**: each **bit** has two possibilities, 1 or 0 (also "on" or "off"). The binary number 1101 equals 13. A **byte** is 8 bits and provides $2^8 = 256$ voltage levels. **Sampling rate** is the number of voltage measurements done on the analog signal per second. The **bit depth** is the number of digital voltage levels available at each sampling. CDs are 44.1 kHz, 16-bit.

[*TV and computer monitors traditionally used a **cathode ray tube** (CRT) which accelerates electrons by high voltage, and sweeps them across the screen in a regular way using magnetic coils or electric deflection plates. **LCD flat screens** contain millions of **pixels**, each with a red, green, and blue **subpixel** whose brightness is addressed every $\frac{1}{60}$ s via a **matrix** of horizontal and vertical wires using a **digital** (**binary**) code.]

[*An **electrocardiogram** (ECG or EKG) records the potential changes of each heart beat as the cells depolarize and repolarize.]

Questions

1. If two points are at the same potential, does this mean that no net work is done in moving a test charge from one point to the other? Does this imply that no force must be exerted? Explain.

2. If a negative charge is initially at rest in an electric field, will it move toward a region of higher potential or lower potential? What about a positive charge? How does the potential energy of the charge change in each instance? Explain.

3. State clearly the difference (a) between electric potential and electric field, (b) between electric potential and electric potential energy.

4. An electron is accelerated from rest by a potential difference of 0.20 V. How much greater would its final speed be if it is accelerated with four times as much voltage? Explain.

5. Is there a point along the line joining two equal positive charges where the electric field is zero? Where the electric potential is zero? Explain.

6. Can a particle ever move from a region of low electric potential to one of high potential and yet have its electric potential energy decrease? Explain.

7. If $V = 0$ at a point in space, must $\vec{E} = 0$? If $\vec{E} = 0$ at some point, must $V = 0$ at that point? Explain. Give examples for each.

8. Can two equipotential lines cross? Explain.

9. Draw in a few equipotential lines in Fig. 16–32b and c.

10. When a battery is connected to a capacitor, why do the two plates acquire charges of the same magnitude? Will this be true if the two plates are different sizes or shapes?

11. A conducting sphere carries a charge Q and a second identical conducting sphere is neutral. The two are initially isolated, but then they are placed in contact. (a) What can you say about the potential of each when they are in contact? (b) Will charge flow from one to the other? If so, how much?

12. The parallel plates of an isolated capacitor carry opposite charges, Q. If the separation of the plates is increased, is a force required to do so? Is the potential difference changed? What happens to the work done in the pulling process?

13. If the electric field $\vec{E}$ is uniform in a region, what can you infer about the electric potential V? If V is uniform in a region of space, what can you infer about $\vec{E}$?

14. Is the electric potential energy of two isolated unlike charges positive or negative? What about two like charges? What is the significance of the sign of the potential energy in each case?

15. If the voltage across a fixed capacitor is doubled, the amount of energy it stores (a) doubles; (b) is halved; (c) is quadrupled; (d) is unaffected; (e) none of these. Explain.

16. How does the energy stored in a capacitor change when a dielectric is inserted if (a) the capacitor is isolated so Q does not change; (b) the capacitor remains connected to a battery so V does not change? Explain.

17. A dielectric is pulled out from between the plates of a capacitor which remains connected to a battery. What changes occur to (a) the capacitance, (b) the charge on the plates, (c) the potential difference, (d) the energy stored in the capacitor, and (e) the electric field? Explain your answers.

18. We have seen that the capacitance C depends on the size and position of the two conductors, as well as on the dielectric constant K. What then did we mean when we said that C is a constant in Eq. 17–7?

MisConceptual Questions

1. A $+0.2 \, \mu C$ charge is in an electric field. What happens if that charge is replaced by a $+0.4 \, \mu C$ charge?
   (a) The electric potential doubles, but the electric potential energy stays the same.
   (b) The electric potential stays the same, but the electric potential energy doubles.
   (c) Both the electric potential and electric potential energy double.
   (d) Both the electric potential and electric potential energy stay the same.

2. Two identical positive charges are placed near each other. At the point halfway between the two charges,
   (a) the electric field is zero and the potential is positive.
   (b) the electric field is zero and the potential is zero.
   (c) the electric field is not zero and the potential is positive.
   (d) the electric field is not zero and the potential is zero.
   (e) None of these statements is true.

3. Four identical point charges are arranged at the corners of a square [*Hint*: Draw a figure]. The electric field E and potential V at the center of the square are
   (a) $E = 0$, $V = 0$.
   (b) $E = 0$, $V \neq 0$.
   (c) $E \neq 0$, $V \neq 0$.
   (d) $E \neq 0$, $V = 0$.
   (e) $E = V$ regardless of the value.

4. Which of the following statements is valid?
   (a) If the potential at a particular point is zero, the field at that point must be zero.
   (b) If the field at a particular point is zero, the potential at that point must be zero.
   (c) If the field throughout a particular region is constant, the potential throughout that region must be zero.
   (d) If the potential throughout a particular region is constant, the field throughout that region must be zero.

5. If it takes an amount of work W to move two $+q$ point charges from infinity to a distance d apart from each other, then how much work should it take to move three $+q$ point charges from infinity to a distance d apart from each other?
   (a) $2W$.
   (b) $3W$.
   (c) $4W$.
   (d) $6W$.

6. A proton $(Q = +e)$ and an electron $(Q = -e)$ are in a constant electric field created by oppositely charged plates. You release the proton from near the positive plate and the electron from near the negative plate. Which feels the larger electric force?
   (a) The proton.
   (b) The electron.
   (c) Neither—there is no force.
   (d) The magnitude of the force is the same for both and in the same direction.
   (e) The magnitude of the force is the same for both but in opposite directions.

7. When the proton and electron in MisConceptual Question 6 strike the opposite plate, which one has more kinetic energy?
   (a) The proton.
   (b) The electron.
   (c) Both acquire the same kinetic energy.
   (d) Neither—there is no change in kinetic energy.
   (e) They both acquire the same kinetic energy but with opposite signs.

8. Which of the following do not affect capacitance?
   (a) Area of the plates.
   (b) Separation of the plates.
   (c) Material between the plates.
   (d) Charge on the plates.
   (e) Energy stored in the capacitor.

9. A battery establishes a voltage V on a parallel-plate capacitor. After the battery is disconnected, the distance between the plates is doubled without loss of charge. Accordingly, the capacitance _____ and the voltage between the plates _____.
   (a) increases; decreases.
   (b) decreases; increases.
   (c) increases; increases.
   (d) decreases; decreases.
   (e) stays the same; stays the same.

10. Which of the following is a vector?
   (a) Electric potential.
   (b) Electric potential energy.
   (c) Electric field.
   (d) Equipotential lines.
   (e) Capacitance.

11. A $+0.2 \, \mu C$ charge is in an electric field. What happens if that charge is replaced by a $-0.2 \, \mu C$ charge?
   (a) The electric potential changes sign, but the electric potential energy stays the same.
   (b) The electric potential stays the same, but the electric potential energy changes sign.
   (c) Both the electric potential and electric potential energy change sign.
   (d) Both the electric potential and electric potential energy stay the same.

Problems

17–1 to 17–4 Electric Potential

1. (I) How much work does the electric field do in moving a $-7.7\,\mu C$ charge from ground to a point whose potential is $+65$ V higher?

2. (I) How much work does the electric field do in moving a proton from a point at a potential of $+125$ V to a point at -45 V? Express your answer both in joules and electron volts.

3. (I) What potential difference is needed to stop an electron that has an initial velocity $v = 6.0 \times 10^5$ m/s?

4. (I) How much kinetic energy will an electron gain (in joules and eV) if it accelerates through a potential difference of $18{,}500$ V?

5. (I) An electron acquires 6.45×10^{-16} J of kinetic energy when it is accelerated by an electric field from plate A to plate B. What is the potential difference between the plates, and which plate is at the higher potential?

6. (I) How strong is the electric field between two parallel plates 6.8 mm apart if the potential difference between them is 220 V?

7. (I) An electric field of 525 V/m is desired between two parallel plates 11.0 mm apart. How large a voltage should be applied?

8. (I) The electric field between two parallel plates connected to a 45-V battery is 1900 V/m. How far apart are the plates?

9. (I) What potential difference is needed to give a helium nucleus ($Q = 2e$) 85.0 keV of kinetic energy?

10. (II) Two parallel plates, connected to a 45-V power supply, are separated by an air gap. How small can the gap be if the air is not to become conducting by exceeding its breakdown value of $E = 3 \times 10^6$ V/m?

11. (II) The work done by an external force to move a $-6.50\,\mu C$ charge from point A to point B is 15.0×10^{-4} J. If the charge was started from rest and had 4.82×10^{-4} J of kinetic energy when it reached point B, what must be the potential difference between A and B?

12. (II) What is the speed of an electron with kinetic energy (a) 850 eV, and (b) 0.50 keV?

13. (II) What is the speed of a proton whose KE is 4.2 keV?

14. (II) An alpha particle (which is a helium nucleus, $Q = +2e$, $m = 6.64 \times 10^{-27}$ kg) is emitted in a radioactive decay with KE $= 5.53$ MeV. What is its speed?

15. (II) An electric field greater than about 3×10^6 V/m causes air to break down (electrons are removed from the atoms and then recombine, emitting light). See Section 17–2 and Table 17–3. If you shuffle along a carpet and then reach for a doorknob, a spark flies across a gap you estimate to be 1 mm between your finger and the doorknob. Estimate the voltage between your finger and the doorknob. Why is no harm done?

16. (II) An electron starting from rest acquires 4.8 keV of KE in moving from point A to point B. (a) How much KE would a proton acquire, starting from rest at B and moving to point A? (b) Determine the ratio of their speeds at the end of their respective trajectories.

17. (II) Draw a conductor in the oblong shape of a football. This conductor carries a net negative charge, $-Q$. Draw in a dozen or so electric field lines and equipotential lines.

17–5 Potential Due to Point Charges

[Let $V = 0$ at $x = \infty$.]

18. (I) What is the electric potential 15.0 cm from a $3.00\,\mu C$ point charge?

19. (I) A point charge Q creates an electric potential of $+165$ V at a distance of 15 cm. What is Q?

20. (II) A $+35\,\mu C$ point charge is placed 46 cm from an identical $+35\,\mu C$ charge. How much work would be required to move a $+0.50\,\mu C$ test charge from a point midway between them to a point 12 cm closer to either of the charges?

21. (II) (a) What is the electric potential 2.5×10^{-15} m away from a proton (charge $+e$)? (b) What is the electric potential energy of a system that consists of two protons 2.5×10^{-15} m apart—as might occur inside a typical nucleus?

22. (II) Three point charges are arranged at the corners of a square of side ℓ as shown in Fig. 17–39. What is the potential at the fourth corner (point A)?

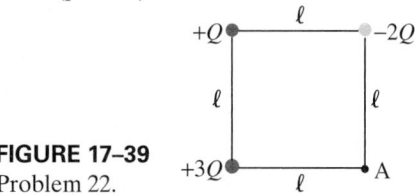

FIGURE 17–39
Problem 22.

23. (II) An electron starts from rest 24.5 cm from a fixed point charge with $Q = -6.50$ nC. How fast will the electron be moving when it is very far away?

24. (II) Two identical $+9.5\,\mu C$ point charges are initially 5.3 cm from each other. If they are released at the same instant from rest, how fast will each be moving when they are very far away from each other? Assume they have identical masses of 1.0 mg.

25. (II) Two point charges, $3.0\,\mu C$ and $-2.0\,\mu C$, are placed 4.0 cm apart on the x axis. At what points along the x axis is (a) the electric field zero and (b) the potential zero?

26. (II) How much work must be done to bring three electrons from a great distance apart to 1.0×10^{-10} m from one another (at the corners of an equilateral triangle)?

27. (II) Point a is 62 cm north of a $-3.8\,\mu C$ point charge, and point b is 88 cm west of the charge (Fig. 17–40). Determine (a) $V_b - V_a$ and (b) $\vec{E}_b - \vec{E}_a$ (magnitude and direction).

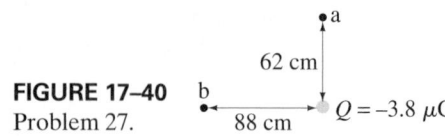

FIGURE 17–40
Problem 27.

28. (II) Many chemical reactions release energy. Suppose that at the beginning of a reaction, an electron and proton are separated by 0.110 nm, and their final separation is 0.100 nm. How much electric potential energy was lost in this reaction (in units of eV)?

29. (III) How much voltage must be used to accelerate a proton (radius 1.2×10^{-15} m) so that it has sufficient energy to just "touch" a silicon nucleus? A silicon nucleus has a charge of $+14e$, and its radius is about 3.6×10^{-15} m. Assume the potential is that for point charges.

30. (III) Two equal but opposite charges are separated by a distance d, as shown in Fig. 17–41. Determine a formula for $V_{BA} = V_B - V_A$ for points B and A on the line between the charges situated as shown.

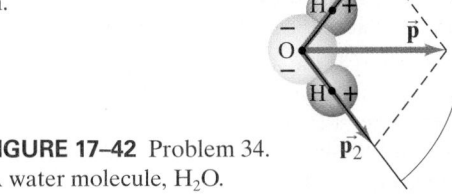

FIGURE 17–41
Problem 30.

31. (III) In the Bohr model of the hydrogen atom, an electron orbits a proton (the nucleus) in a circular orbit of radius 0.53×10^{-10} m. (a) What is the electric potential at the electron's orbit due to the proton? (b) What is the kinetic energy of the electron? (c) What is the total energy of the electron in its orbit? (d) What is the *ionization energy*— that is, the energy required to remove the electron from the atom and take it to $r = \infty$, at rest? Express the results of parts (b), (c), and (d) in joules and eV.

*17–6 Electric Dipoles

*32. (I) An electron and a proton are 0.53×10^{-10} m apart. What is their dipole moment if they are at rest?

*33. (II) Calculate the electric potential due to a dipole whose dipole moment is 4.2×10^{-30} C·m at a point 2.4×10^{-9} m away if this point is (a) along the axis of the dipole nearer the positive charge; (b) 45° above the axis but nearer the positive charge; (c) 45° above the axis but nearer the negative charge.

*34. (III) The dipole moment, considered as a vector, points from the negative to the positive charge. The water molecule, Fig. 17–42, has a dipole moment $\vec{p}$ which can be considered as the vector sum of the two dipole moments, $\vec{p}_1$ and $\vec{p}_2$, as shown. The distance between each H and the O is about 0.96×10^{-10} m. The lines joining the center of the O atom with each H atom make an angle of 104°, as shown, and the net dipole moment has been measured to be $p = 6.1 \times 10^{-30}$ C·m. Determine the charge q on each H atom.

FIGURE 17–42 Problem 34.
A water molecule, H_2O.

17–7 Capacitance

35. (I) The two plates of a capacitor hold $+2500 \, \mu C$ and $-2500 \, \mu C$ of charge, respectively, when the potential difference is 960 V. What is the capacitance?

36. (I) An 8500-pF capacitor holds plus and minus charges of 16.5×10^{-8} C. What is the voltage across the capacitor?

37. (I) How much charge flows from each terminal of a 12.0-V battery when it is connected to a 5.00-μF capacitor?

38. (I) A 0.20-F capacitor is desired. What area must the plates have if they are to be separated by a 3.2-mm air gap?

39. (II) The charge on a capacitor increases by 15 μC when the voltage across it increases from 97 V to 121 V. What is the capacitance of the capacitor?

40. (II) An electric field of 8.50×10^5 V/m is desired between two parallel plates, each of area 45.0 cm^2 and separated by 2.45 mm of air. What charge must be on each plate?

41. (II) If a capacitor has opposite 4.2 μC charges on the plates, and an electric field of 2.0 kV/mm is desired between the plates, what must each plate's area be?

42. (II) It takes 18 J of energy to move a 0.30-mC charge from one plate of a 15-μF capacitor to the other. How much charge is on each plate?

43. (II) To get an idea how big a farad is, suppose you want to make a 1-F air-filled parallel-plate capacitor for a circuit you are building. To make it a reasonable size, suppose you limit the plate area to 1.0 cm^2. What would the gap have to be between the plates? Is this practically achievable?

44. (II) How strong is the electric field between the plates of a 0.80-μF air-gap capacitor if they are 2.0 mm apart and each has a charge of 62 μC?

45. (III) A 2.50-μF capacitor is charged to 746 V and a 6.80-μF capacitor is charged to 562 V. These capacitors are then disconnected from their batteries. Next the positive plates are connected to each other and the negative plates are connected to each other. What will be the potential difference across each and the charge on each? [*Hint*: Charge is conserved.]

46. (III) A 7.7-μF capacitor is charged by a 165-V battery (Fig. 17–43a) and then is disconnected from the battery. When this capacitor (C_1) is then connected (Fig. 17–43b) to a second (initially uncharged) capacitor, C_2, the final voltage on each capacitor is 15 V. What is the value of C_2? [*Hint*: Charge is conserved.]

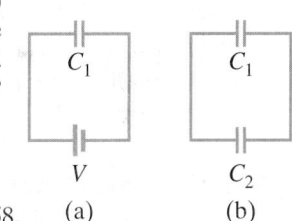

FIGURE 17–43
Problems 46 and 58.　(a)　(b)

17–8 Dielectrics

47. (I) What is the capacitance of two square parallel plates 6.6 cm on a side that are separated by 1.8 mm of paraffin?

48. (I) What is the capacitance of a pair of circular plates with a radius of 5.0 cm separated by 2.8 mm of mica?

49. (II) An uncharged capacitor is connected to a 21.0-V battery until it is fully charged, after which it is disconnected from the battery. A slab of paraffin is then inserted between the plates. What will now be the voltage between the plates?

50. (II) A 3500-pF air-gap capacitor is connected to a 32-V battery. If a piece of mica is placed between the plates, how much charge will flow from the battery?

51. (II) The electric field between the plates of a paper-separated $(K = 3.75)$ capacitor is 8.24×10^4 V/m. The plates are 1.95 mm apart, and the charge on each is 0.675 μC. Determine the capacitance of this capacitor and the area of each plate.

17–9 Electric Energy Storage

52. (I) 650 V is applied to a 2800-pF capacitor. How much energy is stored?

53. (I) A cardiac defibrillator is used to shock a heart that is beating erratically. A capacitor in this device is charged to 5.0 kV and stores 1200 J of energy. What is its capacitance?

54. (II) How much energy is stored by the electric field between two square plates, 8.0 cm on a side, separated by a 1.5-mm air gap? The charges on the plates are equal and opposite and of magnitude 370 μC.

55. (II) A homemade capacitor is assembled by placing two 9-in. pie pans 4 cm apart and connecting them to the opposite terminals of a 9-V battery. Estimate (a) the capacitance, (b) the charge on each plate, (c) the electric field halfway between the plates, and (d) the work done by the battery to charge them. (e) Which of the above values change if a dielectric is inserted?

56. (II) A parallel-plate capacitor has fixed charges $+Q$ and $-Q$. The separation of the plates is then halved. (a) By what factor does the energy stored in the electric field change? (b) How much work must be done to reduce the plate separation from d to $\frac{1}{2}d$? The area of each plate is A.

57. (II) There is an electric field near the Earth's surface whose magnitude is about 150 V/m. How much energy is stored per cubic meter in this field?

58. (III) A 3.70-μF capacitor is charged by a 12.0-V battery. It is disconnected from the battery and then connected to an uncharged 5.00-μF capacitor (Fig. 17–43). Determine the total stored energy (a) before the two capacitors are connected, and (b) after they are connected. (c) What is the change in energy?

17–10 Digital

59. (I) Write the decimal number 116 in binary.

60. (I) Write the binary number 01010101 as a decimal number.

61. (I) Write the binary number 1010101010101010 as a decimal number.

62. (II) Consider a rather coarse 4-bit analog-to-digital conversion where the maximum voltage is 5.0 V. (a) What voltage does 1011 represent? (b) What is the 4-bit representation for 2.0 V?

63. (II) (a) 16-bit sampling provides how many different possible voltages? (b) 24-bit sampling provides how many different possible voltages? (c) For color TV, 3 subpixels, each 8 bits, provides a total of how many different colors?

64. (II) A few extraterrestrials arrived. They had two hands, but claimed that $3 + 2 = 11$. How many fingers did they have on their two hands? Note that our decimal system (and ten characters: 0, 1, 2, ···, 9) surely has its origin because we have ten fingers. [*Hint*: 11 is in their system. In our decimal system, the result would be written as 5.]

*17–11 TV and Computer Monitors

*65. (II) Figure 17–44 is a photograph of a computer screen shot by a camera set at an exposure time of $\frac{1}{4}$ s. During the exposure the cursor arrow was moved around by the mouse, and we see it 15 times. (a) Explain why we see the cursor 15 times. (b) What is the refresh rate of the screen?

FIGURE 17–44
Problem 65.

*66. (III) In a given CRT, electrons are accelerated horizontally by 9.0 kV. They then pass through a uniform electric field E for a distance of 2.8 cm, which deflects them upward so they travel 22 cm to the top of the screen, 11 cm above the center. Estimate the value of E.

*67. (III) Electrons are accelerated by 6.0 kV in a CRT. The screen is 30 cm wide and is 34 cm from the 2.6-cm-long deflection plates. Over what range must the horizontally deflecting electric field vary to sweep the beam fully across the screen?

General Problems

68. A lightning flash transfers 4.0 C of charge and 5.2 MJ of energy to the Earth. (a) Across what potential difference did it travel? (b) How much water could this boil and vaporize, starting from room temperature? (See also Chapter 14.)

69. In an older television tube, electrons are accelerated by thousands of volts through a vacuum. If a television set were laid on its back, would electrons be able to move upward against the force of gravity? What potential difference, acting over a distance of 2.4 cm, would be needed to balance the downward force of gravity so that an electron would remain stationary? Assume that the electric field is uniform.

70. How does the energy stored in a capacitor change, as the capacitor remains connected to a battery, if the separation of the plates is doubled?

71. How does the energy stored in an isolated capacitor change if (a) the potential difference is doubled, or (b) the separation of the plates is doubled?

72. A huge 4.0-F capacitor has enough stored energy to heat 2.8 kg of water from 21°C to 95°C. What is the potential difference across the plates?

73. A proton ($q = +e$) and an alpha particle ($q = +2e$) are accelerated by the same voltage V. Which gains the greater kinetic energy, and by what factor?

74. Dry air will break down if the electric field exceeds 3.0×10^6 V/m. What amount of charge can be placed on a parallel-plate capacitor if the area of each plate is 65 cm^2?

75. Three charges are at the corners of an equilateral triangle (side ℓ) as shown in Fig. 17–45. Determine the potential at the midpoint of each of the sides. Let $V = 0$ at $r = \infty$.

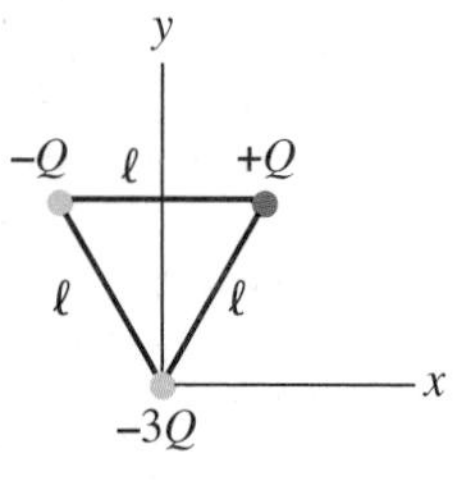

FIGURE 17–45
Problem 75.

76. It takes 15.2 J of energy to move a 13.0-mC charge from one plate of a 17.0-μF capacitor to the other. How much charge is on each plate? Assume constant voltage.

77. A $3.4\,\mu C$ and a $-2.6\,\mu C$ charge are placed 2.5 cm apart. At what points along the line joining them is (*a*) the electric field zero, and (*b*) the electric potential zero?

78. Near the surface of the Earth there is an electric field of about 150 V/m which points downward. Two identical balls with mass $m = 0.670\,\text{kg}$ are dropped from a height of 2.00 m, but one of the balls is positively charged with $q_1 = 650\,\mu C$, and the second is negatively charged with $q_2 = -650\,\mu C$. Use conservation of energy to determine the difference in the speed of the two balls when they hit the ground. (Neglect air resistance.)

79. The power supply for a pulsed nitrogen laser has a $0.050\text{-}\mu F$ capacitor with a maximum voltage rating of 35 kV. (*a*) Estimate how much energy could be stored in this capacitor. (*b*) If 12% of this stored electrical energy is converted to light energy in a pulse that is 6.2 microseconds long, what is the power of the laser pulse?

80. In a **photocell**, ultraviolet (UV) light provides enough energy to some electrons in barium metal to eject them from the surface at high speed. To measure the maximum energy of the electrons, another plate above the barium surface is kept at a negative enough potential that the emitted electrons are slowed down and stopped, and return to the barium surface. See Fig. 17–46. If the plate voltage is −3.02 V (compared to the barium) when the fastest electrons are stopped, what was the speed of these electrons when they were emitted?

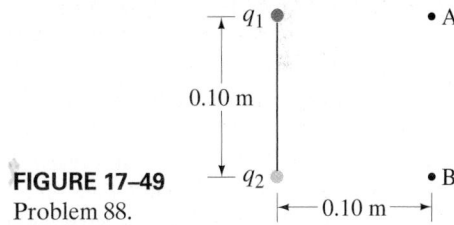

FIGURE 17–46
Problem 80.

81. A $+38\,\mu C$ point charge is placed 36 cm from an identical $+38\,\mu C$ charge. A $-1.5\,\mu C$ charge is moved from point A to point B as shown in Fig. 17–47. What is the change in potential energy?

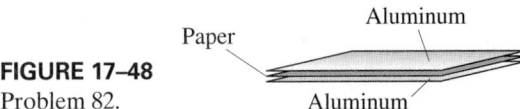

FIGURE 17–47   Problem 81.

82. Paper has a dielectric constant $K = 3.7$ and a dielectric strength of 15×10^6 V/m. Suppose that a typical sheet of paper has a thickness of 0.11 mm. You make a "homemade" capacitor by placing a sheet of 21×14 cm paper between two aluminum foil sheets (Fig. 17–48) of the same size. (*a*) What is the capacitance C of your device? (*b*) About how much charge could you store on your capacitor before it would break down?

FIGURE 17–48
Problem 82.

83. A capacitor is made from two 1.1-cm-diameter coins separated by a 0.10-mm-thick piece of paper ($K = 3.7$). A 12-V battery is connected to the capacitor. How much charge is on each coin?

84. A $+3.5\,\mu C$ charge is 23 cm to the right of a $-7.2\,\mu C$ charge. At the midpoint between the two charges, (*a*) determine the potential and (*b*) the electric field.

85. A parallel-plate capacitor with plate area 3.0 cm^2 and air-gap separation 0.50 mm is connected to a 12-V battery, and fully charged. The battery is then disconnected. (*a*) What is the charge on the capacitor? (*b*) The plates are now pulled to a separation of 0.75 mm. What is the charge on the capacitor now? (*c*) What is the potential difference between the plates now? (*d*) How much work was required to pull the plates to their new separation?

86. A 2.1-μF capacitor is fully charged by a 6.0-V battery. The battery is then disconnected. The capacitor is not ideal and the charge slowly leaks out from the plates. The next day, the capacitor has lost half its stored energy. Calculate the amount of charge lost.

87. Two point charges are fixed 4.0 cm apart from each other. Their charges are $Q_1 = Q_2 = 6.5\,\mu C$, and their masses are $m_1 = 1.5$ mg and $m_2 = 2.5$ mg. (*a*) If Q_1 is released from rest, what will be its speed after a very long time? (*b*) If both charges are released from rest at the same time, what will be the speed of Q_1 after a very long time?

88. Two charges are placed as shown in Fig. 17–49 with $q_1 = 1.2\,\mu C$ and $q_2 = -3.3\,\mu C$. Find the potential difference between points A and B.

FIGURE 17–49
Problem 88.

89. If the electrons in a single raindrop, 3.5 mm in diameter, could be removed from the Earth (without removing the atomic nuclei), by how much would the potential of the Earth increase?

90. Thunderclouds may develop a voltage difference of about 5×10^7 V. Given that an electric field of 3×10^6 V/m is required to produce an electrical spark within a volume of air, estimate the length of a thundercloud lightning bolt. [Can you see why, when lightning strikes from a cloud to the ground, the bolt has to propagate as a sequence of steps?]

91. A manufacturer claims that a carpet will not generate more than 6.0 kV of static electricity. What magnitude of charge would have to be transferred between a carpet and a shoe for there to be a 6.0-kV potential difference between the shoe and the carpet? Approximate the area of the shoe and assume the shoe and carpet are large sheets of charge separated by a small distance $d = 1.0$ mm.

92. Compact "ultracapacitors" with capacitance values up to several thousand farads are now commercially available. One application for ultracapacitors is in providing power for electrical circuits when other sources (such as a battery) are turned off. To get an idea of how much charge can be stored in such a component, assume a 1200-F ultracapacitor is initially charged to 12.0 V by a battery and is then disconnected from the battery. If charge is then drawn off the plates of this capacitor at a rate of 1.0 mC/s, say, to power the backup memory of some electrical device, how long (in days) will it take for the potential difference across this capacitor to drop to 6.0 V?

93. An electron is accelerated horizontally from rest by a potential difference of 2200 V. It then passes between two horizontal plates 6.5 cm long and 1.3 cm apart that have a potential difference of 250 V (Fig. 17–50). At what angle θ will the electron be traveling after it passes between the plates?

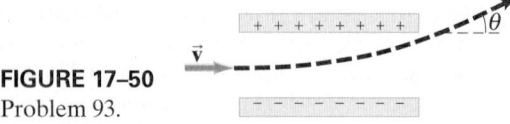

FIGURE 17–50
Problem 93.

94. In the **dynamic random access memory (DRAM)** of a computer, each memory cell contains a capacitor for charge storage. Each of these cells represents a single binary-bit value of "1" when its 35-fF capacitor $\left(1\text{ fF} = 10^{-15}\text{ F}\right)$ is charged at 1.5 V, or "0" when uncharged at 0 V. (a) When fully charged, how many excess electrons are on a cell capacitor's negative plate? (b) After charge has been placed on a cell capacitor's plate, it slowly "leaks" off at a rate of about 0.30 fC/s. How long does it take for the potential difference across this capacitor to decrease by 2.0% from its fully charged value? (Because of this leakage effect, the charge on a DRAM capacitor is "refreshed" many times per second.) Note: A DRAM cell is shown in Fig. 21–29.

95. In the DRAM computer chip of Problem 94, suppose the two parallel plates of one cell's 35-fF capacitor are separated by a 2.0-nm-thick insulating material with dielectric constant $K = 25$. (a) Determine the area A (in μm^2) of the cell capacitor's plates. (b) If the plate area A accounts for half of the area of each cell, estimate how many megabytes of memory can be placed on a 3.0-cm^2 silicon wafer. (1 byte = 8 bits.)

96. A parallel-plate capacitor with plate area $A = 2.0\text{ m}^2$ and plate separation $d = 3.0$ mm is connected to a 35-V battery (Fig. 17–51a). (a) Determine the charge on the capacitor, the electric field, the capacitance, and the energy stored in the capacitor. (b) With the capacitor still connected to the battery, a slab of plastic with dielectric strength $K = 3.2$ is placed between the plates of the capacitor, so that the gap is completely filled with the dielectric (Fig. 17–51b). What are the new values of charge, electric field, capacitance, and the energy stored in the capacitor?

FIGURE 17–51
Problem 96.

Search and Learn

1. Make a list of rules for and properties of equipotential surfaces or lines. You should be able to find eight distinct rules in the text.

2. Figure 17–8 shows contour lines (elevations). Just for fun, assume they are equipotential lines on a flat 2-dimensional surface with the values shown being in volts. Estimate the magnitude and direction of the "electric field" (a) between Iceberg Lake and Cecile Lake and (b) at the Minaret Mine. Assume that up is $+y$, right is $+x$, and that Cecile Lake is about 1.0 km wide in the middle.

3. In lightning storms, the potential difference between the Earth and the bottom of thunderclouds may be 35,000,000 V. The bottoms of the thunderclouds are typically 1500 m above the Earth, and can have an area of 110 km^2. Modeling the Earth–cloud system as a huge capacitor, calculate (a) the capacitance of the Earth–cloud system, (b) the charge stored in the "capacitor," and (c) the energy stored in the "capacitor."

4. The potential energy stored in a capacitor (Section 17–9) can be written as either $CV^2/2$ or $Q^2/2C$. In the first case the energy is proportional to C; in the second case the energy is proportional to $1/C$. (a) Explain how both of these equations can be correct. (b) When might you use the first equation and when might you use the second equation? (c) If a paper dielectric is inserted into a parallel-plate capacitor that is attached to a battery (V does not change), by what factor will the energy stored in the capacitor change? (d) If a quartz dielectric is inserted into a charged parallel-plate capacitor that is isolated from any battery, by what factor will the energy stored in the capacitor change?

5. Suppose it takes 75 kW of power for your car to travel at a constant speed on the highway. (a) What is this in horsepower? (b) How much energy in joules would it take for your car to travel at highway speed for 5.0 hours? (c) Suppose this amount of energy is to be stored in the electric field of a parallel-plate capacitor (Section 17–9). If the voltage on the capacitor is to be 850 V, what is the required capacitance? (d) If this capacitor were to be made from activated carbon (Section 17–7), the voltage would be limited to no more than 10 V. In this case, how many grams of activated carbon would be required? (e) Is this practical?

6. Capacitors can be used as "electric charge counters." Consider an initially uncharged capacitor of capacitance C with its bottom plate grounded and its top plate connected to a source of electrons. (a) If N electrons flow onto the capacitor's top plate, show that the resulting potential difference V across the capacitor is directly proportional to N. (b) Assume the voltage-measuring device can accurately resolve voltage changes of about 1 mV. What value of C would be necessary to resolve the arrival of an individual electron? (c) Using modern semiconductor technology, a micron-size capacitor can be constructed with parallel conducting plates separated by an insulator of dielectric constant $K = 3$ and thickness $d = 100$ nm. What side length ℓ should the square plates have (in μm)?

The glow of the thin wire filament of incandescent lightbulbs is caused by the electric current passing through it. Electric energy is transformed to thermal energy (via collisions between moving electrons and atoms of the wire), which causes the wire's temperature to become so high that it glows. In halogen lamps (tungsten–halogen), shown on the right, the tungsten filament is surrounded by a halogen gas such as bromine or iodine in a clear tube. Halogens, via chemical reactions, restore many of the tungsten atoms that were evaporated from the hot filament, allowing longer life, higher temperature (typically 2900 K versus 2700 K), better efficiency, and whiter light.

Electric current and electric power in electric circuits are of basic importance in everyday life. We examine both dc and ac in this Chapter, and include the microscopic analysis of electric current.

Electric Currents

CHAPTER 18

CHAPTER-OPENING QUESTION—Guess now!

The conductors shown are all made of copper and are at the same temperature. Which conductor would have the greatest resistance to the flow of charge entering from the left? Which would offer the least resistance?

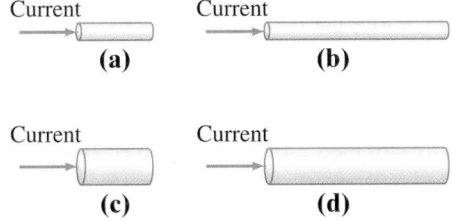

Current
(a)

Current
(b)

Current
(c)

Current
(d)

CONTENTS

I n the previous two Chapters we have been studying static electricity: electric charges at rest. In this Chapter we begin our study of charges in motion, and we call a flow of charge an electric current.

In everyday life we are familiar with electric currents in wires and other conductors. Most practical electrical devices depend on electric current: current through a lightbulb, current in the heating element of a stove, hair dryer, or electric heater, as well as currents in electronic devices. Electric currents can exist in conductors such as wires, but also in semiconductor devices, human cells and their membranes (Section 18–10), and in empty space.

In electrostatic situations, we saw in Section 16–9 that the electric field must be zero inside a conductor (if it weren't, the charges would move). But when charges are *moving* along a conductor, an electric field is needed to set charges into motion, and to keep them in motion against even low resistance in any normal conductor. We can control the flow of charge using electric fields and electric potential (voltage), concepts we have just been discussing. In order to have a current in a wire, a potential difference is needed, which can be provided by a battery.

We first look at electric current from a macroscopic point of view. Later in the Chapter we look at currents from a microscopic (theoretical) point of view as a flow of electrons in a wire.

FIGURE 18–1 Alessandro Volta. In this portrait, Volta demonstrates his battery to Napoleon in 1801.

18–1 The Electric Battery

Until the year 1800, the technical development of electricity consisted mainly of producing a static charge by friction. It all changed in 1800 when Alessandro Volta (1745–1827; Fig. 18–1) invented the electric battery, and with it produced the first steady flow of electric charge—that is, a steady electric current.

The events that led to the discovery of the battery are interesting. Not only was this an important discovery, but it also gave rise to a famous scientific debate.

In the 1780s, Luigi Galvani (1737–1798), professor at the University of Bologna, carried out a series of experiments on the contraction of a frog's leg muscle by using static electricity. Galvani found that the muscle also contracted when dissimilar metals were inserted into the frog. Galvani believed that the source of the electric charge was in the frog muscle or nerve itself, and that the metal merely transmitted the charge to the proper points. When he published his work in 1791, he termed this charge "animal electricity." Many wondered, including Galvani himself, if he had discovered the long-sought "life-force."

Volta, at the University of Pavia 200 km away, was skeptical of Galvani's results, and came to believe that the source of the electricity was not in the animal itself, but rather in the *contact between the dissimilar metals*. Volta realized that a moist conductor, such as a frog muscle or moisture at the contact point of two dissimilar metals, was necessary in the circuit if it was to be effective. He also saw that the contracting frog muscle was a sensitive instrument for detecting electric "tension" or "electromotive force" (his words for what we now call voltage), in fact more sensitive than the best available electroscopes that he and others had developed.[†]

FIGURE 18–2 A voltaic battery, from Volta's original publication.

Volta's research found that certain combinations of metals produced a greater effect than others, and, using his measurements, he listed them in order of effectiveness. (This "electrochemical series" is still used by chemists today.) He also found that carbon could be used in place of one of the metals.

Volta then conceived his greatest contribution to science. Between a disc of zinc and one of silver, he placed a piece of cloth or paper soaked in salt solution or dilute acid and piled a "battery" of such couplings, one on top of another, as shown in Fig. 18–2. This "pile" or "battery" produced a much increased potential difference. Indeed, when strips of metal connected to the two ends of the pile were brought close, a spark was produced. Volta had designed and built the first electric battery. He published his discovery in 1800.

[†]Volta's most sensitive electroscope (Section 16–4) measured about 40 V per degree (angle of leaf separation). Nonetheless, he was able to estimate the potential differences produced by combinations of dissimilar metals in contact. For a silver–zinc contact he got about 0.7 V, remarkably close to today's value of 0.78 V.

Electric Cells and Batteries

A battery produces electricity by transforming chemical energy into electrical energy. Today a great variety of electric cells and batteries are available, from flashlight batteries to the storage battery of a car. The simplest batteries contain two plates or rods made of dissimilar metals (one can be carbon) called **electrodes**. The electrodes are immersed in a solution or paste, such as a dilute acid, called the **electrolyte**. Such a device is properly called an **electric cell**, and several cells connected together is a **battery**, although today even a single cell is called a battery. The chemical reactions involved in most electric cells are quite complicated. Here we describe how one very simple cell works, emphasizing the physical aspects.

The cell shown in Fig. 18–3 uses dilute sulfuric acid as the electrolyte. One of the electrodes is made of carbon, the other of zinc. The part of each electrode outside the solution is called the **terminal**, and connections to wires and circuits are made here. The acid tends to dissolve the zinc electrode. Each zinc atom leaves two electrons behind on the electrode and enters the solution as a positive ion. The zinc electrode thus acquires a negative charge. The electrolyte becomes positively charged, and can pull electrons off the carbon electrode. Thus the carbon electrode becomes positively charged. Because there is an opposite charge on the two electrodes, there is a potential difference between the two terminals.

In a cell whose terminals are not connected, only a small amount of the zinc is dissolved, for as the zinc electrode becomes increasingly negative, any new positive zinc ions produced are attracted back to the electrode. Thus, a particular potential difference (or voltage) is maintained between the two terminals. If charge is allowed to flow between the terminals, say, through a wire (or a lightbulb), then more zinc can be dissolved. After a time, one or the other electrode is used up and the cell becomes "dead."

The voltage that exists between the terminals of a battery depends on what the electrodes are made of and their relative ability to be dissolved or give up electrons.

When two or more cells are connected so that the positive terminal of one is connected to the negative terminal of the next, they are said to be connected in *series* and their voltages add up. Thus, the voltage between the ends of two 1.5-V AA flashlight batteries connected in series is 3.0 V, whereas the six 2-V cells of an automobile storage battery give 12 V. Figure 18–4a shows a diagram of a common "dry cell" or "flashlight battery" used not only in flashlights but in many portable electronic devices, and Fig. 18–4b shows two smaller ones connected in series to a flashlight bulb. An incandescent lightbulb consists of a thin, coiled wire (filament) inside an evacuated glass bulb, as shown in Fig. 18–5 and in the Chapter-Opening Photos, page 501. When charge passes through the filament, it gets very hot (≈ 2800 K) and glows. Other bulb types, such as fluorescent, work differently.

FIGURE 18–3 Simple electric cell.

FIGURE 18–4 (a) Diagram of an ordinary dry cell (like a D-cell or AA). The cylindrical zinc cup is covered on the sides; its flat bottom is the negative terminal. (b) Two dry cells (AA type) connected in series. Note that the positive terminal of one cell pushes against the negative terminal of the other.

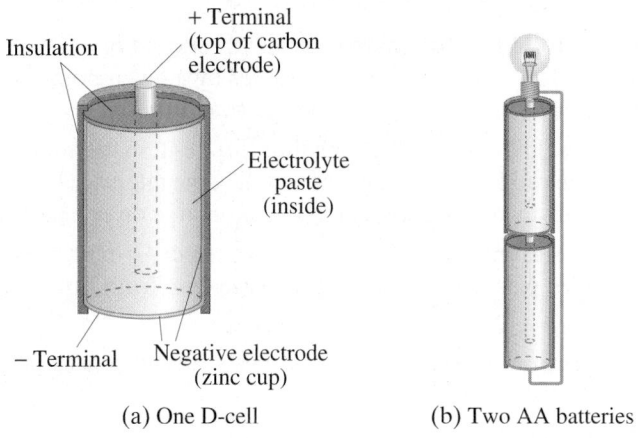

(a) One D-cell   (b) Two AA batteries

FIGURE 18–5 An ordinary incandescent lightbulb: the fine wire of the filament becomes so hot that it glows. Incandescent halogen bulbs enclose the filament in a small quartz tube filled with a halogen gas (bromine or iodine) which allows longer filament life and higher filament temperature for greater efficiency and whiteness.

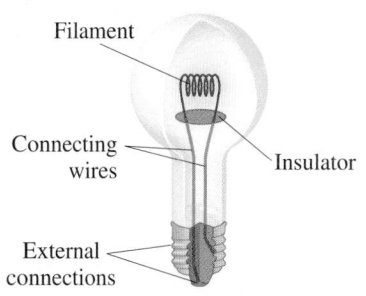

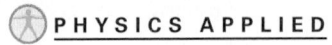
Electric Cars

Considerable research is being done to improve batteries for electric cars and for hybrids (which use both a gasoline internal combustion engine and an electric motor). One type of battery is lithium-ion, in which the anode contains lithium and the cathode is carbon. Electric cars need no gear changes and can develop full torque starting from rest, and so can accelerate quickly and smoothly. The distance an electric car can go between charges of the battery (its "range") is an important parameter because each recharging of an electric car battery may take hours, not minutes like a gas fill-up. Because charging an electric car can draw a large current over a period of several hours, electric power companies may need to upgrade their power grids so they won't fail when many electric cars are being charged at the same time in a small urban area.

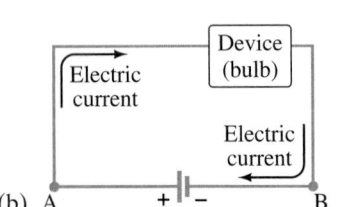

(a)

(b)

A ⊢⊢ B

FIGURE 18–6 (a) A simple electric circuit. (b) Schematic drawing of the same circuit, consisting of a battery, connecting wires (thick gray lines), and a lightbulb or other device.

⚠ CAUTION

A battery does not create charge; a lightbulb does not destroy charge

18–2 Electric Current

The purpose of a battery is to produce a potential difference, which can then make charges move. When a continuous conducting path is connected between the terminals of a battery, we have an electric **circuit**, Fig. 18–6a. On any diagram of a circuit, as in Fig. 18–6b, we use the symbol

$$\text{—}\vert\!\vdash\text{—} \qquad \text{or} \qquad \text{—}\vert\!\vdash\text{—} \qquad\qquad \text{[battery symbol]}$$

to represent a battery. The device connected to the battery could be a lightbulb, a heater, a radio, or some other device. When such a circuit is formed, charge can move (or flow) through the wires of the circuit, from one terminal of the battery to the other, as long as the conducting path is continuous. Any flow of charge such as this is called an **electric current**.

More precisely, the electric current in a wire is defined as the net amount of charge that passes through the wire's full cross section at any point per unit time. Thus, the current I is defined as

$$I = \frac{\Delta Q}{\Delta t}, \tag{18–1}$$

where ΔQ is the amount of charge that passes through the conductor at any location during the time interval Δt.

Electric current is measured in coulombs per second; this is given a special name, the **ampere** (abbreviated amp or A), after the French physicist André Ampère (1775–1836). Thus, 1 A = 1 C/s. Smaller units of current are often used, such as the milliampere $(1\,\text{mA} = 10^{-3}\,\text{A})$ and microampere $(1\,\mu\text{A} = 10^{-6}\,\text{A})$.

A current can flow in a circuit only if there is a *continuous* conducting path. We then have a **complete circuit**. If there is a break in the circuit, say, a cut wire, we call it an **open circuit** and no current flows. In any single circuit, with only a single path for current to follow such as in Fig. 18–6b, a steady current at any instant is the same at one point (say, point A) as at any other point (such as B). This follows from the conservation of electric charge: charge doesn't disappear. A battery does not create (or destroy) any net charge, nor does a lightbulb absorb or destroy charge.

EXAMPLE 18–1 **Current is flow of charge.** A steady current of 2.5 A exists in a wire for 4.0 min. (*a*) How much total charge passes by a given point in the circuit during those 4.0 min? (*b*) How many electrons would this be?

APPROACH (*a*) Current is flow of charge per unit time, Eq. 18–1, so the amount of charge passing a point is the product of the current and the time interval. (*b*) To get the number of electrons, we divide the total charge by the charge on one electron.

SOLUTION (*a*) Since the current was 2.5 A, or 2.5 C/s, then in 4.0 min (= 240 s) the total charge that flowed past a given point in the wire was, from Eq. 18–1,

$$\Delta Q = I\,\Delta t = (2.5\,\text{C/s})(240\,\text{s}) = 600\,\text{C}.$$

(*b*) The charge on one electron is 1.60×10^{-19} C, so 600 C would consist of

$$\frac{600\,\text{C}}{1.6 \times 10^{-19}\,\text{C/electron}} = 3.8 \times 10^{21} \text{ electrons}.$$

EXERCISE A If 1 million electrons per second pass a point in a wire, what is the current?

CONCEPTUAL EXAMPLE 18–2 | **How to connect a battery.** What is wrong with each of the schemes shown in Fig. 18–7 for lighting a flashlight bulb with a flashlight battery and a single wire?

RESPONSE (*a*) There is no closed path for charge to flow around. Charges might briefly start to flow from the battery toward the lightbulb, but there they run into a "dead end," and the flow would immediately come to a stop.

(*b*) Now there is a closed path passing to and from the lightbulb; but the wire touches only one battery terminal, so there is no potential difference in the circuit to make the charge move. Neither here, nor in (*a*), does the bulb light up.

(*c*) Nothing is wrong here. This is a complete circuit: charge can flow out from one terminal of the battery, through the wire and the bulb, and into the other terminal. This scheme will light the bulb.

In many real circuits, wires are connected to a common conductor that provides continuity. This common conductor is called **ground**, usually represented as ⏚ or ⏚, and really is connected to the ground for a building or house. In a car, one terminal of the battery is called "ground," but is not connected to the earth itself—it is connected to the frame of the car, as is one connection to each lightbulb and other devices. Thus the car frame is a conductor in each circuit, ensuring a continuous path for charge flow, and is called "ground" for the car's circuits. (Note that the car frame is well insulated from the earth by the rubber tires.)

We saw in Chapter 16 that conductors contain many free electrons. Thus, if a continuous conducting wire is connected to the terminals of a battery, negatively charged electrons flow in the wire. When the wire is first connected, the potential difference between the terminals of the battery sets up an electric field inside the wire and parallel to it. Free electrons at one end of the wire are attracted into the positive terminal, and at the same time other electrons enter the other end of the wire at the negative terminal of the battery. There is a continuous flow of electrons throughout the wire that begins as soon as the wire is connected to *both* terminals.

When the conventions of positive and negative charge were invented two centuries ago, however, it was assumed that positive charge flowed in a wire. For nearly all purposes, positive charge flowing in one direction is exactly equivalent to negative charge flowing in the opposite direction, as shown in Fig. 18–8. Today, we still use the historical convention of positive charge flow when discussing the direction of a current. So when we speak of the current direction in a circuit, we mean the direction positive charge would flow. This is sometimes referred to as **conventional current**. When we want to speak of the direction of electron flow, we will specifically state it is the electron current. In liquids and gases, both positive and negative charges (ions) can move.

In practical life, such as rating the total charge of a car battery, you may see the unit **ampere-hour** (A·h): from Eq. 18–1, $\Delta Q = I\,\Delta t$.

EXERCISE B How many coulombs is 1.00 A·h?

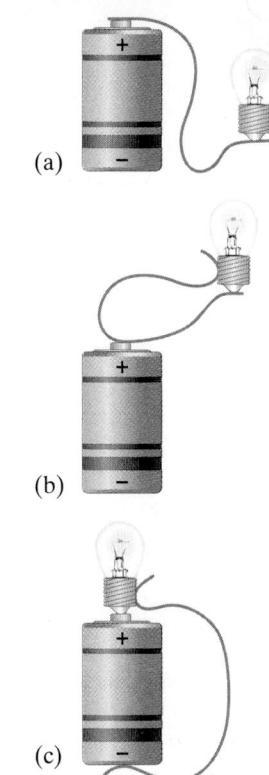

(a)

(b)

(c)

FIGURE 18–7 Example 18–2.

FIGURE 18–8 Conventional current from + to − is equivalent to a negative electron flow from − to +.

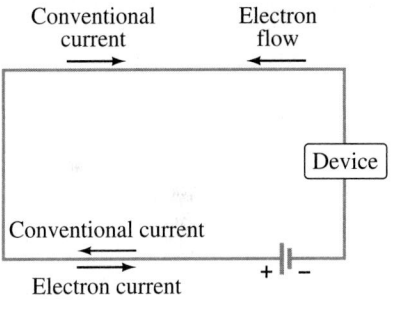

⚠ **CAUTION**

Distinguish conventional current from electron flow

18–3 Ohm's Law: Resistance and Resistors

To produce an electric current in a circuit, a difference in potential is required. One way of producing a potential difference along a wire is to connect its ends to the opposite terminals of a battery. It was Georg Simon Ohm (1787–1854) who established experimentally that the current in a metal wire is proportional to the potential difference V applied to its two ends:

$$I \propto V.$$

If, for example, we connect a wire to the two terminals of a 6-V battery, the current in the wire will be twice what it would be if the wire were connected to a 3-V battery. It is also found that reversing the sign of the voltage does not affect the magnitude of the current.

Exactly how large the current is in a wire depends not only on the voltage between its ends, but also on the resistance the wire offers to the flow of electrons. Electron flow is impeded because of collisions with the atoms of the wire. We define electrical **resistance** R as the proportionality factor between the voltage V (between the ends of the wire) and the current I (passing through the wire):

$$V = IR. \tag{18-2}$$

Ohm found experimentally that in metal conductors R is a constant independent of V, a result known as **Ohm's law**. Equation 18–2, $V = IR$, is itself sometimes called Ohm's law, but only when referring to materials or devices for which R is a constant independent of V. But R is not a constant for many substances other than metals, nor for devices such as diodes, vacuum tubes, transistors, and so on. Even for metals, R is not constant if the temperature changes much: for a lightbulb filament the measured resistance is low for small currents, but is much higher at the filament's normal large operating current that puts it at the high temperature needed to make it glow (≈ 3000 K). Thus Ohm's "law" is not a fundamental law of nature, but rather a description of a certain class of materials: metal conductors, whose temperature does not change much. Such materials are said to be "ohmic." Materials or devices that do not follow Ohm's law are said to be *nonohmic*. See Fig. 18–9.

The unit for resistance is called the **ohm** and is abbreviated Ω (Greek capital letter omega). Because $R = V/I$, we see that $1.0\ \Omega$ is equivalent to 1.0 V/A.

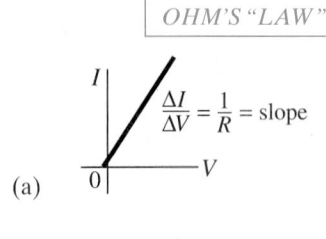

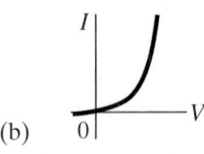

FIGURE 18–9 Graphs of current vs. voltage (a) for a metal conductor which obeys Ohm's law, and (b) for a nonohmic device, in this case a semiconductor diode.

FIGURE 18–10 Flashlight (Example 18–3). Note how the circuit is completed along the side strip.

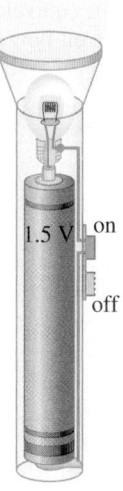

| EXAMPLE 18–3 | **Flashlight bulb resistance.** A small flashlight bulb (Fig. 18–10) draws 300 mA from its 1.5-V battery. (*a*) What is the resistance of the bulb? (*b*) If the battery becomes weak and the voltage drops to 1.2 V, how would the current change? Assume the bulb is approximately ohmic.

APPROACH We apply Ohm's law to the bulb, where the voltage applied across it is the battery voltage.

SOLUTION (*a*) We change 300 mA to 0.30 A and use Eq. 18–2:

$$R = \frac{V}{I} = \frac{1.5\ \text{V}}{0.30\ \text{A}} = 5.0\ \Omega.$$

(*b*) If the resistance stays the same, the current would be

$$I = \frac{V}{R} = \frac{1.2\ \text{V}}{5.0\ \Omega} = 0.24\ \text{A} = 240\ \text{mA},$$

or a decrease of 60 mA.

NOTE With the smaller current in part (*b*), the bulb filament's temperature would be lower and the bulb less bright. Also, resistance does depend on temperature (Section 18–4), so our calculation is only a rough approximation.

EXERCISE C What is the resistance of a lightbulb if 0.50 A flows through it when 120 V is connected across it?

All electric devices, from heaters to lightbulbs to stereo amplifiers, offer resistance to the flow of current. The filaments of lightbulbs (Fig. 18–5) and electric heaters are special types of wires whose resistance results in their becoming very hot. Generally, the connecting wires have very low resistance in comparison to the resistance of the wire filaments or coils, so the connecting wires usually have a minimal effect on the magnitude of the current.[†]

[†]A useful analogy compares the flow of electric charge in a wire to the flow of water in a river, or in a pipe, acted on by gravity. If the river (or pipe) is nearly level, the flow rate is small. But if one end is somewhat higher than the other, the water flow rate—or current—is greater. The greater the difference in height, the swifter the current. We saw in Chapter 17 that electric potential is analogous to the height of a cliff for gravity. Just as an increase in height can cause a greater flow of water, so a greater electric potential difference, or voltage, causes a greater electric current. Resistance in a wire is analogous to rocks in a river that retard water flow.

In many circuits, particularly in electronic devices, **resistors** are used to control the amount of current. Resistors have resistances ranging from less than an ohm to millions of ohms (see Figs. 18–11 and 18–12). The main types are "wire-wound" resistors which consist of a coil of fine wire, "composition" resistors which are usually made of carbon, resistors made of thin carbon or metal films, and (on tiny integrated circuit "chips") undoped semiconductors.

When we draw a diagram of a circuit, we use the symbol

–\/\/\–   [resistor symbol]

to indicate a resistance. Wires whose resistance is negligible, however, are shown simply as straight lines. Figure 18–12 and its Table show one way to specify the resistance of a resistor.

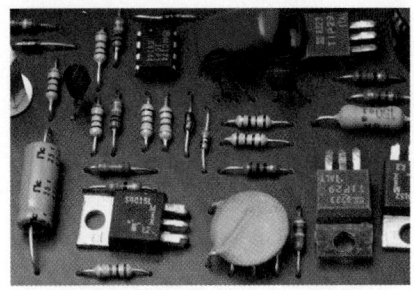

FIGURE 18–11 Photo of resistors (striped), plus other devices on a circuit board.

Resistor Color Code			
Color	**Number**	**Multiplier**	**Tolerance**
Black	0	1	
Brown	1	10^1	1%
Red	2	10^2	2%
Orange	3	10^3	
Yellow	4	10^4	
Green	5	10^5	
Blue	6	10^6	
Violet	7	10^7	
Gray	8	10^8	
White	9	10^9	
Gold		10^{-1}	5%
Silver		10^{-2}	10%
No color			20%

FIGURE 18–12 The resistance value of a given resistor is written on the exterior, or may be given as a color code as shown below and in the Table: the first two colors represent the first two digits in the value of the resistance, the third color represents the power of ten that it must be multiplied by, and the fourth is the manufactured tolerance. For example, a resistor whose four colors are red, green, yellow, and silver has a resistance of $25 \times 10^4 \, \Omega = 250{,}000 \, \Omega = 250 \, k\Omega$, plus or minus 10%. [An alternative code is a number such as 104, which means $R = 1.0 \times 10^4 \, \Omega$.]

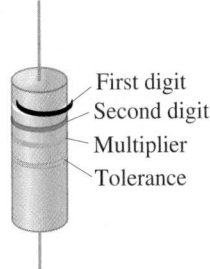

First digit
Second digit
Multiplier
Tolerance

CONCEPTUAL EXAMPLE 18–4 | **Current and potential.** Current I enters a resistor R as shown in Fig. 18–13. (*a*) Is the potential higher at point A or at point B? (*b*) Is the current greater at point A or at point B?

RESPONSE (*a*) Positive charge always flows from + to −, from high potential to low potential. So if current I is conventional (positive) current, point A is at a higher potential than point B.

(*b*) Conservation of charge requires that whatever charge flows into the resistor at point A, an equal amount of charge emerges at point B. Charge or current does not get "used up" by a resistor. So the current is the same at A and B.

FIGURE 18–13 Example 18–4.

A R B

An electric potential decrease, as from point A to point B in Example 18–4, is often called a **potential drop** or a **voltage drop**.

Some Helpful Clarifications

Here we briefly summarize some possible misunderstandings and clarifications. Batteries do not put out a constant current. Instead, batteries are intended to maintain a constant potential difference, or very nearly so. (Details in the next Chapter.) Thus a battery should be considered a source of voltage. The voltage is applied *across* a wire or device.

Electric current passes *through* a wire or device (connected to a battery), and its magnitude depends on that device's resistance. The resistance is a *property* of the wire or device. The voltage, on the other hand, is external to the wire or device, and is applied across the two ends of the wire or device. The current through the device might be called the "response": the current increases if the voltage increases or the resistance decreases, as $I = V/R$.

⚠ **CAUTION**

Voltage is applied across *a device; current passes* through *a device*

Current is *not* a vector, even though current does have a direction. In a thin wire, the direction of the current is always parallel to the wire at each point, no matter how the wire curves, just like water in a pipe. The direction of conventional (positive) current is from high potential $(+)$ toward lower potential $(-)$.

Current and charge do not increase or decrease or get "used up" when going through a wire or other device. The amount of charge that goes in at one end comes out at the other end.

CAUTION

Current is not consumed

18–4 Resistivity

It is found experimentally that the resistance R of a uniform wire is directly proportional to its length ℓ and inversely proportional to its cross-sectional area A. That is,

$$R = \rho \frac{\ell}{A}, \tag{18–3}$$

where ρ (Greek letter "rho"), the constant of proportionality, is called the **resistivity** and depends on the material used. Typical values of ρ, whose units are $\Omega \cdot m$ (see Eq. 18–3), are given for various materials in the middle column of Table 18–1 which is divided into the categories *conductors*, *insulators*, and *semiconductors* (Section 16–3). The values depend somewhat on purity, heat treatment, temperature, and other factors. Notice that silver has the lowest resistivity and is thus the best conductor (although it is expensive). Copper is close, and much less expensive, which is why most wires are made of copper. Aluminum, although it has a higher resistivity, is much less dense than copper; it is thus preferable to copper in some situations, such as for transmission lines, because its resistance for the same weight is less than that for copper.[†]

EXERCISE D Return to the Chapter-Opening Question, page 501, and answer it again now. Try to explain why you may have answered differently the first time.

[†]The reciprocal of the resistivity, called the **electrical conductivity**, is $\sigma = 1/\rho$ and has units of $(\Omega \cdot m)^{-1}$.

TABLE 18–1 Resistivity and Temperature Coefficients (at 20°C)

Material	Resistivity, ρ $(\Omega \cdot m)$	Temperature Coefficient, α $(C°)^{-1}$
Conductors		
Silver	1.59×10^{-8}	0.0061
Copper	1.68×10^{-8}	0.0068
Gold	2.44×10^{-8}	0.0034
Aluminum	2.65×10^{-8}	0.00429
Tungsten	5.6×10^{-8}	0.0045
Iron	9.71×10^{-8}	0.00651
Platinum	10.6×10^{-8}	0.003927
Mercury	98×10^{-8}	0.0009
Nichrome (Ni, Fe, Cr alloy)	100×10^{-8}	0.0004
Semiconductors[‡]		
Carbon (graphite)	$(3-60) \times 10^{-5}$	−0.0005
Germanium	$(1-500) \times 10^{-3}$	−0.05
Silicon	$0.1-60$	−0.07
Insulators		
Glass	10^9-10^{12}	
Hard rubber	$10^{13}-10^{15}$	

[‡] Values depend strongly on the presence of even slight amounts of impurities.

EXERCISE E A copper wire has a resistance of $10\,\Omega$. What would its resistance be if it had the same diameter but was only half as long? (a) $20\,\Omega$, (b) $10\,\Omega$, (c) $5\,\Omega$, (d) $1\,\Omega$, (e) none of these.

EXAMPLE 18–5 **Speaker wires.** Suppose you want to connect your stereo to remote speakers (Fig. 18–14). (a) If each wire must be 20 m long, what diameter copper wire should you use to keep the resistance less than $0.10\,\Omega$ per wire? (b) If the current to each speaker is 4.0 A, what is the potential difference, or voltage drop, across each wire?

APPROACH We solve Eq. 18–3 to get the area A, from which we can calculate the wire's radius using $A = \pi r^2$. The diameter is $2r$. In (b) we can use Ohm's law, $V = IR$.

SOLUTION (a) We solve Eq. 18–3 for the area A and find ρ for copper in Table 18–1:

$$A = \rho\frac{\ell}{R} = \frac{(1.68 \times 10^{-8}\,\Omega\cdot\mathrm{m})(20\,\mathrm{m})}{(0.10\,\Omega)} = 3.4 \times 10^{-6}\,\mathrm{m}^2.$$

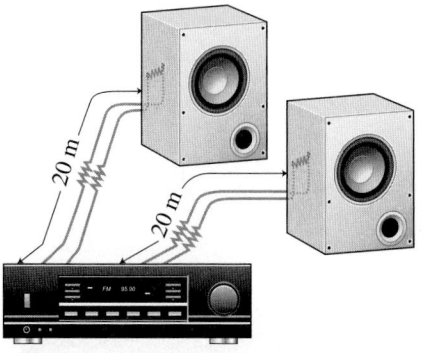

FIGURE 18–14 Example 18–5.

The cross-sectional area A of a circular wire is $A = \pi r^2$. The radius must then be at least

$$r = \sqrt{\frac{A}{\pi}} = 1.04 \times 10^{-3}\,\mathrm{m} = 1.04\,\mathrm{mm}.$$

The diameter is twice the radius and so must be at least $2r = 2.1\,\mathrm{mm}$.

(b) From $V = IR$ we find that the voltage drop across each wire is

$$V = IR = (4.0\,\mathrm{A})(0.10\,\Omega) = 0.40\,\mathrm{V}.$$

NOTE The voltage drop across the wires reduces the voltage that reaches the speakers from the stereo amplifier, thus reducing the sound level a bit.

CONCEPTUAL EXAMPLE 18–6 **Stretching changes resistance.** Suppose a wire of resistance R could be stretched uniformly until it was twice its original length. What would happen to its resistance? Assume the amount of material, and therefore its volume, doesn't change.

RESPONSE If the length ℓ doubles, then the cross-sectional area A is halved, because the volume $(V = A\ell)$ of the wire remains the same. From Eq. 18–3 we see that the resistance would increase by a factor of four $\left(2/\frac{1}{2} = 4\right)$.

EXERCISE F Copper wires in houses typically have a diameter of about 1.5 mm. How long a wire would have a $1.0\text{-}\Omega$ resistance?

Temperature Dependence of Resistivity

The resistivity of a material depends somewhat on temperature. The resistance of metals generally increases with temperature. This is not surprising, because at higher temperatures, the atoms are moving more rapidly and are arranged in a less orderly fashion. So they might be expected to interfere more with the flow of electrons. If the temperature change is not too great, the resistivity of metals usually increases nearly linearly with temperature. That is,

$$\rho_T = \rho_0\big[1 + \alpha(T - T_0)\big] \qquad\qquad \textbf{(18–4)}$$

where ρ_0 is the resistivity at some reference temperature T_0 (such as $0°C$ or $20°C$), ρ_T is the resistivity at a temperature T, and α is the **temperature coefficient of resistivity**. Values for α are given in Table 18–1. Note that the temperature coefficient for semiconductors can be negative. Why? It seems that at higher temperatures, some of the electrons that are normally not free in a semiconductor become free and can contribute to the current. Thus, the resistance of a semiconductor can decrease with an increase in temperature.

EXAMPLE 18–7 **Resistance thermometer.** The variation in electrical resistance with temperature can be used to make precise temperature measurements. Platinum is commonly used since it is relatively free from corrosive effects and has a high melting point. Suppose at 20.0°C the resistance of a platinum resistance thermometer is 164.2 Ω. When placed in a particular solution, the resistance is 187.4 Ω. What is the temperature of this solution?

APPROACH Since the resistance R is directly proportional to the resistivity ρ, we can combine Eq. 18–3 with Eq. 18–4 to find R as a function of temperature T, and then solve that equation for T.

SOLUTION Equation 18–3 tells us $R = \rho\ell/A$, so we multiply Eq. 18–4 by (ℓ/A) to obtain

$$R = R_0\left[1 + \alpha(T - T_0)\right].$$

Here $R_0 = \rho_0\ell/A$ is the resistance of the wire at $T_0 = 20.0°C$. We solve this equation for T and find (see Table 18–1 for α)

$$T = T_0 + \frac{R - R_0}{\alpha R_0} = 20.0°C + \frac{187.4\ \Omega - 164.2\ \Omega}{(3.927 \times 10^{-3}(C°)^{-1})(164.2\ \Omega)} = 56.0°C.$$

NOTE Resistance thermometers have the advantage that they can be used at very high or low temperatures where gas or liquid thermometers would be useless.

NOTE More convenient for some applications is a **thermistor** (Fig. 18–15), which consists of a metal oxide or semiconductor whose resistance also varies in a repeatable way with temperature. Thermistors can be made quite small and respond very quickly to temperature changes.

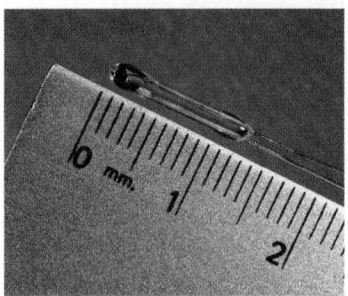

FIGURE 18–15 A thermistor only 13 mm long, shown next to a millimeter ruler.

EXERCISE G The resistance of the tungsten filament of a common incandescent lightbulb is how many times greater at its operating temperature of 2800 K than its resistance at room temperature? (*a*) Less than 1% greater; (*b*) roughly 10% greater; (*c*) about 2 times greater; (*d*) roughly 10 times greater; (*e*) more than 100 times greater.

The value of α in Eq. 18–4 can itself depend on temperature, so it is important to check the temperature range of validity of any value (say, in a handbook of physical data). If the temperature range is wide, Eq. 18–4 is not adequate and terms proportional to the square and cube of the temperature are needed, but these terms are generally very small except when $T - T_0$ is large.

18–5 Electric Power

Electric energy is useful to us because it can be easily transformed into other forms of energy. Motors transform electric energy into mechanical energy, and are examined in Chapter 20.

In other devices such as electric heaters, stoves, toasters, and hair dryers, electric energy is transformed into thermal energy in a wire resistance known as a "heating element." And in an ordinary lightbulb, the tiny wire filament (Fig. 18–5 and Chapter-Opening Photo) becomes so hot it glows; only a few percent of the energy is transformed into visible light, and the rest, over 90%, into thermal energy. Lightbulb filaments and heating elements (Fig. 18–16) in household appliances have resistances typically of a few ohms to a few hundred ohms.

Electric energy is transformed into thermal energy or light in such devices, and there are many collisions between the moving electrons and the atoms of the wire. In each collision, part of the electron's kinetic energy is transferred to the atom with which it collides. As a result, the kinetic energy of the wire's atoms increases and hence the temperature (Section 13–9) of the wire element increases. The increased thermal energy can be transferred as heat by conduction and convection to the air in a heater or to food in a pan, by radiation to bread in a toaster, or radiated as light.

FIGURE 18–16 Hot electric stove burner glows because of energy transformed by electric current.

To find the power transformed by an electric device, recall that the energy transformed when a charge Q moves through a potential difference V is QV (Eq. 17–3). Then the power P, which is the rate energy is transformed, is

$$P = \frac{\text{energy transformed}}{\text{time}} = \frac{QV}{t}.$$

The charge that flows per second, Q/t, is the electric current I. Thus we have

$$P = IV. \tag{18–5}$$

This general relation gives us the power transformed by any device, where I is the current passing through it and V is the potential difference across it. It also gives the power delivered by a source such as a battery. The SI unit of electric power is the same as for any kind of power, the **watt** ($1\text{ W} = 1\text{ J/s}$).

The rate of energy transformation in a resistance R can be written in two other ways, starting with the general relation $P = IV$ and substituting in Ohm's law, $V = IR$:

$$P = IV = I(IR) = I^2R \tag{18–6a}$$

$$P = IV = \left(\frac{V}{R}\right)V = \frac{V^2}{R}. \tag{18–6b}$$

Equations 18–6a and b apply only to resistors, whereas Eq. 18–5, $P = IV$, is more general and applies to any device.

EXAMPLE 18–8 **Headlights.** Calculate the resistance of a 40-W automobile headlight designed for 12 V (Fig. 18–17).

APPROACH We solve for R in Eq. 18–6b, which has the given variables.

SOLUTION From Eq. 18–6b,

$$R = \frac{V^2}{P} = \frac{(12\text{ V})^2}{(40\text{ W})} = 3.6\ \Omega.$$

NOTE This is the resistance when the bulb is burning brightly at 40 W. When the bulb is cold, the resistance is much lower, as we saw in Eq. 18–4 (see also Exercise G). Since the current is high when the resistance is low, lightbulbs burn out most often when first turned on.

It is energy, not power, that you pay for on your electric bill. Since power is the *rate* energy is transformed, the total energy used by any device is simply its power consumption multiplied by the time it is on. If the power is in watts and the time is in seconds, the energy will be in joules since $1\text{ W} = 1\text{ J/s}$. Electric companies usually specify the energy with a much larger unit, the **kilowatt-hour** (kWh). One kWh $= (1000\text{ W})(3600\text{ s}) = 3.60 \times 10^6\text{ J}$.

EXAMPLE 18–9 **Electric heater.** An electric heater draws a steady 15.0 A on a 120-V line. How much power does it require and how much does it cost per month (30 days) if it operates 3.0 h per day and the electric company charges 9.2 cents per kWh?

APPROACH We use Eq. 18–5, $P = IV$, to find the power. We multiply the power (in kW) by the time (h) used in a month and by the cost per energy unit, \$0.092 per kWh, to get the cost per month.

SOLUTION The power is

$$P = IV = (15.0\text{ A})(120\text{ V})$$
$$= 1800\text{ W} = 1.80\text{ kW}.$$

The time (in hours) the heater is used per month is $(3.0\text{ h/d})(30\text{ d}) = 90\text{ h}$, which at 9.2¢/kWh would cost $(1.80\text{ kW})(90\text{ h})(\$0.092/\text{kWh}) = \$15$, just for this heater.

NOTE Household current is actually alternating (ac), but our solution is still valid assuming the given values for V and I are the proper averages (rms) as we discuss in Section 18–7.

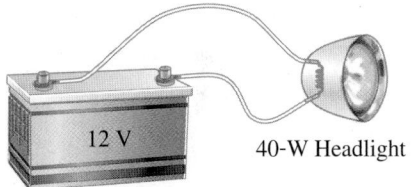
12 V   40-W Headlight

FIGURE 18–17 Example 18–8.

PHYSICS APPLIED
Why lightbulbs burn out when first turned on

CAUTION
You pay for energy, which is power × time, not for power

PHYSICS APPLIED

Lightning

FIGURE 18–18 Example 18–10. A lightning bolt.

EXAMPLE 18–10 ESTIMATE Lightning bolt. Lightning is a spectacular example of electric current in a natural phenomenon (Fig. 18–18). There is much variability to lightning bolts, but a typical event might transfer 10^9 J of energy across a potential difference of perhaps 5×10^7 V during a time interval of about 0.2 s. Use this information to estimate (a) the total amount of charge transferred between cloud and ground, (b) the current in the lightning bolt, and (c) the average power delivered over the 0.2 s.

APPROACH We estimate the charge Q, recalling that potential energy change equals the potential difference ΔV times the charge Q, Eq. 17–3. We equate ΔPE with the energy transferred, ΔPE $\approx 10^9$ J. Next, the current I is Q/t (Eq. 18–1) and the power P is energy/time.

SOLUTION (a) From Eq. 17–3, the energy transformed is ΔPE $= Q\,\Delta V$. We solve for Q:

$$Q = \frac{\Delta \text{PE}}{\Delta V} \approx \frac{10^9 \, \text{J}}{5 \times 10^7 \, \text{V}} = 20 \text{ coulombs.}$$

(b) The current during the 0.2 s is about

$$I = \frac{Q}{t} \approx \frac{20 \, \text{C}}{0.2 \, \text{s}} = 100 \, \text{A.}$$

(c) The average power delivered is

$$P = \frac{\text{energy}}{\text{time}} = \frac{10^9 \, \text{J}}{0.2 \, \text{s}} = 5 \times 10^9 \, \text{W} = 5 \, \text{GW.}$$

We can also use Eq. 18–5:

$$P = IV = (100 \, \text{A})(5 \times 10^7 \, \text{V}) = 5 \, \text{GW.}$$

NOTE Since most lightning bolts consist of several stages, it is possible that individual parts could carry currents much higher than the 100 A calculated above.

EXERCISE H Since 1 kWh $= 3.6 \times 10^6$ J, how much mass must be lifted against gravity through one meter to do the equivalent amount of work?

18–6 Power in Household Circuits

PHYSICS APPLIED

Safety—wires getting hot

PHYSICS APPLIED

Fuses, circuit breakers, and shorts

The electric wires that carry electricity to lights and other electric appliances in houses and buildings have some resistance, although usually it is quite small. Nonetheless, if the current is large enough, the wires will heat up and produce thermal energy at a rate equal to I^2R, where R is the wire's resistance. One possible hazard is that the current-carrying wires in the wall of a building may become so hot as to start a fire. Thicker wires have less resistance (see Eq. 18–3) and thus can carry more current without becoming too hot. When a wire carries more current than is safe, it is said to be "overloaded." To prevent overloading, **fuses** or **circuit breakers** are installed in circuits. They are basically switches (Fig. 18–19, top of next page) that open the circuit when the current exceeds a safe value. A 20-A fuse or circuit breaker, for example, opens when the current passing through it exceeds 20 A. If a circuit repeatedly burns out a fuse or opens a circuit breaker, and no connected device requires more than 20 A, there are two possibilities: there may be too many devices drawing current in that circuit; or there is a fault somewhere, such as a "short." A short, or "short circuit," means that two wires have touched that should not have (perhaps because the insulation has worn through) so the path of the current is shortened through a path of very low resistance. With reduced resistance, the current becomes very large and can make a wire hot enough to start a fire. Short circuits should be remedied immediately.

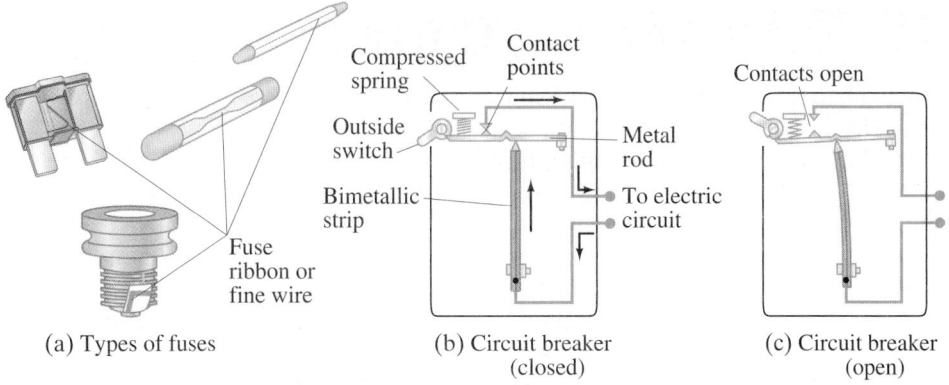

(a) Types of fuses

Compressed spring

Contact points

Outside switch

Bimetallic strip

Metal rod

To electric circuit

(b) Circuit breaker (closed)

Contacts open

(c) Circuit breaker (open)

FIGURE 18–19 (a) Fuses. When current exceeds a certain value, the metallic ribbon or wire inside melts and the circuit opens. Then the fuse must be replaced. (b) One type of circuit breaker. Current passes through a bimetallic strip. When the current exceeds a safe level, the heating of the bimetallic strip causes the strip to bend so far to the left that the notch in the spring-loaded metal rod drops down over the end of the bimetallic strip (c) and the circuit opens at the contact points (one is attached to the rod) and the outside switch is also flipped. When the bimetallic strip cools, it can be reset using the outside switch. Better magnetic-type circuit breakers are discussed in Chapters 20 and 21.

Household circuits are designed with the various devices connected so that each receives the standard voltage (Fig. 18–20) from the electric company (usually 120 V in the United States). Circuits with the devices arranged as in Fig. 18–20 are called *parallel circuits*, as we will discuss in the next Chapter. When a fuse blows or circuit breaker opens, it is important to check the total current being drawn on that circuit, which is the sum of the currents in each device.

FIGURE 18–20 Connection of household appliances.

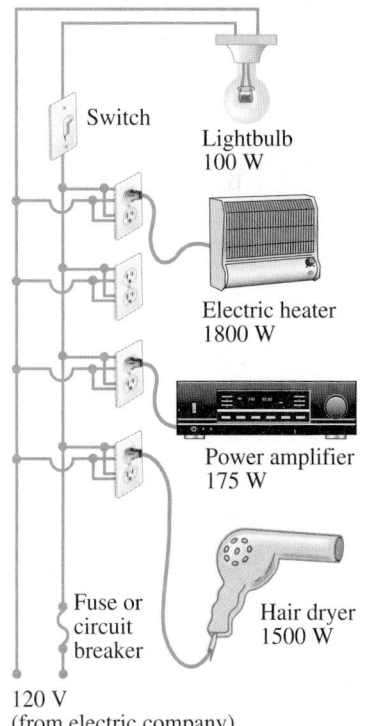

Switch

Lightbulb 100 W

Electric heater 1800 W

Power amplifier 175 W

Fuse or circuit breaker

Hair dryer 1500 W

120 V (from electric company)

EXAMPLE 18–11 **Will a fuse blow?** Determine the total current drawn by all the devices in the circuit of Fig. 18–20.

APPROACH Each device has the same 120-V voltage across it. The current each draws from the source is found from $I = P/V$, Eq. 18–5.

SOLUTION The circuit in Fig. 18–20 draws the following currents: the lightbulb draws $I = P/V = 100\,\text{W}/120\,\text{V} = 0.8\,\text{A}$; the heater draws $1800\,\text{W}/120\,\text{V} = 15.0\,\text{A}$; the power amplifier draws a maximum of $175\,\text{W}/120\,\text{V} = 1.5\,\text{A}$; and the hair dryer draws $1500\,\text{W}/120\,\text{V} = 12.5\,\text{A}$. The total current drawn, if all devices are used at the same time, is

$$0.8\,\text{A} + 15.0\,\text{A} + 1.5\,\text{A} + 12.5\,\text{A} = 29.8\,\text{A}.$$

NOTE The heater draws as much current as 18 100-W lightbulbs. For safety, the heater should probably be on a circuit by itself.

If the circuit in Fig. 18–20 is designed for a 20-A fuse, the fuse should blow, and we hope it will, to prevent overloaded wires from getting hot enough to start a fire. Something will have to be turned off to get this circuit below 20 A. (Houses and apartments usually have several circuits, each with its own fuse or circuit breaker; try moving one of the devices to another circuit.) If the circuit is designed with heavier wire and a 30-A fuse, the fuse shouldn't blow—if it does, a short may be the problem. (The most likely place for a short is in the cord of one of the devices.) Proper fuse size is selected according to the wire used to supply the current. A properly rated fuse should *never* be replaced by a higher-rated one, even in a car. A fuse blowing or a circuit breaker opening is acting like a switch, making an "open circuit." By an open circuit, we mean that there is no longer a complete conducting path, so no current can flow; it is as if $R = \infty$.

 **PHYSICS APPLIED**
Proper fuses and shorts

CONCEPTUAL EXAMPLE 18–12 **A dangerous extension cord.** Your 1800-W portable electric heater is too far from your desk to warm your feet. Its cord is too short, so you plug it into an extension cord rated at 11 A. Why is this dangerous?

RESPONSE 1800 W at 120 V draws a 15-A current. The wires in the extension cord rated at 11 A could become hot enough to melt the insulation and cause a fire.

PHYSICS APPLIED
Extension cords and possible danger

EXERCISE I How many 60-W 120-V lightbulbs can operate on a 20-A line? (*a*) 2; (*b*) 3; (*c*) 6; (*d*) 20; (*e*) 40.

18–7 Alternating Current

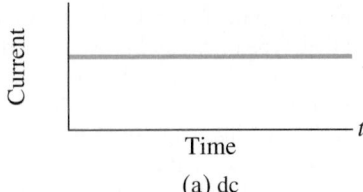

(a) dc

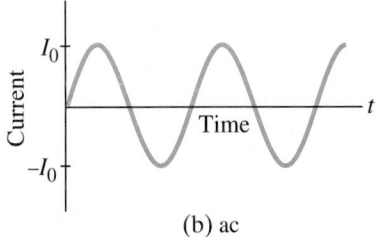

(b) ac

FIGURE 18–21 (a) Direct current, and (b) alternating current, as functions of time.

When a battery is connected to a circuit, the current moves steadily in one direction. This is called a **direct current**, or **dc**. Electric generators at electric power plants, however, produce **alternating current**, or **ac**. (Sometimes capital letters are used, DC and AC.) An alternating current reverses direction many times per second and is commonly sinusoidal, Fig. 18–21. The electrons in a wire first move in one direction and then in the other. The current supplied to homes and businesses by electric companies is ac throughout virtually the entire world. We will discuss and analyze ac circuits in detail in Chapter 21. But because ac circuits are so common in real life, we will discuss some of their basic aspects here.

The voltage produced by an ac electric generator is sinusoidal, as we shall see later. The current it produces is thus sinusoidal (Fig. 18–21b). We can write the voltage as a function of time as

$$V = V_0 \sin 2\pi f t = V_0 \sin \omega t. \qquad \textbf{(18–7a)}$$

The potential V oscillates between $+V_0$ and $-V_0$, and V_0 is referred to as the **peak voltage**. The frequency f is the number of complete oscillations made per second, and $\omega = 2\pi f$. In most areas of the United States and Canada, f is 60 Hz (the unit "hertz," as we saw in Chapters 8 and 11, means cycles per second). In many countries, 50 Hz is used.

Equation 18–2, $V = IR$, works also for ac: if a voltage V exists across a resistance R, then the current I through the resistance is

$$I = \frac{V}{R} = \frac{V_0}{R} \sin \omega t = I_0 \sin \omega t. \qquad \textbf{(18–7b)}$$

The quantity $I_0 = V_0/R$ is the **peak current**. The current is considered positive when the electrons flow in one direction and negative when they flow in the opposite direction. It is clear from Fig. 18–21b that an alternating current is as often positive as it is negative. Thus, the average current is zero. This does not mean, however, that no power is needed or that no heat is produced in a resistor. Electrons do move back and forth, and do produce heat. Indeed, the power transformed in a resistance R at any instant is (Eq. 18–7b)

$$P = I^2 R = I_0^2 R \sin^2 \omega t.$$

Because the current is squared, we see that the power is always positive, as graphed in Fig. 18–22. The quantity $\sin^2 \omega t$ varies between 0 and 1; and it is not too difficult to show[†] that its average value is $\frac{1}{2}$, as indicated in Fig. 18–22. Thus, the *average power* transformed, $\overline{P}$, is

$$\overline{P} = \tfrac{1}{2} I_0^2 R.$$

FIGURE 18–22 Power transformed in a resistor in an ac circuit.

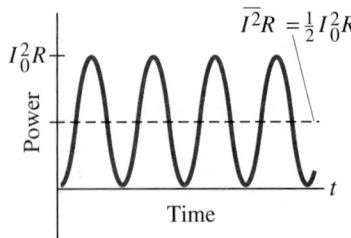

Since power can also be written $P = V^2/R = (V_0^2/R) \sin^2 \omega t$, we also have that the average power is

$$\overline{P} = \tfrac{1}{2} \frac{V_0^2}{R}.$$

The average or mean value of the *square* of the current or voltage is thus what is important for calculating average power: $\overline{I^2} = \tfrac{1}{2} I_0^2$ and $\overline{V^2} = \tfrac{1}{2} V_0^2$. The square root of each of these is the **rms** (root-mean-square) value of the current or voltage:

$$I_{rms} = \sqrt{\overline{I^2}} = \frac{I_0}{\sqrt{2}} = 0.707 I_0, \qquad \textbf{(18–8a)}$$

$$V_{rms} = \sqrt{\overline{V^2}} = \frac{V_0}{\sqrt{2}} = 0.707 V_0. \qquad \textbf{(18–8b)}$$

[†]A graph of $\cos^2 \omega t$ versus t is identical to that for $\sin^2 \omega t$ in Fig. 18–22, except that the points are shifted (by $\frac{1}{4}$ cycle) on the time axis. Thus the average value of $\sin^2$ and $\cos^2$, averaged over one or more full cycles, will be the same. From the trigonometric identity $\sin^2 \theta + \cos^2 \theta = 1$, we can write

$$\overline{(\sin^2 \omega t)} + \overline{(\cos^2 \omega t)} = 2\overline{(\sin^2 \omega t)} = 1.$$

Hence the average value of $\sin^2 \omega t$ is $\frac{1}{2}$.

The rms values of V and I are sometimes called the *effective values*. They are useful because they can be substituted directly into the power formulas, Eqs. 18–5 and 18–6, to get the average power:

$$\overline{P} = I_{rms} V_{rms} \qquad \text{(18–9a)}$$

$$\overline{P} = \tfrac{1}{2} I_0^2 R = I_{rms}^2 R \qquad \text{(18–9b)}$$

$$\overline{P} = \tfrac{1}{2} \frac{V_0^2}{R} = \frac{V_{rms}^2}{R}. \qquad \text{(18–9c)}$$

Thus, a direct current whose values of I and V equal the rms values of I and V for an alternating current will produce the same power. Hence it is usually the rms value of current and voltage that is specified or measured. For example, in the United States and Canada, standard line voltage is 120-V ac. The 120 V is V_{rms}; the peak voltage V_0 is (Eq. 18–8b)

$$V_0 = \sqrt{2} \, V_{rms} = 170 \, \text{V}.$$

In much of the world (Europe, Australia, Asia) the rms voltage is 240 V, so the peak voltage is 340 V. The line voltage can vary, depending on the total load; the frequency of 60 Hz or 50 Hz, however, remains extremely steady.

EXAMPLE 18–13 **Hair dryer.** (*a*) Calculate the resistance and the peak current in a 1500-W hair dryer (Fig. 18–23) connected to a 120-V ac line. (*b*) What happens if it is connected to a 240-V ac line in Britain?

APPROACH We are given $\overline{P}$ and V_{rms}, so $I_{rms} = \overline{P}/V_{rms}$ (Eq. 18–9a or 18–5), and $I_0 = \sqrt{2} \, I_{rms}$. Then we find R from $V = IR$.

SOLUTION (*a*) We solve Eq. 18–9a for the rms current:

$$I_{rms} = \frac{\overline{P}}{V_{rms}} = \frac{1500 \, \text{W}}{120 \, \text{V}} = 12.5 \, \text{A}.$$

Then

$$I_0 = \sqrt{2} \, I_{rms} = 17.7 \, \text{A}.$$

The resistance is

$$R = \frac{V_{rms}}{I_{rms}} = \frac{120 \, \text{V}}{12.5 \, \text{A}} = 9.6 \, \Omega.$$

The resistance could equally well be calculated using peak values:

$$R = \frac{V_0}{I_0} = \frac{170 \, \text{V}}{17.7 \, \text{A}} = 9.6 \, \Omega.$$

(*b*) When connected to a 240-V line, more current would flow and the resistance would change with the increased temperature (Section 18–4). But let us make an estimate of the power transformed based on the same 9.6-Ω resistance. The average power would be

$$\overline{P} = \frac{V_{rms}^2}{R}$$

$$= \frac{(240 \, \text{V})^2}{(9.6 \, \Omega)} = 6000 \, \text{W}.$$

This is four times the dryer's power rating and would undoubtedly melt the heating element or the wire coils of the motor.

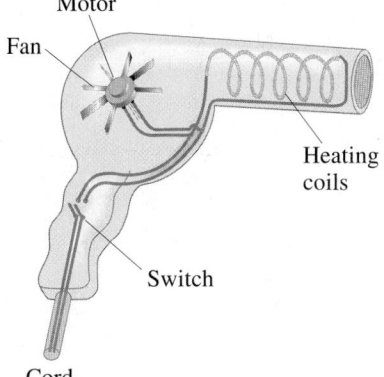

FIGURE 18–23 A hair dryer. Most of the current goes through the heating coils, a pure resistance; a small part goes to the motor to turn the fan. Example 18–13.

This Section has given a brief introduction to the simpler aspects of alternating currents. We will discuss ac circuits in more detail in Chapter 21. In Chapter 19 we will deal with the details of dc circuits only.

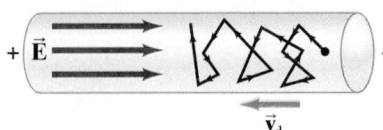

FIGURE 18–24 Electric field $\vec{E}$ in a wire gives electrons in random motion a drift velocity $\vec{v}_d$. Note $\vec{v}_d$ is in the opposite direction of $\vec{E}$ because electrons have a negative charge $(\vec{F} = q\vec{E})$.

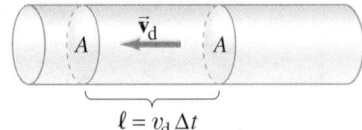

FIGURE 18–25 Electrons in the volume $A\ell$ will all pass through the cross section indicated in a time Δt, where $\ell = v_d \Delta t$.

*18–8 Microscopic View of Electric Current

It can be useful to analyze a simple model of electric current at the microscopic level of atoms and electrons. In a conducting wire, for example, we can imagine the free electrons as moving about randomly at high speeds, bouncing off the atoms of the wire (somewhat like the molecules of a gas—Sections 13–8 to 13–10). When an electric field exists in the wire, Fig. 18–24, the electrons feel a force and initially begin to accelerate. But they soon reach a more or less steady average velocity known as their **drift velocity**, v_d (collisions with atoms in the wire keep them from accelerating further). The drift velocity is normally very much smaller than the electrons' average random speed.

We can relate v_d to the macroscopic current I in the wire. In a time Δt, the electrons will travel a distance $\ell = v_d \Delta t$ on average. Suppose the wire has cross-sectional area A. Then in time Δt, electrons in a volume $V = A\ell = Av_d \Delta t$ will pass through the cross section A of wire, as shown in Fig. 18–25. If there are n free electrons (each with magnitude of charge e) per unit volume, then the total number of electrons is $N = nV$ (V is volume, not voltage) and the total charge ΔQ that passes through the area A in a time Δt is

$$\Delta Q = \text{(number of charges, } N) \times \text{(charge per particle)}$$
$$= (nV)(e) = (nAv_d \Delta t)(e).$$

The magnitude of the current I in the wire is thus

$$I = \frac{\Delta Q}{\Delta t} = neAv_d. \qquad (18\text{–}10)$$

EXAMPLE 18–14 **Electron speed in wire.** A copper wire 3.2 mm in diameter carries a 5.0-A current. Determine the drift velocity of the free electrons. Assume that one electron per Cu atom is free to move (the others remain bound to the atom).

APPROACH We apply Eq. 18–10 to find the drift velocity v_d if we can determine the number n of free electrons per unit volume. Since we assume there is one free electron per atom, the density of free electrons, n, is the same as the number of Cu atoms per unit volume. The atomic mass of Cu is 63.5 u (see Periodic Table inside the back cover), so 63.5 g of Cu contains one mole or 6.02×10^{23} free electrons. To find the volume V of this amount of copper, and then $n = N/V$, we use the mass density of copper (Table 10–1), $\rho_D = 8.9 \times 10^3 \text{ kg/m}^3$, where $\rho_D = m/V$. (We use ρ_D to distinguish it here from ρ for resistivity.)

SOLUTION The number of free electrons per unit volume, $n = N/V$ (where $V = \text{volume} = m/\rho_D$), is

$$n = \frac{N}{V} = \frac{N}{m/\rho_D} = \frac{N(1 \text{ mole})}{m(1 \text{ mole})}\rho_D$$

$$n = \left(\frac{6.02 \times 10^{23} \text{ electrons}}{63.5 \times 10^{-3} \text{ kg}}\right)(8.9 \times 10^3 \text{ kg/m}^3) = 8.4 \times 10^{28} \text{ m}^{-3}.$$

The cross-sectional area of the wire is $A = \pi r^2 = \pi(1.6 \times 10^{-3} \text{ m})^2 = 8.0 \times 10^{-6} \text{ m}^2$. Then, by Eq. 18–10, the drift velocity has magnitude

$$v_d = \frac{I}{neA} = \frac{5.0 \text{ A}}{(8.4 \times 10^{28} \text{ m}^{-3})(1.6 \times 10^{-19} \text{ C})(8.0 \times 10^{-6} \text{ m}^2)}$$

$$= 4.6 \times 10^{-5} \text{ m/s} \approx 0.05 \text{ mm/s}.$$

NOTE The actual speed of electrons bouncing around inside the metal is estimated to be about $1.6 \times 10^6 \text{ m/s}$ at 20°C, very much greater than the drift velocity.

The drift velocity of electrons in a wire is slow, only about 0.05 mm/s in Example 18–14, which means it takes an electron about 20×10^3 s, or $5\frac{1}{2}$ h, to travel only 1 m. This is not how fast "electricity travels": when you flip a light switch, the light—even if many meters away—goes on nearly instantaneously. Why? Because electric fields travel essentially at the speed of light $(3 \times 10^8 \text{ m/s})$. We can think of electrons in a wire as being like a pipe full of water: when a little water enters one end of the pipe, some water immediately comes out the other end.

*18–9 Superconductivity

At very low temperatures, well below 0°C, the resistivity (Section 18–4) of certain metals and certain compounds or alloys becomes zero as measured by the highest-precision techniques. Materials in such a state are said to be **superconducting**. This phenomenon was first observed by H. K. Onnes (1853–1926) in 1911 when he cooled mercury below 4.2 K (−269°C) and found that the resistance of mercury suddenly dropped to zero. In general, superconductors become superconducting only below a certain *transition temperature* or *critical temperature*, T_C, which is usually within a few degrees of absolute zero. Current in a ring-shaped superconducting material has been observed to flow for years in the absence of a potential difference, with no measurable decrease. Measurements show that the resistivity ρ of superconductors is less than $4 \times 10^{-25}\ \Omega\cdot m$, which is over 10^{16} times smaller than that for copper, and is considered to be zero in practice. See Fig. 18–26.

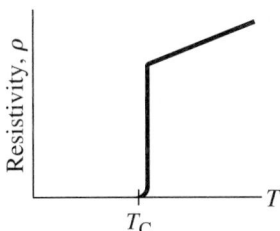

FIGURE 18–26 A superconducting material has zero resistivity when its temperature is below T_C, its "critical temperature." At temperatures above T_C, the resistivity jumps to a "normal" nonzero value and increases with temperature as most materials do (Eq. 18–4).

Before 1986 the highest temperature at which a material was found to superconduct was 23 K, which required liquid helium to keep the material cold. In 1987, a compound of yttrium, barium, copper, and oxygen (YBCO) was developed that can be superconducting at 90 K. Since this is above the boiling temperature of liquid nitrogen, 77 K, liquid nitrogen is sufficiently cold to keep the material superconducting. This was an important breakthrough because liquid nitrogen is much more easily and cheaply obtained than is the liquid helium needed for earlier superconductors. Superconductivity at temperatures as high as 160 K has been reported, though in fragile compounds.

To develop high-T_C superconductors for use as wires (such as for wires in "superconducting electromagnets"—Section 20–7), many applications today utilize a bismuth-strontium-calcium-copper oxide (BSCCO). A major challenge is how to make a useable, bendable wire out of the BSCCO, which is very brittle. (One solution is to embed tiny filaments of the high-T_C superconductor in a metal alloy, which is not resistanceless but has resistance much less than a conventional copper cable.)

*18–10 Electrical Conduction in the Human Nervous System

An interesting example of the flow of electric charge is in the human nervous system, which provides us with the means for being aware of the world, for communication within the body, and for controlling the body's muscles. Although the detailed functioning of the hugely complex nervous system still is not well understood, we do have a reasonable understanding of how messages are transmitted within the nervous system: they are electrical signals passing along the basic element of the nervous system, the **neuron**.

Neurons are living cells of unusual shape (Fig. 18–27). Attached to the main cell body are several small appendages known as *dendrites* and a long tail called the *axon*. Signals are received by the dendrites and are propagated along the axon. When a signal reaches the nerve endings, it is transmitted to the next neuron or to a muscle at a connection called a *synapse*.

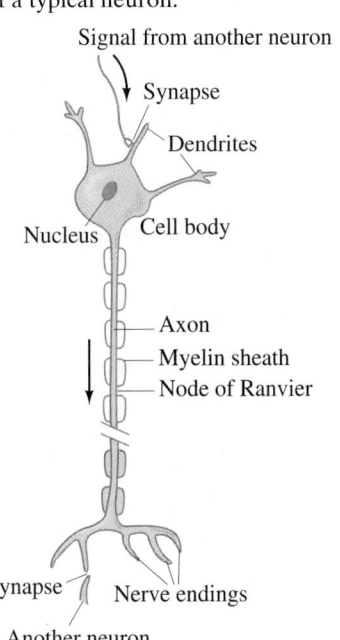

FIGURE 18–27 A simplified sketch of a typical neuron.

TABLE 18–2
Concentrations of Ions Inside and Outside a Typical Axon

	Concentration inside axon (mol/m³)	Concentration outside axon (mol/m³)
K^+	140	5
Na^+	15	140
Cl^-	9	125

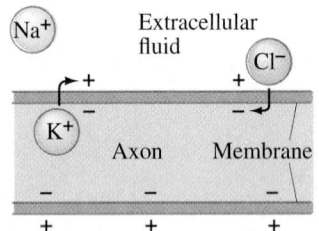

FIGURE 18–28 How a dipole layer of charge forms on a cell membrane.

FIGURE 18–29 Measuring the potential difference between the inside and outside of a nerve cell.

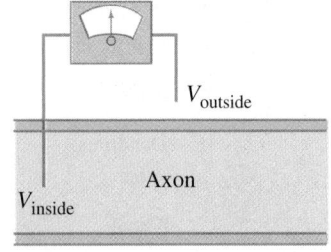

FIGURE 18–30 Action potential.

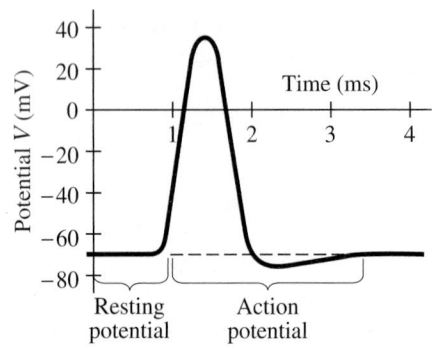

A neuron, before transmitting an electrical signal, is in the so-called "resting state." Like nearly all living cells, neurons have a net positive charge on the outer surface of the cell membrane and a negative charge on the inner surface. This difference in charge, or **dipole layer**, means that a potential difference exists across the cell membrane. When a neuron is not transmitting a signal, this **resting potential**, normally stated as

$$V_{\text{inside}} - V_{\text{outside}},$$

is typically $-60\,\text{mV}$ to $-90\,\text{mV}$, depending on the type of organism. The most common ions in a cell are K^+, Na^+, and Cl^-. There are large differences in the concentrations of these ions inside and outside an axon, as indicated by the typical values given in Table 18–2. Other ions are also present, so the fluids both inside and outside the axon are electrically neutral. Because of the differences in concentration, there is a tendency for ions to diffuse across the membrane (see Section 13–13 on diffusion). However, in the resting state the cell membrane prevents any net flow of Na^+ through a mechanism of active transport[‡] of Na^+ ions out of the cell by a particular protein to which Na^+ attach; energy needed comes from ATP. But it does allow the flow of Cl^- ions, and less so of K^+ ions, and it is these two ions that produce the dipole charge layer on the membrane. Because there is a greater concentration of K^+ inside the cell than outside, more K^+ ions tend to diffuse outward across the membrane than diffuse inward. A K^+ ion that passes through the membrane becomes attached to the outer surface of the membrane, and leaves behind an equal negative charge that lies on the inner surface of the membrane (Fig. 18–28). The fluids themselves remain neutral. What keeps the ions on the membrane is their attraction for each other across the membrane. Independently, Cl^- ions tend to diffuse *into* the cell since their concentration outside is higher. Both K^+ and Cl^- diffusion tends to charge the interior surface of the membrane negative and the outside positive. As charge accumulates on the membrane surface, it becomes increasingly difficult for more ions to diffuse: K^+ ions trying to move outward, for example, are repelled by the positive charge already there. Equilibrium is reached when the tendency to diffuse because of the concentration difference is just balanced by the electrical potential difference across the membrane. The greater the concentration difference, the greater the potential difference across the membrane ($-60\,\text{mV}$ to $-90\,\text{mV}$).

The most important aspect of a neuron is not that it has a resting potential (most cells do), but rather that it can respond to a stimulus and conduct an electrical signal along its length. The stimulus could be thermal (when you touch a hot stove) or chemical (as in taste buds); it could be pressure (as on the skin or at the eardrum), or light (as in the eye); or it could be the electric stimulus of a signal coming from the brain or another neuron. In the laboratory, the stimulus is usually electrical and is applied by a tiny probe at some point on the neuron. If the stimulus exceeds some threshold, a voltage pulse will travel down the axon. This voltage pulse can be detected at a point on the axon using a voltmeter or an oscilloscope connected as in Fig. 18–29. This voltage pulse has the shape shown in Fig. 18–30, and is called an **action potential**. As can be seen, the potential increases from a resting potential of about $-70\,\text{mV}$ and becomes a positive $30\,\text{mV}$ or $40\,\text{mV}$. The action potential lasts for about 1 ms and travels down an axon with a speed of $30\,\text{m/s}$ to $150\,\text{m/s}$. When an action potential is stimulated, the nerve is said to have "fired."

What causes the action potential? At the point where the stimulus occurs, the membrane suddenly alters its permeability, becoming much more permeable to Na^+ than to K^+ and Cl^- ions. Thus, Na^+ ions rush into the cell and the inner surface of the wall becomes positively charged, and the potential difference quickly swings positive ($\approx +30\,\text{mV}$ in Fig. 18–30). Just as suddenly, the membrane returns to its original characteristics; it becomes impermeable to Na^+ and in fact pumps out Na^+ ions. The diffusion of Cl^- and K^+ ions again predominates and the original resting potential is restored ($-70\,\text{mV}$ in Fig. 18–30).

[‡]This transport mechanism is sometimes referred to as the "sodium pump."

What causes the action potential to travel along the axon? The action potential occurs at the point of stimulation, as shown in Fig. 18–31a. The membrane momentarily is positive on the inside and negative on the outside at this point. Nearby charges are attracted toward this region, as shown in Fig. 18–31b. The potential in these adjacent regions then drops, causing an action potential there. Thus, as the membrane returns to normal at the original point, nearby it experiences an action potential, so the action potential moves down the axon (Figs. 18–31c and d).

You may wonder if the number of ions that pass through the membrane would significantly alter the concentrations. The answer is no; and we can show why (and again show the power and usefulness of physics) by treating the axon as a capacitor as we do in Search and Learn Problem 8 (the concentration changes by less than 1 part in 10^4).

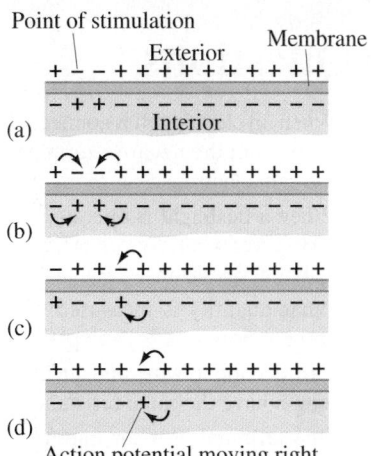

FIGURE 18–31 Propagation of action potential along axon membrane.

Summary

An electric **battery** serves as a source of nearly constant potential difference by transforming chemical energy into electric energy. A simple battery consists of two electrodes made of different metals immersed in a solution or paste known as an electrolyte.

Electric current, I, refers to the rate of flow of electric charge and is measured in **amperes** (A): 1 A equals a flow of 1 C/s past a given point.

The direction of **conventional current** is that of positive charge flow. In a wire, it is actually negatively charged electrons that move, so they flow in a direction opposite to the conventional current. A positive charge flow in one direction is almost always equivalent to a negative charge flow in the opposite direction. Positive conventional current always flows from a high potential to a low potential.

The **resistance** R of a device is defined by the relation

$$V = IR, \tag{18–2}$$

where I is the current in the device when a potential difference V is applied across it. For materials such as metals, R is a constant independent of V (thus $I \propto V$), a result known as **Ohm's law**. Thus, the current I coming from a battery of voltage V depends on the resistance R of the circuit connected to it.

Voltage is applied *across* a device or between the ends of a wire. Current passes *through* a wire or device. Resistance is a property *of* the wire or device.

The unit of resistance is the **ohm** (Ω), where $1\,\Omega = 1\,\text{V/A}$. See Table 18–3.

TABLE 18–3 Summary of Units

Current	$1\,\text{A} = 1\,\text{C/s}$
Potential difference	$1\,\text{V} = 1\,\text{J/C}$
Power	$1\,\text{W} = 1\,\text{J/s}$
Resistance	$1\,\Omega = 1\,\text{V/A}$

The resistance R of a wire is inversely proportional to its cross-sectional area A, and directly proportional to its length ℓ and to a property of the material called its resistivity:

$$R = \frac{\rho \ell}{A}. \tag{18–3}$$

The **resistivity**, ρ, increases with temperature for metals, but for semiconductors it may decrease.

The rate at which energy is transformed in a resistance R from electric to other forms of energy (such as heat and light) is equal to the product of current and voltage. That is, the **power** transformed, measured in watts, is given by

$$P = IV, \tag{18–5}$$

which for resistors can be written as

$$P = I^2R = \frac{V^2}{R}. \tag{18–6}$$

The SI unit of power is the **watt** ($1\,\text{W} = 1\,\text{J/s}$).

The total electric energy transformed in any device equals the product of the power and the time during which the device is operated. In SI units, energy is given in joules ($1\,\text{J} = 1\,\text{W}\cdot\text{s}$), but electric companies use a larger unit, the **kilowatt-hour** ($1\,\text{kWh} = 3.6 \times 10^6\,\text{J}$).

Electric current can be **direct current** (**dc**), in which the current is steady in one direction; or it can be **alternating current** (**ac**), in which the current reverses direction at a particular frequency f, typically 60 Hz. Alternating currents are typically sinusoidal in time,

$$I = I_0 \sin \omega t, \tag{18–7b}$$

where $\omega = 2\pi f$, and are produced by an alternating voltage.

The **rms** values of sinusoidally alternating currents and voltages are given by

$$I_{\text{rms}} = \frac{I_0}{\sqrt{2}} \quad \text{and} \quad V_{\text{rms}} = \frac{V_0}{\sqrt{2}}, \tag{18–8}$$

respectively, where I_0 and V_0 are the **peak** values. The power relationship, $P = IV = I^2R = V^2/R$, is valid for the average power in alternating currents when the rms values of V and I are used.

[*The current in a wire, at the microscopic level, is considered to be a slow **drift velocity** of electrons, $\vec{v}_d$. The current I is given by

$$I = neAv_d, \tag{18–10}$$

where n is the number of free electrons per unit volume, e is the magnitude of the charge on an electron, and A is the cross-sectional area of the wire.]

[*At very low temperatures certain materials become **superconducting**, which means their electrical resistance becomes zero.]

[*The human nervous system operates via electrical conduction: when a nerve "fires," an electrical signal travels as a voltage pulse known as an **action potential**.]

Questions

1. When an electric cell is connected to a circuit, electrons flow away from the negative terminal in the circuit. But within the cell, electrons flow *to* the negative terminal. Explain.

2. When a flashlight is operated, what is being used up: battery current, battery voltage, battery energy, battery power, or battery resistance? Explain.

3. What quantity is measured by a battery rating given in ampere-hours $(A \cdot h)$? Explain.

4. Can a copper wire and an aluminum wire of the same length have the same resistance? Explain.

5. One terminal of a car battery is said to be connected to "ground." Since it is not really connected to the ground, what is meant by this expression?

6. The equation $P = V^2/R$ indicates that the power dissipated in a resistor decreases if the resistance is increased, whereas the equation $P = I^2R$ implies the opposite. Is there a contradiction here? Explain.

7. What happens when a lightbulb burns out?

8. If the resistance of a small immersion heater (to heat water for tea or soup, Fig. 18–32) was increased, would it speed up or slow down the heating process? Explain.

FIGURE 18–32
Question 8.

9. If a rectangular solid made of carbon has sides of lengths a, $2a$, and $3a$, to which faces would you connect the wires from a battery so as to obtain (a) the least resistance, (b) the greatest resistance?

10. Explain why lightbulbs almost always burn out just as they are turned on and not after they have been on for some time.

11. Which draws more current, a 100-W lightbulb or a 75-W bulb? Which has the higher resistance?

12. Electric power is transferred over large distances at very high voltages. Explain how the high voltage reduces power losses in the transmission lines.

13. A 15-A fuse blows out repeatedly. Why is it dangerous to replace this fuse with a 25-A fuse?

14. When electric lights are operated on low-frequency ac (say, 5 Hz), they flicker noticeably. Why?

15. Driven by ac power, the same electrons pass back and forth through your reading lamp over and over again. Explain why the light stays lit instead of going out after the first pass of electrons.

16. The heating element in a toaster is made of Nichrome wire. Immediately after the toaster is turned on, is the current magnitude (I_{rms}) in the wire increasing, decreasing, or staying constant? Explain.

17. Is current used up in a resistor? Explain.

18. Why is it more dangerous to turn on an electric appliance when you are standing outside in bare feet than when you are inside wearing shoes with thick soles?

*19. Compare the drift velocities and electric currents in two wires that are geometrically identical and the density of atoms is similar, but the number of free electrons per atom in the material of one wire is twice that in the other.

*20. A voltage V is connected across a wire of length ℓ and radius r. How is the electron drift speed affected if (a) ℓ is doubled, (b) r is doubled, (c) V is doubled, assuming in each case that other quantities stay the same?

MisConceptual Questions

1. When connected to a battery, a lightbulb glows brightly. If the battery is reversed and reconnected to the bulb, the bulb will glow
   (a) brighter.   (c) with the same brightness.
   (b) dimmer.   (d) not at all.

2. When a battery is connected to a lightbulb properly, current flows through the lightbulb and makes it glow. How much current flows through the battery compared with the lightbulb?
   (a) More.
   (b) Less.
   (c) The same amount.
   (d) No current flows through the battery.

3. Which of the following statements about Ohm's law is true?
   (a) Ohm's law relates the current through a wire to the voltage across the wire.
   (b) Ohm's law holds for all materials.
   (c) Any material that obeys Ohm's law does so independently of temperature.
   (d) Ohm's law is a fundamental law of physics.
   (e) Ohm's law is valid for superconductors.

4. Electrons carry energy from a battery to a lightbulb. What happens to the electrons when they reach the lightbulb?
   (a) The electrons are used up.
   (b) The electrons stay in the lightbulb.
   (c) The electrons are emitted as light.
   (d) Fewer electrons leave the bulb than enter it.
   (e) None of the above.

5. Where in the circuit of Fig. 18–33 is the current the largest, (a), (b), (c), or (d)? Or (e) it is the same at all points?

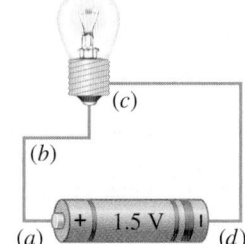

FIGURE 18–33
MisConceptual Question 5.

6. When you double the *voltage* across a certain material or device, you observe that the *current* increases by a factor of 3. What can you conclude?
   (a) Ohm's law is obeyed, because the current increases when V increases.
   (b) Ohm's law is not obeyed in this case.
   (c) This situation has nothing to do with Ohm's law.

7. When current flows through a resistor,
   (a) some of the charge is used up by the resistor.
   (b) some of the current is used up by the resistor.
   (c) Both (a) and (b) are true.
   (d) Neither (a) nor (b) is true.

8. The unit kilowatt-hour is a measure of
   (a) the rate at which energy is transformed.
   (b) power.
   (c) an amount of energy.
   (d) the amount of power used per second.

9. Why might a circuit breaker open if you plug too many electrical devices into a single circuit?
   (a) The voltage becomes too high.
   (b) The current becomes too high.
   (c) The resistance becomes too high.
   (d) A circuit breaker will not "trip" no matter how many electrical devices you plug into the circuit.

10. Nothing happens when birds land on a power line, yet we are warned not to touch a power line with a ladder. What is the difference?
    (a) Birds have extremely high internal resistance compared to humans.
    (b) There is little to no voltage drop between a bird's two feet, but there is a significant voltage drop between the top of a ladder touching a power line and the bottom of the ladder on the ground.
    (c) Dangerous current comes from the ground only.
    (d) Most birds don't understand the situation.

11. When a light switch is turned on, the light comes on immediately because
    (a) the electrons coming from the power source move through the initially empty wires very fast.
    (b) the electrons already in the wire are instantly "pushed" by a voltage difference.
    (c) the lightbulb may be old with low resistance. It would take longer if the bulb were new and had high resistance.
    (d) the electricity bill is paid. The electric company can make it take longer when the bill is unpaid.

For assigned homework and other learning materials, go to the MasteringPhysics website.

Problems

18–2 and 18–3 Electric Current, Resistance, Ohm's Law
(*Note*: The charge on one electron is 1.60×10^{-19} C.)

1. (I) A current of 1.60 A flows in a wire. How many electrons are flowing past any point in the wire per second?

2. (I) A service station charges a battery using a current of 6.7 A for 5.0 h. How much charge passes through the battery?

3. (I) What is the current in amperes if 1200 Na^+ ions flow across a cell membrane in 3.1 μs? The charge on the sodium is the same as on an electron, but positive.

4. (I) What is the resistance of a toaster if 120 V produces a current of 4.6 A?

5. (I) What voltage will produce 0.25 A of current through a 4800-Ω resistor?

6. (I) How many coulombs are there in a 75 ampere-hour car battery?

7. (II) (a) What is the current in the element of an electric clothes dryer with a resistance of 8.6 Ω when it is connected to 240 V? (b) How much charge passes through the element in 50 min? (Assume direct current.)

8. (II) A bird stands on a dc electric transmission line carrying 4100 A (Fig. 18–34). The line has 2.5×10^{-5} Ω resistance per meter, and the bird's feet are 4.0 cm apart. What is the potential difference between the bird's feet?

FIGURE 18–34
Problem 8.

9. (II) A hair dryer draws 13.5 A when plugged into a 120-V line. (a) What is its resistance? (b) How much charge passes through it in 15 min? (Assume direct current.)

10. (II) A 4.5-V battery is connected to a bulb whose resistance is 1.3 Ω. How many electrons leave the battery per minute?

11. (II) An electric device draws 5.60 A at 240 V. (a) If the voltage drops by 15%, what will be the current, assuming nothing else changes? (b) If the resistance of the device were reduced by 15%, what current would be drawn at 240 V?

18–4 Resistivity

12. (I) What is the diameter of a 1.00-m length of tungsten wire whose resistance is 0.32 Ω?

13. (I) What is the resistance of a 5.4-m length of copper wire 1.5 mm in diameter?

14. (II) Calculate the ratio of the resistance of 10.0 m of aluminum wire 2.2 mm in diameter, to 24.0 m of copper wire 1.8 mm in diameter.

15. (II) Can a 2.2-mm-diameter copper wire have the same resistance as a tungsten wire of the same length? Give numerical details.

16. (II) A certain copper wire has a resistance of 15.0 Ω. At what point along its length must the wire be cut so that the resistance of one piece is 4.0 times the resistance of the other? What is the resistance of each piece?

17. (II) Compute the voltage drop along a 21-m length of household no. 14 copper wire (used in 15-A circuits). The wire has diameter 1.628 mm and carries a 12-A current.

18. (II) Two aluminum wires have the same resistance. If one has twice the length of the other, what is the ratio of the diameter of the longer wire to the diameter of the shorter wire?

19. (II) A rectangular solid made of carbon has sides of lengths 1.0 cm, 2.0 cm, and 4.0 cm, lying along the x, y, and z axes, respectively (Fig. 18–35). Determine the resistance for current that passes through the solid in (a) the x direction, (b) the y direction, and (c) the z direction. Assume the resistivity is $\rho = 3.0 \times 10^{-5}\ \Omega\cdot\text{m}$.

FIGURE 18–35
Problem 19.

20. (II) A length of wire is cut in half and the two lengths are wrapped together side by side to make a thicker wire. How does the resistance of this new combination compare to the resistance of the original wire?

21. (II) How much would you have to raise the temperature of a copper wire (originally at 20°C) to increase its resistance by 12%?

22. (II) Determine at what temperature aluminum will have the same resistivity as tungsten does at 20°C.

23. (II) A 100-W lightbulb has a resistance of about 12 Ω when cold (20°C) and 140 Ω when on (hot). Estimate the temperature of the filament when hot assuming an average temperature coefficient of resistivity $\alpha = 0.0045\ (\text{C}°)^{-1}$.

24. (III) A length of aluminum wire is connected to a precision 10.00-V power supply, and a current of 0.4212 A is precisely measured at 23.5°C. The wire is placed in a new environment of unknown temperature where the measured current is 0.3818 A. What is the unknown temperature?

25. (III) For some applications, it is important that the value of a resistance not change with temperature. For example, suppose you made a 3.20-kΩ resistor from a carbon resistor and a Nichrome wire-wound resistor connected together so the total resistance is the sum of their separate resistances. What value should each of these resistors have (at 0°C) so that the combination is temperature independent?

26. (III) A 10.0-m length of wire consists of 5.0 m of copper followed by 5.0 m of aluminum, both of diameter 1.4 mm. A voltage difference of 95 mV is placed across the composite wire. (a) What is the total resistance (sum) of the two wires? (b) What is the current through the wire? (c) What are the voltages across the aluminum part and across the copper part?

18–5 and 18–6 Electric Power

27. (I) What is the maximum power consumption of a 3.0-V portable CD player that draws a maximum of 240 mA of current?

28. (I) The heating element of an electric oven is designed to produce 3.3 kW of heat when connected to a 240-V source. What must be the resistance of the element?

29. (I) What is the maximum voltage that can be applied across a 3.9-kΩ resistor rated at $\frac{1}{4}$ watt?

30. (I) (a) Determine the resistance of, and current through, a 75-W lightbulb connected to its proper source voltage of 110 V. (b) Repeat for a 250-W bulb.

31. (I) An electric car has a battery that can hold 16 kWh of energy (approximately 6×10^7 J). If the battery is designed to operate at 340 V, how many coulombs of charge would need to leave the battery at 340 V and return at 0 V to equal the stored energy of the battery?

32. (I) An electric car uses a 45-kW (160-hp) motor. If the battery pack is designed for 340 V, what current would the motor need to draw from the battery? Neglect any energy losses in getting energy from the battery to the motor.

33. (II) A 120-V hair dryer has two settings: 950 W and 1450 W. (a) At which setting do you guess the resistance to be higher? After making a guess, determine the resistance at (b) the lower setting, and (c) the higher setting.

34. (II) A 12-V battery causes a current of 0.60 A through a resistor. (a) What is its resistance, and (b) how many joules of energy does the battery lose in a minute?

35. (II) A 120-V fish-tank heater is rated at 130 W. Calculate (a) the current through the heater when it is operating, and (b) its resistance.

36. (II) You buy a 75-W lightbulb in Europe, where electricity is delivered at 240 V. If you use the bulb in the United States at 120 V (assume its resistance does not change), how bright will it be relative to 75-W 120-V bulbs? [*Hint*: Assume roughly that brightness is proportional to power consumed.]

37. (II) How many kWh of energy does a 550-W toaster use in the morning if it is in operation for a total of 5.0 min? At a cost of 9.0 cents/kWh, estimate how much this would add to your monthly electric energy bill if you made toast four mornings per week.

38. (II) At $0.095/kWh, what does it cost to leave a 25-W porch light on day and night for a year?

39. (II) What is the total amount of energy stored in a 12-V, 65 A·h car battery when it is fully charged?

40. (II) An ordinary flashlight uses two D-cell 1.5-V batteries connected in series to provide 3.0 V across the bulb, as in Fig. 18–4b (Fig. 18–36). The bulb draws 380 mA when turned on. (a) Calculate the resistance of the bulb and the power dissipated. (b) By what factor would the power increase if four D-cells in series (total 6.0 V) were used with the same bulb? (Neglect heating effects of the filament.) Why shouldn't you try this?

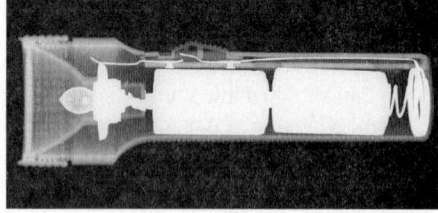

FIGURE 18–36
Problem 40
(X-ray of a
flashlight).

41. (II) How many 75-W lightbulbs, connected to 120 V as in Fig. 18–20, can be used without blowing a 15-A fuse?

42. (II) An extension cord made of two wires of diameter 0.129 cm (no. 16 copper wire) and of length 2.7 m (9 ft) is connected to an electric heater which draws 18.0 A on a 120-V line. How much power is dissipated in the cord?

43. (II) You want to design a portable electric blanket that runs on a 1.5-V battery. If you use a 0.50-mm-diameter copper wire as the heating element, how long should the wire be if you want to generate 18 W of heating power? What happens if you accidentally connect the blanket to a 9.0-V battery?

44. (II) A power station delivers 750 kW of power at 12,000 V to a factory through wires with total resistance 3.0 Ω. How much less power is wasted if the electricity is delivered at 50,000 V rather than 12,000 V?

45. (III) A small immersion heater can be used in a car to heat a cup of water for coffee or tea. If the heater can heat 120 mL of water from 25°C to 95°C in 8.0 min, (a) approximately how much current does it draw from the car's 12-V battery, and (b) what is its resistance? Assume the manufacturer's claim of 85% efficiency.

46. (III) The current in an electromagnet connected to a 240-V line is 21.5 A. At what rate must cooling water pass over the coils for the water temperature to rise no more than 6.50 C°?

18–7 Alternating Current

47. (I) Calculate the peak current in a 2.7-kΩ resistor connected to a 220-V rms ac source.

48. (I) An ac voltage, whose peak value is 180 V, is across a 310-Ω resistor. What are the rms and peak currents in the resistor?

49. (II) Estimate the resistance of the 120-V_{rms} circuits in your house as seen by the power company, when (a) everything electrical is unplugged, and (b) two 75-W lightbulbs are on.

50. (II) The peak value of an alternating current in a 1500-W device is 6.4 A. What is the rms voltage across it?

51. (II) An 1800-W arc welder is connected to a 660-V_{rms} ac line. Calculate (a) the peak voltage and (b) the peak current.

52. (II) Each channel of a stereo receiver is capable of an average power output of 100 W into an 8-Ω loudspeaker (see Fig. 18–14). What are the rms voltage and the rms current fed to the speaker (a) at the maximum power of 100 W, and (b) at 1.0 W when the volume is turned down?

53. (II) Determine (a) the maximum instantaneous power dissipated by a 2.2-hp pump connected to a 240-V_{rms} ac power source, and (b) the maximum current passing through the pump.

54. (II) A heater coil connected to a 240-V_{rms} ac line has a resistance of 38 Ω. (a) What is the average power used? (b) What are the maximum and minimum values of the instantaneous power?

*18–8 Microscopic View of Electric Current

*55. (II) A 0.65-mm-diameter copper wire carries a tiny dc current of 2.7 μA. Estimate the electron drift velocity.

*56. (II) A 4.80-m length of 2.0-mm-diameter wire carries a 750-mA dc current when 22.0 mV is applied to its ends. If the drift velocity is 1.7×10^{-5} m/s, determine (a) the resistance R of the wire, (b) the resistivity ρ, and (c) the number n of free electrons per unit volume.

*57. (III) At a point high in the Earth's atmosphere, He^{2+} ions in a concentration of $2.4 \times 10^{12}/m^3$ are moving due north at a speed of 2.0×10^6 m/s. Also, a $7.0 \times 10^{11}/m^3$ concentration of O_2^- ions is moving due south at a speed of 6.2×10^6 m/s. Determine the magnitude and direction of the net current passing through unit area (A/m^2).

*18–10 Nerve Conduction

*58. (I) What is the magnitude of the electric field across an axon membrane 1.0×10^{-8} m thick if the resting potential is −70 mV?

*59. (II) A neuron is stimulated with an electric pulse. The action potential is detected at a point 3.70 cm down the axon 0.0052 s later. When the action potential is detected 7.20 cm from the point of stimulation, the time required is 0.0063 s. What is the speed of the electric pulse along the axon? (Why are two measurements needed instead of only one?)

*60. (III) During an action potential, Na^+ ions move into the cell at a rate of about 3×10^{-7} mol/m²·s. How much power must be produced by the "active Na^+ pumping" system to produce this flow against a +30-mV potential difference? Assume that the axon is 10 cm long and 20 μm in diameter.

General Problems

61. A person accidentally leaves a car with the lights on. If each of the two headlights uses 40 W and each of the two taillights 6 W, for a total of 92 W, how long will a fresh 12-V battery last if it is rated at 75 A·h? Assume the full 12 V appears across each bulb.

62. A sequence of potential differences V is applied across a wire (diameter = 0.32 mm, length = 11 cm) and the resulting currents I are measured as follows:

V (V)	0.100	0.200	0.300	0.400	0.500
I (mA)	72	142	218	290	357

(a) If this wire obeys Ohm's law, graphing I vs. V will result in a straight-line plot. Explain why this is so and determine the theoretical predictions for the straight line's slope and y-intercept. (b) Plot I vs. V. Based on this plot, can you conclude that the wire obeys Ohm's law (i.e., did you obtain a straight line with the expected y-intercept, within the values of the significant figures)? If so, determine the wire's resistance R. (c) Calculate the wire's resistivity and use Table 18–1 to identify the solid material from which it is composed.

63. What is the average current drawn by a 1.0-hp 120-V motor? (1 hp = 746 W.)

64. The **conductance** G of an object is defined as the reciprocal of the resistance R; that is, $G = 1/R$. The unit of conductance is a *mho* $(= ohm^{-1})$, which is also called the *siemens* (S). What is the conductance (in siemens) of an object that draws 440 mA of current at 3.0 V?

65. The heating element of a 110-V, 1500-W heater is 3.8 m long. If it is made of iron, what must its diameter be?

66. (a) A particular household uses a 2.2-kW heater 2.0 h/day ("on" time), four 100-W lightbulbs 6.0 h/day, a 3.0-kW electric stove element for a total of 1.0 h/day, and miscellaneous power amounting to 2.0 kWh/day. If electricity costs $0.115 per kWh, what will be their monthly bill (30 d)? (b) How much coal (which produces 7500 kcal/kg) must be burned by a 35%-efficient power plant to provide the yearly needs of this household?

67. A small city requires about 15 MW of power. Suppose that instead of using high-voltage lines to supply the power, the power is delivered at 120 V. Assuming a two-wire line of 0.50-cm-diameter copper wire, estimate the cost of the energy lost to heat per hour per meter. Assume the cost of electricity is about 12 cents per kWh.

68. A 1600-W hair dryer is designed for 117 V. (*a*) What will be the percentage change in power output if the voltage drops to 105 V? Assume no change in resistance. (*b*) How would the actual change in resistivity with temperature affect your answer?

69. The wiring in a house must be thick enough so it does not become so hot as to start a fire. What diameter must a copper wire be if it is to carry a maximum current of 35 A and produce no more than 1.5 W of heat per meter of length?

70. Determine the resistance of the tungsten filament in a 75-W 120-V incandescent lightbulb (*a*) at its operating temperature of about 2800 K, (*b*) at room temperature.

71. Suppose a current is given by the equation $I = 1.40 \sin 210t$, where I is in amperes and t in seconds. (*a*) What is the frequency? (*b*) What is the rms value of the current? (*c*) If this is the current through a 24.0-Ω resistor, write the equation that describes the voltage as a function of time.

72. A microwave oven running at 65% efficiency delivers 950 W to the interior. Find (*a*) the power drawn from the source, and (*b*) the current drawn. Assume a source voltage of 120 V.

73. A 1.00-Ω wire is stretched uniformly to 1.50 times its original length. What is its resistance now?

74. 220 V is applied to two different conductors made of the same material. One conductor is twice as long and twice the diameter of the second. What is the ratio of the power transformed in the first relative to the second?

75. An electric power plant can produce electricity at a fixed power P, but the plant operator is free to choose the voltage V at which it is produced. This electricity is carried as an electric current I through a transmission line (resistance R) from the plant to the user, where it provides the user with electric power P'. (*a*) Show that the reduction in power $\Delta P = P - P'$ due to transmission losses is given by $\Delta P = P^2R/V^2$. (*b*) In order to reduce power losses during transmission, should the operator choose V to be as large or as small as possible?

76. A 2800-W oven is connected to a 240-V source. (*a*) What is the resistance of the oven? (*b*) How long will it take to bring 120 mL of 15°C water to 100°C assuming 65% efficiency? (*c*) How much will this cost at 11 cents/kWh?

77. A proposed electric vehicle makes use of storage batteries as its source of energy. It is powered by 24 batteries, each 12 V, 95 A·h. Assume that the car is driven on level roads at an average speed of 45 km/h, and the average friction force is 440 N. Assume 100% efficiency and neglect energy used for acceleration. No energy is consumed when the vehicle is stopped, since the engine doesn't need to idle. (*a*) Determine the horsepower required. (*b*) After approximately how many kilometers must the batteries be recharged?

78. A 15.2-Ω resistor is made from a coil of copper wire whose total mass is 15.5 g. What is the diameter of the wire, and how long is it?

79. A fish-tank heater is rated at 95 W when connected to 120 V. The heating element is a coil of Nichrome wire. When uncoiled, the wire has a total length of 3.5 m. What is the diameter of the wire?

80. A 100-W, 120-V lightbulb has a resistance of 12 Ω when cold (20°C) and 140 Ω when on (hot). Calculate its power consumption (*a*) at the instant it is turned on, and (*b*) after a few moments when it is hot.

81. In an automobile, the system voltage varies from about 12 V when the car is off to about 13.8 V when the car is on and the charging system is in operation, a difference of 15%. By what percentage does the power delivered to the headlights vary as the voltage changes from 12 V to 13.8 V? Assume the headlight resistance remains constant.

82. A tungsten filament used in a flashlight bulb operates at 0.20 A and 3.0 V. If its resistance at 20°C is 1.5 Ω, what is the temperature of the filament when the flashlight is on?

83. Lightbulb A is rated at 120 V and 40 W for household applications. Lightbulb B is rated at 12 V and 40 W for automotive applications. (*a*) What is the current through each bulb? (*b*) What is the resistance of each bulb? (*c*) In one hour, how much charge passes through each bulb? (*d*) In one hour, how much energy does each bulb use? (*e*) Which bulb requires larger diameter wires to connect its power source and the bulb?

84. An air conditioner draws 18 A at 220-V ac. The connecting cord is copper wire with a diameter of 1.628 mm. (*a*) How much power does the air conditioner draw? (*b*) If the length of the cord (containing two wires) is 3.5 m, how much power is dissipated in the wiring? (*c*) If no. 12 wire, with a diameter of 2.053 mm, was used instead, how much power would be dissipated in the wiring? (*d*) Assuming that the air conditioner is run 12 h per day, how much money per month (30 days) would be saved by using no. 12 wire? Assume that the cost of electricity is 12 cents per kWh.

85. An electric wheelchair is designed to run on a single 12-V battery rated to provide 100 ampere-hours (100 A·h). (*a*) How much energy is stored in this battery? (*b*) If the wheelchair experiences an average total retarding force (mainly friction) of 210 N, how far can the wheelchair travel on one charge?

86. If a wire of resistance R is stretched uniformly so that its length doubles, by what factor does the power dissipated in the wire change, assuming it remains hooked up to the same voltage source? Assume the wire's volume and density remain constant.

87. Copper wire of diameter 0.259 cm is used to connect a set of appliances at 120 V, which draw 1450 W of power total. (*a*) What power is wasted in 25.0 m of this wire? (*b*) What is your answer if wire of diameter 0.412 cm is used?

88. Battery-powered electricity is very expensive compared with that available from a wall outlet. Estimate the cost per kWh of (*a*) an alkaline D-cell (cost $1.70) and (*b*) an alkaline AA-cell (cost $1.25). These batteries can provide a continuous current of 25 mA for 820 h and 120 h, respectively, at 1.5 V. (*c*) Compare to the cost of a normal 120-V ac house source at $0.10/kWh.

89. A copper pipe has an inside diameter of 3.00 cm and an outside diameter of 5.00 cm (Fig. 18–37). What is the resistance of a 10.0-m length of this pipe?

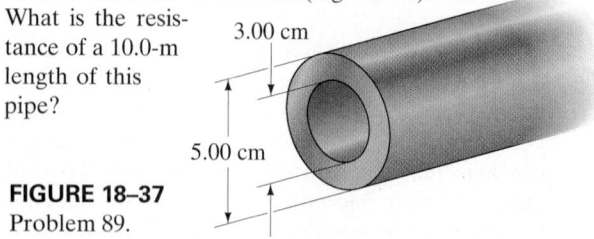

FIGURE 18–37
Problem 89.

*90. The Tevatron accelerator at Fermilab (Illinois) is designed to carry an 11-mA beam of protons ($q = 1.6 \times 10^{-19}$ C) traveling at very nearly the speed of light (3.0×10^8 m/s) around a ring 6300 m in circumference. How many protons are in the beam?

*91. The level of liquid helium (temperature $\approx$ 4 K) in its storage tank can be monitored using a vertically aligned niobium–titanium (NbTi) wire, whose length ℓ spans the height of the tank. In this level-sensing setup, an electronic circuit maintains a constant electrical current I at all times in the NbTi wire and a voltmeter monitors the voltage difference V across this wire. Since the superconducting critical temperature for NbTi is 10 K, the portion of the wire immersed in the liquid helium is in the superconducting state, while the portion above the liquid (in helium vapor with temperature above 10 K) is in the normal state. Define $f = x/\ell$ to be the fraction of the tank filled with liquid helium (Fig. 18–38) and V_0 to be the value of V when the tank is empty ($f = 0$). Determine the relation between f and V (in terms of V_0).

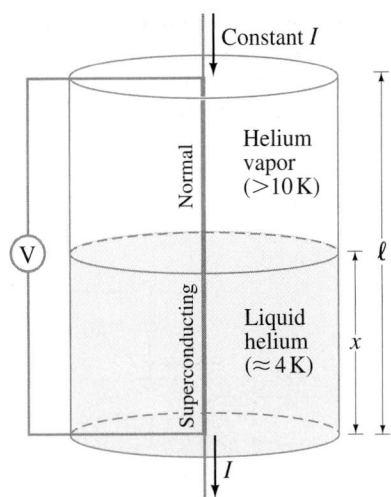

FIGURE 18–38 Problem 91.

Search and Learn

1. Why is Ohm's law less of a law than Newton's laws?

2. A traditional incandescent lamp filament may have been lit to a temperature of 2700 K. A contemporary halogen incandescent lamp filament may be at around 2900 K. (a) Estimate the percent improvement of the halogen bulb over the traditional one. [*Hint:* See Section 14–8.] (b) To produce the same amount of light as a traditional 100-W bulb, estimate what wattage a halogen bulb should use.

3. You find a small cylindrical resistor that measures 9.00 mm in length and 2.15 mm in diameter, and it has a color code of red, yellow, brown, and gold. What is the resistor made of primarily?

4. Small changes in the length of an object can be measured using a **strain gauge** sensor, which is a wire that when undeformed has length ℓ_0, cross-sectional area A_0, and resistance R_0. This sensor is rigidly affixed to the object's surface, aligning its length in the direction in which length changes are to be measured. As the object deforms, the length of the wire sensor changes by $\Delta\ell$, and the resulting change ΔR in the sensor's resistance is measured. Assuming that as the solid wire is deformed to a length ℓ, its density and volume remain constant (only approximately valid), show that the strain $(= \Delta\ell/\ell_0)$ of the wire sensor, and thus of the object to which it is attached, is approximately $\Delta R/2R_0$. [See Sections 18–4 and 9–5.]

5. An electric heater is used to heat a room of volume 65 m³. Air is brought into the room at 5°C and is completely replaced twice per hour. Heat loss through the walls amounts to approximately 850 kcal/h. If the air is to be maintained at 22°C, what minimum wattage must the heater have? (The specific heat of air is about 0.17 kcal/kg·C°. Reread parts of Chapter 14 and Section 18–5.)

6. Household wiring has sometimes used aluminium instead of copper. (a) Using Table 18–1, find the ratio of the resistance of a copper wire to that of an aluminum wire of the same length and diameter. (b) Typical copper wire used for home wiring in the U.S. has a diameter of 1.63 mm. What is the resistance of 125 m of this wire? (c) What would be the resistance of the same wire if it were made of aluminum? (d) How much power would be dissipated in each wire if it carried 18 A of current? (e) What should be the diameter of the aluminum wire for it to have the same resistance as the copper wire? (f) In Section 18–4, a statement is made about the resistance of copper and aluminum wires of the same weight. Using Table 10–1 for the densities of copper and aluminum, find the resistance of an aluminum wire of the same mass and length as the copper wire in part (b). Is the statement true?

*7. How far can an average electron move along the wires of a 650-W toaster during an alternating current cycle? The power cord has copper wires of diameter 1.7 mm and is plugged into a standard 60-Hz 120-V ac outlet. [*Hint:* The maximum current in the cycle is related to the maximum drift velocity. The maximum velocity in an oscillation is related to the maximum displacement; see Chapter 11.]

*8. Capacitance of an axon. (a) Do an order-of-magnitude estimate for the capacitance of an axon 10 cm long of radius 10 μm. The thickness of the membrane is about 10^{-8} m, and the dielectric constant is about 3. (b) By what factor does the concentration (number of ions per volume) of Na^+ ions in the cell change as a result of one action potential?

ANSWERS TO EXERCISES

A: 1.6×10^{-13} A.
B: 3600 C.
C: 240 Ω.
D: (b), (c).
E: (c).

F: 110 m.
G: (d).
H: 370,000 kg, or about 5000 people.
I: (e) 40.

This cell phone has an attachment that measures a person's blood sugar level, and plots it over a period of days. All electronic devices contain circuits that are dc, at least in part. The circuit diagram below shows a possible amplifier circuit for an audio output (cell phone ear piece). We have already met two of the circuit elements shown: resistors and capacitors, and we discuss them in circuits in this Chapter. (The large triangle is an amplifier chip containing transistors.) We also discuss how voltmeters and ammeters work, and how measurements affect the quantity being measured.

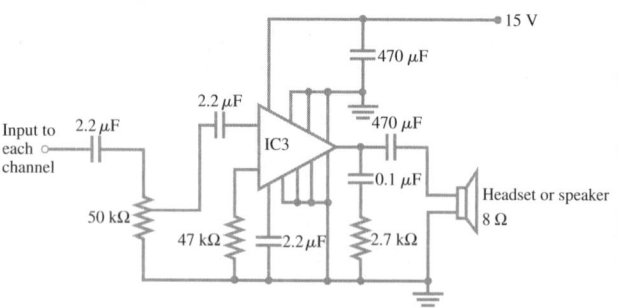

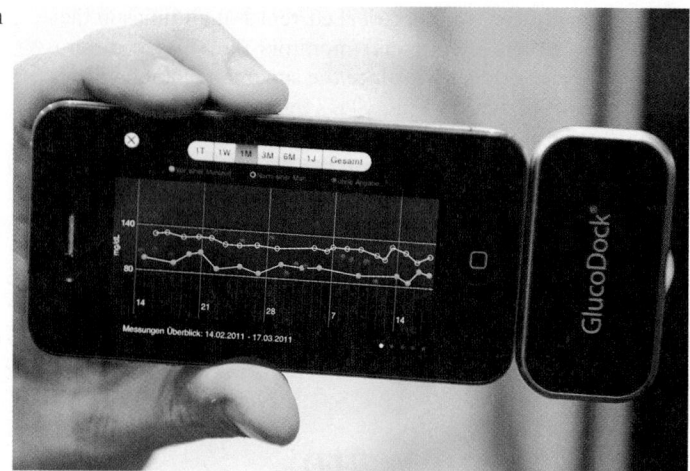

CHAPTER 19

DC Circuits

CONTENTS

TABLE 19–1 Symbols for Circuit Elements

Symbol	Device
⊣⊢	Battery
⊣⊢ or ⊣⊱	Capacitor
-⋀⋁⋀-	Resistor
——	Wire with negligible resistance
⟋—	Switch
⏚ or ↓	Ground

CHAPTER-OPENING QUESTION—Guess now!

The automobile headlight bulbs shown in the circuits here are identical. The battery connection which produces more light is

(a) circuit 1.

(b) circuit 2.

(c) both the same.

(d) not enough information.

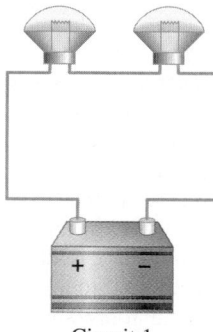

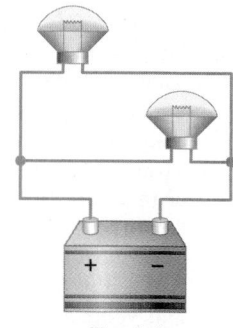

Circuit 1   Circuit 2

Electric circuits are basic parts of all electronic devices from cell phones and TV sets to computers and automobiles. Scientific measurements—whether in physics, biology, or medicine—make use of electric circuits. In Chapter 18, we discussed the basic principles of electric current. Now we apply these principles to analyze dc circuits involving combinations of batteries, resistors, and capacitors. We also study the operation of some useful instruments.[†]

When we draw a diagram for a circuit, we represent batteries, capacitors, and resistors by the symbols shown in Table 19–1. Wires whose resistance is negligible compared with other resistance in the circuit are drawn as straight lines. A ground symbol (⏚ or ↓) may mean a real connection to the ground, perhaps via a metal pipe, or it may mean a common connection, such as the frame of a car.

For the most part in this Chapter, except in Section 19–6 on *RC* circuits, we will be interested in circuits operating in their steady state. We won't be looking at a circuit at the moment a change is made in it, such as when a battery or resistor is connected or disconnected, but only when the currents have reached their steady values.

[†]AC circuits that contain only a voltage source and resistors can be analyzed like the dc circuits in this Chapter. However, ac circuits that contain capacitors and other circuit elements are more complicated, and we discuss them in Chapter 21.

19–1 EMF and Terminal Voltage

To have current in an electric circuit, we need a device such as a battery or an electric generator that transforms one type of energy (chemical, mechanical, or light, for example) into electric energy. Such a device is called a **source** of **electromotive force**[†] or of **emf**. The *potential difference* between the terminals of such a source, when no current flows to an external circuit, is called the **emf** of the source. The symbol $\mathscr{E}$ is usually used for emf (don't confuse $\mathscr{E}$ with E for electric field), and its unit is volts.

A battery is not a source of constant current—the current out of a battery varies according to the resistance in the circuit. A battery *is*, however, a nearly constant voltage source, but not perfectly constant as we now discuss. For example, if you start a car with the headlights on, you may notice the headlights dim. This happens because the starter draws a large current, and the battery voltage drops below its rated emf as a result. The voltage drop occurs because the chemical reactions in a battery cannot supply charge fast enough to maintain the full emf. For one thing, charge must move (within the electrolyte) between the electrodes of the battery, and there is always some hindrance to completely free flow. Thus, a battery itself has some resistance, which is called its **internal resistance**; it is usually designated r.

A real battery is modeled as if it were a perfect emf $\mathscr{E}$ in series with a resistor r, as shown in Fig. 19–1. Since this resistance r is inside the battery, we can never separate it from the battery. The two points a and b in Fig. 19–1 represent the two terminals of the battery. What we measure is the **terminal voltage** $V_{ab} = V_a - V_b$. When no current is drawn from the battery, the terminal voltage equals the emf, which is determined by the chemical reactions in the battery: $V_{ab} = \mathscr{E}$. However, when a current I flows from the battery there is an internal drop in voltage equal to Ir. Thus the terminal voltage (the actual voltage applied to a circuit) is

$$V_{ab} = \mathscr{E} - Ir. \qquad \text{[current } I \text{ flows from battery]} \quad \textbf{(19–1)}$$

For example, if a 12-V battery has an internal resistance of 0.1 Ω, then when 10 A flows from the battery, the terminal voltage is $12\text{ V} - (10\text{ A})(0.1\text{ Ω}) = 11\text{ V}$. The internal resistance of a battery is usually small. For example, an ordinary flashlight battery when fresh may have an internal resistance of perhaps 0.05 Ω. (However, as it ages and the electrolyte dries out, the internal resistance increases to many ohms.)

CAUTION

Why battery voltage isn't perfectly constant

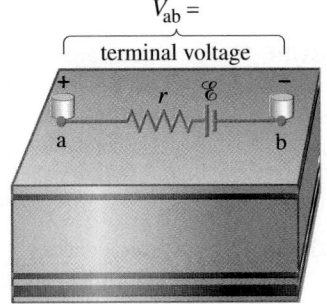

FIGURE 19–1 Diagram for an electric cell or battery.

| **EXAMPLE 19–1** | **Battery with internal resistance.** A 65.0-Ω resistor is connected to the terminals of a battery whose emf is 12.0 V and whose internal resistance is 0.5 Ω, Fig. 19–2. Calculate (*a*) the current in the circuit, (*b*) the terminal voltage of the battery, V_{ab}, and (*c*) the power dissipated in the resistor R and in the battery's internal resistance r.

APPROACH We first consider the battery as a whole, which is shown in Fig. 19–2 as an emf $\mathscr{E}$ and internal resistance r between points a and b. Then we apply $V = IR$ to the circuit itself.

SOLUTION (*a*) From Eq. 19–1, we have $V_{ab} = \mathscr{E} - Ir$. We apply Ohm's law (Eq. 18–2) to this battery and the resistance R of the circuit: $V_{ab} = IR$. Hence

$$\mathscr{E} - Ir = IR$$

or $\mathscr{E} = I(R + r)$. So

$$I = \frac{\mathscr{E}}{R + r} = \frac{12.0\text{ V}}{65.0\text{ Ω} + 0.5\text{ Ω}} = \frac{12.0\text{ V}}{65.5\text{ Ω}} = 0.183\text{ A}.$$

(*b*) The terminal voltage is

$$V_{ab} = \mathscr{E} - Ir = 12.0\text{ V} - (0.183\text{ A})(0.5\text{ Ω}) = 11.9\text{ V}.$$

(*c*) The power dissipated in R (Eq. 18–6) is

$$P_R = I^2 R = (0.183\text{ A})^2(65.0\text{ Ω}) = 2.18\text{ W},$$

and in the battery's resistance r it is

$$P_r = I^2 r = (0.183\text{ A})^2(0.5\text{ Ω}) = 0.02\text{ W}.$$

FIGURE 19–2 Example 19–1.

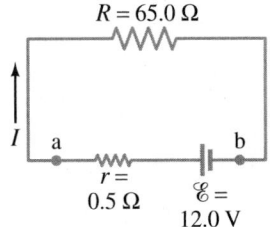

[†]The term "electromotive force" is a misnomer—it does not refer to a "force" that is measured in newtons. To avoid confusion, we use the abbreviation, emf.

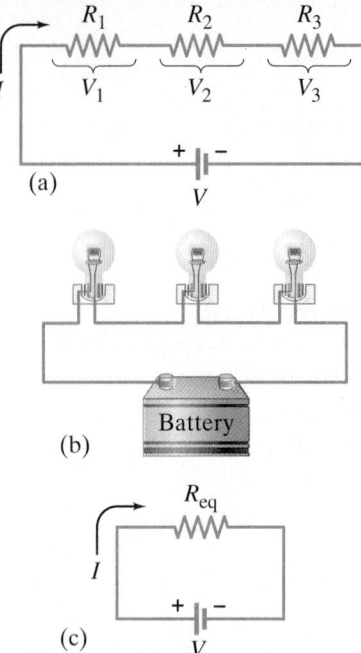

(a)

(b) Battery

(c)

FIGURE 19–3 (a) Resistances connected in series. (b) Resistances could be lightbulbs, or any other type of resistance. (c) Equivalent single resistance R_{eq} that draws the same current: $R_{eq} = R_1 + R_2 + R_3$.

FIGURE 19–4 (a) Resistances connected in parallel. (b) Resistances could be lightbulbs. (c) The equivalent circuit with R_{eq} obtained from Eq. 19–4:
$$\frac{1}{R_{eq}} = \frac{1}{R_1} + \frac{1}{R_2} + \frac{1}{R_3}.$$

(a)

(b)

(c)

Unless stated otherwise, we assume the battery's internal resistance is negligible, and the battery voltage given is its terminal voltage, which we will usually write as V rather than V_{ab}. Do not confuse V (italic) for voltage, with V (not italic) for the volt unit.

19–2 Resistors in Series and in Parallel

When two or more resistors are connected end to end along a single path as shown in Fig. 19–3a, they are said to be connected in **series**. The resistors could be simple resistors as were pictured in Fig. 18–11, or they could be lightbulbs (Fig. 19–3b), or heating elements, or other resistive devices. Any charge that passes through R_1 in Fig. 19–3a will also pass through R_2 and then R_3. Hence the same current I passes through each resistor. (If it did not, this would imply that either charge was not conserved, or that charge was accumulating at some point in the circuit, which does not happen in the steady state.)

We let V represent the potential difference (voltage) across all three resistors in Fig. 19–3a. We assume all other resistance in the circuit can be ignored, so V equals the terminal voltage supplied by the battery. We let V_1, V_2, and V_3 be the potential differences across each of the resistors, R_1, R_2, and R_3, respectively. From Ohm's law, $V = IR$, we can write $V_1 = IR_1$, $V_2 = IR_2$, and $V_3 = IR_3$. Because the resistors are connected end to end, energy conservation tells us that the total voltage V is equal to the sum of the voltages across each resistor:

$$V = V_1 + V_2 + V_3 = IR_1 + IR_2 + IR_3. \qquad \text{[series]} \quad \textbf{(19–2)}$$

Now let us determine the equivalent single resistance R_{eq} that would draw the same current I as our combination of three resistors in series; see Fig. 19–3c. Such a single resistance R_{eq} would be related to V by

$$V = IR_{eq}.$$

We equate this expression with Eq. 19–2, $V = I(R_1 + R_2 + R_3)$, and find

$$R_{eq} = R_1 + R_2 + R_3. \qquad \text{[series]} \quad \textbf{(19–3)}$$

When we put several resistances in series, the total or equivalent resistance is the sum of the separate resistances. (Sometimes we call it "net resistance.") This sum applies to any number of resistances in series. Note that when you add more resistance to the circuit, the current through the circuit will decrease. For example, if a 12-V battery is connected to a 4-Ω resistor, the current will be 3 A. But if the 12-V battery is connected to three 4-Ω resistors in series, the total resistance is 12 Ω and the current through the entire circuit will be only 1 A.

Another way to connect resistors is in **parallel**, so that the current from the source splits into separate branches or paths (Fig. 19–4a). Wiring in houses and buildings is arranged so all electric devices are in parallel, as we saw in Chapter 18, Fig. 18–20. With parallel wiring, if you disconnect one device (say, R_1 in Fig. 19–4a), the current to the other devices is not interrupted. Compare to a series circuit, where if one device (say, R_1 in Fig. 19–3a) is disconnected, the current *is* stopped to all others.

In a parallel circuit, Fig. 19–4a, the total current I that leaves the battery splits into three separate paths. We let I_1, I_2, and I_3 be the currents through each of the resistors, R_1, R_2, and R_3, respectively. Because *electric charge is conserved*, the current I flowing into junction A (where the different wires or conductors meet, Fig. 19–4a) must equal the current flowing out of the junction. Thus

$$I = I_1 + I_2 + I_3. \qquad \text{[parallel]}$$

When resistors are connected in parallel, each has the same voltage across it. (Indeed, any two points in a circuit connected by a wire of negligible resistance are at the same potential.) Hence the full voltage of the battery is applied to each resistor in Fig. 19–4a. Applying Ohm's law to each resistor, we have

$$I_1 = \frac{V}{R_1}, \quad I_2 = \frac{V}{R_2}, \quad \text{and} \quad I_3 = \frac{V}{R_3}.$$

Let us now determine what single resistor R_{eq} (Fig. 19–4c) will draw the same current I as these three resistances in parallel. This equivalent resistance R_{eq} must satisfy Ohm's law too:

$$I = \frac{V}{R_{eq}}.$$

We now combine the equations above:

$$I = I_1 + I_2 + I_3,$$

$$\frac{V}{R_{eq}} = \frac{V}{R_1} + \frac{V}{R_2} + \frac{V}{R_3}.$$

When we divide out the V from each term, we have

$$\frac{1}{R_{eq}} = \frac{1}{R_1} + \frac{1}{R_2} + \frac{1}{R_3}. \qquad \text{[parallel]} \quad (19\text{–}4)$$

For example, suppose you connect two 4-Ω loudspeakers in parallel to a single set of output terminals of an amplifier. The equivalent resistance of the two 4-Ω "resistors" in parallel is

$$\frac{1}{R_{eq}} = \frac{1}{4\,\Omega} + \frac{1}{4\,\Omega} = \frac{2}{4\,\Omega} = \frac{1}{2\,\Omega},$$

and so $R_{eq} = 2\,\Omega$. Thus the net (or equivalent) resistance is *less* than each single resistance. This may at first seem surprising. But remember that when you connect resistors in parallel, you are giving the current additional paths to follow. Hence the net resistance will be less.[†]

Equations 19–3 and 19–4 make good sense. Recalling Eq. 18–3 for resistivity, $R = \rho\ell/A$, we see that placing resistors in series effectively increases the length and therefore the resistance; putting resistors in parallel effectively increases the area through which current flows, thus reducing the overall resistance.

Note that whenever a group of resistors is replaced by the equivalent resistance, current and voltage and power in the rest of the circuit are unaffected.

EXERCISE A You have a 10-Ω and a 15-Ω resistor. What is the smallest and largest equivalent resistance that you can make with these two resistors?

CONCEPTUAL EXAMPLE 19–2 | **Series or parallel?** (a) The lightbulbs in Fig. 19–5 are identical. Which configuration produces more light? (b) Which way do you think the headlights of a car are wired? Ignore change of filament resistance R with current.

RESPONSE (a) The equivalent resistance of the parallel circuit is found from Eq. 19–4, $1/R_{eq} = 1/R + 1/R = 2/R$. Thus $R_{eq} = R/2$. The parallel combination then has lower resistance ($= R/2$) than the series combination ($R_{eq} = R + R = 2R$). There will be more total current in the parallel configuration (2), since $I = V/R_{eq}$ and V is the same for both circuits. The total power transformed, which is related to the light produced, is $P = IV$, so the greater current in (2) means more light is produced.

(b) Headlights are wired in parallel (2), because if one bulb goes out, the other bulb can stay lit. If they were in series (1), when one bulb burned out (the filament broke), the circuit would be open and no current would flow, so neither bulb would light.

EXERCISE B Return to the Chapter-Opening Question, page 526, and answer it again now. Try to explain why you may have answered differently the first time.

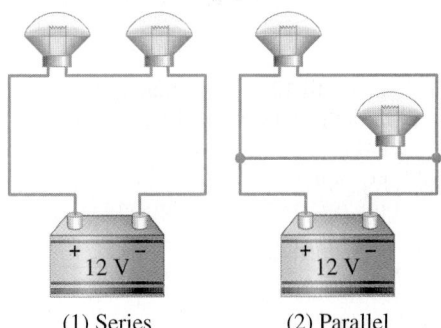

FIGURE 19–5 Example 19–2.

(1) Series   (2) Parallel

[†]An analogy may help. Consider two identical pipes taking in water near the top of a dam and releasing it at the bottom as shown in the figure to the right. If both pipes are open, rather than only one, twice as much water will flow through. That is, the net resistance to the flow of water will be reduced by half with two equal pipes open, just as for electrical resistors in parallel.

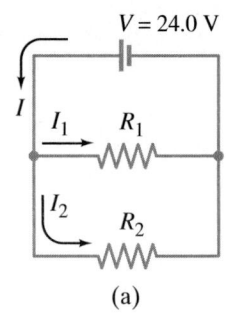

$V = 24.0$ V

I_1 R_1

I_2 R_2

(a)

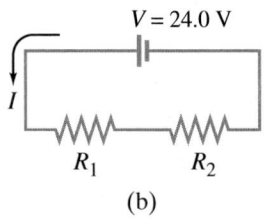

$V = 24.0$ V

I

R_1 R_2

(b)

FIGURE 19–6 Example 19–3.

FIGURE 19–7 (a) Circuit for Examples 19–4 and 19–5. (b) Equivalent circuit, showing the equivalent resistance of 290 Ω for the two parallel resistors in (a).

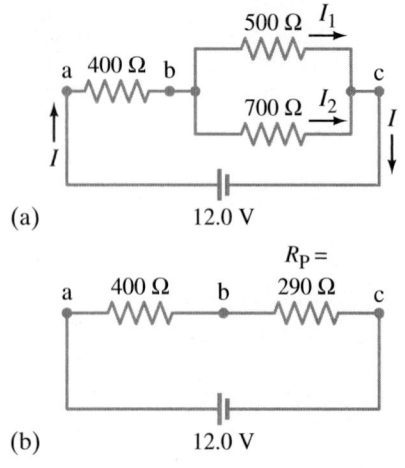

500 Ω I_1

a 400 Ω b c

700 Ω I_2

I

I

(a) 12.0 V

$R_P =$

a 400 Ω b 290 Ω c

(b) 12.0 V

EXAMPLE 19–3 **Series and parallel resistors.** Two 100-Ω resistors are connected (a) in parallel, and (b) in series, to a 24.0-V battery (Fig. 19–6). What is the current through each resistor and what is the equivalent resistance of each circuit?

APPROACH We use Ohm's law and the ideas just discussed for series and parallel connections to get the current in each case. We can also use Eqs. 19–3 and 19–4.

SOLUTION (a) Any given charge (or electron) can flow through only one or the other of the two resistors in Fig. 19–6a. Just as a river may break into two streams when going around an island, here too the total current I from the battery (Fig. 19–6a) splits to flow through each resistor, so I equals the sum of the separate currents through the two resistors:

$$I = I_1 + I_2.$$

The potential difference across each resistor is the battery voltage $V = 24.0$ V. Applying Ohm's law to each resistor gives

$$I = I_1 + I_2 = \frac{V}{R_1} + \frac{V}{R_2} = \frac{24.0 \text{ V}}{100 \text{ } \Omega} + \frac{24.0 \text{ V}}{100 \text{ } \Omega} = 0.24 \text{ A} + 0.24 \text{ A} = 0.48 \text{ A}.$$

The equivalent resistance is

$$R_{eq} = \frac{V}{I} = \frac{24.0 \text{ V}}{0.48 \text{ A}} = 50 \text{ } \Omega.$$

We could also have obtained this result from Eq. 19–4:

$$\frac{1}{R_{eq}} = \frac{1}{100 \text{ } \Omega} + \frac{1}{100 \text{ } \Omega} = \frac{2}{100 \text{ } \Omega} = \frac{1}{50 \text{ } \Omega},$$

so $R_{eq} = 50 \text{ } \Omega$.

(b) All the current that flows out of the battery passes first through R_1 and then through R_2 because they lie along a single path, Fig. 19–6b. So the current I is the same in both resistors; the potential difference V across the battery equals the total change in potential across the two resistors:

$$V = V_1 + V_2.$$

Ohm's law gives $V = IR_1 + IR_2 = I(R_1 + R_2)$. Hence

$$I = \frac{V}{R_1 + R_2} = \frac{24.0 \text{ V}}{100 \text{ } \Omega + 100 \text{ } \Omega} = 0.120 \text{ A}.$$

The equivalent resistance, using Eq. 19–3, is $R_{eq} = R_1 + R_2 = 200 \text{ } \Omega$. We can also get R_{eq} by thinking from the point of view of the battery: the total resistance R_{eq} must equal the battery voltage divided by the current it delivers:

$$R_{eq} = \frac{V}{I} = \frac{24.0 \text{ V}}{0.120 \text{ A}} = 200 \text{ } \Omega.$$

NOTE The voltage across R_1 is $V_1 = IR_1 = (0.120 \text{ A})(100 \text{ } \Omega) = 12.0$ V, and that across R_2 is $V_2 = IR_2 = 12.0$ V, each being half of the battery voltage. A simple circuit like Fig. 19–6b is thus often called a simple **voltage divider**.

EXAMPLE 19–4 **Circuit with series and parallel resistors.** How much current is drawn from the battery shown in Fig. 19–7a?

APPROACH The current I that flows out of the battery all passes through the 400-Ω resistor, but then it splits into I_1 and I_2 passing through the 500-Ω and 700-Ω resistors. The latter two resistors are in parallel with each other. We look for something that we already know how to treat. So let's start by finding the equivalent resistance, R_P, of the parallel resistors, 500 Ω and 700 Ω. Then we can consider this R_P to be in series with the 400-Ω resistor.

SOLUTION The equivalent resistance, R_P, of the 500-Ω and 700-Ω resistors in parallel is

$$\frac{1}{R_P} = \frac{1}{500 \text{ } \Omega} + \frac{1}{700 \text{ } \Omega} = 0.0020 \text{ } \Omega^{-1} + 0.0014 \text{ } \Omega^{-1} = 0.0034 \text{ } \Omega^{-1}.$$

This is $1/R_P$, so we take the reciprocal to find R_P.

It is a common mistake to forget to take this reciprocal. The units of reciprocal ohms, Ω^{-1}, are a reminder. Thus

$$R_P = \frac{1}{0.0034\,\Omega^{-1}} = 290\,\Omega.$$

CAUTION

Remember to take the reciprocal

This $290\,\Omega$ is the equivalent resistance of the two parallel resistors, and is in series with the $400\text{-}\Omega$ resistor (see equivalent circuit, Fig. 19–7b). To find the total equivalent resistance R_{eq}, we add the $400\text{-}\Omega$ and $290\text{-}\Omega$ resistances, since they are in series:

$$R_{eq} = 400\,\Omega + 290\,\Omega = 690\,\Omega.$$

The total current flowing from the battery is then

$$I = \frac{V}{R_{eq}} = \frac{12.0\,\text{V}}{690\,\Omega} = 0.0174\,\text{A} \approx 17\,\text{mA}.$$

NOTE This I is also the current flowing through the $400\text{-}\Omega$ resistor, but not through the $500\text{-}\Omega$ and $700\text{-}\Omega$ resistors (both currents are less—see the next Example).

EXAMPLE 19–5 **Current in one branch.** What is the current I_1 through the $500\text{-}\Omega$ resistor in Fig. 19–7a?

APPROACH We need the voltage across the $500\text{-}\Omega$ resistor, which is the voltage between points b and c in Fig. 19–7a, and we call it V_{bc}. Once V_{bc} is known, we can apply Ohm's law, $V = IR$, to get the current. First we find the voltage across the $400\text{-}\Omega$ resistor, V_{ab}, since we know that $17.4\,\text{mA}$ passes through it (Example 19–4).

SOLUTION V_{ab} can be found using $V = IR$:

$$V_{ab} = (0.0174\,\text{A})(400\,\Omega) = 7.0\,\text{V}.$$

The total voltage across the network of resistors is $V_{ac} = 12.0\,\text{V}$, so V_{bc} must be $12.0\,\text{V} - 7.0\,\text{V} = 5.0\,\text{V}$. Ohm's law gives the current I_1 through the $500\text{-}\Omega$ resistor:

$$I_1 = \frac{5.0\,\text{V}}{500\,\Omega} = 1.0 \times 10^{-2}\,\text{A} = 10\,\text{mA}.$$

This is the answer we wanted. We can also calculate the current I_2 through the $700\text{-}\Omega$ resistor since the voltage across it is also $5.0\,\text{V}$:

$$I_2 = \frac{5.0\,\text{V}}{700\,\Omega} = 7\,\text{mA}.$$

NOTE When I_1 combines with I_2 to form the total current I (at point c in Fig. 19–7a), their sum is $10\,\text{mA} + 7\,\text{mA} = 17\,\text{mA}$. This equals the total current I as calculated in Example 19–4, as it should.

CONCEPTUAL EXAMPLE 19–6 **Bulb brightness in a circuit.** The circuit in Fig. 19–8 has three identical lightbulbs, each of resistance R. (*a*) When switch S is closed, how will the brightness of bulbs A and B compare with that of bulb C? (*b*) What happens when switch S is opened? Use a minimum of mathematics.

RESPONSE (*a*) With switch S closed, the current that passes through bulb C must split into two equal parts when it reaches the junction leading to bulbs A and B because the resistance of bulb A equals that of B. Thus, A and B each receive half of C's current; A and B will be equally bright, but less bright than C ($P = I^2R$). (*b*) When the switch S is open, no current can flow through bulb A, so it will be dark. Now, the same current passes through bulbs B and C, so B and C will be equally bright. The equivalent resistance of this circuit ($= R + R$) is greater than that of the circuit with the switch closed, so the current leaving the battery is reduced. Thus, bulb C will be dimmer when we open the switch, but bulb B will be brighter because it gets more current when the switch is open (you may want to use some mathematics here).

EXERCISE C A 100-W, 120-V lightbulb and a 60-W, 120-V lightbulb are connected in two different ways as shown in Fig. 19–9. In each case, which bulb glows more brightly? Ignore change of filament resistance with current (and temperature).

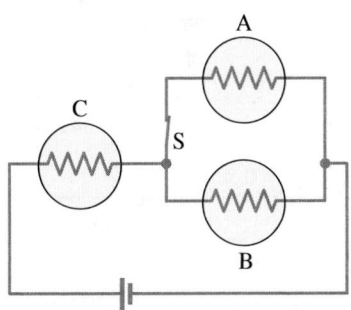

FIGURE 19–8 Example 19–6, three identical lightbulbs. Each yellow circle with —$\bigwedge\!\bigwedge\!\bigwedge$— inside represents a lightbulb and its resistance.

FIGURE 19–9 Exercise C.

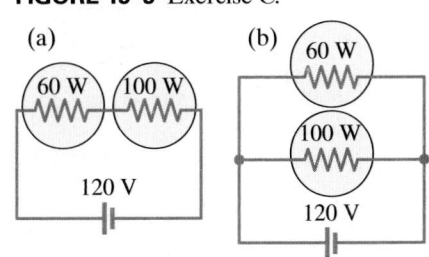

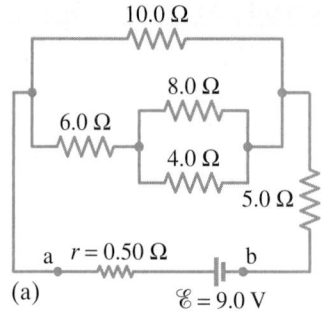

10.0 Ω

8.0 Ω

6.0 Ω

4.0 Ω

5.0 Ω

a $r = 0.50\ \Omega$ b

(a) $\mathscr{E} = 9.0\ \text{V}$

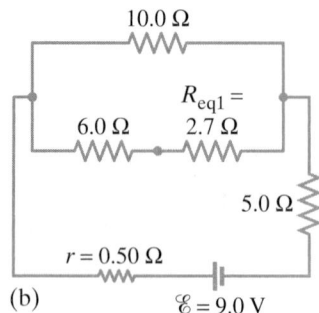

10.0 Ω

$R_{eq1} = 2.7\ \Omega$

6.0 Ω

5.0 Ω

$r = 0.50\ \Omega$

(b) $\mathscr{E} = 9.0\ \text{V}$

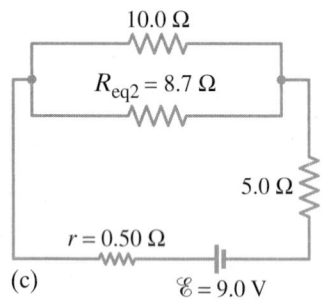

10.0 Ω

$R_{eq2} = 8.7\ \Omega$

5.0 Ω

$r = 0.50\ \Omega$

(c) $\mathscr{E} = 9.0\ \text{V}$

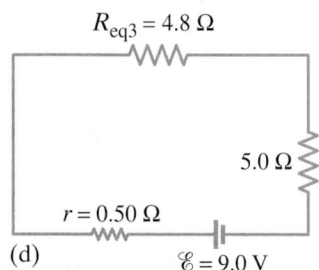

$R_{eq3} = 4.8\ \Omega$

5.0 Ω

$r = 0.50\ \Omega$

(d) $\mathscr{E} = 9.0\ \text{V}$

FIGURE 19–10 Circuit for Example 19–7, where r is the internal resistance of the battery.

FIGURE 19–11 Currents can be calculated using Kirchhoff's rules.

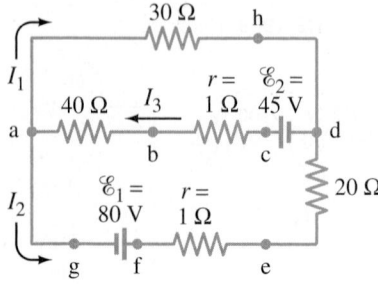

30 Ω   h

I_1

40 Ω I_3 $r =$ $\mathscr{E}_2 =$
        $1\ \Omega$ 45 V

a   b   c   d

$\mathscr{E}_1 =$ $r =$
80 V $1\ \Omega$

I_2

20 Ω

g   f   e

EXAMPLE 19–7 **Analyzing a circuit.** A 9.0-V battery whose internal resistance r is 0.50 Ω is connected in the circuit shown in Fig. 19–10a. (*a*) How much current is drawn from the battery? (*b*) What is the terminal voltage of the battery? (*c*) What is the current in the 6.0-Ω resistor?

APPROACH To find the current out of the battery, we first need to determine the equivalent resistance R_{eq} of the entire circuit, including r, which we do by identifying and isolating simple series or parallel combinations of resistors. Once we find I from Ohm's law, $I = \mathscr{E}/R_{eq}$, we get the terminal voltage using $V_{ab} = \mathscr{E} - Ir$. For (*c*) we apply Ohm's law to the 6.0-Ω resistor.

SOLUTION (*a*) We want to determine the equivalent resistance of the circuit. But where do we start? We note that the 4.0-Ω and 8.0-Ω resistors are in parallel, and so have an equivalent resistance R_{eq1} given by

$$\frac{1}{R_{eq1}} = \frac{1}{8.0\ \Omega} + \frac{1}{4.0\ \Omega} = \frac{3}{8.0\ \Omega};$$

so $R_{eq1} = 2.7\ \Omega$. This 2.7 Ω is in series with the 6.0-Ω resistor, as shown in the equivalent circuit of Fig. 19–10b. The net resistance of the lower arm of the circuit is then

$$R_{eq2} = 6.0\ \Omega + 2.7\ \Omega = 8.7\ \Omega,$$

as shown in Fig. 19–10c. The equivalent resistance R_{eq3} of the 8.7-Ω and 10.0-Ω resistances in parallel is given by

$$\frac{1}{R_{eq3}} = \frac{1}{10.0\ \Omega} + \frac{1}{8.7\ \Omega} = 0.21\ \Omega^{-1},$$

so $R_{eq3} = \left(1/0.21\ \Omega^{-1}\right) = 4.8\ \Omega$. This 4.8 Ω is in series with the 5.0-Ω resistor and the 0.50-Ω internal resistance of the battery (Fig. 19–10d), so the total equivalent resistance R_{eq} of the circuit is $R_{eq} = 4.8\ \Omega + 5.0\ \Omega + 0.50\ \Omega = 10.3\ \Omega$. Hence the current drawn is

$$I = \frac{\mathscr{E}}{R_{eq}} = \frac{9.0\ \text{V}}{10.3\ \Omega} = 0.87\ \text{A}.$$

(*b*) The terminal voltage of the battery is

$$V_{ab} = \mathscr{E} - Ir = 9.0\ \text{V} - (0.87\ \text{A})(0.50\ \Omega) = 8.6\ \text{V}.$$

(*c*) Now we can work back and get the current in the 6.0-Ω resistor. It must be the same as the current through the 8.7 Ω shown in Fig. 19–10c (why?). The voltage across that 8.7 Ω will be the emf of the battery minus the voltage drops across r and the 5.0-Ω resistor: $V_{8.7} = 9.0\ \text{V} - (0.87\ \text{A})(0.50\ \Omega + 5.0\ \Omega)$. Applying Ohm's law, we get the current (call it I')

$$I' = \frac{9.0\ \text{V} - (0.87\ \text{A})(0.50\ \Omega + 5.0\ \Omega)}{8.7\ \Omega} = 0.48\ \text{A}.$$

This is the current through the 6.0-Ω resistor.

19–3 Kirchhoff's Rules

In the last few Examples we have been able to find the currents in circuits by combining resistances in series and parallel, and using Ohm's law. This technique can be used for many circuits. However, some circuits are too complicated for that analysis. For example, we cannot find the currents in each part of the circuit shown in Fig. 19–11 simply by combining resistances as we did before.

To deal with complicated circuits, we use Kirchhoff's rules, devised by G. R. Kirchhoff (1824–1887) in the mid-nineteenth century. There are two rules, and they are simply convenient applications of the laws of conservation of charge and energy.

Kirchhoff's first rule or **junction rule** is based on the conservation of electric charge (we already used it to derive the equation for parallel resistors). It states that

> **at any junction point, the sum of all currents entering the junction must equal the sum of all currents leaving the junction.**

Junction rule (conservation of charge)

That is, whatever charge goes in must come out. For example, at the junction point a in Fig. 19–11, I_3 is entering whereas I_1 and I_2 are leaving. Thus Kirchhoff's junction rule states that $I_3 = I_1 + I_2$. We already saw an instance of this in the NOTE at the end of Example 19–5.

Kirchhoff's second rule or **loop rule** is based on the conservation of energy. It states that

> **the sum of the changes in potential around any closed loop of a circuit must be zero.**

Loop rule (conservation of energy)

To see why this rule should hold, consider a rough analogy with the potential energy of a roller coaster on its track. When the roller coaster starts from the station, it has a particular potential energy. As it is pulled up the first hill, its gravitational potential energy increases and reaches a maximum at the top. As it descends the other side, its potential energy decreases and reaches a local minimum at the bottom of the hill. As the roller coaster continues on its up and down path, its potential energy goes through more changes. But when it arrives back at the starting point, it has exactly as much potential energy as it had when it started at this point. Another way of saying this is that there was as much uphill as there was downhill.

Similar reasoning can be applied to an electric circuit. We will analyze the circuit of Fig. 19–11 shortly, but first we consider the simpler circuit in Fig. 19–12. We have chosen it to be the same as the equivalent circuit of Fig. 19–7b already discussed. The current in this circuit is $I = (12.0\,\text{V})/(690\,\Omega) = 0.0174\,\text{A}$, as we calculated in Example 19–4. (We keep an extra digit in I to reduce rounding errors.) The positive side of the battery, point e in Fig. 19–12a, is at a high potential compared to point d at the negative side of the battery. That is, point e is like the top of a hill for a roller coaster. We follow the current around the circuit starting at any point. We choose to start at point d and follow a small positive test charge completely around this circuit. As we go, we note all changes in potential. When the test charge returns to point d, the potential will be the same as when we started (total change in potential around the circuit is zero). We plot the changes in potential around the circuit in Fig. 19–12b; point d is arbitrarily taken as zero.

As our positive test charge goes from point d, which is the negative or low potential side of the battery, to point e, which is the positive terminal (high potential side) of the battery, the potential increases by 12.0 V. (This is like the roller coaster being pulled up the first hill.) That is,

$$V_{ed} = +12.0\,\text{V}.$$

When our test charge moves from point e to point a, there is no change in potential because there is no source of emf and negligible resistance in the connecting wires.

Next, as the charge passes through the 400-Ω resistor to get to point b, there is a decrease in potential of $V = IR = (0.0174\,\text{A})(400\,\Omega) = 7.0\,\text{V}$. The positive test charge is flowing "downhill" since it is heading toward the negative terminal of the battery, as indicated in the graph of Fig. 19–12b. Because this is a *decrease* in potential, we use a *negative* sign:

$$V_{ba} = V_b - V_a = -7.0\,\text{V}.$$

As the charge proceeds from b to c there is another potential decrease (a "voltage drop") of $(0.0174\,\text{A}) \times (290\,\Omega) = 5.0\,\text{V}$, and this too is a decrease in potential:

$$V_{cb} = -5.0\,\text{V}.$$

There is no change in potential as our test charge moves from c to d as we assume negligible resistance in the wires.

The sum of all the changes in potential around the circuit of Fig. 19–12 is

$$+12.0\,\text{V} - 7.0\,\text{V} - 5.0\,\text{V} = 0.$$

This is exactly what Kirchhoff's loop rule said it would be.

FIGURE 19–12 Changes in potential around the circuit in (a) are plotted in (b).

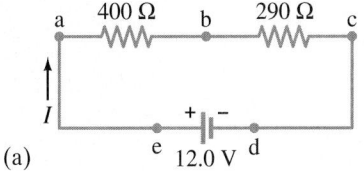
(a)

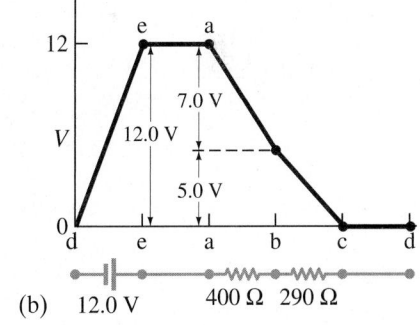
(b)

🖋 **PROBLEM SOLVING**
Be consistent with signs when applying the loop rule

Kirchhoff's Rules

1. **Label the current** in each separate branch of the given circuit with a different subscript, such as I_1, I_2, I_3 (see Fig. 19–11 or 19–13). Each current refers to a segment between two junctions. Choose the direction of each current, using an arrow. The direction can be chosen arbitrarily: if the current is actually in the opposite direction, it will come out with a minus sign in the solution.

2. **Identify the unknowns.** You will need as many independent equations as there are unknowns. You may write down more equations than this, but you will find that some of the equations will be redundant (that is, not be independent in the sense of providing new information). You may use $V = IR$ for each resistor, which sometimes will reduce the number of unknowns.

3. **Apply Kirchhoff's junction rule** at one or more junctions.

4. **Apply Kirchhoff's loop rule** for one or more loops: follow each loop in one direction only. Pay careful attention to subscripts, and to signs:
   (a) For a resistor, apply Ohm's law; the potential difference is negative (a decrease) if your chosen loop direction is the same as the chosen current direction through that resistor. The potential difference is positive (an increase) if your chosen loop direction is *opposite* to the chosen current direction.
   (b) For a battery, the potential difference is positive if your chosen loop direction is from the negative terminal toward the positive terminal; the potential difference is negative if the loop direction is from the positive terminal toward the negative terminal.

5. **Solve the equations** algebraically for the unknowns. Be careful with signs. At the end, check your answers by plugging them into the original equations, or even by using any additional loop or junction rule equations not used previously.

EXAMPLE 19–8 **Using Kirchhoff's rules.** Calculate the currents I_1, I_2, and I_3 in the three branches of the circuit in Fig. 19–13 (which is the same as Fig. 19–11).

APPROACH and SOLUTION

1. **Label the currents** and their directions. Figure 19–13 uses the labels I_1, I_2, and I_3 for the current in the three separate branches. Since (positive) current tends to move away from the positive terminal of a battery, we choose I_2 and I_3 to have the directions shown in Fig. 19–13. The direction of I_1 is not obvious in advance, so we arbitrarily chose the direction indicated. If the current actually flows in the opposite direction, our answer will have a negative sign.

2. **Identify the unknowns.** We have three unknowns (I_1, I_2, and I_3) and therefore we need three equations, which we get by applying Kirchhoff's junction and loop rules.

3. **Junction rule**: We apply Kirchhoff's junction rule to the currents at point a, where I_3 enters and I_2 and I_1 leave:

$$I_3 = I_1 + I_2. \qquad \text{(i)}$$

This same equation holds at point d, so we get no new information by writing an equation for point d.

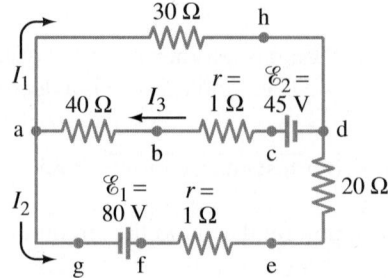

FIGURE 19–13 Currents can be calculated using Kirchhoff's rules. See Example 19–8.

4. Loop rule: We apply Kirchhoff's loop rule to two different closed loops. First we apply it to the upper loop ahdcba. We start (and end) at point a. From a to h we have a potential decrease $V_{ha} = -(I_1)(30\ \Omega)$. From h to d there is no change, but from d to c the potential increases by 45 V: that is, $V_{cd} = +45\ V$. From c to a the potential decreases through the two resistances by an amount $V_{ac} = -(I_3)(40\ \Omega + 1\ \Omega) = -(41\ \Omega)I_3$. Thus we have $V_{ha} + V_{cd} + V_{ac} = 0$, or

$$-30I_1 + 45 - 41I_3 = 0, \qquad \text{(ii)}$$

where we have omitted the units (volts and amps) so we can more easily see the algebra. For our second loop, we take the outer loop ahdefga. (We could have chosen the lower loop abcdefga instead.) Again we start at point a, and going to point h we have $V_{ha} = -(I_1)(30\ \Omega)$. Next, $V_{dh} = 0$. But when we take our positive test charge from d to e, it actually is going uphill, against the current—or at least against the *assumed* direction of the current, which is what counts in this calculation. Thus $V_{ed} = +I_2(20\ \Omega)$ has a *positive* sign. Similarly, $V_{fe} = +I_2(1\ \Omega)$. From f to g there is a decrease in potential of 80 V because we go from the high potential terminal of the battery to the low. Thus $V_{gf} = -80\ V$. Finally, $V_{ag} = 0$, and the sum of the potential changes around this loop is

$$-30I_1 + (20 + 1)I_2 - 80 = 0. \qquad \text{(iii)}$$

Our major work is done. The rest is algebra.

5. Solve the equations. We have three equations—labeled (i), (ii), and (iii)—and three unknowns. From Eq. (iii) we have

$$I_2 = \frac{80 + 30I_1}{21} = 3.8 + 1.4I_1. \qquad \text{(iv)}$$

From Eq. (ii) we have

$$I_3 = \frac{45 - 30I_1}{41} = 1.1 - 0.73I_1. \qquad \text{(v)}$$

We substitute Eqs. (iv) and (v) into Eq. (i):

$$I_1 = I_3 - I_2 = 1.1 - 0.73I_1 - 3.8 - 1.4I_1.$$

We solve for I_1, collecting terms:

$$3.1I_1 = -2.7$$
$$I_1 = -0.87\ A.$$

The negative sign indicates that the direction of I_1 is actually opposite to that initially assumed and shown in Fig. 19–13. The answer automatically comes out in amperes because our voltages and resistances were in volts and ohms. From Eq. (iv) we have

$$I_2 = 3.8 + 1.4I_1 = 3.8 + 1.4(-0.87) = 2.6\ A,$$

and from Eq. (v)

$$I_3 = 1.1 - 0.73I_1 = 1.1 - 0.73(-0.87) = 1.7\ A.$$

This completes the solution.

NOTE The unknowns in different situations are not necessarily currents. It might be that the currents are given and we have to solve for unknown resistance or voltage. The variables are then different, but the technique is the same.

EXERCISE D Write the Kirchhoff equation for the lower loop abcdefga of Example 19–8 and show, assuming the currents calculated in this Example, that the potentials add to zero for this lower loop.

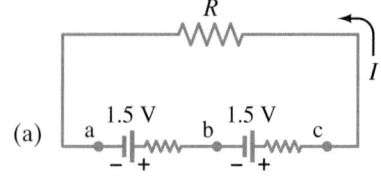

(a)

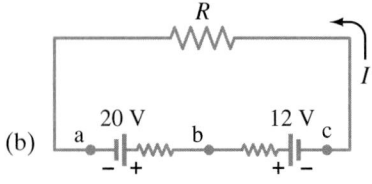

(b)

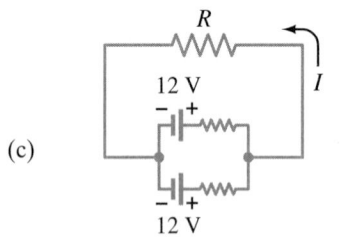

(c)

FIGURE 19–14 Batteries in series, (a) and (b), and in parallel (c).

FIGURE 19–15 Example 19–9, a jump start.

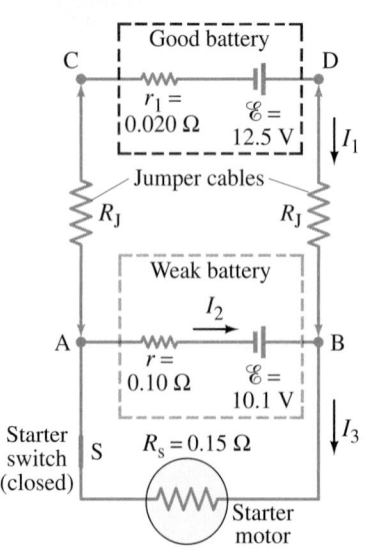

19–4 EMFs in Series and in Parallel; Charging a Battery

When two or more sources of emf, such as batteries, are arranged in series as in Fig. 19–14a, the total voltage is the algebraic sum of their respective voltages. On the other hand, when a 20-V and a 12-V battery are connected oppositely, as shown in Fig. 19–14b, the net voltage V_{ca} is 8 V (ignoring voltage drop across internal resistances). That is, a positive test charge moved from a to b gains in potential by 20 V, but when it passes from b to c it drops by 12 V. So the net change is 20 V − 12 V = 8 V. You might think that connecting batteries in reverse like this would be wasteful. For most purposes that would be true. But such a reverse arrangement is precisely how a battery charger works. In Fig. 19–14b, the 20-V source is charging up the 12-V battery. Because of its greater voltage, the 20-V source is forcing charge back into the 12-V battery: electrons are being forced into its negative terminal and removed from its positive terminal.

An automobile alternator keeps the car battery charged in the same way. A voltmeter placed across the terminals of a (12-V) car battery with the engine running fairly fast can tell you whether or not the alternator is charging the battery. If it is, the voltmeter reads 13 or 14 V. If the battery is not being charged, the voltage will be 12 V, or less if the battery is discharging. Car batteries can be recharged, but other batteries may not be rechargeable because the chemical reactions in many cannot be reversed. In such cases, the arrangement of Fig. 19–14b would simply waste energy.

Sources of emf can also be arranged in parallel, Fig. 19–14c, which—if the emfs are the same—can provide more energy when large currents are needed. Each of the cells in parallel has to produce only a fraction of the total current, so the energy loss due to internal resistance is less than for a single cell; and the batteries will go dead less quickly.

EXAMPLE 19–9 **Jump starting a car.** A good car battery is being used to jump start a car with a weak battery. The good battery has an emf of 12.5 V and internal resistance 0.020 Ω. Suppose the weak battery has an emf of 10.1 V and internal resistance 0.10 Ω. Each copper jumper cable is 3.0 m long and 0.50 cm in diameter, and can be attached as shown in Fig. 19–15. Assume the starter motor can be represented as a resistor $R_s = 0.15$ Ω. Determine the current through the starter motor (a) if only the weak battery is connected to it, and (b) if the good battery is also connected, as shown in Fig. 19–15.

APPROACH We apply Kirchhoff's rules, but in (b) we will first need to determine the resistance of the jumper cables using their dimensions and the resistivity ($\rho = 1.68 \times 10^{-8}$ Ω·m for copper) as discussed in Section 18–4.

SOLUTION (a) The circuit with only the weak battery and no jumper cables is simple: an emf of 10.1 V connected to two resistances in series, 0.10 Ω + 0.15 Ω = 0.25 Ω. Hence the current is $I = V/R = (10.1 \text{ V})/(0.25 \text{ Ω}) = 40$ A.

(b) We need to find the resistance of the jumper cables that connect the good battery to the weak one. From Eq. 18–3, each has resistance

$$R_J = \frac{\rho \ell}{A} = \frac{(1.68 \times 10^{-8} \text{ Ω·m})(3.0 \text{ m})}{(\pi)(0.25 \times 10^{-2} \text{ m})^2} = 0.0026 \text{ Ω}.$$

Kirchhoff's loop rule for the full outside loop gives

$$12.5 \text{ V} - I_1(2R_J + r_1) - I_3 R_s = 0$$
$$12.5 \text{ V} - I_1(0.025 \text{ Ω}) - I_3(0.15 \text{ Ω}) = 0 \quad \text{(i)}$$

since $(2R_J + r) = (0.0052 \text{ Ω} + 0.020 \text{ Ω}) = 0.025$ Ω.

The loop rule for the lower loop, including the weak battery and the starter, gives

$$10.1 \text{ V} - I_3(0.15\ \Omega) - I_2(0.10\ \Omega) = 0. \qquad \textbf{(ii)}$$

The junction rule at point B gives

$$I_1 + I_2 = I_3. \qquad \textbf{(iii)}$$

We have three equations in three unknowns. From Eq. (iii),

$$I_1 = I_3 - I_2$$

and we substitute this into Eq. (i):

$$12.5 \text{ V} - (I_3 - I_2)(0.025\ \Omega) - I_3(0.15\ \Omega) = 0,$$
$$12.5 \text{ V} - I_3(0.175\ \Omega) + I_2(0.025\ \Omega) = 0.$$

Combining this last equation with Eq. (ii) gives

$$12.5 \text{ V} - I_3(0.175\ \Omega) + \left(\frac{10.1 \text{ V} - I_3(0.15\ \Omega)}{0.10\ \Omega} \right)(0.025\ \Omega) = 0$$

or

$$I_3 = \frac{12.5 \text{ V} + 2.5 \text{ V}}{(0.175\ \Omega + 0.0375\ \Omega)} = 71 \text{ A},$$

quite a bit better than in part (a).

The other currents are $I_2 = -5$ A and $I_1 = 76$ A. Note that $I_2 = -5$ A is in the opposite direction from what we assumed in Fig. 19–15. The terminal voltage of the weak 10.1-V battery when being charged is

$$V_{BA} = 10.1 \text{ V} - (-5 \text{ A})(0.10\ \Omega) = 10.6 \text{ V}.$$

NOTE The circuit in Fig. 19–15, without the starter motor, is how a battery can be charged. The stronger battery pushes charge back into the weaker battery.

FIGURE 19–16 Exercise E.

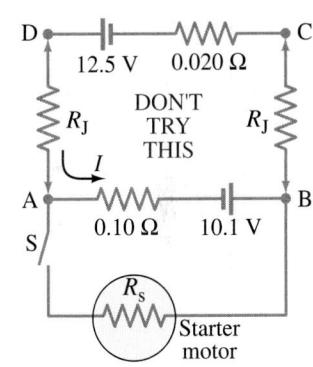

EXERCISE E If the jumper cables of Example 19–9 were mistakenly connected in reverse, the positive terminal of each battery would be connected to the negative terminal of the other battery (Fig. 19–16). What would be the current I even before the starter motor is engaged (the switch S in Fig. 19–16 is open)? Why could this cause the batteries to explode?

Safety when Jump Starting

Before jump starting a car's weak battery, be sure both batteries are 12 V and check the polarity of both batteries. The following (cautious) procedure applies if the negative ($-$) terminal is ground (attached by a cable to the metal car frame and motor), and the "hot" terminal is positive ($+$) on both batteries, as is the case for most modern cars. The $+$ terminal is usually marked by a red color, often a red cover. The safest procedure is to first connect the hot ($+$) terminal of the weak battery to the hot terminal of the good battery (using the cable with red clamps). Spread apart the handles of each clamp to squeeze the contact tightly. Then connect the black cable, first to the ground terminal of the good battery, and the other end to a clean exposed metal part (i.e., at ground) on the car with the weak battery. (This last connection should preferably be not too close to the battery, which in rare cases might leak H_2 gas that could ignite at the spark that may accompany the final connection.) This is safer than connecting directly to the ground terminal. When you are ready to start the disabled car, it helps to have the good car running (to keep its battery fully charged). As soon as the disabled car starts, immediately detach the cables in the exact reverse order (ground cable first).

In the photo of Fig. 19–15, the above procedure is not being followed. Note the safety error: with ground terminals connected, if the red clamp ($+12$ V) touches a metal part ($=$ ground), even if dropped by the person, a short circuit with damaging high electric current can occur (hundreds of amps).

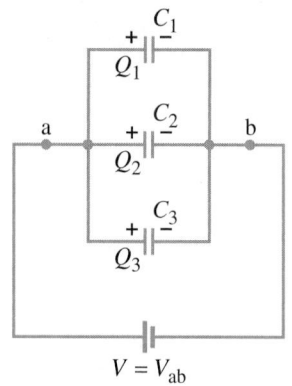

FIGURE 19–17 Capacitors in parallel: $C_{eq} = C_1 + C_2 + C_3$.

19–5 Circuits Containing Capacitors in Series and in Parallel

Just as resistors can be placed in series or in parallel in a circuit, so can capacitors (Chapter 17). We first consider a **parallel** connection as shown in Fig. 19–17. If a battery supplies a potential difference V to points a and b, this same potential difference $V = V_{ab}$ exists across each of the capacitors. That is, since the left-hand plates of all the capacitors are connected by conductors, they all reach the same potential V_a when connected to the battery; and the right-hand plates each reach potential V_b. Each capacitor plate acquires a charge given by $Q_1 = C_1 V$, $Q_2 = C_2 V$, and $Q_3 = C_3 V$. The total charge Q that must leave the battery is then

$$Q = Q_1 + Q_2 + Q_3 = C_1 V + C_2 V + C_3 V.$$

Let us try to find a single equivalent capacitor that will hold the same charge Q at the same voltage $V = V_{ab}$. It will have a capacitance C_{eq} given by

$$Q = C_{eq} V.$$

Combining the two previous equations, we have

$$C_{eq} V = C_1 V + C_2 V + C_3 V = (C_1 + C_2 + C_3)V$$

or

$$C_{eq} = C_1 + C_2 + C_3. \qquad \text{[parallel]} \quad \text{(19–5)}$$

The net effect of connecting capacitors in parallel is thus to *increase* the capacitance. Connecting capacitors in parallel is essentially increasing the area of the plates where charge can accumulate (see, for example, Eq. 17–8).

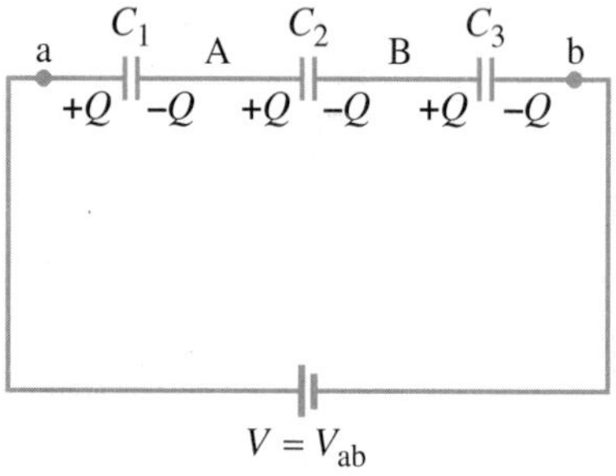

Capacitors can also be connected in **series**: that is, end to end as shown in Fig. 19–18. A charge $+Q$ flows from the battery to one plate of C_1, and $-Q$ flows to one plate of C_3. The regions A and B between the capacitors were originally neutral, so the net charge there must still be zero. The $+Q$ on the left plate of C_1 attracts a charge of $-Q$ on the opposite plate. Because region A must have a zero net charge, there is $+Q$ on the left plate of C_2. The same considerations apply to the other capacitors, so we see that the charge on each capacitor plate has the same magnitude Q. A single capacitor that could replace these three in series without affecting the circuit (that is, Q and V the same) would have a capacitance C_{eq} where

$$Q = C_{eq} V.$$

FIGURE 19–18 Capacitors in series:
$$\frac{1}{C_{eq}} = \frac{1}{C_1} + \frac{1}{C_2} + \frac{1}{C_3}.$$

The total voltage V across the three capacitors in series must equal the sum of the voltages across each capacitor:

$$V = V_1 + V_2 + V_3.$$

We also have for each capacitor $Q = C_1 V_1$, $Q = C_2 V_2$, and $Q = C_3 V_3$, so we substitute for V_1, V_2, V_3, and V into the last equation and get

$$\frac{Q}{C_{eq}} = \frac{Q}{C_1} + \frac{Q}{C_2} + \frac{Q}{C_3} = Q\left(\frac{1}{C_1} + \frac{1}{C_2} + \frac{1}{C_3}\right)$$

or

$$\frac{1}{C_{eq}} = \frac{1}{C_1} + \frac{1}{C_2} + \frac{1}{C_3}. \qquad \text{[series]} \quad \text{(19–6)}$$

⚠️ **CAUTION**

Formula for capacitors in series resembles formula for resistors in parallel

Notice that the equivalent capacitance C_{eq} is *smaller* than the smallest contributing capacitance. Notice also that the forms of the equations for capacitors in series or in parallel are the reverse of their counterparts for resistance. That is, the formula for capacitors in series resembles the formula for resistors in parallel.

EXAMPLE 19–10 **Equivalent capacitance.** Determine the capacitance of a single capacitor that will have the same effect as the combination shown in Fig. 19–19a. Take $C_1 = C_2 = C_3 = C$.

APPROACH First we find the equivalent capacitance of C_2 and C_3 in parallel, and then consider that capacitance in series with C_1.

SOLUTION Capacitors C_2 and C_3 are connected in parallel, so they are equivalent to a single capacitor having capacitance

$$C_{23} = C_2 + C_3 = C + C = 2C.$$

This C_{23} is in series with C_1, Fig. 19–19b, so the equivalent capacitance of the entire circuit, C_{eq}, is given by

$$\frac{1}{C_{eq}} = \frac{1}{C_1} + \frac{1}{C_{23}} = \frac{1}{C} + \frac{1}{2C} = \frac{3}{2C}.$$

Hence the equivalent capacitance of the entire combination is $C_{eq} = \frac{2}{3}C$, and it is smaller than any of the contributing capacitances, $C_1 = C_2 = C_3 = C$.

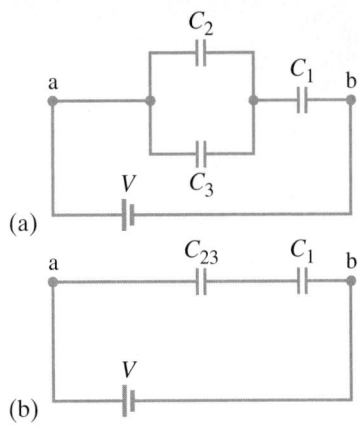

(a)

(b)

FIGURE 19–19
Examples 19–10 and 19–11.

📝 **PROBLEM SOLVING**
Remember to take the reciprocal

EXERCISE F Consider two identical capacitors $C_1 = C_2 = 10\,\mu\text{F}$. What are the smallest and largest capacitances that can be obtained by connecting these in series or parallel combinations? (*a*) $0.2\,\mu\text{F}$, $5\,\mu\text{F}$; (*b*) $0.2\,\mu\text{F}$, $10\,\mu\text{F}$; (*c*) $0.2\,\mu\text{F}$, $20\,\mu\text{F}$; (*d*) $5\,\mu\text{F}$, $10\,\mu\text{F}$; (*e*) $5\,\mu\text{F}$, $20\,\mu\text{F}$; (*f*) $10\,\mu\text{F}$, $20\,\mu\text{F}$.

EXAMPLE 19–11 **Charge and voltage on capacitors.** Determine the charge on each capacitor in Fig. 19–19a of Example 19–10 and the voltage across each, assuming $C = 3.0\,\mu\text{F}$ and the battery voltage is $V = 4.0\,\text{V}$.

APPROACH We have to work "backward" through Example 19–10. That is, we find the charge Q that leaves the battery, using the equivalent capacitance. Then we find the charge on each separate capacitor and the voltage across each. Each step uses Eq. 17–7, $Q = CV$.

SOLUTION The 4.0-V battery behaves as if it is connected to a capacitance $C_{eq} = \frac{2}{3}C = \frac{2}{3}(3.0\,\mu\text{F}) = 2.0\,\mu\text{F}$. Therefore the charge Q that leaves the battery, by Eq. 17–7, is

$$Q = CV = (2.0\,\mu\text{F})(4.0\,\text{V}) = 8.0\,\mu\text{C}.$$

From Fig. 19–19a, this charge arrives at the negative plate of C_1, so $Q_1 = 8.0\,\mu\text{C}$. The charge Q that leaves the positive plate of the battery is split evenly between C_2 and C_3 (*symmetry*: $C_2 = C_3$) and is $Q_2 = Q_3 = \frac{1}{2}Q = 4.0\,\mu\text{C}$. Next, the voltages across C_2 and C_3 have to be the same. The voltage across each capacitor is obtained using $V = Q/C$. So

$$V_1 = Q_1/C_1 = (8.0\,\mu\text{C})/(3.0\,\mu\text{F}) = 2.7\,\text{V}$$
$$V_2 = Q_2/C_2 = (4.0\,\mu\text{C})/(3.0\,\mu\text{F}) = 1.3\,\text{V}$$
$$V_3 = Q_3/C_3 = (4.0\,\mu\text{C})/(3.0\,\mu\text{F}) = 1.3\,\text{V}.$$

19–6 *RC* Circuits—Resistor and Capacitor in Series

Capacitor Charging

Capacitors and resistors are often found together in a circuit. Such ***RC* circuits** are common in everyday life. They are used to control the speed of a car's windshield wipers and the timing of traffic lights; they are used in camera flashes, in heart pacemakers, and in many other electronic devices. In *RC* circuits, we are not so interested in the final "steady state" voltage and charge on the capacitor, but rather in how these variables change in time.

A simple *RC* circuit is shown in Fig. 19–20a. When the switch S is closed, current immediately begins to flow through the circuit. Electrons will flow out from the negative terminal of the battery, through the resistor *R*, and accumulate on the upper plate of the capacitor. And electrons will flow into the positive terminal of the battery, leaving a positive charge on the other plate of the capacitor. As charge accumulates on the capacitor, the potential difference across it increases ($V_C = Q/C$), and the current is reduced until eventually the voltage across the capacitor equals the emf of the battery, $\mathscr{E}$. There is then no further current flow, and no potential difference across the resistor. The potential difference V_C across the capacitor, which is proportional to the charge on it ($V_C = Q/C$, Eq. 17–7), thus increases in time as shown in Fig. 19–20b. The shape of this curve is a type of exponential, and is given by the formula[†]

$$V_C = \mathscr{E}(1 - e^{-t/RC}),\qquad (19\text{--}7a)$$

where we use the subscript c to remind us that V_C is the voltage across the capacitor and is given here as a function of time *t*. [The constant *e*, known as the base for natural logarithms, has the value $e = 2.718\cdots$. Do not confuse this *e* with *e* for the charge on the electron.]

We can write a similar formula for the charge $Q\ (= CV_C)$ on the capacitor:

$$Q = Q_0(1 - e^{-t/RC}),\qquad (19\text{--}7b)$$

where Q_0 represents the maximum charge.

The product of the resistance *R* times the capacitance *C*, which appears in the exponent, is called the **time constant** τ of the circuit:

$$\tau = RC.\qquad (19\text{--}7c)$$

The time constant is a measure of how quickly the capacitor becomes charged. [The units of *RC* are $\Omega \cdot F = (V/A)(C/V) = C/(C/s) = s.$] Specifically, it can be shown that the product *RC* gives the time required for the capacitor's voltage (and charge) to reach 63% of the maximum. This can be checked[‡] using any calculator with an e^x key: $e^{-1} = 0.37$, so for $t = RC$, then $(1 - e^{-t/RC}) = (1 - e^{-1}) = (1 - 0.37) = 0.63$. In a circuit, for example, where $R = 200\,\text{k}\Omega$ and $C = 3.0\,\mu\text{F}$, the time constant is $(2.0 \times 10^5\,\Omega)(3.0 \times 10^{-6}\,\text{F}) = 0.60\,\text{s}$. If the resistance is much smaller, the time constant is much smaller and the capacitor becomes charged much more quickly. This makes sense, because a lower resistance will retard the flow of charge less. All circuits contain some resistance (if only in the connecting wires), so a capacitor can never be charged instantaneously when connected to a battery.

Finally, what is the voltage V_R across the resistor in Fig. 19–20a? The imposed battery voltage is $\mathscr{E}$, so

$$V_R = \mathscr{E} - V_C = \mathscr{E}(1 - 1 + e^{-t/RC}) = \mathscr{E}e^{-t/RC}.$$

This is called an **exponential decay**. The current *I* flowing in the circuit is that flowing through the resistor and is also an exponential decay:

$$I = \frac{V_R}{R} = \frac{\mathscr{E}}{R}e^{-t/RC}.\qquad (19\text{--}7d)$$

When the switch of the circuit in Fig. 19–20a is closed, the current is largest at first because there is no charge on the capacitor to impede it. As charge builds on the capacitor, the current decreases in time. That is exactly what Eq. 19–7d and Fig. 19–20c tell us.

FIGURE 19–20 After the switch S closes in the *RC* circuit shown in (a), the voltage V_C across the capacitor increases with time as shown in (b), and the current through the resistor decreases with time as shown in (c).

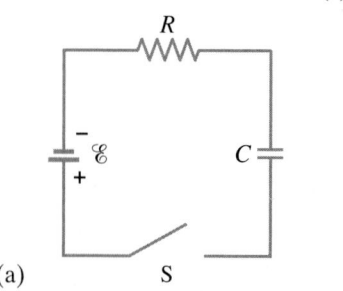

(a)

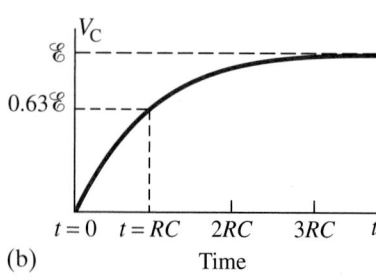

(b)

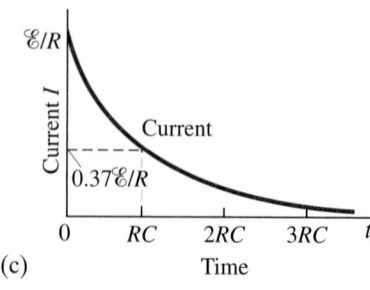

(c)

[†]The derivation uses calculus.

[‡]More simply, since $e = 2.718\cdots$, then $e^{-1} = 1/e = 1/2.718 = 0.37$. Note that *e* is the inverse operation to the natural logarithm ln: $\ln(e) = 1$, and $\ln(e^x) = x$.

EXAMPLE 19–12 *RC circuit, with emf.* The capacitance in the circuit of Fig. 19–20a is $C = 0.30\,\mu\text{F}$, the total resistance is $R = 20\,\text{k}\Omega$, and the battery emf is 12 V. Determine (*a*) the time constant, (*b*) the maximum charge the capacitor could acquire, (*c*) the time it takes for the charge to reach 99% of this value, and (*d*) the maximum current.

APPROACH We use Fig. 19–20 and Eqs. 19–7a, b, c, and d.

SOLUTION (*a*) The time constant is $RC = (2.0 \times 10^4\,\Omega)(3.0 \times 10^{-7}\,\text{F}) = 6.0 \times 10^{-3}\,\text{s} = 6.0\,\text{ms}$.

(*b*) The maximum charge would occur when no further current flows, so $Q_0 = C\mathscr{E} = (3.0 \times 10^{-7}\,\text{F})(12\,\text{V}) = 3.6\,\mu\text{C}$.

(*c*) In Eq. 19–7b, we set $Q = 0.99C\mathscr{E}$:

$$0.99C\mathscr{E} = C\mathscr{E}\left(1 - e^{-t/RC}\right),$$

or

$$e^{-t/RC} = 1 - 0.99 = 0.01.$$

We take the natural logarithm of both sides (Appendix A–8), recalling that $\ln e^x = x$:

$$\frac{t}{RC} = -\ln(0.01) = 4.6$$

so

$$t = 4.6RC = (4.6)(6.0 \times 10^{-3}\,\text{s}) = 28 \times 10^{-3}\,\text{s}$$

or 28 ms (less than $\frac{1}{30}$ s).

(*d*) The current is a maximum at $t = 0$ (the moment when the switch is closed) and there is no charge yet on the capacitor ($Q = 0$):

$$I_{\text{max}} = \frac{\mathscr{E}}{R} = \frac{12\,\text{V}}{2.0 \times 10^4\,\Omega} = 600\,\mu\text{A}.$$

Capacitor Discharging

The circuit just discussed involved the *charging* of a capacitor by a battery through a resistance. Now let us look at another situation: a capacitor is already charged to a voltage V_0 and charge Q_0, and it is then allowed to *discharge* through a resistance R as shown in Fig. 19–21a. In this case there is no battery. When the switch S is closed, charge begins to flow through resistor R from one side of the capacitor toward the other side, until the capacitor is fully discharged. The voltage across the capacitor decreases, as shown in Fig. 19–21b. This "exponential decay" curve is given by

$$V_C = V_0 e^{-t/RC},$$

where V_0 is the initial voltage across the capacitor. The voltage falls 63% of the way to zero (to $0.37V_0$) in a time $\tau = RC$. Because the charge Q on the capacitor is $Q = CV$ (and $Q_0 = CV_0$), we can write

$$Q = Q_0 e^{-t/RC}$$

for a discharging capacitor, where Q_0 is the initial charge. The voltage across the resistor will have the same magnitude as that across the capacitor at any instant, but the opposite sign, because there is zero applied emf: $V_C + V_R = 0$ so $V_R = -V_C = -V_0 e^{-t/RC}$. A graph of V_R vs. time would just be Fig. 19–21b upside down. The current $I = V_R/R = -(V_0/R)e^{-t/RC} = -I_0 e^{-t/RC}$. The current has its greatest magnitude at $t = 0$ and decreases exponentially in time. (The current has a minus sign because in Fig. 19–21a it flows in the opposite direction as compared to the current in Fig. 19–20a.)

FIGURE 19–21 For the *RC* circuit shown in (a), the voltage V_C across the capacitor decreases with time t, as shown in (b), after the switch S is closed at $t = 0$. The charge on the capacitor follows the same curve since $Q \propto V_C$.

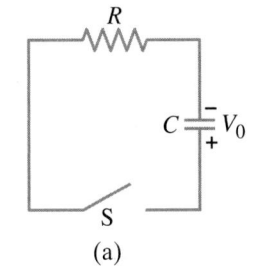

(a)

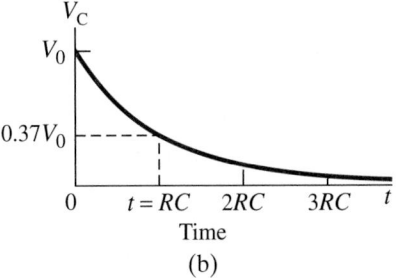

(b)

EXAMPLE 19–13 **A discharging *RC* circuit.** If a charged capacitor, $C = 35\,\mu\text{F}$, is connected to a resistance $R = 120\,\Omega$ as in Fig. 19–21a, how much time will elapse until the voltage falls to 10% of its original (maximum) value?

APPROACH The voltage across the capacitor decreases according to $V_C = V_0 e^{-t/RC}$. We set $V_C = 0.10V_0$ (10% of V_0), but first we need to calculate $\tau = RC$.

SOLUTION The time constant for this circuit is given by

$$\tau = RC = (120\,\Omega)(35 \times 10^{-6}\,\text{F}) = 4.2 \times 10^{-3}\,\text{s}.$$

After a time t the voltage across the capacitor will be

$$V_C = V_0 e^{-t/RC}.$$

We want to know the time t for which $V_C = 0.10V_0$. We substitute into the above equation

$$0.10V_0 = V_0 e^{-t/RC}$$

so

$$e^{-t/RC} = 0.10.$$

The inverse operation to the exponential e is the natural log, ln. Thus

$$\ln\!\left(e^{-t/RC}\right) = -\frac{t}{RC} = \ln 0.10 = -2.3.$$

Solving for t, we find the elapsed time is

$$t = 2.3(RC) = (2.3)(4.2 \times 10^{-3}\,\text{s}) = 9.7 \times 10^{-3}\,\text{s} = 9.7\,\text{ms}.$$

NOTE We can find the time for any specified voltage across a capacitor by using $t = RC \ln(V_0/V_C)$.

CONCEPTUAL EXAMPLE 19–14 **Bulb in *RC* circuit.** In the circuit of Fig. 19–22, the capacitor is originally uncharged. Describe the behavior of the lightbulb from the instant switch S is closed until a long time later.

RESPONSE When the switch is first closed, the current in the circuit is high and the lightbulb burns brightly. As the capacitor charges, the voltage across the capacitor increases, causing the current to be reduced, and the lightbulb dims. As the potential difference across the capacitor approaches the same voltage as the battery, the current decreases toward zero and the lightbulb goes out.

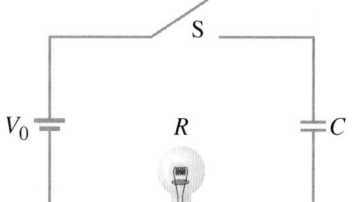

FIGURE 19–22 Example 19–14.

Medical and Other Applications of *RC* Circuits

The charging and discharging in an *RC* circuit can be used to produce voltage pulses at a regular frequency. The charge on the capacitor increases to a particular voltage, and then discharges. One way of initiating the discharge of the capacitor is by the use of a gas-filled tube which has an electrical breakdown when the voltage across it reaches a certain value V_0. After the discharge is finished, the tube no longer conducts current and the recharging process repeats itself, starting at a lower voltage V_0'. Figure 19–23 shows a possible circuit, and the **sawtooth voltage** it produces.

A simple blinking light can be an application of a sawtooth oscillator circuit. Here the emf is supplied by a battery; the neon bulb flashes on at a rate of perhaps 1 cycle per second. The main component of a "flasher unit" is a moderately large capacitor.

PHYSICS APPLIED
Sawtooth voltage

PHYSICS APPLIED
Blinking flashers

FIGURE 19–23 (a) An *RC* circuit, coupled with a gas-filled tube as a switch, can produce (b) a repeating "sawtooth" voltage.

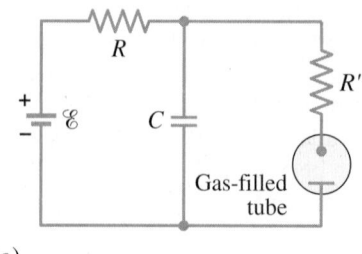
(a)

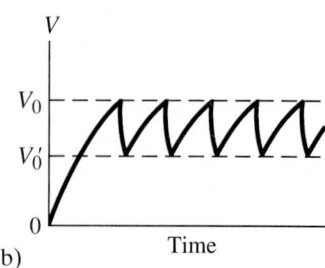
(b)

The intermittent windshield wipers of a car can also use an *RC* circuit. The *RC* time constant, which can be changed using a multi-positioned switch for different values of *R* with fixed *C*, determines the rate at which the wipers come on.

EXERCISE G A typical turn signal flashes perhaps twice per second, so its time constant is on the order of 0.5 s. Estimate the resistance in the circuit, assuming a moderate capacitor of $C = 1\,\mu F$.

An important medical use of an *RC* circuit is the electronic heart pacemaker, which can make a stopped heart start beating again by applying an electric stimulus through electrodes attached to the chest. The stimulus can be repeated at the normal heartbeat rate if necessary. The heart itself contains *pacemaker* cells, which send out tiny electric pulses at a rate of 60 to 80 per minute. These signals induce the start of each heartbeat. In some forms of heart disease, the natural pacemaker fails to function properly, and the heart loses its beat. Such patients use *electronic pacemakers* which produce a regular voltage pulse that starts and controls the frequency of the heartbeat. The electrodes are implanted in or near the heart (Fig. 19–24), and the circuit contains a capacitor and a resistor. The charge on the capacitor increases to a certain point and then discharges a pulse to the heart. Then it starts charging again. The pulsing rate depends on the time constant *RC*.

19–7 Electric Hazards

Excess electric current can overheat wires in buildings and cause fires, as discussed in Section 18–6. Electric current can also damage the human body or even be fatal. Electric current through the human body can cause damage in two ways: (1) heating tissue and causing burns; (2) stimulating nerves and muscles, and we feel a "shock." The severity of a shock depends on the magnitude of the current, how long it acts, and through what part of the body it passes. A current passing through vital organs such as the heart or brain is especially damaging.

A current of about 1 mA or more can be felt and may cause pain. Currents above 10 mA cause severe contraction of the muscles, and a person may not be able to let go of the source of the current (say, a faulty appliance or wire). Death from paralysis of the respiratory system can occur. Artificial respiration can sometimes revive a victim. If a current above about 80 to 100 mA passes across the torso, so that a portion passes through the heart for more than a second or two, the heart muscles will begin to contract irregularly and blood will not be properly pumped. This condition is called **ventricular fibrillation**. If it lasts for long, death results. Strangely enough, if the current is much larger, on the order of 1 A, death by heart failure may be less likely,[†] but such currents can cause serious burns if concentrated through a small area of the body.

It is current that harms, but it is voltage that drives the current. The seriousness of an electric shock depends on the current and thus on the applied voltage and the effective resistance of the body. Living tissue has low resistance because the fluid of cells contains ions that can conduct quite well. However, the outer layer of skin, when dry, offers high resistance and is thus protective. The effective resistance between two points on opposite sides of the body when the skin is dry is on the order of 10^4 to $10^6\,\Omega$. But when the skin is wet, the resistance may be $10^3\,\Omega$ or less. A person who is barefoot or wearing thin-soled shoes will be in good contact with the ground, and touching a 120-V line with a wet hand can result in a current

$$I = \frac{120\,\text{V}}{1000\,\Omega} = 120\,\text{mA}.$$

As we saw, this could be lethal.

[†]Larger currents apparently bring the entire heart to a standstill. Upon release of the current, the heart returns to its normal rhythm. This may not happen when fibrillation occurs because, once started, it can be hard to stop. Fibrillation may also occur as a result of a heart attack or during heart surgery. A device known as a *defibrillator* (described in Section 17–9) can apply a brief high current to the heart, causing complete heart stoppage which is often followed by resumption of normal beating.

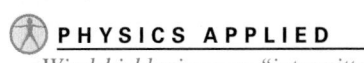

PHYSICS APPLIED
Windshield wipers on "intermittent"

PHYSICS APPLIED
Heart pacemaker

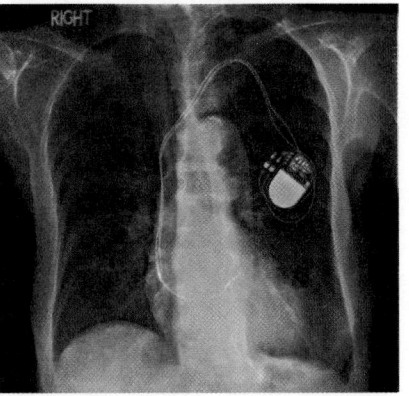

FIGURE 19–24 Electronic battery-powered pacemaker can be seen on the rib cage in this X-ray (color added).

PHYSICS APPLIED
Dangers of electricity

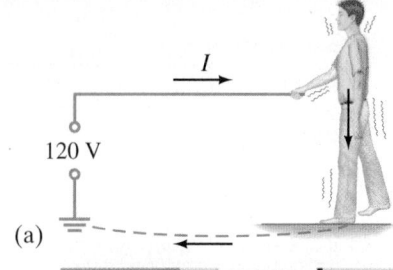

(a)

120 V

"Yikes!"

(b)

FIGURE 19–25 You can receive a shock when the circuit is completed.

⚠ **CAUTION**

Keep one hand in your pocket when other touches electricity

🚶 **PHYSICS APPLIED**

Grounding and shocks

You can get a shock by becoming part of a complete circuit. Figure 19–25 shows two ways the circuit might be completed when you accidentally touch a "hot" electric wire—"hot" meaning a high potential relative to ground such as 120 V (normal U.S. household voltage) or 240 V (many other countries). The other wire of building wiring is connected to ground—either by a wire connected to a buried conductor, or via a metal water pipe into the ground. The current in Fig. 19–25a passes from the high-voltage wire through you to ground through your bare feet, and back along the ground (a fair conductor) to the ground terminal of the source. If you stand on a good insulator—thick rubber-soled shoes or a dry wood floor—there will be much more resistance in the circuit and much less current through you. If you stand with bare feet on the ground, or in a bathtub, there is lethal danger because the resistance is much less and the current greater. In a bathtub (or swimming pool), not only are you wet, which reduces your resistance, but the water is in contact with the drain pipe (typically metal) that leads to the ground. It is strongly recommended that you not touch anything electrical when wet or in bare feet. The use of non-metal pipes would be protective.

In Fig. 19–25b, a person touches a faulty "hot" wire with one hand, and the other hand touches a sink faucet (connected to ground via the pipe or even by water in a non-metal pipe). The current is particularly dangerous because it passes across the chest, through the heart and lungs. A useful rule: if one hand is touching something electrical, keep your other hand in your back pocket (don't use it!), and wear thick rubber-soled shoes. Also remove metal jewelry, especially rings (your finger is usually moist under a ring).

You can come into contact with a hot wire by touching a bare wire whose insulation has worn off, or from a bare wire inside an appliance when you're tinkering with it. (Always unplug an electrical device before investigating its insides!)[†] Also, a wire inside a device may break or lose its insulation and come in contact with the case. If the case is metal, it will conduct electricity. A person could then suffer a severe shock merely by touching the case, as shown in Fig. 19–26b. To prevent

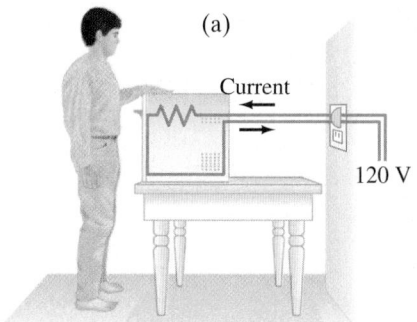

(a)

Current

120 V

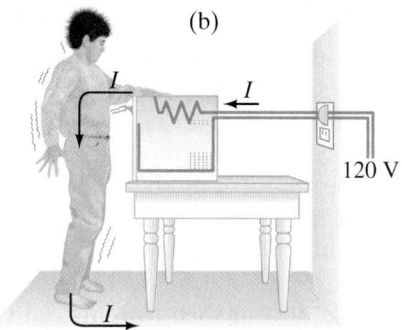

(b)

I   *I*

120 V

I

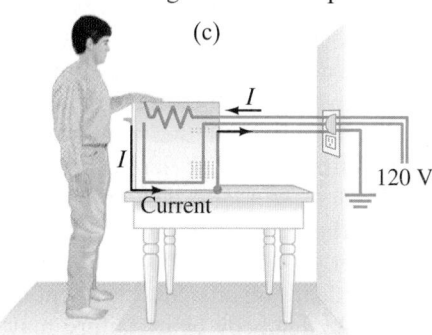

(c)

I   *I*

I

120 V

Current

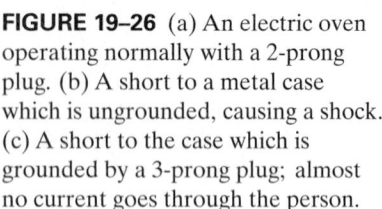

FIGURE 19–26 (a) An electric oven operating normally with a 2-prong plug. (b) A short to a metal case which is ungrounded, causing a shock. (c) A short to the case which is grounded by a 3-prong plug; almost no current goes through the person.

an accident, metal cases are supposed to be connected directly to ground by a separate ground wire. Then if a "hot" wire touches the grounded case, a short circuit to ground immediately occurs internally, as shown in Fig. 19–26c, and most of the current passes through the low-resistance ground wire rather than through the person. Furthermore, the high current should open a fuse or circuit breaker. Grounding a metal case is done by a separate ground wire connected to the third (round) prong of a 3-prong plug. Never cut off the third prong of a plug—it could save your life. A three-prong plug, and an adapter, are shown in Figs. 19–27a and b.

[†]Even then you can get a bad shock from a capacitor that hasn't been discharged until you touch it.

FIGURE 19–27 (a) A 3-prong plug, and (b) an adapter (white) for old-fashioned 2-prong outlets—be sure to screw down the ground tab (green color in photo).

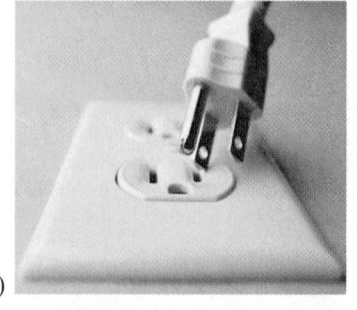

(a)

(b)

Safe Wiring

Why is a third wire needed? The 120 V is carried by the other two wires—one **hot** (120 V ac), the other **neutral**, which is itself grounded. The third "dedicated" ground wire with the round prong may seem redundant. But it is protection for two reasons: (1) It protects against internal wiring that may have been done incorrectly or is faulty as discussed above, Fig. 19–26. (2) The *neutral* wire carries the full normal current ("return" current from the hot 120 V) and it does have resistance—so there can be a voltage drop along the neutral wire, normally small; but if connections are poor or corroded, or the plug is loose, the resistance could be large enough that you might feel that voltage if you touched the neutral wire some distance from its grounding point.

Some electrical devices come with only two wires, and the plug's two prongs are of different widths; the plug can be inserted only one way into the outlet so that the intended neutral (wider prong) in the device is connected to neutral in the wiring (Fig. 19–28). For example, the screw threads of a lightbulb are meant to be connected to neutral (and the base contact to hot), to avoid shocks when changing a bulb in a possibly protruding socket. Devices with 2-prong plugs do *not* have their cases grounded; they are supposed to have double electric insulation (or have a nonmetal case). Take extra care anyway.

The insulation on a wire may be color coded. Hand-held meters (Section 19–8) may have red (hot) and black (ground) lead wires. But in a U.S. house, the hot wire is often black (though it may be red), whereas white is neutral and green (or bare) is the dedicated ground, Fig. 19–29. But beware: these color codes cannot always be trusted.

[In the U.S., three wires normally enter a house: two *hot* wires at 120 V each (which add together to 240 V for appliances or devices that run on 240 V) plus the grounded *neutral* (carrying return current for the two hot wires). See Fig. 19–29. The "dedicated" *ground* wire (non-current carrying) is a fourth wire that does not come from the electric company but enters the house from a nearby heavy stake in the ground or a buried metal pipe. The two hot wires can feed separate 120-V circuits in the house, so each 120-V circuit inside the house has only three wires, including the dedicated ground.]

Normal circuit breakers (Sections 18–6 and 20–7) protect equipment and buildings from overload and fires. They protect humans only in some circumstances, such as the very high currents that result from a short, if they respond quickly enough. *Ground fault circuit interrupters* (GFCI or GFI), described in Section 21–9, are designed to protect people from the much lower currents (10 mA to 100 mA) that are lethal but would not throw a 15-A circuit breaker or blow a 20-A fuse.

Another danger is **leakage current**, by which we mean a current along an unintended path. Leakage currents are often "capacitively coupled." For example, a wire in a lamp forms a capacitor with the metal case; charges moving in one conductor attract or repel charge in the other, so there is a current. Typical electrical codes limit leakage currents to 1 mA for any device, which is usually harmless. It could be dangerous, however, to a hospital patient with implanted electrodes, due to the absence of the protective skin layer and because the current can pass directly through the heart. Although 100 mA may be needed to cause heart fibrillation when entering through the hands and spreading out through the body (very little of it actually passing through the heart), but as little as 0.02 mA can cause fibrillation when passing directly to the heart. Thus, a "wired" patient is in considerable danger from leakage current even from as simple an act as touching a lamp.

Finally, don't touch a downed power line (lethal!) or even get near it. A hot power line is at thousands of volts. A huge current can flow along the ground from the point where the high-voltage wire touches the ground. This current is great enough that the voltage between your two feet could be large and dangerous. Tip: stand on one foot, or run so only one foot touches the ground at a time.

⚠ CAUTION
Necessity of third (ground) wire

FIGURE 19–28 A polarized 2-prong plug.

⚠ CAUTION
Black wire may be either ground or hot. Beware!

FIGURE 19–29 Four wires entering a typical house. The color codes for wires are not always as shown here—be careful!

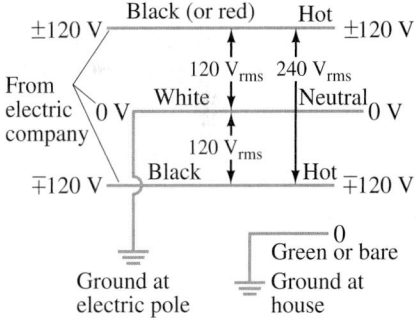

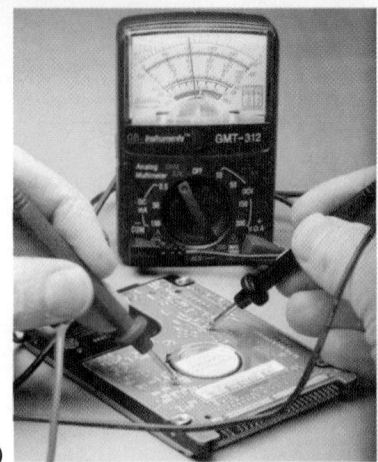

(a)

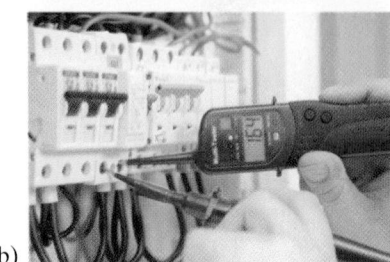

(b)

FIGURE 19–30 (a) An analog multimeter. (b) An electronic digital meter measuring voltage at a circuit breaker.

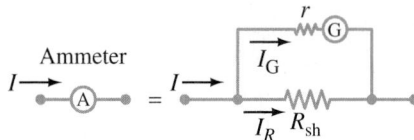

PHYSICS APPLIED

Ammeters use shunt resistor in parallel

FIGURE 19–31 An ammeter is a galvanometer in parallel with a (shunt) resistor with low resistance, R_{sh}.

Ammeter

$$I \xrightarrow{\hspace{0.3cm}} \text{—}(A)\text{—} = I \xrightarrow{\hspace{0.3cm}} \begin{array}{c} r \quad (G) \\ I_G \\ I_R \quad R_{sh} \end{array}$$

PHYSICS APPLIED

Voltmeters use series resistor

19–8 Ammeters and Voltmeters—Measurement Affects the Quantity Being Measured

Measurement is a fundamental part of physics, and is not as simple as you might think. Measuring instruments can not be taken for granted; their results are not perfect and often need to be interpreted. As an illustration of measurement "theory" we examine here how electrical quantities are measured using meters. We also examine how meters affect the quantity they attempt to measure.

An **ammeter** measures current, and a **voltmeter** measures potential difference or voltage. Each can be either: (1) an *analog* meter, which displays numerical values by the position of a pointer that can move across a scale (Fig. 19–30a); or (2) a *digital* meter, which displays the numerical value in numbers (Fig. 19–30b). We now examine how analog meters work.

An analog ammeter or voltmeter, in which the reading is by a pointer on a scale (Fig. 19–30a), uses a *galvanometer*. A galvanometer works on the principle of the force between a magnetic field and a current-carrying coil of wire; it is straightforward to understand and will be discussed in Chapter 20. For now, we only need to know that the deflection of the needle of a galvanometer is proportional to the current flowing through it. The *full-scale current sensitivity* of a galvanometer, I_m, is the electric current needed to make the needle deflect full scale, typically about 50 μA.

A galvanometer whose sensitivity I_m is 50 μA can measure currents from about 1 μA (currents smaller than this would be hard to read on the scale) up to 50 μA. To measure larger currents, a resistor is placed in parallel with the galvanometer. An analog **ammeter**, represented by the symbol •-(A)-•, consists of a galvanometer (•-(G)-•) in parallel with a resistor called the **shunt resistor**, as shown in Fig. 19–31. ("Shunt" is a synonym for "in parallel.") The shunt resistance is R_{sh}, and the resistance of the galvanometer coil is r. The value of R_{sh} is chosen according to the full-scale deflection desired; R_{sh} is normally very small—giving an ammeter a very small net resistance—so most of the current passes through R_{sh} and very little ($\lesssim 50$ μA) passes through the galvanometer to deflect the needle.

EXAMPLE 19–15 **Ammeter design.** Design an ammeter to read 1.0 A at full scale using a galvanometer with a full-scale sensitivity of 50 μA and a resistance $r = 30\ \Omega$. Check if the scale is linear.

APPROACH Only 50 μA $(= I_G = 0.000050\ \text{A})$ of the 1.0-A current passes through the galvanometer to give full-scale deflection. The rest of the current $(I_R = 0.999950\ \text{A})$ passes through the small shunt resistor, R_{sh}, Fig. 19–31. The potential difference across the galvanometer equals that across the shunt resistor (they are in parallel). We apply Ohm's law to find R_{sh}.

SOLUTION Because $I = I_G + I_R$, when $I = 1.0$ A flows into the meter, we want I_R through the shunt resistor to be $I_R = 0.999950$ A. The potential difference across the shunt is the same as across the galvanometer, so

$$I_R R_{sh} = I_G r.$$

Then

$$R_{sh} = \frac{I_G r}{I_R} = \frac{(5.0 \times 10^{-5}\ \text{A})(30\ \Omega)}{(0.999950\ \text{A})} = 1.5 \times 10^{-3}\ \Omega,$$

or 0.0015 Ω. The shunt resistor must thus have a *very* low resistance and most of the current passes through it.

Because $I_G = I_R(R_{sh}/r)$ and (R_{sh}/r) is constant, we see that the scale is linear (needle deflection is proportional to I_G). If the current $I \approx I_R$ into the meter is half of full scale, 0.50 A, the current to the galvanometer will be $I_G = I_R(R_{sh}/r) = (0.50\ \text{A})(1.5 \times 10^{-3}\ \Omega)/(30\ \Omega) = 25$ μA, which would make the needle deflect halfway, as it should.

An analog **voltmeter** (•-(V)-•) consists of a galvanometer and a resistor R_{ser} connected in series, Fig. 19–32. R_{ser} is usually large, giving a voltmeter a high internal resistance.

EXAMPLE 19–16 Voltmeter design.

Using a galvanometer with internal resistance $r = 30\,\Omega$ and full-scale current sensitivity of $50\,\mu A$, design a voltmeter that reads from 0 to 15 V. Is the scale linear?

APPROACH When a potential difference of 15 V exists across the terminals of our voltmeter, we want $50\,\mu A$ to be passing through it so as to give a full-scale deflection.

SOLUTION From Ohm's law, $V = IR$, we have (Fig. 19–32)

$$15\,\text{V} = (50\,\mu A)(r + R_{ser}),$$

so

$$R_{ser} = (15\,\text{V})/(5.0 \times 10^{-5}\,\text{A}) - r = 300\,\text{k}\Omega - 30\,\Omega \approx 300\,\text{k}\Omega.$$

Notice that $r = 30\,\Omega$ is so small compared to the value of R_{ser} that it doesn't influence the calculation significantly. The scale will again be linear: if the voltage to be measured is 6.0 V, the current passing through the voltmeter will be $(6.0\,\text{V})/(3.0 \times 10^5\,\Omega) = 2.0 \times 10^{-5}\,\text{A}$, or $20\,\mu A$. This will produce two-fifths of full-scale deflection, as required $(6.0\,\text{V}/15.0\,\text{V} = 2/5)$.

How to Connect Meters

Suppose you wish to determine the current I in the circuit shown in Fig. 19–33a, and the voltage V across the resistor R_1. How exactly are ammeters and voltmeters connected to the circuit being measured?

Because an ammeter is used to measure the current flowing in the circuit, it must be inserted directly into the circuit, in series with the other elements, as shown in Fig. 19–33b. The smaller its internal resistance, the less it affects the circuit.

A voltmeter is connected "externally," in parallel with the circuit element across which the voltage is to be measured. It measures the potential difference between two points. Its two wire leads (connecting wires) are connected to the two points, as shown in Fig. 19–33c, where the voltage across R_1 is being measured. The larger its internal resistance ($R_{ser} + r$, Fig. 19–32), the less it affects the circuit being measured.

Effects of Meter Resistance

It is important to know the sensitivity of a meter, for in many cases the resistance of the meter can seriously affect your results. Consider the following Example.

EXAMPLE 19–17 Voltage reading vs. true voltage.

An electronic circuit has two 15-kΩ resistors, R_1 and R_2, connected in series, Fig. 19–34a. The battery voltage is 8.0 V and it has negligible internal resistance. A voltmeter has resistance of 50 kΩ on the 5.0-V scale. What voltage does the meter read when connected across R_1, Fig. 19–34b, and what error is caused by the meter's finite resistance?

APPROACH The meter acts as a resistor in parallel with R_1. We use parallel and series resistor analyses and Ohm's law to find currents and voltages.

SOLUTION The voltmeter resistance of 50,000 Ω is in parallel with $R_1 = 15\,\text{k}\Omega$, Fig. 19–34b. The net resistance R_{eq} of these two is

$$\frac{1}{R_{eq}} = \frac{1}{50\,\text{k}\Omega} + \frac{1}{15\,\text{k}\Omega} = \frac{13}{150\,\text{k}\Omega};$$

so $R_{eq} = 11.5\,\text{k}\Omega$. This $R_{eq} = 11.5\,\text{k}\Omega$ is in series with $R_2 = 15\,\text{k}\Omega$, so the total resistance is now 26.5 kΩ (not the original 30 kΩ). Hence the current from the battery is

$$I = \frac{8.0\,\text{V}}{26.5\,\text{k}\Omega} = 3.0 \times 10^{-4}\,\text{A} = 0.30\,\text{mA}.$$

Then the voltage drop across R_1, which is the same as that across the voltmeter, is $(3.0 \times 10^{-4}\,\text{A})(11.5 \times 10^3\,\Omega) = 3.5\,\text{V}$. [The voltage drop across R_2 is $(3.0 \times 10^{-4}\,\text{A})(15 \times 10^3\,\Omega) = 4.5\,\text{V}$, for a total of 8.0 V.] If we assume the meter is accurate, it reads 3.5 V. In the original circuit, without the meter, $R_1 = R_2$ so the voltage across R_1 is half that of the battery, or 4.0 V. Thus the voltmeter, because of its internal resistance, gives a low reading. It is off by 0.5 V, or more than 10%.

NOTE Often the **sensitivity** of a voltmeter is specified on its face as, for example, $10,000\,\Omega/\text{V}$. Then on a 5.0-V scale, the voltmeter would have a resistance given by $(5.0\,\text{V})(10,000\,\Omega/\text{V}) = 50,000\,\Omega$. The meter's resistance depends on the scale used.

Voltmeter

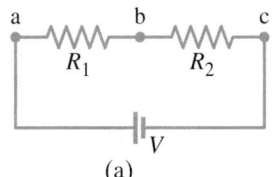

FIGURE 19–32 A voltmeter is a galvanometer in series with a resistor with high resistance, R_{ser}.

FIGURE 19–33 Measuring current and voltage.

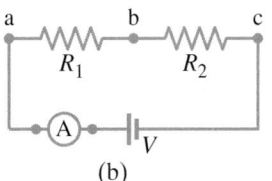

(a)

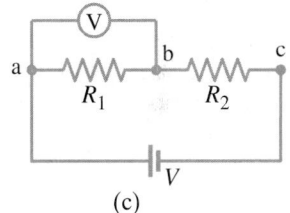

(b)

(c)

FIGURE 19–34 Example 19–17.

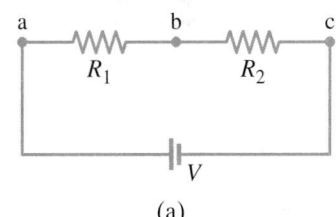

(a)

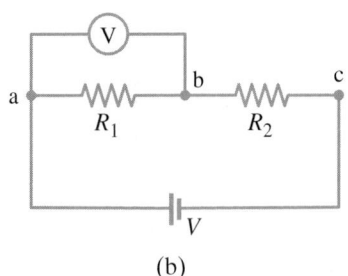

(b)

Example 19–17 illustrates how seriously a meter can affect a circuit and give a misleading reading. If the resistance of a voltmeter is much higher than the resistance of the circuit, however, it will have little effect and its readings can be more accurate, at least to the manufactured precision of the meter, which for analog meters is typically 3% to 4% of full-scale deflection. Even an ammeter can interfere with a circuit, but the effect is minimal if its resistance is much less than that of the circuit as a whole. For both voltmeters and ammeters, the more sensitive the galvanometer, the less effect it will have on the circuit. [A 50,000-Ω/V meter is far better than a 1000-Ω/V meter.]

Whenever we make a measurement on a circuit, to some degree we affect that circuit (Example 19–17). This is true for other types of measurement as well: when we make a measurement on a system, we affect that system in some way. On a temperature measurement, for example, the thermometer can exchange heat with the system, thus altering the temperature it is measuring. It is important to be able to make needed corrections, as we saw in Example 19–17.

Other Meters

The meters described above are for direct current. A dc meter can be modified to measure ac (alternating current, Section 18–7) with the addition of diodes (Chapter 29), which allow current to flow in one direction only. An ac meter can be calibrated to read rms or peak values.

Voltmeters and ammeters can have several series or shunt resistors to offer a choice of range. **Multimeters** can measure voltage, current, and resistance. Sometimes a multimeter is called a VOM (Volt-Ohm-Meter or Volt-Ohm-Milliammeter).

An **ohmmeter** measures resistance, and must contain a battery of known voltage connected in series to a resistor (R_{ser}) and to an ammeter which contains a shunt R_{sh} (Fig. 19–35). The resistor whose resistance is to be measured completes the circuit, and must not be connected in a circuit containing a voltage source. The needle deflection of the meter is inversely proportional to the resistance. The scale calibration depends on the value of its series resistor, which is changeable in a multimeter. Because an ohmmeter sends a current through the device whose resistance is to be measured, it should not be used on very delicate devices that could be damaged by the current.

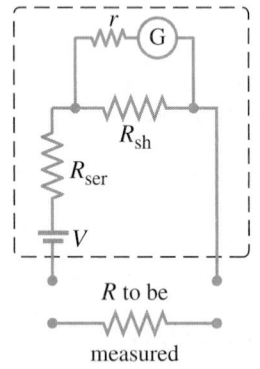

FIGURE 19–35 An ohmmeter.

Digital Meters

Digital meters (see Fig. 19–30b) are used in the same way as analog meters: they are inserted directly into the circuit, in series, to measure current (Fig. 19–33b), and connected "outside," in parallel with the circuit, to measure voltage (Fig. 19–33c).

The internal construction of digital meters is very different from analog meters. First, digital meters do not use a galvanometer, but rather semiconductor devices (Chapter 29). The electronic circuitry and digital readout are more sensitive than a galvanometer, and have less effect on the circuit to be measured. When we measure dc voltages, a digital meter's resistance is very high, commonly on the order of 10 MΩ to 100 MΩ $(10^7–10^8\ \Omega)$, and doesn't change significantly when different voltage scales are selected. A 100-MΩ digital meter draws very little current when connected across even a 1-MΩ resistance.

The precision of digital meters is exceptional, often one part in $10^4\ (= 0.01\%)$ or better. This precision is not the same as accuracy, however. A precise meter of internal resistance $10^8\ \Omega$ will not give accurate results if used to measure a voltage across a 10^8-Ω resistor—in which case it is necessary to do a calculation like that in Example 19–17.

Summary

A device that transforms another type of energy into electrical energy is called a **source** of **emf**. A battery behaves like a source of emf in series with an **internal resistance**. The emf is the potential difference determined by the chemical reactions in the battery and equals the terminal voltage when no current is drawn. When a current is drawn, the voltage at the battery's terminals is less than its emf by an amount equal to the potential decrease Ir across the internal resistance.

When resistances are connected in **series** (end to end in a single linear path), the equivalent resistance is the sum of the individual resistances:

$$R_{eq} = R_1 + R_2 + \cdots. \qquad (19\text{-}3)$$

In a series combination, R_{eq} is greater than any component resistance.

When resistors are connected in **parallel**, it is the reciprocals that add up:

$$\frac{1}{R_{eq}} = \frac{1}{R_1} + \frac{1}{R_2} + \cdots. \qquad (19\text{-}4)$$

In a parallel connection, the net resistance is less than any of the individual resistances.

Kirchhoff's rules are useful in determining the currents and voltages in circuits. Kirchhoff's **junction rule** is based on conservation of electric charge and states that the sum of all currents entering any junction equals the sum of all currents leaving that junction. The second, or **loop rule**, is based on conservation of energy and states that the algebraic sum of the changes in potential around any closed path of the circuit must be zero.

When capacitors are connected in **parallel**, the equivalent capacitance is the sum of the individual capacitances:

$$C_{eq} = C_1 + C_2 + \cdots. \qquad (19\text{-}5)$$

When capacitors are connected in **series**, it is the reciprocals that add up:

$$\frac{1}{C_{eq}} = \frac{1}{C_1} + \frac{1}{C_2} + \cdots. \qquad (19\text{-}6)$$

When an **RC circuit** containing a resistance R in series with a capacitance C is connected to a dc source of emf, the voltage across the capacitor rises gradually in time characterized by an exponential of the form $(1 - e^{-t/RC})$, where the **time constant**

$$\tau = RC \qquad (19\text{-}7)$$

is the time it takes for the voltage to reach 63% of its maximum value.

A capacitor discharging through a resistor is characterized by the same time constant: in a time $\tau = RC$, the voltage across the capacitor drops to 37% of its initial value. The charge on the capacitor, and the voltage across it, decrease as $e^{-t/RC}$.

Electric shocks are caused by current passing through the body. To avoid shocks, the body must not become part of a complete circuit by allowing different parts of the body to touch objects at different potentials. Commonly, shocks are caused by one part of the body touching ground ($V = 0$) and another part touching a nonzero electric potential.

An **ammeter** measures current. An analog ammeter consists of a galvanometer and a parallel **shunt resistor** that carries most of the current. An analog **voltmeter** consists of a galvanometer and a series resistor. An ammeter is inserted *into* the circuit whose current is to be measured. A voltmeter is external, being connected in parallel to the element whose voltage is to be measured. Digital meters have greater internal resistance and affect the circuit to be measured less than do analog meters.

Questions

1. Explain why birds can sit on power lines safely, even though the wires have no insulation around them, whereas leaning a metal ladder up against a power line is extremely dangerous.

2. Discuss the advantages and disadvantages of Christmas tree lights connected in parallel versus those connected in series.

3. If all you have is a 120-V line, would it be possible to light several 6-V lamps without burning them out? How?

4. Two lightbulbs of resistance R_1 and R_2 $(R_2 > R_1)$ and a battery are all connected in series. Which bulb is brighter? What if they are connected in parallel? Explain.

5. Household outlets are often double outlets. Are these connected in series or parallel? How do you know?

6. With two identical lightbulbs and two identical batteries, explain how and why you would arrange the bulbs and batteries in a circuit to get the maximum possible total power to the lightbulbs. (Ignore internal resistance of batteries.)

7. If two identical resistors are connected in series to a battery, does the battery have to supply more power or less power than when only one of the resistors is connected? Explain.

8. You have a single 60-W bulb lit in your room. How does the overall resistance of your room's electric circuit change when you turn on an additional 100-W bulb? Explain.

9. Suppose three identical capacitors are connected to a battery. Will they store more energy if connected in series or in parallel?

10. When applying Kirchhoff's loop rule (such as in Fig. 19–36), does the sign (or direction) of a battery's emf depend on the direction of current through the battery? What about the terminal voltage?

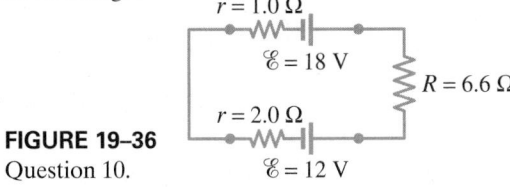

FIGURE 19–36
Question 10.

11. Different lamps might have batteries connected in either of the two arrangements shown in Fig. 19–37. What would be the advantages of each scheme?

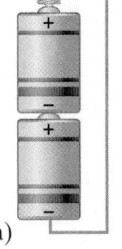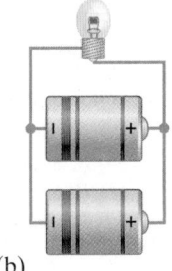

FIGURE 19–37
Question 11. (a) (b)

12. For what use are batteries connected in series? For what use are they connected in parallel? Does it matter if the batteries are nearly identical or not in either case?

13. Can the terminal voltage of a battery ever exceed its emf? Explain.

14. Explain in detail how you could measure the internal resistance of a battery.

15. In an *RC* circuit, current flows from the battery until the capacitor is completely charged. Is the total energy supplied by the battery equal to the total energy stored by the capacitor? If not, where does the extra energy go?

16. Given the circuit shown in Fig. 19–38, use the words "increases," "decreases," or "stays the same" to complete the following statements:
    (a) If R_7 increases, the potential difference between A and E _____. Assume no resistance in Ⓐ and ⅇ.
    (b) If R_7 increases, the potential difference between A and E _____. Assume Ⓐ and ⅇ have resistance.
    (c) If R_7 increases, the voltage drop across R_4 _____.
    (d) If R_2 decreases, the current through R_1 _____.
    (e) If R_2 decreases, the current through R_6 _____.
    (f) If R_2 decreases, the current through R_3 _____.
    (g) If R_5 increases, the voltage drop across R_2 _____.
    (h) If R_5 increases, the voltage drop across R_4 _____.
    (i) If R_2, R_5, and R_7 increase, $\mathcal{E} (r = 0)$ _____.

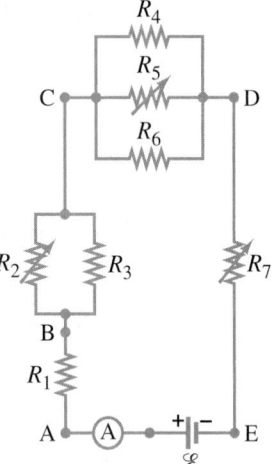

FIGURE 19–38 Question 16. R_2, R_5, and R_7 are *variable* resistors (you can change their resistance), given the symbol ⎓⎓⎓.

17. Design a circuit in which two different switches of the type shown in Fig. 19–39 can be used to operate the same lightbulb from opposite sides of a room.

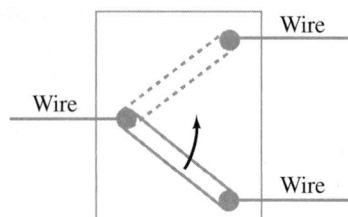

FIGURE 19–39
Question 17.

18. Why is it more dangerous to turn on an electric appliance when you are standing outside in bare feet than when you are inside wearing shoes with thick soles?

19. What is the main difference between an analog voltmeter and an analog ammeter?

20. What would happen if you mistakenly used an ammeter where you needed to use a voltmeter?

21. Explain why an ideal ammeter would have zero resistance and an ideal voltmeter infinite resistance.

22. A voltmeter connected across a resistor always reads *less* than the actual voltage (i.e., when the meter is not present). Explain.

23. A small battery-operated flashlight requires a single 1.5-V battery. The bulb is barely glowing. But when you take the battery out and check it with a digital voltmeter, it registers 1.5 V. How would you explain this?

MisConceptual Questions

1. In which circuits shown in Fig. 19–40 are resistors connected in series?

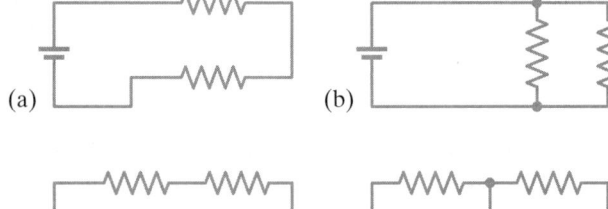

FIGURE 19–40 MisConceptual Question 1.

2. Which resistors in Fig. 19–41 are connected in parallel?
   (a) All three.
   (b) R_1 and R_2.
   (c) R_2 and R_3.
   (d) R_1 and R_3.
   (e) None of the above.

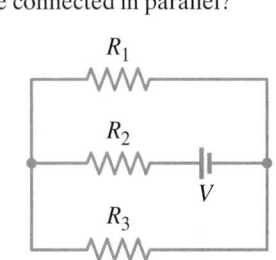

FIGURE 19–41
MisConceptual Question 2.

3. A 10,000-Ω resistor is placed in series with a 100-Ω resistor. The current in the 10,000-Ω resistor is 10 A. If the resistors are swapped, how much current flows through the 100-Ω resistor?
   (a) >10 A.   (b) <10 A.   (c) 10 A.
   (d) Need more information about the circuit.

4. Two identical 10-V batteries and two identical 10-Ω resistors are placed in series as shown in Fig. 19–42. If a 10-Ω lightbulb is connected with one end connected between the batteries and other end between the resistors, how much current will flow through the lightbulb?
   (a) 0 A.
   (b) 1 A.
   (c) 2 A.
   (d) 4 A.

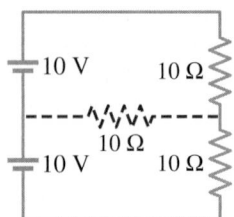

FIGURE 19–42 MisConceptual Question 4.

5. Which resistor shown in Fig. 19–43 has the greatest current going through it? Assume that all the resistors are equal.
(a) R_1.
(d) R_5.
(b) R_1 and R_2.
(e) All of them the same.
(c) R_3 and R_4.

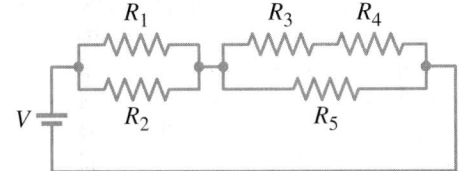

FIGURE 19–43
MisConceptual
Question 5.

6. Figure 19–44 shows three identical bulbs in a circuit. What happens to the brightness of bulb A if you replace bulb B with a short circuit?
(a) Bulb A gets brighter.
(b) Bulb A gets dimmer.
(c) Bulb A's brightness does not change.
(d) Bulb A goes out.

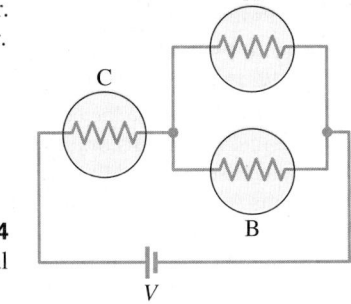

FIGURE 19–44
MisConceptual
Question 6.

7. When the switch shown in Fig. 19–45 is closed, what will happen to the voltage across resistor R_4? It will
(a) increase.
(b) decrease.
(c) stay the same.

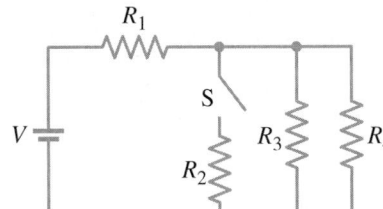

FIGURE 19–45
MisConceptual
Questions 7 and 8.

8. When the switch shown in Fig. 19–45 is closed, what will happen to the voltage across resistor R_1? It will
(a) increase.
(b) decrease.
(c) stay the same.

9. As a capacitor is being charged in an RC circuit, the current flowing through the resistor is
(a) increasing.
(c) constant.
(b) decreasing.
(d) zero.

10. For the circuit shown in Fig. 19–46, what happens when the switch S is closed?
(a) Nothing. Current cannot flow through the capacitor.
(b) The capacitor immediately charges up to the battery emf.
(c) The capacitor eventually charges up to the full battery emf at a rate determined by R and C.
(d) The capacitor charges up to a fraction of the battery emf determined by R and C.
(e) The capacitor charges up to a fraction of the battery emf determined by R only.

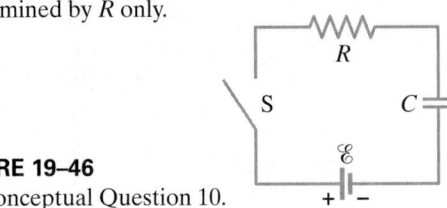

FIGURE 19–46
MisConceptual Question 10.

11. The capacitor in the circuit shown in Fig. 19–47 is charged to an initial value Q. When the switch is closed, it discharges through the resistor. It takes 2.0 seconds for the charge to drop to $\frac{1}{2}Q$. How long does it take to drop to $\frac{1}{4}Q$?
(a) 3.0 seconds.
(b) 4.0 seconds.
(c) Between 2.0 and 3.0 seconds.
(d) Between 3.0 and 4.0 seconds.
(e) More than 4.0 seconds.

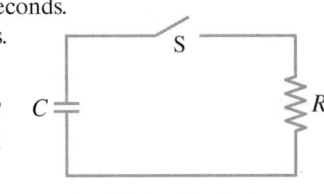

FIGURE 19–47
MisConceptual
Question 11.

12. A resistor and a capacitor are used in series to control the timing in the circuit of a heart pacemaker. To design a pacemaker that can double the heart rate when the patient is exercising, which statement below is true? The capacitor
(a) needs to discharge faster, so the resistance should be decreased.
(b) needs to discharge faster, so the resistance should be increased.
(c) needs to discharge slower, so the resistance should be decreased.
(d) needs to discharge slower, so the resistance should be increased.
(e) does not affect the timing, regardless of the resistance.

13. Why is an appliance cord with a three-prong plug safer than one with two prongs?
(a) The 120 V from the outlet is split among three wires, so it isn't as high a voltage as when it is only split between two wires.
(b) Three prongs fasten more securely to the wall outlet.
(c) The third prong grounds the case, so the case cannot reach a high voltage.
(d) The third prong acts as a ground wire, so the electrons have an easier time leaving the appliance. As a result, fewer electrons build up in the appliance.
(e) The third prong controls the capacitance of the appliance, so it can't build up a high voltage.

14. When capacitors are connected in series, the effective capacitance is _____ the smallest capacitance; when capacitors are connected in parallel, the effective capacitance is _____ the largest capacitance.
(a) greater than; equal to.
(d) equal to; less than.
(b) greater than; less than.
(e) equal to; equal to.
(c) less than; greater than.

15. If ammeters and voltmeters are not to significantly alter the quantities they are measuring,
(a) the resistance of an ammeter and a voltmeter should be much higher than that of the circuit element being measured.
(b) the resistance of an ammeter should be much lower, and the resistance of a voltmeter should be much higher, than those of the circuit being measured.
(c) the resistance of an ammeter should be much higher, and the resistance of a voltmeter should be much lower, than those of the circuit being measured.
(d) the resistance of an ammeter and a voltmeter should be much lower than that of the circuit being measured.
(e) None of the above.

Problems

19–1 Emf and Terminal Voltage

1. (I) Calculate the terminal voltage for a battery with an internal resistance of $0.900\ \Omega$ and an emf of 6.00 V when the battery is connected in series with (a) a $71.0\text{-}\Omega$ resistor, and (b) a $710\text{-}\Omega$ resistor.

2. (I) Four 1.50-V cells are connected in series to a $12.0\text{-}\Omega$ lightbulb. If the resulting current is 0.45 A, what is the internal resistance of each cell, assuming they are identical and neglecting the resistance of the wires?

3. (II) What is the internal resistance of a 12.0-V car battery whose terminal voltage drops to 8.8 V when the starter motor draws 95 A? What is the resistance of the starter?

19–2 Resistors in Series and Parallel

[In these Problems neglect the internal resistance of a battery unless the Problem refers to it.]

4. (I) A $650\text{-}\Omega$ and an $1800\text{-}\Omega$ resistor are connected in series with a 12-V battery. What is the voltage across the $1800\text{-}\Omega$ resistor?

5. (I) Three $45\text{-}\Omega$ lightbulbs and three $65\text{-}\Omega$ lightbulbs are connected in series. (a) What is the total resistance of the circuit? (b) What is the total resistance if all six are wired in parallel?

6. (II) Suppose that you have a $580\text{-}\Omega$, a $790\text{-}\Omega$, and a $1.20\text{-k}\Omega$ resistor. What is (a) the maximum, and (b) the minimum resistance you can obtain by combining these?

7. (II) How many $10\text{-}\Omega$ resistors must be connected in series to give an equivalent resistance to five $100\text{-}\Omega$ resistors connected in parallel?

8. (II) Design a "voltage divider" (see Example 19–3) that would provide one-fifth (0.20) of the battery voltage across R_2, Fig. 19–6. What is the ratio R_1/R_2?

9. (II) Suppose that you have a 9.0-V battery and wish to apply a voltage of only 3.5 V. Given an unlimited supply of $1.0\text{-}\Omega$ resistors, how could you connect them to make a "voltage divider" that produces a 3.5-V output for a 9.0-V input?

10. (II) Three $1.70\text{-k}\Omega$ resistors can be connected together in four different ways, making combinations of series and/or parallel circuits. What are these four ways, and what is the net resistance in each case?

11. (II) A battery with an emf of 12.0 V shows a terminal voltage of 11.8 V when operating in a circuit with two lightbulbs, each rated at 4.0 W (at 12.0 V), which are connected in parallel. What is the battery's internal resistance?

12. (II) Eight identical bulbs are connected in series across a 120-V line. (a) What is the voltage across each bulb? (b) If the current is 0.45 A, what is the resistance of each bulb, and what is the power dissipated in each?

13. (II) Eight bulbs are connected in parallel to a 120-V source by two long leads of total resistance $1.4\ \Omega$. If 210 mA flows through each bulb, what is the resistance of each, and what fraction of the total power is wasted in the leads?

14. (II) A close inspection of an electric circuit reveals that a $480\text{-}\Omega$ resistor was inadvertently soldered in the place where a $350\text{-}\Omega$ resistor is needed. How can this be fixed without removing anything from the existing circuit?

15. (II) Eight 7.0-W Christmas tree lights are connected in series to each other and to a 120-V source. What is the resistance of each bulb?

16. (II) Determine (a) the equivalent resistance of the circuit shown in Fig. 19–48, (b) the voltage across each resistor, and (c) the current through each resistor.

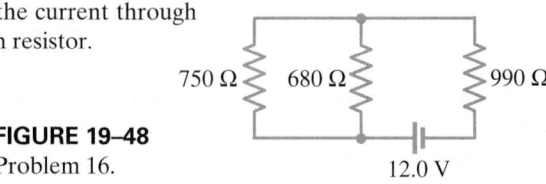

FIGURE 19–48
Problem 16.

750 Ω 680 Ω 990 Ω 12.0 V

17. (II) A 75-W, 120-V bulb is connected in parallel with a 25-W, 120-V bulb. What is the net resistance?

18. (II) (a) Determine the equivalent resistance of the "ladder" of equal $175\text{-}\Omega$ resistors shown in Fig. 19–49. In other words, what resistance would an ohmmeter read if connected between points A and B? (b) What is the current through each of the three resistors on the left if a 50.0-V battery is connected between points A and B?

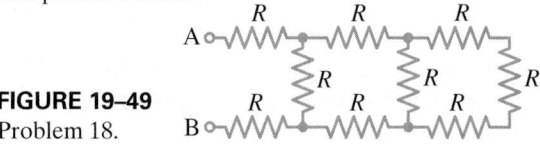

FIGURE 19–49
Problem 18.

19. (II) What is the net resistance of the circuit connected to the battery in Fig. 19–50?

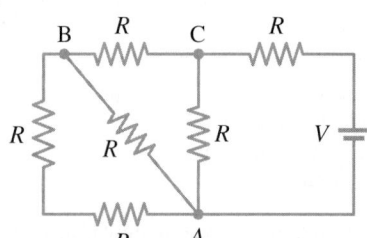

FIGURE 19–50
Problems 19 and 20.

20. (II) Calculate the current through each resistor in Fig. 19–50 if each resistance $R = 3.25\ \text{k}\Omega$ and $V = 12.0$ V. What is the potential difference between points A and B?

21. (III) Two resistors when connected in series to a 120-V line use one-fourth the power that is used when they are connected in parallel. If one resistor is $4.8\ \text{k}\Omega$, what is the resistance of the other?

22. (III) Three equal resistors (R) are connected to a battery as shown in Fig. 19–51. Qualitatively, what happens to (a) the voltage drop across each of these resistors, (b) the current flow through each, and (c) the terminal voltage of the battery, when the switch S is opened, after having been closed for a long time? (d) If the emf of the battery is 9.0 V, what is its terminal voltage when the switch is closed if the internal resistance r is $0.50\ \Omega$ and $R = 5.50\ \Omega$? (e) What is the terminal voltage when the switch is open?

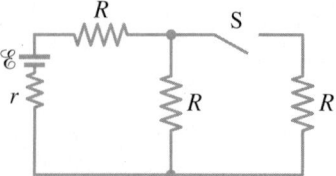

FIGURE 19–51
Problem 22.

23. (III) A 2.5-kΩ and a 3.7-kΩ resistor are connected in parallel; this combination is connected in series with a 1.4-kΩ resistor. If each resistor is rated at 0.5 W (maximum without overheating), what is the maximum voltage that can be applied across the whole network?

24. (III) Consider the network of resistors shown in Fig. 19–52. Answer qualitatively: (*a*) What happens to the voltage across each resistor when the switch S is closed? (*b*) What happens to the current through each when the switch is closed? (*c*) What happens to the power output of the battery when the switch is closed? (*d*) Let $R_1 = R_2 = R_3 = R_4 = 155\,\Omega$ and $V = 22.0$ V. Determine the current through each resistor before and after closing the switch. Are your qualitative predictions confirmed?

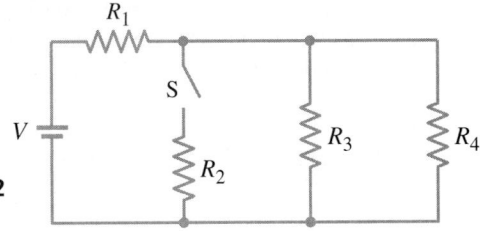

FIGURE 19–52
Problem 24.

19–3 Kirchhoff's Rules

25. (I) Calculate the current in the circuit of Fig. 19–53, and show that the sum of all the voltage changes around the circuit is zero.

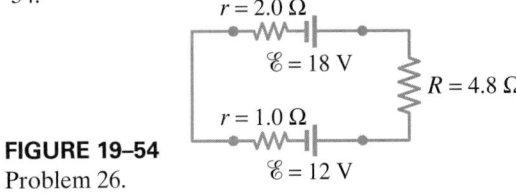

FIGURE 19–53
Problem 25.

26. (II) Determine the terminal voltage of each battery in Fig. 19–54.

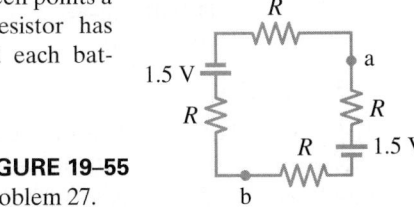

FIGURE 19–54
Problem 26.

27. (II) For the circuit shown in Fig. 19–55, find the potential difference between points a and b. Each resistor has $R = 160\,\Omega$ and each battery is 1.5 V.

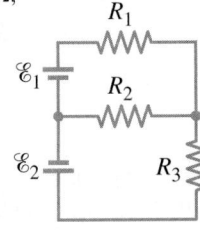

FIGURE 19–55
Problem 27.

28. (II) Determine the magnitudes and directions of the currents in each resistor shown in Fig. 19–56. The batteries have emfs of $\mathscr{E}_1 = 9.0$ V and $\mathscr{E}_2 = 12.0$ V and the resistors have values of $R_1 = 25\,\Omega$, $R_2 = 68\,\Omega$, and $R_3 = 35\,\Omega$. (*a*) Ignore internal resistance of the batteries. (*b*) Assume each battery has internal resistance $r = 1.0\,\Omega$.

FIGURE 19–56
Problem 28.

29. (II) (*a*) What is the potential difference between points a and d in Fig. 19–57 (similar to Fig. 19–13, Example 19–8), and (*b*) what is the terminal voltage of each battery?

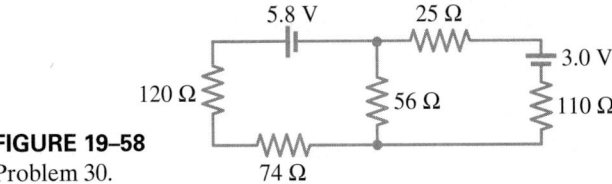

FIGURE 19–57
Problem 29.

30. (II) Calculate the magnitude and direction of the currents in each resistor of Fig. 19–58.

FIGURE 19–58
Problem 30.

31. (II) Determine the magnitudes and directions of the currents through R_1 and R_2 in Fig. 19–59.

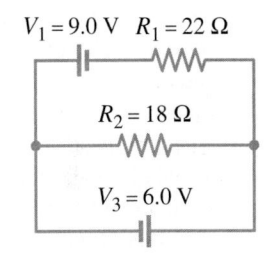

FIGURE 19–59
Problems 31 and 32.

32. (II) Repeat Problem 31, now assuming that each battery has an internal resistance $r = 1.4\,\Omega$.

33. (III) (*a*) A network of five equal resistors R is connected to a battery $\mathscr{E}$ as shown in Fig. 19–60. Determine the current I that flows out of the battery. (*b*) Use the value determined for I to find the single resistor R_{eq} that is equivalent to the five-resistor network.

FIGURE 19–60
Problem 33.

34. (III) (*a*) Determine the currents I_1, I_2, and I_3 in Fig. 19–61. Assume the internal resistance of each battery is $r = 1.0\,\Omega$. (*b*) What is the terminal voltage of the 6.0-V battery?

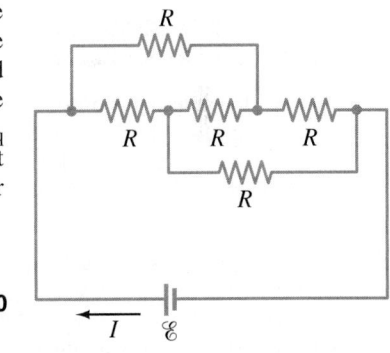

FIGURE 19–61
Problems 34 and 35.

35. (III) What would the current I_1 be in Fig. 19–61 if the 12-Ω resistor is shorted out (resistance $= 0$)? Let $r = 1.0\,\Omega$.

19–4 Emfs Combined, Battery Charging

36. (II) Suppose two batteries, with unequal emfs of 2.00 V and 3.00 V, are connected as shown in Fig. 19–62. If each internal resistance is $r = 0.350\,\Omega$, and $R = 4.00\,\Omega$, what is the voltage across the resistor R?

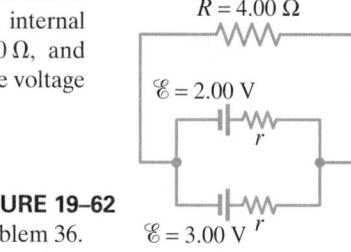

FIGURE 19–62
Problem 36.

37. (II) A battery for a proposed electric car is to have three hundred 3-V lithium ion cells connected such that the total voltage across all of the cells is 300 V. Describe a possible connection configuration (using series and parallel connections) that would meet these battery specifications.

19–5 Capacitors in Series and Parallel

38. (I) (a) Six 4.8-μF capacitors are connected in parallel. What is the equivalent capacitance? (b) What is their equivalent capacitance if connected in series?

39. (I) A 3.00-μF and a 4.00-μF capacitor are connected in series, and this combination is connected in parallel with a 2.00-μF capacitor (see Fig. 19–63). What is the net capacitance?

40. (II) If 21.0 V is applied across the whole network of Fig. 19–63, calculate (a) the voltage across each capacitor and (b) the charge on each capacitor.

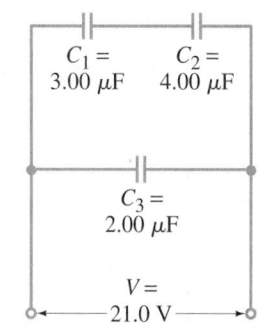

FIGURE 19–63
Problems 39 and 40.

41. (II) The capacitance of a portion of a circuit is to be reduced from 2900 pF to 1200 pF. What capacitance can be added to the circuit to produce this effect without removing existing circuit elements? Must any existing connections be broken to accomplish this?

42. (II) An electric circuit was accidentally constructed using a 7.0-μF capacitor instead of the required 16-μF value. Without removing the 7.0-μF capacitor, what can a technician add to correct this circuit?

43. (II) Consider three capacitors, of capacitance 3200 pF, 5800 pF, and 0.0100 μF. What maximum and minimum capacitance can you form from these? How do you make the connection in each case?

44. (II) Determine the equivalent capacitance between points a and b for the combination of capacitors shown in Fig. 19–64.

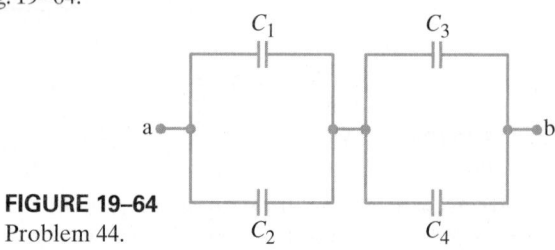

FIGURE 19–64
Problem 44.

45. (II) What is the ratio of the voltage V_1 across capacitor C_1 in Fig. 19–65 to the voltage V_2 across capacitor C_2?

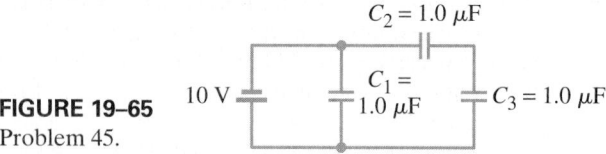

FIGURE 19–65
Problem 45.

46. (II) A 0.50-μF and a 1.4-μF capacitor are connected in series to a 9.0-V battery. Calculate (a) the potential difference across each capacitor and (b) the charge on each. (c) Repeat parts (a) and (b) assuming the two capacitors are in parallel.

47. (II) A circuit contains a single 250-pF capacitor hooked across a battery. It is desired to store four times as much energy in a combination of two capacitors by adding a single capacitor to this one. How would you hook it up, and what would its value be?

48. (II) Suppose three parallel-plate capacitors, whose plates have areas A_1, A_2, and A_3 and separations d_1, d_2, and d_3, are connected in parallel. Show, using only Eq. 17–8, that Eq. 19–5 is valid.

49. (II) Two capacitors connected in parallel produce an equivalent capacitance of 35.0 μF but when connected in series the equivalent capacitance is only 4.8 μF. What is the individual capacitance of each capacitor?

50. (III) Given three capacitors, $C_1 = 2.0\,\mu$F, $C_2 = 1.5\,\mu$F, and $C_3 = 3.0\,\mu$F, what arrangement of parallel and series connections with a 12-V battery will give the minimum voltage drop across the 2.0-μF capacitor? What is the minimum voltage drop?

51. (III) In Fig. 19–66, suppose $C_1 = C_2 = C_3 = C_4 = C$. (a) Determine the equivalent capacitance between points a and b. (b) Determine the charge on each capacitor and the potential difference across each in terms of V.

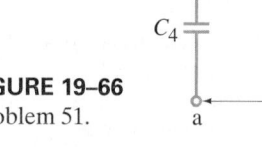

FIGURE 19–66
Problem 51.

19–6 RC Circuits

52. (I) Estimate the value of resistances needed to make a variable timer for intermittent windshield wipers: one wipe every 15 s, 8 s, 4 s, 2 s, 1 s. Assume the capacitor used is on the order of 1 μF. See Fig. 19–67.

FIGURE 19–67
Problem 52.

53. (II) Electrocardiographs are often connected as shown in Fig. 19–68. The lead wires to the legs are said to be capacitively coupled. A time constant of 3.0 s is typical and allows rapid changes in potential to be recorded accurately. If $C = 3.0\,\mu\text{F}$, what value must R have? [*Hint:* Consider each leg as a separate circuit.]

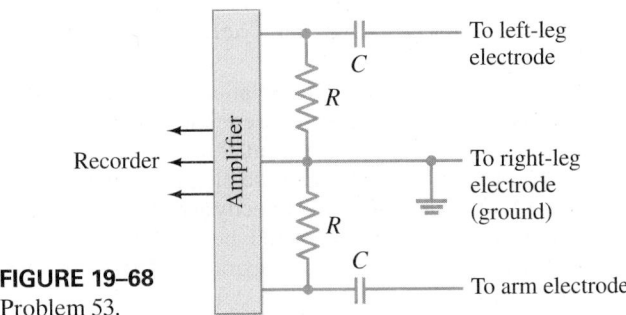

FIGURE 19–68
Problem 53.

54. (II) In Fig. 19–69 (same as Fig. 19–20a), the total resistance is 15.0 kΩ, and the battery's emf is 24.0 V. If the time constant is measured to be 18.0 μs, calculate (*a*) the total capacitance of the circuit and (*b*) the time it takes for the voltage across the resistor to reach 16.0 V after the switch is closed.

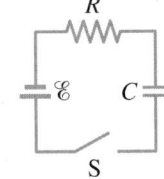

FIGURE 19–69
Problem 54.

55. (II) Two 3.8-μF capacitors, two 2.2-kΩ resistors, and a 16.0-V source are connected in series. Starting from the uncharged state, how long does it take for the current to drop from its initial value to 1.50 mA?

56. (II) The *RC* circuit of Fig. 19–70 (same as Fig. 19–21a) has $R = 8.7\,\text{k}\Omega$ and $C = 3.0\,\mu\text{F}$. The capacitor is at voltage V_0 at $t = 0$, when the switch is closed. How long does it take the capacitor to discharge to 0.25% of its initial voltage?

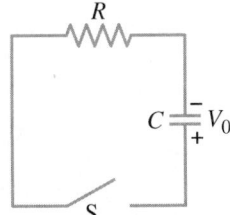

FIGURE 19–70
Problem 56.

57. (III) Consider the circuit shown in Fig. 19–71, where all resistors have the same resistance R. At $t = 0$, with the capacitor C uncharged, the switch is closed. (*a*) At $t = 0$, the three currents can be determined by analyzing a simpler, but equivalent, circuit. Draw this simpler circuit and use it to find the values of I_1, I_2, and I_3 at $t = 0$. (*b*) At $t = \infty$, the currents can be determined by analyzing a simpler, equivalent circuit. Draw this simpler circuit and implement it in finding the values of I_1, I_2, and I_3 at $t = \infty$. (*c*) At $t = \infty$, what is the potential difference across the capacitor?

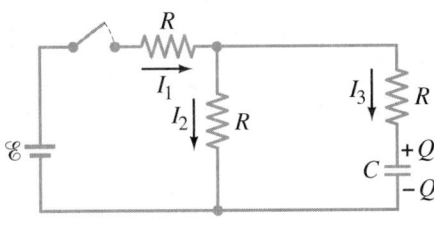

FIGURE 19–71
Problem 57.

58. (III) Two resistors and two uncharged capacitors are arranged as shown in Fig. 19–72. Then a potential difference of 24 V is applied across the combination as shown. (*a*) What is the potential at point a with switch S open? (Let $V = 0$ at the negative terminal of the source.) (*b*) What is the potential at point b with the switch open? (*c*) When the switch is closed, what is the final potential of point b? (*d*) How much charge flows through the switch S after it is closed?

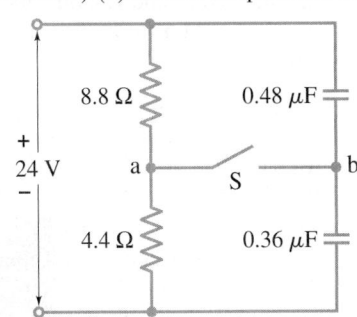

FIGURE 19–72
Problem 58.

19–8 Ammeters and Voltmeters

59. (I) (*a*) An ammeter has a sensitivity of 35,000 Ω/V. What current in the galvanometer produces full-scale deflection? (*b*) What is the resistance of a voltmeter on the 250-V scale if the meter sensitivity is 35,000 Ω/V?

60. (II) An ammeter whose internal resistance is 53 Ω reads 5.25 mA when connected in a circuit containing a battery and two resistors in series whose values are 720 Ω and 480 Ω. What is the actual current when the ammeter is absent?

61. (II) A milliammeter reads 35 mA full scale. It consists of a 0.20-Ω resistor in parallel with a 33-Ω galvanometer. How can you change this ammeter to a voltmeter giving a full-scale reading of 25 V without taking the ammeter apart? What will be the sensitivity (Ω/V) of your voltmeter?

62. (II) A galvanometer has an internal resistance of 32 Ω and deflects full scale for a 55-μA current. Describe how to use this galvanometer to make (*a*) an ammeter to read currents up to 25 A, and (*b*) a voltmeter to give a full-scale deflection of 250 V.

63. (III) A battery with $\mathscr{E} = 12.0$ V and internal resistance $r = 1.0\,\Omega$ is connected to two 7.5-kΩ resistors in series. An ammeter of internal resistance 0.50 Ω measures the current, and at the same time a voltmeter with internal resistance 15 kΩ measures the voltage across one of the 7.5-kΩ resistors in the circuit. What do the ammeter and voltmeter read? What is the % "error" from the current and voltage *without* meters?

64. (III) What internal resistance should the voltmeter of Example 19–17 have to be in error by less than 5%?

65. (III) Two 9.4-kΩ resistors are placed in series and connected to a battery. A voltmeter of sensitivity 1000 Ω/V is on the 3.0-V scale and reads 1.9 V when placed across either resistor. What is the emf of the battery? (Ignore its internal resistance.)

66. (III) When the resistor R in Fig. 19–73 is 35 Ω, the high-resistance voltmeter reads 9.7 V. When R is replaced by a 14.0-Ω resistor, the voltmeter reading drops to 8.1 V. What are the emf and internal resistance of the battery?

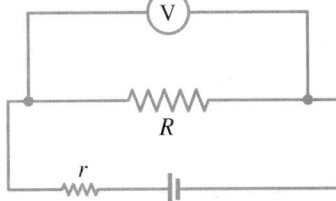

FIGURE 19–73
Problem 66.

General Problems

67. Suppose that you wish to apply a 0.25-V potential difference between two points on the human body. The resistance is about 1800 Ω, and you only have a 1.5-V battery. How can you connect up one or more resistors to produce the desired voltage?

68. A **three-way lightbulb** can produce 50 W, 100 W, or 150 W, at 120 V. Such a bulb contains two filaments that can be connected to the 120 V individually or in parallel (Fig. 19–74). (*a*) Describe how the connections to the two filaments are made to give each of the three wattages. (*b*) What must be the resistance of each filament?

FIGURE 19–74
Problem 68.

69. What are the values of effective capacitance which can be obtained by connecting four identical capacitors, each having a capacitance *C*?

70. Electricity can be a hazard in hospitals, particularly to patients who are connected to electrodes, such as an ECG. Suppose that the motor of a motorized bed shorts out to the bed frame, and the bed frame's connection to a ground has broken (or was not there in the first place). If a nurse touches the bed and the patient at the same time, the nurse becomes a conductor and a complete circuit can be made through the patient to ground through the ECG apparatus. This is shown schematically in Fig. 19–75. Calculate the current through the patient.

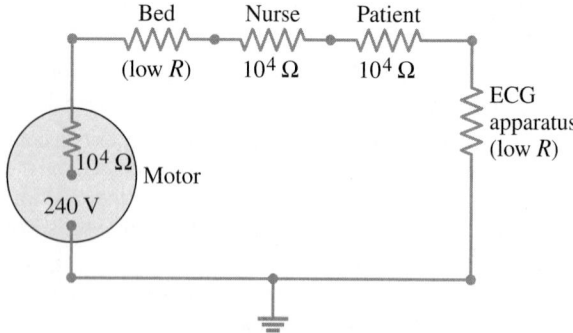

FIGURE 19–75 Problem 70.

71. A heart pacemaker is designed to operate at 72 beats/min using a 6.5-μF capacitor in a simple *RC* circuit. What value of resistance should be used if the pacemaker is to fire (capacitor discharge) when the voltage reaches 75% of maximum and then drops to 0 V (72 times a minute)?

72. Suppose that a person's body resistance is 950 Ω (moist skin). (*a*) What current passes through the body when the person accidentally is connected to 120 V? (*b*) If there is an alternative path to ground whose resistance is 25 Ω, what then is the current through the body? (*c*) If the voltage source can produce at most 1.5 A, how much current passes through the person in case (*b*)?

73. One way a multiple-speed ventilation fan for a car can be designed is to put resistors in series with the fan motor. The resistors reduce the current through the motor and make it run more slowly. Suppose the current in the motor is 5.0 A when it is connected directly across a 12-V battery. (*a*) What series resistor should be used to reduce the current to 2.0 A for low-speed operation? (*b*) What power rating should the resistor have? Assume that the motor's resistance is roughly the same at all speeds.

74. A **Wheatstone bridge** is a type of "bridge circuit" used to make measurements of resistance. The unknown resistance to be measured, R_x, is placed in the circuit with accurately known resistances R_1, R_2, and R_3 (Fig. 19–76). One of these, R_3, is a variable resistor which is adjusted so that when the switch is closed momentarily, the ammeter Ⓐ shows zero current flow. The bridge is then said to be balanced. (*a*) Determine R_x in terms of R_1, R_2, and R_3. (*b*) If a Wheatstone bridge is "balanced" when $R_1 = 590\ \Omega$, $R_2 = 972\ \Omega$, and $R_3 = 78.6\ \Omega$, what is the value of the unknown resistance?

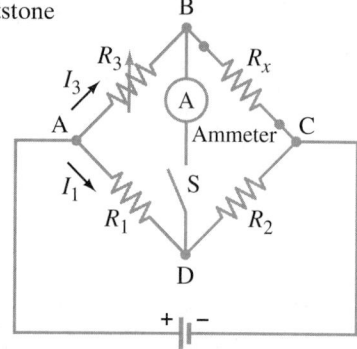

FIGURE 19–76
Problem 74.
Wheatstone bridge.

75. The internal resistance of a 1.35-V mercury cell is 0.030 Ω, whereas that of a 1.5-V dry cell is 0.35 Ω. Explain why three mercury cells can more effectively power a 2.5-W hearing aid that requires 4.0 V than can three dry cells.

76. How many $\frac{1}{2}$-W resistors, each of the same resistance, must be used to produce an equivalent 3.2-kΩ, 3.5-W resistor? What is the resistance of each, and how must they be connected? Do not exceed $P = \frac{1}{2}$ W in each resistor.

77. A **solar cell**, 3.0 cm square, has an output of 350 mA at 0.80 V when exposed to full sunlight. A solar panel that delivers close to 1.3 A of current at an emf of 120 V to an external load is needed. How many cells will you need to create the panel? How big a panel will you need, and how should you connect the cells to one another?

78. The current through the 4.0-kΩ resistor in Fig. 19–77 is 3.10 mA. What is the terminal voltage V_{ba} of the "unknown" battery? (There are two answers. Why?)

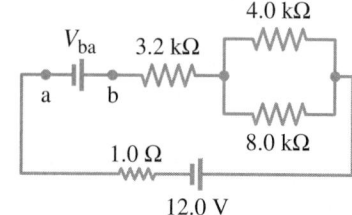

FIGURE 19–77
Problem 78.

79. A power supply has a fixed output voltage of 12.0 V, but you need $V_T = 3.5$ V output for an experiment. (a) Using the voltage divider shown in Fig. 19–78, what should R_2 be if R_1 is 14.5 Ω? (b) What will the terminal voltage V_T be if you connect a load to the 3.5-V output, assuming the load has a resistance of 7.0 Ω?

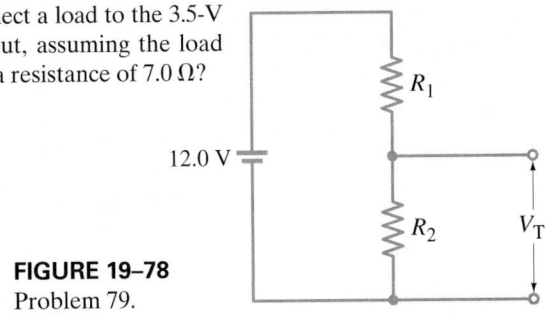

FIGURE 19–78
Problem 79.

80. A battery produces 40.8 V when 8.40 A is drawn from it, and 47.3 V when 2.80 A is drawn. What are the emf and internal resistance of the battery?

81. In the circuit shown in Fig. 19–79, the 33-Ω resistor dissipates 0.80 W. What is the battery voltage?

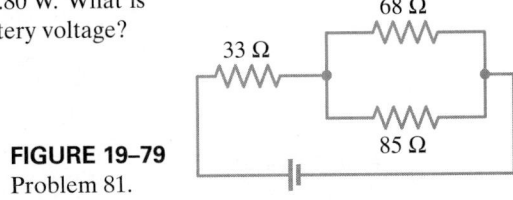

FIGURE 19–79
Problem 81.

82. For the circuit shown in Fig. 19–80, determine (a) the current through the 16-V battery and (b) the potential difference between points a and b, $V_a - V_b$.

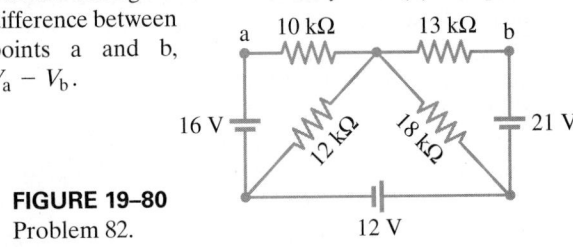

FIGURE 19–80
Problem 82.

83. The current through the 20-Ω resistor in Fig. 19–81 does not change whether the two switches S_1 and S_2 are both open or both closed. Use this clue to determine the value of the unknown resistance R.

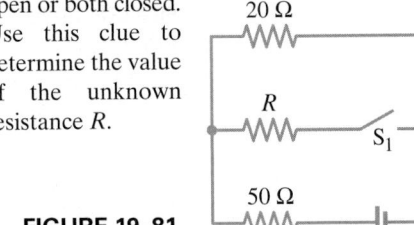

FIGURE 19–81
Problem 83.

84. (a) What is the equivalent resistance of the circuit shown in Fig. 19–82? [Hint: Redraw the circuit to see series and parallel better.] (b) What is the current in the 14-Ω resistor? (c) What is the current in the 12-Ω resistor? (d) What is the power dissipation in the 4.5-Ω resistor?

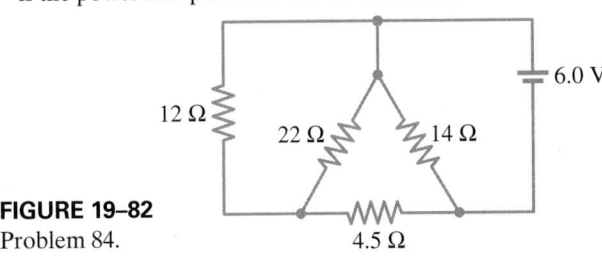

FIGURE 19–82
Problem 84.

85. (a) A voltmeter and an ammeter can be connected as shown in Fig. 19–83a to measure a resistance R. If V is the voltmeter reading, and I is the ammeter reading, the value of R will not quite be V/I (as in Ohm's law) because some current goes through the voltmeter. Show that the actual value of R is

$$\frac{1}{R} = \frac{I}{V} - \frac{1}{R_V},$$

where R_V is the voltmeter resistance. Note that $R \approx V/I$ if $R_V \gg R$. (b) A voltmeter and an ammeter can also be connected as shown in Fig. 19–83b to measure a resistance R. Show in this case that

$$R = \frac{V}{I} - R_A,$$

where V and I are the voltmeter and ammeter readings and R_A is the resistance of the ammeter. Note that $R \approx V/I$ if $R_A \ll R$.

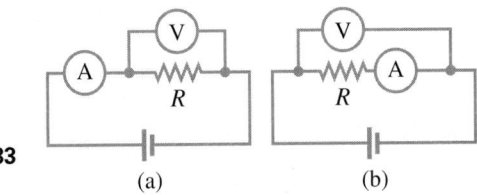

FIGURE 19–83
Problem 85.   (a)   (b)

86. The circuit shown in Fig. 19–84 uses a neon-filled tube as in Fig. 19–23a. This neon lamp has a threshold voltage V_0 for conduction, because no current flows until the neon gas in the tube is ionized by a sufficiently strong electric field. Once the threshold voltage is exceeded, the lamp has negligible resistance. The capacitor stores electrical energy, which can be released to flash the lamp. Assume that $C = 0.150\ \mu$F, $R = 2.35 \times 10^6\ \Omega$, $V_0 = 90.0$ V, and $\mathscr{E} = 105$ V. (a) Assuming the circuit is hooked up to the emf at time $t = 0$, at what time will the light first flash? (b) If the value of R is increased, will the time you found in part (a) increase or decrease? (c) The flashing of the lamp is very brief. Why? (d) Explain what happens after the lamp flashes for the first time.

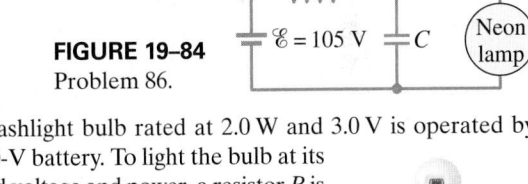

FIGURE 19–84
Problem 86.

87. A flashlight bulb rated at 2.0 W and 3.0 V is operated by a 9.0-V battery. To light the bulb at its rated voltage and power, a resistor R is connected in series as shown in Fig. 19–85. What value should the resistor have?

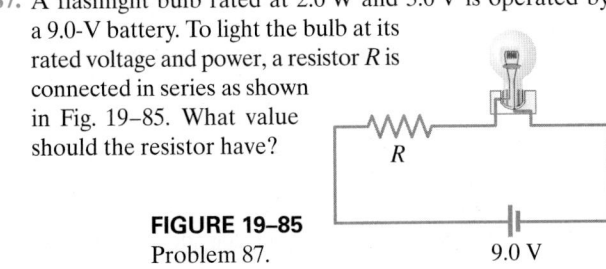

FIGURE 19–85
Problem 87.   9.0 V

88. In Fig. 19–86, let $V = 10.0$ V and $C_1 = C_2 = C_3 = 25.4\ \mu$F. How much energy is stored in the capacitor network (a) as shown, (b) if the capacitors were all in series, and (c) if the capacitors were all in parallel?

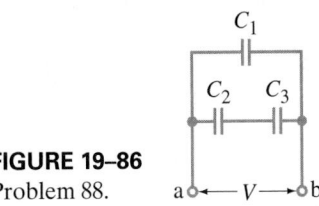

FIGURE 19–86
Problem 88.   a $\circ \!\!\leftarrow\!\! V \!\!\rightarrow\!\! \circ$ b

89. A 12.0-V battery, two resistors, and two capacitors are connected as shown in Fig. 19–87. After the circuit has been connected for a long time, what is the charge on each capacitor?

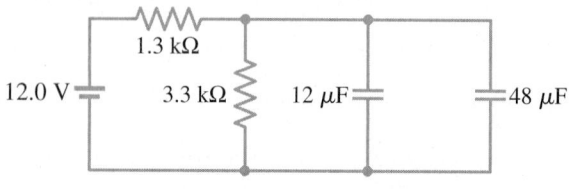

FIGURE 19–87 Problem 89.

90. Determine the current in each resistor of the circuit shown in Fig. 19–88.

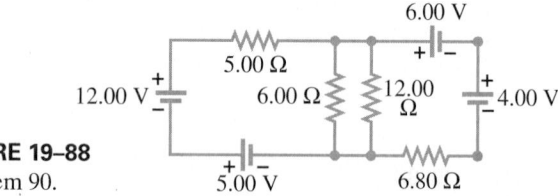

FIGURE 19–88
Problem 90.

91. How much energy must a 24-V battery expend to charge a 0.45-μF and a 0.20-μF capacitor fully when they are placed (a) in parallel, (b) in series? (c) How much charge flowed from the battery in each case?

92. Two capacitors, $C_1 = 2.2\,\mu$F and $C_2 = 1.2\,\mu$F, are connected in parallel to a 24-V source as shown in Fig. 19–89a. After they are charged they are disconnected from the source and from each other, and then reconnected directly to each other with plates of opposite sign connected together (see Fig. 19–89b). Find the charge on each capacitor and the potential across each after equilibrium is established (Fig. 19–89c).

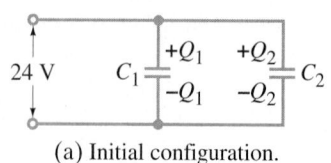

(a) Initial configuration.

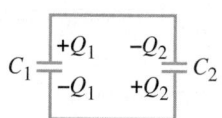

(b) At the instant of reconnection only.

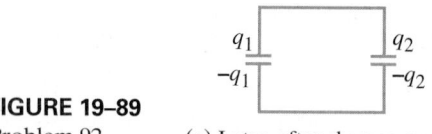

FIGURE 19–89
Problem 92.

(c) Later, after charges move.

93. The switch S in Fig. 19–90 is connected downward so that capacitor C_2 becomes fully charged by the battery of voltage V_0. If the switch is then connected upward, determine the charge on each capacitor after the switching.

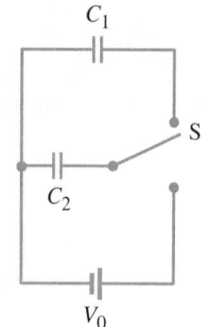

FIGURE 19–90
Problem 93.

94. The performance of the starter circuit in a car can be significantly degraded by a small amount of corrosion on a battery terminal. Figure 19–91a depicts a properly functioning circuit with a battery (12.5-V emf, 0.02-Ω internal resistance) attached via corrosion-free cables to a starter motor of resistance $R_S = 0.15\,\Omega$. Sometime later, corrosion between a battery terminal and a starter cable introduces an extra series resistance of only $R_C = 0.10\,\Omega$ into the circuit as suggested in Fig. 19–91b. Let P_0 be the power delivered to the starter in the circuit free of corrosion, and let P be the power delivered to the circuit with corrosion. Determine the ratio P/P_0.

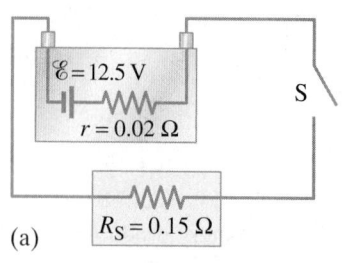

(a)

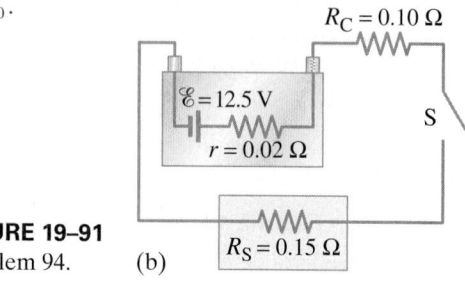

FIGURE 19–91
Problem 94.

(b)

95. The variable capacitance of an old radio tuner consists of four plates connected together placed alternately between four other plates, also connected together (Fig. 19–92). Each plate is separated from its neighbor by 1.6 mm of air. One set of plates can move so that the area of overlap of each plate varies from 2.0 cm^2 to 9.0 cm^2. (a) Are these seven capacitors connected in series or in parallel? (b) Determine the range of capacitance values.

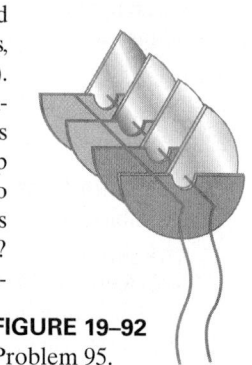

FIGURE 19–92
Problem 95.

96. A 175-pF capacitor is connected in series with an unknown capacitor, and as a series combination they are connected to a 25.0-V battery. If the 175-pF capacitor stores 125 pC of charge on its plates, what is the unknown capacitance?

97. In the circuit shown in Fig. 19–93, $C_1 = 1.0\,\mu$F, $C_2 = 2.0\,\mu$F, $C_3 = 2.4\,\mu$F, and a voltage $V_{ab} = 24$ V is applied across points a and b. After C_1 is fully charged, the switch is thrown to the right. What is the final charge and potential difference on each capacitor?

FIGURE 19–93
Problem 97.

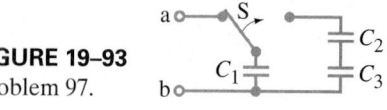

Search and Learn

1. Compare the formulas for resistors and for capacitors when connected in series and in parallel by filling in the Table below. Discuss and explain the differences. Consider the role of voltage V.

	R_{eq}	C_{eq}
Series		
Parallel		

2. Fill in the Table below for a combination of two unequal resistors of resistance R_1 and R_2. Assume the electric potential on the low-voltage end of the combination is V_A volts and the potential at the high-voltage end of the combination is V_B volts. First draw diagrams.

Property	Resistors in Series	Resistors in Parallel
Equivalent resistance		
Current through equivalent resistance		
Voltage across equivalent resistance		
Voltage across the pair of resistors		
Voltage across each resistor	$V_1 =$ $V_2 =$	$V_1 =$ $V_2 =$
Voltage at a point between the resistors		Not applicable
Current through each resistor	$I_1 =$ $I_2 =$	$I_1 =$ $I_2 =$

3. Cardiac defibrillators are discussed in Section 17–9. (a) Choose a value for the resistance so that the 1.0-μF capacitor can be charged to 3000 V in 2.0 seconds. Assume that this 3000 V is 95% of the full source voltage. (b) The effective resistance of the human body is given in Section 19–7. If the defibrillator discharges with a time constant of 10 ms, what is the effective capacitance of the human body?

4. A **potentiometer** is a device to precisely measure potential differences or emf, using a **null** technique. In the simple potentiometer circuit shown in Fig. 19–94, R' represents the total resistance of the resistor from A to B (which could be a long uniform "slide" wire), whereas R represents the resistance of only the part from A to the movable contact at C. When the unknown emf to be measured, $\mathcal{E}_x$, is placed into the circuit as shown, the movable contact C is moved until the galvanometer G gives a null reading (i.e., zero) when the switch S is closed. The resistance between A and C for this situation we call R_x. Next, a standard emf, $\mathcal{E}_s$, which is known precisely, is inserted into the circuit in place of $\mathcal{E}_x$ and again the contact C is moved until zero current flows through the galvanometer when the switch S is closed. The resistance between A and C now is called R_s. Show that the unknown emf is given by

$$\mathcal{E}_x = \left(\frac{R_x}{R_s} \right) \mathcal{E}_s$$

where R_x, R_s, and $\mathcal{E}_s$ are all precisely known. The working battery is assumed to be fresh and to give a constant voltage.

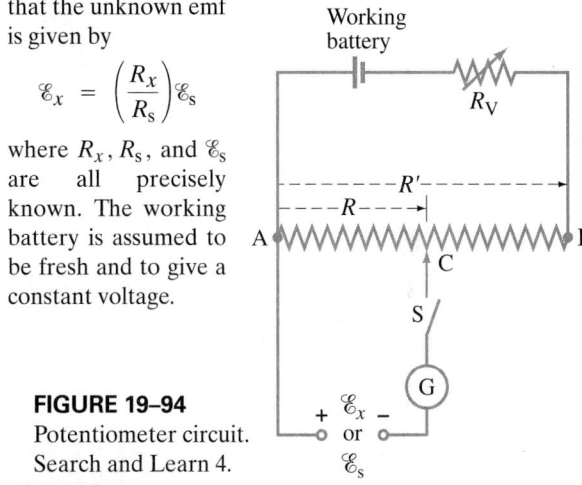

FIGURE 19–94
Potentiometer circuit.
Search and Learn 4.

5. The circuit shown in Fig. 19–95 is a primitive 4-bit **digital-to-analog converter** (**DAC**). In this circuit, to represent each digit (2^n) of a binary number, a "1" has the nth switch closed whereas zero ("0") has the switch open. For example, 0010 is represented by closing switch $n = 1$, while all other switches are open. Show that the voltage V across the 1.0-Ω resistor for the binary numbers 0001, 0010, 0100, and 1001 (which in decimal represent 1, 2, 4, 9) follows the pattern that you expect for a 4-bit DAC. (Section 17–10 may help.)

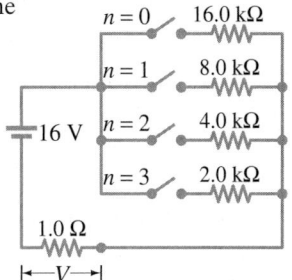

FIGURE 19–95
Search and Learn 5.

ANSWERS TO EXERCISES

A: 6 Ω and 25 Ω.
B: (b).
C: (a) 60-W bulb; (b) 100-W bulb. [Can you explain why? In (a), recall $P = I^2R$.]
D: $41I_3 - 45 + 21I_2 - 80 = 0$.
E: 180 A; this high current through the batteries could cause them to become very hot (and dangerous—possibly exploding): the power dissipated in the weak battery would be $P = I^2r = (180 \text{ A})^2(0.10 \ \Omega) = 3200$ W!

F: (e).
G: ≈ 500 kΩ.

Magnets produce magnetic fields, but so do electric currents. Compass needles are magnets, and they align along the direction of any magnetic field present. Here, the compasses show the presence (and direction) of a magnetic field near a current-carrying wire. We shall see in this Chapter how magnetic field is defined, and how magnetic fields exert forces on electric currents and on charged particles. We also discuss useful applications of the interaction between magnetic fields and electric currents and moving electric charges, such as motors and loudspeakers.

20 CHAPTER

CONTENTS

Magnetism

CHAPTER-OPENING QUESTION—Guess now!

Which of the following can experience a force when placed in the magnetic field of a magnet?

(a) An electric charge at rest.

(b) An electric charge moving.

(c) An electric current in a wire.

(d) Another magnet.

The history of magnetism began thousands of years ago, when in a region of Asia Minor known as Magnesia, rocks were found that could attract each other. These rocks were called "magnets" after their place of discovery.

Not until the nineteenth century, however, was it seen that magnetism and electricity are closely related. A crucial discovery was that electric currents produce magnetic effects (we will say "magnetic fields") like magnets do. All kinds of practical devices depend on magnetism, from compasses to motors, loudspeakers, computer memory, and electric generators.

20–1 Magnets and Magnetic Fields

You probably have observed a magnet attract paper clips, nails, and other objects made of iron, as in Fig. 20–1. Any magnet, whether it is in the shape of a bar or a horseshoe, has two ends or faces, called **poles**, which is where the magnetic effect is strongest. If a bar magnet is suspended from a fine thread, it is found that one pole of the magnet will always point toward the north. It is not known for sure when this fact was discovered, but it is known that the Chinese were making use of it as an aid to navigation by the eleventh century and perhaps earlier.

This is the principle of a compass. A compass needle is simply a bar magnet which is supported at its center of gravity so that it can rotate freely. The pole of a freely suspended magnet that points toward geographic north is called the **north pole** of the magnet. The other pole points toward the south and is called the **south pole**.

It is a familiar observation that when two magnets are brought near one another, each exerts a force on the other. The force can be either attractive or repulsive and can be felt even when the magnets don't touch. If the north pole of one bar magnet is brought near the north pole of a second magnet, the force is repulsive. Similarly, if the south poles are brought close, the force is repulsive. But when the north pole of one magnet is brought near the south pole of another magnet, the force is attractive. These results are shown in Fig. 20–2, and are reminiscent of the forces between electric charges: like poles repel, and unlike poles attract. But *do not confuse magnetic poles with electric charge.* They are very different. One important difference is that a positive or negative electric charge can easily be isolated. But an isolated single magnetic pole has never been observed. If a bar magnet is cut in half, you do not obtain isolated north and south poles. Instead, two new magnets are produced, Fig. 20–3, each with north (N) and south (S) poles. If the cutting operation is repeated, more magnets are produced, each with a north and a south pole. Physicists have searched for isolated single magnetic poles (monopoles), but no **magnetic monopole** has ever been observed.

Besides iron, a few other materials, such as cobalt, nickel, gadolinium, and some of their oxides and alloys, show strong magnetic effects. They are said to be **ferromagnetic** (from the Latin word *ferrum* for iron). Other materials show some slight magnetic effect, but it is very weak and can be detected only with delicate instruments. We will look in more detail at ferromagnetism in Section 20–12.

In Chapter 16, we used the concept of an electric field surrounding an electric charge. In a similar way, we can picture a **magnetic field** surrounding a magnet. The force one magnet exerts on another can then be described as the interaction between one magnet and the magnetic field of the other. Just as we drew electric field lines, we can also draw **magnetic field lines**. They can be drawn, as for electric field lines, so that

1. the direction of the magnetic field is tangent to a field line at any point, and
2. the number of lines per unit area is proportional to the strength of the magnetic field.

The *direction* of the magnetic field at a given location can be defined as the direction that the north pole of a compass needle would point if placed at that location. (We will give a more precise definition of magnetic field shortly.) Figure 20–4a shows how thin iron filings (acting like tiny magnets) reveal the magnetic field lines by lining up like the compass needles. The magnetic field determined in this way for the field surrounding a bar magnet is shown in Fig. 20–4b. Notice that because of our definition, the lines always point out from the north pole and in toward the south pole of a magnet (the north pole of a magnetic compass needle is attracted to the south pole of the magnet).

Magnetic field lines continue inside a magnet, as indicated in Fig. 20–4b. Indeed, given the lack of single magnetic poles, magnetic field lines always form closed loops, unlike electric field lines that begin on positive charges and end on negative charges.

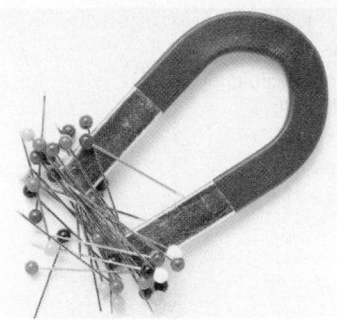

FIGURE 20–1 A horseshoe magnet attracts pins made of iron.

FIGURE 20–2 Like poles of two magnets repel; unlike poles attract.

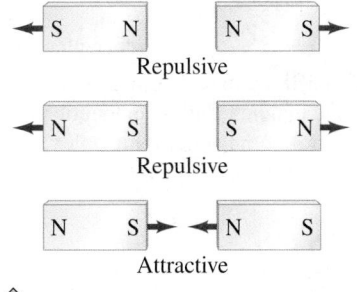

C A U T I O N
Magnets do not attract all metals

FIGURE 20–3 If you split a magnet, you won't get isolated north and south poles; instead, two new magnets are produced, each with a north and a south pole.

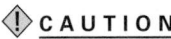
C A U T I O N
Magnetic field lines form closed loops, unlike electric field lines

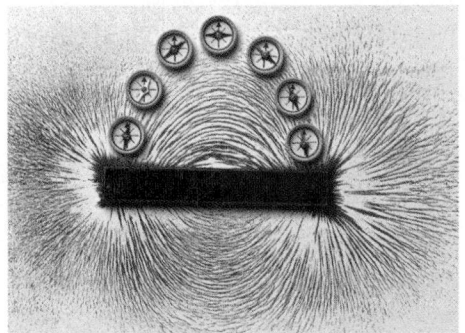

(a)

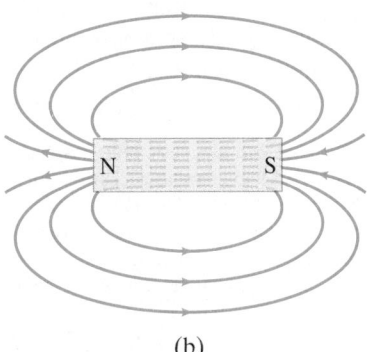

(b)

FIGURE 20–4 (a) Visualizing magnetic field lines around a bar magnet, using iron filings and compass needles. The red end of the bar magnet is its north pole. The N pole of a nearby compass needle points away from the north pole of the magnet. (b) Diagram of magnetic field lines for a bar magnet.

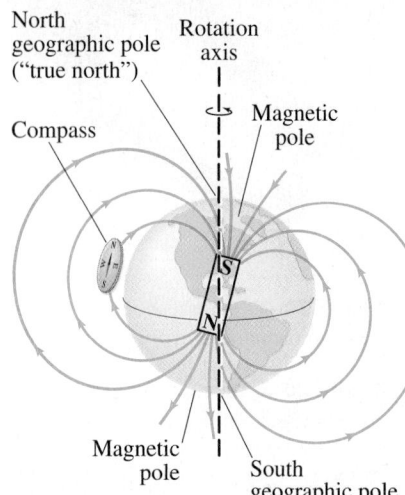

North geographic pole ("true north")

Rotation axis

Magnetic pole

Compass

Magnetic pole

South geographic pole

FIGURE 20–5 The Earth acts like a huge magnet. But its magnetic poles are not at the geographic poles (on the Earth's rotation axis).

Earth's Magnetic Field

The Earth's magnetic field is shown in Fig. 20–5, and is thought to be produced by electric currents in the Earth's molten iron outer core. The pattern of field lines is almost as though there were an imaginary bar magnet inside the Earth. Since the north pole (N) of a compass needle points north, the Earth's **magnetic pole** which is in the geographic north is magnetically a south pole, as indicated in Fig. 20–5 by the S on the schematic bar magnet inside the Earth. Remember that the north pole of one magnet is attracted to the south pole of another magnet. Nonetheless, Earth's pole in the north is still often called the "north magnetic pole," or "geomagnetic north," simply because it is in the north. Similarly, the Earth's southern magnetic pole, which is near the geographic south pole, is magnetically a north pole (N). The Earth's magnetic poles do not coincide with the **geographic poles**, which are on the Earth's axis of rotation. The north magnetic pole, for example, is in the Canadian Arctic, now on the order of 1000 km[†] from the geographic north pole, or **true north**. This difference must be taken into account for accurate use of a compass (Fig. 20–6). The angular difference between the direction of a compass needle (which points along the magnetic field lines) at any location and true (geographical) north is called the **magnetic declination**. In the U.S. it varies from 0° to about 20°, depending on location.

Notice in Fig. 20–5 that the Earth's magnetic field at most locations is not tangent to the Earth's surface. The angle that the Earth's magnetic field makes with the horizontal at any point is referred to as the **angle of dip**, or the "inclination." It is 67° at New York, for example, and 55° at Miami.

EXERCISE A Does the Earth's magnetic field have a greater magnitude near the poles or near the equator? [How can you tell using the field lines in Fig. 20–5?]

PHYSICS APPLIED

Use of a compass

FIGURE 20–6 Using a map and compass in the wilderness. First you align the compass case so the needle points away from true north (N) exactly the number of degrees of declination stated on the map (for this topographic map, it is 17° as shown just to the left of the compass). Then align the map with true north, as shown, *not* with the compass needle. [This is an old map (1953) of a part of California; on new maps (2012) the declination is only 13°, telling us the position of magnetic north has moved—see footnote below.]

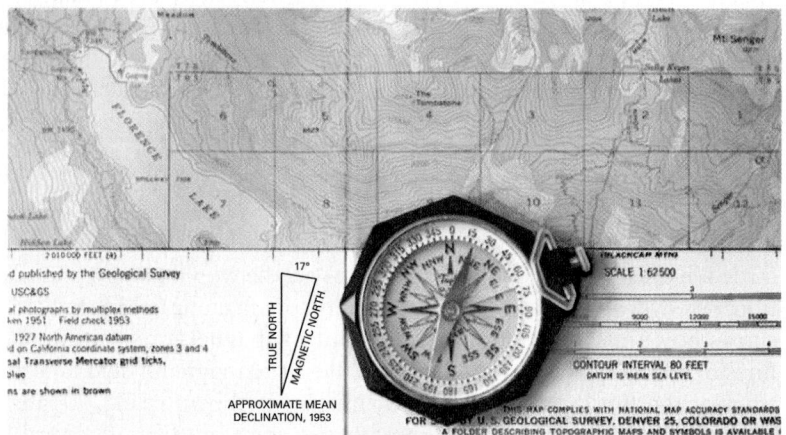

Uniform Magnetic Field

The simplest magnetic field is one that is uniform—it doesn't change in magnitude or direction from one point to another. A perfectly uniform field over a large area is not easy to produce. But the field between two flat parallel pole pieces of a magnet is nearly uniform if the area of the pole faces is large compared to their separation, as shown in Fig. 20–7. At the edges, the field "fringes" out somewhat: the magnetic field lines are no longer quite parallel and uniform. The parallel evenly spaced field lines in the central region of the gap indicate that the field is uniform at points not too near the edges, much like the electric field between two parallel plates (Fig. 17–1).

FIGURE 20–7 Magnetic field between two wide poles of a magnet is nearly uniform, except near the edges.

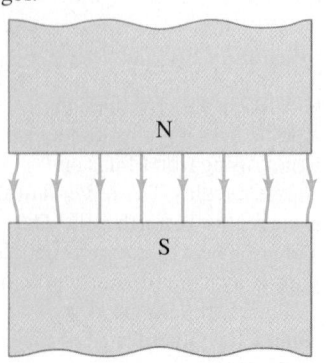

[†]Earth's north magnetic pole has been moving over time, on the order of 10 km per year in recent decades. Magnetism in rocks solidified at various times in the past (age determined by radioactive dating—see Section 30–11) suggests that Earth's magnetic poles have not only moved significantly over geologic time, but have also reversed direction 400 times over the last 330 million years. Also note that a compass gives a false reading if you are standing on rock containing magnetized iron ore (as you move around, the compass needle is inconsistent).

(a)

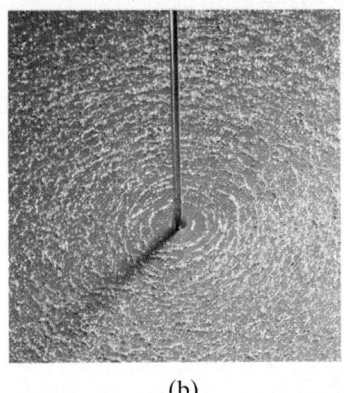
(b)

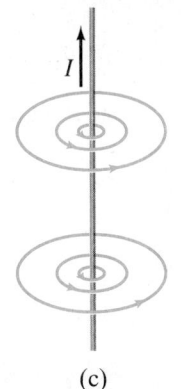
(c)

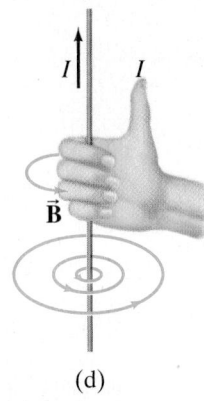
(d)

FIGURE 20–8 (a) Deflection of compass needles near a current-carrying wire, showing the presence and direction of the magnetic field. (b) Iron filings also align along the direction of the magnetic field lines near a straight current-carrying wire. (c) Diagram of the magnetic field lines around an electric current in a straight wire. (d) Right-hand rule for remembering the direction of the magnetic field: when the thumb points in the direction of the conventional current, the fingers wrapped around the wire point in the direction of the magnetic field. ($\vec{B}$ is the symbol for magnetic field.)

20–2 Electric Currents Produce Magnetic Fields

During the eighteenth century, many scientists sought to find a connection between electricity and magnetism. A stationary electric charge and a magnet were shown to have no influence on each other. But in 1820, Hans Christian Oersted (1777–1851) found that when a compass is placed near a wire, the compass needle deflects if (and only if) the wire carries an electric current. As we have seen, a compass needle is deflected by a magnetic field. So Oersted's experiment showed that **an electric current produces a magnetic field**. He had found a connection between electricity and magnetism.

A compass needle placed near a straight section of current-carrying wire experiences a force, causing the needle to align tangent to a circle around the wire, Fig. 20–8a. Thus, the magnetic field lines produced by a current in a straight wire are in the form of circles with the wire at their center, Figs. 20–8b and c. The direction of these lines is indicated by the north pole of the compasses in Fig. 20–8a. There is a simple way to remember the direction of the magnetic field lines in this case. It is called a **right-hand rule**: grasp the wire with your right hand so that your thumb points in the direction of the conventional (positive) current; then your fingers will encircle the wire in the direction of the magnetic field, Fig. 20–8d.

Right-Hand-Rule-1:
Magnetic field direction produced by electric current

The magnetic field lines due to a circular loop of current-carrying wire can be determined in a similar way by placing a compass at various locations near the loop. The result is shown in Fig. 20–9. Again the right-hand rule can be used, as shown in Fig. 20–10. Unlike the uniform field shown in Fig. 20–7, the magnetic fields shown in Figs. 20–8 and 20–9 are *not* uniform—the fields are different in magnitude and direction at different locations.

FIGURE 20–9 Magnetic field lines due to a circular loop of wire.

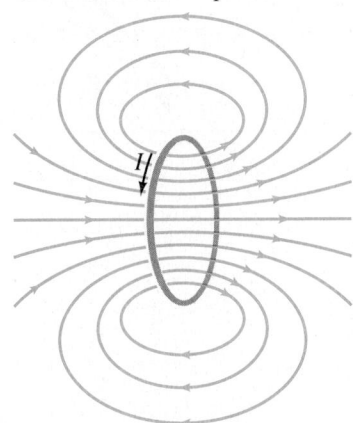

EXERCISE B A straight wire carries a current directly toward you. In what direction are the magnetic field lines surrounding the wire?

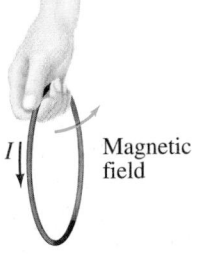

FIGURE 20–10 Right-hand rule for determining the direction of the magnetic field relative to the current in a loop of wire.

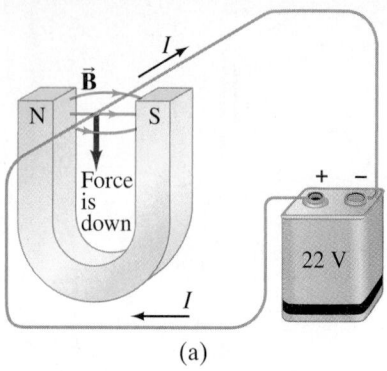

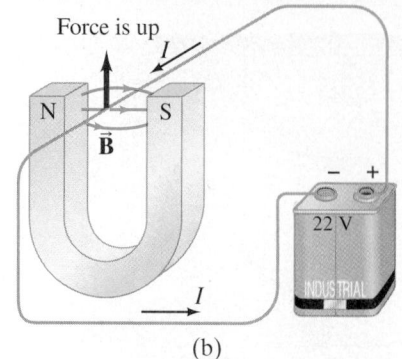

Force is up

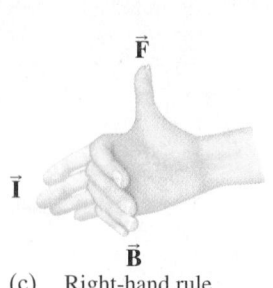

(a)

(b)

(c)   Right-hand rule

FIGURE 20–11 (a) Force on a current-carrying wire placed in a magnetic field $\vec{B}$; (b) same, but current reversed; (c) right-hand rule for setup in (b), with current $\vec{I}$ shown as if a vector with direction.

20–3 Force on an Electric Current in a Magnetic Field; Definition of $\vec{B}$

In Section 20–2 we saw that an electric current exerts a force on a magnet, such as a compass needle. By Newton's third law, we might expect the reverse to be true as well: we should expect that *a magnet exerts a force on a current-carrying wire*. Experiments indeed confirm this effect, and it too was first observed by Oersted.

Suppose a straight wire is placed in the magnetic field between the poles of a horseshoe magnet as shown in Fig. 20–11, where the vector symbol $\vec{B}$ represents the magnitude and direction of the magnetic field. When a current flows in the wire, experiment shows that a force is exerted on the wire. But this force is *not* toward one or the other pole of the magnet. Instead, the force is directed at right angles to the magnetic field direction, downward in Fig. 20–11a. If the current is reversed in direction, the force is in the opposite direction, upward as shown in Fig. 20–11b. Experiments show that *the direction of the force is always perpendicular to the direction of the current and also perpendicular to the direction of the magnetic field, $\vec{B}$.*

Right-Hand-Rule-2:
Force on current exerted by $\vec{B}$

The direction of the force is given by another **right-hand rule**, as illustrated in Fig. 20–11c. Orient your right hand until your outstretched fingers can point in the direction of the conventional current I, and when you bend your fingers they point in the direction of the magnetic field lines, $\vec{B}$. Then your outstretched thumb will point in the direction of the force $\vec{F}$ on the wire.

This right-hand rule describes the direction of the force. What about the magnitude of the force on the wire? It is found experimentally that the magnitude of the force is directly proportional to the current I in the wire, to the magnetic field B (assumed uniform), and to the length ℓ of wire exposed to the magnetic field. The force also depends on the angle θ between the current direction and the magnetic field (Fig. 20–12), being proportional to $\sin \theta$. Thus, the force on a wire carrying a current I with length ℓ in a uniform magnetic field B is given by

$$F \propto I\ell B \sin \theta.$$

FIGURE 20–12 Current-carrying wire in a magnetic field. Force on the wire is directed into the page.

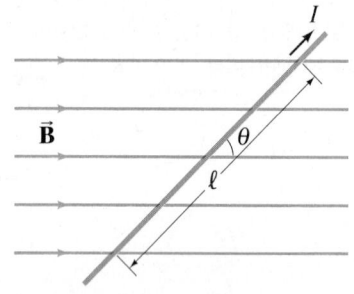

When the current is perpendicular to the field lines ($\theta = 90°$ and $\sin 90° = 1$), the force is strongest. When the wire is parallel to the magnetic field lines ($\theta = 0°$), there is no force at all.

Up to now we have not defined the magnetic field strength precisely. In fact, the magnetic field B can be conveniently defined in terms of the above proportion so that the proportionality constant is precisely 1. Thus we have

$$F = I\ell B \sin \theta. \tag{20–1}$$

If the current's direction is perpendicular to the field $\vec{B}$ ($\theta = 90°$), then the force is

$$F_{\text{max}} = I\ell B. \qquad\qquad \left[\text{current} \perp \vec{B}\right] \tag{20–2}$$

(If B is not uniform, then B in Eqs. 20–1 and 20–2 can be the average field over the length ℓ of the wire.)

The magnitude of $\vec{B}$ can be defined using Eq. 20–2 as $B = F_{\text{max}}/I\ell$, where F_{max} is the magnitude of the force on a straight length ℓ of wire carrying a current I when the wire is perpendicular to $\vec{B}$.

EXERCISE C A wire carrying current I is perpendicular to a magnetic field of strength B. Assuming a fixed length of wire, which of the following changes will result in decreasing the force on the wire by a factor of 2? (*a*) Decrease the angle from 90° to 45°; (*b*) decrease the angle from 90° to 30°; (*c*) decrease the current in the wire to $I/2$; (*d*) decrease the magnetic field to $B/2$; (*e*) none of these will do it.

The SI unit for magnetic field B is the **tesla** (T). From Eq. 20–1 or 20–2, we see that $1\,\text{T} = 1\,\text{N/A}\cdot\text{m}$. An older name for the tesla is the "weber per meter squared" $(1\,\text{Wb/m}^2 = 1\,\text{T})$. Another unit sometimes used to specify magnetic field is a cgs unit, the **gauss** (G): $1\,\text{G} = 10^{-4}\,\text{T}$. A field given in gauss should always be changed to teslas before using with other SI units. To get a "feel" for these units, we note that the magnetic field of the Earth at its surface is about $\frac{1}{2}\,\text{G}$ or $0.5 \times 10^{-4}\,\text{T}$. On the other hand, strong electromagnets can produce fields on the order of 2 T and superconducting magnets can produce over 10 T.

EXAMPLE 20–1 **Magnetic force on a current-carrying wire.** A wire carrying a steady (dc) 30-A current has a length $\ell = 12\,\text{cm}$ between the pole faces of a magnet. The wire is at an angle $\theta = 60°$ to the field (Fig. 20–13). The magnetic field is approximately uniform at 0.90 T. We ignore the field beyond the pole pieces. Determine the magnitude and direction of the force on the wire.

APPROACH We use Eq. 20–1, $F = I\ell B \sin\theta$.

SOLUTION The force F on the 12-cm length of wire within the uniform field B is

$$F = I\ell B \sin\theta = (30\,\text{A})(0.12\,\text{m})(0.90\,\text{T})(\sin 60°) = 2.8\,\text{N}.$$

We use right-hand-rule-2 to find the direction of $\vec{\mathbf{F}}$. Hold your right hand flat, pointing your fingers in the direction of the current. Then bend your fingers (maybe needing to rotate your hand) so they point along $\vec{\mathbf{B}}$, Fig. 20–13. Your thumb then points into the page, which is thus the direction of the force F.

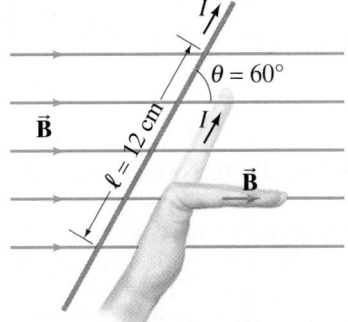

FIGURE 20–13 Example 20–1. For right-hand-rule-2, the thumb points into the page. See Fig. 20–11c.

EXERCISE D A straight power line carries 30 A and is perpendicular to the Earth's magnetic field of $0.50 \times 10^{-4}\,\text{T}$. What magnitude force is exerted on 100 m of this power line?

On a diagram, when we want to represent an electric current or a magnetic field that is pointing out of the page (toward us) or into the page, we use $\odot$ or $\times$, respectively. The $\odot$ is meant to resemble the tip of an arrow pointing directly toward the reader, whereas the $\times$ or $\otimes$ resembles the tail of an arrow pointing away. See Fig. 20–14.

EXAMPLE 20–2 **Measuring a magnetic field.** A rectangular loop of wire hangs vertically as shown in Fig. 20–14. A magnetic field $\vec{\mathbf{B}}$ is directed horizontally, perpendicular to the plane of the loop, and points out of the page as represented by the symbol $\odot$. The magnetic field $\vec{\mathbf{B}}$ is very nearly uniform along the horizontal portion of wire ab (length $\ell = 10.0\,\text{cm}$) which is near the center of the gap of a large magnet producing the field. The top portion of the wire loop is out of the field. The loop hangs from a balance (reads 0 when $B = 0$) which measures a downward magnetic force of $F = 3.48 \times 10^{-2}\,\text{N}$ when the wire carries a current $I = 0.245\,\text{A}$. What is the magnitude of the magnetic field B?

APPROACH Three straight sections of the wire loop are in the magnetic field: a horizontal section and two vertical sections. We apply Eq. 20–1 to each section and use the right-hand rule.

SOLUTION Using right-hand-rule-2 (page 564), we see that the magnetic force on the left vertical section of wire points to the left, and the force on the vertical section on the right points to the right. These two forces are equal and in opposite directions and so add up to zero. Hence, the net magnetic force on the loop is that on the horizontal section ab, whose length is $\ell = 0.100\,\text{m}$. The angle θ between $\vec{\mathbf{B}}$ and the wire is $\theta = 90°$, so $\sin\theta = 1$. Thus Eq. 20–1 gives

$$B = \frac{F}{I\ell} = \frac{3.48 \times 10^{-2}\,\text{N}}{(0.245\,\text{A})(0.100\,\text{m})} = 1.42\,\text{T}.$$

NOTE This technique can be a precise means of determining magnetic field strength.

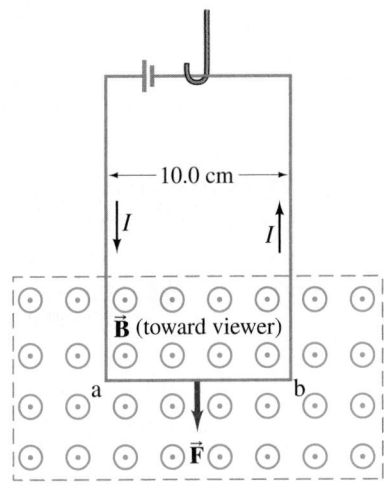

FIGURE 20–14 Measuring a magnetic field $\vec{\mathbf{B}}$. Example 20–2.

20–4 Force on an Electric Charge Moving in a Magnetic Field

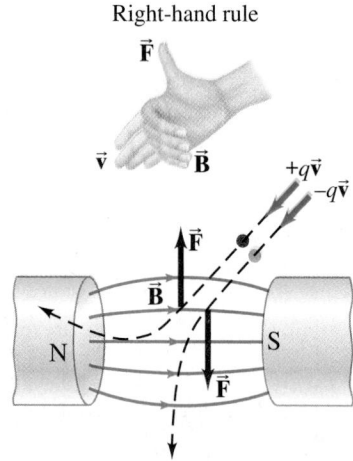

Right-hand rule

$\vec{\mathbf{F}}$

$\vec{\mathbf{v}}$ $\vec{\mathbf{B}}$ $+q\vec{\mathbf{v}}$

$-q\vec{\mathbf{v}}$

$\vec{\mathbf{F}}$

$\vec{\mathbf{B}}$

N   S

$\vec{\mathbf{F}}$

FIGURE 20–15 Force on charged particles due to a magnetic field is perpendicular to the magnetic field direction. If $\vec{\mathbf{v}}$ is horizontal, then $\vec{\mathbf{F}}$ is vertical. The right-hand rule is shown for the force on a positive charge, $+q$.

Right-hand-rule-3:
Force on moving charge exerted by $\vec{\mathbf{B}}$

We have seen that a current-carrying wire experiences a force when placed in a magnetic field. Since a current in a wire consists of moving electric charges, we might expect that freely moving charged particles (not in a wire) would also experience a force when passing through a magnetic field. Free electric charges are not as easy to produce in the lab as a current in a wire, but it can be done, and experiments do show that moving electric charges experience a force in a magnetic field.

From what we already know, we can predict the force on a single electric charge moving in a magnetic field $\vec{\mathbf{B}}$. If N such particles of charge q pass by a given point in time t, they constitute a current $I = Nq/t$. We let t be the time for a charge q to travel a distance ℓ in a magnetic field $\vec{\mathbf{B}}$; then $\ell = vt$ where v is the magnitude of the velocity $\vec{\mathbf{v}}$ of the particle. Thus, the force on these N particles is, by Eq. 20–1, $F = I\ell B \sin\theta = (Nq/t)(vt)B \sin\theta = NqvB \sin\theta$. The force on *one* of the N particles is then

$$F = qvB \sin\theta. \qquad [\theta \text{ between } \vec{\mathbf{v}} \text{ and } \vec{\mathbf{B}}] \quad (20\text{–}3)$$

This equation gives the magnitude of the force exerted by a magnetic field on a particle of charge q moving with velocity v at a point where the magnetic field has magnitude B. The angle between $\vec{\mathbf{v}}$ and $\vec{\mathbf{B}}$ is θ. The force is greatest when the particle moves perpendicular to $\vec{\mathbf{B}}$ ($\theta = 90°$):

$$F_{\text{max}} = qvB. \qquad [\vec{\mathbf{v}} \perp \vec{\mathbf{B}}] \quad (20\text{–}4)$$

The force is *zero* if the particle moves *parallel* to the field lines ($\theta = 0°$). The *direction* of the force is perpendicular to the magnetic field $\vec{\mathbf{B}}$ and to the velocity $\vec{\mathbf{v}}$ of the particle. For a positive charge, the force direction is given by another **right-hand rule**: you orient your right hand so that your outstretched fingers point along the direction of the particle's velocity ($\vec{\mathbf{v}}$), and when you bend your fingers they must point along the direction of $\vec{\mathbf{B}}$. Then your thumb will point in the direction of the force. This is true only for *positively* charged particles, and will be "up" for the positive particle shown in Fig. 20–15. For negatively charged particles, the force is in exactly the opposite direction, "down" in Fig. 20–15.

CONCEPTUAL EXAMPLE 20–3 | **Negative charge near a magnet.** A negative charge $-Q$ is placed at rest near a magnet. Will the charge begin to move? Will it feel a force? What if the charge were positive, $+Q$?

RESPONSE No to all questions. A charge at rest has velocity equal to zero. Magnetic fields exert a force only on moving electric charges (Eq. 20–3).

EXERCISE E Return to the Chapter-Opening Question, page 560, and answer it again now. Try to explain why you may have answered differently the first time.

EXAMPLE 20–4 | **Magnetic force on a proton.** A magnetic field exerts a force of 8.0×10^{-14} N toward the west on a proton moving vertically upward at a speed of 5.0×10^6 m/s (Fig. 20–16a). When moving horizontally in a northerly direction, the force on the proton is zero (Fig. 20–16b). Determine the magnitude and direction of the magnetic field in this region. (The charge on a proton is $q = +e = 1.6 \times 10^{-19}$ C.)

FIGURE 20–16 Example 20–4.

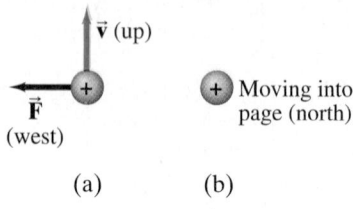

$\vec{\mathbf{v}}$ (up)

$+$

$\vec{\mathbf{F}}$
(west)

$+$ Moving into page (north)

(a)   (b)

APPROACH Since the force on the proton is zero when moving north, the field must be in a north–south direction ($\theta = 0°$ in Eq. 20–3). To produce a force to the west when the proton moves upward, right-hand-rule-3 tells us that $\vec{\mathbf{B}}$ must point toward the north. (Your thumb points west and the outstretched fingers of your right hand point upward only when your bent fingers point north.) The magnitude of $\vec{\mathbf{B}}$ is found using Eq. 20–3.

SOLUTION Equation 20–3 with $\theta = 90°$ gives

$$B = \frac{F}{qv} = \frac{8.0 \times 10^{-14}\,\text{N}}{(1.6 \times 10^{-19}\,\text{C})(5.0 \times 10^6\,\text{m/s})} = 0.10\,\text{T}.$$

EXERCISE F Determine the force on the proton of Example 20–4 if it heads horizontally south.

EXAMPLE 20–5 **ESTIMATE** **Magnetic force on ions during a nerve pulse.**
Estimate the magnitude of the magnetic force due to the Earth's magnetic field on ions crossing a cell membrane during an action potential (Section 18–10). Assume the speed of the ions is $10^{-2}\,\text{m/s}$.

APPROACH Using $F = qvB$, set the magnetic field of the Earth to be roughly $B \approx 10^{-4}\,\text{T}$, and the charge $q \approx e \approx 10^{-19}\,\text{C}$.

SOLUTION $F \approx (10^{-19}\,\text{C})(10^{-2}\,\text{m/s})(10^{-4}\,\text{T}) = 10^{-25}\,\text{N}$.

NOTE This is an extremely small force. Yet it is thought that migrating animals do somehow detect the Earth's magnetic field, and this is an area of active research.

The path of a charged particle moving in a plane perpendicular to a uniform magnetic field is a circle as we shall now show. In Fig. 20–17 the magnetic field is directed *into* the paper, as represented by ×'s. An electron at point P is moving to the right, and the force on it at this point is toward the bottom of the page as shown (use the right-hand rule and reverse the direction for negative charge). The electron is thus deflected toward the page bottom. A moment later, say, when it reaches point Q, the force is still perpendicular to the velocity and is in the direction shown. Because the force is always perpendicular to $\vec{\mathbf{v}}$, the magnitude of $\vec{\mathbf{v}}$ does not change—the electron moves at constant speed. We saw in Chapter 5 that if the force on a particle is always perpendicular to its velocity $\vec{\mathbf{v}}$, the particle moves in a circle and has a centripetal acceleration of magnitude $a = v^2/r$ (Eq. 5–1). Thus a charged particle moves in a circular path with a constant magnitude of centripetal acceleration in a uniform magnetic field (see Fig. 20–18). The electron moves clockwise in Fig. 20–17. A positive particle in this field would feel a force in the opposite direction and would thus move counterclockwise.

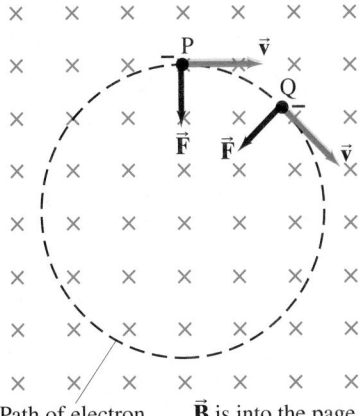

Path of electron   **B** is into the page

FIGURE 20–17 Force exerted by a uniform magnetic field on a moving charged particle (in this case, an electron) produces a circular path.

FIGURE 20–18 The white ring inside the glass tube is the glow of a beam of electrons that ionize the gas molecules. The red coils of current-carrying wire produce a nearly uniform magnetic field, illustrating the circular path of charged particles in a uniform magnetic field.

EXAMPLE 20–6 **Electron's path in a uniform magnetic field.** An electron travels at $1.5 \times 10^7\,\text{m/s}$ in a plane perpendicular to a uniform 0.010-T magnetic field. Describe its path quantitatively. Ignore gravity (= very small in comparison).

APPROACH The electron moves at speed v in a curved path and so must have a centripetal acceleration $a = v^2/r$ (Eq. 5–1). We find the radius of curvature using Newton's second law. The force is given by Eq. 20–3 with $\sin\theta = 1$: $F = qvB$.

SOLUTION We insert F and a into Newton's second law:

$$\Sigma F = ma$$
$$qvB = \frac{mv^2}{r}.$$

We solve for r and find

$$r = \frac{mv}{qB}.$$

Since $\vec{\mathbf{F}}$ is perpendicular to $\vec{\mathbf{v}}$, the magnitude of $\vec{\mathbf{v}}$ doesn't change. From this equation we see that if $\vec{\mathbf{B}} = $ constant, then $r = $ constant, and the curve must be a circle as we claimed above. To get r we put in the numbers:

$$r = \frac{(9.1 \times 10^{-31}\,\text{kg})(1.5 \times 10^7\,\text{m/s})}{(1.6 \times 10^{-19}\,\text{C})(0.010\,\text{T})} = 0.85 \times 10^{-2}\,\text{m} = 8.5\,\text{mm}.$$

NOTE See Fig. 20–18. If the magnetic field B is larger, is the radius larger or smaller?

The time T required for a particle of charge q moving with constant speed v to make one circular revolution in a uniform magnetic field $\vec{\mathbf{B}}$ ($\perp \vec{\mathbf{v}}$) is $T = 2\pi r/v$, where $2\pi r$ is the circumference of its circular path. From Example 20–6, $r = mv/qB$, so

$$T = \frac{2\pi m}{qB}.$$

Since T is the period of rotation, the frequency of rotation is

$$f = \frac{1}{T} = \frac{qB}{2\pi m}. \qquad \textbf{(20–5)}$$

This is often called the **cyclotron frequency** of a particle in a field because this is the frequency at which particles revolve in a cyclotron (see Problem 88).

CONCEPTUAL EXAMPLE 20–7 **Stopping charged particles.** An electric charge q moving in an electric field $\vec{\mathbf{E}}$ can be decelerated to a stop if the force $\vec{\mathbf{F}} = q\vec{\mathbf{E}}$ (Eq. 16–5) acts in the direction opposite to the charge's velocity. Can a magnetic field be used to stop a charged particle?

RESPONSE No, because the force is always *perpendicular* to the velocity of the particle and thus can only change the direction but not the magnitude of its velocity. Also the magnetic force cannot do work on the particle (force and displacement are perpendicular, Eq. 6–1) and so cannot change the kinetic energy of the particle, Eq. 6–4.

Magnetic Fields

Magnetic fields are somewhat analogous to the electric fields of Chapter 16, but there are several important differences to recall:

1. The force experienced by a charged particle moving in a magnetic field is *perpendicular* to the direction of the magnetic field (and to the direction of the velocity of the particle), whereas the force exerted by an electric field is *parallel* to the direction of the field (and independent of the velocity of the particle).

2. The *right-hand rule*, in its different forms, is intended to help you determine the directions of magnetic field, and the forces they exert, and/or the directions of electric current or charged particle velocity. The right-hand rules (Table 20–1) are designed to deal with the "perpendicular" nature of these quantities.

3. The equations in this Chapter are generally not printed as vector equations, but involve magnitudes only. Right-hand rules are to be used to find directions of vector quantities.

TABLE 20–1 Summary of Right-hand Rules (= RHR)

Physical Situation	Example	How to Orient Right Hand	Result
1. Magnetic field produced by current (RHR-1)	I I $\vec{\mathbf{B}}$ **Fig. 20–8d**	Wrap fingers around wire with thumb pointing in direction of current I	Fingers curl in direction of $\vec{\mathbf{B}}$
2. Force on electric current I due to magnetic field (RHR-2)	$\vec{\mathbf{F}}$ $\vec{\mathbf{I}}$ $\vec{\mathbf{B}}$ **Fig. 20–11c**	Fingers first point straight along current I, then bend along magnetic field $\vec{\mathbf{B}}$	Thumb points in direction of the force $\vec{\mathbf{F}}$
3. Force on electric charge $+q$ due to magnetic field (RHR-3)	$\vec{\mathbf{F}}$ $\vec{\mathbf{v}}$ $\vec{\mathbf{B}}$ **Fig. 20–15**	Fingers point along particle's velocity $\vec{\mathbf{v}}$, then along $\vec{\mathbf{B}}$	Thumb points in direction of the force $\vec{\mathbf{F}}$

CONCEPTUAL EXAMPLE 20–8 | **A helical path.** What is the path of a charged particle in a uniform magnetic field if its velocity is *not* perpendicular to the magnetic field?

RESPONSE The velocity vector can be broken down into components parallel and perpendicular to the field. The velocity component parallel to the field lines experiences no force ($\theta = 0$), so this component remains constant. The velocity component perpendicular to the field results in circular motion about the field lines. Putting these two motions together produces a helical (spiral) motion around the field lines as shown in Fig. 20–19.

EXERCISE G What is the sign of the charge in Fig. 20–19? How would you modify the drawing if the charge had the opposite sign?

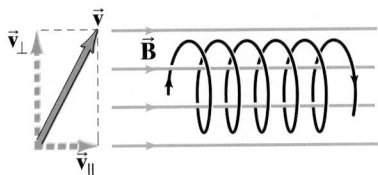

FIGURE 20–19 Example 20–8.

FIGURE 20–20 (a) Diagram showing a charged particle that approaches the Earth and is "captured" by the magnetic field of the Earth. Such particles follow the field lines toward the poles as shown. (b) Photo of aurora borealis.

* Aurora Borealis

Charged ions approach the Earth from the Sun (the "solar wind") and enter the atmosphere mainly near the poles, sometimes causing a phenomenon called the **aurora borealis** or "northern lights" in northern latitudes. To see why, consider Example 20–8 and Fig. 20–20 (see also Fig. 20–19). In Fig. 20–20 we imagine a stream of charged particles approaching the Earth. The velocity component *perpendicular* to the field for each particle becomes a circular orbit around the field lines, whereas the velocity component *parallel* to the field carries the particle along the field lines toward the poles. As a particle approaches the Earth's North Pole, the magnetic field is stronger and the radius of the helical path becomes smaller (see Example 20–6, $r \propto 1/B$).

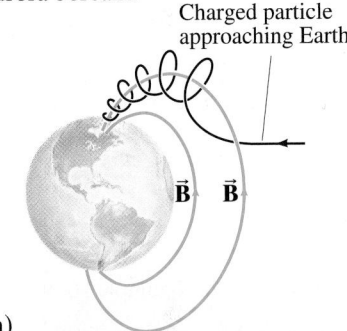

(a)

A high concentration of high-speed charged particles ionizes the air, and as the electrons recombine with atoms, light is emitted (Chapter 27) which is the aurora. Auroras are especially spectacular during periods of high sunspot activity when more charged particles are emitted and more come toward Earth.

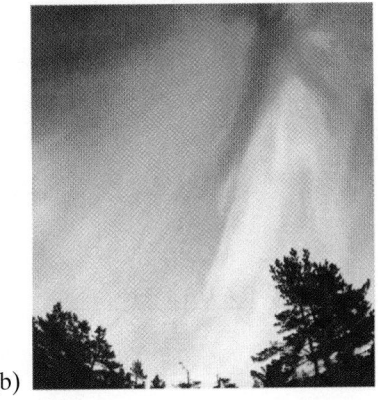

(b)

* The Hall Effect

When a current-carrying conductor is held fixed in a magnetic field, the field exerts a sideways force on the charges moving in the conductor. For example, if electrons move to the right in the rectangular conductor shown in Fig. 20–21a, the inward magnetic field will exert a downward force on the electrons of magnitude $F = ev_dB$, where v_d is the drift velocity of the electrons (Section 18–8). Thus the electrons will tend to move nearer to side D than side C, causing a potential difference between sides C and D of the conductor. This potential difference builds up until the electric field $\vec{E}_H$ that it produces exerts a force ($= e\vec{E}_H$) on the moving charges that is equal and opposite to the magnetic force ($= ev_dB$). This is the **Hall effect**, named after E. H. Hall who discovered it in 1879. The difference of potential produced is called the **Hall emf**. Its magnitude is $V_{Hall} = E_Hd = (F/e)d = v_dBd$, where d is the width of the conductor.

A current of negative charges moving to the right is equivalent to positive charges moving to the left, at least for most purposes. But the Hall effect can distinguish these two. As can be seen in Fig. 20–21b, positive particles moving to the left are deflected downward, so that the bottom surface is positive relative to the top surface. This is the reverse of part (a). Indeed, the direction of the emf in the Hall effect first revealed that it is negative particles that move in metal conductors, and that positive "holes" move in *p*-type semiconductors.

Because the Hall emf is proportional to B, the Hall effect can be used to measure magnetic fields. A device to do so is called a *Hall probe*. When B is known, the Hall emf can be used to determine the drift velocity of charge carriers.

FIGURE 20–21 The Hall effect. (a) Negative charges moving to the right as the current. (b) Positive charges moving to the left as the current.

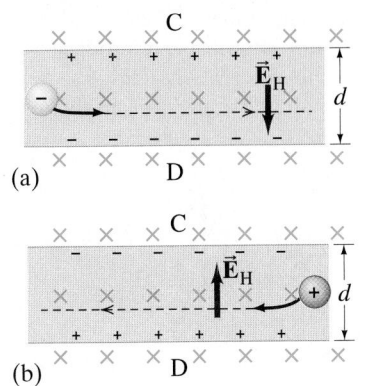

(a)

(b)

20–5 Magnetic Field Due to a Long Straight Wire

We saw in Section 20–2, Fig. 20–8, that the magnetic field lines due to the electric current in a long straight wire form circles with the wire at the center (Fig. 20–22). You might expect that the field strength at a given point would be greater if the current flowing in the wire were greater; and that the field would be less at points farther from the wire. This is indeed the case. Careful experiments show that the magnetic field B due to the current in a long straight wire is directly proportional to the current I in the wire and inversely proportional to the distance r from the wire:

$$B \propto \frac{I}{r}.$$

This relation is valid as long as r, the perpendicular distance to the wire, is much less than the distance to the ends of the wire (i.e., the wire is long).

The proportionality constant is written as $\mu_0/2\pi$, so

$$B = \frac{\mu_0}{2\pi}\frac{I}{r}.$$ [near a long straight wire]   (20–6)

The value of the constant μ_0, which is called the **permeability of free space**, is[†]

$$\mu_0 = 4\pi \times 10^{-7}\,\text{T}\cdot\text{m/A}.$$

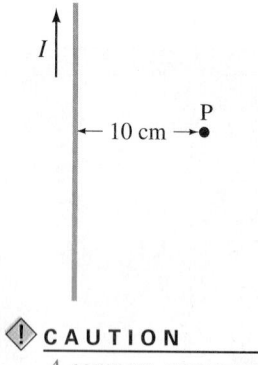

FIGURE 20–22 Same as Fig. 20–8c, magnetic field lines around a long straight wire carrying an electric current I.

FIGURE 20–23 Example 20–9.

EXAMPLE 20–9 **Calculation of $\vec{B}$ near a wire.** An electric wire in the wall of a building carries a dc current of 25 A vertically upward. What is the magnetic field due to this current at a point P, 10 cm due north of the wire (Fig. 20–23)?

APPROACH We assume the wire is much longer than the 10-cm distance to the point P so we can apply Eq. 20–6.

SOLUTION According to Eq. 20–6:

$$B = \frac{\mu_0 I}{2\pi r} = \frac{(4\pi \times 10^{-7}\,\text{T}\cdot\text{m/A})(25\,\text{A})}{(2\pi)(0.10\,\text{m})} = 5.0 \times 10^{-5}\,\text{T},$$

or 0.50 G. By right-hand-rule-1 (page 568), the field points to the west (into the page in Fig. 20–23) at point P.

NOTE The magnetic field at point P produced by the wire has about the same magnitude as Earth's, so a compass at P would not point north but to the northwest.

NOTE Most electrical wiring in buildings consists of cables with two wires in each cable. Since the two wires carry current in opposite directions, their magnetic fields cancel to a large extent, but may still affect sensitive electronic devices.

⚠ **CAUTION**

A compass, near a current, may not point north

FIGURE 20–24 Example 20–10. Wire 1 carrying current I_1 out towards us, and wire 2 carrying current I_2 into the page, produce magnetic fields whose lines are circles around their respective wires.

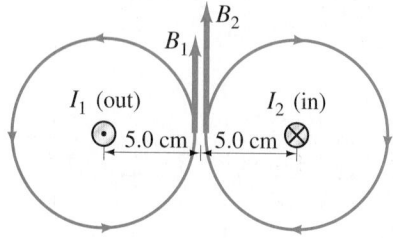

EXAMPLE 20–10 **Magnetic field midway between two currents.** Two parallel straight wires 10.0 cm apart carry currents in opposite directions (Fig. 20–24). Current $I_1 = 5.0$ A is out of the page, and $I_2 = 7.0$ A is into the page. Determine the magnitude and direction of the magnetic field halfway between the two wires.

APPROACH The magnitude of the field produced by each wire is calculated from Eq. 20–6. The direction of *each* wire's field is determined with the right-hand rule. The total field is the vector sum of the two fields at the midway point.

SOLUTION The magnetic field lines due to current I_1 form circles around the wire of I_1, and right-hand-rule-1 (Fig. 20–8d) tells us they point counterclockwise around the wire. The field lines due to I_2 form circles around the wire of I_2 and point clockwise, Fig. 20–24. At the midpoint, both fields point upward in Fig. 20–24 as shown, and so add together. The midpoint is 0.050 m from each wire.

[†]The constant is chosen in this complicated way so that Ampère's law (Section 20–8), which is considered more fundamental, will have a simple and elegant form.

From Eq. 20–6 the magnitudes of B_1 and B_2 are

$$B_1 = \frac{\mu_0 I_1}{2\pi r} = \frac{(4\pi \times 10^{-7}\,\text{T}\cdot\text{m/A})(5.0\,\text{A})}{2\pi(0.050\,\text{m})} = 2.0 \times 10^{-5}\,\text{T};$$

$$B_2 = \frac{\mu_0 I_2}{2\pi r} = \frac{(4\pi \times 10^{-7}\,\text{T}\cdot\text{m/A})(7.0\,\text{A})}{2\pi(0.050\,\text{m})} = 2.8 \times 10^{-5}\,\text{T}.$$

The total field is *up* with a magnitude of

$$B = B_1 + B_2 = 4.8 \times 10^{-5}\,\text{T}.$$

EXERCISE H Suppose both I_1 and I_2 point into the page in Fig. 20–24. What then is the field B midway between the wires?

CONCEPTUAL EXAMPLE 20–11 | **Magnetic field due to four wires.**
Figure 20–25 shows four long parallel wires which carry equal currents into or out of the page as shown. In which configuration, (*a*) or (*b*), is the magnetic field greater at the center of the square?

RESPONSE It is greater in (*a*). The arrows illustrate the directions of the field produced by each wire; check it out, using the right-hand rule to confirm these results. The net field at the center is the superposition of the four fields (which are of equal magnitude), which will point to the left in (*a*) and is zero in (*b*).

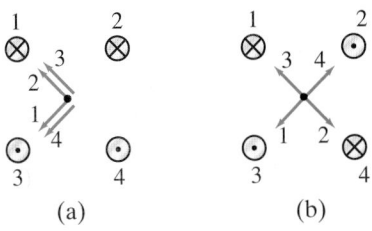

FIGURE 20–25 Example 20–11.

20–6 Force between Two Parallel Wires

We have seen that a wire carrying a current produces a magnetic field (magnitude given by Eq. 20–6 for a long straight wire). Also, a current-carrying wire feels a force when placed in a magnetic field (Section 20–3, Eq. 20–1). Thus, we expect that two current-carrying wires will exert a force on each other.

Consider two long parallel wires separated by a distance d, as in Fig. 20–26a. They carry currents I_1 and I_2, respectively. Each current produces a magnetic field that is "felt" by the other, so each must exert a force on the other. For example, the magnetic field B_1 produced by I_1 in Fig 20–26 is given by Eq. 20–6, which at the location of wire 2 points into the page and has magnitude

$$B_1 = \frac{\mu_0}{2\pi}\frac{I_1}{d}.$$

See Fig. 20–26b, where the field due *only* to I_1 is shown. According to Eq. 20–2, the force F_2 exerted by B_1 on a length ℓ_2 of wire 2, carrying current I_2, has magnitude

$$F_2 = I_2 B_1 \ell_2.$$

Note that the force on I_2 is due only to the field produced by I_1. Of course, I_2 also produces a field, but it does not exert a force on itself. We substitute B_1 into the formula for F_2 and find that the force on a length ℓ_2 of wire 2 is

$$F_2 = \frac{\mu_0}{2\pi}\frac{I_1 I_2}{d}\ell_2. \qquad \text{[parallel wires]} \quad \textbf{(20–7)}$$

If we use right-hand-rule-1 of Fig. 20–8d, we see that the lines of B_1 are as shown in Fig. 20–26b. Then using right-hand-rule-2 of Fig. 20–11c, we see that the force exerted on I_2 will be to the left in Fig. 20–26b. That is, I_1 exerts an attractive force on I_2 (Fig. 20–27a). This is true as long as the currents are in the same direction. If I_2 is in the opposite direction from I_1, right-hand-rule-2 indicates that the force is in the opposite direction. That is, I_1 exerts a repulsive force on I_2 (Fig. 20–27b).

Reasoning similar to that above shows that the magnetic field produced by I_2 exerts an equal but opposite force on I_1. We expect this to be true also from Newton's third law. Thus, as shown in Fig. 20–27, parallel currents in the same direction attract each other, whereas parallel currents in opposite directions repel.

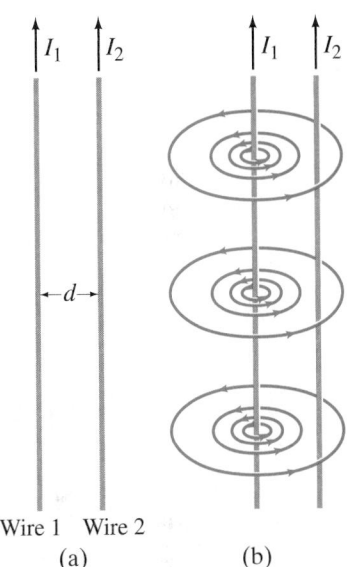

FIGURE 20–26 (a) Two parallel conductors carrying currents I_1 and I_2. (b) Magnetic field $\vec{B}_1$ produced by I_1. (Field produced by I_2 is not shown.) $\vec{B}_1$ points into page at position of I_2.

FIGURE 20–27 (a) Parallel currents in the same direction exert an attractive force on each other. (b) Antiparallel currents (in opposite directions) exert a repulsive force on each other.

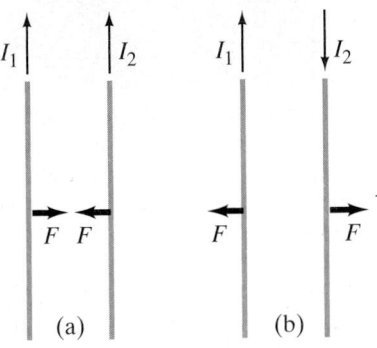

EXAMPLE 20–12 **Force between two current-carrying wires.** The two wires of a 2.0-m-long appliance cord are 3.0 mm apart and carry a current of 8.0 A. Calculate the force one wire exerts on the other.

APPROACH Each wire is in the magnetic field of the other when the current is on, so we can apply Eq. 20–7.

SOLUTION Equation 20–7 gives

$$F = \frac{\mu_0}{2\pi} \frac{I_1 I_2}{d} \ell_2 = \frac{(4\pi \times 10^{-7}\,\text{T·m/A})(8.0\,\text{A})^2(2.0\,\text{m})}{(2\pi)(3.0 \times 10^{-3}\,\text{m})} = 8.5 \times 10^{-3}\,\text{N}.$$

The currents are in opposite directions (one toward the appliance, the other away from it), so the force would be repulsive and tend to spread the wires apart.

Definition of the Ampere and the Coulomb

You may have wondered how the constant μ_0 in Eq. 20–6 could be exactly $4\pi \times 10^{-7}\,\text{T·m/A}$. Here is how it happened. With an older definition of the ampere, μ_0 was measured experimentally to be very close to this value. Today, μ_0 is *defined* to be exactly $4\pi \times 10^{-7}\,\text{T·m/A}$. This could not be done if the ampere were defined independently. The ampere, the unit of current, is now defined in terms of the magnetic field it produces using the defined value of μ_0.

In particular, we use the force between two parallel current-carrying wires, Eq. 20–7, to define the ampere precisely. If $I_1 = I_2 = 1\,\text{A}$ exactly, and the two wires are exactly 1 m apart, then

$$\frac{F}{\ell} = \frac{\mu_0}{2\pi} \frac{I_1 I_2}{d} = \frac{(4\pi \times 10^{-7}\,\text{T·m/A})}{(2\pi)} \frac{(1\,\text{A})(1\,\text{A})}{(1\,\text{m})} = 2 \times 10^{-7}\,\text{N/m}.$$

Thus, *one* **ampere** *is defined as that current flowing in each of two long parallel wires, 1 m apart, which results in a force of exactly 2×10^{-7} N per meter of length of each wire.*

This is the precise definition of the ampere, and because it is readily reproducible, is called an **operational definition**. The **coulomb** is defined in terms of the ampere as being *exactly* one ampere-second: $1\,\text{C} = 1\,\text{A·s}$.

PHYSICS APPLIED
Solenoids and electromagnets

20–7 Solenoids and Electromagnets

A long coil of wire consisting of many loops (or turns) of wire is called a **solenoid**. The current in each loop produces a magnetic field, as we saw in Fig. 20–9. The magnetic field within a solenoid can be fairly large because it is the sum of the fields due to the current in each loop (Fig. 20–28). A solenoid acts like a magnet; one end can be considered the north pole and the other the south pole, depending on the direction of the current in the loops (use the right-hand rule). Since the magnetic field lines leave the north pole of a magnet, the north pole of the solenoid in Fig. 20–28 is on the right. As we will see in the next Section, the magnetic field inside a tightly wrapped solenoid with N turns of wire in a length ℓ, each carrying current I, is

$$B = \frac{\mu_0 N I}{\ell}. \tag{20–8}$$

If a piece of iron is placed inside a solenoid, the magnetic field is increased greatly because the iron becomes a magnet. The resulting magnetic field is the sum of the field due to the current and the field due to the iron, and can be hundreds or thousands of times the field due to the current alone (see Section 20–12). Such an iron-core solenoid is an **electromagnet**.

Electromagnets have many practical applications, from use in motors and generators to producing large magnetic fields for research. Sometimes an iron core is not present—the magnetic field then comes only from the current in the wire coils. A large field B in this case requires a large current I, which produces a large amount of waste heat $(P = I^2 R)$. But if the current-carrying wires are made of superconducting material kept below the transition temperature (Section 18–9), very high fields can be produced, and no electric power is needed to maintain the large current in the superconducting coils. Energy is required, however, to refrigerate the coils at the low temperatures where they superconduct.

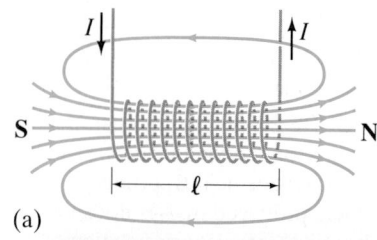

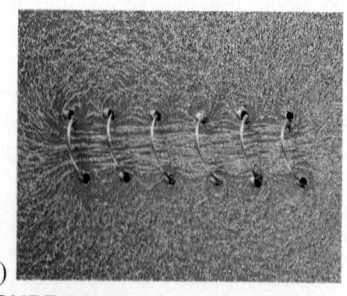

FIGURE 20–28 (a) Magnetic field of a solenoid. The north pole of this solenoid, thought of as a magnet, is on the right, and the south pole is on the left. (b) Photo of iron filings aligning along $\vec{B}$ field lines of a solenoid with loosely spaced loops. The field is smoother if the loops are closely spaced.

Another useful device consists of a solenoid into which a rod of iron is partially inserted. This combination is also referred to as a **solenoid**. One simple use is as a doorbell (Fig. 20–29). When the circuit is closed by pushing the button, the coil effectively becomes a magnet and exerts a force on the iron rod. The rod is pulled into the coil and strikes the bell. A large solenoid is used for the starter of a car: when you engage the starter, you are closing a circuit that not only turns the starter motor, but first activates a solenoid that moves the starter into direct contact with the gears on the engine's flywheel. Solenoids are used a lot as switches in cars and many other devices. They have the advantage of moving mechanical parts quickly and accurately.

Magnetic Circuit Breakers

Modern circuit breakers that protect houses and buildings from overload and fire contain not only a "thermal" part (bimetallic strip as described in Section 18–6, Fig. 18–19) but also a magnetic sensor. If the current is above a certain level, the magnetic field the current produces pulls an iron plate that breaks the same contact points as in Figs. 18–19b and c. Magnetic circuit breakers react quickly (<10 ms), and for buildings are designed to react to the high currents of short circuits (but not shut off for the start-up surges of motors).

In more sophisticated circuit breakers, including ground fault circuit interrupters (GFCIs—discussed in Section 21–9), a solenoid is used. The iron rod of Fig. 20–29, instead of striking a bell, strikes one side of a pair of electric contact points, opening them and opening the circuit. They react very quickly (≈ 1 ms) and to very small currents (≈ 5 mA) and thus protect humans (not just property) and save lives.

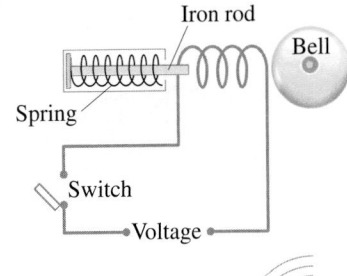
PHYSICS APPLIED
Doorbell, car starter

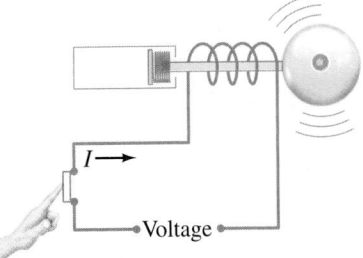

FIGURE 20–29 Solenoid used as a doorbell.

PHYSICS APPLIED
Magnetic circuit breakers

20–8 Ampère's Law

The relation between the current in a long straight wire and the magnetic field it produces is given by Eq. 20–6, Section 20–5. This equation is valid *only* for a long straight wire. Is there a general relation between a current in a wire of any shape and the magnetic field around it? Yes: the French scientist André Marie Ampère (1775–1836) proposed such a relation shortly after Oersted's discovery. Consider any (arbitrary) closed path around a current, as shown in Fig. 20–30, and imagine this path as being made up of short segments each of length $\Delta \ell$. We take the product of the length of each segment times the component of magnetic field $\vec{\mathbf{B}}$ parallel to that segment. If we now sum all these terms, the result (according to Ampère) will be equal to μ_0 times the net current I_{encl} that passes through the surface *enclosed* by the path. This is known as **Ampère's law** and can be written

$$\Sigma B_\parallel \Delta \ell = \mu_0 I_{encl}. \qquad \textbf{(20–9)}$$

AMPÈRE'S LAW

The symbol Σ means "the sum of" and $B_\parallel$ means the component of $\vec{\mathbf{B}}$ parallel to that particular $\Delta \ell$. The lengths $\Delta \ell$ are chosen small enough so that $B_\parallel$ is essentially constant along each length. The sum must be made over a closed path, and I_{encl} is the total net current enclosed by this closed path.

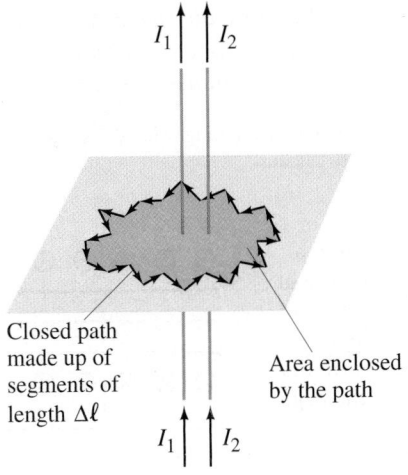

FIGURE 20–30 Arbitrary path enclosing electric currents, for Ampère's law. The path is broken down into segments of equal length $\Delta \ell$. The total current enclosed by the path shown is $I_{encl} = I_1 + I_2$.

Closed path made up of segments of length $\Delta \ell$

Area enclosed by the path

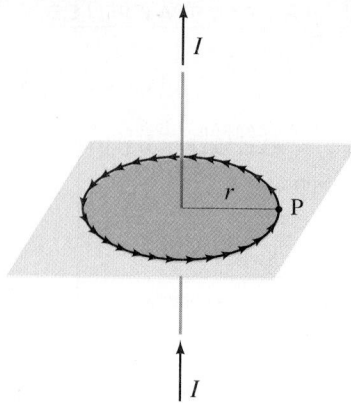

FIGURE 20-31 Circular path of radius r.

Field Due to a Straight Wire

We can check Ampère's law by applying it to the simple case of a long straight wire carrying a current I. Let us find the magnitude of B at point P, a distance r from the wire in Fig. 20–31. The magnetic field lines are circles with the wire at their center (as in Fig. 20–8). As the path to be used in Eq. 20–9, we choose a convenient one: a circle of radius r, because at any point on this path, $\vec{B}$ will be tangent to this circle. For any short segment of the circle (Fig. 20–31), $\vec{B}$ will be parallel to that segment, so $B_{\parallel} = B$. Suppose we break the circular path down into 100 segments.[†] Then Ampère's law states that

$$(B\,\Delta\ell)_1 + (B\,\Delta\ell)_2 + (B\,\Delta\ell)_3 + \cdots + (B\,\Delta\ell)_{100} = \mu_0 I.$$

The dots represent all the terms we did not write down. All the segments are the same distance from the wire, so by *symmetry* we expect B to be the same at each segment. We can then factor out B from the sum:

$$B(\Delta\ell_1 + \Delta\ell_2 + \Delta\ell_3 + \cdots + \Delta\ell_{100}) = \mu_0 I.$$

The sum of the segment lengths $\Delta\ell$ equals the circumference of the circle, $2\pi r$. Thus we have

$$B(2\pi r) = \mu_0 I,$$

or

$$B = \frac{\mu_0 I}{2\pi r}.$$

This is just Eq. 20–6 for the magnetic field near a long straight wire, so Ampère's law agrees with experiment in this case.

A great many experiments indicate that Ampère's law is valid in general. Practically, it can be used to calculate the magnetic field mainly for simple or symmetric situations. Its importance is that it relates the magnetic field to the current in a direct and mathematically elegant way. Ampère's law is considered one of the basic laws of electricity and magnetism. It is valid for any situation where the currents and fields are not changing in time.

Field Inside a Solenoid

FIGURE 20-32 (a) Magnetic field due to several loops of a solenoid. (b) For many closely spaced loops, the field is very nearly uniform.

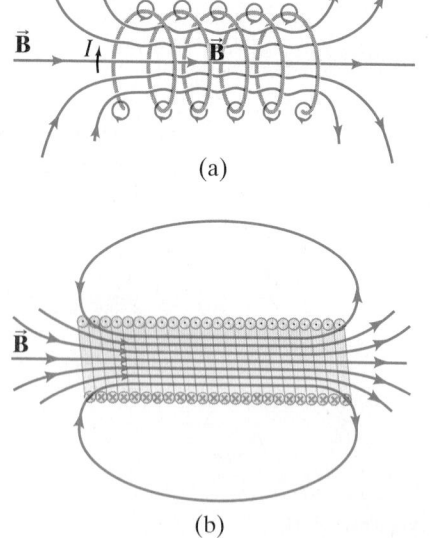

(a)

(b)

We now use Ampère's law to calculate the magnetic field inside a *solenoid* (Section 20–7), a long coil of wire with many loops or turns, Fig. 20–32. Each loop produces a magnetic field as was shown in Fig. 20–9, and the total field inside the solenoid will be the sum of the fields due to each current loop as shown in Fig. 20–32a for a few loops. If the solenoid has many loops and they are close together, the field inside will be nearly uniform and parallel to the solenoid axis except at the ends, as shown in Fig. 20–32b. Outside the solenoid, the field lines spread out in space, so the magnetic field is much weaker than inside. For applying Ampère's law, we choose the path abcd shown in Fig. 20–33 far from either end. We consider this path as made up of four straight segments, the sides of the rectangle: ab, bc, cd, da. Then Ampère's law, Eq. 20–9, becomes

$$(B_{\parallel}\,\Delta\ell)_{ab} + (B_{\parallel}\,\Delta\ell)_{bc} + (B_{\parallel}\,\Delta\ell)_{cd} + (B_{\parallel}\,\Delta\ell)_{da} = \mu_0 I_{encl}.$$

The first term in the sum on the left will be (nearly) zero because the field outside the solenoid is negligible compared to the field inside. Furthermore, $\vec{B}$ is perpendicular to the segments bc and da, so these terms are zero, too.

[†]Actually, Ampère's law is precisely accurate when there is an infinite number of infinitesimally short segments, but that leads into calculus.

FIGURE 20-33 Cross-sectional view into a solenoid. The magnetic field inside is straight except at the ends. Red dashed lines indicate the path chosen for use in Ampère's law. ⊙ and ⊗ are electric current direction (in the wire loops) out of the page and into the page.

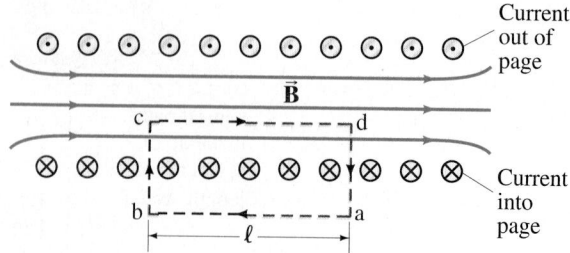

Thus the left side of our Ampère equation we just wrote becomes $(B_{\parallel}\,\Delta\ell)_{cd} = B\ell$, where B is the field inside the solenoid, and ℓ is the length cd. We set $B\ell$ equal to μ_0 times the current enclosed by our chosen rectangular loop: if a current I flows in the wire of the solenoid, the total current enclosed by our path abcd is NI, where N is the number of loops (or turns) our path encircles (five in Fig. 20–33). Thus Ampère's law gives us

$$B\ell = \mu_0 NI,$$

so

$$B = \frac{\mu_0 IN}{\ell},$$ [solenoid]   (20–8 *again*)

as we quoted in the previous Section. This is the magnetic field magnitude inside a solenoid. B depends only on the number of loops per unit length, N/ℓ, and the current I. The field does not depend on the position within the solenoid, so B is uniform inside the solenoid. This is strictly true only for an infinite solenoid, but it is a good approximation for real ones at points not close to the ends.

The direction of the magnetic field inside the solenoid is found by applying right-hand-rule-1, Fig. 20–8d (see also Figs. 20–9 and 20–10), and is as shown in Fig. 20–33.

20–9 Torque on a Current Loop; Magnetic Moment

When an electric current flows in a closed loop of wire placed in an external magnetic field, as shown in Fig. 20–34, the magnetic force on the current can produce a torque. This is the principle behind a number of important practical devices, including motors and analog voltmeters and ammeters, which we discuss in the next Section.

Current flows through the rectangular loop in Fig. 20–34a, whose face we assume is parallel to $\vec{B}$. $\vec{B}$ exerts no force and no torque on the horizontal segments of wire because they are parallel to the field and $\sin\theta = 0$ in Eq. 20–1. But the magnetic field does exert a force on each of the vertical sections of wire as shown, $\vec{F}_1$ and $\vec{F}_2$ (see also top view, Fig. 20–34b). By right-hand-rule-2 (Fig. 20–11c or Table 20–1) the direction of the force on the upward current on the left is in the opposite direction from the equal magnitude force $\vec{F}_2$ on the downward current on the right. These forces give rise to a net torque that acts to rotate the coil about its vertical axis.

Let us calculate the magnitude of this torque. From Eq. 20–2 (current $\perp \vec{B}$), the force $F = IaB$, where a is the length of the vertical arm of the coil (Fig. 20–34a). The lever arm for each force is $b/2$, where b is the width of the coil and the "axis" is at the midpoint. The torques around this axis produced by $\vec{F}_1$ and $\vec{F}_2$ act in the same direction (Fig. 20–34b), so the total torque τ is the sum of the two torques:

$$\tau = IaB\frac{b}{2} + IaB\frac{b}{2} = IabB = IAB,$$

where $A = ab$ is the area of the coil. If the coil consists of N loops of wire, the current is then NI, so the torque becomes

$$\tau = NIAB.$$

If the coil makes an angle with the magnetic field, as shown in Fig. 20–34c, the forces are unchanged, but each lever arm is reduced from $\frac{1}{2}b$ to $\frac{1}{2}b\sin\theta$. Note that the angle θ is taken to be the angle between $\vec{B}$ and the perpendicular to the face of the coil, Fig. 20–34c. So the torque becomes

$$\tau = NIAB \sin\theta.$$ (20–10)

This formula, derived here for a rectangular coil, is valid for any shape of flat coil.

The quantity NIA is called the **magnetic dipole moment** of the coil:

$$M = NIA$$ (20–11)

and is considered a vector perpendicular to the coil.

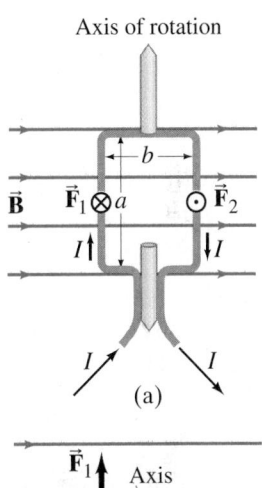

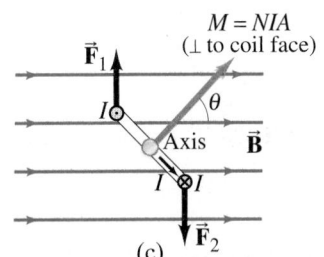

FIGURE 20–34 Calculating the torque on a current loop in a magnetic field $\vec{B}$. (a) Loop face parallel to $\vec{B}$ field lines; (b) top view; (c) loop makes an angle to $\vec{B}$, reducing the torque since the lever arm is reduced.

EXAMPLE 20–13 | **Torque on a coil.** A circular loop of wire has a diameter of 20.0 cm and contains 10 loops. The current in each loop is 3.00 A, and the coil is placed in a 2.00-T external magnetic field. Determine the maximum and minimum torque exerted on the coil by the field.

APPROACH Equation 20–10 is valid for any shape of coil, including circular loops. Maximum and minimum torque are determined by the angle θ the coil makes with the magnetic field.

SOLUTION The area of one loop of the coil is

$$A = \pi r^2 = \pi (0.100\,\text{m})^2 = 3.14 \times 10^{-2}\,\text{m}^2.$$

The maximum torque occurs when the coil's face is parallel to the magnetic field, so $\theta = 90°$ in Fig. 20–34c, and $\sin \theta = 1$ in Eq. 20–10:

$$\tau = NIAB \sin \theta = (10)(3.00\,\text{A})(3.14 \times 10^{-2}\,\text{m}^2)(2.00\,\text{T})(1) = 1.88\,\text{N·m}.$$

The minimum torque occurs if $\sin \theta = 0$, for which $\theta = 0°$, and then $\tau = 0$ from Eq. 20–10.

NOTE If the coil is free to turn, it will rotate toward the orientation with $\theta = 0°$.

20–10 Applications: Motors, Loudspeakers, Galvanometers

There are many practical applications of the forces related to magnetism. Among the most common are motors and loudspeakers. First we look at the galvanometer, which is the easiest to explain, and which you find on the instrument panels of automobiles and other devices whose readout is via a pointer or needle.

Galvanometer

The basic component of analog meters (those with pointer and dial), including analog ammeters, voltmeters, and ohmmeters, including gauges on car dashboards, is a galvanometer. We have already seen how these meters are designed (Section 19–8), and now we can examine how the crucial element, a galvanometer, works. As shown in Fig. 20–35, a **galvanometer** consists of a coil of wire (with attached pointer) suspended in the magnetic field of a permanent magnet. When current flows through the loop of wire, the magnetic field B exerts a torque τ on the loop, as given by Eq. 20–10,

$$\tau = NIAB \sin \theta.$$

This torque is opposed by a spring which exerts a torque τ_s approximately proportional to the angle ϕ through which it is turned (Hooke's law). That is,

$$\tau_s = k\phi,$$

where k is the stiffness constant of the spring. The coil and attached pointer rotate to the angle where the torques balance. When the needle is in equilibrium at rest, the torques have equal magnitude: $k\phi = NIAB \sin \theta$, so

$$\phi = \frac{NIAB \sin \theta}{k}.$$

The deflection of the pointer, ϕ, is directly proportional to the current I flowing in the coil, but also depends on the angle θ the coil makes with $\vec{\mathbf{B}}$. For a useful meter we need ϕ to depend only on the current I, independent of θ. To solve this problem, magnets with curved pole pieces are used and the galvanometer coil is wrapped around a cylindrical iron core as shown in Fig. 20–36. The iron tends to concentrate the magnetic field lines so that $\vec{\mathbf{B}}$ always points parallel to the face of the coil at the wire outside the core. The force is then always perpendicular to the face of the coil, and the torque will not vary with angle. Thus ϕ will be proportional to I, as required for a useful meter.

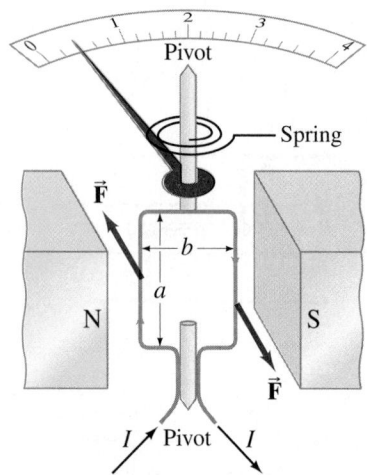

FIGURE 20–35 Galvanometer.

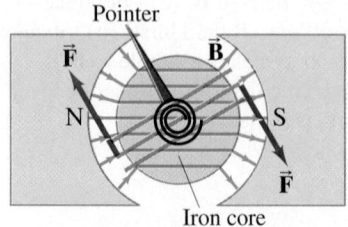

FIGURE 20–36 Galvanometer coil (3 loops shown) wrapped on an iron core.

Electric Motors

An **electric motor** changes electric energy into (rotational) mechanical energy. A motor works on the same principle as a galvanometer (a torque is exerted on a current-carrying loop in a magnetic field) except that the coil must turn continuously in one direction. The coil is mounted on an iron cylinder called the **rotor** or **armature**, Fig. 20–37. Actually, there are several coils, although only one is indicated in Fig. 20–37. The armature is mounted on a shaft or axle. When the armature is in the position shown in Fig. 20–37, the magnetic field exerts forces on the current in the loop as shown (perpendicular to $\vec{\mathbf{B}}$ and to the current direction). However, when the coil, which is rotating clockwise in Fig. 20–37, passes beyond the vertical position, the forces would then act to return the coil back toward the vertical if the current remained the same. But if the current could be reversed at that critical moment, the forces would reverse, and the coil would continue rotating in the same direction. Thus, alternation of the current is necessary if a motor is to turn continuously in one direction. This can be achieved in a **dc motor** with the use of **commutators** and **brushes**: as shown in Fig. 20–38, input current passes through stationary brushes that rub against the conducting commutators mounted on the motor shaft. At every half revolution, each commutator changes its connection over to the other brush. Thus the current in the coil reverses every half revolution as required for continuous rotation.

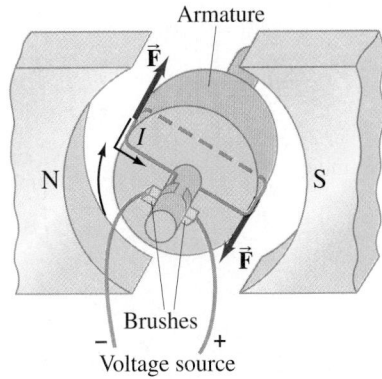

FIGURE 20–37 Diagram of a simple dc motor. (Magnetic field lines are as shown in Fig. 20–36.)

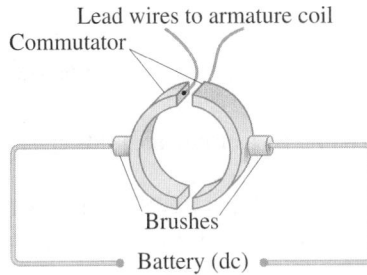

FIGURE 20–38 Commutator-brush arrangement in a dc motor ensures alternation of the current in the armature to keep rotation continuous in one direction. The commutators are attached to the motor shaft and turn with it, whereas the brushes remain stationary.

Most motors contain several coils, called *windings*, each connected to a different portion of the armature, Fig. 20–39. Current flows through each coil only during a small part of a revolution, at the time when its orientation results in the maximum torque. In this way, a motor produces a much steadier torque than can be obtained from a single coil.

An **ac motor**, with ac current as input, can work without commutators since the current itself alternates. Many motors use wire coils to produce the magnetic field (electromagnets) instead of a permanent magnet. Indeed the design of most motors is more complex than described here, but the general principles remain the same.

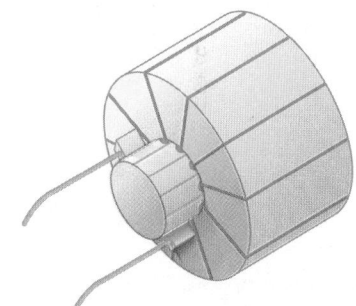

FIGURE 20–39 Motor with many windings.

FIGURE 20–40 Loudspeaker.

Loudspeakers and Headsets

Loudspeakers and audio headsets also work on the principle that a magnet exerts a force on a current-carrying wire. The electrical output of a stereo or TV set is connected to the wire leads of the speaker or earbuds. The speaker leads are connected internally to a coil of wire, which is itself attached to the speaker cone, Fig. 20–40. The speaker cone is usually made of stiffened cardboard and is mounted so that it can move back and forth freely (except at its attachment on the outer edges). A permanent magnet is mounted directly in line with the coil of wire. When the alternating current of an audio signal flows through the wire coil, which is free to move within the magnet, the coil experiences a force due to the magnetic field of the magnet. (The force is to the right at the instant shown in Fig. 20–40, RHR-2, page 568.) As the current alternates at the frequency of the audio signal, the coil and attached speaker cone move back and forth at the same frequency, causing alternate compressions and rarefactions of the adjacent air, and sound waves are produced. A speaker thus changes electrical energy into sound energy, and the frequencies and intensities of the emitted sound waves can be an accurate reproduction of the electrical input.

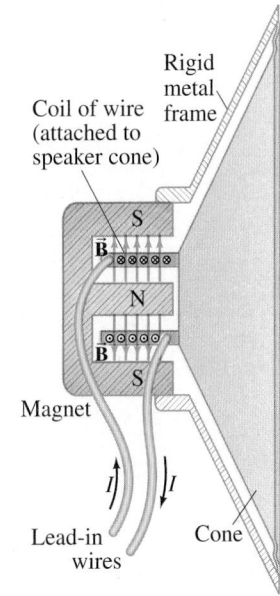

*20–11 Mass Spectrometer

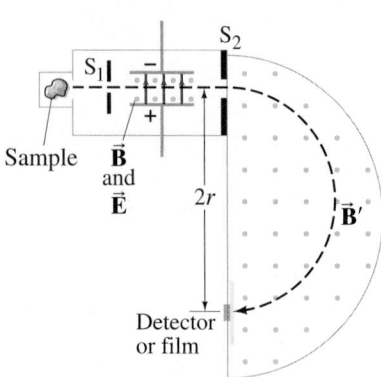

FIGURE 20–41 Bainbridge-type mass spectrometer. The magnetic fields B and B' point out of the paper (indicated by the dots).

A **mass spectrometer** is a device to measure masses of atoms. It is used today not only in physics but also in chemistry, geology, and medicine, often to identify atoms (and their concentration) in given samples. Ions are produced by heating the sample, or by using an electric current. As shown in Fig. 20–41, the ions (mass m, charge q) pass through slit S_1 and enter a region (before S_2) where there are crossed ($\perp$) electric and magnetic fields. Ions follow a straight-line path in this region if the electric force qE (upward on a positive ion) is just balanced by the magnetic force qvB (downward on a positive ion): that is, if $qE = qvB$, or

$$ v = \frac{E}{B}. $$

Only those ions whose speed is $v = E/B$ will pass through undeflected and emerge through slit S_2. (This arrangement is called a **velocity selector**.) In the semicircular region, after S_2, there is only a magnetic field, B', so the ions follow a circular path. The radius of the circular path is found from their mark on film, or by detectors, if B' is fixed. If instead r is fixed by the position of a detector, then B' is varied until detection occurs. Newton's second law, $\Sigma F = ma$, applied to an ion moving in a circle under the influence only of the magnetic field B' gives $qvB' = mv^2/r$. Since $v = E/B$, we have

$$ m = \frac{qB'r}{v} = \frac{qBB'r}{E}. \tag{20–12} $$

All the quantities on the right side are known or can be measured, and thus m can be determined.

Historically, the masses of many atoms were measured this way. When a pure substance was used, it was sometimes found that two or more closely spaced marks would appear on the film. For example, neon produced two marks whose radii corresponded to atoms of 20 and 22 atomic mass units (u). Impurities were ruled out and it was concluded that there must be two types of neon with different masses. These different forms were called **isotopes**. It was soon found that most elements are mixtures of isotopes, and the difference in mass is due to different numbers of neutrons (discussed in Chapter 30).

EXAMPLE 20–14 **Mass spectrometry.** Carbon atoms of atomic mass 12.0 u are found to be mixed with an unknown element. In a mass spectrometer with fixed B', the carbon traverses a path of radius 22.4 cm and the unknown's path has a 26.2-cm radius. What is the unknown element? Assume the ions of both elements have the same charge.

APPROACH The carbon and unknown atoms pass through the same electric and magnetic fields. Hence their masses are proportional to the radius of their respective paths (see Eq. 20–12).

SOLUTION We write a ratio for the masses, using Eq. 20–12:

$$ \frac{m_x}{m_C} = \frac{qBB'r_x/E}{qBB'r_C/E} = \frac{r_x}{r_C} $$

$$ = \frac{26.2 \text{ cm}}{22.4 \text{ cm}} = 1.17. $$

Thus $m_x = 1.17 \times 12.0 \text{ u} = 14.0 \text{ u}$. The other element is probably nitrogen (see the Periodic Table, inside the back cover).

NOTE The unknown could also be an isotope such as carbon-14 $\left(^{14}_{6}\text{C}\right)$. See Appendix B. Further physical or chemical analysis would be needed.

*20–12 Ferromagnetism: Domains and Hysteresis

We saw in Section 20–1 that iron (and a few other materials) can be made into strong magnets. These materials are said to be **ferromagnetic**.

*Sources of Ferromagnetism

Microscopic examination reveals that a piece of iron is made up of tiny regions known as **domains**, less than 1 mm in length or width. Each domain behaves like a tiny magnet with a north and a south pole. In an unmagnetized piece of iron, the domains are arranged randomly, Fig. 20–42a. The magnetic effects of the domains cancel each other out, so this piece of iron is not a magnet. In a magnet, the domains are preferentially aligned in one direction as shown in Fig. 20–42b (downward in this case). A magnet can be made from an unmagnetized piece of iron by placing it in a strong magnetic field. (You can make a needle magnetic, for example, by stroking it with one pole of a strong magnet.) The magnetization direction of domains may actually rotate slightly to be more nearly parallel to the external field, and the borders of domains may move so domains with magnetic orientation parallel to the external field grow larger (compare Figs. 20–42a and b).

We can now explain how a magnet can pick up unmagnetized pieces of iron like paper clips. The magnet's field causes a slight realignment of the domains in the unmagnetized object so that it becomes a temporary magnet with its north pole facing the south pole of the permanent magnet; thus, attraction results. Similarly, elongated iron filings in a magnetic field acquire aligned domains and align themselves to reveal the shape of the magnetic field, Fig. 20–43.

An iron magnet can remain magnetized for a long time, and is referred to as a "permanent magnet." But if you drop a magnet on the floor or strike it with a hammer, you can jar the domains into randomness and the magnet loses some or all of its magnetism. Heating a permanent magnet can also cause loss of magnetism, for raising the temperature increases the random thermal motion of atoms, which tends to randomize the domains. Above a certain temperature known as the **Curie temperature** (1043 K for iron), a magnet cannot be made at all.

The striking similarity between the fields produced by a bar magnet and by a loop of electric current (Figs. 20–4b, 20–9) offers a clue that perhaps magnetic fields produced by electric currents may have something to do with ferromagnetism. According to modern atomic theory, atoms can be roughly visualized as having electrons that orbit around a central nucleus. The electrons are charged, and so constitute an electric current and therefore produce a magnetic field. But the fields due to orbiting electrons end up adding to zero. Electrons themselves produce an additional magnetic field, almost as if they and their electric charge were spinning about their own axes. And it is this magnetic field due to electron **spin**[†] that is believed to produce ferromagnetism in most ferromagnetic materials.

It is believed today that *all* magnetic fields are caused by electric currents. This means that magnetic field lines always form closed loops, unlike electric field lines which begin on positive charges and end on negative charges.

*Magnetic Permeability

If a piece of ferromagnetic material like iron is placed inside a solenoid to form an electromagnet (Section 20–7), the magnetic field increases greatly over that produced by the current in the solenoid coils alone, often by hundreds or thousands of times. This happens because the domains in the iron become aligned by the external field produced by the current in the solenoid coil.

[†]The name "spin" comes from an early suggestion that this intrinsic magnetic field arises from the electron "spinning" on its axis (as well as "orbiting" the nucleus) to produce the extra field. However, this view of a spinning electron is oversimplified and not valid (see Chapter 28).

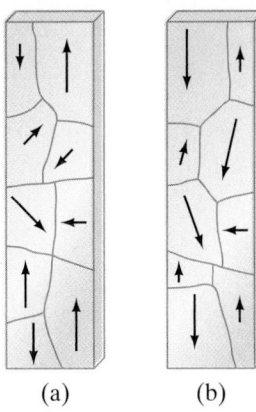

FIGURE 20–42 (a) An unmagnetized piece of iron is made up of domains that are randomly arranged. Each domain is like a tiny magnet; the arrows represent the magnetization direction, with the arrowhead being the N pole. (b) In a magnet, the domains are preferentially aligned in one direction (down in this case), and may be altered in size by the magnetization process.

FIGURE 20–43 Iron filings line up along magnetic field lines due to a permanent magnet.

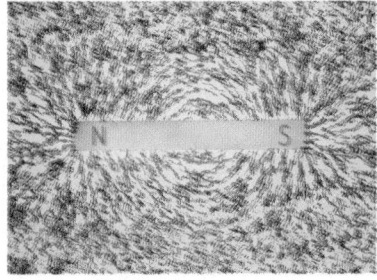

◇ **CAUTION**

$\vec{\mathbf{B}}$ *lines form closed loops,*
$\vec{\mathbf{E}}$ *lines start on* ⊕ *and end on* ⊖

The total magnetic field $\vec{B}$ is then the sum of two terms,

$$\vec{B} = \vec{B}_0 + \vec{B}_M.$$

$\vec{B}_0$ is the field due to the current in the solenoid coil and $\vec{B}_M$ is the additional field due to the iron. Often $B_M \gg B_0$. The total field can also be written by replacing the constant μ_0 in Eq. 20–8 ($B = \mu_0 NI/\ell$ for a solenoid) by another constant called the **magnetic permeability** μ, which is characteristic of the magnetic material inside the coil. Then $B = \mu NI/\ell$. For ferromagnetic materials, μ is much greater than μ_0. For all other materials, its value is very close to μ_0.[†] The value of μ, however, is not constant for ferromagnetic materials; it depends on the strength of the "external" field B_0, as the following experiment shows.

* Hysteresis

Measurements on magnetic materials often use a **torus** or **toroid**, which is like a long solenoid bent into the shape of a donut (Fig. 20–44), so practically all the lines of $\vec{B}$ remain within the toroid. Consider a toroid with an iron core that is initially unmagnetized and there is no current in the wire loops. Then the current I is slowly increased, and B_0 (which is due only to I) increases linearly with I. The total field B also increases, but follows the curved line shown in Fig. 20–45 which is a graph of total B vs. B_0. Initially, point a, the domains are randomly oriented. As B_0 increases, the domains become more and more aligned until at point b, nearly all are aligned. The iron is said to be approaching **saturation**.

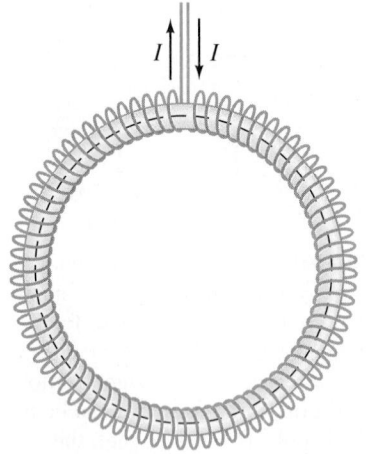

FIGURE 20–44 Iron-core toroid.

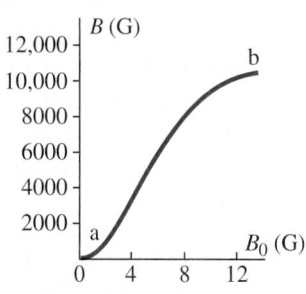

FIGURE 20–45 Total magnetic field B in an iron-core toroid as a function of the external field B_0 (B_0 is caused by the current I in the coil). We use gauss ($1\,G = 10^{-4}\,T$) so that labels are clear.

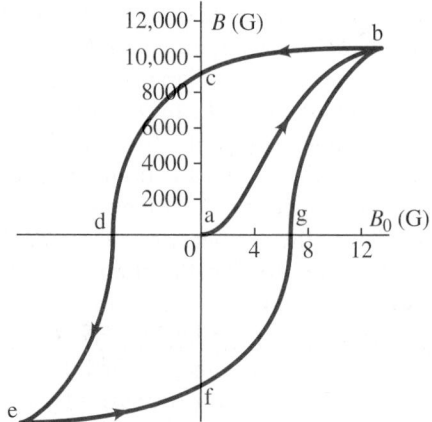

FIGURE 20–46 Hysteresis curve.

Next, suppose current in the coil is reduced, so the field B_0 decreases. If the current (and B_0) is reduced to zero, point c in Fig. 20–46, the domains do *not* become completely random. Instead, some permanent magnetism remains in the iron core. If the current is increased in the opposite direction, enough domains can be turned around so the total B becomes zero at point d. As the reverse current is increased further, the iron approaches saturation in the opposite direction, point e. Finally, if the current is again reduced to zero (point f) and then increased in the original direction, the total field follows the path efgb, again approaching saturation at point b.

Notice that the field did not pass through the origin (point a) in this cycle. The fact that the curve does not retrace itself on the same path is called **hysteresis**. The curve bcdefgb is called a **hysteresis loop**. In such a cycle, much energy is transformed to thermal energy (friction) due to realigning of the domains. Note that at points c and f, the iron core is magnetized even though there is no current in the coils. These points correspond to a permanent magnet.

[†]All materials are slightly magnetic. Nonferromagnetic materials fall into two principal classes: (1) **paramagnetic** materials consist of atoms that have a net magnetic dipole moment which can align slightly with an external field, just as the galvanometer coil in Fig. 20–35 experiences a torque that tends to align it; (2) **diamagnetic** materials have atoms with no net dipole moment, but in the presence of an external field electrons revolving in one direction increase in speed slightly whereas electrons revolving in the opposite direction are reduced in speed; the result is a slight net magnetic effect that opposes the external field.

Summary

A magnet has two **poles**, north and south. The north pole is that end which points toward geographic north when the magnet is freely suspended. Like poles of two magnets repel each other, whereas unlike poles attract.

We can picture that a **magnetic field** surrounds every magnet. The SI unit for magnetic field is the **tesla** (T).

Electric currents produce magnetic fields. For example, the lines of magnetic field due to a current in a straight wire form circles around the wire, and the field exerts a force on magnets (or currents) near it.

A magnetic field exerts a force on an electric current. For a straight wire of length ℓ carrying a current I, the force has magnitude

$$F = I\ell B \sin\theta, \tag{20–1}$$

where θ is the angle between the magnetic field $\vec{\mathbf{B}}$ and the current direction. The direction of the force is perpendicular to the current-carrying wire and to the magnetic field, and is given by a right-hand rule. Equation 20–1 serves as the definition of magnetic field $\vec{\mathbf{B}}$.

Similarly, a magnetic field exerts a force on a charge q moving with velocity v of magnitude

$$F = qvB \sin\theta, \tag{20–3}$$

where θ is the angle between $\vec{\mathbf{v}}$ and $\vec{\mathbf{B}}$. The direction of $\vec{\mathbf{F}}$ is perpendicular to $\vec{\mathbf{v}}$ and to $\vec{\mathbf{B}}$ (again a right-hand rule). The path of a charged particle moving perpendicular to a uniform magnetic field is a circle.

The magnitude of the magnetic field produced by a current I in a long straight wire, at a distance r from the wire, is

$$B = \frac{\mu_0}{2\pi}\frac{I}{r}. \tag{20–6}$$

Two currents exert a force on each other via the magnetic field each produces. Parallel currents in the same direction attract each other; currents in opposite directions repel.

The magnetic field inside a long tightly wound solenoid is

$$B = \mu_0 NI/\ell, \tag{20–8}$$

where N is the number of loops in a length ℓ of coil, and I is the current in each loop.

Ampère's law states that around any chosen closed loop path, the sum of each path segment $\Delta\ell$ times the component of $\vec{\mathbf{B}}$ parallel to the segment equals μ_0 times the current I enclosed by the closed path:

$$\Sigma B_\parallel \Delta\ell = \mu_0 I_{\text{encl}}. \tag{20–9}$$

The torque τ on N loops of current I in a magnetic field $\vec{\mathbf{B}}$ is

$$\tau = NIAB \sin\theta. \tag{20–10}$$

The force or torque exerted on a current-carrying wire by a magnetic field is the basis for operation of many devices, such as **motors**, **loudspeakers**, and **galvanometers** used in analog electric meters.

[*A **mass spectrometer** uses electric and magnetic fields to determine the mass of ions.]

[*Iron and a few other materials that are **ferromagnetic** can be made into strong permanent magnets. Ferromagnetic materials are made up of tiny **domains**—each a tiny magnet—which are preferentially aligned in a permanent magnet. When iron or another ferromagnetic material is placed in a magnetic field B_0 due to a current, the iron becomes magnetized. When the current is turned off, the material remains magnetized; when the current is increased in the opposite direction, a graph of the total field B versus B_0 is a **hysteresis loop**, and the fact that the curve does not retrace itself is called **hysteresis**.]

Questions

1. A compass needle is not always balanced parallel to the Earth's surface, but one end may dip downward. Explain.

2. Explain why the Earth's "north pole" is really a magnetic south pole. Indicate how north and south magnetic poles were defined and how we can tell experimentally that the north pole is really a south magnetic pole.

3. In what direction are the magnetic field lines surrounding a straight wire carrying a current that is moving directly away from you? Explain.

4. A horseshoe magnet is held vertically with the north pole on the left and south pole on the right. A wire passing between the poles, equidistant from them, carries a current directly away from you. In what direction is the force on the wire? Explain.

5. Will a magnet attract any metallic object, such as those made of aluminum or copper? (Try it and see.) Why is this so?

6. Two iron bars attract each other no matter which ends are placed close together. Are both magnets? Explain.

7. The magnetic field due to current in wires in your home can affect a compass. Discuss the effect in terms of currents, including if they are ac or dc.

8. If a negatively charged particle enters a region of uniform magnetic field which is perpendicular to the particle's velocity, will the kinetic energy of the particle increase, decrease, or stay the same? Explain your answer. (Neglect gravity and assume there is no electric field.)

9. In Fig. 20–47, charged particles move in the vicinity of a current-carrying wire. For each charged particle, the arrow indicates the initial direction of motion of the particle, and the + or − indicates the sign of the charge. For each of the particles, indicate the direction of the magnetic force due to the magnetic field produced by the wire. Explain.

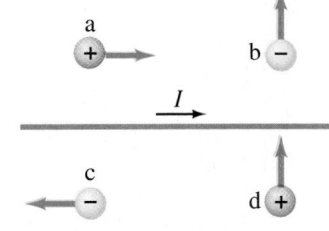

FIGURE 20–47
Question 9.

10. Three particles, a, b, and c, enter a magnetic field and follow paths as shown in Fig. 20–48. What can you say about the charge on each particle? Explain.

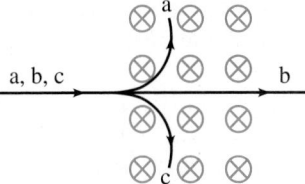

FIGURE 20–48
Question 10.

11. Can an iron rod attract a magnet? Can a magnet attract an iron rod? What must you consider to answer these questions?

12. A positively charged particle in a nonuniform magnetic field follows the trajectory shown in Fig. 20–49. Indicate the direction of the magnetic field at points near the path, assuming the path is always in the plane of the page, and indicate the relative magnitudes of the field in each region. Explain your answers.

FIGURE 20–49
Question 12.

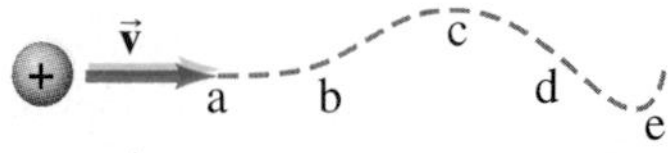

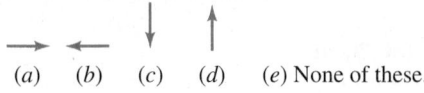

13. Explain why a strong magnet held near a CRT television screen (Section 17–11) causes the picture to become distorted. Also, explain why the picture sometimes goes completely black where the field is the strongest. [But don't risk damage to your TV by trying this.]

14. Suppose you have three iron rods, two of which are magnetized but the third is not. How would you determine which two are the magnets without using any additional objects?

15. Can you set a resting electron into motion with a magnetic field? With an electric field? Explain.

16. A charged particle is moving in a circle under the influence of a uniform magnetic field. If an electric field that points in the same direction as the magnetic field is turned on, describe the path the charged particle will take.

17. A charged particle moves in a straight line through a particular region of space. Could there be a nonzero magnetic field in this region? If so, give two possible situations.

18. If a moving charged particle is deflected sideways in some region of space, can we conclude, for certain, that $\vec{B} \neq 0$ in that region? Explain.

19. Two insulated long wires carrying equal currents I cross at right angles to each other. Describe the magnetic force one exerts on the other.

20. A horizontal current-carrying wire, free to move in Earth's gravitational field, is suspended directly above a parallel, current-carrying wire. (a) In what direction is the current in the lower wire? (b) Can the lower wire be held in stable equilibrium due to the magnetic force of the upper wire? Explain.

21. What would be the effect on B inside a long solenoid if (a) the diameter of all the loops was doubled, (b) the spacing between loops was doubled, or (c) the solenoid's length was doubled along with a doubling in the total number of loops?

22. A type of magnetic switch similar to a solenoid is a **relay** (Fig. 20–50). A relay is an electromagnet (the iron rod inside the coil does not move) which, when activated, attracts a strip of iron on a pivot. Design a relay to close an electrical switch. A relay is used when you need to switch on a circuit carrying a very large current but do not want that large current flowing through the main switch. For example, a car's starter switch is connected to a relay so that the large current needed for the starter doesn't pass to the dashboard switch.

FIGURE 20–50
Question 22.

*23. Two ions have the same mass, but one is singly ionized and the other is doubly ionized. How will their positions on the film of a mass spectrometer (Fig. 20–41) differ? Explain.

*24. Why will either pole of a magnet attract an unmagnetized piece of iron?

*25. An unmagnetized nail will not attract an unmagnetized paper clip. However, if one end of the nail is in contact with a magnet, the other end *will* attract a paper clip. Explain.

MisConceptual Questions

1. Indicate which of the following will produce a magnetic field:
   (a) A magnet.
   (b) The Earth.
   (c) An electric charge at rest.
   (d) A moving electric charge.
   (e) An electric current.
   (f) The voltage of a battery not connected to anything.
   (g) An ordinary piece of iron.
   (h) A piece of any metal.

2. A current in a wire points into the page as shown at the right. In which direction is the magnetic field at point A (choose below)?

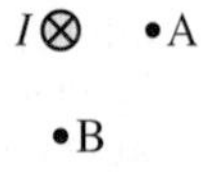

   $I \otimes$ •A

   •B

   → ← ↓ ↑
   (a) (b) (c) (d)   (e) None of these.

3. In which direction (see above) is the magnetic field at point B?

4. When a charged particle moves parallel to the direction of a magnetic field, the particle travels in a
   (a) straight line.   (c) helical path.
   (b) circular path.   (d) hysteresis loop.

5. As a proton moves through space, it creates
   (a) an electric field only.
   (b) a magnetic field only.
   (c) both an electric field and magnetic field.
   (d) nothing; the electric field and magnetic fields cancel each other out.

6. Which statements about the force on a charged particle placed in a magnetic field are true?
   (a) A magnetic force is exerted only if the particle is moving.
   (b) The force is a maximum if the particle is moving in the direction of the field.
   (c) The force causes the particle to gain kinetic energy.
   (d) The direction of the force is along the magnetic field.
   (e) A magnetic field always exerts a force on a charged particle.

7. Which of the following statements is false? The magnetic field of a current-carrying wire
(a) is directed circularly around the wire.
(b) decreases inversely with the distance from the wire.
(c) exists only if the current in the wire is changing.
(d) depends on the magnitude of the current.

8. A wire carries a current directly away from you. Which way do the magnetic field lines produced by this wire point?
(a) They point parallel to the wire in the direction of the current.
(b) They point parallel to the wire opposite the direction of the current.
(c) They point toward the wire.
(d) They point away from the wire.
(e) They make circles around the wire.

9. A proton enters a uniform magnetic field that is perpendicular to the proton's velocity (Fig. 20–51). What happens to the kinetic energy of the proton?
(a) It increases.
(b) It decreases.
(c) It stays the same.
(d) It depends on the velocity direction.
(e) It depends on the B field direction.

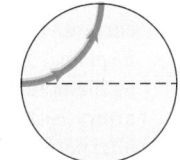

FIGURE 20–51
MisConceptual Question 9.

10. For a charged particle, a constant magnetic field can be used to change
(a) only the direction of the particle's velocity.
(b) only the magnitude of the particle's velocity.
(c) both the magnitude and direction of the particle's velocity.
(d) None of the above.

11. Which of the following statements about the force on a charged particle due to a magnetic field are not valid?
(a) It depends on the particle's charge.
(b) It depends on the particle's velocity.
(c) It depends on the strength of the external magnetic field.
(d) It acts at right angles to the direction of the particle's motion.
(e) None of the above; all of these statements are valid.

12. Two parallel wires are vertical. The one on the left carries a 10-A current upward. The other carries 5-A current downward. Compare the magnitude of the force that each wire exerts on the other.
(a) The wire on the left carries twice as much current, so it exerts twice the force on the right wire as the right one exerts on the left one.
(b) The wire on the left exerts a smaller force. It creates a magnetic field twice that due to the wire on the right; and therefore has less energy to cause a force on the wire on the right.
(c) The two wires exert the same force on each other.
(d) Not enough information; we need the length of the wire.

For assigned homework and other learning materials, go to the MasteringPhysics website.

Problems

20–3 Force on Electric Current in Magnetic Field

1. (I) (a) What is the force per meter of length on a straight wire carrying a 6.40-A current when perpendicular to a 0.90-T uniform magnetic field? (b) What if the angle between the wire and field is 35.0°?

2. (I) How much current is flowing in a wire 4.80 m long if the maximum force on it is 0.625 N when placed in a uniform 0.0800-T field?

3. (I) A 240-m length of wire stretches between two towers and carries a 120-A current. Determine the magnitude of the force on the wire due to the Earth's magnetic field of 5.0×10^{-5} T which makes an angle of 68° with the wire.

4. (I) A 2.6-m length of horizontal wire carries a 4.5-A current toward the south. The dip angle of the Earth's magnetic field makes an angle of 41° to the wire. Estimate the magnitude of the magnetic force on the wire due to the Earth's magnetic field of 5.5×10^{-5} T.

5. (I) The magnetic force per meter on a wire is measured to be only 45% of its maximum possible value. What is the angle between the wire and the magnetic field?

6. (II) The force on a wire carrying 6.45 A is a maximum of 1.28 N when placed between the pole faces of a magnet. If the pole faces are 55.5 cm in diameter, what is the approximate strength of the magnetic field?

7. (II) The force on a wire is a maximum of 8.50×10^{-2} N when placed between the pole faces of a magnet. The current flows horizontally to the right and the magnetic field is vertical. The wire is observed to "jump" toward the observer when the current is turned on. (a) What type of magnetic pole is the top pole face? (b) If the pole faces have a diameter of 10.0 cm, estimate the current in the wire if the field is 0.220 T. (c) If the wire is tipped so that it makes an angle of 10.0° with the horizontal, what force will it now feel? [*Hint*: What length of wire will now be in the field?]

8. (II) Suppose a straight 1.00-mm-diameter copper wire could just "float" horizontally in air because of the force due to the Earth's magnetic field $\vec{\mathbf{B}}$, which is horizontal, perpendicular to the wire, and of magnitude 5.0×10^{-5} T. What current would the wire carry? Does the answer seem feasible? Explain briefly.

20–4 Force on Charge Moving in Magnetic Field

9. (I) Determine the magnitude and direction of the force on an electron traveling 7.75×10^5 m/s horizontally to the east in a vertically upward magnetic field of strength 0.45 T.

10. (I) An electron is projected vertically upward with a speed of 1.70×10^6 m/s into a uniform magnetic field of 0.640 T that is directed horizontally away from the observer. Describe the electron's path in this field.

11. (I) Alpha particles (charge $q = +2e$, mass $m = 6.6 \times 10^{-27}$ kg) move at 1.6×10^6 m/s. What magnetic field strength would be required to bend them into a circular path of radius $r = 0.14$ m?

12. (I) Find the direction of the force on a negative charge for each diagram shown in Fig. 20–52, where $\vec{v}$ (green) is the velocity of the charge and $\vec{B}$ (blue) is the direction of the magnetic field. ($\otimes$ means the vector points inward. $\odot$ means it points outward, toward you.)

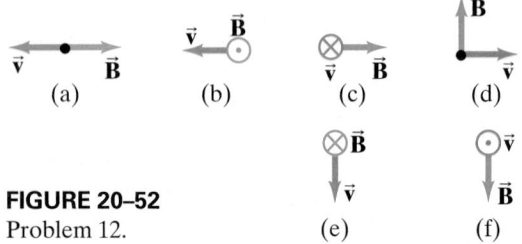

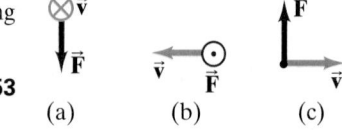

FIGURE 20–52
Problem 12.

13. (I) Determine the direction of $\vec{B}$ for each case in Fig. 20–53, where $\vec{F}$ represents the maximum magnetic force on a positively charged particle moving with velocity $\vec{v}$.

FIGURE 20–53
Problem 13. (a) (b) (c)

14. (II) Determine the velocity of a beam of electrons that goes undeflected when moving perpendicular to an electric and to a magnetic field. $\vec{E}$ and $\vec{B}$ are also perpendicular to each other and have magnitudes 7.7×10^3 V/m and 7.5×10^{-3} T, respectively. What is the radius of the electron orbit if the electric field is turned off?

15. (II) A helium ion ($Q = +2e$) whose mass is 6.6×10^{-27} kg is accelerated by a voltage of 3700 V. (a) What is its speed? (b) What will be its radius of curvature if it moves in a plane perpendicular to a uniform 0.340-T field? (c) What is its period of revolution?

16. (II) For a particle of mass m and charge q moving in a circular path in a magnetic field B, (a) show that its kinetic energy is proportional to r^2, the square of the radius of curvature of its path. (b) Show that its angular momentum is $L = qBr^2$, around the center of the circle.

17. (II) A 1.5-MeV (kinetic energy) proton enters a 0.30-T field, in a plane perpendicular to the field. What is the radius of its path? See Section 17–4.

18. (II) An electron experiences the greatest force as it travels 2.8×10^6 m/s in a magnetic field when it is moving northward. The force is vertically upward and of magnitude 6.2×10^{-13} N. What is the magnitude and direction of the magnetic field?

19. (II) A proton and an electron have the same kinetic energy upon entering a region of constant magnetic field. What is the ratio of the radii of their circular paths?

20. (III) A proton (mass m_p), a deuteron ($m = 2m_p$, $Q = e$), and an alpha particle ($m = 4m_p$, $Q = 2e$) are accelerated by the same potential difference V and then enter a uniform magnetic field $\vec{B}$, where they move in circular paths perpendicular to $\vec{B}$. Determine the radius of the paths for the deuteron and alpha particle in terms of that for the proton.

21. (III) A 3.40-g bullet moves with a speed of 155 m/s perpendicular to the Earth's magnetic field of 5.00×10^{-5} T. If the bullet possesses a net charge of 18.5×10^{-9} C, by what distance will it be deflected from its path due to the Earth's magnetic field after it has traveled 1.50 km?

*22. (III) A **Hall probe**, consisting of a thin rectangular slab of current-carrying material, is calibrated by placing it in a known magnetic field of magnitude 0.10 T. When the field is oriented normal to the slab's rectangular face, a Hall emf of 12 mV is measured across the slab's width. The probe is then placed in a magnetic field of unknown magnitude B, and a Hall emf of 63 mV is measured. Determine B assuming that the angle θ between the unknown field and the plane of the slab's rectangular face is (a) $\theta = 90°$, and (b) $\theta = 60°$.

*23. (III) The Hall effect can be used to measure **blood flow rate** because the blood contains ions that constitute an electric current. (a) Does the sign of the ions influence the emf? Explain. (b) Determine the flow velocity in an artery 3.3 mm in diameter if the measured emf across the width of the artery is 0.13 mV and B is 0.070 T. (In actual practice, an alternating magnetic field is used.)

*24. (III) A long copper strip 1.8 cm wide and 1.0 mm thick is placed in a 1.2-T magnetic field as in Fig. 20–21a. When a steady current of 15 A passes through it, the Hall emf is measured to be 1.02 μV. Determine (a) the drift velocity of the electrons and (b) the density of free (conducting) electrons (number per unit volume) in the copper. [*Hint*: See also Section 18–8.]

20–5 and 20–6 Magnetic Field of Straight Wire, Force between Two Wires

25. (I) Jumper cables used to start a stalled vehicle often carry a 65-A current. How strong is the magnetic field 4.5 cm from one cable? Compare to the Earth's magnetic field (5.0×10^{-5} T).

26. (I) If an electric wire is allowed to produce a magnetic field no larger than that of the Earth (0.50×10^{-4} T) at a distance of 12 cm from the wire, what is the maximum current the wire can carry?

27. (I) Determine the magnitude and direction of the force between two parallel wires 25 m long and 4.0 cm apart, each carrying 25 A in the same direction.

28. (I) A vertical straight wire carrying an upward 28-A current exerts an attractive force per unit length of 7.8×10^{-4} N/m on a second parallel wire 9.0 cm away. What current (magnitude and direction) flows in the second wire?

29. (II) In Fig. 20–54, a long straight wire carries current I out of the page toward you. Indicate, with appropriate arrows, the direction and (relative) magnitude of $\vec{B}$ at each of the points C, D, and E in the plane of the page.

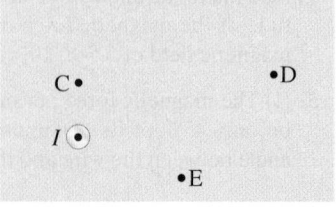

FIGURE 20–54
Problem 29.

30. (II) An experiment on the Earth's magnetic field is being carried out 1.00 m from an electric cable. What is the maximum allowable current in the cable if the experiment is to be accurate to ± 3.0%?

31. (II) A rectangular loop of wire is placed next to a straight wire, as shown in Fig. 20–55. There is a current of 3.5 A in both wires. Determine the magnitude and direction of the net force on the loop.

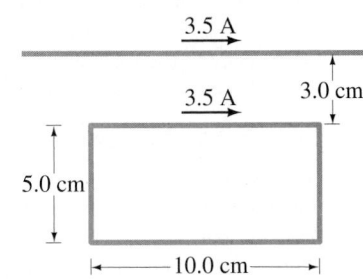

FIGURE 20–55
Problem 31.

32. (II) A horizontal compass is placed 18 cm due south from a straight vertical wire carrying a 48-A current downward. In what direction does the compass needle point at this location? Assume the horizontal component of the Earth's field at this point is 0.45×10^{-4} T and the magnetic declination is $0°$.

33. (II) A long horizontal wire carries 24.0 A of current due north. What is the net magnetic field 20.0 cm due west of the wire if the Earth's field there points downward, $44°$ below the horizontal, and has magnitude 5.0×10^{-5} T?

34. (II) A straight stream of protons passes a given point in space at a rate of 2.5×10^9 protons/s. What magnetic field do they produce 1.5 m from the beam?

35. (II) Determine the magnetic field midway between two long straight wires 2.0 cm apart in terms of the current I in one when the other carries 25 A. Assume these currents are (a) in the same direction, and (b) in opposite directions.

36. (II) Two straight parallel wires are separated by 7.0 cm. There is a 2.0-A current flowing in the first wire. If the magnetic field strength is found to be zero between the two wires at a distance of 2.2 cm from the first wire, what is the magnitude and direction of the current in the second wire?

37. (II) Two long straight wires each carry a current I out of the page toward the viewer, Fig. 20–56. Indicate, with appropriate arrows, the direction of $\vec{B}$ at each of the points 1 to 6 in the plane of the page. State if the field is zero at any of the points.

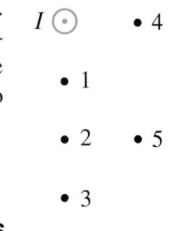

FIGURE 20–56
Problem 37.

38. (II) A power line carries a current of 95 A west along the tops of 8.5-m-high poles. (a) What is the magnitude and direction of the magnetic field produced by this wire at the ground directly below? How does this compare with the Earth's magnetic field of about $\frac{1}{2}$ G? (b) Where would the wire's magnetic field cancel the Earth's field?

39. (II) A compass needle points $17°$ E of N outdoors. However, when it is placed 12.0 cm to the east of a vertical wire inside a building, it points $32°$ E of N. What is the magnitude and direction of the current in the wire? The Earth's field there is 0.50×10^{-4} T and is horizontal.

40. (II) A long pair of insulated wires serves to conduct 24.5 A of dc current to and from an instrument. If the wires are of negligible diameter but are 2.8 mm apart, what is the magnetic field 10.0 cm from their midpoint, in their plane (Fig. 20–57)? Compare to the magnetic field of the Earth.

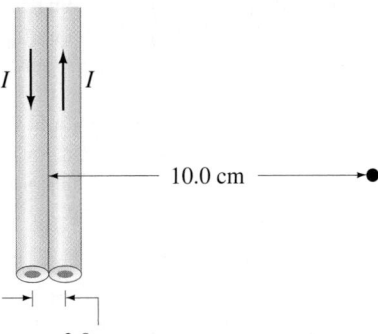

FIGURE 20–57
Problems 40 and 41.

41. (II) A third wire is placed in the plane of the two wires shown in Fig. 20–57 parallel and just to the right. If it carries 25.0 A upward, what force per meter of length does it exert on each of the other two wires? Assume it is 2.8 mm from the nearest wire, center to center.

42. (III) Two long thin parallel wires 13.0 cm apart carry 28-A currents in the same direction. Determine the magnetic field vector at a point 10.0 cm from one wire and 6.0 cm from the other (Fig. 20–58).

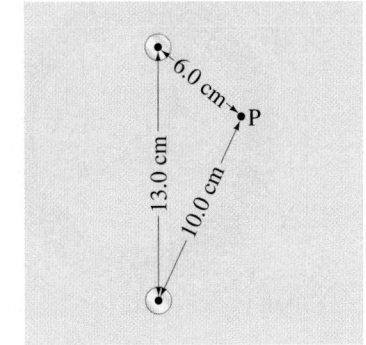

FIGURE 20–58
Problem 42.

43. (III) Two long wires are oriented so that they are perpendicular to each other. At their closest, they are 20.0 cm apart (Fig. 20–59). What is the magnitude of the magnetic field at a point midway between them if the top one carries a current of 20.0 A and the bottom one carries 12.0 A?

FIGURE 20–59
Problem 43.

20–7 Solenoids and Electromagnets

44. (I) A thin 12-cm-long solenoid has a total of 460 turns of wire and carries a current of 2.0 A. Calculate the field inside the solenoid near the center.

45. (I) A 30.0-cm-long solenoid 1.25 cm in diameter is to produce a field of 4.65 mT at its center. How much current should the solenoid carry if it has 935 turns of the wire?

46. (I) A 42-cm-long solenoid, 1.8 cm in diameter, is to produce a 0.030-T magnetic field at its center. If the maximum current is 4.5 A, how many turns must the solenoid have?

47. (II) A 550-turn horizontal solenoid is 15 cm long. The current in its coils is 38 A. A straight wire cuts through the center of the solenoid, along a 3.0-cm diameter. This wire carries a 22-A current downward (and is connected by other wires that don't concern us). What is the force on this wire assuming the solenoid's magnetic field points due east?

48. (III) You have 1.0 kg of copper and want to make a practical solenoid that produces the greatest possible magnetic field for a given voltage. Should you make your copper wire long and thin, short and fat, or something else? Consider other variables, such as solenoid diameter, length, and so on. Explain your reasoning.

20–8 Ampère's Law

49. (III) A *toroid* is a solenoid in the shape of a donut (Fig. 20–60). Use Ampère's law along the circular paths, shown dashed in Fig. 20–60a, to determine that the magnetic field (a) inside the toroid is $B = \mu_0 NI/2\pi R$, where N is the total number of turns, and (b) outside the toroid is $B = 0$. (c) Is the field inside a toroid uniform like a solenoid's? If not, how does it vary?

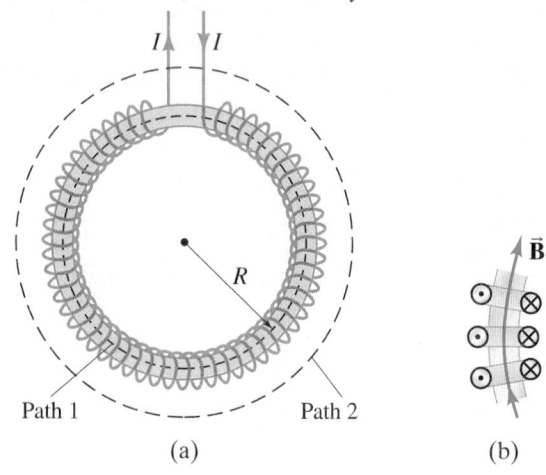

Path 1

R

Path 2

(a)

$\vec{B}$

(b)

FIGURE 20–60 Problem 49. (a) A toroid or torus. (b) A section of the toroid showing direction of the current for three loops: ⊙ means current toward you, and ⊗ means current away from you.

50. (III) (a) Use Ampère's law to show that the magnetic field between the conductors of a **coaxial cable** (Fig. 20–61) is $B = \mu_0 I/2\pi r$ if r (distance from center) is greater than the radius of the inner wire and less than the radius of the outer cylindrical braid (= ground). (b) Show that $B = 0$ outside the coaxial cable.

FIGURE 20–61
Coaxial cable.
Problem 50.

Insulating sleeve

I r

Cylindrical braid

I Solid wire

20–9 and 20–10 Torque on Current Loop, Motors, Galvanometers

51. (I) A single square loop of wire 22.0 cm on a side is placed with its face parallel to the magnetic field as in Fig. 20–34b. When 5.70 A flows in the coil, the torque on it is 0.325 m·N. What is the magnetic field strength?

52. (I) If the current to a motor drops by 12%, by what factor does the output torque change?

53. (I) A galvanometer needle deflects full scale for a 53.0-μA current. What current will give full-scale deflection if the magnetic field weakens to 0.760 of its original value?

54. (II) A circular coil 12.0 cm in diameter and containing nine loops lies flat on the ground. The Earth's magnetic field at this location has magnitude 5.50×10^{-5} T and points into the Earth at an angle of 56.0° below a line pointing due north. If a 7.20-A clockwise current passes through the coil, (a) determine the torque on the coil, and (b) which edge of the coil rises up: north, east, south, or west?

*20–11 Mass Spectrometer

*55. (I) Protons move in a circle of radius 6.10 cm in a 0.566-T magnetic field. What value of electric field could make their paths straight? In what direction must the electric field point?

*56. (I) In a mass spectrometer, germanium atoms have radii of curvature equal to 21.0, 21.6, 21.9, 22.2, and 22.8 cm. The largest radius corresponds to an atomic mass of 76 u. What are the atomic masses of the other isotopes?

*57. (II) Suppose the electric field between the electric plates in the mass spectrometer of Fig. 20–41 is 2.88×10^4 V/m and the magnetic fields are $B = B' = 0.68$ T. The source contains carbon isotopes of mass numbers 12, 13, and 14 from a long-dead piece of a tree. (To estimate masses of the atoms, multiply by 1.67×10^{-27} kg.) How far apart are the lines formed by the singly charged ions of each type on the photographic film? What if the ions were doubly charged?

*58. (II) One form of mass spectrometer accelerates ions by a voltage V before they enter a magnetic field B. The ions are assumed to start from rest. Show that the mass of an ion is $m = qB^2R^2/2V$, where R is the radius of the ions' path in the magnetic field and q is their charge.

*59. (II) An unknown particle moves in a straight line through crossed electric and magnetic fields with $E = 1.5$ kV/m and $B = 0.034$ T. If the electric field is turned off, the particle moves in a circular path of radius $r = 2.7$ cm. What might the particle be?

*60. (III) A mass spectrometer is monitoring air pollutants. It is difficult, however, to separate molecules of nearly equal mass such as CO (28.0106 u) and N_2 (28.0134 u). How large a radius of curvature must a spectrometer have (Fig. 20–41) if these two molecules are to be separated on the film by 0.50 mm?

*20–12 Ferromagnetism, Hysteresis

*61. (I) A long thin iron-core solenoid has 380 loops of wire per meter, and a 350-mA current flows through the wire. If the permeability of the iron is $3000\mu_0$, what is the total field B inside the solenoid?

*62. (II) An iron-core solenoid is 38 cm long and 1.8 cm in diameter, and has 780 turns of wire. The magnetic field inside the solenoid is 2.2 T when 48 A flows in the wire. What is the permeability μ at this high field strength?

*63. (II) The following are some values of B and B_0 for a piece of iron as it is being magnetized (note different units):

$B_0(10^{-4}$ T)	0.0	0.13	0.25	0.50	0.63	0.78	1.0	1.3
B(T)	0.0	0.0042	0.010	0.028	0.043	0.095	0.45	0.67

$B_0(10^{-4}$ T)	1.9	2.5	6.3	13.0	130	1300	10,000
B(T)	1.01	1.18	1.44	1.58	1.72	2.26	3.15

Determine the magnetic permeability μ for each value and plot a graph of μ versus B_0.

General Problems

64. Two long straight parallel wires are 15 cm apart. Wire A carries 2.0-A current. Wire B's current is 4.0 A in the same direction. (a) Determine the magnetic field magnitude due to wire A at the position of wire B. (b) Determine the magnetic field due to wire B at the position of wire A. (c) Are these two magnetic fields equal and opposite? Why or why not? (d) Determine the force on wire A due to wire B, and the force on wire B due to wire A. Are these two forces equal and opposite? Why or why not?

65. Protons with momentum 4.8×10^{-21} kg·m/s are magnetically steered clockwise in a circular path 2.2 m in diameter. Determine the magnitude and direction of the field in the magnets surrounding the beam pipe.

66. A small but rigid ∪-shaped wire carrying a 5.0-A current (Fig. 20–62) is placed inside a solenoid. The solenoid is 15.0 cm long and has 700 loops of wire, and the current in each loop is 7.0 A. What is the net force on the ∪-shaped wire?

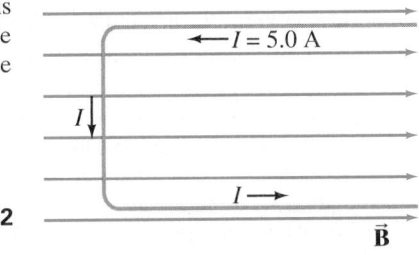

FIGURE 20–62
Problem 66.

67. The power cable for an electric trolley (Fig. 20–63) carries a horizontal current of 330 A toward the east. The Earth's magnetic field has a strength 5.0×10^{-5} T and makes an angle of dip of 22° at this location. Calculate the magnitude and direction of the magnetic force on an 18-m length of this cable.

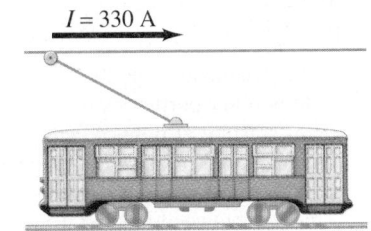

FIGURE 20–63
Problem 67.

68. A particle of charge q moves in a circular path of radius r perpendicular to a uniform magnetic field B. Determine its linear momentum in terms of the quantities given.

69. An airplane has acquired a net charge of 1280 μC. If the Earth's magnetic field of 5.0×10^{-5} T is perpendicular to the airplane's velocity of magnitude 120 m/s, determine the force on the airplane.

70. A 32-cm-long solenoid, 1.8 cm in diameter, is to produce a 0.050-T magnetic field at its center. If the maximum current is 6.4 A, how many turns must the solenoid have?

71. Near the equator, the Earth's magnetic field points almost horizontally to the north and has magnitude $B = 0.50 \times 10^{-4}$ T. What should be the magnitude and direction for the velocity of an electron if its weight is to be exactly balanced by the magnetic force?

72. A doubly charged helium atom, whose mass is 6.6×10^{-27} kg, is accelerated by a voltage of 3200 V. (a) What will be its radius of curvature in a uniform 0.240-T field? (b) What is its period of revolution?

73. Four very long straight parallel wires, located at the corners of a square of side ℓ, carry equal currents I_0 perpendicular to the page as shown in Fig. 20–64. Determine the magnitude and direction of $\vec{B}$ at the center C of the square.

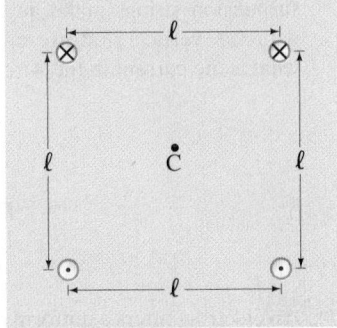

FIGURE 20–64
Problem 73.

74. (a) What value of magnetic field would make a beam of electrons, traveling to the west at a speed of 4.8×10^6 m/s, go undeflected through a region where there is a uniform electric field of 12,000 V/m pointing south? (b) What is the direction of the magnetic field if it is perpendicular to the electric field? (c) What is the frequency of the circular orbit of the electrons if the electric field is turned off?

75. Magnetic fields are very useful in particle accelerators for "beam steering"; that is, the magnetic fields can be used to change the direction of the beam of charged particles without altering their speed (Fig. 20–65). Show how this could work with a beam of protons. What happens to protons that are not moving with the speed for which the magnetic field was designed? If the field extends over a region 5.0 cm wide and has a magnitude of 0.41 T, by approximately what angle θ will a beam of protons traveling at 2.5×10^6 m/s be bent?

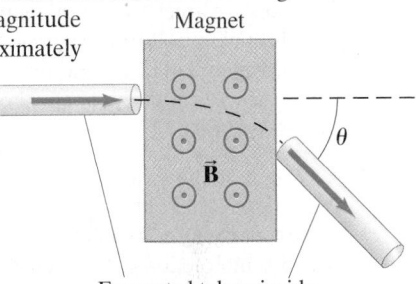

FIGURE 20–65
Problem 75.

Evacuated tubes, inside of which the protons move with velocity indicated by the green arrows

76. The magnetic field B at the center of a circular coil of wire carrying a current I (as in Fig. 20–9) is

$$B = \frac{\mu_0 NI}{2r},$$

where N is the number of loops in the coil and r is its radius. Imagine a simple model in which the Earth's magnetic field of about 1 G $(= 1 \times 10^{-4}$ T) near the poles is produced by a single current loop around the equator. Roughly estimate the current this loop would carry.

77. A proton follows a spiral path through a gas in a uniform magnetic field of 0.010 T, perpendicular to the plane of the spiral, as shown in Fig. 20–66. In two successive loops, at points P and Q, the radii are 10.0 mm and 8.5 mm, respectively. Calculate the change in the kinetic energy of the proton as it travels from P to Q.

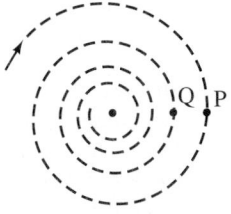

FIGURE 20–66
Problem 77.

78. Two long straight aluminum wires, each of diameter 0.42 mm, carry the same current but in opposite directions. They are suspended by 0.50-m-long strings as shown in Fig. 20–67. If the suspension strings make an angle of 3.0° with the vertical and are hanging freely, what is the current in the wires?

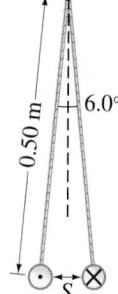

FIGURE 20–67
Problem 78.

79. An electron enters a uniform magnetic field $B = 0.23$ T at a 45° angle to $\vec{B}$. Determine the radius r and pitch p (distance between loops) of the electron's helical path assuming its speed is 3.0×10^6 m/s. See Fig. 20–68.

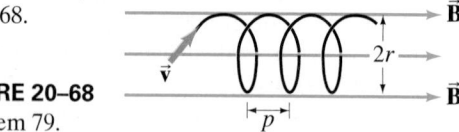

FIGURE 20–68
Problem 79.

80. A motor run by a 9.0-V battery has a 20-turn square coil with sides of length 5.0 cm and total resistance 28 Ω. When spinning, the magnetic field felt by the wire in the coil is 0.020 T. What is the maximum torque on the motor?

81. Electrons are accelerated horizontally by 2.2 kV. They then pass through a uniform magnetic field B for a distance of 3.8 cm, which deflects them upward so they reach the top of a screen 22 cm away, 11 cm above the center. Estimate the value of B.

82. A 175-g model airplane charged to 18.0 mC and traveling at 3.4 m/s passes within 8.6 cm of a wire, nearly parallel to its path, carrying a 25-A current. What acceleration (in g's) does this interaction give the airplane?

83. A uniform conducting rod of length ℓ and mass m sits atop a fulcrum, which is placed a distance $\ell/4$ from the rod's left-hand end and is immersed in a uniform magnetic field of magnitude B directed into the page (Fig. 20–69). An object whose mass M is 6.0 times greater than the rod's mass is hung from the rod's left-hand end. What current (direction and magnitude) should flow through the rod in order for it to be "balanced" (i.e., be at rest horizontally) on the fulcrum? (Flexible connecting wires which exert negligible force on the rod are not shown.)

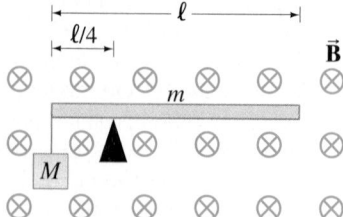

FIGURE 20–69
Problem 83.

84. Suppose the Earth's magnetic field at the equator has magnitude 0.50×10^{-4} T and a northerly direction at all points. Estimate the speed a singly ionized uranium ion ($m = 238$ u, $q = +e$) would need to circle the Earth 6.0 km above the equator. Can you ignore gravity? [Ignore relativity.]

85. A particle with charge q and momentum p, initially moving along the x axis, enters a region where a uniform magnetic field B_0 extends over a width $x = \ell$ as shown in Fig. 20–70. The particle is deflected a distance d in the $+y$ direction as it traverses the field. Determine (a) whether q is positive or negative, and (b) the magnitude of its momentum p in terms of q, B_0, ℓ, and d.

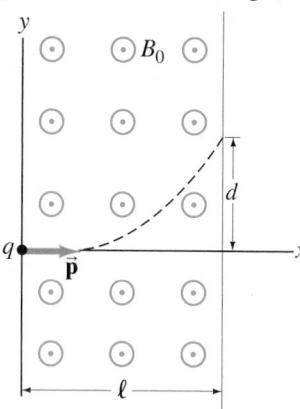

FIGURE 20–70
Problem 85.

86. A bolt of lightning strikes a metal flag pole, one end of which is anchored in the ground. Estimate the force the Earth's magnetic field can exert on the flag pole while the lightning-induced current flows. See Example 18–10.

87. A sort of "projectile launcher" is shown in Fig. 20–71. A large current moves in a closed loop composed of fixed rails, a power supply, and a very light, almost frictionless bar (pale green) touching the rails. A magnetic field is perpendicular to the plane of the circuit. If the bar has a length $\ell = 28$ cm, a mass of 1.5 g, and is placed in a field of 1.7 T, what constant current flow is needed to accelerate the bar from rest to 28 m/s in a distance of 1.0 m? In what direction must the magnetic field point?

FIGURE 20–71
Problem 87.

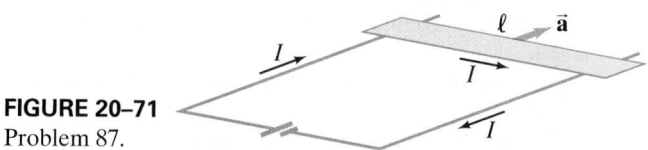

88. The **cyclotron** (Fig. 20–72) is a device used to accelerate elementary particles such as protons to high speeds. Particles starting at point A with some initial velocity travel in semicircular orbits in the magnetic field B. The particles are accelerated to higher speeds each time they pass through the gap between the metal "dees," where there is an electric field E. (There is no electric field inside the hollow metal dees where the electrons move in circular paths.) The electric field changes direction each half-cycle, owing to an ac voltage $V = V_0 \sin 2\pi ft$, so that the particles are increased in speed at each passage through the gap. (a) Show that the frequency f of the voltage must be $f = Bq/2\pi m$, where q is the charge on the particles and m their mass. (b) Show that the kinetic energy of the particles increases by $2qV_0$ each revolution, assuming that the gap is small. (c) If the radius of the cyclotron is 2.0 m and the magnetic field strength is 0.50 T, what will be the maximum kinetic energy of accelerated protons in MeV?

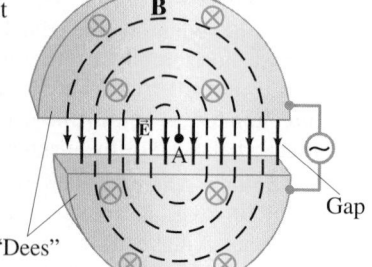

FIGURE 20–72
A cyclotron.   "Dees"
Problem 88.

588 **CHAPTER 20 Magnetism**

89. Three long parallel wires are 3.8 cm from one another. (Looking along them, they are at three corners of an equilateral triangle.) The current in each wire is 8.00 A, but its direction in wire M is opposite to that in wires N and P (Fig. 20–73). (*a*) Determine the magnetic force per unit length on each wire due to the other two. (*b*) In Fig. 20–73, determine the magnitude and direction of the magnetic field at the midpoint of the line between wire M and wire N.

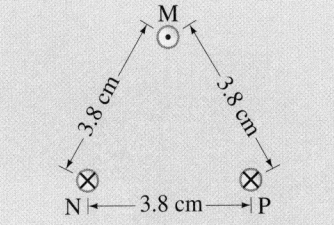

FIGURE 20–73
Problems 89 and 90.

90. In Fig. 20–73 the top wire is 1.00-mm-diameter copper wire and is suspended in air due to the two magnetic forces from the bottom two wires. The current flow through the two bottom wires is 75 A in each. Calculate the required current flow in the suspended wire (M).

91. You want to get an idea of the magnitude of magnetic fields produced by overhead power lines. You estimate that a transmission wire is about 13 m above the ground. The local power company tells you that the lines operate at 240 kV and provide a maximum power of 46 MW. Estimate the magnetic field you might experience walking under one such power line, and compare to the Earth's field.

92. Two long parallel wires 8.20 cm apart carry 19.2-A currents in the same direction. Determine the magnetic field vector at a point P, 12.0 cm from one wire and 13.0 cm from the other (Fig. 20–74). [*Hint*: Use the law of cosines; see Appendix A or inside rear cover.]

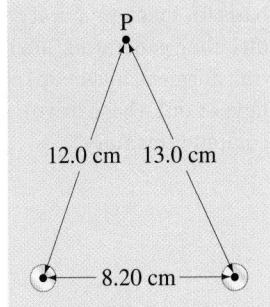

FIGURE 20–74
Problem 92.

Search and Learn

1. How many magnetic force equations are there in Chapter 20? List each one and explain when it applies. For each magnetic force equation, show how the units work out to give force in newtons.

2. An electron is moving north at a constant speed of 3.0×10^4 m/s. (*a*) In what direction should an electric field point if the electron is to be accelerated to the east? (*b*) In what direction should a magnetic field point if the electron is to be accelerated to the west? (*c*) If the electric field of part *a* has a strength of 330 V/m, what magnetic field (magnitude and direction) will produce zero net force on the electron? (*d*) If the electron in part *c* is moving faster than 3.0×10^4 m/s, in which direction will it be accelerated? What if it is moving slower than 3.0×10^4 m/s? (*e*) Now consider electrons that move perpendicular to both a magnetic field and to an electric field, which are perpendicular to each other. If only electrons with speeds of 5.5×10^4 m/s go straight through undeflected, what is the ratio of the magnitudes of electric field to magnetic field? Without knowing the value of the electric field, can you know the value of the magnetic field?

3. (*a*) A particle of charge *q* moves in a circular path of radius *r* in a uniform magnetic field $\vec{B}$. If the magnitude of the magnetic field is double, and the kinetic energy of the particle is the same, how does the angular momentum of the particle differ? (*b*) Show that the magnetic dipole moment *M* (Section 20–9) of an electron orbiting the proton nucleus of a hydrogen atom is related to the orbital angular momentum *L* of the electron by

$$M = \frac{e}{2m}L.$$

4. (*a*) Two long parallel wires, each 2.0 mm in diameter and 9.00 cm apart, carry equal 1.0-A currents in the same direction, Fig. 20–75. Determine $\vec{B}$ along the *x* axis between the wires as a function of *x*. (*b*) Graph *B* vs. *x* from $x = 1.0$ mm to $x = 89.0$ mm.

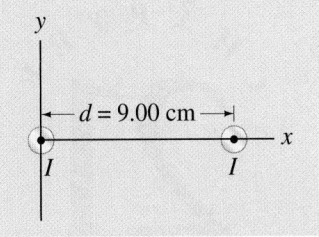

FIGURE 20–75
Search and Learn 4.

5. The force on a moving particle in a magnetic field is the idea behind **electromagnetic pumping**. It can be used to pump metallic fluids (such as sodium) and to pump blood in artificial heart machines. A basic design is shown in Fig. 20–76. For blood, an electric field is applied perpendicular to a blood vessel and to the magnetic field. Explain in detail how ions in the blood are caused to move. Do positive and negative ions feel a force in the same direction?

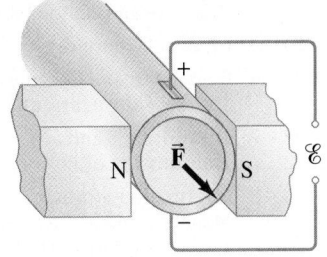

FIGURE 20–76
Electromagnetic pumping in a blood vessel.
Search and Learn 5.

ANSWERS TO EXERCISES

A: Near the poles, where the field lines are closer together.
B: Circles, pointing counterclockwise.
C: (*b*), (*c*), (*d*).
D: 0.15 N.
E: (*b*), (*c*), (*d*).

F: Zero.
G: Negative; the helical path would rotate in the opposite direction (still going to the right).
H: 0.8×10^{-5} T, up.

One of the great laws of physics is Faraday's law of induction, which says that a changing magnetic flux produces an induced emf. This photo shows a bar magnet moving into (or out of) a coil of wire, and the galvanometer registers an induced current. This phenomenon of electromagnetic induction is the basis for many practical devices, including generators, alternators, transformers, magnetic recording on tape or disk (hard drive), and computer memory.

CHAPTER 21

Electromagnetic Induction and Faraday's Law

CHAPTER-OPENING QUESTION—Guess now!

In the photograph above, the bar magnet is inserted down into the coil of wire, and is left there for 1 minute; then it is pulled up and out from the coil. What would an observer watching the galvanometer see?

(a) No change (pointer stays on zero): without a battery there is no current to detect.

(b) A small current flows while the magnet is inside the coil of wire.

(c) A current spike as the magnet enters the coil, and then nothing.

(d) A current spike as the magnet enters the coil, and then a steady small current.

(e) A current spike as the magnet enters the coil, then nothing (pointer at zero), then a current spike in the opposite direction as the magnet exits the coil.

In Chapter 20, we discussed two ways in which electricity and magnetism are related: (1) an electric current produces a magnetic field; and (2) a magnetic field exerts a force on an electric current or on a moving electric charge. These discoveries were made in 1820–1821. Scientists then began to wonder: if electric currents produce a magnetic field, is it possible that a magnetic field can produce an electric current? Ten years later the American Joseph Henry (1797–1878) and the Englishman Michael Faraday (1791–1867) independently found that it was possible. Henry actually made the discovery first. But Faraday published his results earlier and investigated the subject in more detail. We now discuss this phenomenon and some of its world-changing applications including the electric generator.

21–1 Induced EMF

In his attempt to produce an electric current from a magnetic field, Faraday used an apparatus like that shown in Fig. 21–1. A coil of wire, X, was connected to a battery. The current that flowed through X produced a magnetic field that was intensified by the ring-shaped iron core around which the wire was wrapped. Faraday hoped that a strong steady current in X would produce a great enough magnetic field to produce a current in a second coil Y wrapped on the same iron ring.

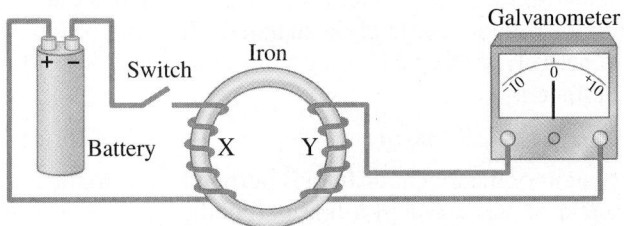

FIGURE 21–1 Faraday's experiment to induce an emf.

This second circuit, Y, contained a galvanometer to detect any current but contained no battery. He met no success with constant currents. But the long-sought effect was finally observed when Faraday noticed the galvanometer in circuit Y deflect strongly at the moment he closed the switch in circuit X. And the galvanometer deflected strongly in the opposite direction when he opened the switch in X. A constant current in X produced a constant magnetic field which produced *no* current in Y. Only when the current in X was starting or stopping was a current produced in Y.

Faraday concluded that although a constant magnetic field produces no current in a conductor, a *changing* magnetic field can produce an electric current. Such a current is called an **induced current**. When the magnetic field through coil Y changes, a current occurs in Y as if there were a source of emf in circuit Y. We therefore say that

a changing magnetic field induces an emf.

Faraday did further experiments on **electromagnetic induction**, as this phenomenon is called. For example, Fig. 21–2 shows that if a magnet is moved quickly into a coil of wire, a current is induced in the wire. If the magnet is quickly removed, a current is induced in the opposite direction ($\vec{\mathbf{B}}$ through the coil decreases). Furthermore, if the magnet is held steady and the coil of wire is moved toward or away from the magnet, again an emf is induced and a current flows. Motion or change is required to induce an emf. It doesn't matter whether the magnet or the coil moves. It is their *relative motion* that counts.

⚠ **CAUTION**

Changing $\vec{\mathbf{B}}$, *not* $\vec{\mathbf{B}}$ *itself, induces current*

⚠ **CAUTION**

Relative motion—magnet or coil moving induces current

FIGURE 21–2 (a) A current is induced when a magnet is moved toward a coil, momentarily increasing the magnetic field through the coil. (b) The induced current is opposite when the magnet is moved away from the coil ($\vec{\mathbf{B}}$ decreases). Note that the galvanometer zero is at the center of the scale and the needle deflects left or right, depending on the direction of the current. In (c), no current is induced if the magnet does not move relative to the coil. It is the relative motion that counts here: the magnet can be held steady and the coil moved, which also induces an emf.

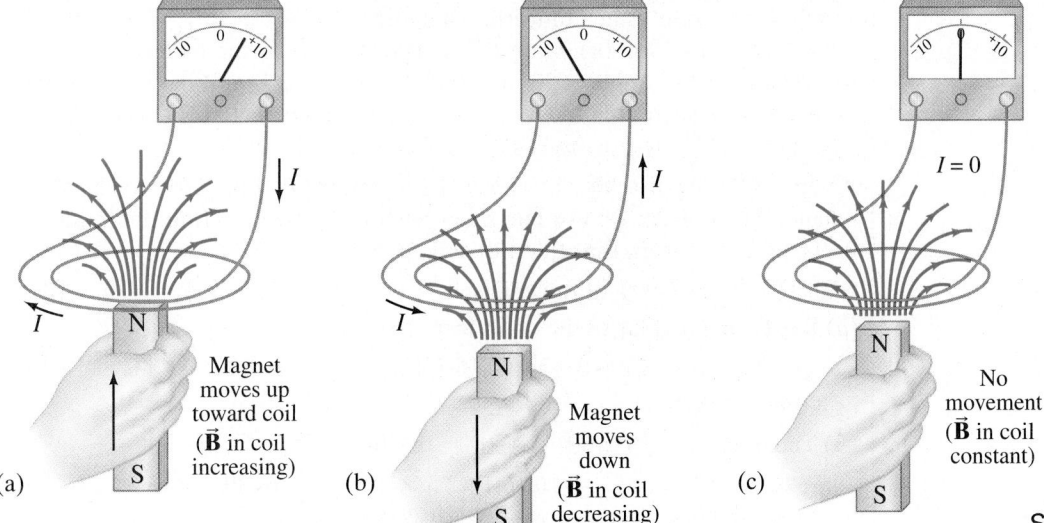

21–2 Faraday's Law of Induction; Lenz's Law

Faraday investigated quantitatively what factors influence the magnitude of the emf induced. He found first of all that the more rapidly the magnetic field changes, the greater the induced emf. He also found that the induced emf depends on the area of the circuit loop (and also the angle it makes with $\vec{\mathbf{B}}$). In fact, it is found that the emf is proportional to the rate of change of the **magnetic flux**, Φ_B, passing through the circuit or loop of area A. Magnetic flux for a uniform magnetic field through a loop of area A is defined as

$$\Phi_B = B_\perp A = BA \cos\theta. \qquad [B \text{ uniform}] \quad \textbf{(21–1)}$$

Here $B_\perp$ is the component of the magnetic field $\vec{\mathbf{B}}$ perpendicular to the face of the loop, and θ is the angle between $\vec{\mathbf{B}}$ and a line perpendicular to the face of the loop. These quantities are shown in Fig. 21–3 for a square loop of side ℓ whose area is $A = \ell^2$. When the face of the loop is parallel to $\vec{\mathbf{B}}$, $\theta = 90°$ and $\Phi_B = 0$. When $\vec{\mathbf{B}}$ is perpendicular to the face of the loop, $\theta = 0°$, and

$$\Phi_B = BA. \qquad [\text{uniform } \vec{\mathbf{B}} \perp \text{loop face}]$$

As we saw in Chapter 20, the lines of $\vec{\mathbf{B}}$ (like lines of $\vec{\mathbf{E}}$) can be drawn such that the number of lines per unit area is proportional to the field strength. Then the flux Φ_B can be thought of as being proportional to the *total number of lines passing through the area enclosed by the loop.* This is illustrated in Fig. 21–4, where three wire loops of a coil are viewed from the side (on edge). For $\theta = 90°$, no magnetic field lines pass through the loops and $\Phi_B = 0$, whereas Φ_B is a maximum when $\theta = 0°$. The unit of magnetic flux is the tesla-meter²; this is called a **weber**: $1\ \text{Wb} = 1\ \text{T} \cdot \text{m}^2$.

With our definition of flux, Eq. 21–1, we can write down the results of Faraday's investigations: The emf $\mathscr{E}$ induced in a circuit is equal to the rate of change of magnetic flux through the circuit:

$$\mathscr{E} = -\frac{\Delta \Phi_B}{\Delta t}. \qquad [1 \text{ loop}] \quad \textbf{(21–2a)}$$

This fundamental result is known as **Faraday's law of induction**, and it is one of the basic laws of electromagnetism.

If the circuit contains N loops that are closely wrapped so the same flux passes through each, the emfs induced in each loop add together, so the total emf is

$$\mathscr{E} = -N\frac{\Delta \Phi_B}{\Delta t}. \qquad [N \text{ loops}] \quad \textbf{(21–2b)}$$

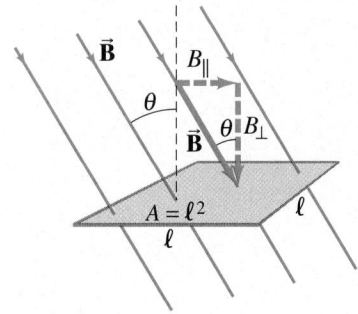

FIGURE 21–3 Determining the flux through a flat loop of wire. This loop is square, of side ℓ and area $A = \ell^2$.

FIGURE 21–4 Magnetic flux Φ_B is proportional to the number of lines of $\vec{\mathbf{B}}$ that pass through the loops of a coil (here with 3 loops).

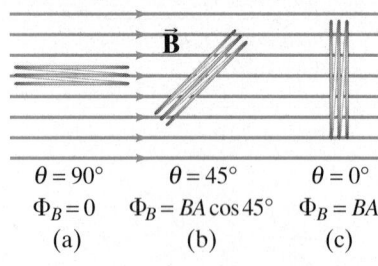

$\theta = 90°$	$\theta = 45°$	$\theta = 0°$
$\Phi_B = 0$	$\Phi_B = BA\cos 45°$	$\Phi_B = BA$
(a)	(b)	(c)

> *FARADAY'S LAW*
> *OF INDUCTION*

> *FARADAY'S LAW*
> *OF INDUCTION*

EXAMPLE 21–1 **A loop of wire in a magnetic field.** A square loop of wire of side $\ell = 5.0\ \text{cm}$ is in a uniform magnetic field $B = 0.16\ \text{T}$. What is the magnetic flux in the loop (a) when $\vec{\mathbf{B}}$ is perpendicular to the face of the loop and (b) when $\vec{\mathbf{B}}$ is at an angle of 30° to the area of the loop? (c) What is the magnitude of the average current in the loop if it has a resistance of $0.012\ \Omega$ and it is rotated from position (b) to position (a) in 0.14 s?

APPROACH We use the definition $\Phi_B = BA\cos\theta$, Eq. 21–1, to calculate the magnetic flux. Then we use Faraday's law of induction to find the induced emf in the coil, and from that the induced current ($I = \mathscr{E}/R$).

SOLUTION The area of the coil is $A = \ell^2 = (5.0 \times 10^{-2}\ \text{m})^2 = 2.5 \times 10^{-3}\ \text{m}^2$. (a) $\vec{\mathbf{B}}$ is perpendicular to the coil's face, so $\theta = 0°$ and

$$\Phi_B = BA\cos 0° = (0.16\ \text{T})(2.5 \times 10^{-3}\ \text{m}^2)(1) = 4.0 \times 10^{-4}\ \text{T} \cdot \text{m}^2$$

or $4.0 \times 10^{-4}\ \text{Wb}$.

(b) The angle θ is 30° and $\cos 30° = 0.866$, so

$$\Phi_B = BA\cos\theta = (0.16\ \text{T})(2.5 \times 10^{-3}\ \text{m}^2)\cos 30° = 3.5 \times 10^{-4}\ \text{T} \cdot \text{m}^2$$

or $3.5 \times 10^{-4}\ \text{Wb}$, a bit less than in part (a).

(c) The magnitude of the induced emf (Eq. 21–2a) during the 0.14-s time interval is

$$\mathcal{E} = \frac{\Delta \Phi_B}{\Delta t} = \frac{(4.0 \times 10^{-4}\,\text{T} \cdot \text{m}^2) - (3.5 \times 10^{-4}\,\text{T} \cdot \text{m}^2)}{0.14\,\text{s}} = 3.6 \times 10^{-4}\,\text{V}.$$

Before and after the loop rotates, when it is at rest, the emf is zero. The current in the wire loop (Ohm's law) while it is rotating is

$$I = \frac{\mathcal{E}}{R} = \frac{3.6 \times 10^{-4}\,\text{V}}{0.012\,\Omega} = 0.030\,\text{A} = 30\,\text{mA}.$$

The minus signs in Eqs. 21–2a and b are there to remind us in which direction the induced emf acts. Experiments show that

> **a current produced by an induced emf moves in a direction so that the magnetic field created by that current opposes the original change in flux.**

This is known as **Lenz's law.** Be aware that we are now discussing two distinct magnetic fields: (1) the changing magnetic field or flux that induces the current, and (2) the magnetic field produced by the induced current (all currents produce a magnetic field). The second (induced) field opposes the *change* in the first.

Lenz's law can be said another way, valid even if no current can flow (as when a circuit is not complete):

> **An induced emf is always in a direction that opposes the original change in flux that caused it.**

Let us apply Lenz's law to the relative motion between a magnet and a coil, Fig. 21–2. The changing flux through the coil induces an emf in the coil, producing a current. This induced current produces its own magnetic field. In Fig. 21–2a the distance between the coil and the magnet decreases. The magnet's magnetic field (and number of field lines) through the coil increases, and therefore the flux increases. The magnetic field of the magnet points upward. To oppose the upward increase, the magnetic field produced by the induced current needs to point *downward* inside the coil. Thus, Lenz's law tells us the current moves as shown in Fig. 21–2a (use the right-hand rule). In Fig. 21–2b, the flux *decreases* (because the magnet is moved away and *B* decreases), so the induced current in the coil produces an *upward* magnetic field through the coil that is "trying" to maintain the status quo. Thus the current in Fig. 21–2b is in the opposite direction from Fig. 21–2a.

It is important to note that an emf is induced whenever there is a change in *flux* through the coil, and we now consider some more possibilities.

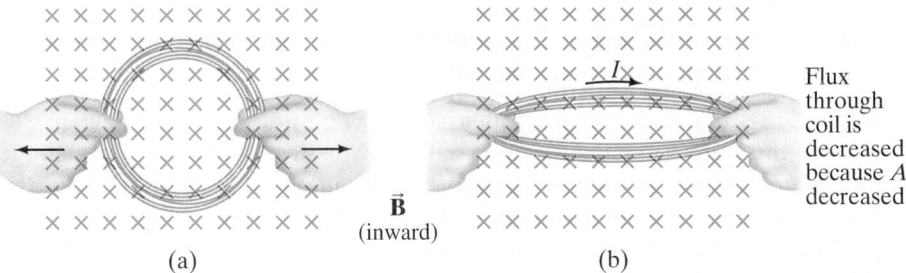

(a)

(b)

Flux through coil is decreased because *A* decreased

Magnetic flux $\Phi_B = BA \cos\theta$, so an emf can be induced in three ways: (1) by a changing magnetic field *B*; (2) by changing the area *A* of the loop in the field; or (3) by changing the loop's orientation θ with respect to the field. Figures 21–1 and 21–2 showed case 1. Cases 2 and 3 are illustrated in Figs. 21–5 and 21–6.

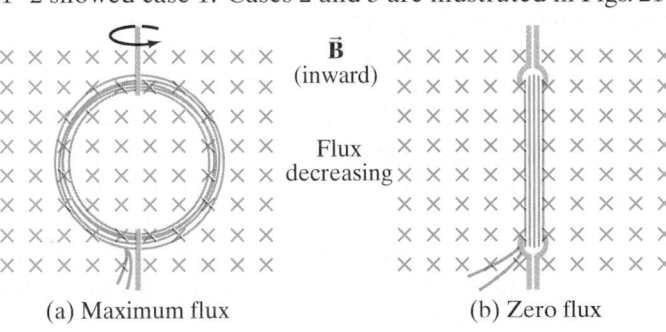

(a) Maximum flux

(b) Zero flux

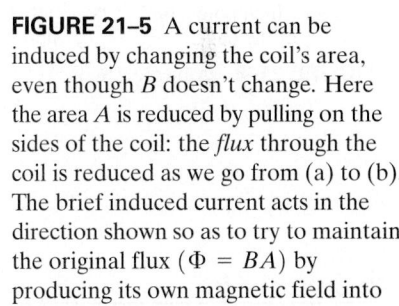

FIGURE 21–5 A current can be induced by changing the coil's area, even though *B* doesn't change. Here the area *A* is reduced by pulling on the sides of the coil: the *flux* through the coil is reduced as we go from (a) to (b). The brief induced current acts in the direction shown so as to try to maintain the original flux ($\Phi = BA$) by producing its own magnetic field into the page. That is, as area *A* decreases, the current acts to increase *B* in the original (inward) direction.

FIGURE 21–6 A current can be induced by rotating a coil in a magnetic field. The flux through the coil changes from (a) to (b) because θ (in Eq. 21–1, $\Phi = BA \cos\theta$) went from 0° ($\cos\theta = 1$) to 90° ($\cos\theta = 0$).

FIGURE 21–7 Example 21–2: An induction stove.

CONCEPTUAL EXAMPLE 21–2 | **Induction stove.** In an induction stove (Fig. 21–7), an ac current exists in a coil that is the "burner" (a burner that never gets hot). Why will it heat a metal pan, usually iron, but not a glass container?

RESPONSE The ac current sets up a changing magnetic field that passes through the pan bottom. This changing magnetic field induces a current in the pan, and since the pan offers resistance, electric energy is transformed to thermal energy which heats the pan and its contents. If the pan is iron, magnetic hysteresis due to the changing current produces additional heating. A glass container offers such high resistance that little current is induced and little energy is transferred $(P = V^2/R)$.

Lenz's Law

Lenz's law is used to determine the direction of the (conventional) electric current induced in a loop due to a change in magnetic flux inside the loop. To produce an induced current you need

(a) a closed conducting loop, and

(b) an external magnetic flux through the loop that is changing in time.

1. Determine whether the magnetic flux $(\Phi_B = BA \cos\theta)$ inside the loop is decreasing, increasing, or unchanged.

2. The magnetic field due to the induced current: (a) points in the same direction as the external field if the flux is decreasing; (b) points in the opposite direction from the external field if the flux is increasing; or (c) is zero if the flux is not changing.

3. Once you know the direction of the induced magnetic field, use right-hand-rule-1 (page 563, Chapter 20) to find the direction of the induced current.

4. Always keep in mind that there are two magnetic fields: (1) an external field whose flux must be changing if it is to induce an electric current, and (2) a magnetic field produced by the induced current.

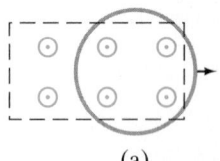
(a)
Pulling a round loop to the right out of a magnetic field which points out of the page

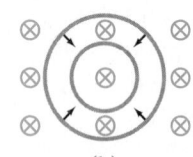
(b)
Shrinking a loop in a magnetic field pointing into the page

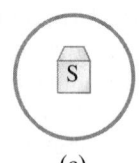
(c)
S magnetic pole moving from below, up toward the loop

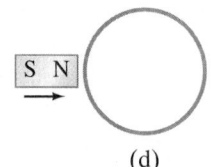
(d)
N magnetic pole moving toward loop in the plane of the loop

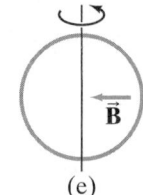
(e)
Rotating the loop by pulling the left side toward us and pushing the right side in; the magnetic field points from right to left

FIGURE 21–8 Example 21–3.

⬦ **CAUTION**

Magnetic field created by induced current opposes change in external flux, not necessarily opposing the external field

CONCEPTUAL EXAMPLE 21–3 | **Practice with Lenz's law.** In which direction is the current, induced in the circular loop for each situation in Fig. 21–8?

RESPONSE In (*a*), the magnetic field initially pointing out of the page passes through the loop. If you pull the loop out of the field, magnetic flux through the loop decreases; so the induced current will be in a direction to maintain the decreasing flux through the loop: the current will be counterclockwise to produce a magnetic field outward (toward the reader).

(*b*) The external field is into the page. The coil area gets smaller, so the flux will decrease; hence the induced current will be clockwise, producing its own field into the page to make up for the flux decrease.

(*c*) Magnetic field lines point into the S pole of a magnet, so as the magnet moves toward us and the loop, the magnet's field points into the page and is getting stronger. The current in the loop will be induced in the counterclockwise direction in order to produce a field $\vec{B}$ *out* of the page.

(*d*) The field is in the plane of the loop, so no magnetic field lines pass through the loop and the flux through the loop is zero throughout the process; hence there is no change in flux and no induced emf or current in the loop.

(*e*) Initially there is no flux through the loop. When you start to rotate the loop, the external field through the loop begins increasing to the left. To counteract this change in flux, the loop will have current induced in a counterclockwise direction so as to produce its own field to the right.

EXAMPLE 21–4 **Pulling a coil from a magnetic field.** A 100-loop square coil of wire, with side $\ell = 5.00$ cm and total resistance $R = 100\,\Omega$, is positioned perpendicular to a uniform magnetic field $B = 0.600$ T, as shown in Fig. 21–9. It is quickly pulled from the field at constant speed (moving perpendicular to $\vec{B}$) to a region where B drops abruptly to zero. At $t = 0$, the right edge of the coil is at the edge of the field. It takes 0.100 s for the whole coil to reach the field-free region. Determine (a) the rate of change in flux through one loop of the coil, and (b) the total emf and current induced in the 100-loop coil. (c) How much energy is dissipated in the coil? (d) What was the average force required (F_{ext})?

APPROACH We start by finding how the magnetic flux, $\Phi_B = BA\cos 0° = BA$, changes during the time interval $\Delta t = 0.100$ s. Faraday's law then gives the induced emf and Ohm's law gives the current.

SOLUTION (a) The coil's area is $A = \ell^2 = (5.00 \times 10^{-2}\,\text{m})^2 = 2.50 \times 10^{-3}\,\text{m}^2$. The flux through one loop is initially $\Phi_B = BA = (0.600\,\text{T})(2.50 \times 10^{-3}\,\text{m}^2) = 1.50 \times 10^{-3}$ Wb. After 0.100 s, the flux is zero. The rate of change in flux is constant (because the coil is square), and for one loop is equal to

$$\frac{\Delta \Phi_B}{\Delta t} = \frac{0 - (1.50 \times 10^{-3}\,\text{Wb})}{0.100\,\text{s}} = -1.50 \times 10^{-2}\,\text{Wb/s}.$$

(b) The emf induced (Eq. 21–2) in the 100-loop coil during this 0.100-s interval is

$$\mathcal{E} = -N\frac{\Delta \Phi_B}{\Delta t} = -(100)(-1.50 \times 10^{-2}\,\text{Wb/s}) = 1.50\,\text{V}.$$

The current is found by applying Ohm's law to the 100-Ω coil:

$$I = \frac{\mathcal{E}}{R} = \frac{1.50\,\text{V}}{100\,\Omega} = 1.50 \times 10^{-2}\,\text{A} = 15.0\,\text{mA}.$$

By Lenz's law, the current must be clockwise to produce more $\vec{B}$ into the page and thus oppose the decreasing flux into the page.

(c) The total energy dissipated in the coil is the product of the power $(= I^2R)$ and the time:

$$E = Pt = I^2Rt = (1.50 \times 10^{-2}\,\text{A})^2(100\,\Omega)(0.100\,\text{s}) = 2.25 \times 10^{-3}\,\text{J}.$$

(d) We can use the result of part (c) and apply the work-energy principle: the energy dissipated E is equal to the work W needed to pull the coil out of the field (Chapter 6). Because $W = \bar{F}_{ext}\,d$ where $d = 5.00$ cm, then

$$\bar{F}_{ext} = \frac{W}{d} = \frac{2.25 \times 10^{-3}\,\text{J}}{5.00 \times 10^{-2}\,\text{m}} = 0.0450\,\text{N}.$$

Alternate Solution (d) We can also calculate the force directly using Eq. 20–2 for constant $\vec{B}$, $F = I\ell B$. The force the magnetic field exerts on the top and bottom sections of the square coil of Fig. 21–9 are in opposite directions and cancel each other. The magnetic force $\vec{F}_M$ exerted on the left vertical section of the square coil acts to the left as shown because the current is up (clockwise). The right side of the loop is in the region where $\vec{B} = 0$. Hence the external force to the right, $\vec{F}_{ext}$, needed to just overcome the magnetic force to the left (on $N = 100$ loops), is

$$F_{ext} = NI\ell B = (100)(0.0150\,\text{A})(0.0500\,\text{m})(0.600\,\text{T}) = 0.0450\,\text{N},$$

which is the same answer, confirming our use of energy conservation above.

EXERCISE B What is the direction of the induced current in the circular loop due to the current shown in each part of Fig. 21–10?

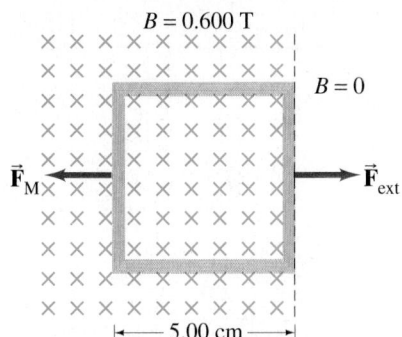

FIGURE 21–9 Example 21–4. The square coil in a magnetic field $B = 0.600$ T is pulled abruptly to the right to a region where $B = 0$. (The forces shown are discussed in the alternate solution at the end of Example 21–4.)

FIGURE 21–10 Exercise B.

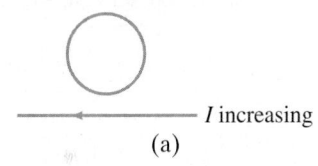

(a)

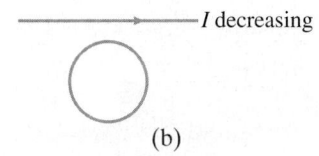

(b)

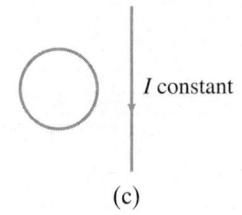

(c)

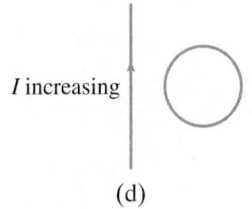

(d)

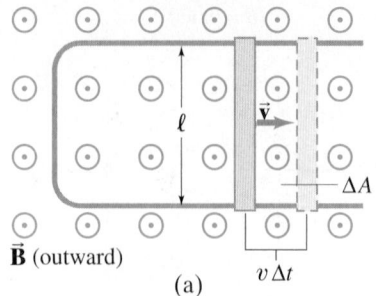

B (outward)

(a)

$v \, \Delta t$

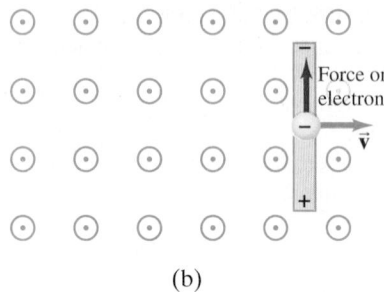

Force on electron

v

(b)

FIGURE 21–11 (a) A conducting rod is moved to the right on a U-shaped conductor in a uniform magnetic field **B** that points out of the page. The induced current is clockwise. (b) Upward force on an electron in the metal rod (moving to the right) due to **B** pointing out of the page; hence electrons can collect at the top of the rod, leaving + charge at the bottom.

FIGURE 21–12 Example 21–5.

🛦 **PHYSICS APPLIED**
Blood-flow measurement

FIGURE 21–13 Measurement of blood velocity from the induced emf. Example 21–6.

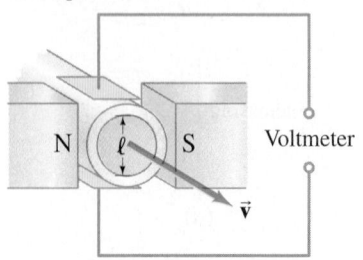

N   S   Voltmeter

v

21–3 EMF Induced in a Moving Conductor

Another way to induce an emf is shown in Fig. 21–11a, and this situation helps illuminate the nature of the induced emf. Assume that a uniform magnetic field **B** is perpendicular to the area bounded by the U-shaped conductor and the movable rod resting on it. If the rod is made to move at a speed v to the right, it travels a distance $\Delta x = v \, \Delta t$ in a time Δt. Therefore, the area of the loop increases by an amount $\Delta A = \ell \, \Delta x = \ell v \, \Delta t$ in a time Δt. By Faraday's law there is an induced emf $\mathscr{E}$ whose magnitude is given by

$$\mathscr{E} = \frac{\Delta \Phi_B}{\Delta t} = \frac{B \, \Delta A}{\Delta t} = \frac{B \ell v \, \Delta t}{\Delta t} = B \ell v. \qquad \textbf{(21–3)}$$

The induced current is clockwise (to counter the increasing flux).

Equation 21–3 is valid as long as B, ℓ, and v are mutually perpendicular. (If they are not, we use only the components of each that are mutually perpendicular.) An emf induced on a conductor moving in a magnetic field is sometimes called **motional emf**.

We can also obtain Eq. 21–3 without using Faraday's law. We saw in Chapter 20 that a charged particle moving with speed v perpendicular to a magnetic field B experiences a force $F = qvB$ (Eq. 20–4). When the rod of Fig. 21–11a moves to the right with speed v, the electrons in the rod also move with this speed. Therefore, since $\vec{v} \perp \vec{B}$, each electron feels a force $F = qvB$, which acts up the page as the red arrow in Fig. 21–11b shows. If the rod is not in contact with the U-shaped conductor, electrons would collect at the upper end of the rod, leaving the lower end positive (see signs in Fig. 21–11b). There must thus be an induced emf. If the rod is in contact with the U-shaped conductor (Fig. 21–11a), the electrons will flow into the U. There will then be a clockwise (conventional) current in the loop. To calculate the emf, we determine the work W needed to move a charge q from one end of the rod to the other against this potential difference: $W = \text{force} \times \text{distance} = (qvB)(\ell)$. The emf equals the work done per unit charge, so $\mathscr{E} = W/q = qvB\ell/q = B\ell v$, the same result as from Faraday's law above, Eq. 21–3.

EXERCISE C In what direction will the electrons flow in Fig. 21–11 if the rod moves to the left, decreasing the area of the current loop?

EXAMPLE 21–5 | ESTIMATE **Does a moving airplane develop a large emf?** An airplane travels 1000 km/h in a region where the Earth's magnetic field is about 5×10^{-5} T and is nearly vertical (Fig. 21–12). What is the potential difference induced between the wing tips that are 70 m apart?

APPROACH We consider the wings to be a 70-m-long conductor moving through the Earth's magnetic field. We use Eq. 21–3 to get the emf.

SOLUTION Since $v = 1000$ km/h $= 280$ m/s, and $\vec{v} \perp \vec{B}$, we have

$$\mathscr{E} = B\ell v = (5 \times 10^{-5} \, \text{T})(70 \, \text{m})(280 \, \text{m/s}) \approx 1 \, \text{V}.$$

NOTE Not much to worry about.

EXAMPLE 21–6 **Electromagnetic blood-flow measurement.** The rate of blood flow in our body's vessels can be measured using the apparatus shown in Fig. 21–13, since blood contains charged ions. Suppose that the blood vessel is 2.0 mm in diameter, the magnetic field is 0.080 T, and the measured emf is 0.10 mV. What is the flow velocity v of the blood?

APPROACH The magnetic field **B** points horizontally from left to right (N pole toward S pole). The induced emf acts over the width $\ell = 2.0$ mm of the blood vessel, perpendicular to **B** and **v** (Fig. 21–13), just as in Fig. 21–11. We can then use Eq. 21–3 to get v.

SOLUTION We solve for v in Eq. 21–3:

$$v = \frac{\mathscr{E}}{B\ell} = \frac{(1.0 \times 10^{-4} \, \text{V})}{(0.080 \, \text{T})(2.0 \times 10^{-3} \, \text{m})} = 0.63 \, \text{m/s}.$$

NOTE In actual practice, an alternating current is used to produce an alternating magnetic field. The induced emf is then alternating.

21–4 Changing Magnetic Flux Produces an Electric Field

We have seen that a changing magnetic flux induces an emf. In a closed loop of wire there will also be an induced current, which implies there is an electric field in the wire causing the electrons to start moving. Indeed, this and other results suggest the important conclusion that

a changing magnetic flux produces an electric field.

This result applies not only to wires and other conductors, but is a general result that applies to any region in space. Indeed, an electric field will be produced (= induced) at any point in space where there is a changing magnetic field.

We can get a simple formula for E in terms of B for the case of electrons in a moving conductor, as in Fig. 21–11. The electrons feel a force (upwards in Fig. 21–11b); and if we put ourselves in the reference frame of the conductor, this force accelerating the electrons implies that there is an electric field in the conductor. Electric field is defined as the force per unit charge, $E = F/q$, where here $F = qvB$ (Eq. 20–4). Thus the effective field E in the rod must be

$$E = \frac{F}{q} = \frac{qvB}{q} = vB, \qquad (21\text{–}4)$$

which is a useful result.

21–5 Electric Generators

We discussed alternating currents (ac) briefly in Section 18–7. Now we examine how ac is generated: by an **electric generator** or **dynamo**. A generator transforms mechanical energy into electric energy, just the opposite of what a motor does (Section 20–10). A simplified diagram of an **ac generator** is shown in Fig. 21–14. A generator consists of many loops of wire (only one is shown) wound on an **armature** that can rotate in a magnetic field. The axle is turned by some mechanical means (falling water, steam turbine, car motor belt), and an emf is induced in the rotating coil. An electric current is thus the *output* of a generator. Suppose in Fig. 21–14 that the armature is rotating clockwise; then right-hand-rule-3 (p. 568) applied to charged particles in the wire (or Lenz's law) tells us that the (conventional) current in the wire labeled b on the armature is outward towards us; therefore the current is outward through brush b. (Each brush is fixed and presses against a continuous slip ring that rotates with the armature.) After one-half revolution, wire b will be where wire a is now in Fig. 21–14, and the current then at brush b will be inward. Thus the current produced is alternating.

The frequency f is 60 Hz for general use in the United States and Canada, whereas 50 Hz is used in many countries. Most of the power generated in the United States is done at steam plants, where the burning of fossil fuels (coal, oil, natural gas) boils water to produce high-pressure steam that turns a turbine connected to the generator axle (Fig. 15–21). Turbines can also be turned by water pressure at a dam (hydroelectric). At nuclear power plants, the nuclear energy released is used to produce steam to turn turbines. Indeed, a heat engine (Chapter 15) connected to a generator is the principal means of generating electric power. The frequency of 60 Hz or 50 Hz is maintained very precisely by power companies.

A **dc generator** is much like an ac generator, except the slip rings are replaced by split-ring commutators, Fig. 21–15a, just as in a dc motor (Figs. 20–37 and 20–38). The output of such a generator is as shown and can be smoothed out by placing a capacitor in parallel with the output.[†] More common is the use of many armature windings, as in Fig. 21–15b, which produces a smoother output.

[†]A capacitor tends to store charge and, if the time constant RC is long enough, helps to smooth out the voltage as shown in the figure to the right.

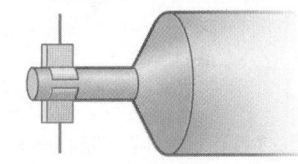

FIGURE 21–14 An ac generator.

FIGURE 21–15 (a) A dc generator with one set of commutators, and (b) a dc generator with many sets of commutators and windings.

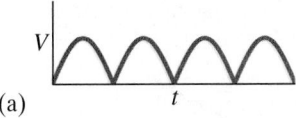

(a)

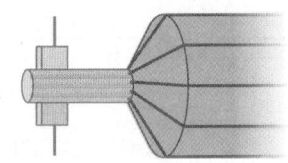

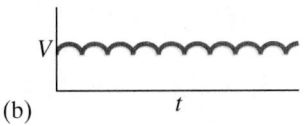

(b)

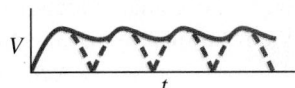

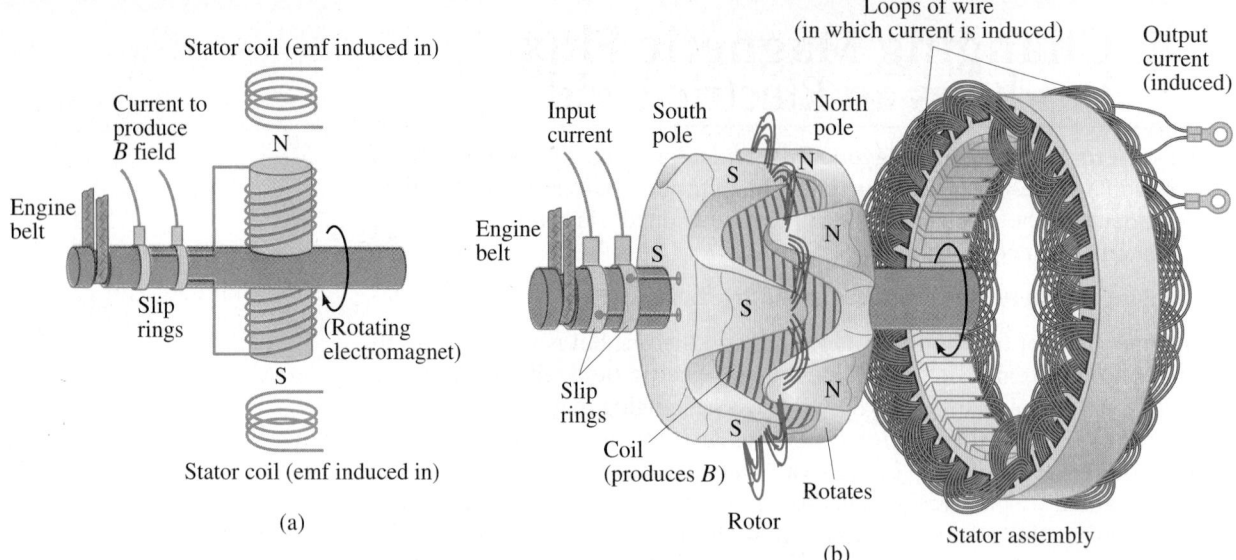

FIGURE 21–16 (a) Simplified schematic diagram of an alternator. The input current to the rotor from the battery is connected through continuous slip rings. Sometimes the rotor electromagnet is replaced by a permanent magnet (no input current). (b) Actual shape of an alternator. The rotor is made to turn by a belt from the engine. The current in the wire coil of the rotor produces a magnetic field inside it on its axis that points horizontally from left to right (not shown), thus making north and south poles of the plates attached at either end. These end plates are made with triangular fingers that are bent over the coil—hence there are alternating N and S poles quite close to one another, with magnetic field lines between them as shown by the blue lines. As the rotor turns, these field lines pass through the fixed stator coils (shown on the right for clarity, but in operation the rotor rotates within the stator), inducing a current in them, which is the output.

*Alternators

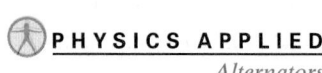

PHYSICS APPLIED
Alternators

Automobiles used to use dc generators. Today they mainly use **alternators**, which avoid the problems of wear and electrical arcing (sparks) across the split-ring commutators of dc generators. Alternators differ from generators in that an electromagnet, called the **rotor**, is fed by current from the battery and is made to rotate by a belt from the engine. The magnetic field of the turning rotor passes through a surrounding set of stationary coils called the **stator** (Fig. 21–16), inducing an alternating current in the stator coils, which is the output. This ac output is changed to dc for charging the battery by the use of semiconductor diodes, which allow current flow in one direction only.

Deriving the Generator Equation

FIGURE 21–17 The emf is induced in the segments ab and cd, whose velocity components perpendicular to the field $\vec{B}$ are $v \sin \theta$.

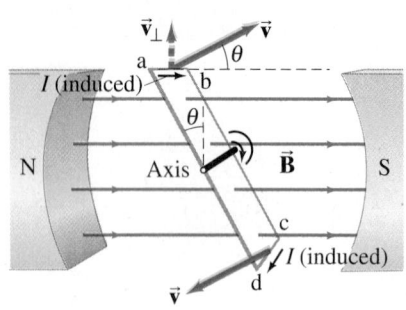

Figure 21–17 shows the wire loop on a generator armature. The loop is being made to rotate clockwise in a uniform magnetic field $\vec{B}$. The velocity of the two lengths ab and cd at this instant are shown. Although the sections of wire bc and da are moving, the force on electrons in these sections is toward the side of the wire, not along the wire's length. The emf generated is thus due only to the force on charges in the sections ab and cd. From right-hand-rule-3, we see that the direction of the induced current in ab is from a toward b. And in the lower section, it is from c to d; so the flow is continuous in the loop. The magnitude of the emf generated in ab is given by Eq. 21–3, except that we must take the component of the velocity perpendicular to B:

$$\mathscr{E} = B\ell v_\perp,$$

where ℓ is the length of ab. From Fig. 21–17 we see that $v_\perp = v \sin \theta$, where θ is the angle the loop's face makes with the vertical. The emf induced in cd has the same magnitude and is in the same direction. Therefore their emfs add, and the total emf is

$$\mathscr{E} = 2NB\ell v \sin \theta,$$

where we have multiplied by N, the number of loops in the coil.

If the coil is rotating with constant angular velocity ω, then the angle $\theta = \omega t$. From the angular equations (Eq. 8–4), $v = \omega r = \omega(h/2)$, where r is the distance from the rotation axis and h is the length of bc or ad. Thus $\mathscr{E} = 2NB\omega\ell(h/2) \sin \omega t$, or

$$\mathscr{E} = NB\omega A \sin \omega t, \tag{21–5}$$

where $A = \ell h$ is the area of the loop. This equation holds for any shape coil, not just for a rectangle as derived. Thus, the output emf of the generator is sinusoidally alternating (see Fig. 21–18 and Section 18–7). Since ω is expressed in radians per second, we can write $\omega = 2\pi f$, where f is the frequency (in Hz = s^{-1}). The rms output (see Section 18–7, Eq. 18–8b) is

$$V_{rms} = \frac{NB\omega A}{\sqrt{2}}.$$

EXAMPLE 21–7 **An ac generator.** The armature of a 60-Hz ac generator rotates in a 0.15-T magnetic field. If the area of the coil is $2.0 \times 10^{-2}\,m^2$, how many loops must the coil contain if the peak output is to be $\mathscr{E}_0 = 170\,V$?

APPROACH From Eq. 21–5 we see that the maximum emf is $\mathscr{E}_0 = NBA\omega$.

SOLUTION We solve Eq. 21–5 for N with $\omega = 2\pi f = (6.28)(60\,s^{-1}) = 377\,s^{-1}$:

$$N = \frac{\mathscr{E}_0}{BA\omega} = \frac{170\,V}{(0.15\,T)(2.0 \times 10^{-2}\,m^2)(377\,s^{-1})} = 150\,\text{turns}.$$

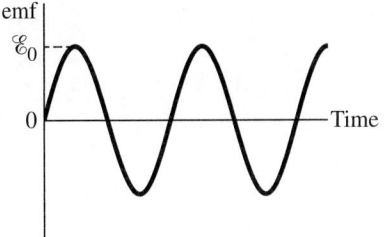

FIGURE 21–18 An ac generator produces an alternating current. The output emf $\mathscr{E} = \mathscr{E}_0 \sin \omega t$, where $\mathscr{E}_0 = NA\omega B$ (Eq. 21–5).

21–6 Back EMF and Counter Torque; Eddy Currents

Back EMF, in a Motor

A motor turns and produces mechanical energy when a current is made to flow in it. From our description in Section 20–10 of a simple dc motor, you might expect that the armature would accelerate indefinitely due to the torque on it. However, as the armature of the motor turns, the magnetic flux through the coil changes and an emf is generated. This induced emf acts to oppose the motion (Lenz's law) and is called the **back emf** or **counter emf**. The greater the speed of the motor, the greater the back emf. A motor normally turns and does work on something, but if there were no load to push (or rotate), the motor's speed would increase until the back emf equaled the input voltage. When there is a mechanical load, the speed of the motor may be limited also by the load. The back emf will then be less than the external applied voltage. The greater the mechanical load, the slower the motor rotates and the lower is the back emf ($\mathscr{E} \propto \omega$, Eq. 21–5).

EXAMPLE 21–8 **Back emf in a motor.** The armature windings of a dc motor have a resistance of 5.0 Ω. The motor is connected to a 120-V line, and when the motor reaches full speed against its normal load, the back emf is 108 V. Calculate (*a*) the current into the motor when it is just starting up, and (*b*) the current when the motor reaches full speed.

APPROACH As the motor is just starting up, it is turning very slowly, so there is negligible back emf. The only voltage is the 120-V line. The current is given by Ohm's law with $R = 5.0\,Ω$. At full speed, we must include as emfs both the 120-V applied emf and the opposing back emf.

SOLUTION (*a*) At start up, the current is controlled by the 120 V applied to the coil's 5.0-Ω resistance. By Ohm's law,

$$I = \frac{V}{R} = \frac{120\,V}{5.0\,Ω} = 24\,A.$$

(*b*) When the motor is at full speed, the back emf must be included in the equivalent circuit shown in Fig. 21–19. In this case, Ohm's law (or Kirchhoff's rule) gives

$$120\,V - 108\,V = I(5.0\,Ω).$$

Therefore

$$I = \frac{12\,V}{5.0\,Ω} = 2.4\,A.$$

NOTE This result shows that the current can be very high when a motor first starts up. This is why the lights in your house may dim when the motor of the refrigerator (or other large motor) starts up. The large initial refrigerator current causes the voltage to the lights to drop because the house wiring has resistance and there is some voltage drop across it when large currents are drawn.

FIGURE 21–19 Circuit of a motor showing induced back emf. Example 21–8.

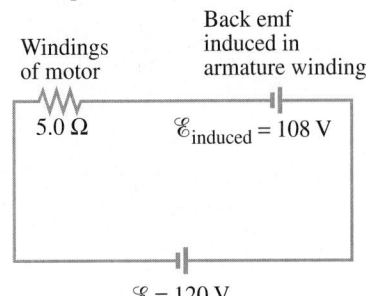

CONCEPTUAL EXAMPLE 21–9 **Motor overload.** When using an appliance such as a blender, electric drill, or sewing machine, if the appliance is overloaded or jammed so that the motor slows appreciably or stops while the power is still connected, the motor can burn out and be ruined. Explain why this happens.

RESPONSE The motors are designed to run at a certain speed for a given applied voltage, and the designer must take the expected back emf into account. If the rotation speed is reduced, the back emf will not be as high as expected ($\mathscr{E} \propto \omega$, Eq. 21–5). The current will increase and may become large enough that the windings of the motor heat up and may melt, ruining the motor.

Counter Torque, in a Generator

In a generator, the situation is the reverse of that for a motor. As we saw, the mechanical turning of the armature induces an emf in the loops, which is the output. If the generator rotates but is not connected to an external circuit, the emf exists at the terminals but there is no current. In this case, it takes little effort to turn the armature. But if the generator *is* connected to a device that draws current, then a current flows in the coils of the armature. Because this current-carrying coil is in an external magnetic field, there will be a torque exerted on it (as in a motor), and this torque opposes the motion (use right-hand-rule-2, page 568, for the force on a wire in Fig. 21–14 or 21–17). This is called a **counter torque**. The greater the electrical load—that is, the more current that is drawn—the greater will be the counter torque. Hence the external applied torque will have to be greater to keep the generator turning. This makes sense from the conservation of energy principle. More mechanical energy input is needed to produce more electric energy output.

EXERCISE D A bicycle headlight is powered by a generator that is turned by the bicycle wheel. (*a*) If you speed up, how does the power to the light change? (*b*) Does the generator resist being turned as the bicycle's speed increases, and if so how?

Eddy Currents

Induced currents are not always confined to well-defined paths such as in wires. Consider, for example, the rotating metal wheel in Fig. 21–20a. An external magnetic field is applied to a limited area of the wheel as shown and points into the page. The section of wheel in the magnetic field has an emf induced in it because the conductor is moving, carrying electrons with it. The flow of induced (conventional) current in the wheel is upward in the region of the magnetic field (Fig. 21–20b), and the current follows a downward return path outside that region. Why? According to Lenz's law, the induced currents oppose the change that causes them. Consider the part of the rotating wheel labeled c in Fig. 21–20b, where the magnetic field is zero but is just about to enter a region where $\vec{\mathbf{B}}$ points into the page. To oppose this inward increase in magnetic field, the induced current is counterclockwise to produce a field pointing out of the page (right-hand-rule-1). Similarly, region d is about to move to e, where $\vec{\mathbf{B}}$ is zero; hence the current is clockwise to produce an inward field opposed to this decreasing flux inward. These currents are referred to as **eddy currents**. They can be present in any conductor that is moving across a magnetic field or through which the magnetic flux is changing.

In Fig. 21–20b, the magnetic field exerts a force $\vec{\mathbf{F}}$ on the induced currents it has created, and that force opposes the rotational motion. Eddy currents can be used in this way as a smooth braking device on, say, a rapid-transit car. In order to stop the car, an electromagnet can be turned on that applies its field either to the wheels or to the moving steel rail below. Eddy currents can also be used to dampen (reduce) the oscillation of a vibrating system, which is referred to as **magnetic damping**.

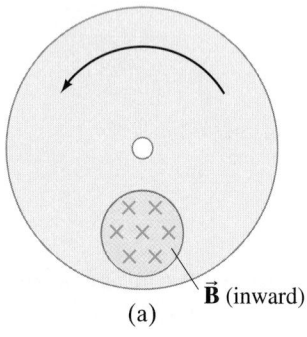

(a) $\vec{\mathbf{B}}$ (inward)

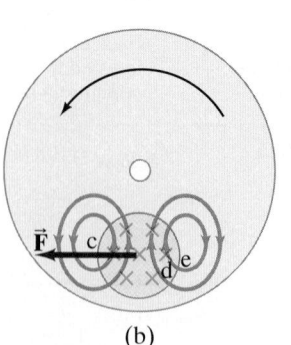

(b)

FIGURE 21–20 Production of eddy currents in a rotating wheel. The gray lines in (b) indicate induced current.

Eddy currents, however, can be a problem. For example, eddy currents induced in the armature of a motor or generator produce heat ($P = I\mathscr{E}$) and waste energy. To reduce the eddy currents, the armatures are *laminated*; that is, they are made of very thin sheets of iron that are well insulated from one another (used also in transformers, Fig. 21–23). The total path length of the eddy currents is confined to each slab, which increases the total resistance; hence the current is less and there is less wasted energy.

Walk-through metal detectors (Fig. 21–21) use electromagnetic induction and eddy currents to detect metal objects. Several coils are situated in the walls of the walk-through at different heights. In one technique, the coils are given brief pulses of current, hundreds or thousands of times per second. When a person passes through the walk-through, any metal object being carried will have eddy currents induced in it, and the small magnetic field produced by that eddy current can be detected, setting off an alert or alarm.

FIGURE 21–21 Metal detector.

PHYSICS APPLIED
Metal detector

21–7 Transformers and Transmission of Power

A transformer is a device for increasing or decreasing an ac voltage. Transformers are found everywhere: on utility poles (Fig. 21–22) to reduce the high voltage from the electric company to a usable voltage in houses (120 V or 240 V), in chargers for cell phones, laptops, and other electrical devices, in your car to give the needed high voltage to the spark plugs, and in many other applications. A **transformer** consists of two coils of wire known as the **primary** and **secondary** coils. The two coils can be interwoven (with insulated wire); or they can be linked by an iron core which is laminated to minimize eddy-current losses (Section 21–6), as shown in Fig. 21–23. Transformers are designed so that (nearly) all the magnetic flux produced by the current in the primary coil also passes through the secondary coil, and we assume this is true in what follows. We also assume that energy losses (in resistance and hysteresis) can be ignored—a good approximation for real transformers, which are often better than 99% efficient.

When an ac voltage is applied to the primary coil, the changing magnetic field it produces will induce an ac voltage of the same frequency in the secondary coil. However, the voltage will be different according to the number of "turns" or loops in each coil. From Faraday's law, the voltage or emf induced in the secondary coil is

$$V_S = N_S \frac{\Delta \Phi_B}{\Delta t},$$

where N_S is the number of turns in the secondary coil, and $\Delta \Phi_B / \Delta t$ is the rate at which the magnetic flux changes.

The input primary voltage, V_P, is related to the rate at which the flux changes through it,

$$V_P = N_P \frac{\Delta \Phi_B}{\Delta t},$$

where N_P is the number of turns in the primary coil. This follows because the changing flux produces a back emf, $N_P \Delta \Phi_B / \Delta t$, in the primary that balances the applied voltage V_P if the resistance of the primary can be ignored (Kirchhoff's rules). We divide these two equations, assuming little or no flux is lost, to find

$$\frac{V_S}{V_P} = \frac{N_S}{N_P}. \tag{21–6}$$

This *transformer equation* tells how the secondary (output) voltage is related to the primary (input) voltage; V_S and V_P in Eq. 21–6 can be the rms values (Section 18–7) for both, or peak values for both. Steady dc voltages don't work in a transformer because there would be no changing magnetic flux.

FIGURE 21–22 Repairing a step-down transformer on a utility pole.

FIGURE 21–23 Step-up transformer ($N_P = 4$, $N_S = 12$).

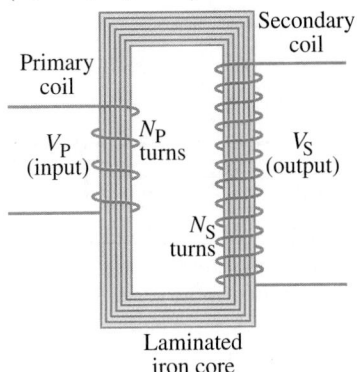

Primary coil

Secondary coil

V_P (input)

N_P turns

V_S (output)

N_S turns

Laminated iron core

If the secondary coil contains more loops than the primary coil ($N_S > N_P$), we have a **step-up transformer**. The secondary voltage is greater than the primary voltage. For example, if the secondary coil has twice as many turns as the primary coil, then the secondary voltage will be twice that of the primary voltage. If N_S is less than N_P, we have a **step-down transformer**.

Although ac voltage can be increased (or decreased) with a transformer, we don't get something for nothing. Energy conservation tells us that the power output can be no greater than the power input. A well-designed transformer can be greater than 99% efficient, so little energy is lost to heat. The power output thus essentially equals the power input. Since power $P = IV$ (Eq. 18–5), we have

$$I_P V_P = I_S V_S,$$

or (remembering Eq. 21–6),

$$\frac{I_S}{I_P} = \frac{N_P}{N_S}. \tag{21-7}$$

EXAMPLE 21–10 **Cell phone charger.** The charger for a cell phone contains a transformer that reduces 120-V (or 240-V) ac to 5.0-V ac to charge the 3.7-V battery (Section 19–4). (It also contains diodes to change the 5.0-V ac to 5.0-V dc.) Suppose the secondary coil contains 30 turns and the charger supplies 700 mA. Calculate (*a*) the number of turns in the primary coil, (*b*) the current in the primary, and (*c*) the power transformed.

APPROACH We assume the transformer is ideal, with no flux loss, so we can use Eq. 21–6 and then Eq. 21–7.

SOLUTION (*a*) This is a step-down transformer, and from Eq. 21–6 we have

$$N_P = N_S \frac{V_P}{V_S} = \frac{(30)(120\,\text{V})}{(5.0\,\text{V})} = 720\ \text{turns}.$$

(*b*) From Eq. 21–7

$$I_P = I_S \frac{N_S}{N_P} = (0.70\,\text{A})\left(\frac{30}{720}\right) = 29\ \text{mA}.$$

(*c*) The power transformed is

$$P = I_S V_S = (0.70\,\text{A})(5.0\,\text{V}) = 3.5\ \text{W}.$$

NOTE The power in the primary coil, $P = (0.029\,\text{A})(120\,\text{V}) = 3.5$ W, is the same as the power in the secondary coil. There is 100% efficiency in power transfer for our ideal transformer.

EXERCISE E How many turns would you want in the secondary coil of a transformer having $N_P = 400$ turns if it were to reduce the voltage from 120-V ac to 3.0-V ac?

A transformer operates only on ac. A dc current in the primary coil does not produce a changing flux and therefore induces no emf in the secondary. However, if a dc voltage is applied to the primary through a switch, at the instant the switch is opened or closed there will be an induced voltage in the secondary. For example, if the dc is turned on and off as shown in Fig. 21–24a, the voltage induced in the secondary is as shown in Fig. 21–24b. Notice that the secondary voltage drops to zero when the dc voltage is steady. This is basically how, in the **ignition system** of an automobile, the high voltage is created to produce the spark across the gap of a spark plug that ignites the gas-air mixture. The transformer is referred to as an "ignition coil," and transforms the 12 V dc of the battery (when switched off in the primary) into a spike of as much as 30 kV in the secondary.

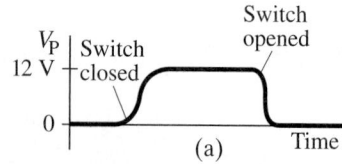

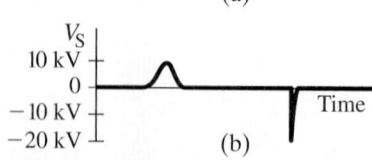

FIGURE 21–24 A dc voltage turned on and off as shown in (a) produces voltage pulses in the secondary (b). Voltage scales in (a) and (b) are not the same.

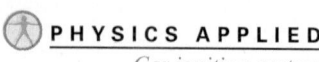

PHYSICS APPLIED

Car ignition system

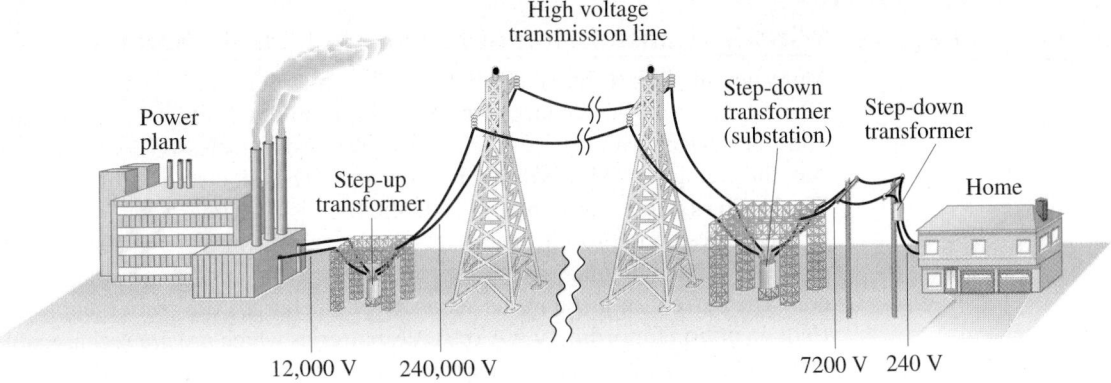

High voltage
transmission line

Power
plant

Step-up
transformer

Step-down
transformer
(substation)

Step-down
transformer

Home

12,000 V   240,000 V

7200 V   240 V

FIGURE 21–25 The transmission of electric power from power plants to homes makes use of transformers at various stages.

Transformers play an important role in the transmission of electricity. Power plants are often situated some distance from metropolitan areas, so electricity must then be transmitted over long distances (Fig. 21–25). There is always some power loss in the transmission lines, and this loss can be minimized if the power is transmitted at high voltage, using transformers, as the following Example shows.

PHYSICS APPLIED
Transformers help power transmission

EXAMPLE 21–11 **Transmission lines.** An average of 120 kW of electric power is sent to a small town from a power plant 10 km away. The transmission lines have a total resistance of 0.40 Ω. Calculate the power loss if the power is transmitted at (*a*) 240 V and (*b*) 24,000 V.

APPROACH We cannot use $P = V^2/R$ because if R is the resistance of the transmission lines, we don't know the voltage drop along them. The given voltages are applied across the lines plus the load (the town). But we can determine the current I in the lines $(= P/V)$, and then find the power loss from $P_L = I^2R$, for both cases (*a*) and (*b*).

SOLUTION (*a*) If 120 kW is sent at 240 V, the total current will be

$$I = \frac{P}{V} = \frac{1.2 \times 10^5 \text{ W}}{2.4 \times 10^2 \text{ V}} = 500 \text{ A}.$$

The power loss in the lines, P_L, is then

$$P_L = I^2R = (500 \text{ A})^2(0.40 \text{ Ω}) = 100 \text{ kW}.$$

Thus, over 80% of all the power would be wasted as heat in the power lines!

(*b*) If 120 kW is sent at 24,000 V, the total current will be

$$I = \frac{P}{V} = \frac{1.2 \times 10^5 \text{ W}}{2.4 \times 10^4 \text{ V}} = 5.0 \text{ A}.$$

The power loss in the lines is then

$$P_L = I^2R = (5.0 \text{ A})^2(0.40 \text{ Ω}) = 10 \text{ W},$$

which is less than $\frac{1}{100}$ of 1%: a far better efficiency.

NOTE We see that the higher voltage results in less current, and thus less power is wasted as heat in the transmission lines. It is for this reason that power is usually transmitted at very high voltages, as high as 700 kV.

The great advantage of ac, and a major reason it is in nearly universal use[†], is that the voltage can easily be stepped up or down by a transformer. The output voltage of an electric generating plant is stepped up prior to transmission. Upon arrival in a city, it is stepped down in stages at electric substations prior to distribution. The voltage in lines along city streets is typically 2400 V or 7200 V and is stepped down to 240 V or 120 V for home use by transformers (Figs. 21–22 and 21–25).

[†]DC transmission along wires does exist, and has some advantages (if the current is constant, there is no induced current in nearby conductors as there is with ac). But boosting to high voltage and down again at the receiving end requires more complicated electronics.

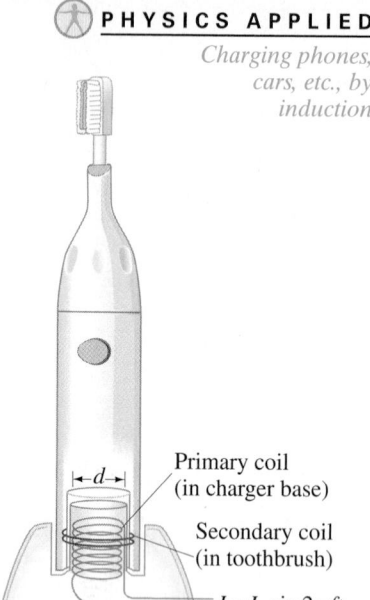

FIGURE 21–26 This electric toothbrush contains rechargeable batteries which are being recharged as it sits on its base. Charging occurs from a primary coil in the base to a secondary coil in the toothbrush. The toothbrush can be lifted from its base when you want to brush your teeth.

Primary coil (in charger base)

Secondary coil (in toothbrush)

$I = I_0 \sin 2\pi ft$

Wireless Transmission of Power—Inductive Charging

Many devices with rechargeable batteries, like cell phones, cordless phones, and even electric cars, can be recharged using a direct metal contact between the device and the charger. But devices can also be charged "wirelessly" by induction, without the need for exposed electric contacts. The electric toothbrush shown in Fig. 21–26 sits on a plastic base. Inside the base is a "primary coil" connected to an ac outlet. Inside the toothbrush is a "secondary coil" in which a current is induced due to the changing magnetic field produced by the changing current in the primary coil. The current induced in the secondary coil charges the rechargeable batteries. (Not an option for ordinary AA or AAA batteries which are *not* rechargeable.) The effect is like a transformer—except here there is no iron to contain the field lines, so there is less efficiency. But you can separate the two parts (toothbrush and charger) and brush your teeth. Many heart pacemakers are given power inductively: power in an external coil is transmitted to a secondary coil in the pacemaker (Fig. 19–25) inside the person's body near the heart. Inductive charging is also a possible means for recharging an electric car's batteries.

Wireless transmission of power must be done over short distances to maintain a reasonable efficiency. Wireless transmission of signals (information) can be done over great distances (Section 22–7) because even fairly low power signals can be detected, and it is the information in the signal voltages that counts, not power.

*21–8 Information Storage: Magnetic and Semiconductor; Tape, Hard Drive, RAM

*Magnetic Storage: Read/Write on Tape and Disks

Recording and playback on tape or disk is done by magnetic **heads**. Magnetic tapes contain a thin layer of ferromagnetic oxide on a thin plastic tape. Computer **hard drives** (HD) store digital information (applications and data): they have a thin layer of ferromagnetic material on the surface of each rotating disk or platter, Fig. 21–27a. During recording of an audio or video signal on tape, or "writing" on a hard drive, the voltage is sent to the recording head which acts as a tiny electromagnet (Fig. 21–27b) that magnetizes the tiny section of tape or disk passing the narrow gap in the head at each instant. During playback, or "reading" of an HD, the changing magnetism of the moving tape or disk at the gap causes corresponding changes in the magnetic field within the soft-iron head, which in turn induces an emf in the coil (Faraday's law). This induced emf is the output signal that can be processed by the computer, or for audio can be amplified and sent to a loudspeaker (for video to a monitor or TV).

FIGURE 21–27 (a) Photo of a hard drive showing several platters and read/write heads that can quickly move from the edge of the disk to the center. (b) Read/Write (playback/recording) head for disk or tape. In writing or recording, the electric input signal to the head, which acts as an electromagnet, magnetizes the passing tape or disk. In reading or playback, the changing magnetic field of the passing tape or disk induces a changing magnetic field in the head, which in turn induces in the coil an emf that is the output signal.

(a)

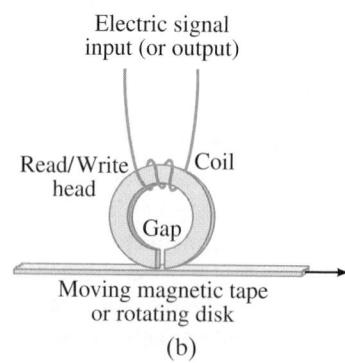

Electric signal input (or output)

Read/Write head

Coil

Gap

Moving magnetic tape or rotating disk

(b)

Audio and video signals may be **analog**, varying continuously in amplitude over time: the variation in degree of magnetization at sequential points reflects the variation in amplitude and frequency of the audio or video signal. In modern equipment, analog signals (say from a microphone) are electronically converted to **digital**—which means a series of **bits**, each of which is a "1" or a "0", that forms a **binary code** as discussed in Section 17–10. (Recall also, 8 bits in a row = 1 **byte**.) Computers process only digital information.

CD-ROMs, CDs (audio compact discs), and DVDs (digital video discs) are read by an **optical drive** (not magnetic): a laser emits a narrow beam of light that reflects off the "grooves" of the rotating disc containing "pits" as described in Section 28–11.

*Semiconductor Memory: DRAM, Flash

Basic to your computer is its **random access memory** (**RAM**). This is where the information you are working with at any given time is temporarily stored and manipulated by you. Each data storage location can be accessed and read (or written) directly and quickly, so you don't have to wait. In contrast, hard drives, tape, flash and external devices are more permanent **storage**, and they are much slower to access because the data must be searched for, sequentially, such as along the circular tracks of hard drives (Fig. 21–27a). Programs, applications, and data that you want to use are imported by the computer into the RAM from their more permanent (and more slowly accessed) storage area.[†]

RAM is based on semiconductor technology, storing the binary bits ("0" or "1") as electric charge or voltage. Some computers may use semiconductors also for long-term storage ("flash memory") in place of a hard drive.

A common type of RAM is **dynamic random access memory** or **DRAM**, which uses arrays of transistors known as MOSFETs (metal-oxide semiconductor field-effect transistors). Transistors will be discussed in Section 29–10, but we already encountered them in Section 17–11 about TV screen addressing, Fig. 17–34, which we show again here, Fig. 21–28. A MOSFET transistor in RAM serves basically as an on–off switch: the voltage on the **gate** terminal acts to control the conductivity between the **source** and the **drain** terminals, thus allowing current to flow (or not) between them.

Each memory "cell," which in DRAM consists of one transistor and a capacitor, stores one bit (= a "0" or a "1"). Each cell is extremely small physically, less than 100 nm across.[‡] Typical DRAM chips (integrated circuits) contain billions of these memory cells. To see how they work, we look at a tiny part, the simple four-cell array shown in Fig. 21–29. One side of each cell capacitor is grounded; the other side is connected to the transistor source. The drain of each transistor is connected to a very thin conducting wire or "line," a **bit-line**, that runs across the array of cells. Each gate is connected to a **word-line**. A particular bit is a "1" or a "0" depending on whether the capacitor of the cell is charged to a voltage V (maybe 5 V) or is at zero (uncharged, or at a very small voltage).

To **write** data, say on the upper left cell in Fig. 21–29, word-line-1 is given a high enough pulse of voltage to "turn on" the transistor. That is, the high gate voltage attracts charge and allows bit-line-1 and the capacitor to be connected. Thus charge can flow from bit-line-1 to the capacitor, charging it either to V or to 0, depending on the bit-line-1 voltage at that moment, thus writing a "1" or a "0".

The lower left cell in Fig. 21–29 can be written at the same time by setting bit-line-2 voltage to V or zero.

Now let us see a simple way to **read** a cell. In order to read the data stored ("1" or "0") on the upper left cell, a voltage of about $\frac{1}{2}V$ is given to bit-line-1. Then word-line-1 is given enough voltage to turn on the transistor and connect bit-line-1 to the capacitor. The capacitor, if uncharged (= "0"), will now drag charge from bit-line-1 and the bit-line voltage will drop *below* $\frac{1}{2}V$. If the capacitor is already charged to V (= "1"), the connection to bit-line-1 will raise bit-line-1's voltage to *above* $\frac{1}{2}V$. A sensor at the end of bit-line-1 will detect either change in voltage (increase means it reads a "1", decrease a "0"). All cells connected to one word-line are read at the same moment. The capacitor voltage has been altered by the small charge flow during the reading process. So that cell or bit which has just been read needs to be written again, or "refreshed."

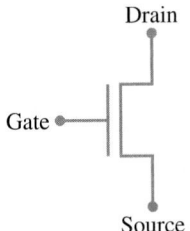

FIGURE 21–28 Symbol for a MOSFET transistor. The gate acts to attract or repel charge, and thus open or close the connection along the semiconductor that connects source and drain.

FIGURE 21–29 A tiny 2 × 2 cell, part of a simple DRAM array. The word-lines and bit-lines do not touch each other where they cross.

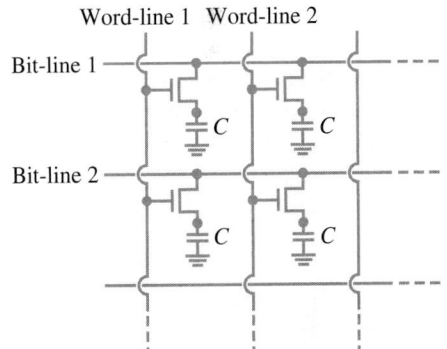

[†]Computer specifications may use "memory" for the random access (fast) memory, and "storage" for the long-term (and slower access) information on hard drives, flash drives, and related devices.

[‡]At 100 nm, 10^5 bits can fit along a 1-cm line, $(1\,\text{cm}/100\,\text{nm}) = 10^{-2}\,\text{m}/10^{-7}\,\text{m}$. So a square, 1 cm on a side, can hold $10^5 \times 10^5 = 10^{10} = 10$ Gbits ≈ 1 GB (gigabyte). Today, cells are even smaller than 100 nm: a $(30\,\text{nm})^2$ cell can hold ≈ 10 GB in a 1 cm^2 area.

The transistors are imperfect switches and allow the charge on the tiny capacitors in each cell to be "leaky" and lose charge fairly quickly, so every cell has to be read and rewritten (refreshed) many times per second. The D in DRAM stands for this "dynamic" refreshing action. If the power is turned off, the capacitors lose their charge and the data are lost. DRAM is thus referred to as being **volatile** memory, whereas a hard drive keeps its (magnetic) memory even when the electric power is off and is called **nonvolatile** memory (doesn't "evaporate").

Flash memory is also made of semiconductor material on tiny "chips." The transistor structures are more complicated, and are able to keep the data even without power so they are nonvolatile. Each MOSFET contains a second gate (the **floating gate**) insulated on both faces, and can hold charge for many years. Charged or not corresponds to a "1" or a "0" bit. Figure 21–30 is a diagram of such an **NVM** (nonvolatile memory) cell. The floating gate is insulated from the standard gate and the semiconductor connecting source and drain. A high positive voltage on the gate (+20 V) forces electrons in the semiconductor (at 0 V) to pass through the thin insulator into the floating gate by a process of quantum mechanical **tunneling** (discussed in Section 30–12). This charge is stored on the floating gate as a "1" bit. The erase process is done by applying the opposite (−20 V) voltage to force electrons to tunnel out of the floating gate, returning it to the uncharged state (= a "0" bit). The erase process is slow (milliseconds vs. ns for DRAM), so erasure is done in large blocks of memory. Flash memory[†] is slower to read or write, and is too slow to use as RAM. Instead, flash memory can be used in place of a hard drive as general storage in computers and tablets, and may be called a "solid state device" (**SSD**). Flash is also used for flash drives, memory cards (such as SD cards), thumb drives, cell phone and portable player memory, and external computer memory.

Magnetoresistive RAM (**MRAM**) is a recent development, involving (again) magnetic properties. One cell (storing one bit) consists of two tiny ferromagnetic plates (separated by an insulator), one of which is permanently magnetized. The other plate can be magnetized in one direction or the other, for a "1" or a "0", by current in nearby wires. Cell size is a bit large, but MRAM is fast and nonvolatile (no power and no refresh needed) and therefore has the potential to be used as any type of memory.

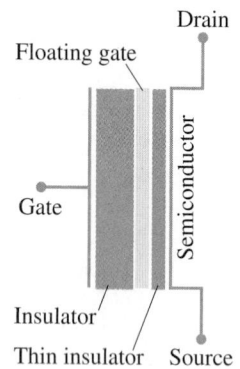

FIGURE 21–30 A floating gate nonvolatile memory cell (NVM).

*21–9 Applications of Induction: Microphone, Seismograph, GFCI

*Microphone

The condenser microphone was discussed in Section 17–7. Many other types operate on the principle of induction. In one form, a microphone is just the inverse of a loudspeaker (Section 20–10). A small coil connected to a membrane is suspended close to a small permanent magnet, as shown in Fig. 21–31. The coil moves in the magnetic field when sound waves strike the membrane, and this motion induces an emf in the moving coil. The frequency of the induced emf will be just that of the impinging sound waves, and this emf is the "signal" that can be amplified and sent to loudspeakers or recorder.

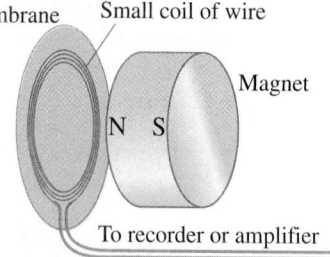

FIGURE 21–31 Diagram of a microphone that works by induction.

PHYSICS APPLIED
Credit card

*Credit Card Reader

When you pass a credit card through a reader at a store, the magnetic stripe on the back of the card passes over a read head just as for a computer hard drive. The magnetic stripe contains personal information about your account and connects by telephone line for approval from your credit card company. Newer cards use semiconductor chips that are more difficult to fraudulently copy.

[†]Why the name "Flash"? It may come from the erase process: large blocks erased "in a flash," and/or because the earliest floating gate memories were erased by a flash of UV light which ejected the stored electrons.

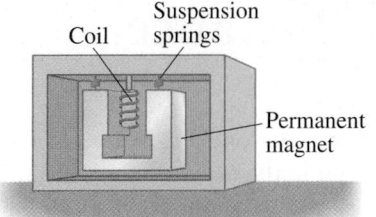

Coil
Suspension springs
Permanent magnet

FIGURE 21–32 One type of seismograph, in which the coil is fixed to the case and moves with the Earth's surface. The magnet, suspended by springs, has inertia and does not move instantaneously with the coil (and case), so there is relative motion between magnet and coil.

*Seismograph

In geophysics, a **seismograph** measures the intensity of earthquake waves using a magnet and a coil of wire. Either the magnet or the coil is fixed to the case, and the other is inertial (suspended by a spring; Fig. 21–32). The relative motion of magnet and coil when the surface of the Earth shakes induces an emf output.

PHYSICS APPLIED
Seismograph

*Ground Fault Circuit Interrupter (GFCI)

Fuses and circuit breakers (Sections 18–6 and 20–7) protect buildings from electricity-induced fire, and apparatus from damage, due to undesired high currents. But they do not turn off the current until it is very much greater than that which can cause permanent damage to humans or death (≈ 100 mA). If fast enough, they may protect humans in some cases, such as very high currents due to short circuits. A **ground fault circuit interrupter** (GFCI) is meant above all to protect humans.

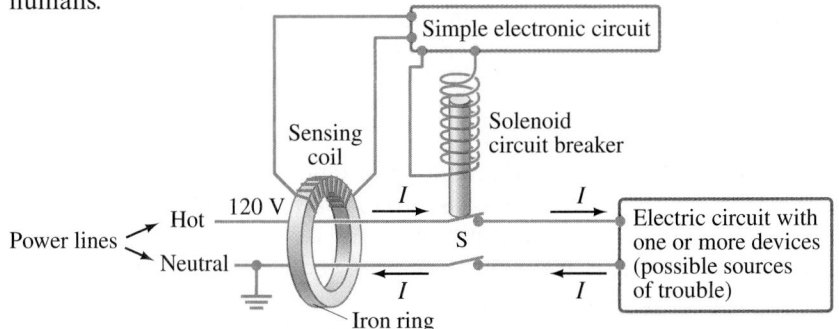

Simple electronic circuit
Sensing coil
Solenoid circuit breaker
Hot 120 V
Power lines
Neutral
I
S
I
Electric circuit with one or more devices (possible sources of trouble)
I
I
Iron ring

FIGURE 21–33 A ground fault circuit interrupter (GFCI).

PHYSICS APPLIED
GFCI

FIGURE 21–34 (a) A GFCI wall outlet. GFCIs can be recognized by their "test" and "reset" buttons. (b) Add-on GFCI that plugs into outlet.

Electromagnetic induction is the physical basis of a GFCI. As shown in Fig. 21–33, the two conductors of a power line connected to an electric circuit or device (red) pass through a small iron ring. Around the ring are many loops of thin wire that serve as a sensing coil. Under normal conditions (no ground fault), the current moving in the hot power wire is exactly balanced by the returning current in the neutral wire. If something goes wrong and the hot wire touches the ungrounded metal case of the device or appliance, some of the entering current can pass through a person who touches the case and then to ground (a **ground fault**). Then the return current in the neutral wire will be less than the entering current in the hot wire, so there is a *net current* passing through the GFCI's iron ring. Because the current is ac, it is changing and that current difference produces a changing magnetic field in the iron, thus inducing an emf in the sensing coil wrapped around the iron. For example, if a device draws 8.0 A, and there is a ground fault through a person of 100 mA ($= 0.1$ A), then 7.9 A will appear in the neutral wire. The emf induced in the sensing coil by this 100-mA difference is amplified by a simple transistor circuit and sent to its own solenoid circuit breaker that opens the circuit at the switch S, thus protecting your life.

If the case of the faulty device is grounded, the difference in current is even higher when there is a fault, and the GFCI trips very quickly.

GFCIs can sense current differences as low as 5 mA and react in 1 ms, saving lives. They can be small to fit as a wall outlet (Fig. 21–34a), or as a plug-in unit into which you plug a hair dryer or toaster (Fig. 21–34b). It is especially important to have GFCIs installed in kitchens, in bathrooms, outdoors, and near swimming pools, where people are most in danger of touching ground. GFCIs always have a "test" button (to be sure the GFCI itself works) and a "reset" button (after it goes off).

(a)

(b)

*21–10 Inductance

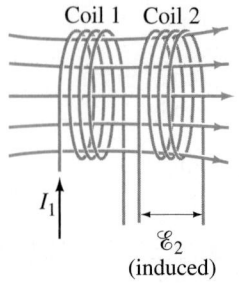

Coil 1 Coil 2

I_1

$\mathcal{E}_2$
(induced)

FIGURE 21–35 A changing current in one coil will induce a current in the second coil.

*Mutual Inductance

If two coils of wire are near one another, as in Fig. 21–35, a changing current in one will induce an emf in the other. We apply Faraday's law to coil 2: the emf $\mathcal{E}_2$ induced in coil 2 is proportional to the rate of change of magnetic flux passing through it. A changing flux in coil 2 is produced by a changing current I_1 in coil 1. So $\mathcal{E}_2$ is proportional to the rate of change of the current in coil 1:

$$\mathcal{E}_2 = -M \frac{\Delta I_1}{\Delta t}, \qquad \text{(21–8a)}$$

where we assume the time interval Δt is very small, and the constant of proportionality, M, is called the **mutual inductance**. (The minus sign is because of Lenz's law, the induced emf opposes the changing flux.) Mutual inductance has units of $V \cdot s/A = \Omega \cdot s$, which is called the **henry** (H), after Joseph Henry: $1\,H = 1\,\Omega \cdot s$.

The mutual inductance M is a "constant" in that it does not depend on I_1; M depends on "geometric" factors such as the size, shape, number of turns, and relative positions of the two coils, and also on whether iron (or other ferromagnetic material) is present. For example, the farther apart the two coils are in Fig. 21–35, the fewer lines of flux can pass through coil 2, so M will be less. If we consider the inverse situation—a changing current in coil 2 inducing an emf in coil 1—the proportionality constant, M, turns out to have the same value,

$$\mathcal{E}_1 = -M \frac{\Delta I_2}{\Delta t}. \qquad \text{(21–8b)}$$

A transformer is an example of mutual inductance in which the coupling is maximized so that nearly all flux lines pass through both coils. Mutual inductance has other uses as well, including inductive charging of cell phones, electric cars, and other devices with rechargeable batteries, as we discussed in Section 21–7. Some types of pacemakers used to maintain blood flow in heart patients (Section 19–6) receive their power from an external coil which is transmitted via mutual inductance to a second coil in the pacemaker near the heart. This type has the advantage over battery-powered pacemakers in that surgery is not needed to replace a battery when it wears out.

PHYSICS APPLIED
Pacemaker

*Self-Inductance

The concept of inductance applies also to an isolated single coil. When a changing current passes through a coil or solenoid, a changing magnetic flux is produced inside the coil, and this in turn induces an emf. This induced emf opposes the change in flux (Lenz's law); it is much like the back emf generated in a motor. (For example, if the current through the coil is increasing, the increasing magnetic flux induces an emf that opposes the original current and tends to retard its increase.) The induced emf $\mathcal{E}$ is proportional to the rate of change in current (and is in the direction opposed to the change, hence the minus sign):

$$\mathcal{E} = -L \frac{\Delta I}{\Delta t}. \qquad \text{(21–9)}$$

The constant of proportionality L is called the **self-inductance**, or simply the **inductance** of the coil. It, too, is measured in henrys. The magnitude of L depends on the size and shape of the coil and on the presence of an iron core.

An ac circuit (Section 18–7) always contains some inductance, but often it is quite small unless the circuit contains a coil of many loops or turns. A coil that has significant self-inductance L is called an **inductor**. It is shown on circuit diagrams by the symbol

. 　　　　　　　　　　　[inductor symbol]

CONCEPTUAL EXAMPLE 21–12 | **Direction of emf in inductor.** Current passes through the coil in Fig. 21–36 from left to right as shown. (*a*) If the current is increasing with time, in which direction is the induced emf? (*b*) If the current is decreasing in time, what then is the direction of the induced emf?

RESPONSE (*a*) From Lenz's law we know that the induced emf must oppose the change in magnetic flux. If the current is increasing, so is the magnetic flux. The induced emf acts to oppose the increasing flux, which means it acts like a source of emf that opposes the outside source of emf driving the current. So the induced emf in the coil acts to oppose *I* in Fig. 21–36a. In other words, the inductor might be thought of as a battery with a positive terminal at point A (tending to block the current entering at A), and negative at point B.

(*b*) If the current is decreasing, then by Lenz's law the induced emf acts to bolster the flux—like a source of emf reinforcing the external emf. The induced emf acts to increase *I* in Fig. 21–36b, so in this situation you can think of the induced emf as a battery with its negative terminal at point A to attract more current (conventional, +) to move to the right.

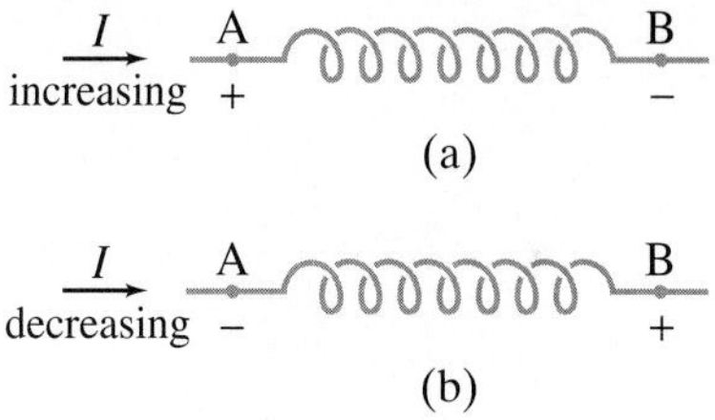

FIGURE 21–36 Example 21–12.

EXAMPLE 21–13 | **Solenoid inductance.** (*a*) Determine a formula for the self-inductance L of a long tightly wrapped solenoid coil of length ℓ and cross-sectional area A, that contains N turns (or loops) of wire. (*b*) Calculate the value of L if $N = 100$, $\ell = 5.0\,\text{cm}$, $A = 0.30\,\text{cm}^2$, and the solenoid is air filled.

APPROACH The induced emf in a coil can be determined either from Faraday's law $\left(\mathscr{E} = -N\,\Delta\Phi_B/\Delta t\right)$ or the self-inductance $\left(\mathscr{E} = -L\,\Delta I/\Delta t\right)$. If we equate these two expressions, we can solve for the inductance L since we know how to calculate the flux Φ_B for a solenoid using Eq. 20–8 $(B = \mu_0\,IN/\ell)$.

SOLUTION (*a*) We equate Faraday's law (Eq. 21–2b) and Eq. 21–9 for the inductance:

$$\mathscr{E} = -N\frac{\Delta\Phi_B}{\Delta t} = -L\frac{\Delta I}{\Delta t},$$

and solve for L:

$$L = N\frac{\Delta\Phi_B}{\Delta I}.$$

We know $\Phi_B = BA$ (Eq. 21–1), and Eq. 20–8 gives us the magnetic field B for a solenoid, $B = \mu_0 NI/\ell$, so the magnetic flux inside the solenoid is

$$\Phi_B = \frac{\mu_0\,NIA}{\ell}.$$

Any change in current, ΔI, causes a change in flux

$$\Delta\Phi_B = \frac{\mu_0\,N\,\Delta I\,A}{\ell}.$$

We put this into our equation above for L:

$$L = N\frac{\Delta\Phi_B}{\Delta I} = \frac{\mu_0\,N^2 A}{\ell}.$$

(*b*) Using $\mu_0 = 4\pi \times 10^{-7}\,\text{T}\cdot\text{m/A}$, and putting in values given,

$$L = \frac{\left(4\pi \times 10^{-7}\,\text{T}\cdot\text{m/A}\right)(100)^2\left(3.0 \times 10^{-5}\,\text{m}^2\right)}{\left(5.0 \times 10^{-2}\,\text{m}\right)} = 7.5\,\mu\text{H}.$$

*21–11 Energy Stored in a Magnetic Field

In Section 17–9 we saw that the energy stored in a capacitor is equal to $\frac{1}{2}CV^2$. By using a similar argument, it can be shown that the energy U stored in an inductance L, carrying a current I, is

$$U = \text{energy} = \tfrac{1}{2}LI^2.$$

Just as the energy stored in a capacitor can be considered to reside in the electric field between its plates, so the energy in an inductor can be considered to be stored in its magnetic field.

To write the energy in terms of the magnetic field, we quote the result of Example 21–13 that the inductance of a solenoid is $L = \mu_0 N^2 A/\ell$. The magnetic field B in a solenoid is related to the current I (see Eq. 20–8) by $B = \mu_0 NI/\ell$. Thus, $I = B\ell/\mu_0 N$, and

$$U = \text{energy} = \tfrac{1}{2}LI^2 = \frac{1}{2}\left(\frac{\mu_0 N^2 A}{\ell}\right)\left(\frac{B\ell}{\mu_0 N}\right)^2 = \frac{1}{2}\frac{B^2}{\mu_0}A\ell.$$

We can think of this energy as residing in the volume enclosed by the windings, which is $A\ell$. Then the energy per unit volume, or **energy density**, is

$$u = \text{energy density} = \frac{1}{2}\frac{B^2}{\mu_0}. \qquad \textbf{(21–10)}$$

This formula, which was derived for the special case of a solenoid, can be shown to be valid for any region of space where a magnetic field exists. If a ferromagnetic material is present, μ_0 is replaced by μ. This equation is analogous to that for an electric field, $\frac{1}{2}\epsilon_0 E^2$, Section 17–9.

*21–12 LR Circuit

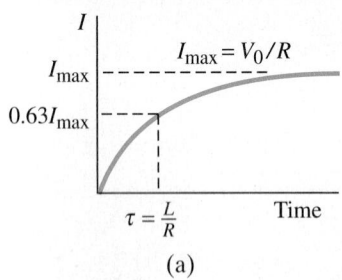

FIGURE 21–37 LR circuit.

FIGURE 21–38 (a) Growth of current in an LR circuit when connected to a battery. (b) Decay of current when the LR circuit is shorted out (battery is out of the circuit).

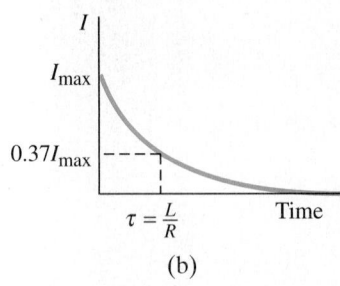

Any inductor will have some resistance. We represent an inductor by drawing the inductance L and its resistance R separately, as in Fig. 21–37. The resistance R could also include any other resistance in the circuit. Now we ask, what happens when a battery of voltage V_0 is connected in series to such an LR circuit? At the instant the switch connecting the battery is closed, the current starts to flow. It is opposed by the induced emf in the inductor because of the changing current. However, as soon as current starts to flow, there is a voltage drop across the resistance ($V = IR$). Hence, the voltage drop across the inductance is reduced and the current increases less rapidly. The current thus rises gradually, as shown in Fig. 21–38a, and approaches the steady value $I_{\max} = V_0/R$ when there is no more emf in the inductor (I is no longer changing) so all the voltage drop is across the resistance. The shape of the curve for I as a function of time is

$$I = \left(\frac{V_0}{R}\right)(1 - e^{-t/\tau}), \qquad [LR \text{ circuit with emf}]$$

where e is the number $e = 2.718\cdots$ (see Section 19–6) and

$$\tau = \frac{L}{R}$$

is the **time constant** of the circuit. When $t = \tau$, then $(1 - e^{-1}) = 0.63$, so τ is the time required for the current to reach $0.63 I_{\max}$.

Next, if the battery is suddenly switched out of the circuit (dashed line in Fig. 21–37), it takes time for the current to drop to zero, as shown in Fig. 21–38b. This is an exponential decay curve given by

$$I = I_{\max} e^{-t/\tau}. \qquad [LR \text{ circuit without emf}]$$

The time constant τ is the time for the current to decrease to $0.37 I_{\max}$ (37% of the original value), and again equals L/R.

These graphs show that there is always some "lag time" or "reaction time" when an electromagnet, for example, is turned on or off. We also see that an LR circuit has properties similar to an RC circuit (Section 19–6). Unlike the capacitor case, however, the time constant here is *inversely* proportional to R.

*21–13 AC Circuits and Reactance

We have previously discussed circuits that contain combinations of resistor, capacitor, and inductor, but only when they are connected to a dc source of emf or to zero voltage. Now we discuss these circuit elements when they are connected to a source of alternating voltage that produces an alternating current (ac).

First we examine, one at a time, how a resistor, a capacitor, and an inductor behave when connected to a source of alternating voltage, represented by the symbol

●—(∼)—●   [alternating voltage]

which produces a sinusoidal voltage of frequency f. We assume in each case that the emf gives rise to a current

$$I = I_0 \cos 2\pi f t,$$

where t is time and I_0 is the peak current. Remember (Section 18–7) that $V_{rms} = V_0/\sqrt{2}$ and $I_{rms} = I_0/\sqrt{2}$ (Eq. 18–8).

*Resistor

When an ac source is connected to a resistor as in Fig. 21–39a, the current increases and decreases with the alternating voltage according to Ohm's law,

$$V = IR = I_0 R \cos 2\pi f t = V_0 \cos 2\pi f t$$

where $V_0 = I_0 R$ is the peak voltage. Figure 21–39b shows the voltage (red curve) and the current (blue curve) as a function of time. Because the current is zero when the voltage is zero and the current reaches a peak when the voltage does, we say that the current and voltage are **in phase**. Energy is transformed into heat (Section 18–7), at an average rate $\overline{P} = \overline{IV} = I_{rms}^2 R = V_{rms}^2/R$.

*Inductor

In Fig. 21–40a an inductor of inductance L (symbol ‑⟞⟞⟞⟞‑) is connected to the ac source. We ignore any resistance it might have (it is usually small). The voltage applied to the inductor will be equal to the "back" emf generated in the inductor by the changing current as given by Eq. 21–9. This is because the sum of the electric potential changes around any closed circuit must add up to zero, by Kirchhoff's rule. Thus

$$V - L \frac{\Delta I}{\Delta t} = 0$$

or

$$\frac{\Delta I}{\Delta t} = \frac{V}{L}$$

where V is the sinusoidally varying voltage of the source and $L\, \Delta I/\Delta t$ is the voltage induced in the inductor. According to the last equation, I is increasing most rapidly when V has its maximum value, $V = V_0$. And I will be decreasing most rapidly when $V = -V_0$. These two instants correspond to points d and b on the graph of voltage versus time in Fig. 21–40b. By going point by point in this manner, the curve of V versus t as compared to that for I versus t can be constructed, and they are shown by the blue and red lines, respectively, in Fig. 21–40b. Notice that the current reaches its peaks (and troughs) $\frac{1}{4}$ cycle after the voltage does. We say that the

current lags the voltage by 90° for an inductor.

Because the current and voltage in an inductor are *out of phase* by 90°, the product IV (= power) is as often positive as it is negative (Fig. 21–40b). So no energy is transformed in an inductor on average; and no energy is dissipated as thermal energy.

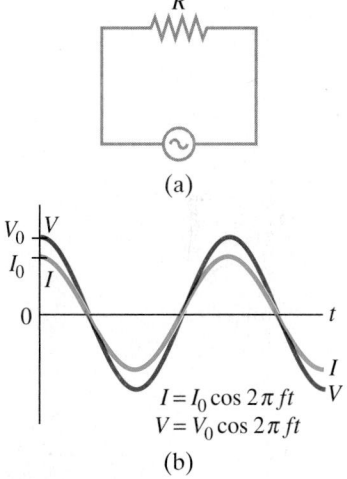

(a)

(b)

$$I = I_0 \cos 2\pi f t$$
$$V = V_0 \cos 2\pi f t$$

FIGURE 21–39 (a) Resistor connected to an ac source. (b) Current (blue curve) is in phase with the voltage (red) across a resistor.

FIGURE 21–40 (a) Inductor connected to an ac source. (b) Current (blue curve) lags voltage (red curve) by a quarter cycle, or 90°.

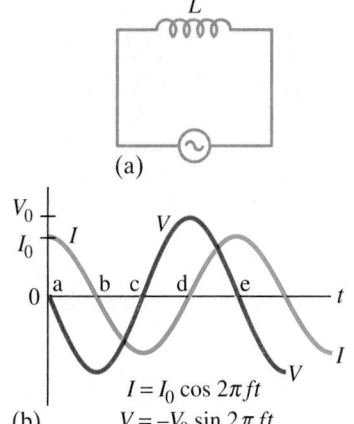

(a)

(b)

$$I = I_0 \cos 2\pi f t$$
$$V = -V_0 \sin 2\pi f t$$

Just as a resistor impedes the flow of charge, so too an inductor impedes the flow of charge in an alternating current due to the back emf produced. For a resistor R, the current and voltage are related by $V = IR$. We can write a similar relation for an inductor:

$$V = IX_L, \qquad \left[\begin{array}{l}\text{rms or peak values,}\\ \text{not at any instant}\end{array}\right] \quad \textbf{(21–11a)}$$

where X_L is called the **inductive reactance**. X_L has units of ohms. The quantities V and I in Eq. 21–11a can refer either to rms for both, or to peak values for both (see Section 18–7). Although this equation can relate the peak values, the peak current and voltage are not reached at the same time; so Eq. 21–11a is *not valid at a particular instant*, as is the case for a resistor ($V = IR$). Careful calculation (using calculus), as well as experiment, shows that

$$X_L = \omega L = 2\pi f L, \qquad \textbf{(21–11b)}$$

where $\omega = 2\pi f$ and f is the frequency of the ac.

For example, the inductive reactance of a 0.300-H inductor at 120 V and 60.0 Hz is $X_L = 2\pi f L = (6.28)(60.0 \text{ s}^{-1})(0.300 \text{ H}) = 113 \ \Omega$.

*Capacitor

When a capacitor is connected to a battery, the capacitor plates quickly acquire equal and opposite charges; but no steady current flows in the circuit. A capacitor prevents the flow of a dc current. But if a capacitor is connected to an alternating source of voltage, as in Fig. 21–41a, an alternating current will flow continuously. This can happen because when the ac voltage is first turned on, charge begins to flow and one plate acquires a negative charge and the other a positive charge. But when the voltage reverses itself, the charges flow in the opposite direction. Thus, for an alternating applied voltage, an ac current is present in the circuit continuously.

FIGURE 21–41 (a) Capacitor connected to an ac source. (b) Current leads voltage by a quarter cycle, or 90°.

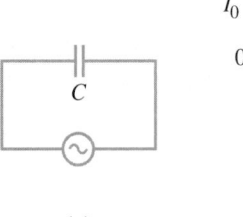
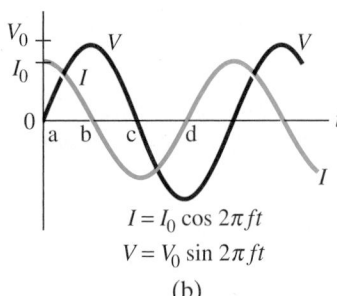

$I = I_0 \cos 2\pi f t$

$V = V_0 \sin 2\pi f t$

(a)   (b)

The applied voltage must equal the voltage across the capacitor: $V = Q/C$, where C is the capacitance and Q the charge on the plates. Thus the charge Q on the plates is in phase with the voltage. But what about the current I? At point a in Fig. 21–41b, when the voltage is zero and starts increasing, the charge on the plates is zero. Thus charge flows readily toward the plates and the current I is large. As the voltage approaches its maximum of V_0 (point b in Fig. 21–41b), the charge that has accumulated on the plates tends to prevent more charge from flowing, so the current I drops to zero at point b. Thus the current follows the blue curve in Fig. 21–41b. Like an inductor, the voltage and current are out of phase by 90°. But for a capacitor, the current reaches its peaks $\frac{1}{4}$ cycle before the voltage does, so we say that the

current leads the voltage by 90° for a capacitor.

⚠ **CAUTION**

Only resistance dissipates energy

Because the current and voltage are out of phase, the average power dissipated is zero, just as for an inductor. Thus *only a resistance will dissipate energy* as thermal energy in an ac circuit.

A relationship between the applied voltage and the current in a capacitor can be written just as for an inductance:

$$V = IX_C,$$

$$\left[\begin{array}{c}\text{rms or peak} \\ \text{values}\end{array}\right] \quad \textbf{(21–12a)}$$

where X_C is called the **capacitive reactance** and has units of ohms. V and I can both be rms or both maximum (V_0 and I_0); X_C depends on both the capacitance C and the frequency f:

$$X_C = \frac{1}{\omega C} = \frac{1}{2\pi f C}, \quad \textbf{(21–12b)}$$

where $\omega = 2\pi f$. For dc conditions, $f = 0$ and X_C becomes infinite, as it should because a capacitor does not pass dc current.

EXAMPLE 21–14 **Capacitor reactance.** What is the rms current in the circuit of Fig. 21–41a if $C = 1.00\ \mu\text{F}$ and $V_{rms} = 120\ \text{V}$? Calculate for (a) $f = 60.0\ \text{Hz}$, and then for (b) $f = 6.00 \times 10^5\ \text{Hz}$.

APPROACH We find the reactance using Eq. 21–12b, and solve for current in the equivalent form of Ohm's law, Eq. 21–12a.

SOLUTION (a) $X_C = 1/2\pi f C = 1/(2\pi)(60.0\ \text{s}^{-1})(1.00 \times 10^{-6}\ \text{F}) = 2.65\ \text{k}\Omega$. The rms current is (Eq. 21–12a):

$$I_{rms} = \frac{V_{rms}}{X_C} = \frac{120\ \text{V}}{2.65 \times 10^3\ \Omega} = 45.2\ \text{mA}.$$

(b) For $f = 6.00 \times 10^5\ \text{Hz}$, X_C will be $0.265\ \Omega$ and $I_{rms} = 452\ \text{A}$, vastly larger!

NOTE The dependence on f is dramatic. For high frequencies, the capacitive reactance is very small.

Two common applications of capacitors are illustrated in Figs. 21–42a and b. In Fig. 21–42a, circuit A is said to be capacitively coupled to circuit B. The purpose of the capacitor is to prevent a dc voltage from passing from A to B but allowing an ac signal to pass relatively unimpeded (if C is sufficiently large). In Fig. 21–42b, the

PHYSICS APPLIED
Capacitors as filters

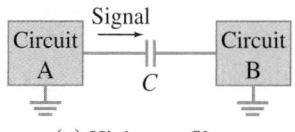

(a) High-pass filter

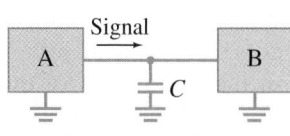

(b) Low-pass filter

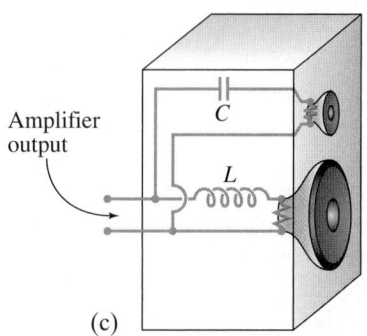

(c)

FIGURE 21–42 (a) and (b): Two common uses for a capacitor as a filter. (c) Simple loudspeaker cross-over.

capacitor also passes ac and not dc. In this case, a dc voltage can be maintained between circuits A and B, but an ac signal leaving A passes to ground instead of into B. Thus the capacitor in Fig. 21–42b acts like a **filter** when a constant dc voltage is required; any sharp variation in voltage passes to ground instead of into circuit B.

EXERCISE F The capacitor C in Fig. 21–42a is often called a "high-pass" filter, and the one in Fig. 21–42b a "low-pass" filter. Explain why.

Loudspeakers having separate "woofer" (low-frequency speaker) and "tweeter" (high-frequency speaker) may use a simple "cross-over" that consists of a capacitor in the tweeter circuit to impede low-frequency signals, and an inductor in the woofer circuit to impede high-frequency signals $(X_L = 2\pi f L)$. Hence mainly low-frequency sounds reach and are emitted by the woofer. See Fig. 21–42c.

PHYSICS APPLIED
Loudspeaker cross-over

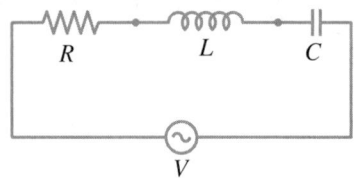

FIGURE 21–43 An *LRC* circuit.

*21–14 *LRC* Series AC Circuit

Let us examine a circuit containing all three elements in series: a resistor R, an inductor L, and a capacitor C, Fig. 21–43. If a given circuit contains only two of these elements, we can still use the results of this Section by setting $R = 0$, $X_L = 0$, or $X_C = 0$, as needed. We let V_R, V_L, and V_C represent the voltage across each element at a *given instant* in time; and V_{R0}, V_{L0}, and V_{C0} represent the *maximum* (peak) values of these voltages. The voltage across each of the elements will follow the phase relations we discussed in the previous Section. At any instant the voltage V supplied by the source will be, by Kirchhoff's loop rule,

$$V = V_R + V_L + V_C. \tag{21–13a}$$

Because the various voltages are not in phase, they do not reach their peak values at the same time, so the peak voltage of the source V_0 will *not* equal $V_{R0} + V_{L0} + V_{C0}$.

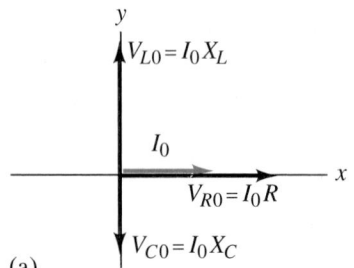

(a)

⚠ **CAUTION**

Peak voltages do not add to yield source voltage

*Phasor Diagrams

The current in an *LRC* circuit at any instant is the same at all points in the circuit (charge does not pile up in the wires). Thus the currents in each element are in phase with each other, even though the voltages are not. We choose our origin in time ($t = 0$) so that the current I at any time t is

$$I = I_0 \cos 2\pi f t.$$

AC circuits are complicated to analyze. The easiest approach is to use a sort of vector device known as a **phasor diagram**. Arrows (treated like vectors) are drawn in an xy coordinate system to represent each voltage. The length of each arrow represents the magnitude of the peak voltage across each element:

$$V_{R0} = I_0 R, \quad V_{L0} = I_0 X_L, \quad \text{and} \quad V_{C0} = I_0 X_C. \tag{21–13b}$$

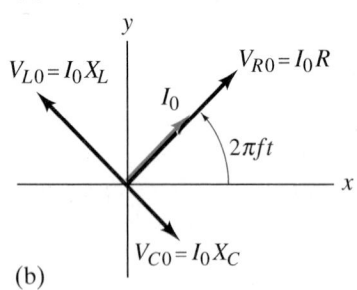

(b)

V_{R0} is in phase with the current and is initially ($t = 0$) drawn along the positive x axis, as is the current I_0. V_{L0} leads the current by 90°, so it leads V_{R0} by 90° and is initially drawn along the positive y axis. V_{C0} lags the current by 90°, so V_{C0} is drawn initially along the negative y axis. See Fig. 21–44a. If we let the vector diagram rotate counterclockwise at frequency f, we get the diagram shown in Fig. 21–44b; after a time, t, each arrow has rotated through an angle $2\pi f t$. Then the projections of each arrow on the x axis represent the voltages across each element at the instant t, as can be seen in Fig. 21–44c. For example $I = I_0 \cos 2\pi f t$.

The sum of the projections of the three voltage vectors represents the instantaneous voltage across the whole circuit, V. Therefore, the vector sum of these vectors will be the vector that represents the peak source voltage, V_0, as shown in Fig. 21–45 where it is seen that V_0 makes an angle ϕ with I_0 and V_{R0}. As time passes, V_0 rotates with the other vectors, so the instantaneous voltage V (projection of V_0 on the x axis) is (see Fig. 21–45)

$$V = V_0 \cos(2\pi f t + \phi).$$

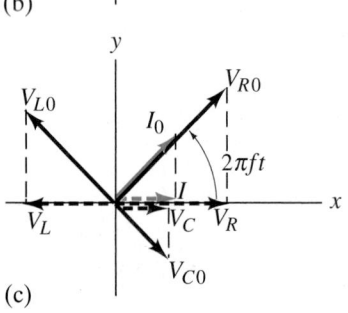

(c)

FIGURE 21–44 Phasor diagram for a series *LRC* circuit at (a) $t = 0$, (b) time t later. (c) Projections on the x axis which give I, V_R, V_C, V_L at time t.

FIGURE 21–45 Phasor diagram for a series *LRC* circuit showing the sum vector, V_0.

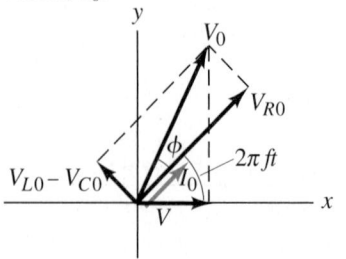

The voltage V across the whole circuit must equal the source voltage (Fig. 21–43). Thus the voltage from the source is out of phase with the current by an angle ϕ.

From this analysis we can now determine the total **impedance** Z of the circuit, which is defined in analogy to resistance and reactance as

$$V_{\text{rms}} = I_{\text{rms}} Z, \quad \text{or} \quad V_0 = I_0 Z. \tag{21–14}$$

From Fig. 21–45 we see, using the Pythagorean theorem (V_0 is the hypotenuse of a

right triangle), that (use Eq. 21–13b)

$$V_0 = \sqrt{V_{R0}^2 + (V_{L0} - V_{C0})^2}$$
$$= I_0 \sqrt{R^2 + (X_L - X_C)^2}.$$

Thus, from Eq. 21–14, the total impedance Z is

$$Z = \sqrt{R^2 + (X_L - X_C)^2}. \qquad (21\text{–}15)$$

Also from Fig. 21–45, we can find the phase angle ϕ between voltage and current:

$$\tan \phi = \frac{V_{L0} - V_{C0}}{V_{R0}} = \frac{I_0(X_L - X_C)}{I_0 R} = \frac{X_L - X_C}{R} \qquad (21\text{–}16a)$$

and

$$\cos \phi = \frac{V_{R0}}{V_0} = \frac{I_0 R}{I_0 Z} = \frac{R}{Z}. \qquad (21\text{–}16b)$$

Figure 21–45 was drawn for the case $X_L > X_C$, and the current lags the source voltage by ϕ. When the reverse is true, $X_L < X_C$, then ϕ in Eqs. 21–16 is less than zero, and the current leads the source voltage.

We saw earlier that power is dissipated only by a resistance; none is dissipated by inductance or capacitance. Therefore, the average power is given by $\overline{P} = I_{\text{rms}}^2 R$. But from Eq. 21–16b, $R = Z \cos \phi$. Therefore

$$\overline{P} = I_{\text{rms}}^2 Z \cos \phi = I_{\text{rms}} V_{\text{rms}} \cos \phi. \qquad (21\text{–}17)$$

The factor $\cos \phi$ is referred to as the **power factor** of the circuit.

EXAMPLE 21–15 **_LRC circuit._** Suppose $R = 25.0\ \Omega$, $L = 30.0\ \text{mH}$, and $C = 12.0\ \mu\text{F}$ in Fig. 21–43, and they are connected in series to a 90.0-V ac (rms) 500-Hz source. Calculate (_a_) the current in the circuit, and (_b_) the voltmeter readings (rms) across each element.

APPROACH To obtain the current, we determine the impedance (Eq. 21–15 plus Eqs. 21–11b and 21–12b), and then use $I_{\text{rms}} = V_{\text{rms}}/Z$ (Eq. 21–14). Voltage drops across each element are found using Ohm's law or equivalent for each element: $V_R = IR$, $V_L = IX_L$, and $V_C = IX_C$.

SOLUTION (_a_) First, we find the reactance of the inductor and capacitor at $f = 500\ \text{Hz} = 500\ \text{s}^{-1}$:

$$X_L = 2\pi f L = 94.2\ \Omega, \qquad X_C = \frac{1}{2\pi f C} = 26.5\ \Omega.$$

Then the total impedance is

$$Z = \sqrt{R^2 + (X_L - X_C)^2} = \sqrt{(25.0\ \Omega)^2 + (94.2\ \Omega - 26.5\ \Omega)^2} = 72.2\ \Omega.$$

From the impedance version of Ohm's law, Eq. 21–14,

$$I_{\text{rms}} = \frac{V_{\text{rms}}}{Z} = \frac{90.0\ \text{V}}{72.2\ \Omega} = 1.25\ \text{A}.$$

(_b_) The rms voltage across each element is

$$(V_R)_{\text{rms}} = I_{\text{rms}} R = (1.25\ \text{A})(25.0\ \Omega) = 31.2\ \text{V}$$
$$(V_L)_{\text{rms}} = I_{\text{rms}} X_L = (1.25\ \text{A})(94.2\ \Omega) = 118\ \text{V}$$
$$(V_C)_{\text{rms}} = I_{\text{rms}} X_C = (1.25\ \text{A})(26.5\ \Omega) = 33.1\ \text{V}.$$

CAUTION

Individual peak or rms voltages do NOT add up to source voltage (due to phase differences)

NOTE These voltages do _not_ add up to the source voltage, 90.0 V (rms). Indeed, the rms voltage across the inductance _exceeds_ the source voltage. This can happen because the different voltages are out of phase with each other: so, at any instant the capacitor's voltage might be negative which compensates for a large positive inductor voltage. The rms voltages, however, are always positive by definition. Although the rms voltages need not add up to the source voltage, the instantaneous voltages at any time must add up to the source voltage at that instant.

*21–15 Resonance in AC Circuits

The rms current in an LRC series circuit is given by (see Eqs. 21–14, 21–15, 21–11b, and 21–12b):

$$I_{rms} = \frac{V_{rms}}{Z} = \frac{V_{rms}}{\sqrt{R^2 + \left(2\pi f L - \frac{1}{2\pi f C}\right)^2}}. \qquad \textbf{(21–18)}$$

Because the reactance of inductors and capacitors depends on the frequency f of the source, the current in an LRC circuit depends on frequency. From Eq. 21–18 we see that the current will be maximum at a frequency that satisfies

$$2\pi f L - \frac{1}{2\pi f C} = 0.$$

We solve this for f, and call the solution f_0:

$$f_0 = \frac{1}{2\pi}\sqrt{\frac{1}{LC}}. \qquad \text{[resonance]} \quad \textbf{(21–19)}$$

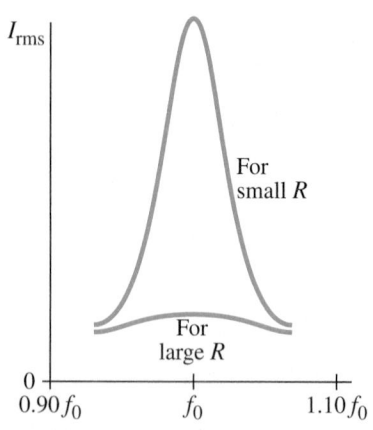

FIGURE 21–46 Current in an LRC circuit as a function of frequency, showing resonance peak at $f = f_0 = 1/(2\pi\sqrt{LC})$.

When $f = f_0$, the circuit is in **resonance**, and f_0 is the **resonant frequency** of the circuit. At this frequency, $X_C = X_L$, so the impedance is purely resistive. A graph of I_{rms} versus f is shown in Fig. 21–46 for particular values of R, L, and C. For smaller R compared to X_L and X_C, the resonance peak will be higher and sharper.

When R is very small, we speak of an **LC circuit**. The energy in an LC circuit oscillates, at frequency f_0, between the inductor and the capacitor, with some being dissipated in R (some resistance is unavoidable). This is called an **LC oscillation** or an **electromagnetic oscillation**. Not only does the charge oscillate back and forth, but so does the energy, which oscillates between being stored in the electric field of the capacitor and in the magnetic field of the inductor.

Electric resonance is used in many circuits. Radio and TV sets, for example, use resonant circuits for tuning in a station. Many frequencies reach the circuit from the antenna, but a significant current flows only for frequencies at or near the resonant frequency chosen (the station you want). Either L or C is variable so that different stations can be tuned in (more on this in Chapter 22).

▌Summary

The **magnetic flux** passing through a loop is equal to the product of the area of the loop times the perpendicular component of the magnetic field:

$$\Phi_B = B_\perp A = BA\cos\theta. \qquad \textbf{(21–1)}$$

If the magnetic flux through a coil of wire changes in time, an emf is induced in the coil. The magnitude of the induced emf equals the time rate of change of the magnetic flux through the loop times the number N of loops in the coil:

$$\mathcal{E} = -N\frac{\Delta\Phi_B}{\Delta t}. \qquad \textbf{(21–2b)}$$

This is **Faraday's law of induction**.

The induced emf can produce a current whose magnetic field opposes the original change in flux (**Lenz's law**).

Faraday's law also tells us that a changing magnetic field produces an electric field; and that a straight wire of length ℓ moving with speed v perpendicular to a magnetic field of strength B has an emf induced between its ends equal to

$$\mathcal{E} = B\ell v. \qquad \textbf{(21–3)}$$

An electric **generator** changes mechanical energy into electric energy. Its operation is based on Faraday's law: a coil of wire is made to rotate uniformly by mechanical means in a magnetic field, and the changing flux through the coil induces a sinusoidal current, which is the output of the generator.

A motor, which operates in the reverse of a generator, acts like a generator in that a **back emf** is induced in its rotating coil. Because this back emf opposes the input voltage, it can act to limit the current in a motor coil. Similarly, a generator acts somewhat like a motor in that a **counter torque** acts on its rotating coil.

A **transformer**, which is a device to change the magnitude of an ac voltage, consists of a primary coil and a secondary coil. The changing flux due to an ac voltage in the primary coil induces an ac voltage in the secondary coil. In a 100% efficient transformer, the ratio of output to input voltages (V_S/V_P) equals the ratio of the number of turns N_S in the secondary to the number N_P in the primary:

$$\frac{V_S}{V_P} = \frac{N_S}{N_P}. \qquad \textbf{(21–6)}$$

The ratio of secondary to primary current is in the inverse ratio of turns:

$$\frac{I_S}{I_P} = \frac{N_P}{N_S}. \qquad \textbf{(21–7)}$$

[*Read/write heads for computer hard drives and tape, as well as microphones, ground fault circuit interrupters, and seismographs, are all applications of electromagnetic induction.]

[*A changing current in a coil of wire will produce a changing magnetic field that induces an emf in a second coil placed nearby. The **mutual inductance**, M, is defined by

$$\mathcal{E}_2 = -M \frac{\Delta I_1}{\Delta t}. \quad (21\text{–}8)]$$

[*Within a single coil, the changing B due to a changing current induces an opposing emf, $\mathcal{E}$, so a coil has a **self-inductance** L defined by

$$\mathcal{E} = -L \frac{\Delta I}{\Delta t}. \quad (21\text{–}9)]$$

[*The energy stored in an inductance L carrying current I is given by $U = \frac{1}{2}LI^2$. This energy can be thought of as being stored in the magnetic field of the inductor. The energy density u in any magnetic field B is given by

$$u = \frac{1}{2} \frac{B^2}{\mu_0}. \quad (21\text{–}10)]$$

[*When an inductance L and resistor R are connected in series to a source of emf, V_0, the current rises as

$$I = \frac{V_0}{R}(1 - e^{-t/\tau}),$$

where $\tau = L/R$ is the **time constant**. If the battery is suddenly switched out of the LR circuit, the current drops exponentially, $I = I_{\max}e^{-t/\tau}$.]

[*Inductive and capacitive **reactance**, X, defined as for resistors, is the proportionality constant between voltage and current (either the rms or peak values). Across an inductor,

$$V = IX_L, \quad (21\text{–}11a)$$

and across a capacitor,

$$V = IX_C. \quad (21\text{–}12a)$$

The reactance of an inductor increases with frequency f,

$$X_L = 2\pi f L, \quad (21\text{–}11b)$$

whereas the reactance of a capacitor decreases with frequency f,

$$X_C = \frac{1}{2\pi f C}. \quad (21\text{–}12b)$$

The current through a resistor is always in phase with the voltage across it, but in an inductor the current lags the voltage by 90°, and in a capacitor the current leads the voltage by 90°.]

[*In an LRC series circuit, the total **impedance** Z is defined by the equivalent of $V = IR$ for resistance, namely,

$$V_0 = I_0 Z \quad \text{or} \quad V_{\text{rms}} = I_{\text{rms}} Z; \quad (21\text{–}14)$$

Z is given by

$$Z = \sqrt{R^2 + (X_L - X_C)^2}. \quad (21\text{–}15)]$$

[*An LRC series circuit **resonates** at a frequency given by

$$f_0 = \frac{1}{2\pi}\sqrt{\frac{1}{LC}}. \quad (21\text{–}19)$$

The rms current in the circuit is largest when the applied voltage has a frequency equal to f_0.]

Questions

1. What would be the advantage, in Faraday's experiments (Fig. 21–1), of using coils with many turns?
2. What is the difference between magnetic flux and magnetic field?
3. Suppose you are holding a circular ring of wire in front of you and (a) suddenly thrust a magnet, south pole first, away from you toward the center of the circle. Is a current induced in the wire? (b) Is a current induced when the magnet is held steady within the ring? (c) Is a current induced when you withdraw the magnet? For each yes answer, specify the direction. Explain your answers.
4. (a) A wire loop is pulled away from a current-carrying wire (Fig. 21–47). What is the direction of the induced current in the loop: clockwise or counterclockwise? (b) What if the wire loop stays fixed as the current I decreases? Explain your answers.

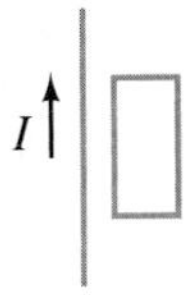

FIGURE 21–47
Question 4.

5. (a) If the north pole of a thin flat magnet moves on a table toward a loop also on the table (Fig. 21–48), in what direction is the induced current in the loop? Assume the magnet is the same thickness as the wire. (b) What if the magnet is four times thicker than the wire loop? Explain your answers.

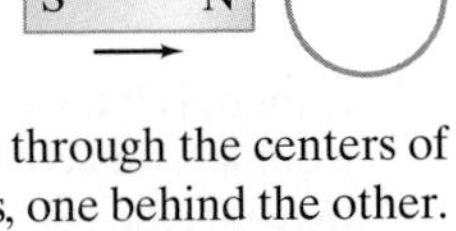

FIGURE 21–48
Question 5.

6. Suppose you are looking along a line through the centers of two circular (but separate) wire loops, one behind the other. A battery is suddenly connected to the front loop, establishing a clockwise current. (a) Will a current be induced in the second loop? (b) If so, when does this current start? (c) When does it stop? (d) In what direction is this current? (e) Is there a force between the two loops? (f) If so, in what direction?

7. The battery mentioned in Question 6 is disconnected. Will a current be induced in the second loop? If so, when does it start and stop? In what direction is this current?
8. In Fig. 21–49, determine the direction of the induced current in resistor R_A (a) when coil B is moved toward coil A, (b) when coil B is moved away from A, (c) when the resistance R_B is increased but the coils remain fixed. Explain your answers.

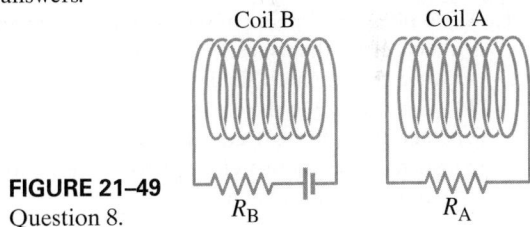

FIGURE 21–49
Question 8.

Coil B   Coil A

R_B R_A

9. In situations where a small signal must travel over a distance, a *shielded cable* is used in which the signal wire is surrounded by an insulator and then enclosed by a cylindrical conductor (shield) carrying the return current. Why is a "shield" necessary?
10. What is the advantage of placing the two insulated electric wires carrying ac close together or even twisted about each other?
11. Explain why, exactly, the lights may dim briefly when a refrigerator motor starts up. When an electric heater is turned on, the lights may stay dimmed as long as the heater is on. Explain the difference.
12. Use Figs. 21–14 and 21–17 plus the right-hand rules to show why the counter torque in a generator *opposes* the motion.
13. Will an eddy current brake (Fig. 21–20) work on a copper or aluminum wheel, or must the wheel be ferromagnetic? Explain.

14. A bar magnet falling inside a vertical metal tube reaches a terminal velocity even if the tube is evacuated so that there is no air resistance. Explain.

15. It has been proposed that eddy currents be used to help sort solid waste for recycling. The waste is first ground into tiny pieces and iron removed with a magnet. The waste then is allowed to slide down an incline over permanent magnets. How will this aid in the separation of nonferrous metals (Al, Cu, Pb, brass) from nonmetallic materials?

16. The pivoted metal bar with slots in Fig. 21–50 falls much more quickly through a magnetic field than does a solid bar. Explain.

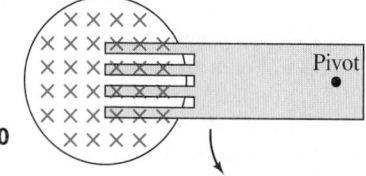

FIGURE 21–50
Question 16.

17. If an aluminum sheet is held between the poles of a large bar magnet, it requires some force to pull it out of the magnetic field even though the sheet is not ferromagnetic and does not touch the pole faces. Explain.

18. A bar magnet is held above the floor and dropped (Fig. 21–51). In case (a), the magnet falls through a wire loop. In case (b), there is nothing between the magnet and the floor. How will the speeds of the magnets compare? Explain.

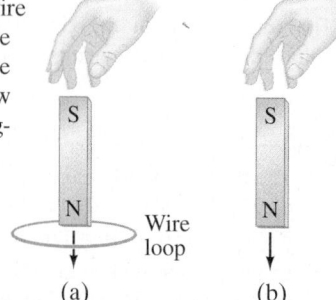

FIGURE 21–51
Question 18 and
MisConceptual
Question 5.   (a)   (b)

19. A metal bar, pivoted at one end, oscillates freely in the absence of a magnetic field; but in a magnetic field, its oscillations are quickly damped out. Explain. (This *magnetic damping* is used in a number of practical devices.)

20. An enclosed transformer has four wire leads coming from it. How could you determine the ratio of turns on the two coils without taking the transformer apart? How would you know which wires paired with which?

21. The use of higher-voltage lines in homes—say, 600 V or 1200 V—would reduce energy waste. Why are they not used?

22. A transformer designed for a 120-V ac input will often "burn out" if connected to a 120-V dc source. Explain. [*Hint*: The resistance of the primary coil is usually very low.]

***23.** How would you arrange two flat circular coils so that their mutual inductance was (*a*) greatest, (*b*) least (without separating them by a great distance)? Explain.

***24.** Does the emf of the battery in Fig. 21–37 affect the time needed for the *LR* circuit to reach (*a*) a given fraction of its maximum possible current, (*b*) a given value of current? Explain.

***25.** In an *LRC* circuit, can the rms voltage across (*a*) an inductor, (*b*) a capacitor, be greater than the rms voltage of the ac source? Explain.

***26.** Describe briefly how the frequency of the source emf affects the impedance of (*a*) a pure resistance, (*b*) a pure capacitance, (*c*) a pure inductance, (*d*) an *LRC* circuit near resonance (*R* small), (*e*) an *LRC* circuit far from resonance (*R* small).

***27.** Describe how to make the impedance in an *LRC* circuit a minimum.

***28.** An *LRC* resonant circuit is often called an *oscillator* circuit. What is it that oscillates?

***29.** Is the ac current in the inductor always the same as the current in the resistor of an *LRC* circuit? Explain.

MisConceptual Questions

1. A coil rests in the plane of the page while a magnetic field is directed into the page. A clockwise current is induced
(*a*) when the magnetic field gets stronger.
(*b*) when the size of the coil decreases.
(*c*) when the coil is moved sideways across the page.
(*d*) when the magnetic field is tilted so it is no longer perpendicular to the page.

2. A wire loop moves at constant velocity without rotation through a constant magnetic field. The induced current in the loop will be
(*a*) clockwise.   (*b*) counterclockwise.   (*c*) zero.
(*d*) We need to know the orientation of the loop relative to the magnetic field.

3. A square loop moves to the right from an area where $\vec{B} = 0$, completely through a region containing a uniform magnetic field directed into the page (Fig. 21–52), and then out to $B = 0$ after point L. A current is induced in the loop
(*a*) only as it passes line J.   (*d*) as it passes line J or line L.
(*b*) only as it passes line K.   (*e*) as it passes all three lines.
(*c*) only as it passes line L.

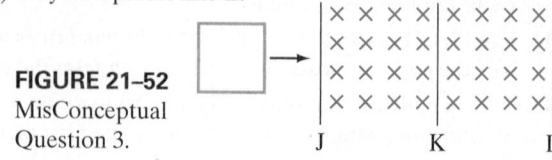

FIGURE 21–52
MisConceptual
Question 3.   J   K   L

4. Two loops of wire are moving in the vicinity of a very long straight wire carrying a steady current (Fig. 21–53). Find the direction of the induced current in each loop.

For C: For D:
 (*a*) clockwise.   (*a*) clockwise.
 (*b*) counterclockwise.   (*b*) counterclockwise.
 (*c*) zero.   (*c*) zero.
 (*d*) alternating (ac).   (*d*) alternating (ac).

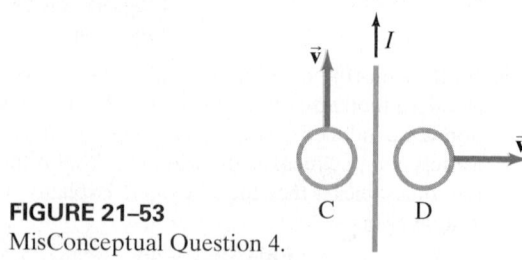

FIGURE 21–53   C   D
MisConceptual Question 4.

5. If there is induced current in Question 18 (see Fig. 21–51), wouldn't that cost energy? Where would that energy come from in case (a)?
(*a*) Induced current doesn't need energy.
(*b*) Energy conservation is violated.
(*c*) There is less kinetic energy.
(*d*) There is more gravitational potential energy.

6. A nonconducting plastic hoop is held in a magnetic field that points out of the page (Fig. 21–54). As the strength of the field increases,
   (a) an induced emf will be produced that causes a clockwise current.
   (b) an induced emf will be produced that causes a counterclockwise current.
   (c) an induced emf will be produced but no current.
   (d) no induced emf will be produced.

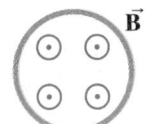

FIGURE 21–54
MisConceptual Question 6.

7. A long straight wire carries a current I as shown in Fig. 21–55. A small loop of wire rests in the plane of the page. Which of the following will *not* induce a current in the loop?
   (a) Increasing the current in the straight wire.
   (b) Moving the loop in a direction parallel to the wire.
   (c) Rotating the loop so that it becomes perpendicular to the plane of the page.
   (d) Moving the loop farther from the wire without rotating it.
   (e) Moving the loop farther from the wire while rotating it.

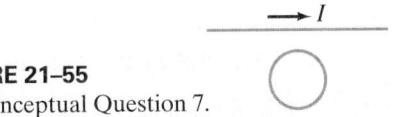

FIGURE 21–55
MisConceptual Question 7.

8. Two separate but nearby coils are mounted along the same axis. A power supply controls the flow of current in the first coil, and thus the magnetic field it produces. The second coil is connected only to an ammeter. The ammeter will indicate that a current is flowing in the second coil
   (a) whenever a current flows in the first coil.
   (b) only when a steady current flows in the first coil.
   (c) only when the current in the first coil changes.
   (d) only if the second coil is connected to the power supply by rewiring it to be in series with the first coil.

9. When a generator is used to produce electric current, the resulting electric energy originates from which source?
   (a) The generator's magnetic field.
   (b) Whatever rotates the generator's axle.
   (c) The resistance of the generator's coil.
   (d) Back emf.
   (e) Empty space.

10. Which of the following will *not* increase a generator's voltage output?
    (a) Rotating the generator faster.
    (b) Increasing the area of the coil.
    (c) Rotating the magnetic field so that it is more closely parallel to the generator's rotation axis.
    (d) Increasing the magnetic field through the coil.
    (e) Increasing the number of turns in the coil.

11. Which of the following can a transformer accomplish?
    (a) Changing voltage but not current.
    (b) Changing current but not voltage.
    (c) Changing power.
    (d) Changing both current and voltage.

12. A laptop computer's charger unit converts 120 V from a wall power outlet to the lower voltage required by the laptop. Inside the charger's plastic case is a diode or rectifier (discussed in Chapter 29) that changes ac to dc plus a
    (a) battery.
    (b) motor.
    (c) generator.
    (d) transformer.
    (e) transmission line.

13. Which of the following statements about transformers is false?
    (a) Transformers work using ac current or dc current.
    (b) If the current in the secondary is higher, the voltage is lower.
    (c) If the voltage in the secondary is higher, the current is lower.
    (d) If no flux is lost, the product of the voltage and the current is the same in the primary and secondary coils.

14. A 10-V, 1.0-A dc current is run through a step-up transformer that has 10 turns on the input side and 20 turns on the output side. What is the output?
    (a) 10 V, 0.5 A.
    (b) 20 V, 0.5 A.
    (c) 20 V, 1 A.
    (d) 10 V, 1 A.
    (e) 0 V, 0 A.

15. The alternating electric current at a wall outlet is most commonly produced by
    (a) a connection to rechargeable batteries.
    (b) a rotating coil that is immersed in a magnetic field.
    (c) accelerating electrons between oppositely charged capacitor plates.
    (d) using an electric motor.
    (e) alternately heating and cooling a wire.

*16. When you swipe a credit card, the machine sometimes fails to read the card. What can you do differently?
    (a) Swipe the card more slowly so that the reader has more time to read the magnetic stripe.
    (b) Swipe the card more quickly so that the induced emf is higher.
    (c) Swipe the card more quickly so that the induced currents are reduced.
    (d) Swipe the card more slowly so that the magnetic fields don't change so fast.

*17. Which of the following is true about all series ac circuits?
    (a) The voltage across any circuit element is a maximum when the current is a maximum in that circuit element.
    (b) The current at any point in the circuit is always the same as the current at any other point in the circuit.
    (c) The current in the circuit is a maximum when the source ac voltage is a maximum.
    (d) Resistors, capacitors, and inductors can all change the phase of the current.

Problems

21–1 to 21–4 Faraday's Law of Induction

1. (I) The magnetic flux through a coil of wire containing two loops changes at a constant rate from -58 Wb to $+38$ Wb in 0.34 s. What is the emf induced in the coil?

2. (I) The north pole of the magnet in Fig. 21–57 is being inserted into the coil. In which direction is the induced current flowing through resistor R? Explain.

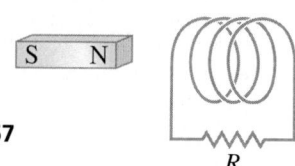

FIGURE 21–57
Problem 2.

R

3. (I) The rectangular loop in Fig. 21–58 is being pushed to the right, where the magnetic field points inward. In what direction is the induced current? Explain your reasoning.

FIGURE 21–58
Problem 3.

4. (I) If the solenoid in Fig. 21–59 is being pulled away from the loop shown, in what direction is the induced current in the loop? Explain.

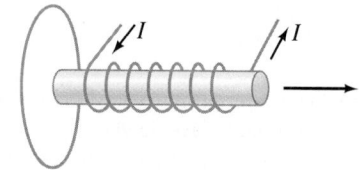

FIGURE 21–59
Problem 4.

5. (II) An 18.5-cm-diameter loop of wire is initially oriented perpendicular to a 1.5-T magnetic field. The loop is rotated so that its plane is parallel to the field direction in 0.20 s. What is the average induced emf in the loop?

6. (II) A fixed 10.8-cm-diameter wire coil is perpendicular to a magnetic field 0.48 T pointing up. In 0.16 s, the field is changed to 0.25 T pointing down. What is the average induced emf in the coil?

7. (II) A 16-cm-diameter circular loop of wire is placed in a 0.50-T magnetic field. (*a*) When the plane of the loop is perpendicular to the field lines, what is the magnetic flux through the loop? (*b*) The plane of the loop is rotated until it makes a 42° angle with the field lines. What is the angle θ in Eq. 21–1 for this situation? (*c*) What is the magnetic flux through the loop at this angle?

8. (II) (*a*) If the resistance of the resistor in Fig. 21–60 is slowly increased, what is the direction of the current induced in the small circular loop inside the larger loop? (*b*) What would it be if the small loop were placed outside the larger one, to the left? Explain your answers.

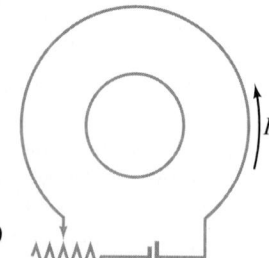

FIGURE 21–60
Problem 8.

9. (II) The moving rod in Fig. 21–11 is 12.0 cm long and is pulled at a speed of 18.0 cm/s. If the magnetic field is 0.800 T, calculate (*a*) the emf developed, and (*b*) the electric field felt by electrons in the rod.

10. (II) A circular loop in the plane of the paper lies in a 0.65-T magnetic field pointing into the paper. The loop's diameter changes from 20.0 cm to 6.0 cm in 0.50 s. What is (*a*) the direction of the induced current, (*b*) the magnitude of the average induced emf, and (*c*) the average induced current if the coil resistance is 2.5 Ω?

11. (II) What is the direction of the induced current in the circular loop due to the current shown in each part of Fig. 21–61? Explain why.

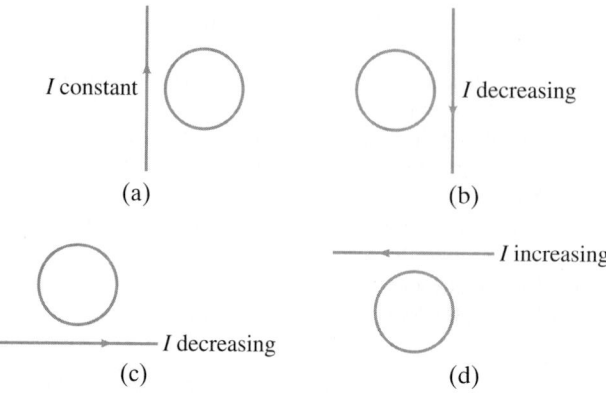

I constant

I decreasing

(a)   (b)

I increasing

I decreasing

(c)   (d)

FIGURE 21–61 Problem 11.

12. (II) A 600-turn solenoid, 25 cm long, has a diameter of 2.5 cm. A 14-turn coil is wound tightly around the center of the solenoid. If the current in the solenoid increases uniformly from 0 to 5.0 A in 0.60 s, what will be the induced emf in the short coil during this time?

13. (II) When a car drives through the Earth's magnetic field, an emf is induced in its vertical 55-cm-long radio antenna. If the Earth's field $(5.0 \times 10^{-5}\,\text{T})$ points north with a dip angle of 38°, what is the maximum emf induced in the antenna and which direction(s) will the car be moving to produce this maximum value? The car's speed is 30.0 m/s on a horizontal road.

14. (II) Part of a single rectangular loop of wire with dimensions shown in Fig. 21–62 is situated inside a region of uniform magnetic field of 0.550 T. The total resistance of the loop is 0.230 Ω. Calculate the force required to pull the loop from the field (to the right) at a constant velocity of 3.10 m/s. Neglect gravity.

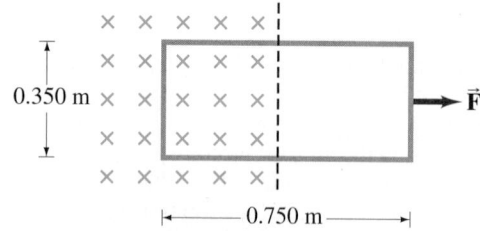

0.350 m

$\vec{\mathbf{F}}$

0.750 m

FIGURE 21–62 Problem 14.

15. (II) In order to make the rod of Fig. 21–11a move to the right at speed v, you need to apply an external force on the rod to the right. (*a*) Explain and determine the magnitude of the required force. (*b*) What external power is needed to move the rod? (Do not confuse this external force on the rod with the upward force on the electrons shown in Fig. 21–11b.)

16. (II) In Fig. 21–11, the moving rod has a resistance of 0.25 Ω and moves on rails 20.0 cm apart. The stationary U-shaped conductor has negligible resistance. When a force of 0.350 N is applied to the rod, it moves to the right at a constant speed of 1.50 m/s. What is the magnetic field?

17. (III) In Fig. 21–11, the rod moves with a speed of 1.6 m/s on rails 30.0 cm apart. The rod has a resistance of 2.5 Ω. The magnetic field is 0.35 T, and the resistance of the U-shaped conductor is 21.0 Ω at a given instant. Calculate (a) the induced emf, (b) the current in the U-shaped conductor, and (c) the external force needed to keep the rod's velocity constant at that instant.

18. (III) A 22.0-cm-diameter coil consists of 30 turns of circular copper wire 2.6 mm in diameter. A uniform magnetic field, perpendicular to the plane of the coil, changes at a rate of 8.65×10^{-3} T/s. Determine (a) the current in the loop, and (b) the rate at which thermal energy is produced.

19. (III) The magnetic field perpendicular to a single 13.2-cm-diameter circular loop of copper wire decreases uniformly from 0.670 T to zero. If the wire is 2.25 mm in diameter, how much charge moves past a point in the coil during this operation?

21–5 Generators

20. (II) The generator of a car idling at 1100 rpm produces 12.7 V. What will the output be at a rotation speed of 2500 rpm, assuming nothing else changes?

21. (II) A 550-loop circular armature coil with a diameter of 8.0 cm rotates at 120 rev/s in a uniform magnetic field of strength 0.55 T. (a) What is the rms voltage output of the generator? (b) What would you do to the rotation frequency in order to double the rms voltage output?

22. (II) A generator rotates at 85 Hz in a magnetic field of 0.030 T. It has 950 turns and produces an rms voltage of 150 V and an rms current of 70.0 A. (a) What is the peak current produced? (b) What is the area of each turn of the coil?

23. (III) A simple generator has a square armature 6.0 cm on a side. The armature has 85 turns of 0.59-mm-diameter copper wire and rotates in a 0.65-T magnetic field. The generator is used to power a lightbulb rated at 12.0 V and 25.0 W. At what rate should the generator rotate to provide 12.0 V to the bulb? Consider the resistance of the wire on the armature.

21–6 Back EMF and Torque

24. (I) A motor has an armature resistance of 3.65 Ω. If it draws 8.20 A when running at full speed and connected to a 120-V line, how large is the back emf?

25. (I) The back emf in a motor is 72 V when operating at 1800 rpm. What would be the back emf at 2300 rpm if the magnetic field is unchanged?

26. (II) What will be the current in the motor of Example 21–8 if the load causes it to run at half speed?

21–7 Transformers

[Assume 100% efficiency, unless stated otherwise.]

27. (I) A transformer is designed to change 117 V into 13,500 V, and there are 148 turns in the primary coil. How many turns are in the secondary coil?

28. (I) A transformer has 360 turns in the primary coil and 120 in the secondary coil. What kind of transformer is this, and by what factor does it change the voltage? By what factor does it change the current?

29. (I) A step-up transformer increases 25 V to 120 V. What is the current in the secondary coil as compared to the primary coil?

30. (I) Neon signs require 12 kV for their operation. To operate from a 240-V line, what must be the ratio of secondary to primary turns of the transformer? What would the voltage output be if the transformer were connected in reverse?

31. (II) A model-train transformer plugs into 120-V ac and draws 0.35 A while supplying 6.8 A to the train. (a) What voltage is present across the tracks? (b) Is the transformer step-up or step-down?

32. (II) The output voltage of a 95-W transformer is 12 V, and the input current is 25 A. (a) Is this a step-up or a step-down transformer? (b) By what factor is the voltage multiplied?

33. (II) A transformer has 330 primary turns and 1240 secondary turns. The input voltage is 120 V and the output current is 15.0 A. What are the output voltage and input current?

34. (II) If 35 MW of power at 45 kV (rms) arrives at a town from a generator via 4.6-Ω transmission lines, calculate (a) the emf at the generator end of the lines, and (b) the fraction of the power generated that is wasted in the lines.

35. (II) For the transmission of electric power from power plant to home, as depicted in Fig. 21–25, where the electric power sent by the plant is 100 kW, about how far away could the house be from the power plant before power loss is 50%? Assume the wires have a resistance per unit length of 5×10^{-5} Ω/m.

36. (II) For the electric power transmission system shown in Fig. 21–25, what is the ratio N_S/N_P for (a) the step-up transformer, (b) the step-down transformer next to the home?

37. (III) Suppose 2.0 MW is to arrive at a large shopping mall over two 0.100-Ω lines. Estimate how much power is saved if the voltage is stepped up from 120 V to 1200 V and then down again, rather than simply transmitting at 120 V. Assume the transformers are each 99% efficient.

38. (III) Design a dc transmission line that can transmit 925 MW of electricity 185 km with only a 2.5% loss. The wires are to be made of aluminum and the voltage is 660 kV.

*21–10 Inductance

*39. (I) If the current in a 160-mH coil changes steadily from 25.0 A to 10.0 A in 350 ms, what is the magnitude of the induced emf?

*40. (I) What is the inductance of a coil if the coil produces an emf of 2.50 V when the current in it changes from −28.0 mA to +31.0 mA in 14.0 ms?

*41. (I) Determine the inductance L of a 0.60-m-long air-filled solenoid 2.9 cm in diameter containing 8500 loops.

*42. (I) How many turns of wire would be required to make a 130-mH inductor out of a 30.0-cm-long air-filled solenoid with a diameter of 5.8 cm?

*43. (II) An air-filled cylindrical inductor has 2600 turns, and it is 2.5 cm in diameter and 28.2 cm long. (a) What is its inductance? (b) How many turns would you need to generate the same inductance if the core were iron-filled instead? Assume the magnetic permeability of iron is about 1200 times that of free space.

*44. (II) A coil has 2.25-Ω resistance and 112-mH inductance. If the current is 3.00 A and is increasing at a rate of 3.80 A/s, what is the potential difference across the coil at this moment?

*45. (III) A physics professor wants to demonstrate the large size of the henry unit. On the outside of a 12-cm-diameter plastic hollow tube, she wants to wind an air-filled solenoid with self-inductance of 1.0 H using copper wire with a 0.81-mm diameter. The solenoid is to be tightly wound with each turn touching its neighbor (the wire has a thin insulating layer on its surface so the neighboring turns are not in electrical contact). How long will the plastic tube need to be and how many kilometers of copper wire will be required? What will be the resistance of this solenoid?

*46. (III) A long thin solenoid of length ℓ and cross-sectional area A contains N_1 closely packed turns of wire. Wrapped tightly around it is an insulated coil of N_2 turns, Fig. 21–63. Assume all the flux from coil 1 (the solenoid) passes through coil 2, and calculate the mutual inductance.

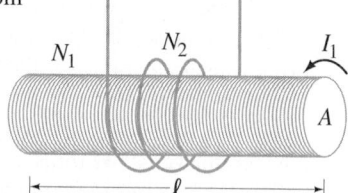

FIGURE 21–63
Problem 46.

*21–11 Magnetic Energy Storage

*47. (I) The magnetic field inside an air-filled solenoid 36 cm long and 2.0 cm in diameter is 0.72 T. Approximately how much energy is stored in this field?

*48. (II) At $t = 0$, the current through a 45.0-mH inductor is 50.0 mA and is increasing at the rate of 115 mA/s. What is the initial energy stored in the inductor, and how long does it take for the energy to increase by a factor of 5.0 from the initial value?

*49. (II) Assuming the Earth's magnetic field averages about 0.50×10^{-4} T near Earth's surface, estimate the total energy stored in this field in the first 10 km above Earth's surface.

*21–12 LR Circuit

*50. (II) It takes 2.56 ms for the current in an LR circuit to increase from zero to 0.75 its maximum value. Determine (a) the time constant of the circuit, (b) the resistance of the circuit if $L = 31.0$ mH.

*51. (II) How many time constants does it take for the potential difference across the resistor in an LR circuit like that in Fig. 21–37 to drop to 2.5% of its original value, after the switch is moved to the upper position, removing V_0 from the circuit?

*52. (III) Determine $\Delta I / \Delta t$ at $t = 0$ (when the battery is connected) for the LR circuit of Fig. 21–37 and show that if I continued to increase at this rate, it would reach its maximum value in one time constant.

*53. (III) After how many time constants does the current in Fig. 21–37 reach within (a) 10%, (b) 1.0%, and (c) 0.1% of its maximum value?

*21–13 AC Circuits and Reactance

*54. (I) What is the reactance of a 6.20-μF capacitor at a frequency of (a) 60.0 Hz, (b) 1.00 MHz?

*55. (I) At what frequency will a 32.0-mH inductor have a reactance of 660 Ω?

*56. (I) At what frequency will a 2.40-μF capacitor have a reactance of 6.10 kΩ?

*57. (II) Calculate the reactance of, and rms current in, a 260-mH radio coil connected to a 240-V (rms) 10.0-kHz ac line. Ignore resistance.

*58. (II) An inductance coil operates at 240 V and 60.0 Hz. It draws 12.2 A. What is the coil's inductance?

*59. (II) (a) What is the reactance of a well-insulated 0.030-μF capacitor connected to a 2.0-kV (rms) 720-Hz line? (b) What will be the peak value of the current?

*21–14 LRC Circuits

*60. (II) For a 120-V rms 60-Hz voltage, an rms current of 70 mA passing through the human body for 1.0 s could be lethal. What must be the impedance of the body for this to occur?

*61. (II) A 36-kΩ resistor is in series with a 55-mH inductor and an ac source. Calculate the impedance of the circuit if the source frequency is (a) 50 Hz, and (b) 3.0×10^4 Hz.

*62. (II) A 3.5-kΩ resistor and a 3.0-μF capacitor are connected in series to an ac source. Calculate the impedance of the circuit if the source frequency is (a) 60 Hz, and (b) 60,000 Hz.

*63. (II) Determine the resistance of a coil if its impedance is 235 Ω and its reactance is 115 Ω.

*64. (II) Determine the total impedance, phase angle, and rms current in an LRC circuit connected to a 10.0-kHz, 725-V (rms) source if $L = 28.0$ mH, $R = 8.70$ kΩ, and $C = 6250$ pF.

*65. (II) An ac voltage source is connected in series with a 1.0-μF capacitor and a 650-Ω resistor. Using a digital ac voltmeter, the amplitude of the voltage source is measured to be 4.0 V rms, while the voltages across the resistor and across the capacitor are found to be 3.0 V rms and 2.7 V rms, respectively. Determine the frequency of the ac voltage source. Why is the voltage measured across the voltage source not equal to the sum of the voltages measured across the resistor and across the capacitor?

*66. (III) (a) What is the rms current in an LR circuit when a 60.0-Hz 120-V rms ac voltage is applied, where $R = 2.80$ kΩ and $L = 350$ mH? (b) What is the phase angle between voltage and current? (c) How much power is dissipated? (d) What are the rms voltage readings across R and L?

*67. (III) (a) What is the rms current in an RC circuit if $R = 6.60$ kΩ, $C = 1.80$ μF, and the rms applied voltage is 120 V at 60.0 Hz? (b) What is the phase angle between voltage and current? (c) What are the voltmeter readings across R and C?

*68. (III) Suppose circuit B in Fig. 21–42a consists of a resistance $R = 520$ Ω. The filter capacitor has capacitance $C = 1.2$ μF. Will this capacitor act to eliminate 60-Hz ac but pass a high-frequency signal of frequency 6.0 kHz? To check this, determine the voltage drop across R for a 130-mV signal of frequency (a) 60 Hz; (b) 6.0 kHz.

*21–15 Resonance in AC Circuits

*69. (I) A 3500-pF capacitor is connected in series to a 55.0-μH coil of resistance 4.00 Ω. What is the resonant frequency of this circuit?

*70. (II) The variable capacitor in the tuner of an AM radio has a capacitance of 2800 pF when the radio is tuned to a station at 580 kHz. (a) What must be the capacitance for a station at 1600 kHz? (b) What is the inductance (assumed constant)?

*71. (II) An LRC circuit has $L = 14.8$ mH and $R = 4.10$ Ω. (a) What value must C have to produce resonance at 3600 Hz? (b) What will be the maximum current at resonance if the peak external voltage is 150 V?

*72. (III) A resonant circuit using a 260-nF capacitor is to resonate at 18.0 kHz. The air-core inductor is to be a solenoid with closely packed coils made from 12.0 m of insulated wire 1.1 mm in diameter. How many loops will the inductor contain?

*73. (III) A 2200-pF capacitor is charged to 120 V and then quickly connected to an inductor. The frequency of oscillation is observed to be 19 kHz. Determine (a) the inductance, (b) the peak value of the current, and (c) the maximum energy stored in the magnetic field of the inductor.

General Problems

74. Suppose you are looking at two wire loops in the plane of the page as shown in Fig. 21–64. When switch S is closed in the left-hand coil, (a) what is the direction of the induced current in the other loop? (b) What is the situation after a "long" time? (c) What is the direction of the induced current in the right-hand loop if that loop is quickly pulled horizontally to the right? (d) Suppose the right-hand loop also has a switch like the left-hand loop. The switch in the left-hand loop has been closed a long time when the switch in the right-hand loop is closed. What happens in this case? Explain each answer.

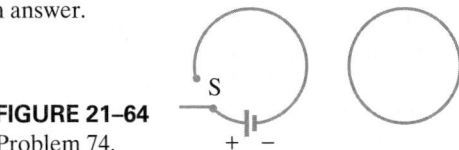

FIGURE 21–64
Problem 74.

75. A square loop 24.0 cm on a side has a resistance of 6.10 Ω. It is initially in a 0.665-T magnetic field, with its plane perpendicular to $\vec{B}$, but is removed from the field in 40.0 ms. Calculate the electric energy dissipated in this process.

76. A high-intensity desk lamp is rated at 45 W but requires only 12 V. It contains a transformer that converts 120-V household voltage. (a) Is the transformer step-up or step-down? (b) What is the current in the secondary coil when the lamp is on? (c) What is the current in the primary coil? (d) What is the resistance of the bulb when on?

77. A flashlight can be made that is powered by the induced current from a magnet moving through a coil of wire. The coil and magnet are inside a plastic tube that can be shaken causing the magnet to move back and forth through the coil. Assume the magnet has a maximum field strength of 0.05 T. Make reasonable assumptions and specify the size of the coil and the number of turns necessary to light a standard 1-watt, 3-V flashlight bulb.

78. Conceptual Example 21–9 states that an overloaded motor may burn out due to high currents. Suppose you have a blender with an internal resistance of 3.0 Ω. (a) At 120 V, what is the initial current through the blender? (b) The blender is rated at 2.0 A for continuous use. What is the back emf of the blender? (c) At what rate is heat dissipated in the blender during normal use? (d) If the blender jams and stops turning, at what rate is heat dissipated in the motor coils?

79. Power is generated at 24 kV at a generating plant located 56 km from a town that requires 55 MW of power at 12 kV. Two transmission lines from the plant to the town each have a resistance of 0.10 Ω/km. What should the output voltage of the transformer at the generating plant be for an overall transmission efficiency of 98.5%, assuming a perfect transformer?

80. The primary windings of a transformer which has an 88% efficiency are connected to 110-V ac. The secondary windings are connected across a 2.4-Ω, 75-W lightbulb. (a) Calculate the current through the primary windings of the transformer. (b) Calculate the ratio of the number of primary windings of the transformer to the number of secondary windings of the transformer.

81. A pair of power transmission lines each have a 0.95-Ω resistance and carry 740 A over 9.0 km. If the rms input voltage is 42 kV, calculate (a) the voltage at the other end, (b) the power input, (c) power loss in the lines, and (d) the power output.

82. Two resistanceless rails rest 32 cm apart on a 6.0° ramp. They are joined at the bottom by a 0.60-Ω resistor. At the top a copper bar of mass 0.040 kg (ignore its resistance) is laid across the rails. Assuming a vertical 0.45-T magnetic field, what is the terminal (steady) velocity of the bar as it slides frictionlessly down the rails?

83. Show that the power loss in transmission lines, P_L, is given by $P_L = (P_T)^2 R_L / V^2$, where P_T is the power transmitted to the user, V is the delivered voltage, and R_L is the resistance of the power lines.

84. A coil with 190 turns, a radius of 5.0 cm, and a resistance of 12 Ω surrounds a solenoid with 230 turns/cm and a radius of 4.5 cm (Fig. 21–65). The current in the solenoid changes at a constant rate from 0 to 2.0 A in 0.10 s. Calculate the magnitude and direction of the induced current in the outer coil.

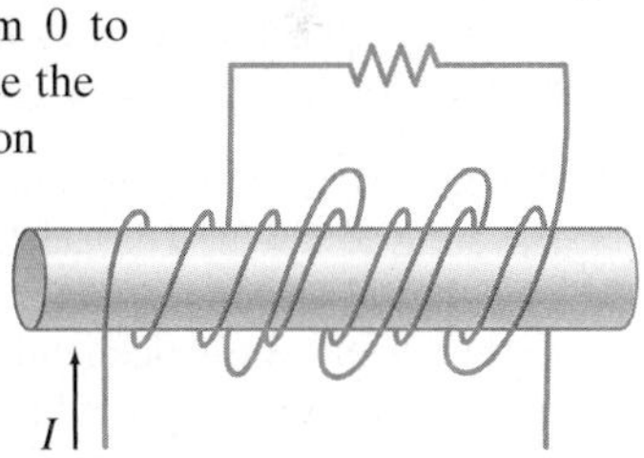

FIGURE 21–65
Problem 84.

85. A certain electronic device needs to be protected against sudden surges in current. In particular, after the power is turned on, the current should rise no more than 7.5 mA in the first 120 μs. The device has resistance 120 Ω and is designed to operate at 55 mA. How would you protect this device?

86. A 35-turn 12.5-cm-diameter coil is placed between the pole pieces of an electromagnet. When the electromagnet is turned on, the flux through the coil changes, inducing an emf. At what rate (in T/s) must the magnetic field change if the emf is to be 120 V?

87. Calculate the peak output voltage of a simple generator whose square armature windings are 6.60 cm on a side; the armature contains 125 loops and rotates in a field of 0.200 T at a rate of 120 rev/s.

*88. Typical large values for electric and magnetic fields attained in laboratories are about 1.0×10^4 V/m and 2.0 T. (a) Determine the energy density for each field and compare. (b) What magnitude electric field would be needed to produce the same energy density as the 2.0-T magnetic field?

*89. Determine the inductance L of the primary of a transformer whose input is 220 V at 60.0 Hz if the current drawn is 6.3 A. Assume no current in the secondary.

*90. A 130-mH coil whose resistance is 15.8 Ω is connected to a capacitor C and a 1360-Hz source voltage. If the current and voltage are to be in phase, what value must C have?

*91. The wire of a tightly wound solenoid is unwound and used to make another tightly wound solenoid of twice the diameter. By what factor does the inductance change?

*92. The **Q factor** of a resonant ac circuit (Section 21–15) can be defined as the ratio of the voltage across the capacitor (or inductor) to the voltage across the resistor, at resonance. The larger the Q factor, the sharper the resonance curve will be and the sharper the tuning. (a) Show that the Q factor is given by the equation $Q = (1/R)\sqrt{L/C}$. (b) At a resonant frequency $f_0 = 1.0$ MHz, what must be the values of L and R to produce a Q factor of 650? Assume that $C = 0.010\,\mu$F.

Search and Learn

1. (a) Sections 19–7 and 21–9 discuss conditions when and where it is especially important to have ground fault circuit interrupters (GFCIs) installed. What is it about those places that makes "touching ground" especially risky? (b) Describe how a GFCI works and compare to fuses and circuit breakers (see also Section 18–6).

2. While demonstrating Faraday's law to her class, a physics professor inadvertently moves the gold ring on her finger from a location where a 0.68-T magnetic field points along her finger to a zero-field location in 45 ms. The 1.5-cm-diameter ring has a resistance and mass of 55 μΩ and 15 g, respectively. (a) Estimate the thermal energy produced in the ring due to the flow of induced current. (b) Find the temperature rise of the ring, assuming all of the thermal energy produced goes into increasing the ring's temperature. The specific heat of gold is 129 J/kg·C°.

3. A small electric car overcomes a 250-N friction force when traveling 35 km/h. The electric motor is powered by ten 12-V batteries connected in series and is coupled directly to the wheels whose diameters are 58 cm. The 290 armature coils are rectangular, 12 cm by 15 cm, and rotate in a 0.65-T magnetic field. (a) How much current does the motor draw to produce the required torque? (b) What is the back emf? (c) How much power is dissipated in the coils? (d) What percent of the input power is used to drive the car? [Hint: Check Sections 6–10, 18–5, 20–9, 20–10, and 21–6.]

4. Explain the advantage of using ac rather than dc current when electric power needs to be transported long distances. (See Section 21–7.)

5. A power line carrying a sinusoidally varying current with frequency $f = 60$ Hz and peak value $I_0 = 155$ A runs at a height of 7.0 m across a farmer's land (Fig. 21–66). The farmer constructs a vertical 2.0-m-high 2000-turn rectangular wire coil below the power line. The farmer hopes to use the induced voltage in this coil to power 120-V electrical equipment, which requires a sinusoidally varying voltage with frequency $f = 60$ Hz and peak value $V_0 = 170$ V. Estimate the length ℓ of the coil needed. Would this be stealing? [Hint: Consider ΔB over one-quarter of a cycle $(\frac{1}{240}$ s$)$. See Sections 20–5 and 18–7.]

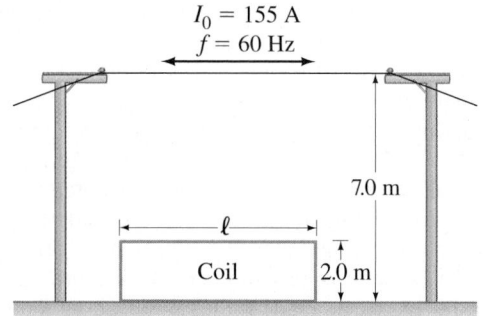

FIGURE 21–66 Search and Learn 5.

6. A **ballistic galvanometer** is a device that measures the total charge Q that passes through it in a short time. It is connected to a **search coil** that measures B (also called a **flip coil**) which is a small coil with N turns, each of cross-sectional area A. The flip coil is placed in the magnetic field to be measured with its face perpendicular to the field. It is then quickly rotated 180° about a diameter. Show that the total charge Q that flows in the induced current during this short "flip" time is proportional to the magnetic field B. In particular, show that

$$B = \frac{QR}{2NA}$$

where R is the total resistance of the circuit including the coil and ballistic galvanometer which measures charge Q.

ANSWERS TO EXERCISES

A: (e).
B: (a) Counterclockwise; (b) clockwise; (c) zero; (d) counterclockwise.
C: Clockwise (conventional current counterclockwise).
D: (a) Increase (brighter); (b) yes; resists more (counter torque).

E: 10 turns.
F: From Eq. 21–11b, the higher the frequency the lower the reactance, so in (a) more high frequency current flows to circuit B. In (b) higher frequencies pass to ground whereas lower frequencies pass more easily to circuit B.

Wireless technology is all around us: radio and television, cell phones, wi-fi, Bluetooth, and all wireless communication. These devices work by electromagnetic waves traveling through space. Wireless devices are applications of Marconi's development of long-distance transmission of information a century ago.

In this photo we see the first humans to land on the Moon. In the background is a television camera that sent live moving images through empty space to Earth where it was shown live.

We will see in this Chapter that Maxwell predicted the existence of EM waves from his famous equations. Maxwell's equations themselves are a magnificent summary of electromagnetism. We will also see that EM waves carry energy and momentum, and that light itself is an electromagnetic wave.

Electromagnetic Waves

CHAPTER

22

CHAPTER-OPENING QUESTION—Guess now!

Which of the following best describes the difference between radio waves and X-rays?

(a) X-rays are radiation whereas radio waves are electromagnetic waves.

(b) Both can be thought of as electromagnetic waves. They differ only in wavelength and frequency.

(c) X-rays are pure energy. Radio waves are made of fields, not energy.

(d) Radio waves come from electric currents in an antenna. X-rays are not related to electric charge.

(e) X-rays are made up of particles called photons whereas radio waves are oscillations in space.

The culmination of electromagnetic theory in the nineteenth century was the prediction, and the experimental verification, that waves of electromagnetic fields could travel through space. This achievement opened a whole new world of communication: first the wireless telegraph, then radio and television, and more recently cell phones, remote-control devices, wi-fi, and Bluetooth. Most important was the spectacular prediction that light is an electromagnetic wave.

The theoretical prediction of electromagnetic waves was the work of the Scottish physicist James Clerk Maxwell (1831–1879; Fig. 22–1), who unified, in one magnificent theory, all the phenomena of electricity and magnetism.

FIGURE 22–1 James Clerk Maxwell.

22–1 Changing Electric Fields Produce Magnetic Fields; Maxwell's Equations

The development of electromagnetic theory in the early part of the nineteenth century by Oersted, Ampère, and others was not actually done in terms of electric and magnetic fields. The idea of the field was introduced somewhat later by Faraday, and was not generally used until Maxwell showed that all electric and magnetic phenomena could be described using only four equations involving electric and magnetic fields. These equations, known as **Maxwell's equations**, are the basic equations for all electromagnetism. They are fundamental in the same sense that Newton's three laws of motion and the law of universal gravitation are for mechanics. In a sense, they are even more fundamental, because they are consistent with the theory of relativity (Chapter 26), whereas Newton's laws are not. Because all of electromagnetism is contained in this set of four equations, Maxwell's equations are considered one of the great triumphs of the human intellect.

Although we will not present Maxwell's equations in mathematical form since they involve calculus, we will summarize them here in words. They are:

(1) a generalized form of Coulomb's law that relates electric field to its source, electric charge (= Gauss's law, Section 16–12);

(2) a similar law for the magnetic field, except that magnetic field lines are always continuous—they do not begin or end (as electric field lines do, on charges);

(3) an electric field is produced by a changing magnetic field (Faraday's law);

(4) a magnetic field is produced by an electric current (Ampère's law), or by a changing electric field.

Law (3) is Faraday's law (see Chapter 21, especially Section 21–4). The first part of law (4), that a magnetic field is produced by an electric current, was discovered by Oersted, and the mathematical relation is given by Ampère's law (Section 20–8). But the second part of law (4) is an entirely new aspect predicted by Maxwell: Maxwell argued that if a changing magnetic field produces an electric field, as given by Faraday's law, then the reverse might be true as well: **a changing electric field will produce a magnetic field**. This was an *hypothesis* by Maxwell. It is based on the idea of *symmetry* in nature. Indeed, the size of the effect in most cases is so small that Maxwell recognized it would be difficult to detect it experimentally.

*Maxwell's Fourth Equation (Ampère's Law Extended)

To back up the idea that a changing electric field might produce a magnetic field, we use an indirect argument that goes something like this. According to Ampère's law (Section 20–8), $\Sigma B_{\parallel} \, \Delta \ell = \mu_0 I$. That is, divide any closed path you choose into short segments $\Delta \ell$, multiply each segment by the parallel component of the magnetic field B at that segment, and then sum all these products over the complete closed path. That sum will then equal μ_0 times the total current I that passes through a surface bounded by the path. When we applied Ampère's law to the field around a straight wire (Section 20–8), we imagined the current as passing through the circular area enclosed by our circular loop. That area is the flat surface 1 shown in Fig. 22–2. However, we could just as well use the sack-shaped surface 2 in Fig. 22–2 as the surface for Ampère's law because the same current I passes through it.

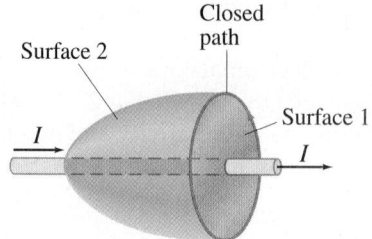

FIGURE 22–2 Ampère's law applied to two different surfaces bounded by the same closed path.

Surface 2

Closed path

Surface 1

I

I

Now consider the closed path for the situation of Fig. 22–3 where a capacitor is being discharged. Ampère's law works for surface 1 (current I passes through surface 1), but it does not work for surface 2 because no current passes through surface 2. There is a magnetic field around the wire, so the left side of Ampère's law

$$\Sigma B_{\parallel}\,\Delta\ell = \mu_0 I$$

is not zero around the circular closed path; yet no current flows through surface 2, so the right side *is* zero for surface 2. We seem to have a contradiction of Ampère's law. There is a magnetic field present in Fig. 22–3, however, only if charge is flowing to or away from the capacitor plates. The changing charge on the plates means that the electric field between the plates is changing in time. Maxwell resolved the problem of no current through surface 2 in Fig. 22–3 by proposing that the changing electric field between the plates is *equivalent to* an electric current. He called it a **displacement current**, I_D. An ordinary current I is then called a "conduction current," and Ampère's law, as generalized by Maxwell, becomes

$$\Sigma B_{\parallel}\,\Delta\ell = \mu_0(I + I_D).$$

Ampère's law will now apply also for surface 2 in Fig. 22–3, where I_D refers to the changing electric field.

Combining Eq. 17–7 for the charge on a capacitor, $Q = CV$, with Eq. 17–8, $C = \epsilon_0 A/d$, and with the magnitudes in Eq. 17–4a, $V = Ed$, we can write $Q = CV = (\epsilon_0 A/d)(Ed) = \epsilon_0 AE$. Then the current I_D becomes

$$I_D = \frac{\Delta Q}{\Delta t} = \epsilon_0 \frac{\Delta\Phi_E}{\Delta t},$$

where $\Phi_E = EA$ is the **electric flux**, defined in analogy to magnetic flux (Section 21–2). Then, Ampère's law becomes

$$\Sigma B_{\parallel}\,\Delta\ell = \mu_0 I + \mu_0\epsilon_0 \frac{\Delta\Phi_E}{\Delta t}. \qquad \textbf{(22–1)}$$

Ampère's law (general form)

This equation embodies Maxwell's idea that a magnetic field can be caused not only by a normal electric current, but also by a changing electric field or changing electric flux.

22–2 Production of Electromagnetic Waves

According to Maxwell, a magnetic field will be produced in empty space if there is a changing electric field. From this, Maxwell derived another startling conclusion. If a changing magnetic field produces an electric field, that electric field is itself changing. This changing electric field will, in turn, produce a magnetic field, which will be changing, and so it too will produce a changing electric field; and so on. When Maxwell worked with his equations, he found that the net result of these interacting changing fields was a *wave* of electric and magnetic fields that can propagate (travel) through space! We now examine, in a simplified way, how such **electromagnetic waves** can be produced.

Consider two conducting rods that will serve as an "antenna" (Fig. 22–4a). Suppose these two rods are connected by a switch to the opposite terminals of a battery. When the switch is closed, the upper rod quickly becomes positively charged and the lower one negatively charged. Electric field lines are formed as indicated by the lines in Fig. 22–4b. While the charges are flowing, a current exists whose direction is indicated by the black arrows. A magnetic field is therefore produced near the antenna. The magnetic field lines encircle the rod-like antenna and therefore, in Fig. 22–4, $\vec{B}$ points into the page ($\otimes$) on the right and out of the page ($\odot$) on the left. Now we ask, how far out do these electric and magnetic fields extend? In the static case, the fields would extend outward indefinitely far. However, when the switch in Fig. 22–4 is closed, the fields quickly appear nearby, but it takes time for them to reach distant points. Both electric and magnetic fields store energy, and this energy cannot be transferred to distant points at infinite speed.

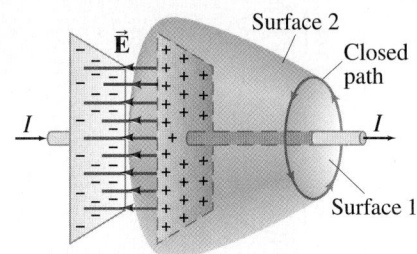

FIGURE 22–3 A capacitor discharging. A conduction current passes through surface 1, but no conduction current passes through the sacklike surface 2. An extra term is needed in Ampère's law.

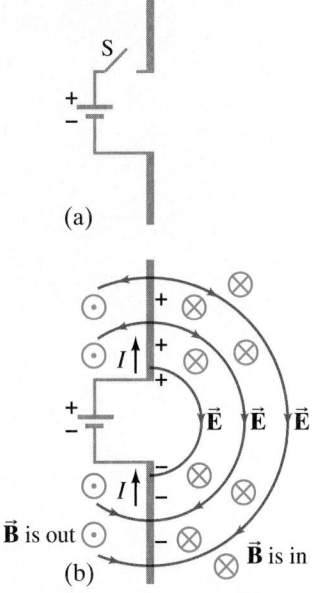

FIGURE 22–4 Fields produced by charge flowing into conductors. It takes time for the $\vec{E}$ and $\vec{B}$ fields to travel outward to distant points. The fields are shown to the right of the antenna, but they move out in all directions, symmetrically about the (vertical) antenna.

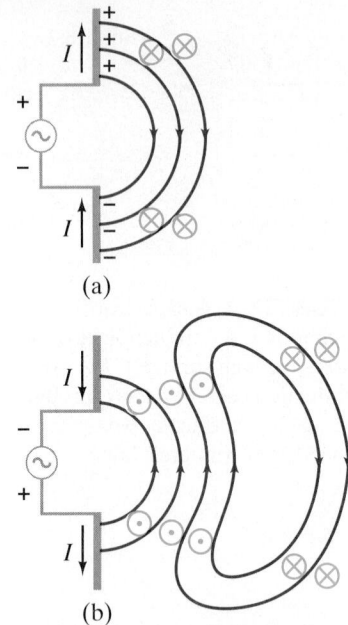

FIGURE 22–5 Sequence showing electric and magnetic fields that spread outward from oscillating charges on two conductors (the antenna) connected to an ac source (see the text).

Now we look at the situation of Fig. 22–5, where our antenna is connected to an ac generator. In Fig. 22–5a, the connection has just been completed. Charge starts building up, and fields form just as in Fig. 22–4b. The + and − signs in Fig. 22–5a indicate the net charge on each rod at a given instant. The black arrows indicate the direction of the current. The electric field is represented by red lines in the plane of the page; and the magnetic field, according to the right-hand rule, is into (⊗) or out of (⊙) the page in blue. In Fig. 22–5b, the voltage of the ac generator has reversed in direction; the current is reversed and the new magnetic field is in the opposite direction. Because the new fields have changed direction, the old lines fold back to connect up to some of the new lines and form closed loops as shown.[†] The old fields, however, don't suddenly disappear; they are on their way to distant points. Indeed, because a changing magnetic field produces an electric field, and a changing electric field produces a magnetic field, this combination of changing electric and magnetic fields moving outward is self-supporting, no longer depending on the antenna charges.

The fields not far from the antenna, referred to as the *near field*, become quite complicated, but we are not so interested in them. We are mainly interested in the fields far from the antenna (they are generally what we detect), which we refer to as the **radiation field**. The electric field lines form loops, as shown in Fig. 22–6a, and continue moving outward. The magnetic field lines also form closed loops, but are not shown because they are perpendicular to the page. Although the lines are shown only on the right of the source, fields also travel in other directions. The field strengths are greatest in directions perpendicular to the oscillating charges; and they drop to zero along the direction of oscillation—above and below the antenna in Fig. 22–6a.

FIGURE 22–6 (a) The radiation fields (far from the antenna) produced by a sinusoidal signal on the antenna. The red closed loops represent electric field lines. The magnetic field lines, perpendicular to the page and represented by blue ⊗ and ⊙, also form closed loops. (b) Very far from the antenna, the wave fronts (field lines) are essentially flat over a fairly large area, and are referred to as *plane waves*.

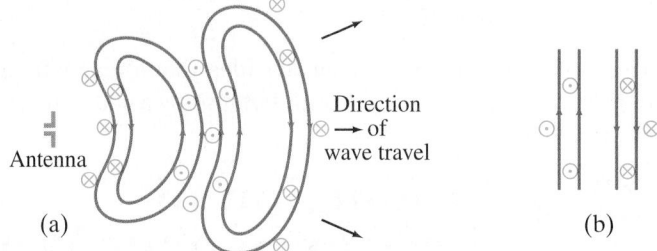

The magnitudes of both $\vec{\mathbf{E}}$ and $\vec{\mathbf{B}}$ in the radiation field are found to decrease with distance as $1/r$. (Compare this to the static electric field given by Coulomb's law where $\vec{\mathbf{E}}$ decreases as $1/r^2$.) The energy carried by the electromagnetic wave is proportional (as for any wave, Chapter 11) to the square of the amplitude, E^2 or B^2, as will be discussed further in Section 22–7, so the intensity of the wave decreases as $1/r^2$. Thus the energy carried by EM waves follows the **inverse square law** just as for sound waves (Eqs. 11–16).

Several things about the radiation field can be noted from Fig. 22–6. First, *the electric and magnetic fields at any point are perpendicular to each other, and to the direction of wave travel.* Second, we can see that the fields alternate in direction ($\vec{\mathbf{B}}$ is into the page at some points and out of the page at others; $\vec{\mathbf{E}}$ points up at some points and down at others). Thus, the field strengths vary from a maximum in one direction, to zero, to a maximum in the other direction. The electric and magnetic fields are "in phase": that is, they each are zero at the same points and reach their maxima at the same points in space. Finally, very far from the antenna (Fig. 22–6b) the field lines are quite flat over a reasonably large area, and the waves are referred to as **plane waves**.

[†]We are considering waves traveling through empty space. There are no charges for lines of $\vec{\mathbf{E}}$ to start or stop on, so they form closed loops. Magnetic field lines always form closed loops.

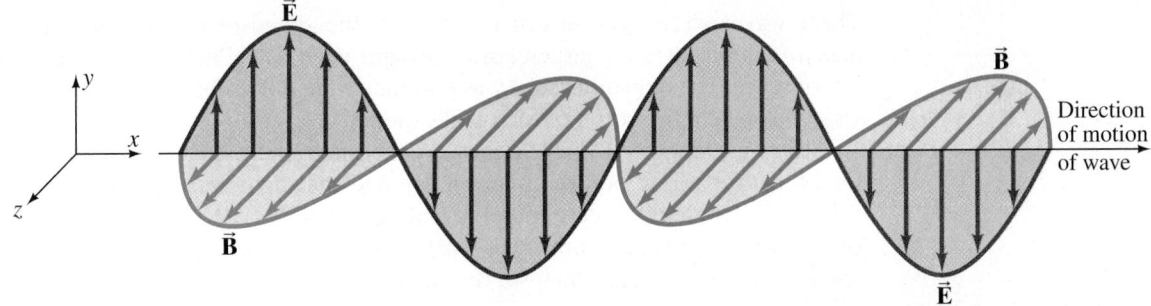

If the source voltage varies sinusoidally, then the electric and magnetic field strengths in the radiation field will also vary sinusoidally. The sinusoidal character of the waves is shown in Fig. 22–7, which displays the field directions and magnitudes plotted as a function of position along the direction of wave travel. Notice that $\vec{B}$ and $\vec{E}$ are perpendicular to each other and to the direction of wave travel.

We call these waves electromagnetic (EM) waves. They are *transverse* waves because the amplitude is perpendicular to the direction of wave travel. However, EM waves are always waves of *fields*, not of matter (like waves on water or a rope). Because they are fields, EM waves can propagate in empty space.

As we have seen, EM waves are produced by electric charges that are oscillating, and hence are undergoing acceleration. In fact, we can say in general that

accelerating electric charges give rise to electromagnetic waves.

Maxwell derived a formula for the speed of EM waves:

$$v = c = \frac{E}{B}, \qquad \qquad (22\text{–}2)$$

where c is the special symbol for the speed of electromagnetic waves in empty space, and E and B are the magnitudes of electric and magnetic fields at the same point in space. More specifically, it was also shown that

$$c = \frac{1}{\sqrt{\epsilon_0 \mu_0}}. \qquad \qquad \text{[speed of EM waves]} \quad (22\text{–}3)$$

When Maxwell put in the values for ϵ_0 and μ_0, he found

$$c = \frac{1}{\sqrt{\epsilon_0 \mu_0}} = \frac{1}{\sqrt{(8.85 \times 10^{-12}\,\text{C}^2/\text{N}\cdot\text{m}^2)(4\pi \times 10^{-7}\,\text{N}\cdot\text{s}^2/\text{C}^2)}} = 3.00 \times 10^8\,\text{m/s},$$

which is exactly equal to the measured speed of light in vacuum (Section 22–4).

EXERCISE A At a particular instant in time, a wave has its electric field pointing north and its magnetic field pointing up. In which direction is the wave traveling? (*a*) South, (*b*) west, (*c*) east, (*d*) down, (*e*) not enough information. [See Fig. 22–7.]

22–3 Light as an Electromagnetic Wave and the Electromagnetic Spectrum

Maxwell's prediction that EM waves should exist was startling. Equally remarkable was the speed at which EM waves were predicted to travel—$3.00 \times 10^8\,\text{m/s}$, the same as the measured speed of light.

Light had been shown some 60 years before Maxwell's work to behave like a wave (we'll discuss this in Chapter 24). But nobody knew what kind of wave it was. What is it that is oscillating in a light wave? Maxwell, on the basis of the calculated speed of EM waves, argued that light must be an electromagnetic wave. This idea soon came to be generally accepted by scientists, but not fully until after EM waves were experimentally detected. EM waves were first generated and detected experimentally by Heinrich Hertz (1857–1894) in 1887, eight years after Maxwell's death. Hertz used a spark-gap apparatus in which charge was made to rush back and forth for a short time, generating waves whose frequency was about 10^9 Hz. He detected them some distance away using a loop of wire in which an emf was induced when a changing magnetic field passed through.

These waves were later shown to travel at the speed of light, 3.00×10^8 m/s, and to exhibit all the characteristics of light such as reflection, refraction, and interference. The only difference was that they were not visible. Hertz's experiment was a strong confirmation of Maxwell's theory.

The wavelengths of visible light were measured in the first decade of the nineteenth century, long before anyone imagined that light was an electromagnetic wave. The wavelengths were found to lie between 4.0×10^{-7} m and 7.5×10^{-7} m, or 400 nm to 750 nm $(1 \text{ nm} = 10^{-9} \text{ m})$. The frequencies of visible light can be found using Eq. 11–12, which we rewrite here:

$$c = \lambda f, \tag{22–4}$$

where f and λ are the frequency and wavelength, respectively, of the wave. Here, c is the speed of light, 3.00×10^8 m/s; it gets the special symbol c because of its universality for all EM waves in free space. Equation 22–4 tells us that the frequencies of visible light are between 4.0×10^{14} Hz and 7.5×10^{14} Hz. (Recall that 1 Hz = 1 cycle per second = 1 s^{-1}.)

But visible light is only one kind of EM wave. As we have seen, Hertz produced EM waves of much lower frequency, about 10^9 Hz. These are now called **radio waves**, because frequencies in this range are used to transmit radio and TV signals. Electromagnetic waves, or EM radiation as we sometimes call it, have been produced or detected over a wide range of frequencies. They are usually categorized as shown in Fig. 22–8, which is known as the **electromagnetic spectrum**.

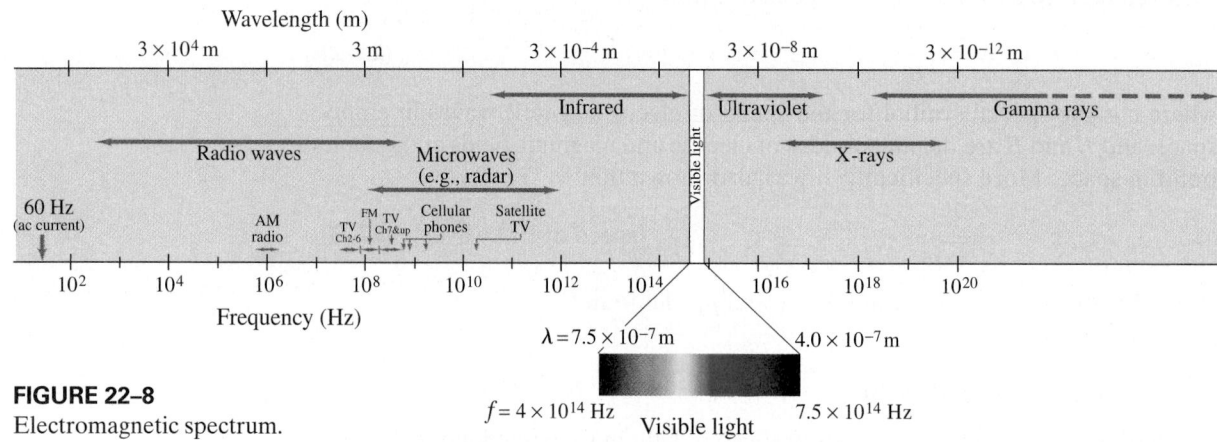

FIGURE 22–8
Electromagnetic spectrum.

Radio waves and microwaves can be produced in the laboratory using electronic equipment (Fig. 22–5). Higher-frequency waves are very difficult to produce electronically. These and other types of EM waves are produced in natural processes, as emission from atoms, molecules, and nuclei (more on this later). EM waves can be produced by the acceleration of electrons or other charged particles, such as electrons in the antenna of Fig. 22–5. X-rays have very short wavelengths (and very high frequencies), and they are produced (Chapters 25 and 28) when fast-moving electrons are rapidly decelerated upon striking a metal target. Even the visible light emitted by an ordinary incandescent bulb is due to electrons undergoing acceleration within the hot filament.

EXERCISE B Return to the Chapter-Opening Question, page 625, and answer it again now. Try to explain why you may have answered differently the first time.

We will meet various types of EM waves later. However, it is worth mentioning here that infrared (IR) radiation (EM waves whose frequency is just less than that of visible light) is mainly responsible for the heating effect of the Sun. The Sun emits not only visible light but substantial amounts of IR and UV (ultraviolet) as well. The molecules of our skin tend to "resonate" at infrared frequencies, so it is these that are preferentially absorbed and thus warm us. We humans experience EM waves differently depending on their wavelengths: Our eyes detect wavelengths between about 4×10^{-7} m and 7.5×10^{-7} m (visible light), whereas our skin detects longer wavelengths (IR). Many EM wavelengths we don't detect directly at all.

Light and other electromagnetic waves travel at a speed of 3×10^8 m/s. Compare this to sound, which travels (see Chapter 12) at a speed of about 300 m/s in air, a million times slower; or to typical freeway speeds of a car, 30 m/s (100 km/h, or 60 mi/h), 10 million times slower than light. EM waves differ from sound waves in another big way: sound waves travel in a medium such as air, and involve motion of air molecules; EM waves do not involve any material—only fields, and they can travel in empty space.

⚠ CAUTION

Sound and EM waves are different

EXAMPLE 22–1 **Wavelengths of EM waves.** Calculate the wavelength (a) of a 60-Hz EM wave, (b) of a 93.3-MHz FM radio wave, and (c) of a beam of visible red light from a laser at frequency 4.74×10^{14} Hz.

APPROACH All of these waves are electromagnetic waves, so their speed is $c = 3.00 \times 10^8$ m/s. We solve for λ in Eq. 22–4: $\lambda = c/f$.

SOLUTION (a)
$$\lambda = \frac{c}{f} = \frac{3.00 \times 10^8 \text{ m/s}}{60 \text{ s}^{-1}} = 5.0 \times 10^6 \text{ m},$$

or 5000 km. 60 Hz is the frequency of ac current in the United States, and, as we see here, one wavelength stretches all the way across the continental USA.

(b)
$$\lambda = \frac{3.00 \times 10^8 \text{ m/s}}{93.3 \times 10^6 \text{ s}^{-1}} = 3.22 \text{ m}.$$

The length of an FM radio antenna is often about half this $\left(\frac{1}{2}\lambda\right)$, or $1\frac{1}{2}$ m.

(c)
$$\lambda = \frac{3.00 \times 10^8 \text{ m/s}}{4.74 \times 10^{14} \text{ s}^{-1}} = 6.33 \times 10^{-7} \text{ m} \ (= 633 \text{ nm}).$$

EXERCISE C What are the frequencies of (a) an 80-m-wavelength radio wave, and (b) an X-ray of wavelength 5.5×10^{-11} m?

EXAMPLE 22–2 ESTIMATE **Cell phone antenna.** The antenna of a cell phone is often $\frac{1}{4}$ wavelength long. A particular cell phone has an 8.5-cm-long straight rod for its antenna. Estimate the operating frequency of this phone.

APPROACH The basic equation relating wave speed, wavelength, and frequency is $c = \lambda f$; the wavelength λ equals four times the antenna's length.

SOLUTION The antenna is $\frac{1}{4}\lambda$ long, so $\lambda = 4(8.5 \text{ cm}) = 34 \text{ cm} = 0.34$ m. Then $f = c/\lambda = (3.0 \times 10^8 \text{ m/s})/(0.34 \text{ m}) = 8.8 \times 10^8 \text{ Hz} = 880$ MHz.

NOTE Radio antennas are not always straight conductors. The conductor may be a round loop to save space. See Fig. 22–18b.

EXERCISE D How long should a $\frac{1}{4}$-λ antenna be for an aircraft radio operating at 165 MHz?

Electromagnetic waves can travel along transmission lines as well as in empty space. When a source of emf is connected to a transmission line—be it two parallel wires or a coaxial cable (Fig. 22–9)—the electric field within the wire is not set up immediately at all points along the wires. This is based on the same argument we used in Section 22–2 with reference to Fig. 22–5. Indeed, it can be shown that if the wires are separated by empty space or air, the electrical signal travels along the wires at the speed $c = 3.0 \times 10^8$ m/s. For example, when you flip a light switch, the light actually goes on a tiny fraction of a second later. If the wires are in a medium whose electric permittivity is ϵ and magnetic permeability is μ (Sections 17–8 and 20–12, respectively), the speed is not given by Eq. 22–3, but by

$$v = \frac{1}{\sqrt{\epsilon\mu}}$$

instead.

FIGURE 22–9 Coaxial cable.

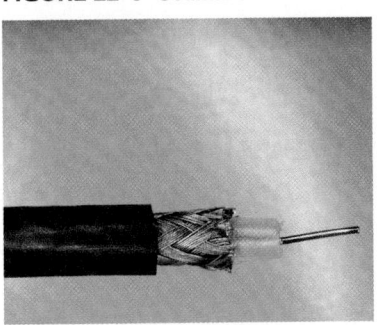

EXAMPLE 22-3 ESTIMATE **Phone call time lag.** You make a telephone call from New York to a friend in London. Estimate how long it will take the electrical signal generated by your voice to reach London, assuming the signal is (*a*) carried on a telephone cable under the Atlantic Ocean, and (*b*) sent via satellite 36,000 km above the ocean. Would there be a noticeable delay in either case?

APPROACH The signal is carried on a telephone wire or in the air via satellite. In either case it is an electromagnetic wave. Electronics as well as the wire or cable slow things down, but as a rough estimate we take the speed to be $c = 3.0 \times 10^8$ m/s.

SOLUTION The distance from New York to London is about 5000 km.

(*a*) The time delay via the cable is $t = d/c \approx (5 \times 10^6 \text{ m})/(3.0 \times 10^8 \text{ m/s}) = 0.017$ s.

(*b*) Via satellite the time would be longer because communications satellites, which are usually geosynchronous (Example 5–12), move at a height of 36,000 km. The signal would have to go up to the satellite and back down, or about 72,000 km. The actual distance the signal would travel would be a little more than this as the signal would go up and down on a diagonal (5000 km New York to London, small compared to the distance up to the satellite). Thus $t = d/c \approx (7.2 \times 10^7 \text{ m})/(3 \times 10^8 \text{ m/s}) \approx 0.24$ s, one way. Both directions $\approx \frac{1}{2}$ s.

NOTE When the signal travels via the underwater cable, there is only a hint of a delay and conversations are fairly normal. When the signal is sent via satellite, the delay *is* noticeable. The length of time between the end of when you speak and your friend receives it and replies, and then you hear the reply, would be about a half second beyond the normal time in a conversation, as we just calculated. This is enough to be noticeable, and you have to adjust for it so you don't start talking again while your friend's reply is on the way back to you.

EXERCISE E If you are on the phone via satellite to someone only 100 km away, would you notice the same effect discussed in the NOTE above?

EXERCISE F If your voice traveled as a sound wave, how long would it take to go from New York to London?

22–4 Measuring the Speed of Light

Galileo attempted to measure the speed of light by trying to measure the time required for light to travel a known distance between two hilltops. He stationed an assistant on one hilltop and himself on another, and ordered the assistant to lift the cover from a lamp the instant he saw a flash from Galileo's lamp. Galileo measured the time between the flash of his lamp and when he received the light from his assistant's lamp. The time was so short that Galileo concluded it merely represented human reaction time, and that the speed of light must be extremely high.

The first successful determination that the speed of light is finite was made by the Danish astronomer Ole Roemer (1644–1710). Roemer had noted that the carefully measured orbital period of Io, a moon of Jupiter with an average period of 42.5 h, varied slightly, depending on the relative position of Earth and Jupiter. He attributed this variation in the apparent period to the change in distance between the Earth and Jupiter during one of Io's periods, and the time it took light to travel the extra distance. Roemer concluded that the speed of light—though great—is finite.

Since then a number of techniques have been used to measure the speed of light. Among the most important were those carried out by the American Albert A. Michelson (1852–1931). Michelson used the rotating mirror apparatus diagrammed in Fig. 22–10 for a series of high-precision experiments carried out from 1880 to the 1920s. Light from a source would hit one face of a rotating eight-sided mirror. The reflected light traveled to a stationary mirror a large distance away and back again as shown. If the rotating mirror was turning at just the right rate, the returning beam of light would reflect from one face of the mirror into a small telescope through which the observer looked. If the speed of rotation was only slightly different, the beam would be deflected to one side and would not be seen by the observer. From the required speed of the rotating mirror and the known distance to the stationary mirror, the speed of light could be calculated. In the 1920s, Michelson set up the rotating mirror on the top of Mt. Wilson in southern California and the stationary mirror on Mt. Baldy (Mt. San Antonio) 35 km away. He later measured the speed of light in vacuum using a long evacuated tube.

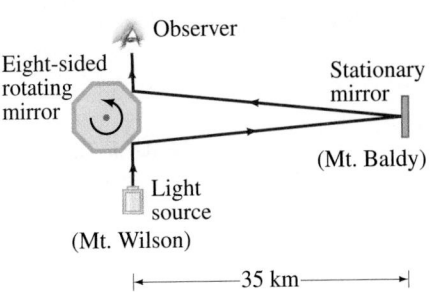

FIGURE 22–10 Michelson's speed-of-light apparatus (not to scale).

Today the speed of light, c, in vacuum is taken as

$$c = 2.99792458 \times 10^8 \text{ m/s},$$

and is *defined* to be this value. This means that the standard for length, the meter, is no longer defined separately. Instead, as we noted in Section 1–5, the meter is now formally defined as the distance light travels in vacuum in $1/299{,}792{,}458$ of a second.

We usually round off c to

$$c = 3.00 \times 10^8 \text{ m/s}$$

when extremely precise results are not required. In air, the speed is only slightly less.

22–5 Energy in EM Waves

Electromagnetic waves carry energy from one region of space to another. This energy is associated with the moving electric and magnetic fields. In Section 17–9, we saw that the energy density u_E (J/m^3) stored in an electric field E is $u_E = \frac{1}{2}\epsilon_0 E^2$ (Eq. 17–11). The energy density stored in a magnetic field B, as we discussed in Section 21–11, is given by $u_B = \frac{1}{2}B^2/\mu_0$ (Eq. 21–10). Thus, the total energy stored per unit volume in a region of space where there is an electromagnetic wave is

$$u = u_E + u_B = \frac{1}{2}\epsilon_0 E^2 + \frac{1}{2}\frac{B^2}{\mu_0}. \qquad \textbf{(22–5)}$$

In this equation, E and B represent the electric and magnetic field strengths of the wave at any instant in a small region of space. We can write Eq. 22–5 in terms of the E field only using Eqs. 22–2 ($B = E/c$) and 22–3 ($c = 1/\sqrt{\epsilon_0\mu_0}$) to obtain

$$u = \frac{1}{2}\epsilon_0 E^2 + \frac{1}{2}\frac{\epsilon_0\mu_0 E^2}{\mu_0} = \epsilon_0 E^2. \qquad \textbf{(22–6a)}$$

Note here that the energy density associated with the B field equals that due to the E field, and each contributes half to the total energy. We can also write the energy density in terms of the B field only:

$$u = \epsilon_0 E^2 = \epsilon_0 c^2 B^2 = \frac{B^2}{\mu_0}, \qquad \textbf{(22–6b)}$$

or in one term containing both E and B,

$$u = \epsilon_0 E^2 = \epsilon_0 EcB = \frac{\epsilon_0 EB}{\sqrt{\epsilon_0\mu_0}}$$

or

$$u = \sqrt{\frac{\epsilon_0}{\mu_0}}\, EB. \qquad \textbf{(22–6c)}$$

Equations 22–6 give the energy density of EM waves in any region of space at any instant.

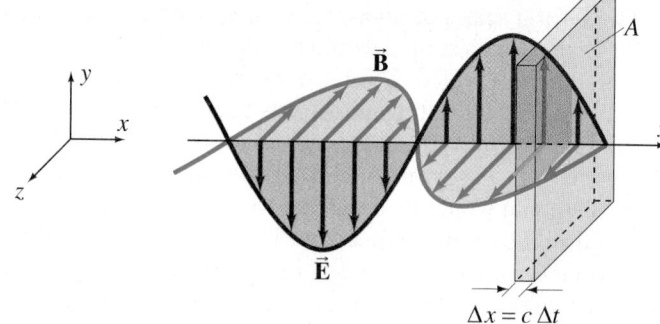

FIGURE 22–11 Electromagnetic wave carrying energy through area A.

$$\Delta x = c\,\Delta t$$

The energy a wave transports per unit time per unit area is the **intensity** I, as defined in Sections 11–9 and 12–2.[†] The units of I are W/m². The energy ΔU is the energy density (Eq. 22–6a) times the volume V. Hence the energy passing through an area A in a time Δt (see Fig. 22–11) is

$$\Delta U = u\,\Delta V = (u)(A\,\Delta x) = (\epsilon_0 E^2)(Ac\,\Delta t)$$

because $\Delta x = c\,\Delta t$. Therefore, the magnitude of the intensity (energy per unit area per time Δt, or power per unit area) is

$$I = \frac{\Delta U}{A\,\Delta t} = \frac{(\epsilon_0 E^2)(Ac\,\Delta t)}{A\,\Delta t} = \epsilon_0 c E^2.$$

From Eqs. 22–2 and 22–3, this can also be written

$$I = \epsilon_0 c E^2 = \frac{c}{\mu_0} B^2 = \frac{EB}{\mu_0}. \tag{22–7}$$

We can also find the *average intensity* over an extended period of time, if E and B are sinusoidal. Then $\overline{E^2} = E_0^2/2$, just as for electric currents and voltages, Section 18–7, Eqs. 18–8. Thus

$$\overline{I} = \frac{1}{2}\epsilon_0 c E_0^2 = \frac{1}{2}\frac{c}{\mu_0} B_0^2 = \frac{E_0 B_0}{2\mu_0}. \tag{22–8}$$

Here E_0 and B_0 are the maximum values of E and B. We can also write

$$\overline{I} = \frac{E_{\text{rms}} B_{\text{rms}}}{\mu_0},$$

where E_{rms} and B_{rms} are the rms values $(E_{\text{rms}} = \sqrt{\overline{E^2}} = E_0/\sqrt{2}$, and $B_{\text{rms}} = \sqrt{\overline{B^2}} = B_0/\sqrt{2})$.

EXAMPLE 22–4 E **and** B **from the Sun.** Radiation from the Sun reaches the Earth (above the atmosphere) with an intensity of about 1350 W/m² = 1350 J/s·m². Assume that this is a single EM wave, and calculate the maximum values of E and B.

APPROACH We solve Eq. 22–8 $\left(\overline{I} = \frac{1}{2}\epsilon_0 c E_0^2\right)$ for E_0 in terms of $\overline{I}$ and use $\overline{I} = 1350$ J/s·m².

SOLUTION $E_0 = \sqrt{\dfrac{2\overline{I}}{\epsilon_0 c}} = \sqrt{\dfrac{2(1350\ \text{J/s·m}^2)}{(8.85 \times 10^{-12}\ \text{C}^2/\text{N·m}^2)(3.00 \times 10^8\ \text{m/s})}}$

$\qquad\qquad = 1.01 \times 10^3$ V/m.

From Eq. 22–2, $B = E/c$, so

$$B_0 = \frac{E_0}{c} = \frac{1.01 \times 10^3\ \text{V/m}}{3.00 \times 10^8\ \text{m/s}} = 3.37 \times 10^{-6}\ \text{T}.$$

NOTE Although B has a small numerical value compared to E (because of the way the different units for E and B are defined), B contributes the same energy to the wave as E does, as we saw earlier.

⚠️ **CAUTION**

E and B have very different values (due to how units are defined), but E and B contribute equal energy

---

[†]The intensity I for EM waves is often called the **Poynting vector** and given the symbol $\vec{\mathbf{S}}$. Its direction is that in which the energy is being transported, which is the direction the wave is traveling, and its magnitude is the intensity $(S = I)$.

22–6 Momentum Transfer and Radiation Pressure

If electromagnetic waves carry energy, then we would expect them to also carry linear momentum. When an electromagnetic wave encounters the surface of an object, a force will be exerted on the surface as a result of the momentum transfer ($F = \Delta p/\Delta t$) just as when a moving object strikes a surface. The force per unit area exerted by the waves is called **radiation pressure**, and its existence was predicted by Maxwell. He showed that if a beam of EM radiation (light, for example) is completely absorbed by an object, then the momentum transferred is

$$\Delta p = \frac{\Delta U}{c}, \qquad \begin{bmatrix} \text{radiation} \\ \text{fully} \\ \text{absorbed} \end{bmatrix} \quad \textbf{(22–9a)}$$

where ΔU is the energy absorbed by the object in a time Δt and c is the speed of light. If, instead, the radiation is fully reflected (suppose the object is a mirror), then the momentum transferred is twice as great, just as when a ball bounces elastically off a surface:

$$\Delta p = \frac{2\,\Delta U}{c}. \qquad \begin{bmatrix} \text{radiation} \\ \text{fully} \\ \text{reflected} \end{bmatrix} \quad \textbf{(22–9b)}$$

If a surface absorbs some of the energy, and reflects some of it, then $\Delta p = a\,\Delta U/c$, where a has a value between 1 and 2.

Using Newton's second law we can calculate the force and the pressure exerted by EM radiation on an object. The force F is given by

$$F = \frac{\Delta p}{\Delta t}.$$

The radiation pressure P (assuming full absorption) is given by (see Eq. 22–9a)

$$P = \frac{F}{A} = \frac{1}{A}\frac{\Delta p}{\Delta t} = \frac{1}{Ac}\frac{\Delta U}{\Delta t}.$$

We discussed in Section 22–5 that the average intensity $\bar{I}$ is defined as energy per unit time per unit area:

$$\bar{I} = \frac{\Delta U}{A\,\Delta t}.$$

Hence the radiation pressure is

$$P = \frac{\bar{I}}{c}. \qquad \begin{bmatrix} \text{fully} \\ \text{absorbed} \end{bmatrix} \quad \textbf{(22–10a)}$$

If the light is fully reflected, the radiation pressure is twice as great (Eq. 22–9b):

$$P = \frac{2\bar{I}}{c}. \qquad \begin{bmatrix} \text{fully} \\ \text{reflected} \end{bmatrix} \quad \textbf{(22–10b)}$$

EXAMPLE 22–5 | ESTIMATE | Solar pressure. Radiation from the Sun that reaches the Earth's surface (after passing through the atmosphere) transports energy at a rate of about 1000 W/m². Estimate the pressure and force exerted by the Sun on your outstretched hand.

APPROACH The radiation is partially reflected and partially absorbed, so let us estimate simply $P = \bar{I}/c$.

SOLUTION $P \approx \dfrac{\bar{I}}{c} = \dfrac{1000 \text{ W/m}^2}{3 \times 10^8 \text{ m/s}} \approx 3 \times 10^{-6} \text{ N/m}^2.$

An estimate of the area of your outstretched hand might be about 10 cm by 20 cm, so $A \approx 0.02 \text{ m}^2$. Then the force is

$$F = PA \approx (3 \times 10^{-6} \text{ N/m}^2)(0.02 \text{ m}^2) \approx 6 \times 10^{-8} \text{ N}.$$

NOTE These numbers are tiny. The force of gravity on your hand, for comparison, is maybe a half pound, or with $m = 0.2$ kg, $mg \approx (0.2 \text{ kg})(9.8 \text{ m/s}^2) \approx 2$ N. The radiation pressure on your hand is imperceptible compared to gravity.

EXAMPLE 22–6 **ESTIMATE** **A solar sail.** Proposals have been made to use the radiation pressure from the Sun to help propel spacecraft around the solar system. (a) About how much force would be applied on a 1 km × 1 km highly reflective sail when about the same distance from the Sun as the Earth is? (b) By how much would this increase the speed of a 5000-kg spacecraft in one year? (c) If the spacecraft started from rest, about how far would it travel in a year?

APPROACH (a) Pressure P is force per unit area, so $F = PA$. We use the estimate of Example 22–5, doubling it for a reflecting surface $P = 2\bar{I}/c$. (b) We find the acceleration from Newton's second law, and assume it is constant, and then find the speed from $v = v_0 + at$. (c) The distance traveled is given by $x = \frac{1}{2}at^2$.

SOLUTION (a) Doubling the result of Example 22–5, we get a solar pressure that is about $2\bar{I}/c \approx 10^{-5}\,\text{N/m}^2$, rounding off. Then the force is $F \approx PA = (10^{-5}\,\text{N/m}^2)(10^3\,\text{m})(10^3\,\text{m}) \approx 10\,\text{N}$.

(b) The acceleration is $a \approx F/m \approx (10\,\text{N})/(5000\,\text{kg}) \approx 2 \times 10^{-3}\,\text{m/s}^2$. One year has $(365\,\text{days})(24\,\text{h/day})(3600\,\text{s/h}) \approx 3 \times 10^7\,\text{s}$. The speed increase is $v - v_0 = at \approx (2 \times 10^{-3}\,\text{m/s}^2)(3 \times 10^7\,\text{s}) \approx 6 \times 10^4\,\text{m/s}$ ($\approx 200{,}000\,\text{km/h}!$).

(c) Starting from rest, this acceleration would result in a distance traveled of about $d = \frac{1}{2}at^2 \approx \frac{1}{2}(2 \times 10^{-3}\,\text{m/s}^2)(3 \times 10^7\,\text{s})^2 \approx 10^{12}\,\text{m}$ in a year, about seven times the Sun–Earth distance. This result would apply if the spacecraft was far from the Earth so the Earth's gravitational force is small compared to 10 N.

NOTE A large sail providing a small force over a long time could result in a lot of motion. [Gravity due to the Sun and planets has been ignored, but in reality would have to be considered.]

Although you cannot directly feel the effects of radiation pressure, the phenomenon is quite dramatic when applied to atoms irradiated by a finely focused laser beam. An atom has a mass on the order of $10^{-27}\,\text{kg}$, and a laser beam can deliver energy at a rate of $1000\,\text{W/m}^2$. This is the same intensity used in Example 22–5, but here a radiation pressure of $10^{-6}\,\text{N/m}^2$ would be very significant on a molecule whose mass might be 10^{-23} to $10^{-26}\,\text{kg}$. It is possible to move atoms and molecules around by steering them with a laser beam, in a device called **optical tweezers**. Optical tweezers have some remarkable applications. They are of great interest to biologists, especially since optical tweezers can manipulate live microorganisms, and components within a cell, without damaging them. Optical tweezers have been used to measure the elastic properties of DNA by pulling each end of the molecule with such a laser "tweezers."

PHYSICS APPLIED
Optical tweezers
(move cell parts,
DNA elasticity)

22–7 Radio and Television; Wireless Communication

PHYSICS APPLIED
Wireless transmission

Electromagnetic waves offer the possibility of transmitting information over long distances. Among the first to realize this and put it into practice was Guglielmo Marconi (1874–1937) who, in the 1890s, invented and developed wireless communication. With it, messages could be sent at the speed of light without the use of wires. The first signals were merely long and short pulses that could be translated into words by a code, such as the "dots" and "dashes" of the Morse code: they were digital wireless, believe it or not. In 1895 Marconi sent wireless signals a kilometer or two in Italy. By 1901 he had sent test signals 3000 km across the ocean from Newfoundland, Canada, to Cornwall, England (Fig. 22–12). In 1903 he sent the first practical commercial messages from Cape Cod, Massachusetts, to England: the London *Times* printed news items sent from its New York correspondent. 1903 was also the year of the first powered airplane flight by the Wright brothers. The hallmarks of the modern age—wireless communication and flight—date from the same year. Our modern world of wireless communication, including radio, television, cordless phones, cell phones, Bluetooth, wi-fi, and satellite communication, are based on Marconi's pioneering work.

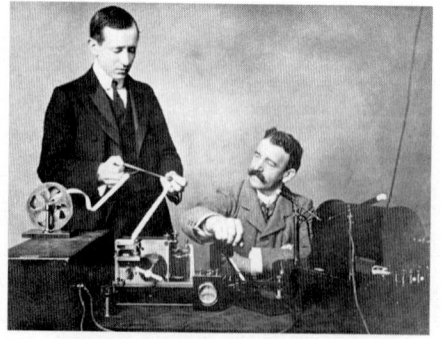
FIGURE 22–12 Guglielmo Marconi (1874–1937), on the left, receiving signals in Cornwall, 1901.

The next decade saw the development of vacuum tubes. Out of this early work radio and television were born. We now discuss briefly (1) how radio and TV signals are transmitted, and (2) how they are received at home.

The process by which a radio station transmits information (words and music) is outlined in Fig. 22–13. The audio (sound) information is changed into an electrical signal of the same frequencies by, say, a microphone, a laser, or a magnetic read/write head. This electrical signal is called an audiofrequency (AF) signal, because the frequencies are in the audio range (20 to 20,000 Hz). The signal is amplified electronically and is then mixed with a radio-frequency (RF) signal called its **carrier frequency**, which represents that station. AM radio stations have carrier frequencies from about 530 kHz to 1700 kHz. For example, "710 on your dial" means a station whose carrier frequency is 710 kHz. FM radio stations have much higher carrier frequencies, between 88 MHz and 108 MHz. The carrier frequencies for broadcast TV stations in the United States lie between 54 MHz and 72 MHz, between 76 MHz and 88 MHz, between 174 MHz and 216 MHz, and between 470 MHz and 698 MHz. Today's digital broadcasting (see Sections 17–10 and 17–11) uses the same frequencies as the pre-2009 analog transmission.

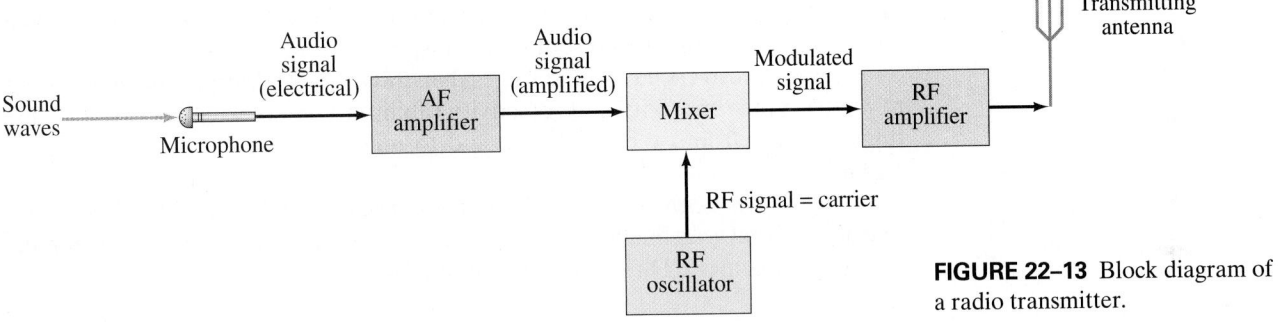

FIGURE 22–13 Block diagram of a radio transmitter.

The mixing of the audio and carrier frequencies is done in two ways. In **amplitude modulation** (AM), the amplitude of the high-frequency carrier wave is made to vary in proportion to the amplitude of the audio signal, as shown in Fig. 22–14. It is called "amplitude modulation" because the *amplitude* of the carrier is altered ("modulate" means to change or alter). In **frequency modulation** (FM), the *frequency* of the carrier wave is made to change in proportion to the audio signal's amplitude, as shown in Fig. 22–15. The mixed signal is amplified further and sent to the transmitting antenna (Fig. 22–13), where the complex mixture of frequencies is sent out in the form of EM waves. In digital communication, the signal is put into digital form (Section 17–10) which modulates the carrier.

A television transmitter works in a similar way, using FM for audio and AM for video; both audio and video signals are mixed with carrier frequencies.

PHYSICS APPLIED
AM and FM

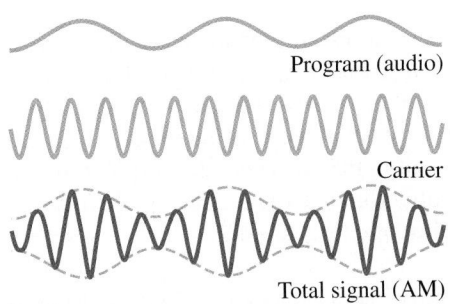

FIGURE 22–14 In amplitude modulation (AM), the amplitude of the carrier signal is made to vary in proportion to the audio signal's amplitude.

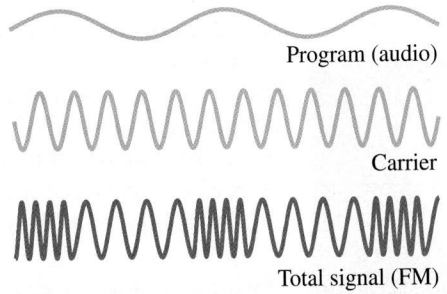

FIGURE 22–15 In frequency modulation (FM), the frequency of the carrier signal is made to change in proportion to the audio signal's amplitude. This method is used by FM radio and television.

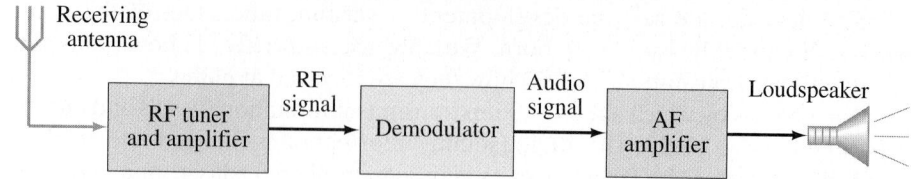

FIGURE 22–16 Block diagram of a simple radio receiver.

PHYSICS APPLIED

Radio and TV receivers

FIGURE 22–17 Simple tuning stage of a radio.

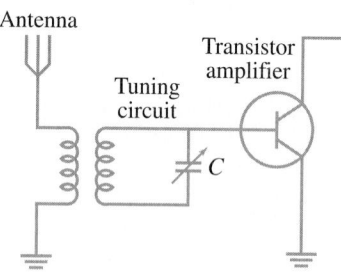

Now let us look at the other end of the process, the reception of radio and TV programs at home. A simple radio receiver is diagrammed in Fig. 22–16. The EM waves sent out by all stations are received by the antenna. The signals the antenna detects and sends to the receiver are very small and contain frequencies from many different stations. The receiver uses a resonant LC circuit (Section 21–15) to select out a particular RF frequency (actually a narrow range of frequencies) corresponding to a particular station. A simple way of tuning a station is shown in Fig. 22–17. A particular station is "tuned in" by adjusting C and/or L so that the resonant frequency of the circuit ($f_0 = 1/(2\pi\sqrt{LC})$, Eq. 21–19) equals that of the station's carrier frequency. The signal, containing both audio and carrier frequencies, next goes to the *demodulator*, or *detector* (Fig. 22–16), where "demodulation" takes place—that is, the audio signal is separated from the RF carrier frequency. The audio signal is amplified and sent to a loudspeaker or headphones.

Modern receivers have more stages than those shown. Various means are used to increase the sensitivity and selectivity (ability to detect weak signals and distinguish them from other stations), and to minimize distortion of the original signal.[†]

A television receiver does similar things to both the audio and the video signals. The audio signal goes finally to the loudspeaker, and the video signal to the monitor screen, such as an LCD (Sections 17–11 and 24–11).

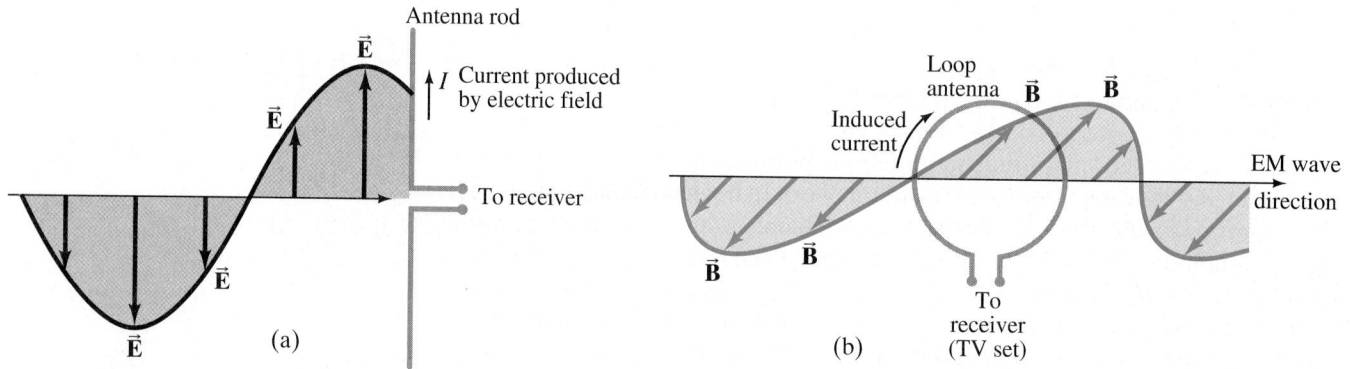

FIGURE 22–18 Antennas. (a) Electric field of EM wave produces a current in an antenna consisting of straight wire or rods. (b) The moving and changing magnetic field induces an emf and current in a loop antenna.

FIGURE 22–19 A satellite dish.

One kind of antenna consists of one or more conducting rods; the electric field in the EM waves exerts a force on the electrons in the conductor, causing them to move back and forth at the frequencies of the waves (Fig. 22–18a). A second type of antenna consists of a tubular coil of wire which detects the magnetic field of the wave: the changing B field induces an emf in the coil (Fig. 22–18b). A satellite dish (Fig. 22–19) consists of a parabolic reflector that focuses the EM waves onto a "horn," similar to a concave mirror telescope (Fig. 25–22).

[†]For *FM stereo broadcasting*, two signals are carried by the carrier wave. One signal contains frequencies up to about 15 kHz, which includes most audio frequencies. The other signal includes the same range of frequencies, but 19 kHz is added to it. A stereo receiver subtracts this 19,000-Hz signal and distributes the two signals to the left and right channels. The first signal consists of the sum of left and right channels (L + R), so monophonic radios (one speaker) detect all the sound. The second signal is the difference between left and right (L − R). Hence a stereo receiver must add and subtract the two signals to get pure left and right signals for each channel.

EXAMPLE 22–7 **Tuning a station.** Calculate the transmitting wavelength of an FM radio station that transmits at 100.1 MHz.

APPROACH Radio is transmitted as an EM wave, so the speed is $c = 3.0 \times 10^8$ m/s. The wavelength is found from Eq. 22–4, $\lambda = c/f$.

SOLUTION The carrier frequency is $f = 100.1$ MHz $\approx 1.0 \times 10^8$ s^{-1}, so

$$\lambda = \frac{c}{f} = \frac{(3.0 \times 10^8 \text{ m/s})}{(1.0 \times 10^8 \text{ s}^{-1})} = 3.0 \text{ m}.$$

NOTE The wavelengths of other FM signals (88 MHz to 108 MHz) are close to the 3.0-m wavelength of this station. FM antennas are typically 1.5 m long, or about a half wavelength. This length is chosen so that the antenna reacts in a resonant fashion and thus is more sensitive to FM frequencies. AM radio antennas would have to be very long and impractical to be either $\frac{1}{2}\lambda$ or $\frac{1}{4}\lambda$.

Other EM Wave Communications

The various regions of the radio-wave spectrum are assigned by governmental agencies for various purposes. Besides those mentioned above, there are "bands" assigned for use by ships, airplanes, police, military, amateurs, satellites and space, and radar. Cell phones, for example, are complete radio transmitters and receivers. In the U.S., CDMA cell phones function on two different bands: 800 MHz and 1900 MHz ($= 1.9$ GHz). Europe, Asia, and much of the rest of the world use a different system: the international standard called GSM (Global System for Mobile Communication), on 900-MHz and 1800-MHz bands. The U.S. now also has the GSM option (at 850 MHz and 1.9 GHz), as does much of the rest of the Americas. A 700-MHz band is being made available for cell phones (it used to carry TV broadcast channels, no longer used). Radio-controlled toys (cars, sailboats, robotic animals, etc.) can use various frequencies from 27 MHz to 75 MHz. Automobile remote entry (keyless) may operate around 300 MHz or 400 MHz.

Cable TV channels are carried as electromagnetic waves along a coaxial cable (Fig. 22–9) rather than being broadcast and received through the "air." The channels are in the same part of the EM spectrum, hundreds of MHz, but some are at frequencies not available for TV broadcast. Digital satellite TV and radio are carried in the microwave portion of the spectrum (12 to 14 GHz and 2.3 GHz, respectively).

Wireless from the Moon

In 1969, astronauts first landed on the Moon. It was shown live on television (Fig. 22–20). The transmitting TV camera can be seen in the Chapter-Opening photo, page 625. At that time, someone pointed out that Columbus and other early navigators could have imagined that humans might one day reach the Moon. But they would never have believed possible that moving images could be sent from the Moon to the Earth through empty space.

FIGURE 22–20 The first person on the Moon, Neil Armstrong, July 20, 1969, pointed out "One small step for a man, one giant leap for mankind."

Summary

James Clerk Maxwell synthesized an elegant theory in which all electric and magnetic phenomena could be described using four equations, now called **Maxwell's equations**. They are based on earlier ideas, but Maxwell added one more—that a changing electric field produces a magnetic field.

Maxwell's theory predicted that transverse **electromagnetic (EM) waves** would be produced by accelerating electric charges, and these waves would propagate (move) through space at the speed of light:

$$c = \frac{1}{\sqrt{\epsilon_0 \mu_0}} = 3.00 \times 10^8 \text{ m/s}. \qquad \textbf{(22–3)}$$

The oscillating electric and magnetic fields in an EM wave are perpendicular to each other and to the direction of propagation. These EM waves are waves of fields, not matter, and can propagate in empty space.

The wavelength λ and frequency f of EM waves are related to their speed c by

$$c = \lambda f \qquad \textbf{(22–4)}$$

just as for other waves.

After EM waves were experimentally detected, it became generally accepted that light is an EM wave. The **electromagnetic spectrum** includes EM waves of a wide variety of wavelengths, from microwaves and radio waves to visible light to X-rays and gamma rays, all of which travel through space at a speed $c = 3.0 \times 10^8$ m/s.

The average *intensity* (W/m^2) of an EM wave is

$$\bar{I} = \frac{1}{2}\epsilon_0 c E_0^2 = \frac{1}{2}\frac{c}{\mu_0} B_0^2 = \frac{1}{2}\frac{E_0 B_0}{\mu_0}, \qquad \textbf{(22–8)}$$

where E_0 and B_0 are the peak values of the electric and magnetic fields, respectively, in the wave.

EM waves carry momentum and exert a **radiation pressure** proportional to the intensity I of the wave.

Radio, TV, cell phone, and other wireless signals are transmitted through space in the radio-wave or microwave part of the EM spectrum.

Questions

1. The electric field in an EM wave traveling north oscillates in an east–west plane. Describe the direction of the magnetic field vector in this wave. Explain.

2. Is sound an EM wave? If not, what kind of wave is it?

3. Can EM waves travel through a perfect vacuum? Can sound waves?

4. When you flip a light switch on, does the light go on immediately? Explain.

5. Are the wavelengths of radio and television signals longer or shorter than those detectable by the human eye?

6. When you connect two loudspeakers to the output of a stereo amplifier, should you be sure the lead-in wires are equal in length to avoid a time lag between speakers? Explain.

7. In the electromagnetic spectrum, what type of EM wave would have a wavelength of 10^3 km? 1 km? 1 m? 1 cm? 1 mm? 1 μm?

8. Can radio waves have the same frequencies as sound waves (20 Hz–20,000 Hz)?

9. If a radio transmitter has a vertical antenna, should a receiver's antenna (rod type) be vertical or horizontal to obtain best reception?

10. The carrier frequencies of FM broadcasts are much higher than for AM broadcasts. On the basis of what you learned about diffraction in Chapter 11, explain why AM signals can be detected more readily than FM signals behind low hills or buildings.

11. Discuss how cordless telephones make use of EM waves. What about cell phones?

12. A lost person may signal by switching a flashlight on and off using Morse code. This is actually a modulated EM wave. Is it AM or FM? What is the frequency of the carrier, approximately?

MisConceptual Questions

1. In a vacuum, what is the difference between a radio wave and an X-ray?
   (a) Wavelength. (b) Frequency. (c) Speed.

2. The radius of an atom is on the order of 10^{-10} m. In comparison, the wavelength of visible light is
   (a) much smaller. (b) about the same size. (c) much larger.

3. Which of the following travel at the same speed as light? (Choose all that apply.)
   (a) Radio waves. (d) Ultrasonic waves. (g) Gamma rays.
   (b) Microwaves. (e) Infrared radiation. (h) X-rays.
   (c) Radar. (f) Cell phone signals.

4. Which of the following types of electromagnetic radiation travels the fastest?
   (a) Radio waves.
   (b) Visible light waves.
   (c) X-rays.
   (d) Gamma rays.
   (e) All the above travel at the same speed.

5. In empty space, which quantity is always larger for X-ray radiation than for a radio wave?
   (a) Amplitude.   (c) Frequency.
   (b) Wavelength.   (d) Speed.

6. If electrons in a wire vibrate up and down 1000 times per second, they will create an electromagnetic wave having
   (a) a wavelength of 1000 m.   (c) a speed of 1000 m/s.
   (b) a frequency of 1000 Hz.   (d) an amplitude of 1000 m.

7. If the Earth–Sun distance were doubled, the intensity of radiation from the Sun that reaches the Earth's surface would
   (a) quadruple. (b) double. (c) drop to $\frac{1}{2}$. (d) drop to $\frac{1}{4}$.

8. An electromagnetic wave is traveling straight down toward the center of the Earth. At a certain moment in time the electric field points west. In which direction does the magnetic field point at this moment?
   (a) North.   (d) West.   (g) Either (a) or (b).
   (b) South.   (e) Up.   (h) Either (c) or (d).
   (c) East.   (f) Down.   (i) Either (e) or (f).

9. If the intensity of an electromagnetic wave doubles,
   (a) the electric field must also double.
   (b) the magnetic field must also double.
   (c) both the magnetic field and the electric field must increase by a factor of $\sqrt{2}$.
   (d) Any of the above.

10. If all else is the same, for which surface would the radiation pressure from light be the greatest?
    (a) A black surface.
    (b) A gray surface.
    (c) A yellow surface.
    (d) A white surface.
    (e) All experience the same radiation pressure, because they are exposed to the same light.

11. Starting in 2009, TV stations in the U.S. switched to digital signals. [See Sections 22–7, 17–10, and 17–11.] To watch today's digital broadcast TV, could you use a pre-2009 TV antenna meant for analog? Explain.
    (a) No; analog antennas do not receive digital signals.
    (b) No; digital signals are broadcast at different frequencies, so you need a different antenna.
    (c) Yes; digital signals are broadcast with the same carrier frequencies, so your old antenna will be fine.
    (d) No; you cannot receive digital signals through an antenna and need to switch to cable or satellite.

Problems

22–1 $\vec{B}$ Produced by Changing $\vec{E}$

*1. (II) Determine the rate at which the electric field changes between the round plates of a capacitor, 8.0 cm in diameter, if the plates are spaced 1.1 mm apart and the voltage across them is changing at a rate of 120 V/s.

*2. (II) Calculate the displacement current I_D between the square plates, 5.8 cm on a side, of a capacitor if the electric field is changing at a rate of 1.6×10^6 V/m·s.

*3. (II) At a given instant, a 3.8-A current flows in the wires connected to a parallel-plate capacitor. What is the rate at which the electric field is changing between the plates if the square plates are 1.60 cm on a side?

*4. (III) A 1500-nF capacitor with circular parallel plates 2.0 cm in diameter is accumulating charge at the rate of 32.0 mC/s at some instant in time. What will be the induced magnetic field strength 10.0 cm radially outward from the center of the plates? What will be the value of the field strength after the capacitor is fully charged?

22–2 EM Waves

5. (I) If the electric field in an EM wave has a peak magnitude of 0.72×10^{-4} V/m, what is the peak magnitude of the magnetic field strength?

6. (I) If the magnetic field in a traveling EM wave has a peak magnitude of 10.5 nT, what is the peak magnitude of the electric field?

7. (I) In an EM wave traveling west, the B field oscillates up and down vertically and has a frequency of 90.0 kHz and an rms strength of 7.75×10^{-9} T. Determine the frequency and rms strength of the electric field. What is the direction of its oscillations?

8. (I) How long does it take light to reach us from the Sun, 1.50×10^8 km away?

9. (II) How long should it take the voices of astronauts on the Moon to reach the Earth? Explain in detail.

22–3 Electromagnetic Spectrum

10. (I) An EM wave has a wavelength of 720 nm. What is its frequency, and how would we classify it?

11. (I) An EM wave has frequency 7.14×10^{14} Hz. What is its wavelength, and how would we classify it?

12. (I) A widely used "short-wave" radio broadcast band is referred to as the 49-m band. What is the frequency of a 49-m radio signal?

13. (I) What is the frequency of a microwave whose wavelength is 1.50 cm?

14. (II) Electromagnetic waves and sound waves can have the same frequency. (a) What is the wavelength of a 1.00-kHz electromagnetic wave? (b) What is the wavelength of a 1.00-kHz sound wave? (The speed of sound in air is 341 m/s.) (c) Can you hear a 1.00-kHz electromagnetic wave?

15. (II) (a) What is the wavelength of a 22.75×10^9 Hz radar signal? (b) What is the frequency of an X-ray with wavelength 0.12 nm?

16. (II) How long would it take a message sent as radio waves from Earth to reach Mars when Mars is (a) nearest Earth, (b) farthest from Earth? Assume that Mars and Earth are in the same plane and that their orbits around the Sun are circles (Mars is $\approx 230 \times 10^6$ km from the Sun).

17. (II) Our nearest star (other than the Sun) is 4.2 light-years away. That is, it takes 4.2 years for the light it emits to reach Earth. How far away is it in meters?

18. (II) A light-year is a measure of distance (not time). How many meters does light travel in a year?

19. (II) Pulsed lasers used for science and medicine produce very brief bursts of electromagnetic energy. If the laser light wavelength is 1062 nm (Neodymium–YAG laser), and the pulse lasts for 34 picoseconds, how many wavelengths are found within the laser pulse? How brief would the pulse need to be to fit only one wavelength?

22–4 Measuring the Speed of Light

20. (II) What is the minimum angular speed at which Michelson's eight-sided mirror would have had to rotate to reflect light into an observer's eye by succeeding mirror faces (1/8 of a revolution, Fig. 22–10)?

21. (II) A student wants to scale down Michelson's light-speed experiment to a size that will fit in one room. An eight-sided mirror is available, and the stationary mirror can be mounted 12 m from the rotating mirror. If the arrangement is otherwise as shown in Fig. 22–10, at what minimum rate must the mirror rotate?

22–5 Energy in EM Wave

22. (I) The $\vec{E}$ field in an EM wave has a peak of 22.5 mV/m. What is the average rate at which this wave carries energy across unit area per unit time?

23. (II) The magnetic field in a traveling EM wave has an rms strength of 22.5 nT. How long does it take to deliver 365 J of energy to 1.00 cm² of a wall that it hits perpendicularly?

24. (II) How much energy is transported across a 1.00-cm² area per hour by an EM wave whose E field has an rms strength of 30.8 mV/m?

25. (II) A spherically spreading EM wave comes from an 1800-W source. At a distance of 5.0 m, what is the intensity, and what is the rms value of the electric field?

26. (II) If the amplitude of the B field of an EM wave is 2.2×10^{-7} T, (a) what is the amplitude of the E field? (b) What is the average power transported across unit area by the EM wave?

27. (II) What is the average energy contained in a 1.00-m³ volume near the Earth's surface due to radiant energy from the Sun? See Example 22–4.

28. (II) A 15.8-mW laser puts out a narrow beam 2.40 mm in diameter. What are the rms values of E and B in the beam?

29. (II) Estimate the average power output of the Sun, given that about 1350 W/m² reaches the upper atmosphere of the Earth.

30. (II) A high-energy pulsed laser emits a 1.0-ns-long pulse of average power 1.5×10^{11} W. The beam is nearly a cylinder 2.2×10^{-3} m in radius. Determine (a) the energy delivered in each pulse, and (b) the rms value of the electric field.

22–6 Radiation Pressure

31. (II) Estimate the radiation pressure due to a bulb that emits 25 W of EM radiation at a distance of 9.5 cm from the center of the bulb. Estimate the force exerted on your fingertip if you place it at this point.

32. (II) What size should the solar panel on a satellite orbiting Jupiter be if it is to collect the same amount of radiation from the Sun as a 1.0-m^2 solar panel on a satellite orbiting Earth? [Hint: Assume the inverse square law (Eq. 11–16b).]

33. (III) Suppose you have a car with a 100-hp engine. How large a solar panel would you need to replace the engine with solar power? Assume that the solar panels can utilize 20% of the maximum solar energy that reaches the Earth's surface (1000 W/m^2).

22–7 Radio, TV

34. (I) What is the range of wavelengths for (a) FM radio (88 MHz to 108 MHz) and (b) AM radio (535 kHz to 1700 kHz)?

35. (I) Estimate the wavelength for a 1.9-GHz cell phone transmitter.

36. (I) Compare 980 on the AM dial to 98.1 on FM. Which has the longer wavelength, and by what factor is it larger?

37. (I) What are the wavelengths for two TV channels that broadcast at 54.0 MHz (Channel 2) and 692 MHz (Channel 51)?

38. (I) The variable capacitor in the tuner of an AM radio has a capacitance of 2500 pF when the radio is tuned to a station at 550 kHz. What must the capacitance be for a station near the other end of the dial, 1610 kHz?

39. (I) The oscillator of a 98.3-MHz FM station has an inductance of 1.8 μH. What value must the capacitance be?

40. (II) A certain FM radio tuning circuit has a fixed capacitor $C = 810$ pF. Tuning is done by a variable inductance. What range of values must the inductance have to tune stations from 88 MHz to 108 MHz?

41. (II) An amateur radio operator wishes to build a receiver that can tune a range from 14.0 MHz to 15.0 MHz. A variable capacitor has a minimum capacitance of 86 pF. (a) What is the required value of the inductance? (b) What is the maximum capacitance used on the variable capacitor?

42. (II) A satellite beams microwave radiation with a power of 13 kW toward the Earth's surface, 550 km away. When the beam strikes Earth, its circular diameter is about 1500 m. Find the rms electric field strength of the beam.

43. (III) A 1.60-m-long FM antenna is oriented parallel to the electric field of an EM wave. How large must the electric field be to produce a 1.00-mV (rms) voltage between the ends of the antenna? What is the rate of energy transport per m^2?

General Problems

44. Who will hear the voice of a singer first: a person in the balcony 50.0 m away from the stage (see Fig. 22–21), or a person 1200 km away at home whose ear is next to the radio listening to a live broadcast? Roughly how much sooner? Assume the microphone is a few centimeters from the singer and the temperature is 20°C.

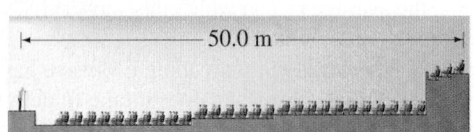

FIGURE 22–21 Problem 44.

45. A global positioning system (GPS) functions by determining the travel times for EM waves from various satellites to a land-based GPS receiver. If the receiver is to detect a change in travel distance on the order of 3 m, what is the associated change in travel time (in ns) that must be measured?

46. Light is emitted from an ordinary lightbulb filament in wavetrain bursts about 10^{-8} s in duration. What is the length in space of such wave trains?

47. The voice from an astronaut on the Moon (Fig. 22–22) was beamed to a listening crowd on Earth. If you were standing 28 m from the loudspeaker on Earth, what was the total time lag between when you heard the sound and when the sound entered a microphone on the Moon? Explain whether the microphone was inside the space helmet, or outside, and why.

FIGURE 22–22
Problem 47.

48. Radio-controlled clocks throughout the United States receive a radio signal from a transmitter in Fort Collins, Colorado, that accurately (within a microsecond) marks the beginning of each minute. A slight delay, however, is introduced because this signal must travel from the transmitter to the clocks. Assuming Fort Collins is no more than 3000 km from any point in the U.S., what is the longest travel-time delay?

49. If the Sun were to disappear or radically change its output, how long would it take for us on Earth to learn about it?

50. Cosmic microwave background radiation fills space with an average energy density of about 4×10^{-14} J/m^3. (a) Find the rms value of the electric field associated with this radiation. (b) How far from a 7.5-kW radio transmitter emitting uniformly in all directions would you find a comparable value?

51. What are E_0 and B_0 at a point 2.50 m from a light source whose output is 18 W? Assume the bulb emits radiation of a single frequency uniformly in all directions.

52. Estimate the rms electric field in the sunlight that hits Mars, knowing that the Earth receives about 1350 W/m^2 and that Mars is 1.52 times farther from the Sun (on average) than is the Earth.

53. The average intensity of a particular TV station's signal is 1.0×10^{-13} W/m^2 when it arrives at a 33-cm-diameter satellite TV antenna. (a) Calculate the total energy received by the antenna during 4.0 hours of viewing this station's programs. (b) Estimate the amplitudes of the E and B fields of the EM wave.

54. What length antenna would be appropriate for a portable device that could receive satellite TV?

55. A radio station is allowed to broadcast at an average power not to exceed 25 kW. If an electric field amplitude of 0.020 V/m is considered to be acceptable for receiving the radio transmission, estimate how many kilometers away you might be able to detect this station.

56. The radiation pressure (Section 22–6) created by electromagnetic waves might someday be used to power spacecraft through the use of a "solar sail," Example 22–6. (a) Assuming total reflection, what would be the pressure on a solar sail located at the same distance from the Sun as the Earth (where $I = 1350 \, \text{W/m}^2$)? (b) Suppose the sail material has a mass of $1 \, \text{g/m}^2$. What would be the acceleration of the sail due to solar radiation pressure? (c) A realistic solar sail would have a payload. How big a sail would you need to accelerate a 100-kg payload at $1 \times 10^{-3} \, \text{m/s}^2$?

57. Suppose a 35-kW radio station emits EM waves uniformly in all directions. (a) How much energy per second crosses a 1.0-m^2 area 1.0 km from the transmitting antenna? (b) What is the rms magnitude of the $\vec{\mathbf{E}}$ field at this point, assuming the station is operating at full power? What is the rms voltage induced in a 1.0-m-long vertical car antenna (c) 1.0 km away, (d) 50 km away?

58. A point source emits light energy uniformly in all directions at an average rate P_0 with a single frequency f. Show that the peak electric field in the wave is given by

$$E_0 = \sqrt{\frac{\mu_0 c P_0}{2\pi r^2}}.$$

[Hint: The surface area of a sphere is $4\pi r^2$.]

59. What is the maximum power level of a radio station so as to avoid electrical breakdown of air at a distance of 0.65 m from the transmitting antenna? Assume the antenna is a point source. Air breaks down in an electric field of about $3 \times 10^6 \, \text{V/m}$.

60. Estimate how long an AM antenna would have to be if it were (a) $\frac{1}{2}\lambda$ or (b) $\frac{1}{4}\lambda$. AM radio is roughly 1 MHz (530 kHz to 1.7 MHz).

61. 12 km from a radio station's transmitting antenna, the amplitude of the electric field is 0.12 V/m. What is the average power output of the radio station?

Search and Learn

1. How practical is solar power for various devices? Assume that on a sunny day, sunlight has an intensity of $1000 \, \text{W/m}^2$ at the surface of Earth and that a solar-cell panel can convert 20% of that sunlight into electric power. Calculate the area A of solar panel needed to power (a) a calculator that consumes 50 mW, (b) a hair dryer that consumes 1500 W, (c) a car that would require 40 hp. (d) In each case, would the area A be small enough to be mounted on the device itself, or in the case of (b) on the roof of a house?

2. A powerful laser portrayed in a movie provides a 3-mm diameter beam of green light with a power of 3 W. A good agent inside the Space Shuttle aims the laser beam at an enemy astronaut hovering outside. The mass of the enemy astronaut is 120 kg and the Space Shuttle 103,000 kg. (a) Determine the "radiation-pressure" force exerted on the enemy by the laser beam assuming her suit is perfectly reflecting. (b) If the enemy is 30 m from the Shuttle's center of mass, estimate the gravitational force the Shuttle exerts on the enemy. (c) Which of the two forces is larger, and by what factor?

3. The Arecibo radio telescope in Puerto Rico can detect a radio wave with an intensity as low as $1 \times 10^{-23} \, \text{W/m}^2$. Consider a "best-case" scenario for communication with extraterrestrials: suppose an advanced civilization a distance x away from Earth is able to transform the entire power output of a Sun-like star completely into a radio-wave signal which is transmitted uniformly in all directions. (a) In order for Arecibo to detect this radio signal, what is the maximum value for x in light-years (1 ly $\approx 10^{16}$ m)? (b) How does this maximum value compare with the 100,000-ly size of our Milky Way galaxy? The intensity of sunlight at Earth's orbital distance from the Sun is $1350 \, \text{W/m}^2$. [Hint: Assume the inverse square law (Eq. 11–16b).]

4. Laser light can be focused (at best) to a spot with a radius r equal to its wavelength λ. Suppose a 1.0-W beam of green laser light $(\lambda = 5 \times 10^{-7} \, \text{m})$ forms such a spot and illuminates a cylindrical object of radius r and length r (Fig. 22–23). Estimate (a) the radiation pressure and force on the object, and (b) its acceleration, if its density equals that of water and it absorbs all the radiation. [This order-of-magnitude calculation convinced researchers of the feasibility of "optical tweezers," page 636.]

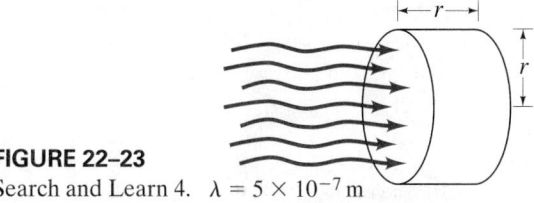

FIGURE 22–23
Search and Learn 4. $\lambda = 5 \times 10^{-7} \, \text{m}$

Reflection from still water, as from a glass mirror, can be analyzed using the ray model of light.

Is this picture right side up, or upside down? How can you tell? What are the clues? Notice the people and position of the Sun. Ray diagrams, which we will learn to draw in this Chapter, can provide the answer. See Example 23–3.

In this first Chapter on light and optics, we use the ray model of light to understand the formation of images by mirrors, both plane and curved (spherical). We also study refraction—how light rays bend when they go from one medium to another—and how, via refraction, images are formed by lenses, which are the crucial part of so many optical instruments.

CHAPTER

23

Light: Geometric Optics

CONTENTS

CHAPTER-OPENING QUESTIONS—Guess now!

1. A 2.0-m-tall person is standing 2.0 m from a flat vertical mirror staring at her image. What minimum height must the mirror's reflecting glass have if the person is to see her entire body, from the top of her head to her feet?

(a) 0.50 m. **(b)** 1.0 m. **(c)** 1.5 m. **(d)** 2.0 m. **(e)** 2.5 m.

2. The focal length of a lens is

(a) the diameter of the lens.

(b) the thickness of the lens.

(c) the distance from the lens at which incoming parallel rays bend to intersect at a point.

(d) the distance from the lens at which all real images are formed.

The sense of sight is extremely important to us, for it provides us with a large part of our information about the world. How do we see? What is the something called *light* that enters our eyes and causes the sensation of sight? How does light behave so that we can see everything that we do? We saw in Chapter 22 that light can be considered a form of electromagnetic radiation. We now examine the subject of light in detail in the next three Chapters.

We see an object in one of two ways: (1) the object may be a *source* of light, such as a lightbulb, a flame, or a star, in which case we see the light emitted directly from the source; or, more commonly, (2) we see an object by light *reflected* from it.

In the latter case, the light may have originated from the Sun, artificial lights, or a campfire. An understanding of how objects *emit* light was not achieved until the 1920s, and will be discussed in Chapter 27. How light is *reflected* from objects was understood much earlier, and will be discussed in Section 23–2.

23–1 The Ray Model of Light

A great deal of evidence suggests that *light travels in straight lines* under a wide variety of circumstances.[†] For example, a source of light like the Sun (which at its great distance from us is nearly a "point source") casts distinct shadows, and the beam from a laser pointer appears to be a straight line. In fact, we infer the positions of objects in our environment by assuming that light moves from the object to our eyes in straight-line paths. Our orientation to the physical world is based on this assumption.

This reasonable assumption is the basis of the **ray model** of light. This model assumes that light travels in straight-line paths called light **rays**. Actually, a ray is an idealization; it is meant to represent an extremely narrow beam of light. When we see an object, according to the ray model, light reaches our eyes from each point on the object. Although light rays leave each point in many different directions, normally only a small bundle of these rays can enter the pupil of an observer's eye, as shown in Fig. 23–1. If the person's head moves to one side, a different bundle of rays will enter the eye from each point.

We saw in Chapter 22 that light can be considered as an electromagnetic wave. Although the ray model of light does not deal with this aspect of light (we discuss the wave nature of light in Chapter 24), the ray model has been very successful in describing many aspects of light such as reflection, refraction, and the formation of images by mirrors and lenses. Because these explanations involve straight-line rays at various angles, this subject is referred to as **geometric optics**.

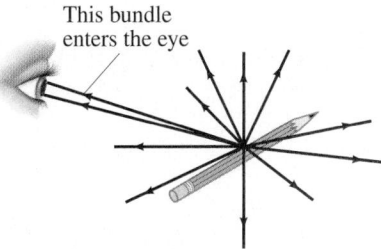

FIGURE 23–1 Light rays come from each single point on an object. A small bundle of rays leaving one point is shown entering a person's eye.

23–2 Reflection; Image Formation by a Plane Mirror

When light strikes the surface of an object, some of the light is reflected. The rest can be absorbed by the object (and transformed to thermal energy) or, if the object is transparent like glass or water, part can be transmitted through. For a very smooth shiny object such as a silvered mirror, over 95% of the light may be reflected.

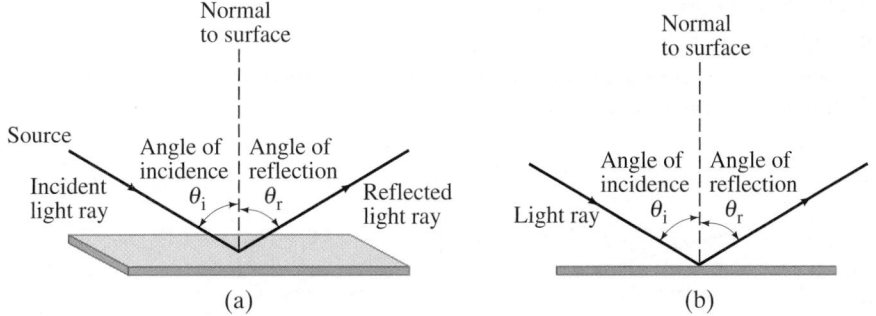

(a)   (b)

FIGURE 23–2 Law of reflection: (a) shows a 3-D view of an incident ray being reflected at the top of a flat surface; (b) shows a side or "end-on" view, which we will usually use because of its clarity.

When a narrow beam of light strikes a flat surface (Fig. 23–2), we define the **angle of incidence**, θ_i, to be the angle an incident ray makes with the normal (perpendicular) to the surface, and the **angle of reflection**, θ_r, to be the angle the reflected ray makes with the normal. It is found that the *incident and reflected rays lie in the same plane with the normal to the surface*, and that

the angle of reflection equals the angle of incidence, $\theta_r = \theta_i$.

LAW OF REFLECTION

This is the **law of reflection**, and it is depicted in Fig. 23–2. It was known to the ancient Greeks, and you can confirm it yourself by shining a narrow flashlight beam or a laser pointer at a mirror in a darkened room.

[†]In a uniform transparent medium such as air or glass: But not always, such as for nonuniform air that allows optical illusions and mirages which we discuss in Section 24–2 (Fig. 24–4).

FIGURE 23–3 Diffuse reflection from a rough surface.

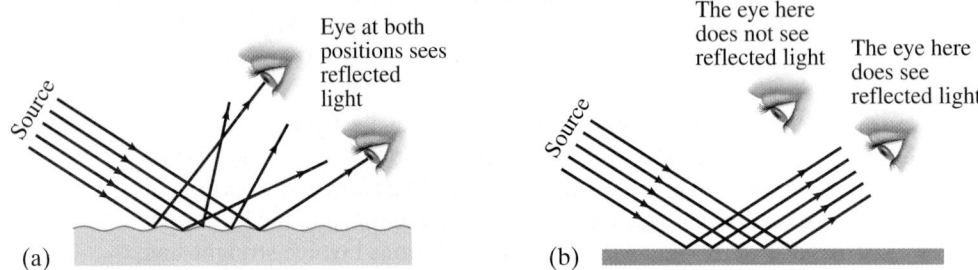

When light is incident upon a rough surface, even microscopically rough such as this page, it is reflected in many directions, as shown in Fig. 23–3. This is called **diffuse reflection**. The law of reflection still holds, however, at each small section of the surface. Because of diffuse reflection in all directions, an ordinary object can be seen at many different angles by the light reflected from it. When you move your head to the side, different reflected rays reach your eye from each point on the object (such as this page), Fig. 23–4a. Let us compare diffuse reflection to reflection from a mirror, which is known as **specular reflection**. ("Speculum" is Latin for mirror.) When a narrow beam of light shines on a mirror, the light will not reach your eye unless your eye is positioned at just the right place where the law of reflection is satisfied, as shown in Fig. 23–4b. This is what gives rise to the special image-forming properties of mirrors.

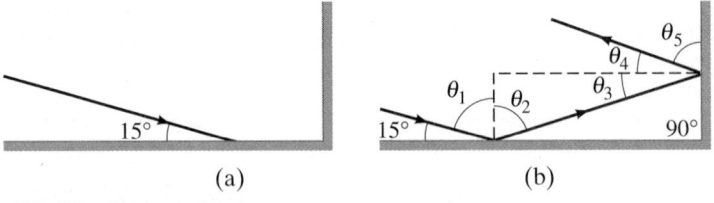

FIGURE 23–4 A narrow beam of light shines on (a) white paper, and (b) a mirror. In part (a), you can see with your eye the white light (and printed words) reflected at various positions because of diffuse reflection. But in part (b), you see the reflected light only when your eye is placed correctly $(\theta_r = \theta_i)$; mirror reflection is also known as specular reflection. (Galileo, using similar arguments, showed that the Moon must have a rough surface rather than a highly polished surface like a mirror, as some people thought.)

EXAMPLE 23–1 **Reflection from flat mirrors.** Two flat mirrors are perpendicular to each other. An incoming beam of light makes an angle of 15° with the first mirror as shown in Fig. 23–5a. What angle will the outgoing beam make with the second mirror?

APPROACH We sketch the path of the beam as it reflects off the two mirrors, and draw the two normals to the mirrors for the two reflections. We use geometry and the law of reflection to find the various angles.

FIGURE 23–5 Example 23–1.

SOLUTION In Fig. 23–5b, $\theta_1 + 15° = 90°$, so $\theta_1 = 75°$; by the law of reflection $\theta_2 = \theta_1 = 75°$ too. Using the fact that the sum of the three angles of a triangle is always 180°, and noting that the two normals to the two mirrors are perpendicular to each other, we have $\theta_2 + \theta_3 + 90° = 180°$. Thus $\theta_3 = 180° - 90° - 75° = 15°$. By the law of reflection, $\theta_4 = \theta_3 = 15°$, so $\theta_5 = 75°$ is the angle the reflected ray makes with the second mirror surface.

NOTE The outgoing ray is parallel to the incoming ray. Reflectors on bicycles, cars, and other applications use this principle.

When you look straight into a mirror, you see what appears to be yourself as well as various objects around and behind you, Fig. 23–6. Your face and the other objects look as if they are in front of you, beyond the mirror. But what you see in the mirror is an **image** of the objects, including yourself, that are in front of the mirror. Also, you don't see yourself as others see you, because left and right appear reversed in the image.

A **plane mirror** is one with a smooth flat reflecting surface. Figure 23–7 shows how an image is formed by a plane mirror according to the ray model. We are viewing the mirror, on edge, in the diagram of Fig. 23–7, and the rays are shown reflecting from the front surface. (Good mirrors are generally made by putting a highly reflective metallic coating on one surface of a very flat piece of glass.) Rays from two different points on an object (the bottle on the left in Fig. 23–7) are shown: two rays are shown leaving from a point on the top of the bottle, and two more from a point on the bottom. Rays leave each point on the object going in many directions (as in Fig. 23–1), but only those that enclose the bundle of rays that enter the eye from each of the two points are shown. Each set of diverging rays that reflect from the mirror and enter the eye *appear to come from a single point* behind the mirror, called the **image point**, as shown by the dashed lines. That is, our eyes and brain interpret any rays that enter an eye as having traveled straight-line paths. The point from which each bundle of rays seems to come is one point on the image. For each point on the object, there is a corresponding image point. (This analysis of how a plane mirror forms an image was published by Kepler in 1604.)

FIGURE 23–6 When you look in a mirror, you see an image of yourself and objects around you. You don't see yourself as others see you, because left and right appear reversed in the image.

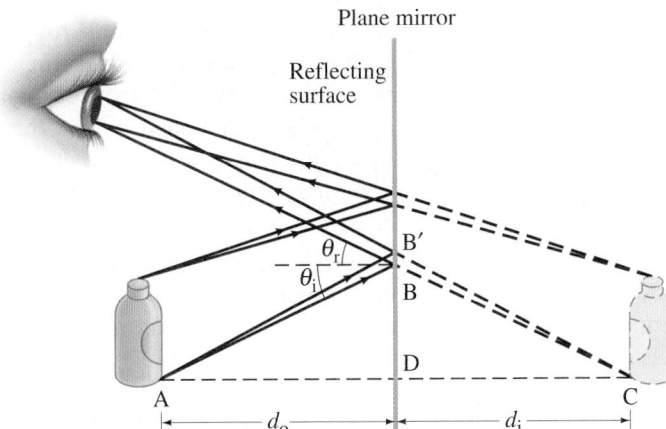

Plane mirror

Reflecting surface

FIGURE 23–7 Formation of a virtual image by a plane mirror. Only the bundle of rays from the top and bottom of the object which reach the eye is shown.

Let us concentrate on the two rays that leave point A on the object in Fig. 23–7, and strike the mirror at points B and B'. We use geometry now, for the rays at B. The angles ADB and CDB are right angles; and because of the law of reflection, $\theta_i = \theta_r$ at point B. Therefore, by geometry, angles ABD and CBD are also equal. The two triangles ABD and CBD are thus congruent, and the length AD = CD. That is, the image appears as far behind the mirror as the object is in front. The **image distance**, d_i (perpendicular distance from mirror to image, Fig. 23–7), equals the **object distance**, d_o (perpendicular distance from object to mirror). From the geometry, we also can see that the height of the image is the same as that of the object.

The light rays do not actually pass through the image location itself in Fig. 23–7. (Note where the red lines are dashed to show they are our projections, not rays.) The image would not appear on paper or film placed at the location of the image. Therefore, it is called a **virtual image**. This is to distinguish it from a **real image** in which the light does pass through the image and which therefore could appear on a white surface, or on film or on an electronic sensor placed at the image position. Our eyes can see both real and virtual images, as long as the diverging rays enter our pupils. We will see that curved mirrors and lenses can form real images, as well as virtual. A movie projector lens, for example, produces a real image that is visible on the screen.

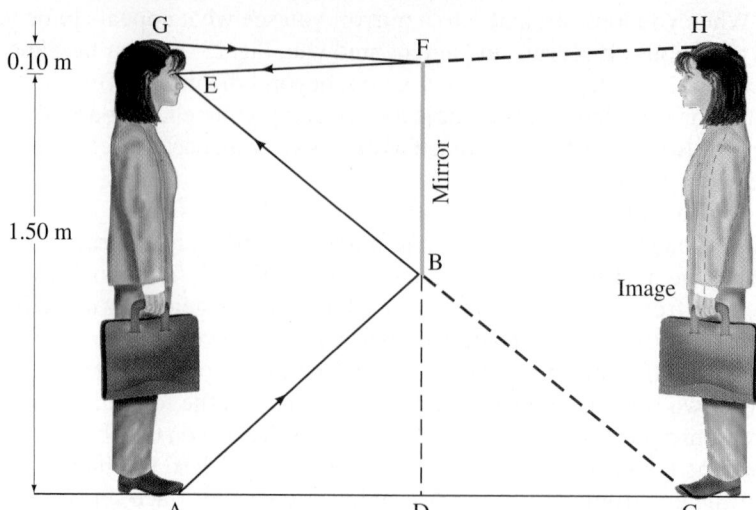

FIGURE 23–8 Seeing oneself in a mirror. Example 23–2.

G

0.10 m

1.50 m

E

F

Mirror

B

Image

H

A

D

C

EXAMPLE 23–2 **How tall must a full-length mirror be?** A woman 1.60 m tall stands in front of a vertical plane mirror. What is the minimum height of the mirror, and how high must its lower edge be above the floor, if she is to be able to see her whole body? Assume that her eyes are 10 cm below the top of her head.

APPROACH For her to see her whole body, light rays from the top of her head (point G) and from the bottom of her foot (A) must reflect from the mirror and enter her eye, Fig. 23–8. We don't show two rays diverging from each point as we did in Fig. 23–7, where we wanted to find where the image is. Now that we know the image is the same distance behind a plane mirror as the object is in front, we only need to show one ray leaving point G (top of head) and one ray leaving point A (her toe), and then use geometry.

SOLUTION First consider the ray that leaves her foot at A, reflects at B, and enters the eye at E. The mirror needs to extend no lower than B. The angle of reflection equals the angle of incidence, so the height BD is half of the height AE. Because AE = 1.60 m − 0.10 m = 1.50 m, then BD = 0.75 m. Similarly, if the woman is to see the top of her head, the top edge of the mirror only needs to reach point F, which is 5 cm below the top of her head (half of GE = 10 cm). Thus, DF = 1.55 m, and the mirror needs to have a vertical height of only (1.55 m − 0.75 m) = 0.80 m. And the mirror's bottom edge must be 0.75 m above the floor.

NOTE We see that a mirror, if positioned at the correct height (as in Fig. 23–8), need be only half as tall as a person for that person to be able to see all of himself or herself.

EXERCISE A Does the result of Example 23–2 depend on your distance from the mirror? (Try it and see, it's fun.)

EXERCISE B Return to Chapter-Opening Question 1, page 644, and answer it again now. Try to explain why you may have answered differently the first time.

CONCEPTUAL EXAMPLE 23–3 **Is the photo upside down?** Close examination of the photograph on the first page of this Chapter reveals that in the top portion, the image of the Sun is seen clearly, whereas in the lower portion, the image of the Sun is partially blocked by the tree branches. Show why the reflection is not the same as the real scene by drawing a sketch of this situation, showing the Sun, the camera, the branch, and two rays going from the Sun to the camera (one direct and one reflected). Is the photograph right side up?

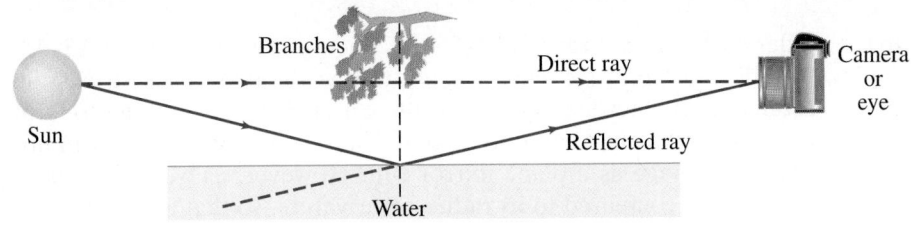
Branches

Sun

Direct ray

Reflected ray

Water

Camera or eye

FIGURE 23–9 Example 23–3.

RESPONSE We need to draw two diagrams, one assuming the photo on p. 644 is right side up, and another assuming it is upside down. Figure 23–9 is drawn assuming the photo is upside down. In this case, the Sun blocked by the tree would be the direct view, and the full view of the Sun the reflection: the ray which reflects off the water and into the camera travels at an angle below the branch, whereas the ray that travels directly to the camera passes through the branches. This works. Try to draw a diagram assuming the photo is right side up (thus assuming that the image of the Sun in the reflection is higher above the horizon than it is as viewed directly). It won't work. The photo on p. 644 is upside down.

Also, what about the people in the photo? Try to draw a diagram showing why they don't appear in the reflection. [*Hint:* Assume they are not sitting at the edge of the pool, but back from the edge.] Then try to draw a diagram of the reverse (i.e., assume the photo is right side up so the people are visible only in the reflection). Reflected images are not perfect replicas when different planes (distances) are involved.

23–3 Formation of Images by Spherical Mirrors

Reflecting surfaces can also be *curved*, usually *spherical*, which means they form a section of a sphere. A **spherical mirror** is called **convex** if the reflection takes place on the outer surface of the spherical shape so that the center of the mirror surface bulges out toward the viewer, Fig. 23–10a. A mirror is called **concave** if the reflecting surface is on the inner surface of the sphere so that the mirror surface curves away from the viewer (like a "cave"), Fig. 23–10b. Concave mirrors are used as shaving or cosmetic mirrors (**magnifying mirrors**), Fig. 23–11a, because they magnify. Convex mirrors are sometimes used on cars and trucks (rearview mirrors) and in shops (to watch for theft), because they take in a wide field of view, Fig. 23–11b.

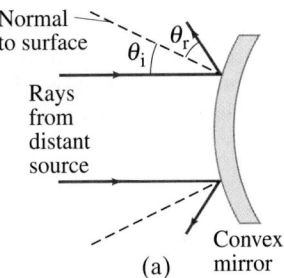

Normal to surface

θ_i θ_r

Rays from distant source

(a) Convex mirror

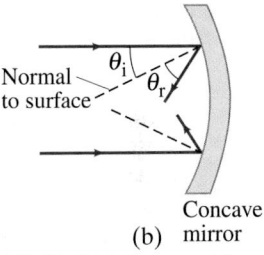

Normal to surface

θ_i θ_r

(b) Concave mirror

FIGURE 23–10 Mirrors with convex and concave spherical surfaces. Note that $\theta_r = \theta_i$ for each ray. (The dashed lines are perpendicular to the mirror surface at each point shown.)

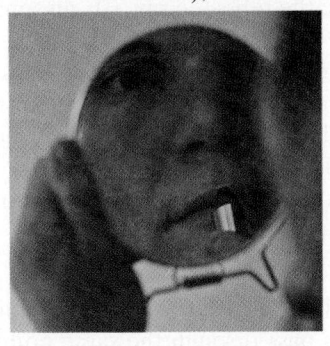

(a)

(b)

FIGURE 23–11 (a) A concave cosmetic mirror gives a magnified image. (b) A convex mirror in a store reduces image size and so includes a wide field of view.

Focal Point and Focal Length

To see how spherical mirrors form images, we first consider an object that is very far from a concave mirror. For a distant object, as shown in Fig. 23–12, the rays from each point on the object that strike the mirror will be nearly parallel. *For an object infinitely far away* (the Sun and stars approach this), *the rays would be precisely parallel.*

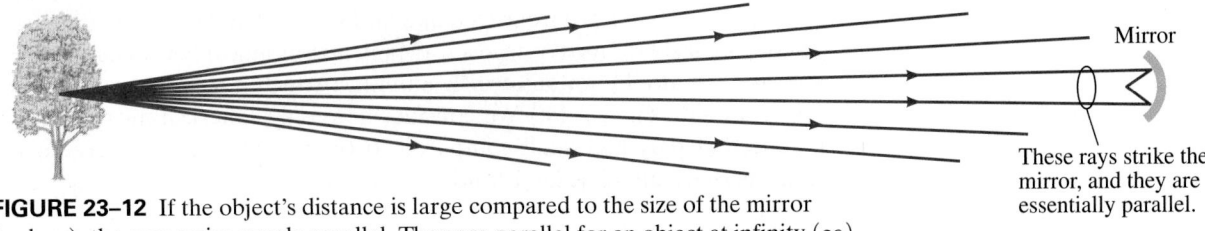

Mirror

These rays strike the mirror, and they are essentially parallel.

FIGURE 23–12 If the object's distance is large compared to the size of the mirror (or lens), the rays arrive nearly parallel. They are parallel for an object at infinity (∞).

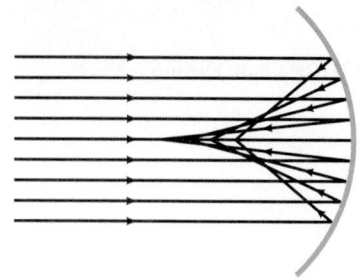

FIGURE 23–13 Parallel rays striking a concave spherical mirror do not intersect (or focus) at precisely a single point. (This "defect" is referred to as "spherical aberration.")

Now consider such parallel rays falling on a concave mirror as in Fig. 23–13. The law of reflection holds for each of these rays at the point each strikes the mirror. As can be seen, they are not all brought to a single point. In order to form a sharp image, the rays must come to a point. Thus a spherical mirror will not make as sharp an image as a plane mirror will. However, as we show below, if the mirror is small compared to its radius of curvature, so that a reflected ray makes only a *small angle* with the incident ray (2θ in Fig. 23–14), then the rays will cross each other at very nearly a single point, or **focus**. In the case shown in Fig. 23–14, the incoming rays are parallel to the **principal axis**, which is defined as the straight line perpendicular to the curved surface at its center (line CA in Fig. 23–14). The point F, where incident parallel rays come to a focus after reflection, is called the **focal point** of the mirror. The distance between F and the center of the mirror, length FA, is called the **focal length**, f, of the mirror. The focal point is also the *image point for an object infinitely far away* along the principal axis. The image of the Sun, for example, would be at F.

FIGURE 23–14 Rays parallel to the principal axis of a concave spherical mirror come to a focus at F, the focal point, as long as the mirror is small in width as compared to its radius of curvature, r, so that the rays are "paraxial"— that is, make only small angles with the horizontal axis.

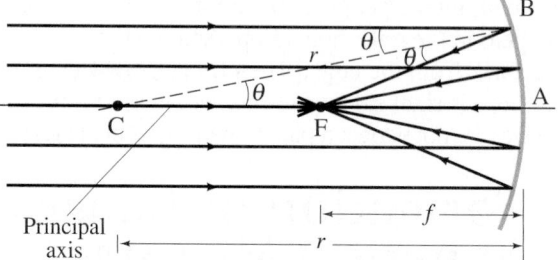

Now we will show, for a mirror whose reflecting surface is small compared to its radius of curvature, that the rays very nearly meet at a common point, F, and we will also determine the focal length f. In this approximation, we consider only rays that make a small angle with the principal axis; such rays are called **paraxial rays**, and their angles are exaggerated in Fig. 23–14 to make the labels clear. First we consider a ray that strikes the mirror at B in Fig. 23–14. The point C is the center of curvature of the mirror (the center of the sphere of which the mirror is a part). So the dashed line CB is equal to r, the radius of curvature, and CB is normal to the mirror's surface at B. The incoming ray that hits the mirror at B makes an angle θ with this normal, and hence the reflected ray, BF, also makes an angle θ with the normal (law of reflection). The angle BCF is also θ, as shown. The triangle CBF is isosceles because two of its angles are equal. Thus length CF = FB. We assume the mirror surface is small compared to the mirror's radius of curvature, so the angles are small, and the length FB is nearly equal to length FA. In this approximation, FA = FC. But FA = f, the focal length, and CA = $2 \times$ FA = r. Thus the focal length is half the radius of curvature:

$$f = \frac{r}{2}. \qquad \text{[spherical mirror]} \quad \textbf{(23–1)}$$

We assumed only that the angle θ was small, so this result applies for all other incident paraxial rays. Thus all paraxial rays pass through the same point F, the focal point.

Since it is only approximately true that the rays come to a perfect focus at F, the more curved the mirror, the worse the approximation (Fig. 23–13) and the more blurred the image. This "defect" of spherical mirrors is called **spherical aberration**; we will discuss it more with regard to lenses in Chapter 25. A **parabolic reflector**, on the other hand, will reflect the rays to a perfect focus. However, because parabolic shapes are much harder to make and thus much more expensive, spherical mirrors are used for most purposes. (Many astronomical telescopes use parabolic reflectors, as do TV satellite dish antennas which concentrate radio waves to nearly a point, Fig. 22–19.) We consider here only spherical mirrors and we will assume that they are small compared to their radius of curvature so that the image is sharp and Eq. 23–1 holds.

Image Formation—Ray Diagrams

We saw that for an object at infinity, the image is located at the focal point of a concave spherical mirror, where $f = r/2$. But where does the image lie for an object not at infinity? First consider the object shown as an arrow in Fig. 23–15a, which is placed between F and C at point O (O for object). Let us determine where the image will be for a given point O′ at the top of the object, by finding the point where rays drawn from the tip of the arrow converge after reflecting from the mirror. To do this we can draw several rays and make sure these reflect from the mirror such that the angle of reflection equals the angle of incidence.

(a) Ray 1 goes out from O′ parallel to the axis and reflects through F.

(b) Ray 2 goes through F and then reflects back parallel to the axis.

(c) Ray 3 is perpendicular to mirror, and so must reflect back on itself and go through C (center of curvature).

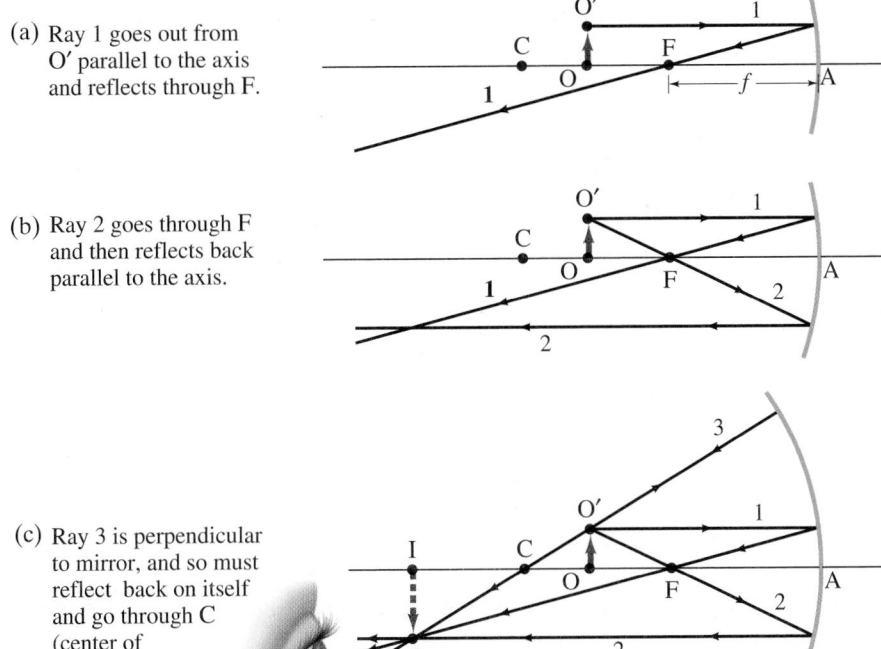

Diverging rays heading toward eye

FIGURE 23–15 Rays leave point O′ on the object (an arrow). Shown are the three most useful rays for determining where the image I′ is formed. [Note that our mirror is not small compared to f, so our diagram will not give the precise position of the image.]

Many rays could be drawn leaving any point on an object, but determining the image position is faster if we deal with three particular rays. These are the rays labeled 1, 2, and 3 in Fig. 23–15 and we draw them leaving object point O′ as follows:

Ray 1 leaving O′ is drawn parallel to the axis; therefore after reflection it must pass along a line through F, Fig. 23–15a (just as parallel rays did in Fig. 23–14).

Ray 2 leaves O′ and is made to pass through F (Fig. 23–15b); therefore it must reflect so it is parallel to the axis. (In reverse, a parallel ray passes through F.)

Ray 3 is drawn along a radius of the spherical surface (Fig. 23–15c) and is perpendicular to the mirror, so it is reflected back on itself and passes through C, the center of curvature.

All three rays leave a single point O′ on the object. After reflection from a (small) mirror, the point at which these rays cross is the image point I′. All other rays from the same object point will also pass through this image point. To find the image point for any object point, only these three types of rays need to be drawn. Only two of these rays are needed, but the third serves as a check.

We have shown the image point in Fig. 23–15 only for a single point on the object. Other points on the object are imaged nearby. For instance, the bottom of the arrow, on the principal axis at point O, is imaged on the axis at point I. So a complete image of the object is formed (dashed arrow in Fig. 23–15c). Because the light actually passes through the image, this is a **real image** that will appear on a white surface or film placed there. This can be compared to the virtual image formed by a plane mirror (the light does not pass through that image, Fig. 23–7).

The image in Fig. 23–15 can be seen by the eye only when the eye is placed to the left of the image, so that some of the rays *diverging* from each point on the image (as point I′) can enter the eye as shown in Fig. 23–15c (just as in Figs. 23–1 and 23–7).

RAY DIAGRAM

Finding the image position for a curved mirror

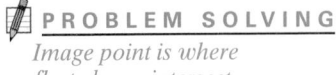 **PROBLEM SOLVING**

Image point is where reflected rays intersect

Mirror Equation and Magnification

Image points can be determined, roughly, by drawing the three rays as just described, Fig. 23–15. But it is difficult to draw small angles for the "paraxial" rays as we assumed. For more accurate results, we now derive an equation that gives the image distance if the object distance and radius of curvature of the mirror are known. To do this, we refer to Fig. 23–16. The **object distance**, d_o, is the distance of the object (point O) from the center of the mirror. The **image distance**, d_i, is the distance of the image (point I) from the center of the mirror. The height of the object OO′ is called h_o and the height of the image, I′I, is h_i. Two rays leaving O′ are shown: O′FBI′ (same as ray 2 in Fig. 23–15) and O′AI′, which is a fourth type of ray that reflects at the center of the mirror and can also be used to find an image point.

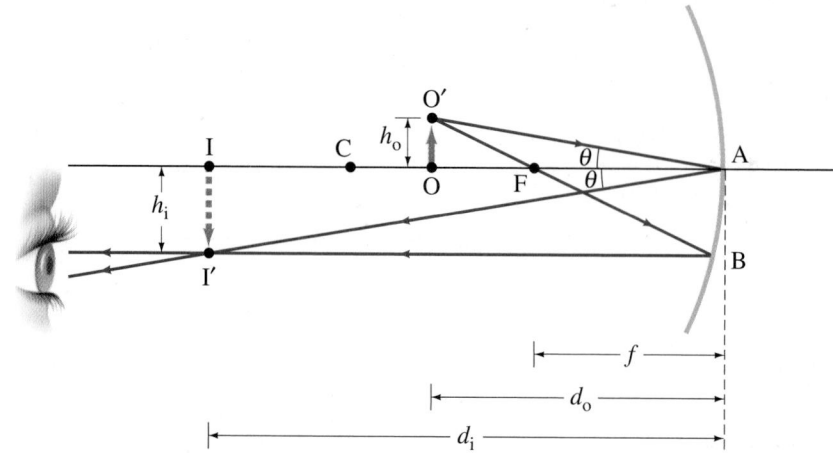

FIGURE 23–16 Diagram for deriving the mirror equation. For the derivation, we assume the mirror size is small compared to its radius of curvature.

The ray O′AI′ obeys the law of reflection, so the two right triangles O′AO and I′AI are similar. Therefore, we have

$$\frac{h_o}{h_i} = \frac{d_o}{d_i}.$$

For the other ray shown, O′FBI′, the triangles O′FO and AFB are also similar because the angles at F are equal and we use the approximation AB = h_i (mirror small compared to its radius). Furthermore FA = f, the focal length of the mirror, so

$$\frac{h_o}{h_i} = \frac{OF}{FA} = \frac{d_o - f}{f}.$$

The left sides of the two preceding expressions are the same, so we can equate the right sides:

$$\frac{d_o}{d_i} = \frac{d_o - f}{f}.$$

We now divide both sides by d_o and rearrange to obtain

Mirror equation
$$\frac{1}{d_o} + \frac{1}{d_i} = \frac{1}{f}. \qquad\qquad \textbf{(23–2)}$$

This is the equation we were seeking. It is called the **mirror equation** and relates the object and image distances to the focal length f (where $f = r/2$).

The mirror equation also holds for a plane mirror: the focal length is $f = r/2 = \infty$ (Eq. 23–1), and Eq. 23–2 gives $d_i = -d_o$.

The **magnification**, m, of a mirror is defined as the height of the image divided by the height of the object. From our first set of similar triangles in Fig. 23–16, or the first equation just below Fig. 23–16, we can write:

$$m = \frac{h_i}{h_o} = -\frac{d_i}{d_o}. \qquad (23\text{–}3)$$

The minus sign in Eq. 23–3 is inserted as a convention. Indeed, we must be careful about the signs of all quantities in Eqs. 23–2 and 23–3. Sign conventions are chosen so as to give the correct locations and orientations of images, as predicted by ray diagrams. The **sign conventions** we use are:

1. the image height h_i is positive if the image is upright, and negative if inverted, relative to the object (assuming h_o is taken as positive);

2. d_i or d_o is positive if image or object is in front of the mirror (as in Fig. 23–16); if either image or object is behind the mirror, the corresponding distance is negative. [An example of $d_i < 0$ can be seen in Fig. 23–17, Example 23–6.]†

Thus the magnification (Eq. 23–3) is positive for an upright image and negative for an inverted image (upside down). We summarize sign conventions more fully in the Problem Solving Strategy following our discussion of convex mirrors later in this Section.

PROBLEM SOLVING
Sign conventions for mirrors

Concave Mirror Examples

EXAMPLE 23–4 | **Image in a concave mirror.** A 1.50-cm-high object is placed 20.0 cm from a concave mirror with radius of curvature 30.0 cm. Determine (*a*) the position of the image, and (*b*) its size.

APPROACH We determine the focal length from the radius of curvature (Eq. 23–1), $f = r/2 = 15.0\,\text{cm}$. The ray diagram is basically the same as Fig. 23–16, since the object is between F and C. The position and size of the image are found from Eqs. 23–2 and 23–3.

SOLUTION Referring to Fig. 23–16, we have CA = r = 30.0 cm, FA = f = 15.0 cm, and OA = d_o = 20.0 cm.

(*a*) We start with the mirror equation, Eq. 23–2, rearranging it (subtracting $(1/d_o)$ from both sides):

$$\frac{1}{d_i} = \frac{1}{f} - \frac{1}{d_o} = \frac{1}{15.0\,\text{cm}} - \frac{1}{20.0\,\text{cm}} = 0.0167\,\text{cm}^{-1}.$$

So $d_i = 1/(0.0167\,\text{cm}^{-1}) = 60.0\,\text{cm}$. Because d_i is positive, the image is 60.0 cm in front of the mirror, on the same side as the object.

CAUTION
Remember to take the reciprocal

(*b*) From Eq. 23–3, the magnification is

$$m = -\frac{d_i}{d_o} = -\frac{60.0\,\text{cm}}{20.0\,\text{cm}} = -3.00.$$

The image is 3.0 times larger than the object, and its height is

$$h_i = mh_o = (-3.00)(1.5\,\text{cm}) = -4.5\,\text{cm}.$$

The minus sign reminds us that the image is inverted, as shown in Fig. 23–16.

NOTE When an object is further from a concave mirror than the focal point, we can see from Fig. 23–15 or 23–16 that the image is always inverted and real.

CONCEPTUAL EXAMPLE 23–5 | **Reversible rays.** If the object in Example 23–4 is placed instead where the image is (see Fig. 23–16), where will the new image be?

RESPONSE The mirror equation is *symmetric* in d_o and d_i. Thus the new image will be where the old object was. Indeed, in Fig. 23–16 we need only reverse the direction of the rays to get our new situation.

---

†d_o is always positive for a real object; $d_o < 0$ can happen only if the object is an image formed by another mirror or lens—see Example 23–16.

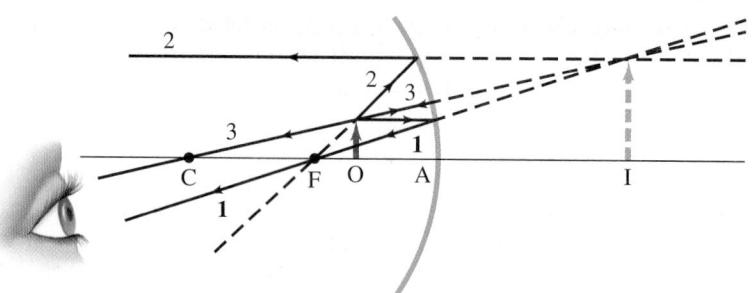

FIGURE 23–17 Object placed within the focal point F. The image is *behind* the mirror and is *virtual*, Example 23–6. [Note that the vertical scale (height of object = 1.0 cm) is different from the horizontal (OA = 10.0 cm) for ease of drawing, and reduces the precision of the drawing.]

EXAMPLE 23–6 **Object closer to concave mirror than focal point.**
A 1.00-cm-high object is placed 10.0 cm from a concave mirror whose radius of curvature is 30.0 cm. (*a*) Draw a ray diagram to locate (approximately) the position of the image. (*b*) Determine the position of the image and the magnification analytically.

APPROACH We draw the ray diagram using the rays as in Fig. 23–15, page 651. An analytic solution uses Eqs. 23–1, 23–2, and 23–3.

SOLUTION (*a*) Since $f = r/2 = 15.0$ cm, the object is between the mirror and the focal point. We draw the three rays as described earlier (Fig. 23–15); they are shown leaving the tip of the object in Fig. 23–17. Ray 1 leaves the tip of our object heading toward the mirror parallel to the axis, and reflects through F. Ray 2 cannot head toward F because it would not strike the mirror; so ray 2 must point as if it started at F (dashed line in Fig. 23–17) and heads to the mirror, and then is reflected parallel to the principal axis. Ray 3 is perpendicular to the mirror and reflects back on itself. The rays reflected from the mirror diverge and so never meet at a point. They appear to be coming from a point behind the mirror (dashed lines). This point locates the image of the tip of the arrow. The image is thus behind the mirror and is *virtual*.

(*b*) We use Eq. 23–2 to find d_i when $d_o = 10.0$ cm:

$$\frac{1}{d_i} = \frac{1}{f} - \frac{1}{d_o} = \frac{1}{15.0 \text{ cm}} - \frac{1}{10.0 \text{ cm}} = \frac{2 - 3}{30.0 \text{ cm}} = -\frac{1}{30.0 \text{ cm}}.$$

Therefore, $d_i = -30.0$ cm. The minus sign means the image is behind the mirror, which our diagram also showed us. The magnification is $m = -d_i/d_o = -(-30.0 \text{ cm})/(10.0 \text{ cm}) = +3.00$. So the image is 3.00 times larger than the object. The plus sign indicates that the image is upright (same as object), which is consistent with the ray diagram, Fig. 23–17.

NOTE The image distance cannot be obtained accurately by measuring on Fig. 23–17, because our diagram violates the paraxial ray assumption (we draw rays at steeper angles to make them clearly visible).

NOTE When the object is located inside the focal point of a concave mirror $(d_o < f)$, the image is always upright and virtual. If the object O in Fig. 23–17 is you, you see yourself clearly, because the reflected rays at point O (you) are diverging. Your image is upright and enlarged. This is how a shaving or cosmetic mirror is used—you must place your head closer to the mirror than the focal point if you are to see yourself right-side up (see the photograph, Fig. 23–11a). [If the object is *beyond* the focal point, as in Fig. 23–15, the image is real and inverted: upside down—and hard to use!]

Seeing the Image; Seeing Yourself

For a person's eye to see a sharp image, the eye must be at a place where it intercepts diverging rays from points on the image, as is the case for the eye's position in Figs. 23–15, 23–16, and 23–17. When we look at normal objects, we always detect rays diverging toward the eye as shown in Fig. 23–1. (Or, for very distant objects like stars, the rays become essentially parallel, as in Fig. 23–12.)

If you placed your eye between points O and I in Fig. 23–16, for example, *converging* rays from the object OO′ would enter your eye and the lens of your eye could not bring them to a focus; you would see a blurry image or no perceptible image at all. [We will discuss the eye more in Chapter 25.]

If *you* are the object OO′ in Fig. 23–16, situated between F and C, and are trying to see yourself in the mirror, you would see a blur; but the person whose eye is shown in Fig. 23–16 could see you clearly. If you are to the left of C in Fig. 23–16, where $d_o > 2f$, you can see yourself clearly, but upside down. Why? Because then the rays arriving from the image will be *diverging* at your position (Fig. 23–18), and your eye can then focus them. You can also see yourself clearly, and right side up, if you are closer to the mirror than its focal point $(d_o < f)$, as we saw in Example 23–6, Fig. 23–17.

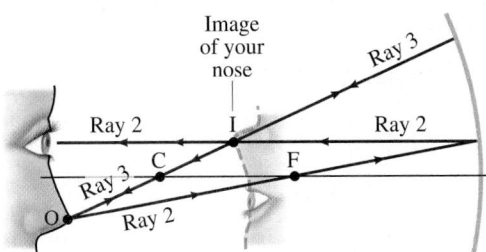

FIGURE 23–18 You can see a clear inverted image of your face in a concave mirror when you are beyond C $(d_o > 2f)$, because the rays that arrive at your eye are *diverging*. Standard rays 2 and 3 are shown leaving point O on your nose. Ray 2 (and other nearby rays) enters your eye. Notice that rays are diverging as they move to the left of image point I.

Convex Mirrors

The analysis used for concave mirrors can be applied to **convex** mirrors. Even the mirror equation (Eq. 23–2) holds for a convex mirror, although the quantities involved must be carefully defined. Figure 23–19a shows parallel rays falling on a convex mirror. Again spherical aberration is significant (Fig. 23–13), unless we assume the mirror is small compared to its radius of curvature. The reflected rays diverge, but seem to come from point F behind the mirror, Fig. 23–19a. This is the **focal point**, and its distance from the center of the mirror (point A) is the **focal length**, f. The equation $f = r/2$ is valid also for a convex mirror. We see that an object at infinity produces a virtual image in a convex mirror. Indeed, no matter where the object is placed on the reflecting side of a convex mirror, the image will be virtual and upright, as indicated in Fig. 23–19b. To find the image we draw rays 1 and 3 according to the rules used before on the concave mirror, as shown in Fig. 23–19b. Note that although rays 1 and 3 don't actually pass through points F and C, the line along which each is drawn does (shown dashed).

The mirror equation, Eq. 23–2, holds for convex mirrors but the focal length f and radius of curvature must be considered negative. The proof is left as a Problem. It is also left as a Problem to show that Eq. 23–3 for the magnification is also valid.

FIGURE 23–19 Convex mirror: (a) the focal point is at F, behind the mirror; (b) the image I of the object at O is virtual, upright, and smaller than the object. [Not to scale for Example 23–7.]

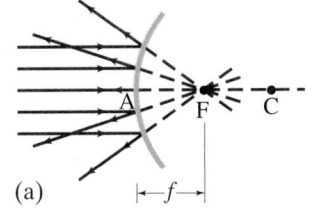

(a)

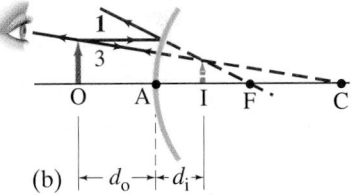

(b)

FIGURE 23–20 Example 23–7.

EXAMPLE 23–7 | **Convex rearview mirror.** An external rearview car mirror is convex with a radius of curvature of 16.0 m (Fig. 23–20). Determine the location of the image and its magnification for an object 10.0 m from the mirror.

APPROACH We follow the steps of the Problem Solving Strategy explicitly.

SOLUTION

1. **Draw a ray diagram.** The ray diagram will be like Fig. 23–19b, but the large object distance $(d_o = 10.0 \text{ m})$ makes a precise drawing difficult. We have a convex mirror, so r is negative by convention.

2. **Mirror and magnification equations.** The center of curvature of a convex mirror is behind the mirror, as is its focal point, so we set $r = -16.0$ m so that the focal length is $f = r/2 = -8.0$ m. The object is in front of the mirror, $d_o = 10.0$ m. Solving the mirror equation, Eq. 23–2, for $1/d_i$ gives

$$\frac{1}{d_i} = \frac{1}{f} - \frac{1}{d_o} = \frac{1}{-8.0 \text{ m}} - \frac{1}{10.0 \text{ m}} = \frac{-10.0 - 8.0}{80.0 \text{ m}} = -\frac{18}{80.0 \text{ m}}.$$

Thus $d_i = -80.0 \text{ m}/18 = -4.4$ m. Equation 23–3 gives the magnification

$$m = -\frac{d_i}{d_o} = -\frac{(-4.4 \text{ m})}{(10.0 \text{ m})} = +0.44.$$

3. **Sign conventions.** The image distance is negative, -4.4 m, so the image is *behind* the mirror. The magnification is $m = +0.44$, so the image is *upright* (same orientation as object, which is useful) and about half what it would be in a plane mirror.

4. **Check.** Our results are consistent with Fig. 23–19b.

Convex rearview mirrors on vehicles sometimes come with a warning that objects are closer than they appear in the mirror. The fact that d_i may be smaller than d_o (as in Example 23–7) seems to contradict this observation. The real reason the object seems farther away is that its image in the convex mirror is *smaller* than it would be in a plane mirror, and we judge distance of ordinary objects such as other cars mostly by their size.

23–4 Index of Refraction

We saw in Chapter 22 that the speed of light in vacuum (like other EM waves) is

$$c = 2.99792458 \times 10^8 \text{ m/s},$$

which is usually rounded off to

$$3.00 \times 10^8 \text{ m/s}$$

when extremely precise results are not required.

In air, the speed is only slightly less. In other transparent materials, such as glass and water, the speed is always less than that in vacuum. For example, in water light travels at about $\frac{3}{4}c$. The ratio of the speed of light in vacuum to the speed v in a given material is called the **index of refraction**, n, of that material:

$$n = \frac{c}{v}. \tag{23–4}$$

The index of refraction is never less than 1, and values for various materials are given in Table 23–1. For example, since $n = 1.33$ for water, the speed of light in water is

$$v = \frac{c}{n} = \frac{(3.00 \times 10^8 \text{ m/s})}{1.33} = 2.26 \times 10^8 \text{ m/s}.$$

As we shall see later, n varies somewhat with the wavelength of the light—except in vacuum—so a particular wavelength is specified in Table 23–1, that of yellow light with wavelength $\lambda = 589$ nm.

That light travels more slowly in matter than in vacuum can be explained at the atomic level as being due to the absorption and reemission of light by atoms and molecules of the material.

TABLE 23–1 Indices of Refraction†

Material	$n = \dfrac{c}{v}$
Vacuum	1.0000
Air (at STP)	1.0003
Water	1.33
Ethyl alcohol	1.36
Glass	
Fused quartz	1.46
Crown glass	1.52
Light flint	1.58
Plastic	
Acrylic, Lucite, CR-39	1.50
Polycarbonate	1.59
"High-index"	1.6–1.7
Sodium chloride	1.53
Diamond	2.42

†$\lambda = 589$ nm.

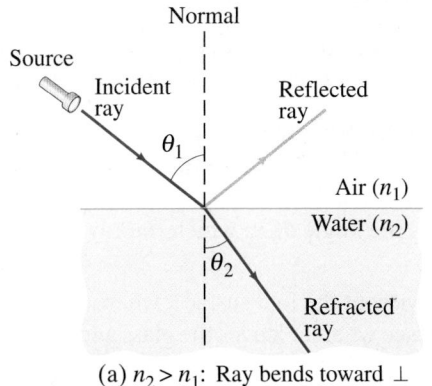

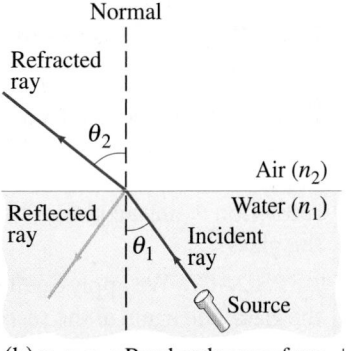

(a) $n_2 > n_1$: Ray bends toward ⊥ (b) $n_1 > n_2$: Ray bends away from ⊥

FIGURE 23–21 Refraction.
(a) Light refracted when passing from air (n_1) into water (n_2): $n_2 > n_1$.
(b) Light refracted when passing from water (n_1) into air (n_2): $n_1 > n_2$.

23–5 Refraction: Snell's Law

When light passes from one transparent medium into another with a different index of refraction, some or all of the incident light is reflected at the boundary. The rest passes into the new medium. If a ray of light is incident at an angle to the surface (other than perpendicular), the ray changes direction as it enters the new medium. This change in direction, or bending, of the light ray is called **refraction**.

Figure 23–21a shows a ray passing from air into water. Angle θ_1 is the angle the incident ray makes with the normal (perpendicular) to the surface and is called the **angle of incidence**. Angle θ_2 is the **angle of refraction**, the angle the refracted ray makes with the normal to the surface. Notice that the ray bends toward the normal when entering the water. This is always the case when the ray enters a medium where the speed of light is *less* (and the index of refraction is greater, Eq. 23–4). If light travels from one medium into a second where its speed is *greater*, the ray bends away from the normal; this is shown in Fig. 23–21b for a ray traveling from water to air.

> ◇ **CAUTION**
> *Angles of incidence and refraction are measured from the perpendicular, not from the surface*

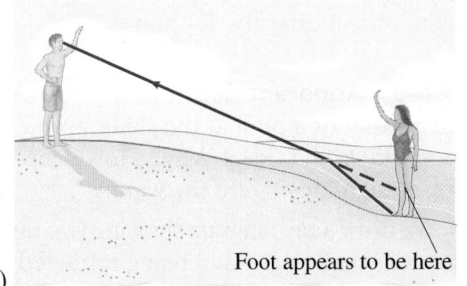

Foot appears to be here

(a)   (b)

FIGURE 23–22 (a) Photograph, and (b) ray diagram showing why a person's legs look shorter standing in water: a ray from the bather's foot to the observer's eye bends at the water's surface, and our brain interprets the light as traveling in a straight line, from higher up (dashed line).

Refraction is responsible for a number of common optical illusions. For example, a person standing in waist-deep water appears to have shortened legs (Fig. 23–22). The rays leaving the person's foot are bent at the surface. The observer's brain assumes the rays to have traveled a straight-line path (dashed red line), and so the feet appear to be higher than they really are. Similarly, when you put a straw in water, it appears to be bent (Fig. 23–23). This also means that water is deeper than it appears.

FIGURE 23–23 A straw in water looks bent even when it isn't.

Snell's Law

The angle of refraction depends on the speed of light in the two media and on the incident angle. An analytic relation between θ_1 and θ_2 in Fig. 23–21 was arrived at experimentally about 1621 by Willebrord Snell (1591–1626). Known as **Snell's law**, it is written:

$$n_1 \sin \theta_1 = n_2 \sin \theta_2. \qquad \textbf{(23–5)}$$

θ_1 is the angle of incidence and θ_2 is the angle of refraction; n_1 and n_2 are the respective indices of refraction in the materials. See Fig. 23–21. The incident and refracted rays lie in the same plane, which also includes the perpendicular to the surface. Snell's law is the **law of refraction**. (Snell's law was derived in Section 11–13 for water waves where Eq. 11–20 is just a combination of Eqs. 23–5 and 23–4, and we derive it again in Chapter 24 using the wave theory of light.)

Snell's law shows that if $n_2 > n_1$, then $\theta_2 < \theta_1$. Thus, if light enters a medium where n is greater (and its speed is less), the ray is bent toward the normal. And if $n_2 < n_1$, then $\theta_2 > \theta_1$, so the ray bends away from the normal. See Fig. 23–21.

> SNELL'S LAW
> (LAW OF REFRACTION)

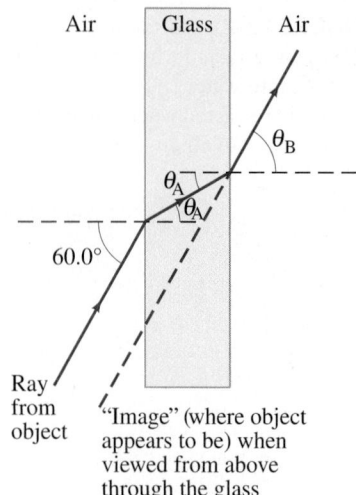

Air Glass Air

θ_B

θ_A
θ_A

60.0°

Ray from object

"Image" (where object appears to be) when viewed from above through the glass

FIGURE 23–24 Light passing through a piece of glass (Example 23–8).

EXERCISE C Light passes from a medium with $n = 1.3$ (water) into a medium with $n = 1.5$ (glass). Is the light bent toward or away from the perpendicular to the interface?

EXAMPLE 23–8 **Refraction through flat glass.** Light traveling in air strikes a flat piece of uniformly thick glass at an incident angle of 60.0°, as shown in Fig. 23–24. If the index of refraction of the glass is 1.50, (a) what is the angle of refraction θ_A in the glass; (b) what is the angle θ_B at which the ray emerges from the glass?

APPROACH We apply Snell's law twice: at the first surface, where the light enters the glass, and again at the second surface where it leaves the glass and enters the air.

SOLUTION (a) The incident ray is in air, so $n_1 = 1.00$ and $n_2 = 1.50$. Applying Snell's law where the light enters the glass $(\theta_1 = 60.0°,\ \theta_2 = \theta_A)$ gives

$$(1.00)\sin 60.0° = (1.50)\sin\theta_A$$

or

$$\sin\theta_A = \frac{1.00}{1.50}\sin 60.0° = 0.5774,$$

and $\theta_A = 35.3°$.

(b) Since the faces of the glass are parallel, the incident angle at the second surface is also θ_A (geometry), so $\sin\theta_A = 0.5774$. At this second interface, $n_1 = 1.50$ and $n_2 = 1.00$. Thus the ray re-enters the air at an angle θ_B given by

$$\sin\theta_B = \frac{1.50}{1.00}\sin\theta_A = 0.866,$$

and $\theta_B = 60.0°$. The direction of a light ray is thus unchanged by passing through a flat piece of glass of uniform thickness.

NOTE This result is valid for any angle of incidence. The ray is displaced slightly to one side, however. You can observe this by looking through a piece of glass (near its edge) at some object and then moving your head to the side slightly so that you see the object directly. It "jumps."

⚠ **CAUTION** **(real life)**

Water is deeper than it looks

FIGURE 23–25 Example 23–9.

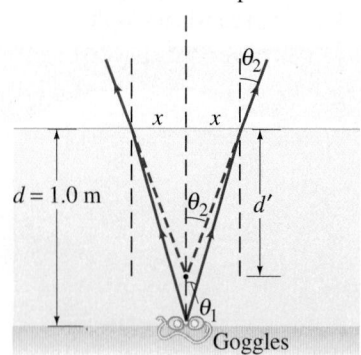

$d = 1.0$ m

θ_2

x | x

θ_2

d'

θ_1

Goggles

EXAMPLE 23–9 **Apparent depth of a pool.** A swimmer has dropped her goggles to the bottom of a pool at the shallow end, marked as 1.0 m deep. But the goggles don't look that deep. Why? How deep do the goggles appear to be when you look straight down into the water?

APPROACH We draw a ray diagram showing two rays going upward from a point on the goggles at a small angle, and being refracted at the water's (flat) surface, Fig. 23–25. The two rays traveling upward from the goggles are refracted *away* from the normal as they exit the water, and so appear to be diverging from a point above the goggles (dashed lines), which is why the water seems less deep than it actually is. We are looking straight down, so all angles are small (but exaggerated in Fig. 23–25 for clarity).

SOLUTION To calculate the apparent depth d' (Fig. 23–25), given a real depth $d = 1.0$ m, we use Snell's law with $n_1 = 1.33$ for water and $n_2 = 1.0$ for air:

$$\sin\theta_2 = n_1\sin\theta_1.$$

We are considering only small angles, so $\sin\theta \approx \tan\theta \approx \theta$, with θ in radians. So Snell's law becomes

$$\theta_2 \approx n_1\theta_1.$$

From Fig. 23–25, we see that $\theta_2 \approx \tan\theta_2 = x/d'$ and $\theta_1 \approx \tan\theta_1 = x/d$. Putting these into Snell's law, $\theta_2 \approx n_1\theta_1$, we get

$$\frac{x}{d'} \approx n_1\frac{x}{d}$$

or

$$d' \approx \frac{d}{n_1} = \frac{1.0 \text{ m}}{1.33} = 0.75 \text{ m}.$$

The pool seems only three-fourths as deep as it actually is.

NOTE Water in general is deeper than it looks—a useful safety guideline.

23–6 Total Internal Reflection; Fiber Optics

When light passes from one material into a second material where the index of refraction is less (say, from water into air), the refracted light ray bends away from the normal, as for rays I and J in Fig. 23–26. At a particular incident angle, the angle of refraction will be 90°, and the refracted ray would skim the surface (ray K).

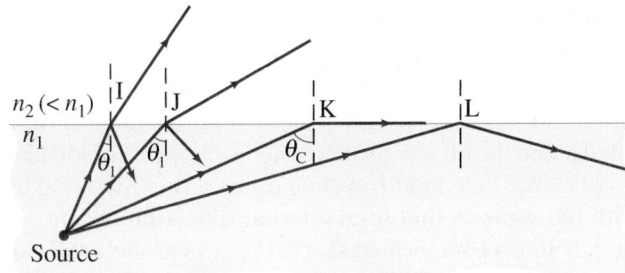

Source

FIGURE 23–26 Since $n_2 < n_1$, light rays are totally internally reflected if the incident angle $\theta_1 > \theta_C$, as for ray L. If $\theta_1 < \theta_C$, as for rays I and J, only a part of the light is reflected, and the rest is refracted.

The incident angle at which this occurs is called the **critical angle**, θ_C. From Snell's law, θ_C is given by

$$\sin \theta_C = \frac{n_2}{n_1} \sin 90° = \frac{n_2}{n_1}. \qquad (23\text{--}6)$$

For any incident angle less than θ_C, there will be a refracted ray, although part of the light will also be reflected at the boundary. However, for incident angles θ_1 greater than θ_C, Snell's law would tell us that $\sin \theta_2$ ($= n_1 \sin \theta_1 / n_2$) would be greater than 1.00 when $n_2 < n_1$. Yet the sine of an angle can never be greater than 1.00. In this case there is no refracted ray at all, and *all of the light is reflected*, as for ray L in Fig. 23–26. This effect is called **total internal reflection**. Total internal reflection occurs only when light strikes a boundary where the medium beyond has a *lower* index of refraction.

⚠ **CAUTION**

Total internal reflection (occurs only if refractive index is smaller beyond boundary)

CONCEPTUAL EXAMPLE 23–10 View up from under water. Describe what a person would see who looked up at the world from beneath the perfectly smooth surface of a lake or swimming pool.

RESPONSE For an air–water interface, the critical angle is given by

$$\sin \theta_C = \frac{1.00}{1.33} = 0.750.$$

Therefore, $\theta_C = 49°$. Thus the person would see the outside world compressed into a circle whose edge makes a 49° angle with the vertical. Beyond this angle, the person would see reflections from the sides and bottom of the lake or pool (Fig. 23–27).

| **EXERCISE D** Light traveling in air strikes a glass surface with $n = 1.48$. For what range of angles will total internal reflection occur?

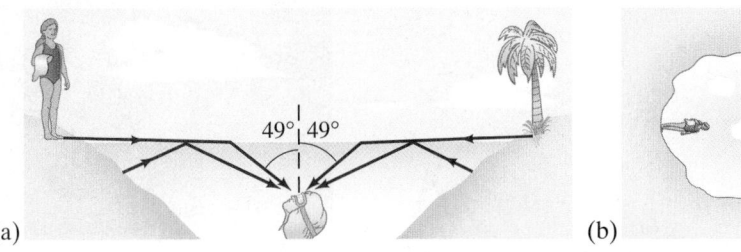

(a) 49° 49° (b)

FIGURE 23–27 (a) Light rays entering submerged person's eye, and (b) view looking upward from beneath the water (the surface of the water must be very smooth). Example 23–10.

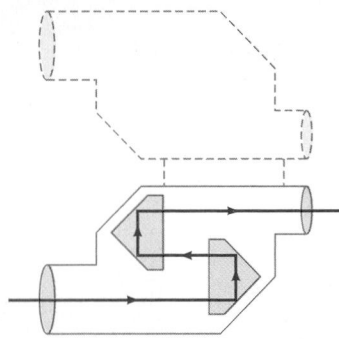

FIGURE 23–28 Total internal reflection of light by prisms in binoculars.

FIGURE 23–29 Light reflected totally at the interior surface of a glass or transparent plastic fiber.

Many optical instruments, such as binoculars, use total internal reflection within a prism to reflect light. The advantage is that very nearly 100% of the light is reflected, whereas even the best mirrors reflect somewhat less than 100%. Thus the image is brighter, especially after several reflections. For glass with $n = 1.50$, $\theta_C = 41.8°$. Therefore, 45° prisms will reflect all the light internally, if oriented as shown in the binoculars of Fig. 23–28.

| **EXERCISE E** What would happen if we immersed the 45° glass prisms in Fig. 23–28 in water?

Fiber Optics; Medical Instruments

Total internal reflection is the principle behind **fiber optics**. Glass and plastic fibers as thin as a few micrometers in diameter are commonly used. A bundle of such slender transparent fibers is called a **light pipe** or **fiber-optic cable**. Light[†] can be transmitted along the fiber with almost no loss because of total internal reflection. Figure 23–29 shows how light traveling down a thin fiber makes only glancing collisions with the walls so that total internal reflection occurs. Even if the light pipe is bent gently into a complicated shape, the critical angle still won't be exceeded, so light is transmitted practically undiminished to the other end. Very small losses do occur, mainly by reflection at the ends and absorption within the fiber.

Important applications of fiber-optic cables are in communications and medicine. They are used in place of wire to carry telephone calls, video signals, and computer data. The signal is a modulated light beam (a light beam whose intensity can be varied) and data is transmitted at a much higher rate and with less loss and less interference than an electrical signal in a copper wire. Fibers have been developed that can support over one hundred separate wavelengths, each modulated to carry more than 10 gigabits (10^{10} bits) of information per second. That amounts to a terabit (10^{12} bits) per second for one hundred wavelengths.

The use of fiber optics to transmit a clear picture is particularly useful in medicine, Fig. 23–30. For example, a patient's lungs can be examined by inserting a fiber-optic cable known as a bronchoscope through the mouth and down the bronchial tube. Light is sent down an outer set of fibers to illuminate the lungs. The reflected light returns up a central core set of fibers. Light directly in front of each fiber travels up that fiber. At the opposite end, a viewer sees a series of bright and dark spots, much like a TV screen—that is, a picture of what lies at the opposite end. Lenses are used at each end of the cable. The image may be viewed directly or on a monitor screen or film. The fibers must be optically insulated from one another, usually by a thin coating of material with index of refraction less than that of the fiber. The more fibers there are, and the smaller they are, the more detailed the picture. Such instruments, including bronchoscopes, colonoscopes (for viewing the colon), and endoscopes (stomach or other organs), are extremely useful for examining hard-to-reach places.

[†]Fiber-optic devices use not only visible light but also infrared light, ultraviolet light, and microwaves.

FIGURE 23–30 (a) How a fiber-optic image is made. (b) Example of a fiber-optic device inserted through the mouth to view the vocal cords, with the image on screen.

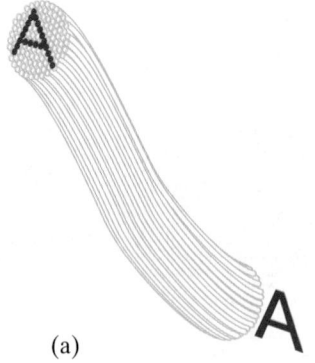

(a)

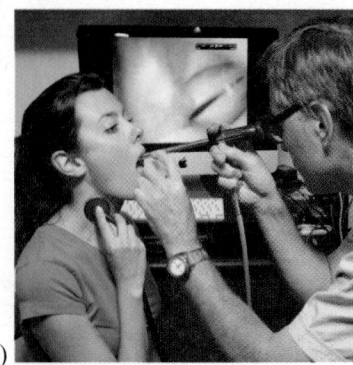

(b)

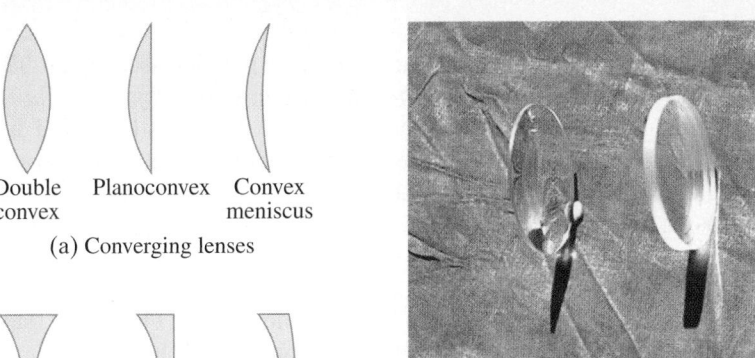

Double convex   Planoconvex   Convex meniscus

(a) Converging lenses

Double concave   Planoconcave   Concave meniscus

(b) Diverging lenses

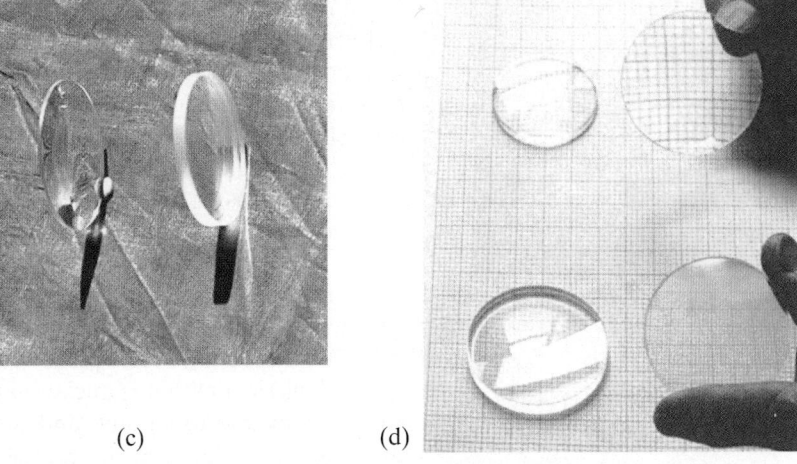

(c)   (d)

FIGURE 23–31 (a) Converging lenses and (b) diverging lenses, shown in cross section. Converging lenses are thicker at the center whereas diverging lenses are thicker at the edges. (c) Photo of a converging lens (on the left) and a diverging lens (right). (d) Converging lenses (above), and diverging lenses (below), lying flat, and raised off the paper to form images.

23–7 Thin Lenses; Ray Tracing

The most important simple optical device is the thin lens. The development of optical devices using lenses dates to the sixteenth and seventeenth centuries, although the earliest record of eyeglasses dates from the late thirteenth century. Today we find lenses in eyeglasses, cameras, magnifying glasses, telescopes, binoculars, microscopes, and medical instruments. A thin lens is usually circular, and its two faces are portions of a sphere. (Cylindrical faces are also possible, but we will concentrate on spherical.) The two faces can be concave, convex, or plane. Several types are shown in Figs. 23–31a and b in cross section. The importance of lenses is that they form images of objects—see Fig. 23–32.

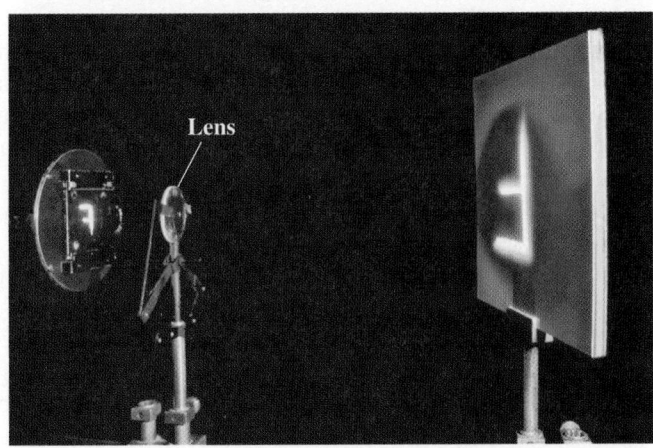

Lens

FIGURE 23–32 Converging lens (in holder) forms an image (large "F" on screen at right) of a bright object (illuminated "F" at the left).

Consider parallel rays striking the double convex lens shown in cross section in Fig. 23–33. We assume the lens is made of transparent material such as glass or transparent plastic with index of refraction greater than that of the air outside. The **axis** of a lens is a straight line passing through the center of the lens and perpendicular to its two surfaces (Fig. 23–33). From Snell's law, we can see that each ray in Fig. 23–33 is bent toward the axis when the ray enters the lens and again when it leaves the lens at the back surface. (Note the dashed lines indicating the normals to each surface for the top ray.) If rays parallel to the axis fall on a thin lens, they will be focused to a point called the **focal point**, F. This will not be precisely true for a lens with spherical surfaces. But it will be very nearly true—that is, parallel rays will be focused to a tiny region that is nearly a point—if the diameter of the lens is small compared to the radii of curvature of the two lens surfaces. This criterion is satisfied by a **thin lens**, one that is very thin compared to its diameter, and we consider only thin lenses here.

FIGURE 23–33 Parallel rays are brought to a focus by a converging thin lens.

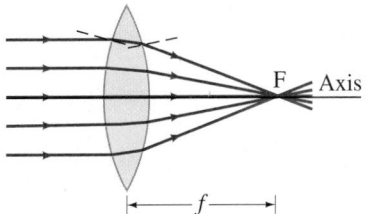

F   Axis

f

FIGURE 23–34 Image of the Sun burning wood.

FIGURE 23–35 Parallel rays at an angle are focused on the focal plane.

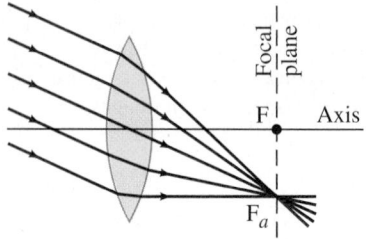

FIGURE 23–36 Diverging lens.

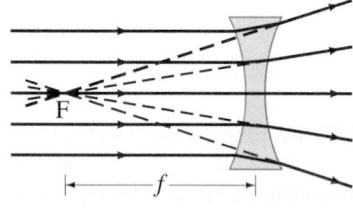

The rays from a point on a distant object are essentially parallel—see Fig. 23–12. Therefore we can say that *the focal point is the image point for an object at infinity on the lens axis*, Fig. 23–33. Thus, the focal point of a lens can be found by locating the point where the Sun's rays (or those from some other distant object) are brought to a sharp image, Fig. 23–34. The distance of the focal point from the center of the lens is called the **focal length**, *f*, Fig. 23–33. A lens can be turned around so that light can pass through it from the opposite side. The *focal length is the same on both sides*, as we shall see later, even if the curvatures of the two lens surfaces are different. If parallel rays fall on a lens at an angle, as in Fig. 23–35, they focus at a point F_a. The plane containing all focus points, such as F and F_a in Fig. 23–35, is called the **focal plane** of the lens.

Any lens (in air) that is thicker in the center than at the edges will make parallel rays converge to a point, and is called a **converging lens** (see Fig. 23–31a). Lenses that are thinner in the center than at the edges (Fig. 23–31b) are called **diverging lenses** because they make parallel light diverge, as shown in Fig. 23–36. The focal point, F, of a diverging lens is defined as that point from which refracted rays, originating from parallel incident rays, seem to emerge as shown in Fig. 23–36. And the distance from F to the center of the lens is called the **focal length**, *f*, just as for a converging lens.

EXERCISE F Return to Chapter-Opening Question 2, page 644, and answer it again now. Try to explain why you may have answered differently the first time.

Optometrists and ophthalmologists, instead of using the focal length, use the reciprocal of the focal length to specify the strength of eyeglass (or contact) lenses. This is called the **power**, *P*, of a lens:

$$P = \frac{1}{f}. \tag{23–7}$$

The unit for lens power is the **diopter** (D), which is an inverse meter: $1\,\text{D} = 1\,\text{m}^{-1}$. For example, a 20-cm-focal-length lens has a power $P = 1/(0.20\,\text{m}) = 5.0\,\text{D}$. We will mainly use the focal length, but we will refer again to the power of a lens when we discuss eyeglass lenses in Chapter 25.

The most important parameter of a lens is its focal length *f*, which is the same on both sides of the lens. For a converging lens, *f* can be measured by finding the image point for the Sun or other distant objects. Once *f* is known, the image position can be determined for any object. To find the image point by drawing rays would be difficult if we had to determine the refractive angles at the front surface of the lens and again at the back surface where the ray exits. We can save ourselves a lot of effort by making use of certain facts we already know, such as that a ray parallel to the axis of the lens passes (after refraction) through the focal point. To determine an image point, we can consider only the three rays indicated in Fig. 23–37, which uses an arrow (on the left) as the object, and a converging lens forming an image (dashed arrow) to the right. These rays, emanating from a single point on the object, are drawn as if the lens were infinitely thin, and we show only a single sharp bend at the center line of the lens instead of the refractions at each surface. These three rays are drawn as follows:

RAY DIAGRAM

Finding the image position formed by a thin lens

Ray 1 is drawn parallel to the axis, Fig. 23–37a; therefore it is refracted by the lens so that it passes along a line through the focal point F behind the lens.

Ray 2 is drawn to pass through the other focal point F′ (front side of lens in Fig. 23–37) and emerge from the lens parallel to the axis, Fig. 23–37b. (In reverse it would be a parallel ray going left and passing through F′.)

Ray 3 is directed toward the very center of the lens, where the two surfaces are essentially parallel to each other, Fig. 23–37c. This ray therefore emerges from the lens at the same angle as it entered. The ray would be displaced slightly to one side, as we saw in Example 23–8; but since we assume the lens is thin, we draw ray 3 straight through as shown.

The point where these three rays cross is the image point for that object point. Actually, any two of these rays will suffice to locate the image point, but drawing the third ray can serve as a check.

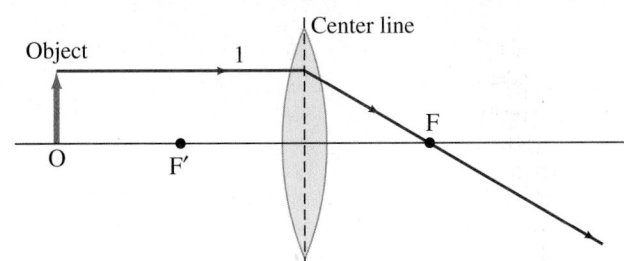

(a) Ray 1 leaves one point on object going parallel to the axis, then refracts through focal point behind the lens.

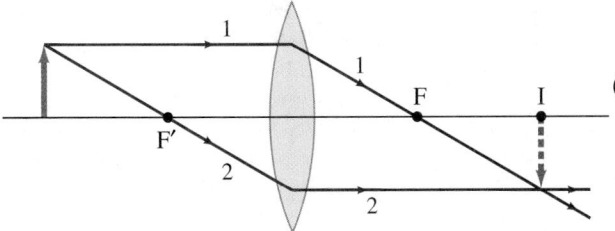

(b) Ray 2 passes through F' in front of the lens; therefore it is parallel to the axis behind the lens.

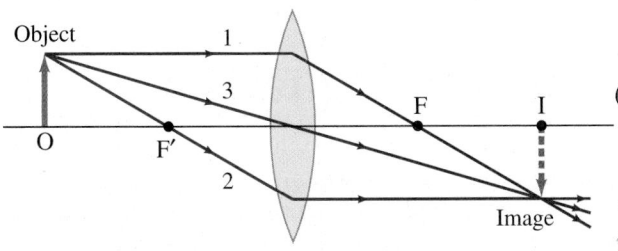

(c) Ray 3 passes straight through the center of the lens (assumed very thin).

FIGURE 23–37 Finding the image by ray tracing for a converging lens. Rays are shown leaving one point on the object (an arrow). Shown are the three most useful rays, leaving the tip of the object, for determining where the image of that point is formed. (Note that the focal points F and F' on either side of the lens are the same distance *f* from the center of the lens.)

Using these three rays for one object point, we can find the image point for that point of the object (the top of the arrow in Fig. 23–37). The image points for all other points on the object can be found similarly to determine the complete image of the object. Because the rays actually pass through the image for the case shown in Fig. 23–37, it is a **real image** (see pages 647 and/or 651). The image could be detected by film or electronic sensor, and actually be seen on a white surface or screen placed at the position of the image (Fig. 23–38).

FIGURE 23–38 (a) A converging lens can form a real image (here of a distant building, upside down) on a white wall. (b) That same real image is also directly visible to the eye. [Figure 23–31d shows images (graph paper) seen by the eye made by both diverging and converging lenses.]

| CONCEPTUAL EXAMPLE 23–11 | **Half-blocked lens.** What happens to the image of an object if the top half of a lens is covered by a piece of cardboard?

RESPONSE Let us look at the rays in Fig. 23–37. If the top half (or any half of the lens) is blocked, you might think that half the image is blocked. But in Fig. 23–37c, we see how the rays used to create the "top" of the image pass through both the top and the bottom of the lens. Only three of many rays are shown—many more rays pass through the lens, and they can form the image. You don't lose the image. But covering part of the lens cuts down on the total light received and reduces the brightness of the image.

NOTE If the lens is partially blocked by your thumb, you may notice an out of focus image of part of that thumb.

(a)

Seeing the Image

The image can also be seen directly by the eye when the eye is placed behind the image, as shown in Fig. 23–37c, so that some of the rays diverging from each point on the image can enter the eye. We can see a sharp image only for rays *diverging* from each point on the image, because we see normal objects when diverging rays from each point enter the eye as shown in Fig. 23–1. A normal eye cannot focus converging rays; if your eye was positioned between points F and I in Fig. 23–37c, it would not see a clear image. (More about our eyes in Section 25–2.) Figure 23–38 shows an image seen (a) on a white surface and (b) directly by the eye (and a camera) placed behind the image. The eye can see both real and virtual images (see next page) as long as the eye is positioned so rays diverging from the image enter it.

(b)

Diverging Lens

By drawing the same three rays emerging from a single object point, we can determine the image position formed by a diverging lens, as shown in Fig. 23–39. Note that ray 1 is drawn parallel to the axis, but does not pass through the focal point F′ behind the lens. Instead it seems to come (dashed line) from the focal point F in front of the lens. Ray 2 is directed toward F′ and is refracted parallel to the lens axis by the lens. Ray 3 passes directly through the center of the lens. The three refracted rays seem to emerge from a point on the left of the lens. This is the image point, I. Because the rays do not pass through the image, it is a **virtual image**. Note that the eye does not distinguish between real and virtual images—both are visible.

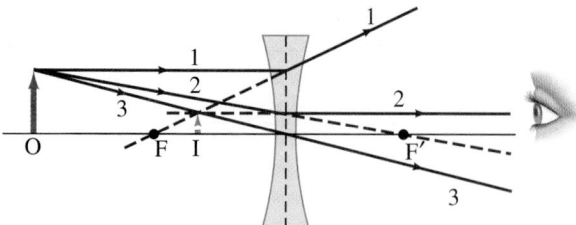

FIGURE 23–39 Finding the image by ray tracing for a diverging lens.

23–8 The Thin Lens Equation

We now derive an equation that relates the image distance to the object distance and the focal length of a thin lens. This equation will make the determination of image position quicker and more accurate than doing ray tracing. Let d_o be the object distance, the distance of the object from the center of the lens, and d_i be the image distance, the distance of the image from the center of the lens, Fig. 23–40.

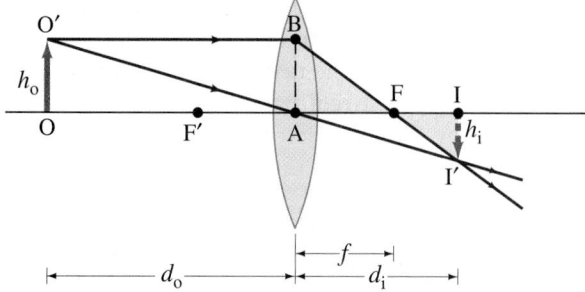

FIGURE 23–40 Deriving the lens equation for a converging lens.

Let h_o and h_i refer to the heights of the object and image. Consider the two rays shown in Fig. 23–40 for a converging lens, assumed to be very thin. The right triangles FI′I and FBA (highlighted in yellow) are similar because angle AFB equals angle IFI′; so

$$\frac{h_i}{h_o} = \frac{d_i - f}{f},$$

since length $AB = h_o$. Triangles OAO′ and IAI′ are similar as well. Therefore,

$$\frac{h_i}{h_o} = \frac{d_i}{d_o}.$$

We equate the right sides of these two equations (the left sides are the same), and divide by d_i to obtain

$$\frac{1}{f} - \frac{1}{d_i} = \frac{1}{d_o}$$

or

THIN LENS EQUATION

$$\frac{1}{d_o} + \frac{1}{d_i} = \frac{1}{f}. \tag{23–8}$$

This is called the **thin lens equation**. It relates the image distance d_i to the object distance d_o and the focal length f. It is the most useful equation in geometric optics. (Interestingly, it is exactly the same as the mirror equation, Eq. 23–2.)

If the object is at infinity, then $1/d_o = 0$, so $d_i = f$. Thus the focal length is the image distance for an object at infinity, as mentioned earlier.

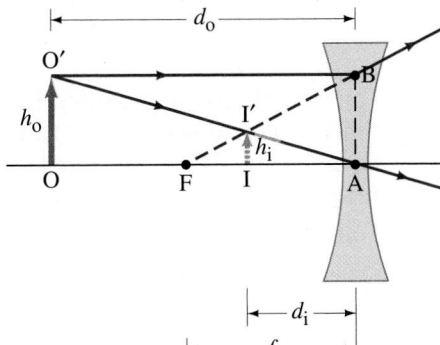

FIGURE 23–41 Deriving the lens equation for a diverging lens.

We can derive the lens equation for a diverging lens using Fig. 23–41. Triangles IAI′ and OAO′ are similar; and triangles IFI′ and AFB are similar. Thus (noting that length $AB = h_o$)

$$\frac{h_i}{h_o} = \frac{d_i}{d_o}$$

and

$$\frac{h_i}{h_o} = \frac{f - d_i}{f}.$$

When we equate the right sides of these two equations and simplify, we obtain

$$\frac{1}{d_o} - \frac{1}{d_i} = -\frac{1}{f}.$$

This equation becomes the same as Eq. 23–8 if we make f and d_i negative. That is, we take f to be *negative for a diverging lens*, and d_i negative when the image is on the same side of the lens as the light comes from. Thus Eq. 23–8 will be valid for both converging and diverging lenses, and for *all* situations, if we use the following **sign conventions**:

⚠ **CAUTION**

Focal length is negative for diverging lens

📝 **PROBLEM SOLVING**

SIGN CONVENTIONS for lenses

1. The focal length is positive for converging lenses and negative for diverging lenses.

2. The object distance is positive if the object is on the side of the lens from which the light is coming (this is always the case for real objects; but when lenses are used in combination, it might not be so: see Example 23–16); otherwise, it is negative.

3. The image distance is positive if the image is on the opposite side of the lens from where the light is coming; if it is on the same side, d_i is negative. Equivalently, the image distance is positive for a real image (Fig. 23–40) and negative for a virtual image (Fig. 23–41).

4. The height of the image, h_i, is positive if the image is upright, and negative if the image is inverted relative to the object. (h_o is always taken as upright and positive.)

The **magnification**, m, of a lens is defined as the ratio of the image height to object height, $m = h_i/h_o$. From Figs. 23–40 and 23–41 and the conventions just stated (for which we will need a minus sign), we have

$$m = \frac{h_i}{h_o} = -\frac{d_i}{d_o}. \tag{23–9}$$

For an upright image the magnification is positive, and for an inverted image the magnification is negative.

From sign convention 1, it follows that the power (Eq. 23–7) of a converging lens, in diopters, is positive, whereas the power of a diverging lens is negative. A converging lens is sometimes referred to as a **positive lens**, and a diverging lens as a **negative lens**.

Diverging lenses (see Fig. 23–41) always produce an upright virtual image for any real object, no matter where that object is. Converging lenses can produce real (inverted) images as in Fig. 23–40, or virtual (upright) images, depending on object position, as we shall see.

Thin Lenses

1. Draw a **ray diagram**, as precise as possible, but even a rough one can serve as confirmation of analytic results. Choose one point on the object and draw at least two, or preferably three, of the easy-to-draw rays described in Figs. 23–37 and 23–39. The image point is where the rays intersect.

2. For analytic solutions, solve for unknowns in the **thin lens equation** (Eq. 23–8) and the **magnification equation** (Eq. 23–9). The thin lens equation involves reciprocals—don't forget to take the reciprocal.

3. Follow the **sign conventions** listed just above.

4. Check that your analytic answers are **consistent** with your ray diagram.

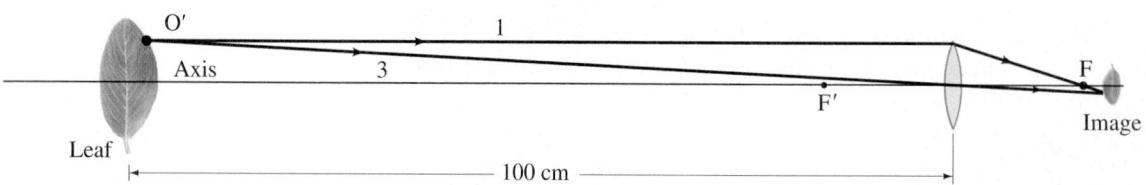

FIGURE 23–42 Example 23–12. (Not to scale.)

EXAMPLE 23–12 **Image formed by converging lens.** What is (a) the position, and (b) the size, of the image of a 7.6-cm-high leaf placed 1.00 m from a +50.0-mm-focal-length camera lens?

APPROACH We follow the steps of the Problem Solving Strategy explicitly.

SOLUTION

1. **Ray diagram.** Figure 23–42 is an approximate ray diagram, showing only rays 1 and 3 for a single point on the leaf. We see that the image ought to be a little behind the focal point F, to the right of the lens.

2. **Thin lens and magnification equations.** (a) We find the image position analytically using the thin lens equation, Eq. 23–8. The camera lens is converging, with $f = +5.00$ cm, and $d_o = 100$ cm, and so the thin lens equation gives

$$\frac{1}{d_i} = \frac{1}{f} - \frac{1}{d_o} = \frac{1}{5.00 \text{ cm}} - \frac{1}{100 \text{ cm}} = \frac{20.0 - 1.0}{100 \text{ cm}} = \frac{19.0}{100 \text{ cm}}.$$

Then, taking the reciprocal,

$$d_i = \frac{100 \text{ cm}}{19.0} = 5.26 \text{ cm},$$

or 52.6 mm behind the lens.

(b) The magnification is

$$m = -\frac{d_i}{d_o} = -\frac{5.26 \text{ cm}}{100 \text{ cm}} = -0.0526,$$

so

$$h_i = mh_o = (-0.0526)(7.6 \text{ cm}) = -0.40 \text{ cm}.$$

The image is 4.0 mm high.

3. **Sign conventions.** The image distance d_i came out positive, so the image is behind the lens. The image height is $h_i = -0.40$ cm; the minus sign means the image is inverted.

4. **Consistency.** The analytic results of steps 2 and 3 are consistent with the ray diagram, Fig. 23–42: the image is behind the lens and inverted.

NOTE Part (a) tells us that the image is 2.6 mm farther from the lens than the image for an object at infinity, which would equal the focal length, 50.0 mm. Indeed, when focusing a camera lens, the closer the object is to the camera, the farther the lens must be from the sensor or film.

EXERCISE G If the leaf (object) of Example 23–12 is moved farther from the lens, does the image move closer to or farther from the lens? (Don't calculate!)

EXAMPLE 23–13 **Object close to converging lens.** An object is placed 10 cm from a 15-cm-focal-length converging lens. Determine the image position and size (a) analytically, and (b) using a ray diagram.

APPROACH The object is within the focal point—closer to the lens than the focal point F as $d_o < f$. We first use Eqs. 23–8 and 23–9 to obtain an analytic solution, and then confirm with a ray diagram using the special rays 1, 2, and 3 for a single object point.

SOLUTION (a) Given $f = 15$ cm and $d_o = 10$ cm, then

$$\frac{1}{d_i} = \frac{1}{15 \text{ cm}} - \frac{1}{10 \text{ cm}} = \frac{2-3}{30 \text{ cm}} = -\frac{1}{30 \text{ cm}},$$

and $d_i = -30$ cm. (Remember to take the reciprocal!) Because d_i is negative, the image must be virtual and on the same side of the lens as the object (sign convention 3, page 665). The magnification

$$m = -\frac{d_i}{d_o} = -\frac{-30 \text{ cm}}{10 \text{ cm}} = 3.0.$$

⚠ CAUTION

Don't forget to take the reciprocal

The image is three times as large as the object and is upright. This lens is being used as a magnifying glass, which we discuss in more detail in Section 25–3.

(b) The ray diagram is shown in Fig. 23–43 and confirms the result in part (a). We choose point O′ on the top of the object and draw ray 1. Ray 2, however, may take some thought: if we draw it heading toward F′, it is going the wrong way—so we have to draw it as if coming from F′ (and so dashed), striking the lens, and then going out parallel to the lens axis. We project it backward, with a dashed line, as we must do also for ray 1, in order to find where they cross. Ray 3 is drawn through the lens center, and it crosses the other two rays at the image point, I′.

NOTE From Fig. 23–43 we can see that, when an object is placed between a converging lens and its focal point, the image is virtual.

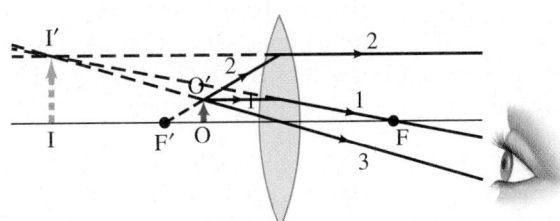

FIGURE 23–43 An object placed within the focal point of a converging lens produces a virtual image. Example 23–13.

EXAMPLE 23–14 **Diverging lens.** Where must a small insect be placed if a 25-cm-focal-length diverging lens is to form a virtual image 20 cm from the lens, on the same side as the object?

APPROACH The ray diagram is basically that of Fig. 23–41 because our lens here is diverging and our image is given as in front of the lens within the focal distance. (It would be a valuable exercise to draw the ray diagram to scale, precisely, now.) The insect's distance, d_o, can be calculated using the thin lens equation.

SOLUTION The lens is diverging, so f is negative: $f = -25$ cm. The image distance must be negative too because the image is in front of the lens (sign conventions), so $d_i = -20$ cm. The lens equation, Eq. 23–8, gives

$$\frac{1}{d_o} = \frac{1}{f} - \frac{1}{d_i} = -\frac{1}{25 \text{ cm}} + \frac{1}{20 \text{ cm}} = \frac{-4+5}{100 \text{ cm}} = \frac{1}{100 \text{ cm}}.$$

So the object must be 100 cm in front of the lens.

*23–9 Combinations of Lenses

Many optical instruments use lenses in combination. When light passes through more than one lens, we find the image formed by the first lens as if it were alone. Then this image becomes the *object* for the second lens. Next we find the image formed by this second lens using the first image as object. This second image is the final image if there are only two lenses. The total magnification will be the product of the separate magnifications of each lens. Even if the second lens intercepts the light from the first lens before it forms an image, this technique still works.

EXAMPLE 23–15 **A two-lens system.** Two converging lenses, A and B, with focal lengths $f_A = 20.0$ cm and $f_B = 25.0$ cm, are placed 80.0 cm apart, as shown in Fig. 23–44a. An object is placed 60.0 cm in front of the first lens as shown in Fig. 23–44b. Determine (a) the position, and (b) the magnification, of the final image formed by the combination of the two lenses.

APPROACH Starting at the tip of our object O, we draw rays 1, 2, and 3 for the first lens, A, and also a ray 4 which, after passing through lens A, acts for the second lens, B, as ray 3′ (through the center). We use primes now for the standard rays relative to lens B. Ray 2 for lens A exits parallel, and so is ray 1′ for lens B. To determine the position of the image I_A formed by lens A, we use Eq. 23–8 with $f_A = 20.0$ cm and $d_{oA} = 60.0$ cm. The distance of I_A (lens A's image) from lens B is the object distance d_{oB} for lens B. The final image is found using the thin lens equation, this time with all distances relative to lens B. For (b) the magnifications are found from Eq. 23–9 for each lens in turn.

SOLUTION (a) The object is a distance $d_{oA} = +60.0$ cm from the first lens, A, and this lens forms an image whose position can be calculated using the thin lens equation:

$$\frac{1}{d_{iA}} = \frac{1}{f_A} - \frac{1}{d_{oA}} = \frac{1}{20.0 \text{ cm}} - \frac{1}{60.0 \text{ cm}} = \frac{3-1}{60.0 \text{ cm}} = \frac{1}{30.0 \text{ cm}}.$$

⬥ CAUTION

*Object distance for second lens is **not** equal to the image distance for first lens*

So the first image I_A is at $d_{iA} = 30.0$ cm behind the first lens. This image becomes the object for the second lens, B. It is a distance $d_{oB} = 80.0$ cm $-$ 30.0 cm $= 50.0$ cm in front of lens B (Fig. 23–44b). The image formed by lens B, again using the thin lens equation, is at a distance d_{iB} from the lens B:

$$\frac{1}{d_{iB}} = \frac{1}{f_B} - \frac{1}{d_{oB}} = \frac{1}{25.0 \text{ cm}} - \frac{1}{50.0 \text{ cm}} = \frac{2-1}{50.0 \text{ cm}} = \frac{1}{50.0 \text{ cm}}.$$

Hence $d_{iB} = 50.0$ cm behind lens B. This is the final image—see Fig. 23–44b.

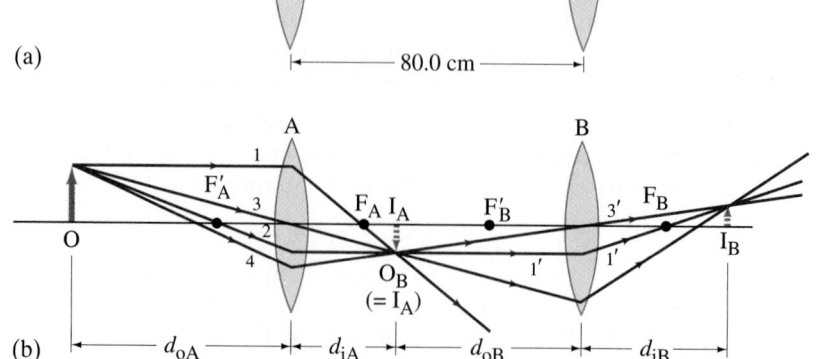

FIGURE 23–44 Two lenses, A and B, used in combination, Example 23–15. The small numbers refer to the easily drawn rays.

(b) Lens A has a magnification (Eq. 23–9)

$$m_A = -\frac{d_{iA}}{d_{oA}} = -\frac{30.0\,\text{cm}}{60.0\,\text{cm}} = -0.500.$$

Thus, the first image is inverted and is half as high as the object (again Eq. 23–9):

$$h_{iA} = m_A h_{oA} = -0.500 h_{oA}.$$

Lens B takes this first image as object and changes its height by a factor

$$m_B = -\frac{d_{iB}}{d_{oB}} = -\frac{50.0\,\text{cm}}{50.0\,\text{cm}} = -1.000.$$

The second lens reinverts the image (the minus sign) but doesn't change its size. The final image height is (remember h_{oB} is the same as h_{iA})

$$h_{iB} = m_B h_{oB} = m_B h_{iA} = m_B m_A h_{oA} = (m_{total}) h_{oA}.$$

The total magnification is the product of m_A and m_B, which here equals $m_{total} = m_A m_B = (-1.000)(-0.500) = +0.500$, or half the original height, and the final image is upright.

PROBLEM SOLVING

Total magnification is
$m_{total} = m_A m_B$

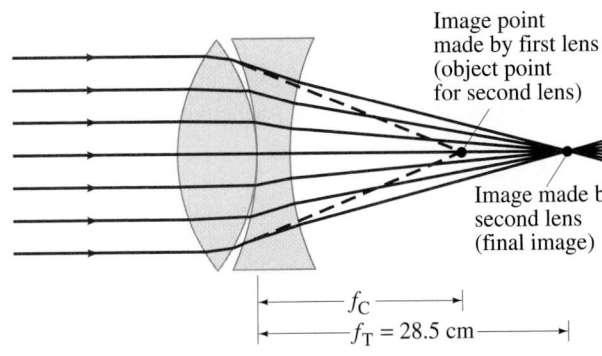

Image point made by first lens (object point for second lens)

Image made by second lens (final image)

f_C

$f_T = 28.5\,\text{cm}$

FIGURE 23–45 Determining the focal length of a diverging lens. Example 23–16.

EXAMPLE 23–16 **Measuring f for a diverging lens.** To measure the focal length of a diverging lens, a converging lens is placed in contact with it, as shown in Fig. 23–45. The Sun's rays are focused by this combination at a point 28.5 cm behind the lenses as shown. If the converging lens has a focal length f_C of 16.0 cm, what is the focal length f_D of the diverging lens? Assume both lenses are thin and the space between them is negligible.

APPROACH The image distance for the first lens equals its focal length (16.0 cm) since the object distance is infinity (∞). The position of this image, even though it is never actually formed, acts as the object for the second (diverging) lens. We apply the thin lens equation to the diverging lens to find its focal length, given that the final image is at $d_i = 28.5\,\text{cm}$.

SOLUTION Rays from the Sun are focused 28.5 cm behind the combination, so the focal length of the total combination is $f_T = 28.5\,\text{cm}$. If the diverging lens was absent, the converging lens would form the image at its focal point—that is, at a distance $f_C = 16.0\,\text{cm}$ behind it (dashed lines in Fig. 23–45). When the diverging lens is placed next to the converging lens, we treat the image formed by the first lens as the *object* for the second lens. Since this object lies to the right of the diverging lens, this is a situation where d_o is negative (see the sign conventions, page 665). Thus, for the diverging lens, the object is virtual and $d_o = -16.0\,\text{cm}$. The diverging lens forms the image of this virtual object at a distance $d_i = 28.5\,\text{cm}$ away (given). Thus,

CAUTION
$d_o < 0$

$$\frac{1}{f_D} = \frac{1}{d_o} + \frac{1}{d_i} = \frac{1}{-16.0\,\text{cm}} + \frac{1}{28.5\,\text{cm}} = -0.0274\,\text{cm}^{-1}.$$

We take the reciprocal to find $f_D = -1/(0.0274\,\text{cm}^{-1}) = -36.5\,\text{cm}$.

NOTE If this technique is to work, the converging lens must be "stronger" than the diverging lens—that is, it must have a focal length whose magnitude is less than that of the diverging lens.

*23–10 Lensmaker's Equation

A useful equation, called the **lensmaker's equation**, relates the focal length of a lens to the radii of curvature R_1 and R_2 of its two surfaces and its index of refraction n:

$$\frac{1}{f} = (n - 1)\left(\frac{1}{R_1} + \frac{1}{R_2}\right). \qquad \text{(23–10)}$$

Lensmaker's equation

If both surfaces are convex, R_1 and R_2 are considered positive.[†] For a concave surface, the radius must be considered *negative*.

Notice that Eq. 23–10 is *symmetrical* in R_1 and R_2. Thus, if a lens is turned around so that light impinges on the other surface, the focal length is the same even if the two lens surfaces are different. This confirms what we said earlier: a lens' focal length is the same on both sides of the lens.

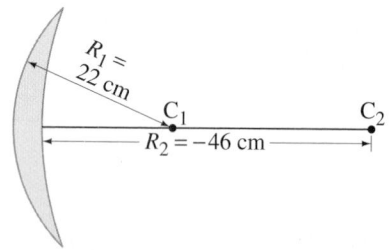

FIGURE 23–46 Example 23–17. The left surface is convex (center bulges outward); the right surface is concave.

EXAMPLE 23–17 | **Calculating f for a converging lens.** A convex meniscus lens (Figs. 23–31a and 23–46) is made from glass with $n = 1.50$. The radius of curvature of the convex surface (left in Fig. 23–46) is 22 cm. The surface on the right is concave with radius of curvature 46 cm. What is the focal length?

APPROACH We use the lensmaker's equation, Eq. 23–10, to find f.

SOLUTION $R_1 = 0.22$ m and $R_2 = -0.46$ m (concave surface). Then

$$\frac{1}{f} = (1.50 - 1.00)\left(\frac{1}{0.22\text{ m}} - \frac{1}{0.46\text{ m}}\right) = 1.19\text{ m}^{-1}.$$

So

$$f = \frac{1}{1.19\text{ m}^{-1}} = 0.84\text{ m},$$

and the lens is converging since $f > 0$.

NOTE If we turn the lens around so that $R_1 = -46$ cm and $R_2 = +22$ cm, we get the same result.

NOTE Because Eq. 23–10 gives $1/f$, it gives directly the power of a lens in diopters, Eq. 23–7. The power of this lens is about 1.2 D.

[†]Some books use a different convention: R_1 and R_2 may be considered positive if their centers of curvature are to the right of the lens; then a minus sign replaces the + sign in their version of Eq. 23–10.

Summary

Light appears to travel along straight-line paths, called **rays**, through uniform transparent materials including air and glass. When light reflects from a flat surface, the *angle of reflection equals the angle of incidence*. This **law of reflection** explains why mirrors can form **images**.

In a **plane mirror**, the image is virtual, upright, the same size as the object, and as far behind the mirror as the object is in front.

A **spherical mirror** can be concave or convex. A **concave** spherical mirror focuses parallel rays of light (light from a very distant object) to a point called the **focal point**. The distance of this point from the mirror is the **focal length** f of the mirror and

$$f = \frac{r}{2} \qquad \text{(23–1)}$$

where r is the radius of curvature of the mirror.

Parallel rays falling on a **convex mirror** reflect from the mirror as if they diverged from a common point behind the mirror. The distance of this point from the mirror is the focal length and is considered negative for a convex mirror.

For a given object, the approximate position and size of the image formed by a mirror can be found by ray tracing. Algebraically, the relation between image and object distances, d_i and d_o, and the focal length f, is given by the **mirror equation**:

$$\frac{1}{d_o} + \frac{1}{d_i} = \frac{1}{f}. \qquad \text{(23–2)}$$

The ratio of image height h_i to object height h_o, which equals the magnification m of a mirror, is

$$m = \frac{h_i}{h_o} = -\frac{d_i}{d_o}. \qquad \text{(23–3)}$$

If the rays that converge to form an image actually pass through the image, so the image would appear on a screen or film placed there, the image is said to be a **real image**. If the light rays do not actually pass through the image, the image is a **virtual image**.

The speed of light v depends on the **index of refraction**, n, of the material:

$$n = \frac{c}{v}, \qquad \text{(23–4)}$$

where c is the speed of light in vacuum.

When light passes from one transparent medium into another, the rays bend or refract. The **law of refraction** (**Snell's law**) states that

$$n_1 \sin \theta_1 = n_2 \sin \theta_2, \qquad \text{(23–5)}$$

where n_1 and θ_1 are the index of refraction and angle with the normal (perpendicular) to the surface for the incident ray, and n_2 and θ_2 are for the refracted ray.

When light rays reach the boundary of a material where the index of refraction decreases, the rays will be **totally internally reflected** if the incident angle, θ_1, is such that Snell's law would predict $\sin \theta_2 > 1$. This occurs if θ_1 exceeds the critical angle θ_C given by

$$\sin \theta_C = \frac{n_2}{n_1}. \qquad (23\text{–}6)$$

A lens uses refraction to produce a real or virtual image. Parallel rays of light are focused to a point, the **focal point**, by a **converging** lens. The distance of the focal point from the lens is the **focal length** f of the lens. It is the same on both sides of the lens.

After parallel rays pass through a **diverging** lens, they appear to diverge from a point in front of the lens, which is its focal point; and the corresponding focal length is considered negative.

The **power** P of a lens, which is $P = 1/f$ (Eq. 23–7), is given in diopters, which are units of inverse meters (m^{-1}).

For a given object, the position and size of the image formed by a lens can be found approximately by ray tracing. Algebraically, the relation between image and object distances, d_i and d_o,

and the focal length f, is given by the **thin lens equation**:

$$\frac{1}{d_o} + \frac{1}{d_i} = \frac{1}{f}. \qquad (23\text{–}8)$$

The ratio of image height to object height, which equals the **magnification** m for a lens, is

$$m = \frac{h_i}{h_o} = -\frac{d_i}{d_o}. \qquad (23\text{–}9)$$

When using the various equations of geometric optics, you must remember the **sign conventions** for all quantities involved: carefully review them (pages 655 and 665) when doing Problems.

[*When two (or more) thin lenses are used in combination to produce an image, the thin lens equation can be used for each lens in sequence. The image produced by the first lens acts as the object for the second lens.]

[*The **lensmaker's equation** relates the radii of curvature of the lens surfaces and the lens' index of refraction to the focal length of the lens.]

Questions

1. Archimedes is said to have burned the whole Roman fleet in the harbor of Syracuse, Italy, by focusing the rays of the Sun with a huge spherical mirror. Is this[†] reasonable?

2. What is the focal length of a plane mirror? What is the magnification of a plane mirror?

3. Although a plane mirror appears to reverse left and right, it doesn't reverse up and down. Discuss why this happens, noting that front to back is also reversed. Also discuss what happens if, while standing, you look up vertically at a horizontal mirror on the ceiling.

4. An object is placed along the principal axis of a spherical mirror. The magnification of the object is -2.0. Is the image real or virtual, inverted or upright? Is the mirror concave or convex? On which side of the mirror is the image located?

5. If a concave mirror produces a real image, is the image necessarily inverted? Explain.

6. How might you determine the speed of light in a solid, rectangular, transparent object?

7. When you look at the Moon's reflection from a ripply sea, it appears elongated (Fig. 23–47). Explain.

FIGURE 23–47
Question 7.

8. What is the angle of refraction when a light ray is incident perpendicular to the boundary between two transparent materials?

9. When you look down into a swimming pool or a lake, are you likely to overestimate or underestimate its depth? Explain. How does the apparent depth vary with the viewing angle? (Use ray diagrams.)

10. Draw a ray diagram to show why a stick or straw looks bent when part of it is under water (Fig. 23–23).

11. When a wide beam of parallel light enters water at an angle, the beam broadens. Explain.

12. You look into an aquarium and view a fish inside. One ray of light from the fish is shown emerging from the tank in Fig. 23–48. The apparent position of the fish is also shown (dashed ray). In the drawing, indicate the approximate position of the actual fish. Briefly justify your answer.

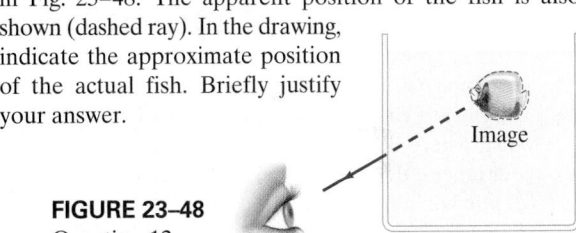

FIGURE 23–48
Question 12.

13. How can you "see" a round drop of water on a table even though the water is transparent and colorless?

14. A ray of light is refracted through three different materials (Fig. 23–49). Which material has (a) the largest index of refraction, (b) the smallest?

FIGURE 23–49
Question 14.

15. A child looks into a pool to see how deep it is. She then drops a small toy into the pool to help decide how deep the pool is. After this careful investigation, she decides it is safe to jump in—only to discover the water is over her head. What went wrong with her interpretation of her experiment?

16. Can a light ray traveling in air be totally reflected when it strikes a smooth water surface if the incident angle is chosen correctly? Explain.

†Students at MIT did a feasibility study. See
www.mit.edu/2.009/www/experiments/deathray/10_ArchimedesResult.html.

17. What type of mirror is shown in Fig. 23–50? Explain.

FIGURE 23–50
Question 17 and Problem 15.

18. Light rays from stars (including our Sun) always bend toward the vertical direction as they pass through the Earth's atmosphere. (*a*) Why does this make sense? (*b*) What can you conclude about the apparent positions of stars as viewed from Earth? Draw a circle for Earth, a dot for you, and 3 or 4 stars at different angles.

19. Where must the film be placed if a camera lens is to make a sharp image of an object far away? Explain.

20. A photographer moves closer to his subject and then refocuses. Does the camera lens move farther away from or closer to the camera film or sensor? Explain.

21. Can a diverging lens form a real image under any circumstances? Explain.

22. Light rays are said to be "reversible." Is this consistent with the thin lens equation? Explain.

23. Can real images be projected on a screen? Can virtual images? Can either be photographed? Discuss carefully.

24. A thin converging lens is moved closer to a nearby object. Does the real image formed change (*a*) in position, (*b*) in size? If yes, describe how.

25. If a glass converging lens is placed in water, its focal length in water will be (*a*) longer, (*b*) shorter, or (*c*) the same as in air. Explain.

26. Compare the mirror equation with the thin lens equation. Discuss similarities and differences, especially the sign conventions for the quantities involved.

27. A lens is made of a material with an index of refraction $n = 1.25$. In air, it is a converging lens. Will it still be a converging lens if placed in water? Explain, using a ray diagram.

28. (*a*) Does the focal length of a lens depend on the fluid in which it is immersed? (*b*) What about the focal length of a spherical mirror? Explain.

29. An underwater lens consists of a carefully shaped thin-walled plastic container filled with air. What shape should it have in order to be (*a*) converging, (*b*) diverging? Use ray diagrams to support your answer.

30. The thicker a double convex lens is in the center as compared to its edges, the shorter its focal length for a given lens diameter. Explain.

*31. A non-symmetrical lens (say, planoconvex) forms an image of a nearby object. Use the lensmaker's equation to explain if the image point changes when the lens is turned around.

*32. Example 23–16 shows how to use a converging lens to measure the focal length of a diverging lens. (*a*) Why can't you measure the focal length of a diverging lens directly? (*b*) It is said that for this to work, the converging lens must be stronger than the diverging lens. What is meant by "stronger," and why is this statement true?

MisConceptual Questions

1. Suppose you are standing about 3 m in front of a mirror. You can see yourself just from the top of your head to your waist, where the bottom of the mirror cuts off the rest of your image. If you walk one step closer to the mirror
(*a*) you will not be able to see any more of your image.
(*b*) you will be able to see more of your image, below your waist.
(*c*) you will see less of your image, with the cutoff rising to be above your waist.

2. When the reflection of an object is seen in a flat mirror, the image is
(*a*) real and upright.
(*b*) real and inverted.
(*c*) virtual and upright.
(*d*) virtual and inverted.

3. You want to create a spotlight that will shine a bright beam of light with all of the light rays parallel to each other. You have a large concave spherical mirror and a small light-bulb. Where should you place the lightbulb?
(*a*) At the focal point of the mirror.
(*b*) At the radius of curvature of the mirror.
(*c*) At any point, because all rays bouncing off the mirror will be parallel.
(*d*) None of the above; you can't make parallel rays with a concave mirror.

4. When you look at a fish in a still stream from the bank, the fish appears shallower than it really is due to refraction. From directly above, it appears
(*a*) deeper than it really is.
(*b*) at its actual depth.
(*c*) shallower than its real depth.
(*d*) It depends on your height above the water.

5. Parallel light rays cross interfaces from medium 1 into medium 2 and then into medium 3 as shown in Fig. 23–51. What can we say about the relative sizes of the indices of refraction of these media?
(*a*) $n_1 > n_2 > n_3$.
(*b*) $n_3 > n_2 > n_1$.
(*c*) $n_2 > n_3 > n_1$.
(*d*) $n_1 > n_3 > n_2$.
(*e*) $n_2 > n_1 > n_3$.
(*f*) None of the above.

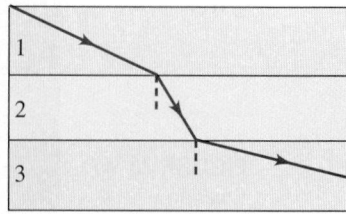

FIGURE 23–51
MisConceptual Question 5.

6. To shoot a swimming fish with an intense light beam from a *laser gun*, you should aim
(*a*) directly at the image.
(*b*) slightly above the image.
(*c*) slightly below the image.

7. When moonlight strikes the surface of a calm lake, what happens to this light?
   (a) All of it reflects from the water surface back to the air.
   (b) Some of it reflects back to the air; some enters the water.
   (c) All of it enters the water.
   (d) All of it disappears via absorption by water molecules.

8. If you shine a light through an optical fiber, why does it come out the end but not out the sides?
   (a) It does come out the sides, but this effect is not obvious because the sides are so much longer than the ends.
   (b) The sides are mirrored, so the light reflects.
   (c) Total internal reflection makes the light reflect from the sides.
   (d) The light flows along the length of the fiber, never touching the sides.

9. A converging lens, such as a typical magnifying glass,
   (a) always produces a magnified image (taller than object).
   (b) always produces an image smaller than the object.
   (c) always produces an upright image.
   (d) always produces an inverted image (upside down).
   (e) None of these statements are true.

10. Virtual images can be formed by
    (a) only mirrors.
    (b) only lenses.
    (c) only plane mirrors.
    (d) only curved mirrors or lenses.
    (e) plane and curved mirrors, and lenses.

11. A lens can be characterized by its *power*, which
    (a) is the same as the magnification.
    (b) tells how much light the lens can focus.
    (c) depends on where the object is located.
    (d) is the reciprocal of the focal length.

12. You cover half of a lens that is forming an image on a screen. Compare what happens when you cover the top half of the lens versus the bottom half.
    (a) When you cover the top half of the lens, the top half of the image disappears; when you cover the bottom half of the lens, the bottom half of the image disappears.
    (b) When you cover the top half of the lens, the bottom half of the image disappears; when you cover the bottom half of the lens, the top half of the image disappears.
    (c) The image becomes half as bright in both cases.
    (d) Nothing happens in either case.
    (e) The image disappears in both cases.

13. Which of the following can form an image?
    (a) A plane mirror.
    (b) A curved mirror.
    (c) A lens curved on both sides.
    (d) A lens curved on only one side.
    (e) All of the above.

14. As an object moves from just outside the focal point of a converging lens to just inside it, the image goes from _____ and _____ to _____ and _____.
    (a) large; inverted; large; upright.
    (b) large; upright; large; inverted.
    (c) small; inverted; small; upright.
    (d) small; upright; small; inverted.

For assigned homework and other learning materials, go to the MasteringPhysics website.

Problems

23–2 Reflection; Plane Mirrors

1. (I) When you look at yourself in a 60-cm-tall plane mirror, you see the same amount of your body whether you are close to the mirror or far away. (Try it and see.) Use ray diagrams to show why this should be true.

2. (I) Suppose that you want to take a photograph of yourself as you look at your image in a mirror 3.1 m away. For what distance should the camera lens be focused?

3. (II) Two plane mirrors meet at a 135° angle, Fig. 23–52. If light rays strike one mirror at 34° as shown, at what angle ϕ do they leave the second mirror?

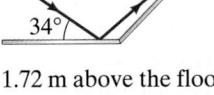

FIGURE 23–52
Problem 3.

4. (II) A person whose eyes are 1.72 m above the floor stands 2.20 m in front of a vertical plane mirror whose bottom edge is 38 cm above the floor, Fig. 23–53. What is the horizontal distance x to the base of the wall supporting the mirror of the nearest point on the floor that can be seen reflected in the mirror?

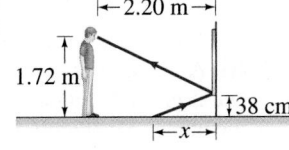

FIGURE 23–53
Problem 4.

5. (II) Stand up two plane mirrors so they form a 90.0° angle as in Fig. 23–54. When you look into this double mirror, you see yourself as others see you, instead of reversed as in a single mirror. Make a ray diagram to show how this occurs.

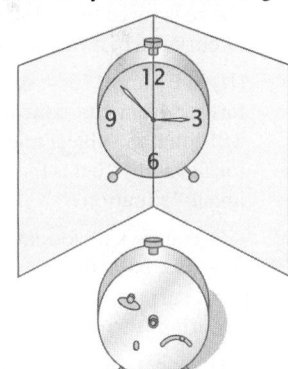

FIGURE 23–54
Problem 5.

6. (II) Two plane mirrors, nearly parallel, are facing each other 2.3 m apart as in Fig. 23–55. You stand 1.6 m away from one of these mirrors and look into it. You will see multiple images of yourself. (a) How far away from you are the first three images of yourself in the mirror in front of you? (b) Are these first three images facing toward you or away from you?

FIGURE 23–55
Problem 6.

7. (III) Suppose you are 94 cm from a plane mirror. What area of the mirror is used to reflect the rays entering one eye from a point on the tip of your nose if your pupil diameter is 4.5 mm?

23–3 Spherical Mirrors

8. (I) A solar cooker, really a concave mirror pointed at the Sun, focuses the Sun's rays 18.8 cm in front of the mirror. What is the radius of the spherical surface from which the mirror was made?

9. (I) How far from a concave mirror (radius 21.0 cm) must an object be placed if its image is to be at infinity?

10. (II) A small candle is 38 cm from a concave mirror having a radius of curvature of 24 cm. (a) What is the focal length of the mirror? (b) Where will the image of the candle be located? (c) Will the image be upright or inverted?

11. (II) An object 3.0 mm high is placed 16 cm from a convex mirror of radius of curvature 16 cm. (a) Show by ray tracing that the image is virtual, and estimate the image distance. (b) Show that the (negative) image distance can be computed from Eq. 23–2 using a focal length of −8.0 cm. (c) Compute the image size, using Eq. 23–3.

12. (II) A dentist wants a small mirror that, when 2.00 cm from a tooth, will produce a 4.0× upright image. What kind of mirror must be used and what must its radius of curvature be?

13. (II) You are standing 3.4 m from a convex security mirror in a store. You estimate the height of your image to be half of your actual height. Estimate the radius of curvature of the mirror.

14. (II) The image of a distant tree is virtual and very small when viewed in a curved mirror. The image appears to be 19.0 cm behind the mirror. What kind of mirror is it, and what is its radius of curvature?

15. (II) A mirror at an amusement park shows an upright image of any person who stands 1.9 m in front of it. If the image is three times the person's height, what is the radius of curvature of the mirror? (See Fig. 23–50.)

16. (II) In Example 23–4, show that if the object is moved 10.0 cm farther from the concave mirror, the object's image size will equal the object's actual size. Stated as a multiple of the focal length, what is the object distance for this "actual-sized image" situation?

17. (II) You look at yourself in a shiny 8.8-cm-diameter Christmas tree ball. If your face is 25.0 cm away from the ball's front surface, where is your image? Is it real or virtual? Is it upright or inverted?

18. (II) Some rearview mirrors produce images of cars to your rear that are smaller than they would be if the mirror were flat. Are the mirrors concave or convex? What is a mirror's radius of curvature if cars 16.0 m away appear 0.33 their normal size?

19. (II) When walking toward a concave mirror you notice that the image flips at a distance of 0.50 m. What is the radius of curvature of the mirror?

20. (II) (a) Where should an object be placed in front of a concave mirror so that it produces an image at the same location as the object? (b) Is the image real or virtual? (c) Is the image inverted or upright? (d) What is the magnification of the image?

21. (II) A shaving or makeup mirror is designed to magnify your face by a factor of 1.40 when your face is placed 20.0 cm in front of it. (a) What type of mirror is it? (b) Describe the type of image that it makes of your face. (c) Calculate the required radius of curvature for the mirror.

22. (II) Use two techniques, (a) a ray diagram, and (b) the mirror equation, to show that the magnitude of the magnification of a concave mirror is less than 1 if the object is beyond the center of curvature C ($d_o > r$), and is greater than 1 if the object is within C ($d_o < r$).

23. (III) Show, using a ray diagram, that the magnification m of a convex mirror is $m = -d_i/d_o$, just as for a concave mirror. [Hint: Consider a ray from the top of the object that reflects at the center of the mirror.]

24. (III) An object is placed a distance r in front of a wall, where r exactly equals the radius of curvature of a certain concave mirror. At what distance from the wall should this mirror be placed so that a real image of the object is formed on the wall? What is the magnification of the image?

23–4 Index of Refraction

25. (I) The speed of light in ice is 2.29×10^8 m/s. What is the index of refraction of ice?

26. (I) What is the speed of light in (a) ethyl alcohol, (b) lucite, (c) crown glass?

27. (II) The speed of light in a certain substance is 82% of its value in water. What is the index of refraction of that substance?

23–5 Refraction; Snell's Law

28. (I) A flashlight beam strikes the surface of a pane of glass ($n = 1.56$) at a 67° angle to the normal. What is the angle of refraction?

29. (I) A diver shines a flashlight upward from beneath the water at a 35.2° angle to the vertical. At what angle does the light leave the water?

30. (I) A light beam coming from an underwater spotlight exits the water at an angle of 56.0°. At what angle of incidence did it hit the air–water interface from below the surface?

31. (I) Rays of the Sun are seen to make a 36.0° angle to the vertical beneath the water. At what angle above the horizon is the Sun?

32. (II) An aquarium filled with water has flat glass sides whose index of refraction is 1.54. A beam of light from outside the aquarium strikes the glass at a 43.5° angle to the perpendicular (Fig. 23–56). What is the angle of this light ray when it enters (a) the glass, and then (b) the water? (c) What would be the refracted angle if the ray entered the water directly?

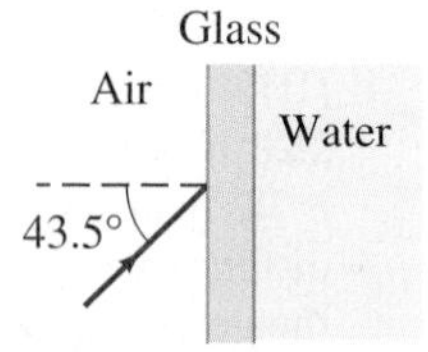

FIGURE 23–56
Problem 32.

33. (II) A beam of light in air strikes a slab of glass ($n = 1.51$) and is partially reflected and partially refracted. Determine the angle of incidence if the angle of reflection is twice the angle of refraction.

34. (II) In searching the bottom of a pool at night, a watchman shines a narrow beam of light from his flashlight, 1.3 m above the water level, onto the surface of the water at a point 2.5 m from his foot at the edge of the pool (Fig. 23–57). Where does the spot of light hit the bottom of the 2.1-m-deep pool? Measure from the bottom of the wall beneath his foot.

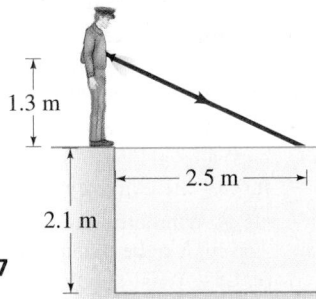

FIGURE 23–57
Problem 34.

23–6 Total Internal Reflection

35. (I) What is the critical angle for the interface between water and crown glass? To be internally reflected, the light must start in which material?

36. (I) The critical angle for a certain liquid–air surface is 47.2°. What is the index of refraction of the liquid?

37. (II) A beam of light is emitted in a pool of water from a depth of 82.0 cm. Where must it strike the air–water interface, relative to the spot directly above it, in order that the light does *not* exit the water?

38. (II) A beam of light is emitted 8.0 cm beneath the surface of a liquid and strikes the air surface 7.6 cm from the point directly above the source. If total internal reflection occurs, what can you say about the index of refraction of the liquid?

39. (III) (*a*) What is the minimum index of refraction for a glass or plastic prism to be used in binoculars (Fig. 23–28) so that total internal reflection occurs at 45°? (*b*) Will binoculars work if their prisms (assume $n = 1.58$) are immersed in water? (*c*) What minimum n is needed if the prisms are immersed in water?

40. (III) A beam of light enters the end of an optic fiber as shown in Fig. 23–58. (*a*) Show that we can guarantee total internal reflection at the side surface of the material (at point A), if the index of refraction is greater than about 1.42. In other words, regardless of the angle α, the light beam reflects back into the material at point A, assuming air outside. (*b*) What if the fiber were immersed in water?

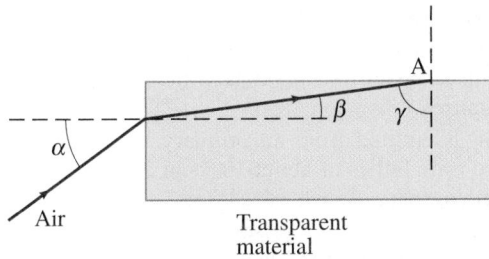

FIGURE 23–58 Problem 40.

23–7 and 23–8 Thin Lenses

41. (I) A sharp image is located 391 mm behind a 215-mm-focal-length converging lens. Find the object distance (*a*) using a ray diagram, (*b*) by calculation.

42. (I) Sunlight is observed to focus at a point 16.5 cm behind a lens. (*a*) What kind of lens is it? (*b*) What is its power in diopters?

43. (I) (*a*) What is the power of a 32.5-cm-focal-length lens? (*b*) What is the focal length of a −6.75-D lens? Are these lenses converging or diverging?

44. (II) A certain lens focuses light from an object 1.55 m away as an image 48.3 cm on the other side of the lens. What type of lens is it and what is its focal length? Is the image real or virtual?

45. (II) A 105-mm-focal-length lens is used to focus an image on the sensor of a camera. The maximum distance allowed between the lens and the sensor plane is 132 mm. (*a*) How far in front of the sensor should the lens (assumed thin) be positioned if the object to be photographed is 10.0 m away? (*b*) 3.0 m away? (*c*) 1.0 m away? (*d*) What is the closest object this lens could photograph sharply?

46. (II) Use ray diagrams to show that a real image formed by a thin lens is always inverted, whereas a virtual image is always upright if the object is real.

47. (II) A stamp collector uses a converging lens with focal length 28 cm to view a stamp 16 cm in front of the lens. (*a*) Where is the image located? (*b*) What is the magnification?

48. (II) It is desired to magnify reading material by a factor of 3.0× when a book is placed 9.0 cm behind a lens. (*a*) Draw a ray diagram and describe the type of image this would be. (*b*) What type of lens is needed? (*c*) What is the power of the lens in diopters?

49. (II) A −7.00-D lens is held 12.5 cm from an ant 1.00 mm high. Describe the position, type, and height of the image.

50. (II) An object is located 1.50 m from a 6.5-D lens. By how much does the image move if the object is moved (*a*) 0.90 m closer to the lens, and (*b*) 0.90 m farther from the lens?

51. (II) (*a*) How far from a 50.0-mm-focal-length lens must an object be placed if its image is to be magnified 2.50× and be real? (*b*) What if the image is to be virtual and magnified 2.50×?

52. (II) Repeat Problem 51 for a −50.0-mm-focal-length lens. [*Hint*: Consider objects real or virtual (formed by some other piece of optics).]

53. (II) How far from a converging lens with a focal length of 32 cm should an object be placed to produce a real image which is the same size as the object?

54. (II) (*a*) A 2.40-cm-high insect is 1.30 m from a 135-mm-focal-length lens. Where is the image, how high is it, and what type is it? (*b*) What if $f = -135$ mm?

55. (III) A bright object and a viewing screen are separated by a distance of 86.0 cm. At what location(s) between the object and the screen should a lens of focal length 16.0 cm be placed in order to produce a sharp image on the screen? [*Hint*: First draw a diagram.]

56. (III) How far apart are an object and an image formed by an 85-cm-focal-length converging lens if the image is 3.25× larger than the object and is real?

57. (III) In a film projector, the film acts as the object whose image is projected on a screen (Fig. 23–59). If a 105-mm-focal-length lens is to project an image on a screen 25.5 m away, how far from the lens should the film be? If the film is 24 mm wide, how wide will the picture be on the screen?

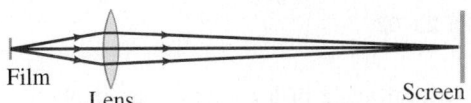

FIGURE 23–59 Film projector, Problem 57.

*58. (II) A diverging lens with $f = -36.5$ cm is placed 14.0 cm behind a converging lens with $f = 20.0$ cm. Where will an object at infinity be focused?

*59. (II) Two 25.0-cm-focal-length converging lenses are placed 16.5 cm apart. An object is placed 35.0 cm in front of one lens. Where will the final image formed by the second lens be located? What is the total magnification?

*60. (II) A 38.0-cm-focal-length converging lens is 28.0 cm behind a diverging lens. Parallel light strikes the diverging lens. After passing through the converging lens, the light is again parallel. What is the focal length of the diverging lens? [Hint: First draw a ray diagram.]

*61. (II) Two lenses, one converging with focal length 20.0 cm and one diverging with focal length -10.0 cm, are placed 25.0 cm apart. An object is placed 60.0 cm in front of the converging lens. Determine (a) the position and (b) the magnification of the final image formed. (c) Sketch a ray diagram for this system.

*62. (II) A lighted candle is placed 36 cm in front of a converging lens of focal length $f_1 = 13$ cm, which in turn is 56 cm in front of another converging lens of focal length $f_2 = 16$ cm (see Fig. 23–60). (a) Draw a ray diagram and estimate the location and the relative size of the final image. (b) Calculate the position and relative size of the final image.

$f_1 = 13$ cm $\qquad$ $f_2 = 16$ cm

FIGURE 23–60
Problem 62.

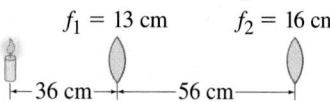

|←—36 cm—→|←——56 cm——→|

*63. (I) A double concave lens has surface radii of 33.4 cm and 28.8 cm. What is the focal length if $n = 1.52$?

*64. (I) Both surfaces of a double convex lens have radii of 34.1 cm. If the focal length is 28.9 cm, what is the index of refraction of the lens material?

*65. (I) A planoconvex lens (Fig. 23–31a) with $n = 1.55$ is to have a focal length of 16.3 cm. What is the radius of curvature of the convex surface?

*66. (II) A symmetric double convex lens with a focal length of 22.0 cm is to be made from glass with an index of refraction of 1.52. What should be the radius of curvature for each surface?

*67. (II) A prescription for an eyeglass lens calls for $+3.50$ diopters. The lensmaker grinds the lens from a "blank" with $n = 1.56$ and convex front surface of radius of curvature of 30.0 cm. What should be the radius of curvature of the other surface?

*68. (III) An object is placed 96.5 cm from a glass lens ($n = 1.52$) with one concave surface of radius 22.0 cm and one convex surface of radius 18.5 cm. Where is the final image? What is the magnification?

General Problems

69. Sunlight is reflected off the Moon. How long does it take that light to reach us from the Moon?

70. You hold a small flat mirror 0.50 m in front of you and can see your reflection twice in that mirror because there is a full-length mirror 1.0 m behind you (Fig. 23–61). Determine the distance of each image from you.

1.0 m $\qquad$ 0.50 m

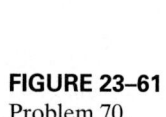

FIGURE 23–61
Problem 70.

71. We wish to determine the depth of a swimming pool filled with water by measuring the width ($x = 6.50$ m) and then noting that the far bottom edge of the pool is just visible at an angle of 13.0° above the horizontal as shown in Fig. 23–62. Calculate the depth of the pool.

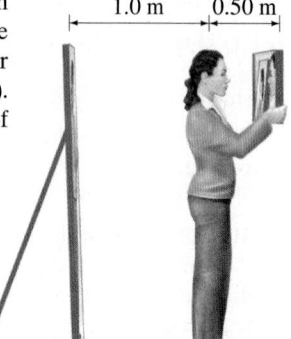

13.0° $\qquad$ |←—6.50 m—→|

Water

Depth ?

FIGURE 23–62
Problem 71.

72. The critical angle of a certain piece of plastic in air is $\theta_C = 37.8°$. What is the critical angle of the same plastic if it is immersed in water?

73. A pulse of light takes 2.63 ns (see Table 1–4) to travel 0.500 m in a certain material. Determine the material's index of refraction, and identify this material.

74. When an object is placed 60.0 cm from a certain converging lens, it forms a real image. When the object is moved to 40.0 cm from the lens, the image moves 10.0 cm farther from the lens. Find the focal length of this lens.

75. A 4.5-cm-tall object is placed 32 cm in front of a spherical mirror. It is desired to produce a virtual image that is upright and 3.5 cm tall. (a) What type of mirror should be used? (b) Where is the image located? (c) What is the focal length of the mirror? (d) What is the radius of curvature of the mirror?

76. Light is emitted from an ordinary lightbulb filament in wave-train bursts of about 10^{-8} s in duration. What is the length in space of such wave trains?

77. If the apex angle of a prism is $\phi = 75°$ (see Fig. 23–63), what is the minimum incident angle for a ray if it is to emerge from the opposite side (i.e., not be totally internally reflected), given $n = 1.58$?

ϕ

FIGURE 23–63
Problem 77.

78. (a) A plane mirror can be considered a limiting case of a spherical mirror. Specify what this limit is. (b) Determine an equation that relates the image and object distances in this limit of a plane mirror. (c) Determine the magnification of a plane mirror in this same limit. (d) Are your results in parts (b) and (c) consistent with the discussion of Section 23–2 on plane mirrors?

79. An object is placed 18 cm from a certain mirror. The image is half the height of the object, inverted, and real. How far is the image from the mirror, and what is the radius of curvature of the mirror?

80. Light is incident on an equilateral glass prism at a 45.0° angle to one face, Fig. 23–64. Calculate the angle at which light emerges from the opposite face. Assume that $n = 1.54$.

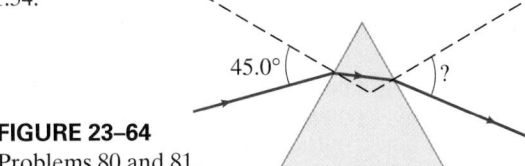

FIGURE 23–64
Problems 80 and 81.

81. Suppose a ray strikes the left face of the prism in Fig. 23–64 at 45.0° as shown, but is totally internally reflected at the opposite side. If the apex angle (at the top) is $\theta = 65.0°$, what can you say about the index of refraction of the prism?

82. (a) An object 37.5 cm in front of a certain lens is imaged 8.20 cm in front of that lens (on the same side as the object). What type of lens is this, and what is its focal length? Is the image real or virtual? (b) If the image were located, instead, 44.5 cm in front of the lens, what type of lens would it be and what focal length would it have?

83. How large is the image of the Sun on a camera sensor with (a) a 35-mm-focal-length lens, (b) a 50-mm-focal-length lens, and (c) a 105-mm-focal-length lens? The Sun has diameter 1.4×10^6 km, and it is 1.5×10^8 km away.

84. Figure 23–65 is a photograph of an eyeball with the image of a boy in a doorway. (a) Is the eye here acting as a lens or as a mirror? (b) Is the eye being viewed right side up or is the camera taking this photo upside down? (c) Explain, based on all possible images made by a convex mirror or lens.

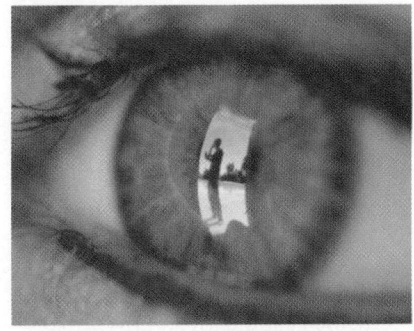

FIGURE 23–65
Problem 84.

85. Which of the two lenses shown in Fig. 23–66 is converging, and which is diverging? Explain using ray diagrams and show how each image is formed.

FIGURE 23–66 Problem 85.

86. Figure 23–67 shows a liquid-detecting prism device that might be used inside a washing machine. If no liquid covers the prism's hypotenuse, total internal reflection of the beam from the light source produces a large signal in the light sensor. If liquid covers the hypotenuse, some light escapes from the prism into the liquid and the light sensor's signal decreases. Thus a large signal from the light sensor indicates the absence of liquid in the reservoir. Determine the allowable range for the prism's index of refraction n.

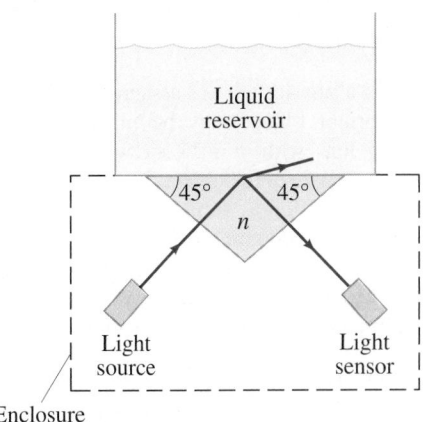

FIGURE 23–67 Problem 86.

*87. (a) Show that if two thin lenses of focal lengths f_1 and f_2 are placed in contact with each other, the focal length of the combination is given by $f_T = f_1 f_2/(f_1 + f_2)$. (b) Show that the power P of the combination of two lenses is the sum of their separate powers, $P = P_1 + P_2$.

*88. Two converging lenses are placed 30.0 cm apart. The focal length of the lens on the right is 20.0 cm, and the focal length of the lens on the left is 15.0 cm. An object is placed to the left of the 15.0-cm-focal-length lens. A final image from both lenses is inverted and located halfway between the two lenses. How far to the left of the 15.0-cm-focal-length lens is the original object?

*89. An object is placed 30.0 cm from a +5.0-D lens. A spherical mirror with focal length 25 cm is placed 75 cm behind the lens. Where is the final image? (Note that the mirror reflects light back through the lens.) Be sure to draw a diagram.

*90. A small object is 25.0 cm from a diverging lens as shown in Fig. 23–68. A converging lens with a focal length of 12.0 cm is 30.0 cm to the right of the diverging lens. The two-lens system forms a real inverted image 17.0 cm to the right of the converging lens. What is the focal length of the diverging lens?

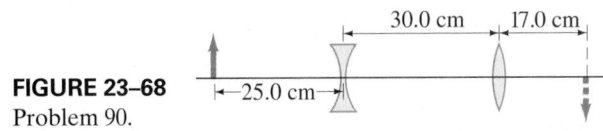

FIGURE 23–68
Problem 90.

Search and Learn

1. (*a*) Describe the difference between a real image and a virtual image? (*b*) Can your eyes tell the difference? (*c*) How can you tell the difference on a ray diagram? (*d*) How could you tell the difference between a virtual image and a real image experimentally? (*e*) If you were to take a photograph of a virtual image, would you see the image in the photograph? (*f*) If you were to put a piece of photographic film at the location of a virtual image, would the image be captured on the film? (*g*) Explain any differences in your answers to parts (*e*) and (*f*).

2. Students in a physics lab are assigned to find the location where a bright object may be placed in order that a converging lens with $f = 12$ cm will produce an image three times the size of the object. Two students complete the assignment at different times using identical equipment, but when they compare notes later, they discover that their answers for the object distance are not the same. Explain why they do not necessarily need to repeat the lab, and justify your response with a calculation.

3. Both a converging lens and a concave mirror can produce virtual images that are larger than the object. Concave mirrors can be used as makeup mirrors, but converging lenses cannot be. (*a*) Draw ray diagrams to explain why not. (*b*) If a concave mirror has the same focal length as a converging lens, and an object is placed first at a distance of $\frac{1}{2}f$ from the lens and then at a distance of $\frac{1}{2}f$ from the mirror, how will the magnification of the object compare in the two cases?

4. (*a*) Did the person we see in Fig. 23–69 shoot the picture we are looking at? We see her in three different mirrors. Describe (*b*) what type of mirror each is, and (*c*) her position relative to the focal point and center of curvature.

FIGURE 23–69 Search and Learn 4.

5. Justify the second part of sign convention 3, page 665, starting "Equivalently." Use ray diagrams for all possible situations. Cite Figures already in the text and draw any others needed.

6. The only means to create a real image with a single lens would be to place
   (*a*) the object inside the focal length of a converging lens;
   (*b*) the object inside the focal length of a diverging lens;
   (*c*) the object outside the focal length of a converging lens;
   (*d*) the object outside the focal length of a diverging lens;
   (*e*) any of the above, given the correct distance from the focal point.

7. Make a table showing the sign conventions for mirrors and lenses. Include the sign convention for the mirrors and lenses themselves and for the image and object heights and distances for each.

8. Figure 23–70 shows a converging lens held above three equal-sized letters A. In (a) the lens is 5 cm from the paper, and in (b) the lens is 15 cm from the paper. Estimate the focal length of the lens. What is the image position for each case?

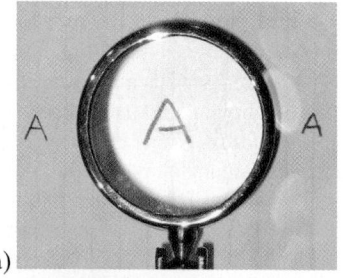

(a)

FIGURE 23–70
Search and Learn 8.

(b)

A: No.
B: (*b*).
C: Toward.
D: None.

E: No total internal reflection, $\theta_C > 45°$.
F: (*c*).
G: Closer to it.

The beautiful colors from the surface of this soap bubble can be nicely explained by the wave theory of light. A soap bubble is a very thin spherical film filled with air. Light reflected from the outer and inner surfaces of this thin film of soapy water interferes constructively to produce the bright colors. Which color we see at any point depends on the thickness of the soapy water film at that point and also on the viewing angle. Near the top of the bubble, we see a small black area surrounded by a silver or white area. The bubble's thickness is smallest at that black spot, perhaps only about 30 nm thick, and is fully transparent (we see the black background).

We cover fundamental aspects of the wave nature of light, including two-slit interference and interference in thin films.

The Wave Nature of Light

CHAPTER 24

CHAPTER-OPENING QUESTION—Guess now!

When a thin layer of oil lies on top of water or wet pavement, you can often see swirls of color. We also see swirls of color on the soap bubble shown above. What causes these colors?

(a) Additives in the oil or soap reflect various colors.

(b) Chemicals in the oil or soap absorb various colors.

(c) Dispersion due to differences in index of refraction in the oil or soap.

(d) The interactions of the light with a thin boundary layer where the oil (or soap) and the water have mixed irregularly.

(e) Light waves reflected from the top and bottom surfaces of the thin oil or soap film can add up constructively for particular wavelengths.

Light carries energy. Evidence for this can come from focusing the Sun's rays with a magnifying glass on a piece of paper and burning a hole in it. But how does light travel, and in what form is this energy carried? In our discussion of waves in Chapter 11, we noted that energy can be carried from place to place in basically two ways: by particles or by waves. In the first case, material objects or particles can carry energy, such as an avalanche of rocks or rushing water. In the second case, water waves and sound waves, for example, can carry energy over long distances even though the oscillating particles of the medium do not travel these distances. In view of this, what can we say about the nature of light: does light travel as a stream of particles away from its source, or does light travel in the form of waves that spread outward from the source?

CONTENTS

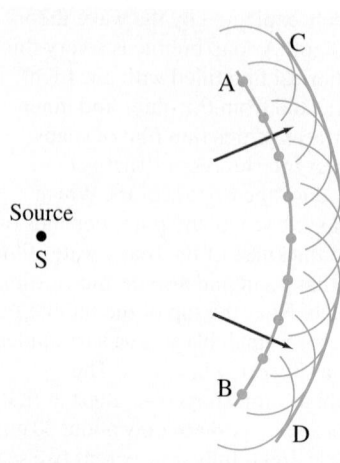

FIGURE 24–1 Huygens' principle, used to determine wave front CD when wave front AB is given.

FIGURE 24–2 Huygens' principle is consistent with diffraction (a) around the edge of an obstacle, (b) through a large hole, (c) through a small hole whose size is on the order of the wavelength of the wave.

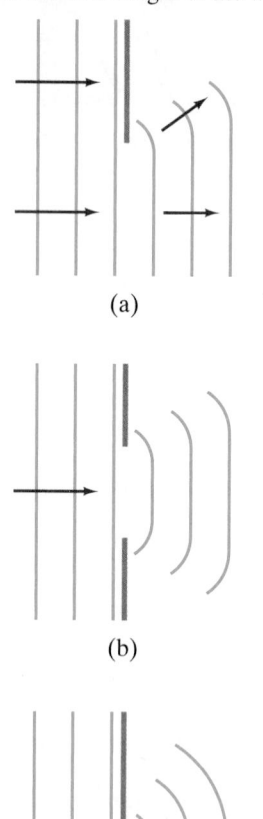

(a)

(b)

(c)

Historically, this question has turned out to be a difficult one. For one thing, light does not reveal itself in any obvious way as being made up of tiny particles; nor do we see tiny light waves passing by as we do water waves. The evidence seemed to favor first one side and then the other until about 1830, when most physicists had accepted the wave theory. By the end of the nineteenth century, light was considered to be an *electromagnetic wave* (Chapter 22). In the early twentieth century, light was shown to have a particle nature as well, as we shall discuss in Chapter 27. We now speak of the wave–particle duality of light. The wave theory of light remains valid and has proved very successful. In this Chapter we investigate the evidence for the wave theory and how it has been used to explain a wide range of phenomena.

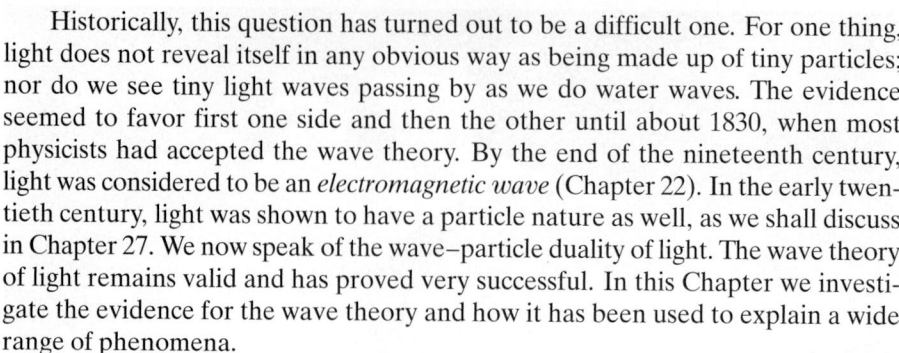

24–1 Waves vs. Particles; Huygens' Principle and Diffraction

The Dutch scientist Christian Huygens (1629–1695), a contemporary of Newton, proposed a wave theory of light that had much merit. Still useful today is a technique Huygens developed for predicting the future position of a wave front when an earlier position is known. By a **wave front**, we mean all the points along a two- or three-dimensional wave that form a wave crest—what we simply call a "wave" as seen on the ocean. Wave fronts are perpendicular to rays as discussed in Chapter 11 (Fig. 11–35). **Huygens' principle** can be stated as follows: *Every point on a wave front can be considered as a source of tiny wavelets that spread out in the forward direction at the speed of the wave itself. The new wave front is the envelope of all the wavelets—that is, the tangent to all of them.*

As an example of the use of Huygens' principle, consider the wave front AB in Fig. 24–1, which is traveling away from a source S. We assume the medium is *isotropic*—that is, the speed v of the waves is the same in all directions. To find the wave front a short time t after it is at AB, tiny circles are drawn at points along AB with radius $r = vt$. The centers of these tiny circles are shown as blue dots on the original wave front AB, and the circles represent Huygens' (imaginary) wavelets. The tangent to all these wavelets, the curved line CD, is the new position of the wave front after a time t.

Huygens' principle is particularly useful for analyzing what happens when waves run into an obstacle and the wave fronts are partially interrupted. Huygens' principle predicts that waves bend in behind an obstacle, as shown in Fig. 24–2. This is just what water waves do, as we saw in Chapter 11 (Figs. 11–45 and 11–46). The bending of waves behind obstacles into the "shadow region" is known as **diffraction**. Since diffraction occurs for waves, but not for particles, it can serve as one means for distinguishing the nature of light.

Note, as shown in Fig. 24–2, that diffraction is most prominent when the size of the opening is on the order of the wavelength of the wave. If the opening is much larger than the wavelength, diffraction may go unnoticed.

Does light exhibit diffraction? In the mid-seventeenth century, the Jesuit priest Francesco Grimaldi (1618–1663) had observed that when sunlight entered a darkened room through a tiny hole in a screen, the spot on the opposite wall was larger than would be expected from geometric rays. He also observed that the border of the image was not clear but was surrounded by colored fringes. Grimaldi attributed this to the diffraction of light.

The wave model of light nicely accounts for diffraction. But the ray model (Chapter 23) cannot account for diffraction, and it is important to be aware of such limitations to the ray model. Geometric optics using rays is successful in a wide range of situations only because normal openings and obstacles are much larger than the wavelength of the light, and so relatively little diffraction or bending occurs.

*24–2 Huygens' Principle and the Law of Refraction

The laws of reflection and refraction were well known in Newton's time. The law of reflection could not distinguish between the two theories we just discussed: waves versus particles. When waves reflect from an obstacle, the angle of incidence equals the angle of reflection (Fig. 11–36). The same is true of particles—think of a tennis ball without spin striking a flat surface.

The law of refraction is another matter. Consider a ray of light entering a medium where it is bent toward the normal, as when traveling from air into water. As shown in Fig. 24–3, this bending can be constructed using Huygens' principle if we assume the speed of light is less in the second medium ($v_2 < v_1$). In time t, point B on wave front AB (perpendicular to the incoming ray) travels a distance $v_1 t$ to reach point D. Point A on the wave front, traveling in the second medium, goes a distance $v_2 t$ to reach point C, and $v_2 t < v_1 t$. Huygens' principle is applied to points A and B to obtain the curved wavelets shown at C and D. The wave front is tangent to these two wavelets, so the new wave front is the line CD. Hence the rays, which are perpendicular to the wave fronts, bend toward the normal if $v_2 < v_1$, as drawn. (This is basically the same discussion as we used around Fig. 11–42.)

Newton favored a particle theory of light which predicted the opposite result, that the speed of light would be greater in the second medium ($v_2 > v_1$). Thus the wave theory predicts that the speed of light in water, for example, is less than in air; and Newton's particle theory predicts the reverse. An experiment to actually measure the speed of light in water was performed in 1850 by the French physicist Jean Foucault, and it confirmed the wave-theory prediction. By then, however, the wave theory was already fully accepted, as we shall see in the next Section.

Snell's law of refraction follows directly from Huygens' principle, given that the speed of light v in any medium is related to the speed in a vacuum, c, and the index of refraction, n, by Eq. 23–4: that is, $v = c/n$. From the Huygens' construction of Fig. 24–3, angle ADC is equal to θ_2 and angle BAD is equal to θ_1. Then for the two triangles that have the common side AD, we have

$$\sin\theta_1 = \frac{v_1 t}{AD}, \qquad \sin\theta_2 = \frac{v_2 t}{AD}.$$

We divide these two equations and obtain

$$\frac{\sin\theta_1}{\sin\theta_2} = \frac{v_1}{v_2}.$$

Then, by Eq. 23–4, $v_1 = c/n_1$ and $v_2 = c/n_2$, so we have

$$n_1 \sin\theta_1 = n_2 \sin\theta_2,$$

which is Snell's law of refraction, Eq. 23–5. (The law of reflection can be derived from Huygens' principle in a similar way.)

When a light wave travels from one medium to another, its frequency does not change, but its wavelength does. This can be seen from Fig. 24–3, where each of the blue lines representing a wave front corresponds to a crest (peak) of the wave. Then

$$\frac{\lambda_2}{\lambda_1} = \frac{v_2 t}{v_1 t} = \frac{v_2}{v_1} = \frac{n_1}{n_2},$$

where, in the last step, we used Eq. 23–4, $v = c/n$. If medium 1 is a vacuum (or air), so $n_1 = 1$, $v_1 = c$, and we call λ_1 simply λ, then the wavelength in another medium of index of refraction $n \,(= n_2)$ will be

$$\lambda_n = \frac{\lambda}{n}. \qquad (24\text{–}1)$$

This result is consistent with the frequency f being unchanged no matter what medium the wave is traveling in, since $c = f\lambda$.

> EXERCISE A A light beam in air with wavelength = 500 nm, frequency = 6.0×10^{14} Hz, and speed = 3.0×10^8 m/s goes into glass which has an index of refraction = 1.5. What are the wavelength, frequency, and speed of the light in the glass?

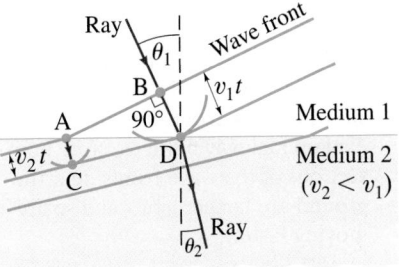

FIGURE 24–3 Refraction explained, using Huygens' principle. Wave fronts are perpendicular to the rays.

⚠️ CAUTION

Frequency is fixed, wavelength can change

(a)

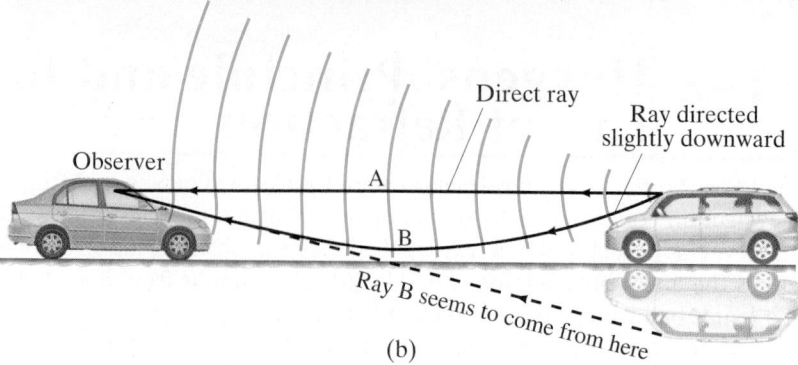

Direct ray

Ray directed slightly downward

Observer

A

B

Ray B seems to come from here

(b)

FIGURE 24–4 (a) A highway mirage. (b) Drawing (greatly exaggerated) showing wave fronts and rays to explain highway mirages. Note how sections of the wave fronts near the ground are farther apart and so are moving faster.

🔵 **PHYSICS APPLIED**

Highway mirages

FIGURE 24–5 (a) Young's double-slit experiment. (b) If light consists of particles, we would expect to see two bright lines on the screen behind the slits. (c) In fact, many lines are observed. The slits and their separation need to be very thin.

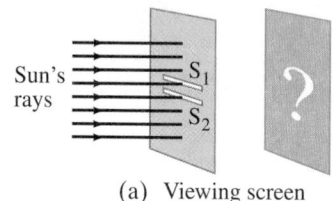

Sun's rays

S_1
S_2

?

(a)   Viewing screen

(b)   Viewing screen (particle theory prediction)

(c)   Viewing screen (actual)

Wave fronts can be used to explain how mirages are produced by refraction of light. For example, on a hot day motorists sometimes see a mirage of water on the highway ahead of them, with distant vehicles seemingly reflected in it (Fig. 24–4a). On a hot day, there can be a layer of very hot air next to the roadway (made hot by the Sun beating down on the road). Hot air is less dense than cooler air, so the index of refraction is slightly lower in the hot air. In Fig. 24–4b, we see a diagram of light coming from one point on a distant car (on the right) heading left toward the observer. Wave fronts and two rays (perpendicular to the wave fronts) are shown. Ray A heads directly at the observer and follows a straight-line path, and represents the normal view of the distant car. Ray B is a ray initially directed slightly downward but, instead of hitting the road, it bends slightly as it moves through layers of air of different index of refraction. The wave fronts, shown in blue in Fig. 24–4b, move slightly faster in the layers of (less dense) air nearer the ground. Thus ray B is bent as shown, and seems to the observer to be coming from below (dashed line) as if reflected off the road. Hence the mirage.

24–3 Interference—Young's Double-Slit Experiment

In 1801, the Englishman Thomas Young (1773–1829) obtained convincing evidence for the wave nature of light and was even able to measure wavelengths for visible light. Figure 24–5a shows a schematic diagram of Young's famous double-slit experiment. To have light from a single source, Young used the sunlight passing through a very narrow slit in a window covering. This beam of parallel rays falls on a screen containing two closely spaced slits, S_1 and S_2. (The slits and their separation are very narrow, not much larger than the wavelength of the light.) If light consists of tiny particles, we would expect to see two bright lines on a screen placed behind the slits as in (b). But instead, a series of bright lines are seen as in (c). Young was able to explain this result as a **wave-interference** phenomenon.

To understand why, consider **plane waves**[†] of light of a single wavelength—called **monochromatic**, meaning "one color"—falling on the two slits as shown in Fig. 24–6. Because of diffraction, the waves leaving the two small slits spread out as shown. This is equivalent to the interference pattern produced when two rocks are thrown into a lake (Fig. 11–38), or when sound from two loudspeakers interferes (Fig. 12–16). Recall Section 11–11 on wave interference.

[†]See pages 312 and 628.

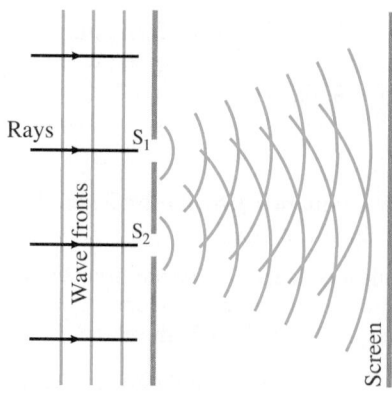

Rays

S_1

Wave fronts

S_2

Screen

FIGURE 24–6 Plane waves (parallel flat wave fronts) fall on two slits. If light is a wave, light passing through one of two slits should interfere with light passing through the other slit.

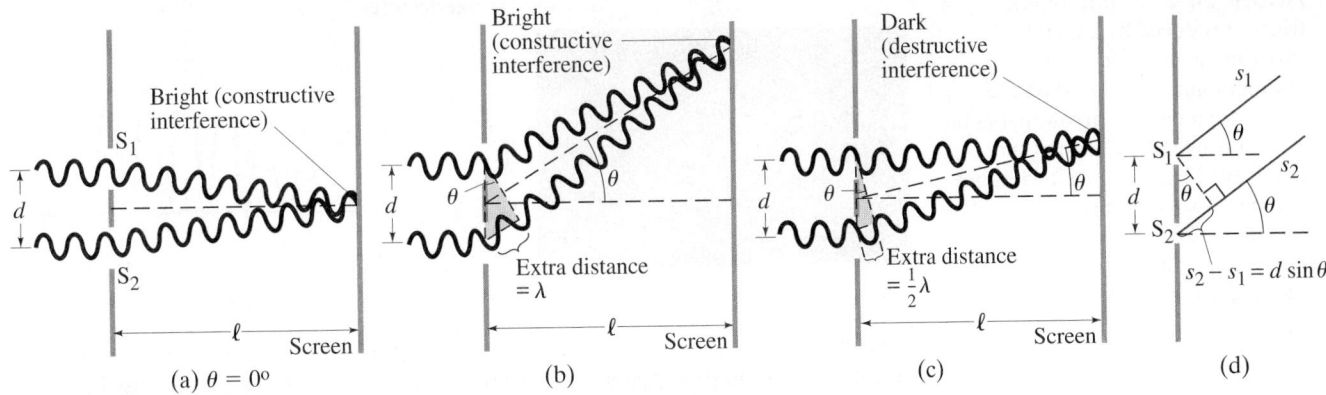

(a) $\theta = 0^\circ$ (b) (c) (d)

FIGURE 24–7 How the wave theory explains the pattern of lines seen in the double-slit experiment. (a) At the center of the screen, waves from each slit travel the same distance and are in phase. [Assume $\ell \gg d$.] (b) At this angle θ, the lower wave travels an extra distance of one whole wavelength, and the waves are in phase; note from the shaded triangle that the path difference equals $d \sin \theta$. (c) For this angle θ, the lower wave travels an extra distance equal to one-half wavelength, so the two waves arrive at the screen fully out of phase. (d) A more detailed diagram showing the geometry for parts (b) and (c).

To see how an interference pattern is produced on the screen, we make use of Fig. 24–7. Waves of wavelength λ are shown entering the slits S_1 and S_2, which are a distance d apart. The waves spread out in all directions after passing through the slits (Fig. 24–6), but they are shown in Figs. 24–7a, b, and c only for three different angles θ. In Fig. 24–7a, the waves reaching the center of the screen are shown ($\theta = 0^\circ$). Waves from the two slits travel the same distance, so they are **in phase**: a crest of one wave arrives at the same time as a crest of the other wave. Hence the amplitudes of the two waves add to form a larger amplitude as shown in Fig. 24–8a. This is **constructive interference**, and there is a bright line at the center of the screen. Constructive interference also occurs when the paths of the two rays differ by one wavelength (or any whole number of wavelengths), as shown in Fig. 24–7b; also here there will be a bright line on the screen. But if one ray travels an extra distance of one-half wavelength (or $\frac{3}{2}\lambda$, $\frac{5}{2}\lambda$, and so on), the two waves are exactly **out of phase** (Section 11–11) when they reach the screen: the crests of one wave arrive at the same time as the troughs of the other wave, and so they add to produce zero amplitude (Fig. 24–8b). This is **destructive interference**, and the screen is dark, Fig. 24–7c. Thus, there will be a series of bright and dark lines (or **fringes**) on the viewing screen.

FIGURE 24–8 Two traveling waves are shown undergoing (a) constructive interference, (b) destructive interference. (See also Section 11–11.)

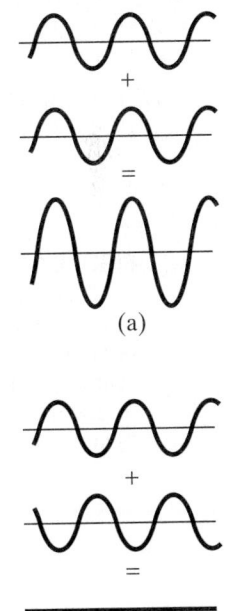

(a)

(b)

To determine exactly where the bright lines fall, first note that Fig. 24–7 is somewhat exaggerated; in real situations, the distance d between the slits is very small compared to the distance ℓ to the screen. The rays from each slit for each case will therefore be essentially parallel, and θ is the angle they make with the horizontal as shown in Fig. 24–7d. From the shaded right triangles shown in Figs. 24–7b and c, we can see that the extra distance traveled by the lower ray is $d \sin \theta$ (seen more clearly in Fig. 24–7d). Constructive interference will occur, and a bright fringe will appear on the screen, when the *path difference*, $d \sin \theta$, equals a whole number of wavelengths:

$$d \sin \theta = m\lambda, \qquad m = 0, 1, 2, \cdots. \qquad \begin{bmatrix} \text{constructive} \\ \text{interference} \\ \text{(bright)} \end{bmatrix} \quad \textbf{(24–2a)}$$

The value of m is called the **order** of the interference fringe. The first order ($m = 1$), for example, is the first fringe on each side of the central fringe (which is at $\theta = 0$, $m = 0$). Destructive interference occurs when the path difference $d \sin \theta$ is $\frac{1}{2}\lambda$, $\frac{3}{2}\lambda$, and so on:

$$d \sin \theta = \left(m + \tfrac{1}{2}\right)\lambda, \qquad m = 0, 1, 2, \cdots. \qquad \begin{bmatrix} \text{destructive} \\ \text{interference} \\ \text{(dark)} \end{bmatrix} \quad \textbf{(24–2b)}$$

The bright fringes are peaks or maxima of light intensity, the dark fringes are minima.

FIGURE 24–9 (a) Interference fringes produced by a double-slit experiment and detected by photographic film placed on the viewing screen. The arrow marks the central fringe. (b) Graph of the intensity of light in the interference pattern. Also shown are values of m for Eq. 24–2a (constructive interference) and Eq. 24–2b (destructive interference).

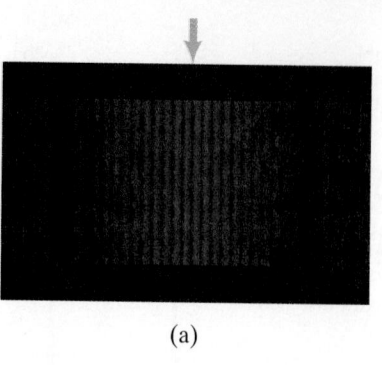

(a)

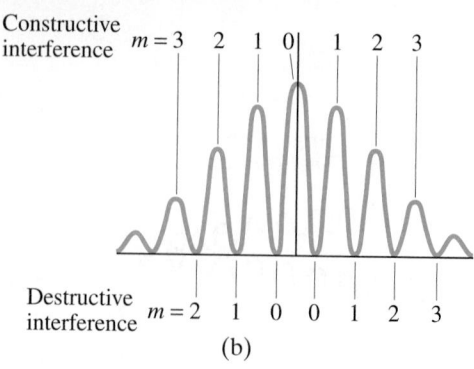

(b)

The intensity of the bright fringes is greatest for the central fringe ($m = 0$) and decreases for higher orders, as shown in Fig. 24–9. How much the intensity decreases with increasing order depends on the width of the two slits.

EXAMPLE 24–1 **Line spacing for double-slit interference.** A screen containing two slits 0.100 mm apart is 1.20 m from the viewing screen. Light of wavelength $\lambda = 500$ nm falls on the slits from a distant source. Approximately how far apart will adjacent bright interference fringes be on the screen?

APPROACH The angular position of bright (constructive interference) fringes is found using Eq. 24–2a. The distance between the first two fringes (say) can be found using right triangles as shown in Fig. 24–10.

SOLUTION Given $d = 0.100$ mm $= 1.00 \times 10^{-4}$ m, $\lambda = 500 \times 10^{-9}$ m, and $\ell = 1.20$ m, the first-order fringe ($m = 1$) occurs at an angle θ given by

$$\sin\theta_1 = \frac{m\lambda}{d} = \frac{(1)(500 \times 10^{-9}\,\text{m})}{1.00 \times 10^{-4}\,\text{m}} = 5.00 \times 10^{-3}.$$

This is a very small angle, so we can take $\sin\theta \approx \theta$, with θ in radians. The first-order fringe will occur a distance x_1 above the center of the screen (see Fig. 24–10), given by $x_1/\ell = \tan\theta_1 \approx \theta_1$, so

$$x_1 \approx \ell\theta_1 = (1.20\,\text{m})(5.00 \times 10^{-3}) = 6.00\,\text{mm}.$$

The second-order fringe ($m = 2$) will occur at

$$x_2 \approx \ell\theta_2 = \ell\frac{2\lambda}{d} = 12.0\,\text{mm}$$

above the center, and so on. Thus the lower-order fringes are 6.00 mm apart.

NOTE The spacing between fringes is essentially uniform until the approximation $\sin\theta \approx \theta$ is no longer valid.

> **⚠ CAUTION**
>
> *Use the approximation*
> $\theta \approx \tan\theta$ *or* $\theta \approx \sin\theta$
> *only if θ is small*
> *and in radians*

FIGURE 24–10 Examples 24–1 and 24–2. For small angles θ (give θ in radians), the interference fringes occur at distance $x = \theta\ell$ above the center fringe ($m = 0$); θ_1 and x_1 are for the first-order fringe ($m = 1$); θ_2 and x_2 are for $m = 2$.

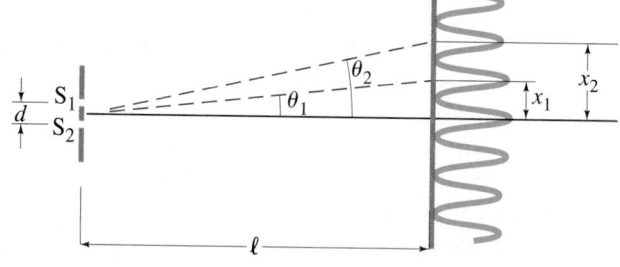

CONCEPTUAL EXAMPLE 24–2 **Changing the wavelength.** (*a*) What happens to the interference pattern shown in Fig. 24–10, Example 24–1, if the incident light (500 nm) is replaced by light of wavelength 700 nm? (*b*) What happens instead if the wavelength stays at 500 nm but the slits are moved farther apart?

RESPONSE (*a*) When λ increases in Eq. 24–2a but d stays the same, the angle θ for bright fringes increases and the interference pattern spreads out. (*b*) Increasing the slit spacing d reduces θ for each order, so the lines are closer together.

From Eqs. 24–2 we can see that, except for the zeroth-order fringe at the center, the position of the fringes depends on wavelength. When white light falls on the two slits, as Young found in his experiments, the central fringe is white, but the first (and higher) order fringes contain a spectrum of colors like a rainbow.

Using Eq. 24–2a, we can see that θ is smallest for violet light and largest for red (Fig. 24–11). By measuring the position of these fringes, Young was the first to determine the wavelengths of visible light. In doing so, he showed that what distinguishes different colors physically is their wavelength (or frequency), an idea put forward earlier by Grimaldi in 1665.

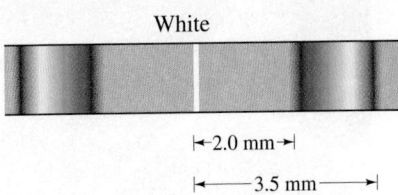

White

|← 2.0 mm →|

|←——— 3.5 mm ———→|

FIGURE 24–11 First-order fringes for a double slit are a full spectrum, like a rainbow. Also Example 24–3.

EXAMPLE 24–3 **Wavelengths from double-slit interference.** White light passes through two slits 0.50 mm apart, and an interference pattern is observed on a screen 2.5 m away. The first-order fringe resembles a rainbow with violet and red light at opposite ends. The violet light is about 2.0 mm and the red 3.5 mm from the center of the central white fringe (Fig. 24–11). Estimate the wavelengths for the violet and red light.

APPROACH We find the angles for violet and red light from the distances given and the diagram of Fig. 24–10. Then we use Eq. 24–2a to obtain the wavelengths. Because 3.5 mm is much less than 2.5 m, we can use the small-angle approximation.

SOLUTION We use Eq. 24–2a ($d \sin \theta = m\lambda$) with $m = 1$, $d = 5.0 \times 10^{-4}$ m, and $\sin \theta \approx \tan \theta \approx \theta$. Also $\theta \approx x/\ell$ (Fig. 24–10), so for violet light, $x = 2.0$ mm, and

$$\lambda = \frac{d \sin \theta}{m} \approx \frac{d\theta}{m} \approx \frac{d}{m}\frac{x}{\ell} = \left(\frac{5.0 \times 10^{-4}\,\text{m}}{1}\right)\left(\frac{2.0 \times 10^{-3}\,\text{m}}{2.5\,\text{m}}\right) = 4.0 \times 10^{-7}\,\text{m},$$

or 400 nm. For red light, $x = 3.5$ mm, so

$$\lambda \approx \frac{d}{m}\frac{x}{\ell} = \left(\frac{5.0 \times 10^{-4}\,\text{m}}{1}\right)\left(\frac{3.5 \times 10^{-3}\,\text{m}}{2.5\,\text{m}}\right) = 7.0 \times 10^{-7}\,\text{m} = 700\,\text{nm}.$$

EXERCISE B For the setup in Example 24–3, how far from the central white fringe is the first-order fringe for green light $\lambda = 500$ nm?

Coherence

The two slits in Figs. 24–6 and 24–7 act as if they were two sources of radiation. They are called **coherent sources** because the waves leaving them have the same wavelength and frequency, and bear the same phase relationship to each other at all times. This happens because the waves come from a single source to the left of the two slits. An interference pattern is observed only when the sources are coherent. If two tiny lightbulbs replaced the two slits, an interference pattern would not be seen. The light emitted by one lightbulb would have a random phase with respect to the second bulb, and the screen would be more or less uniformly illuminated. Two such sources, whose output waves have phases that bear no fixed relationship to each other over time, are called **incoherent sources**.

24–4 The Visible Spectrum and Dispersion

Two of the most important properties of light are readily describable in terms of the wave theory of light: intensity (or brightness) and color. The **intensity** of light is the energy it carries per unit area per unit time, and is related to the square of the amplitude of the wave, as for any wave (see Section 11–9, or Eqs. 22–7 and 22–8). The **color** of light is related to the frequency f or wavelength λ of the light. (Recall $\lambda f = c = 3.0 \times 10^8$ m/s, Eq. 22–4.) Visible light—that to which our eyes are sensitive—consists of frequencies from 4×10^{14} Hz to 7.5×10^{14} Hz, corresponding to wavelengths in air of about 400 nm to 750 nm.[†] This is the **visible spectrum**, and within it lie the different colors from violet to red, as shown in Fig. 24–12.

[†]Sometimes the angstrom (Å) unit is used when referring to light: $1\,\text{Å} = 1 \times 10^{-10}$ m. Visible light has wavelengths in air of 4000 Å to 7500 Å.

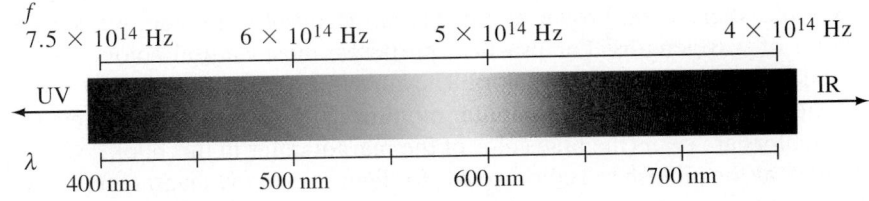

FIGURE 24–12 The spectrum of visible light, showing the range of frequencies and wavelengths in air for the various colors. Many colors, such as brown, do not appear in the spectrum; they are made from a mixture of wavelengths.

FIGURE 24–13 White light passing through a prism is spread out into its constituent colors.

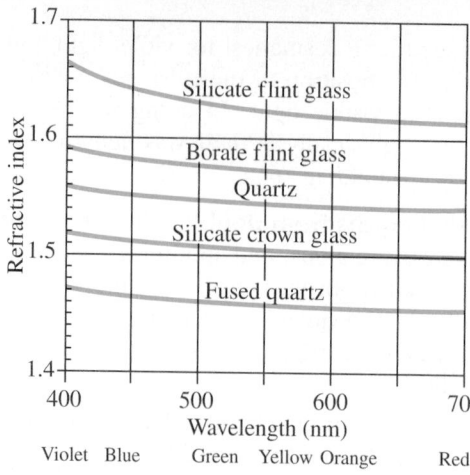

FIGURE 24–14 Index of refraction as a function of wavelength for various transparent solids.

FIGURE 24–15 White light dispersed by a prism into the visible spectrum.

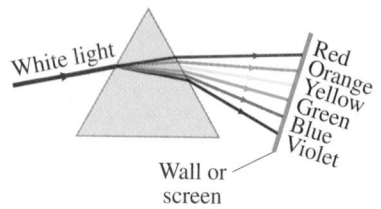

PHYSICS APPLIED

Rainbows

Light with wavelength (in air) shorter than 400 nm (= violet) is called **ultraviolet** (UV), and light with wavelength longer than 750 nm (= red) is called **infrared** (IR).[†] Although human eyes are not sensitive to UV or IR, some types of photographic film and other detectors do respond to them.

A prism can separate white light into a rainbow of colors, as shown in Fig. 24–13. This happens when the index of refraction of a material depends on the wavelength, as shown for several materials in Fig. 24–14. White light is a mixture of all visible wavelengths, and when incident on a prism, as in Fig. 24–15, the different wavelengths are bent to varying degrees. Because the index of refraction is greater for the shorter wavelengths, violet light is bent the most and red the least, as shown in Fig. 24–15. This spreading of white light into the full spectrum is called **dispersion**.

Rainbows are a spectacular example of dispersion—by drops of water. You can see rainbows when you look at falling water droplets with the Sun behind you. Figure 24–16 shows how red and violet rays are bent by spherical water droplets and are reflected off the back surface of the droplet. Red is bent the least and so reaches the observer's eyes from droplets higher in the sky, as shown in Fig. 24–16a. Thus the top of the rainbow is red.

FIGURE 24–16 (a) Ray diagram showing how a rainbow (b) is formed.

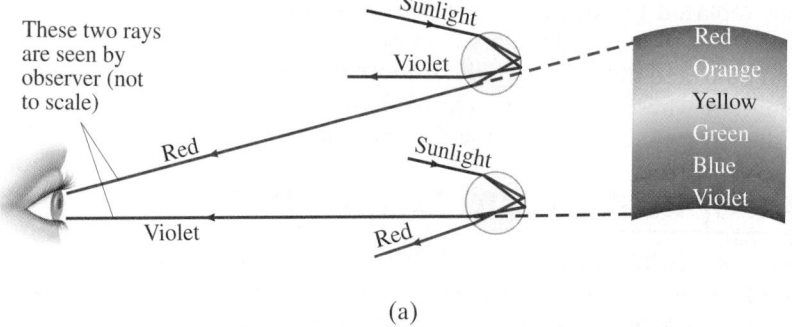

(a)

(b)

FIGURE 24–17 Diamond.

Diamonds achieve their brilliance (Fig. 24–17) from a combination of dispersion and total internal reflection. Because diamonds have a very high index of refraction of about 2.4, the critical angle for total internal reflection is only 25°. The light dispersed into a spectrum inside the diamond therefore strikes many of the internal surfaces of the diamond before it strikes one at less than 25° and emerges. After many such reflections, the light has traveled far enough that the colors have become sufficiently separated to be seen individually and brilliantly by the eye after leaving the diamond.

The visible spectrum, Fig. 24–12, does not show all the colors seen in nature. For example, there is no brown in Fig. 24–12. Many of the colors we see are a mixture of wavelengths. For practical purposes, most natural colors can be reproduced using three primary colors. They are red, green, and blue for direct source viewing such as TV and computer monitors. For inks used in printing, the primary colors are cyan (the blue color of the margin notes in this book), yellow, and magenta (the pinkish red color we use for light rays in ray diagrams).

[†]The complete electromagnetic spectrum is illustrated in Fig. 22–8.

| CONCEPTUAL EXAMPLE 24–4 | **Observed color of light under water.** We |

said that color depends on wavelength. For example, light of wavelength $\lambda_0 = 650$ nm in air, we see red. If we observe the same object when under water, it still looks red. But the wavelength in water λ_w is (Eq. 24–1) $\lambda_w = \lambda_0/n_w = 650$ nm$/1.33 = 489$ nm. Light with wavelength 489 nm in air would appear blue in air. Can you explain why the light appears red rather than blue when observed under water?

RESPONSE Today we have little doubt that it is our brains that express colors, based on the wavelengths of light that strike the receptor cells within the retina (at the rear of the eyeball, as diagrammed in the next Chapter, Fig. 25–9). For objects under water, the water does nothing to change the frequency, but does change the wavelength to λ_0/n_w. When that light enters the eye, the frequency is still unchanged, but the speed is changed to c/n_{eye} where n_{eye} is the index of refraction of the fluid that fills the interior of the eye and is in contact with the retina. The wavelength of light that reaches the retina is $\lambda_{eye} = \lambda_0/n_{eye}$, and is the same whether the light enters from the air or from water.

24–5 Diffraction by a Single Slit or Disk

Young's double-slit experiment put the wave theory of light on a firm footing. But full acceptance came only with studies on diffraction (Section 24–1) more than a decade later, in the 1810s and 1820s.

We have already discussed diffraction briefly with regard to water waves (Section 11–14) as well as for light (Section 24–1). We have seen that diffraction refers to the spreading or bending of waves around edges. Let's look in more detail.

In 1819 Augustin Fresnel (1788–1827) presented to the French Academy a wave theory of light that predicted and explained interference and diffraction effects. Almost immediately Siméon Poisson (1781–1840) pointed out a counter-intuitive inference: according to Fresnel's wave theory, if light from a point source were to fall on a solid disk, part of the incident light would be diffracted around the edges and would constructively interfere at the center of the shadow (Fig. 24–18). That prediction seemed very unlikely. But when the experiment was actually carried out by Francois Arago, the bright spot was seen at the very center of the shadow (Fig. 24–19a). This was strong evidence for the wave theory.

Figure 24–19a is a photograph of the shadow cast by a coin using a coherent point source of light, a laser in this case. The bright spot is clearly present at the center. Note also the bright and dark fringes beyond the shadow. These resemble the interference fringes of a double slit. Indeed, they are due to interference of waves diffracted around the outer edge of the disk, and the group of fringes is referred to as a **diffraction pattern**. A diffraction pattern exists around any sharp-edged object illuminated by a point source, as shown in Figs. 24–19b and c. We are not always aware of diffraction because most sources of light in everyday life are not points, so light from different parts of the source washes out the pattern.

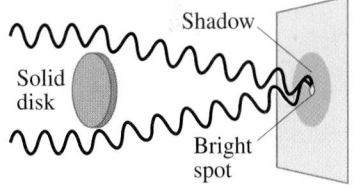

FIGURE 24–18 If light is a wave, a bright spot will appear at the center of the shadow of a solid disk illuminated by a point source of monochromatic light.

FIGURE 24–19 Diffraction pattern of (a) a circular disk (a coin), (b) scissors, (c) a single slit, each illuminated by a (nearly) point source of coherent monochromatic light.

(a)

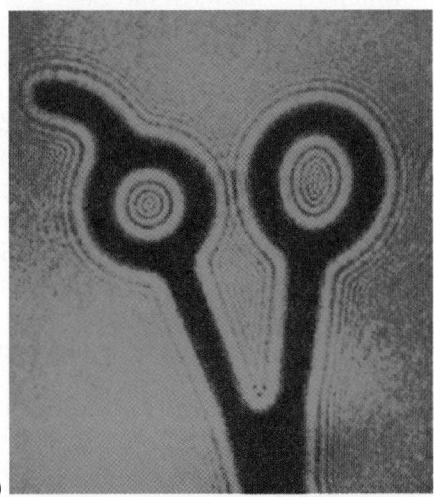

(b)

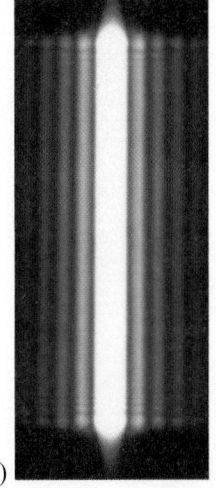

(c)

687

To see how a diffraction pattern arises, we analyze the important case of monochromatic light passing through a narrow slit (as for Fig. 24–19c). We assume that parallel rays (plane waves) of light pass straight through a slit of width D to a viewing screen very far away.[†] As we know from studying water waves and from Huygens' principle, waves passing through a slit spread out in all directions. We will now examine how the waves passing through different parts of the slit interfere with each other.

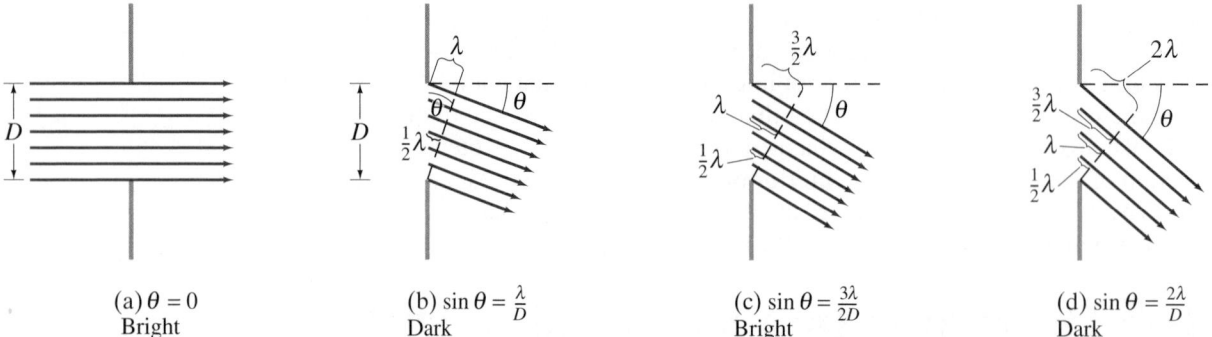

(a) $\theta = 0$
Bright

(b) $\sin\theta = \frac{\lambda}{D}$
Dark

(c) $\sin\theta = \frac{3\lambda}{2D}$
Bright

(d) $\sin\theta = \frac{2\lambda}{D}$
Dark

FIGURE 24–20 Analysis of diffraction pattern formed by light passing through a narrow slit of width D.

Parallel rays of monochromatic light pass through the narrow slit as shown in Fig. 24–20a. The slit width D is on the order of the wavelength λ of the light, but the slit's length (into and out of page) may be large compared to λ. The light falls on a screen which is assumed to be very far away, so the rays heading toward any point are very nearly parallel before they meet at the screen. First we consider rays that pass straight through as in Fig. 24–20a. They are all in phase, so there will be a central bright spot on the screen (see Fig. 24–19c). In Fig. 24–20b, we consider rays moving at an angle θ such that the ray from the top of the slit travels exactly one wavelength farther than the ray from the bottom edge to reach the screen. The ray passing through the very center of the slit will travel one-half wavelength farther than the ray at the bottom of the slit. These two rays will be exactly out of phase with one another and so will destructively interfere when they overlap at the screen. Similarly, a ray slightly above the bottom one will cancel a ray that is the same distance above the central one. Indeed, each ray passing through the lower half of the slit will cancel with a corresponding ray passing through the upper half. Thus, all the rays destructively interfere in pairs, and so the light intensity will be zero on the viewing screen at this angle. The angle θ at which this takes place can be seen from Fig. 24–20b to occur when $\lambda = D \sin\theta$, so

$$\sin\theta = \frac{\lambda}{D}. \qquad \text{[first minimum]} \quad \textbf{(24–3a)}$$

The light intensity is a maximum at $\theta = 0°$ and decreases to a minimum (intensity = zero) at the angle θ given by Eq. 24–3a.

Now consider a larger angle θ such that the top ray travels $\frac{3}{2}\lambda$ farther than the bottom ray, as in Fig. 24–20c. In this case, the rays from the bottom third of the slit will cancel in pairs with those in the middle third because they will be $\lambda/2$ out of phase. However, light from the top third of the slit will still reach the screen, so there will be a bright spot (or fringe) centered near $\sin\theta \approx 3\lambda/2D$, but it will not be nearly as bright as the central spot at $\theta = 0°$. For an even larger angle θ such that the top ray travels 2λ farther than the bottom ray, Fig. 24–20d, rays from the bottom quarter of the slit will cancel with those in the quarter just above it because the path lengths differ by $\lambda/2$. And the rays through the quarter of the slit just above center will cancel with those through the top quarter. At this angle there will again be a minimum of zero intensity in the diffraction pattern.

[†]If the viewing screen is not far away, lenses can be used to make the rays parallel.

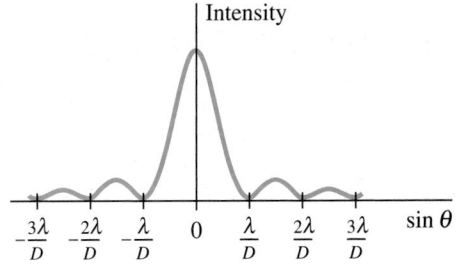

Intensity

$$-\frac{3\lambda}{D} \quad -\frac{2\lambda}{D} \quad -\frac{\lambda}{D} \quad 0 \quad \frac{\lambda}{D} \quad \frac{2\lambda}{D} \quad \frac{3\lambda}{D} \quad \sin\theta$$

FIGURE 24–21 Intensity in the diffraction pattern of a single slit as a function of $\sin\theta$. Note that the central maximum is not only much higher than the maxima to each side, but it is also twice as wide ($2\lambda/D$ wide) as any of the others (each only λ/D wide).

A plot of the intensity as a function of angle is shown in Fig. 24–21. This corresponds well with the photo of Fig. 24–19c. Notice that minima (zero intensity) occur on both sides of center at

$$D \sin\theta = m\lambda, \qquad m = \pm 1, \pm 2, \pm 3, \cdots, \qquad \text{[minima]} \quad \textbf{(24–3b)}$$

but *not* at $m = 0$ where there is the strongest maximum. Between the minima, smaller intensity maxima occur at approximately (not exactly) $m \approx \frac{3}{2}, \frac{5}{2}, \cdots$.

Note that the *minima* for a diffraction pattern, Eq. 24–3b, satisfy a criterion that looks very similar to that for the *maxima* (bright spots or fringes) for double-slit interference, Eq. 24–2a. Also note that D is a single slit width, whereas d in Eqs. 24–2 is the distance between two slits.

◇ **CAUTION**
Don't confuse Eqs. 24–2 for interference with Eqs. 24–3 for diffraction; note the differences

EXAMPLE 24–5 **Single-slit diffraction maximum.** Light of wavelength 750 nm passes through a slit 1.0×10^{-3} mm wide. How wide is the central maximum (a) in degrees, and (b) in centimeters, on a screen 20 cm away?

APPROACH The width of the central maximum goes from the first minimum on one side to the first minimum on the other side. We use Eq. 24–3a to find the angular position of the first single-slit diffraction minimum.

SOLUTION (a) The first minimum occurs at

$$\sin\theta = \frac{\lambda}{D} = \frac{7.5 \times 10^{-7}\,\text{m}}{1.0 \times 10^{-6}\,\text{m}} = 0.75.$$

So $\theta = 49°$. This is the angle between the center and the first minimum, Fig. 24–22. The angle subtended by the whole central maximum, between the minima above and below the center, is twice this, or 98°.

(b) The width of the central maximum is $2x$, where $\tan\theta = x/20$ cm. So $2x = 2(20\,\text{cm})(\tan 49°) = 46\,\text{cm}$.

NOTE A large width of the screen will be illuminated, but it will not normally be very bright since the amount of light that passes through such a small slit will be small and it is spread over a large area. Note also that we *cannot* use the small-angle approximation here ($\theta \approx \sin\theta \approx \tan\theta$) because θ is large.

FIGURE 24–22 Example 24–5.

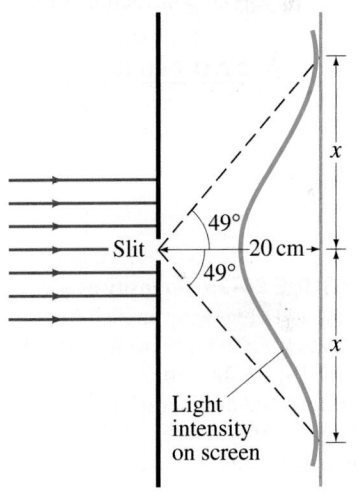

Slit — 49° — 20 cm — 49°
x
x
Light intensity on screen

EXERCISE C In Example 24–5, red light ($\lambda = 750$ nm) was used. If instead yellow light ($\lambda = 550$ nm) had been used, would the central maximum be wider or narrower?

CONCEPTUAL EXAMPLE 24–6 **Diffraction spreads.** Light shines through a small rectangular slit that is narrower in the vertical direction than the horizontal, Fig. 24–23. (a) Would you expect the diffraction pattern to be more spread out in the vertical direction or in the horizontal direction? (b) Should a rectangular loudspeaker horn at a stadium be tall and narrow, or wide and flat?

RESPONSE (a) From Eq. 24–3a we can see that if we make the slit width D smaller, the pattern spreads out more (θ will be larger in Eq. 24–3a). This is consistent with our study of waves in Chapter 11. The diffraction through the rectangular hole will be wider vertically, since the opening is smaller in that direction.

(b) For a stadium loudspeaker, the sound pattern desired is one spread out horizontally, so the horn should be tall and narrow (rotate Fig. 24–23 by 90°).

FIGURE 24–23 Example 24–6.

24–6 Diffraction Grating

A large number of equally spaced parallel slits is called a **diffraction grating**, although the term "interference grating" might be as appropriate. Gratings can be made by precision machining of very fine parallel lines on a glass plate. The untouched spaces between the lines serve as the slits. Photographic transparencies of an original grating serve as inexpensive gratings. Gratings containing 10,000 lines or slits per centimeter are common, and are very useful for precise measurements of wavelengths. A diffraction grating containing slits is called a **transmission grating**. Another type of diffraction grating is the **reflection grating**, made by ruling fine lines on a metallic or glass surface from which light is reflected and analyzed. The analysis is basically the same as for a transmission grating, which we now discuss.

The analysis of a diffraction grating is much like that of Young's double-slit experiment. We assume parallel rays of light are incident on the grating as shown in Fig. 24–24. We also assume that the slits are narrow enough so that diffraction by each of them spreads light over a very wide angle on a distant screen beyond the grating, and interference can occur with light from all the other slits. Light rays that pass through each slit without deviation ($\theta = 0°$) interfere constructively to produce a bright maximum at the center of the screen. Constructive interference also occurs at an angle θ such that rays from adjacent slits travel an extra distance of $\Delta\ell = m\lambda$, where m is an integer. If d is the distance *between* slits, then we see from Fig. 24–24 that $\Delta\ell = d \sin\theta$, and

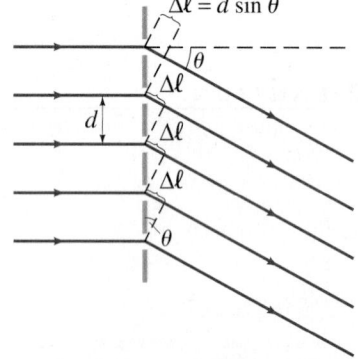

FIGURE 24–24 Diffraction grating.

$$\sin\theta = \frac{m\lambda}{d}, \qquad m = 0, 1, 2, \cdots \qquad \left[\begin{array}{l}\text{diffraction grating}\\\text{principal maxima}\end{array}\right] \quad \textbf{(24–4)}$$

is the criterion to have a brightness maximum. This is the same equation as for the double-slit situation, and again m is called the **order** of the pattern.

There is an important difference between a double-slit and a multiple-slit pattern. The bright maxima are much *sharper* and *narrower* for a grating. Why? Suppose the angle θ in Fig. 24–24 is increased just slightly beyond θ required for a maximum. For only two slits, the two waves will be only slightly out of phase, so nearly full constructive interference occurs. This means the maxima are wide (see Fig. 24–9). For a grating, the waves from two adjacent slits will also not be significantly out of phase. But waves from one slit and those from a second one a few hundred slits away may be exactly out of phase; all or nearly all the light can cancel in pairs in this way. For example, suppose the angle θ is very slightly different from its first-order maximum, so that the extra path length for a pair of adjacent slits is not exactly λ but rather 1.0010λ. The wave through one slit and another one 500 slits below will have a path difference of $1\lambda + (500)(0.0010\lambda) = 1.5000\lambda$, or $1\frac{1}{2}$ wavelengths, so the two will be out of phase and cancel. A pair of slits, one below each of these, will also cancel. That is, the light from slit 1 cancels with light from slit 501; light from slit 2 cancels with light from slit 502, and so on. Thus even for a tiny angle[†] corresponding to an extra path length of $\frac{1}{1000}\lambda$, there is much destructive interference, and so the maxima of a diffraction grating are very narrow. The more slits there are in a grating, the sharper will be the peaks (see Fig. 24–25). Because a grating produces much sharper maxima than two slits alone, and also much brighter maxima because there are many more slits, a grating is a far more precise device for measuring wavelengths.

FIGURE 24–25 Intensity as a function of viewing angle θ (or position on the screen) for (a) two slits, (b) six slits. For a diffraction grating, the number of slits is very large ($\approx 10^4$) and the peaks are narrower still.

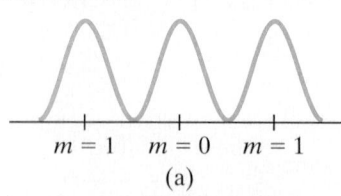

$m = 1$ $m = 0$ $m = 1$

(a)

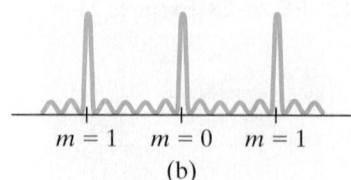

$m = 1$ $m = 0$ $m = 1$

(b)

Suppose the light striking a diffraction grating is not monochromatic, but consists of two or more distinct wavelengths. Then for all orders other than $m = 0$, each wavelength will produce a maximum at a different angle (Eq. 24–4), forming a line on the screen as shown in Fig. 24–26a.

[†]Depending on the total number of slits, there may or may not be complete cancellation for such an angle, so there will be very tiny peaks between the main maxima (see Fig. 24–25b), but they are usually much too small to be seen.

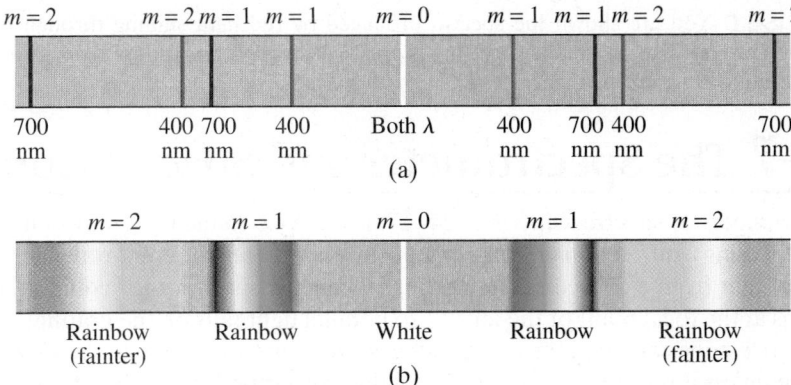

FIGURE 24–26 Spectra produced by a grating: (a) two wavelengths, 400 nm and 700 nm; (b) white light. The second order will normally be dimmer than the first order. (Higher orders are not shown.) If the grating spacing is small enough, the second and higher orders will be missing.

If white light strikes a grating, the central ($m = 0$) maximum will be a sharp white line. But for all other orders, there will be a distinct spectrum of colors spread out over a certain angular width, Fig. 24–26b. Because a diffraction grating spreads out light into its component wavelengths, the resulting pattern is called a **spectrum**.

EXAMPLE 24–7 **Diffraction grating: line positions.** Determine the angular positions of the first- and second-order lines (maxima) for light of wavelength 400 nm and 700 nm incident on a grating containing 10,000 slits per centimeter.

APPROACH First we find the distance d between grating slits: if the grating has N slits in 1 m, then the distance between slits is $d = 1/N$ meters. Then we use Eq. 24–4, $\sin\theta = m\lambda/d$, to get the angles for the two wavelengths for $m = 1$ and 2.

SOLUTION The grating contains 1.00×10^4 slits/cm $= 1.00 \times 10^6$ slits/m, which means the distance between slits is $d = (1/1.00 \times 10^6)\,\text{m} = 1.00 \times 10^{-6}\,\text{m} = 1.00\ \mu\text{m}$. In first order ($m = 1$), the angles are

$$\sin\theta_{400} = \frac{m\lambda}{d} = \frac{(1)(4.00 \times 10^{-7}\,\text{m})}{1.00 \times 10^{-6}\,\text{m}} = 0.400$$

$$\sin\theta_{700} = \frac{(1)(7.00 \times 10^{-7}\,\text{m})}{1.00 \times 10^{-6}\,\text{m}} = 0.700$$

so $\theta_{400} = 23.6°$ and $\theta_{700} = 44.4°$. In second order,

$$\sin\theta_{400} = \frac{2\lambda}{d} = \frac{(2)(4.00 \times 10^{-7}\,\text{m})}{1.00 \times 10^{-6}\,\text{m}} = 0.800$$

$$\sin\theta_{700} = \frac{(2)(7.00 \times 10^{-7}\,\text{m})}{1.00 \times 10^{-6}\,\text{m}} = 1.40$$

so $\theta_{400} = 53.1°$. But the second order does not exist for $\lambda = 700\ \text{nm}$ because $\sin\theta$ cannot exceed 1. No higher orders will appear.

EXAMPLE 24–8 **Spectra overlap.** White light containing wavelengths from 400 nm to 750 nm strikes a grating containing 4000 slits/cm. Show that the blue at $\lambda = 450\ \text{nm}$ of the third-order spectrum overlaps the red at 700 nm of the second order.

APPROACH We use $\sin\theta = m\lambda/d$ to calculate the angular positions of the $m = 3$ blue maximum and the $m = 2$ red one.

SOLUTION The grating spacing is $d = (1/4000)\,\text{cm} = 2.50 \times 10^{-6}\,\text{m}$. The blue of the third order occurs at an angle θ given by

$$\sin\theta = \frac{m\lambda}{d} = \frac{(3)(4.50 \times 10^{-7}\,\text{m})}{(2.50 \times 10^{-6}\,\text{m})} = 0.540.$$

Red in second order occurs at

$$\sin\theta = \frac{(2)(7.00 \times 10^{-7}\,\text{m})}{(2.50 \times 10^{-6}\,\text{m})} = 0.560,$$

which is a greater angle; so the second order overlaps into the beginning of the third-order spectrum.

24–7 The Spectrometer and Spectroscopy

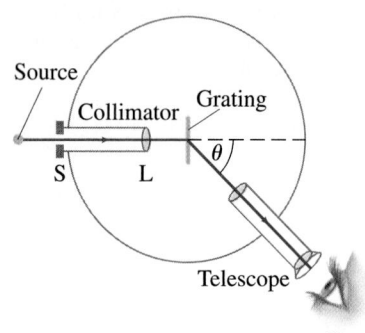

FIGURE 24–27 Spectrometer or spectroscope.

A **spectrometer** or **spectroscope**, Fig. 24–27, is a device to measure wavelengths accurately using a diffraction grating (or a prism) to separate different wavelengths of light. Light from a source passes through a narrow slit S in the "collimator." The slit is at the focal point of the lens L, so parallel light falls on the grating. The movable telescope can bring the rays to a focus. Nothing will be seen in the viewing telescope unless it is positioned at an angle θ that corresponds to a diffraction peak (first order is usually used) of a wavelength emitted by the source. The angle θ can be measured to very high accuracy, so the wavelength can be determined to high accuracy using Eq. 24–4:

$$\lambda = \frac{d}{m}\sin\theta,$$

where m is an integer representing the order, and d is the distance between grating slits. The bright line you see in a spectrometer corresponding to a discrete particular wavelength is actually an image of the slit S. A narrower slit results in dimmer light, but we can measure the angular position more precisely. If the light contains a continuous range of wavelengths, then a continuous spectrum is seen in the spectroscope.

The spectrometer in Fig. 24–27 uses a transmission grating. Others may use a reflection grating, or sometimes a prism. A prism works because of dispersion (Section 24–4), bending light of different wavelengths into different angles. A prism is not a linear device and must be calibrated because λ is not $\propto \sin\theta$; see Fig. 24–14.

An important use of a spectrometer is for the identification of atoms or molecules. When a gas is heated or an electric current is passed through it, the gas emits a characteristic **line spectrum**. That is, only certain discrete wavelengths of light are emitted, and these are different for different elements and compounds.[†] Figure 24–28 shows the line spectra for a number of elements in the gas state. Line spectra occur only for gases at high temperatures and low pressure and density. The light from heated solids, such as a lightbulb filament, and even from a dense gaseous object such as the Sun, produces a **continuous spectrum** including a wide range of wavelengths.

Figure 24–28 also shows the Sun's "continuous spectrum," which contains a number of *dark* lines (only the most prominent are shown), called **absorption lines**. Atoms and molecules can absorb light at the same wavelengths at which they emit light.

FIGURE 24–28 Line spectra for the gases indicated, and the spectrum from the Sun showing absorption lines.

[†]Why atoms and molecules emit line spectra was a great mystery for many years and played a central role in the development of modern quantum theory, as we shall see in Chapter 27.

The Sun's absorption lines are due to absorption by atoms and molecules in the cooler outer atmosphere of the Sun, as well as by atoms and molecules in the Earth's atmosphere. A careful analysis of all the Sun's thousands of absorption lines reveals that at least two-thirds of all elements are present in the Sun's atmosphere. The presence of elements in the atmosphere of nearby planets, in interstellar space, and in stars, is also determined by spectroscopy.

Spectroscopy is useful for determining the presence of certain types of molecules in laboratory specimens where chemical analysis would be difficult. For example, biological DNA and different types of protein absorb light in particular regions of the spectrum (such as in the UV). The material to be examined, which is often in solution, is placed in a monochromatic light beam whose wavelength is selected by the placement angle of a diffraction grating or prism. The amount of absorption, as compared to a standard solution without the specimen, can reveal not only the presence of a particular type of molecule, but also its concentration.

Light emission and absorption also occur outside the visible part of the spectrum, such as in the UV and IR regions. Glass absorbs light in these regions, so reflection gratings and mirrors (in place of lenses) are used. Special types of film or sensors are used for detection.

PHYSICS APPLIED
Chemical and biochemical analysis by spectroscopy

EXAMPLE 24–9 **Hydrogen spectrum.** Light emitted by hot hydrogen gas is observed with a spectroscope using a diffraction grating having 1.00×10^4 slits/cm. The spectral lines nearest to the center (0°) are a violet line at 24.2°, a blue line at 25.7°, a blue-green line at 29.1°, and a red line at 41.0° from the center. What are the wavelengths of these spectral lines of hydrogen?

APPROACH We get the wavelengths from the angles by using $\lambda = (d/m) \sin \theta$ where d is the spacing between slits, and m is the order of the spectrum (Eq. 24–4).

SOLUTION Since these are the closest lines to $\theta = 0°$, this is the first-order spectrum ($m = 1$). The slit spacing is $d = 1/(1.00 \times 10^4 \, \text{cm}^{-1}) = 1.00 \times 10^{-6} \, \text{m}$. The wavelength of the violet line is

$$\lambda = \left(\frac{d}{m}\right) \sin \theta = \left(\frac{1.00 \times 10^{-6} \, \text{m}}{1}\right) \sin 24.2° = 4.10 \times 10^{-7} \, \text{m} = 410 \, \text{nm}.$$

The other wavelengths are:

blue:	$\lambda = (1.00 \times 10^{-6} \, \text{m}) \sin 25.7° = 434 \, \text{nm},$
blue-green:	$\lambda = (1.00 \times 10^{-6} \, \text{m}) \sin 29.1° = 486 \, \text{nm},$
red:	$\lambda = (1.00 \times 10^{-6} \, \text{m}) \sin 41.0° = 656 \, \text{nm}.$

NOTE In an unknown mixture of gases, these four spectral lines need to be seen to identify that the mixture contains hydrogen.

24–8 Interference in Thin Films

Interference of light gives rise to many everyday phenomena such as the bright colors reflected from soap bubbles and from thin oil or gasoline films on water, Fig. 24–29. In these and other cases, the colors are a result of constructive interference between light reflected from the two surfaces of the thin film. The effect is observed only if the thickness of the film is on the order of the wavelength of the light. If the film thickness is greater than a few wavelengths, the effect gets washed out.

FIGURE 24–29 Thin-film interference patterns seen in (a) a soap bubble, (b) a thin film of soapy water, and (c) a thin layer of oil on wet pavement.

(a)

(b)

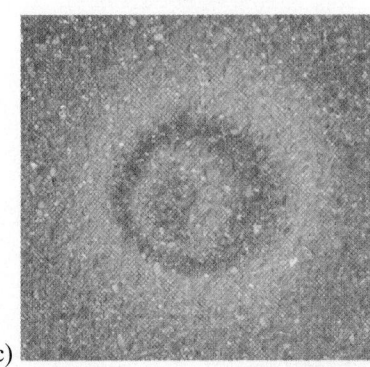
(c)

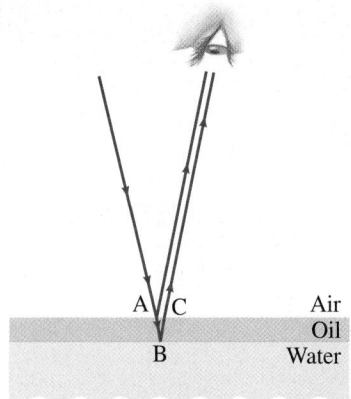

FIGURE 24–30 Light reflected from the upper and lower surfaces of a thin film of oil lying on water.

To see how this **thin-film interference** happens, consider a smooth surface of water on top of which is a thin uniform layer of another substance, say an oil whose index of refraction is less than that of water (we'll see why we assume this shortly); see Fig. 24–30. Assume for now that the incident light is of a single wavelength. Part of the incident light is reflected at A on the top surface, and part of the light transmitted is reflected at B on the lower surface. The part reflected at the lower surface must travel the extra distance ABC. If this *path difference* ABC equals one or a whole number of wavelengths in the film (λ_n), the two waves will reach the eye in phase and interfere constructively. Hence the region AC on the surface film will appear bright. But if ABC equals $\frac{1}{2}\lambda_n, \frac{3}{2}\lambda_n$, and so on, the two waves will be exactly out of phase and destructive interference occurs: the area AC on the film will show no reflection—it will be dark (transparent to the dark material below). The wavelength λ_n is *the wavelength in the film:* $\lambda_n = \lambda/n$, where n is the index of refraction in the film and λ is the wavelength in vacuum. See Eq. 24–1.

When white light falls on such a film, the path difference ABC will equal λ_n (or $m\lambda_n$, with $m =$ an integer) for only one wavelength at a given viewing angle. The color corresponding to λ (λ in air) will be seen as very bright. For light viewed at a slightly different angle, the path difference ABC will be longer or shorter and a different color will undergo constructive interference. Thus, for an extended (nonpoint) source emitting white light, a series of bright colors will be seen next to one another. Variations in thickness of the film will also alter the path difference ABC and therefore affect the color of light that is most strongly reflected.

EXERCISE E Return to the Chapter-Opening Question, page 679, and answer it again now. Try to explain why you may have answered differently the first time.

When a curved glass surface is placed in contact with a flat glass surface, Fig. 24–31, a series of concentric rings is seen when illuminated from above by either white light (as shown) or by monochromatic light. These are called **Newton's rings**[†] and they are due to interference between waves reflected by the top and bottom surfaces of the very thin *air gap* between the two pieces of glass. Because this gap (which is equivalent to a thin film) increases in width from the central contact point out to the edges, the extra path length for the lower ray (equal to BCD) varies. Where it equals 0, $\frac{1}{2}\lambda$, λ, $\frac{3}{2}\lambda$, 2λ, and so on, it corresponds to constructive and destructive interference; and this gives rise to the series of bright colored circles seen in Fig. 24–31b. The color you see at a given radius corresponds to constructive interference; at that radius, other colors partially or fully destructively interfere. (If monochromatic light is used, the rings are alternately bright and dark.)

The point of contact of the two glass surfaces (A in Fig. 24–31a) is not bright in Fig. 24–31b. Since the path difference is zero here, our previous analysis would suggest that the waves reflected from each surface are in phase—so this central area ought to be bright. But it is dark, which tells us the two waves must be completely

[†]Although Newton gave an elaborate description of them, they had been first observed and described by his contemporary, Robert Hooke.

FIGURE 24–31 Newton's rings. (a) Light rays reflected from upper and lower surfaces of the thin air gap can interfere. (b) Photograph of interference patterns using white light.

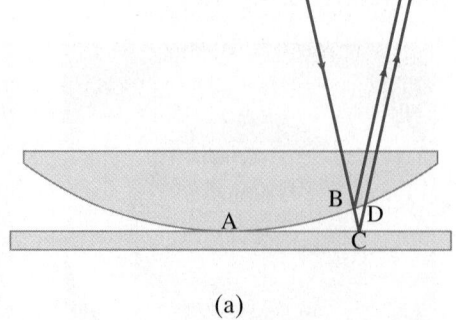

(a)

(b)

out of phase. This can happen only if one of the waves, upon reflection, flips over—a crest becomes a trough—see Fig. 24–32. We say that the reflected wave has undergone a **phase shift** of 180°, or of half a wave cycle ($\frac{1}{2}\lambda$). Indeed, this and other experiments reveal that, at normal incidence,

> **a beam of light, reflected by a material with index of refraction greater than that of the material in which it is traveling, changes phase by 180° or $\frac{1}{2}$ cycle;**

see Fig. 24–32. This phase shift acts just like a path difference of $\frac{1}{2}\lambda$. If the index of refraction of the reflecting material is less than that of the material in which the light is traveling, no phase shift occurs.[†]

Thus the wave reflected at the curved surface above the air gap in Fig. 24–31a undergoes no change in phase. But the wave reflected at the lower surface, where the beam in air strikes the glass, undergoes a $\frac{1}{2}$-cycle phase shift, equivalent to a $\frac{1}{2}\lambda$ path difference. Thus the two waves reflected near the point of contact A of the two glass surfaces (where the air gap approaches zero thickness) will be a half cycle (or 180°) out of phase, and a dark spot occurs. Bright colored rings will occur when the path difference is $\frac{1}{2}\lambda$, $\frac{3}{2}\lambda$, and so on, because the phase shift at one surface effectively adds a path difference of $\frac{1}{2}\lambda$ ($=\frac{1}{2}$cycle). (If monochromatic light is used, the bright Newton's rings will be separated by dark bands which occur when the path difference BCD in Fig. 24–31a is equal to an integral number of wavelengths.)

Returning for a moment to Fig. 24–30, the light reflecting at both interfaces, air–oil and oil–water, *each* underwent a phase shift of 180° equivalent to a path difference of $\frac{1}{2}\lambda$, since we assumed $n_{\text{water}} > n_{\text{oil}} > n_{\text{air}}$. Because the two phase shifts were equal, they didn't affect our analysis.

EXAMPLE 24–10 | Thin film of air, wedge-shaped. A very fine wire 7.35×10^{-3} mm in diameter is placed between two flat glass plates as in Fig. 24–33a. Light whose wavelength in air is 600 nm falls (and is viewed) perpendicular to the plates and a series of bright and dark bands is seen, Fig. 24–33b. How many light and dark bands will there be in this case? Will the area next to the wire be bright or dark?

APPROACH We need to consider two effects: (1) path differences for rays reflecting from the two close surfaces (thin wedge of air between the two glass plates), and (2) the $\frac{1}{2}$-cycle phase shift at the lower surface (point E in Fig. 24–33a), where rays in air can enter glass (or be reflected). Because there is a phase shift only at the lower surface, there will be a dark band (no reflection) when the path difference is 0, λ, 2λ, 3λ, and so on. Since the light rays are perpendicular to the plates, the extra path length (DEF) equals $2t$, where t is the thickness of the air gap at any point.

SOLUTION Dark bands will occur where

$$2t = m\lambda, \qquad m = 0, 1, 2, \cdots.$$

Bright bands occur when $2t = (m + \frac{1}{2})\lambda$, where m is an integer. At the position of the wire, $t = 7.35 \times 10^{-6}$ m. At this point there will be $2t/\lambda = (2)(7.35 \times 10^{-6}\,\text{m})/(6.00 \times 10^{-7}\,\text{m}) = 24.5$ wavelengths. This is a "half integer," so the area next to the wire will be bright. There will be a total of 25 dark lines along the plates, corresponding to path lengths DEF of 0λ, 1λ, 2λ, 3λ, $\cdots$, 24λ, including the one at the point of contact A ($m = 0$). Between them, there will be 24 bright lines plus the one at the end, or 25.

NOTE The bright and dark bands will be straight only if the glass plates are extremely flat. If they are not, the pattern is uneven, as in Fig. 24–33c. Thus we see a very precise way of testing a glass surface for flatness. Spherical lens surfaces can be tested for precision by placing the lens on a flat glass surface and observing Newton's rings (Fig. 24–31b) for perfect circularity.

[†]This result corresponds to the reflection of a wave traveling along a cord when it reaches the end. As we saw in Fig. 11–33, if the end is tied down, the wave changes phase and the pulse flips over, but if the end is free, no phase shift occurs.

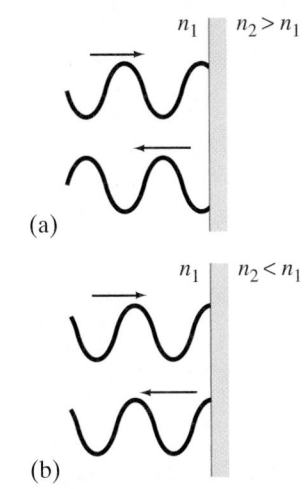

FIGURE 24–32 (a) Reflected ray changes phase by 180° or $\frac{1}{2}$ cycle if $n_2 > n_1$, but (b) does not if $n_2 < n_1$.

FIGURE 24–33 (a) Light rays reflected from the upper and lower surfaces of a thin wedge of air (between two glass plates) interfere to produce bright and dark bands. (b) Pattern observed when glass plates are optically flat; (c) pattern when plates are not so flat. See Example 24–10.

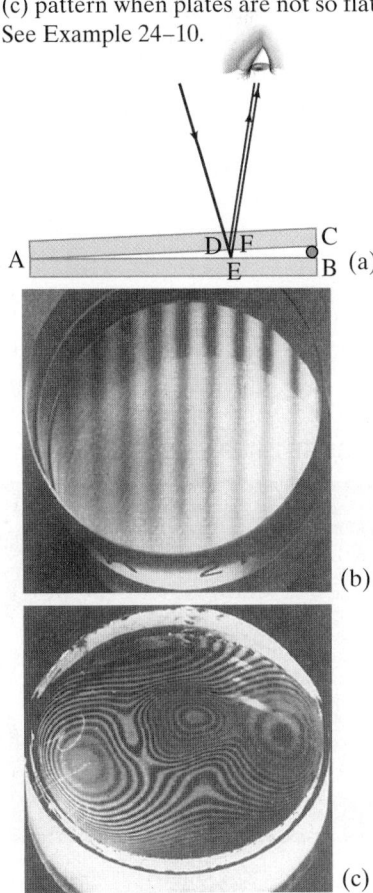

PHYSICS APPLIED
Testing glass for flatness

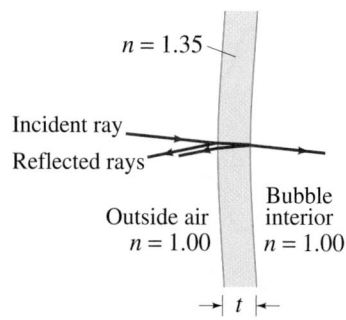

n = 1.35

Incident ray

Reflected rays

Outside air
n = 1.00

Bubble interior
n = 1.00

→| t |←

FIGURE 24–34 Soap bubble, Example 24–11. The incident and reflected rays are assumed to be perpendicular to the bubble's surface. They are shown at a slight angle so we can distinguish them.

◆ **C A U T I O N**

A formula is not enough: you must also check for phase changes at surfaces

When white light (rather than monochromatic light) is incident on the thin wedge of air in Fig. 24–31a or 24–33a, a colorful series of fringes is seen because constructive interference occurs for different wavelengths in the reflected light at different thicknesses along the wedge.

A soap bubble (Fig. 24–29a and Chapter-Opening Photo) is a thin spherical shell (or film) with air inside. The variations in thickness of a soap bubble film give rise to bright colors reflected from the soap bubble. (There is air on both sides of the bubble film.) Similar variations in film thickness produce the bright colors seen reflecting from a thin layer of oil or gasoline on a puddle or lake (Fig. 24–29c). Which wavelengths appear brightest also depends on the viewing angle.

EXAMPLE 24–11 **Thickness of soap bubble skin.** A soap bubble appears green ($\lambda = 540$ nm) at the point on its front surface nearest the viewer. What is the smallest thickness the soap bubble film could have? Assume $n = 1.35$.

APPROACH Assume the light is reflected perpendicularly from the point on a spherical surface nearest the viewer, Fig. 24–34. The light rays also reflect from the inner surface of the soap bubble film as shown. The path difference of these two reflected rays is $2t$, where t is the thickness of the soap film. Light reflected from the first (outer) surface undergoes a 180° phase change (index of refraction of soap is greater than that of air), whereas reflection at the second (inner) surface does not. To determine the thickness t for an interference maximum, we must use the wavelength of light in the soap ($n = 1.35$).

SOLUTION The 180° phase change at only one surface is equivalent to a $\frac{1}{2}\lambda$ path difference. Therefore, green light is bright when the minimum path difference equals $\frac{1}{2}\lambda_n$. Thus, $2t = \lambda_n/2$, so

$$ t = \frac{\lambda_n}{4} = \frac{\lambda}{4n} = \frac{(540 \text{ nm})}{(4)(1.35)} = 100 \text{ nm}. $$

This is the smallest thickness.

NOTE At this small thickness, blue (450 nm) and red (600 nm) also would reflect fairly constructively, so the bubble would appear almost white. The green color is more likely to be seen at the *next* thickness that gives constructive interference, $2t = 3\lambda/2n$, because other colors would be more fully cancelled by destructive interference. Then t would be $t = 3\lambda/4n = 300$ nm. Note that green is seen in air, so $\lambda = 540$ nm (not λ/n).

*Colors in a Thin Soap Film

FIGURE 24–29b (Repeated.)

The thin film of soapy water (in a plastic loop) shown in Fig. 24–29b (repeated here) has stood vertically for a long time. Gravity has pulled the soapy water downward, so the film increases in thickness going toward the bottom. The top section is so thin (perhaps 30 nm thick $\ll \lambda$) that light reflected from the front and back surfaces have almost zero path difference. Thus the 180° phase change at the front surface assures that the two reflected waves are 180° out of phase for all wavelengths of visible light. The white light incident on this thin film does not reflect at the top part of the film, so the top is transparent and we see the background which is black.

Below the black area at the top, there is a thin blue line, and then a white band. The film has thickened to perhaps 75 to 100 nm, so the shortest wavelength (blue) light begins to partially interfere constructively. But just below, where the thickness is slightly greater (100 nm), the path difference is reasonably close to $\lambda/2$ for much of the spectrum and we see white or silver.[†]

Immediately below the white band in this Figure we see a brown band, where $t \approx 200$ nm, and many wavelengths (not all) are close to λ—and those colors destructively interfere, leaving only a few colors to partially interfere constructively, giving us murky brown.

[†]Why? Recall that red starts at 600 nm in air; so most colors in the spectrum lie between 450 nm and 600 nm in air; but in water the wavelengths are $n = 1.33$ times smaller, 340 nm to 450 nm, so a 100-nm thickness is a 200-nm path difference, not far from $\lambda/2$ for most colors.

Farther down in Fig. 24–29b, with increasing thickness t, a path difference $2t = 510$ nm corresponds nicely to $\frac{3}{2}\lambda$ for blue, but not for other colors, so we see blue ($\frac{3}{2}\lambda$ path difference plus $\frac{1}{2}\lambda$ phase change = constructive interference). Other colors experience constructive interference (at $\frac{3}{2}\lambda$ and then at $\frac{5}{2}\lambda$) at still greater thicknesses, so going down we see a series of separated colors something like a rainbow.

In the soap bubble of our Chapter-Opening Photo (page 679), similar things happen: at the top (where the film is thinnest) we see black and then silver-white, just as in the soap film shown in Fig. 24–29b.

Also examine the oil film on wet pavement shown in Fig. 24–29c (repeated here). The oil film is thickest at the center and thins out toward the edges. Notice the whitish outer ring where most colors constructively interfere, which would suggest a thickness on the order of 100 nm as discussed above for the white band in the soap film. Beyond the outer white band of the oil film, Fig. 24–29c, there is still some oil, but the film is so thin that reflected light from upper and lower surfaces destructively interfere and you can see right through this very thin oil film.

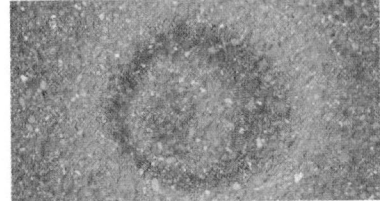

FIGURE 24–29c (Repeated.)

Lens Coatings

An important application of thin-film interference is in the coating of glass to make it "nonreflecting," particularly for lenses. A glass surface reflects about 4% of the light incident upon it. Good-quality cameras, microscopes, and other optical devices may contain six to ten thin lenses. Reflection from all these surfaces can reduce the light level considerably, and multiple reflections produce a background haze that reduces the quality of the image. By reducing reflection, transmission and sharpness are increased.

PHYSICS APPLIED
Lens coatings

A very thin coating on the lens surfaces can reduce reflections considerably. The thickness of the coating is chosen so that light (at least for one wavelength) reflecting from the front and rear surfaces of the film destructively interferes. Destructive interference can occur nearly completely for one particular wavelength depending on the thickness of the coating. Nearby wavelengths will at least partially destructively interfere, but a single coating cannot eliminate reflections for all wavelengths. Nonetheless, a single coating can reduce total reflection from 4% to 1% of the incident light. Often the coating is designed to eliminate the center of the reflected spectrum (around 550 nm). The extremes of the spectrum—red and violet—will not be reduced as much. Since a mixture of red and violet produces purple, the light seen reflected from such coated lenses is purple (Fig. 24–35). Lenses containing two or three separate coatings can more effectively reduce a wider range of reflecting wavelengths.

FIGURE 24–35 A coated lens. Note color of light reflected from the front lens surface.

EXAMPLE 24–12 **Nonreflective coating.** What is the thickness of an optical coating of MgF$_2$ whose index of refraction is $n = 1.38$ and which is designed to eliminate reflected light at wavelengths (in air) around 550 nm when incident normally on glass for which $n = 1.50$?

APPROACH We explicitly follow the procedure outlined in the Problem Solving Strategy on page 697.

SOLUTION

1. **Interference effects.** Consider two rays reflected from the front and rear surfaces of the coating on the lens as shown in Fig. 24–36. The rays are drawn not quite perpendicular to the lens so we can see each of them. These two reflected rays will interfere with each other.

2. **Constructive interference.** We want to eliminate reflected light, so we do not consider constructive interference.

3. **Destructive interference.** To eliminate reflection, we want reflected rays 1 and 2 to be $\frac{1}{2}$ cycle out of phase with each other so that they destructively interfere. The phase difference is due to the path difference $2t$ traveled by ray 2, as well as any phase change in either ray due to reflection.

4. **Reflection phase shift.** Rays 1 and 2 *both* undergo a change of phase by $\frac{1}{2}$ cycle when they reflect from the coating's front and rear surfaces, respectively (at both surfaces the index of refraction increases). Thus there is no net change in phase due to the reflections. The net phase difference will be due to the extra path $2t$ taken by ray 2 in the coating, where $n = 1.38$. We want $2t$ to equal $\frac{1}{2}\lambda_n$ so that destructive interference occurs, where $\lambda_n = \lambda/n$ is the wavelength in the coating. With $2t = \lambda_n/2 = \lambda/2n$, then

$$t = \frac{\lambda_n}{4} = \frac{\lambda}{4n} = \frac{(550 \text{ nm})}{(4)(1.38)} = 99.6 \text{ nm}.$$

NOTE We could have set $2t = \left(m + \frac{1}{2}\right)\lambda_n$, where m is an integer. The smallest thickness ($m = 0$) is usually chosen because destructive interference will occur over the widest angle.

NOTE Complete destructive interference occurs only for the given wavelength of visible light. Longer and shorter wavelengths will have only partial cancellation.

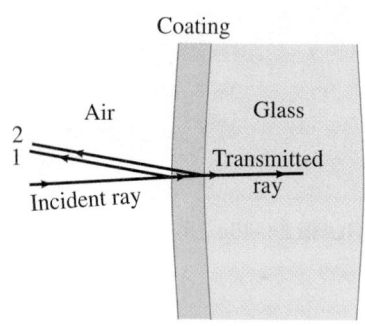

FIGURE 24–36 Lens coating, Example 24–12. Incident ray of light is partially reflected at the front surface of a lens coating (ray 1) and again partially reflected at the rear surface of the coating (ray 2), with most of the energy passing through as the transmitted ray into the glass.

*24–9 Michelson Interferometer

A useful instrument involving wave interference is the **Michelson interferometer** (Fig. 24–37),[†] invented by the American Albert A. Michelson (Section 22–4). Monochromatic light from a single point on an extended source is shown striking a half-silvered mirror M$_S$. This **beam splitter** mirror M$_S$ has a thin layer of silver that reflects only half the light that hits it, so that half of the beam passes through to a fixed mirror M$_2$, where it is reflected back. The other half is reflected by M$_S$ to a mirror M$_1$ that is movable (by a fine-thread screw), where it is also reflected back. Upon its return, part of beam 1 passes through M$_S$ and reaches a sensor or the eye; and part of beam 2, on its return, is reflected by M$_S$ into the eye. If the two path lengths are identical, the two coherent beams entering the eye constructively interfere and brightness will be seen. If the movable mirror is moved a distance $\lambda/4$, one beam will travel an extra distance equal to $\lambda/2$ (because it travels back and forth over the distance $\lambda/4$). In this case, the two beams will destructively interfere and darkness will be seen. As M$_1$ is moved farther, brightness will recur (when the path difference is λ), then darkness, and so on.

Very precise length measurements can be made with an interferometer. The motion of mirror M$_1$ by only $\frac{1}{4}\lambda$ produces a clear difference between brightness and darkness. For $\lambda = 400$ nm, this means a precision of 100 nm, or 10^{-4} mm! If mirror M$_1$ is tilted very slightly, the bright or dark spots are seen instead as a series of bright and dark lines or "fringes" that move as M$_1$ moves. By counting the number of fringes (or fractions thereof) that pass a reference line, extremely precise length measurements can be made.

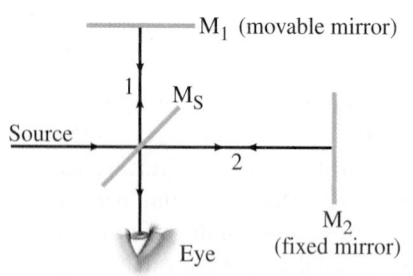

FIGURE 24–37 Michelson interferometer.

[†]There are other types of interferometer, but Michelson's is the best known.

24–10 Polarization

An important and useful property of light is that it can be *polarized*. To see what this means, let us examine waves traveling on a rope. A rope can oscillate in a vertical plane, Fig. 24–38a, or in a horizontal plane, Fig. 24–38b. In either case, the wave is said to be **linearly polarized** or **plane-polarized**—*the oscillations are in a plane.*

If we now place an obstacle containing a vertical slit in the path of the wave, Fig. 24–39, a vertically polarized wave passes through the vertical slit, but a horizontally polarized wave will not. If a horizontal slit were used, the vertically polarized wave would be stopped. If both types of slit were used, both types of wave would be stopped by one slit or the other. Note that polarization can exist *only* for *transverse waves*, and not for longitudinal waves such as sound. The latter oscillate only along the direction of motion, and neither orientation of slit would stop them.

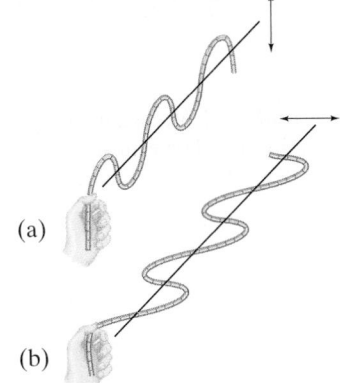

FIGURE 24–38 Transverse waves on a rope polarized (a) in a vertical plane and (b) in a horizontal plane.

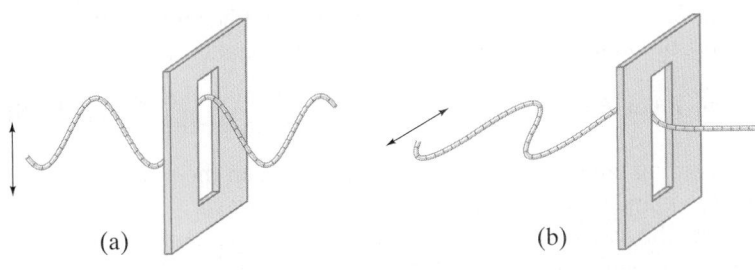

(a)   (b)

FIGURE 24–39 (a) A vertically polarized wave passes through a vertical slit, but (b) a horizontally polarized wave will not.

Maxwell's theory of light as electromagnetic (EM) waves predicted that light can be polarized since an EM wave is a transverse wave. The direction of polarization in a plane-polarized EM wave is taken as the direction of the electric field vector $\vec{E}$.

Light is not necessarily polarized. It can also be **unpolarized**, which means that the source has oscillations in many planes at once, as shown in Fig. 24–40. Ordinary lightbulbs emit unpolarized light, as does the Sun.

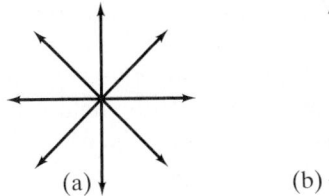

(a)   (b)

FIGURE 24–40 (a) Oscillation of the electric field vectors in unpolarized light. The light is traveling into or out of the page. (b) Electric field in linear polarized light.

Polaroids (Polarization by Absorption)

Plane-polarized light can be obtained from unpolarized light using certain crystals such as tourmaline. Or, more commonly, we use a **Polaroid sheet**. (Polaroid materials were invented in 1929 by Edwin Land.) A Polaroid sheet consists of long complex molecules arranged parallel to one another. Such a Polaroid acts like a series of parallel slits to allow one orientation of polarization to pass through nearly undiminished. This direction is called the *transmission axis* of the Polaroid. Polarization perpendicular to this direction is absorbed almost completely by the Polaroid.

Absorption by a Polaroid can be explained at the molecular level. An electric field $\vec{E}$ that oscillates parallel to the long molecules can set electrons into motion along the molecules, thus doing work on them and transferring energy. Hence, if $\vec{E}$ is parallel to the molecules, it gets absorbed. An electric field $\vec{E}$ perpendicular to the long molecules does not have this possibility of doing work and transferring its energy, and so passes through freely. When we speak of the *transmission axis* of a Polaroid, we mean the direction for which $\vec{E}$ is passed, so a Polaroid axis is *perpendicular* to the long molecules. [If we want to think of there being slits between the parallel molecules in the sense of Fig. 24–39, then Fig. 24–39 would apply for the $\vec{B}$ field in the EM wave, not the $\vec{E}$ field.]

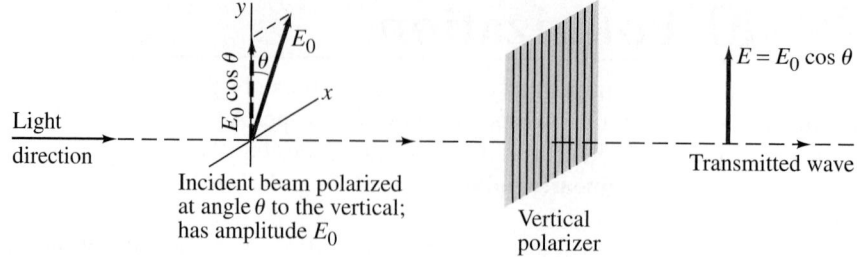

FIGURE 24–41 Vertical Polaroid transmits only the vertical component of a wave (electric field) incident upon it.

Light direction

Incident beam polarized at angle θ to the vertical; has amplitude E_0

Vertical polarizer

$E = E_0 \cos \theta$

Transmitted wave

If a beam of plane-polarized light strikes a Polaroid whose transmission axis is at an angle θ to the incident polarization direction, the beam will emerge plane-polarized parallel to the Polaroid transmission axis, and the amplitude of E will be reduced to $E \cos \theta$, Fig. 24–41. Thus, a Polaroid passes only that component of polarization (the electric field vector, $\vec{E}$) that is parallel to its transmission axis. Because the intensity of a light beam is proportional to the square of the amplitude (Sections 11–9 and 22–5), the intensity of a plane-polarized beam transmitted by a polarizer is proportional to $(E_0 \cos \theta)^2$, a relation called Malus' law,

$$I = I_0 \cos^2 \theta, \qquad \left[\begin{array}{c}\text{intensity of plane-polarized} \\ \text{wave passed by polarizer}\end{array}\right] \quad \textbf{(24–5)}$$

where I_0 is the incoming intensity and θ is the angle between the polarizer transmission axis and the plane of polarization of the incoming wave.

A Polaroid can be used as a **polarizer** to *produce* plane-polarized light from unpolarized light, since only the component of light parallel to the axis is transmitted. A Polaroid can also be used as an **analyzer** to determine (1) if light is polarized and (2) the plane of polarization. A Polaroid acting as an analyzer will pass the same amount of light independent of the orientation of its axis if the light is unpolarized; try rotating one lens of a pair of Polaroid sunglasses while looking through it at a lightbulb. If the light is polarized, however, when you rotate the Polaroid the transmitted light will be a maximum when the plane of polarization is parallel to the Polaroid's transmission axis, and a minimum when perpendicular to it. If you do this while looking at the sky, preferably at right angles to the Sun's direction, you will see that skylight is polarized. (Direct sunlight is unpolarized, but don't look directly at the Sun, even through a polarizer, for damage to the eye may occur.) If the light transmitted by an analyzer Polaroid falls to zero at one orientation, then the light is 100% plane-polarized. If it merely reaches a minimum, the light is *partially polarized*.

Unpolarized light consists of light with random directions of polarization. Each of these polarization directions can be resolved into components along two mutually perpendicular directions. On average, an unpolarized beam can be thought of as two plane-polarized beams of equal magnitude perpendicular to one another. When unpolarized light passes through a polarizer, one component is eliminated. So the intensity of the light passing through is reduced by half because half the light is eliminated: $I = \frac{1}{2} I_0$ (Fig. 24–42).

When two Polaroids are *crossed*—that is, their polarizing axes are perpendicular to one another—unpolarized light can be entirely stopped. As shown in Fig. 24–43, unpolarized light is made plane-polarized by the first Polaroid (the polarizer). The second Polaroid, the analyzer, then eliminates this component since its transmission axis is perpendicular to the first.

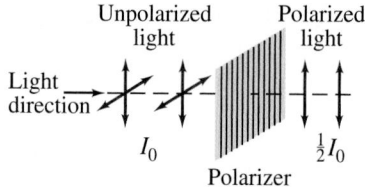

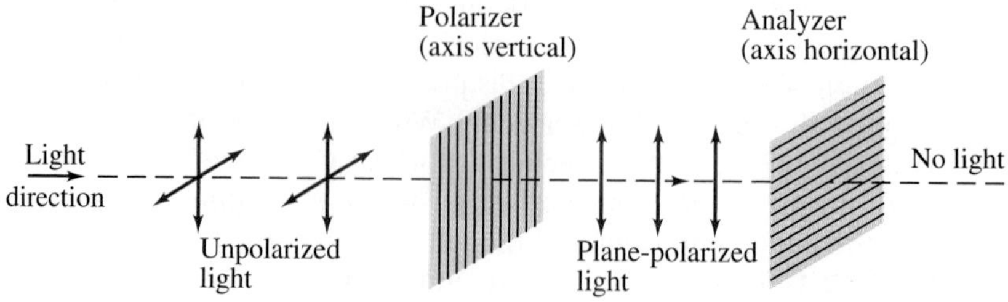

Unpolarized light

Polarized light

Light direction

I_0

Polarizer

$\frac{1}{2} I_0$

FIGURE 24–42 Unpolarized light has equal intensity vertical and horizontal components. After passing through a polarizer, one of these components is eliminated. The intensity of the light is reduced to half.

FIGURE 24–43 Crossed Polaroids completely eliminate light.

Polarizer (axis vertical)

Analyzer (axis horizontal)

Light direction

Unpolarized light

Plane-polarized light

No light

You can try this with Polaroid sunglasses (Fig. 24–44). Note that Polaroid sunglasses eliminate 50% of unpolarized light because of their polarizing property; they absorb even more because they are colored. Plane-polarized light in any direction is also stopped by crossed Polaroids.

EXAMPLE 24–13 | **Two Polaroids at 60°.** Unpolarized light passes through two Polaroids; the axis of the first is vertical and that of the second is at 60° to the vertical. Describe the orientation and intensity of the transmitted light.

APPROACH Half of the unpolarized light is absorbed by the first Polaroid, and the remaining light emerges plane-polarized vertically. When that light passes through the second Polaroid, the intensity is further reduced according to Eq. 24–5, and the plane of polarization is then along the axis of the second Polaroid.

FIGURE 24–44 Crossed Polaroids. When the two polarized sunglass lenses overlap, with axes perpendicular, almost no light passes through.

SOLUTION The first Polaroid eliminates half the light, so the intensity is reduced by half: $I_1 = \frac{1}{2}I_0$. The light reaching the second polarizer is vertically polarized and so is reduced in intensity (Eq. 24–5) to

$$I_2 = I_1(\cos 60°)^2 = \frac{1}{4}I_1.$$

Thus, $I_2 = \frac{1}{8}I_0$. The transmitted light has an intensity one-eighth that of the original and is plane-polarized at a 60° angle to the vertical.

CONCEPTUAL EXAMPLE 24–14 | **Three Polaroids.** We saw in Fig. 24–43 that when unpolarized light falls on two crossed Polaroids (axes at 90°), no light passes through. What happens if a third Polaroid, with axis at 45° to each of the other two, is placed between them (Fig. 24–45a)?

RESPONSE We start just as in Example 24–13 and recall again that light emerging from each Polaroid is polarized parallel to that Polaroid's axis. Thus the angle in Eq. 24–5 is that between the transmission axes of each pair of Polaroids taken in turn. The first Polaroid changes the unpolarized light to plane-polarized and reduces the intensity from I_0 to $I_1 = \frac{1}{2}I_0$. The second polarizer further reduces the intensity by $(\cos 45°)^2$, Eq. 24–5:

$$I_2 = I_1(\cos 45°)^2 = \frac{1}{2}I_1 = \frac{1}{4}I_0.$$

FIGURE 24–45 Example 24–14.

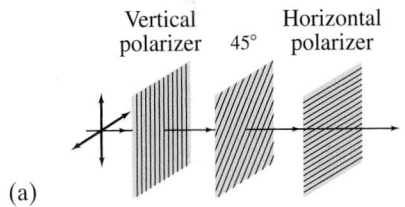

(a)

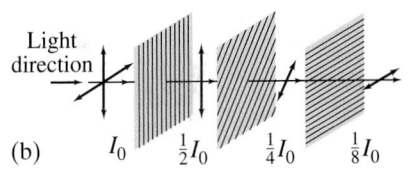

(b)

The light leaving the second polarizer is plane-polarized at 45° (Fig. 24–45b) relative to the third polarizer, so the third one reduces the intensity to

$$I_3 = I_2(\cos 45°)^2 = \frac{1}{2}I_2,$$

or $I_3 = \frac{1}{8}I_0$. Thus $\frac{1}{8}$ of the original intensity gets transmitted.

NOTE If we don't insert the 45° Polaroid, zero intensity results (Fig. 24–43).

EXERCISE F How much light would pass through if the 45° polarizer in Example 24–14 was placed not between the other two polarizers but (a) before the vertical (first) polarizer, or (b) after the horizontal polarizer?

Polarization by Reflection

Another means of producing polarized light from unpolarized light is by reflection. When light strikes a nonmetallic surface at any angle other than perpendicular, the reflected beam is polarized preferentially in the plane parallel to the surface, Fig. 24–46. In other words, the component with polarization in the plane perpendicular to the surface is preferentially transmitted or absorbed. You can check this by rotating Polaroid sunglasses while looking through them at a flat surface of a lake or road. Since most outdoor surfaces are horizontal, Polaroid sunglasses are made with their axes vertical to eliminate the more strongly reflected horizontal component, and thus reduce glare.

FIGURE 24–46 Light reflected from a nonmetallic surface, such as the smooth surface of water in a lake, is partially polarized parallel to the surface.

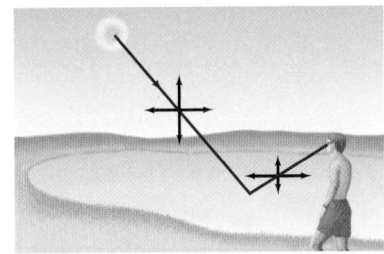

FIGURE 24–47 Photographs of a lake, (a) allowing all light into the camera lens, and (b) using a polarizer. The polarizer is adjusted to absorb most of the (polarized) light reflected from the water's surface, allowing the dimmer light from the bottom of the lake, and any fish lying there, to be seen more readily.

 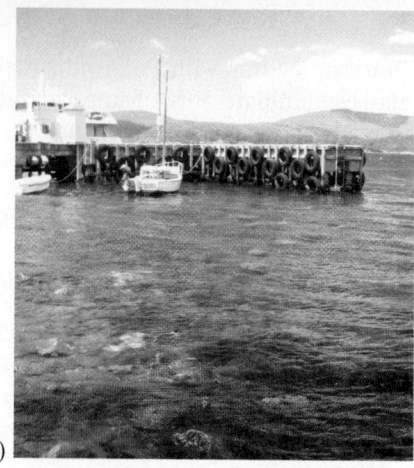

(a)   (b)

People who go fishing wear Polaroids to eliminate reflected glare from the surface of a lake or stream and thus see beneath the water more clearly (Fig. 24–47).

The amount of polarization in the reflected beam depends on the angle, varying from no polarization at normal incidence to 100% polarization at an angle known as the **polarizing angle** θ_p.[†] This angle is related to the index of refraction of the two materials on either side of the boundary by the equation

$$\tan \theta_p = \frac{n_2}{n_1}, \tag{24–6a}$$

where n_1 is the index of refraction of the material in which the incident beam is traveling, and n_2 is that of the medium beyond the reflecting boundary. If the beam is traveling in air, $n_1 = 1$, and Eq. 24–6a becomes

$$\tan \theta_p = n. \tag{24–6b}$$

The polarizing angle θ_p is also called **Brewster's angle**, and Eqs. 24–6 *Brewster's law*, after the Scottish physicist David Brewster (1781–1868), who worked it out experimentally in 1812. Equations 24–6 can be derived from the electromagnetic wave theory of light. It is interesting that at Brewster's angle, the reflected ray and the transmitted (refracted) ray make a 90° angle to each other; that is, $\theta_p + \theta_r = 90°$, where θ_r is the refraction angle (Fig. 24–48). This can be seen

FIGURE 24–48 At θ_p the reflected light is plane-polarized parallel to the surface, and $\theta_p + \theta_r = 90°$, where θ_r is the refraction angle. (The large dots represent vibrations perpendicular to the page.)

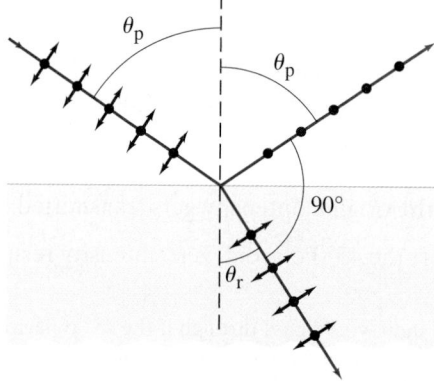

by substituting Eq. 24–6a, $n_2 = n_1 \tan \theta_p = n_1 \sin \theta_p / \cos \theta_p$, into Snell's law, $n_1 \sin \theta_p = n_2 \sin \theta_r$, which gives $\cos \theta_p = \sin \theta_r$ which can only hold if $\theta_p = 90° - \theta_r$ (see Trigonometric identities inside back cover or Appendix page A-8).

EXAMPLE 24–15 **Polarizing angle.** (*a*) At what incident angle is sunlight reflected from a lake perfectly plane-polarized? (*b*) What is the refraction angle?

APPROACH The polarizing angle at the surface is Brewster's angle, Eq. 24–6b. We find the angle of refraction from Snell's law.

SOLUTION (*a*) We use Eq. 24–6b with $n = 1.33$, so $\tan \theta_p = 1.33$ giving $\theta_p = 53.1°$.
(*b*) From Snell's law, $\sin \theta_r = \sin \theta_p / n = \sin 53.1° / 1.33 = 0.601$ giving $\theta_r = 36.9°$.

NOTE $\theta_p + \theta_r = 53.1° + 36.9° = 90.0°$, as expected.

[†]Only a fraction of the incident light is reflected at the surface of a transparent medium. Although this reflected light is 100% polarized (if $\theta = \theta_p$), the remainder of the light, which is transmitted into the new medium, is only partially polarized.

*24–11 Liquid Crystal Displays (LCD)

A wonderful use of polarization is in a **liquid crystal display** (LCD). LCDs are used as the display in cell phones, other hand-held electronic devices, and flat-panel computer and television screens.

A liquid crystal display is made up of many tiny rectangles called **pixels**, or "picture elements." The picture you see depends on which pixels are dark or light and of what color, as suggested in Fig. 24–49 for a simple black and white picture.

Liquid crystals are organic materials that at room temperature exist in a phase that is neither fully solid nor fully liquid. They are sort of gooey, and their molecules display a randomness of position characteristic of liquids, as discussed in Section 13–1 and Fig. 13–2b. They also show some of the orderliness of a solid crystal (Fig. 13–2a), but only in one dimension.

The liquid crystals we find useful are made up of relatively rigid rod-like molecules that interact weakly with each other and tend to align parallel to each other, as shown in Fig. 24–50.

In a simple LCD, each pixel (picture element) contains liquid crystal material sandwiched between two glass plates whose inner surfaces have been brushed to form nanometer-wide parallel scratches. The rod-like liquid crystal molecules in contact with the scratches tend to line up along the scratches. The two plates typically have their scratches at 90° to each other, and the weak electric forces between the rod-like molecules tend to keep them nearly aligned with their nearest neighbors, resulting in the twisted pattern shown in Fig. 24–51a.

The outer surfaces of the glass plates each have a thin film polarizer, they too oriented at 90° to each other. Unpolarized light incident from the left becomes plane-polarized, and the liquid crystal molecules keep this polarization aligned with their rod-like shape. That is, the plane of polarization of the light rotates with the molecules as the light passes through the liquid crystal. The light emerges with its plane of polarization rotated by 90°, and readily passes through the second polarizer, Fig. 24–51a. A tiny LCD pixel in this situation will appear bright.

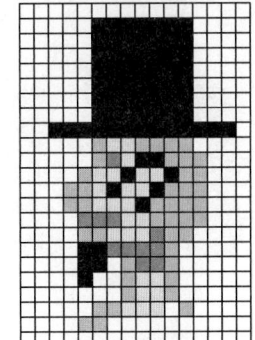

FIGURE 24–49 Example of an image made up of many small squares or *pixels* (picture elements).

FIGURE 24–50 Liquid crystal molecules tend to align in one dimension (parallel to each other) but have random positions (left-right, up-down).

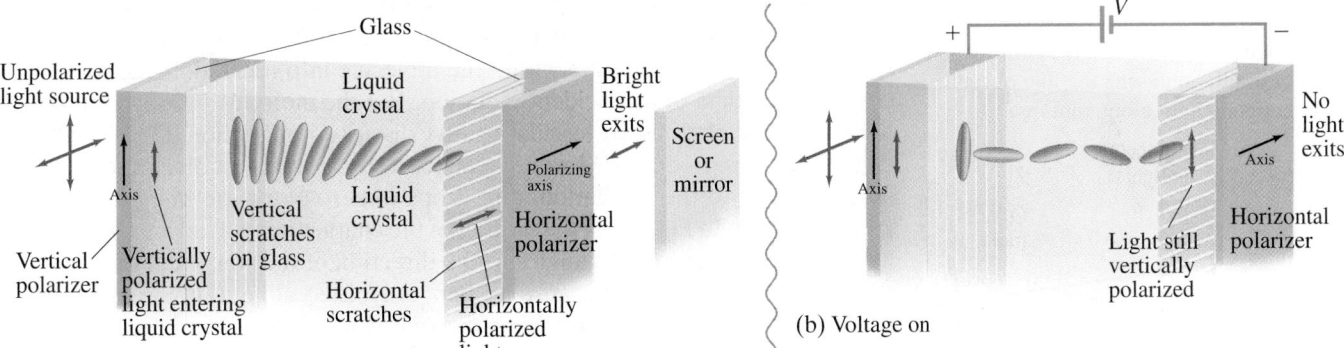

(a) Voltage off

(b) Voltage on

FIGURE 24–51 (a) "Twisted" form of liquid crystal. Light polarization plane is rotated 90°, and so is transmitted by the horizontal polarizer. Only one line of molecules is shown. (b) Molecules disoriented by electric field. The plane of polarization is not changed, so light does not pass through the horizontal polarizer. (The transparent electrodes are not shown.)

Now suppose a voltage is applied to transparent electrodes on each glass plate of the pixel. The rod-like molecules are polar (or can acquire an internal separation of charge due to the applied electric field). The applied voltage tends to align the molecules end-to-end, and they no longer follow the careful twisted pattern shown in Fig. 24–51a. Instead the applied electric field tends to align the molecules end-to-end, left to right (perpendicular to the glass plates), Fig. 24–51b, and then they don't affect the light polarization significantly. The entering plane-polarized light no longer has its plane of polarization rotated as it passes through the liquid crystal, and no light can exit through the second (horizontal) polarizer (Fig. 24–51b). With the voltage on, the pixel appears dark.[†]

[†]Some displays use an opposite system: the polarizers are parallel to each other (the scratches remain at 90° to maintain the twist). Then voltage *off* results in *black* (no light), and voltage *on* results in bright light.

FIGURE 24–52 Watch-face LCD display with altimeter. The black segments or pixels have a voltage applied to them. Note that the 8 uses all seven segments (pixels); other numbers use fewer.

FIGURE 24–53 Arrangement of subpixels on a TV or computer display (enlarged).

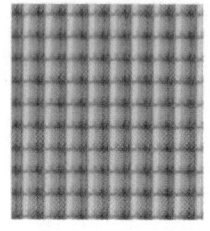

FIGURE 24–54 Unpolarized sunlight scattered by molecules of the air. An observer at right angles sees plane-polarized light, since the component of oscillation along the line of sight emits no light along that line.

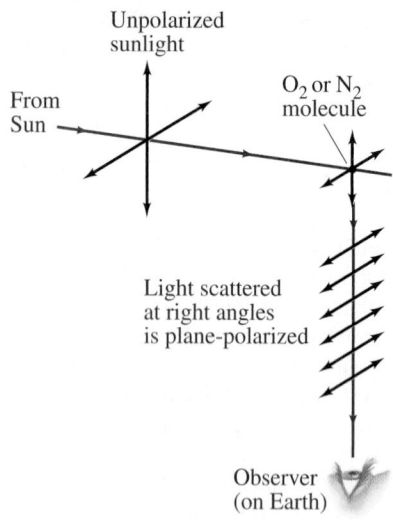

PHYSICS APPLIED

Why the sky is blue
Why sunsets are red
Why clouds are white

Simple display screens, such as for watches and calculators, use ambient light as the source (you can't see the display in the dark), with a mirror behind the LCD to reflect the light back. There are only a few pixels, corresponding to the elongated segments needed to form the numbers from 0 to 9 (and letters in some displays), as seen in Fig. 24–52. Any pixels to which a voltage is applied appear dark and form part of a number. With no voltage, pixels pass light through the polarizers to the mirror and back out, which forms a bright background. Displays with white numbers on a dark background have the voltages reversed.

Television, cell phone, and computer LCDs are more sophisticated. A color pixel consists of three cells, or subpixels, each covered with a red, green, or blue filter (Fig. 24–53). Varying brightnesses of these three primary colors can yield almost any natural color. A good-quality screen consists of millions of pixels. Behind this array of pixels is a light source, often thin fluorescent tubes the diameter of a straw, or light-emitting diodes (LEDs). The light passes through the liquid crystal subpixels, or not, depending on the voltage applied to each, as we discussed in detail in Section 17–11. See especially Figs. 17–31 and 17–33.

[To obtain a range of gray scale or range of color brightness, each subpixel cannot simply go on or off as in Fig. 24–51. Several techniques can be used depending on the construction of the LCD. If the voltage applied in Fig. 24–51b is small enough, the disorientation of the molecules may be small, allowing some rotation of the polarization vector and thus some light can pass through, the actual amount depending on the voltage. Alternatively, each subpixel can be pulsed—the length of time it is *on* affects the perceived brightness. The effect of stronger or weaker brightness can instead be provided by the number of nearby subpixels of the same color that are turned on or off; this third system lets the eye "average" over many pixels, but reduces the sharpness or resolution of the picture.]

*24–12 Scattering of Light by the Atmosphere

Sunsets are red, the sky is blue, and skylight is polarized (at least partially). These phenomena can be explained on the basis of the *scattering* of light by the molecules of the atmosphere. In Fig. 24–54 we see unpolarized light from the Sun impinging on a molecule of the Earth's atmosphere. The electric field of the EM wave sets the electric charges within the molecule into oscillation, and the molecule absorbs some of the incident radiation. But the molecule quickly reemits this light since the charges are oscillating. As discussed in Section 22–2, oscillating electric charges produce EM waves. The intensity is strongest along the direction perpendicular to the oscillation, and drops to zero along the line of oscillation (Section 22–2). In Fig. 24–54 the motion of the charges is resolved into two components. An observer at right angles to the direction of the sunlight, as shown, will see plane-polarized light because no light is emitted along the line of the other component of the oscillation. (When viewing along the line of an oscillation, you don't see that oscillation, and hence see no waves made by it.) At other viewing angles, both components will be present; one will be stronger, however, so the light appears partially polarized. Thus, the process of scattering explains the polarization of skylight.

Scattering of light by the Earth's atmosphere depends on wavelength λ. For particles much smaller than the wavelength of light (such as molecules of air), the particles will be less of an obstruction to long wavelengths than to short ones. The scattering decreases, in fact, as $1/\lambda^4$. Blue and violet light are thus scattered much more than red and orange, which is why the sky looks blue. At sunset, the Sun's rays pass through a maximum length of atmosphere. Much of the blue has been taken out by scattering. The light that reaches us at this low angle where the Sun is near the horizon, and reflects off clouds and haze, is thus lacking in blue. That is why sunsets appear reddish.

The dependence of scattering on $1/\lambda^4$ is valid only if the scattering objects are much smaller than the wavelength of the light. This is valid for oxygen and nitrogen molecules whose diameters are about 0.2 nm. Clouds, however, contain water droplets or crystals that are much larger than λ. They scatter all frequencies of light nearly uniformly. Hence clouds appear white (or gray, if shadowed).

Summary

The wave theory of light is strongly supported by observations that light exhibits **interference** and **diffraction**. Wave theory also explains the refraction of light and the fact that light travels more slowly in transparent solids and liquids than it does in air.

[*An aid to predicting wave behavior is **Huygens' principle**, which states that every point on a wave front can be considered as a source of tiny wavelets that spread out in the forward direction at the speed of the wave itself. The new wave front is the envelope (the common tangent) of all the wavelets.]

The wavelength of light in a medium with index of refraction n is

$$\lambda_n = \frac{\lambda}{n}, \qquad \text{(24–1)}$$

where λ is the wavelength in vacuum; the frequency is not changed.

Young's double-slit experiment demonstrated the interference of light. The observed bright spots of the interference pattern are explained as constructive interference between the beams coming through the two slits, where the beams differ in path length by an integral number of wavelengths. The dark areas in between are due to destructive interference when the path lengths differ by $\frac{1}{2}\lambda$, $\frac{3}{2}\lambda$, and so on. The angles θ at which **constructive interference** occurs are given by

$$\sin\theta = m\frac{\lambda}{d}, \qquad \text{(24–2a)}$$

where λ is the wavelength of the light, d is the separation of the slits, and m is an integer (0, 1, 2, $\cdots$). **Destructive interference** occurs at angles θ given by

$$\sin\theta = \left(m + \tfrac{1}{2}\right)\frac{\lambda}{d}, \qquad \text{(24–2b)}$$

where m is an integer (0, 1, 2, $\cdots$).

Two sources of light are perfectly **coherent** if the waves leaving them are of the same single frequency and maintain the same phase relationship at all times. If the light waves from the two sources have a random phase with respect to each other over time (as for two lightbulbs), the two sources are **incoherent**.

The frequency or wavelength of light determines its color. The **visible spectrum** in air extends from about 400 nm (violet) to about 750 nm (red).

Glass prisms spread white light into its constituent colors because the index of refraction varies with wavelength, a phenomenon known as **dispersion**.

The formula $\sin\theta = m\lambda/d$ for constructive interference also holds for a **diffraction grating**, which consists of many parallel slits or lines, separated from each other by a distance d.

The peaks of constructive interference are much brighter and sharper for a diffraction grating than for a two-slit apparatus.

A diffraction grating (or a prism) is used in a **spectrometer** to separate different colors and observe **line spectra**. For a given order m, θ depends on λ. Precise determination of wavelength can be done with a spectrometer by careful measurement of θ.

Diffraction refers to the fact that light, like other waves, bends around objects it passes, and spreads out after passing through narrow slits. This bending gives rise to a **diffraction pattern** due to interference between rays of light that travel different distances.

Light passing through a very narrow slit of width D (on the order of the wavelength λ) will produce a pattern with a bright central maximum of half-width θ given by

$$\sin\theta = \frac{\lambda}{D}, \qquad \text{(24–3a)}$$

flanked by fainter lines to either side.

Light reflected from the front and rear surfaces of a thin film of transparent material can interfere constructively or destructively, depending on the path difference. A phase change of 180° or $\frac{1}{2}\lambda$ occurs when the light reflects at a surface where the index of refraction increases. Such **thin-film interference** has many practical applications, such as lens coatings and using Newton's rings to check uniformity of glass surfaces.

In **unpolarized light**, the electric field vectors oscillate in all transverse directions. If the electric vector oscillates only in one plane, the light is said to be **plane-polarized**. Light can also be partially polarized.

When an unpolarized light beam passes through a **Polaroid** sheet, the emerging beam is plane-polarized. When a light beam is polarized and passes through a Polaroid, the intensity varies as the Polaroid is rotated. Thus a Polaroid can act as a **polarizer** or as an **analyzer**.

The intensity I_0 of a plane-polarized light beam incident on a Polaroid is reduced to

$$I = I_0 \cos^2\theta \qquad \text{(24–5)}$$

where θ is the angle between the axis of the Polaroid and the initial plane of polarization.

Light can also be partially or fully **polarized by reflection**. If light traveling in air is reflected from a medium of index of refraction n, the reflected beam will be *completely* plane-polarized if the incident angle θ_p is given by

$$\tan\theta_p = n. \qquad \text{(24–6b)}$$

The fact that light can be polarized shows that it must be a transverse wave.

Questions

1. Does Huygens' principle apply to sound waves? To water waves? Explain how Huygens' principle makes sense for water waves, where each point vibrates up and down.

2. Why is light sometimes described as rays and sometimes as waves?

3. We can hear sounds around corners but we cannot see around corners; yet both sound and light are waves. Explain the difference.

4. Two rays of light from the same source destructively interfere if their path lengths differ by how much?

5. Monochromatic red light is incident on a double slit, and the interference pattern is viewed on a screen some distance away. Explain how the fringe pattern would change if the red light source is replaced by a blue light source.

6. If Young's double-slit experiment were submerged in water, how would the fringe pattern be changed?

7. Why doesn't the light from the two headlights of a distant car produce an interference pattern?

8. Why are interference fringes noticeable only for a *thin* film like a soap bubble and not for a thick piece of glass?

9. Why are the fringes of Newton's rings (Fig. 24–31) closer together as you look farther from the center?

10. Some coated lenses appear greenish yellow when seen by reflected light. What reflected wavelengths do you suppose the coating is designed to eliminate completely?

11. A drop of oil on a pond appears bright at its edges, where its thickness is much less than the wavelengths of visible light. What can you say about the index of refraction of the oil compared to that of water?

12. Radio waves and visible light are both electromagnetic waves. Why can a radio receive a signal behind a hill when we cannot see the transmitting antenna?

13. Hold one hand close to your eye and focus on a distant light source through a narrow slit between two fingers. (Adjust your fingers to obtain the best pattern.) Describe the pattern that you see.

14. For diffraction by a single slit, what is the effect of increasing (a) the slit width, (b) the wavelength?

15. Describe the single-slit diffraction pattern produced when white light falls on a slit having a width of (a) 60 nm, (b) 60,000 nm.

16. What happens to the diffraction pattern of a single slit if the whole apparatus is immersed in (a) water, (b) a vacuum, instead of in air.

17. What is the difference in the interference patterns formed by two slits 10^{-4} cm apart as compared to a diffraction grating containing 10^4 slits/cm?

18. For a diffraction grating, what is the advantage of (a) many slits, (b) closely spaced slits?

19. White light strikes (a) a diffraction grating and (b) a prism. A rainbow appears on a wall just below the direction of the horizontal incident beam in each case. What is the color of the top of the rainbow in each case? Explain.

20. What does polarization tell us about the nature of light?

21. Explain the advantage of polarized sunglasses over plain tinted sunglasses.

22. How can you tell if a pair of sunglasses is polarizing or not?

*23. What would be the color of the sky if the Earth had no atmosphere?

*24. If the Earth's atmosphere were 50 times denser than it is, would sunlight still be white, or would it be some other color?

MisConceptual Questions

1. Light passing through a double-slit arrangement is viewed on a distant screen. The interference pattern observed on the screen would have the widest spaced fringes for the case of
   (a) red light and a small slit spacing.
   (b) blue light and a small slit spacing.
   (c) red light and a large slit spacing.
   (d) blue light and a large slit spacing.

2. Light from a green laser of wavelength 530 nm passes through two slits that are 400 nm apart. The resulting pattern formed on a screen in front of the slits is shown in Fig. 24–55. If point A is the same distance from both slits, how much closer is point B to one slit than to the other?
   (a) 530 nm.
   (b) 265 nm.
   (c) 400 nm.
   (d) 0 nm.
   (e) It depends on the distance to the screen.

FIGURE 24–55
MisConceptual
Question 2.

A   B

3. The colors in a rainbow are caused by
   (a) the interaction of the light reflected from different raindrops.
   (b) different amounts of absorption for light of different colors by the water in the raindrops.
   (c) different amounts of refraction for light of different colors by the water in the raindrops.
   (d) the downward motion of the raindrops.

4. A double-slit experiment yields an interference pattern due to the path length difference from light traveling through one slit versus the other. Why does a single slit show a diffraction pattern?
   (a) There is a path length difference from waves originating at different parts of the slit.
   (b) The wavelength of the light is shorter than the slit.
   (c) The light passing through the slit interferes with light that does not pass through.
   (d) The single slit must have something in the middle of it, causing it to act like a double slit.

5. If you hold two fingers very close together and look at a bright light, you see lines between the fingers. What is happening?
   (a) You are holding your fingers too close to your eye to be able to focus on it.
   (b) You are seeing a diffraction pattern.
   (c) This is a quantum-mechanical tunneling effect.
   (d) The brightness of the light is overwhelming your eye.

6. Light passes through a slit that is about 5×10^{-3} m high and 5×10^{-7} m wide. The central bright light visible on a distant screen will be
   (a) about 5×10^{-3} m high and about 5×10^{-7} m wide.
   (b) about 5×10^{-3} m high and wider than 5×10^{-7} m.
   (c) about 5×10^{-3} m high and narrower than 5×10^{-7} m.
   (d) taller than 5×10^{-3} m high and wider than 5×10^{-7} m.
   (e) taller than 5×10^{-3} m high and about 5×10^{-7} m wide.

7. Blue light of wavelength λ passes through a single slit of width d and forms a diffraction pattern on a screen. If we replace the blue light by red light of wavelength 2λ, we can retain the original diffraction pattern if we change the slit width
(*a*) to $d/4$.
(*b*) to $d/2$.
(*c*) not at all.
(*d*) to $2d$.
(*e*) to $4d$.

8. Imagine holding a circular disk in a beam of monochromatic light (Fig. 24–56). If diffraction occurs at the edge of the disk, the center of the shadow is
(*a*) darker than the rest of the shadow.
(*b*) a bright spot.
(*c*) bright or dark, depending on the wavelength.
(*d*) bright or dark, depending on the distance to the screen.

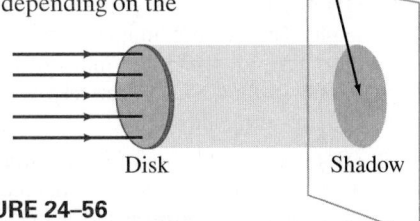

FIGURE 24–56
MisConceptual Question 8.
Disk   Shadow   Screen

9. If someone is around a corner from you, what is the main reason you can hear him speaking but can't see him?
(*a*) Sound travels farther in air than light does.
(*b*) Sound can travel through walls, but light cannot.
(*c*) Sound waves have long enough wavelengths to bend around a corner; light wavelengths are too short to bend much.
(*d*) Sound waves reflect off walls, but light cannot.

10. When a CD is held at an angle, the reflected light contains many colors. What causes these colors?
(*a*) An anti-theft encoding intended to prevent copying of the CD.
(*b*) The different colors correspond to different data bits.
(*c*) Light reflected from the closely spaced grooves adds constructively for different wavelengths at different angles.
(*d*) It is part of the decorative label on the CD.

11. If a thin film has a thickness that is
(*a*) $\frac{1}{4}$ of a wavelength, constructive interference will always occur.
(*b*) $\frac{1}{4}$ of a wavelength, destructive interference will always occur.
(*c*) $\frac{1}{2}$ of a wavelength, constructive interference will always occur.
(*d*) $\frac{1}{2}$ of a wavelength, destructive interference will always occur.
(*e*) None of the above is always true.

12. If unpolarized light is incident from the left on three polarizers as shown in Fig. 24–57, in which case will some light get through?
(*a*) Case 1 only.
(*b*) Case 2 only.
(*c*) Case 3 only.
(*d*) Cases 1 and 3.
(*e*) All three cases.

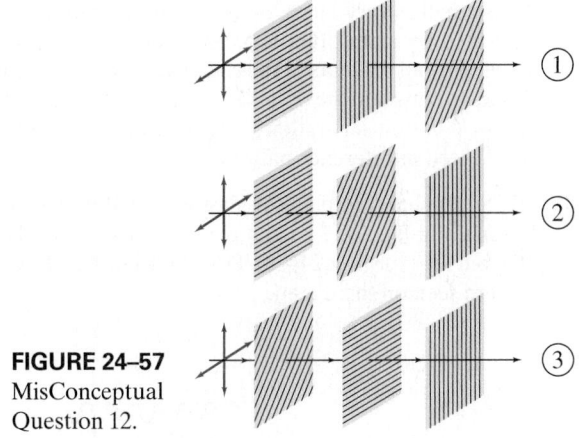

FIGURE 24–57
MisConceptual
Question 12.

For assigned homework and other learning materials, go to the MasteringPhysics website.

Problems

24–3 Double-Slit Interference

1. (I) Monochromatic light falling on two slits 0.018 mm apart produces the fifth-order bright fringe at an 8.6° angle. What is the wavelength of the light used?

2. (I) The third-order bright fringe of 610-nm light is observed at an angle of 31° when the light falls on two narrow slits. How far apart are the slits?

3. (II) Monochromatic light falls on two very narrow slits 0.048 mm apart. Successive fringes on a screen 6.50 m away are 8.5 cm apart near the center of the pattern. Determine the wavelength and frequency of the light.

4. (II) If 720-nm and 660-nm light passes through two slits 0.62 mm apart, how far apart are the second-order fringes for these two wavelengths on a screen 1.0 m away?

5. (II) Water waves having parallel crests 4.5 cm apart pass through two openings 7.5 cm apart in a board. At a point 3.0 m beyond the board, at what angle relative to the "straight-through" direction would there be little or no wave action?

6. (II) A red laser from the physics lab is marked as producing 632.8-nm light. When light from this laser falls on two closely spaced slits, an interference pattern formed on a wall several meters away has bright red fringes spaced 5.00 mm apart near the center of the pattern. When the laser is replaced by a small laser pointer, the fringes are 5.14 mm apart. What is the wavelength of light produced by the laser pointer?

7. (II) Light of wavelength 680 nm falls on two slits and produces an interference pattern in which the third-order bright red fringe is 38 mm from the central fringe on a screen 2.8 m away. What is the separation of the two slits?

8. (II) Light of wavelength λ passes through a pair of slits separated by 0.17 mm, forming a double-slit interference pattern on a screen located a distance 37 cm away. Suppose that the image in Fig. 24–9a is an actual-size reproduction of this interference pattern. Use a ruler to measure a pertinent distance on this image; then utilize this measured value to determine λ (nm).

9. (II) A parallel beam of light from a He–Ne laser, with a wavelength 633 nm, falls on two very narrow slits 0.068 mm apart. How far apart are the fringes in the center of the pattern on a screen 3.3 m away?

10. (II) A physics professor wants to perform a lecture demonstration of Young's double-slit experiment for her class using the 633-nm light from a He–Ne laser. Because the lecture hall is very large, the interference pattern will be projected on a wall that is 5.0 m from the slits. For easy viewing by all students in the class, the professor wants the distance between the $m = 0$ and $m = 1$ maxima to be 35 cm. What slit separation is required in order to produce the desired interference pattern?

11. (II) Suppose a thin piece of glass is placed in front of the lower slit in Fig. 24–7 so that the two waves enter the slits 180° out of phase (Fig. 24–58). Draw in detail the interference pattern seen on the screen.

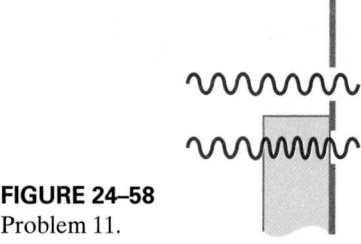

FIGURE 24–58
Problem 11.

12. (II) In a double-slit experiment it is found that blue light of wavelength 480 nm gives a second-order maximum at a certain location on the screen. What wavelength of visible light would have a minimum at the same location?

13. (II) Two narrow slits separated by 1.0 mm are illuminated by 544-nm light. Find the distance between adjacent bright fringes on a screen 4.0 m from the slits.

14. (II) Assume that light of a single color, rather than white light, passes through the two-slit setup described in Example 24–3. If the distance from the central fringe to a first-order fringe is measured to be 2.9 mm on the screen, determine the light's wavelength (in nm) and color (see Fig. 24–12).

15. (II) In a double-slit experiment, the third-order maximum for light of wavelength 480 nm is located 16 mm from the central bright spot on a screen 1.6 m from the slits. Light of wavelength 650 nm is then projected through the same slits. How far from the central bright spot will the second-order maximum of this light be located?

16. (II) Light of wavelength 470 nm in air shines on two slits 6.00×10^{-2} mm apart. The slits are immersed in water, as is a viewing screen 40.0 cm away. How far apart are the fringes on the screen?

17. (III) A very thin sheet of plastic ($n = 1.60$) covers one slit of a double-slit apparatus illuminated by 680-nm light. The center point on the screen, instead of being a maximum, is dark. What is the (minimum) thickness of the plastic?

24–4 Visible Spectrum; Dispersion

18. (I) By what percent is the speed of blue light (450 nm) less than the speed of red light (680 nm), in silicate flint glass (see Fig. 24–14)?

19. (II) A light beam strikes a piece of glass at a 65.00° incident angle. The beam contains two wavelengths, 450.0 nm and 700.0 nm, for which the index of refraction of the glass is 1.4831 and 1.4754, respectively. What is the angle between the two refracted beams?

20. (III) A parallel beam of light containing two wavelengths, $\lambda_1 = 455$ nm and $\lambda_2 = 642$ nm, enters the silicate flint glass of an equilateral prism as shown in Fig. 24–59. At what angles, θ_1 and θ_2, does each beam leave the prism (give angle with normal to the face)? See Fig. 24–14.

FIGURE 24–59
Problem 20.

24–5 Single-Slit Diffraction

21. (I) If 680-nm light falls on a slit 0.0425 mm wide, what is the angular width of the central diffraction peak?

22. (I) Monochromatic light falls on a slit that is 2.60×10^{-3} mm wide. If the angle between the first dark fringes on either side of the central maximum is 28.0° (dark fringe to dark fringe), what is the wavelength of the light used?

23. (II) When blue light of wavelength 440 nm falls on a single slit, the first dark bands on either side of center are separated by 51.0°. Determine the width of the slit.

24. (II) A single slit 1.0 mm wide is illuminated by 450-nm light. What is the width of the central maximum (in cm) in the diffraction pattern on a screen 6.0 m away?

25. (II) How wide is the central diffraction peak on a screen 2.30 m behind a 0.0348-mm-wide slit illuminated by 558-nm light?

26. (II) Consider microwaves which are incident perpendicular to a metal plate which has a 1.6-cm slit in it. Discuss the angles at which there are diffraction minima for wavelengths of (a) 0.50 cm, (b) 1.0 cm, and (c) 3.0 cm.

27. (II) (a) For a given wavelength λ, what is the minimum slit width for which there will be no diffraction minima? (b) What is the minimum slit width so that no visible light exhibits a diffraction minimum?

28. (II) Light of wavelength 620 nm falls on a slit that is 3.80×10^{-3} mm wide. Estimate how far the first bright diffraction fringe is from the strong central maximum if the screen is 10.0 m away.

29. (II) Monochromatic light of wavelength 633 nm falls on a slit. If the angle between the first two bright fringes on either side of the central maximum is 32°, estimate the slit width.

30. (II) Coherent light from a laser diode is emitted through a rectangular area $3.0\,\mu m \times 1.5\,\mu m$ (horizontal-by-vertical). If the laser light has a wavelength of 780 nm, determine the angle between the first diffraction minima (a) above and below the central maximum, (b) to the left and right of the central maximum.

31. (III) If parallel light falls on a single slit of width D at a 28.0° angle to the normal, describe the diffraction pattern.

24–6 and 24–7 Diffraction Gratings

32. (I) At what angle will 510-nm light produce a second-order maximum when falling on a grating whose slits are 1.35×10^{-3} cm apart?

33. (I) A grating that has 3800 slits per cm produces a third-order fringe at a 22.0° angle. What wavelength of light is being used?

34. (I) A grating has 7400 slits/cm. How many spectral orders can be seen (400 to 700 nm) when it is illuminated by white light?

35. (II) Red laser light from a He–Ne laser ($\lambda = 632.8$ nm) creates a second-order fringe at 53.2° after passing through the grating. What is the wavelength λ of light that creates a first-order fringe at 20.6°?

36. (II) How many slits per centimeter does a grating have if the third order occurs at a 15.0° angle for 620-nm light?

37. (II) A source produces first-order lines when incident normally on a 9800-slit/cm diffraction grating at angles 28.8°, 36.7°, 38.6°, and 41.2°. What are the wavelengths?

38. (II) White light containing wavelengths from 410 nm to 750 nm falls on a grating with 7800 slits/cm. How wide is the first-order spectrum on a screen 3.40 m away?

39. (II) A diffraction grating has 6.5×10^5 slits/m. Find the angular spread in the second-order spectrum between red light of wavelength 7.0×10^{-7} m and blue light of wavelength 4.5×10^{-7} m.

40. (II) Two first-order spectrum lines are measured by a 9650-slit/cm spectroscope at angles, on each side of center, of $+26°38'$, $+41°02'$ and $-26°18'$, $-40°27'$. Calculate the wavelengths based on these data.

41. (II) What is the highest spectral order that can be seen if a grating with 6500 slits per cm is illuminated with 633-nm laser light? Assume normal incidence.

42. (II) The first-order line of 589-nm light falling on a diffraction grating is observed at a 14.5° angle. How far apart are the slits? At what angle will the third order be observed?

43. (II) Two (and only two) full spectral orders can be seen on either side of the central maximum when white light is sent through a diffraction grating. What is the maximum number of slits per cm for the grating?

24–8 Thin-Film Interference

44. (I) If a soap bubble is 120 nm thick, what wavelength is most strongly reflected at the center of the outer surface when illuminated normally by white light? Assume that $n = 1.32$.

45. (I) How far apart are the dark bands in Example 24–10 if the glass plates are each 21.5 cm long?

46. (II) (a) What is the smallest thickness of a soap film ($n = 1.33$) that would appear black if illuminated with 480-nm light? Assume there is air on both sides of the soap film. (b) What are two other possible thicknesses for the film to appear black? (c) If the thickness t was much less than λ, why would the film also appear black?

47. (II) A lens appears greenish yellow ($\lambda = 570$ nm is strongest) when white light reflects from it. What minimum thickness of coating ($n = 1.25$) do you think is used on such a glass lens ($n = 1.52$), and why?

48. (II) A thin film of oil ($n_o = 1.50$) with varying thickness floats on water ($n_w = 1.33$). When it is illuminated from above by white light, the reflected colors are as shown in Fig. 24–60. In air, the wavelength of yellow light is 580 nm. (a) Why are there no reflected colors at point A? (b) What is the oil's thickness t at point B?

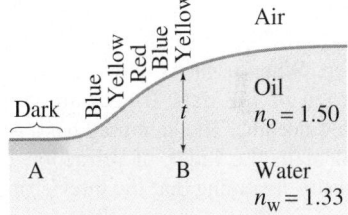

FIGURE 24–60 Problem 48.

49. (II) How many uncoated thin lenses in an optical instrument would reduce the amount of light passing through the instrument to 50% or less? (Assume the same transmission percent at each of the two surfaces—see page 697.)

50. (II) A total of 35 bright and 35 dark Newton's rings (not counting the dark spot at the center) are observed when 560-nm light falls normally on a planoconvex lens resting on a flat glass surface (Fig. 24–31). How much thicker is the lens at the center than the edges?

51. (II) If the wedge between the glass plates of Example 24–10 is filled with some transparent substance other than air—say, water—the pattern shifts because the wavelength of the light changes. In a material where the index of refraction is n, the wavelength is $\lambda_n = \lambda/n$, where λ is the wavelength in vacuum (Eq. 24–1). How many dark bands would there be if the wedge of Example 24–10 were filled with water?

52. (II) A fine metal foil separates one end of two pieces of optically flat glass, as in Fig. 24–33. When light of wavelength 670 nm is incident normally, 24 dark bands are observed (with one at each end). How thick is the foil?

53. (II) How thick (minimum) should the air layer be between two flat glass surfaces if the glass is to appear bright when 450-nm light is incident normally? What if the glass is to appear dark?

54. (III) A thin oil slick (n_o = 1.50) floats on water (n_w = 1.33). When a beam of white light strikes this film at normal incidence from air, the only enhanced reflected colors are red (650 nm) and violet (390 nm). From this information, deduce the (minimum) thickness t of the oil slick.

55. (III) A uniform thin film of alcohol (n = 1.36) lies on a flat glass plate (n = 1.56). When monochromatic light, whose wavelength can be changed, is incident normally, the reflected light is a minimum for λ = 525 nm and a maximum for λ = 655 nm. What is the minimum thickness of the film?

*24–9 Michelson Interferometer

*56. (II) How far must the mirror M_1 in a Michelson interferometer be moved if 680 fringes of 589-nm light are to pass by a reference line?

*57. (II) What is the wavelength of the light entering an interferometer if 362 bright fringes are counted when the movable mirror moves 0.125 mm?

*58. (II) A micrometer is connected to the movable mirror of an interferometer. When the micrometer is tightened down on a thin metal foil, the net number of bright fringes that move, compared to closing the empty micrometer, is 296. What is the thickness of the foil? The wavelength of light used is 589 nm.

*59. (III) One of the beams of an interferometer (Fig. 24–61) passes through a small evacuated glass container 1.155 cm deep. When a gas is allowed to slowly fill the container, a total of 158 dark fringes are counted to move past a reference line. The light used has a wavelength of 632.8 nm. Calculate the index of refraction of the gas at its final density, assuming that the interferometer is in vacuum.

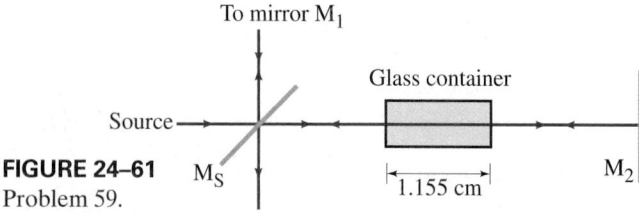

To mirror M_1

Glass container

Source

FIGURE 24–61
Problem 59.

M_S 1.155 cm M_2

24–10 Polarization

60. (I) Two polarizers are oriented at 72° to one another. Unpolarized light falls on them. What fraction of the light intensity is transmitted?

61. (I) What is Brewster's angle for an air–glass (n = 1.56) surface?

62. (II) At what angle should the axes of two Polaroids be placed so as to reduce the intensity of the incident unpolarized light to (a) $\frac{1}{3}$, (b) $\frac{1}{10}$?

63. (II) Two polarizers are oriented at 42.0° to one another. Light polarized at a 21.0° angle to each polarizer passes through both. What is the transmitted intensity (%)?

64. (II) Three perfectly polarizing sheets are spaced 2 cm apart and in parallel planes. The transmission axis of the second sheet is 30° relative to the first one. The transmission axis of the third sheet is 90° relative to the *first* one. Unpolarized light impinges on the first polarizing sheet. What percent of this light is transmitted out through the third polarizer?

65. (II) A piece of material, suspected of being a stolen diamond (n = 2.42), is submerged in oil of refractive index 1.43 and illuminated by unpolarized light. It is found that the reflected light is completely polarized at an angle of 62°. Is it diamond? Explain.

66. (II) Two Polaroids are aligned so that the initially unpolarized light passing through them is a maximum. At what angle should one of them be placed so the transmitted intensity is subsequently reduced by half?

67. (II) What is Brewster's angle for a diamond submerged in water?

68. (II) The critical angle for total internal reflection at a boundary between two materials is 58°. What is Brewster's angle at this boundary? Give two answers, one for each material.

69. (II) What would Brewster's angle be for reflections off the surface of water for light coming from beneath the surface? Compare to the angle for total internal reflection, and to Brewster's angle from above the surface.

70. (II) Unpolarized light of intensity I_0 passes through six successive Polaroid sheets each of whose axis makes a 35° angle with the previous one. What is the intensity of the transmitted beam?

71. (III) Two polarizers are oriented at 48° to each other and plane-polarized light is incident on them. If only 35% of the light gets through both of them, what was the initial polarization direction of the incident light?

72. (III) Four polarizers are placed in succession with their axes vertical, at 30.0° to the vertical, at 60.0° to the vertical, and at 90.0° to the vertical. (a) Calculate what fraction of the incident unpolarized light is transmitted by the four polarizers. (b) Can the transmitted light be *decreased* by removing one of the polarizers? If so, which one? (c) Can the transmitted light intensity be extinguished by removing polarizers? If so, which one(s)?

General Problems

73. Light of wavelength 5.0×10^{-7} m passes through two parallel slits and falls on a screen 5.0 m away. Adjacent bright bands of the interference pattern are 2.0 cm apart. (a) Find the distance between the slits. (b) The same two slits are next illuminated by light of a different wavelength, and the fifth-order minimum for this light occurs at the same point on the screen as the fourth-order minimum for the previous light. What is the wavelength of the second source of light?

74. Television and radio waves reflecting from mountains or airplanes can interfere with the direct signal from the station. (a) What kind of interference will occur when 75-MHz television signals arrive at a receiver directly from a distant station, and are reflected from a nearby airplane 122 m directly above the receiver? Assume $\frac{1}{2}\lambda$ change in phase of the signal upon reflection. (b) What kind of interference will occur if the plane is 22 m closer to the receiver?

75. Red light from three separate sources passes through a diffraction grating with 3.60×10^5 slits/m. The wavelengths of the three lines are 6.56×10^{-7} m (hydrogen), 6.50×10^{-7} m (neon), and 6.97×10^{-7} m (argon). Calculate the angles for the first-order diffraction line of each source.

76. What is the index of refraction of a clear material if a minimum thickness of 125 nm, when laid on glass, is needed to reduce reflection to nearly zero when light of 675 nm is incident normally upon it? Do you have a choice for an answer?

77. Light of wavelength 650 nm passes through two narrow slits 0.66 mm apart. The screen is 2.40 m away. A second source of unknown wavelength produces its second-order fringe 1.23 mm closer to the central maximum than the 650-nm light. What is the wavelength of the unknown light?

78. Monochromatic light of variable wavelength is incident normally on a thin sheet of plastic film in air. The reflected light is a maximum only for $\lambda = 491.4$ nm and $\lambda = 688.0$ nm in the visible spectrum. What is the thickness of the film ($n = 1.58$)? [*Hint*: Assume successive values of m.]

79. Show that the second- and third-order spectra of white light produced by a diffraction grating always overlap. What wavelengths overlap?

80. A radio station operating at 90.3 MHz broadcasts from two identical antennas at the same elevation but separated by a 9.0-m horizontal distance d, Fig. 24–62. A maximum signal is found along the midline, perpendicular to d at its midpoint and extending horizontally in both directions. If the midline is taken as $0°$, at what other angle(s) θ is a maximum signal detected? A minimum signal? Assume all measurements are made much farther than 9.0 m from the antenna towers.

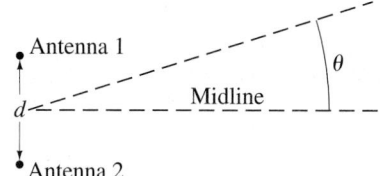

FIGURE 24–62 Problem 80.

81. Calculate the minimum thickness needed for an antireflective coating ($n = 1.38$) applied to a glass lens in order to eliminate (*a*) blue (450 nm), or (*b*) red (720 nm) reflections for light at normal incidence.

82. Stealth aircraft are designed to not reflect radar, whose wavelength is typically 2 cm, by using an antireflecting coating. Ignoring any change in wavelength in the coating, estimate its thickness.

83. A laser beam passes through a slit of width 1.0 cm and is pointed at the Moon, which is approximately 380,000 km from the Earth. Assume the laser emits waves of wavelength 633 nm (the red light of a He–Ne laser). Estimate the width of the beam when it reaches the Moon due to diffraction.

84. A thin film of soap ($n = 1.34$) coats a piece of flat glass ($n = 1.52$). How thick is the film if it reflects 643-nm red light most strongly when illuminated normally by white light?

85. When violet light of wavelength 415 nm falls on a single slit, it creates a central diffraction peak that is 8.20 cm wide on a screen that is 3.15 m away. How wide is the slit?

86. A series of polarizers are each rotated $10°$ from the previous polarizer. Unpolarized light is incident on this series of polarizers. How many polarizers does the light have to go through before it is $\frac{1}{5}$ of its original intensity?

87. The wings of a certain beetle have a series of parallel lines across them. When normally incident 480-nm light is reflected from the wing, the wing appears bright when viewed at an angle of $56°$. How far apart are the lines?

88. A teacher stands well back from an outside doorway 0.88 m wide, and blows a whistle of frequency 950 Hz. Ignoring reflections, estimate at what angle(s) it is *not* possible to hear the whistle clearly on the playground outside the doorway. Assume 340 m/s for the speed of sound.

89. Light is incident on a diffraction grating with 7200 slits/cm and the pattern is viewed on a screen located 2.5 m from the grating. The incident light beam consists of two wavelengths, $\lambda_1 = 4.4 \times 10^{-7}$ m and $\lambda_2 = 6.8 \times 10^{-7}$ m. Calculate the linear distance between the first-order bright fringes of these two wavelengths on the screen.

90. How many slits per centimeter must a grating have if there is to be no second-order spectrum for any visible wavelength?

91. When yellow sodium light, $\lambda = 589$ nm, falls on a diffraction grating, its first-order peak on a screen 72.0 cm away falls 3.32 cm from the central peak. Another source produces a line 3.71 cm from the central peak. What is its wavelength? How many slits/cm are on the grating?

92. Two of the lines of the atomic hydrogen spectrum have wavelengths of 656 nm and 410 nm. If these fall at normal incidence on a grating with 7700 slits/cm, what will be the angular separation of the two wavelengths in the first-order spectrum?

93. A tungsten–halogen bulb emits a continuous spectrum of ultraviolet, visible, and infrared light in the wavelength range 360 nm to 2000 nm. Assume that the light from a tungsten–halogen bulb is incident on a diffraction grating with slit spacing d and that the first-order brightness maximum for the wavelength of 1200 nm occurs at angle θ. What other wavelengths within the spectrum of incident light will produce a brightness maximum at this same angle θ? [Optical filters are used to deal with this bothersome effect when a continuous spectrum of light is measured by a spectrometer.]

94. At what angle above the horizon is the Sun when light reflecting off a smooth lake is polarized most strongly?

95. Unpolarized light falls on two polarizer sheets whose axes are at right angles. (*a*) What fraction of the incident light intensity is transmitted? (*b*) What fraction is transmitted if a third polarizer is placed between the first two so that its axis makes a $56°$ angle with the axis of the first polarizer? (*c*) What if the third polarizer is in front of the other two?

96. At what angle should the axes of two Polaroids be placed so as to reduce the intensity of the incident unpolarized light by an additional factor (after the first Polaroid cuts it in half) of (*a*) 4, (*b*) 10, (*c*) 100?

Search and Learn

1. Compare Figs. 24–5, 24–6, and 24–7, which are different representations of the double-slit experiment. For each figure state the direction the light is traveling. Where are the wave crests in terms of this direction? How are they represented in each figure? Give one advantage of each figure in helping you understand the double-slit experiment and interference.

2. Discuss the similarities, and differences, of double-slit interference and single-slit diffraction.

3. Describe why the various colors of visible light appear as they do in Fig. 24–16, where red is at the top and violet at the bottom, and in Fig. 24–26, where violet is closest to the central maximum and red is farthest from the central maximum.

4. When can we use geometric optics as in Chapter 23, and when do we need to use the more complicated wave model of light discussed in Chapter 24? In particular, what are the physical characteristics that matter in making this decision?

5. A parallel beam of light containing two wavelengths, 420 nm and 650 nm, enters a borate flint glass equilateral prism (Fig. 24–63). (a) What is the angle between the two beams leaving the prism? (b) Repeat part (a) for a diffraction grating with 5800 slits/cm. (c) Discuss two advantages of a diffraction grating, including one that you see from your results.

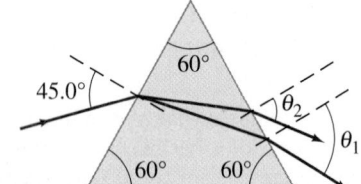

FIGURE 24–63
Search and Learn 5.

6. Suppose you viewed the light *transmitted* through a thin coating layered on a flat piece of glass. Draw a diagram, similar to Fig. 24–30 or 24–36, and describe the conditions required for maxima and minima. Consider all possible values of index of refraction. Discuss the relative intensity of the minima compared to the maxima and to zero.

7. What percent of visible light is reflected from plain glass? Assume your answer refers to transmission through each surface, front and back. How does the presence of multiple lenses in a good camera degrade the image? What is suggested in Section 24–8 to reduce this reflection? Explain in words, and sketch how this solution works. For a glass lens in air, about how much improvement does this solution provide?

ANSWERS TO EXERCISES

A: 333 nm; 6.0×10^{14} Hz; 2.0×10^8 m/s.
B: 2.5 mm.
C: Narrower.
D: A.

E: (e).
F: Zero for both (a) and (b), because the two successive polarizers at 90° cancel all light. The 45° Polaroid must be inserted *between* the other two if transmission is to occur.

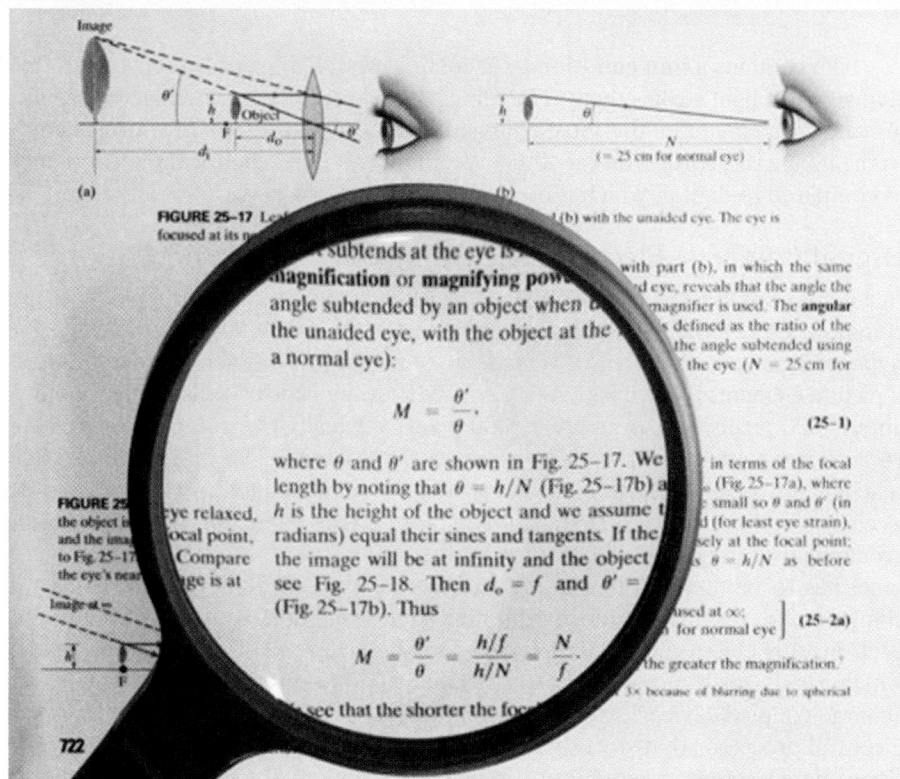

FIGURE 25–17 ...
focused at its ...

(text obscured by magnifying glass in the image)

FIGURE 25 ...
the object is ...
and the ima... ...ocal point,
to Fig. 25–17 ...Compare
the eye's near... ...ge is at

Of the many optical devices we discuss in this Chapter, the magnifying glass is the simplest. Here it is magnifying part of page 722 of this Chapter, which describes how the magnifying glass works according to the ray model. In this Chapter we examine film and digital cameras, the human eye, eyeglasses, telescopes, and microscopes as well as image resolution, X-rays, and CT scans.

Optical Instruments

C H A P T E R

25

CHAPTER-OPENING QUESTION—Guess now!

Because of diffraction, a light microscope has a useful magnification of about

(a) 50×; (b) 100×; (c) 500×; (d) 2000×; (e) 5000×;

and the smallest objects it can resolve have a size of about

(a) 10 nm; (b) 100 nm; (c) 500 nm; (d) 2500 nm; (e) 5500 nm.

I n our discussion of the behavior of light in the two previous Chapters, we also described a few instruments such as the spectrometer and the Michelson interferometer. In this Chapter, we will discuss some more common instruments, most of which use lenses, including the camera, telescope, microscope, and the human eye. To describe their operation, we will use ray diagrams as we did in Chapter 23. However, we will see that understanding some aspects of their operation will require the wave nature of light.

25–1 Cameras: Film and Digital

The basic elements of a **camera** are a lens, a light-tight box, a shutter to let light pass through the lens only briefly, and in a digital camera an electronic sensor or in a traditional camera a piece of film (Fig. 25–1). When the shutter is opened for a brief "exposure," light from external objects in the field of view is focused by the lens as an image on the sensor or film.

You can see the image yourself if you remove the back of a conventional camera, keeping the shutter open, and view through a piece of tissue paper (on which an image can form) placed where the film should be.

CONTENTS

FIGURE 25–1 A simple camera.

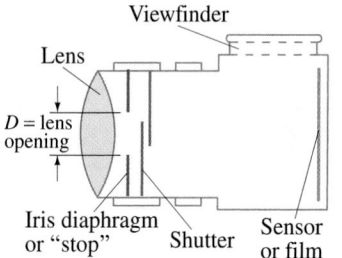

Film contains a thin **emulsion** (= a coating) with light-sensitive chemicals that change when light strikes them. The film is then developed by chemicals dissolved in water, which causes the most changed areas (brightest light) to turn opaque, so the image is recorded on the film.[†] We might call film "chemical photography" as compared to digital, which is electronic.

Digital Cameras, Electronic Sensors (CCD, CMOS)

In a **digital camera**, the film is replaced by a semiconductor sensor. Two types are common: **CCD** (*charge-coupled device*) and **CMOS** (*complementary metal oxide semiconductor*). A CCD sensor is made up of millions of tiny semiconductor **pixels** ("picture elements")—see Fig. 24–49. A 12-MP (12-megapixel) sensor might contain about 4000 pixels horizontally by 3000 pixels vertically over an area of perhaps 16×12 mm, and preferably larger such as 36×24 mm like 35-mm film. Light reaching any pixel liberates electrons within the semiconductor[‡] which are stored as charge in that pixel's capacitance. The more intense the light, the more charge accumulates during the brief exposure time. After exposure, the charge on each pixel has to be "read" (measured) and stored. A reader circuit first reads the charge on the pixel capacitance right next to it. Immediately after, the charge on each pixel is electronically transferred to its adjacent pixel, towards the reader which reads each pixel charge in sequence, one-by-one. Hence the name "charge-coupled device." All this information (the brightness of each pixel) goes to a central processor that stores it and allows re-formation of the image later onto the camera's screen, a computer screen, or a printer. After all the pixel charge information is transferred to memory (Section 21–8), a new picture can be taken.

A CMOS sensor also uses a silicon semiconductor, and incorporates transistor electronics within each pixel, allowing parallel readout, somewhat like the similar MOSFET array that was shown in Fig. 21–29.

Sensor sizes are typically in the ratio 4:3 or 3:2. A larger sensor is better because it can hold more pixels, and/or each pixel can be larger and hold more charge (free electrons) to provide a wider range of brightness, better color accuracy, and better sensitivity in low-light conditions.

In the most common array of pixels, referred to as a **Bayer mosaic**, color is achieved by red, green, and blue filters over alternating pixels as shown in Fig. 25–2, similar to what a color LCD or CRT screen does (Sections 17–11 and 24–11). The sensor type shown in Fig. 25–2 contains twice as many green pixels as red or blue (because green seems to have a stronger influence on the human eye's sensation of sharpness). The computer-analyzed color at many pixels is often an average with nearest-neighbor colors to reduce memory size (= compression, see page 489).

Each different color of pixel in a Bayer array is counted as a *separate* pixel. In contrast, in an LCD screen (Sections 17–11 and 24–11), a group of three subpixels is counted as one pixel, a more conservative count.

An alternative technology, called "Foveon," uses a semiconductor layer system. Different wavelengths of light penetrate silicon to different depths, as shown in Fig. 25–3: blue wavelengths are absorbed in the top layer, allowing green and red light to pass through. Longer wavelengths (green) are absorbed in the second layer, and the bottom layer detects the longest wavelengths (red). All three colors are detected by each "tri-pixel" site, resulting in better color resolution and fewer artifacts.

Digital Artifacts

Digital cameras can produce image artifacts (artificial effects in the image not present in the original) resulting from the electronic sensing of the image. One example using the Bayer mosaic of pixels (Fig. 25–2) is described in Fig. 25–4.

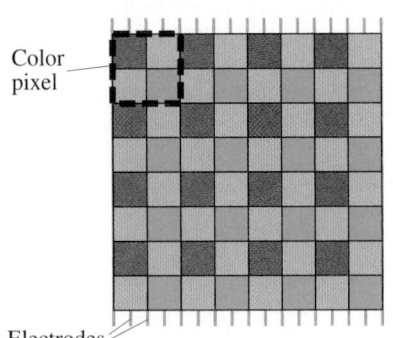

Color pixel

Electrodes

FIGURE 25–2 Portion of a typical Bayer array sensor. A square group of four pixels $^{RG}_{GB}$ is sometimes called a "color pixel."

FIGURE 25–3 A layered or "Foveon" tri-pixel that includes all three colors, arranged vertically so light can pass through all three subpixels.

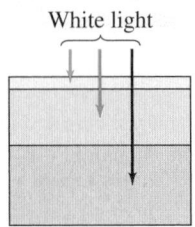

White light

FIGURE 25–4 Suppose we take a picture that includes a thin black line (our object) on a white background. The *image* of this black line has a colored "halo" (red above, blue below) due to the mosaic arrangement of color filter pixels, as shown by the colors transmitted to the image. Computer averaging minimizes such color problems (the green at top and bottom of image may average with nearby pixels to give white or nearly so) but the image is consequently "softened" or blurred.

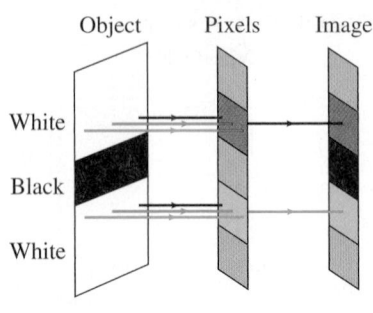

Object   Pixels   Image

White

Black

White

[†]This is called a *negative*, because the black areas correspond to bright objects and vice versa. The same process occurs during printing to produce a black-and-white "positive" picture from the negative. Color film has three emulsion layers (or dyes) corresponding to the three primary colors.

[‡]Specifically, photons of light knock electrons in the valence band up to the conduction band. This material on semiconductors is covered in Chapter 29.

Camera Adjustments

There are three main adjustments on good-quality cameras: shutter speed, *f*-stop, and focusing. Although most cameras today make these adjustments automatically, it is valuable to understand these adjustments to use a camera effectively. For special or top-quality work, a manual camera is indispensable (Fig. 25–5).

Exposure time or shutter speed This refers to how quickly the digital sensor can make an accurate reading of the images, or how long the shutter of a camera is open and the film or sensor is exposed. It could vary from a second or more ("time exposures") to $\frac{1}{1000}$ s or faster. To avoid blurring from camera movement, exposure times shorter than $\frac{1}{100}$ s are normally needed. If the object is moving, even shorter exposure times are needed to "stop" the action ($\frac{1}{1000}$ s or less). If the exposure time is not fast enough, the image will be blurred by camera shake. Blurring in low light conditions is more of a problem with cell-phone cameras whose inexpensive sensors need to have the shutter open longer to collect enough light. Digital still cameras or cell phones that take short videos must have a fast enough "sampling" time and fast "clearing" (of the charge) so as to take pictures at least 15 frames per second; preferable is 24 fps (like film) or 30, 60, or 120 fps like TV refresh rates (25, 50, or 100 in areas like Europe where 50 Hz is the normal line voltage frequency).

f-stop The amount of light reaching the sensor or film depends on the area of the **lens opening** as well as shutter speed, and must be carefully controlled to avoid **underexposure** (too little light so the picture is dark and only the brightest objects show up) or **overexposure** (too much light, so that bright objects have a lack of contrast and "washed-out" appearance). A high quality camera controls the exposure with a "stop" or iris diaphragm, whose opening is of variable diameter, placed behind the lens (Fig. 25–1). The lens opening is controlled (automatically or manually) to compensate for bright or dark lighting conditions, the sensitivity of the sensor or film,[†] and for different shutter speeds. The size of the opening is specified by the **f-stop** or **f-number**, defined as

$$f\text{-stop} = \frac{f}{D},$$

where f is the focal length of the lens and D is the diameter of the lens opening (see Fig. 25–1). For example, when a 50-mm-focal-length lens has an opening $D = 25$ mm, then $f/D = 50$ mm/25 mm $= 2$, so we say it is set at $f/2$. When this lens is set at $f/5.6$, the opening is only 9 mm $(50/9 = 5.6)$. For faster shutter speeds, or low light conditions, a wider lens opening must be used to get a proper exposure, which corresponds to a smaller f-stop number. The smaller the f-stop number, the larger the opening and the more light passes through the lens to the sensor or film. The smallest f-number of a lens (largest opening) is referred to as the *speed* of the lens. The best lenses may have a speed of $f/2.0$, or even faster. The advantage of a fast lens is that it allows pictures to be taken under poor lighting conditions. Good quality lenses consist of several elements to reduce the defects present in simple thin lenses (Section 25–6). Standard f-stops are

$$1.0, \ 1.4, \ 2.0, \ 2.8, \ 4.0, \ 5.6, \ 8, \ 11, \ 16, \ 22, \text{ and } 32$$

(Fig. 25–5). Each of these stops corresponds to a diameter reduction by a factor of $\sqrt{2} \approx 1.4$. Because the amount of light reaching the film is proportional to the *area* of the opening, and therefore proportional to the diameter squared, each standard f-stop corresponds to a factor of 2 change in light intensity reaching the film.

FIGURE 25–5 On this camera, the *f*-stops and the focusing ring are on the camera lens. Shutter speeds are selected on the small wheel on top of the camera body.

[†]Different films have different sensitivities to light, referred to as the "film speed" and specified as an "ISO (or ASA) number." A "faster" film is more sensitive and needs less light to produce a good image, but is grainier which you see when the image is enlarged. Digital cameras may have a "gain" or "ISO" adjustment for sensitivity. A typical everyday ISO might be 200 or so. Adjusting a CCD to be "faster" (high ISO like 3200) for low light conditions results in "noise," resulting in graininess just as in film cameras.

(a)

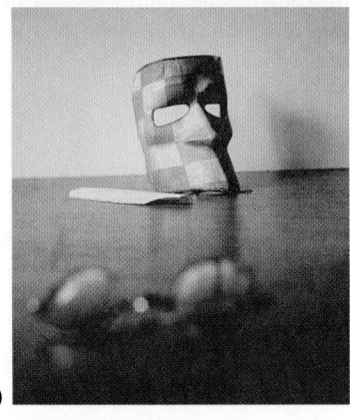

(b)

FIGURE 25–6 Photos taken with a camera lens (a) focused on a nearby object with distant object blurry, and (b) focused on a more distant object with nearby object blurry.

Focusing Focusing is the operation of placing the lens at the correct position relative to the sensor or film for the sharpest image. The image distance is smallest for objects at infinity (the symbol ∞ is used for infinity) and is equal to the focal length, as we saw in Section 23–7. For closer objects, the image distance is greater than the focal length, as can be seen from the lens equation, $1/f = 1/d_o + 1/d_i$ (Eq. 23–8). To focus on nearby objects, the lens must therefore be moved away from the sensor or film, and this is usually done on a manual camera by turning a ring on the lens.

If the lens is focused on a nearby object, a sharp image of it will be formed, but the image of distant objects may be blurry (Fig. 25–6). The rays from a point on the distant object will be out of focus—instead of a point, they will form a circle on the sensor or film as shown (exaggerated) in Fig. 25–7. The distant object will thus produce an image consisting of overlapping circles and will be blurred. These circles are called **circles of confusion**. To have near and distant objects sharp in the same photo, you (or the camera) can try setting the lens focus at an intermediate position. For a given distance setting, there is a range of distances over which the circles of confusion will be small enough that the images will be reasonably sharp. This is called the **depth of field**. For a sensor or film width of 36 mm (including 35-mm film cameras), the depth of field is usually based on a maximum circle of confusion diameter of 0.030 mm, even 0.02 mm or 0.01 mm for critical work or very large photographs. The depth of field varies with the lens opening. If the lens opening is smaller, only rays through the central part of the lens are accepted, and these form smaller circles of confusion for a given object distance. Hence, at smaller lens openings, a greater range of object distances will fit within the circle of confusion criterion, so the depth of field is greater. Smaller lens openings, however, result in reduced resolution due to diffraction (discussed later in this Chapter). Best resolution is typically found around $f/5.6$ or $f/8$.

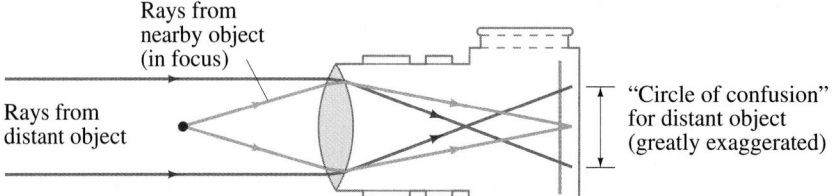

FIGURE 25–7 When the lens is positioned to focus on a nearby object, points on a distant object produce circles and are therefore blurred. (The effect is shown greatly exaggerated.)

EXAMPLE 25–1 **Camera focus.** How far must a 50.0-mm-focal-length camera lens be moved from its infinity setting to sharply focus an object 3.00 m away?

APPROACH For an object at infinity, the image is at the focal point, by definition (Section 23–7). For an object distance of 3.00 m, we use the thin lens equation, Eq. 23–8, to find the image distance (distance of lens to film or sensor).

SOLUTION When focused at infinity, the lens is 50.0 mm from the film. When focused at $d_o = 3.00$ m, the image distance is given by the lens equation,

$$\frac{1}{d_i} = \frac{1}{f} - \frac{1}{d_o} = \frac{1}{50.0\ \text{mm}} - \frac{1}{3000\ \text{mm}} = \frac{3000 - 50}{(3000)(50.0)\ \text{mm}} = \frac{2950}{150{,}000\ \text{mm}}.$$

We solve for d_i and find $d_i = 50.8$ mm, so the lens needs to move 0.8 mm away from the film or digital sensor.

EXERCISE A If the lens of Example 25–1 is 50.4 mm from the film or sensor, what is the object distance for sharp focus?

Shutter speed. To improve the depth of field, you "stop down" your camera lens by two f-stops from $f/4$ to $f/8$. What should you do to the shutter speed to maintain the same exposure?

RESPONSE The amount of light admitted by the lens is proportional to the area of the lens opening. Reducing the lens opening by two f-stops reduces the diameter by a factor of 2, and the area by a factor of 4. To maintain the same exposure, the shutter must be open four times as long. If the shutter speed had been $\frac{1}{500}$ s, you would have to increase the exposure time to $\frac{1}{125}$ s.

*Picture Sharpness

The sharpness of a picture depends not only on accurate focusing and short exposure times, but also on the graininess of the film, or number of pixels on a digital sensor. Fine-grained films and tiny pixels are "slower," meaning they require longer exposures for a given light level. All pixels are rarely used because digital cameras have averaging (or "compression") programs, such as JPEG, which reduce memory size by averaging over pixels where little contrast is detected. But some detail is inevitably lost. For example, a small blue lake may seem uniform, and coding 600 pixels as identical takes less memory than specifying all 600. Any slight variation of the water surface is lost. Full "RAW" data uses more memory. Film records everything (down to its grain size), as does RAW if your camera offers it. The processor also averages over pixels in low light conditions, resulting in a less sharp photo.

The quality of the lens strongly affects the image quality, and we discuss lens resolution and diffraction effects in Sections 25–6 and 25–7. The sharpness, or **resolution**, of a lens is often given as so many lines per millimeter, measured by photographing a standard set of parallel black lines on a white background (sometimes said as "line pairs/mm") on fine-grain film or high quality sensor, or as so many dots per inch (dpi). The minimum spacing of distinguishable lines or dots gives the resolution. A lens that can give 50 lines/mm is reasonable, 100 lines/mm is very good (= 100 dots/mm ≈ 2500 dpi). Electronic sensors also have a resolution and it is sometimes given as line pairs across the full sensor width.

A "full" Bayer pixel (upper left in Fig. 25–2) is 4 regular pixels: for example, to make a white dot as part of a white line (between two black lines when determining lens resolution), all 4 Bayer pixels (RGGB) would have to be bright. For a Foveon, all three colors of one pixel need to be bright to produce a white dot.[†]

Pixels and resolution. A digital camera offers a maximum resolution of 4000×3000 pixels on a 32 mm × 24 mm sensor. How sharp should the lens be to make use of this sensor resolution in RAW?

APPROACH We find the number of pixels per millimeter and require the lens to be at least that good.

SOLUTION We can either take the image height (3000 pixels in 24 mm) or the width (4000 pixels in 32 mm):

$$\frac{3000 \text{ pixels}}{24 \text{ mm}} = 125 \text{ pixels/mm}.$$

We would like the lens to match this resolution of 125 lines or dots per mm, which would be a quite good lens. If the lens is not this good, fewer pixels and less memory could be used.

NOTE Increasing lens resolution is a tougher problem today than is squeezing more pixels on a CCD or CMOS sensor. The sensor for high quality cameras must also be physically larger for better image accuracy and greater light sensitivity in low light conditions.

[†]Consider a 4000×3000 pixel array. For a Foveon, each "full pixel" (Fig. 25–3) has all 3 colors, each of which can be counted as a pixel, so it may be considered as $4000 \times 3000 \times 3 = 36$ MP. For a Bayer sensor, Fig. 25–2, 4000×3000 is 12 MP (6 MP of green, 3 MP each of red and blue). There are more green pixels because they are most important in our eyes' ability to note resolution. So the distance between green pixels is a rough guide to the sharpness of a Bayer. To match a 4000×3000 Foveon (36 MP, or 12 MP of tri-pixel sites), a Bayer would need to have about 24 MP (because it would then have 12 MP of green). This "equivalence" is only a rough approximation.

EXAMPLE 25–4 **Blown-up photograph.** A photograph looks sharp at normal viewing distances if the dots or lines are resolved to perhaps 10 dots/mm. Would an 8×10-inch enlargement of a photo taken by the camera in Example 25–3 seem sharp?

APPROACH We assume the image is 4000×3000 pixels on a 32×24-mm sensor as in Example 25–3, or 125 pixels/mm. We make an enlarged photo 8×10 in. $= 20$ cm $\times 25$ cm.

SOLUTION The short side of the sensor is 24 mm $= 2.4$ cm long, and that side of the photograph is 8 inches or 20 cm. Thus the size is increased by a factor of 20 cm/2.4 cm $\approx 8\times$ (or 25 cm/3.2 cm $\approx 8\times$). To fill the 8×10-in. paper, we assume the enlargement is $8\times$. The pixels are thus enlarged $8\times$. So the pixel count of 125/mm on the sensor becomes $125/8 = 15$ per mm on the print. Hence an 8×10-inch print would be a sharp photograph. We could go 50% larger—11×14 or maybe even 12×18 inches.

In order to make very large photographic prints, large-format cameras are used such as 6 cm $\times$ 6 cm $\left(2\frac{1}{4}\text{ inch square}\right)$—either film or sensor—and even 4×5 inch and 8×10 inch (using sheet film or glass plates).

EXERCISE B The criterion of 0.030 mm as the diameter of a circle of confusion as acceptable sharpness is how many dots per mm on the sensor?

Telephotos and Wide-angles

Camera lenses are categorized into normal, telephoto, and wide angle, according to focal length and film size. A **normal lens** covers the sensor or film with a field of view that corresponds approximately to that of normal vision. A "normal" lens for 35-mm film has a focal length of 50 mm. The best digital cameras aim for a sensor of the same size[†] (24 mm $\times$ 36 mm). (If the sensor is smaller, digital cameras sometimes specify focal lengths to correspond with classic 35-mm cameras.) **Telephoto lenses** act like telescopes to magnify images. They have longer focal lengths than a normal lens: as we saw in Section 23–8 (Eq. 23–9), the height of the image for a given object distance is proportional to the image distance, and the image distance will be greater for a lens with longer focal length. For distant objects, the image height is very nearly proportional to the focal length. Thus a 200-mm telephoto lens for use with a 35-mm camera gives a $4\times$ magnification over the normal 50-mm lens. A **wide-angle lens** has a shorter focal length than normal: a wider field of view is included, and objects appear smaller. A **zoom lens** is one whose focal length can be changed (by changing the distance between the thin lenses that make up the compound lens) so that you seem to zoom up to, or away from, the subject as you change the focal length.

Digital cameras may have an **optical zoom** meaning the lens can change focal length and maintain resolution. But an "electronic" or **digital zoom** just enlarges the dots (pixels) with loss of sharpness.

Different types of viewing systems are used in cameras. In some cameras, you view through a small window just above the lens as in Fig. 25–1. In a **single-lens reflex** camera (SLR), you actually view through the lens with the use of prisms and mirrors (Fig. 25–8). A mirror hangs at a 45° angle behind the lens and flips up out of the way just before the shutter opens. SLRs have the advantage that you can see almost exactly what you will get. Digital cameras use an LCD display, and it too can show what you will get on the photo if it is carefully designed.

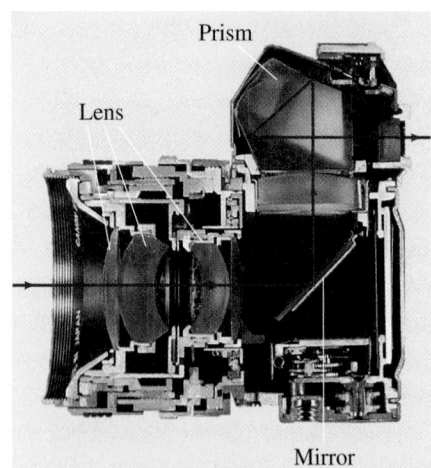

FIGURE 25–8 Single-lens reflex (SLR) camera, showing how the image is viewed through the lens with the help of a movable mirror and prism.

Prism

Lens

Mirror

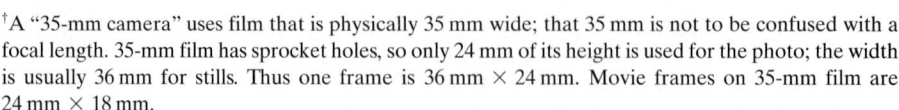

[†]A "35-mm camera" uses film that is physically 35 mm wide; that 35 mm is not to be confused with a focal length. 35-mm film has sprocket holes, so only 24 mm of its height is used for the photo; the width is usually 36 mm for stills. Thus one frame is 36 mm $\times$ 24 mm. Movie frames on 35-mm film are 24 mm $\times$ 18 mm.

25–2 The Human Eye; Corrective Lenses

The human eye resembles a camera in its basic structure (Fig. 25–9), but is far more sophisticated. The interior of the eye is filled with a transparent gel-like substance called the *vitreous humor* with index of refraction $n = 1.337$. Light enters this enclosed volume through the cornea and lens. Between the cornea and lens is a watery fluid, the aqueous humor (*aqua* is "water" in Latin) with $n = 1.336$. A diaphragm, called the **iris** (the colored part of your eye), adjusts automatically to control the amount of light entering the eye, similar to a camera. The hole in the iris through which light passes (the **pupil**) is black because no light is reflected from it (it's a hole), and very little light is reflected back out from the interior of the eye. The **retina**, which plays the role of the film or sensor in a camera, is on the curved rear surface of the eye. The retina consists of a complex array of nerves and receptors known as *rods* and *cones* which act to change light energy into electrical signals that travel along the nerves. The reconstruction of the image from all these tiny receptors is done mainly in the brain, although some analysis may also be done in the complex interconnected nerve network at the retina itself. At the center of the retina is a small area called the **fovea**, about 0.25 mm in diameter, where the cones are very closely packed and the sharpest image and best color discrimination are found.

Unlike a camera, the eye contains no shutter. The equivalent operation is carried out by the nervous system, which analyzes the signals to form images at the rate of about 30 per second. This can be compared to motion picture or television cameras, which operate by taking a series of still pictures at a rate of 24 (movies) and 60 or 30 (U.S. television) per second. Their rapid projection on the screen gives the appearance of motion.

The lens of the eye ($n = 1.386$ to 1.406) does little of the bending of the light rays. Most of the refraction is done at the front surface of the **cornea** ($n = 1.376$) at its interface with air ($n = 1.0$). The lens acts as a fine adjustment for focusing at different distances. This is accomplished by the ciliary muscles (Fig. 25–9), which change the curvature of the lens so that its focal length is changed. To focus on a distant object, the ciliary muscles of the eye are relaxed and the lens is thin, as shown in Fig. 25–10a, and parallel rays focus at the focal point (on the retina). To focus on a nearby object, the muscles contract, causing the center of the lens to thicken, Fig. 25–10b, thus shortening the focal length so that images of nearby objects can be focused on the retina, behind the new focal point. This focusing adjustment is called **accommodation**.

The closest distance at which the eye can focus clearly is called the **near point** of the eye. For young adults it is typically 25 cm, although younger children can often focus on objects as close as 10 cm. As people grow older, the ability to accommodate is reduced and the near point increases. A given person's **far point** is the farthest distance at which an object can be seen clearly. For some purposes it is useful to speak of a **normal eye** (a sort of average over the population), defined as an eye having a near point of 25 cm and a far point of infinity. To check your own near point, place this book close to your eye and slowly move it away until the type is sharp.

The "normal" eye is sort of an ideal. Many people have eyes that do not accommodate within the "normal" range of 25 cm to infinity, or have some other defect. Two common defects are nearsightedness and farsightedness. Both can be corrected to a large extent with lenses—either eyeglasses or contact lenses.

In **nearsightedness**, or **myopia**, the human eye can focus only on nearby objects. The far point is not infinity but some shorter distance, so distant objects are not seen clearly. Nearsightedness is usually caused by an eyeball that is too long, although sometimes it is the curvature of the cornea that is too great. In either case, images of distant objects are focused in front of the retina.

PHYSICS APPLIED
The eye

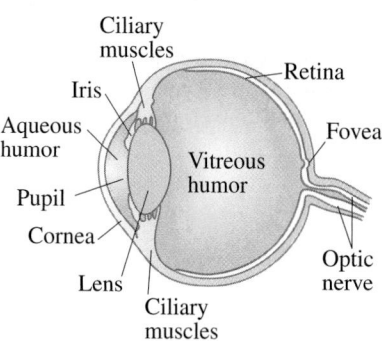

FIGURE 25–9 Diagram of a human eye.

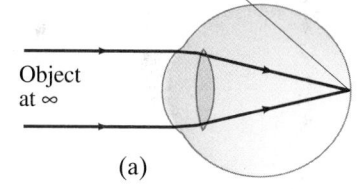

FIGURE 25–10 Accommodation by a normal eye: (a) lens relaxed, focused at infinity; (b) lens thickened, focused on a nearby object.

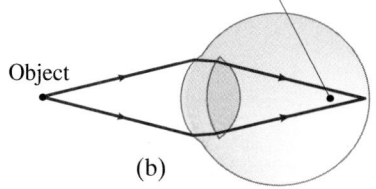

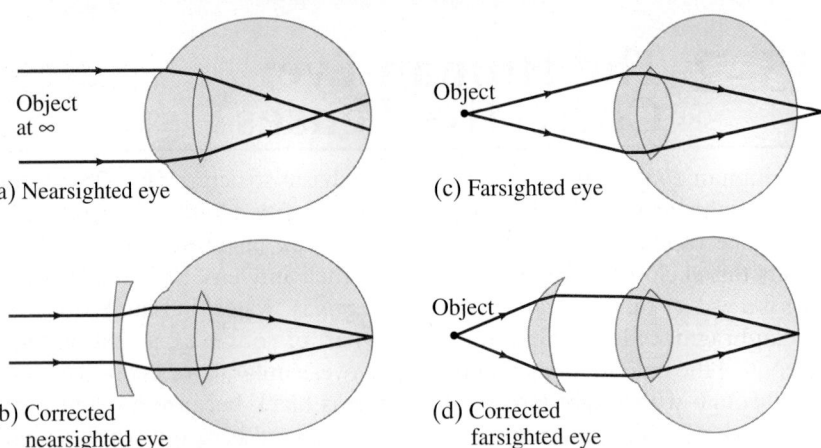

FIGURE 25–11 Correcting eye defects with lenses. (a) A nearsighted eye, which cannot focus clearly on distant objects (focal point is in front of retina), can be corrected (b) by use of a diverging lens. (c) A farsighted eye, which cannot focus clearly on nearby objects (focus point behind retina), can be corrected (d) by use of a converging lens.

(a) Nearsighted eye

(b) Corrected nearsighted eye

(c) Farsighted eye

(d) Corrected farsighted eye

PHYSICS APPLIED

Corrective lenses

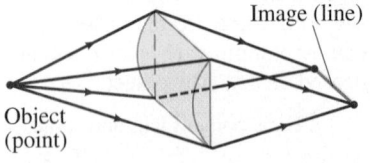

FIGURE 25–12 A cylindrical lens forms a line image of a point object because it is converging in one plane only.

Image (line)

Object (point)

A diverging lens, because it causes parallel rays to diverge, allows the rays to be focused at the retina (Figs. 25–11a and b) and thus can correct nearsightedness.

In **farsightedness**, or *hyperopia*, the eye cannot focus on nearby objects. Although distant objects are usually seen clearly, the near point is somewhat greater than the "normal" 25 cm, which makes reading difficult. This defect is caused by an eyeball that is too short or (less often) by a cornea that is not sufficiently curved. It is corrected by a converging lens, Figs. 25–11c and d. Similar to hyperopia is *presbyopia*, which is the lessening ability of the eye to accommodate as a person ages, and the near point moves out. Converging lenses also compensate for this.

Astigmatism is usually caused by an out-of-round cornea or lens so that point objects are focused as short lines, which blurs the image. It is as if the cornea were spherical with a cylindrical section superimposed. As shown in Fig. 25–12, a cylindrical lens focuses a point into a line parallel to its axis. An astigmatic eye may focus rays in one plane, such as the vertical plane, at a shorter distance than it does for rays in a horizontal plane. Astigmatism is corrected with the use of a compensating cylindrical lens. Lenses for eyes that are nearsighted or farsighted as well as astigmatic are ground with superimposed spherical and cylindrical surfaces, so that the radius of curvature of the correcting lens is different in different planes.

EXAMPLE 25–5 **Farsighted eye.** A farsighted eye has a near point of 100 cm. Reading glasses must have what lens power so that a newspaper can be read at a distance of 25 cm? Assume the lens is very close to the eye.

APPROACH When the object is placed 25 cm from the lens (= d_o), we want the image to be 100 cm away on the *same* side of the lens (so the eye can focus it), and so the image is virtual, as shown in Fig. 25–13, and $d_i = -100$ cm will be negative. We use the thin lens equation (Eq. 23–8) to determine the needed focal length. Optometrists' prescriptions specify the power ($P = 1/f$, Eq. 23–7) given in diopters $(1\,D = 1\,m^{-1})$.

SOLUTION Given that $d_o = 25$ cm and $d_i = -100$ cm, the thin lens equation gives

$$\frac{1}{f} = \frac{1}{d_o} + \frac{1}{d_i} = \frac{1}{25\text{ cm}} + \frac{1}{-100\text{ cm}} = \frac{4-1}{100\text{ cm}} = \frac{1}{33\text{ cm}}.$$

So $f = 33$ cm $= 0.33$ m. The power P of the lens is $P = 1/f = +3.0\,D$. The plus sign indicates that it is a converging lens.

NOTE We chose the image position to be where the eye can actually focus. The lens needs to put the image there, given the desired placement of the object (newspaper).

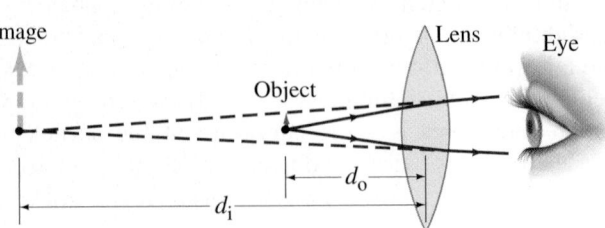

FIGURE 25–13 Lens of reading glasses (Example 25–5).

EXAMPLE 25–6 **Nearsighted eye.** A nearsighted eye has near and far points of 12 cm and 17 cm, respectively. (a) What lens power is needed for this person to see distant objects clearly, and (b) what then will be the near point? Assume that the lens is 2.0 cm from the eye (typical for eyeglasses).

APPROACH For a distant object $(d_o = \infty)$, the lens must put the image at the far point of the eye as shown in Fig. 25–14a, 17 cm in front of the eye. We can use the thin lens equation to find the focal length of the lens, and from this its lens power. The new near point (as shown in Fig. 25–14b) can be calculated for the lens by again using the thin lens equation.

SOLUTION (a) For an object at infinity $(d_o = \infty)$, the image must be in front of the lens 17 cm from the eye or $(17\,\text{cm} - 2\,\text{cm}) = 15\,\text{cm}$ from the lens; hence $d_i = -15\,\text{cm}$. We use the thin lens equation to solve for the focal length of the needed lens:

$$\frac{1}{f} = \frac{1}{d_o} + \frac{1}{d_i} = \frac{1}{\infty} + \frac{1}{-15\,\text{cm}} = -\frac{1}{15\,\text{cm}}.$$

So $f = -15\,\text{cm} = -0.15\,\text{m}$ or $P = 1/f = -6.7\,\text{D}$. The minus sign indicates that it must be a diverging lens for the myopic eye.

(b) The near point when glasses are worn is where an object is placed (d_o) so that the lens forms an image at the "near point of the naked eye," namely 12 cm from the eye. That image point is $(12\,\text{cm} - 2\,\text{cm}) = 10\,\text{cm}$ in front of the lens, so $d_i = -0.10\,\text{m}$ and the thin lens equation gives

$$\frac{1}{d_o} = \frac{1}{f} - \frac{1}{d_i} = -\frac{1}{0.15\,\text{m}} + \frac{1}{0.10\,\text{m}} = \frac{-2 + 3}{0.30\,\text{m}} = \frac{1}{0.30\,\text{m}}.$$

So $d_o = 30\,\text{cm}$, which means the near point when the person is wearing glasses is 30 cm in front of the lens, or 32 cm from the eye.

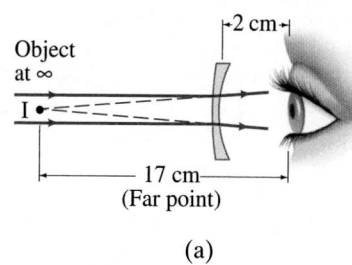

(a)

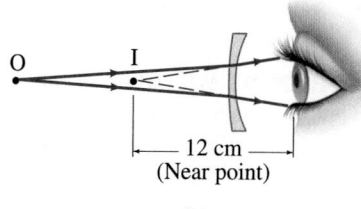

(b)

FIGURE 25–14 Example 25–6.

Contact Lenses

Suppose contact lenses are used to correct the eye in Example 25–6. Since contacts are placed directly on the cornea, we would not subtract out the 2.0 cm for the image distances. That is, for distant objects $d_i = f = -17\,\text{cm}$, so $P = 1/f = -5.9\,\text{D}$. The new near point would be 41 cm. Thus we see that a contact lens and an eyeglass lens will require slightly different powers, or focal lengths, for the same eye because of their different placements relative to the eye. We also see that glasses in this case give a better near point than contacts.

PHYSICS APPLIED
Contact lenses—different f and P

EXERCISE C What power of contact lens is needed for an eye to see distant objects if its far point is 25 cm?

Underwater Vision

When your eyes are under water, distant underwater objects look blurry because at the water–cornea interface, the difference in indices of refraction is very small: $n = 1.33$ for water, 1.376 for the cornea. Hence light rays are bent very little and are focused far behind the retina, Fig. 25–15a. If you wear goggles or a face mask, you restore an air–cornea interface ($n = 1.0$ and 1.376, respectively) and the rays can be focused, Fig. 25–15b.

PHYSICS APPLIED
Underwater vision

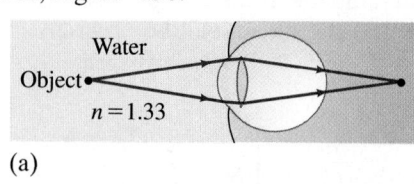

(a)

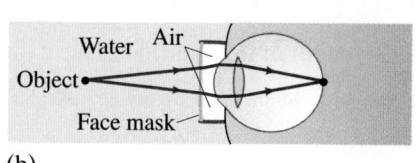

(b)

FIGURE 25–15 (a) Under water, we see a blurry image because light rays are bent much less than in air. (b) If we wear goggles, we again have an air–cornea interface and can see clearly.

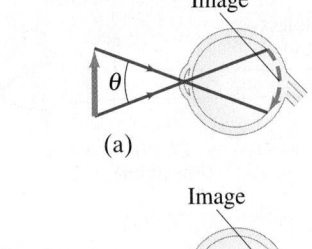

(a)

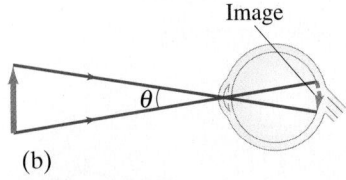

(b)

FIGURE 25–16 When the same object is viewed at a shorter distance, the image on the retina is greater, so the object appears larger and more detail can be seen. The angle θ that the object subtends in (a) is greater than in (b). *Note*: This is not a normal ray diagram because we are showing only one ray from each point.

25–3 Magnifying Glass

Much of the remainder of this Chapter will deal with optical devices that are used to produce magnified images of objects. We first discuss the **simple magnifier**, or **magnifying glass**, which is simply a converging lens (see Chapter-Opening Photo, page 713).

How large an object appears, and how much detail we can see on it, depends on the size of the image it makes on the retina. This, in turn, depends on the angle subtended by the object at the eye. For example, a penny held 30 cm from the eye looks twice as tall as one held 60 cm away because the angle it subtends is twice as great (Fig. 25–16). When we want to examine detail on an object, we bring it up close to our eyes so that it subtends a greater angle. However, our eyes can accommodate only up to a point (the near point), and we will assume a standard distance of $N = 25$ cm as the near point in what follows.

A magnifying glass allows us to place the object closer to our eye so that it subtends a greater angle. As shown in Fig. 25–17a, the object is placed at the focal point or just within it. Then the converging lens produces a virtual image, which must be at least 25 cm from the eye if the eye is to focus on it. If the eye is relaxed, the image will be at infinity, and in this case the object is exactly at the focal point. (You make this slight adjustment yourself when you "focus" on the object by moving the magnifying glass.)

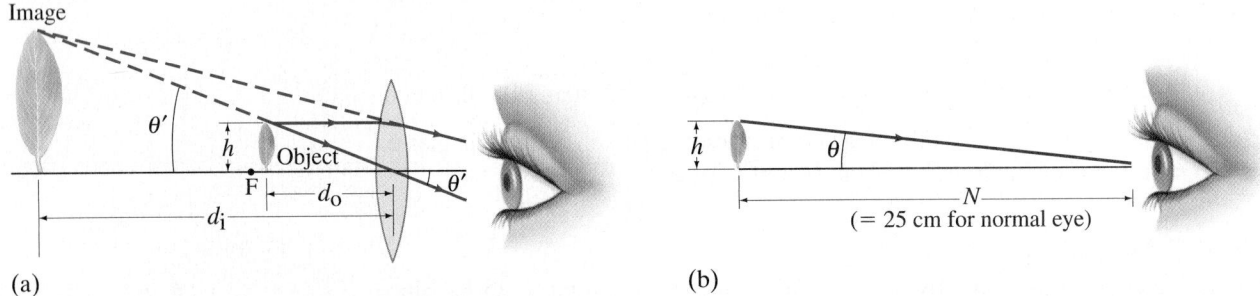

(a)

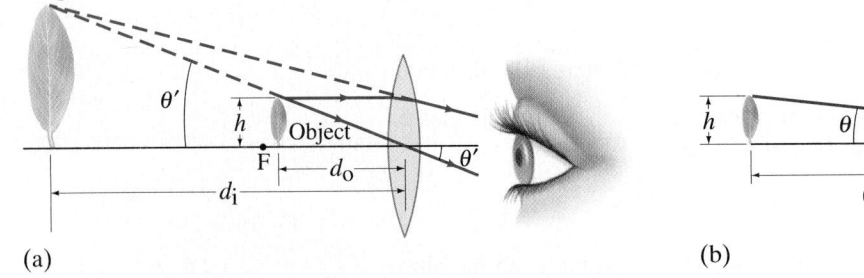

(b)

FIGURE 25–17 Leaf viewed (a) through a magnifying glass, and (b) with the unaided eye. The eye is focused at its near point in both cases.

A comparison of part (a) of Fig. 25–17 with part (b), in which the same object is viewed at the near point with the unaided eye, reveals that the angle the object subtends at the eye is much larger when the magnifier is used. The **angular magnification** or **magnifying power**, M, of the lens is defined as the ratio of the angle subtended by an object when using the lens, to the angle subtended using the unaided eye, with the object at the near point N of the eye ($N = 25$ cm for a normal eye):

$$M = \frac{\theta'}{\theta}, \tag{25–1}$$

where θ and θ' are shown in Fig. 25–17. We can write M in terms of the focal length by noting that $\theta = h/N$ (Fig. 25–17b) and $\theta' = h/d_o$ (Fig. 25–17a), where h is the height of the object and we assume the angles are small so θ and θ' (in radians) equal their sines and tangents. If the eye is relaxed (for least eye strain), the image will be at infinity and the object will be precisely at the focal point; see Fig. 25–18. Then $d_o = f$ and $\theta' = h/f$, whereas $\theta = h/N$ as before (Fig. 25–17b). Thus

$$M = \frac{\theta'}{\theta} = \frac{h/f}{h/N} = \frac{N}{f}. \quad \left[\begin{array}{l} \text{eye focused at } \infty; \\ N = 25 \text{ cm for normal eye} \end{array}\right] \tag{25–2a}$$

We see that the shorter the focal length of the lens, the greater the magnification.[†]

[†]Simple single-lens magnifiers are limited to about 2 or 3× because of blurring due to spherical aberration (Section 25–6).

FIGURE 25–18 With the eye relaxed, the object is placed at the focal point, and the image is at infinity. Compare to Fig. 25–17a where the image is at the eye's near point.

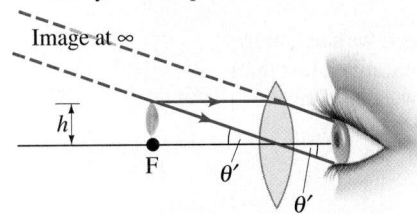

The magnification of a given lens can be increased a bit by moving the lens and adjusting your eye so it focuses on the image at the eye's near point. In this case, $d_i = -N$ (see Fig. 25–17a) if your eye is very near the magnifier. Then the object distance d_o is given by

$$\frac{1}{d_o} = \frac{1}{f} - \frac{1}{d_i} = \frac{1}{f} + \frac{1}{N}.$$

We see from this equation that $d_o = fN/(f + N) < f$, as shown in Fig. 25–17a, since $N/(f + N)$ must be less than 1. With $\theta' = h/d_o$ the magnification is

$$M = \frac{\theta'}{\theta} = \frac{h/d_o}{h/N}$$

$$= N\left(\frac{1}{d_o}\right) = N\left(\frac{1}{f} + \frac{1}{N}\right)$$

or

$$M = \frac{N}{f} + 1. \qquad \left[\begin{array}{c} \text{eye focused at near point, } N; \\ N = 25 \text{ cm for normal eye} \end{array}\right] \quad \textbf{(25–2b)}$$

We see that the magnification is slightly greater when the eye is focused at its near point, as compared to when it is relaxed.

EXAMPLE 25–7 ESTIMATE | **A jeweler's "loupe."** An 8-cm-focal-length converging lens is used as a "jeweler's loupe," which is a magnifying glass. Estimate (a) the magnification when the eye is relaxed, and (b) the magnification if the eye is focused at its near point $N = 25$ cm.

APPROACH The magnification when the eye is relaxed is given by Eq. 25–2a. When the eye is focused at its near point, we use Eq. 25–2b and we assume the lens is near the eye.

SOLUTION (a) With the relaxed eye focused at infinity,

$$M = \frac{N}{f} = \frac{25 \text{ cm}}{8 \text{ cm}} \approx 3\times.$$

(b) The magnification when the eye is focused at its near point ($N = 25$ cm), and the lens is near the eye, is

$$M = 1 + \frac{N}{f} = 1 + \frac{25}{8} \approx 4\times.$$

25–4 Telescopes

A telescope is used to magnify objects that are very far away. In most cases, the object can be considered to be at infinity.

Galileo, although he did not invent it,[†] developed the telescope into a usable and important instrument. He was the first to examine the heavens with the telescope (Fig. 25–19), and he made world-shaking discoveries, including the moons of Jupiter, the phases of Venus, sunspots, the structure of the Moon's surface, and that the Milky Way is made up of a huge number of individual stars.

[†]Galileo built his first telescope in 1609 after having heard of such an instrument existing in Holland. The first telescopes magnified only three to four times, but Galileo soon made a 30-power instrument. The first Dutch telescopes date from about 1604 and probably were copies of an Italian telescope built around 1590. Kepler (see Chapter 5) gave a ray description (in 1611) of the Keplerian telescope, which is named for him because he first described it, although he did not build it.

FIGURE 25–19 (a) Objective lens (mounted now in an ivory frame) from the telescope with which Galileo made his world-shaking discoveries, including the moons of Jupiter. (b) Telescopes made by Galileo (1609).

(a)

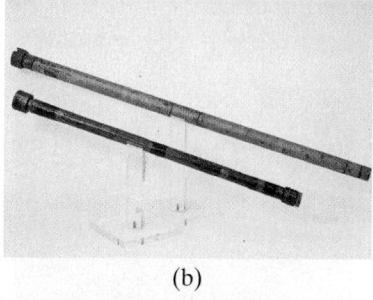

(b)

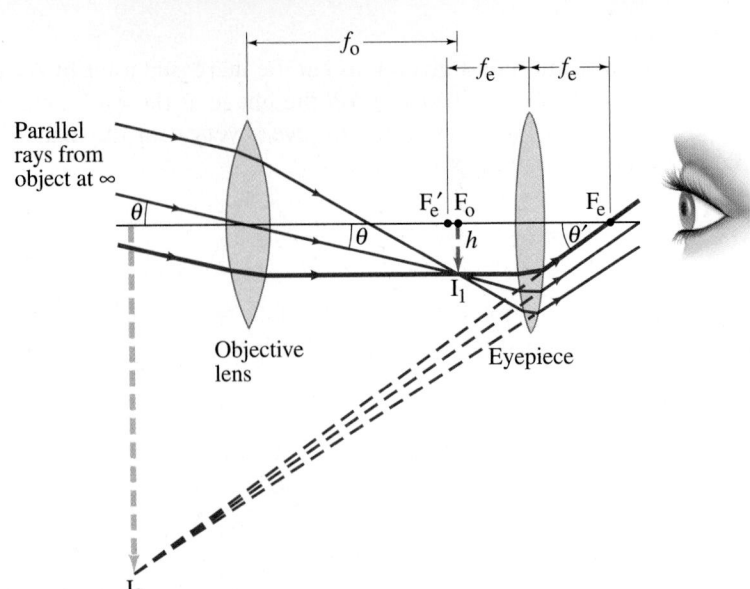

FIGURE 25-20 Astronomical telescope (refracting). Parallel light from one point on a distant object $(d_o = \infty)$ is brought to a focus by the objective lens in its focal plane. This image (I_1) is magnified by the eyepiece to form the final image I_2. Only two of the rays shown entering the objective are standard rays (2 and 3) as described in Fig. 23-37.

Several types of **astronomical telescope** exist. The common **refracting** type, sometimes called **Keplerian**, contains two converging lenses located at opposite ends of a long tube, as illustrated in Fig. 25-20. The lens closest to the object is called the **objective lens** (focal length f_o) and forms a real image I_1 of the distant object in the plane of its focal point F_o (or near it if the object is not at infinity). The second lens, called the **eyepiece** (focal length f_e), acts as a magnifier. That is, the eyepiece magnifies the image I_1 formed by the objective lens to produce a second, greatly magnified image, I_2, which is virtual and inverted. If the viewing eye is relaxed, the eyepiece is adjusted so the image I_2 is at infinity. Then the real image I_1 is at the focal point F'_e of the eyepiece, and the distance between the lenses is $f_o + f_e$ for an object at infinity.

To find the total angular magnification of this telescope, we note that the angle an object subtends as viewed by the unaided eye is just the angle θ subtended at the telescope objective. From Fig. 25-20 we can see that $\theta \approx h/f_o$, where h is the height of the image I_1 and we assume θ is small so that $\tan\theta \approx \theta$. Note, too, that the thickest of the three rays drawn in Fig. 25-20 is parallel to the axis before it strikes the eyepiece and therefore is refracted through the eyepiece focal point F_e on the far side. Thus, $\theta' \approx h/f_e$ and the **total magnifying power** (that is, angular magnification, which is what is always quoted) of this telescope is

$$M = \frac{\theta'}{\theta} = \frac{(h/f_e)}{(h/f_o)} = -\frac{f_o}{f_e}, \qquad \left[\begin{array}{c}\text{telescope}\\\text{magnification}\end{array}\right] \quad \textbf{(25-3)}$$

where we used Eq. 23-1 and we inserted a minus sign to indicate that the image is inverted. To achieve a large magnification, the objective lens should have a long focal length and the eyepiece a short focal length.

FIGURE 25-21 This large refracting telescope was built in 1897 and is housed at Yerkes Observatory in Wisconsin. The objective lens is 102 cm (40 inches) in diameter, and the telescope tube is about 19 m long. Example 25-8.

EXAMPLE 25-8 **Telescope magnification.** The largest optical refracting telescope in the world is located at the Yerkes Observatory in Wisconsin, Fig. 25-21. It is referred to as a "40-inch" telescope, meaning that the diameter of the objective is 40 in., or 102 cm. The objective lens has a focal length of 19 m, and the eyepiece has a focal length of 10 cm. (a) Calculate the total magnifying power of this telescope. (b) Estimate the length of the telescope.

APPROACH Equation 25-3 gives the magnification. The length of the telescope is the distance between the two lenses.

SOLUTION (a) From Eq. 25-3 we find

$$M = -\frac{f_o}{f_e} = -\frac{19\,\text{m}}{0.10\,\text{m}} = -190\times.$$

(b) For a relaxed eye, the image I_1 is at the focal point of both the eyepiece and the objective lenses. The distance between the two lenses is thus $f_o + f_e \approx 19\,\text{m}$, which is essentially the length of the telescope.

EXERCISE D A 40× telescope has a 1.2-cm focal length eyepiece. What is the focal length of the objective lens?

For an astronomical telescope to produce bright images of faint stars, the objective lens must be large to allow in as much light as possible. Indeed, the diameter of the objective lens (and hence its "light-gathering power") is an important parameter for an astronomical telescope, which is why the largest ones are specified by giving the objective diameter (such as the 10-meter Keck telescope in Hawaii). The construction and grinding of large lenses is very difficult. Therefore, the largest telescopes are **reflecting telescopes** which use a curved mirror as the objective, Fig. 25–22. A mirror has only one surface to be ground and can be supported along its entire surface[†] (a large lens, supported at its edges, would sag under its own weight). Often, the eyepiece lens or mirror (see Fig. 25–22) is removed so that the real image formed by the objective mirror can be recorded directly on film or on an electronic sensor (CCD or CMOS, Section 25–1).

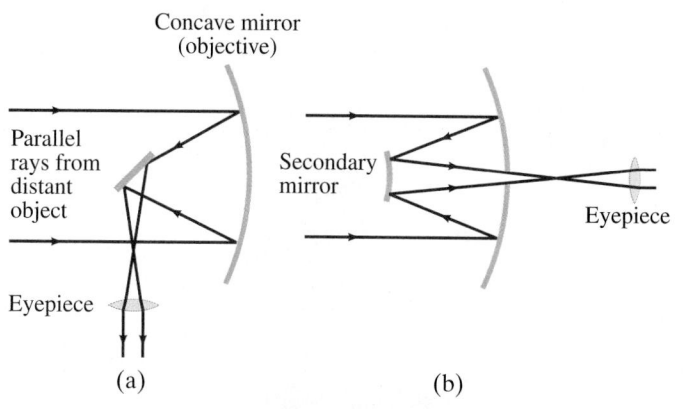

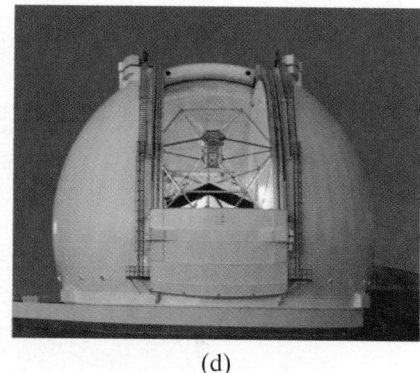

(a) (b) (c) (d)

FIGURE 25–22 A concave mirror can be used as the objective of an astronomical telescope. Arrangement (a) is called the Newtonian focus, and (b) the Cassegrainian focus. Other arrangements are also possible. (c) The 200-inch (mirror diameter) Hale telescope on Palomar Mountain in California. (d) The 10-meter Keck telescope on Mauna Kea, Hawaii. The Keck combines thirty-six 1.8-meter six-sided mirrors into the equivalent of a very large single reflector, 10 m in diameter.

A **terrestrial telescope**, for viewing objects on Earth, must provide an upright image—seeing normal objects upside down would be difficult (much less important for viewing stars). Two designs are shown in Fig. 25–23. The **Galilean** type, which Galileo used for his great astronomical discoveries, has a diverging lens as eyepiece which intercepts the converging rays from the objective lens before they reach a focus, and acts to form a virtual upright image, Fig. 25–23a. This design is still used in opera glasses. The tube is reasonably short, but the field of view is small. The second type, shown in Fig. 25–23b, is often called a **spyglass** and makes use of a third convex lens that acts to make the image upright as shown. A spyglass must be quite long. The most practical design today is the **prism binocular** which was shown in Fig. 23–28. The objective and eyepiece are converging lenses. The prisms reflect the rays by total internal reflection and shorten the physical size of the device, and they also act to produce an upright image. One prism reinverts the image in the vertical plane, the other in the horizontal plane.

[†]Another advantage of mirrors is that they exhibit no chromatic aberration because the light doesn't pass through them; and they can be ground into a parabolic shape to correct for spherical aberration (Section 25–6). The reflecting telescope was first proposed by Newton.

FIGURE 25–23 Terrestrial telescopes that produce an upright image: (a) Galilean; (b) spyglass, or erector type.

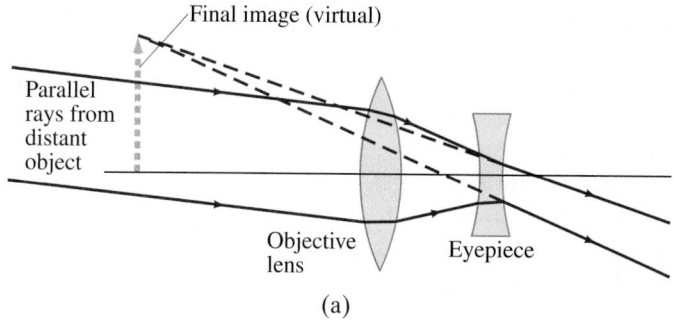
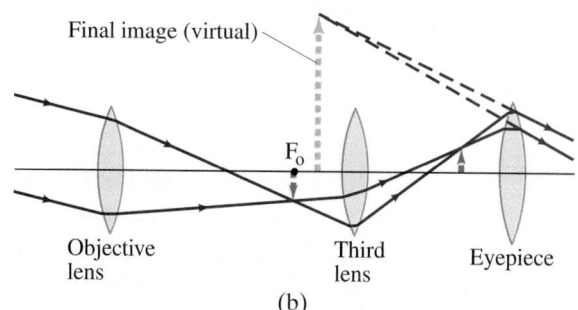

(a) (b)

25–5 Compound Microscope

The compound **microscope**, like the telescope, has both objective and eyepiece (or ocular) lenses, Fig. 25–24. The design is different from that for a telescope because a microscope is used to view objects that are very close, so the object distance is very small. The object is placed just beyond the objective's focal point as shown in Fig. 25–24a. The image I_1 formed by the objective lens is real, quite far from the objective lens, and much enlarged. The eyepiece is positioned so that this image is near the eyepiece focal point F_e. The image I_1 is magnified by the eyepiece into a very large virtual image, I_2, which is seen by the eye and is inverted. Modern microscopes use a third "tube" lens behind the objective, but we will analyze the simpler arrangement shown in Fig. 25–24a.

FIGURE 25–24 Compound microscope: (a) ray diagram, (b) photograph (illumination comes from below, outlined in red, then up through the slide holding the sample or object).

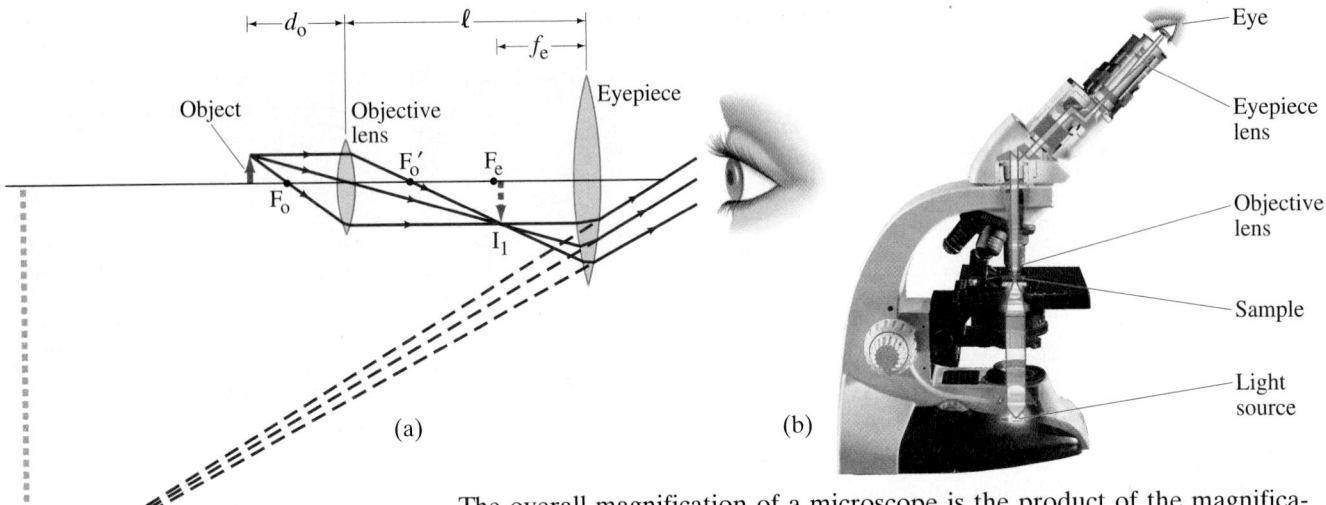

(a)

(b)

The overall magnification of a microscope is the product of the magnifications produced by the two lenses. The image I_1 formed by the objective lens is a factor m_o greater than the object itself. From Fig. 25–24a and Eq. 23–9 for the magnification of a simple lens, we have

$$m_o = \frac{h_i}{h_o} = \frac{d_i}{d_o} = \frac{\ell - f_e}{d_o}, \tag{25–4}$$

where d_o and d_i are the object and image distances for the objective lens, ℓ is the distance between the lenses (equal to the length of the barrel), and we ignored the minus sign in Eq. 23–9 which only tells us that the image is inverted. We set $d_i = \ell - f_e$, which is exact only if the eye is relaxed, so that the image I_1 is at the eyepiece focal point F_e. The eyepiece acts like a simple magnifier. If we assume that the eye is relaxed, the eyepiece angular magnification M_e is (from Eq. 25–2a)

$$M_e = \frac{N}{f_e}, \tag{25–5}$$

where the near point $N = 25\ \text{cm}$ for the normal eye. Since the eyepiece enlarges the image formed by the objective, the overall angular magnification M is the product of the magnification of the objective lens, m_o, times the angular magnification, M_e, of the eyepiece lens (Eqs. 25–4 and 25–5):

$$M = M_e m_o = \left(\frac{N}{f_e}\right)\left(\frac{\ell - f_e}{d_o}\right) \qquad \left[\begin{array}{c}\text{microscope}\\ \text{magnification}\end{array}\right] \tag{25–6a}$$

$$\approx \frac{N\ell}{f_e f_o}. \qquad [f_o \text{ and } f_e \ll \ell] \tag{25–6b}$$

The approximation, Eq. 25–6b, is accurate when f_e and f_o are small compared to ℓ, so $\ell - f_e \approx \ell$, and the object is near F_o so $d_o \approx f_o$ (Fig. 25–24a). This is a good approximation for large magnifications, which are obtained when f_o and f_e are very small (they are in the denominator of Eq. 25–6b). To make lenses of very short focal length, compound lenses involving several elements must be used to avoid serious aberrations, as discussed in the next Section.

EXAMPLE 25–9 **Microscope.** A compound microscope consists of a 10× eyepiece and a 50× objective 17.0 cm apart. Determine (a) the overall magnification, (b) the focal length of each lens, and (c) the position of the object when the final image is in focus with the eye relaxed. Assume a normal eye, so $N = 25$ cm.

APPROACH The overall magnification is the product of the eyepiece magnification and the objective magnification. The focal length of the eyepiece is found from Eq. 25–2a or 25–5 for the magnification of a simple magnifier. For the objective lens, it is easier to next find d_o (part c) using Eq. 25–4 before we find f_o.

SOLUTION (a) The overall magnification is $(10\times)(50\times) = 500\times$.

(b) The eyepiece focal length is (Eq. 25–5) $f_e = N/M_e = 25$ cm/10 = 2.5 cm. Next we solve Eq. 25–4 for d_o, and find

$$d_o = \frac{\ell - f_e}{m_o} = \frac{(17.0\,\text{cm} - 2.5\,\text{cm})}{50} = 0.29\,\text{cm}.$$

Then, from the thin lens equation for the objective with $d_i = \ell - f_e = 14.5$ cm (see Fig. 25–24a),

$$\frac{1}{f_o} = \frac{1}{d_o} + \frac{1}{d_i} = \frac{1}{0.29\,\text{cm}} + \frac{1}{14.5\,\text{cm}} = 3.52\,\text{cm}^{-1};$$

so $f_o = 1/(3.52\,\text{cm}^{-1}) = 0.28$ cm.

(c) We just calculated $d_o = 0.29$ cm, which is very close to f_o.

25–6 Aberrations of Lenses and Mirrors

In Chapter 23 we developed a theory of image formation by a thin lens. We found, for example, that all rays from each point on an object are brought to a single point as the image point. This result, and others, were based on approximations for a thin lens, mainly that all rays make small angles with the axis and that we can use $\sin \theta \approx \theta$. Because of these approximations, we expect deviations from the simple theory, which are referred to as **lens aberrations**. There are several types of aberration; we will briefly discuss each of them separately, but all may be present at one time.

Consider an object at any point (even at infinity) on the axis of a lens with spherical surfaces. Rays from this point that pass through the outer regions of the lens are brought to a focus at a different point from those that pass through the center of the lens. This is called **spherical aberration**, and is shown exaggerated in Fig. 25–25. Consequently, the image seen on a screen or film will not be a point but a tiny circular patch of light. If the sensor or film is placed at the point C, as indicated, the circle will have its smallest diameter, which is referred to as the **circle of least confusion**. Spherical aberration is present whenever spherical surfaces are used. It can be reduced by using nonspherical (= aspherical) lens surfaces, but grinding such lenses is difficult and expensive. Spherical aberration can be reduced by the use of several lenses in combination, and by using primarily the central part of lenses.

For object points off the lens axis, additional aberrations occur. Rays passing through the different parts of the lens cause spreading of the image that is noncircular. There are two effects: **coma** (because the image of a point is comet-shaped rather than a tiny circle) and **off-axis astigmatism**.[†] Furthermore, the image points for objects off the axis but at the same distance from the lens do not fall on a flat plane but on a curved surface—that is, the focal plane is not flat. (We expect this because the points on a flat plane, such as the film in a camera, are not equidistant from the lens.) This aberration is known as **curvature of field** and is a problem in cameras and other devices where the sensor or film is a flat plane. In the eye, however, the retina is curved, which compensates for this effect.

PHYSICS APPLIED
Lens aberrations

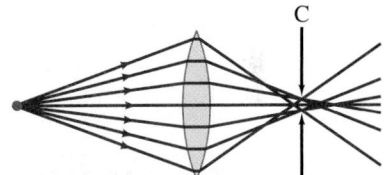

FIGURE 25–25 Spherical aberration (exaggerated). Circle of least confusion is at C.

[†]Although the effect is the same as for astigmatism in the eye (Section 25–2), the cause is different. Off-axis astigmatism is no problem in the eye because objects are clearly seen only at the fovea, on the lens axis.

Another aberration, **distortion**, is a result of variation of magnification at different distances from the lens axis. Thus a straight-line object some distance from the axis may form a curved image. A square grid of lines may be distorted to produce "barrel distortion," or "pincushion distortion," Fig. 25–26. The former is common in extreme wide-angle lenses.

FIGURE 25–26 Distortion: lenses may image a square grid of perpendicular lines to produce (a) barrel distortion or (b) pincushion distortion. These distortions can be seen in the photograph of Fig. 23–31d for a simple lens.

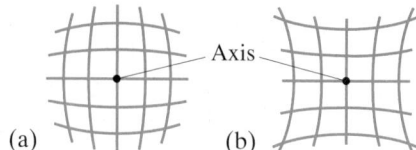

(a) (b)

FIGURE 25–27 Chromatic aberration. Different colors are focused at different points.

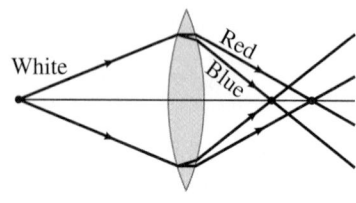

FIGURE 25–28 Achromatic doublet.

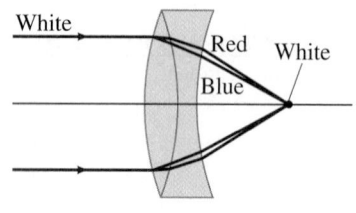

All the above aberrations occur for monochromatic light and hence are referred to as *monochromatic aberrations*. Normal light is not monochromatic, and there will also be **chromatic aberration**. This aberration arises because of dispersion—the variation of index of refraction of transparent materials with wavelength (Section 24–4). For example, blue light is bent more than red light by glass. So if white light is incident on a lens, the different colors are focused at different points, Fig. 25–27, and have slightly different magnifications resulting in colored fringes in the image. Chromatic aberration can be eliminated for any two colors (and reduced greatly for all others) by the use of two lenses made of different materials with different indices of refraction and dispersion. Normally one lens is converging and the other diverging, and they are often cemented together (Fig. 25–28). Such a lens combination is called an **achromatic doublet** (or "color-corrected" lens).

To reduce aberrations, high-quality lenses are **compound lenses** consisting of many simple lenses, referred to as **elements**. A typical high-quality camera lens may contain six to eight (or more) elements. For simplicity we will usually indicate lenses in diagrams as if they were simple lenses.

The human eye is also subject to aberrations, but they are minimal. Spherical aberration, for example, is minimized because (1) the cornea is less curved at the edges than at the center, and (2) the lens is less dense at the edges than at the center. Both effects cause rays at the outer edges to be bent less strongly, and thus help to reduce spherical aberration. Chromatic aberration is partially compensated for because the lens absorbs the shorter wavelengths appreciably and the retina is less sensitive to the blue and violet wavelengths. This is just the region of the spectrum where dispersion—and thus chromatic aberration—is greatest (Fig. 24–14).

Spherical mirrors (Section 23–3) also suffer aberrations including spherical aberration (see Fig. 23–13). Mirrors can be ground in a parabolic shape to correct for aberrations, but they are much harder to make and therefore very expensive. Spherical mirrors do not, however, exhibit chromatic aberration because the light does not pass through them (no refraction, no dispersion).

25–7 Limits of Resolution; Circular Apertures

The ability of a lens to produce distinct images of two point objects very close together is called the **resolution** of the lens. The closer the two images can be and still be seen as distinct (rather than overlapping blobs), the higher the resolution. The resolution of a camera lens, for example, is often specified as so many dots or lines per millimeter, as mentioned in Section 25–1.

Two principal factors limit the resolution of a lens. The first is lens aberrations. As we just saw, because of spherical and other aberrations, a point object is not a point on the image but a tiny blob. Careful design of compound lenses can reduce aberrations significantly, but they cannot be eliminated entirely. The second factor that limits resolution is *diffraction*, which cannot be corrected for because it is a natural result of the wave nature of light. We discuss it now.

In Section 24–5, we saw that because light travels as a wave, light from a point source passing through a slit is spread out into a diffraction pattern (Figs. 24–19 and 24–21). A lens, because it has edges, acts like a round slit. When a lens forms the image of a point object, the image is actually a tiny diffraction pattern. Thus *an image would be blurred even if aberrations were absent.*

In the analysis that follows, we assume that the lens is free of aberrations, so we can concentrate on diffraction effects and how much they limit the resolution of a lens. In Fig. 24–21 we saw that the diffraction pattern produced by light passing through a rectangular slit has a central maximum in which most of the light falls. This central peak falls to a minimum on either side of its center at an angle θ given by

$$\sin \theta = \lambda/D$$

(this is Eq. 24–3a), where D is the slit width and λ the wavelength of light used. θ is the angular half-width of the central maximum, and for small angles (in radians) can be written

$$\theta \approx \sin \theta = \frac{\lambda}{D}.$$

There are also low-intensity fringes beyond.

For a lens, or any circular hole, the image of a point object will consist of a *circular* central peak (called the *diffraction spot* or *Airy disk*) surrounded by faint circular fringes, as shown in Fig. 25–29a. The central maximum has an angular half-width given by

$$\theta = \frac{1.22\lambda}{D},$$

where D is the diameter of the circular opening. [This is a theoretical result for a perfect circle or lens. For real lenses or circles, the factor is on the order of 1 to 2.] This formula differs from that for a slit (Eq. 24–3) by the factor 1.22. This factor appears because the width of a circular hole is not uniform (like a rectangular slit) but varies from its diameter D to zero. A mathematical analysis shows that the "average" width is $D/1.22$. Hence we get the equation above rather than Eq. 24–3. The intensity of light in the diffraction pattern from a point source of light passing through a circular opening is shown in Fig. 25–30. The image for a non-point source is a superposition of such patterns. For most purposes we need consider only the central spot, since the concentric rings are so much dimmer.

If two point objects are very close, the diffraction patterns of their images will overlap as shown in Fig. 25–29b. As the objects are moved closer, a separation is reached where you can't tell if there are two overlapping images or a single image. The separation at which this happens may be judged differently by different observers. However, a generally accepted criterion is that proposed by Lord Rayleigh (1842–1919). This **Rayleigh criterion** states that *two images are just resolvable when the center of the diffraction disk of one image is directly over the first minimum in the diffraction pattern of the other.* This is shown in Fig. 25–31. Since the first minimum is at an angle $\theta = 1.22\lambda/D$ from the central maximum, Fig. 25–31 shows that two objects can be considered *just resolvable* if they are separated by at least an angle θ given by

$$\theta = \frac{1.22\lambda}{D}. \qquad \begin{bmatrix} \text{2 points just resolvable;} \\ \theta \text{ in radians} \end{bmatrix} \quad \textbf{(25-7)}$$

In this equation, D is the diameter of the lens, and applies also to a mirror diameter. This is the limit on resolution set by the wave nature of light due to diffraction. A smaller angle means better resolution: you can make out closer objects. We see from Eq. 25–7 that using a shorter wavelength λ can reduce θ and thus increase resolution.

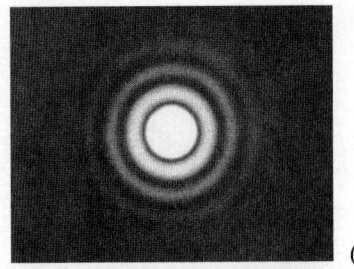
(a)

(b)

FIGURE 25–29 Photographs of images (greatly magnified) formed by a lens, showing the diffraction pattern of an image for: (a) a single point object; (b) two point objects whose images are barely resolved.

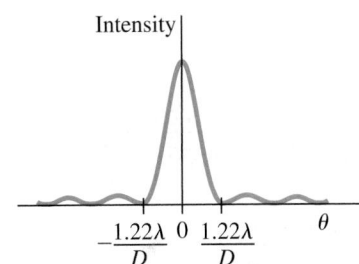

FIGURE 25–30 Intensity of light across the diffraction pattern of a circular hole.

FIGURE 25–31 The *Rayleigh criterion.* Two images are just resolvable when the center of the diffraction peak of one is directly over the first minimum in the diffraction pattern of the other. The two point objects O and O′ subtend an angle θ at the lens; only one ray (it passes through the center of the lens) is drawn for each object, to indicate the center of the diffraction pattern of its image.

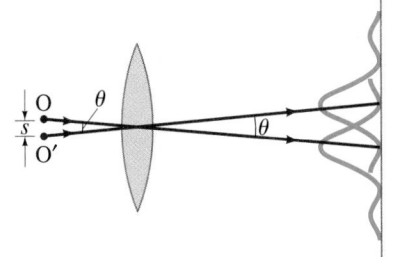

FIGURE 25–32 Hubble Space Telescope, with Earth in the background. The flat orange panels are solar cells that collect energy from the Sun to power the equipment.

PHYSICS APPLIED
How well the eye can see

EXAMPLE 25–10 **Hubble Space Telescope.** The Hubble Space Telescope (HST) is a reflecting telescope that was placed in orbit above the Earth's atmosphere, so its resolution would not be limited by turbulence in the atmosphere (Fig. 25–32). Its objective diameter is 2.4 m. For visible light, say $\lambda = 550$ nm, estimate the improvement in resolution the Hubble offers over Earth-bound telescopes, which are limited in resolution by movement of the Earth's atmosphere to about half an arc second. (Each degree is divided into 60 minutes each containing 60 seconds, so $1° = 3600$ arc seconds.)

APPROACH Angular resolution for the Hubble is given (in radians) by Eq. 25–7. The resolution for Earth telescopes is given, and we first convert it to radians so we can compare.

SOLUTION Earth-bound telescopes are limited to an angular resolution of

$$\theta = \frac{1}{2}\left(\frac{1}{3600}\right)°\left(\frac{2\pi \text{ rad}}{360°}\right) = 2.4 \times 10^{-6} \text{ rad}.$$

The Hubble, on the other hand, is limited by diffraction (Eq. 25–7) which for $\lambda = 550$ nm is

$$\theta = \frac{1.22\lambda}{D} = \frac{1.22(550 \times 10^{-9} \text{ m})}{2.4 \text{ m}} = 2.8 \times 10^{-7} \text{ rad},$$

thus giving almost ten times better resolution $(2.4 \times 10^{-6} \text{ rad}/2.8 \times 10^{-7} \text{ rad} \approx 9\times)$.

EXAMPLE 25–11 ESTIMATE **Eye resolution.** You are in an airplane at an altitude of 10,000 m. If you look down at the ground, estimate the minimum separation s between objects that you could distinguish. Could you count cars in a parking lot? Consider only diffraction, and assume your pupil is about 3.0 mm in diameter and $\lambda = 550$ nm.

APPROACH We use the Rayleigh criterion, Eq. 25–7, to estimate θ. The separation s of objects is $s = \ell\theta$, where $\ell = 10^4$ m and θ is in radians.

SOLUTION In Eq. 25–7, we set $D = 3.0$ mm for the opening of the eye:

$$s = \ell\theta = \ell\frac{1.22\lambda}{D}$$

$$= (10^4 \text{ m})\frac{(1.22)(550 \times 10^{-9} \text{ m})}{3.0 \times 10^{-3} \text{ m}} = 2.2 \text{ m}.$$

Yes, you could just resolve a car (roughly 2 m wide by 3 or 4 m long) and so could count the number of cars in the lot.

EXERCISE E Someone claims a spy satellite camera can see 3-cm-high newspaper headlines from an altitude of 100 km. If diffraction were the only limitation ($\lambda = 550$ nm), use Eq. 25–7 to determine what diameter lens the camera would have.

25–8 Resolution of Telescopes and Microscopes; the λ Limit

You might think that a microscope or telescope could be designed to produce any desired magnification, depending on the choice of focal lengths and quality of the lenses. But this is not possible, because of diffraction. An increase in magnification above a certain point merely results in magnification of the diffraction patterns. This can be highly misleading since we might think we are seeing details of an object when we are really seeing details of the diffraction pattern.

To examine this problem, we apply the Rayleigh criterion: two objects (or two nearby points on one object) are just resolvable if they are separated by an angle θ (Fig. 25–31) given by Eq. 25–7:

$$\theta = \frac{1.22\lambda}{D}.$$

This formula is valid for either a microscope or a telescope, where D is the diameter of the objective lens or mirror. For a telescope, the resolution is specified by stating θ as given by this equation.[†]

EXAMPLE 25–12 **Telescope resolution (radio wave vs. visible light).**
What is the theoretical minimum angular separation of two stars that can just be resolved by (a) the 200-inch telescope on Palomar Mountain (Fig. 25–22c); and (b) the Arecibo radiotelescope (Fig. 25–33), whose diameter is 300 m and whose radius of curvature is also 300 m. Assume $\lambda = 550$ nm for the visible-light telescope in part (a), and $\lambda = 4$ cm (the shortest wavelength at which the radio-telescope has operated) in part (b).

APPROACH We apply the Rayleigh criterion (Eq. 25–7) for each telescope.

SOLUTION (a) With $D = 200$ in. $= 5.1$ m, we have from Eq. 25–7 that

$$\theta = \frac{1.22\lambda}{D} = \frac{(1.22)(5.50 \times 10^{-7}\,\text{m})}{(5.1\,\text{m})} = 1.3 \times 10^{-7}\,\text{rad},$$

or 0.75×10^{-5} deg. (Note that this is equivalent to resolving two points less than 1 cm apart from a distance of 100 km!)
(b) For radio waves with $\lambda = 0.04$ m emitted by stars, the resolution is

$$\theta = \frac{(1.22)(0.04\,\text{m})}{(300\,\text{m})} = 1.6 \times 10^{-4}\,\text{rad}.$$

The resolution is less because the wavelength is so much larger, but the larger objective collects more radiation and thus detects fainter objects.

NOTE In both cases, we determined the limit set by diffraction. The resolution for a visible-light Earth-bound telescope is not this good because of aberrations and, more importantly, turbulence in the atmosphere. In fact, large-diameter objectives are not justified by increased resolution, but by their greater light-gathering ability—they allow more light in, so fainter objects can be seen. Radiotelescopes are not hindered by atmospheric turbulence, and the resolution found in (b) is a good estimate.

FIGURE 25–33 The 300-meter radiotelescope in Arecibo, Puerto Rico, uses radio waves (Fig. 22–8) instead of visible light.

PHYSICS APPLIED
Why large-diameter objectives

For a microscope, it is more convenient to specify the actual distance, s, between two points that are just barely resolvable: see Fig. 25–31. Since objects are normally placed near the focal point of the microscope objective, the angle subtended by two objects is $\theta = s/f$, so $s = f\theta$. If we combine this with Eq. 25–7, we obtain the **resolving power (RP)** of a microscope

$$\text{RP} = s = f\theta = \frac{1.22\lambda f}{D}, \qquad \text{[microscope]} \quad \textbf{(25–8)}$$

where f is the objective lens' focal length (not frequency) and D its diameter. The distance s is called the resolving power of the lens because it is the minimum separation of two object points that can just be resolved—assuming the highest quality lens since this limit is imposed by the wave nature of light. A smaller RP means better resolution, better detail.

[†]Earth-bound telescopes with large-diameter objectives are usually limited not by diffraction but by other effects such as turbulence in the atmosphere. The resolution of a high-quality microscope, on the other hand, normally *is* limited by diffraction; microscope objectives are complex compound lenses containing many elements of small diameter (since f is small), thus reducing aberrations.

Diffraction sets an ultimate limit on the detail that can be seen on any object. In Eq. 25–8 for the resolving power of a microscope, the focal length of the lens cannot practically be made less than (approximately) the radius of the lens (= $D/2$), and even that is very difficult (see the lensmaker's equation, Eq. 23–10). In this best case, Eq. 25–8 gives, with $f \approx D/2$,

$$\text{RP} \approx \frac{\lambda}{2}. \qquad \qquad \textbf{(25–9)}$$

Thus we can say, to within a factor of 2 or so, that

it is not possible to resolve detail of objects smaller than the wavelength of the radiation being used.

This is an important and useful rule of thumb.

Compound lenses in microscopes are now designed so well that the actual limit on resolution is often set by diffraction—that is, by the wavelength of the light used. To obtain greater detail, one must use radiation of shorter wavelength. The use of UV radiation can increase the resolution by a factor of perhaps 2. Far more important, however, was the discovery in the early twentieth century that electrons have wave properties (Chapter 27) and that their wavelengths can be very small. The wave nature of electrons is utilized in the electron microscope (Section 27–9), which can magnify 100 to 1000 times more than a visible-light microscope because of the much shorter wavelengths. X-rays, too, have very short wavelengths and are often used to study objects in great detail (Section 25–11).

25–9 Resolution of the Human Eye and Useful Magnification

The resolution of the human eye is limited by several factors, all of roughly the same order of magnitude. The resolution is best at the fovea, where the cone spacing is smallest, about 3 μm (= 3000 nm). The diameter of the pupil varies from about 0.1 cm to about 0.8 cm. So for $\lambda = 550$ nm (where the eye's sensitivity is greatest), the diffraction limit is about $\theta \approx 1.22\lambda/D \approx 8 \times 10^{-5}$ rad to 6×10^{-4} rad. The eye is about 2 cm long, giving a resolving power (Eq. 25–8) of $s \approx (2 \times 10^{-2} \text{ m})(8 \times 10^{-5} \text{ rad}) \approx 2 \mu$m at best, to about 10 μm at worst (pupil small). Spherical and chromatic aberration also limit the resolution to about 10 μm. The net result is that the eye can just resolve objects whose angular separation is around

$$5 \times 10^{-4} \text{ rad.} \qquad \qquad \begin{bmatrix} \text{best eye} \\ \text{resolution} \end{bmatrix}$$

This corresponds to objects separated by 1 cm at a distance of about 20 m.

The typical near point of a human eye is about 25 cm. At this distance, the eye can just resolve objects that are $(25 \text{ cm})(5 \times 10^{-4} \text{ rad}) \approx 10^{-4} \text{ m} = \frac{1}{10}$ mm apart.[†] Since the best light microscopes can resolve objects no smaller than about 200 nm at best (Eq. 25–9 for violet light, $\lambda = 400$ nm), the useful magnification [= (resolution by naked eye)/(resolution by microscope)] is limited to about

$$\frac{10^{-4} \text{ m}}{200 \times 10^{-9} \text{ m}} \approx 500\times. \qquad \begin{bmatrix} \text{maximum useful} \\ \text{microscope magnification} \end{bmatrix}$$

In practice, magnifications of about 1000× are often used to minimize eyestrain. Any greater magnification would simply make visible the diffraction pattern produced by the microscope objective lens.

EXERCISE F Return to the Chapter-Opening Question, page 713, and answer it again now. Try to explain why you may have answered differently the first time.

[†]A nearsighted eye that needs −8 or −10 D lenses can have a near point of 8 or 10 cm, and a higher resolution up close (without glasses) of a factor of $2\frac{1}{2}$ or 3, or $\approx \frac{1}{25}$ mm $\approx 40 \mu$m.

*25–10 Specialty Microscopes and Contrast

All the resolving power a microscope can attain will be useless if the object to be seen cannot be distinguished from the background. The difference in brightness between the image of an object and the image of the surroundings is called **contrast**. Achieving high contrast is an important problem in microscopy and other forms of imaging. The problem arises in biology, for example, because cells consist largely of water and are almost uniformly transparent to light. We now briefly discuss two special types of microscope that can increase contrast: the interference and phase-contrast microscopes.

An **interference microscope** makes use of the wave properties of light in a direct way to increase contrast in a transparent object. Consider a transparent object—say, a bacterium in water (Fig. 25–34). Coherent light enters uniformly from the left and is in phase at all points such as a and b. If the object is as transparent as the water, the beam leaving at d will be as bright as that at c. There will be no contrast and the object will not be seen. However, if the object's refractive index is slightly different from that of the surrounding medium, the wavelength within the object will be altered as shown. Hence light waves at points c and d will differ in phase, if not in amplitude. The interference microscope changes this difference in phase into a difference of amplitude which our eyes can detect. Light that passes through the sample is superimposed onto a reference beam that does not pass through the object, so that they interfere. One way of doing this is shown in Fig. 25–35. Light from a source is split into two equal beams by a half-silvered mirror, MS_1. One beam passes through the object, and the second (comparison beam) passes through an identical system without the object. The two meet again and are superposed by the half-silvered mirror MS_2 before entering the eyepiece and eye. The path length (and amplitude) of the comparison beam is adjustable so that the background can be dark; that is, full destructive interference occurs. Light passing through the object (beam bd in Fig. 25–34) will also interfere with the comparison beam. But because of its different phase, the interference will not be completely destructive. Thus it will appear brighter than the background. Where the object varies in thickness, the phase difference between beams ac and bd in Fig. 25–34 will be different, thus affecting the amount of interference. Hence *variation in the thickness of the object will appear as variations in brightness in the image.*

A **phase-contrast microscope** also makes use of interference and differences in phase to produce a high-contrast image. Contrast is achieved by a circular glass *phase plate* that has a groove (or a raised portion) in the shape of a ring, positioned so undeviated source rays pass through it, but rays deviated by the object do not pass through this ring. Because the rays deviated by the object travel through a different thickness of glass than the undeviated source rays, the two can be out of phase and can interfere destructively at the object image plane. Thus the image of the object can contrast sharply with the background. Phase-contrast microscope images tend to have "halos" around them (as a result of diffraction from the phase-plate opening), so care must be taken in the interpretation of images.

25–11 X-Rays and X-Ray Diffraction

In 1895, W. C. Roentgen (1845–1923) discovered that when electrons were accelerated by a high voltage in a vacuum tube and allowed to strike a glass or metal surface inside the tube, fluorescent minerals some distance away would glow, and photographic film would become exposed. Roentgen attributed these effects to a new type of radiation (different from cathode rays). They were given the name **X-rays** after the algebraic symbol x, meaning an unknown quantity. He soon found that X-rays penetrated through some materials better than through others, and within a few weeks he presented the first X-ray photograph (of his wife's hand). The production of X-rays today is usually done in a tube (Fig. 25–36) similar to Roentgen's, using voltages of typically 30 kV to 150 kV.

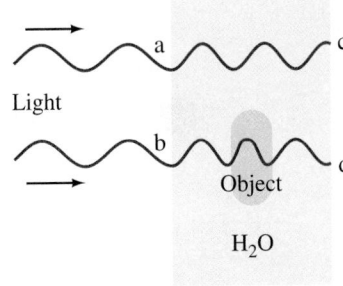

FIGURE 25–34 Object—say, a bacterium—in a water solution.

PHYSICS APPLIED
Interference microscope

FIGURE 25–35 Diagram of an interference microscope.

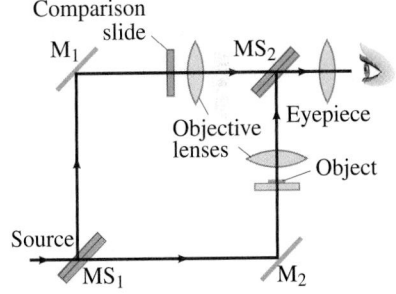

PHYSICS APPLIED
Phase-contrast microscope

FIGURE 25–36 X-ray tube. Electrons emitted by a heated filament in a vacuum tube are accelerated by a high voltage. When they strike the surface of the anode, the "target," X-rays are emitted.

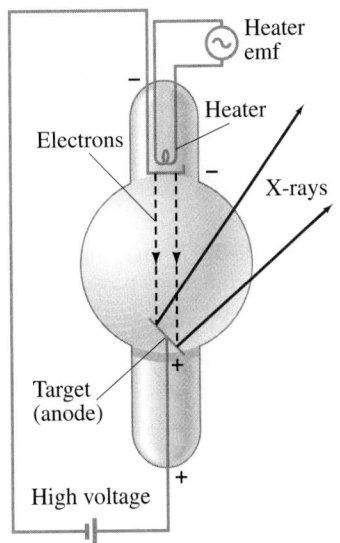

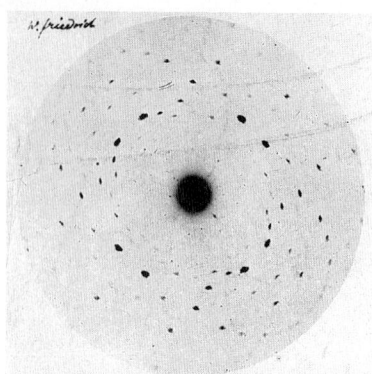

FIGURE 25–37 This X-ray diffraction pattern is one of the first observed by Max von Laue in 1912 when he aimed a beam of X-rays at a zinc sulfide crystal. The diffraction pattern was detected directly on a photographic plate.

Investigations into the nature of X-rays indicated they were not charged particles (such as electrons) since they could not be deflected by electric or magnetic fields. It was suggested that they might be a form of invisible light. However, they showed no diffraction or interference effects using ordinary gratings. Indeed, if their wavelengths were much smaller than the typical grating spacing of 10^{-6} m ($= 10^3$ nm), no effects would be expected. Around 1912, Max von Laue (1879–1960) suggested that if the atoms in a crystal were arranged in a regular array (see Fig. 13–2a), such a crystal might serve as a diffraction grating for very short wavelengths on the order of the spacing between atoms, estimated to be about 10^{-10} m ($= 10^{-1}$ nm). Experiments soon showed that X-rays scattered from a crystal did indeed show the peaks and valleys of a diffraction pattern (Fig. 25–37). Thus it was shown, in a single blow, that X-rays have a wave nature and that atoms are arranged in a regular way in crystals. Today, X-rays are recognized as electromagnetic radiation with wavelengths in the range of about 10^{-2} nm to 10 nm, the range readily produced in an X-ray tube.

*X-Ray Diffraction

We saw in Sections 25–7 and 25–8 that light of shorter wavelength provides greater resolution when we are examining an object microscopically. Since X-rays have much shorter wavelengths than visible light, they should in principle offer much greater resolution. However, there seems to be no effective material to use as lenses for the very short wavelengths of X-rays. Instead, the clever but complicated technique of **X-ray diffraction** (or **crystallography**) has proved very effective for examining the microscopic world of atoms and molecules. In a simple crystal such as NaCl, the atoms are arranged in an orderly cubical fashion, Fig. 25–38, with atoms spaced a distance d apart. Suppose that a beam of X-rays is incident on the crystal at an angle ϕ to the surface, and that the two rays shown are reflected from two subsequent planes of atoms as shown. The two rays will constructively interfere if the extra distance ray I travels is a whole number of wavelengths farther than the distance ray II travels. This extra distance is $2d \sin \phi$. Therefore, constructive interference will occur when

$$m\lambda = 2d \sin \phi, \qquad m = 1, 2, 3, \cdots, \qquad \textbf{(25–10)}$$

where m can be any integer. (Notice that ϕ is *not* the angle with respect to the normal to the surface.) This is called the **Bragg equation** after W. L. Bragg (1890–1971), who derived it and who, together with his father, W. H. Bragg (1862–1942), developed the theory and technique of X-ray diffraction by crystals in 1912–1913. If the X-ray wavelength is known and the angle ϕ is measured, the distance d between atoms can be obtained. This is the basis for X-ray crystallography.

FIGURE 25–38 X-ray diffraction by a crystal.

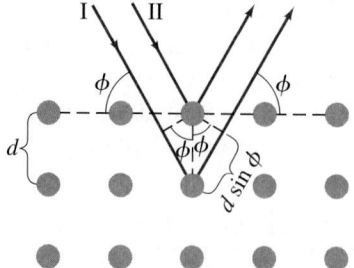

FIGURE 25–39 X-rays can be diffracted from many possible planes within a crystal.

EXERCISE G When X-rays of wavelength 0.10×10^{-9} m are scattered from a sodium chloride crystal, a second-order diffraction peak is observed at 21°. What is the spacing between the planes of atoms for this scattering?

Actual X-ray diffraction patterns are quite complicated. First of all, a crystal is a three-dimensional object, and X-rays can be diffracted from different planes at different angles within the crystal, as shown in Fig. 25–39. Although the analysis is complex, a great deal can be learned from X-ray diffraction about any substance that can be put in crystalline form.

X-ray diffraction has been very useful in determining the structure of biologically important molecules, such as the double helix structure of DNA, worked out by James Watson and Francis Crick in 1953. See Fig. 25–40, and for models of the double helix, Figs. 16–39a and 16–40. Around 1960, the first detailed structure of a protein molecule, myoglobin, was elucidated with the aid of X-ray diffraction. Soon the structure of an important constituent of blood, hemoglobin, was worked out, and since then the structures of a great many molecules have been determined with the help of X-rays.

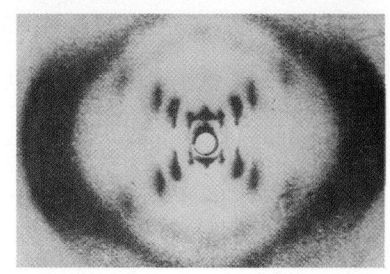

FIGURE 25–40 X-ray diffraction photo of DNA molecules taken by Rosalind Franklin in the early 1950s. The cross of spots suggested that DNA is a helix.

*25–12 X-Ray Imaging and Computed Tomography (CT Scan)

*Normal X-Ray Image

For a conventional medical or dental X-ray photograph, the X-rays emerging from the tube (Fig. 25–36) pass through the body and are detected on photographic film, a digital sensor, or a fluorescent screen, Fig. 25–41. The rays travel in very nearly straight lines through the body with minimal deviation since at X-ray wavelengths there is little diffraction or refraction. There is absorption (and scattering), however; and the difference in absorption by different structures in the body is what gives rise to the image produced by the transmitted rays. The less the absorption, the greater the transmission and the darker the film. The image is, in a sense, a "shadow" of what the rays have passed through. The X-ray image is *not* produced by focusing rays with lenses as for the instruments discussed earlier in this Chapter.

PHYSICS APPLIED
Normal X-ray image

CAUTION
Normal X-ray image is a sort of shadow (no lenses are involved)

FIGURE 25–41 Conventional X-ray imaging, which is essentially shadowing.

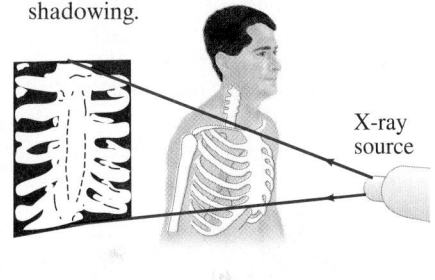

X-ray source

*Tomography Images (CT)

In conventional X-ray images, a body's thickness is projected onto film or a sensor; structures overlap and in many cases are difficult to distinguish. In the 1970s, a revolutionary X-ray technique was developed called **computed tomography** (CT), which produces an image of a *slice* through the body. (The word **tomography** comes from the Greek: *tomos* = slice, *graph* = picture.) Structures and lesions previously impossible to visualize can now be seen with remarkable clarity. The principle behind CT is shown in Fig. 25–42: a thin collimated (parallel) beam of X-rays passes through the body to a detector that measures the transmitted intensity. Measurements are made at a large number of points as the source and detector are moved past the body together. The apparatus is then rotated slightly about the body axis and again scanned; this is repeated at (perhaps) 1° intervals for 180°. The intensity of the transmitted beam for the many points of each scan, and for each angle, is sent to a computer that reconstructs the image of the slice. Note that the imaged slice is perpendicular to the long axis of the body. For this reason, CT is sometimes called **computerized axial tomography** (CAT), although the abbreviation CAT, as in CAT scan, can also be read as **computer-assisted tomography**.

PHYSICS APPLIED
Computed tomography images (CT or CAT scans)

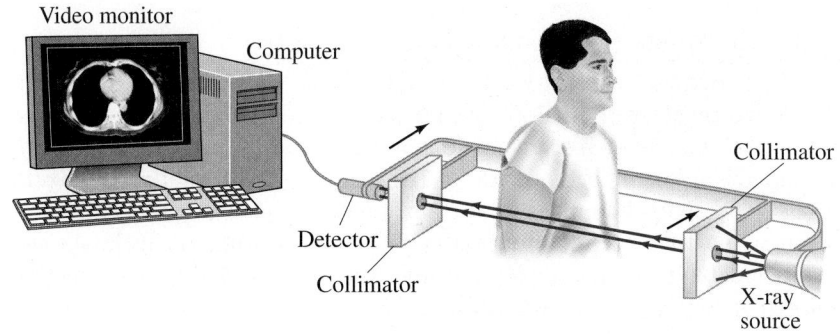

Video monitor
Computer
Detector
Collimator
Collimator
X-ray source

FIGURE 25–42 Tomographic imaging: the X-ray source and detector move together across the body, the transmitted intensity being measured at a large number of points. Then the "source-detector" assembly is rotated slightly (say, 1°) around a vertical axis, and another scan is made. This process is repeated for perhaps 180°. The computer reconstructs the image of the slice and it is presented on a TV or computer monitor.

FIGURE 25–43 (a) Fan-beam scanner. Rays transmitted through the entire body are measured simultaneously at each angle. The source and detector rotate to take measurements at different angles. In another type of fan-beam scanner, there are detectors around the entire 360° of the circle which remain fixed as the source moves. (b) In still another type, a beam of electrons from a source is directed by magnetic fields at tungsten targets surrounding the patient.

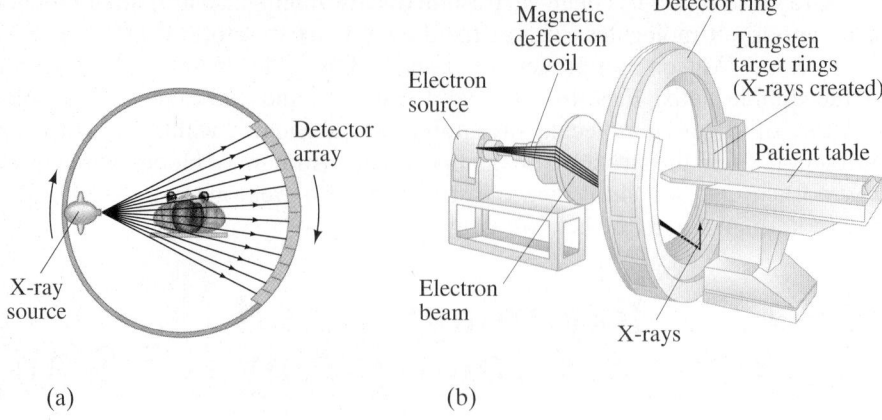

(a)   (b)

The use of a single detector as in Fig. 25–42 would require a few minutes for the many scans needed to form a complete image. Much faster scanners use a fan beam, Fig. 25–43a, in which beams passing through the entire cross section of the body are detected simultaneously by many detectors. The source and detectors are then rotated about the patient, and an image requires only a few seconds. Even faster, and therefore useful for heart scans, are fixed source machines wherein an electron beam is directed (by magnetic fields) to tungsten targets surrounding the patient, creating the X-rays. See Fig. 25–43b.

*Image Formation

But how is the image formed? We can think of the slice to be imaged as being divided into many tiny picture elements (or **pixels**), which could be squares (as in Fig. 24–49). For CT, the width of each pixel is chosen according to the width of the detectors and/or the width of the X-ray beams, and this determines the resolution of the image, which might be 1 mm. An X-ray detector measures the intensity of the transmitted beam. Subtracting this value from the intensity of the beam at the source yields the total absorption (called a "projection") along that beam line. Complicated mathematical techniques are used to analyze all the absorption projections for the huge number of beam scans measured (see the next Subsection), obtaining the absorption at each pixel and assigning each a "grayness value" according to how much radiation was absorbed. The image is made up of tiny spots (pixels) of varying shades of gray. Often the amount of absorption is color-coded. The colors in the resulting **false-color** image have nothing to do, however, with the actual color of the object. The actual images are monochromatic (various shades of gray, depending on the absorption). Only *visible* light has color; X-rays do not.

Figure 25–44 illustrates what actual CT images look like. It is generally agreed that CT scanning has revolutionized some areas of medicine by providing much less invasive, and/or more accurate, diagnosis.

Computed tomography can also be applied to ultrasound imaging (Section 12–9) and to emissions from radioisotopes and nuclear magnetic resonance (Sections 31–8 and 31–9).

*Tomographic Image Reconstruction

How can the "grayness" of each pixel be determined even though all we can measure is the total absorption along each beam line in the slice? It can be done only by using the many beam scans made at a great many different angles. Suppose the image is to be an array of 100×100 elements for a total of 10^4 pixels. If we have 100 detectors and measure the absorption projections at 100 different angles, then we get 10^4 pieces of information. From this information, an image can be reconstructed, but not precisely. If more angles are measured, the reconstruction of the image can be done more accurately.

FIGURE 25–44 Two CT images, with different resolutions, each showing a cross section of a brain. Photo (a) is of low resolution; photo (b), of higher resolution, shows a brain tumor, and uses false color to highlight it.

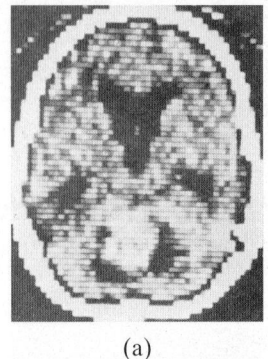

(a)

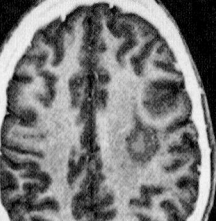

(b)

To suggest how mathematical reconstruction is done, we consider a very simple case using the **iterative** technique ("to iterate" is from the Latin "to repeat"). Suppose our sample slice is divided into the simple 2 × 2 pixels as shown in Fig. 25–45. The number inside each pixel represents the amount of absorption by the material in that area (say, in tenths of a percent): that is, 4 represents twice as much absorption as 2. But we cannot directly measure these values—they are the unknowns we want to solve for. All we can measure are the projections—the total absorption along each beam line—and these are shown in Fig. 25–45 outside the yellow squares as the sum of the absorptions for the pixels along each line at four different angles. These projections (given at the tip of each arrow) are what we can measure, and we now want to work back from them to see how close we can get to the true absorption value for each pixel. We start our analysis with each pixel being assigned a zero value, Fig. 25–46a. In the iterative technique, we use the projections to estimate the absorption value in each square, and repeat for each angle. The angle 1 projections are 7 and 13 (Fig. 25–45). We divide each of these equally between their two squares: each square in the left column of Fig. 25–46a gets $3\frac{1}{2}$ (half of 7), and each square in the right column gets $6\frac{1}{2}$ (half of 13).

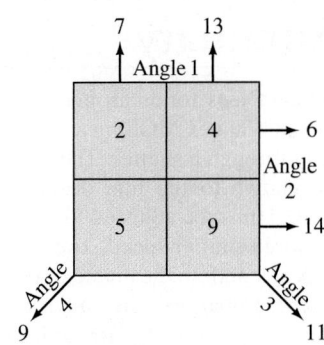

FIGURE 25–45 A simple 2 × 2 image showing true absorption values (inside the squares) and measured projections.

FIGURE 25–46 Reconstructing the image using projections in an iterative procedure.

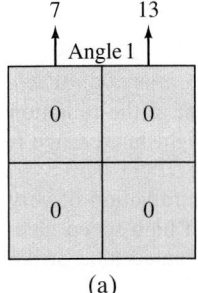 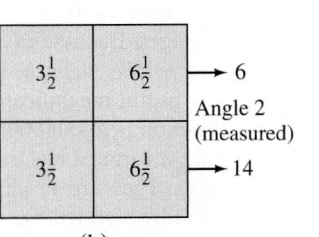 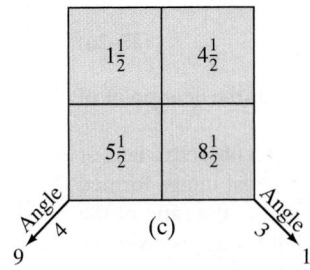 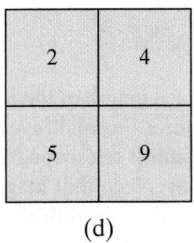

(a) (b) (c) (d)

Next we use the projections at angle 2. We calculate the difference between the measured projections at angle 2 (6 and 14) and the projections based on the previous estimate (top row: $3\frac{1}{2} + 6\frac{1}{2} = 10$; same for bottom row). Then we distribute this difference equally to the squares in that row. For the top row, we have

$$3\frac{1}{2} + \frac{6 - 10}{2} = 1\frac{1}{2} \quad \text{and} \quad 6\frac{1}{2} + \frac{6 - 10}{2} = 4\frac{1}{2};$$

and for the bottom row,

$$3\frac{1}{2} + \frac{14 - 10}{2} = 5\frac{1}{2} \quad \text{and} \quad 6\frac{1}{2} + \frac{14 - 10}{2} = 8\frac{1}{2}.$$

These values are inserted as shown in Fig. 25–46c. Next, the projection at angle 3 (= 11), combined with the difference as above, gives

(upper left) $1\frac{1}{2} + \dfrac{11 - 10}{2} = 2$ and (lower right) $8\frac{1}{2} + \dfrac{11 - 10}{2} = 9$;

and then for angle 4 we have

(lower left) $5\frac{1}{2} + \dfrac{9 - 10}{2} = 5$ and (upper right) $4\frac{1}{2} + \dfrac{9 - 10}{2} = 4.$

The result, shown in Fig. 25–46d, corresponds exactly to the true values. (In real situations, the true values are not known, which is why these computer techniques are required.) To obtain these numbers exactly, we used six pieces of information (two each at angles 1 and 2, one each at angles 3 and 4). For the much larger number of pixels used for actual images, exact values are generally not attained. Many iterations may be needed, and the calculation is considered sufficiently precise when the difference between calculated and measured projections is sufficiently small. The above example illustrates the "convergence" of the process: the first iteration (b to c in Fig. 25–46) changed the values by 2, the last iteration (c to d) by only $\frac{1}{2}$.

Summary

A **camera** lens forms an image on film, or on an electronic sensor (CCD or CMOS) in a digital camera. Light is allowed in briefly through a shutter. The image is focused by moving the lens relative to the film or sensor, and the **f-stop** (or lens opening) must be adjusted for the brightness of the scene and the chosen shutter speed. The f-stop is defined as the ratio of the focal length to the diameter of the lens opening.

The human **eye** also adjusts for the available light—by opening and closing the iris. It focuses not by moving the lens, but by adjusting the shape of the lens to vary its focal length. The image is formed on the retina, which contains an array of receptors known as rods and cones.

Diverging eyeglass or contact lenses are used to correct the defect of a nearsighted eye, which cannot focus well on distant objects. Converging lenses are used to correct for defects in which the eye cannot focus on close objects.

A **simple magnifier** is a converging lens that forms a virtual image of an object placed at (or within) the focal point. The **angular magnification**, when viewed by a relaxed normal eye, is

$$M = \frac{N}{f}, \qquad (25\text{--}2a)$$

where f is the focal length of the lens and N is the near point of the eye (25 cm for a "normal" eye).

An **astronomical telescope** consists of an **objective lens** or mirror, and an **eyepiece** that magnifies the real image formed by the objective. The **magnification** is equal to the ratio of the objective and eyepiece focal lengths, and the image is inverted:

$$M = -\frac{f_o}{f_e}. \qquad (25\text{--}3)$$

A compound **microscope** also uses objective and eyepiece lenses, and the final image is inverted. The total magnification is the product of the magnifications of the two lenses and is approximately

$$M \approx \frac{N\ell}{f_e f_o}, \qquad (25\text{--}6b)$$

where ℓ is the distance between the lenses, N is the near point of the eye, and f_o and f_e are the focal lengths of objective and eyepiece, respectively.

Microscopes, telescopes, and other optical instruments are limited in the formation of sharp images by **lens aberrations**. These include **spherical aberration**, in which rays passing through the edge of a lens are not focused at the same point as those that pass near the center; and **chromatic aberration**, in which different colors are focused at different points. Compound lenses, consisting of several elements, can largely correct for aberrations.

The wave nature of light also limits the sharpness, or **resolution**, of images. Because of diffraction, it is *not possible to discern details smaller than the wavelength* of the radiation being used. The useful magnification of a light microscope is limited by diffraction to about 500×.

[*X-rays are a form of electromagnetic radiation of very short wavelength. They are produced when high-speed electrons, accelerated by high voltage in an evacuated tube, strike a glass or metal target.]

[*Computed tomography (CT or CAT scan) uses many narrow X-ray beams through a section of the body to construct an image of that section.]

Questions

1. Why must a camera lens be moved farther from the sensor or film to focus on a closer object?

2. Why is the depth of field greater, and the image sharper, when a camera lens is "stopped down" to a larger f-number? Ignore diffraction.

3. Describe how diffraction affects the statement of Question 2. [*Hint*: See Eq. 24–3 or 25–7.]

4. Why are bifocals needed mainly by older persons and not generally by younger people?

5. Will a nearsighted person who wears corrective lenses in her glasses be able to see clearly underwater when wearing those glasses? Use a diagram to show why or why not.

6. You can tell whether people are nearsighted or farsighted by looking at the width of their face through their glasses. If a person's face appears narrower through the glasses (Fig. 25–47), is the person farsighted or nearsighted? Try to explain, but also check experimentally with friends who wear glasses.

FIGURE 25–47
Question 6.

7. In attempting to discern distant details, people will sometimes squint. Why does this help?

8. Is the image formed on the retina of the human eye upright or inverted? Discuss the implications of this for our perception of objects.

9. The human eye is much like a camera—yet, when a camera shutter is left open and the camera is moved, the image will be blurred. But when you move your head with your eyes open, you still see clearly. Explain.

10. Reading glasses use converging lenses. A simple magnifier is also a converging lens. Are reading glasses therefore magnifiers? Discuss the similarities and differences between converging lenses as used for these two different purposes.

11. Nearsighted people often look over (or under) their glasses when they want to see something small up close, like a cell phone screen. Why?

12. Spherical aberration in a thin lens is minimized if rays are bent equally by the two surfaces. If a planoconvex lens is used to form a real image of an object at infinity, which surface should face the object? Use ray diagrams to show why.

13. Explain why chromatic aberration occurs for thin lenses but not for mirrors.

14. Inexpensive microscopes for children's use usually produce images that are colored at the edges. Why?

15. Which aberrations present in a simple lens are not present (or are greatly reduced) in the human eye?

16. By what factor can you improve resolution, other things being equal, if you use blue light ($\lambda = 450$ nm) rather than red (700 nm)?

17. Atoms have diameters of about 10^{-8} cm. Can visible light be used to "see" an atom? Explain.

18. Which color of visible light would give the best resolution in a microscope? Explain.

19. For both converging and diverging lenses, discuss how the focal length for red light differs from that for violet light.

20. The 300-meter radiotelescope in Arecibo, Puerto Rico (Fig. 25–33), is the world's largest radiotelescope, but many other radiotelescopes are also very large. Why are radiotelescopes so big? Why not make optical telescopes that are equally large? (The largest optical telescopes have diameters of about 10 meters.)

MisConceptual Questions

1. The image of a nearby object formed by a camera lens is
(a) at the lens' focal point.
(b) always blurred.
(c) at the same location as the image of an object at infinity.
(d) farther from the lens than the lens' focal point.

2. What is a megapixel in a digital camera?
(a) A large spot on the detector where the image is focused.
(b) A special kind of lens that gives a sharper image.
(c) A number related to how many photographs the camera can store.
(d) A million light-sensitive spots on the detector.
(e) A number related to how fast the camera can take pictures.

3. When a nearsighted person looks at a distant object through her glasses, the image produced by the glasses should be
(a) about 25 cm from her eye.
(b) at her eye's far point.
(c) at her eye's near point.
(d) at the far point for a normal eye.

4. If the distance from your eye's lens to the retina is shorter than for a normal eye, you will struggle to see objects that are
(a) nearby.   (c) colorful.
(b) far away.   (d) moving fast.

5. The image produced on the retina of the eye is _____ compared to the object being viewed.
(a) inverted.   (c) sideways.
(b) upright.   (d) enlarged.

6. How do eyeglasses help a nearsighted person see more clearly?
(a) Diverging lenses bend light entering the eye, so the image focuses farther from the front of the eye.
(b) Diverging lenses bend light entering the eye, so the image focuses closer to the front of the eye.
(c) Converging lenses bend light entering the eye, so the image focuses farther from the front of the eye.
(d) Converging lenses bend light entering the eye, so the image focuses closer to the front of the eye.
(e) Lenses adjust the distance from the cornea to the back of the eye.

7. When you closely examine an object through a magnifying glass, the magnifying glass
(a) makes the object bigger.
(b) makes the object appear closer than it actually is.
(c) makes the object appear farther than it actually is.
(d) causes additional light rays to be emitted by the object.

8. It would be impossible to build a microscope that could use visible light to see the molecular structure of a crystal because.
(a) lenses with enough magnification cannot be made.
(b) lenses cannot be ground with fine enough precision.
(c) lenses cannot be placed in the correct place with enough precision.
(d) diffraction limits the resolving power to about the size of the wavelength of the light used.
(e) More than one of the above is correct.

9. Why aren't white-light microscopes made with a magnification of 3000×?
(a) Lenses can't be made large enough.
(b) Lenses can't be made small enough.
(c) Lenses can't be made with short enough focal lengths.
(d) Lenses can't be made with long enough focal lengths.
(e) Diffraction limits useful magnification to several times less than this.

10. The resolving power of a microscope is greatest when the object being observed is illuminated by
(a) ultraviolet light.   (c) visible light.
(b) infrared light.   (d) radio waves.

11. Which of the following statements is true?
(a) A larger-diameter lens can better resolve two distant points.
(b) Red light can better resolve two distant points than blue light can.
(c) It is easier to resolve distant objects than nearer objects.
(d) Objects that are closer together are easier to resolve than objects that are farther apart.

12. While you are photographing a dog, it begins to move away. What must you do to keep it in focus?
(a) Increase the f-stop value.
(b) Decrease the f-stop value.
(c) Move the lens away from the sensor or film.
(d) Move the lens closer to the sensor or film.
(e) None of the above.

13. A converging lens, like the type used in a magnifying glass,
(a) always produces a magnified image (image taller than the object).
(b) can also produce an image smaller than the object.
(c) always produces an upright image.
(d) can also produce an inverted image (upside down).
(e) None of these statements are true.

Problems

25–1 Camera

1. (I) A properly exposed photograph is taken at $f/16$ and $\frac{1}{100}$ s. What lens opening is required if the shutter speed is $\frac{1}{400}$ s?

2. (I) A television camera lens has a 17-cm focal length and a lens diameter of 6.0 cm. What is its f-number?

3. (I) A 65-mm-focal-length lens has f-stops ranging from $f/1.4$ to $f/22$. What is the corresponding range of lens diaphragm diameters?

4. (I) A light meter reports that a camera setting of $\frac{1}{500}$ s at $f/5.6$ will give a correct exposure. But the photographer wishes to use $f/11$ to increase the depth of field. What should the shutter speed be?

5. (II) For a camera equipped with a 55-mm-focal-length lens, what is the object distance if the image height equals the object height? How far is the object from the image on the film?

6. (II) A nature photographer wishes to shoot a 34-m-tall tree from a distance of 65 m. What focal-length lens should be used if the image is to fill the 24-mm height of the sensor?

7. (II) A 200-mm-focal-length lens can be adjusted so that it is 200.0 mm to 208.2 mm from the film. For what range of object distances can it be adjusted?

8. (II) How large is the image of the Sun on film used in a camera with (a) a 28-mm-focal-length lens, (b) a 50-mm-focal-length lens, and (c) a 135-mm-focal-length lens? (d) If the 50-mm lens is considered normal for this camera, what relative magnification does each of the other two lenses provide? The Sun has diameter 1.4×10^6 km, and it is 1.5×10^8 km away.

9. (II) If a 135-mm telephoto lens is designed to cover object distances from 1.30 m to ∞, over what distance must the lens move relative to the plane of the sensor or film?

10. (III) Show that for objects very far away (assume infinity), the magnification of any camera lens is proportional to its focal length.

25–2 Eye and Corrective Lenses

11. (I) A human eyeball is about 2.0 cm long and the pupil has a maximum diameter of about 8.0 mm. What is the "speed" of this lens?

12. (II) A person struggles to read by holding a book at arm's length, a distance of 52 cm away. What power of reading glasses should be prescribed for her, assuming they will be placed 2.0 cm from the eye and she wants to read at the "normal" near point of 25 cm?

13. (II) Reading glasses of what power are needed for a person whose near point is 125 cm, so that he can read a computer screen at 55 cm? Assume a lens–eye distance of 1.8 cm.

14. (II) An eye is corrected by a -5.50-D lens, 2.0 cm from the eye. (a) Is this eye near- or farsighted? (b) What is this eye's far point without glasses?

15. (II) A person's right eye can see objects clearly only if they are between 25 cm and 85 cm away. (a) What power of contact lens is required so that objects far away are sharp? (b) What will be the near point with the lens in place?

16. (II) About how much longer is the nearsighted eye in Example 25–6 than the 2.0 cm of a normal eye?

17. (II) A person has a far point of 14 cm. What power glasses would correct this vision if the glasses were placed 2.0 cm from the eye? What power contact lenses, placed on the eye, would the person need?

18. (II) One lens of a nearsighted person's eyeglasses has a focal length of -26.0 cm and the lens is 1.8 cm from the eye. If the person switches to contact lenses placed directly on the eye, what should be the focal length of the corresponding contact lens?

19. (II) What is the focal length of the eye–lens system when viewing an object (a) at infinity, and (b) 34 cm from the eye? Assume that the lens–retina distance is 2.0 cm.

20. (III) The closely packed cones in the fovea of the eye have a diameter of about 2 μm. For the eye to discern two images on the fovea as distinct, assume that the images must be separated by at least one cone that is not excited. If these images are of two point-like objects at the eye's 25-cm near point, how far apart are these barely resolvable objects? Assume the eye's diameter (cornea-to-fovea distance) is 2.0 cm.

21. (III) A nearsighted person has near and far points of 10.6 and 20.0 cm, respectively. If she puts on contact lenses with power $P = -4.00$ D, what are her new near and far points?

25–3 Magnifying Glass

22. (I) What is the focal length of a magnifying glass of 3.2× magnification for a relaxed normal eye?

23. (I) What is the magnification of a lens used with a relaxed eye if its focal length is 16 cm?

24. (I) A magnifier is rated at 3.5× for a normal eye focusing on an image at the near point. (a) What is its focal length? (b) What is its focal length if the 3.5× refers to a relaxed eye?

25. (II) Sherlock Holmes is using an 8.20-cm-focal-length lens as his magnifying glass. To obtain maximum magnification, where must the object be placed (assume a normal eye), and what will be the magnification?

26. (II) A small insect is placed 4.85 cm from a $+5.00$-cm-focal-length lens. Calculate (a) the position of the image, and (b) the angular magnification.

27. (II) A 3.80-mm-wide bolt is viewed with a 9.60-cm-focal-length lens. A normal eye views the image at its near point. Calculate (a) the angular magnification, (b) the width of the image, and (c) the object distance from the lens.

28. (II) A magnifying glass with a focal length of 9.2 cm is used to read print placed at a distance of 8.0 cm. Calculate (a) the position of the image; (b) the angular magnification.

29. (III) A writer uses a converging lens of focal length $f = 12$ cm as a magnifying glass to read fine print on his book contract. Initially, the writer holds the lens above the fine print so that its image is at infinity. To get a better look, he then moves the lens so that the image is at his 25-cm near point. How far, and in what direction (toward or away from the fine print) did the writer move the lens? Assume his eye is adjusted to remain always very near the magnifying glass.

30. (III) A magnifying glass is rated at 3.0× for a normal eye that is relaxed. What would be the magnification for a relaxed eye whose near point is (a) 75 cm, and (b) 15 cm? Explain the differences.

25–4 Telescopes

31. (I) What is the magnification of an astronomical telescope whose objective lens has a focal length of 82 cm, and whose eyepiece has a focal length of 2.8 cm? What is the overall length of the telescope when adjusted for a relaxed eye?

32. (I) The overall magnification of an astronomical telescope is desired to be 25×. If an objective of 88-cm focal length is used, what must be the focal length of the eyepiece? What is the overall length of the telescope when adjusted for use by the relaxed eye?

33. (II) A 7.0× binocular has 3.5-cm-focal-length eyepieces. What is the focal length of the objective lenses?

34. (II) An astronomical telescope has an objective with focal length 75 cm and a +25-D eyepiece. What is the total magnification?

35. (II) An astronomical telescope has its two lenses spaced 82.0 cm apart. If the objective lens has a focal length of 78.5 cm, what is the magnification of this telescope? Assume a relaxed eye.

36. (II) A Galilean telescope adjusted for a relaxed eye is 36.8 cm long. If the objective lens has a focal length of 39.0 cm, what is the magnification?

37. (II) What is the magnifying power of an astronomical telescope using a reflecting mirror whose radius of curvature is 6.1 m and an eyepiece whose focal length is 2.8 cm?

38. (II) The Moon's image appears to be magnified 150× by a reflecting astronomical telescope with an eyepiece having a focal length of 3.1 cm. What are the focal length and radius of curvature of the main (objective) mirror?

39. (II) A 120× astronomical telescope is adjusted for a relaxed eye when the two lenses are 1.10 m apart. What is the focal length of each lens?

40. (II) An astronomical telescope longer than about 50 cm is not easy to hold by hand. Estimate the maximum angular magnification achievable for a telescope designed to be handheld. Assume its eyepiece lens, if used as a magnifying glass, provides a magnification of 5× for a relaxed eye with near point $N = 25$ cm.

41. (III) A reflecting telescope (Fig. 25–22b) has a radius of curvature of 3.00 m for its objective mirror and a radius of curvature of −1.50 m for its eyepiece mirror. If the distance between the two mirrors is 0.90 m, how far in front of the eyepiece should you place the electronic sensor to record the image of a star?

42. (III) A 6.5× pair of binoculars has an objective focal length of 26 cm. If the binoculars are focused on an object 4.0 m away (from the objective), what is the magnification? (The 6.5× refers to objects at infinity; Eq. 25–3 holds only for objects at infinity and not for nearby ones.)

25–5 Microscopes

43. (I) A microscope uses an eyepiece with a focal length of 1.70 cm. Using a normal eye with a final image at infinity, the barrel length is 17.5 cm and the focal length of the objective lens is 0.65 cm. What is the magnification of the microscope?

44. (I) A 720× microscope uses a 0.40-cm-focal-length objective lens. If the barrel length is 17.5 cm, what is the focal length of the eyepiece? Assume a normal eye and that the final image is at infinity.

45. (I) A 17-cm-long microscope has an eyepiece with a focal length of 2.5 cm and an objective with a focal length of 0.33 cm. What is the approximate magnification?

46. (II) A microscope has a 14.0× eyepiece and a 60.0× objective lens 20.0 cm apart. Calculate (a) the total magnification, (b) the focal length of each lens, and (c) where the object must be for a normal relaxed eye to see it in focus.

47. (II) Repeat Problem 46 assuming that the final image is located 25 cm from the eyepiece (near point of a normal eye).

48. (II) A microscope has a 1.8-cm-focal-length eyepiece and a 0.80-cm objective. Assuming a relaxed normal eye, calculate (a) the position of the object if the distance between the lenses is 14.8 cm, and (b) the total magnification.

49. (II) The eyepiece of a compound microscope has a focal length of 2.80 cm and the objective lens has $f = 0.740$ cm. If an object is placed 0.790 cm from the objective lens, calculate (a) the distance between the lenses when the microscope is adjusted for a relaxed eye, and (b) the total magnification.

50. (III) An inexpensive instructional lab microscope allows the user to select its objective lens to have a focal length of 32 mm, 15 mm, or 3.9 mm. It also has two possible eyepieces with magnifications 5× and 15×. Each objective forms a real image 160 mm beyond its focal point. What are the largest and smallest overall magnifications obtainable with this instrument?

25–6 Lens Aberrations

51. (II) An achromatic lens is made of two very thin lenses, placed in contact, that have focal lengths $f_1 = -27.8$ cm and $f_2 = +25.3$ cm. (a) Is the combination converging or diverging? (b) What is the net focal length?

*52. (III) A planoconvex lens (Fig. 23–31a) has one flat surface and the other has $R = 14.5$ cm. This lens is used to view a red and yellow object which is 66.0 cm away from the lens. The index of refraction of the glass is 1.5106 for red light and 1.5226 for yellow light. What are the locations of the red and yellow images formed by the lens? [Hint: See Section 23–10.]

25–7 to 25–9 Resolution Limits

53. (I) What is the angular resolution limit (degrees) set by diffraction for the 100-inch (254-cm mirror diameter) Mt. Wilson telescope ($\lambda = 560$ nm)?

54. (I) What is the resolving power of a microscope ($\lambda = 550$ nm) with a 5-mm-diameter objective which has $f = 9$ mm?

55. (II) Two stars 18 light-years away are barely resolved by a 66-cm (mirror diameter) telescope. How far apart are the stars? Assume $\lambda = 550$ nm and that the resolution is limited by diffraction.

56. (II) The nearest neighboring star to the Sun is about 4 light-years away. If a planet happened to be orbiting this star at an orbital radius equal to that of the Earth–Sun distance, what minimum diameter would an Earth-based telescope's aperture have to be in order to obtain an image that resolved this star–planet system? Assume the light emitted by the star and planet has a wavelength of 550 nm.

57. (II) If you could shine a very powerful flashlight beam toward the Moon, estimate the diameter of the beam when it reaches the Moon. Assume that the beam leaves the flashlight through a 5.0-cm aperture, that its white light has an average wavelength of 550 nm, and that the beam spreads due to diffraction only.

58. (II) The normal lens on a 35-mm camera has a focal length of 50.0 mm. Its aperture diameter varies from a maximum of 25 mm ($f/2$) to a minimum of 3.0 mm ($f/16$). Determine the resolution limit set by diffraction for ($f/2$) and ($f/16$). Specify as the number of lines per millimeter resolved on the detector or film. Take $\lambda = 560$ nm.

59. (III) Suppose that you wish to construct a telescope that can resolve features 6.5 km across on the Moon, 384,000 km away. You have a 2.0-m-focal-length objective lens whose diameter is 11.0 cm. What focal-length eyepiece is needed if your eye can resolve objects 0.10 mm apart at a distance of 25 cm? What is the resolution limit set by the size of the objective lens (that is, by diffraction)? Use $\lambda = 560$ nm.

*25–11 X-Ray Diffraction

*60. (II) X-rays of wavelength 0.138 nm fall on a crystal whose atoms, lying in planes, are spaced 0.285 nm apart. At what angle ϕ (relative to the surface, Fig. 25–38) must the X-rays be directed if the first diffraction maximum is to be observed?

*61. (II) First-order Bragg diffraction is observed at 23.8° relative to the crystal surface, with spacing between atoms of 0.24 nm. (a) At what angle will second order be observed? (b) What is the wavelength of the X-rays?

*62. (II) If X-ray diffraction peaks corresponding to the first three orders ($m = 1$, 2, and 3) are measured, can both the X-ray wavelength λ and lattice spacing d be determined? Prove your answer.

*25–12 Imaging by Tomography

*63. (II) (a) Suppose for a conventional X-ray image that the X-ray beam consists of parallel rays. What would be the magnification of the image? (b) Suppose, instead, that the X-rays come from a point source (as in Fig. 25–41) that is 15 cm in front of a human body which is 25 cm thick, and the film is pressed against the person's back. Determine and discuss the range of magnifications that result.

General Problems

64. A **pinhole** camera uses a tiny pinhole instead of a lens. Show, using ray diagrams, how reasonably sharp images can be formed using such a pinhole camera. In particular, consider two point objects 2.0 cm apart that are 1.0 m from a 1.0-mm-diameter pinhole. Show that on a piece of film 7.0 cm behind the pinhole the two objects produce two separate circles that do not overlap.

65. Suppose that a correct exposure is $\frac{1}{250}$ s at $f/11$. Under the same conditions, what exposure time would be needed for a *pinhole* camera (Problem 64) if the pinhole diameter is 1.0 mm and the film is 7.0 cm from the hole?

66. An astronomical telescope has a magnification of 7.5×. If the two lenses are 28 cm apart, determine the focal length of each lens.

67. (a) How far away can a human eye distinguish two car headlights 2.0 m apart? Consider only diffraction effects and assume an eye pupil diameter of 6.0 mm and a wavelength of 560 nm. (b) What is the minimum angular separation an eye could resolve when viewing two stars, considering only diffraction effects? In reality, it is about 1′ of arc. Why is it not equal to your answer in (b)?

68. Figure 25–48 was taken from the NIST Laboratory (National Institute of Standards and Technology) in Boulder, CO, 2.0 km from the hiker in the photo. The Sun's image was 15 mm across on the film. Estimate the focal length of the camera lens (actually a telescope). The Sun has diameter 1.4×10^6 km, and it is 1.5×10^8 km away.

FIGURE 25–48
Problem 68.

69. A 1.0-cm-diameter lens with a focal length of 35 cm uses blue light to image two objects 15 m away that are very close together. What is the closest those objects can be to each other and still be imaged as separate objects?

70. A movie star catches a reporter shooting pictures of her at home. She claims the reporter was trespassing. To prove her point, she gives as evidence the film she seized. Her 1.65-m height is 8.25 mm high on the film, and the focal length of the camera lens was 220 mm. How far away from the subject was the reporter standing?

71. As early morning passed toward midday, and the sunlight got more intense, a photographer noted that, if she kept her shutter speed constant, she had to change the f-number from $f/5.6$ to $f/16$. By what factor had the sunlight intensity increased during that time?

72. A child has a near point of 15 cm. What is the maximum magnification the child can obtain using a 9.5-cm-focal-length magnifier? What magnification can a normal eye obtain with the same lens? Which person sees more detail?

73. A woman can see clearly with her right eye only when objects are between 45 cm and 135 cm away. Prescription bifocals should have what powers so that she can see distant objects clearly (upper part) and be able to read a book 25 cm away (lower part) with her right eye? Assume that the glasses will be 2.0 cm from the eye.

74. What is the magnifying power of a +4.0-D lens used as a magnifier? Assume a relaxed normal eye.

75. A physicist lost in the mountains tries to make a telescope using the lenses from his reading glasses. They have powers of +2.0 D and +5.5 D, respectively. (a) What maximum magnification telescope is possible? (b) Which lens should be used as the eyepiece?

76. A person with normal vision adjusts a microscope for a good image when her eye is relaxed. She then places a camera where her eye was. For what object distance should the camera be set? Explain.

77. A 50-year-old man uses +2.5-D lenses to read a newspaper 25 cm away. Ten years later, he must hold the paper 38 cm away to see clearly with the same lenses. What power lenses does he need now in order to hold the paper 25 cm away? (Distances are measured from the lens.)

78. Two converging lenses, one with $f = 4.0$ cm and the other with $f = 48$ cm, are made into a telescope. (a) What are the length and magnification? Which lens should be the eyepiece? (b) Assume these lenses are now combined to make a microscope; if the magnification needs to be 25×, how long would the microscope be?

79. An X-ray tube operates at 95 kV with a current of 25 mA and nearly all the electron energy goes into heat. If the specific heat of the 0.065-kg anode plate is 0.11 kcal/kg·C°, what will be the temperature rise per minute if no cooling water is used? (See Fig. 25–36.)

80. Human vision normally covers an angle of roughly 40° horizontally. A "normal" camera lens then is defined as follows: When focused on a distant horizontal object which subtends an angle of 40°, the lens produces an image that extends across the full horizontal extent of the camera's light-recording medium (film or electronic sensor). Determine the focal length f of the "normal" lens for the following types of cameras: (a) a 35-mm camera that records images on film 36 mm wide; (b) a digital camera that records images on a charge-coupled device (CCD) 1.60 cm wide.

81. The objective lens and the eyepiece of a telescope are spaced 85 cm apart. If the eyepiece is +19 D, what is the total magnification of the telescope?

82. Sam purchases +3.50-D eyeglasses which correct his faulty vision to put his near point at 25 cm. (Assume he wears the lenses 2.0 cm from his eyes.) Calculate (a) the focal length of Sam's glasses, (b) Sam's near point without glasses. (c) Pam, who has normal eyes with near point at 25 cm, puts on Sam's glasses. Calculate Pam's near point with Sam's glasses on.

83. Spy planes fly at extremely high altitudes (25 km) to avoid interception. If their cameras are to discern features as small as 5 cm, what is the minimum aperture of the camera lens to afford this resolution? (Use $\lambda = 580$ nm.)

84. X-rays of wavelength 0.0973 nm are directed at an unknown crystal. The second diffraction maximum is recorded when the X-rays are directed at an angle of 21.2° relative to the crystal surface. What is the spacing between crystal planes?

85. The Hubble Space Telescope, with an objective diameter of 2.4 m, is viewing the Moon. Estimate the minimum distance between two objects on the Moon that the Hubble can distinguish. Consider diffraction of light with wavelength 550 nm. Assume the Hubble is near the Earth.

86. The Earth and Moon are separated by about 400×10^6 m. When Mars is 8×10^{10} m from Earth, could a person standing on Mars resolve the Earth and its Moon as two separate objects without a telescope? Assume a pupil diameter of 5 mm and $\lambda = 550$ nm.

87. You want to design a spy satellite to photograph license plate numbers. Assuming it is necessary to resolve points separated by 5 cm with 550-nm light, and that the satellite orbits at a height of 130 km, what minimum lens aperture (diameter) is required?

88. Given two 12-cm-focal-length lenses, you attempt to make a crude microscope using them. While holding these lenses a distance 55 cm apart, you position your microscope so that its objective lens is distance d_o from a small object. Assume your eye's near point $N = 25$ cm. (a) For your microscope to function properly, what should d_o be? (b) Assuming your eye is relaxed when using it, what magnification M does your microscope achieve? (c) Since the length of your microscope is not much greater than the focal lengths of its lenses, the approximation $M \approx N\ell/f_e f_o$ is not valid. If you apply this approximation to your microscope, what % error do you make in your microscope's true magnification?

*89. The power of one lens in a pair of eyeglasses is -3.5 D. The radius of curvature of the outside surface is 16.0 cm. What is the radius of curvature of the inside surface? The lens is made of plastic with $n = 1.62$.

Search and Learn

1. Digital cameras may offer an optical zoom or a digital zoom. An optical zoom uses a variable focal-length lens, so only the central part of the field of view fills the entire sensor; a digital zoom electronically includes only the central pixels of the sensor, so objects are larger in the final picture. Discuss which is better, and why.

2. Which of the following statements is true? (See Section 25–2.) Write a brief explanation why each is true or false. (a) Contact lenses and eyeglasses for the same person would have the same power. (b) Farsighted people can see far clearly but not near. (c) Nearsighted people cannot see near or far clearly. (d) Astigmatism in vision is corrected by using different spherical lenses for each eye.

3. Redo Examples 25–3 and 25–4 assuming the sensor has only 6 MP. Explain the different results and their impact on finished photographs.

4. Describe at least four advantages of using mirrors rather than lenses for an astronomical telescope.

5. An astronomical telescope, Fig. 25–20, produces an inverted image. One way to make a telescope that produces an upright image is to insert a third lens between the objective and the eyepiece, Fig. 25–23b. To have the same magnification, the non-inverting telescope will be longer. Suppose lenses of focal length 150 cm, 1.5 cm, and 10 cm are available. Where should these three lenses be placed to make a non-inverting telescope with magnification 100×?

6. Mizar, the second star from the end of the Big Dipper's handle, appears to have a companion star, Alcor. From Earth, Mizar and Alcor have an angular separation of 12 arc minutes (1 arc min = $\frac{1}{60}$ of 1°). Using Examples 25–10 and 25–11, estimate the angular resolution of the human eye (in arc min). From your estimate, explain if these two stars can be resolved by the naked eye.

ANSWERS TO EXERCISES

A: 6.3 m.
B: 33 dots/mm.
C: $P = -4.0$ D.
D: 48 cm.

E: 2 m.
F: (c) as stated on page 732; (c) by the λ rule.
G: 0.28 nm.

A science fantasy book called *Mr Tompkins in Wonderland* (1940), by physicist George Gamow, imagined a world in which the speed of light was only 10 m/s (20 mi/h). Mr Tompkins had studied relativity and when he began "speeding" on a bicycle, he "expected that he would be immediately shortened, and was very happy about it as his increasing figure had lately caused him some anxiety. To his great surprise, however, nothing happened to him or to his cycle. On the other hand, the picture around him completely changed. The streets grew shorter, the windows of the shops began to look like narrow slits, and the policeman on the corner became the thinnest man he had ever seen. 'By Jove!' exclaimed Mr Tompkins excitedly, 'I see the trick now. This is where the word *relativity* comes in.'"

Relativity does indeed predict that objects moving relative to us at high speed, close to the speed of light c, are shortened in length. We don't notice it as Mr Tompkins did, because $c = 3 \times 10^8$ m/s is incredibly fast. We will study length contraction, time dilation, simultaneity non-agreement, and how energy and mass are equivalent ($E = mc^2$).

26 The Special Theory of Relativity

CHAPTER-OPENING QUESTION—Guess now!

A rocket is headed away from Earth at a speed of $0.80c$. The rocket fires a small payload at a speed of $0.70c$ (relative to the rocket) aimed away from Earth. How fast is the payload moving relative to Earth?

(a) $1.50c$;

(b) a little less than $1.50c$;

(c) a little over c;

(d) a little under c;

(e) $0.75c$.

Physics at the end of the nineteenth century looked back on a period of great progress. The theories developed over the preceding three centuries had been very successful in explaining a wide range of natural phenomena. Newtonian mechanics beautifully explained the motion of objects on Earth and in the heavens. Furthermore, it formed the basis for successful treatments of fluids, wave motion, and sound. Kinetic theory explained the behavior of gases and other materials. Maxwell's theory of electromagnetism embodied all of electric and magnetic phenomena, and it predicted the existence of electromagnetic waves that would behave just like light—so light came to be thought of as an electromagnetic wave. Indeed, it seemed that the natural world, as seen through the eyes of physicists, was very well explained. A few puzzles remained, but it was felt that these would soon be explained using already known principles.

It did not turn out so simply. Instead, these puzzles were to be solved only by the introduction, in the early part of the twentieth century, of two revolutionary new theories that changed our whole conception of nature: the *theory of relativity* and *quantum theory*.

Physics as it was known at the end of the nineteenth century (what we've covered up to now in this book) is referred to as **classical physics**. The new physics that grew out of the great revolution at the turn of the twentieth century is now called **modern physics**. In this Chapter, we present the special theory of relativity, which was first proposed by Albert Einstein (1879–1955; Fig. 26–1) in 1905. In Chapter 27, we introduce the equally momentous quantum theory.

FIGURE 26–1 Albert Einstein (1879–1955), one of the great minds of the twentieth century, was the creator of the special and general theories of relativity.

26–1 Galilean–Newtonian Relativity

Einstein's special theory of relativity deals with how we observe events, particularly how objects and events are observed from different frames of reference.[†] This subject had already been explored by Galileo and Newton.

The special theory of relativity deals with events that are observed and measured from so-called **inertial reference frames** (Section 4–2 and Appendix C), which are reference frames in which Newton's first law is valid: if an object experiences no net force, the object either remains at rest or continues in motion with constant speed in a straight line. It is usually easiest to analyze events when they are observed and measured by observers at rest in an inertial frame. The Earth, though not quite an inertial frame (it rotates), is close enough that for most purposes we can approximate it as an inertial frame. Rotating or otherwise accelerating frames of reference are noninertial frames,[‡] and won't concern us in this Chapter (they are dealt with in Einstein's general theory of relativity, as we will see in Chapter 33).

A reference frame that moves with constant velocity with respect to an inertial frame is itself also an inertial frame, since Newton's laws hold in it as well. When we say that we observe or make measurements from a certain reference frame, it means that we are at rest in that reference frame.

[†]A reference frame is a set of coordinate axes fixed to some object such as the Earth, a train, or the Moon. See Section 2–1.

[‡]On a rotating platform (say a merry-go-round), for example, a ball at rest starts moving outward even though no object exerts a force on it. This is therefore not an inertial frame. See Appendix C, Fig. C–1.

(a)
Reference frame = car

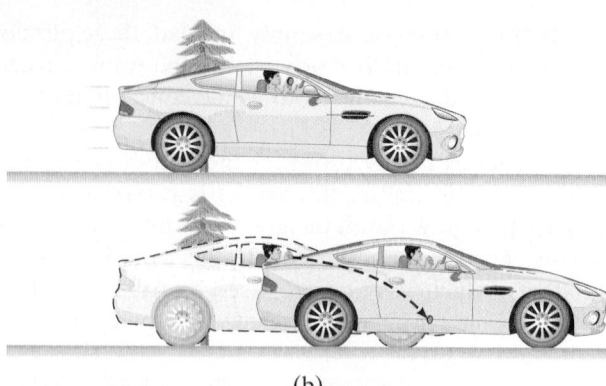

(b)
Reference frame = Earth

FIGURE 26–2 A coin is dropped by a person in a moving car. The upper views show the moment of the coin's release, the lower views are a short time later. (a) In the reference frame of the car, the coin falls straight down (and the tree moves to the left). (b) In a reference frame fixed on the Earth, the coin has an initial velocity (= to car's) and follows a curved (parabolic) path.

Both Galileo and Newton were aware of what we now call the **relativity principle** applied to mechanics: that *the basic laws of physics are the same in all inertial reference frames*. You may have recognized its validity in everyday life. For example, objects move in the same way in a smoothly moving (constant-velocity) train or airplane as they do on Earth. (This assumes no vibrations or rocking which would make the reference frame noninertial.) When you walk, drink a cup of soup, play pool, or drop a pencil on the floor while traveling in a train, airplane, or ship moving at constant velocity, the objects move just as they do when you are at rest on Earth. Suppose you are in a car traveling rapidly at constant velocity. If you drop a coin from above your head inside the car, how will it fall? It falls straight downward with respect to the car, and hits the floor directly below the point of release, Fig. 26–2a. This is just how objects fall on the Earth—straight down—and thus our experiment in the moving car is in accord with the relativity principle. (If you drop the coin out the car's window, this won't happen because the moving air drags the coin backward relative to the car.)

Note in this example, however, that to an observer on the Earth, the coin follows a curved path, Fig. 26–2b. The actual path followed by the coin is different as viewed from different frames of reference. This does not violate the relativity principle because this principle states that the *laws* of physics are the same in all inertial frames. The same law of gravity, and the same laws of motion, apply in both reference frames. The acceleration of the coin is the same in both reference frames. The difference in Figs. 26–2a and b is that in the Earth's frame of reference, the coin has an initial velocity (equal to that of the car). The laws of physics therefore predict it will follow a parabolic path like any projectile (Chapter 3). In the car's reference frame, there is no initial velocity, and the laws of physics predict that the coin will fall straight down. The laws are the same in both reference frames, although the specific paths are different.

Galilean–Newtonian relativity involves certain unprovable assumptions that make sense from everyday experience. It is assumed that the lengths of objects are the same in one reference frame as in another, and that time passes at the same rate in different reference frames. In classical mechanics, then, space and time intervals are considered to be **absolute**: their measurement does not change from one reference frame to another. The mass of an object, as well as all forces, are assumed to be unchanged by a change in inertial reference frame.

The position of an object, however, is different when specified in different reference frames, and so is velocity. For example, a person may walk inside a bus toward the front with a speed of 2 m/s. But if the bus moves 10 m/s with respect to the Earth, the person is then moving with a speed of 12 m/s with respect to the Earth. The acceleration of an object, however, is the same in any inertial reference frame according to classical mechanics. This is because the change in velocity, and the time interval, will be the same. For example, the person in the bus may accelerate from 0 to 2 m/s in 1.0 seconds, so $a = 2 \text{ m/s}^2$ in the reference frame of the bus. With respect to the Earth, the acceleration is

$$(12 \text{ m/s} - 10 \text{ m/s})/(1.0 \text{ s}) = 2 \text{ m/s}^2,$$

which is the same.

Since neither F, m, nor a changes from one inertial frame to another, Newton's second law, $F = ma$, does not change. Thus Newtons' second law satisfies the relativity principle. The other laws of mechanics also satisfy the relativity principle.

That the laws of mechanics are the same in all inertial reference frames implies that no one inertial frame is special in any sense. We express this important conclusion by saying that **all inertial reference frames are equivalent** for the description of mechanical phenomena. No one inertial reference frame is any better than another. A reference frame fixed to a car or an aircraft traveling at constant velocity is as good as one fixed on the Earth. When you travel smoothly at constant velocity in a car or airplane, it is just as valid to say you are at rest and the Earth is moving as it is to say the reverse.[†] There is no experiment you can do to tell which frame is "really" at rest and which is moving. Thus, there is no way to single out one particular reference frame as being at absolute rest.

A complication arose, however, in the last half of the nineteenth century. Maxwell's comprehensive and successful theory of electromagnetism (Chapter 22) predicted that light is an electromagnetic wave. Maxwell's equations gave the velocity of light c as $3.00 \times 10^8\,\text{m/s}$; and this is just what is measured. The question then arose: in what reference frame does light have precisely the value predicted by Maxwell's theory? It was assumed that light would have a different speed in different frames of reference. For example, if observers could travel on a rocket ship at a speed of $1.0 \times 10^8\,\text{m/s}$ away from a source of light, we might expect them to measure the speed of the light reaching them to be $(3.0 \times 10^8\,\text{m/s}) - (1.0 \times 10^8\,\text{m/s}) = 2.0 \times 10^8\,\text{m/s}$. But Maxwell's equations have no provision for relative velocity. They predicted the speed of light to be $c = 3.0 \times 10^8\,\text{m/s}$, which seemed to imply that there must be some preferred reference frame where c would have this value.

We discussed in Chapters 11 and 12 that waves can travel on water and along ropes or strings, and sound waves travel in air and other materials. Nineteenth-century physicists viewed the material world in terms of the laws of mechanics, so it was natural for them to assume that light too must travel in some *medium*. They called this transparent medium the **ether** and assumed it permeated all space.[‡] It was therefore assumed that the velocity of light given by Maxwell's equations must be with respect to the ether.[§]

Scientists soon set out to determine the speed of the Earth relative to this absolute frame, whatever it might be. A number of clever experiments were designed. The most direct were performed by A. A. Michelson and E. W. Morley in the 1880s. They measured the difference in the speed of light in different directions using Michelson's interferometer (Section 24–9). They expected to find a difference depending on the orientation of their apparatus with respect to the ether. For just as a boat has different speeds relative to the land when it moves upstream, downstream, or across the stream, so too light would be expected to have different speeds depending on the velocity of the ether past the Earth.

Strange as it may seem, they detected no difference at all. This was a great puzzle. A number of explanations were put forth over a period of years, but they led to contradictions or were otherwise not generally accepted. This **null result** was one of the great puzzles at the end of the nineteenth century.

Then in 1905, Albert Einstein proposed a radical new theory that reconciled these many problems in a simple way. But at the same time, as we shall see, it completely changed our ideas of space and time.

[†]We use the reasonable approximation that Earth is an inertial reference frame.

[‡]The medium for light waves could not be air, since light travels from the Sun to Earth through nearly empty space. Therefore, another medium was postulated, the ether. The ether was not only transparent but, because of difficulty in detecting it, was assumed to have zero density.

[§]Also, it appeared that Maxwell's equations did *not* satisfy the relativity principle: They were simplest in the frame where $c = 3.00 \times 10^8\,\text{m/s}$, in a reference frame at rest in the ether. In any other reference frame, extra terms were needed to account for relative velocity. Although other laws of physics obeyed the relativity principle, the laws of electricity and magnetism apparently did not. Einstein's second postulate (next Section) resolved this problem: Maxwell's equations do satisfy relativity.

26–2 Postulates of the Special Theory of Relativity

The problems that existed at the start of the twentieth century with regard to electromagnetic theory and Newtonian mechanics were beautifully resolved by Einstein's introduction of the special theory of relativity in 1905. Unaware of the Michelson–Morley null result, Einstein was motivated by certain questions regarding electromagnetic theory and light waves. For example, he asked himself: "What would I see if I rode a light beam?" The answer was that instead of a traveling electromagnetic wave, he would see alternating electric and magnetic fields at rest whose magnitude changed in space, but did not change in time. Such fields, he realized, had never been detected and indeed were not consistent with Maxwell's electromagnetic theory. He argued, therefore, that it was unreasonable to think that the speed of light relative to any observer could be reduced to zero, or in fact reduced at all. This idea became the second postulate of his theory of relativity.

In his famous 1905 paper, Einstein proposed doing away with the idea of the ether and the accompanying assumption of a preferred or absolute reference frame at rest. This proposal was embodied in two postulates. The first was an extension of the Galilean–Newtonian relativity principle to include not only the laws of mechanics but also those of the rest of physics, including electricity and magnetism:

First postulate (the relativity principle): **The laws of physics have the same form in all inertial reference frames.**

The first postulate can also be stated as: *there is no experiment you can do in an inertial reference frame to determine if you are at rest or moving uniformly at constant velocity.*

The second postulate is consistent with the first:

Second postulate (constancy of the speed of light): **Light propagates through empty space with a definite speed c independent of the speed of the source or observer.**

These two postulates form the foundation of Einstein's **special theory of relativity**. It is called "special" to distinguish it from his later "general theory of relativity," which deals with noninertial (accelerating) reference frames (Chapter 33). The special theory, which is what we discuss here, deals only with inertial frames.

The second postulate may seem hard to accept, for it seems to violate common sense. First of all, we have to think of light traveling through empty space. Giving up the ether is not too hard, however, since it had never been detected. But the second postulate also tells us that the speed of light in vacuum is always the same, $3.00 \times 10^8 \, \text{m/s}$, no matter what the speed of the observer or the source. Thus, a person traveling toward or away from a source of light will measure the same speed for that light as someone at rest with respect to the source. This conflicts with our everyday experience: we would expect to have to add in the velocity of the observer. On the other hand, perhaps we can't expect our everyday experience to be helpful when dealing with the high velocity of light. Furthermore, the null result of the Michelson–Morley experiment is fully consistent with the second postulate.[†]

Einstein's proposal has a certain beauty. By doing away with the idea of an absolute reference frame, it was possible to reconcile classical mechanics with Maxwell's electromagnetic theory. The speed of light predicted by Maxwell's equations *is* the speed of light in vacuum in *any* reference frame.

Einstein's theory required us to give up common sense notions of space and time, and in the following Sections we will examine some strange but interesting consequences of special relativity. Our arguments for the most part will be simple ones.

[†]The Michelson–Morley experiment can also be considered as evidence for the first postulate, since it was intended to measure the motion of the Earth relative to an absolute reference frame. Its failure to do so implies the absence of any such preferred frame.

We will use a technique that Einstein himself did: we will imagine very simple experimental situations in which little mathematics is needed. In this way, we can see many of the consequences of relativity theory without getting involved in detailed calculations. Einstein called these **thought experiments**.

26–3 Simultaneity

An important consequence of the theory of relativity is that we can no longer regard time as an absolute quantity. No one doubts that time flows onward and never turns back. But according to relativity, the time interval between two events, and even whether or not two events are simultaneous, depends on the observer's reference frame. By an **event**, which we use a lot here, we mean something that happens at a particular place and at a particular time.

Two events are said to occur simultaneously if they occur at exactly the same time. But how do we know if two events occur precisely at the same time? If they occur at the same point in space—such as two apples falling on your head at the same time—it is easy. But if the two events occur at widely separated places, it is more difficult to know whether the events are simultaneous since we have to take into account the time it takes for the light from them to reach us. Because light travels at finite speed, a person who sees two events must calculate back to find out when they actually occurred. For example, if two events are *observed* to occur at the same time, but one actually took place farther from the observer than the other, then the more distant one must have occurred earlier, and the two events were not simultaneous.

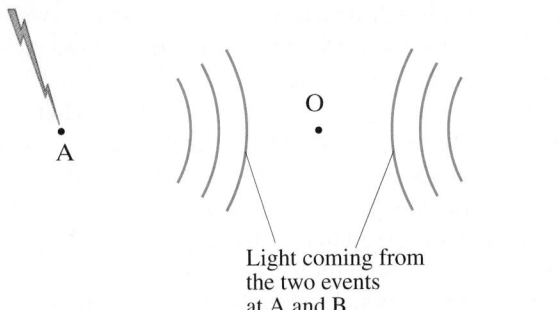

Light coming from
the two events
at A and B

FIGURE 26–3 A moment after lightning strikes at points A and B, the pulses of light (shown as blue waves) are traveling toward the observer O, but O "sees" the lightning only when the light reaches O.

We now imagine a simple thought experiment. Assume an observer, called O, is located exactly halfway between points A and B where two events occur, Fig. 26–3. Suppose the two events are lightning that strikes the points A and B, as shown. For brief events like lightning, only short pulses of light (blue in Fig. 26–3) will travel outward from A and B and reach O. Observer O "sees" the events when the pulses of light reach point O. If the two pulses reach O at the same time, then the two events had to be simultaneous. This is because (i) the two light pulses travel at the same speed (postulate 2), and (ii) the distance OA equals OB, so the time for the light to travel from A to O and from B to O must be the same. Observer O can then definitely state that the two events occurred simultaneously. On the other hand, if O sees the light from one event before that from the other, then the former event occurred first.

The question we really want to examine is this: if two events are simultaneous to an observer in one reference frame, are they also simultaneous to another observer moving with respect to the first? Let us call the observers O_1 and O_2 and assume they are fixed in reference frames 1 and 2 that move with speed v relative to one another. These two reference frames can be thought of as two rockets or two trains (Fig. 26–4). O_2 says that O_1 is moving to the right with speed v, as in Fig. 26–4a; and O_1 says O_2 is moving to the left with speed v, as in Fig. 26–4b. Both viewpoints are legitimate according to the relativity principle. [There is no third point of view that will tell us which one is "really" moving.]

FIGURE 26–4 Observers O_1 and O_2, on two different trains (two different reference frames), are moving with relative speed v. (a) O_2 says that O_1 is moving to the right. (b) O_1 says that O_2 is moving to the left. Both viewpoints are legitimate: it all depends on your reference frame.

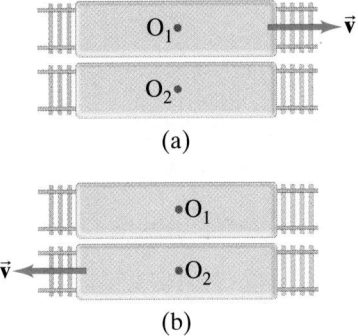

Now suppose that observers O_1 and O_2 observe and measure two lightning strikes. The lightning bolts mark both trains where they strike: at A_1 and B_1 on O_1's train, and at A_2 and B_2 on O_2's train, Fig. 26–5a. For simplicity, we assume that O_1 is exactly halfway between A_1 and B_1, and O_2 is halfway between A_2 and B_2. Let us first put ourselves in O_2's reference frame, so we observe O_1 moving to the right with speed v. Let us also assume that the two events occur *simultaneously* in O_2's frame, and just at the instant when O_1 and O_2 are opposite each other, Fig. 26–5a. A short time later, Fig. 26–5b, light from A_2 and from B_2 reach O_2 at the same time (we assumed this). Since O_2 knows (or measures) the distances O_2A_2 and O_2B_2 as equal, O_2 knows the two events are simultaneous in the O_2 reference frame.

FIGURE 26–5 Thought experiment on simultaneity. In both (a) and (b) we are in the reference frame of observer O_2, who sees the reference frame of O_1 moving to the right. In (a), one lightning bolt strikes the two reference frames at A_1 and A_2, and a second lightning bolt strikes at B_1 and B_2. (b) A moment later, the light (shown in blue) from the two events reaches O_2 at the same time. So according to observer O_2, the two bolts of lightning struck simultaneously. But in O_1's reference frame, the light from B_1 has already reached O_1, whereas the light from A_1 has not yet reached O_1. So in O_1's reference frame, the event at B_1 must have preceded the event at A_1. Simultaneity in time is not absolute.

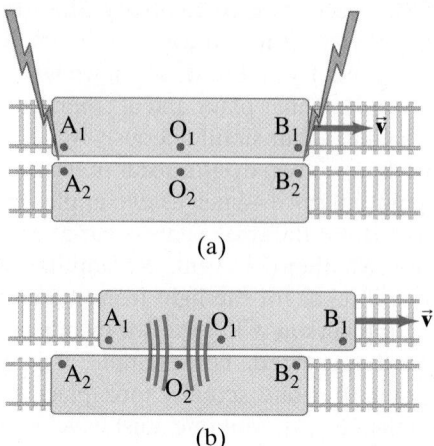

(a)

(b)

But what does observer O_1 observe and measure? From our (O_2) reference frame, we can predict what O_1 will observe. We see that O_1 moves to the right during the time the light is traveling to O_1 from A_1 and B_1. As shown in Fig. 26–5b, we can see from our O_2 reference frame that the light from B_1 has already passed O_1, whereas the light from A_1 has not yet reached O_1. That is, O_1 observes the light coming from B_1 before observing the light coming from A_1. Given (i) that light travels at the same speed c in any direction and in any reference frame, and (ii) that the distance O_1A_1 equals O_1B_1, then observer O_1 can only conclude that the event at B_1 occurred before the event at A_1. The two events are *not* simultaneous for O_1, even though they are for O_2.

We thus find that two events which take place at different locations and are simultaneous to one observer, are actually not simultaneous to a second observer who moves relative to the first.

It may be tempting to ask: "Which observer is right, O_1 or O_2?" The answer, according to relativity, is that they are *both* right. There is no "best" reference frame we can choose to determine which observer is right. Both frames are equally good. We can only conclude that *simultaneity is not an absolute concept*, but is relative. We are not aware of this lack of agreement on simultaneity in everyday life because the effect is noticeable only when the relative speed of the two reference frames is very large (near c), or the distances involved are very large.

26–4 Time Dilation and the Twin Paradox

The fact that two events simultaneous to one observer may not be simultaneous to a second observer suggests that time itself is not absolute. Could it be that time passes differently in one reference frame than in another? This is, indeed, just what Einstein's theory of relativity predicts, as the following thought experiment shows.

Figure 26–6 shows a spaceship traveling past Earth at high speed. The point of view of an observer on the spaceship is shown in part (a), and that of an observer on Earth in part (b). Both observers have accurate clocks. The person on the spaceship (Fig. 26–6a) flashes a light and measures the time it takes the light to travel directly across the spaceship and return after reflecting from a mirror (the rays are drawn at a slight angle for clarity). In the reference frame of the spaceship, the

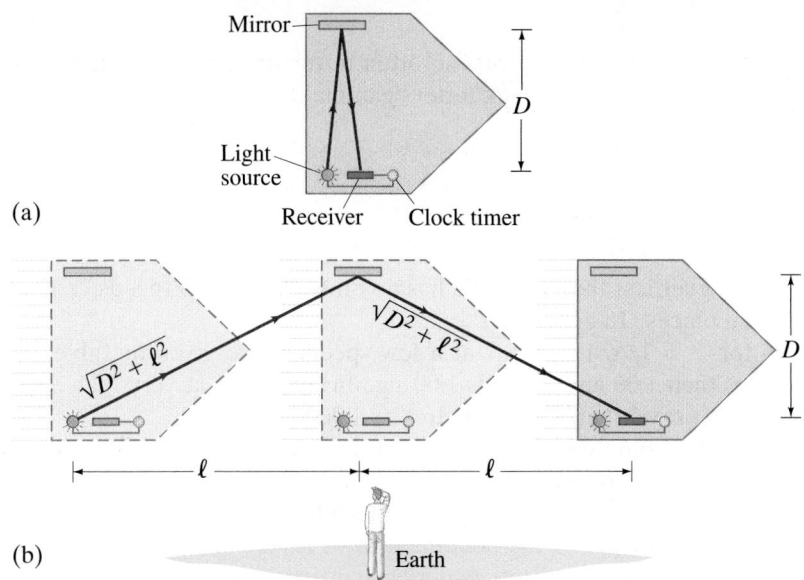

(a)

(b) Earth

FIGURE 26–6 Time dilation can be shown by a thought experiment: the time it takes for light to travel across a spaceship and back is longer for the observer on Earth (b) than for the observer on the spaceship (a).

light travels a distance $2D$ at speed c, Fig. 26–6a; so the time required to go across and back, Δt_0, is

$$\Delta t_0 = \frac{2D}{c}.$$

The observer on Earth, Fig. 26–6b, observes the same process. But to this observer, the spaceship is moving. So the light travels the diagonal path shown going across the spaceship, reflecting off the mirror, and returning to the sender. Although the light travels at the same speed to this observer (the second postulate), it travels a greater distance. Hence the time required, as measured by the observer on Earth, will be *greater* than that measured by the observer on the spaceship.

Let us determine the time interval Δt measured by the observer on Earth between sending and receiving the light. In time Δt, the spaceship travels a distance $2\ell = v\,\Delta t$ where v is the speed of the spaceship (Fig. 26–6b). The light travels a total distance on its diagonal path (Pythagorean theorem) of $2\sqrt{D^2 + \ell^2} = c\,\Delta t$, where $\ell = v\,\Delta t/2$. Therefore

$$c\,\Delta t = 2\sqrt{D^2 + \ell^2} = 2\sqrt{D^2 + v^2(\Delta t)^2/4}.$$

We square both sides to find $c^2(\Delta t)^2 = 4D^2 + v^2(\Delta t)^2$, and solve for $(\Delta t)^2$:

$$(\Delta t)^2 = 4D^2/(c^2 - v^2)$$

so

$$\Delta t = \frac{2D}{c\sqrt{1 - v^2/c^2}}.$$

We combine this equation for Δt with the formula for Δt_0 above, $\Delta t_0 = 2D/c$:

$$\Delta t = \frac{\Delta t_0}{\sqrt{1 - v^2/c^2}}.$$ **(26–1a)**   *TIME DILATION*

Since $\sqrt{1 - v^2/c^2}$ is always less than 1, we see that $\Delta t > \Delta t_0$. That is, the time interval between the two events (the sending of the light, and its reception on the spaceship) is *greater* for the observer on Earth than for the observer on the spaceship. This is a general result of the theory of relativity, and is known as **time dilation**. The time dilation effect can be stated as

clocks moving relative to an observer are measured to run more slowly, as compared to clocks at rest.

However, we should not think that the clocks are somehow at fault.

Time is actually measured to pass more slowly in any moving reference frame as compared to your own.

This remarkable result is an inevitable outcome of the two postulates of the special theory of relativity.

The factor $1/\sqrt{1 - v^2/c^2}$ occurs so often in relativity that we often give it the shorthand symbol γ (the Greek letter "gamma"), and write Eq. 26–1a as

$$\Delta t = \gamma \, \Delta t_0 \tag{26–1b}$$

where

$$\gamma = \frac{1}{\sqrt{1 - v^2/c^2}}. \tag{26–2}$$

Note that γ is never less than one, and has no units. At normal speeds, $\gamma = 1$ to many decimal places. In general, $\gamma \geq 1$.

Values for $\gamma = 1/\sqrt{1 - v^2/c^2}$ at a few speeds v are given in Table 26–1. γ is never less than 1.00 and exceeds 1.00 significantly only at very high speeds, much above let's say 10^6 m/s (for which $\gamma = 1.000006$).

The concept of time dilation may be hard to accept, for it contradicts our experience. We can see from Eq. 26–1 that the time dilation effect is indeed negligible unless v is reasonably close to c. If v is much less than c, then the term v^2/c^2 is much smaller than the 1 in the denominator of Eq. 26–1, and then $\Delta t \approx \Delta t_0$ (see Example 26–2). The speeds we experience in everyday life are much smaller than c, so it is little wonder we don't ordinarily notice time dilation. But experiments that have tested the time dilation effect have confirmed Einstein's predictions. In 1971, for example, extremely precise atomic clocks were flown around the Earth in jet planes. The speed of the planes $(10^3$ km/h$)$ was much less than c, so the clocks had to be accurate to nanoseconds $(10^{-9}$ s$)$ in order to detect any time dilation. They were this accurate, and they confirmed Eqs. 26–1 to within experimental error. Time dilation had been confirmed decades earlier, however, by observations on "elementary particles" which have very small masses (typically 10^{-30} to 10^{-27} kg) and so require little energy to be accelerated to speeds close to the speed of light, c. Many of these elementary particles are not stable and decay after a time into lighter particles. One example is the muon, whose mean lifetime is 2.2 μs when at rest. Careful experiments showed that when a muon is traveling at high speeds, its lifetime is measured to be longer than when it is at rest, just as predicted by the time dilation formula.

TABLE 26–1 Values of γ

v	γ
0	1.00000 …
0.01c	1.00005
0.10c	1.005
0.50c	1.15
0.90c	2.3
0.99c	7.1

EXAMPLE 26–1 **Lifetime of a moving muon.** (a) What will be the mean lifetime of a muon as measured in the laboratory if it is traveling at $v = 0.60c = 1.80 \times 10^8$ m/s with respect to the laboratory? A muon's mean lifetime at rest is 2.20 μs $= 2.20 \times 10^{-6}$ s. (b) How far does a muon travel in the laboratory, on average, before decaying?

APPROACH If an observer were to move along with the muon (the muon would be at rest to this observer), the muon would have a mean life of 2.20×10^{-6} s. To an observer in the lab, the muon lives longer because of time dilation. We find the mean lifetime using Eq. 26–1 and the average distance using $d = v \, \Delta t$.

SOLUTION (a) From Eq. 26–1 with $v = 0.60c$, we have

$$\Delta t = \frac{\Delta t_0}{\sqrt{1 - v^2/c^2}}$$

$$= \frac{2.20 \times 10^{-6}\,\text{s}}{\sqrt{1 - 0.36c^2/c^2}} = \frac{2.20 \times 10^{-6}\,\text{s}}{\sqrt{0.64}} = 2.8 \times 10^{-6}\,\text{s}.$$

(b) Relativity predicts that a muon with speed 1.80×10^8 m/s would travel an average distance $d = v \, \Delta t = (1.80 \times 10^8\,\text{m/s})(2.8 \times 10^{-6}\,\text{s}) = 500$ m, and this is the distance that is measured experimentally in the laboratory.

NOTE At a speed of 1.8×10^8 m/s, classical physics would tell us that with a mean life of 2.2 μs, an average muon would travel $d = vt = (1.8 \times 10^8\,\text{m/s})(2.2 \times 10^{-6}\,\text{s}) = 400$ m. This is shorter than the distance measured.

EXERCISE A What is the muon's mean lifetime (Example 26–1) if it is traveling at $v = 0.90c$? (a) 0.42 μs; (b) 2.3 μs; (c) 5.0 μs; (d) 5.3 μs; (e) 12.0 μs.

We need to clarify how to use Eq. 26–1, $\Delta t = \gamma \Delta t_0$, and the meaning of Δt and Δt_0. The equation is true only when Δt_0 represents the time interval between the two events *in a reference frame where an observer at rest sees the two events occur at the same point in space* (as in Fig. 26–6a where the two events are the light flash being sent and being received). This time interval, Δt_0, is called the **proper time**. Then Δt in Eqs. 26–1 represents the time interval between the two events as measured in a reference frame *moving* with speed v with respect to the first. In Example 26–1 above, Δt_0 (and not Δt) was set equal to 2.2×10^{-6} s because it is only in the rest frame of the muon that the two events ("birth" and "decay") occur at the same point in space. The proper time Δt_0 is the shortest time between the events any observer can measure. In any other moving reference frame, the time Δt is greater.

 CAUTION

Proper time Δt_0 is for 2 events at the same point in space

 CAUTION

Proper time is shortest:
$\Delta t > \Delta t_0$

EXAMPLE 26–2 Time dilation at 100 km/h. Let us check time dilation for everyday speeds. A car traveling 100 km/h covers a certain distance in 10.00 s according to the driver's watch. What does an observer at rest on Earth measure for the time interval?

APPROACH The car's speed relative to Earth, written in meters per second, is $100 \text{ km/h} = (1.00 \times 10^5 \text{ m})/(3600 \text{ s}) = 27.8 \text{ m/s}$. The driver is at rest in the reference frame of the car, so we set $\Delta t_0 = 10.00$ s in the time dilation formula.

SOLUTION We use Eq. 26–1a:

$$\Delta t = \frac{\Delta t_0}{\sqrt{1 - \dfrac{v^2}{c^2}}} = \frac{10.00 \text{ s}}{\sqrt{1 - \left(\dfrac{27.8 \text{ m/s}}{3.00 \times 10^8 \text{ m/s}}\right)^2}}$$

$$= \frac{10.00 \text{ s}}{\sqrt{1 - (8.59 \times 10^{-15})}}.$$

If you put these numbers into a calculator, you will obtain $\Delta t = 10.00$ s, because the denominator differs from 1 by such a tiny amount. The time measured by an observer fixed on Earth would show no difference from that measured by the driver, even with the best instruments. A computer that could calculate to a large number of decimal places would reveal a slight difference between Δt and Δt_0.

NOTE We can estimate the difference using the binomial expansion (Appendix A–5),

 PROBLEM SOLVING

Use of the binomial expansion

$$(1 \pm x)^n \approx 1 \pm nx. \qquad [\text{for } x \ll 1]$$

In our time dilation formula, we have the factor $\gamma = \left(1 - v^2/c^2\right)^{-\frac{1}{2}}$. Thus[†]

$$\Delta t = \gamma \Delta t_0 = \Delta t_0 \left(1 - \frac{v^2}{c^2}\right)^{-\frac{1}{2}} \approx \Delta t_0 \left(1 + \frac{1}{2} \frac{v^2}{c^2}\right)$$

$$\approx 10.00 \text{ s} \left[1 + \frac{1}{2}\left(\frac{27.8 \text{ m/s}}{3.00 \times 10^8 \text{ m/s}}\right)^2\right]$$

$$\approx 10.00 \text{ s} + 4 \times 10^{-14} \text{ s}.$$

So the difference between Δt and Δt_0 is predicted to be 4×10^{-14} s, an extremely small amount.

EXERCISE B A certain atomic clock keeps precise time on Earth. If the clock is taken on a spaceship traveling at a speed $v = 0.60c$, does this clock now run slow according to the people (*a*) on the spaceship, (*b*) on Earth?

[†]Recall that $1/x^n$ is written as x^{-n}, such as $1/x^2 = x^{-2}$, Appendix A–2.

EXAMPLE 26–3 **Reading a magazine on a spaceship.** A passenger on a fictional high-speed spaceship traveling between Earth and Jupiter at a steady speed of 0.75c reads a magazine which takes 10.0 min according to her watch. (a) How long does this take as measured by Earth-based clocks? (b) How much farther is the spaceship from Earth at the end of reading the article than it was at the beginning?

APPROACH (a) The time interval in one reference frame is related to the time interval in the other by Eq. 26–1a or b. (b) At constant speed, distance is speed × time. Because there are two time intervals (Δt and Δt_0) we will get two distances, one for each reference frame. [This surprising result is explored in the next Section (26–5).]

SOLUTION (a) The given 10.0-min time interval is the proper time Δt_0—starting and finishing the magazine happen at the same place on the spaceship. Earth clocks measure

$$\Delta t = \frac{\Delta t_0}{\sqrt{1 - (v^2/c^2)}} = \frac{10.00 \text{ min}}{\sqrt{1 - (0.75)^2}} = 15.1 \text{ min.}$$

(b) In the Earth frame, the rocket travels a distance $D = v \, \Delta t = (0.75c)(15.1 \text{ min}) = (0.75)(3.0 \times 10^8 \text{ m/s})(15.1 \text{ min} \times 60 \text{ s/min}) = 2.04 \times 10^{11}$ m. In the spaceship's frame, the Earth is moving away from the spaceship at 0.75c, but the time is only 10.0 min, so the distance is measured to be $D_0 = v \, \Delta t_0 = (2.25 \times 10^8 \text{ m/s})(600 \text{ s}) = 1.35 \times 10^{11}$ m.

Space Travel?

Time dilation has aroused interesting speculation about space travel. According to classical (Newtonian) physics, to reach a star 100 light-years away would not be possible for ordinary mortals (1 light-year is the distance light can travel in 1 year = 3.0×10^8 m/s $\times 3.16 \times 10^7$ s $= 9.5 \times 10^{15}$ m). Even if a spaceship could travel at close to the speed of light, it would take over 100 years to reach such a star. But time dilation tells us that the time involved could be less. In a spaceship traveling at $v = 0.999c$, the time for such a trip would be only about $\Delta t_0 = \Delta t \sqrt{1 - v^2/c^2} = (100 \text{ yr})\sqrt{1 - (0.999)^2} = 4.5 \text{ yr}$. Thus time dilation allows such a trip, but the enormous practical problems of achieving such speeds may not be possible to overcome, certainly not in the near future.

When we talk in this Chapter and in the Problems about spaceships moving at speeds close to c, it is for understanding and for fun, but not realistic, although for tiny elementary particles such high speeds *are* realistic.

In this example, 100 years would pass on Earth, whereas only 4.5 years would pass for the astronaut on the trip. Is it just the clocks that would slow down for the astronaut? No.

All processes, including aging and other life processes, run more slowly for the astronaut as measured by an Earth observer. But to the astronaut, time would pass in a normal way.

The astronaut would experience 4.5 years of normal sleeping, eating, reading, and so on. And people on Earth would experience 100 years of ordinary activity.

Twin Paradox

Not long after Einstein proposed the special theory of relativity, an apparent paradox was pointed out. According to this **twin paradox**, suppose one of a pair of 20-year-old twins takes off in a spaceship traveling at very high speed to a distant star and back again, while the other twin remains on Earth. According to the Earth twin, the astronaut twin will age less. Whereas 20 years might pass for the Earth twin, perhaps only 1 year (depending on the spacecraft's speed) would pass for the traveler. Thus, when the traveler returns, the earthbound twin could expect to be 40 years old whereas the traveling twin would be only 21.

This is the viewpoint of the twin on the Earth. But what about the traveling twin? If all inertial reference frames are equally good, won't the traveling twin make all the claims the Earth twin does, only in reverse? Can't the astronaut twin claim that since the Earth is moving away at high speed, time passes more slowly on Earth and the twin on Earth will age less? This is the opposite of what the Earth twin predicts. They cannot both be right, for after all the spacecraft returns to Earth and a direct comparison of ages and clocks can be made.

There is, however, no contradiction here. One of the viewpoints is indeed incorrect. The consequences of the special theory of relativity—in this case, time dilation—can be applied only by observers in an inertial reference frame. The Earth is such a frame (or nearly so), whereas the spacecraft is not. The spacecraft accelerates at the start and end of its trip and when it turns around at the far point of its journey. Part of the time, the astronaut twin may be in an inertial frame (and is justified in saying the Earth twin's clocks run slow). But during the accelerations, the twin on the spacecraft is not in an inertial frame. So she cannot use special relativity to predict their relative ages when she returns to Earth. The Earth twin stays in the same inertial frame, and we can thus trust her predictions based on special relativity. Thus, there is no paradox. The prediction of the Earth twin that the traveling twin ages less is the correct one.

*Global Positioning System (GPS)

Airplanes, cars, boats, and hikers use **global positioning system** (**GPS**) receivers to tell them quite accurately where they are at a given moment (Fig. 26–7). There are more than 30 global positioning system satellites that send out precise time signals using atomic clocks. Your receiver compares the times received from at least four satellites, all of whose times are carefully synchronized to within 1 part in 10^{13}. By comparing the time differences with the known satellite positions and the fixed speed of light, the receiver can determine how far it is from each satellite and thus where it is on the Earth. It can do this to an accuracy of a few meters, if it has been constructed to make corrections such as the one below due to relativity.

CONCEPTUAL EXAMPLE 26–4 | **A relativity correction to GPS.** GPS satellites move at about $4 \text{ km/s} = 4000 \text{ m/s}$. Show that a good GPS receiver needs to correct for time dilation if it is to produce results consistent with atomic clocks accurate to 1 part in 10^{13}.

RESPONSE Let us calculate the magnitude of the time dilation effect by inserting $v = 4000 \text{ m/s}$ into Eq. 26–1a:

$$\Delta t = \frac{1}{\sqrt{1 - \dfrac{v^2}{c^2}}} \Delta t_0 = \frac{1}{\sqrt{1 - \left(\dfrac{4 \times 10^3 \text{ m/s}}{3 \times 10^8 \text{ m/s}}\right)^2}} \Delta t_0$$

$$= \frac{1}{\sqrt{1 - 1.8 \times 10^{-10}}} \Delta t_0.$$

We use the binomial expansion: $(1 \pm x)^n \approx 1 \pm nx$ for $x \ll 1$ (see Appendix A–5) which here is $(1 - x)^{-\frac{1}{2}} \approx 1 + \frac{1}{2}x$. That is

$$\Delta t = \left(1 + \tfrac{1}{2}(1.8 \times 10^{-10})\right) \Delta t_0 = \left(1 + 9 \times 10^{-11}\right) \Delta t_0.$$

The time "error" divided by the time interval is

$$\frac{(\Delta t - \Delta t_0)}{\Delta t_0} = 1 + 9 \times 10^{-11} - 1 = 9 \times 10^{-11} \approx 1 \times 10^{-10}.$$

Time dilation, if not accounted for, would introduce an error of about 1 part in 10^{10}, which is 1000 times greater than the precision of the atomic clocks. Not correcting for time dilation means a receiver could give much poorer position accuracy.

NOTE GPS devices must make other corrections as well, including effects associated with general relativity.

PHYSICS APPLIED
Global positioning system (GPS)

FIGURE 26–7 A visiting professor of physics uses the GPS on her smart phone to find a restaurant (red dot). Her location in the physics department is the blue dot. Traffic on some streets is also shown (green = good, orange = slow, red = heavy traffic) which comes in part by tracking cell phone movements.

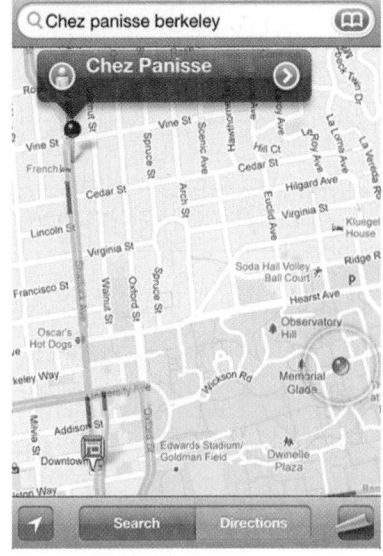

26–5 Length Contraction

Time intervals are not the only things different in different reference frames. Space intervals—lengths and distances—are different as well, according to the special theory of relativity, and we illustrate this with a thought experiment.

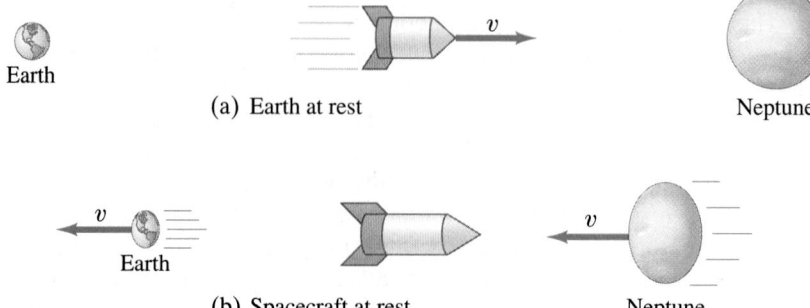

FIGURE 26–8 (a) A spaceship traveling at very high speed v from Earth to the planet Neptune, as seen from Earth's frame of reference. (b) According to an observer on the spaceship, Earth and Neptune are moving at the very high speed v: Earth leaves the spaceship, and a time Δt_0 later Neptune arrives at the spaceship.

(a) Earth at rest

(b) Spacecraft at rest

Observers on Earth watch a spacecraft traveling at speed v from Earth to, say, Neptune, Fig. 26–8a. The distance between the planets, as measured by the Earth observers, is ℓ_0. The time required for the trip, measured from Earth, is

$$\Delta t = \frac{\ell_0}{v}. \qquad \text{[Earth observer]}$$

In Fig. 26–8b we see the point of view of observers on the spacecraft. In this frame of reference, the spaceship is at rest; Earth and Neptune move[†] with speed v. The time between departure of Earth and arrival of Neptune, as observed from the spacecraft, is the "proper time" Δt_0 (page 753), because these two events occur at the same point in space (i.e., at the spacecraft). Therefore the time interval is less for the spacecraft observers than for the Earth observers. That is, because of time dilation (Eq. 26–1a), the time for the trip as viewed by the spacecraft is

$$\Delta t_0 = \Delta t \sqrt{1 - v^2/c^2}$$
$$= \Delta t / \gamma. \qquad \text{[spacecraft observer]}$$

Because the spacecraft observers measure the same speed but less time between these two events, they also measure the distance as less. If we let ℓ be the distance between the planets as viewed by the spacecraft observers, then $\ell = v\,\Delta t_0$, which we can rewrite as $\ell = v\,\Delta t_0 = v\,\Delta t \sqrt{1 - v^2/c^2} = \ell_0\sqrt{1 - v^2/c^2}$. Thus we have the important result that

LENGTH CONTRACTION

$$\ell = \ell_0\sqrt{1 - v^2/c^2} \qquad (26\text{–}3a)$$

or, using γ (Eq. 26–2),

$$\ell = \frac{\ell_0}{\gamma}. \qquad (26\text{–}3b)$$

This is a general result of the special theory of relativity and applies to lengths of objects as well as to distance between objects. The result can be stated most simply in words as:

> **the length of an object moving relative to an observer is measured to be shorter along its direction of motion than when it is at rest.**

⚠ **CAUTION**

Proper length is measured in reference frame where the two positions are at rest

This is called **length contraction**. The length ℓ_0 in Eqs. 26–3 is called the **proper length**. It is the length of the object (or distance between two points whose positions are measured at the same time) as determined by *observers at rest* with respect to the object. Equations 26–3 give the length ℓ that will be measured by observers when the object travels past them at speed v.

[†]We assume v is much greater than the relative speed of Neptune and Earth (which we thus ignore).

It is important to note that length contraction occurs *only along the direction of motion*. For example, the moving spaceship in Fig. 26–8a is shortened in length, but its height is the same as when it is at rest.

Length contraction, like time dilation, is not noticeable in everyday life because the factor $\sqrt{1 - v^2/c^2}$ in Eq. 26–3a differs significantly from 1.00 only when v is very large.

EXAMPLE 26–5 **Painting's contraction.** A rectangular painting measures 1.00 m tall and 1.50 m wide, Fig. 26–9a. It is hung on the side wall of a spaceship which is moving past the Earth at a speed of $0.90c$. (*a*) What are the dimensions of the picture according to the captain of the spaceship? (*b*) What are the dimensions as seen by an observer on the Earth?

APPROACH We apply the length contraction formula, Eq. 26–3a, to the dimension parallel to the motion; v is the speed of the painting relative to the Earth observer.

SOLUTION (*a*) The painting is at rest ($v = 0$) on the spaceship so it (as well as everything else in the spaceship) looks perfectly normal to everyone on the spaceship. The captain sees a 1.00-m by 1.50-m painting.

(*b*) Only the dimension in the direction of motion is shortened, so the height is unchanged at 1.00 m, Fig. 26–9b. The length, however, is contracted to

$$\ell = \ell_0 \sqrt{1 - \frac{v^2}{c^2}}$$
$$= (1.50 \text{ m})\sqrt{1 - (0.90)^2} = 0.65 \text{ m}.$$

So the picture has dimensions $1.00 \text{ m} \times 0.65 \text{ m}$ to an observer on Earth.

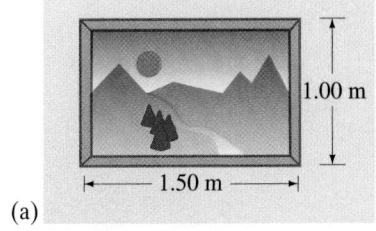
(a)

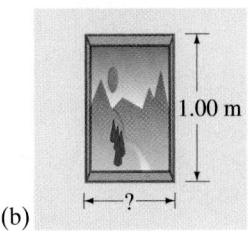
(b)

FIGURE 26–9 Example 26–5.

EXAMPLE 26–6 **A fantasy supertrain.** A very fast train with a "proper length" of $\ell_0 = 500$ m (measured by people at rest on the train) is passing through a tunnel that is 200 m long according to observers on the ground. Let us imagine the train's speed to be so great that the train fits completely within the tunnel as seen by observers on the ground. That is, the engine is just about to emerge from one end of the tunnel at the time the last car disappears into the other end. What is the train's speed?

APPROACH Since the train just fits inside the tunnel, its length measured by the person on the ground is $\ell = 200$ m. The length contraction formula, Eq. 26–3a or b, can thus be used to solve for v.

SOLUTION Substituting $\ell = 200$ m and $\ell_0 = 500$ m into Eq. 26–3a gives

$$200 \text{ m} = 500 \text{ m} \sqrt{1 - \frac{v^2}{c^2}};$$

dividing both sides by 500 m and squaring, we get

$$(0.40)^2 = 1 - \frac{v^2}{c^2}$$

or

$$\frac{v}{c} = \sqrt{1 - (0.40)^2}$$

and

$$v = 0.92c.$$

NOTE No real train could go this fast. But it is fun to think about.

NOTE An observer on the *train* would *not* see the two ends of the train inside the tunnel at the same time. Recall that observers moving relative to each other do not agree about simultaneity. (See Example 26–7, next.)

EXERCISE C What is the length of the tunnel as measured by observers on the train in Example 26–6?

Observers at rest on the Earth see a very fast 200-m-long train pass through a 200-m-long tunnel (as in Example 26–6) so that the train momentarily disappears from view inside the tunnel. Observers on the train measure the train's length to be 500 m and the tunnel's length to be only 80 m (Exercise C, using Eq. 26–3a). Clearly a 500-m-long train cannot fit inside an 80-m-long tunnel. How is this apparent inconsistency explained?

RESPONSE Events simultaneous in one reference frame may not be simultaneous in another. Let the engine emerging from one end of the tunnel be "event A," and the last car disappearing into the other end of the tunnel "event B." To observers in the Earth frame, events A and B are simultaneous. To observers on the train, however, the events are not simultaneous. In the train's frame, event A occurs before event B. As the engine emerges from the tunnel, observers on the train observe the last car as still 500 m − 80 m = 420 m from the entrance to the tunnel.

26–6 Four-Dimensional Space–Time

Let us imagine a person is on a train moving at a very high speed, say 0.65c, Fig. 26–10. This person begins a meal at 7:00 and finishes at 7:15, according to a clock on the train. The two events, beginning and ending the meal, take place at the same point on the train, so the "proper time" between these two events is 15 min. To observers on Earth, the plate is moving and the meal will take longer—20 min according to Eqs. 26–1. Let us assume that the meal was served on a 20-cm-diameter plate (its "proper length"). To observers on the Earth, the plate is moving and is only 15 cm wide (length contraction). Thus, to observers on the Earth, the meal looks smaller but lasts longer.

FIGURE 26–10 According to an accurate clock on a fast-moving train, a person (a) begins dinner at 7:00 and (b) finishes at 7:15. At the beginning of the meal, two observers on Earth set their watches to correspond with the clock on the train. These observers measure the eating time as 20 minutes.

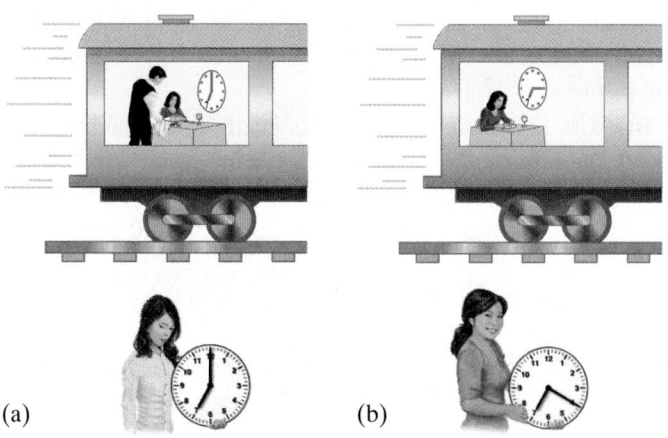

(a)   (b)

In a sense the two effects, time dilation and length contraction, balance each other. When viewed from the Earth, what an object seems to lose in size it gains in length of time it lasts. Space, or length, is exchanged for time.

Considerations like this led to the idea of **four-dimensional space–time**: space takes up three dimensions and time is a fourth dimension. Space and time are intimately connected. Just as when we squeeze a balloon we make one dimension larger and another smaller, so when we examine objects and events from different reference frames, a certain amount of space is exchanged for time, or vice versa.

Although the idea of four dimensions may seem strange, it refers to the idea that any object or event is specified by four quantities—three to describe where in space, and one to describe when in time. The really unusual aspect of four-dimensional space–time is that space and time can intermix: a little of one can be exchanged for a little of the other when the reference frame is changed.

[In Galilean–Newtonian relativity, the time interval between two events, Δt, and the distance between two events or points, Δx, are invariant quantities no matter what inertial reference frame they are viewed from. Neither of these quantities is invariant according to Einstein's relativity. But there is an invariant quantity in four-dimensional space–time, called the **space–time interval**, which is $(\Delta s)^2 = (c\,\Delta t)^2 - (\Delta x)^2$.]

26–7 Relativistic Momentum

So far in this Chapter, we have seen that two basic mechanical quantities, length and time intervals, need modification because they are relative—their value depends on the reference frame from which they are measured. We might expect that other physical quantities might need some modification according to the theory of relativity, such as momentum and energy.

The analysis of collisions between two particles shows that if we want to preserve the law of conservation of momentum in relativity, we must redefine momentum as

$$p = \frac{mv}{\sqrt{1 - v^2/c^2}} = \gamma mv. \qquad (26\text{–}4)$$

Here γ is shorthand for $1/\sqrt{1 - v^2/c^2}$ as before (Eq. 26–2). For speeds much less than the speed of light, Eq. 26–4 gives the classical momentum, $p = mv$.

Relativistic momentum has been tested many times on tiny elementary particles (such as muons), and it has been found to behave in accord with Eq. 26–4.

EXAMPLE 26–8 **Momentum of moving electron.** Compare the momentum of an electron to its classical value when it has a speed of (a) 4.00×10^7 m/s in the CRT of an old TV set, and (b) $0.98c$ in an accelerator used for cancer therapy.

APPROACH We use Eq. 26–4 for the momentum of a moving electron.

SOLUTION (a) At $v = 4.00 \times 10^7$ m/s, the electron's momentum is

$$p = \frac{mv}{\sqrt{1 - \dfrac{v^2}{c^2}}} = \frac{mv}{\sqrt{1 - \dfrac{(4.00 \times 10^7\,\text{m/s})^2}{(3.00 \times 10^8\,\text{m/s})^2}}} = 1.01mv.$$

The factor $\gamma = 1/\sqrt{1 - v^2/c^2} \approx 1.01$, so the momentum is only about 1% greater than the classical value. (If we put in the mass of an electron, $m = 9.11 \times 10^{-31}$ kg, the momentum is $p = 1.01mv = 3.68 \times 10^{-23}$ kg·m/s, compared to 3.64×10^{-23} kg·m/s classically.)

(b) With $v = 0.98c$, the momentum is

$$p = \frac{mv}{\sqrt{1 - \dfrac{v^2}{c^2}}} = \frac{mv}{\sqrt{1 - \dfrac{(0.98c)^2}{c^2}}} = \frac{mv}{\sqrt{1 - (0.98)^2}} = 5.0mv.$$

An electron traveling at 98% the speed of light has $\gamma = 5.0$ and a momentum 5.0 times its classical value.

The relativistic definition of momentum, Eq. 26–4, has sometimes been interpreted as an increase in the mass of an object. In this interpretation, a particle can have a **relativistic mass**, m_{rel}, which increases with speed according to

$$m_{rel} = \frac{m}{\sqrt{1 - v^2/c^2}}.$$

In this "mass-increase" formula, m is referred to as the **rest mass** of the object. With this interpretation, *the mass of an object appears to increase as its speed increases*. But there are problems with relativistic mass. If we plug it into formulas like $F = ma$ or $KE = \frac{1}{2}mv^2$, we obtain formulas that do not agree with experiment. (If we write Newton's second law in its more general form, $\vec{F} = \Delta\vec{p}/\Delta t$, that would get a correct result.) Also, be careful *not* to think a mass acquires more particles or more molecules as its speed becomes very large. It doesn't. Today, most physicists prefer not to use relativistic mass, so an object has only one mass (its rest mass), and it is only the momentum that increases with speed.

Whenever we talk about the mass of an object, we will always mean its rest mass (a fixed value). [But see Problem 46.]

26–8 The Ultimate Speed

A basic result of the special theory of relativity is that the speed of an object cannot equal or exceed the speed of light. That the speed of light is a natural speed limit in the universe can be seen from any of Eqs. 26–1, 26–3, or 26–4. It is perhaps easiest to see from Eq. 26–4. As an object is accelerated to greater and greater speeds, its momentum becomes larger and larger. Indeed, if v were to equal c, the denominator in this equation would be zero, and the momentum would be infinite. To accelerate an object up to $v = c$ would thus require infinite energy, and so is not possible.

26–9 $E = mc^2$; Mass and Energy

If momentum needs to be modified to fit with relativity as we just saw in Eq. 26–4, then we might expect that energy would also need to be rethought. Indeed, Einstein not only developed a new formula for kinetic energy, but also found a new relation between mass and energy, and the startling idea that mass is a form of energy.

We start with the work-energy principle (Chapter 6), hoping it is still valid in relativity and will give verifiable results. That is, we assume the net work done on a particle is equal to its change in kinetic energy (KE). Using this principle, Einstein showed that at high speeds the formula $KE = \frac{1}{2}mv^2$ is not correct. Instead, Einstein showed that the kinetic energy of a particle of mass m traveling at speed v is given by

$$KE = \frac{mc^2}{\sqrt{1 - v^2/c^2}} - mc^2. \qquad \textbf{(26–5a)}$$

In terms of $\gamma = 1/\sqrt{1 - v^2/c^2}$ we can rewrite Eq. 26–5a as

$$KE = \gamma mc^2 - mc^2 = (\gamma - 1)mc^2. \qquad \textbf{(26–5b)}$$

Equation 26–5a requires some interpretation. The first term increases with the speed v of the particle. The second term, mc^2, is constant; it is called the **rest energy** of the particle, and represents a form of energy that a particle has even when at rest. Note that if a particle is at rest ($v = 0$) the first term in Eq. 26–5a becomes mc^2, so $KE = 0$ as it should.

We can rearrange Eq. 26–5b to get

$$\gamma mc^2 = mc^2 + \text{KE}.$$

We call γmc^2 the *total energy* E of the particle (assuming no potential energy), because it equals the rest energy plus the kinetic energy:

$$E = \text{KE} + mc^2. \tag{26–6a}$$

The total energy† can also be written, using Eqs. 26–5, as

$$E = \gamma mc^2 = \frac{mc^2}{\sqrt{1 - v^2/c^2}}. \tag{26–6b}$$

For a particle at rest in a given reference frame, KE is zero in Eq. 26–6a, so the total energy is its rest energy:

$$E = mc^2. \tag{26–7}$$

MASS RELATED TO ENERGY

Here we have Einstein's famous formula, $E = mc^2$. This formula mathematically relates the concepts of energy and mass. But if this idea is to have any physical meaning, then mass ought to be convertible to other forms of energy and vice versa. Einstein suggested that this might be possible, and indeed changes of mass to other forms of energy, and vice versa, have been experimentally confirmed countless times in nuclear and elementary particle physics. For example, an electron and a positron (= a positive electron, see Section 32–3) have often been observed to collide and disappear, producing pure electromagnetic radiation. The amount of electromagnetic energy produced is found to be exactly equal to that predicted by Einstein's formula, $E = mc^2$. The reverse process is also commonly observed in the laboratory: electromagnetic radiation under certain conditions can be converted into material particles such as electrons (see Section 27–6 on pair production). On a larger scale, the energy produced in nuclear power plants is a result of the loss in mass of the uranium fuel as it undergoes the process called fission (Chapter 31). Even the radiant energy we receive from the Sun is an example of $E = mc^2$; the Sun's mass is continually decreasing as it radiates electromagnetic energy outward.

The relation $E = mc^2$ is now believed to apply to all processes, although the changes are often too small to measure. That is, when the energy of a system changes by an amount ΔE, the mass of the system changes by an amount Δm given by

$$\Delta E = (\Delta m)(c^2). \tag{26–8}$$

In a nuclear reaction where an energy E is required or released, the masses of the reactants and the products will be different by $\Delta m = \Delta E/c^2$.

EXAMPLE 26–9 **Pion's kinetic energy.** A π^0 meson $(m = 2.4 \times 10^{-28}\,\text{kg})$ travels at a speed $v = 0.80c = 2.4 \times 10^8\,\text{m/s}$. What is its kinetic energy? Compare to a classical calculation.

APPROACH We use Eq. 26–5 and compare to $\frac{1}{2}mv^2$.

SOLUTION We substitute values into Eq. 26–5a

$$\begin{aligned}
\text{KE} &= mc^2\left(\frac{1}{\sqrt{1 - v^2/c^2}} - 1\right) \\
&= (2.4 \times 10^{-28}\,\text{kg})(3.0 \times 10^8\,\text{m/s})^2\left(\frac{1}{(1 - 0.64)^{\frac{1}{2}}} - 1\right) \\
&= 1.4 \times 10^{-11}\,\text{J}.
\end{aligned}$$

Notice that the units of mc^2 are $\text{kg} \cdot \text{m}^2/\text{s}^2$, which is the joule.

NOTE Classically $\text{KE} = \frac{1}{2}mv^2 = \frac{1}{2}(2.4 \times 10^{-28}\,\text{kg})(2.4 \times 10^8\,\text{m/s})^2 = 6.9 \times 10^{-12}\,\text{J}$, about half as much, but this is not a correct result. Note that $\frac{1}{2}\gamma mv^2$ also does not work.

PROBLEM SOLVING
Relativistic kinetic energy

†This is for a "free particle," without forces and potential energy. Potential energy terms can be added.

EXAMPLE 26–10 **Energy from nuclear decay.** The energy required or released in nuclear reactions and decays comes from a change in mass between the initial and final particles. In one type of radioactive decay (Chapter 30), an atom of uranium ($m = 232.03716\,\mathrm{u}$) decays to an atom of thorium ($m = 228.02874\,\mathrm{u}$) plus an atom of helium ($m = 4.00260\,\mathrm{u}$) where the masses given are in atomic mass units ($1\,\mathrm{u} = 1.6605 \times 10^{-27}\,\mathrm{kg}$). Calculate the energy released in this decay.

APPROACH The initial mass minus the total final mass gives the mass loss in atomic mass units (u); we convert that to kg, and multiply by c^2 to find the energy released, $\Delta E = \Delta m\, c^2$.

SOLUTION The initial mass is 232.03716 u, and after the decay the mass is $228.02874\,\mathrm{u} + 4.00260\,\mathrm{u} = 232.03134\,\mathrm{u}$, so there is a loss of mass of 0.00582 u. This mass, which equals $(0.00582\,\mathrm{u})(1.66 \times 10^{-27}\,\mathrm{kg}) = 9.66 \times 10^{-30}\,\mathrm{kg}$, is changed into energy. By $\Delta E = \Delta m\, c^2$, we have

$$\Delta E = (9.66 \times 10^{-30}\,\mathrm{kg})(3.0 \times 10^8\,\mathrm{m/s})^2$$

$$= 8.70 \times 10^{-13}\,\mathrm{J}.$$

Since $1\,\mathrm{MeV} = 1.60 \times 10^{-13}\,\mathrm{J}$ (Section 17–4), the energy released is 5.4 MeV.

In the tiny world of atoms and nuclei, it is common to quote energies in eV (electron volts) or multiples such as MeV ($10^6\,\mathrm{eV}$). Momentum (see Eq. 26–4) can be quoted in units of eV/c (or MeV/c). And mass can be quoted (from $E = mc^2$) in units of eV/c^2 (or MeV/c^2). Note the use of c to keep the units correct. The masses of the electron and the proton can be shown to be $0.511\,\mathrm{MeV}/c^2$ and $938\,\mathrm{MeV}/c^2$, respectively. For example, for the electron, $mc^2 = (9.11 \times 10^{-31}\,\mathrm{kg})(2.998 \times 10^8\,\mathrm{m/s})^2/(1.602 \times 10^{-13}\,\mathrm{J/MeV}) = 0.511\,\mathrm{MeV}$. See also the Table inside the front cover.

EXAMPLE 26–11 **A 1-TeV proton.** The Tevatron accelerator at Fermilab in Illinois can accelerate protons to a kinetic energy of $1.0\,\mathrm{TeV}$ ($10^{12}\,\mathrm{eV}$). What is the speed of such a proton?

APPROACH We solve the kinetic energy formula, Eq. 26–5a, for v.

SOLUTION The rest energy of a proton is $mc^2 = 938\,\mathrm{MeV}$ or $9.38 \times 10^8\,\mathrm{eV}$. Compared to the kinetic energy of $10^{12}\,\mathrm{eV}$, the rest energy can be neglected, so we simplify Eq. 26–5a to

$$\mathrm{KE} \approx \frac{mc^2}{\sqrt{1 - v^2/c^2}}.$$

We solve this for v in the following steps:

$$\sqrt{1 - \frac{v^2}{c^2}} = \frac{mc^2}{\mathrm{KE}};$$

$$1 - \frac{v^2}{c^2} = \left(\frac{mc^2}{\mathrm{KE}}\right)^2;$$

$$\frac{v^2}{c^2} = 1 - \left(\frac{mc^2}{\mathrm{KE}}\right)^2 = 1 - \left(\frac{9.38 \times 10^8\,\mathrm{eV}}{1.0 \times 10^{12}\,\mathrm{eV}}\right)^2;$$

$$v = \sqrt{1 - (9.38 \times 10^{-4})^2}\, c$$

$$= 0.99999956\, c.$$

So the proton is traveling at a speed very nearly equal to c.

At low speeds, $v \ll c$, the relativistic formula for kinetic energy reduces to the classical one, as we now show by using the binomial expansion (Appendix A): $(1 \pm x)^n = 1 \pm nx + \cdots$, keeping only two terms because $x = v/c$ is very much less than 1. With $n = -\frac{1}{2}$ we expand the square root in Eq. 26–5a

$$\text{KE} = mc^2 \left(\frac{1}{\sqrt{1 - v^2/c^2}} - 1 \right)$$

so that

$$\text{KE} \approx mc^2 \left(1 + \frac{1}{2} \frac{v^2}{c^2} + \cdots - 1 \right) \approx \tfrac{1}{2} mv^2.$$

The dots in the first expression represent very small terms in the expansion which we neglect since we assumed that $v \ll c$. Thus at low speeds, the relativistic form for kinetic energy reduces to the classical form, $\text{KE} = \frac{1}{2} mv^2$. This makes relativity a viable theory in that it can predict accurate results at low speed as well as at high. Indeed, the other equations of special relativity also reduce to their classical equivalents at ordinary speeds: length contraction, time dilation, and modifications to momentum as well as kinetic energy, all disappear for $v \ll c$ since $\sqrt{1 - v^2/c^2} \approx 1$.

A useful relation between the total energy E of a particle and its momentum p can also be derived. The momentum of a particle of mass m and speed v is given by Eq. 26–4

$$p = \gamma mv = \frac{mv}{\sqrt{1 - v^2/c^2}}.$$

The total energy is

$$E = \text{KE} + mc^2$$

or

$$E = \gamma mc^2 = \frac{mc^2}{\sqrt{1 - v^2/c^2}}.$$

We square this equation (and we insert "$v^2 - v^2$" which is zero, but will help us):

$$E^2 = \frac{m^2 c^2 c^2}{1 - v^2/c^2} = \frac{m^2 c^2 (c^2 - v^2 + v^2)}{1 - v^2/c^2} = \frac{m^2 c^2 v^2}{1 - v^2/c^2} + \frac{m^2 c^2 (c^2 - v^2)}{1 - v^2/c^2}$$

$$= p^2 c^2 + \frac{m^2 c^4 (1 - v^2/c^2)}{1 - v^2/c^2}$$

or

$$E^2 = p^2 c^2 + m^2 c^4. \tag{26–9}$$

Thus, the total energy can be written in terms of the momentum p, or in terms of the kinetic energy (Eq. 26–6a), where we have assumed there is no potential energy.

*Invariant Energy–Momentum

We can rewrite Eq. 26–9 as $E^2 - p^2 c^2 = m^2 c^4$. Since the mass m of a given particle is the same in any reference frame, we see that the quantity $E^2 - p^2 c^2$ must also be the same in any reference frame. Thus, at any given moment the total energy E and momentum p of a particle will be different in different reference frames, but the quantity $E^2 - p^2 c^2$ will have the same value in all inertial reference frames. We say that the quantity $E^2 - p^2 c^2$ is **invariant**.

When Do We Use Relativistic Formulas?

From a practical point of view, we do not have much opportunity in our daily lives to use the mathematics of relativity. For example, the γ factor, $\gamma = 1/\sqrt{1 - v^2/c^2}$, has a value of 1.005 when $v = 0.10c$. Thus, for speeds even as high as $0.10c = 3.0 \times 10^7$ m/s, the factor $\sqrt{1 - v^2/c^2}$ in relativistic formulas gives a numerical correction of less than 1%. For speeds less than $0.10c$, or unless mass and energy are interchanged, we don't usually need the more complicated relativistic formulas, and can use the simpler classical formulas.

If you are given a particle's mass m and its kinetic energy KE, you can do a quick calculation to determine if you need to use relativistic formulas or if classical ones are good enough. You simply compute the ratio KE/mc^2 because (Eq. 26–5b)

$$\frac{KE}{mc^2} = \gamma - 1 = \frac{1}{\sqrt{1 - v^2/c^2}} - 1.$$

If this ratio comes out to be less than, say, 0.01, then $\gamma \le 1.01$ and relativistic equations will correct the classical ones by about 1%. If your expected precision is no better than 1%, classical formulas are good enough. But if your precision is 1 part in 1000 (0.1%) then you would want to use relativistic formulas. If your expected precision is only 10%, you need relativity if $(KE/mc^2) \gtrsim 0.1$.

EXERCISE D For 1% accuracy, does an electron with $KE = 100$ eV need to be treated relativistically? [*Hint*: The mass of an electron is 0.511 MeV.]

26–10 Relativistic Addition of Velocities

Consider a rocket ship that travels away from the Earth with speed v, and assume that this rocket has fired off a second rocket that travels at speed u' with respect to the first (Fig. 26–11). We might expect that the speed u of rocket 2 with respect to Earth is $u = v + u'$, which in the case shown in Fig. 26–11 is $u = 0.60c + 0.60c = 1.20c$. But, as discussed in Section 26–8, no object can travel faster than the speed of light in any reference frame. Indeed, Einstein showed that since length and time are different in different reference frames, the classical addition-of-velocities formula is no longer valid. Instead, the correct formula is

$$u = \frac{v + u'}{1 + vu'/c^2} \qquad \left[\begin{array}{l}\bar{\mathbf{u}} \text{ and } \bar{\mathbf{v}} \text{ along} \\ \text{the same direction}\end{array}\right] \quad (26\text{--}10)$$

for motion along a straight line. We derive this formula in Appendix E. If u' is in the opposite direction from v, then u' must have a minus sign in the above equation so $u = (v - u')/(1 - vu'/c^2)$.

EXAMPLE 26–12 **Relative velocity, relativistically.** Calculate the speed of rocket 2 in Fig. 26–11 with respect to Earth.

APPROACH We combine the speed of rocket 2 relative to rocket 1 with the speed of rocket 1 relative to Earth, using the relativistic Eq. 26–10 because the speeds are high and they are along the same line.

SOLUTION Rocket 2 moves with speed $u' = 0.60c$ with respect to rocket 1. Rocket 1 has speed $v = 0.60c$ with respect to Earth. The speed of rocket 2 with respect to Earth is (Eq. 26–10)

$$u = \frac{0.60c + 0.60c}{1 + \dfrac{(0.60c)(0.60c)}{c^2}} = \frac{1.20c}{1.36} = 0.88c.$$

NOTE The speed of rocket 2 relative to Earth is less than c, as it must be.

We can see that Eq. 26–10 reduces to the classical form for velocities small compared to the speed of light since $1 + vu'/c^2 \approx 1$ for v and $u' \ll c$. Thus, $u \approx v + u'$, as in classical physics (Chapter 3).

Let us test our formula at the other extreme, that of the speed of light. Suppose that rocket 1 in Fig. 26–11 sends out a beam of light so that $u' = c$. Equation 26–10 tells us that the speed of this light relative to Earth is

$$u = \frac{0.60c + c}{1 + \dfrac{(0.60c)(c)}{c^2}} = \frac{1.60c}{1.60} = c,$$

which is fully consistent with the second postulate of relativity.

EXERCISE E Use Eq. 26–10 to calculate the speed of rocket 2 in Fig. 26–11 relative to Earth if it was shot from rocket 1 at a speed $u' = 3000$ km/s = $0.010c$. Assume rocket 1 had a speed $v = 6000$ km/s = $0.020c$.

EXERCISE F Return to the Chapter-Opening Question, page 744, and answer it again now. Try to explain why you may have answered differently the first time.

⚠ CAUTION

Relative velocities do not add simply, as in classical mechanics ($v \ll c$)

Relativistic addition of velocities formula ($\bar{\mathbf{u}}$ and $\bar{\mathbf{v}}$ along same line)

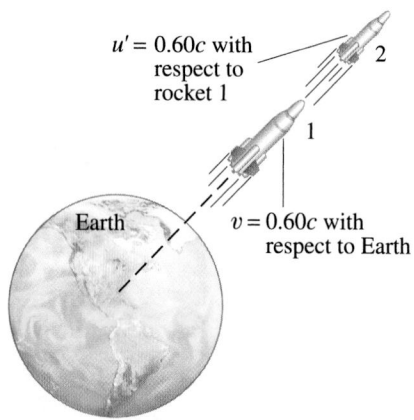

$u' = 0.60c$ with respect to rocket 1

Earth

$v = 0.60c$ with respect to Earth

FIGURE 26–11 Rocket 1 leaves Earth at $v = 0.60c$. Rocket 2 is fired from rocket 1 with speed $u' = 0.60c$. What is the speed of rocket 2 with respect to the Earth? Example 26–12.

26–11 The Impact of Special Relativity

A great many experiments have been performed to test the predictions of the special theory of relativity. Within experimental error, no contradictions have been found. Scientists have therefore accepted relativity as an accurate description of nature.

At speeds much less than the speed of light, the relativistic formulas reduce to the old classical ones, as we have discussed. We would, of course, hope—or rather, insist—that this be true since Newtonian mechanics works so well for objects moving with speeds $v \ll c$. This insistence that a more general theory (such as relativity) give the same results as a more restricted theory (such as classical mechanics which works for $v \ll c$) is called the **correspondence principle**. The two theories must correspond where their realms of validity overlap. Relativity thus does not contradict classical mechanics. Rather, it is a more general theory, of which classical mechanics is now considered to be a limiting case.

The importance of relativity is not simply that it gives more accurate results, especially at very high speeds. Much more than that, it has changed the way we view the world. The concepts of space and time are now seen to be relative, and intertwined with one another, whereas before they were considered absolute and separate. Even our concepts of matter and energy have changed: either can be converted to the other. The impact of relativity extends far beyond physics. It has influenced the other sciences, and even the world of art and literature; it has, indeed, entered the general culture.

The special theory of relativity we have studied in this Chapter deals with inertial (nonaccelerating) reference frames. In Chapter 33 we will discuss briefly the more complicated "general theory of relativity" which can deal with non-inertial reference frames.

Summary

An **inertial reference frame** is one in which Newton's law of inertia holds. Inertial reference frames move at constant velocity relative to one another. Accelerating reference frames are **noninertial**.

The **special theory of relativity** is based on two principles: the **relativity principle**, which states that the laws of physics are the same in all inertial reference frames, and the principle of the **constancy of the speed of light**, which states that the speed of light in empty space has the same value in all inertial reference frames.

One consequence of relativity theory is that two events that are simultaneous in one reference frame may not be simultaneous in another. Other effects are **time dilation**: moving clocks are measured to run slow; and **length contraction**: the length of a moving object is measured to be shorter (in its direction of motion) than when it is at rest. Quantitatively,

$$\Delta t = \frac{\Delta t_0}{\sqrt{1 - v^2/c^2}} = \gamma \, \Delta t_0 \qquad \text{(26–1)}$$

$$\ell = \ell_0 \sqrt{1 - v^2/c^2} = \frac{\ell_0}{\gamma} \qquad \text{(26–3)}$$

where ℓ and Δt are the length and time interval of objects (or events) observed as they move by at the speed v; ℓ_0 and Δt_0 are the **proper length** and **proper time**—that is, the same quantities as measured in the rest frame of the objects or events. The quantity γ is shorthand for

$$\gamma = \frac{1}{\sqrt{1 - v^2/c^2}}. \qquad \text{(26–2)}$$

The theory of relativity has changed our notions of space and time, and of momentum, energy, and mass. Space and time are seen to be intimately connected, with time being the fourth dimension in addition to the three dimensions of space.

The **momentum** of an object is given by

$$p = \gamma m v = \frac{mv}{\sqrt{1 - v^2/c^2}}. \qquad \text{(26–4)}$$

Mass and energy are interconvertible. The equation

$$E = mc^2 \qquad \text{(26–7)}$$

tells how much energy E is needed to create a mass m, or vice versa. Said another way, $E = mc^2$ is the amount of energy an object has because of its mass m. The law of conservation of energy must include mass as a form of energy.

The kinetic energy KE of an object moving at speed v is given by

$$\text{KE} = \frac{mc^2}{\sqrt{1 - v^2/c^2}} - mc^2 = (\gamma - 1)mc^2 \qquad \text{(26–5)}$$

where m is the mass of the object. The total energy E, if there is no potential energy, is

$$\begin{aligned} E &= \text{KE} + mc^2 \\ &= \gamma mc^2. \end{aligned} \qquad \text{(26–6)}$$

The momentum p of an object is related to its total energy E (assuming no potential energy) by

$$E^2 = p^2 c^2 + m^2 c^4. \qquad \text{(26–9)}$$

Velocity addition also must be done in a special way. All these relativistic effects are significant only at high speeds, close to the speed of light, which itself is the ultimate speed in the universe.

Questions

1. You are in a windowless car in an exceptionally smooth train moving at constant velocity. Is there any physical experiment you can do in the train car to determine whether you are moving? Explain.

2. You might have had the experience of being at a red light when, out of the corner of your eye, you see the car beside you creep forward. Instinctively you stomp on the brake pedal, thinking that you are rolling backward. What does this say about absolute and relative motion?

3. A worker stands on top of a railroad car moving at constant velocity and throws a heavy ball straight up (from his point of view). Ignoring air resistance, explain whether the ball will land back in his hand or behind him.

4. Does the Earth really go around the Sun? Or is it also valid to say that the Sun goes around the Earth? Discuss in view of the relativity principle (that there is no best reference frame). Explain. See Section 5–8.

5. If you were on a spaceship traveling at $0.6c$ away from a star, at what speed would the starlight pass you?

6. The time dilation effect is sometimes expressed as "moving clocks run slowly." Actually, this effect has nothing to do with motion affecting the functioning of clocks. What then does it deal with?

7. Does time dilation mean that time actually passes more slowly in moving reference frames or that it only *seems* to pass more slowly?

8. A young-looking woman astronaut has just arrived home from a long trip. She rushes up to an old gray-haired man and in the ensuing conversation refers to him as her son. How might this be possible?

9. If you were traveling away from Earth at speed $0.6c$, would you notice a change in your heartbeat? Would your mass, height, or waistline change? What would observers on Earth using telescopes say about you?

10. Do time dilation and length contraction occur at ordinary speeds, say 90 km/h?

11. Suppose the speed of light were infinite. What would happen to the relativistic predictions of length contraction and time dilation?

12. Explain how the length contraction and time dilation formulas might be used to indicate that c is the limiting speed in the universe.

13. Discuss how our everyday lives would be different if the speed of light were only 25 m/s.

14. The drawing at the start of this Chapter shows the street as seen by Mr Tompkins, for whom the speed of light is $c = 20$ mi/h. What does Mr Tompkins look like to the people standing on the street (Fig. 26–12)? Explain.

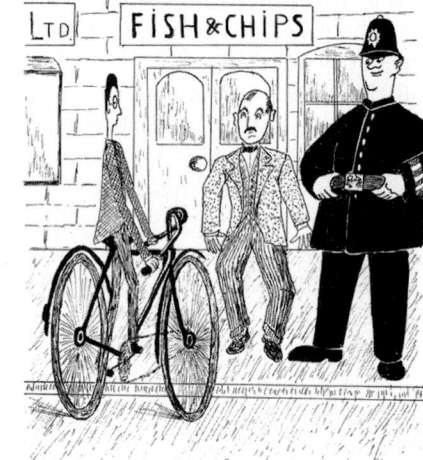

FIGURE 26–12
Question 14.
Mr Tompkins as seen by people on the sidewalk. See also Chapter-Opening Figure on page 744.

15. An electron is limited to travel at speeds less than c. Does this put an upper limit on the momentum of an electron? If so, what is this upper limit? If not, explain.

16. Can a particle of nonzero mass attain the speed of light? Explain.

17. Does the equation $E = mc^2$ conflict with the conservation of energy principle? Explain.

18. If mass is a form of energy, does this mean that a spring has more mass when compressed than when relaxed? Explain.

19. It is not correct to say that "matter can neither be created nor destroyed." What must we say instead?

20. Is our intuitive notion that velocities simply add, as in Section 3–8, completely wrong?

MisConceptual Questions

1. The fictional rocket ship *Adventure* is measured to be 50 m long by the ship's captain inside the rocket. When the rocket moves past a space dock at $0.5c$, space-dock personnel measure the rocket ship to be 43.3 m long. What is its proper length?
   (*a*) 50 m.   (*b*) 43.3 m.   (*c*) 93.3 m.   (*d*) 13.3 m.

2. As rocket ship *Adventure* (MisConceptual Question 1) passes by the space dock, the ship's captain flashes a flashlight at 1.00-s intervals as measured by space-dock personnel. How often does the flashlight flash relative to the captain?
   (*a*) Every 1.15 s.   (*b*) Every 1.00 s.   (*c*) Every 0.87 s.
   (*d*) We need to know the distance between the ship and the space dock.

3. For the flashing of the flashlight in MisConceptual Question 2, what time interval is the proper time interval?
   (*a*) 1.15 s.   (*b*) 1.00 s.   (*c*) 0.87 s.   (*d*) 0.13 s.

4. The rocket ship of MisConceptual Question 1 travels to a star many light-years away, then turns around and returns at the same speed. When it returns to the space dock, who would have aged less: the space-dock personnel or ship's captain?
   (*a*) The space-dock personnel.
   (*b*) The ship's captain.
   (*c*) Both the same amount, because both sets of people were moving relative to each other.
   (*d*) We need to know how far away the star is.

5. An Earth observer notes that clocks on a passing spacecraft run slowly. The person on the spacecraft
   (*a*) agrees her clocks move slower than those on Earth.
   (*b*) feels normal, and her heartbeat and eating habits are normal.
   (*c*) observes that Earth clocks are moving slowly.
   (*d*) The real time is in between the times measured by the two observers.
   (*e*) Both (*a*) and (*b*).
   (*f*) Both (*b*) and (*c*).

6. Spaceships A and B are traveling directly toward each other at a speed $0.5c$ relative to the Earth, and each has a headlight aimed toward the other ship. What value do technicians on ship B get by measuring the speed of the light emitted by ship A's headlight?
   (*a*) $0.5c$. (*b*) $0.75c$. (*c*) $1.0c$. (*d*) $1.5c$.

7. Relativistic formulas for time dilation, length contraction, and mass are valid
   (*a*) only for speeds less than $0.10c$.
   (*b*) only for speeds greater than $0.10c$.
   (*c*) only for speeds very close to c.
   (*d*) for all speeds.

8. Which of the following will two observers in inertial reference frames always agree on? (Choose all that apply.)
   (*a*) The time an event occurred.
   (*b*) The distance between two events.
   (*c*) The time interval between the occurence of two events.
   (*d*) The speed of light.
   (*e*) The validity of the laws of physics.
   (*f*) The simultaneity of two events.

9. Two observers in different inertial reference frames moving relative to each other at nearly the speed of light see the same two events but, using precise equipment, record different time intervals between the two events. Which of the following is true of their measurements?
   (*a*) One observer is incorrect, but it is impossible to tell which one.
   (*b*) One observer is incorrect, and it is possible to tell which one.
   (*c*) Both observers are incorrect.
   (*d*) Both observers are correct.

10. You are in a rocket ship going faster and faster. As your speed increases and your velocity gets closer to the speed of light, which of the following do you observe in your frame of reference?
    (*a*) Your mass increases.
    (*b*) Your length shortens in the direction of motion.
    (*c*) Your wristwatch slows down.
    (*d*) All of the above.
    (*e*) None of the above.

11. You are in a spaceship with no windows, radios, or other means to check outside. How could you determine whether your spaceship is at rest or moving at constant velocity?
    (*a*) By determining the apparent velocity of light in the spaceship.
    (*b*) By checking your precision watch. If it's running slow, then the ship is moving.
    (*c*) By measuring the lengths of objects in the spaceship. If they are shortened, then the ship is moving.
    (*d*) Give up, because you can't tell.

12. The period of a pendulum attached in a spaceship is 2 s while the spaceship is parked on Earth. What is the period to an observer on Earth when the spaceship moves at $0.6c$ with respect to the Earth?
    (*a*) Less than 2 s.
    (*b*) More than 2 s.
    (*c*) 2 s.

13. Two spaceships, each moving at a speed $0.75c$ relative to the Earth, are headed directly toward each other. What do occupants of one ship measure the speed of other ship to be?
    (*a*) $0.96c$. (*b*) $1.0c$. (*c*) $1.5c$. (*d*) $1.75c$. (*e*) $0.75c$.

For assigned homework and other learning materials, go to the MasteringPhysics website.

Problems

26–4 and 26–5 Time Dilation, Length Contraction

1. (I) A spaceship passes you at a speed of $0.850c$. You measure its length to be 44.2 m. How long would it be when at rest?

2. (I) A certain type of elementary particle travels at a speed of 2.70×10^8 m/s. At this speed, the average lifetime is measured to be 4.76×10^{-6} s. What is the particle's lifetime at rest?

3. (II) You travel to a star 135 light-years from Earth at a speed of 2.90×10^8 m/s. What do you measure this distance to be?

4. (II) What is the speed of a pion if its average lifetime is measured to be 4.40×10^{-8} s? At rest, its average lifetime is 2.60×10^{-8} s.

5. (II) In an Earth reference frame, a star is 49 light-years away. How fast would you have to travel so that to you the distance would be only 35 light-years?

6. (II) At what speed v will the length of a 1.00-m stick look 10.0% shorter (90.0 cm)?

7. (II) At what speed do the relativistic formulas for (*a*) length and (*b*) time intervals differ from classical values by 1.00%? (This is a reasonable way to estimate when to use relativistic calculations rather than classical.)

8. (II) You decide to travel to a star 62 light-years from Earth at a speed that tells you the distance is only 25 light-years. How many years would it take you to make the trip?

9. (II) A friend speeds by you in her spacecraft at a speed of $0.720c$. It is measured in your frame to be 4.80 m long and 1.35 m high. (*a*) What will be its length and height at rest? (*b*) How many seconds elapsed on your friend's watch when 20.0 s passed on yours? (*c*) How fast did you appear to be traveling according to your friend? (*d*) How many seconds elapsed on your watch when she saw 20.0 s pass on hers?

10. (II) A star is 21.6 light-years from Earth. How long would it take a spacecraft traveling $0.950c$ to reach that star as measured by observers: (*a*) on Earth, (*b*) on the spacecraft? (*c*) What is the distance traveled according to observers on the spacecraft? (*d*) What will the spacecraft occupants compute their speed to be from the results of (*b*) and (*c*)?

11. (II) A fictional news report stated that starship *Enterprise* had just returned from a 5-year voyage while traveling at $0.70c$. (*a*) If the report meant 5.0 years of *Earth time*, how much time elapsed on the ship? (*b*) If the report meant 5.0 years of *ship time*, how much time passed on Earth?

12. (II) A box at rest has the shape of a cube 2.6 m on a side. This box is loaded onto the flat floor of a spaceship and the spaceship then flies past us with a horizontal speed of $0.80c$. What is the volume of the box as we observe it?

13. (III) Escape velocity from the Earth is 11.2 km/s. What would be the percent decrease in length of a 68.2-m-long spacecraft traveling at that speed as seen from Earth?

14. (III) An unstable particle produced in an accelerator experiment travels at constant velocity, covering 1.00 m in 3.40 ns in the lab frame before changing ("decaying") into other particles. In the rest frame of the particle, determine (a) how long it lived before decaying, (b) how far it moved before decaying.

15. (III) How fast must a pion be moving on average to travel 32 m before it decays? The average lifetime, at rest, is 2.6×10^{-8} s.

26–7 Relativistic Momentum

16. (I) What is the momentum of a proton traveling at $v = 0.68c$?

17. (II) (a) A particle travels at $v = 0.15c$. By what percentage will a calculation of its momentum be wrong if you use the classical formula? (b) Repeat for $v = 0.75c$.

18. (II) A particle of mass m travels at a speed $v = 0.22c$. At what speed will its momentum be doubled?

19. (II) An unstable particle is at rest and suddenly decays into two fragments. No external forces act on the particle or its fragments. One of the fragments has a speed of $0.60c$ and a mass of 6.68×10^{-27} kg, while the other has a mass of 1.67×10^{-27} kg. What is the speed of the less massive fragment?

20. (II) What is the percent change in momentum of a proton that accelerates from (a) $0.45c$ to $0.85c$, (b) $0.85c$ to $0.98c$?

26–9 $E = mc^2$; Mass and Energy

21. (I) Calculate the rest energy of an electron in joules and in MeV $(1 \text{ MeV} = 1.60 \times 10^{-13} \text{ J})$.

22. (I) When a uranium nucleus at rest breaks apart in the process known as *fission* in a nuclear reactor, the resulting fragments have a total kinetic energy of about 200 MeV. How much mass was lost in the process?

23. (I) The total annual energy consumption in the United States is about 1×10^{20} J. How much mass would have to be converted to energy to fuel this need?

24. (I) Calculate the mass of a proton $(1.67 \times 10^{-27} \text{ kg})$ in MeV/c^2.

25. (I) A certain chemical reaction requires 4.82×10^4 J of energy input for it to go. What is the increase in mass of the products over the reactants?

26. (II) Calculate the kinetic energy and momentum of a proton traveling 2.90×10^8 m/s.

27. (II) What is the momentum of a 950-MeV proton (that is, its kinetic energy is 950 MeV)?

28. (II) What is the speed of an electron whose kinetic energy is 1.12 MeV?

29. (II) (a) How much work is required to accelerate a proton from rest up to a speed of $0.985c$? (b) What would be the momentum of this proton?

30. (II) At what speed will an object's kinetic energy be 33% of its rest energy?

31. (II) Determine the speed and the momentum of an electron $(m = 9.11 \times 10^{-31} \text{ kg})$ whose KE equals its rest energy.

32. (II) A proton is traveling in an accelerator with a speed of 1.0×10^8 m/s. By what factor does the proton's kinetic energy increase if its speed is doubled?

33. (II) How much energy can be obtained from conversion of 1.0 gram of mass? How much mass could this energy raise to a height of 1.0 km above the Earth's surface?

34. (II) To accelerate a particle of mass m from rest to speed $0.90c$ requires work W_1. To accelerate the particle from speed $0.90c$ to $0.99c$ requires work W_2. Determine the ratio W_2/W_1.

35. (II) Suppose there was a process by which two photons, each with momentum $0.65 \text{ MeV}/c$, could collide and make a single particle. What is the maximum mass that the particle could possess?

36. (II) What is the speed of a proton accelerated by a potential difference of 165 MV?

37. (II) What is the speed of an electron after being accelerated from rest by 31,000 V?

38. (II) The kinetic energy of a particle is 45 MeV. If the momentum is 121 MeV/c, what is the particle's mass?

39. (II) Calculate the speed of a proton $(m = 1.67 \times 10^{-27} \text{ kg})$ whose kinetic energy is exactly half (a) its total energy, (b) its rest energy.

40. (II) Calculate the kinetic energy and momentum of a proton $(m = 1.67 \times 10^{-27} \text{ kg})$ traveling 8.65×10^7 m/s. By what percentages would your calculations have been in error if you had used classical formulas?

41. (II) Suppose a spacecraft of mass 17,000 kg is accelerated to $0.15c$. (a) How much kinetic energy would it have? (b) If you used the classical formula for kinetic energy, by what percentage would you be in error?

42. (II) A negative muon traveling at 53% the speed of light collides head on with a positive muon traveling at 65% the speed of light. The two muons (each of mass 105.7 MeV/c^2) annihilate, and produce how much electromagnetic energy?

43. (II) Two identical particles of mass m approach each other at equal and opposite speeds, v. The collision is completely inelastic and results in a single particle at rest. What is the mass of the new particle? How much energy was lost in the collision? How much kinetic energy was lost in this collision?

44. (III) The americium nucleus, $^{241}_{95}$Am, decays to a neptunium nucleus, $^{237}_{93}$Np, by emitting an alpha particle of mass 4.00260 u and kinetic energy 5.5 MeV. Estimate the mass of the neptunium nucleus, ignoring its recoil, given that the americium mass is 241.05682 u.

45. (III) Show that the kinetic energy KE of a particle of mass m is related to its momentum p by the equation

$$p = \sqrt{\text{KE}^2 + 2\text{KE } mc^2}/c.$$

*46. (III) What magnetic field B is needed to keep 998-GeV protons revolving in a circle of radius 1.0 km? Use the relativistic mass. The proton's "rest mass" is 0.938 GeV/c^2. $(1 \text{ GeV} = 10^9 \text{ eV}.)$ [Hint: In relativity, $m_{\text{rel}} v^2/r = qvB$ is still valid in a magnetic field, where $m_{\text{rel}} = \gamma m$.]

26–10 Relativistic Addition of Velocities

47. (I) A person on a rocket traveling at $0.40c$ (with respect to the Earth) observes a meteor come from behind and pass her at a speed she measures as $0.40c$. How fast is the meteor moving with respect to the Earth?

48. (II) Two spaceships leave Earth in opposite directions, each with a speed of $0.60c$ with respect to Earth. (a) What is the velocity of spaceship 1 relative to spaceship 2? (b) What is the velocity of spaceship 2 relative to spaceship 1?

49. (II) A spaceship leaves Earth traveling at 0.65c. A second spaceship leaves the first at a speed of 0.82c with respect to the first. Calculate the speed of the second ship with respect to Earth if it is fired (a) in the same direction the first spaceship is already moving, (b) directly backward toward Earth.

50. (II) An observer on Earth sees an alien vessel approach at a speed of 0.60c. The fictional starship *Enterprise* comes to the rescue (Fig. 26–13), overtaking the aliens while moving directly toward Earth at a speed of 0.90c relative to Earth. What is the relative speed of one vessel as seen by the other?

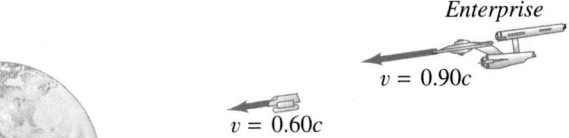

Enterprise

$v = 0.90c$

$v = 0.60c$

FIGURE 26–13 Problem 50.

51. (II) A spaceship in distress sends out two escape pods in opposite directions. One travels at a speed $v_1 = +0.70c$ in one direction, and the other travels at a speed $v_2 = -0.80c$ in the other direction, as observed from the spaceship. What speed does the first escape pod measure for the second escape pod?

52. (II) Rocket A passes Earth at a speed of 0.65c. At the same time, rocket B passes Earth moving 0.95c relative to Earth in the same direction as A. How fast is B moving relative to A when it passes A?

53. (II) Your spaceship, traveling at 0.90c, needs to launch a probe out the forward hatch so that its speed relative to the planet that you are approaching is 0.95c. With what speed must it leave your ship?

General Problems

54. What is the speed of a particle when its kinetic energy equals its rest energy? Does the mass of the particle affect the result?

55. The nearest star to Earth is Proxima Centauri, 4.3 light-years away. (a) At what constant velocity must a spacecraft travel from Earth if it is to reach the star in 4.9 years, as measured by travelers on the spacecraft? (b) How long does the trip take according to Earth observers?

56. According to the special theory of relativity, the factor γ that determines the length contraction and the time dilation is given by $\gamma = 1/\sqrt{1 - v^2/c^2}$. Determine the numerical values of γ for an object moving at speed $v = 0.01c, 0.05c, 0.10c, 0.20c, 0.30c, 0.40c, 0.50c, 0.60c, 0.70c, 0.80c, 0.90c, 0.95c,$ and $0.99c$. Make a graph of γ versus v.

57. A healthy astronaut's heart rate is 60 beats/min. Flight doctors on Earth can monitor an astronaut's vital signs remotely while in flight. How fast would an astronaut be flying away from Earth if the doctor measured her having a heart rate of 25 beats/min?

58. (a) What is the speed v of an electron whose kinetic energy is 14,000 times its rest energy? You can state the answer as the difference $c - v$. Such speeds are reached in the Stanford Linear Accelerator, SLAC. (b) If the electrons travel in the lab through a tube 3.0 km long (as at SLAC), how long is this tube in the electrons' reference frame? [*Hint:* Use the binomial expansion.]

59. What minimum amount of electromagnetic energy is needed to produce an electron and a positron together? A positron is a particle with the same mass as an electron, but has the opposite charge. (Note that electric charge is conserved in this process. See Section 27–6.)

60. How many grams of matter would have to be totally destroyed to run a 75-W lightbulb for 1.0 year?

61. A free neutron can decay into a proton, an electron, and a neutrino. Assume the neutrino's mass is zero; the other masses can be found in the Table inside the front cover. Determine the total kinetic energy shared among the three particles when a neutron decays at rest.

62. An electron $(m = 9.11 \times 10^{-31}\,\text{kg})$ is accelerated from rest to speed v by a conservative force. In this process, its potential energy decreases by $6.20 \times 10^{-14}\,\text{J}$. Determine the electron's speed, v.

63. The Sun radiates energy at a rate of about 4×10^{26} W. (a) At what rate is the Sun's mass decreasing? (b) How long does it take for the Sun to lose a mass equal to that of Earth? (c) Estimate how long the Sun could last if it radiated constantly at this rate.

64. How much energy would be required to break a helium nucleus into its constituents, two protons and two neutrons? The masses of a proton (including an electron), a neutron, and neutral helium are, respectively, 1.00783 u, 1.00867 u, and 4.00260 u. (This energy difference is called the *total binding energy* of the $^4_2$He nucleus.)

65. Show analytically that a particle with momentum p and energy E has a speed given by

$$v = \frac{pc^2}{E} = \frac{pc}{\sqrt{m^2c^2 + p^2}}.$$

66. Two protons, each having a speed of 0.990c in the laboratory, are moving toward each other. Determine (a) the momentum of each proton in the laboratory, (b) the total momentum of the two protons in the laboratory, and (c) the momentum of one proton as seen by the other proton.

67. When two moles of hydrogen molecules (H_2) and one mole of oxygen molecules (O_2) react to form two moles of water (H_2O), the energy released is 484 kJ. How much does the mass decrease in this reaction? What % of the total original mass is this?

68. The fictional starship *Enterprise* obtains its power by combining matter and antimatter, achieving complete conversion of mass into energy. If the mass of the *Enterprise* is approximately 6×10^9 kg, how much mass must be converted into kinetic energy to accelerate it from rest to one-tenth the speed of light?

69. Make a graph of the kinetic energy versus momentum for (a) a particle of nonzero mass, and (b) a particle with zero mass.

70. A spaceship and its occupants have a total mass of 160,000 kg. The occupants would like to travel to a star that is 35 light-years away at a speed of $0.70c$. To accelerate, the engine of the spaceship changes mass directly to energy. (a) Estimate how much mass will be converted to energy to accelerate the spaceship to this speed. (b) Assuming the acceleration is rapid, so the speed for the entire trip can be taken to be $0.70c$, determine how long the trip will take according to the astronauts on board.

71. In a nuclear reaction two identical particles are created, traveling in opposite directions. If the speed of each particle is $0.82c$, relative to the laboratory frame of reference, what is one particle's speed relative to the other particle?

72. A 36,000-kg spaceship is to travel to the vicinity of a star 6.6 light-years from Earth. Passengers on the ship want the (one-way) trip to take no more than 1.0 year. How much work must be done on the spaceship to bring it to the speed necessary for this trip?

73. Suppose a 14,500-kg spaceship left Earth at a speed of $0.90c$. What is the spaceship's kinetic energy? Compare with the total U.S. annual energy consumption (about 10^{20} J).

74. A pi meson of mass m_π decays at rest into a muon (mass m_μ) and a neutrino of negligible or zero mass. Show that the kinetic energy of the muon is $\mathrm{KE}_\mu = (m_\pi - m_\mu)^2 c^2/(2m_\pi)$.

75. An astronaut on a spaceship traveling at $0.75c$ relative to Earth measures his ship to be 23 m long. On the ship, he eats his lunch in 28 min. (a) What length is the spaceship according to observers on Earth? (b) How long does the astronaut's lunch take to eat according to observers on Earth?

76. Astronomers measure the distance to a particular star to be 6.0 light-years (1 ly = distance light travels in 1 year). A spaceship travels from Earth to the vicinity of this star at steady speed, arriving in 3.50 years as measured by clocks on the spaceship. (a) How long does the trip take as measured by clocks in Earth's reference frame? (b) What distance does the spaceship travel as measured in its own reference frame?

77. An electron is accelerated so that its kinetic energy is greater than its rest energy mc^2 by a factor of (a) 5.00, (b) 999. What is the speed of the electron in each case?

78. You are traveling in a spaceship at a speed of $0.70c$ away from Earth. You send a laser beam toward the Earth traveling at velocity c relative to you. What do observers on the Earth measure for the speed of the laser beam?

79. A farm boy studying physics believes that he can fit a 13.0-m-long pole into a 10.0-m-long barn if he runs fast enough, carrying the pole. Can he do it? Explain in detail. How does this fit with the idea that when he is running the barn looks even shorter than 10.0 m?

80. An atomic clock is taken to the North Pole, while another stays at the Equator. How far will they be out of synchronization after 2.0 years has elapsed? [Hint: Use the binomial expansion, Appendix A.]

81. An airplane travels 1300 km/h around the Earth in a circle of radius essentially equal to that of the Earth, returning to the same place. Using special relativity, estimate the difference in time to make the trip as seen by Earth and by airplane observers. [Hint: Use the binomial expansion, Appendix A.]

Search and Learn

1. Determine about how fast Mr Tompkins is traveling in the Chapter-Opening Photograph. Do you agree with the picture in terms of the way Mr Tompkins would see the world? Explain. [Hint: Assume the bank clock and Stop sign facing us are round according to the people on the sidewalk.]

2. Examine the experiment of Fig. 26–5 from O_1's reference frame. In this case, O_1 will be at rest and will see the lightning bolt at B_1 and B_2, before the lightning bolt at A_1 and A_2. Will O_1 recognize that O_2, who is moving with speed v to the left, will see the two events as simultaneous? Explain in detail, drawing diagrams equivalent to Fig. 26–5. [Hint: Include length contraction.]

3. Using Example 26–2 as a guide, show that for objects that move slowly in comparison to c, the length contraction formula is roughly $\ell \approx \ell_0(1 - \frac{1}{2}v^2/c^2)$. Use this approximation to find the "length shortening" $\Delta\ell = \ell_0 - \ell$ of the train in Example 26–6 if the train travels at 100 km/h (rather than $0.92c$).

4. In Example 26–5, the spaceship is moving at $0.90c$ in the horizontal direction relative to an observer on the Earth. If instead the spaceship moved at $0.90c$ directed at $30°$ above the horizontal, what would be the painting's dimensions as seen by the observer on Earth?

5. Protons from outer space crash into the Earth's atmosphere at a high rate. These protons create particles that eventually decay into other particles called *muons*. This cosmic debris travels through the atmosphere. Every second, dozens of muons pass through your body. If a muon is created 30 km above the Earth's surface, what minimum speed and kinetic energy must the muon have in order to hit Earth's surface? A muon's mean lifetime (at rest) is $2.20\ \mu s$ and its mass is $105.7\ \mathrm{MeV}/c^2$.

6. As a rough rule, anything traveling faster than about $0.1c$ is called *relativistic*—that is, special relativity is a significant effect. Determine the speed of an electron in a hydrogen atom (radius 0.53×10^{-10} m) and state whether or not it is relativistic. (Treat the electron as though it were in a circular orbit around the proton. See hint for Problem 46.)

ANSWERS TO EXERCISES

A: (c).
B: (a) No; (b) yes.
C: 80 m.
D: No: $\mathrm{KE}/mc^2 \approx 2 \times 10^{-4}$.

E: $0.030c$, same as classical, to an accuracy of better than 0.1%.
F: (d).

Electron microscopes (EM) produce images using electrons which have wave properties just as light does. Because the wavelength of electrons can be much smaller than that of visible light, much greater resolution and magnification can be obtained. A scanning electron microscope (SEM) can produce images with a three-dimensional quality.

All EM images are monochromatic (black and white). Artistic coloring has been added here, as is common. On the left is an SEM image of a blood clot forming (yellow-color web) due to a wound. White blood cells are colored green here for visibility. On the right, red blood cells in a small artery. A red blood cell travels about 15 km a day inside our bodies and lives roughly 4 months before damage or rupture. Humans contain 4 to 6 liters of blood, and 2 to 3×10^{13} red blood cells.

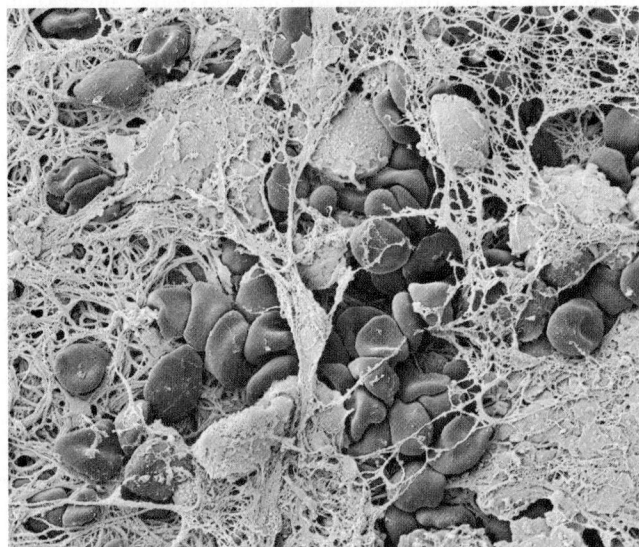 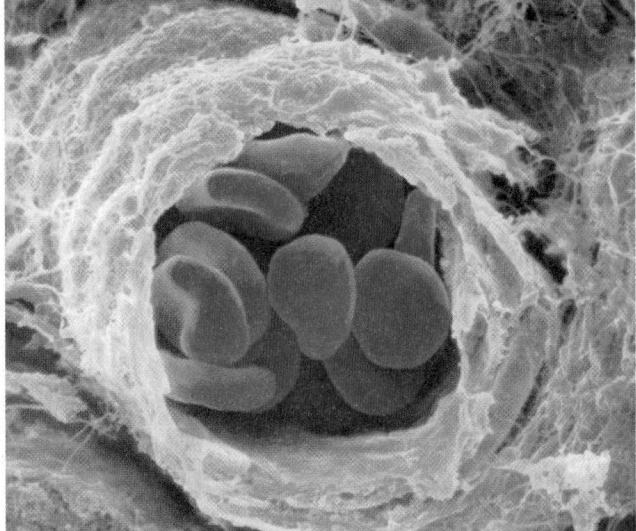

Early Quantum Theory and Models of the Atom

CHAPTER 27

CHAPTER-OPENING QUESTION—Guess now!

It has been found experimentally that
- **(a)** light behaves as a wave.
- **(b)** light behaves as a particle.
- **(c)** electrons behave as particles.
- **(d)** electrons behave as waves.
- **(e)** all of the above are true.
- **(f)** only (a) and (b) are true.
- **(g)** only (a) and (c) are true.
- **(h)** none of the above are true.

The second aspect of the revolution that shook the world of physics in the early part of the twentieth century was the quantum theory (the other was Einstein's theory of relativity). Unlike the special theory of relativity, the revolution of quantum theory required almost three decades to unfold, and many scientists contributed to its development. It began in 1900 with Planck's quantum hypothesis, and culminated in the mid-1920s with the theory of quantum mechanics of Schrödinger and Heisenberg which has been so effective in explaining the structure of matter. The discovery of the electron in the 1890s, with which we begin this Chapter, might be said to mark the beginning of modern physics, and is a sort of precursor to the quantum theory.

CONTENTS

27–1 Discovery and Properties of the Electron

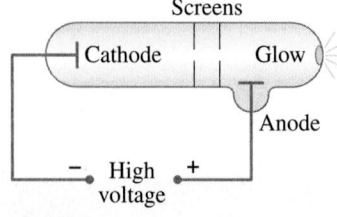

FIGURE 27–1 Discharge tube. In some models, one of the screens is the anode (positive plate).

Toward the end of the nineteenth century, studies were being done on the discharge of electricity through rarefied gases. One apparatus, diagrammed in Fig. 27–1, was a glass tube fitted with electrodes and evacuated so only a small amount of gas remained inside. When a very high voltage was applied to the electrodes, a dark space seemed to extend outward from the cathode (negative electrode) toward the opposite end of the tube; and that far end of the tube would glow. If one or more screens containing a small hole were inserted as shown, the glow was restricted to a tiny spot on the end of the tube. It seemed as though something being emitted by the cathode traveled across to the opposite end of the tube. These "somethings" were named **cathode rays**.

There was much discussion at the time about what these rays might be. Some scientists thought they might resemble light. But the observation that the bright spot at the end of the tube could be deflected to one side by an electric or magnetic field suggested that cathode rays were charged particles; and the direction of the deflection was consistent with a negative charge. Furthermore, if the tube contained certain types of rarefied gas, the path of the cathode rays was made visible by a slight glow.

Estimates of the charge e of the cathode-ray particles, as well as of their charge-to-mass ratio e/m, had been made by 1897. But in that year, J. J. Thomson (1856–1940) was able to measure e/m directly, using the apparatus shown in Fig. 27–2. Cathode rays are accelerated by a high voltage and then pass between a pair of parallel plates built into the tube. Another voltage applied to the parallel plates produces an electric field $\vec{\mathbf{E}}$, and a pair of coils produces a magnetic field $\vec{\mathbf{B}}$. If $E = B = 0$, the cathode rays follow path b in Fig. 27–2.

FIGURE 27–2 Cathode rays deflected by electric and magnetic fields. (See also Section 17–11 on the CRT.)

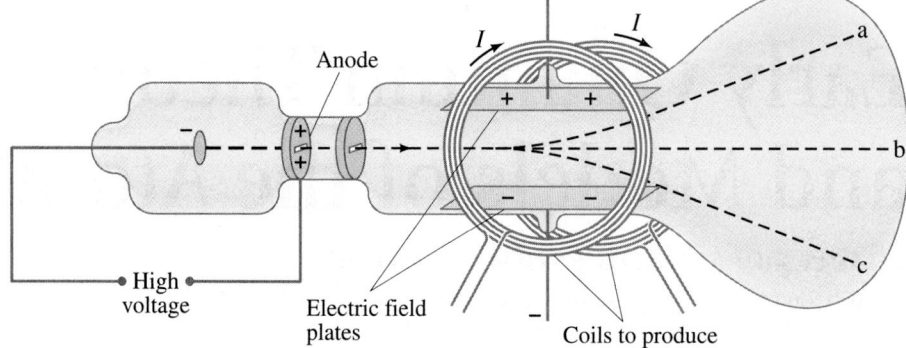

When only the electric field is present, say with the upper plate positive, the cathode rays are deflected upward as in path a in Fig. 27–2. If only a magnetic field exists, say inward, the rays are deflected downward along path c. These observations are just what is expected for a negatively charged particle. The force on the rays due to the magnetic field is $F = evB$, where e is the charge and v is the velocity of the cathode rays (Eq. 20–4). In the absence of an electric field, the rays are bent into a curved path, and applying Newton's second law $F = ma$ with a = centripetal acceleration gives

$$evB = m\frac{v^2}{r},$$

and thus

$$\frac{e}{m} = \frac{v}{Br}.$$

The radius of curvature r can be measured and so can B. The velocity v can be found by applying an electric field in addition to the magnetic field. The electric field E is adjusted so that the cathode rays are undeflected and follow path b in Fig. 27–2. In this situation the upward force due to the electric field, $F = eE$, is balanced by the downward force due to the magnetic field, $F = evB$. We equate the two forces, $eE = evB$, and find

$$v = \frac{E}{B}.$$

Combining this with the above equation we have

$$\frac{e}{m} = \frac{E}{B^2 r}. \qquad \text{(27–1)}$$

The quantities on the right side can all be measured, and although e and m could not be determined separately, the ratio e/m could be determined. The accepted value today is $e/m = 1.76 \times 10^{11}$ C/kg. Cathode rays soon came to be called **electrons**.

Discovery in Science

The "discovery" of the electron, like many others in science, is not quite so obvious as discovering gold or oil. Should the discovery of the electron be credited to the person who first saw a glow in the tube? Or to the person who first called them cathode rays? Perhaps neither one, for they had no conception of the electron as we know it today. In fact, the credit for the discovery is generally given to Thomson, but not because he was the first to see the glow in the tube. Rather it is because he believed that this phenomenon was due to tiny negatively charged particles and made careful measurements on them. Furthermore he argued that these particles were constituents of atoms, and not ions or atoms themselves as many thought, and he developed an electron theory of matter. His view is close to what we accept today, and this is why Thomson is credited with the "discovery." Note, however, that neither he nor anyone else ever actually saw an electron itself. We discuss this briefly, for it illustrates the fact that discovery in science is not always a clear-cut matter. In fact some philosophers of science think the word "discovery" is often not appropriate, such as in this case.

Electron Charge Measurement

Thomson believed that an electron was not an atom, but rather a constituent, or part, of an atom. Convincing evidence for this came soon with the determination of the charge and the mass of the cathode rays. Thomson's student J. S. Townsend made the first direct (but rough) measurements of e in 1897. But it was the more refined **oil-drop experiment** of Robert A. Millikan (1868–1953) that yielded a precise value for the charge on the electron and showed that charge comes in discrete amounts. In this experiment, tiny droplets of mineral oil carrying an electric charge were allowed to fall under gravity between two parallel plates, Fig. 27–3. The electric field E between the plates was adjusted until the drop was suspended in midair. The downward pull of gravity, mg, was then just balanced by the upward force due to the electric field. Thus $qE = mg$ so the charge $q = mg/E$. The mass of the droplet was determined by measuring its terminal velocity in the absence of the electric field. Often the droplet was charged negatively, but sometimes it was positive, suggesting that the droplet had acquired or lost electrons (by friction, leaving the atomizer). Millikan's painstaking observations and analysis presented convincing evidence that any charge was an integral multiple of a smallest charge, e, that was ascribed to the electron, and that the value of e was 1.6×10^{-19} C. This value of e, combined with the measurement of e/m, gives the mass of the electron to be $(1.6 \times 10^{-19}\,\text{C})/(1.76 \times 10^{11}\,\text{C/kg}) = 9.1 \times 10^{-31}$ kg. This mass is less than a thousandth the mass of the smallest atom, and thus confirmed the idea that the electron is only a part of an atom. The accepted value today for the mass of the electron is

$$m_e = 9.11 \times 10^{-31}\,\text{kg}.$$

The experimental result that any charge is an integral multiple of e means that electric charge is *quantized* (exists only in discrete amounts).

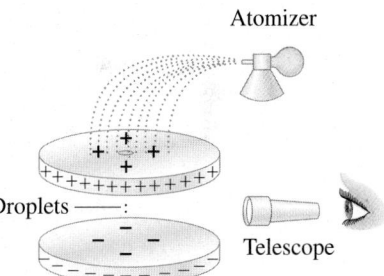

FIGURE 27–3 Millikan's oil-drop experiment.

27–2 Blackbody Radiation; Planck's Quantum Hypothesis

Blackbody Radiation

One of the observations that was unexplained at the end of the nineteenth century was the spectrum of light emitted by hot objects. We saw in Section 14–8 that all objects emit radiation whose total intensity is proportional to the fourth power of the Kelvin (absolute) temperature (T^4). At normal temperatures (≈ 300 K), we are not aware of this electromagnetic radiation because of its low intensity. At higher temperatures, there is sufficient infrared radiation that we can feel heat if we are close to the object. At still higher temperatures (on the order of 1000 K), objects actually glow, such as a red-hot electric stove burner or the heating element in a toaster. At temperatures above 2000 K, objects glow with a yellow or whitish color, such as white-hot iron and the filament of a lightbulb. The light emitted contains a continuous range of wavelengths or frequencies, and the spectrum is a plot of intensity vs. wavelength or frequency. As the temperature increases, the electromagnetic radiation emitted by objects not only increases in total intensity but has its peak intensity at higher and higher frequencies.

The spectrum of light emitted by a hot dense object is shown in Fig. 27–4 for an idealized **blackbody**. A blackbody is a body that, when cool, would absorb all the radiation falling on it (and so would appear black under reflection when illuminated by other sources). The radiation such an idealized blackbody would emit when hot and luminous, called **blackbody radiation** (though not necessarily black in color), approximates that from many real objects. The 6000-K curve in Fig. 27–4, corresponding to the temperature of the surface of the Sun, peaks in the visible part of the spectrum. For lower temperatures, the total intensity drops considerably and the peak occurs at longer wavelengths (or lower frequencies). This is why objects glow with a red color at around 1000 K. It is found experimentally that the wavelength at the peak of the spectrum, λ_P, is related to the Kelvin temperature T by

$$\lambda_P T = 2.90 \times 10^{-3} \,\text{m} \cdot \text{K}. \tag{27–2}$$

This is known as **Wien's law**.

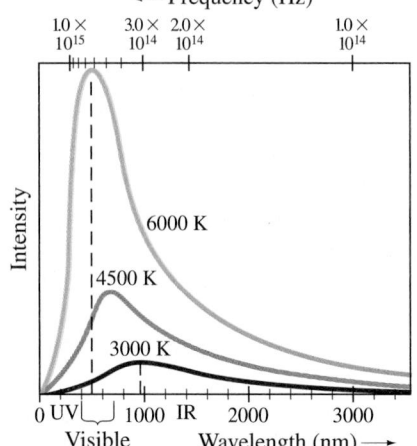

FIGURE 27–4 Measured spectra of wavelengths and frequencies emitted by a blackbody at three different temperatures.

EXAMPLE 27–1 **The Sun's surface temperature.** Estimate the temperature of the surface of our Sun, given that the Sun emits light whose peak intensity occurs in the visible spectrum at around 500 nm.

APPROACH We assume the Sun acts as a blackbody, and use $\lambda_P = 500$ nm in Wien's law (Eq. 27–2).

SOLUTION Wien's law gives

$$T = \frac{2.90 \times 10^{-3} \,\text{m} \cdot \text{K}}{\lambda_P} = \frac{2.90 \times 10^{-3} \,\text{m} \cdot \text{K}}{500 \times 10^{-9} \,\text{m}} \approx 6000 \,\text{K}.$$

EXAMPLE 27–2 **Star color.** Suppose a star has a surface temperature of 32,500 K. What color would this star appear?

APPROACH We assume the star emits radiation as a blackbody, and solve for λ_P in Wien's law, Eq. 27–2.

SOLUTION From Wien's law we have

$$\lambda_P = \frac{2.90 \times 10^{-3} \,\text{m} \cdot \text{K}}{T} = \frac{2.90 \times 10^{-3} \,\text{m} \cdot \text{K}}{3.25 \times 10^4 \,\text{K}} = 89.2 \,\text{nm}.$$

The peak is in the UV range of the spectrum, and will be way to the left in Fig. 27–4. In the visible region, the curve will be descending, so the shortest visible wavelengths will be strongest. Hence the star will appear bluish (or blue-white).

NOTE This example helps us to understand why stars have different colors (reddish for the coolest stars; orangish, yellow, white, bluish for "hotter" stars.)

| **EXERCISE A** What is the color of an object at 4000 K?

Planck's Quantum Hypothesis

In the year 1900, Max Planck (1858–1947) proposed a theory that was able to reproduce the graphs of Fig. 27–4. His theory, still accepted today, made a new and radical assumption: that the energy of the oscillations of atoms within molecules cannot have just any value; instead each has energy which is a multiple of a minimum value related to the frequency of oscillation by

$$E = hf.$$

Here h is a new constant, now called **Planck's constant**, whose value was estimated by Planck by fitting his formula for the blackbody radiation curve to experiment. The value accepted today is

$$h = 6.626 \times 10^{-34} \, \text{J} \cdot \text{s}.$$

Planck's assumption suggests that the energy of any molecular vibration could be only a whole number multiple of hf:

$$E = nhf, \qquad n = 1, 2, 3, \cdots, \qquad \textbf{(27–3)}$$

where n is called a **quantum number** ("quantum" means "discrete amount" as opposed to "continuous"). This idea is often called **Planck's quantum hypothesis**, although little attention was brought to this point at the time. In fact, it appears that Planck considered it more as a mathematical device to get the "right answer" rather than as an important discovery. Planck himself continued to seek a classical explanation for the introduction of h. The recognition that this was an important and radical innovation did not come until later, after about 1905 when others, particularly Einstein, entered the field.

The quantum hypothesis, Eq. 27–3, states that the energy of an oscillator can be $E = hf$, or $2hf$, or $3hf$, and so on, but there cannot be vibrations with energies between these values. That is, energy would not be a continuous quantity as had been believed for centuries; rather it is **quantized**—it exists only in discrete amounts. The smallest amount of energy possible (hf) is called the **quantum of energy**. Recall from Chapter 11 that the energy of an oscillation is proportional to the amplitude squared. Another way of expressing the quantum hypothesis is that not just any amplitude of vibration is possible. The possible values for the amplitude are related to the frequency f.

A simple analogy may help. Compare a ramp, on which a box can be placed at any height, to a flight of stairs on which the box can have only certain discrete amounts of potential energy, as shown in Fig. 27–5.

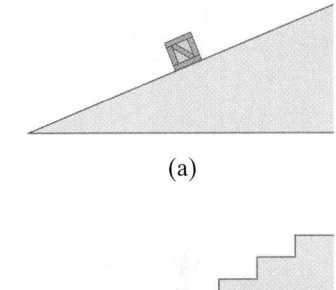

(a)

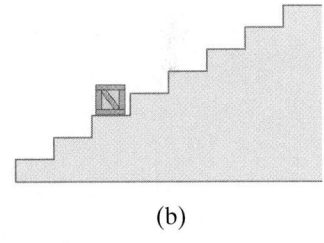

(b)

FIGURE 27–5 Ramp versus stair analogy. (a) On a ramp, a box can have continuous values of potential energy. (b) But on stairs, the box can have only discrete (quantized) values of energy.

27–3 Photon Theory of Light and the Photoelectric Effect

In 1905, the same year that he introduced the special theory of relativity, Einstein made a bold extension of the quantum idea by proposing a new theory of light. Planck's work had suggested that the vibrational energy of molecules in a radiating object is quantized with energy $E = nhf$, where n is an integer and f is the frequency of molecular vibration. Einstein argued that when light is emitted by a molecular oscillator, the molecule's vibrational energy of nhf must decrease by an amount hf (or by $2hf$, etc.) to another integer times hf, such as $(n - 1)hf$. Then to conserve energy, the light ought to be emitted in packets, or *quanta*, each with an energy

$$E = hf, \qquad \textbf{(27–4)} \qquad \textit{Photon energy}$$

where f is here the frequency of the emitted light. Again h is Planck's constant. Because all light ultimately comes from a radiating source, this idea suggests that *light is transmitted as tiny particles*, or **photons** as they are now called, as well as via the waves predicted by Maxwell's electromagnetic theory. The photon theory of light was also a radical departure from classical ideas. Einstein proposed a test of the quantum theory of light: quantitative measurements on the photoelectric effect.

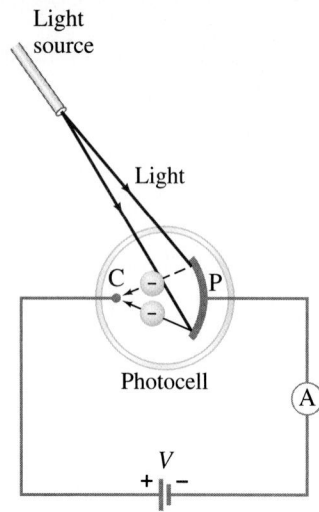

FIGURE 27–6 The photoelectric effect.

When light shines on a metal surface, electrons are found to be emitted from the surface. This effect is called the **photoelectric effect** and it occurs in many materials, but is most easily observed with metals. It can be observed using the apparatus shown in Fig. 27–6. A metal plate P and a smaller electrode C are placed inside an evacuated glass tube, called a **photocell**. The two electrodes are connected to an ammeter and a source of emf, as shown. When the photocell is in the dark, the ammeter reads zero. But when light of sufficiently high frequency illuminates the plate, the ammeter indicates a current flowing in the circuit. We explain completion of the circuit by imagining that electrons, ejected from the plate by the impinging light, flow across the tube from the plate to the "collector" C as indicated in Fig. 27–6.

That electrons should be emitted when light shines on a metal is consistent with the electromagnetic (EM) wave theory of light: the electric field of an EM wave could exert a force on electrons in the metal and eject some of them. Einstein pointed out, however, that the wave theory and the photon theory of light give very different predictions on the details of the photoelectric effect. For example, one thing that can be measured with the apparatus of Fig. 27–6 is the maximum kinetic energy (KE_{max}) of the emitted electrons. This can be done by using a variable voltage source and reversing the terminals so that electrode C is negative and P is positive. The electrons emitted from P will be repelled by the negative electrode, but if this reverse voltage is small enough, the fastest electrons will still reach C and there will be a current in the circuit. If the reversed voltage is increased, a point is reached where the current reaches zero—no electrons have sufficient kinetic energy to reach C. This is called the *stopping potential*, or *stopping voltage*, V_0, and from its measurement, KE_{max} can be determined using conservation of energy (loss of kinetic energy = gain in potential energy):

$$\text{KE}_{max} = eV_0.$$

Now let us examine the details of the photoelectric effect from the point of view of the wave theory versus Einstein's particle theory.

First the wave theory, assuming monochromatic light. The two important properties of a light wave are its intensity and its frequency (or wavelength). When these two quantities are varied, the wave theory makes the following predictions:

Wave

theory

predictions

1. If the light intensity is increased, the number of electrons ejected and their maximum kinetic energy should be increased because the higher intensity means a greater electric field amplitude, and the greater electric field should eject electrons with higher speed.

2. The frequency of the light should not affect the kinetic energy of the ejected electrons. Only the intensity should affect KE_{max}.

The photon theory makes completely different predictions. First we note that in a monochromatic beam, all photons have the same energy $(= hf)$. Increasing the intensity of the light beam means increasing the number of photons in the beam, but does not affect the energy of each photon as long as the frequency is not changed. According to Einstein's theory, an electron is ejected from the metal by a collision with a single photon. In the process, all the photon energy is transferred to the electron and the photon ceases to exist. Since electrons are held in the metal by attractive forces, some minimum energy W_0 is required just to get an electron out through the surface. W_0 is called the **work function**, and is a few electron volts $(1\,\text{eV} = 1.6 \times 10^{-19}\,\text{J})$ for most metals. If the frequency f of the incoming light is so low that hf is less than W_0, then the photons will not have enough energy to eject any electrons at all. If $hf > W_0$, then electrons will be ejected and energy will be conserved in the process. That is, the input energy (of the photon), hf, will equal the outgoing kinetic energy KE of the electron plus the energy required to get it out of the metal, W:

$$hf = \text{KE} + W. \tag{27–5a}$$

The least tightly held electrons will be emitted with the most kinetic energy (KE_{max}), in which case W in this equation becomes the work function W_0, and KE becomes KE_{max}:

$$hf = KE_{max} + W_0. \qquad \text{[least bound electrons]} \quad \textbf{(27-5b)}$$

Many electrons will require more energy than the bare minimum (W_0) to get out of the metal, and thus the kinetic energy of such electrons will be less than the maximum.

From these considerations, the photon theory makes the following predictions:

1. An increase in intensity of the light beam means more photons are incident, so more electrons will be ejected; but since the energy of each photon is not changed, the maximum kinetic energy of electrons is not changed by an increase in intensity.

2. If the frequency of the light is increased, the maximum kinetic energy of the electrons increases linearly, according to Eq. 27–5b. That is,

$$KE_{max} = hf - W_0.$$

This relationship is plotted in Fig. 27–7.

3. If the frequency f is less than the "cutoff" frequency f_0, where $hf_0 = W_0$, no electrons will be ejected, no matter how great the intensity of the light.

Photon

theory

predictions

These predictions of the photon theory are very different from the predictions of the wave theory. In 1913–1914, careful experiments were carried out by R. A. Millikan. The results were fully in agreement with Einstein's photon theory.

One other aspect of the photoelectric effect also confirmed the photon theory. If extremely low light intensity is used, the wave theory predicts a time delay before electron emission so that an electron can absorb enough energy to exceed the work function. The photon theory predicts no such delay—it only takes one photon (if its frequency is high enough) to eject an electron—and experiments showed no delay. This too confirmed Einstein's photon theory.

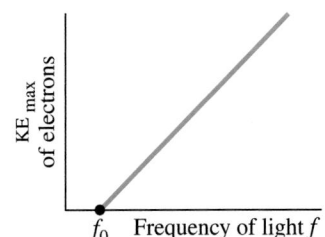

FIGURE 27–7 Photoelectric effect: the maximum kinetic energy of ejected electrons increases linearly with the frequency of incident light. No electrons are emitted if $f < f_0$.

EXAMPLE 27–3 **Photon energy.** Calculate the energy of a photon of blue light, $\lambda = 450\,nm$ in air (or vacuum).

APPROACH The photon has energy $E = hf$ (Eq. 27–4) where $f = c/\lambda$ (Eq. 22–4).

SOLUTION Since $f = c/\lambda$, we have

$$E = hf = \frac{hc}{\lambda} = \frac{(6.63 \times 10^{-34}\,J \cdot s)(3.00 \times 10^8\,m/s)}{(4.5 \times 10^{-7}\,m)} = 4.4 \times 10^{-19}\,J,$$

or $(4.4 \times 10^{-19}\,J)/(1.60 \times 10^{-19}\,J/eV) = 2.8\,eV$. (See definition of eV in Section 17–4, $1\,eV = 1.60 \times 10^{-19}\,J$.)

EXAMPLE 27–4 **ESTIMATE** **Photons from a lightbulb.** Estimate how many visible light photons a 100-W lightbulb emits per second. Assume the bulb has a typical efficiency of about 3% (that is, 97% of the energy goes to heat).

APPROACH Let's assume an average wavelength in the middle of the visible spectrum, $\lambda \approx 500\,nm$. The energy of each photon is $E = hf = hc/\lambda$. Only 3% of the 100-W power is emitted as visible light, or $3\,W = 3\,J/s$. The number of photons emitted per second equals the light output of $3\,J/s$ divided by the energy of each photon.

SOLUTION The energy emitted in one second ($= 3\,J$) is $E = Nhf$ where N is the number of photons emitted per second and $f = c/\lambda$. Hence

$$N = \frac{E}{hf} = \frac{E\lambda}{hc} = \frac{(3\,J)(500 \times 10^{-9}\,m)}{(6.63 \times 10^{-34}\,J \cdot s)(3.00 \times 10^8\,m/s)} \approx 8 \times 10^{18}$$

per second, or almost 10^{19} photons emitted per second, an enormous number.

EXERCISE B A beam contains infrared light of a single wavelength, 1000 nm, and monochromatic UV at 100 nm, both of the same intensity. Are there more 100-nm photons or more 1000-nm photons?

EXAMPLE 27–5 **Photoelectron speed and energy.** What is the kinetic energy and the speed of an electron ejected from a sodium surface whose work function is $W_0 = 2.28$ eV when illuminated by light of wavelength (a) 410 nm, (b) 550 nm?

APPROACH We first find the energy of the photons ($E = hf = hc/\lambda$). If the energy is greater than W_0, then electrons will be ejected with varying amounts of KE, with a maximum of $\text{KE}_{max} = hf - W_0$.

SOLUTION (a) For $\lambda = 410$ nm,

$$hf = \frac{hc}{\lambda} = 4.85 \times 10^{-19} \text{ J} \quad \text{or} \quad 3.03 \text{ eV}.$$

The maximum kinetic energy an electron can have is given by Eq. 27–5b, $\text{KE}_{max} = 3.03 \text{ eV} - 2.28 \text{ eV} = 0.75 \text{ eV}$, or $(0.75 \text{ eV})(1.60 \times 10^{-19} \text{ J/eV}) = 1.2 \times 10^{-19}$ J. Since $\text{KE} = \frac{1}{2}mv^2$ where $m = 9.1 \times 10^{-31}$ kg,

$$v_{max} = \sqrt{\frac{2\text{KE}}{m}} = 5.1 \times 10^5 \text{ m/s}.$$

Most ejected electrons will have less KE and less speed than these maximum values.

(b) For $\lambda = 550$ nm, $hf = hc/\lambda = 3.61 \times 10^{-19} \text{ J} = 2.26$ eV. Since this photon energy is less than the work function, no electrons are ejected.

NOTE In (a) we used the nonrelativistic equation for kinetic energy. If v had turned out to be more than about $0.1c$, our calculation would have been inaccurate by more than a percent or so, and we would probably prefer to redo it using the relativistic form (Eq. 26–5).

EXERCISE C Determine the lowest frequency and the longest wavelength needed to emit electrons from sodium.

By converting units, we can show that the energy of a photon in electron volts, when given the wavelength λ in nm, is

$$E \text{ (eV)} = \frac{1.240 \times 10^3 \text{ eV} \cdot \text{nm}}{\lambda \text{ (nm)}}. \qquad \text{[photon energy in eV]}$$

Applications of the Photoelectric Effect

The photoelectric effect, besides playing an important historical role in confirming the photon theory of light, also has many practical applications. Burglar alarms and automatic doors often make use of the photocell circuit of Fig. 27–6. When a person interrupts the beam of light, the sudden drop in current in the circuit activates a switch—often a solenoid—which operates a bell or opens the door. UV or IR light is sometimes used in burglar alarms because of its invisibility. Many smoke detectors use the photoelectric effect to detect tiny amounts of smoke that interrupt the flow of light and so alter the electric current. Photographic light meters use this circuit as well. Photocells are used in many other devices, such as absorption spectrophotometers, to measure light intensity. One type of film sound track is a variably shaded narrow section at the side of the film, Fig. 27–8. Light passing through the film is thus "modulated," and the output electrical signal of the photocell detector follows the frequencies on the sound track. For many applications today, the vacuum-tube photocell of Fig. 27–6 has been replaced by a semiconductor device known as a **photodiode** (Section 29–9). In these semiconductors, the absorption of a photon liberates a bound electron so it can move freely, which changes the conductivity of the material and the current through a photodiode is altered.

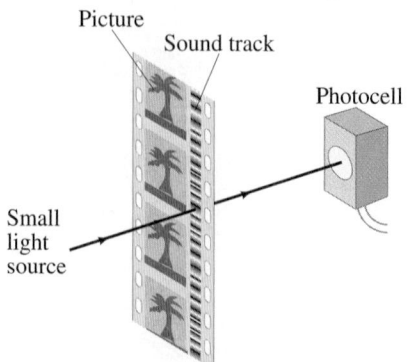

FIGURE 27–8 Optical sound track on movie film. In the projector, light from a small source (different from that for the picture) passes through the sound track on the moving film.

27–4 Energy, Mass, and Momentum of a Photon

We have just seen (Eq. 27–4) that the total energy of a single photon is given by $E = hf$. Because a photon always travels at the speed of light, it is truly a relativistic particle. Thus we must use relativistic formulas for dealing with its mass, energy, and momentum. The momentum of any particle of mass m is given by $p = mv/\sqrt{1 - v^2/c^2}$. Since $v = c$ for a photon, the denominator is zero. To avoid having an infinite momentum, we conclude that the photon's mass must be zero: $m = 0$. This makes sense too because a photon can never be at rest (it always moves at the speed of light). A photon's kinetic energy is its total energy:

$$\text{KE} = E = hf. \qquad \text{[photon]}$$

The momentum of a photon can be obtained from the relativistic formula (Eq. 26–9) $E^2 = p^2c^2 + m^2c^4$ where we set $m = 0$, so $E^2 = p^2c^2$ or

$$p = \frac{E}{c}. \qquad \text{[photon]}$$

⚠ **CAUTION**

Momentum of photon is not mv

Since $E = hf$ for a photon, its momentum is related to its wavelength by

$$p = \frac{E}{c} = \frac{hf}{c} = \frac{h}{\lambda}. \qquad \textbf{(27–6)}$$

EXAMPLE 27–6 | ESTIMATE **Photon momentum and force.** Suppose the 10^{19} photons emitted per second from the 100-W lightbulb in Example 27–4 were all focused onto a piece of black paper and absorbed. (*a*) Calculate the momentum of one photon and (*b*) estimate the force all these photons could exert on the paper.

APPROACH Each photon's momentum is obtained from Eq. 27–6, $p = h/\lambda$. Next, each absorbed photon's momentum changes from $p = h/\lambda$ to zero. We use Newton's second law, $F = \Delta p/\Delta t$, to get the force. Let $\lambda = 500\,\text{nm}$.

SOLUTION (*a*) Each photon has a momentum

$$p = \frac{h}{\lambda} = \frac{6.63 \times 10^{-34}\,\text{J}\cdot\text{s}}{500 \times 10^{-9}\,\text{m}} = 1.3 \times 10^{-27}\,\text{kg}\cdot\text{m/s}.$$

(*b*) Using Newton's second law for $N = 10^{19}$ photons (Example 27–4) whose momentum changes from h/λ to 0, we obtain

$$F = \frac{\Delta p}{\Delta t} = \frac{Nh/\lambda - 0}{1\,\text{s}} = N\frac{h}{\lambda} \approx (10^{19}\,\text{s}^{-1})(10^{-27}\,\text{kg}\cdot\text{m/s}) \approx 10^{-8}\,\text{N}.$$

NOTE This is a tiny force, but we can see that a very strong light source could exert a measurable force, and near the Sun or a star the force due to photons in electromagnetic radiation could be considerable. See Section 22–6.

EXAMPLE 27–7 **Photosynthesis.** In *photosynthesis*, pigments such as chlorophyll in plants capture the energy of sunlight to change CO_2 to useful carbohydrate. About nine photons are needed to transform one molecule of CO_2 to carbohydrate and O_2. Assuming light of wavelength $\lambda = 670\,\text{nm}$ (chlorophyll absorbs most strongly in the range 650 nm to 700 nm), how efficient is the photosynthetic process? The reverse chemical reaction releases an energy of $4.9\,\text{eV}$/molecule of CO_2, so 4.9 eV is needed to transform CO_2 to carbohydrate.

 **PHYSICS APPLIED**

Photosynthesis

APPROACH The efficiency is the minimum energy required (4.9 eV) divided by the actual energy absorbed, nine times the energy (hf) of one photon.

SOLUTION The energy of nine photons, each of energy $hf = hc/\lambda$, is $(9)(6.63 \times 10^{-34}\,\text{J}\cdot\text{s})(3.00 \times 10^8\,\text{m/s})/(6.7 \times 10^{-7}\,\text{m}) = 2.7 \times 10^{-18}\,\text{J}$ or 17 eV. Thus the process is about $(4.9\,\text{eV}/17\,\text{eV}) = 29\%$ efficient.

*27–5 Compton Effect

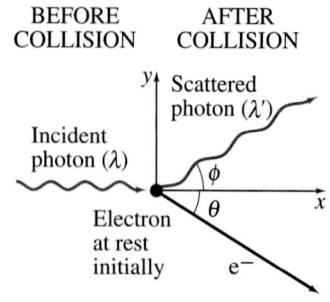

BEFORE COLLISION   AFTER COLLISION

FIGURE 27–9 The Compton effect. A single photon of wavelength λ strikes an electron in some material, knocking it out of its atom. The scattered photon has less energy (some energy is given to the electron) and hence has a longer wavelength λ' (shown exaggerated). Experiments found scattered X-rays of just the wavelengths predicted by conservation of energy and momentum using the photon model.

Besides the photoelectric effect, a number of other experiments were carried out in the early twentieth century which also supported the photon theory. One of these was the **Compton effect** (1923) named after its discoverer, A. H. Compton (1892–1962). Compton aimed short-wavelength light (actually X-rays) at various materials, and detected light scattered at various angles. He found that the scattered light had a slightly longer wavelength than did the incident light, and therefore a slightly lower frequency indicating a loss of energy. He explained this result on the basis of the photon theory as incident photons colliding with electrons of the material, Fig. 27–9. Using Eq. 27–6 for momentum of a photon, Compton applied the laws of conservation of momentum and energy to the collision of Fig. 27–9 and derived the following equation for the wavelength of the scattered photons:

$$\lambda' = \lambda + \frac{h}{m_{\mathrm{e}}c}(1 - \cos\phi), \tag{27–7}$$

where m_{e} is the mass of the electron. (The quantity $h/m_{\mathrm{e}}c$, which has the dimensions of length, is called the **Compton wavelength** of the electron.) We see that the predicted wavelength of scattered photons depends on the angle ϕ at which they are detected. Compton's measurements of 1923 were consistent with this formula. The wave theory of light predicts no such shift: an incoming electromagnetic wave of frequency f should set electrons into oscillation at frequency f; and such oscillating electrons would reemit EM waves of this same frequency f (Section 22–2), which would not change with angle (ϕ). Hence the Compton effect adds to the firm experimental foundation for the photon theory of light.

EXERCISE D When a photon scatters off an electron by the Compton effect, which of the following increases: its energy, frequency, wavelength?

EXAMPLE 27–8 **X-ray scattering.** X-rays of wavelength 0.140 nm are scattered from a very thin slice of carbon. What will be the wavelengths of X-rays scattered at (a) 0°, (b) 90°, (c) 180°?

APPROACH This is an example of the Compton effect, and we use Eq. 27–7 to find the wavelengths.

SOLUTION (a) For $\phi = 0°$, $\cos\phi = 1$ and $1 - \cos\phi = 0$. Then Eq. 27–7 gives $\lambda' = \lambda = 0.140\,\mathrm{nm}$. This makes sense since for $\phi = 0°$, there really isn't any collision as the photon goes straight through without interacting.

(b) For $\phi = 90°$, $\cos\phi = 0$, and $1 - \cos\phi = 1$. So

$$\lambda' = \lambda + \frac{h}{m_{\mathrm{e}}c} = 0.140\,\mathrm{nm} + \frac{6.63 \times 10^{-34}\,\mathrm{J\cdot s}}{(9.11 \times 10^{-31}\,\mathrm{kg})(3.00 \times 10^8\,\mathrm{m/s})}$$

$$= 0.140\,\mathrm{nm} + 2.4 \times 10^{-12}\,\mathrm{m} = 0.142\,\mathrm{nm};$$

that is, the wavelength is longer by one Compton wavelength ($= h/m_{\mathrm{e}}c^2 = 0.0024\,\mathrm{nm}$ for an electron).

(c) For $\phi = 180°$, which means the photon is scattered backward, returning in the direction from which it came (a direct "head-on" collision), $\cos\phi = -1$, and $1 - \cos\phi = 2$. So

$$\lambda' = \lambda + 2\frac{h}{m_{\mathrm{e}}c} = 0.140\,\mathrm{nm} + 2(0.0024\,\mathrm{nm}) = 0.145\,\mathrm{nm}.$$

NOTE The maximum shift in wavelength occurs for backward scattering, and it is twice the Compton wavelength.

PHYSICS APPLIED

Measuring bone density

The Compton effect has been used to diagnose bone disease such as osteoporosis. Gamma rays, which are photons of even shorter wavelength than X-rays, coming from a radioactive source are scattered off bone material. The total intensity of the scattered radiation is proportional to the density of electrons, which is in turn proportional to the bone density. A low bone density may indicate osteoporosis.

27-6 Photon Interactions; Pair Production

When a photon passes through matter, it interacts with the atoms and electrons. There are four important types of interactions that a photon can undergo:

1. The *photoelectric effect*: A photon may knock an electron out of an atom and in the process the photon disappears.

2. The photon may knock an atomic electron to a higher energy state in the atom if its energy is not sufficient to knock the electron out altogether. In this process the photon also disappears, and all its energy is given to the atom. Such an atom is then said to be in an *excited state*, and we shall discuss it more later.

3. The photon can be scattered from an electron (or a nucleus) and in the process lose some energy; this is the *Compton effect* (Fig. 27–9). But notice that the photon is not slowed down. It still travels with speed c, but its frequency will be lower because it has lost some energy.

4. *Pair production*: A photon can actually create matter, such as the production of an electron and a positron, Fig. 27–10. (A positron has the same mass as an electron, but the opposite charge, $+e$.)

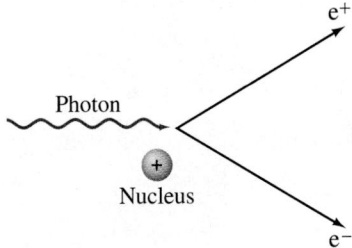

FIGURE 27-10 Pair production: a photon disappears and produces an electron and a positron.

In process 4, **pair production**, the photon disappears in the process of creating the electron–positron pair. This is an example of mass being created from pure energy, and it occurs in accord with Einstein's equation $E = mc^2$. Notice that a photon cannot create an electron alone since electric charge would not then be conserved. The inverse of pair production also occurs: if a positron comes close to an electron, the two quickly **annihilate** each other and their energy, including their mass, appears as electromagnetic energy of photons. Because positrons are not as plentiful in nature as electrons, they usually do not last long.

Electron–positron annihilation is the basis for the type of medical imaging known as PET, as discussed in Section 31–8.

EXAMPLE 27-9 **Pair production.** (*a*) What is the minimum energy of a photon that can produce an electron–positron pair? (*b*) What is this photon's wavelength?

APPROACH The minimum photon energy E equals the rest energy (mc^2) of the two particles created, via Einstein's famous equation $E = mc^2$ (Eq. 26–7). There is no energy left over, so the particles produced will have zero kinetic energy. The wavelength is $\lambda = c/f$ where $E = hf$ for the original photon.

SOLUTION (*a*) Because $E = mc^2$, and the mass created is equal to two electron masses, the photon must have energy

$$E = 2(9.11 \times 10^{-31}\,\text{kg})(3.00 \times 10^8\,\text{m/s})^2 = 1.64 \times 10^{-13}\,\text{J} = 1.02\,\text{MeV}$$

$(1\,\text{MeV} = 10^6\,\text{eV} = 1.60 \times 10^{-13}\,\text{J})$. A photon with less energy cannot undergo pair production.

(*b*) Since $E = hf = hc/\lambda$, the wavelength of a 1.02-MeV photon is

$$\lambda = \frac{hc}{E} = \frac{(6.63 \times 10^{-34}\,\text{J}\cdot\text{s})(3.00 \times 10^8\,\text{m/s})}{(1.64 \times 10^{-13}\,\text{J})} = 1.2 \times 10^{-12}\,\text{m},$$

which is 0.0012 nm. Such photons are in the gamma-ray (or very short X-ray) region of the electromagnetic spectrum (Fig. 22–8).

NOTE Photons of higher energy (shorter wavelength) can also create an electron–positron pair, with the excess energy becoming kinetic energy of the particles.

Pair production cannot occur in empty space, for momentum could not be conserved. In Example 27–9, for instance, energy is conserved, but only enough energy was provided to create the electron–positron pair at rest and thus with zero momentum, which could not equal the initial momentum of the photon. Indeed, it can be shown that at any energy, an additional massive object, such as an atomic nucleus (Fig. 27–10), must take part in the interaction to carry off some of the momentum.

27–7 Wave–Particle Duality; the Principle of Complementarity

The photoelectric effect, the Compton effect, and other experiments have placed the particle theory of light on a firm experimental basis. But what about the classic experiments of Young and others (Chapter 24) on interference and diffraction which showed that the wave theory of light also rests on a firm experimental basis?

We seem to be in a dilemma. Some experiments indicate that light behaves like a wave; others indicate that it behaves like a stream of particles. These two theories seem to be incompatible, but both have been shown to have validity. Physicists finally came to the conclusion that this duality of light must be accepted as a fact of life. It is referred to as the **wave–particle duality**. Apparently, light is a more complex phenomenon than just a simple wave or a simple beam of particles.

To clarify the situation, the great Danish physicist Niels Bohr (1885–1962, Fig. 27–11) proposed his famous **principle of complementarity**. It states that to understand an experiment, sometimes we find an explanation using wave theory and sometimes using particle theory. Yet we must be aware of both the wave and particle aspects of light if we are to have a full understanding of light. Therefore these two aspects of light complement one another.

It is not easy to "visualize" this duality. We cannot readily picture a combination of wave and particle. Instead, we must recognize that the two aspects of light are different "faces" that light shows to experimenters.

Part of the difficulty stems from how we think. Visual pictures (or models) in our minds are based on what we see in the everyday world. We apply the concepts of waves and particles to light because in the macroscopic world we see that energy is transferred from place to place by these two methods. We cannot see directly whether light is a wave or particle, so we do indirect experiments. To explain the experiments, we apply the models of waves or of particles to the nature of light. But these are abstractions of the human mind. When we try to conceive of what light really "is," we insist on a visual picture. Yet there is no reason why light should conform to these models (or visual images) taken from the macroscopic world. The "true" nature of light—if that means anything—is not possible to visualize. The best we can do is recognize that our knowledge is limited to the indirect experiments, and that in terms of everyday language and images, light reveals both wave and particle properties.

It is worth noting that Einstein's equation $E = hf$ itself links the particle and wave properties of a light beam. In this equation, E refers to the energy of a particle; and on the other side of the equation, we have the frequency f of the corresponding wave.

27–8 Wave Nature of Matter

In 1923, Louis de Broglie (1892–1987) extended the idea of the wave–particle duality. He appreciated the *symmetry* in nature, and argued that if light sometimes behaves like a wave and sometimes like a particle, then perhaps those things in nature thought to be particles—such as electrons and other material objects—might also have wave properties. De Broglie proposed that the wavelength of a material particle would be related to its momentum in the same way as for a photon, Eq. 27–6, $p = h/\lambda$. That is, for a particle having linear momentum $p = mv$, the wavelength λ is given by

de Broglie wavelength

$$\lambda = \frac{h}{p},$$

(27–8)

and is valid classically ($p = mv$ for $v \ll c$) and relativistically $\left(p = \gamma mv = mv/\sqrt{1 - v^2/c^2}\right)$. This is sometimes called the **de Broglie wavelength** of a particle.

FIGURE 27–11 Niels Bohr (right), walking with Enrico Fermi along the Appian Way outside Rome. This photo shows one important way physics is done.

⚠ **CAUTION**

*Not correct to say light is a wave and/or a particle. Light can **act** like a wave or like a particle*

EXAMPLE 27–10 **Wavelength of a ball.** Calculate the de Broglie wavelength of a 0.20-kg ball moving with a speed of 15 m/s.

APPROACH We use Eq. 27–8.

SOLUTION $\lambda = \dfrac{h}{p} = \dfrac{h}{mv} = \dfrac{(6.6 \times 10^{-34}\,\text{J} \cdot \text{s})}{(0.20\,\text{kg})(15\,\text{m/s})} = 2.2 \times 10^{-34}\,\text{m}.$

Ordinary objects, such as the ball of Example 27–10, have unimaginably small wavelengths. Even if the speed is extremely small, say 10^{-4} m/s, the wavelength would be about 10^{-29} m. Indeed, the wavelength of any ordinary object is much too small to be measured and detected. The problem is that the properties of waves, such as interference and diffraction, are significant only when the size of objects or slits is not much larger than the wavelength. And there are no known objects or slits to diffract waves only 10^{-30} m long, so the wave properties of ordinary objects go undetected.

But tiny elementary particles, such as electrons, are another matter. Since the mass m appears in the denominator of Eq. 27–8, a very small mass should have a much larger wavelength.

EXAMPLE 27–11 **Wavelength of an electron.** Determine the wavelength of an electron that has been accelerated through a potential difference of 100 V.

APPROACH If the kinetic energy is much less than the rest energy, we can use the classical formula, $\text{KE} = \frac{1}{2}mv^2$ (see end of Section 26–9). For an electron, $mc^2 = 0.511$ MeV. We then apply conservation of energy: the kinetic energy acquired by the electron equals its loss in potential energy. After solving for v, we use Eq. 27–8 to find the de Broglie wavelength.

SOLUTION The gain in kinetic energy equals the loss in potential energy: $\Delta\text{PE} = eV - 0$. Thus $\text{KE} = eV$, so $\text{KE} = 100\,\text{eV}$. The ratio $\text{KE}/mc^2 = 100\,\text{eV}/(0.511 \times 10^6\,\text{eV}) \approx 10^{-4}$, so relativity is not needed. Thus

$$\frac{1}{2}mv^2 = eV$$

and

$$v = \sqrt{\frac{2\,eV}{m}} = \sqrt{\frac{(2)(1.6 \times 10^{-19}\,\text{C})(100\,\text{V})}{(9.1 \times 10^{-31}\,\text{kg})}} = 5.9 \times 10^6\,\text{m/s}.$$

Then

$$\lambda = \frac{h}{mv} = \frac{(6.63 \times 10^{-34}\,\text{J} \cdot \text{s})}{(9.1 \times 10^{-31}\,\text{kg})(5.9 \times 10^6\,\text{m/s})} = 1.2 \times 10^{-10}\,\text{m},$$

or 0.12 nm.

EXERCISE E As a particle travels faster, does its de Broglie wavelength decrease, increase, or remain the same?

EXERCISE F Return to the Chapter-Opening Question, page 771, and answer it again now. Try to explain why you may have answered differently the first time.

Electron Diffraction

From Example 27–11, we see that electrons can have wavelengths on the order of 10^{-10} m, and even smaller. Although small, this wavelength can be detected: the spacing of atoms in a crystal is on the order of 10^{-10} m and the orderly array of atoms in a crystal could be used as a type of diffraction grating, as was done earlier for X-rays (see Section 25–11). C. J. Davisson and L. H. Germer performed the crucial experiment: they scattered electrons from the surface of a metal crystal and, in early 1927, observed that the electrons were scattered into a pattern of regular peaks. When they interpreted these peaks as a diffraction pattern, the wavelength of the diffracted electron wave was found to be just that predicted by de Broglie, Eq. 27–8. In the same year, G. P. Thomson (son of J. J. Thomson) used a different experimental arrangement and also detected diffraction of electrons. (See Fig. 27–12. Compare it to X-ray diffraction, Section 25–11.) Later experiments showed that protons, neutrons, and other particles also have wave properties.

FIGURE 27–12 Diffraction pattern of electrons scattered from aluminum foil, as recorded on film.

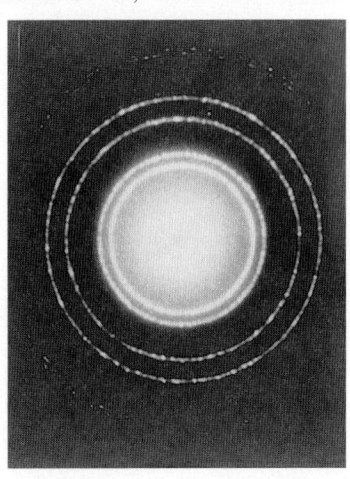

Thus the wave–particle duality applies to material objects as well as to light. The principle of complementarity applies to matter as well. That is, we must be aware of both the particle and wave aspects in order to have an understanding of matter, including electrons. But again we must recognize that a visual picture of a "wave–particle" is not possible.

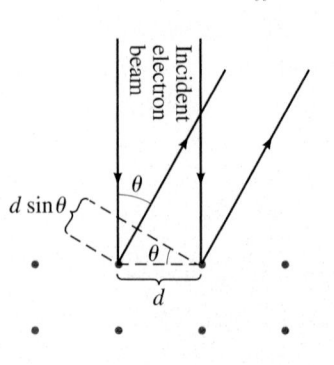

FIGURE 27–13 Example 27–12. The red dots represent atoms in an orderly array in a solid.

EXAMPLE 27–12 **Electron diffraction.** The wave nature of electrons is manifested in experiments where an electron beam interacts with the atoms on the surface of a solid, especially crystals. By studying the angular distribution of the diffracted electrons, one can indirectly measure the geometrical arrangement of atoms. Assume that the electrons strike perpendicular to the surface of a solid (see Fig. 27–13), and that their energy is low, KE = 100 eV, so that they interact only with the surface layer of atoms. If the smallest angle at which a diffraction maximum occurs is at 24°, what is the separation d between the atoms on the surface?

SOLUTION Treating the electrons as waves, we need to determine the condition where the difference in path traveled by the wave diffracted from adjacent atoms is an integer multiple of the de Broglie wavelength, so that constructive interference occurs. The path length difference is $d \sin \theta$ (Fig. 27–13); so for the smallest value of θ we must have

$$d \sin \theta = \lambda.$$

However, λ is related to the (non-relativistic) kinetic energy KE by

$$\text{KE} = \frac{p^2}{2m_e} = \frac{h^2}{2m_e \lambda^2}.$$

Thus

$$\lambda = \frac{h}{\sqrt{2m_e \, \text{KE}}}$$

$$= \frac{\left(6.63 \times 10^{-34} \, \text{J} \cdot \text{s}\right)}{\sqrt{2\left(9.11 \times 10^{-31} \, \text{kg}\right)\left(100 \, \text{eV}\right)\left(1.6 \times 10^{-19} \, \text{J/eV}\right)}} = 0.123 \, \text{nm}.$$

The surface inter-atomic spacing is

$$d = \frac{\lambda}{\sin \theta} = \frac{0.123 \, \text{nm}}{\sin 24°} = 0.30 \, \text{nm}.$$

NOTE Experiments of this type verify both the wave nature of electrons and the orderly array of atoms in crystalline solids.

What Is an Electron?

We might ask ourselves: "What is an electron?" The early experiments of J. J. Thomson (Section 27–1) indicated a glow in a tube, and that glow moved when a magnetic field was applied. The results of these and other experiments were best interpreted as being caused by tiny negatively charged particles which we now call electrons. No one, however, has actually seen an electron directly. The drawings we sometimes make of electrons as tiny spheres with a negative charge on them are merely convenient pictures (now recognized to be inaccurate). Again we must rely on experimental results, some of which are best interpreted using the particle model and others using the wave model. These models are mere pictures that we use to extrapolate from the macroscopic world to the tiny microscopic world of the atom. And there is no reason to expect that these models somehow reflect the reality of an electron. We thus use a wave or a particle model (whichever works best in a situation) so that we can talk about what is happening. But we should not be led to believe that an electron *is* a wave or a particle. Instead we could say that an electron is the set of its properties that we can measure. Bertrand Russell said it well when he wrote that an electron is "a logical construction."

27–9 Electron Microscopes

The idea that electrons have wave properties led to the development of the **electron microscope** (EM), which can produce images of much greater magnification than a light microscope. Figures 27–14 and 27–15 are diagrams of two types, developed around the middle of the twentieth century: the **transmission electron microscope** (TEM), which produces a two-dimensional image, and the **scanning electron microscope** (SEM), which produces images with a three-dimensional quality.

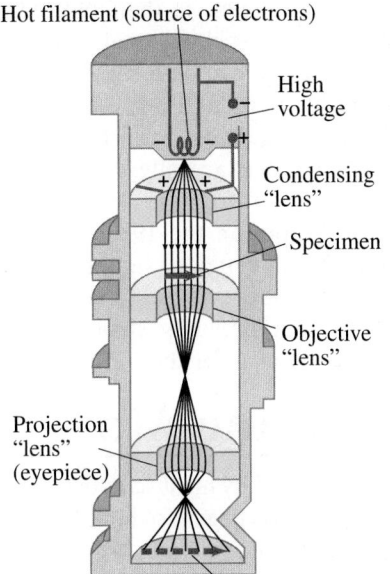

Hot filament (source of electrons)
High voltage
Condensing "lens"
Specimen
Objective "lens"
Projection "lens" (eyepiece)
Image (on screen, film, or semiconductor detector)

FIGURE 27–14 Transmission electron microscope. The magnetic field coils are designed to be "magnetic lenses," which bend the electron paths and bring them to a focus, as shown. The sensors of the image measure electron intensity only, no color.

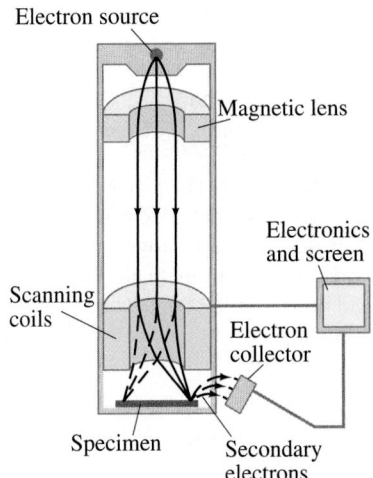

Electron source
Magnetic lens
Electronics and screen
Scanning coils
Electron collector
Specimen
Secondary electrons

FIGURE 27–15 Scanning electron microscope. Scanning coils move an electron beam back and forth across the specimen. Secondary electrons produced when the beam strikes the specimen are collected and their intensity affects the brightness of pixels in a monitor to produce a picture.

In both types, the objective and eyepiece lenses are actually magnetic fields that exert forces on the electrons to bring them to a focus. The fields are produced by carefully designed current-carrying coils of wire. Photographs using each type are shown in Fig. 27–16. EMs measure the intensity of electrons, producing monochromatic photos. Color is often added artificially to highlight.

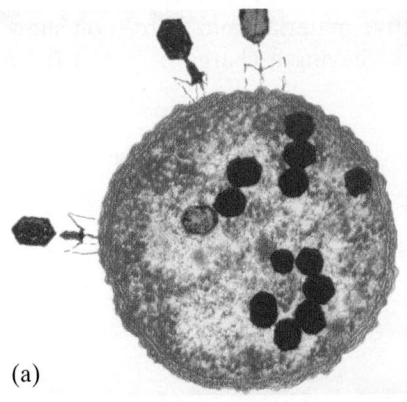

(a)

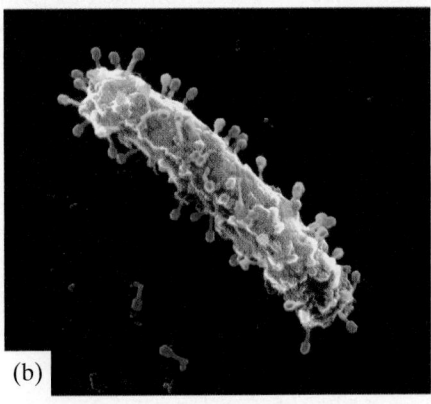

(b)

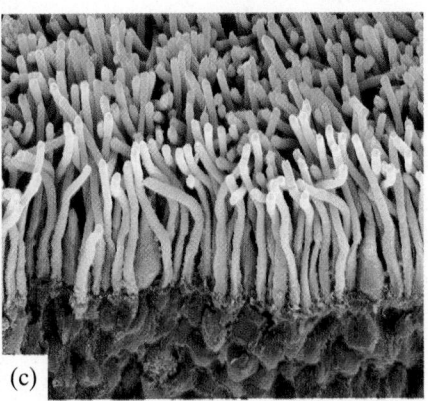

(c)

FIGURE 27–16 Electron micrographs, in false color, of (a) viruses attacking a cell of the bacterium *Escherichia coli* (TEM, $\approx 50,000\times$). (b) Same subject by an SEM ($\approx 35,000\times$). (c) SEM image of an eye's retina (Section 25–2); the rods and cones have been colored beige and green, respectively. Part (c) is also on the cover of this book.

As discussed in Sections 25–7 and 25–8, the maximum resolution of details on an object is about the size of the wavelength of the radiation used to view it. Electrons accelerated by voltages on the order of $10^5\,$V have wavelengths of about 0.004 nm. The maximum resolution obtainable would be on this order, but in practice, aberrations in the magnetic lenses limit the resolution in transmission electron microscopes to about 0.1 to 0.5 nm. This is still 1000 times better than a visible-light microscope, and corresponds to a useful magnification of about a million. Such magnifications are difficult to achieve, and more common magnifications are 10^4 to 10^5. The maximum resolution of a scanning electron microscope is less, typically 5 to 10 nm although new high-resolution SEMs approach 1 nm.

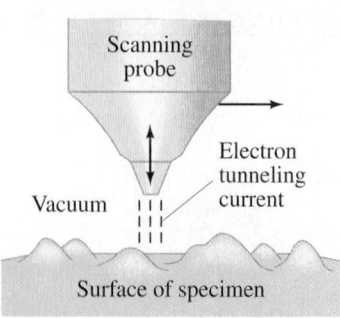

FIGURE 27–17 The probe tip of a scanning tunneling electron microscope, as it is moved horizontally, automatically moves up and down to maintain a constant tunneling current, and this motion is translated into an image of the surface.

FIGURE 27–18 Plum-pudding model of the atom.

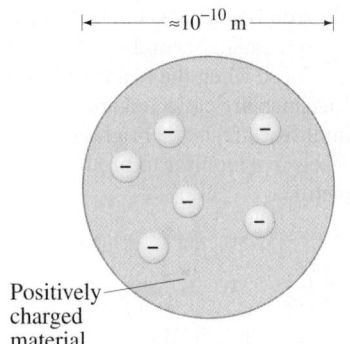

The **scanning tunneling electron microscope** (STM), developed in the 1980s, contains a tiny probe, whose tip may be only one (or a few) atoms wide, that is moved across the specimen to be examined in a series of linear passes. The tip, as it scans, remains very close to the surface of the specimen, about 1 nm above it, Fig. 27–17. A small voltage applied between the probe and the surface causes electrons to leave the surface and pass through the vacuum to the probe, by a process known as *tunneling* (discussed in Section 30–12). This "tunneling" current is very sensitive to the gap width, so a feedback mechanism can be used to raise and lower the probe to maintain a constant electron current. The probe's vertical motion, following the surface of the specimen, is then plotted as a function of position, scan after scan, producing a three-dimensional image of the surface. Surface features as fine as the size of an atom can be resolved: a resolution better than 50 pm (0.05 nm) laterally and 0.01 to 0.001 nm vertically. This kind of resolution has given a great impetus to the study of the surface structure of materials. The "topographic" image of a surface actually represents the distribution of electron charge.

The **atomic force microscope** (AFM), developed in the 1980s, is in many ways similar to an STM, but can be used on a wider range of sample materials. Instead of detecting an electric current, the AFM measures the force between a cantilevered tip and the sample, a force which depends strongly on the tip–sample separation at each point. The tip is moved as for the STM.

27–10 Early Models of the Atom

The idea that matter is made up of atoms was accepted by most scientists by 1900. With the discovery of the electron in the 1890s, scientists began to think of the atom itself as having a structure with electrons as part of that structure. We now discuss how our modern view of the atom developed, and the quantum theory with which it is intertwined.[†]

A typical model of the atom in the 1890s visualized the atom as a homogeneous sphere of positive charge inside of which there were tiny negatively charged electrons, a little like plums in a pudding, Fig. 27–18.

Around 1911, Ernest Rutherford (1871–1937) and his colleagues performed experiments whose results contradicted the plum-pudding model of the atom. In these experiments a beam of positively charged alpha (α) particles was directed at a thin sheet of metal foil such as gold, Fig. 27–19. (These newly discovered α particles were emitted by certain radioactive materials and were soon shown to be doubly ionized helium atoms—that is, having a charge of $+2e$.) It was

FIGURE 27–19 Experimental setup for Rutherford's experiment: α particles emitted by radon are deflected by the atoms of a thin metal foil and a few rebound backward.

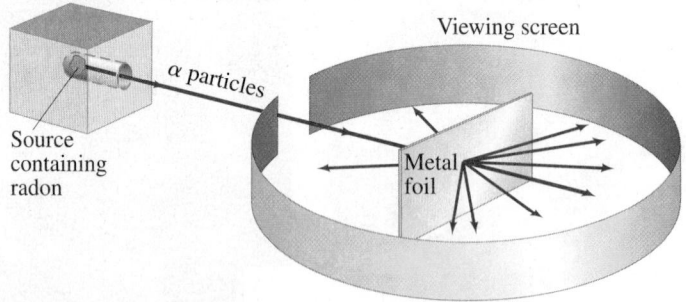

expected from the plum-pudding model that the alpha particles would not be deflected significantly because electrons are so much lighter than alpha particles, and the alpha particles should not have encountered any massive concentration of positive charge to strongly repel them. The experimental results completely contradicted these predictions. It was found that most of the alpha particles passed through the foil unaffected, as if the foil were mostly empty space.

[†]Some readers may say: "Tell us the facts as we know them today, and don't bother us with the historical background and its outmoded theories." Such an approach would ignore the creative aspect of science and thus give a false impression of how science develops. Moreover, it is not really possible to understand today's view of the atom without insight into the concepts that led to it.

And of those deflected, a few were deflected at very large angles—some even backward, nearly in the direction from which they had come. This could happen, Rutherford reasoned, only if the positively charged alpha particles were being repelled by a massive positive charge concentrated in a very small region of space (see Fig. 27–20). He hypothesized that the atom must consist of a tiny but massive positively charged nucleus, containing over 99.9% of the mass of the atom, surrounded by much lighter electrons some distance away. The electrons would be moving in orbits about the nucleus—much as the planets move around the Sun—because if they were at rest, they would fall into the nucleus due to electrical attraction. See Fig. 27–21. Rutherford's experiments suggested that the nucleus must have a radius of about 10^{-15} to 10^{-14} m. From kinetic theory, and especially Einstein's analysis of Brownian motion (see Section 13–1), the radius of atoms was estimated to be about 10^{-10} m. Thus the electrons would seem to be at a distance from the nucleus of about 10,000 to 100,000 times the radius of the nucleus itself. (If the nucleus were the size of a baseball, the atom would have the diameter of a big city several kilometers across.) So an atom would be mostly empty space.

Rutherford's **planetary model** of the atom (also called the **nuclear model** of the atom) was a major step toward how we view the atom today. It was not, however, a complete model and presented some major problems, as we shall see.

27–11 Atomic Spectra: Key to the Structure of the Atom

Earlier in this Chapter we saw that heated solids (as well as liquids and dense gases) emit light with a continuous spectrum of wavelengths. This radiation is assumed to be due to oscillations of atoms and molecules, which are largely governed by the interaction of each atom or molecule with its neighbors.

Rarefied gases can also be excited to emit light. This is done by intense heating, or more commonly by applying a high voltage to a "discharge tube" containing the gas at low pressure, Fig. 27–22. The radiation from excited gases had been observed early in the nineteenth century, and it was found that the spectrum was not continuous. Rather, excited gases emit light of only certain wavelengths, and when this light is analyzed through the slit of a spectroscope or spectrometer, a **line spectrum** is seen rather than a continuous spectrum. The line spectra emitted by a number of elements in the visible region are shown below in Fig. 27–23, and in Chapter 24, Fig. 24–28. The **emission spectrum** is characteristic of the material and can serve as a type of "fingerprint" for identification of the gas.

We also saw (Chapter 24) that if a continuous spectrum passes through a rarefied gas, dark lines are observed in the emerging spectrum, at wavelengths corresponding to lines normally emitted by the gas. This is called an **absorption spectrum** (Fig. 27–23c), and it became clear that gases can absorb light at the same frequencies at which they emit. Using film sensitive to ultraviolet and to infrared light, it was found that gases emit and absorb discrete frequencies in these regions as well as in the visible.

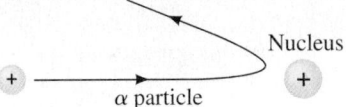

FIGURE 27–20 Backward rebound of α particles in Fig. 27–19 explained as the repulsion from a heavy positively charged nucleus.

FIGURE 27–21 Rutherford's model of the atom: electrons orbit a tiny positive nucleus (not to scale). The atom is visualized as mostly empty space.

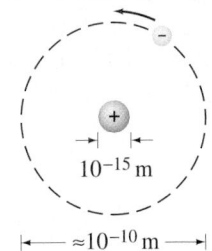

FIGURE 27–22 Gas-discharge tube: (a) diagram; (b) photo of an actual discharge tube for hydrogen.

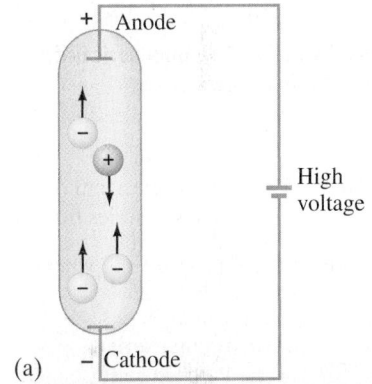

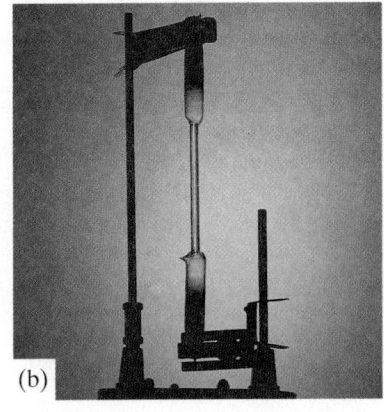

(a)

(b)

(c)

FIGURE 27–23 Emission spectra of the gases (a) atomic hydrogen, (b) helium, and (c) the *solar absorption* spectrum.

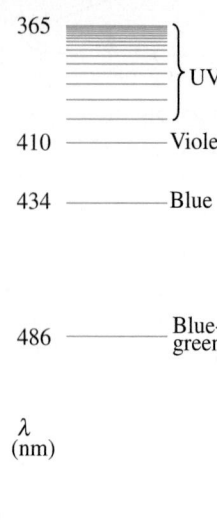

365 ⎤
   ⎬ UV
410 ——————— Violet

434 ——————— Blue

486 ——————— Blue-
                green

λ
(nm)

656 ——————— Red

FIGURE 27–24 Balmer series of lines for hydrogen.

FIGURE 27–25 Line spectrum of atomic hydrogen. Each series fits the formula $\frac{1}{\lambda} = R\left(\frac{1}{n'^2} - \frac{1}{n^2}\right)$ where $n' = 1$ for the Lyman series, $n' = 2$ for the Balmer series, $n' = 3$ for the Paschen series, and so on; n can take on all integer values from $n = n' + 1$ up to infinity. The only lines in the visible region of the electromagnetic spectrum are part of the Balmer series.

In low-density gases, the atoms are far apart on average and hence the light emitted or absorbed is assumed to be by *individual atoms* rather than through interactions between atoms, as in a solid, liquid, or dense gas. Thus the line spectra serve as a key to the structure of the atom: any theory of atomic structure must be able to explain why atoms emit light only of discrete wavelengths, and it should be able to predict what these wavelengths are.

Hydrogen is the simplest atom—it has only one electron orbiting its nucleus. It also has the simplest spectrum. The spectrum of most atoms shows little apparent regularity. But the spacing between lines in the hydrogen spectrum decreases in a regular way, Fig. 27–24. Indeed, in 1885, J. J. Balmer (1825–1898) showed that the four lines in the visible portion of the hydrogen spectrum (with measured wavelengths 656 nm, 486 nm, 434 nm, and 410 nm) have wavelengths that fit the formula

$$\frac{1}{\lambda} = R\left(\frac{1}{2^2} - \frac{1}{n^2}\right), \qquad n = 3, 4, \cdots. \qquad \textbf{(27–9)}$$

Here n takes on the values 3, 4, 5, 6 for the four visible lines, and R, called the **Rydberg constant**, has the value $R = 1.0974 \times 10^7 \text{ m}^{-1}$. Later it was found that this **Balmer series** of lines extended into the UV region, ending at $\lambda = 365 \text{ nm}$, as shown in Fig. 27–24. Balmer's formula, Eq. 27–9, also worked for these lines with higher integer values of n. The lines near 365 nm become too close together to distinguish, but the limit of the series at 365 nm corresponds to $n = \infty$ (so $1/n^2 = 0$ in Eq. 27–9).

Later experiments on hydrogen showed that there were similar series of lines in the UV and IR regions, and each series had a pattern just like the Balmer series, but at different wavelengths, Fig. 27–25. Each of these series was found to

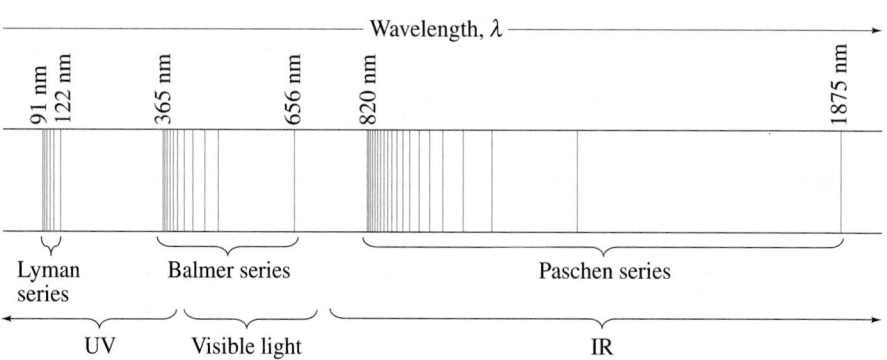

fit a formula with the same form as Eq. 27–9 but with the $1/2^2$ replaced by $1/1^2$, $1/3^2$, $1/4^2$, and so on. For example, the **Lyman series** contains lines with wavelengths from 91 nm to 122 nm (in the UV region) and fits the formula

$$\frac{1}{\lambda} = R\left(\frac{1}{1^2} - \frac{1}{n^2}\right), \qquad n = 2, 3, \cdots.$$

The wavelengths of the **Paschen series** (in the IR region) fit

$$\frac{1}{\lambda} = R\left(\frac{1}{3^2} - \frac{1}{n^2}\right), \qquad n = 4, 5, \cdots.$$

The Rutherford model was unable to explain why atoms emit line spectra. It had other difficulties as well. According to the Rutherford model, electrons orbit the nucleus, and since their paths are curved the electrons are accelerating. Hence they should give off light like any other accelerating electric charge (Chapter 22).

Since light carries off energy and energy is conserved, the electron's own energy must decrease to compensate. Hence electrons would be expected to spiral into the nucleus. As they spiraled inward, their frequency would increase in a short time and so too would the frequency of the light emitted. Thus the two main difficulties of the Rutherford model are these: (1) it predicts that light of a continuous range of frequencies will be emitted, whereas experiment shows line spectra; (2) it predicts that atoms are unstable—electrons would quickly spiral into the nucleus—but we know that atoms in general are stable, because there is stable matter all around us.

Clearly Rutherford's model was not sufficient. Some sort of modification was needed, and Niels Bohr provided it in a model that included the quantum hypothesis. Although the Bohr model has been superseded, it did provide a crucial stepping stone to our present understanding. And some aspects of the Bohr model are still useful today, so we examine it in detail in the next Section.

27–12 The Bohr Model

Bohr had studied in Rutherford's laboratory for several months in 1912 and was convinced that Rutherford's planetary model of the atom had validity. But in order to make it work, he felt that the newly developing quantum theory would somehow have to be incorporated in it. The work of Planck and Einstein had shown that in heated solids, the energy of oscillating electric charges must change discontinuously—from one discrete energy state to another, with the emission of a quantum of light. Perhaps, Bohr argued, the electrons in an atom also cannot lose energy continuously, but must do so in quantum "jumps." In working out his model during the next year, Bohr postulated that electrons move about the nucleus in circular orbits, but that only certain orbits are allowed. He further postulated that an electron in each orbit would have a definite energy and would move in the orbit *without radiating energy* (even though this violated classical ideas since accelerating electric charges are supposed to emit EM waves; see Chapter 22). He thus called the possible orbits **stationary states**. In this **Bohr model**, light is emitted only when an electron jumps from a higher (upper) stationary state to another of lower energy, Fig. 27–26. When such a transition occurs, a single photon of light is emitted whose energy, by energy conservation, is given by

$$hf = E_u - E_\ell, \qquad (27\text{–}10)$$

where E_u refers to the energy of the upper state and E_ℓ the energy of the lower state.

In 1912–13, Bohr set out to determine what energies these orbits would have in the simplest atom, hydrogen; the spectrum of light emitted could then be predicted from Eq. 27–10. In the Balmer formula he had the key he was looking for. Bohr quickly found that his theory would agree with the Balmer formula if he assumed that the electron's angular momentum L is quantized and equal to an integer n times $h/2\pi$. As we saw in Chapter 8 angular momentum is given by $L = I\omega$, where I is the moment of inertia and ω is the angular velocity. For a single particle of mass m moving in a circle of radius r with speed v, $I = mr^2$ and $\omega = v/r$; hence, $L = I\omega = (mr^2)(v/r) = mvr$. Bohr's **quantum condition** is

$$L = mvr_n = n\frac{h}{2\pi}, \qquad n = 1, 2, 3, \cdots, \qquad (27\text{–}11)$$

where n is an integer and r_n is the radius of the nth possible orbit. The allowed orbits are numbered 1, 2, 3, $\cdots$, according to the value of n, which is called the **principal quantum number** of the orbit.

Equation 27–11 did not have a firm theoretical foundation. Bohr had searched for some "quantum condition," and such tries as $E = hf$ (where E represents the energy of the electron in an orbit) did not give results in accord with experiment. Bohr's reason for using Eq. 27–11 was simply that it worked; and we now look at how. In particular, let us determine what the Bohr theory predicts for the measurable wavelengths of emitted light.

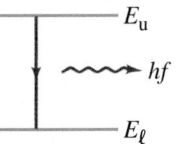

FIGURE 27–26 An atom emits a photon (energy $= hf$) when its energy changes from E_u to a lower energy E_ℓ.

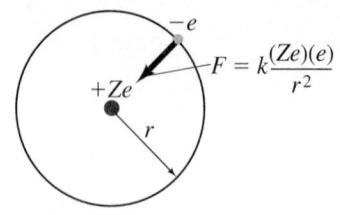

FIGURE 27–27 Electric force (Coulomb's law) keeps the negative electron in orbit around the positively charged nucleus.

An electron in a circular orbit of radius r (Fig. 27–27) would have a centripetal acceleration v^2/r produced by the electrical force of attraction between the negative electron and the positive nucleus. This force is given by Coulomb's law,

$$F = k\frac{(Ze)(e)}{r^2},$$

where $k = 1/4\pi\epsilon_0 = 8.99 \times 10^9 \,\text{N} \cdot \text{m}^2/\text{C}^2$. The charge on the electron is $q_1 = -e$, and that on the nucleus is $q_2 = +Ze$, where Ze is the charge on the nucleus: $+e$ is the charge on a proton, Z is the number of protons in the nucleus (called "atomic number," Section 28–7).[†] For the hydrogen atom, $Z = +1$.

In Newton's second law, $F = ma$, we substitute Coulomb's law for F and $a = v^2/r_n$ for a particular allowed orbit of radius r_n, and obtain

$$F = ma$$
$$k\frac{Ze^2}{r_n^2} = \frac{mv^2}{r_n}.$$

We solve this for r_n,

$$r_n = \frac{kZe^2}{mv^2},$$

and then substitute for v from Eq. 27–11 (which says $v = nh/2\pi mr_n$):

$$r_n = \frac{kZe^2 4\pi^2 mr_n^2}{n^2h^2}.$$

We solve for r_n (it appears on both sides, so we cancel one of them) and find

$$r_n = \frac{n^2h^2}{4\pi^2 mkZe^2} = \frac{n^2}{Z}r_1 \qquad n = 1, 2, 3 \cdots, \qquad \textbf{(27–12)}$$

where n is an integer (Eq. 27–11), and

$$r_1 = \frac{h^2}{4\pi^2 mke^2}.$$

Equation 27–12 gives the radii of all possible orbits. The smallest orbit is for $n = 1$, and for hydrogen ($Z = 1$) has the value

$$r_1 = \frac{(1)^2(6.626 \times 10^{-34}\,\text{J}\cdot\text{s})^2}{4\pi^2(9.11 \times 10^{-31}\,\text{kg})(8.99 \times 10^9\,\text{N}\cdot\text{m}^2/\text{C}^2)(1.602 \times 10^{-19}\,\text{C})^2}$$
$$r_1 = 0.529 \times 10^{-10}\,\text{m}. \qquad \textbf{(27–13)}$$

The radius of the smallest orbit in hydrogen, r_1, is sometimes called the **Bohr radius**. From Eq. 27–12, we see that the radii of the larger orbits[‡] increase as n^2, so

$$r_2 = 4r_1 = 2.12 \times 10^{-10}\,\text{m},$$
$$r_3 = 9r_1 = 4.76 \times 10^{-10}\,\text{m},$$
$$\vdots$$
$$r_n = n^2 r_1, \qquad n = 1, 2, 3, \cdots.$$

FIGURE 27–28 The four smallest orbits in the Bohr model of hydrogen; $r_1 = 0.529 \times 10^{-10}$ m.

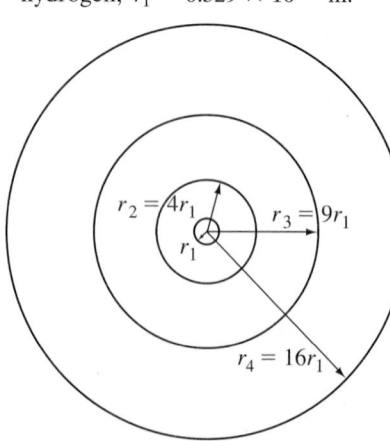

The first four orbits are shown in Fig. 27–28. Notice that, according to Bohr's model, an electron can exist only in the orbits given by Eq. 27–12. There are no allowable orbits in between.

For an atom with $Z \neq 1$, we can write the orbital radii, r_n, using Eq. 27–12:

$$r_n = \frac{n^2}{Z}(0.529 \times 10^{-10}\,\text{m}), \qquad n = 1, 2, 3, \cdots. \qquad \textbf{(27–14)}$$

[†]We include Z in our derivation so that we can treat other single-electron ("hydrogenlike") atoms such as the ions He^+ ($Z = 2$) and Li^{2+} ($Z = 3$). Helium in the neutral state has two electrons; if one electron is missing, the remaining He^+ ion consists of one electron revolving around a nucleus of charge $+2e$. Similarly, doubly ionized lithium, Li^{2+}, also has a single electron, and in this case $Z = 3$.

[‡]Be careful not to believe that these well-defined orbits actually exist. Today electrons are better thought of as forming "clouds," as discussed in Chapter 28.

In each of its possible orbits, the electron in a Bohr model atom would have a definite energy, as the following calculation shows. The total energy equals the sum of the kinetic and potential energies. The potential energy of the electron is given by $\text{PE} = qV = -eV$, where V is the potential due to a point charge $+Ze$ as given by Eq. 17–5: $V = kQ/r = kZe/r$. So

$$\text{PE} = -eV = -k\frac{Ze^2}{r}.$$

The total energy E_n for an electron in the nth orbit of radius r_n is the sum of the kinetic and potential energies:

$$E_n = \tfrac{1}{2}mv^2 - \frac{kZe^2}{r_n}.$$

When we substitute v from Eq. 27–11 and r_n from Eq. 27–12 into this equation, we obtain

$$E_n = -\frac{2\pi^2 Z^2 e^4 m k^2}{h^2}\frac{1}{n^2} \qquad n = 1, 2, 3, \cdots. \qquad \textbf{(27–15a)}$$

If we evaluate the constant term in Eq. 27–15a and convert it to electron volts, as is customary in atomic physics, we obtain

$$E_n = -(13.6\,\text{eV})\frac{Z^2}{n^2}, \qquad n = 1, 2, 3, \cdots. \qquad \textbf{(27–15b)}$$

The lowest energy level $(n = 1)$ for hydrogen $(Z = 1)$ is

$$E_1 = -13.6\,\text{eV}.$$

Since n^2 appears in the denominator of Eq. 27–15b, the energies of the larger orbits in hydrogen $(Z = 1)$ are given by

$$E_n = \frac{-13.6\,\text{eV}}{n^2}.$$

For example,

$$E_2 = \frac{-13.6\,\text{eV}}{4} = -3.40\,\text{eV},$$

$$E_3 = \frac{-13.6\,\text{eV}}{9} = -1.51\,\text{eV}.$$

We see that not only are the orbit radii quantized, but from Eqs. 27–15, so is the energy. The quantum number n that labels the orbit radii also labels the energy levels. The lowest **energy level** or **energy state** has energy E_1, and is called the **ground state**. The higher states, E_2, E_3, and so on, are called **excited states**. The fixed energy levels are also called **stationary states**.

Notice that although the energy for the larger orbits has a smaller numerical value, all the energies are less than zero. Thus, $-3.4\,\text{eV}$ is a higher energy than $-13.6\,\text{eV}$. Hence the orbit closest to the nucleus (r_1) has the lowest energy (the most negative). The reason the energies have negative values has to do with the way we defined the zero for potential energy. For two point charges, $\text{PE} = kq_1 q_2/r$ corresponds to zero potential energy when the two charges are infinitely far apart (Section 17–5). Thus, an electron that can just barely be free from the atom by reaching $r = \infty$ (or, at least, far from the nucleus) with zero kinetic energy will have $E = \text{KE} + \text{PE} = 0 + 0 = 0$, corresponding to $n = \infty$ in Eqs. 27–15. If an electron is free and has kinetic energy, then $E > 0$. To remove an electron that is part of an atom requires an energy input (otherwise atoms would not be stable). Since $E \geq 0$ for a free electron, then an electron bound to an atom needs to have $E < 0$. That is, energy must be added to bring its energy up, from a negative value to at least zero in order to free it.

The minimum energy required to remove an electron from an atom initially in the ground state is called the **binding energy** or **ionization energy**. The ionization energy for hydrogen has been measured to be 13.6 eV, and this corresponds precisely to removing an electron from the lowest state, $E_1 = -13.6\,\text{eV}$, up to $E = 0$ where it can be free.

Spectra Lines Explained

It is useful to show the various possible energy values as horizontal lines on an energy-level diagram. This is shown for hydrogen in Fig. 27–29. The electron in a hydrogen atom can be in any one of these levels according to Bohr theory. But it could never be in between, say at -9.0 eV. At room temperature, nearly all H atoms will be in the ground state ($n = 1$). At higher temperatures, or during an electric discharge when there are many collisions between free electrons and atoms, many atoms can be in excited states ($n > 1$). Once in an excited state, an atom's electron can jump down to a lower state, and give off a photon in the process. This is, according to the Bohr model, the origin of the emission spectra of excited gases.

Note that above $E = 0$, an electron is free and can have any energy (E is not quantized). Thus there is a continuum of energy states above $E = 0$, as indicated in the energy-level diagram of Fig. 27–29.

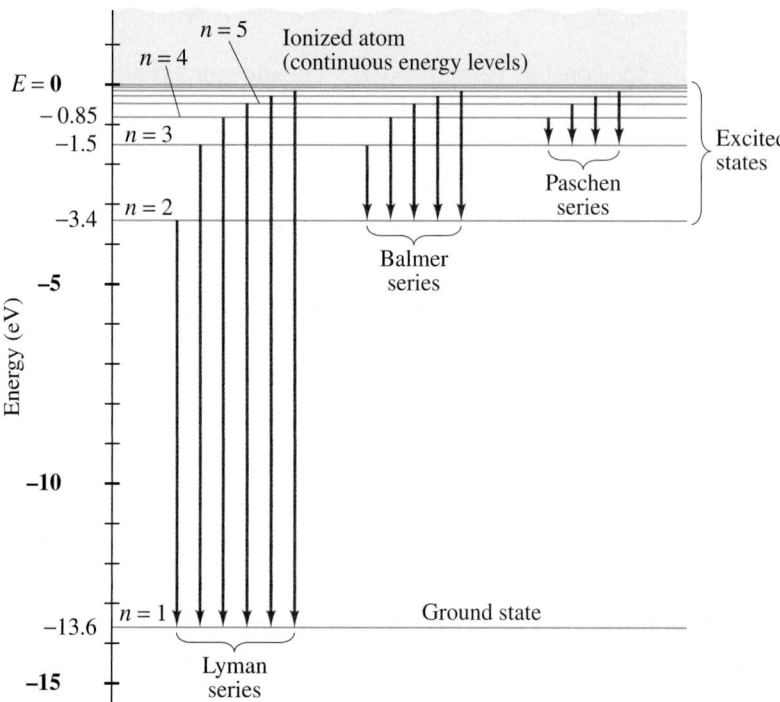

FIGURE 27–29 Energy-level diagram for the hydrogen atom, showing the transitions for the spectral lines of the Lyman, Balmer, and Paschen series (Fig. 27–25). Each vertical arrow represents an atomic transition that gives rise to the photons of one spectral line (a single wavelength or frequency).

The vertical arrows in Fig. 27–29 represent the transitions or jumps that correspond to the various observed spectral lines. For example, an electron jumping from the level $n = 3$ to $n = 2$ would give rise to the 656-nm line in the Balmer series, and the jump from $n = 4$ to $n = 2$ would give rise to the 486-nm line (see Fig. 27–24). We can predict wavelengths of the spectral lines emitted according to Bohr theory by combining Eq. 27–10 with Eq. 27–15. Since $hf = hc/\lambda$, we have from Eq. 27–10

$$\frac{1}{\lambda} = \frac{hf}{hc} = \frac{1}{hc}(E_n - E_{n'}),$$

where n refers to the upper state and n' to the lower state. Then using Eq. 27–15,

$$\frac{1}{\lambda} = \frac{2\pi^2 Z^2 e^4 m k^2}{h^3 c}\left(\frac{1}{n'^2} - \frac{1}{n^2}\right). \qquad (27\text{–}16)$$

This theoretical formula has the same form as the experimental Balmer formula, Eq. 27–9, with $n' = 2$. Thus we see that the Balmer series of lines corresponds to transitions or "jumps" that bring the electron down to the second energy level. Similarly, $n' = 1$ corresponds to the Lyman series and $n' = 3$ to the Paschen series (see Fig. 27–29).

When the constant in Eq. 27–16 is evaluated with $Z = 1$, it is found to have the measured value of the Rydberg constant, $R = 1.0974 \times 10^7$ m^{-1} in Eq. 27–9, in accord with experiment (see Problem 54).

The great success of Bohr's model is that it gives an explanation for why atoms emit line spectra, and accurately predicts the wavelengths of emitted light for hydrogen. The Bohr model also explains absorption spectra: photons of just the right wavelength can knock an electron from one energy level to a higher one. To conserve energy, only photons that have just the right energy will be absorbed. This explains why a continuous spectrum of light entering a gas will emerge with dark (absorption) lines at frequencies that correspond to emission lines (Fig. 27–23c).

The Bohr theory also ensures the stability of atoms. It establishes stability by decree: the ground state is the lowest state for an electron and there is no lower energy level to which it can go and emit more energy. Finally, as we saw above, the Bohr theory accurately predicts the ionization energy of 13.6 eV for hydrogen. However, the Bohr model was not so successful for other atoms, and has been superseded as we shall discuss in the next Chapter. We discuss the Bohr model because it *was* an important start and because we still use the concept of stationary states, the ground state, and transitions between states. Also, the terminology used in the Bohr model is still used by chemists and spectroscopists.

EXAMPLE 27–13 **Wavelength of a Lyman line.** Use Fig. 27–29 to determine the wavelength of the first Lyman line, the transition from $n = 2$ to $n = 1$. In what region of the electromagnetic spectrum does this lie?

APPROACH We use Eq. 27–10, $hf = E_u - E_\ell$, with the energies obtained from Fig. 27–29 to find the energy and the wavelength of the transition. The region of the electromagnetic spectrum is found using the EM spectrum in Fig. 22–8.

SOLUTION In this case, $hf = E_2 - E_1 = \{-3.4\,\text{eV} - (-13.6\,\text{eV})\} = 10.2\,\text{eV} = (10.2\,\text{eV})(1.60 \times 10^{-19}\,\text{J/eV}) = 1.63 \times 10^{-18}\,\text{J}$. Since $\lambda = c/f$, we have

$$\lambda = \frac{c}{f} = \frac{hc}{E_2 - E_1} = \frac{(6.63 \times 10^{-34}\,\text{J·s})(3.00 \times 10^8\,\text{m/s})}{1.63 \times 10^{-18}\,\text{J}} = 1.22 \times 10^{-7}\,\text{m},$$

or 122 nm, which is in the UV region of the EM spectrum, Fig. 22–8. See also Fig. 27–25, where this value is confirmed experimentally.

NOTE An alternate approach: use Eq. 27–16 to find λ, and get the same result.

EXAMPLE 27–14 **Wavelength of a Balmer line.** Use the Bohr model to determine the wavelength of light emitted when a hydrogen atom makes a transition from the $n = 6$ to the $n = 2$ energy level.

APPROACH We can use Eq. 27–16 or its equivalent, Eq. 27–9, with $R = 1.097 \times 10^7\,\text{m}^{-1}$.

SOLUTION We find

$$\frac{1}{\lambda} = (1.097 \times 10^7\,\text{m}^{-1})\left(\frac{1}{4} - \frac{1}{36}\right) = 2.44 \times 10^6\,\text{m}^{-1}.$$

So $\lambda = 1/(2.44 \times 10^6\,\text{m}^{-1}) = 4.10 \times 10^{-7}\,\text{m}$ or 410 nm. This is the fourth line in the Balmer series, Fig. 27–24, and is violet in color.

EXAMPLE 27–15 **Absorption wavelength.** Use Fig. 27–29 to determine the maximum wavelength that hydrogen in its ground state can absorb. What would be the next smaller wavelength that would work?

APPROACH Maximum wavelength corresponds to minimum energy, and this would be the jump from the ground state up to the first excited state (Fig. 27–29). The next smaller wavelength occurs for the jump from the ground state to the second excited state.

SOLUTION The energy needed to jump from the ground state to the first excited state is $13.6\,\text{eV} - 3.4\,\text{eV} = 10.2\,\text{eV}$; the required wavelength, as we saw in Example 27–13, is 122 nm. The energy to jump from the ground state to the second excited state is $13.6\,\text{eV} - 1.5\,\text{eV} = 12.1\,\text{eV}$, which corresponds to a wavelength

$$\lambda = \frac{c}{f} = \frac{hc}{hf} = \frac{hc}{E_3 - E_1} = \frac{(6.63 \times 10^{-34}\,\text{J·s})(3.00 \times 10^8\,\text{m/s})}{(12.1\,\text{eV})(1.60 \times 10^{-19}\,\text{J/eV})} = 103\,\text{nm}.$$

EXAMPLE 27–16 **He⁺ ionization energy.** (*a*) Use the Bohr model to determine the ionization energy of the He⁺ ion, which has a single electron. (*b*) Also calculate the maximum wavelength a photon can have to cause ionization. The helium atom is the second atom, after hydrogen, in the Periodic Table (next Chapter); its nucleus contains 2 protons and normally has 2 electrons circulating around it, so $Z = 2$.

APPROACH We want to determine the minimum energy required to lift the electron from its ground state and to barely reach the free state at $E = 0$. The ground state energy of He⁺ is given by Eq. 27–15b with $n = 1$ and $Z = 2$.

SOLUTION (*a*) Since all the symbols in Eq. 27–15b are the same as for the calculation for hydrogen, except that Z is 2 instead of 1, we see that E_1 will be $Z^2 = 2^2 = 4$ times the E_1 for hydrogen:

$$E_1 = 4(-13.6 \text{ eV}) = -54.4 \text{ eV}.$$

Thus, to ionize the He⁺ ion should require 54.4 eV, and this value agrees with experiment.

(*b*) The maximum wavelength photon that can cause ionization will have energy $hf = 54.4$ eV and wavelength

$$\lambda = \frac{c}{f} = \frac{hc}{hf} = \frac{(6.63 \times 10^{-34} \text{ J} \cdot \text{s})(3.00 \times 10^8 \text{ m/s})}{(54.4 \text{ eV})(1.60 \times 10^{-19} \text{ J/eV})} = 22.8 \text{ nm}.$$

If $\lambda > 22.8$ nm, ionization can not occur.

NOTE If the atom absorbed a photon of greater energy (wavelength shorter than 22.8 nm), the atom could still be ionized and the freed electron would have kinetic energy of its own.

In this Example 27–16, we saw that E_1 for the He⁺ ion is four times more negative than that for hydrogen. Indeed, the energy-level diagram for He⁺ looks just like that for hydrogen, Fig. 27–29, except that the numerical values for each energy level are four times larger. Note, however, that we are talking here about the He⁺ *ion*. Normal (neutral) helium has two electrons and its energy level diagram is entirely different.

CONCEPTUAL EXAMPLE 27–17 **Hydrogen at 20°C.** (*a*) Estimate the average kinetic energy of whole hydrogen atoms (not just the electrons) at room temperature. (*b*) Use the result to explain why, at room temperature, very few H atoms are in excited states and nearly all are in the ground state, and hence emit no light.

RESPONSE According to kinetic theory (Chapter 13), the average kinetic energy of atoms or molecules in a gas is given by Eq. 13–8:

$$\overline{\text{KE}} = \tfrac{3}{2}kT,$$

where $k = 1.38 \times 10^{-23}$ J/K is Boltzmann's constant, and T is the kelvin (absolute) temperature. Room temperature is about $T = 300$ K, so

$$\overline{\text{KE}} = \tfrac{3}{2}(1.38 \times 10^{-23} \text{ J/K})(300 \text{ K}) = 6.2 \times 10^{-21} \text{ J},$$

or, in electron volts:

$$\overline{\text{KE}} = \frac{6.2 \times 10^{-21} \text{ J}}{1.6 \times 10^{-19} \text{ J/eV}} = 0.04 \text{ eV}.$$

The average KE of an atom as a whole is thus very small compared to the energy between the ground state and the next higher energy state (13.6 eV − 3.4 eV = 10.2 eV). Any atoms in excited states quickly fall to the ground state and emit light. Once in the ground state, collisions with other atoms can transfer energy of only 0.04 eV on the average. A small fraction of atoms can have much more energy (see Section 13–10 on the distribution of molecular speeds), but even a kinetic energy that is 10 times the average is not nearly enough to excite atoms into states above the ground state. Thus, at room temperature, practically all atoms are in the ground state. Atoms can be excited to upper states by very high temperatures, or by applying a high voltage so a current of high energy electrons passes through the gas as in a discharge tube (Fig. 27–22).

Correspondence Principle

We should note that Bohr made some radical assumptions that were at variance with classical ideas. He assumed that electrons in fixed orbits do not radiate light even though they are accelerating (moving in a circle), and he assumed that angular momentum is quantized. Furthermore, he was not able to say how an electron moved when it made a transition from one energy level to another. On the other hand, there is no real reason to expect that in the tiny world of the atom electrons would behave as ordinary-sized objects do. Nonetheless, he felt that where quantum theory overlaps with the macroscopic world, it should predict classical results. This is the **correspondence principle**, already mentioned in regard to relativity (Section 26–11). This principle does work for Bohr's theory of the hydrogen atom. The orbit sizes and energies are quite different for $n = 1$ and $n = 2$, say. But orbits with $n = 100,000,000$ and $100,000,001$ would be very close in radius and energy (see Fig. 27–29). Indeed, transitions between such large orbits, which would approach macroscopic sizes, would be imperceptible. Such orbits would thus appear to be continuously spaced, which is what we expect in the everyday world.

Finally, it must be emphasized that the well-defined orbits of the Bohr model do not actually exist. The Bohr model is only a model, not reality. The idea of electron orbits was rejected a few years later, and today electrons are thought of (Chapter 28) as forming "probability clouds."

27–13 de Broglie's Hypothesis Applied to Atoms

Bohr's theory was largely of an *ad hoc* nature. Assumptions were made so that theory would agree with experiment. But Bohr could give no reason why the orbits were quantized, nor why there should be a stable ground state. Finally, ten years later, a reason was proposed by Louis de Broglie. We saw in Section 27–8 that in 1923, de Broglie proposed that material particles, such as electrons, have a wave nature; and that this hypothesis was confirmed by experiment several years later.

One of de Broglie's original arguments in favor of the wave nature of electrons was that it provided an explanation for Bohr's theory of the hydrogen atom. According to de Broglie, a particle of mass m moving with a nonrelativistic speed v would have a wavelength (Eq. 27–8) of

$$\lambda = \frac{h}{mv}.$$

Each electron orbit in an atom, he proposed, is actually a standing wave. As we saw in Chapter 11, when a violin or guitar string is plucked, a vast number of wavelengths are excited. But only certain ones—those that have nodes at the ends—are sustained. These are the *resonant* modes of the string. Waves with other wavelengths interfere with themselves upon reflection and their amplitudes quickly drop to zero. With electrons moving in circles, according to Bohr's theory, de Broglie argued that the electron wave was a *circular* standing wave that closes on itself, Fig. 27–30a. If the wavelength of a wave does not close on itself, as in Fig. 27–30b, destructive interference takes place as the wave travels around the loop, and the wave quickly dies out. Thus, the only waves that persist are those for which the circumference of the circular orbit contains a whole number of wavelengths, Fig. 27–31. The circumference of a Bohr orbit of radius r_n is $2\pi r_n$, so to have constructive interference, we need

$$2\pi r_n = n\lambda, \qquad n = 1, 2, 3, \cdots.$$

When we substitute $\lambda = h/mv$, we get $2\pi r_n = nh/mv$, or

$$mvr_n = \frac{nh}{2\pi}.$$

This is just the *quantum condition* proposed by Bohr on an *ad hoc* basis, Eq. 27–11. It is from this equation that the discrete orbits and energy levels were derived.

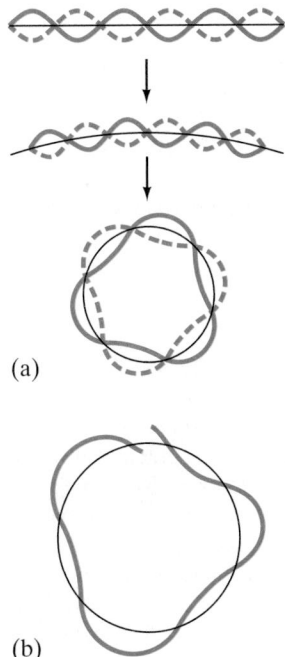

(a)

(b)

FIGURE 27–30 (a) An ordinary standing wave compared to a circular standing wave. (b) When a wave does not close (and hence interferes destructively with itself), it rapidly dies out.

FIGURE 27–31 Standing circular waves for two, three, and five wavelengths on the circumference; n, the number of wavelengths, is also the quantum number.

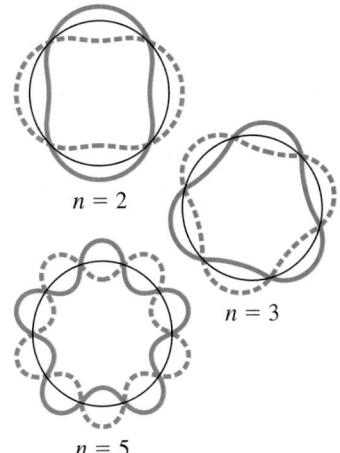

$n = 2$

$n = 3$

$n = 5$

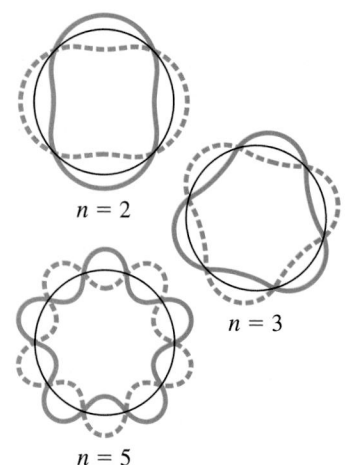

$n = 2$

$n = 3$

$n = 5$

FIGURE 27–31 (Repeated.) Standing circular waves for two, three, and five wavelengths on the circumference; n, the number of wavelengths, is also the quantum number.

Now we have a first explanation for the quantized orbits and energy states in the Bohr model: they are due to the wave nature of the electron, and only resonant "standing" waves can persist.[†] This implies that the *wave–particle duality* is at the root of atomic structure.

In viewing the circular electron waves of Fig. 27–31, the electron is not to be thought of as following the oscillating wave pattern. In the Bohr model of hydrogen, the electron moves in a circle. The circular wave, on the other hand, represents the *amplitude* of the electron "matter wave," and in Fig. 27–31 the wave amplitude is shown superimposed on the circular path of the particle orbit for convenience.

Bohr's theory worked well for hydrogen and for one-electron ions. But it did not prove successful for multi-electron atoms. Bohr theory could not predict line spectra even for the next simplest atom, helium. It could not explain why some emission lines are brighter than others, nor why some lines are split into two or more closely spaced lines ("fine structure"). A new theory was needed and was indeed developed in the 1920s. This new and radical theory is called *quantum mechanics*. It finally solved the problem of atomic structure, but it gives us a very different view of the atom: the idea of electrons in well-defined orbits was replaced with the idea of electron "clouds." This new theory of quantum mechanics has given us a wholly different view of the basic mechanisms underlying physical processes.

[†]We note, however, that Eq. 27–11 is no longer considered valid, as discussed in the next Chapter.

Summary

The electron was discovered using an evacuated cathode ray tube. The measurement of the charge-to-mass ratio (e/m) of the electron was done using magnetic and electric fields. The charge e on the electron was first measured in the Millikan oil-drop experiment and then its mass was obtained from the measured value of the e/m ratio.

Quantum theory has its origins in **Planck's quantum hypothesis** that molecular oscillations are **quantized**: their energy E can only be integer (n) multiples of hf, where h is Planck's constant and f is the natural frequency of oscillation:

$$E = nhf. \qquad (27-3)$$

This hypothesis explained the spectrum of radiation emitted by a **blackbody** at high temperature.

Einstein proposed that for some experiments, light could be pictured as being emitted and absorbed as **quanta** (particles), which we now call **photons**, each with energy

$$E = hf \qquad (27-4)$$

and momentum

$$p = \frac{E}{c} = \frac{hf}{c} = \frac{h}{\lambda}. \qquad (27-6)$$

He proposed the photoelectric effect as a test for the photon theory of light. In the **photoelectric effect**, the photon theory says that each incident photon can strike an electron in a material and eject it if the photon has sufficient energy. The maximum energy of ejected electrons is then linearly related to the frequency of the incident light.

The photon theory is also supported by the **Compton effect** and the observation of electron–positron **pair production**.

The **wave–particle duality** refers to the idea that light and matter (such as electrons) have both wave and particle properties. The wavelength of an object is given by

$$\lambda = \frac{h}{p}, \qquad (27-8)$$

where p is the momentum of the object ($p = mv$ for a particle of mass m and speed v).

The **principle of complementarity** states that we must be aware of both the particle and wave properties of light and of matter for a complete understanding of them.

Electron microscopes (EM) make use of the wave properties of electrons to form an image: their "lenses" are magnetic. Various types of EM exist: some can magnify 100,000× (1000× better than a light microscope); others can give a 3-D image.

Early models of the atom include Rutherford's planetary (or nuclear) model of an atom which consists of a tiny but massive positively charged nucleus surrounded (at a relatively great distance) by electrons.

To explain the **line spectra** emitted by atoms, as well as the stability of atoms, the **Bohr model** postulated that: (1) electrons bound in an atom can only occupy orbits for which the angular momentum is quantized, which results in discrete values for the radius and energy; (2) an electron in such a **stationary state** emits no radiation; (3) if an electron jumps to a lower state, it emits a photon whose energy equals the difference in energy between the two states; (4) the angular momentum L of atomic electrons is quantized by the rule $L = nh/2\pi$, where n is an integer called the **quantum number**. The $n = 1$ state is the **ground state**, which in hydrogen has an energy $E_1 = -13.6$ eV. Higher values of n correspond to **excited states**, and their energies are

$$E_n = -(13.6 \text{ eV}) \frac{Z^2}{n^2}, \qquad (27-15b)$$

where Ze is the charge on the nucleus. Atoms are excited to these higher states by collisions with other atoms or electrons, or by absorption of a photon of just the right frequency.

De Broglie's hypothesis that electrons (and other matter) have a wavelength $\lambda = h/mv$ gave an explanation for Bohr's quantized orbits by bringing in the wave–particle duality: the orbits correspond to circular standing waves in which the circumference of the orbit equals a whole number of wavelengths.

Questions

1. Does a lightbulb at a temperature of 2500 K produce as white a light as the Sun at 6000 K? Explain.

2. If energy is radiated by all objects, why can we not see them in the dark? (See also Section 14–8.)

3. What can be said about the relative temperatures of whitish-yellow, reddish, and bluish stars? Explain.

4. Darkrooms for developing black-and-white film were sometimes lit by a red bulb. Why red? Explain if such a bulb would work in a darkroom for developing color film.

5. If the threshold wavelength in the photoelectric effect increases when the emitting metal is changed to a different metal, what can you say about the work functions of the two metals?

6. Explain why the existence of a cutoff frequency in the photoelectric effect more strongly favors a particle theory rather than a wave theory of light.

7. UV light causes sunburn, whereas visible light does not. Suggest a reason.

8. The work functions for sodium and cesium are 2.28 eV and 2.14 eV, respectively. For incident photons of a given frequency, which metal will give a higher maximum kinetic energy for the electrons? Explain.

9. Explain how the photoelectric circuit of Fig. 27–6 could be used in (a) a burglar alarm, (b) a smoke detector, (c) a photographic light meter.

10. (a) Does a beam of infrared photons always have less energy than a beam of ultraviolet photons? Explain. (b) Does a single photon of infrared light always have less energy than a single photon of ultraviolet light? Why?

11. Light of 450-nm wavelength strikes a metal surface, and a stream of electrons emerges from the metal. If light of the same intensity but of wavelength 400 nm strikes the surface, are more electrons emitted? Does the energy of the emitted electrons change? Explain.

*12. If an X-ray photon is scattered by an electron, does the photon's wavelength change? If so, does it increase or decrease? Explain.

*13. In both the photoelectric effect and in the Compton effect, a photon collides with an electron causing the electron to fly off. What is the difference between the two processes?

14. Why do we say that light has wave properties? Why do we say that light has particle properties?

15. Why do we say that electrons have wave properties? Why do we say that electrons have particle properties?

16. What are the differences between a photon and an electron? Be specific: make a list.

17. If an electron and a proton travel at the same speed, which has the shorter wavelength? Explain.

18. An electron and a proton are accelerated through the same voltage. Which has the longer wavelength? Explain why.

19. In Rutherford's planetary model of the atom, what keeps the electrons from flying off into space?

20. When a wide spectrum of light passes through hydrogen gas at room temperature, absorption lines are observed that correspond only to the Lyman series. Why don't we observe the other series?

21. How can you tell if there is oxygen near the surface of the Sun?

22. (a) List at least three successes of the Bohr model of the atom, according to Section 27–12. (b) List at least two observations that the Bohr model could not explain, according to Section 27–13.

23. According to Section 27–11, what were the two main difficulties of the Rutherford model of the atom?

24. Is it possible for the de Broglie wavelength of a "particle" to be greater than the dimensions of the particle? To be smaller? Is there any direct connection? Explain.

25. How can the spectrum of hydrogen contain so many lines when hydrogen contains only one electron?

26. Explain how the closely spaced energy levels for hydrogen near the top of Fig. 27–29 correspond to the closely spaced spectral lines at the top of Fig. 27–24.

27. In a helium atom, which contains two electrons, do you think that on average the electrons are closer to the nucleus or farther away than in a hydrogen atom? Why?

28. The Lyman series is brighter than the Balmer series, because this series of transitions ends up in the most common state for hydrogen, the ground state. Why then was the Balmer series discovered first?

29. Use conservation of momentum to explain why photons emitted by hydrogen atoms have slightly less energy than that predicted by Eq. 27–10.

30. State if a continuous or a line spectrum is produced by each of the following: (a) a hot solid object; (b) an excited, rarefied gas; (c) a hot liquid; (d) light from a hot solid that passes through a cooler rarefied gas; (e) a hot dense gas. For each, if a line spectrum is produced, is it an emission or an absorption spectrum?

31. Suppose we obtain an emission spectrum for hydrogen at very high temperature (when some of the atoms are in excited states), and an absorption spectrum at room temperature, when all atoms are in the ground state. Will the two spectra contain identical lines?

MisConceptual Questions

1. Which of the following statements is true regarding how blackbody radiation changes as the temperature of the radiating object increases?
   (a) Both the maximum intensity and the peak wavelength increase.
   (b) The maximum intensity increases, and the peak wavelength decreases.
   (c) Both the maximum intensity and the peak wavelength decrease.
   (d) The maximum intensity decreases, and the peak wavelength increases.

2. As red light shines on a piece of metal, no electrons are released. When the red light is slowly changed to shorter-wavelength light (basically progressing through the rainbow), nothing happens until yellow light shines on the metal, at which point electrons are released from the metal. If this metal is replaced with a metal having a higher work function, which light would have the best chance of releasing electrons from the metal?
   (a) Blue.
   (b) Red.
   (c) Yellow would still work fine.
   (d) We need to know more about the metals involved.

3. A beam of red light and a beam of blue light have equal intensities. Which statement is true?
   (a) There are more photons in the blue beam.
   (b) There are more photons in the red beam.
   (c) Both beams contain the same number of photons.
   (d) The number of photons is not related to intensity.

4. Which of the following is necessarily true?
   (a) Red light has more energy than violet light.
   (b) Violet light has more energy than red light.
   (c) A single photon of red light has more energy than a single photon of violet light.
   (d) A single photon of violet light has more energy than a single photon of red light.
   (e) None of the above.
   (f) A combination of the above (specify).

5. If a photon of energy E ejects electrons from a metal with kinetic energy KE, then a photon with energy $E/2$
   (a) will eject electrons with kinetic energy KE/2.
   (b) will eject electrons with an energy greater than KE/2.
   (c) will eject electrons with an energy less than KE/2.
   (d) might not eject any electrons.

6. If the momentum of an electron were doubled, how would its wavelength change?
   (a) No change.
   (b) It would be halved.
   (c) It would double.
   (d) It would be quadrupled.
   (e) It would be reduced to one-fourth.

7. Which of the following can be thought of as either a wave or a particle?
   (a) Light.
   (b) An electron.
   (c) A proton.
   (d) All of the above.

8. When you throw a baseball, its de Broglie wavelength is
   (a) the same size as the ball.
   (b) about the same size as an atom.
   (c) about the same size as an atom's nucleus.
   (d) much smaller than the size of an atom's nucleus.

9. Electrons and photons of light are similar in that
   (a) both have momentum given by h/λ.
   (b) both exhibit wave–particle duality.
   (c) both are used in diffraction experiments to explore structure.
   (d) All of the above.
   (e) None of the above.

10. In Rutherford's famous set of experiments described in Section 27–10, the fact that some alpha particles were deflected at large angles indicated that (choose all that apply)
    (a) the nucleus was positive.
    (b) charge was quantized.
    (c) the nucleus was concentrated in a small region of space.
    (d) most of the atom is empty space.
    (e) None of the above.

11. Which of the following electron transitions between two energy states (n) in the hydrogen atom corresponds to the emission of a photon with the longest wavelength?
    (a) $2 \rightarrow 5$.
    (b) $5 \rightarrow 2$.
    (c) $5 \rightarrow 8$.
    (d) $8 \rightarrow 5$.

12. If we set the potential energy of an electron and a proton to be zero when they are an infinite distance apart, then the lowest energy a bound electron in a hydrogen atom can have is
    (a) 0.
    (b) $-13.6\,\text{eV}$.
    (c) any possible value.
    (d) any value between $-13.6\,\text{eV}$ and 0.

13. Which of the following is the currently accepted model of the atom?
    (a) The plum-pudding model.
    (b) The Rutherford atom.
    (c) The Bohr atom.
    (d) None of the above.

14. Light has all of the following except:
    (a) mass.
    (b) momentum.
    (c) kinetic energy.
    (d) frequency.
    (e) wavelength.

Problems

27–1 Discovery of the Electron

1. (I) What is the value of e/m for a particle that moves in a circle of radius 14 mm in a 0.86-T magnetic field if a perpendicular 640-V/m electric field will make the path straight?

2. (II) (a) What is the velocity of a beam of electrons that go undeflected when passing through crossed (perpendicular) electric and magnetic fields of magnitude 1.88×10^4 V/m and 2.60×10^{-3} T, respectively? (b) What is the radius of the electron orbit if the electric field is turned off?

3. (II) An oil drop whose mass is 2.8×10^{-15} kg is held at rest between two large plates separated by 1.0 cm (Fig. 27–3), when the potential difference between the plates is 340 V. How many excess electrons does this drop have?

27–2 Blackbodies; Planck's Quantum Hypothesis

4. (I) How hot is a metal being welded if it radiates most strongly at 520 nm?

5. (I) Estimate the peak wavelength for radiation emitted from (a) ice at 0°C, (b) a floodlamp at 3100 K, (c) helium at 4 K, assuming blackbody emission. In what region of the EM spectrum is each?

6. (I) (a) What is the temperature if the peak of a blackbody spectrum is at 18.0 nm? (b) What is the wavelength at the peak of a blackbody spectrum if the body is at a temperature of 2200 K?

7. (I) An HCl molecule vibrates with a natural frequency of 8.1×10^{13} Hz. What is the difference in energy (in joules and electron volts) between successive values of the oscillation energy?

8. (II) The steps of a flight of stairs are 20.0 cm high (vertically). If a 62.0-kg person stands with both feet on the same step, what is the gravitational potential energy of this person, relative to the ground, on (a) the first step, (b) the second step, (c) the third step, (d) the nth step? (e) What is the change in energy as the person descends from step 6 to step 2?

9. (II) Estimate the peak wavelength of light emitted from the pupil of the human eye (which approximates a blackbody) assuming normal body temperature.

27–3 and 27–4 Photons and the Photoelectric Effect

10. (I) What is the energy of photons (joules) emitted by a 91.7-MHz FM radio station?

11. (I) What is the energy range (in joules and eV) of photons in the visible spectrum, of wavelength 400 nm to 750 nm?

12. (I) A typical gamma ray emitted from a nucleus during radioactive decay may have an energy of 320 keV. What is its wavelength? Would we expect significant diffraction of this type of light when it passes through an everyday opening, such as a door?

13. (I) Calculate the momentum of a photon of yellow light of wavelength 5.80×10^{-7} m.

14. (I) What is the momentum of a $\lambda = 0.014$ nm X-ray photon?

15. (I) For the photoelectric effect, make a table that shows expected observations for a particle theory of light and for a wave theory of light. Circle the actual observed effects. (See Section 27–3.)

16. (II) About 0.1 eV is required to break a "hydrogen bond" in a protein molecule. Calculate the minimum frequency and maximum wavelength of a photon that can accomplish this.

17. (II) What minimum frequency of light is needed to eject electrons from a metal whose work function is 4.8×10^{-19} J?

18. (II) The human eye can respond to as little as 10^{-18} J of light energy. For a wavelength at the peak of visual sensitivity, 550 nm, how many photons lead to an observable flash?

19. (II) What is the longest wavelength of light that will emit electrons from a metal whose work function is 2.90 eV?

20. (II) The work functions for sodium, cesium, copper, and iron are 2.3, 2.1, 4.7, and 4.5 eV, respectively. Which of these metals will not emit electrons when visible light shines on it?

21. (II) In a photoelectric-effect experiment it is observed that no current flows unless the wavelength is less than 550 nm. (a) What is the work function of this material? (b) What stopping voltage is required if light of wavelength 400 nm is used?

22. (II) What is the maximum kinetic energy of electrons ejected from barium $(W_0 = 2.48$ eV$)$ when illuminated by white light, $\lambda = 400$ to 750 nm?

23. (II) Barium has a work function of 2.48 eV. What is the maximum kinetic energy of electrons if the metal is illuminated by UV light of wavelength 365 nm? What is their speed?

24. (II) When UV light of wavelength 255 nm falls on a metal surface, the maximum kinetic energy of emitted electrons is 1.40 eV. What is the work function of the metal?

25. (II) The threshold wavelength for emission of electrons from a given surface is 340 nm. What will be the maximum kinetic energy of ejected electrons when the wavelength is changed to (a) 280 nm, (b) 360 nm?

26. (II) A certain type of film is sensitive only to light whose wavelength is less than 630 nm. What is the energy (eV and kcal/mol) needed for the chemical reaction to occur which causes the film to change?

27. (II) When 250-nm light falls on a metal, the current through a photoelectric circuit (Fig. 27–6) is brought to zero at a stopping voltage of 1.64 V. What is the work function of the metal?

28. (II) In a photoelectric experiment using a clean sodium surface, the maximum energy of the emitted electrons was measured for a number of different incident frequencies, with the following results.

Frequency ($\times 10^{14}$ Hz)	Energy (eV)
11.8	2.60
10.6	2.11
9.9	1.81
9.1	1.47
8.2	1.10
6.9	0.57

Plot the graph of these results and find: (a) Planck's constant; (b) the cutoff frequency of sodium; (c) the work function.

29. (II) Show that the energy E (in electron volts) of a photon whose wavelength is λ (nm) is given by

$$E = \frac{1.240 \times 10^3 \text{ eV} \cdot \text{nm}}{\lambda \text{ (nm)}}.$$

Use at least 4 significant figures for values of h, c, e (see inside front cover).

*27–5 Compton Effect

*30. (I) A high-frequency photon is scattered off of an electron and experiences a change of wavelength of 1.7×10^{-4} nm. At what angle must a detector be placed to detect the scattered photon (relative to the direction of the incoming photon)?

*31. (II) The quantity h/mc, which has the dimensions of length, is called the *Compton wavelength*. Determine the Compton wavelength for (a) an electron, (b) a proton. (c) Show that if a photon has wavelength equal to the Compton wavelength of a particle, the photon's energy is equal to the rest energy of the particle, mc^2.

*32. (II) X-rays of wavelength $\lambda = 0.140$ nm are scattered from carbon. What is the expected Compton wavelength shift for photons detected at angles (relative to the incident beam) of exactly (a) 45°, (b) 90°, (c) 180°?

27–6 Pair Production

33. (I) How much total kinetic energy will an electron–positron pair have if produced by a 3.64-MeV photon?

34. (II) What is the longest wavelength photon that could produce a proton–antiproton pair? (Each has a mass of 1.67×10^{-27} kg.)

35. (II) What is the minimum photon energy needed to produce a $\mu^+\mu^-$ pair? The mass of each μ (muon) is 207 times the mass of an electron. What is the wavelength of such a photon?

36. (II) An electron and a positron, each moving at 3.0×10^5 m/s, collide head on, disappear, and produce two photons, each with the same energy and momentum moving in opposite directions. Determine the energy and momentum of each photon.

37. (II) A gamma-ray photon produces an electron and a positron, each with a kinetic energy of 285 keV. Determine the energy and wavelength of the photon.

27–8 Wave Nature of Matter

38. (I) Calculate the wavelength of a 0.21-kg ball traveling at 0.10 m/s.

39. (I) What is the wavelength of a neutron $\left(m = 1.67 \times 10^{-27} \text{ kg}\right)$ traveling at 8.5×10^4 m/s?

40. (II) Through how many volts of potential difference must an electron, initially at rest, be accelerated to achieve a wavelength of 0.27 nm?

41. (II) Calculate the ratio of the kinetic energy of an electron to that of a proton if their wavelengths are equal. Assume that the speeds are nonrelativistic.

42. (II) An electron has a de Broglie wavelength $\lambda = 4.5 \times 10^{-10}$ m. (a) What is its momentum? (b) What is its speed? (c) What voltage was needed to accelerate it from rest to this speed?

43. (II) What is the wavelength of an electron of energy (a) 10 eV, (b) 100 eV, (c) 1.0 keV?

44. (II) Show that if an electron and a proton have the same nonrelativistic kinetic energy, the proton has the shorter wavelength.

45. (II) Calculate the de Broglie wavelength of an electron if it is accelerated from rest by 35,000 V as in Fig. 27–2. Is it relativistic? How does its wavelength compare to the size of the "neck" of the tube, typically 5 cm? Do we have to worry about diffraction problems blurring the picture on the CRT screen?

46. (III) A Ferrari with a mass of 1400 kg approaches a freeway underpass that is 12 m across. At what speed must the car be moving, in order for it to have a wavelength such that it might somehow "diffract" after passing through this "single slit"? How do these conditions compare to normal freeway speeds of 30 m/s?

27–9 Electron Microscope

47. (II) What voltage is needed to produce electron wavelengths of 0.26 nm? (Assume that the electrons are nonrelativistic.)

48. (II) Electrons are accelerated by 2850 V in an electron microscope. Estimate the maximum possible resolution of the microscope.

27–11 and 27–12 Spectra and the Bohr Model

49. (I) For the three hydrogen transitions indicated below, with n being the initial state and n' being the final state, is the transition an absorption or an emission? Which is higher, the initial state energy or the final state energy of the atom? Finally, which of these transitions involves the largest energy photon? (a) $n = 1$, $n' = 3$; (b) $n = 6$, $n' = 2$; (c) $n = 4$, $n' = 5$.

50. (I) How much energy is needed to ionize a hydrogen atom in the $n = 3$ state?

51. (I) The second longest wavelength in the Paschen series in hydrogen (Fig. 27–29) corresponds to what transition?

52. (I) Calculate the ionization energy of doubly ionized lithium, Li^{2+}, which has $Z = 3$ (and is in the ground state).

53. (I) (a) Determine the wavelength of the second Balmer line ($n = 4$ to $n = 2$ transition) using Fig. 27–29. Determine likewise (b) the wavelength of the second Lyman line and (c) the wavelength of the third Balmer line.

54. (I) Evaluate the Rydberg constant R using the Bohr model (compare Eqs. 27–9 and 27–16) and show that its value is $R = 1.0974 \times 10^7 \text{ m}^{-1}$. (Use values inside front cover to 5 or 6 significant figures.)

55. (II) What is the longest wavelength light capable of ionizing a hydrogen atom in the ground state?

56. (II) What wavelength photon would be required to ionize a hydrogen atom in the ground state and give the ejected electron a kinetic energy of 11.5 eV?

57. (II) In the Sun, an ionized helium (He^+) atom makes a transition from the $n = 6$ state to the $n = 2$ state, emitting a photon. Can that photon be absorbed by hydrogen atoms present in the Sun? If so, between what energy states will the hydrogen atom transition occur?

58. (II) Construct the energy-level diagram for the He^+ ion (like Fig. 27–29).

59. (II) Construct the energy-level diagram for doubly ionized lithium, Li^{2+}.

60. (II) Determine the electrostatic potential energy and the kinetic energy of an electron in the ground state of the hydrogen atom.

61. (II) A hydrogen atom has an angular momentum of 5.273×10^{-34} kg·m²/s. According to the Bohr model, what is the energy (eV) associated with this state?

62. (II) An excited hydrogen atom could, in principle, have a radius of 1.00 cm. What would be the value of n for a Bohr orbit of this size? What would its energy be?

63. (II) Is the use of nonrelativistic formulas justified in the Bohr atom? To check, calculate the electron's velocity, v, in terms of c, for the ground state of hydrogen, and then calculate $\sqrt{1 - v^2/c^2}$.

64. (III) Show that the magnitude of the electrostatic potential energy of an electron in any Bohr orbit of a hydrogen atom is twice the magnitude of its kinetic energy in that orbit.

65. (III) Suppose an electron was bound to a proton, as in the hydrogen atom, but by the gravitational force rather than by the electric force. What would be the radius, and energy, of the first Bohr orbit?

General Problems

66. The Big Bang theory (Chapter 33) states that the beginning of the universe was accompanied by a huge burst of photons. Those photons are still present today and make up the so-called cosmic microwave background radiation. The universe radiates like a blackbody with a temperature today of about 2.7 K. Calculate the peak wavelength of this radiation.

67. At low temperatures, nearly all the atoms in hydrogen gas will be in the ground state. What minimum frequency photon is needed if the photoelectric effect is to be observed?

68. A beam of 72-eV electrons is scattered from a crystal, as in X-ray diffraction, and a first-order peak is observed at $\theta = 38°$. What is the spacing between planes in the diffracting crystal? (See Section 25–11.)

69. A microwave oven produces electromagnetic radiation at $\lambda = 12.2$ cm and produces a power of 720 W. Calculate the number of microwave photons produced by the microwave oven each second.

70. Sunlight reaching the Earth's atmosphere has an intensity of about 1300 W/m^2. Estimate how many photons per square meter per second this represents. Take the average wavelength to be 550 nm.

71. A beam of red laser light ($\lambda = 633$ nm) hits a black wall and is fully absorbed. If this light exerts a total force $F = 5.8$ nN on the wall, how many photons per second are hitting the wall?

72. A flashlight emits 2.5 W of light. As the light leaves the flashlight in one direction, a reaction force is exerted on the flashlight in the opposite direction. Estimate the size of this reaction force.

73. A **photomultiplier tube** (a very sensitive light sensor), is based on the photoelectric effect: incident photons strike a metal surface and the resulting ejected electrons are collected. By counting the number of collected electrons, the number of incident photons (i.e., the incident light intensity) can be determined. (a) If a photomultiplier tube is to respond properly for incident wavelengths throughout the visible range (410 nm to 750 nm), what is the maximum value for the work function W_0 (eV) of its metal surface? (b) If W_0 for its metal surface is above a certain threshold value, the photomultiplier will only function for incident ultraviolet wavelengths and be unresponsive to visible light. Determine this threshold value (eV).

74. If a 100-W lightbulb emits 3.0% of the input energy as visible light (average wavelength 550 nm) uniformly in all directions, estimate how many photons per second of visible light will strike the pupil (4.0 mm diameter) of the eye of an observer, (a) 1.0 m away, (b) 1.0 km away.

75. An electron and a positron collide head on, annihilate, and create two 0.85-MeV photons traveling in opposite directions. What were the initial kinetic energies of electron and positron?

76. By what potential difference must (a) a proton $\left(m = 1.67 \times 10^{-27} \text{ kg}\right)$, and (b) an electron $\left(m = 9.11 \times 10^{-31} \text{ kg}\right)$, be accelerated from rest to have a wavelength $\lambda = 4.0 \times 10^{-12}$ m?

77. In some of Rutherford's experiments (Fig. 27–19) the α particles (mass = 6.64×10^{-27} kg) had a kinetic energy of 4.8 MeV. How close could they get to the surface of a gold nucleus (radius $\approx 7.0 \times 10^{-15}$ m, charge = $+79e$)? Ignore the recoil motion of the nucleus.

78. By what fraction does the mass of an H atom decrease when it makes an $n = 3$ to $n = 1$ transition?

79. Calculate the ratio of the gravitational force to the electric force for the electron in the ground state of a hydrogen atom. Can the gravitational force be reasonably ignored?

80. Electrons accelerated from rest by a potential difference of 12.3 V pass through a gas of hydrogen atoms at room temperature. What wavelengths of light will be emitted?

81. In a particular photoelectric experiment, a stopping potential of 2.10 V is measured when ultraviolet light of wavelength 270 nm is incident on the metal. Using the same setup, what will the new stopping potential be if blue light of wavelength 440 nm is used, instead?

82. Neutrons can be used in diffraction experiments to probe the lattice structure of crystalline solids. Since the neutron's wavelength needs to be on the order of the spacing between atoms in the lattice, about 0.3 nm, what should the speed of the neutrons be?

83. In Chapter 22, the intensity of light striking a surface was related to the electric field of the associated electromagnetic wave. For photons, the intensity is the number of photons striking a 1-m^2 area per second. Suppose 1.0×10^{12} photons of 497-nm light are incident on a 1-m^2 surface every second. What is the intensity of the light? Using the wave model of light, what is the maximum electric field of the electromagnetic wave?

84. The intensity of the Sun's light in the vicinity of the Earth is about 1350 W/m^2. Imagine a spacecraft with a mirrored square sail of dimension 1.0 km. Estimate how much thrust (in newtons) this craft will experience due to collisions with the Sun's photons. [*Hint*: Assume the photons bounce off the sail with no change in the magnitude of their momentum.]

85. Light of wavelength 280 nm strikes a metal whose work function is 2.2 eV. What is the shortest de Broglie wavelength for the electrons that are produced as photoelectrons?

86. Photons of energy 6.0 eV are incident on a metal. It is found that current flows from the metal until a stopping potential of 3.8 V is applied. If the wavelength of the incident photons is doubled, what is the maximum kinetic energy of the ejected electrons? What would happen if the wavelength of the incident photons was tripled?

87. What would be the theoretical limit of resolution for an electron microscope whose electrons are accelerated through 110 kV? (Relativistic formulas should be used.)

88. Assume hydrogen atoms in a gas are initially in their ground state. If free electrons with kinetic energy 12.75 eV collide with these atoms, what photon wavelengths will be emitted by the gas?

89. Visible light incident on a diffraction grating with slit spacing of 0.010 mm has the first maximum at an angle of 3.6° from the central peak. If electrons could be diffracted by the same grating, what electron velocity would produce the same diffraction pattern as the visible light?

90. (a) Suppose an unknown element has an absorption spectrum with lines corresponding to 2.5, 4.7, and 5.1 eV above its ground state and an ionization energy of 11.5 eV. Draw an energy level diagram for this element. (b) If a 5.1-eV photon is absorbed by an atom of this substance, in which state was the atom before absorbing the photon? What will be the energies of the photons that can subsequently be emitted by this atom?

91. A photon of momentum 3.53×10^{-28} kg·m/s is emitted from a hydrogen atom. To what spectrum series does this photon belong, and from what energy level was it ejected?

92. Light of wavelength 464 nm falls on a metal which has a work function of 2.28 eV. (a) How much voltage should be applied to bring the current to zero? (b) What is the maximum speed of the emitted electrons? (c) What is the de Broglie wavelength of these electrons?

93. An electron accelerated from rest by a 96-V potential difference is injected into a 3.67×10^{-4} T magnetic field where it travels in an 18-cm-diameter circle. Calculate e/m from this information.

94. Estimate the number of photons emitted by the Sun in a year. (Take the average wavelength to be 550 nm and the intensity of sunlight reaching the Earth (outer atmosphere) as 1350 W/m^2.)

95. Apply Bohr's assumptions to the Earth–Moon system to calculate the allowed energies and radii of motion. Given the known distance between the Earth and Moon, is the quantization of the energy and radius apparent?

96. At what temperature would the average kinetic energy (Chapter 13) of a molecule of hydrogen gas (H$_2$) be sufficient to excite a hydrogen atom out of the ground state?

Search and Learn

1. Name the person or people who did each of the following: (a) made the first direct measurement of the charge-to-mass ratio of the electron (Section 27–1); (b) measured the charge on the electron and showed that it is quantized (Section 27–1); (c) proposed the radical assumption that the vibrational energy of molecules in a radiating object is quantized (Sections 27–2, 27–3); (d) found that light (X-rays) scattered off electrons in a material will decrease the energy of the photons (Section 27–5); (e) proposed that the wavelength of a material particle would be related to its momentum in the same way as for a photon (Section 27–8); (f) performed the first crucial experiment illustrating electron diffraction (Section 27–8); (g) deciphered the nuclear model of the atom by aiming α particles at gold foil (Section 27–10).

2. State the principle of complementarity, and give at least two experimental results that support this principle for electrons and for photons. (See Section 27–7 and also Sections 27–3 and 27–8.)

3. Imagine the following Young's double-slit experiment using matter rather than light: electrons are accelerated through a potential difference of 12 V, pass through two closely spaced slits separated by a distance d, and create an interference pattern. (a) Using Example 27–11 and Section 24–3 as guides, find the required value for d if the first-order interference fringe is to be produced at an angle of 10°. (b) Given the approximate size of atoms, would it be possible to construct the required two-slit set-up for this experiment?

4. Does each of the following support the wave nature or the particle nature of light? (a) The existence of the cutoff frequency in the photoelectric effect; (b) Young's double-slit experiment; (c) the shift in the photon frequency in Compton scattering; (d) the diffraction of light.

5. (a) From Sections 22–3, 24–4, and 27–3, estimate the minimum energy (eV) that initiates the chemical process on the retina responsible for vision. (b) Estimate the threshold photon energy above which the eye registers no sensation of sight.

6. (a) A rubidium atom ($m = 85$ u) is at rest with one electron in an excited energy level. When the electron jumps to the ground state, the atom emits a photon of wavelength $\lambda = 780$ nm. Determine the resulting (nonrelativistic) recoil speed v of the atom. (b) The recoil speed sets the lower limit on the temperature to which an ideal gas of rubidium atoms can be cooled in a laser-based **atom trap**. Using the kinetic theory of gases (Chapter 13), estimate this "lowest achievable" temperature.

7. Suppose a particle of mass m is confined to a one-dimensional box of width L. According to quantum theory, the particle's wave (with $\lambda = h/mv$) is a standing wave with nodes at the edges of the box. (a) Show the possible modes of vibration on a diagram. (b) Show that the kinetic energy of the particle has quantized energies given by KE $= n^2h^2/8mL^2$, where n is an integer. (c) Calculate the ground-state energy ($n = 1$) for an electron confined to a box of width 0.50×10^{-10} m. (d) What is the ground-state energy, and speed, of a baseball ($m = 140$ g) in a box 0.65 m wide? (e) An electron confined to a box has a ground-state energy of 22 eV. What is the width of the box? [Hint: See Sections 27–8, 27–13, and 11–12.]

ANSWERS TO EXERCISES

A: $\lambda_p = 725$ nm, so red.
B: More 1000-nm photons (each has lower energy).
C: 5.50×10^{14} Hz, 545 nm.

D: Only λ.
E: Decrease.
F: (e).

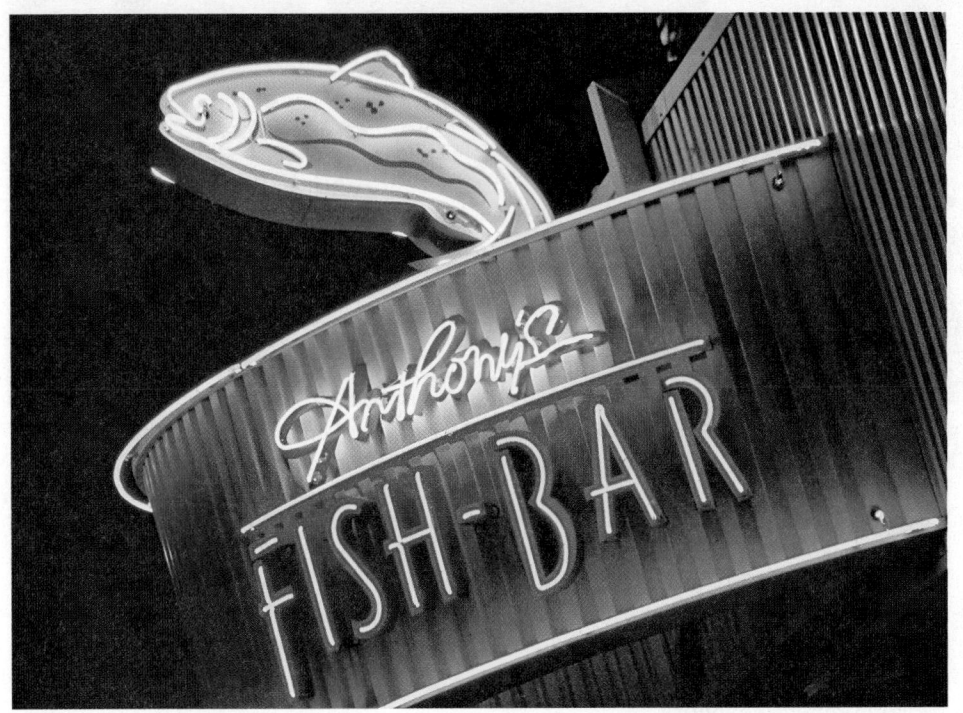

A neon tube is a thin glass tube, moldable into various shapes, filled with neon (or other) gas that glows with a particular color when a current at high voltage passes through it. Gas atoms, excited to upper energy levels, jump down to lower energy levels and emit light (photons) whose wavelengths (color) are characteristic of the type of gas.

In this Chapter we study what quantum mechanics tells us about atoms, their energy levels, and the effect of the exclusion principle for atoms with more than one electron. We also discuss interesting applications such as lasers and holography.

Quantum Mechanics of Atoms

CONTENTS

CHAPTER-OPENING QUESTION—Guess now!

The uncertainty principle states that

(a) no measurement can be perfect because it is technologically impossible to make perfect measuring instruments.

(b) it is impossible to measure exactly where a particle is, unless it is at rest.

(c) it is impossible to simultaneously know both the position and the momentum of a particle with complete certainty.

(d) a particle cannot actually have a completely certain value of momentum.

Bohr's model of the atom gave us a first (though rough) picture of what an atom is like. It proposed explanations for why there is emission and absorption of light by atoms at only certain wavelengths. The wavelengths of the line spectra and the ionization energy for hydrogen (and one-electron ions) are in excellent agreement with experiment. But the Bohr model had important limitations. It was not able to predict line spectra for more complex atoms—atoms with more than one electron—not even for the neutral helium atom, which has only two electrons. Nor could it explain why emission lines, when viewed with great precision, consist of two or more very closely spaced lines (referred to as *fine structure*). The Bohr model also did not explain why some spectral lines were brighter than others. And it could not explain the bonding of atoms in molecules or in solids and liquids.

From a theoretical point of view, too, the Bohr model was not satisfactory: it was a strange mixture of classical and quantum ideas. Moreover, the wave–particle duality was not really resolved.

FIGURE 28–1 Erwin Schrödinger with Lise Meitner (codiscoverers of nuclear fission, Chapter 31).

FIGURE 28–2 Werner Heisenberg (center) on Lake Como (Italy) with Enrico Fermi (left) and Wolfgang Pauli (right).

We mention these limitations of the Bohr model not to disparage it—for it was a landmark in the history of science. Rather, we mention them to show why, in the early 1920s, it became increasingly evident that a new, more comprehensive theory was needed. It was not long in coming. Less than two years after de Broglie gave us his matter–wave hypothesis, Erwin Schrödinger (1887–1961; Fig. 28–1) and Werner Heisenberg (1901–1976; Fig. 28–2) independently developed a new comprehensive theory.

28–1 Quantum Mechanics—A New Theory

The new theory, called **quantum mechanics**, has been extremely successful. It unifies the wave–particle duality into a single consistent theory and has successfully dealt with the spectra emitted by complex atoms, even the fine details. It explains the relative brightness of spectral lines and how atoms form molecules. It is also a much more general theory that covers all quantum phenomena from blackbody radiation to atoms and molecules. It has explained a wide range of natural phenomena and from its predictions many new practical devices have become possible. Indeed, it has been so successful that it is accepted today by nearly all physicists as the fundamental theory underlying physical processes.

Quantum mechanics deals mainly with the microscopic world of atoms and light. But this new theory, when it is applied to macroscopic phenomena, must be able to produce the old classical laws. This, the **correspondence principle** (already mentioned in Section 27–12), is satisfied fully by quantum mechanics.

This doesn't mean we should throw away classical theories such as Newton's laws. In the everyday world, classical laws are far easier to apply and they give sufficiently accurate descriptions. But when we deal with high speeds, close to the speed of light, we must use the theory of relativity; and when we deal with the tiny world of the atom, we use quantum mechanics.

Although we won't go into the detailed mathematics of quantum mechanics, we will discuss the main ideas and how they involve the wave and particle properties of matter to explain atomic structure and other applications.

28–2 The Wave Function and Its Interpretation; the Double-Slit Experiment

The important properties of any wave are its wavelength, frequency, and amplitude. For an electromagnetic wave, the frequency (or wavelength) determines whether the light is in the visible spectrum or not, and if so, what color it is. We also have seen that the frequency is a measure of the energy of the corresponding photon, $E = hf$ (Eq. 27–4). The amplitude or displacement of an electromagnetic wave at any point is the strength of the electric (or magnetic) field at that point, and is related to the intensity of the wave (the brightness of the light).

For material particles such as electrons, quantum mechanics relates the wavelength to momentum according to de Broglie's formula, $\lambda = h/p$, Eq. 27–8. But what corresponds to the *amplitude* or *displacement* of a matter wave? The amplitude of an electromagnetic wave is represented by the electric and magnetic fields, E and B. In quantum mechanics, this role is played by the **wave function**, which is given the symbol Ψ (the Greek capital letter psi, pronounced "sigh"). Thus Ψ represents the wave displacement, as a function of time and position, of a new kind of field which we might call a "matter" field or a matter wave.

To understand how to interpret the wave function Ψ, we make an analogy with light using the wave–particle duality.

We saw in Chapter 11 that the intensity I of any wave is proportional to the square of the amplitude. This holds true for light waves as well, as we saw in Chapter 22. That is,

$$I \propto E^2,$$

where E is the electric field strength. From the *particle* point of view, the intensity of a light beam (of given frequency) is proportional to the number of photons, N, that pass through a given area per unit time. The more photons there are, the greater the intensity. Thus

$$I \propto E^2 \propto N.$$

This proportion can be turned around so that we have

$$N \propto E^2.$$

That is, the number of photons (striking a page of this book, say) is proportional to the square of the electric field strength.

If the light beam is very weak, only a few photons will be involved. Indeed, it is possible to "build up" a photograph in a camera using very weak light so the effect of photons arriving can be seen. If we are dealing with only one photon, the relationship above $\left(N \propto E^2\right)$ can be interpreted in a slightly different way. At any point, the square of the electric field strength E^2 is a measure of the *probability* that a photon will be at that location. At points where E^2 is large, there is a high probability the photon will be there; where E^2 is small, the probability is low.

We can interpret matter waves in the same way, as was first suggested by Max Born (1882–1970) in 1927. The wave function Ψ may vary in magnitude from point to point in space and time. If Ψ describes a collection of many electrons, then Ψ^2 at any point will be proportional to the number of electrons expected to be found at that point. When dealing with small numbers of electrons we can't make very exact predictions, so Ψ^2 takes on the character of a probability. If Ψ, which depends on time and position, represents a single electron (say, in an atom), then Ψ^2 is interpreted like this: Ψ^2 *at a certain point in space and time represents the probability of finding the electron at the given position and time.* Thus Ψ^2 is often referred to as the **probability density** or **probability distribution**.

Double-Slit Interference Experiment for Electrons

To understand this better, we take as a thought experiment the familiar double-slit experiment, and consider it both for light and for electrons.

Consider two slits whose size and separation are on the order of the wavelength of whatever we direct at them, either light or electrons, Fig. 28–3. We know very well what would happen in this case for light, since this is just Young's double-slit experiment (Section 24–3): an interference pattern would be seen on the screen behind. If light were replaced by electrons with wavelength comparable to the slit size, they too would produce an interference pattern (recall Fig. 27–12). In the case of light, the pattern would be visible to the eye or could be recorded on film, semiconductor sensor, or screen. For electrons, a fluorescent screen could be used (it glows where an electron strikes).

FIGURE 28–3 Parallel beam, of light or electrons, falls on two slits whose sizes are comparable to the wavelength. An interference pattern is observed.

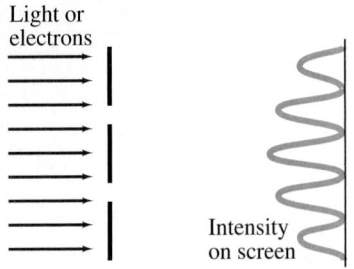

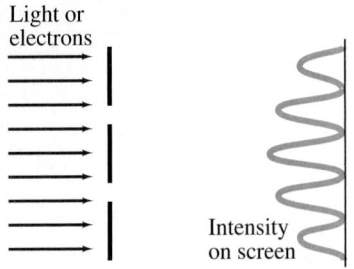Light or electrons → → → → → → → → | | | | Intensity on screen

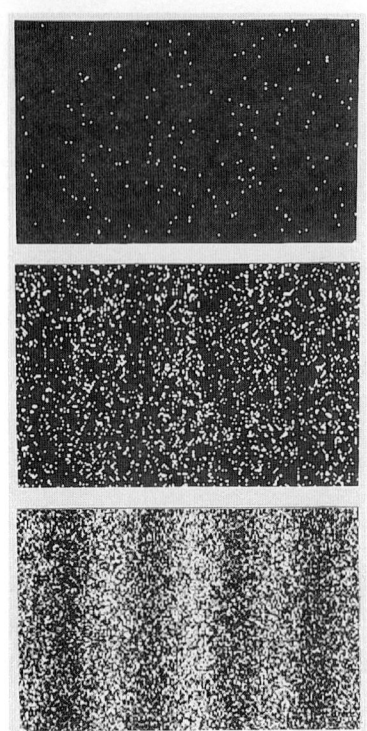

FIGURE 28–4 Young's double-slit experiment done with electrons—note that the pattern is not evident with only a few electrons (top photo), but with more and more electrons (second and third photos), the familiar double-slit interference pattern (Chapter 24) is seen.

If we reduced the flow of electrons (or photons) so they passed through the slits one at a time, we would see a flash each time one struck the screen. At first, the flashes would seem random. Indeed, there is no way to predict just where any one electron would hit the screen. If we let the experiment run for a long time, and kept track of where each electron hit the screen, we would soon see a pattern emerging—the interference pattern predicted by the wave theory; see Fig. 28–4. Thus, although we could not predict where a given electron would strike the screen, we could predict probabilities. (The same can be said for photons.) The probability, as we saw, is proportional to Ψ^2. Where Ψ^2 is zero, we would get a minimum in the interference pattern. And where Ψ^2 is a maximum, we would get a peak in the interference pattern.

The interference pattern would thus occur even when electrons (or photons) passed through the slits one at a time. So the interference pattern could not arise from the interaction of one electron with another. It is as if an electron passed through both slits at the same time, interfering with itself. This is possible because an electron is not precisely a particle. It is as much a wave as it is a particle, and a wave could travel through both slits at once. But what would happen if we covered one of the slits so we knew that the electron passed through the other slit, and a little later we covered the second slit so the electron had to have passed through the first slit? The result would be that no interference pattern would be seen. We would see, instead, two bright areas (or diffraction patterns) on the screen behind the slits.

If both slits are open, the screen shows an interference pattern as if each electron passed through both slits, like a wave. Yet each electron would make a tiny spot on the screen as if it were a particle.

The main point of this discussion is this: if we treat electrons (and other particles) as if they were waves, then Ψ represents the wave amplitude. If we treat them as particles, then we must treat them on a *probabilistic* basis. The square of the wave function, Ψ^2, gives the probability of finding a given electron at a given point. We cannot predict—or even follow—the path of a single electron precisely through space and time.

28–3 The Heisenberg Uncertainty Principle

Whenever a measurement is made, some uncertainty is always involved. For example, you cannot make an absolutely exact measurement of the length of a table. Even with a measuring stick that has markings 1 mm apart, there will be an inaccuracy of perhaps $\frac{1}{2}$ mm or so. More precise instruments will produce more precise measurements. But there is always some uncertainty involved in a measurement, no matter how good the measuring device. We expect that by using more precise instruments, the uncertainty in a measurement can be made indefinitely small.

But according to quantum mechanics, there is actually a limit to the precision of certain measurements. This limit is not a restriction on how well instruments can be made; rather, it is inherent in nature. It is the result of two factors: the wave–particle duality, and the unavoidable interaction between the thing observed and the observing instrument. Let us look at this in more detail.

To make a measurement on an object without disturbing it, at least a little, is not possible. Consider trying to locate a lost Ping-pong ball in a dark room: you could probe about with your hand or a stick, or you could shine a light and detect the photons reflecting off the ball. When you search with your hand or a stick, you find the ball's position when you touch it, but at the same time you unavoidably bump it, and give it some momentum. Thus you won't know its *future* position. If you search for the Ping-pong ball using light, in order to "see" the ball at least one photon (really, quite a few) must scatter from it, and the reflected photon must enter your eye or some other detector. When a photon strikes an ordinary-sized object, it only slightly alters the motion or position of the object.

But a photon striking a tiny object like an electron transfers enough momentum to greatly change the electron's motion and position in an unpredictable way. The mere act of measuring the position of an object at one time makes our knowledge of its future position imprecise.

Now let us see where the wave–particle duality comes in. Imagine a thought experiment in which we are trying to measure the position of an object, say an electron, with photons, Fig. 28–5. (The arguments would be similar if we were using, instead, an electron microscope.) As we saw in Chapter 25, objects can be seen to a precision at best of about the wavelength of the radiation used due to diffraction. If we want a precise position measurement, we must use a short wavelength. But a short wavelength corresponds to high frequency and large momentum ($p = h/\lambda$); and the more momentum the photons have, the more momentum they can give the object when they strike it. If we use photons of longer wavelength, and correspondingly smaller momentum, the object's motion when struck by the photons will not be affected as much. But the longer wave-length means lower resolution, so the object's position will be less accurately known. Thus the act of observing produces an uncertainty in both the *position* and the *momentum* of the electron. This is the essence of the *uncertainty principle* first enunciated by Heisenberg in 1927.

Quantitatively, we can make an approximate calculation of the magnitude of the uncertainties. If we use light of wavelength λ, the position can be measured at best to a precision of about λ. That is, the uncertainty in the position measurement, Δx, is approximately

$$\Delta x \approx \lambda.$$

Suppose that the object can be detected by a single photon. The photon has a momentum $p_x = h/\lambda$ (Eq. 27–6). When the photon strikes our object, it will give some or all of this momentum to the object, Fig. 28–5. Therefore, the final x momentum of our object will be uncertain in the amount

$$\Delta p_x \approx \frac{h}{\lambda}$$

since we can't tell how much momentum will be transferred. The product of these uncertainties is

$$(\Delta x)(\Delta p_x) \approx (\lambda)\left(\frac{h}{\lambda}\right) \approx h.$$

The uncertainties could be larger than this, depending on the apparatus and the number of photons needed for detection. A more careful mathematical calcula-tion shows the product of the uncertainties as, at best, about

$$(\Delta x)(\Delta p_x) \gtrsim \frac{h}{2\pi}. \qquad \textbf{(28–1)}$$

This is a mathematical statement of the **Heisenberg uncertainty principle**, or, as it is sometimes called, the **indeterminacy principle**. It tells us that we cannot measure both the position *and* momentum of an object precisely at the same time. The more accurately we try to measure the position, so that Δx is small, the greater will be the uncertainty in momentum, Δp_x. If we try to measure the momentum very accurately, then the uncertainty in the position becomes large. The uncertainty principle does not forbid individual precise measurements, however. For example, in principle we could measure the position of an object exactly. But then its momentum would be completely unknown. Thus, although we might know the position of the object exactly at one instant, we could have no idea at all where it would be a moment later. The uncertainties expressed here are inherent in nature, and reflect the best precision theoretically attainable even with the best instruments.

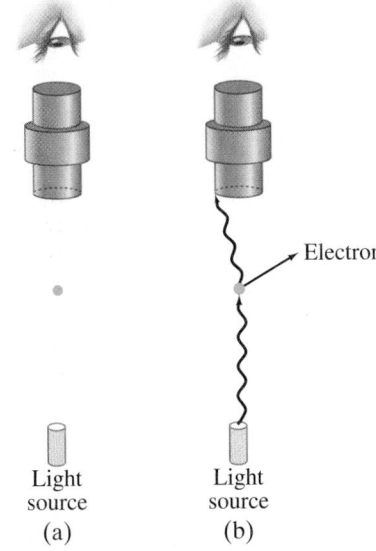

FIGURE 28–5 Thought experiment for observing an electron with a powerful light microscope. At least one photon must scatter from the electron (transferring some momentum to it) and enter the microscope.

UNCERTAINTY PRINCIPLE
(position and momentum)

 C A U T I O N

Uncertainties not due
to instrument deficiency,
but inherent in nature (wave–particle)

EXERCISE A Return to the Chapter-Opening Question, page 803, and answer it again now. Try to explain why you may have answered differently the first time.

Another useful form of the uncertainty principle relates energy and time, and we examine this as follows. The object to be detected has an uncertainty in position $\Delta x \approx \lambda$. The photon that detects it travels with speed c, and it takes a time $\Delta t \approx \Delta x/c \approx \lambda/c$ to pass through the distance of uncertainty. Hence, the measured time when our object is at a given position is uncertain by about

$$\Delta t \approx \frac{\lambda}{c}.$$

Since the photon can transfer some or all of its energy $(= hf = hc/\lambda)$ to our object, the uncertainty in energy of our object as a result is

$$\Delta E \approx \frac{hc}{\lambda}.$$

The product of these two uncertainties is

$$(\Delta E)(\Delta t) \approx \left(\frac{hc}{\lambda}\right)\left(\frac{\lambda}{c}\right) \approx h.$$

A more careful calculation gives

UNCERTAINTY PRINCIPLE *(energy and time)*	$$(\Delta E)(\Delta t) \gtrsim \frac{h}{2\pi}. \qquad (28\text{--}2)$$

This form of the uncertainty principle tells us that the energy of an object can be uncertain (or can be interpreted as briefly nonconserved) by an amount ΔE for a time $\Delta t \approx h/(2\pi \, \Delta E)$.

The quantity $(h/2\pi)$ appears so often in quantum mechanics that for convenience it is given the symbol $\hbar$ ("h-bar"). That is,

$$\hbar = \frac{h}{2\pi} = \frac{6.626 \times 10^{-34} \, \text{J} \cdot \text{s}}{2\pi} = 1.055 \times 10^{-34} \, \text{J} \cdot \text{s}.$$

By using this notation, Eqs. 28–1 and 28–2 for the uncertainty principle can be written

$$(\Delta x)(\Delta p_x) \gtrsim \hbar$$

and

$$(\Delta E)(\Delta t) \gtrsim \hbar.$$

We have been discussing the position and velocity of an electron as if it were a particle. But it isn't simply a particle. Indeed, we have the uncertainty principle because an electron—and matter in general—has wave as well as particle properties. What the uncertainty principle really tells us is that if we insist on thinking of the electron as a particle, then there are certain limitations on this simplified view—namely, that the position and velocity cannot both be known precisely at the same time; and even that the electron does not *have* a precise position and momentum at the same time (because it is not simply a particle). Similarly, the energy can be uncertain (or nonconserved) by an amount ΔE for a time $\Delta t \approx \hbar/\Delta E$.

Because Planck's constant, h, is so small, the uncertainties expressed in the uncertainty principle are usually negligible on the macroscopic level. But at the level of atomic sizes, the uncertainties are significant. Because we consider ordinary objects to be made up of atoms containing nuclei and electrons, the uncertainty principle is relevant to our understanding of all of nature. The uncertainty principle expresses, perhaps most clearly, the probabilistic nature of quantum mechanics. It thus is often used as a basis for philosophic discussion.

EXAMPLE 28–1 **Position uncertainty of electron.** An electron moves in a straight line with a constant speed $v = 1.10 \times 10^6$ m/s which has been measured to a precision of 0.10%. What is the maximum precision with which its position could be simultaneously measured?

APPROACH The momentum is $p = mv$, and the uncertainty in p is $\Delta p = 0.0010p$. The uncertainty principle (Eq. 28–1) gives us the smallest uncertainty in position Δx using the equals sign.

SOLUTION The momentum of the electron is

$$p = mv = (9.11 \times 10^{-31}\,\text{kg})(1.10 \times 10^6\,\text{m/s}) = 1.00 \times 10^{-24}\,\text{kg} \cdot \text{m/s}.$$

The uncertainty in the momentum is 0.10% of this, or $\Delta p = 1.0 \times 10^{-27}\,\text{kg} \cdot \text{m/s}$. From the uncertainty principle, the best simultaneous position measurement will have an uncertainty of

$$\Delta x \approx \frac{\hbar}{\Delta p} = \frac{1.055 \times 10^{-34}\,\text{J} \cdot \text{s}}{1.0 \times 10^{-27}\,\text{kg} \cdot \text{m/s}} = 1.1 \times 10^{-7}\,\text{m},$$

or 110 nm.

NOTE This is about 1000 times the diameter of an atom.

EXERCISE B An electron's position is measured with a precision of 0.50×10^{-10} m. Find the minimum uncertainty in its momentum and velocity.

EXAMPLE 28–2 **Position uncertainty of a baseball.** What is the uncertainty in position, imposed by the uncertainty principle, on a 150-g baseball thrown at $(93 \pm 2)\,\text{mi/h} = (42 \pm 1)\,\text{m/s}$?

APPROACH The uncertainty in the speed is $\Delta v = 1$ m/s. We multiply Δv by m to get Δp and then use the uncertainty principle, solving for Δx.

SOLUTION The uncertainty in the momentum is

$$\Delta p = m\,\Delta v = (0.150\,\text{kg})(1\,\text{m/s}) = 0.15\,\text{kg} \cdot \text{m/s}.$$

Hence the uncertainty in a position measurement could be as small as

$$\Delta x = \frac{\hbar}{\Delta p} = \frac{1.055 \times 10^{-34}\,\text{J} \cdot \text{s}}{0.15\,\text{kg} \cdot \text{m/s}} = 7 \times 10^{-34}\,\text{m}.$$

NOTE This distance is far smaller than any we could imagine observing or measuring. It is trillions of trillions of times smaller than an atom. Indeed, the uncertainty principle sets no relevant limit on measurement for macroscopic objects.

EXAMPLE 28–3 **ESTIMATE** **J/ψ lifetime calculated.** The J/ψ meson, discovered in 1974, was measured to have an average mass of 3100 MeV/c^2 (note the use of energy units since $E = mc^2$) and a mass "width" of 63 keV/c^2. By this we mean that the masses of different J/ψ mesons were actually measured to be slightly different from one another. This mass "width" is related to the very short lifetime of the J/ψ before it decays into other particles. From the uncertainty principle, if the particle exists for only a time Δt, its mass (or rest energy) will be uncertain by $\Delta E \approx \hbar/\Delta t$. Estimate the J/$\psi$ lifetime.

APPROACH We use the energy–time version of the uncertainty principle, Eq. 28–2.

SOLUTION The uncertainty of 63 keV/c^2 in the J/ψ's mass is an uncertainty in its rest energy, which in joules is

$$\Delta E = (63 \times 10^3\,\text{eV})(1.60 \times 10^{-19}\,\text{J/eV}) = 1.01 \times 10^{-14}\,\text{J}.$$

Then we expect its lifetime τ ($= \Delta t$ using Eq. 28–2) to be

$$\tau \approx \frac{\hbar}{\Delta E} = \frac{1.055 \times 10^{-34}\,\text{J} \cdot \text{s}}{1.01 \times 10^{-14}\,\text{J}} \approx 1 \times 10^{-20}\,\text{s}.$$

Lifetimes this short are difficult to measure directly, and the assignment of very short lifetimes depends on this use of the uncertainty principle.

28–4 Philosophic Implications; Probability versus Determinism

The classical Newtonian view of the world is a deterministic one (see Section 5–8). One of its basic ideas is that once the position and velocity of an object are known at a particular time, its future position can be predicted if the forces on it are known. For example, if a stone is thrown a number of times with the same initial velocity and angle, and the forces on it remain the same, the path of the projectile will always be the same. If the forces are known (gravity and air resistance, if any), the stone's path can be precisely predicted. This mechanistic view implies that the future unfolding of the universe, assumed to be made up of particulate objects, is completely determined.

This classical deterministic view of the physical world has been radically altered by quantum mechanics. As we saw in the analysis of the double-slit experiment (Section 28–2), electrons all treated in the same way will not all end up in the same place. According to quantum mechanics, certain probabilities exist that an electron will arrive at different points. This is very different from the classical view, in which the path of a particle is precisely predictable from the initial position and velocity and the forces exerted on it. According to quantum mechanics, the position and velocity of an object cannot even be known accurately at the same time. This is expressed in the uncertainty principle, and arises because basic entities, such as electrons, are not considered simply as particles: they have wave properties as well. Quantum mechanics allows us to calculate only the probability[†] that, say, an electron (when thought of as a particle) will be observed at various places. Quantum mechanics says there is some inherent unpredictability in nature. This is very different from the deterministic view of classical mechanics.

Because matter is considered to be made up of atoms, even ordinary-sized objects are expected to be governed by probability, rather than by strict determinism. For example, quantum mechanics predicts a finite (but negligibly small) probability that when you throw a stone, its path might suddenly curve upward instead of following the downward-curved parabola of normal projectile motion. Quantum mechanics predicts with extremely high probability that ordinary objects will behave just as the classical laws of physics predict. But these predictions are considered probabilities, not absolute certainties. The reason that macroscopic objects behave in accordance with classical laws with such high probability is due to the large number of molecules involved: when large numbers of objects are present in a statistical situation, deviations from the average (or most probable) approach zero. It is the average configuration of vast numbers of molecules that follows the so-called fixed laws of classical physics with such high probability, and gives rise to an apparent "determinism." Deviations from classical laws are observed when small numbers of molecules are dealt with. We can say, then, that although there are no precise deterministic laws in quantum mechanics, there are statistical laws based on probability.

It is important to note that there is a difference between the probability imposed by quantum mechanics and that used in the nineteenth century to understand thermodynamics and the behavior of gases in terms of molecules (Chapters 13 and 15). In thermodynamics, probability is used because there are far too many particles to keep track of. But the molecules are still assumed to move and interact in a deterministic way following Newton's laws. Probability in quantum mechanics is quite different; it is seen as *inherent* in nature, and not as a limitation on our abilities to calculate or to measure.

[†]Note that these probabilities can be calculated precisely, just like predictions of probabilities at rolling dice or dealing cards; but they are unlike predictions of probabilities at sporting events, which are only estimates.

The view presented here is the generally accepted one and is called the **Copenhagen interpretation** of quantum mechanics in honor of Niels Bohr's home, since it was largely developed there through discussions between Bohr and other prominent physicists.

Because electrons are not simply particles, they cannot be thought of as following particular paths in space and time. This suggests that a description of matter in space and time may not be completely correct. This deep and far-reaching conclusion has been a lively topic of discussion among philosophers. Perhaps the most important and influential philosopher of quantum mechanics was Bohr. He argued that a space–time description of actual atoms and electrons is not possible. Yet a description of experiments on atoms or electrons must be given in terms of space and time and other concepts familiar to ordinary experience, such as waves and particles. We must not let our *descriptions* of experiments lead us into believing that atoms or electrons themselves actually move in space and time as classical particles.

28–5 Quantum-Mechanical View of Atoms

At the beginning of this Chapter, we discussed the limitations of the Bohr model of atomic structure. Now we examine the quantum-mechanical theory of atoms, which is a far more complete theory than the old Bohr model. Although the Bohr model has been discarded as an accurate description of nature, nonetheless, quantum mechanics reaffirms certain aspects of the older theory, such as that electrons in an atom exist only in discrete states of definite energy, and that a photon of light is emitted (or absorbed) when an electron makes a transition from one state to another. But quantum mechanics is a much deeper theory, and has provided us with a very different view of the atom. According to quantum mechanics, electrons do not exist in well-defined circular orbits as in the Bohr model. Rather, the electron (because of its wave nature) can be thought of as spread out in space as a "**cloud**." The size and shape of the electron cloud can be calculated for a given state of an atom. For the ground state in the hydrogen atom, the electron cloud is spherically symmetric, as shown in Fig. 28–6. The electron cloud at its higher densities roughly indicates the "size" of an atom. But just as a cloud may not have a distinct border, atoms do not have a precise boundary or a well-defined size. Not all electron clouds have a spherical shape, as we shall see later in this Chapter.

The electron cloud can be interpreted from either the particle or the wave viewpoint. Remember that by a particle we mean something that is localized in space—it has a definite position at any given instant. By contrast, a wave is spread out in space. The electron cloud, spread out in space as in Fig. 28–6, is a result of the wave nature of electrons. Electron clouds can also be interpreted as **probability distributions** (or **probability density**) for a particle. As we saw in Section 28–3, we cannot predict the path an electron will follow (thinking of it as a particle). After one measurement of its position we cannot predict exactly where it will be at a later time. We can only calculate the probability that it will be found at different points. If you were to make 500 different measurements of the position of an electron in a hydrogen atom, the majority of the results would show the electron at points where the probability is high (dark area in Fig. 28–6). Only occasionally would the electron be found where the probability is low. The electron cloud or probability distribution becomes small (or thin) at places, especially far away, but never becomes zero. So quantum mechanics suggests that an atom is *not* mostly empty space, and that there is no truly empty space in the universe.

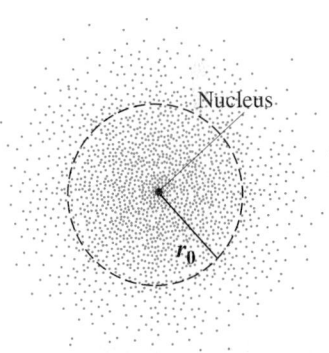

FIGURE 28–6 Electron cloud or "probability distribution" for the ground state of the hydrogen atom, as seen from afar. The dots represent a hypothetical detection of an electron at each point: dots closer together represent more probable presence of an electron (denser cloud). The dashed circle represents the Bohr radius r_0.

28–6 Quantum Mechanics of the Hydrogen Atom; Quantum Numbers

We now look more closely at what quantum mechanics tells us about the hydrogen atom. Much of what we say here also applies to more complex atoms, which are discussed in the next Section.

Quantum mechanics is a much more sophisticated and successful theory than Bohr's. Yet in a few details they agree. Quantum mechanics predicts the same basic energy levels (Fig. 27–29) for the hydrogen atom as does the Bohr model. That is,

$$E_n = -\frac{13.6\,\text{eV}}{n^2}, \qquad n = 1, 2, 3, \cdots,$$

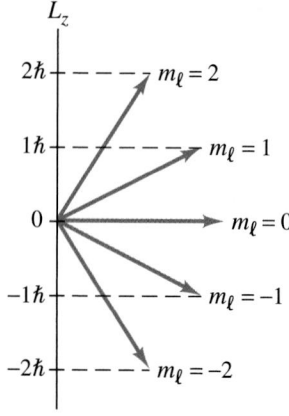

L_z

$2\hbar$ ----- $m_\ell = 2$

$1\hbar$ ----- $m_\ell = 1$

0 ----- $m_\ell = 0$

$-1\hbar$ ----- $m_\ell = -1$

$-2\hbar$ ----- $m_\ell = -2$

FIGURE 28–7 Quantization of angular momentum direction for $\ell = 2$. (Magnitude of $\vec{L}$ is $L = \sqrt{6}\,\hbar$.)

FIGURE 28–8 Energy levels (not to scale). When a magnetic field is applied, the $n = 3$, $\ell = 2$ energy level is split into five separate levels, corresponding to the five values of m_ℓ (2, 1, 0, −1, −2). An $n = 2$, $\ell = 1$ level is split into three levels ($m_\ell = 1, 0, -1$). Transitions can occur between levels (not all transitions are shown), with photons of several slightly different frequencies being given off (the Zeeman effect).

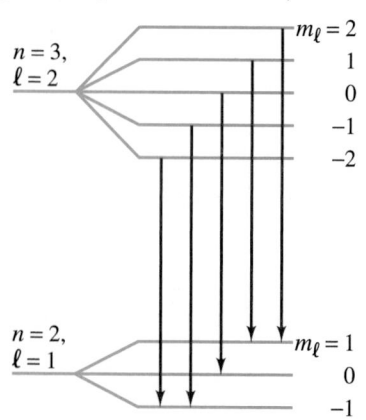

$n = 3,$
$\ell = 2$

$m_\ell = 2$
1
0
-1
-2

$n = 2,$
$\ell = 1$

$m_\ell = 1$
0
-1

where n is an integer. In the simple Bohr model, there was only one quantum number, n. In quantum mechanics, four different quantum numbers are needed to specify each state in the atom:

(1) The quantum number, n, from the Bohr model is found also in quantum mechanics and is called the **principal quantum number**. It can have any integer value from 1 to ∞. The total energy of a state in the hydrogen atom depends on n, as we saw above.

(2) The **orbital quantum number**, ℓ, is related to the magnitude of the angular momentum of the electron; ℓ can take on integer values from 0 to $(n-1)$. For the ground state, $n = 1$, ℓ can only be zero.† For $n = 3$, ℓ can be 0, 1, or 2. The actual magnitude of the angular momentum L is related to the quantum number ℓ by

$$L = \sqrt{\ell(\ell + 1)}\,\hbar \qquad\qquad \textbf{(28–3)}$$

(where again $\hbar = h/2\pi$). The value of ℓ has almost no effect on the total energy in the hydrogen atom; only n does to any appreciable extent (but see *fine structure* below). In atoms with two or more electrons, the energy does depend on ℓ as well as n, as we shall see.

(3) The **magnetic quantum number**, m_ℓ, is related to the *direction* of the electron's angular momentum, and it can take on integer values ranging from $-\ell$ to $+\ell$. For example, if $\ell = 2$, then m_ℓ can be $-2, -1, 0, +1$, or $+2$. Since angular momentum is a vector, it is not surprising that both its magnitude and its direction would be quantized. For $\ell = 2$, the five different directions allowed can be represented by the diagram of Fig. 28–7. This limitation on the direction of $\vec{L}$ is often called **space quantization**. In quantum mechanics, the direction of the angular momentum is usually specified by giving its component along the z axis (this choice is arbitrary). Then L_z is related to m_ℓ by the equation

$$L_z = m_\ell \hbar.$$

The values of L_x and L_y are not definite, however. The name for m_ℓ derives not from theory (which relates it to L_z), but from experiment. It was found that when a gas-discharge tube was placed in a magnetic field, the spectral lines were split into several very closely spaced lines. This splitting, known as the **Zeeman effect**, implies that the energy levels must be split (Fig. 28–8), and thus that the energy of a state depends not only on n but also on m_ℓ when a magnetic field is applied—hence the name "magnetic quantum number."

---

†This replaces Bohr theory, which assigned $\ell = 1$ to the ground state (Eq. 27–11).

(4) Finally, there is the **spin quantum number**, m_s, which for an electron can have only two values, $m_s = +\frac{1}{2}$ and $m_s = -\frac{1}{2}$. The existence of this quantum number did not come out of Schrödinger's original wave theory, as did n, ℓ, and m_ℓ. Instead, a subsequent modification by P. A. M. Dirac (1902–1984) explained its presence as a relativistic effect. The first hint that m_s was needed, however, came from experiment. A careful study of the spectral lines of hydrogen showed that each actually consisted of two (or more) very closely spaced lines even in the absence of an external magnetic field. It was at first hypothesized that this tiny splitting of energy levels, called **fine structure**, was due to angular momentum associated with a spinning of the electron. That is, the electron might spin on its axis as well as orbit the nucleus, just as the Earth spins on its axis as it orbits the Sun. The interaction between the tiny current of the spinning electron could then interact with the magnetic field due to the orbiting charge and cause the small observed splitting of energy levels. (The energy thus depends slightly on m_ℓ and m_s.)[†] Today we consider the picture of a spinning electron as not legitimate. We cannot even view an electron as a localized object, much less a spinning one. What is important is that the electron can have two different states due to some intrinsic property that behaves like an angular momentum, and we still call this property "spin." The two possible values of m_s ($+\frac{1}{2}$ and $-\frac{1}{2}$) are often said to be "spin up" and "spin down," referring to the two possible directions of the spin angular momentum.

The possible values of the four quantum numbers for an electron in the hydrogen atom are summarized in Table 28–1.

TABLE 28–1 Quantum Numbers for an Electron

Name	Symbol	Possible Values
Principal	n	$1, 2, 3, \cdots, \infty$.
Orbital	ℓ	For a given n: ℓ can be $0, 1, 2, \cdots, n-1$.
Magnetic	m_ℓ	For given n and ℓ: m_ℓ can be $\ell, \ell-1, \cdots, 0, \cdots, -\ell$.
Spin	m_s	For each set of n, ℓ, and m_ℓ: m_s can be $+\frac{1}{2}$ or $-\frac{1}{2}$.

CONCEPTUAL EXAMPLE 28–4 **Possible states for $n = 3$.** How many different states are possible for an electron with principal quantum number $n = 3$?

RESPONSE For $n = 3$, ℓ can have the values $\ell = 2, 1, 0$. For $\ell = 2$, m_ℓ can be $2, 1, 0, -1, -2$, which is five different possibilities. For each of these, m_s can be either up or down ($+\frac{1}{2}$ or $-\frac{1}{2}$); so for $\ell = 2$, there are $2 \times 5 = 10$ states. For $\ell = 1$, m_ℓ can be $1, 0, -1$, and since m_s can be $+\frac{1}{2}$ or $-\frac{1}{2}$ for each of these, we have 6 more possible states. Finally, for $\ell = 0$, m_ℓ can only be 0, and there are only 2 states corresponding to $m_s = +\frac{1}{2}$ and $-\frac{1}{2}$. The total number of states is $10 + 6 + 2 = 18$, as detailed in the following Table:

n	ℓ	m_ℓ	m_s	n	ℓ	m_ℓ	m_s
3	2	2	$\frac{1}{2}$	3	1	1	$\frac{1}{2}$
3	2	2	$-\frac{1}{2}$	3	1	1	$-\frac{1}{2}$
3	2	1	$\frac{1}{2}$	3	1	0	$\frac{1}{2}$
3	2	1	$-\frac{1}{2}$	3	1	0	$-\frac{1}{2}$
3	2	0	$\frac{1}{2}$	3	1	-1	$\frac{1}{2}$
3	2	0	$-\frac{1}{2}$	3	1	-1	$-\frac{1}{2}$
3	2	-1	$\frac{1}{2}$	3	0	0	$\frac{1}{2}$
3	2	-1	$-\frac{1}{2}$	3	0	0	$-\frac{1}{2}$
3	2	-2	$\frac{1}{2}$				
3	2	-2	$-\frac{1}{2}$				

EXERCISE C An electron has $n = 4$, $\ell = 2$. Which of the following values of m_ℓ are possible: $4, 3, 2, 1, 0, -1, -2, -3, -4$?

[†]Fine structure is said to be due to a **spin–orbit interaction**.

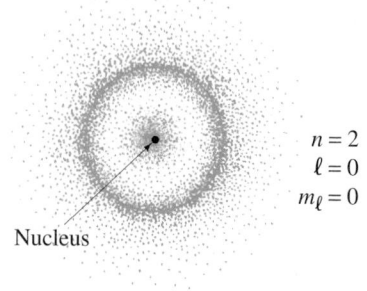

Nucleus

$n = 2$
$\ell = 0$
$m_\ell = 0$

(a)

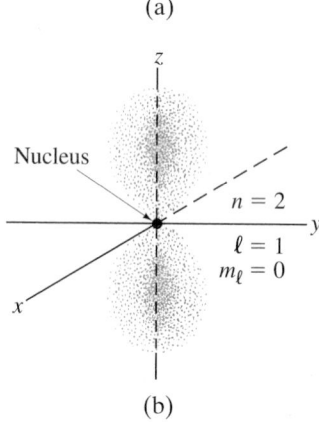

Nucleus

$n = 2$
$\ell = 1$
$m_\ell = 0$

(b)

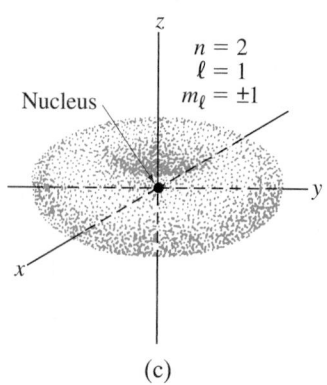

$n = 2$
$\ell = 1$
$m_\ell = \pm 1$

Nucleus

(c)

FIGURE 28–9 Electron cloud, or probability distribution, for $n = 2$ states in hydrogen. [The donut-shaped orbit in (c) is the sum of two dumbbell-shaped orbits, as in (b), along the x and y axes added together.]

EXAMPLE 28–5 **E and L for $n = 3$.** Determine (a) the energy and (b) the orbital angular momentum for an electron in each of the hydrogen atom states with $n = 3$, as in Example 28–4.

APPROACH The energy of a state depends only on n, except for the very small corrections mentioned above, which we will ignore. Energy is calculated as in the Bohr model, $E_n = -13.6\,\text{eV}/n^2$. For angular momentum we use Eq. 28–3.

SOLUTION (a) Since $n = 3$ for all these states, they all have the same energy,

$$ E_3 = -\frac{13.6\,\text{eV}}{(3)^2} = -1.51\,\text{eV}. $$

(b) For $\ell = 0$, Eq. 28–3 gives

$$ L = \sqrt{\ell(\ell + 1)}\,\hbar = 0. $$

For $\ell = 1$,

$$ L = \sqrt{1(1 + 1)}\,\hbar = \sqrt{2}\,\hbar = 1.49 \times 10^{-34}\,\text{J}\cdot\text{s}. $$

For $\ell = 2$, $L = \sqrt{2(2 + 1)}\,\hbar = \sqrt{6}\,\hbar$.

NOTE Atomic angular momenta are generally given as a multiple of $\hbar$ ($\sqrt{2}\,\hbar$ or $\sqrt{6}\,\hbar$ in this case), rather than in SI units.

EXERCISE D What are the energy and angular momentum of the electron in a hydrogen atom with $n = 6$, $\ell = 4$?

Although ℓ and m_ℓ do not significantly affect the energy levels in hydrogen, they do affect the electron probability distribution in space. For $n = 1$, ℓ and m_ℓ can only be zero and the electron distribution is as shown in Fig. 28–6. For $n = 2$, ℓ can be 0 or 1. The distribution for $n = 2$, $\ell = 0$ is shown in Fig. 28–9a, and it is seen to differ from that for the ground state (Fig. 28–6), although it is still spherically symmetric. For $n = 2$, $\ell = 1$, the distributions are not spherically symmetric as shown in Figs. 28–9b (for $m_\ell = 0$) and 28–9c (for $m_\ell = +1$ or -1).

Although the spatial distributions of the electron can be calculated for the various states, it is difficult to measure them experimentally. Most of the experimental information about atoms has come from a careful examination of the emission spectra under various conditions as in Figs. 27–23 and 24–28.

[Chemists refer to atomic states, and especially the shape in space of their probability distributions, as **orbitals**. Each atomic orbital is characterized by its quantum numbers n, ℓ, and m_ℓ, and can hold one or two electrons ($m_s = +\frac{1}{2}$ or $m_s = -\frac{1}{2}$); s-orbitals ($\ell = 0$) are spherically symmetric, Figs. 28–6 and 28–9a; p-orbitals ($\ell = 1$) can be dumbbell shaped with lobes, Fig. 28–9b, or donut shaped if combining $m_\ell = +1$ and $m_\ell = -1$, Fig. 28–9c.]

Selection Rules: Allowed and Forbidden Transitions

Another prediction of quantum mechanics is that when a photon is emitted or absorbed, transitions can occur only between states with values of ℓ that differ by exactly one unit:

$$ \Delta \ell = \pm 1. $$

According to this **selection rule**, an electron in an $\ell = 2$ state can jump only to a state with $\ell = 1$ or $\ell = 3$. It cannot jump to a state with $\ell = 2$ or $\ell = 0$. A transition such as $\ell = 2$ to $\ell = 0$ is called a **forbidden transition**. Actually, such a transition is not absolutely forbidden and can occur, but only with very low probability compared to **allowed transitions**—those that satisfy the selection rule $\Delta \ell = \pm 1$. Since the orbital angular momentum of an H atom must change by one unit when it emits a photon, conservation of angular momentum tells us that the photon must carry off angular momentum. Indeed, experimental evidence of many sorts shows that the photon can be assigned a spin angular momentum of $1\hbar$.

28–7 Multielectron Atoms; the Exclusion Principle

We have discussed the hydrogen atom in detail because it is the simplest to deal with. Now we briefly discuss more complex atoms, those that contain more than one electron. Their energy levels can be determined experimentally from an analysis of their emission spectra. The energy levels are *not* the same as in the H atom, because the electrons interact with each other as well as with the nucleus. Each electron in a complex atom still occupies a particular state characterized by the quantum numbers n, ℓ, m_ℓ, and m_s. For atoms with more than one electron, the energy levels depend on both n and ℓ.

The number of electrons in a neutral atom is called its **atomic number**, Z; Z is also the number of positive charges (protons) in the nucleus, and determines what kind of atom it is. That is, Z determines the fundamental properties that distinguish one type of atom from another.

To understand the possible arrangements of electrons in an atom, a new principle was needed. It was introduced by Wolfgang Pauli (1900–1958; Fig. 28–2) and is called the **Pauli exclusion principle**. It states:

No two electrons in an atom can occupy the same quantum state.

Thus, no two electrons in an atom can have exactly the same set of the quantum numbers n, ℓ, m_ℓ, and m_s. The Pauli exclusion principle forms the basis not only for understanding atoms, but also for understanding molecules and bonding, and other phenomena as well. (See also note at end of this Section.)

Let us now look at the structure of some of the simpler atoms when they are in the ground state. After hydrogen, the next simplest atom is *helium* with two electrons. Both electrons can have $n = 1$, because one can have spin up $\left(m_s = +\frac{1}{2}\right)$ and the other spin down $\left(m_s = -\frac{1}{2}\right)$, thus satisfying the exclusion principle. Since $n = 1$, then ℓ and m_ℓ must be zero (Table 28–1, page 813). Thus the two electrons have the quantum numbers indicated at the top of Table 28–2.

Lithium has three electrons, two of which can have $n = 1$. But the third cannot have $n = 1$ without violating the exclusion principle. Hence the third electron must have $n = 2$. It happens that the $n = 2$, $\ell = 0$ level has a lower energy than $n = 2$, $\ell = 1$. So the electrons in the ground state have the quantum numbers indicated in Table 28–2. The quantum numbers of the third electron could also be, say, $(n, \ell, m_\ell, m_s) = (3, 1, -1, \frac{1}{2})$. But the atom in this case would be in an excited state, because it would have greater energy. It would not be long before it jumped to the ground state with the emission of a photon. At room temperature, unless extra energy is supplied (as in a discharge tube), the vast majority of atoms are in the ground state.

We can continue in this way to describe the quantum numbers of each electron in the ground state of larger and larger atoms. The quantum numbers for sodium, with its eleven electrons, are shown in Table 28–2.

EXERCISE E Construct a Table of the ground-state quantum numbers for beryllium, $Z = 4$ (like those in Table 28–2).

Figure 28–10 shows a simple energy level diagram where occupied states are shown as up or down arrows $\left(m_s = +\frac{1}{2} \text{ or } -\frac{1}{2}\right)$, and possible empty states are shown as a small circle.

TABLE 28–2 Ground-State Quantum Numbers

Helium, $Z = 2$

n	ℓ	m_ℓ	m_s
1	0	0	$\frac{1}{2}$
1	0	0	$-\frac{1}{2}$

Lithium, $Z = 3$

n	ℓ	m_ℓ	m_s
1	0	0	$\frac{1}{2}$
1	0	0	$-\frac{1}{2}$
2	0	0	$\frac{1}{2}$

Sodium, $Z = 11$

n	ℓ	m_ℓ	m_s
1	0	0	$\frac{1}{2}$
1	0	0	$-\frac{1}{2}$
2	0	0	$\frac{1}{2}$
2	0	0	$-\frac{1}{2}$
2	1	1	$\frac{1}{2}$
2	1	1	$-\frac{1}{2}$
2	1	0	$\frac{1}{2}$
2	1	0	$-\frac{1}{2}$
2	1	-1	$\frac{1}{2}$
2	1	-1	$-\frac{1}{2}$
3	0	0	$\frac{1}{2}$

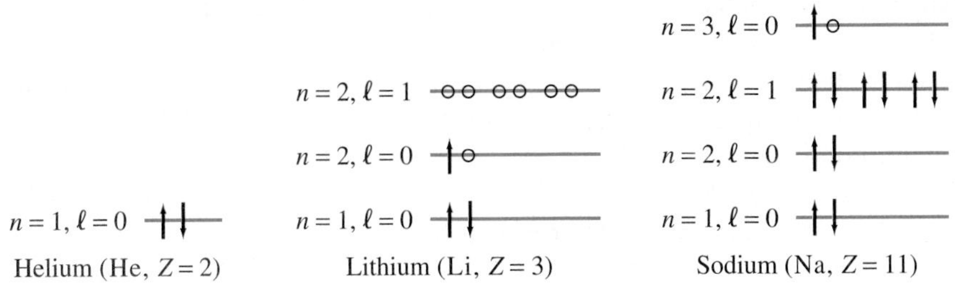

FIGURE 28–10 Energy level diagrams (not to scale) showing occupied states (arrows) and unoccupied states (o) for the ground states of He, Li, and Na. Note that we have shown the $n = 2$, $\ell = 1$ level of Li even though it is empty.

TABLE 28–3 Value of ℓ

Value of ℓ	Letter Symbol	Maximum Number of Electrons in Subshell
0	s	2
1	p	6
2	d	10
3	f	14
4	g	18
5	h	22
⋮	⋮	⋮

TABLE 28–4 Electron Configuration of Some Elements

Z (Number of Electrons)	Element†	Ground State Configuration (outer electrons)
1	H	$1s^1$
2	He	$1s^2$
3	Li	$2s^1$
4	Be	$2s^2$
5	B	$2s^2 2p^1$
6	C	$2s^2 2p^2$
7	N	$2s^2 2p^3$
8	O	$2s^2 2p^4$
9	F	$2s^2 2p^5$
10	Ne	$2s^2 2p^6$
11	Na	$3s^1$
12	Mg	$3s^2$
13	Al	$3s^2 3p^1$
14	Si	$3s^2 3p^2$
15	P	$3s^2 3p^3$
16	S	$3s^2 3p^4$
17	Cl	$3s^2 3p^5$
18	Ar	$3s^2 3p^6$
19	K	$3d^0 4s^1$
20	Ca	$3d^0 4s^2$
21	Sc	$3d^1 4s^2$
22	Ti	$3d^2 4s^2$
23	V	$3d^3 4s^2$
24	Cr	$3d^5 4s^1$
25	Mn	$3d^5 4s^2$
26	Fe	$3d^6 4s^2$

†Names of elements can be found in Appendix B.

The ground-state configuration for all atoms is given in the **Periodic Table**, which is displayed inside the back cover of this book, and discussed in the next Section.

[The *exclusion principle* applies to identical particles whose spin quantum number is a half-integer ($\frac{1}{2}$, $\frac{3}{2}$, and so on), including electrons, protons, and neutrons; such particles are called **fermions**, after Enrico Fermi who derived a statistical theory describing them. A basic assumption is that all electrons are **identical**, indistinguishable one from another. Similarly, all protons are identical, all neutrons are identical, and so on. The exclusion principle does not apply to particles with integer spin (0, 1, 2, and so on), such as the photon and π meson, all of which are referred to as **bosons** (after Satyendranath Bose, who derived a statistical theory for them).]

28–8 The Periodic Table of Elements

More than a century ago, Dmitri Mendeleev (1834–1907) arranged the (then) known elements into what we now call the **Periodic Table** of the elements. The atoms were arranged according to increasing mass, but also so that elements with similar chemical properties would fall in the same column. Today's version is shown inside the back cover of this book. Each square contains the atomic number Z, the symbol for the element, and the atomic mass (in atomic mass units). Finally, the lower left corner shows the configuration of the ground state of the atom. This requires some explanation. Electrons with the same value of n are referred to as being in the same **shell**. Electrons with $n = 1$ are in one shell (the K shell), those with $n = 2$ are in a second shell (the L shell), those with $n = 3$ are in the third (M) shell, and so on. Electrons with the same values of n and ℓ are referred to as being in the same **subshell**. Letters are often used to specify the value of ℓ as shown in Table 28–3. That is, $\ell = 0$ is the s subshell; $\ell = 1$ is the p subshell; $\ell = 2$ is the d subshell; beginning with $\ell = 3$, the letters follow the alphabet, f, g, h, i, and so on. (The first letters s, p, d, and f were originally abbreviations of "sharp," "principal," "diffuse," and "fundamental," terms referring to the experimental spectra.)

The Pauli exclusion principle limits the number of electrons possible in each shell and subshell. For any value of ℓ, there are $2\ell + 1$ possible m_ℓ values (m_ℓ can be any integer from 1 to ℓ, from -1 to $-\ell$, or zero), and two possible m_s values. There can be, therefore, at most $2(2\ell + 1)$ electrons in any ℓ subshell. For example, for $\ell = 2$, five m_ℓ values are possible $(2, 1, 0, -1, -2)$, and for each of these, m_s can be $+\frac{1}{2}$ or $-\frac{1}{2}$ for a total of $2(5) = 10$ states. Table 28–3 lists the maximum number of electrons that can occupy each subshell.

Because the energy levels depend almost entirely on the values of n and ℓ, it is customary to specify the electron configuration simply by giving the n value and the appropriate letter for ℓ, with the number of electrons in each subshell given as a superscript. The ground-state configuration of sodium, for example, is written as $1s^2 2s^2 2p^6 3s^1$. This is simplified in the Periodic Table by specifying the configuration only of the outermost electrons and any other nonfilled subshells (see Table 28–4 here, and the Periodic Table inside the back cover).

CONCEPTUAL EXAMPLE 28–6 **Electron configurations.** Which of the following electron configurations are possible, and which are not: (a) $1s^2 2s^2 2p^6 3s^3$; (b) $1s^2 2s^2 2p^6 3s^2 3p^5 4s^2$; (c) $1s^2 2s^2 2p^6 2d^1$?

RESPONSE (a) This is not allowed, because too many electrons (three) are shown in the s subshell of the M ($n = 3$) shell. The s subshell has $m_\ell = 0$, with two slots only, for "spin up" and "spin down" electrons.

(b) This is allowed, but it is an excited state. One of the electrons from the $3p$ subshell has jumped up to the $4s$ subshell. Since there are 19 electrons, the element is potassium.

(c) This is not allowed, because there is no d ($\ell = 2$) subshell in the $n = 2$ shell (Table 28–1). The outermost electron will have to be (at least) in the $n = 3$ shell.

EXERCISE F Write the complete ground-state configuration for gallium, with its 31 electrons.

The grouping of atoms in the Periodic Table is according to increasing atomic number, Z. It was designed to also show regularity according to chemical properties. Although this is treated in chemistry textbooks, we discuss it here briefly because it is a result of quantum mechanics. See the Periodic Table inside the back cover.

All the **noble gases** (in column VIII of the Periodic Table) have completely filled shells or subshells. That is, their outermost subshell is completely full, and the electron distribution is spherically symmetric. With such full spherical symmetry, other electrons are not attracted nor are electrons readily lost (ionization energy is high). This is why the noble gases are chemically inert (more on this when we discuss molecules and bonding in Chapter 29). Column VII contains the **halogens**, which lack one electron from a filled shell. Because of the shapes of the orbits (see Section 29–1), an additional electron can be accepted from another atom, and hence these elements are quite reactive. They have a valence of −1, meaning that when an extra electron is acquired, the resulting ion has a net charge of −1e. Column I of the Periodic Table contains the **alkali metals**, all of which have a single outer s electron. This electron spends most of its time outside the inner closed shells and subshells which shield it from most of the nuclear charge. Indeed, it is relatively far from the nucleus and is attracted to it by a net charge of only about +1e, because of the shielding effect of the other electrons. Hence this outer electron is easily removed and can spend much of its time around another atom, forming a molecule. This is why the alkali metals are very chemically reactive and have a valence of +1. The other columns of the Periodic Table can be treated similarly.

The presence of the **transition elements** in the center of the Periodic Table, as well as the lanthanides (rare earths) and actinides below, is a result of incomplete inner shells. For the lowest Z elements, the subshells are filled in a simple order: first 1s, then 2s, followed by 2p, 3s, and 3p. You might expect that 3d ($n = 3$, $\ell = 2$) would be filled next, but it isn't. Instead, the 4s level actually has a slightly lower energy than the 3d (due to electrons interacting with each other), so it fills first (K and Ca). Only then does the 3d shell start to fill up, beginning with Sc, as can be seen in Table 28–4. (The 4s and 3d levels are close, so some elements have only one 4s electron, such as Cr.) Most of the chemical properties of these transition elements are governed by the relatively loosely held 4s electrons, and hence they usually have valences of +1 or +2. A similar effect is responsible for the *lanthanides* and *actinides*, which are shown at the bottom of the Periodic Table for convenience. All have very similar chemical properties, which are determined by their two outer 6s or 7s electrons, whereas the different numbers of electrons in the unfilled inner shells have little effect.

CAUTION
Subshells are not always filled in "order"

*28–9 X-Ray Spectra and Atomic Number

The line spectra of atoms in the visible, UV, and IR regions of the EM spectrum are mainly due to transitions between states of the outer electrons. Much of the positive charge of the nucleus is shielded from these electrons by the negative charge on the inner electrons. But the innermost electrons in the $n = 1$ shell "see" the full charge of the nucleus. Since the energy of a level is proportional to Z^2 (see Eq. 27–15), for an atom with $Z = 50$, we would expect wavelengths about $50^2 = 2500$ times shorter than those found in the Lyman series of hydrogen (around 100 nm), or $(100 \text{ nm})/(2500) \approx 10^{-2}$ to 10^{-1} nm. Such short wavelengths lie in the X-ray region of the spectrum.

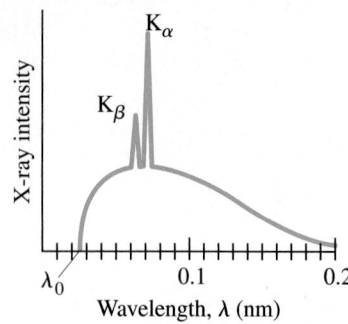

FIGURE 28–11 Spectrum of X-rays emitted from a molybdenum target in an X-ray tube operated at 50 kV.

FIGURE 28–12 Plot of $\sqrt{1/\lambda}$ vs. Z for K_α X-ray lines.

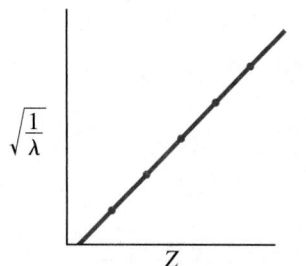

X-rays are produced when electrons accelerated by a high voltage strike the metal target inside an X-ray tube (Section 25–11). If we look at the spectrum of wavelengths emitted by an X-ray tube, we see that the spectrum consists of two parts: a continuous spectrum with a cutoff at some λ_0 which depends only on the voltage across the tube, and a series of peaks superimposed. A typical example is shown in Fig. 28–11. The smooth curve and the cutoff wavelength λ_0 move to the left as the voltage across the tube increases. The sharp lines or peaks (labeled K_α and K_β in Fig. 28–11), however, remain at the same wavelength when the voltage is changed, although they are located at different wavelengths when different target materials are used. This observation suggests that the peaks are characteristic of the target material used. Indeed, we can explain the peaks by imagining that the electrons accelerated by the high voltage of the tube can reach sufficient energies that, when they collide with the atoms of the target, they can knock out one of the very tightly held inner electrons. Then we explain these **characteristic X-rays** (the peaks in Fig. 28–11) as photons emitted when an electron in an upper state drops down to fill the vacated lower state. The K lines result from transitions *into* the K shell ($n = 1$). The K_α line consists of photons emitted in a transition that originates from the $n = 2$ (L) shell and drops to the $n = 1$ (K) shell. On the other hand, the K_β line reflects a transition from the $n = 3$ (M) shell down to the K shell. An L line is due to a transition into the L shell, and so on.

Measurement of the characteristic X-ray spectra has allowed a determination of the inner energy levels of atoms. It has also allowed the determination of Z values for many atoms, because (as we have seen) the wavelength of the shortest characteristic X-rays emitted will be inversely proportional to Z^2. Actually, for an electron jumping from, say, the $n = 2$ to the $n = 1$ level (K_α line), the wavelength is inversely proportional to $(Z - 1)^2$ because the nucleus is shielded by the one electron that still remains in the $1s$ level. In 1914, H. G. J. Moseley (1887–1915) found that a plot of $\sqrt{1/\lambda}$ vs. Z produced a straight line, Fig. 28–12, where λ is the wavelength of the K_α line. The Z values of a number of elements were determined by fitting them to such a **Moseley plot**. The work of Moseley put the concept of atomic number on a firm experimental basis.

EXAMPLE 28–7 **X-ray wavelength.** Estimate the wavelength for an $n = 2$ to $n = 1$ transition in molybdenum ($Z = 42$). What is the energy of such a photon?

APPROACH We use the Bohr formula, Eq. 27–16 for $1/\lambda$, with Z^2 replaced by $(Z - 1)^2 = (41)^2$.

SOLUTION Equation 27–16 gives

$$\frac{1}{\lambda} = \left(\frac{2\pi^2 e^4 m k^2}{h^3 c}\right)(Z - 1)^2\left(\frac{1}{n'^2} - \frac{1}{n^2}\right)$$

where $n = 2$, $n' = 1$, and $k = 8.99 \times 10^9\,\text{N·m}^2/\text{c}^2$. We substitute in values:

$$\frac{1}{\lambda} = (1.097 \times 10^7\,\text{m}^{-1})(41)^2\left(\frac{1}{1} - \frac{1}{4}\right) = 1.38 \times 10^{10}\,\text{m}^{-1}.$$

So

$$\lambda = \frac{1}{1.38 \times 10^{10}\,\text{m}^{-1}} = 0.072\,\text{nm}.$$

This is close to the measured value (Fig. 28–11) of 0.071 nm. Each of these photons would have energy (in eV) of:

$$E = hf = \frac{hc}{\lambda} = \frac{(6.63 \times 10^{-34}\,\text{J·s})(3.00 \times 10^8\,\text{m/s})}{(7.2 \times 10^{-11}\,\text{m})(1.60 \times 10^{-19}\,\text{J/eV})} = 17\,\text{keV}.$$

The denominator includes the conversion factor from joules to eV.

EXAMPLE 28–8 **Determining atomic number.** High-energy electrons are used to bombard an unknown material. The strongest peak is found for X-rays emitted with an energy of 66.3 keV. Guess what the material is.

APPROACH The highest intensity X-rays are generally for the K_α line (see Fig. 28–11) which occurs when high-energy electrons knock out K shell electrons (the innermost orbit, $n = 1$) and their place is taken by electrons from the L shell ($n = 2$). We use the Bohr model, and assume the electrons of the unknown atoms (Z) "see" a nuclear charge of $Z - 1$ (screened by one electron).

SOLUTION The hydrogen transition $n = 2$ to $n = 1$ would yield $E_H = 13.6 \text{ eV} - 3.4 \text{ eV} = 10.2 \text{ eV}$ (see Fig. 27–29 or Example 27–13). Energy of our unknown E_Z is proportional to Z^2 (Eq. 27–15), or rather $(Z - 1)^2$ because the nucleus is shielded by the one electron in a $1s$ state (see above), so we can use ratios:

$$\frac{E_Z}{E_H} = \frac{(Z - 1)^2}{1^2} = \frac{66.3 \times 10^3 \text{ eV}}{10.2 \text{ eV}} = 6.50 \times 10^3,$$

so $Z - 1 = \sqrt{6500} = 81$, and $Z = 82$, which makes it lead.

Now we briefly analyze the continuous part of an X-ray spectrum (Fig. 28–11) based on the photon theory of light. When electrons strike the target, they collide with atoms of the material and give up most of their energy as heat (about 99%, so X-ray tubes must be cooled). Electrons can also give up energy by emitting a photon of light: an electron decelerated by interaction with atoms of the target (Fig. 28–13) emits radiation because of its deceleration (Chapter 22), and in this case it is called **bremsstrahlung** (German for "braking radiation"). Because energy is conserved, the energy of the emitted photon, hf, equals the loss of kinetic energy of the electron, $\Delta \text{KE} = \text{KE} - \text{KE}'$, so

$$hf = \Delta \text{KE}.$$

An electron may lose all or a part of its energy in such a collision. The continuous X-ray spectrum (Fig. 28–11) is explained as being due to such bremsstrahlung collisions in which varying amounts of energy are lost by the electrons. The shortest-wavelength X-ray (the highest frequency) must be due to an electron that gives up *all* its kinetic energy to produce one photon in a single collision. Since the initial kinetic energy of an electron is equal to the energy given it by the accelerating voltage, V, then $\text{KE} = eV$. In a single collision in which the electron is brought to rest ($\text{KE}' = 0$), then $\Delta \text{KE} = eV$ and

$$hf_0 = eV.$$

We set $f_0 = c/\lambda_0$ where λ_0 is the cutoff wavelength (Fig. 28–11) and find

$$\lambda_0 = \frac{hc}{eV}. \qquad \textbf{(28–4)}$$

This prediction for λ_0 corresponds precisely with that observed experimentally. This result is further evidence that X-rays are a form of electromagnetic radiation (light) and that the photon theory of light is valid.

EXAMPLE 28–9 **Cutoff wavelength.** What is the shortest-wavelength X-ray photon emitted in an X-ray tube subjected to 50 kV?

APPROACH The electrons striking the target will have a KE of 50 keV. The shortest-wavelength photons are due to collisions in which all of the electron's KE is given to the photon so $\text{KE} = eV = hf_0$.

SOLUTION From Eq. 28–4,

$$\lambda_0 = \frac{hc}{eV} = \frac{(6.63 \times 10^{-34} \text{ J·s})(3.0 \times 10^8 \text{ m/s})}{(1.6 \times 10^{-19} \text{ C})(5.0 \times 10^4 \text{ V})} = 2.5 \times 10^{-11} \text{ m},$$

or 0.025 nm.

NOTE This result agrees well with experiment, Fig. 28–11.

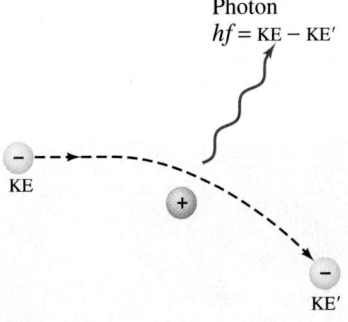

FIGURE 28–13 Bremsstrahlung photon produced by an electron decelerated by interaction with a target atom.

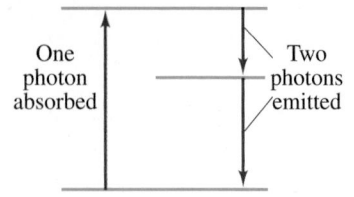

One photon absorbed — Two photons emitted

FIGURE 28–14 Fluorescence.

PHYSICS APPLIED
Fluorescence analysis and fluorescent lightbulbs

FIGURE 28–15 When UV light (a range of wavelengths) illuminates these various "fluorescent" rocks, they fluoresce in the visible region of the spectrum.

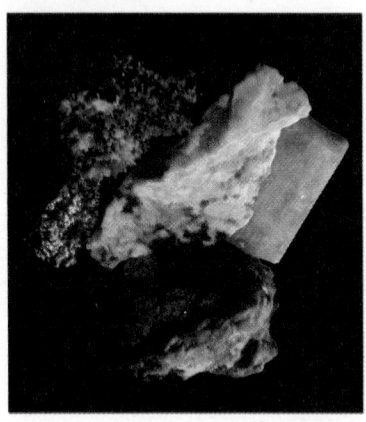

FIGURE 28–16 (a) Absorption of a photon. (b) Stimulated emission. E_u and E_ℓ refer to "upper" and "lower" energy states.

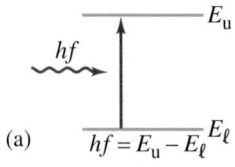

(a) $hf = E_u - E_\ell$

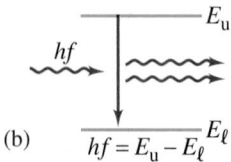

(b) $hf = E_u - E_\ell$

*28–10 Fluorescence and Phosphorescence

When an atom is excited from one energy state to a higher one by the absorption of a photon, it may return to the lower level in a series of two (or more) transitions if there is at least one energy level in between (Fig. 28–14). The photons emitted will consequently have lower energy and frequency than the absorbed photon. When the absorbed photon is in the UV and the emitted photons are in the visible region of the spectrum, this phenomenon is called **fluorescence** (Fig. 28–15).

The wavelength for which fluorescence will occur depends on the energy levels of the particular atoms. Because the frequencies are different for different substances, and because many substances fluoresce readily, fluorescence is a powerful tool for identification of compounds. It is also used for assaying—determining how much of a substance is present—and for following substances along a natural *metabolic pathway* in biological organisms. For detection of a given compound, the stimulating light must be monochromatic, and solvents or other materials present must not fluoresce in the same region of the spectrum. Sometimes the observation of fluorescent light being emitted is sufficient to detect a compound. In other cases, spectrometers are used to measure the wavelengths and intensities of the emitted light.

Fluorescent lightbulbs work in a two-step process. The applied voltage accelerates electrons that strike atoms of the gas in the tube and cause them to be excited. When the excited atoms jump down to their normal levels, they emit UV photons which strike a fluorescent coating on the inside of the tube. The light we see is a result of this material fluorescing in response to the UV light striking it.

Materials such as those used for luminous watch dials, and other glow-in-the-dark products, are said to be **phosphorescent**. When an atom is raised to a normal excited state, it drops back down within about 10^{-8} s. In phosphorescent substances, atoms can be excited by photon absorption to energy levels called **metastable**, which are states that last much longer because to jump down is a "forbidden" transition (Section 28–6). Metastable states can last even a few seconds or longer. In a collection of such atoms, many of the atoms will descend to the lower state fairly soon, but many will remain in the excited state for over an hour. Hence light will be emitted even after long periods. When you put a luminous watch dial close to a bright lamp, many atoms are excited to metastable states, and you can see the glow for a long time afterward.

28–11 Lasers

A **laser** is a device that can produce a very narrow intense beam of monochromatic coherent light. (By **coherent**, we mean that across any cross section of the beam, all parts have the same phase.[†]) The emitted beam is a nearly perfect plane wave. An ordinary light source, on the other hand, emits light in all directions (so the intensity decreases rapidly with distance), and the emitted light is incoherent (the different parts of the beam are not in phase with each other). The excited atoms that emit the light in an ordinary lightbulb act independently, so each photon emitted can be considered as a short wave train lasting about 10^{-8} s. Different wave trains bear no phase relation to one another. Just the opposite is true of lasers.

The action of a laser is based on quantum theory. We have seen that a photon can be absorbed by an atom if (and only if) the photon energy hf corresponds to the energy difference between an occupied energy level of the atom and an available excited state, Fig. 28–16a. If the atom is already in the excited state, it may jump down spontaneously (i.e., no stimulus) to the lower state with the emission of a photon. However, if a photon with this same energy strikes the excited atom, it can stimulate the atom to make the transition sooner to the lower state, Fig. 28–16b. This phenomenon is called **stimulated emission**: not only do we still have the original photon, but also a second one of the same frequency as a result

[†]See also Section 24–3.

of the atom's transition. These two photons are exactly *in phase*, and they are moving in the same direction. This is how coherent light is produced in a laser. The name "laser" is an acronym for **L**ight **A**mplification by **S**timulated **E**mission of **R**adiation.

Normally, most atoms are in the lower state, so the majority of incident photons will be absorbed. To obtain the coherent light from stimulated emission, two conditions must be satisfied. First, the atoms must be excited to the higher state so that an **inverted population** is produced in which more atoms are in the upper state than in the lower one (Fig. 28–17). Then *emission* of photons will dominate over absorption. And second, the higher state must be a **metastable state**—a state in which the electrons remain longer than usual[†] so that the transition to the lower state occurs by stimulated emission rather than spontaneously.

Figure 28–18 is a schematic diagram of a laser: the "lasing" material is placed in a long narrow tube at the ends of which are two mirrors, one of which is partially transparent (transmitting perhaps 1 or 2%). Some of the excited atoms drop down fairly soon after being excited. One of these is the blue atom shown on the far left in Fig. 28–18. If the emitted photon strikes another atom in the excited state, it stimulates this atom to emit a photon of the *same* frequency, moving in the *same* direction, and *in phase* with it. These two photons then move on to strike other atoms causing more stimulated emission. As the process continues, the number of photons multiplies. When the photons strike the end mirrors, most are reflected back, and as they move in the opposite direction, they continue to stimulate more atoms to emit photons. As the photons move back and forth between the mirrors, a small percentage passes through the partially transparent mirror at one end. These photons make up the narrow coherent external laser beam. (Inside the tube, some spontaneously emitted photons will be emitted at an angle to the axis, and these will merely go out the side of the tube and not affect the narrow width of the main beam.)

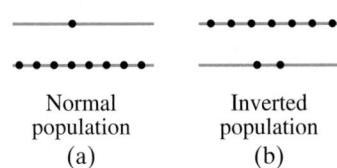

FIGURE 28–17 Two energy levels for a collection of atoms. Each dot represents the energy state of one atom. (a) A normal situation; (b) an inverted population.

**CAUTION**

Laser: photons have same frequency and direction, and are in phase

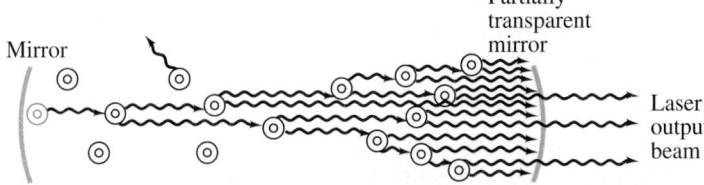

FIGURE 28–18 Laser diagram, showing excited atoms stimulated to emit light.

In a well-designed laser, the spreading of the beam is limited only by diffraction, so the angular spread is $\approx \lambda/D$ (see Eq. 24–3 or 25–7) where D is the diameter of the end mirror. The diffraction spreading can be incredibly small. The light energy, instead of spreading out in space as it does for an ordinary light source, can be a pencil-thin beam.

Creating an Inverted Population

The excitation of the atoms in a laser can be done in several ways to produce the necessary inverted population. In a **ruby laser**, the lasing material is a ruby rod consisting of Al_2O_3 with a small percentage of aluminum (Al) atoms replaced by chromium (Cr) atoms. The Cr atoms are the ones involved in lasing. In a process called **optical pumping**, the atoms are excited by strong flashes of light of wavelength 550 nm, which corresponds to a photon energy of 2.2 eV. As shown in Fig. 28–19, the atoms are excited from state E_0 to state E_2. The atoms quickly decay either back to E_0 or to the intermediate state E_1, which is metastable with a lifetime of about 3×10^{-3} s (compared to 10^{-8} s for ordinary levels). With strong pumping action, more atoms can be found in the E_1 state than are in the E_0 state. Thus we have the inverted population needed for lasing. As soon as a few atoms in the E_1 state jump down to E_0, they emit photons that produce stimulated emission of the other atoms, and the lasing action begins. A ruby laser thus emits a beam whose photons have energy 1.8 eV and a wavelength of 694.3 nm (or "ruby-red" light).

FIGURE 28–19 Energy levels of chromium in a ruby crystal. Photons of energy 2.2 eV "pump" atoms from E_0 to E_2, which then decay to metastable state E_1. Lasing action occurs by stimulated emission of photons in transition from E_1 to E_0.

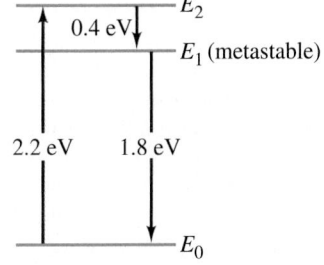

[†]An excited atom may land in such a state and can jump to a lower state only by a so-called forbidden transition (Section 28–6), which is why its lifetime is longer than normal.

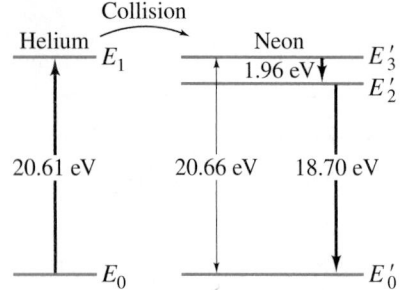

FIGURE 28–20 Energy levels for He and Ne. He is excited in the electric discharge to the E_1 state. This energy is transferred to the E_3' level of the Ne by collision. E_3' is metastable and decays to E_2' by stimulated emission.

FIGURE 28–21 (a) Reading a CD (or DVD). The fine beam of a laser, focused even more finely with lenses, is directed at the undersurface of a rotating compact disc. The beam is reflected back from the areas between pits but reflects much less from pits. The reflected light is detected as shown, reflected by a half-reflecting mirror MS. The strong and weak reflections correspond to the 0s and 1s of the binary code representing the audio or video signal. (b) A laser follows the CD track which starts near the center and spirals outward.

In a **helium–neon laser** (He–Ne), the lasing material is a gas, a mixture of about 85% He and 15% Ne. The atoms are excited by applying a high voltage to the tube so that an electric discharge takes place within the gas. In the process, some of the He atoms are raised to the metastable state E_1 shown in Fig. 28–20, which corresponds to a jump of 20.61 eV, almost exactly equal to an excited state in neon, 20.66 eV. The He atoms do not quickly return to the ground state by spontaneous emission, but instead often give their excess energy to a Ne atom when they collide—see Fig. 28–20. In such a collision, the He drops to the ground state and the Ne atom is excited to the state E_3' (the prime refers to neon states). The slight difference in energy (0.05 eV) is supplied by the kinetic energy of the moving atoms. In this manner, the E_3' state in Ne—which is metastable—becomes more populated than the E_2' level. This inverted population between E_3' and E_2' is what is needed for lasing.

Very common now are **semiconductor diode lasers**, also called **pn junction lasers**, which utilize an inverted population of electrons between the conduction band and the lower-energy valence band (Section 29–9). When an electron jumps down, a photon can be emitted, which in turn can stimulate another electron to make the transition and emit another photon, in phase. The needed mirrors (as in Fig. 28–18) are made by the polished ends of the *pn* crystal. Semiconductor lasers are used in CD and DVD players (see below), and in many other applications.

Other types of laser include: *chemical lasers*, in which the energy input comes from the chemical reaction of highly reactive gases; *dye lasers*, whose frequency is tunable; CO_2 *gas lasers*, capable of high power output in the infrared; and *rare-earth solid-state lasers* such as the high-power Nd:YAG laser.

The excitation of the atoms in a laser can be done continuously or in pulses. In a **pulsed laser**, the atoms are excited by periodic inputs of energy. In a **continuous laser**, the energy input is continuous: as atoms are stimulated to jump down to the lower level, they are soon excited back up to the upper level so the output is a continuous laser beam.

No laser is a source of energy. Energy must be put in, and the laser converts a part of it into an intense narrow beam output.

*Applications

The unique feature of light from a laser, that it is a coherent narrow beam, has found many applications. In everyday life, lasers are used as bar-code readers (at store checkout stands) and in compact disc (CD) and digital video disc (DVD) players. The laser beam reflects off the stripes and spaces of a bar code, or off the tiny pits of a CD or DVD as shown in Fig. 28–21a. The recorded information on a CD or DVD is a series of pits and spaces representing 0s and 1s (or "off" and "on") of a binary code (Section 17–10) that is decoded electronically before being sent to the audio or video system. The laser of a CD player starts reading at the inside of the disc which rotates at about 500 rpm at the start. As the disc rotates, the laser follows the spiral track (Fig. 28–21b), and as it moves outward the disc must slow down because each successive circumference ($C = 2\pi r$) is slightly longer as *r* increases; at the outer edge, the disc is rotating about 200 rpm. A 1-hour CD has a track roughly 5 km long; the track width is about 1600 nm ($= 1.6\ \mu m$) and the distance between pits is about 800 nm. DVDs contain much more information.

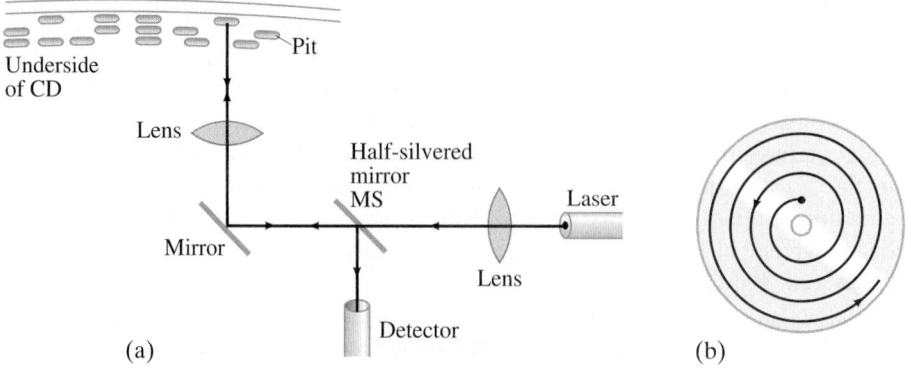

(a)

(b)

Standard DVDs use a thinner track $(0.7 \mu m)$ and shorter pit length $(400 \, nm)$. Blu-ray discs use a "blue" laser with a short wavelength $(405 \, nm)$ and narrower beam, allowing a narrower track $(0.3 \mu m)$ that can store much more data for high definition. DVDs can also have two layers, one below the other. When the laser focuses on the second layer, the light passes through the semitransparent surface layer. The second layer may start reading at the outer edge instead of inside. DVDs can also have a single or double layer on *both* surfaces of the disc.

Lasers are a useful surgical tool. The narrow intense beam can be used to destroy tissue in a localized area, or to break up gallstones and kidney stones. Because of the heat produced, a laser beam can be used to "weld" broken tissue, such as a detached retina, Fig. 28–22, or to mold the cornea of the eye (by vaporizing tiny bits of material) to correct myopia and other eye defects (LASIK surgery). The laser beam can be carried by an optical fiber (Section 23–6) to the surgical point, sometimes as an additional fiber-optic path on an endoscope (again Section 23–6). An example is the removal of plaque clogging human arteries. Lasers have been used to destroy tiny organelles within a living cell by researchers studying how the absence of that organelle affects the behavior of the cell. Laser beams are used to destroy cancerous and precancerous cells; and the heat seals off capillaries and lymph vessels, thus "cauterizing" the wound to prevent spread of the disease.

The intense heat produced in a small area by a laser beam is used for welding and machining metals and for drilling tiny holes in hard materials. Because a laser beam is coherent, monochromatic, narrow, and essentially parallel, lenses can be used to focus the light into even smaller areas. The precise straightness of a laser beam is also useful to surveyors for lining up equipment accurately, especially in inaccessible places.

PHYSICS APPLIED
Medical and other uses of lasers

FIGURE 28–22 Laser being used in eye surgery.

PHYSICS APPLIED
Holography

*28–12 Holography

One of the most interesting applications of laser light is the production of three-dimensional images called **holograms** (see Fig. 28–23). In an ordinary photograph, the film simply records the intensity of light reaching it at each point. When the photograph or transparency is viewed, light reflecting from it or passing through it gives us a two-dimensional picture. In holography, the images are formed by interference, without lenses. A laser hologram is typically made on a photographic emulsion (film). A broadened laser beam is split into two parts by a half-silvered mirror, Fig. 28–24. One part goes directly to the film; the rest passes to the object to be photographed, from which it is reflected to the film. Light from every point on the object reaches each point on the film, and the interference of the two beams allows the film to record both the intensity and relative phase of the light at each point. It is crucial that the incident light be coherent—that is, in phase at all points—which is why a laser is used. After the film is developed, it is placed again in a laser beam and a three-dimensional image of the object is created. You can walk around such an image and see it from different sides as if it were the original object. Yet, if you try to touch it with your hand, there will be nothing material there.

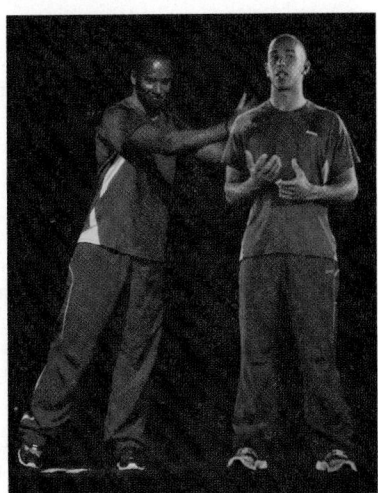

FIGURE 28–23 An athlete (race-car driver) puts his hand on, or through, a holographic image of himself.

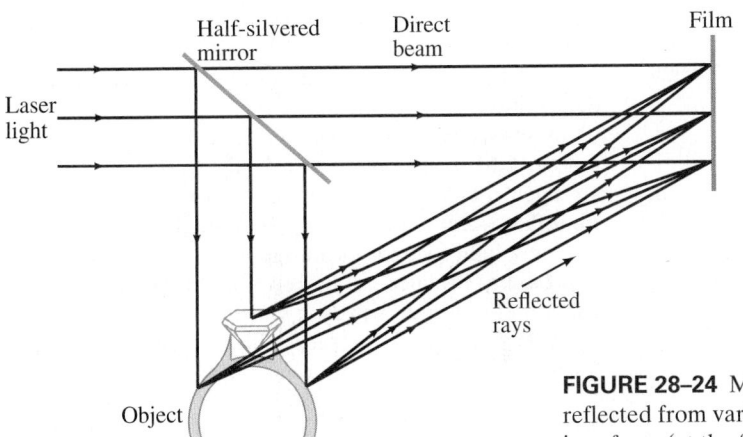

FIGURE 28–24 Making a hologram. Light reflected from various points on the object interferes (at the film) with light from the direct beam.

Volume or **white-light holograms** do not require a laser to see the image, but can be viewed with ordinary white light (preferably a nearly point source, such as the Sun or a clear bulb with a small bright filament). Such holograms must be made, however, with a laser. They are made not on thin film, but on a *thick* emulsion. The interference pattern in the film emulsion can be thought of as an array of bands or ribbons where constructive interference occurred. This array, and the reconstruction of the image, can be compared to Bragg scattering of X-rays from the atoms in a crystal (see Section 25–11). White light can reconstruct the image because the Bragg condition $(m\lambda = 2d \sin\theta)$ selects out the appropriate single wavelength. If the hologram is originally produced by lasers emitting the three additive primary colors (red, green, and blue), the three-dimensional image can be seen in full color when viewed with white light.

Summary

In 1925, Schrödinger and Heisenberg separately worked out a new theory, **quantum mechanics**, which is now considered to be the fundamental theory at the atomic level. It is a statistical theory rather than a deterministic one.

An important aspect of quantum mechanics is the Heisenberg **uncertainty principle**. It results from the wave–particle duality and the unavoidable interaction between an observed object and the observer.

One form of the uncertainty principle states that the position x and momentum p_x of an object cannot both be measured precisely at the same time. The products of the uncertainties, $(\Delta x)(\Delta p_x)$, can be no less than $\hbar\ (= h/2\pi)$:

$$(\Delta p_x)(\Delta x) \gtrsim \hbar. \qquad \textbf{(28–1)}$$

Another form of the uncertainty principle states that the energy can be uncertain (or nonconserved) by an amount ΔE for a time Δt, where

$$(\Delta E)(\Delta t) \gtrsim \hbar. \qquad \textbf{(28–2)}$$

According to quantum mechanics, the state of an electron in an atom is specified by four **quantum numbers**: n, ℓ, m_ℓ, and m_s:

(1) n, the **principal quantum number**, can take on any integer value $(1, 2, 3, \cdots)$ and corresponds to the quantum number of the old Bohr model;

(2) ℓ, the **orbital quantum number**, can take on values from 0 up to $n - 1$;

(3) m_ℓ, the **magnetic quantum number**, can take on integer values from $-\ell$ to $+\ell$;

(4) m_s, the **spin quantum number**, can be $+\frac{1}{2}$ or $-\frac{1}{2}$.

The energy levels in the hydrogen atom depend on n, whereas in other atoms they depend on n and ℓ.

The orbital angular momentum of an atom has magnitude $L = \sqrt{\ell(\ell + 1)}\ \hbar$ and z component $L_z = m_\ell \hbar$.

When an external magnetic field is applied, the spectral lines are split (the **Zeeman effect**), indicating that the energy depends also on m_ℓ in this case.

Even in the absence of a magnetic field, precise measurements of spectral lines show a tiny splitting of the lines called **fine structure**, whose explanation is that the energy depends very slightly on m_ℓ and m_s.

Transitions between states that obey the **selection rule** $\Delta\ell = \pm 1$ are far more probable than other so-called **forbidden transitions**.

The arrangement of electrons in multi-electron atoms is governed by the Pauli **exclusion principle**, which states that no two electrons can occupy the same quantum state—that is, they cannot have the same set of quantum numbers n, ℓ, m_ℓ, and m_s.

As a result, electrons in multi-electron atoms are grouped into **shells** (according to the value of n) and **subshells** (according to ℓ).

Electron configurations are specified using the numerical values of n, and using letters for ℓ: s, p, d, f, etc., for $\ell = 0, 1, 2, 3$, and so on, plus a superscript for the number of electrons in that subshell. Thus, the ground state of hydrogen is $1s^1$, whereas that for oxygen is $1s^2 2s^2 2p^4$.

In the **Periodic Table**, the elements are arranged in horizontal rows according to increasing **atomic number** (= number of electrons in the neutral atom). The shell structure gives rise to a periodicity in the properties of the elements, so that each vertical column can contain elements with similar chemical properties.

X-rays, which are a form of electromagnetic radiation of very short wavelength, are produced when high-speed electrons strike a target. The spectrum of X-rays so produced consists of two parts, a *continuous* spectrum produced when the electrons are decelerated by atoms of the target, and *peaks* representing photons emitted by atoms of the target after being excited by collision with the high-speed electrons. Measurement of these peaks allows determination of inner energy levels of atoms and determination of atomic number Z.

[*Fluorescence** occurs when absorbed UV photons are followed by emission of visible light, due to the special arrangement of energy levels of atoms of the material. **Phosphorescent** materials have **metastable** states (long-lived) that emit light seconds or minutes after absorption of light.]

Lasers produce a narrow beam of monochromatic coherent light (light waves *in phase*).

[*Holograms** are images with a 3-dimensional quality, formed by interference of laser light.]

Questions

1. Compare a matter wave Ψ to (a) a wave on a string, (b) an EM wave. Discuss similarities and differences.
2. Explain why Bohr's theory of the atom is not compatible with quantum mechanics, particularly the uncertainty principle.
3. Explain why it is that the more massive an object is, the easier it becomes to predict its future position.
4. In view of the uncertainty principle, why does a baseball seem to have a well-defined position and speed, whereas an electron does not?
5. Would it ever be possible to balance a very sharp needle precisely on its point? Explain.
6. A cold thermometer is placed in a hot bowl of soup. Will the temperature reading of the thermometer be the same as the temperature of the hot soup before the measurement was made? Explain.
7. Does the uncertainty principle set a limit to how well you can make any single measurement of position? Explain.
8. If you knew the position of an object precisely, with no uncertainty, how well would you know its momentum?
9. When you check the pressure in a tire, doesn't some air inevitably escape? Is it possible to avoid this escape of air altogether? What is the relation to the uncertainty principle?
10. It has been said that the ground-state energy in the hydrogen atom can be precisely known but the excited states have some uncertainty in their values (an "energy width"). Is this consistent with the uncertainty principle in its energy form? Explain.
11. Which model of the hydrogen atom, the Bohr model or the quantum-mechanical model, predicts that the electron spends more time near the nucleus? Explain.
12. The size of atoms varies by only a factor of three or so, from largest to smallest, yet the number of electrons varies from one to over 100. Explain.
13. Excited hydrogen and excited helium atoms both radiate light as they jump down to the $n = 1$, $\ell = 0$, $m_\ell = 0$ state. Why do the two elements have very different emission spectra?
14. How would the Periodic Table look if there were no electron spin but otherwise quantum mechanics were valid? Consider the first 20 elements or so.

15. Which of the following electron configurations are not allowed: (a) $1s^2 2s^2 2p^4 3s^2 4p^2$; (b) $1s^2 2s^2 2p^8 3s^1$; (c) $1s^2 2s^2 2p^6 3s^2 3p^5 4s^2 4d^5 4f^1$? If not allowed, explain why.
16. In what column of the Periodic Table would you expect to find the atom with each of the following configurations: (a) $1s^2 2s^2 2p^5$; (b) $1s^2 2s^2 2p^6 3s^2$; (c) $1s^2 2s^2 2p^6 3s^2 3p^6$; (d) $1s^2 2s^2 2p^6 3s^2 3p^6 4s^1$?
17. Why do chlorine and iodine exhibit similar properties?
18. Explain why potassium and sodium exhibit similar properties.
19. Why are the chemical properties of the rare earths so similar? [Hint: Examine the Periodic Table.]
20. The ionization energy for neon ($Z = 10$) is 21.6 eV, and that for sodium ($Z = 11$) is 5.1 eV. Explain the large difference.
21. Why do we expect electron transitions deep within an atom to produce shorter wavelengths than transitions by outer electrons?
22. Does the Bohr model of the atom violate the uncertainty principle? Explain.
23. Briefly explain why noble gases are nonreactive and why alkali metals are highly reactive. (See Section 28–8.)
24. Compare spontaneous emission to stimulated emission.
25. How does laser light differ from ordinary light? How is it the same?
26. Explain how a 0.0005-W laser beam, photographed at a distance, can seem much stronger than a 1000-W street lamp at the same distance.
27. Does the intensity of light from a laser fall off as the inverse square of the distance? Explain.
*28. Why does the cutoff wavelength in Fig. 28–11 imply a photon nature for light?
*29. Why do we not expect perfect agreement between measured values of characteristic X-ray line wavelengths and those calculated using the Bohr model, as in Example 28–7?
*30. How would you figure out which lines in an X-ray spectrum correspond to K_α, K_β, L, etc., transitions?

MisConceptual Questions

1. An atom has the electron configuration $1s^2 2s^2 2p^6 3s^2 3p^6 4s^1$. How many electrons does this atom have?
   (a) 15. (b) 19. (c) 30. (d) 46.
2. For the electron configuration of MisConceptual Question 1, what orbital quantum numbers do the electrons have?
   (a) 0.
   (b) 0 and 1.
   (c) 0 and 1 and 2.
   (d) 0 and 1 and 2 and 3.
   (e) 0 and 1 and 2 and 3 and 4.
3. If a beam of electrons is fired through a slit,
   (a) the electrons can be deflected because of their wave properties.
   (b) only electrons that hit the edge of the slit are deflected.
   (c) electrons can interact with electromagnetic waves in the slit, forming a diffraction pattern.
   (d) the probability of an electron making it through the slit depends on the uncertainty principle.

4. What is meant by the ground state of an atom?
   (a) All of the quantum numbers have their lowest values $(n = 1, \ell = m_\ell = 0)$.
   (b) The principal quantum number of the electrons in the outer shell is 1.
   (c) All of the electrons are in the lowest energy state, consistent with the exclusion principle.
   (d) The electrons are in the lowest state allowed by the uncertainty principle.

5. The Pauli exclusion principle applies to all electrons
   (a) in the same shell, but not electrons in different shells.
   (b) in the same container of atoms.
   (c) in the same column of the Periodic Table.
   (d) in incomplete shells.
   (e) in the same atom.

6. Which of the following is the best paraphrasing of the Heisenberg uncertainty principle?
  (*a*) Only if you know the exact position of a particle can you know the exact momentum of the particle.
  (*b*) The larger the momentum of a particle, the smaller the position of the particle.
  (*c*) The more precisely you know the position of a particle, the less well you can know the momentum of the particle.
  (*d*) The better you know the position of a particle, the better you can know the momentum of the particle.
  (*e*) How well you can determine the position and momentum of a particle depends on the particle's quantum numbers.

7. Which of the following is required by the Pauli exclusion principle?
  (*a*) No electron in an atom can have the same set of quantum numbers as any other electron in that atom.
  (*b*) Each electron in an atom must have the same *n* value.
  (*c*) Each electron in an atom must have different m_ℓ values.
  (*d*) Only two electrons can be in any particular shell of an atom.
  (*e*) No two electrons in a collection of atoms can have the exact same set of quantum numbers.

8. Under what condition(s) can the exact location and velocity of an electron be measured at the same time?
  (*a*) The electron is in the ground state of the atom.
  (*b*) The electron is in an excited state of the atom.
  (*c*) The electron is free (not bound to an atom).
  (*d*) Both (*a*) and (*b*).
  (*e*) Never.

9. According to the uncertainty principle,
  (*a*) there is always an uncertainty in a measurement of the position of a particle.
  (*b*) there is always an uncertainty in a measurement of the momentum of a particle.
  (*c*) there is always an uncertainty in a simultaneous measurement of both the position and momentum of a particle.
  (*d*) All of the above.

10. Which of the following is *not* always a property of lasers?
  (*a*) All of the photons in laser light have the same phase.
  (*b*) All laser photons have nearly identical frequencies.
  (*c*) Laser light moves as a beam, spreading out very slowly.
  (*d*) Laser light is always brighter than other sources of light.
  (*e*) Lasers depend on an inverted population of atoms where more atoms occupy a higher energy state than some lower energy state.

For assigned homework and other learning materials, go to the MasteringPhysics website.

Problems

28–2 Wave Function, Double-Slit

1. (II) The neutrons in a parallel beam, each having kinetic energy 0.025 eV, are directed through two slits 0.40 mm apart. How far apart will the interference peaks be on a screen 1.0 m away? [*Hint*: First find the wavelength of the neutron.]

2. (II) Pellets of mass 2.0 g are fired in parallel paths with speeds of 120 m/s through a hole 3.0 mm in diameter. How far from the hole must you be to detect a 1.0-cm-diameter spread in the beam of pellets?

28–3 Uncertainty Principle

3. (I) A proton is traveling with a speed of $(8.660 \pm 0.012) \times 10^5$ m/s. With what maximum precision can its position be ascertained? [*Hint*: $\Delta p = m \, \Delta v$.]

4. (I) If an electron's position can be measured to a precision of 2.4×10^{-8} m, how precisely can its speed be known?

5. (I) An electron remains in an excited state of an atom for typically 10^{-8} s. What is the minimum uncertainty in the energy of the state (in eV)?

6. (II) The Z^0 boson, discovered in 1985, is the mediator of the weak nuclear force, and it typically decays very quickly. Its average rest energy is 91.19 GeV, but its short lifetime shows up as an intrinsic width of 2.5 GeV. What is the lifetime of this particle? [*Hint*: See Example 28–3.]

7. (II) What is the uncertainty in the mass of a muon $(m = 105.7 \text{ MeV}/c^2)$, specified in eV/$c^2$, given its lifetime of 2.20 μs?

8. (II) A free neutron $(m = 1.67 \times 10^{-27} \text{ kg})$ has a mean life of 880 s. What is the uncertainty in its mass (in kg)?

9. (II) An electron and a 140-g baseball are each traveling 120 m/s measured to a precision of 0.065%. Calculate and compare the uncertainty in position of each.

10. (II) A radioactive element undergoes an alpha decay with a lifetime of 12 μs. If alpha particles are emitted with 5.5-MeV kinetic energy, find the percent uncertainty $\Delta E/E$ in the particle energy.

11. (II) If an electron's position can be measured to a precision of 15 nm, what is the uncertainty in its speed? Assuming the minimum speed must be at least equal to its uncertainty, what is the electron's minimum kinetic energy?

12. (II) Estimate the lowest possible energy of a neutron contained in a typical nucleus of radius 1.2×10^{-15} m. [*Hint*: Assume a particle can have an energy as large as its uncertainty.]

13. (III) How precisely can the position of a 5.00-keV electron be measured assuming its energy is known to 1.00%?

14. (III) Use the uncertainty principle to show that if an electron were present in the nucleus $(r \approx 10^{-15} \text{ m})$, its kinetic energy (use relativity) would be hundreds of MeV. (Since such electron energies are not observed, we conclude that electrons are not present in the nucleus.) [*Hint*: Assume a particle can have an energy as large as its uncertainty.]

15. (I) For $n = 6$, what values can ℓ have?

16. (I) For $n = 6$, $\ell = 3$, what are the possible values of m_ℓ and m_s?

17. (I) How many electrons can be in the $n = 5$, $\ell = 3$ subshell?

18. (I) How many different states are possible for an electron whose principal quantum number is $n = 4$? Write down the quantum numbers for each state.

19. (I) List the quantum numbers for each electron in the ground state of (a) carbon ($Z = 6$), (b) aluminum ($Z = 13$).

20. (I) List the quantum numbers for each electron in the ground state of oxygen ($Z = 8$).

21. (I) Calculate the magnitude of the angular momentum of an electron in the $n = 5$, $\ell = 3$ state of hydrogen.

22. (I) If a hydrogen atom has $\ell = 4$, what are the possible values for n, m_ℓ, and m_s?

23. (II) If a hydrogen atom has $m_\ell = -3$, what are the possible values of n, ℓ, and m_s?

24. (II) Show that there can be 18 electrons in a "g" subshell.

25. (II) What is the full electron configuration in the ground state for elements with Z equal to (a) 26, (b) 34, (c) 38? [Hint: See the Periodic Table inside the back cover.]

26. (II) What is the full electron configuration for (a) silver (Ag), (b) gold (Au), (c) uranium (U)? [Hint: See the Periodic Table inside the back cover.]

27. (II) A hydrogen atom is in the $5d$ state. Determine (a) the principal quantum number, (b) the energy of the state, (c) the orbital angular momentum and its quantum number ℓ, and (d) the possible values for the magnetic quantum number.

28. (II) Estimate the binding energy of the third electron in lithium using the Bohr model. [Hint: This electron has $n = 2$ and "sees" a net charge of approximately $+1e$.] The measured value is 5.36 eV.

29. (II) Show that the total angular momentum is zero for a filled subshell.

30. (II) For each of the following atomic transitions, state whether the transition is allowed or forbidden, and why: (a) $4p \rightarrow 3p$; (b) $3p \rightarrow 1s$; (c) $4d \rightarrow 2d$; (d) $5d \rightarrow 3s$; (e) $4s \rightarrow 2p$.

31. (II) An electron has $m_\ell = 2$ and is in its lowest possible energy state. What are the values of n and ℓ for this electron?

32. (II) An excited H atom is in a $6d$ state. (a) Name all the states (n, ℓ) to which the atom is "allowed" to make a transition with the emission of a photon. (b) How many different wavelengths are there (ignoring fine structure)?

*33. (I) What are the shortest-wavelength X-rays emitted by electrons striking the face of a 28.5-kV TV picture tube? What are the longest wavelengths?

*34. (I) If the shortest-wavelength bremsstrahlung X-rays emitted from an X-ray tube have $\lambda = 0.035$ nm, what is the voltage across the tube?

*35. (I) Show that the cutoff wavelength λ_0 in an X-ray spectrum is given by
$$\lambda_0 = \frac{1240 \text{ nm}}{V},$$
where V is the X-ray tube voltage in volts.

*36. (I) For the spectrum of X-rays emitted from a molybdenum target (Fig. 28–11), determine the maximum and minimum energy.

*37. (II) Use the result of Example 28–7 ($Z = 42$) to estimate the X-ray wavelength emitted when a cobalt atom ($Z = 27$) makes a transition from $n = 2$ to $n = 1$.

*38. (II) Estimate the wavelength for an $n = 3$ to $n = 2$ transition in iron ($Z = 26$).

*39. (II) Use the Bohr model to estimate the wavelength for an $n = 3$ to $n = 1$ transition in molybdenum ($Z = 42$). The measured value is 0.063 nm. Why do we not expect perfect agreement?

*40. (II) A mixture of iron and an unknown material is bombarded with electrons. The wavelengths of the K_α lines are 194 pm for iron and 229 pm for the unknown. What is the unknown material?

28–11 Lasers

41. (II) A laser used to weld detached retinas puts out 25-ms-long pulses of 640-nm light which average 0.68-W output during a pulse. How much energy can be deposited per pulse and how many photons does each pulse contain? [Hint: See Example 27–4.]

42. (II) A low-power laser used in a physics lab might have a power of 0.50 mW and a beam diameter of 3.0 mm. Calculate (a) the average light intensity of the laser beam, and (b) compare it to the intensity of a lightbulb producing 100-W light viewed from 2.0 m.

43. (II) Calculate the wavelength of the He–Ne laser (see Fig. 28–20).

44. (II) Estimate the angular spread of a laser beam due to diffraction if the beam emerges through a 3.0-mm-diameter mirror. Assume that $\lambda = 694$ nm. What would be the diameter of this beam if it struck (a) a satellite 340 km above the Earth, or (b) the Moon? [Hint: See Sections 24–5 and 25–7.]

General Problems

45. The magnitude of the orbital angular momentum in an excited state of hydrogen is 6.84×10^{-34} J·s and the z component is 2.11×10^{-34} J·s. What are all the possible values of n, ℓ, and m_ℓ for this state?

46. An electron in the $n = 2$ state of hydrogen remains there on average about 10^{-8} s before jumping to the $n = 1$ state. (a) Estimate the uncertainty in the energy of the $n = 2$ state. (b) What fraction of the transition energy is this? (c) What is the wavelength, and width (in nm), of this line in the spectrum of hydrogen?

47. What are the largest and smallest possible values for the angular momentum L of an electron in the $n = 6$ shell?

48. A 12-g bullet leaves a rifle at a speed of 150 m/s. (a) What is the wavelength of this bullet? (b) If the position of the bullet is known to a precision of 0.60 cm (radius of the barrel), what is the minimum uncertainty in its momentum?

49. If an electron's position can be measured to a precision of 2.0×10^{-8} m, what is the uncertainty in its momentum? Assuming its momentum must be at least equal to its uncertainty, estimate the electron's wavelength.

50. The ionization (binding) energy of the outermost electron in boron is 8.26 eV. (a) Use the Bohr model to estimate the "effective charge," Z_{eff}, seen by this electron. (b) Estimate the average orbital radius.

51. Using the Bohr formula for the radius of an electron orbit, estimate the average distance from the nucleus for an electron in the innermost ($n = 1$) orbit of a uranium atom ($Z = 92$). Approximately how much energy would be required to remove this innermost electron?

52. Protons are accelerated from rest across 480 V. They are then directed at two slits 0.70 mm apart. How far apart will the interference peaks be on a screen 28 m away?

53. How many electrons can there be in an "h" subshell?

54. (a) Show that the number of different states possible for a given value of ℓ is equal to $2(2\ell + 1)$. (b) What is this number for $\ell = 0, 1, 2, 3, 4, 5$, and 6?

55. Show that the number of different electron states possible for a given value of n is $2n^2$. (See Problem 54.)

56. A beam of electrons with kinetic energy 45 keV is shot through two narrow slits in a barrier. The slits are a distance 2.0×10^{-6} m apart. If a screen is placed 45.0 cm behind the barrier, calculate the spacing between the "bright" fringes of the interference pattern produced on the screen.

57. The angular momentum in the hydrogen atom is given both by the Bohr model and by quantum mechanics. Compare the results for $n = 2$.

58. The lifetime of a typical excited state in an atom is about 10 ns. Suppose an atom falls from one such excited state to a lower one, and emits a photon of wavelength about 500 nm. Find the fractional energy uncertainty $\Delta E/E$ and wavelength uncertainty $\Delta \lambda / \lambda$ of this photon.

59. A 1300-kg car is traveling with a speed of (22 ± 0.22) m/s. With what maximum precision can its position be determined?

60. An atomic spectrum contains a line with a wavelength centered at 488 nm. Careful measurements show the line is really spread out between 487 and 489 nm. Estimate the lifetime of the excited state that produced this line.

61. An electron and a proton, each initially at rest, are accelerated across the same voltage. Assuming that the uncertainty in their position is given by their de Broglie wavelength, find the ratio of the uncertainty in their momentum.

62. If the principal quantum number n were limited to the range from 1 to 6, how many elements would we find in nature?

63. If your de Broglie wavelength were 0.50 m, how fast would you be moving if your mass is 68.0 kg? Would you notice diffraction effects as you walk through a doorway? Approximately how long would it take you to walk through the doorway?

64. Suppose that the spectrum of an unknown element shows a series of lines with one out of every four matching a line from the Lyman series of hydrogen. Assuming that the unknown element is an ion with Z protons and one electron, determine Z and the element in question.

*65. Photons of wavelength 0.154 nm are emitted from the surface of a certain metal when it is bombarded with high-energy radiation. If this photon wavelength corresponds to the K_α line, what is the element?

Search and Learn

1. Use the uncertainty principle to estimate the position uncertainty for the electron in the ground state of the hydrogen atom. [Hint: Determine the momentum using the Bohr model of Section 27–12 and assume the momentum can be anywhere between this value and zero.] How does this result compare to the Bohr radius?

2. On what factors does the periodicity of the Periodic Table depend? Consider the exclusion principle, quantization of angular momentum, spin, and any others you can think of.

3. As discussed in Section 28–5: (a) List two aspects of the Bohr model that the quantum-mechanical theory of the atom retained. (b) Give one major difference between the Bohr model and the quantum-mechanical theory of the atom.

4. Estimate (a) the quantum number ℓ for the orbital angular momentum of the Earth about the Sun, and (b) the number of possible orientations for the plane of Earth's orbit.

5. Show that the diffraction spread of a laser beam, $\approx \lambda/D$ (Section 28–11), is precisely what you might expect from the uncertainty principle. See also Chapters 24 and 25. [Hint: Since the beam's width is constrained by the dimension of the aperture D, the component of the light's momentum perpendicular to the laser axis is uncertain.]

6. For noble gases, the halogens, and the alkali metals, explain the atomic structure that is common to each group and how that structure explains a common property of the group. (See Section 28–8.)

7. Imagine a line whose length equals the diameter of the smallest orbit in the Bohr model. If we are told only that an electron is located somewhere on this line, then the electron's position can be specified as $x = 0 \pm r_1$, where the origin is at the line's center and r_1 is the Bohr radius. Such an electron can never be observed at rest, but instead at a minimum will have a speed somewhere in the range from $v = 0 - \Delta v$ to $0 + \Delta v$. Determine Δv.

8. What is uncertain in the Heisenberg uncertainty principle? Explain. (See Section 28–3.)

ANSWERS TO EXERCISES

A: (c).
B: 2.1×10^{-24} kg·m/s, 2.3×10^6 m/s.
C: 2, 1, 0, −1, −2.

D: −0.38 eV, $\sqrt{20} \, \hbar$.
E: Add one line to Li in Table 28–2: 2, 0, 0, $-\frac{1}{2}$.
F: $1s^2 2s^2 2p^6 3s^2 3p^6 3d^{10} 4s^2 4p^1$.

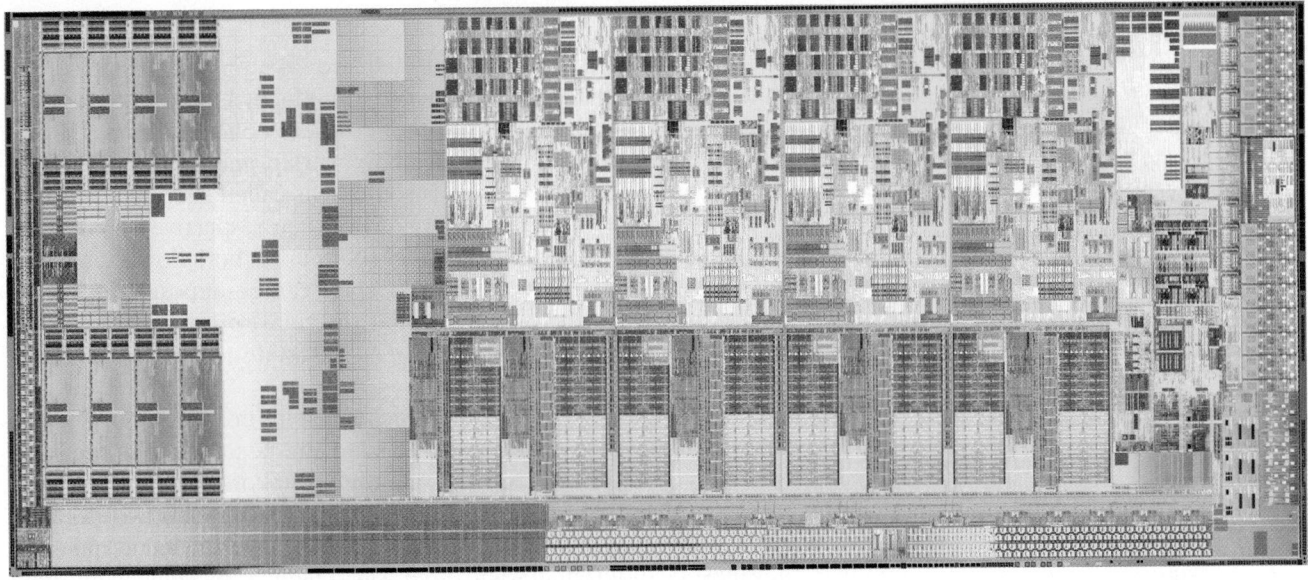

This computer processor chip contains over 1.4 billion transistors, plus diodes and other semiconductor electronic elements, all in a space of about 1 cm². It uses 22-nm technology, meaning the "wires" (conducting lines) are 22 nm wide.

Before discussing semiconductors and their applications, we study the quantum theory description of bonding between atoms to form molecules, and how it explains molecular behavior. We then examine how atoms and molecules form solids, with emphasis on metals as well as on semiconductors and their use in electronics.

Molecules and Solids

C H A P T E R

29

CHAPTER-OPENING QUESTION—Guess now!

What holds a solid together?
- **(a)** Gravitational forces.
- **(b)** Magnetic forces.
- **(c)** Electric forces.
- **(d)** Glue.
- **(e)** Nuclear forces.

Since its development in the 1920s, quantum mechanics has had a profound influence on our lives, both intellectually and technologically. Even the way we view the world has changed, as we have seen in the last few Chapters. Now we discuss how quantum mechanics has given us an understanding of the structure of molecules and matter in bulk, as well as a number of important applications including semiconductor devices and applications to biology. Semiconductor devices, like transistors, now may be only a few atoms thick, which is the realm of quantum mechanics.

*29–1 Bonding in Molecules

One of the great successes of quantum mechanics was to give scientists, at last, an understanding of the nature of chemical bonds. Because it is based in physics, and because this understanding is so important in many fields, we discuss it here.

By a molecule, we mean a group of two or more atoms that are strongly held together so as to function as a single unit. When atoms make such an attachment, we say that a chemical **bond** has been formed. There are two main types of strong chemical bond: covalent and ionic. Many bonds are actually intermediate between these two types.

829

*Covalent Bonds

To understand how *covalent bonds* are formed, we take the simplest case, the bond that holds two hydrogen atoms together to form the hydrogen molecule, H_2. The mechanism is basically the same for other covalent bonds. As two H atoms approach each other, the electron clouds begin to overlap, and the electrons from each atom can "orbit" both nuclei. (This is sometimes called **sharing** electrons.) If both electrons are in the ground state ($n = 1$) of their respective atoms, there are two possibilities: their spins can be parallel (both up or both down), in which case the total spin is $S = \frac{1}{2} + \frac{1}{2} = 1$; or their spins can be opposite ($m_s = +\frac{1}{2}$ for one, and $m_s = -\frac{1}{2}$ for the other), so that the total spin $S = 0$. We shall now see that a bond is formed only for the $S = 0$ state, when the spins are opposite.

First we consider the $S = 1$ state, for which the spins are the same. The two electrons cannot both be in the lowest energy state and be attached to the same atom, for then they would have identical quantum numbers in violation of the exclusion principle. The exclusion principle tells us that, because no two electrons can occupy the same quantum state, if two electrons have the same quantum numbers, they must be different in some other way—namely, by being in different places in space (for example, attached to different atoms). Thus, for $S = 1$, when the two atoms approach each other, the electrons will stay away from each other as shown by the probability distribution of Fig. 29–1. The electrons spend very little time between the two nuclei, so the positively charged nuclei repel each other and no bond is formed.

For the $S = 0$ state, on the other hand, the spins are opposite and the two electrons are consequently in different quantum states (m_s is different, $+\frac{1}{2}$ for one, $-\frac{1}{2}$ for the other). Hence the two electrons can come close together, and the probability distribution looks like Fig. 29–2: the electrons can spend much of their time between the two nuclei. The two positively charged nuclei are attracted to the negatively charged electron cloud between them, and this is the attraction that holds the two hydrogen atoms together to form a hydrogen molecule. This is a **covalent bond**.

The probability distributions of Figs. 29–1 and 29–2 can perhaps be better understood on the basis of waves. What the exclusion principle requires is that when the spins are the same, there is destructive interference of the electron wave functions in the region between the two atoms. But when the spins are opposite, constructive interference occurs in the region between the two atoms, resulting in a large amount of negative charge there. Thus a covalent bond can be said to be the result of constructive interference of the electron wave functions in the space between the two atoms, and of the electrostatic attraction of the two positive nuclei for the negative charge concentration between them.

Why a bond is formed can also be understood from the energy point of view. When the two H atoms approach close to one another, if the spins of their electrons are opposite, the electrons can occupy the same space, as discussed above. This means that each electron can now move about in the space of two atoms instead of in the volume of only one. Because each electron now occupies more space, it is less well localized. From the uncertainty principle with Δx larger, we see that Δp and the minimum momentum can be less. With less momentum, each electron has less energy when the two atoms combine than when they are separate. That is, the molecule has less energy than the two separate atoms, and so is more stable. An energy input is required to break the H_2 molecule into two separate H atoms, so the H_2 molecule is a stable entity. This is what we mean by a *bond*. The energy required to break a bond is called the **bond energy**, the **binding energy**, or the **dissociation energy**. For the hydrogen molecule, H_2, the bond energy is 4.5 eV.

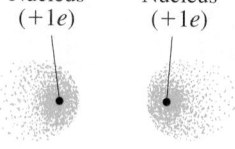

FIGURE 29–1 Electron probability distribution (electron cloud) for two H atoms when the spins are the same: $S = \frac{1}{2} + \frac{1}{2} = 1$.

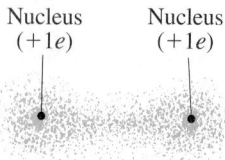

FIGURE 29–2 Electron probability distribution (cloud) around two H atoms when the spins are opposite ($S = 0$): in this case, a bond is formed because the positive nuclei are attracted to the concentration of the electron cloud's negative charge between them. This is a hydrogen molecule, H_2.

*Ionic Bonds

An *ionic bond* is, in a sense, a special case of the covalent bond. Instead of the electrons being shared equally, they are shared unequally. For example, in sodium chloride (NaCl), the outer electron of the sodium spends nearly all its time around the chlorine (Fig. 29–3). The chlorine atom acquires a net negative charge as a result of the extra electron, whereas the sodium atom is left with a net positive charge. The electrostatic attraction between these two charged atoms holds them together. The resulting bond is called an **ionic bond** because it is created by the attraction between the two ions (Na$^+$ and Cl$^-$). But to understand the ionic bond, we must understand why the extra electron from the sodium spends so much of its time around the chlorine. After all, the chlorine atom is neutral; why should it attract another electron?

The answer lies in the probability distributions of the electrons in the two neutral atoms. Sodium contains 11 electrons, 10 of which are in spherically symmetric closed shells (Fig. 29–4). The last electron spends most of its time beyond these closed shells. Because the closed shells have a total charge of $-10e$ and the nucleus has charge $+11e$, the outermost electron in sodium "feels" a net attraction due to $+1e$. It is not held very strongly. On the other hand, 12 of chlorine's 17 electrons form closed shells, or subshells (corresponding to $1s^2 2s^2 2p^6 3s^2$). These 12 electrons form a spherically symmetric shield around the nucleus. The other five electrons are in $3p$ states whose probability distributions are not spherically symmetric and have a form similar to those for the $2p$ states in hydrogen shown in Figs. 28–9b and c. Four of these $3p$ electrons can have "doughnut-shaped" distributions symmetric about the z axis, as shown in Fig. 29–5. The fifth can have a "barbell-shaped" distribution (as for $m_\ell = 0$ in Fig. 28–9b), which in Fig. 29–5 is shown only in dashed outline because it is half empty. That is, the exclusion principle allows one more electron to be in this state (it will have spin opposite to that of the electron already there). If an extra electron—say from a Na atom—happens to be in the vicinity, it can be in this state, perhaps at point x in Fig. 29–5. It could experience an attraction due to as much as $+5e$ because the $+17e$ of the nucleus is partly shielded at this point by the 12 inner electrons. Thus, the outer electron of a sodium atom will be more strongly attracted by the $+5e$ of the chlorine atom than by the $+1e$ of its own atom. This, combined with the strong attraction between the two ions when the extra electron stays with the Cl$^-$, produces the charge distribution of Fig. 29–3, and hence the ionic bond.

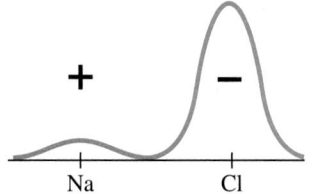

FIGURE 29–3 Probability distribution for the outermost electron of Na in NaCl.

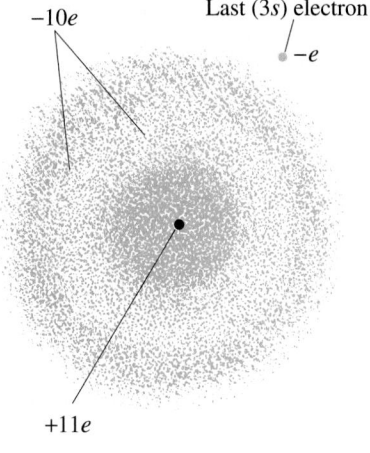

FIGURE 29–4 In a neutral sodium atom, the 10 inner electrons shield the nucleus, so the single outer electron is attracted by a net charge of $+1e$.

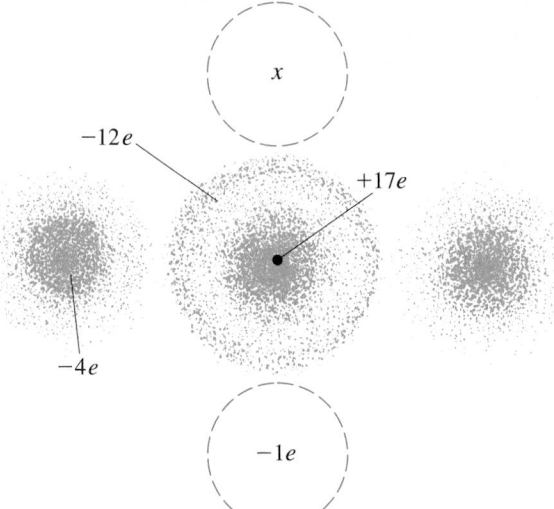

FIGURE 29–5 Neutral chlorine atom. The $+17e$ of the nucleus is shielded by the 12 electrons in the inner shells and subshells. Four of the five $3p$ electrons are shown in doughnut-shaped clouds (seen in cross section at left and right), and the fifth is in the dashed-line cloud concentrated about the z axis (vertical). An extra electron at x will be attracted by a net charge that can be as much as $+5e$.

*Partial Ionic Character of Covalent Bonds

A pure covalent bond in which the electrons are shared equally occurs mainly in symmetrical molecules such as H_2, O_2, and Cl_2. When the atoms involved are different from each other, usually the shared electrons are more likely to be in the vicinity of one atom than the other. The extreme case is an ionic bond. In intermediate cases the *covalent bond* is said to have a **partial ionic character**.

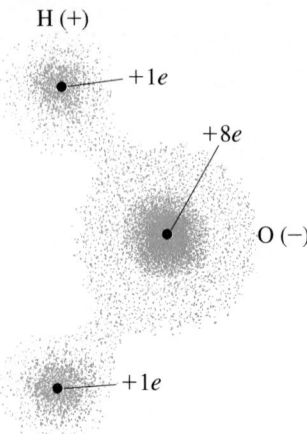

H (+)

+1e

+8e

O (−)

+1e

H (+)

FIGURE 29–6 The water molecule H_2O is polar.

FIGURE 29–7 Potential energy PE as a function of separation r for two point charges of (a) like sign and (b) opposite sign.

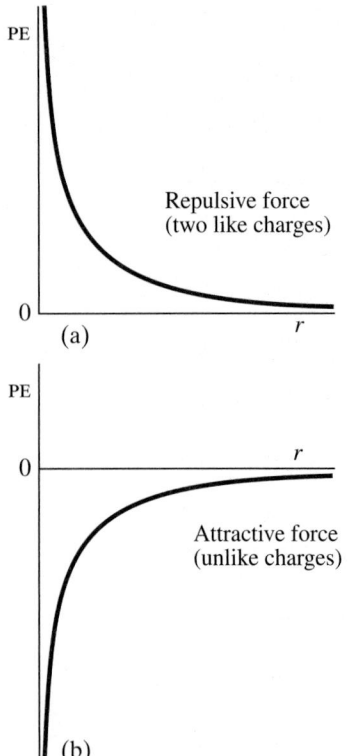

The molecules themselves are **polar**—that is, one part (or parts) of the molecule has a net positive charge and other parts a net negative charge. An example is the water molecule, H_2O (Fig. 29–6). Covalent bonds have shared electrons, which in H_2O are more likely to be found around the oxygen atom than around the two hydrogens. The reason is similar to that discussed above in connection with ionic bonds. Oxygen has eight electrons $(1s^2 2s^2 2p^4)$, of which four form a spherically symmetric core and the other four could have, for example, a doughnut-shaped distribution. The barbell-shaped distribution on the z axis (like that shown dashed in Fig. 29–5) could be empty, so electrons from hydrogen atoms can be attracted by a net charge of $+4e$. They are also attracted by the H nuclei, so they partly orbit the H atoms as well as the O atom. The net effect is that there is a net positive charge on each H atom (less than $+1e$), because the electrons spend only part of their time there. And, there is a net negative charge on the O atom.

*29–2 Potential-Energy Diagrams for Molecules

It is useful to analyze the interaction between two objects—say, between two atoms or molecules—with the use of a potential-energy diagram, which is a plot of the potential energy versus the separation distance.

For the simple case of two point charges, q_1 and q_2, the potential energy PE is given by (we combine Eqs. 17–2a and 17–5)

$$\text{PE} = k\frac{q_1 q_2}{r},$$

where r is the distance between the charges, and the constant $k \, (= 1/4\pi\epsilon_0)$ is equal to $9.0 \times 10^9 \, \text{N} \cdot \text{m}^2/\text{C}^2$. If the two charges have the same sign, the potential energy is positive for all values of r, and a graph of PE versus r in this case is shown in Fig. 29–7a. The force is repulsive (the charges have the *same* sign) and the curve rises as r decreases; this makes sense because if one particle moves freely toward the other (r getting smaller), the repulsion slows it down so its KE gets smaller, meaning PE gets larger. If, on the other hand, the two charges are of the *opposite* sign, the potential energy is negative because the product $q_1 q_2$ is negative. The force is attractive in this case, and the graph of PE ($\propto -1/r$) versus r looks like Fig. 29–7b. The potential energy becomes more *negative* as r decreases.

Now let us look at the potential-energy diagram for the formation of a covalent bond, such as for the hydrogen molecule, H_2. The potential energy PE of one H atom in the presence of the other is plotted in Fig. 29–8. Starting at large r, the PE decreases as the atoms approach, because the electrons concentrate between the two nuclei (Fig. 29–2), so attraction occurs. However, at very short distances, the electrons would be "squeezed out"—there is no room for them between the two nuclei. Without the electrons between them, each nucleus would feel a repulsive force due to the other, so the curve rises as r decreases further.

FIGURE 29–8 Potential-energy diagram for the H_2 molecule; r is the separation of the two H atoms. The binding energy (the energy difference between PE = 0 and the lowest energy state near the bottom of the well) is 4.5 eV, and $r_0 = 0.074$ nm.

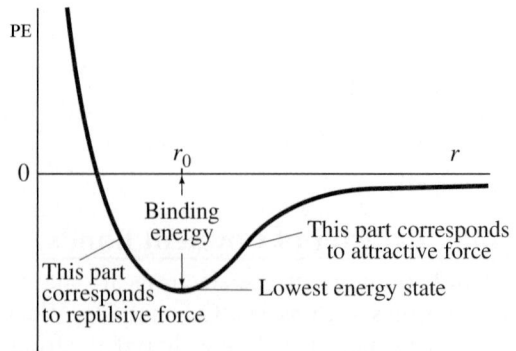

There is an optimum separation of the atoms, r_0 in Fig. 29–8, at which the energy is lowest. This is the point of greatest stability for the hydrogen molecule, and r_0 is the average separation of atoms in the H_2 molecule. The depth of this "well" is the *binding energy*,[†] as shown. This is how much energy must be put into the system to separate the two atoms to infinity, where the PE = 0. For the H_2 molecule, the binding energy is about 4.5 eV and $r_0 = 0.074$ nm.

For many bonds, the potential-energy curve has the shape shown in Fig. 29–9. There is still an optimum distance r_0 at which the molecule is stable. But when the atoms approach from a large distance, the force is initially repulsive rather than attractive. The atoms thus do not form a bond spontaneously. Some additional energy must be injected into the system to get it over the "hump" (or barrier) in the potential-energy diagram. This required energy is called the **activation energy**.

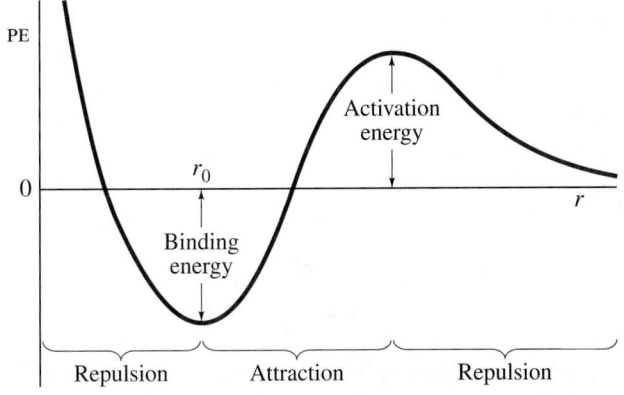

FIGURE 29–9 Potential-energy diagram for a bond requiring an activation energy.

The curve of Fig. 29–9 is much more common than that of Fig. 29–8. The activation energy often reflects a need to break other bonds, before the one under discussion can be made. For example, to make water from O_2 and H_2, the H_2 and O_2 molecules must first be broken into H and O atoms by an input of energy; this is what the activation energy represents. Then the H and O atoms can combine to form H_2O with the release of a great deal more energy than was put in initially. The initial activation energy can be provided by applying an electric spark to a mixture of H_2 and O_2, breaking a few of these molecules into H and O atoms. When these atoms combine to form H_2O, a lot of energy is released (the ground state is near the bottom of the well) which provides the activation energy needed for further reactions: additional H_2 and O_2 molecules are broken up and recombined to form H_2O.

The potential-energy diagrams for ionic bonds, such as NaCl, may be more like Fig. 29–8: the Na^+ and Cl^- ions attract each other at distances a bit larger than some r_0, but at shorter distances the overlapping of inner electron shells gives rise to repulsion. The two atoms thus are most stable at some intermediate separation, r_0. For partially ionic bonds, there is usually an activation energy, Fig. 29–9.

Sometimes the potential energy of a bond looks like that of Fig. 29–10. In this case, the energy of the bonded molecule, at a separation r_0, is greater than when there is no bond ($r = \infty$). That is, an energy *input* is required to make the bond (hence the binding energy is negative), and there is energy release when the bond is broken. Such a bond is stable only because there is the barrier of the activation energy. This type of bond is important in living cells, for it is in such bonds that energy can be stored efficiently in certain molecules, particularly ATP (adenosine triphosphate). The bond that connects the last phosphate group (designated Ⓟ in Fig. 29–10) to the rest of the molecule (ADP, meaning adenosine diphosphate, since it contains only two phosphates) has potential energy of the shape shown in Fig. 29–10. Energy is stored in this bond. When the bond is broken (ATP → ADP + Ⓟ), energy is released and this energy can be used to make other chemical reactions "go."

[†]The binding energy corresponds not quite to the bottom of the potential-energy curve, but to the lowest quantum energy state, slightly above the bottom, as shown in Fig. 29–8.

PHYSICS APPLIED
ATP and energy in the cell

FIGURE 29–10 Potential-energy diagram for the formation of ATP from ADP and phosphate (Ⓟ).

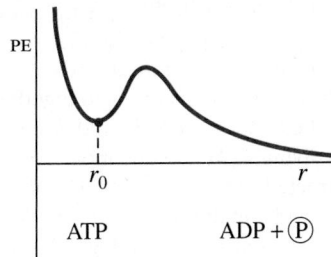

In living cells, many chemical reactions have activation energies that are often on the order of several eV. Such energy barriers are not easy to overcome in the cell. This is where enzymes come in. They act as *catalysts*, which means they act to lower the activation energy so that reactions can occur that otherwise would not. Enzymes act via the electrostatic force to distort the bonding electron clouds, so that the initial bonds are easily broken.

*29–3 Weak (van der Waals) Bonds

Once a bond between two atoms or ions is made, energy must normally be supplied to break the bond and separate the atoms. As mentioned in Section 29–1, this energy is called the *bond energy* or *binding energy*. The binding energy for covalent and ionic bonds is typically 2 to 5 eV. These bonds, which hold atoms together to *form* molecules, are often called **strong bonds** to distinguish them from so-called "weak bonds." The term **weak bond**, as we use it here, refers to an attachment *between* molecules due to simple electrostatic attraction—such as *between* polar molecules (and not *within* a polar molecule, which is a strong bond). The strength of the attachment is much less than for the strong bonds. Binding energies are typically in the range 0.04 to 0.3 eV—hence their name "weak bonds."

Weak bonds are generally the result of attraction between dipoles. (A pair of equal point charges q of opposite sign, separated by a distance ℓ, is called an **electric dipole**, as we saw in Chapter 17.) For example, Fig. 29–11 shows two molecules, which have permanent dipole moments, attracting one another. Besides such **dipole–dipole bonds**, there can also be **dipole–induced dipole bonds**, in which a polar molecule with a permanent dipole moment can induce a dipole moment in an otherwise electrically balanced (nonpolar) molecule, just as a single charge can induce a separation of charge in a nearby object (see Fig. 16–7). There can even be an attraction between two nonpolar molecules, because their electrons are moving about: at any instant there may be a transient separation of charge, creating a brief dipole moment and weak attraction. All these weak bonds are referred to as **van der Waals bonds**, and the forces involved **van der Waals forces**. The potential energy has the general shape shown in Fig. 29–8, with the attractive van der Waals potential energy varying as $1/r^6$. The force decreases greatly with increased distance.

When one of the atoms in a dipole–dipole bond is hydrogen, as in Fig. 29–11, it is called a **hydrogen bond**. A hydrogen bond is generally the strongest of the weak bonds, because the hydrogen atom is the smallest atom and can be approached more closely. Hydrogen bonds also have a partial "covalent" character: that is, electrons between the two dipoles may be shared to a small extent, making a stronger, more lasting bond.

Weak bonds are very important for understanding the activities of cells, such as the double helix shape of DNA (Fig. 29–12), and DNA replication

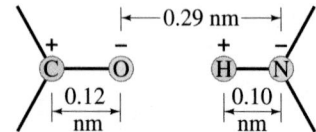

FIGURE 29–11 The C^+—O^- and H^+—N^- dipoles attract each other. (These dipoles may be part of, for example, the nucleotide bases cytosine and guanine in DNA molecules. See Fig. 29–12.) The $+$ and $-$ charges typically have magnitudes of a fraction of e.

PHYSICS APPLIED
DNA

FIGURE 29–12 (a) Model of part of a DNA double helix. The red dots represent hydrogen bonds between the two strands. (b) "Close-up" view: cytosine (C) and guanine (G) molecules on separate strands of a DNA double helix are held together by the hydrogen bonds (red dots) involving an H^+ on one molecule attracted to an N^- or C^+—O^- of a molecule on the adjacent chain. See also Section 16–10 and Figs. 16–39 and 16–40.

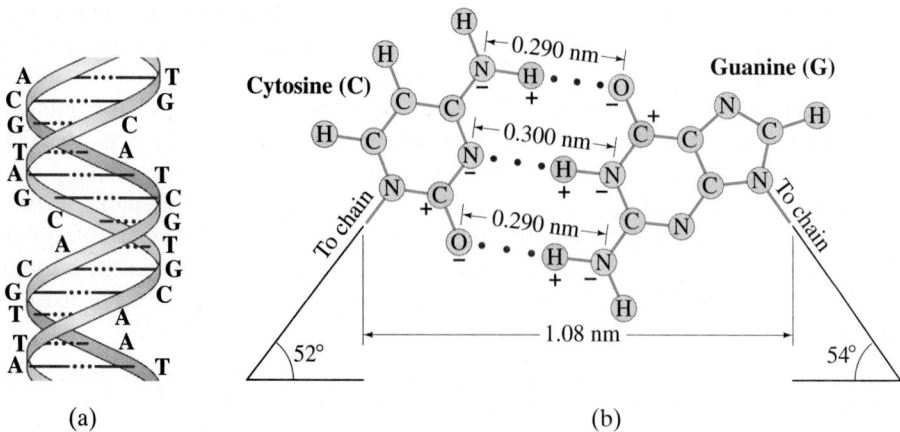

(a)   (b)

(see Section 16–10). The average kinetic energy of molecules in a living cell at normal temperatures $(T \approx 300\,\text{K})$ is around $\frac{3}{2}kT \approx 0.04\,\text{eV}$ (kinetic theory, Chapter 13), about the magnitude of weak bonds. This means that a weak bond can readily be broken just by a molecular collision. Hence weak bonds are not very permanent—they are, instead, brief attachments. This helps them play particular roles in the cell. On the other hand, strong bonds—those that hold molecules together—are almost never broken simply by molecular collision because their binding energies are much higher (≈ 2 to $5\,\text{eV}$). Thus they are relatively permanent. They can be broken by chemical action (the making of even stronger bonds), and this usually happens in the cell with the aid of an enzyme, which is a protein molecule.

EXAMPLE 29–1 **Nucleotide energy.** Calculate the potential energy between a C^+–O^- dipole of the nucleotide base cytosine and the nearby H^+–N^- dipole of guanine, assuming that the two dipoles are lined up as shown in Fig. 29–11. Dipole moment $(= q\ell)$ measurements (see Table 17–2 and Fig. 29–11) give

$$q_\text{H} = -q_\text{N} = \frac{3.0 \times 10^{-30}\,\text{C}\cdot\text{m}}{0.10 \times 10^{-9}\,\text{m}} = 3.0 \times 10^{-20}\,\text{C} = 0.19e,$$

and

$$q_\text{C} = -q_\text{O} = \frac{8.0 \times 10^{-30}\,\text{C}\cdot\text{m}}{0.12 \times 10^{-9}\,\text{m}} = 6.7 \times 10^{-20}\,\text{C} = 0.42e.$$

APPROACH We want to find the potential energy of the two charges in one dipole due to the two charges in the other, because this will be equal to the work needed to pull the two dipoles infinitely far apart. The potential energy of a charge q_1 in the presence of a charge q_2 is $\text{PE} = k(q_1 q_2 / r_{12})$ where $k = 9.0 \times 10^9\,\text{N}\cdot\text{m}^2/\text{C}^2$ and r_{12} is the distance between the two charges. (See Eqs. 17–2 and 17–5.)

SOLUTION The potential energy consists of four terms:

$$\text{PE} = \text{PE}_\text{CH} + \text{PE}_\text{CN} + \text{PE}_\text{OH} + \text{PE}_\text{ON}$$

where PE_CH means the potential energy of C in the presence of H, and similarly for the other terms. We do not have terms corresponding to C and O, or N and H, because the two dipoles are assumed to be stable entities. Then, using the distances shown in Fig. 29–11, we get:

$$\text{PE} = k\left[\frac{q_\text{C} q_\text{H}}{r_\text{CH}} + \frac{q_\text{C} q_\text{N}}{r_\text{CN}} + \frac{q_\text{O} q_\text{H}}{r_\text{OH}} + \frac{q_\text{O} q_\text{N}}{r_\text{ON}}\right]$$

$$= (9.0 \times 10^9\,\text{N}\cdot\text{m}^2/\text{C}^2)(6.7 \times 10^{-20}\,\text{C})(3.0 \times 10^{-20}\,\text{C})\left(\frac{1}{r_\text{CH}} - \frac{1}{r_\text{CN}} - \frac{1}{r_\text{OH}} + \frac{1}{r_\text{ON}}\right)$$

$$= (9.0 \times 10^9\,\text{N}\cdot\text{m}^2/\text{C}^2)(6.7)(3.0)\frac{(10^{-20}\,\text{C})^2}{(10^{-9}\,\text{m})}\left(\frac{1}{0.31} - \frac{1}{0.41} - \frac{1}{0.19} + \frac{1}{0.29}\right)$$

$$= -1.86 \times 10^{-20}\,\text{J} = -0.12\,\text{eV}.$$

The potential energy is negative, meaning 0.12 eV of work (or energy input) is required to separate the dipoles. That is, the binding energy of this "weak" or hydrogen bond is 0.12 eV. This is only an estimate, of course, since other charges in the vicinity would have an influence too.

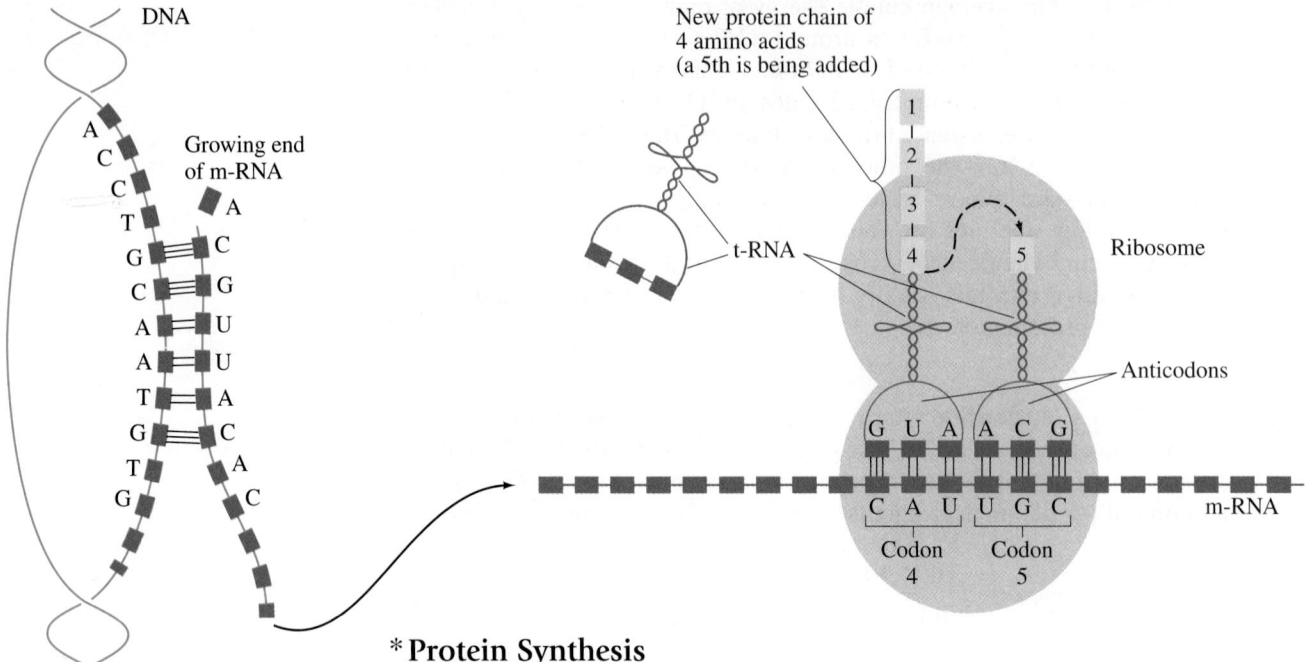

FIGURE 29–13 Protein synthesis. The yellow rectangles represent amino acids. See text for details.

PHYSICS APPLIED

Protein synthesis

*Protein Synthesis

Weak bonds, especially hydrogen bonds, are crucial to the process of protein synthesis. Proteins serve as structural parts of the cell and as enzymes to catalyze chemical reactions needed for the growth and survival of the organism. A protein molecule consists of one or more chains of small molecules known as *amino acids*. There are 20 different amino acids, and a single protein chain may contain hundreds of them in a specific order. The standard model for how amino acids are connected together in the correct order to form a protein molecule is shown schematically in Fig. 29–13.

We begin at the DNA double helix: each gene on a chromosome contains the information for producing one protein. The ordering of the four bases, A, C, G, and T, provides the "code," the **genetic code**, for the order of amino acids in the protein. First, the DNA double helix unwinds and a new molecule called *messenger*-RNA (m-RNA) is synthesized using one strand of the DNA as a "template." m-RNA is a chain molecule containing four different bases, like those of DNA (Section 16–10) except that thymine (T) is replaced by the similar uracil molecule (U). Near the top left in Fig. 29–13, a C has just been added to the growing m-RNA chain in much the same way that DNA replicates (Fig. 16–40); and an A, attracted and held close to the T on the DNA chain by the electrostatic force, will soon be attached to the C by an enzyme. The order of the bases, and thus the genetic information, is preserved in the m-RNA because the shapes of the molecules only allow the "proper" one to get close enough so the electrostatic force can act to form weak bonds.

Next, the m-RNA is buffeted about in the cell (recall kinetic theory, Chapter 13) until it gets close to a tiny organelle known as a *ribosome*, to which it can attach by electrostatic attraction (on the right in Fig. 29–13), because their shapes allow the charged parts to get close enough to form weak bonds. (Recall that force decreases greatly with separation distance.) Also held by the electrostatic force to the ribosome are one or two *transfer*-RNA (t-RNA) molecules. These t-RNA molecules "translate" the genetic code of nucleotide bases into amino acids in the following way. There is a different t-RNA molecule for each amino acid and each combination of three bases. On one end of a t-RNA molecule is an amino acid. On the other end of the t-RNA molecule is the appropriate "anticodon," a set of three nucleotide bases that "code" for that amino acid. If all three bases of an anticodon match the three bases of the "codon" on the m-RNA (in the sense of G to C and A to U), the anticodon is attracted electrostatically to the m-RNA codon and that t-RNA molecule is held there briefly. The ribosome has two particular attachment sites which hold two t-RNA molecules while enzymes bond the two amino acids together to lengthen the amino acid chain (yellow in Fig. 29–13). As each amino acid is connected by an enzyme (four are already connected in Fig. 29–13, top right, and a fifth is about to be connected), the old t-RNA molecule is removed—perhaps by a random collision with some molecule in the cellular fluid. A new one soon becomes attracted as the ribosome moves along the m-RNA.

This process of protein synthesis is often presented as if it occurred in clockwork fashion—as if each molecule knew its role and went to its assigned place. But this is not the case. The forces of attraction between the electric charges of the molecules are rather weak and become significant only when the molecules can come close together, and when several weak bonds can be made. Indeed, if the shapes are not just right, the electrostatic attraction is nearly zero, which is why there are few mistakes. The fact that weak bonds are weak is very important. If they were strong, collisions with other molecules would not allow a t-RNA molecule to be released from the ribosome, or the m-RNA to be released from the DNA. If they were not temporary encounters, metabolism would grind to a halt.

As each amino acid is added to the next, the protein molecule grows in length until it is complete. Even as it is being made, this chain is being buffeted about in the cell—we might think of a wiggling worm. But a protein molecule has electrically charged polar groups along its length. And as it takes on various shapes, the electric forces of attraction between different parts of the molecule will eventually lead to a particular shape of the protein which is quite stable. Each type of protein has its own special shape, depending on the location of charged atoms. In the last analysis, the final shape depends on the order of the amino acids.

*29–4 Molecular Spectra

When atoms combine to form molecules, the probability distributions of the outer electrons overlap and this interaction alters the energy levels. Nonetheless, molecules can undergo transitions between electron energy levels just as atoms do. For example, the H_2 molecule can absorb a photon of just the right frequency to excite one of its ground-state electrons to an excited state. The excited electron can then return to the ground state, emitting a photon. The energy of photons emitted by molecules can be of the same order of magnitude as for atoms, typically 1 to 10 eV, or less.

Additional energy levels become possible for molecules (but not for atoms) because the molecule as a whole can rotate, and the atoms of the molecule can vibrate relative to each other. The energy levels for both rotational and vibrational levels are quantized, and are generally spaced much more closely (10^{-3} to 10^{-1} eV) than the electronic levels. Each atomic energy level thus becomes a set of closely spaced levels corresponding to the vibrational and rotational motions, Fig. 29–14. Transitions from one level to another appear as many very closely spaced lines. In fact, the lines are not always distinguishable, and these spectra are called **band spectra**. Each type of molecule has its own characteristic spectrum, which can be used for identification and for determination of structure. We now look in more detail at rotational and vibrational states in molecules.

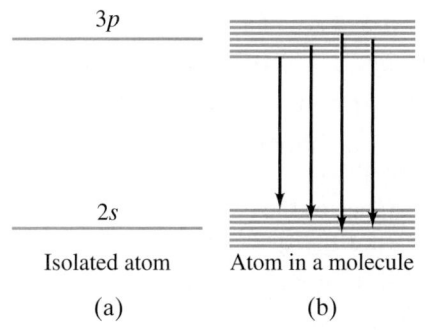

3p

2s

Isolated atom   Atom in a molecule

(a)   (b)

FIGURE 29–14 (a) The individual energy levels of an isolated atom become (b) bands of closely spaced levels in molecules, as well as in solids and liquids.

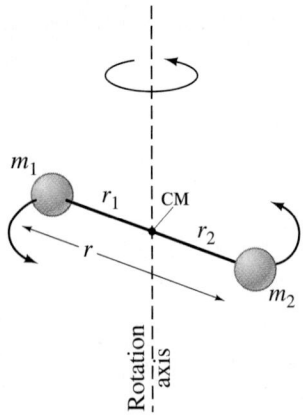

FIGURE 29–15 Diatomic molecule rotating about a vertical axis.

FIGURE 29–16 Rotational energy levels and allowed transitions (emission and absorption) for a diatomic molecule. Upward-pointing arrows represent absorption of a photon, and downward arrows represent emission of a photon.

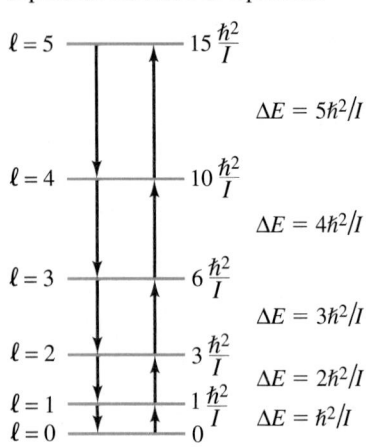

*Rotational Energy Levels in Molecules

We consider only diatomic molecules, although the analysis can be extended to polyatomic molecules. When a diatomic molecule rotates about its center of mass as shown in Fig. 29–15, its kinetic energy of rotation (see Section 8–7) is

$$E_{rot} = \frac{1}{2} I\omega^2 = \frac{(I\omega)^2}{2I},$$

where $I\omega$ is the angular momentum (Section 8–8). Quantum mechanics predicts quantization of angular momentum just as in atoms (see Eq. 28–3):

$$I\omega = \sqrt{\ell(\ell + 1)}\,\hbar, \qquad \ell = 0, 1, 2, \cdots,$$

where ℓ is an integer called the **rotational angular momentum quantum number**. Thus the rotational energy is quantized:

$$E_{rot} = \frac{(I\omega)^2}{2I} = \ell(\ell + 1)\frac{\hbar^2}{2I}. \qquad \ell = 0, 1, 2, \cdots. \quad \textbf{(29–1)}$$

Transitions between rotational energy levels are subject to the **selection rule** (as in Section 28–6):

$$\Delta\ell = \pm 1.$$

The energy of a photon emitted or absorbed for a transition between rotational states with angular momentum quantum number ℓ and $\ell - 1$ will be

$$\Delta E_{rot} = E_\ell - E_{\ell-1} = \frac{\hbar^2}{2I}\ell(\ell + 1) - \frac{\hbar^2}{2I}(\ell - 1)(\ell)$$

$$= \frac{\hbar^2}{I}\ell. \qquad \begin{bmatrix} \ell \text{ is for upper} \\ \text{energy state} \end{bmatrix} \quad \textbf{(29–2)}$$

We see that the transition energy increases directly with ℓ. Figure 29–16 shows some of the allowed rotational energy levels and transitions. Measured absorption lines fall in the microwave or far-infrared regions of the spectrum (energies $\approx 10^{-3}$ eV), and their frequencies are generally 2, 3, 4, $\cdots$ times higher than the lowest one, as predicted by Eq. 29–2.

EXERCISE A Determine the three lowest rotational energy states (in eV) for a nitrogen molecule which has a moment of inertia $I = 1.39 \times 10^{-46}$ kg·m².

EXAMPLE 29–2 **Rotational transition.** A rotational transition $\ell = 1$ to $\ell = 0$ for the molecule CO has a measured absorption wavelength $\lambda_1 = 2.60$ mm (microwave region). Use this to calculate (a) the moment of inertia of the CO molecule, and (b) the CO bond length, r.

APPROACH The absorption wavelength is used to find the energy of the absorbed photon, and we can then calculate the moment of inertia, I, from Eq. 29–2. The moment of inertia is related to the CO separation (bond length r).

SOLUTION (a) The photon energy, $E = hf = hc/\lambda$, equals the rotational energy level difference, ΔE_{rot}. From Eq. 29–2, we can write

$$\frac{\hbar^2}{I}\ell = \Delta E_{rot} = hf = \frac{hc}{\lambda_1}.$$

With $\ell = 1$ (the upper state) in this case, we solve for I:

$$I = \frac{\hbar^2\ell}{hc}\lambda_1 = \frac{h\lambda_1}{4\pi^2 c} = \frac{(6.63 \times 10^{-34}\,\text{J·s})(2.60 \times 10^{-3}\,\text{m})}{4\pi^2(3.00 \times 10^8\,\text{m/s})}$$

$$= 1.46 \times 10^{-46}\,\text{kg·m}^2.$$

(b) The molecule rotates about its center of mass (CM) as shown in Fig. 29–15. Let m_1 be the mass of the C atom, $m_1 = 12$ u, and let m_2 be the mass of the O, $m_2 = 16$ u. The distance of the CM from the C atom, which is r_1 in Fig. 29–15, is given by the CM formula, Eq. 7–9:

$$r_1 = \frac{0 + m_2 r}{m_1 + m_2} = \frac{16}{12 + 16}r = 0.57r.$$

The O atom is a distance $r_2 = r - r_1 = 0.43r$ from the CM. The moment of inertia of the CO molecule about its CM is then (see Example 8–9)

$$
\begin{aligned}
I &= m_1 r_1^2 + m_2 r_2^2 \\
&= \left[(12\,u)(0.57r)^2 + (16\,u)(0.43r)^2 \right]\left[1.66 \times 10^{-27}\,kg/u \right] \\
&= \left(1.14 \times 10^{-26}\,kg \right) r^2.
\end{aligned}
$$

We solve for r and use the result of part (a) for I:

$$
r = \sqrt{\frac{1.46 \times 10^{-46}\,kg\cdot m^2}{1.14 \times 10^{-26}\,kg}} = 1.13 \times 10^{-10}\,m = 0.113\,nm \approx 0.11\,nm.
$$

| **EXERCISE B** What are the wavelengths of the next three rotational transitions for CO?

*Vibrational Energy Levels in Molecules

The potential energy of the two atoms in a typical diatomic molecule has the shape shown in Fig. 29–8 or 29–9, and Fig. 29–17 again shows the PE for the H_2 molecule (solid curve). This PE curve, at least in the vicinity of the equilibrium separation r_0, closely resembles the potential energy of a harmonic oscillator, $PE = \frac{1}{2}kx^2$, which is shown superposed in dashed lines. Thus, for small displacements from r_0, each atom experiences a restoring force approximately proportional to the displacement, and the molecule vibrates as a simple harmonic oscillator (SHO)—see Chapter 11. According to quantum mechanics, the possible quantized energy levels are

$$
E_{vib} = \left(\nu + \tfrac{1}{2} \right)hf, \qquad \nu = 0, 1, 2, \cdots, \tag{29–3}
$$

where f is the classical frequency (see Chapter 11—f depends on the mass of the atoms and on the bond strength or "stiffness") and ν is an integer called the **vibrational quantum number**. The lowest energy state ($\nu = 0$) is not zero (as for rotation), but has $E = \frac{1}{2}hf$. This is called the **zero-point energy**. Higher states have energy $\frac{3}{2}hf$, $\frac{5}{2}hf$, and so on, as shown in Fig. 29–18. Transitions between vibrational energy levels are subject to the **selection rule**

$$
\Delta\nu = \pm 1,
$$

so allowed transitions occur only between adjacent states[†], and all give off (or absorb) photons of energy

$$
\Delta E_{vib} = hf. \tag{29–4}
$$

This is very close to experimental values for small ν. But for higher energies, the PE curve (Fig. 29–17) begins to deviate from a perfect SHO curve, which affects the wavelengths and frequencies of the transitions. Typical transition energies are on the order of $10^{-1}\,eV$, roughly 10 to 100 times larger than for rotational transitions, with wavelengths in the infrared region of the spectrum ($\approx 10^{-5}\,m$).

| **EXAMPLE 29–3** | **Vibrational energy levels in hydrogen.** Hydrogen molecule vibrations emit infrared radiation of wavelength around 2300 nm. (a) What is the separation in energy between adjacent vibrational levels? (b) What is the lowest vibrational energy state?

APPROACH The energy separation between adjacent vibrational levels is (Eq. 29–4) $\Delta E_{vib} = hf = hc/\lambda$. The lowest energy (Eq. 29–3) has $\nu = 0$.

SOLUTION

(a) $\Delta E_{vib} = hf = \dfrac{hc}{\lambda} = \dfrac{\left(6.63 \times 10^{-34}\,J\cdot s \right)\left(3.00 \times 10^8\,m/s \right)}{\left(2300 \times 10^{-9}\,m \right)\left(1.60 \times 10^{-19}\,J/eV \right)} = 0.54\,eV,$

where the denominator includes the conversion factor from joules to eV.

(b) The lowest vibrational energy has $\nu = 0$ in Eq. 29–3:

$$
E_{vib} = \left(\nu + \tfrac{1}{2} \right)hf = \tfrac{1}{2}hf = 0.27\,eV.
$$

| **EXERCISE C** What is the energy of the first vibrational state above the ground state in the hydrogen molecule?

[†]Forbidden transitions with $\Delta\nu = 2$ are emitted with much lower probability, but their observation can be important in some cases, such as in astronomy.

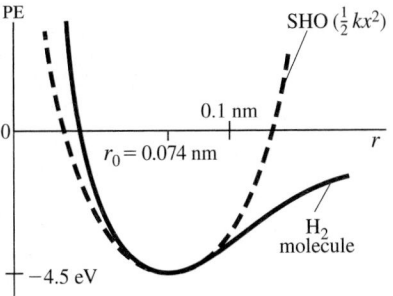

FIGURE 29–17 Potential energy for the H_2 molecule and for a simple harmonic oscillator (PE $= \frac{1}{2}kx^2$, with $|x| = |r - r_0|$).

FIGURE 29–18 Allowed vibrational energies for a diatomic molecule, where f is the fundamental frequency of vibration (see Chapter 11). The energy levels are equally spaced. Transitions are allowed only between adjacent levels ($\Delta\nu = \pm 1$).

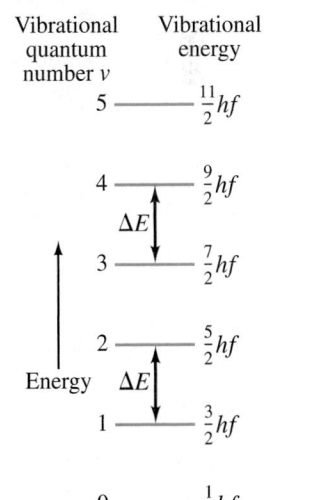

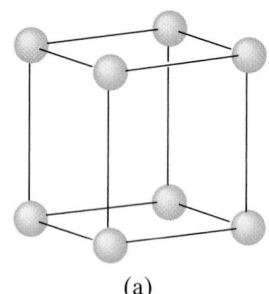

(a)

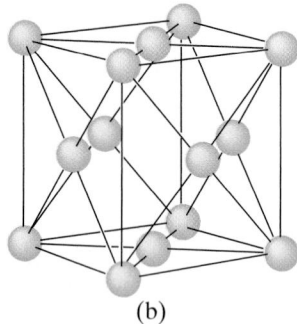

(b)

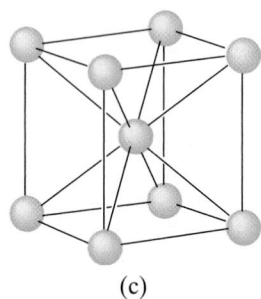

(c)

FIGURE 29–19 Arrangement of atoms in (a) a simple cubic crystal, (b) face-centered cubic crystal (note the atom at the center of each face), and (c) body-centered cubic crystal. Each of these "cells" is repeated in three dimensions to the edges of the macroscopic crystal.

FIGURE 29–20 Diagram of an NaCl crystal, showing the "packing" of atoms.

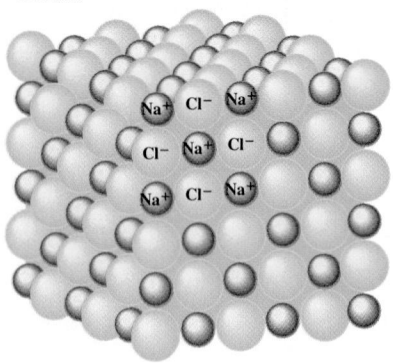

*29–5 Bonding in Solids

Quantum mechanics has been a great tool for understanding the structure of solids. This active field of research today is called **solid-state physics**, or **condensed-matter physics** so as to include liquids as well. The rest of this Chapter is devoted to this subject, and we begin with a brief look at the structure of solids and the bonds that hold them together.

Although some solid materials are *amorphous* in structure (such as glass), in that the atoms and molecules show no long-range order, we are interested here in the large class of *crystalline* substances whose atoms, ions, or molecules are generally accepted to form an orderly array known as a **lattice**. Figure 29–19 shows three of the possible arrangements of atoms in a crystal: simple cubic, face-centered cubic, and body-centered cubic. The NaCl crystal lattice is shown in Fig. 29–20.

The molecules of a solid are held together in a number of ways. The most common are by *covalent* bonding (such as between the carbon atoms of the diamond crystal) and by *ionic* bonding (as in a NaCl crystal). Often the bonds are partially covalent and partially ionic. Our discussion of these bonds earlier in this Chapter for molecules applies equally well to solids.

Let us look for a moment at the NaCl crystal of Fig. 29–20. Each Na^+ ion feels an attractive Coulomb potential due to each of the six "nearest neighbor" Cl^- ions surrounding it. Note that one Na^+ does not "belong" exclusively to one Cl^-, so we must not think of ionic solids as consisting of individual molecules. Each Na^+ also feels a repulsive Coulomb potential due to other Na^+ ions, although this is weaker since the Na^+ ions are farther away.

A different type of bond occurs in metals. Metal atoms have relatively loosely held outer electrons. **Metallic bond** theories propose that in a metallic solid, these outer electrons roam rather freely among all the metal atoms which, without their outer electrons, act like positive ions. According to the theory, the electrostatic attraction between the metal ions and this negative electron "gas" is responsible, at least in part, for holding the solid together. The binding energy of metal bonds is typically 1 to 3 eV, somewhat weaker than ionic or covalent bonds (5 to 10 eV in solids). The "free electrons" are responsible for the high electrical and thermal conductivity of metals. This theory also nicely accounts for the shininess of smooth metal surfaces: the free electrons can vibrate at any frequency, so when light of a range of frequencies falls on a metal, the electrons can vibrate in response and re-emit light of those same frequencies. Hence, the reflected light will consist largely of the same frequencies as the incident light. Compare this to nonmetallic materials that have a distinct color—the atomic electrons exist only in certain energy states, and when white light falls on them, the atoms absorb at certain frequencies, and reflect other frequencies which make up the color we see.

Here is a brief comparison of important strong bonds:

- ionic: an electron is "grabbed" from one atom by another;
- covalent: electrons are shared by atoms within a single molecule;
- metallic: electrons are shared by all atoms in the metal.

The atoms or molecules of some materials, such as the noble gases, can form only **weak bonds** with each other. As we saw in Section 29–3, weak bonds have very low binding energies and would not be expected to hold atoms together as a liquid or solid at room temperature. The noble gases condense only at very low temperatures, where the atomic (thermal) kinetic energy is small and the weak attraction can then hold the atoms together.

EXERCISE D Return to the Chapter-Opening Question, page 829, and answer it again now. Try to explain why you may have answered differently the first time.

*29–6 Free-Electron Theory of Metals; Fermi Energy

The free-electron theory of metals considers electrons in a metal as being in constant motion like an ideal gas, which we discussed in Chapter 13. For a classical ideal gas, at very low temperatures near absolute zero, $T = 0$ K, all the particles would be in the lowest state, with zero kinetic energy $(= \frac{3}{2}kT = 0)$. But the situation is vastly different for an electron gas because, according to quantum mechanics, electrons obey the exclusion principle and can be only in certain possible energy levels or states. Electrons also obey a quantum statistics called **Fermi–Dirac statistics**[†] that takes into account the exclusion principle. All particles that have spin $\frac{1}{2}$ (or other half-integral spin: $\frac{3}{2}$, $\frac{5}{2}$, etc.), such as electrons, protons, and neutrons, obey Fermi–Dirac statistics and are referred to as **fermions** (see Section 28–7). The electron gas in a metal is often called a **Fermi gas**. According to the exclusion principle, no two electrons in the metal can have the same set of quantum numbers. Therefore, in each of the energy states available for the electrons in our "gas," there can be at most two electrons: one with spin up $\left(m_s = +\frac{1}{2}\right)$ and one with spin down $\left(m_s = -\frac{1}{2}\right)$. Thus, at $T = 0$ K, the possible energy levels will be filled, two electrons each, up to a maximum level called the **Fermi level**. This is shown in Fig. 29–21, where the vertical axis is the "density of occupied states," whose meaning is similar to the Maxwell distribution for a classical gas (Section 13–10). The energy of the state at the Fermi level is called the **Fermi energy**, E_F. For copper, $E_F = 7.0$ eV. This is very much greater than the energy of thermal motion at room temperature $(\overline{\text{KE}} = \frac{3}{2}kT \approx 0.04$ eV, Eq. 13–8). Clearly, all motion does not stop at absolute zero.

At $T = 0$, all states with energy below E_F are occupied, and all states above E_F are empty. What happens for $T > 0$? We expect that at least some of the electrons will increase in energy due to thermal motion. Figure 29–22 shows the density of occupied states for $T = 1200$ K, a temperature at which a metal is so hot it would glow. We see that the distribution differs very little from that at $T = 0$. We see also that the changes that do occur are concentrated about the Fermi level. A few electrons from slightly below the Fermi level move to energy states slightly above it. The average energy of the electrons increases only very slightly when the temperature is increased from $T = 0$ K to $T = 1200$ K. This is very different from the behavior of an ideal gas, for which kinetic energy increases directly with T. Nonetheless, this behavior is readily understood as follows. Energy of thermal motion at $T = 1200$ K is about $\frac{3}{2}kT \approx 0.1$ eV. The Fermi level, on the other hand, is on the order of several eV: for copper it is $E_F \approx 7.0$ eV. An electron at $T = 1200$ K may have 7 eV of energy, but it can acquire at most only a few times 0.1 eV of energy by a (thermal) collision with the lattice. Only electrons very near the Fermi level would find vacant states close enough to make such a transition. Essentially none of the electrons could increase in energy by, say, 3 eV, so electrons farther down in the electron gas are unaffected. Only electrons near the top of the energy distribution can be thermally excited to higher states. And their new energy is on the average only slightly higher than their old energy. This model of free electrons in a metal as a "gas," though incomplete, provides good explanations for the thermal and electrical conductivity of metals.

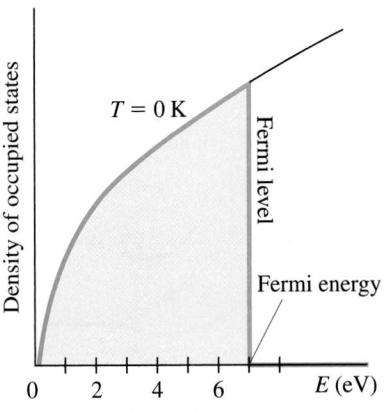

FIGURE 29–21 At $T = 0$ K, all states up to energy E_F, called the Fermi energy, are filled. (Shown here for copper.)

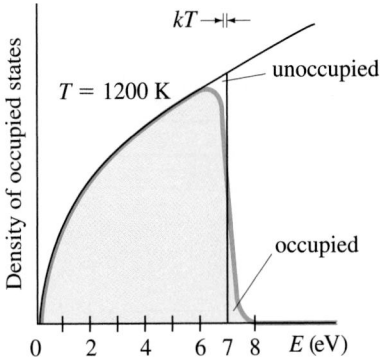

FIGURE 29–22 The density of occupied states for the electron gas in copper. The width kT shown above the graph represents thermal energy at $T = 1200$ K.

[†]Developed independently by Enrico Fermi (Figs. 1–13, 27–11, 28–2, 30–7) in early 1926 and by P. A. M. Dirac a few months later. See Section 28–7.

*29–7 Band Theory of Solids

We saw in Section 29–1 that when two hydrogen atoms approach each other, the wave functions overlap, and the two 1s states (one for each atom) divide into two states of different energy. (As we saw, only one of these states, $S = 0$, has low enough energy to give a bound H_2 molecule.) Figure 29–23a shows this situation for 1s and 2s states for two atoms: as the two atoms get closer (toward the left in Fig. 29–23a), the 1s and 2s states split into two levels. If six atoms come together, as in Fig. 29–23b, each of the states splits into six levels. If a large number of atoms come together to form a solid, then each of the original atomic levels becomes a **band** as shown in Fig. 29–23c. The energy levels are so close together in each band that they seem essentially continuous. This is why the spectrum of heated solids (Section 27–2) appears continuous. (See also Fig. 29–14 and its discussion at start of Section 29–4.)

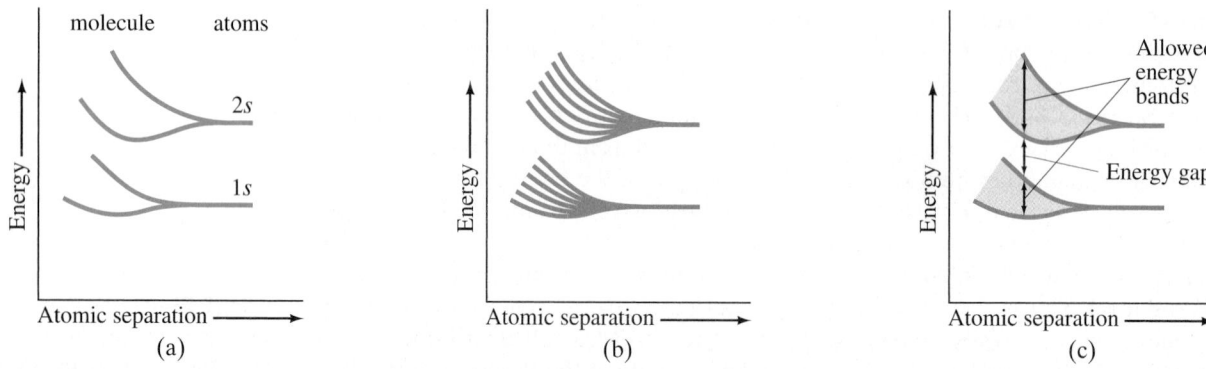

FIGURE 29–23 The splitting of 1s and 2s atomic energy levels as (a) two atoms approach each other (the atomic separation decreases toward the left on the graph); (b) the same for six atoms, and (c) for many atoms when they come together to form a solid.

FIGURE 29–24 Energy bands for sodium (Na).

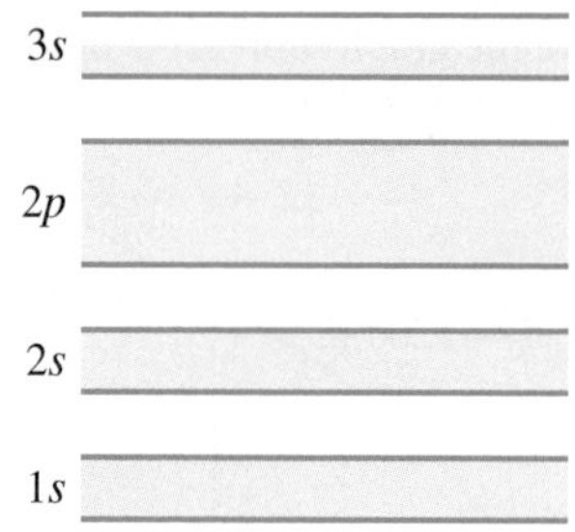

The crucial aspect of a good **conductor** is that the highest energy band containing electrons is only partially filled. Consider sodium metal, for example, whose energy bands are shown in Fig. 29–24. The 1s, 2s, and 2p bands are full (just as in a sodium atom) and don't concern us. The 3s band, however, is only half full. To see why, recall that the exclusion principle stipulates that in an atom, only two electrons can be in the 3s state, one with spin up and one with spin down. These two states have slightly different energy. For a solid consisting of N atoms, the 3s band will contain 2N possible energy states. A sodium atom has a single 3s electron, so in a sample of sodium metal containing N atoms, there are N electrons in the 3s band, and N unoccupied states. When a potential difference is applied across the metal, electrons can respond by accelerating and increasing their energy, since there are plenty of unoccupied states of slightly higher energy available. Hence, a current flows readily and sodium is a good conductor. The characteristic of all good conductors is that the highest energy band is only partially filled, or two bands overlap so that unoccupied states are available. An example of the latter is magnesium, which has two 3s electrons, so its 3s band is filled. But the unfilled 3p band overlaps the 3s band in energy, so there are lots of available states for the electrons to move into. Thus magnesium, too, is a good conductor.

In a material that is a good **insulator**, on the other hand, the highest band containing electrons, called the **valence band**, is completely filled. The next highest energy band, called the **conduction band**, is separated from the valence band by a "forbidden" **energy gap** (or **band gap**), E_g, of typically 5 to 10 eV. So at room temperature (300 K), where thermal energies (that is, average kinetic energy—see Chapter 13) are on the order of $\frac{3}{2}kT \approx 0.04$ eV, almost no electrons can acquire the 5 eV needed to reach the conduction band. When a potential difference is applied across the material, no available states are accessible to the electrons, and no current flows. Hence, the material is a good insulator.

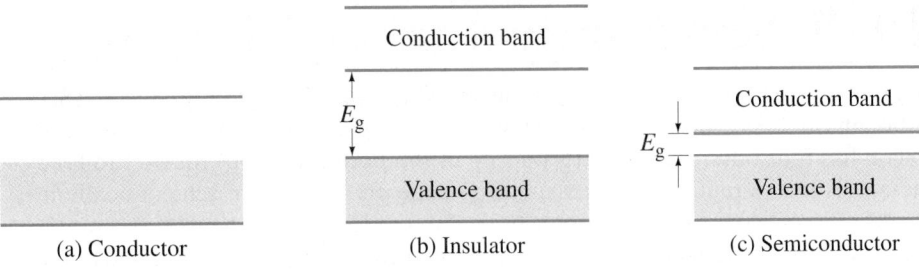

(a) Conductor   (b) Insulator   (c) Semiconductor

FIGURE 29–25 Energy bands for (a) a conductor, (b) an insulator, which has a large energy gap E_g, and (c) a semiconductor, which has a small energy gap E_g. Shading represents occupied states. Pale shading in (c) represents electrons that can pass from the top of the valence band to the bottom of the conduction band due to thermal agitation at room temperature (exaggerated).

Figure 29–25 compares the relevant energy bands (a) for conductors, (b) for insulators, and also (c) for the important class of materials known as **semiconductors**. The bands for a pure (or **intrinsic**) semiconductor, such as silicon or germanium, are like those for an insulator, except that the unfilled conduction band is separated from the filled valence band by a much smaller energy gap, E_g, which for silicon is $E_g = 1.12\ \text{eV}$. At room temperature, electrons are moving about with varying amounts of kinetic energy $\left(\overline{\text{KE}} = \frac{3}{2}kT\right)$, according to kinetic theory, Chapter 13. A few electrons can acquire enough thermal energy to reach the conduction band, and so a very small current may flow when a voltage is applied. At higher temperatures, more electrons have enough energy to jump the gap (top end of thermal distribution—see Fig. 13–20). Often this effect can more than offset the effects of more frequent collisions due to increased disorder at higher temperature, so the resistivity of semiconductors can *decrease* with increasing temperature (see Table 18–1). But this is not the whole story of semiconductor conduction. When a potential difference is applied to a semiconductor, the few electrons in the conduction band move toward the positive electrode. Electrons in the valence band try to do the same thing, and a few can because there are a small number of unoccupied states which were left empty by the electrons reaching the conduction band. Such unfilled electron states are called **holes**. Each electron in the valence band that fills a hole in this way as it moves toward the positive electrode leaves behind its own hole, so the holes migrate toward the negative electrode. As the electrons tend to accumulate at one side of the material, the holes tend to accumulate on the opposite side. We will look at this phenomenon in more detail in the next Section.

EXAMPLE 29–4 Calculating the energy gap. It is found that the conductivity of a certain semiconductor increases when light of wavelength 345 nm or shorter strikes it, suggesting that electrons are being promoted from the valence band to the conduction band. What is the energy gap, E_g, for this semiconductor?

APPROACH The longest wavelength (lowest energy) photon to cause an increase in conductivity has $\lambda = 345\ \text{nm}$, and its energy $(= hf)$ equals the energy gap.

SOLUTION The gap energy equals the energy of a $\lambda = 345$-nm photon:

$$E_g = hf = \frac{hc}{\lambda} = \frac{\left(6.63 \times 10^{-34}\ \text{J} \cdot \text{s}\right)\left(3.00 \times 10^{8}\ \text{m/s}\right)}{\left(345 \times 10^{-9}\ \text{m}\right)\left(1.60 \times 10^{-19}\ \text{J/eV}\right)} = 3.6\ \text{eV}.$$

CONCEPTUAL EXAMPLE 29–5 Which is transparent? The energy gap for silicon is 1.12 eV at room temperature, whereas that of zinc sulfide (ZnS) is 3.6 eV. Which one of these is opaque to visible light, and which is transparent?

RESPONSE Visible-light photons span energies from roughly 1.8 eV to 3.1 eV. ($E = hf = hc/\lambda$ where $\lambda = 400\ \text{nm}$ to 700 nm and $1\ \text{eV} = 1.6 \times 10^{-19}\ \text{J}$.) Light is absorbed by the electrons in a material. Silicon's energy gap is small enough to absorb these photons, thus bumping electrons well up into the conduction band, so silicon is opaque. On the other hand, zinc sulfide's energy gap is so large that no visible-light photons would be absorbed; they would pass right through the material which would thus be transparent.

PHYSICS APPLIED
Transparency

*29–8 Semiconductors and Doping

Nearly all electronic devices today use semiconductors—mainly silicon (Si), although the first transistor (1948) was made with germanium (Ge). An atom of silicon has four outer electrons (group IV of the Periodic Table) that act to hold the atoms in the regular lattice structure of the crystal, shown schematically in Fig. 29–26a. Silicon acquires properties useful for electronics when a tiny amount of impurity is introduced into the crystal structure (perhaps 1 part in 10^6 or 10^7). This is called **doping** the semiconductor. Two kinds of doped semiconductor can be made, depending on the type of impurity used. The impurity can be an element whose atoms have five outer electrons (group V in the Periodic Table), such as arsenic. Then we have the situation shown in Fig. 29–26b, with a few arsenic atoms holding positions in the crystal lattice where normally silicon atoms are. Only four of arsenic's electrons fit into the bonding structure. The fifth does not fit in and can move relatively freely, somewhat like the electrons in a conductor. Because of this small number of extra electrons, a doped semiconductor becomes slightly conducting. The density of conduction electrons in an **intrinsic** (= undoped) semiconductor at room temperature is very low, usually less than 1 per 10^9 atoms. With an impurity concentration of 1 in 10^6 or 10^7 when doped, the conductivity will be much higher and it can be controlled with great precision. An arsenic-doped silicon crystal is an **n-type semiconductor** because *negative* charges (electrons) carry the electric current.

FIGURE 29–26 Two-dimensional representation of a silicon crystal. (a) Four (outer) electrons surround each silicon atom. (b) Silicon crystal doped with a small percentage of arsenic atoms: the extra electron doesn't fit into the crystal lattice and so is free to move about. This is an *n*-type semiconductor.

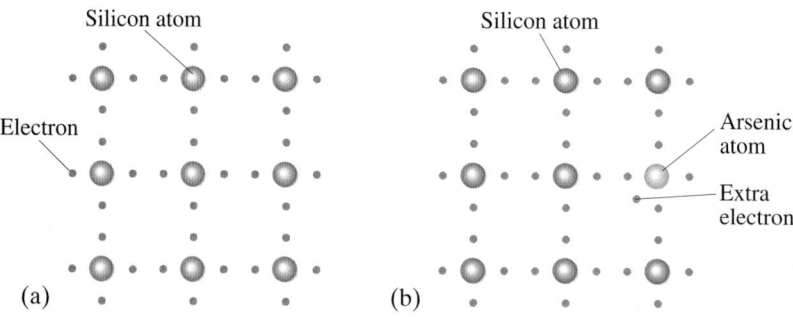

In a **p-type semiconductor**, a small percentage of semiconductor atoms are replaced by atoms with three outer electrons (group III in the Periodic Table), such as boron. As shown in Fig. 29–27a, there is a **hole** in the lattice structure near a boron atom because it has only three outer electrons. Electrons from nearby silicon atoms can jump into this hole and fill it. But this leaves a hole where that electron had previously been, Fig. 29–27b. The vast majority of atoms are silicon, so holes are almost always next to a silicon atom. Since silicon atoms require four outer electrons to be neutral, this means there is a net positive charge at the hole. Whenever an electron moves to fill a hole, the positive hole is then at the previous position of that electron. Another electron can then fill this hole, and the hole thus moves to a new location; and so on. This type of semiconductor is called **p-type** because it is the positive holes that carry the electric current.[†] Note, however, that both *p-type* and *n-type* semiconductors have *no net charge* on them.

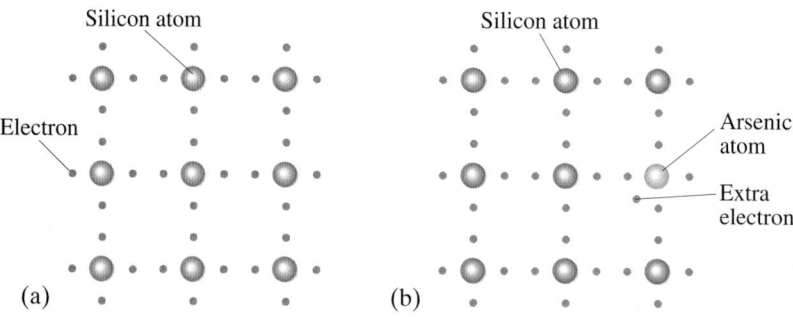
> **CAUTION**
>
> *p-type semiconductors act as though + charges move—but electrons actually do the moving*

[†]Each electron that fills a hole moves a very short distance (~1 atom < 1 nm) whereas holes move much larger distances and so are the real carriers of the current. We can tell the current is carried by positive charges (holes) by using the Hall effect, Section 20–4.

FIGURE 29–27 A *p*-type semiconductor, boron-doped silicon. (a) Boron has only three outer electrons, so there is an empty spot, or *hole* in the structure. (b) Electrons from silicon atoms can jump into the hole and fill it. As a result, the hole moves to a new location (to the right in this diagram), to where the electron used to be.

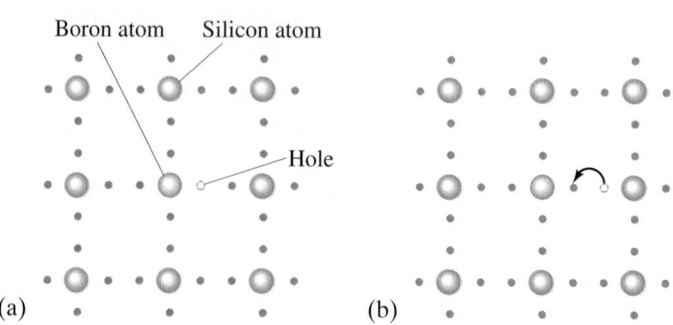

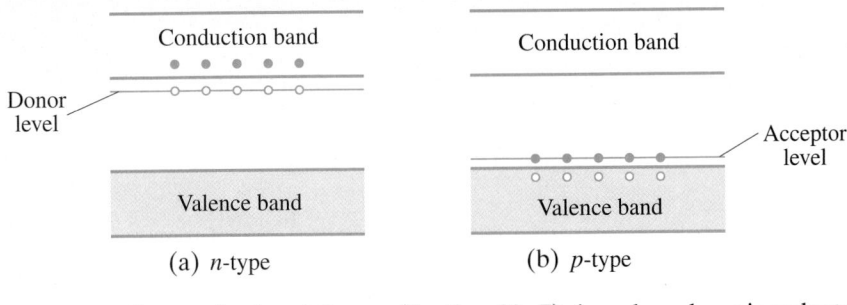

Conduction band

Donor
level

Valence band

(a) *n*-type

Conduction band

Acceptor
level

Valence band

(b) *p*-type

FIGURE 29–28 Impurity energy
levels in doped semiconductors.

According to the band theory (Section 29–7), in a doped semiconductor the impurity provides additional energy states between the bands as shown in Fig. 29–28. In an *n*-type semiconductor, the impurity energy level lies just below the conduction band, Fig. 29–28a. Electrons in this energy level need only about 0.05 eV in Si to reach the conduction band which is on the order of the thermal energy, $\frac{3}{2}kT$ (≈ 0.04 eV at 300 K). At room temperature, the small % of electrons in this donor level (~ 1 in 10^6) can readily make the transition upward. This energy level can thus supply electrons to the conduction band, so it is called a **donor** level. In *p*-type semiconductors, the impurity energy level is just above the valence band (Fig. 29–28b). It is called an **acceptor** level because electrons from the valence band can jump into it with only average thermal energy. Positive holes are left behind in the valence band, and as other electrons move into these holes, the holes move as discussed earlier.

EXERCISE E Which of the following impurity atoms in silicon would produce a *p*-type semiconductor? (*a*) Ge; (*b*) Ne; (*c*) Al; (*d*) As; (*e*) Ga; (*f*) none of the above.

*29–9 Semiconductor Diodes, LEDs, OLEDs

Semiconductor diodes and transistors are essential components of modern electronic devices. The miniaturization achieved today allows many millions of diodes, transistors, resistors, etc., to be fabricated (adding doping atoms) on a single *chip* less than a millimeter on a side.

At the interface between an *n*-type and a *p*-type semiconductor, a ***pn* junction diode** is formed. Separately, the two semiconductors are electrically neutral. But near the junction, a few electrons diffuse from the *n*-type into the *p*-type semiconductor, where they fill a few of the holes. The *n*-type is left with a positive charge, and the *p*-type acquires a net negative charge. Thus an "intrinsic" *potential difference* is established, with the *n* side positive relative to the *p* side, and this prevents further diffusion of electrons. The "junction" is actually a very thin layer between the charged *n* and *p* semiconductors where all holes are filled with electrons. This junction region is called the **depletion layer** (depleted of electrons and holes).[†]

If a battery is connected to a diode with the positive terminal to the *p* side and the negative terminal to the *n* side as in Fig. 29–29a, the externally applied voltage opposes the intrinsic potential difference and the diode is said to be **forward biased**. If the voltage is great enough, about 0.6 V for Si at room temperature, it overcomes that intrinsic potential difference and a large current can flow. The positive holes in the *p*-type semiconductor are repelled by the positive terminal of the battery, and the electrons in the *n*-type are repelled by the negative terminal of the battery. The holes and electrons meet at the junction, and the electrons cross over and fill the holes. A current is flowing. The positive terminal of the battery is continually pulling electrons off the *p* end, forming new holes, and electrons are being supplied by the negative terminal at the *n* end.

When the diode is **reverse biased**, as in Fig. 29–29b, the holes in the *p* end are attracted to the battery's negative terminal and the electrons in the *n* end are attracted to the positive terminal. Almost no current carriers meet near the junction and, ideally, no current flows.

[†]One way to form the *pn* boundary at the **nanometer** thicknesses on chips is to implant (or diffuse) *n*-type donor atoms into the surface of a *p*-type semiconductor, converting a layer of the *p*-type semiconductor into *n*-type.

FIGURE 29–29 Schematic diagram showing how a semiconductor diode operates. Current flows when the voltage is connected in forward bias, as in (a), but not when connected in reverse bias, as in (b).

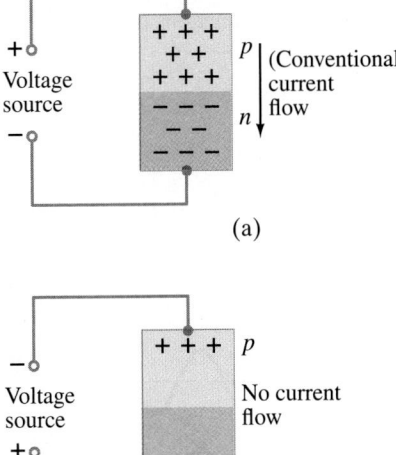

(a)

(b)

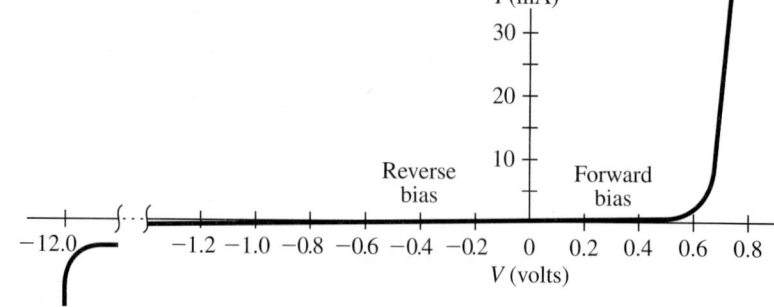

FIGURE 29–30 Current through a silicon *pn* diode as a function of applied voltage.

A graph of current versus voltage for a typical diode is shown in Fig. 29–30. A forward bias greater than 0.6 V allows a large current to flow. In reverse bias, a real diode allows a small amount of reverse current to flow; for most practical purposes, it is negligible.[†]

The symbol for a diode is

$$\longrightarrow\!\!\!\!\vdash\longleftarrow$$ [diode]

where the arrow represents the direction conventional (+) current flows readily.

EXAMPLE 29–6 A diode. The diode whose current–voltage characteristics are shown in Fig. 29–30 is connected in series with a 4.0-V battery in forward bias and a resistor. If a current of 15 mA is to pass through the diode, what resistance must the resistor have?

APPROACH We use Fig. 29–30, where we see that the voltage drop across the diode is about 0.7 V when the current is 15 mA. Then we use simple circuit analysis and Ohm's law (Chapters 18 and 19).

SOLUTION The voltage drop across the resistor is $4.0\,V - 0.7\,V = 3.3\,V$, so $R = V/I = (3.3\,V)/(1.5 \times 10^{-2}\,A) = 220\,\Omega$.

If the voltage across a diode connected in reverse bias is increased greatly, breakdown occurs. The electric field across the junction becomes so large that ionization of atoms results. The electrons thus pulled off their atoms contribute to a larger and larger current as breakdown continues. The voltage remains constant over a wide range of currents. This is shown on the far left in Fig. 29–30. This property of diodes can be used to accurately regulate a voltage supply. A diode designed for this purpose is called a **zener diode**. When placed across the output of an unregulated power supply, a zener diode can maintain the voltage at its own breakdown voltage as long as the supply voltage is always above this point. Zener diodes can be obtained corresponding to voltages of a few volts to hundreds of volts.

A diode is called a **nonlinear device** because the current is not proportional to the voltage. That is, a graph of current versus voltage (Fig. 29–30) is not a straight line, as it is for a resistor (which ideally *is* linear).

*Rectifiers

Since a *pn* junction diode allows current to flow only in one direction (as long as the voltage is not too high), it can serve as a **rectifier**—to change ac into dc. A simple rectifier circuit is shown in Fig. 29–31a. The ac source applies a voltage across the diode alternately positive and negative. Only during half of each cycle will a current pass through the diode; only then is there a current through the resistor R. Hence, a graph of the voltage V_{ab} across R as a function of time looks like the output voltage shown in Fig. 29–31b. This **half-wave rectification** is not exactly dc, but it is unidirectional. More useful is a **full-wave rectifier** circuit, which uses two diodes (or sometimes four) as shown in Fig. 29–32a (top of next page). At any given instant, either one diode or the other will conduct current to the right.

FIGURE 29–31 (a) A simple (half-wave) rectifier circuit using a semiconductor diode. (b) AC source input voltage, and output voltage across R, as functions of time.

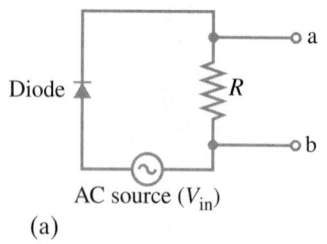

(a)

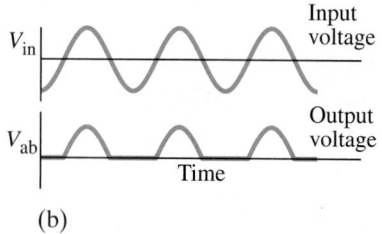

(b)

[†]At room temperature, the reverse current is a few pA in Si; but it increases rapidly with temperature, and may render a diode ineffective above 200°C.

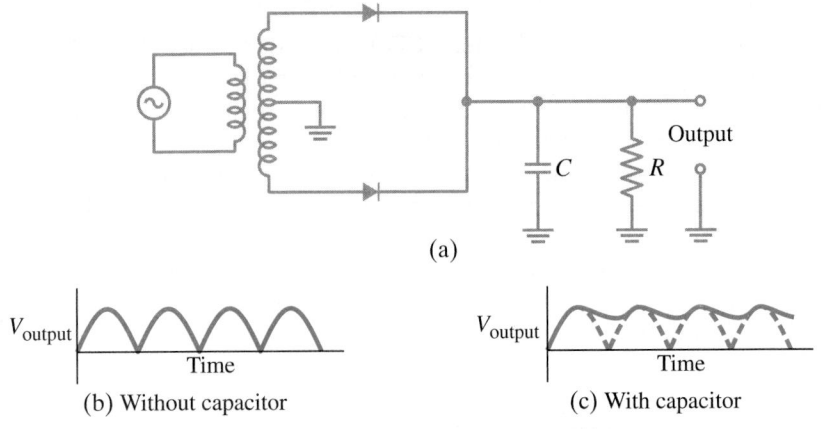

FIGURE 29–32 (a) Full-wave rectifier circuit (including a transformer so the magnitude of the voltage can be changed). (b) Output voltage in the absence of capacitor C. (c) Output voltage with the capacitor in the circuit.

(a)

(b) Without capacitor

(c) With capacitor

Therefore, the output across the load resistor R will be as shown in Fig. 29–32b. Actually this is the voltage if the capacitor C were not in the circuit. The capacitor tends to store charge and, if the time constant RC is sufficiently long, helps to smooth out the current as shown in Fig. 29–32c. (The variation in output shown in Fig. 29–32c is called **ripple voltage**.)

Rectifier circuits are important because most line voltage in buildings is ac, and most electronic devices require a dc voltage for their operation. Hence, diodes are found in nearly all electronic devices including radios, TV sets, computers, and chargers for cell phones and other devices.

*Photovoltaic Cells

Solar cells, also called **photovoltaic cells**, are rather heavily doped *pn* junction diodes used to convert sunlight into electric energy. Photons are absorbed, creating electron–hole pairs if the photon energy is greater than the band gap energy, E_g (see Figs. 29–25c and 29–28). That is, the absorbed photon excites an electron from the valence band up to the conduction band, leaving behind a hole in the valence band. The created electrons and holes produce a current that, when connected to an external circuit, becomes a source of emf and power. A typical silicon *pn* junction may produce about 0.6 V. Many are connected in series to produce a higher voltage. Such series strings are connected in parallel within a **photovoltaic panel**. Research includes experimenting with combinations of semiconductors. A good photovoltaic panel can have an output of perhaps 50 W/m², averaged over day and night, sunny and cloudy. The world's total electricity demand is on the order of 10^{12} W, which could be met with solar cells covering an area of only about 200 km × 200 km of Earth's surface.[†]

Photodiodes (Section 27–3) and **semiconductor particle detectors** (Section 30–13) operate similarly.

*LEDs

A **light-emitting diode (LED)** is sort of the reverse of a photovoltaic cell. When a *pn* junction is forward biased, a current begins to flow. Electrons cross from the *n*-region into the *p*-region, recombining with holes, and a photon can be emitted with an energy about equal to the band gap energy, E_g. This does not work well with silicon diodes.[‡] But high light-emission is achieved with **compound semiconductors**, typically involving a group III and a group V element such as gallium and arsenic (= gallium arsenide = GaAs). Remarkably, GaAs has a crystal structure very similar to Si. See Fig. 29–33. For doping of GaAs, group VI atoms (like Se) can serve as donors, and group II atoms (valence +2, such as Zn) as acceptors. The energy gap for GaAs is $E_g = 1.42$ eV, corresponding to near-infrared photons with wavelength 870 nm (almost visible). Such infrared LEDs are suitable for use in remote-control devices for TVs, DVD players, stereos, car door locks, and so on.

The first visible-light LED, developed in the early 1960s, was made of a semiconductor compound of gallium, arsenic, and phosphorus (= GaAsP) which emitted red light. The red LED soon found use as the familiar indicator lights (on–off) on electronic devices, and as the bright red read-out on calculators and

PHYSICS APPLIED

*LEDs and applications
Car safety (brakes)*

FIGURE 29–33 (a) Two Si atoms forming the covalent bond showing the electrons in different colors for each of the two separate atoms. (In Fig. 29–26a we showed each atom separately to emphasize the four outer electrons in each.) (b) A gallium–arsenic pair, also covalently bonded.

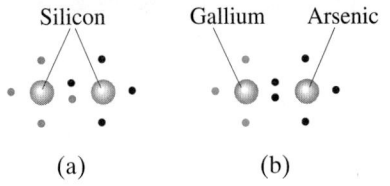

Silicon   Gallium   Arsenic

(a)   (b)

[†]Electricity makes up about 5% of total global energy use.
[‡]Electron-hole recombination in silicon results mostly in heat, as lattice vibrations called **phonons**.

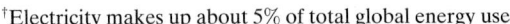

digital clocks (brighter than the dimmer LCD readouts). Further development led to LEDs with higher E_g and shorter wavelengths: first yellow, then finally in 1995, blue (InGaN). A blue LED was important because it gave the possibility of a **white-light LED**. White light can be approximated by LEDs in two ways: (1) using a red, a yellow–green, and a blue LED; (2) using a blue LED with coatings of "powders" or "phosphors" that are fluorescent (Section 28–10). For the latter, the high-energy blue LED photons are themselves emitted, plus they can excite the various phosphors to excited states which decay in two or more steps, emitting light of lower energy and longer wavelengths. Figure 29–34 shows typical spectra of both types.

FIGURE 29–34 (a) A combination of three LEDs of three different colors gives a sort of white color, but there are large wavelength gaps, so some colors would not be reflected and would appear black; this type is rarely used now. (b) A blue LED with fluorescent phosphors or powders gives a better approximation of white light. (Thanks to M. Vannoni and G. Molesini for (b).)

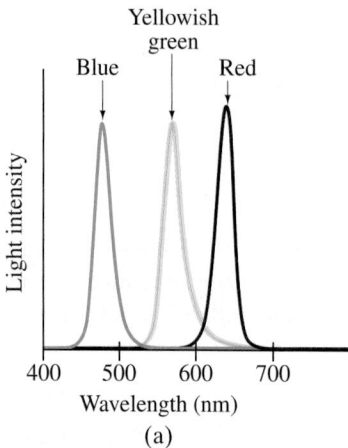
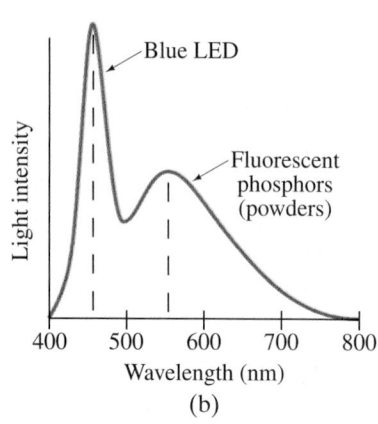

(a)

(b)

FIGURE 29–35 LED flashlights. Note the tiny LEDs, each maybe $\frac{1}{2}$ cm in diameter.

LED "bulbs" are available to replace other types of lighting in applications such as flashlights (Fig. 29–35), street lighting, traffic signals, car brake lights, billboards, backlighting for LCD screens, and large display screens at stadiums. LED lights, sometimes called **solid-state** lighting, are longer-lived (50,000 hours vs. 1000–2000 for ordinary bulbs), more efficient (up to 5 times), and rugged. A small town in Italy, Torraca, was the first to have all its street lighting be LED (2007). LEDs can be as small as 1 or 2 mm wide, and are individual units with wires connected directly to them. They can be used for large TV screens in stadiums, but a home TV would require much smaller LED size, meaning fabrication of many on a crystalline semiconductor, and the pixels would be addressed as discussed in Section 17–11 for LCD screens.

FIGURE 29–36 A pulse oximeter.

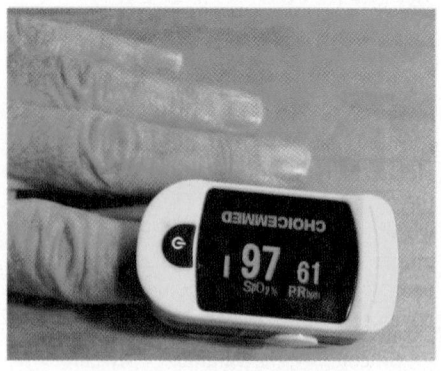

*Pulse Oximeter

A **pulse oximeter** uses two LEDs to measure the % oxygen (O_2) saturation in your blood. One LED is red, 660 nm, and the other IR (900–940 nm). The LED beams pass through a finger (Fig. 29–36) or earlobe and are detected by a photodiode. Oxygenated red blood cells absorb less red and more infrared light than deoxygenated cells. A ratio of absorbed light (red/IR) of 0.5 corresponds to nearly 100% O_2 saturation; a ratio of 1.0 is about 85% and 2.0 corresponds to about 50% (bad). The LED measures during complete pulses, including blood surges, and the device can also count your heartbeat rate.

*pn Diode Lasers

Diode lasers, using a *pn*-junction in forward bias like an LED, are the most compact of lasers and are very common: they read CDs and DVDs and are used as pointers and in laser printers. They emit photons like an LED but, like all lasers (Section 28–11), need to have an *inverted population* of states for the lasing frequency. This is achieved by applying a high forward-bias voltage. The large current brings many electrons into the conduction band at the junction layer, and holes into the valence band, and before the electrons have time to combine with holes, they form an inverted population. When one electron drops down into a hole and emits a photon, that photon stimulates other electrons to drop down as well, *in phase*, creating coherent laser light. Opposite ends of the crystal are made parallel and very smooth so they act as the mirrors needed for lasing, as shown in our laser diagram, Fig. 28–18.

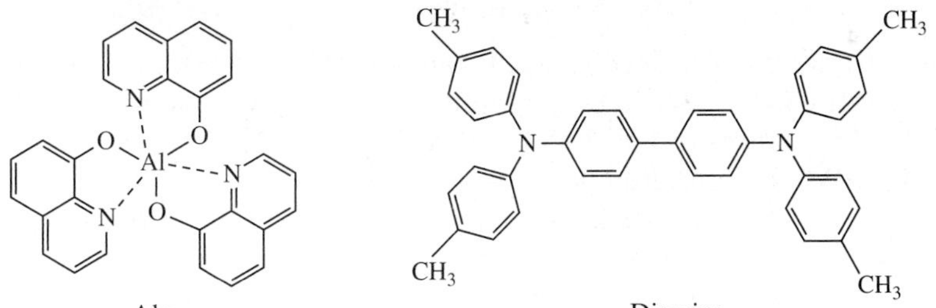

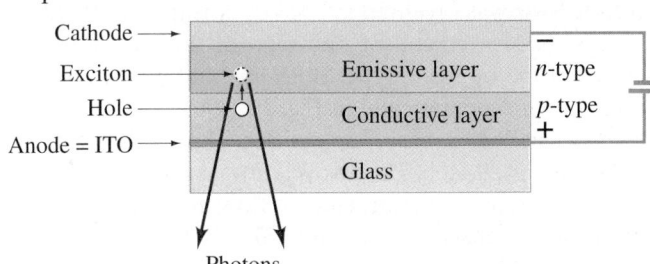

FIGURE 29–37 These two organic molecules were used in the first OLEDs (1987). The hexagons have carbon at each corner, and an attached hydrogen, unless otherwise noted.

Alq₃

Diamine

*OLED (Organic LED)

Many organic compounds have semiconductor properties. Useful ones can have mobile electrons and holes. A practical organic **electroluminescent** (**EL**) device, an **organic light-emitting diode** (**OLED**) was first described in the late 1980s.

Organic compounds contain carbon (C), hydrogen (H), often nitrogen and oxygen, and sometimes other atoms. We usually think of them as coming from life—plants and animals. They are also found in petroleum, and some can be synthesized in the lab. Organic compounds can be complex, and often contain the familiar hexagonal "benzene ring" with C atoms at all (or most) of the six corners. The two organic compounds shown in Fig. 29–37 were used as *n*-type and *p*-type layers in the earliest useful OLED. **Polymers**, long organic molecules with repeating structural units, can also be used for an OLED.

The simplest OLED consists of two organic layers, the **emissive layer** and the **conductive layer**, each 20 to 50 nm thick, sandwiched between two electrodes, Fig. 29–38. The anode is typically transparent, to let the light out. It can be made of a very thin layer of indium–tin oxide (**ITO**), which is transparent and conductive, coated on a glass slab. The cathode is often metallic, but could also be made of transparent material.

Cathode

Exciton

Hole

Anode = ITO

Emissive layer — *n*-type

Conductive layer — *p*-type

Glass

Photons

FIGURE 29–38 An OLED with two organic layers. Hole–electron recombination into an exciton (dashed circle) occurs in the emissive layer, followed by photon emission. Photons emitted in the wrong direction (upward in the diagram) reduce efficiency.

OLEDs can be smaller and thinner than ordinary inorganic LEDs. They can be more easily constructed as a unit for a screen display (i.e., more cheaply, but still quite expensive) than for inorganic LEDs. Their use as screens on cell phones, cameras, and TVs produces brighter light and greater contrast, and they need less power (important for battery life of portable devices) than LCD screens. Why? They need no backlight (like LCDs) because they emit the light themselves. OLEDs can be fabricated as a matrix, usually active matrix (**AMOLED**), using the same type of addressing described in Section 17–11 for LCDs. OLED displays are much thinner than LCDs and retain brightness at larger viewing angles. They can even be fabricated on curved or flexible substrates—try the windshield of your car (Fig. 29–39). The array may be RGBG (similar to a Bayer mosaic, Fig. 25–2) or RGBW where W = white is meant to give greater brightness. The subpixels can also be stacked, one above the other (similar to the Foveon, Fig. 25–3).

FIGURE 29–39 Head up displays on curved windshields can use curved OLEDs to show, for example, your speed without having to look down at the speedometer.

*OLED Functioning (advanced)

According to band theory, when a voltage is applied (≈ 2 to 5 V), electrons are "injected" (engineering term) into energy states of the **lowest unoccupied molecular orbitals** (**LUMO**) of the emissive layer. At the same time, electrons are withdrawn from the **highest occupied molecular orbitals** (**HOMO**) of the conductive layer at the cathode—which is equivalent to **holes** being "injected" into the conductive layer. The LUMO and HOMO energy levels are analogous to the conduction and valence bands of inorganic silicon diodes (Fig. 29–28). Holes travel in the HOMO, electrons in the LUMO. ("Orbital" is a chemistry word for the states occupied by the electrons in a molecule.)

When electrons and holes meet near the junction (Fig. 29–38), they can form a sort of bound state (like in the hydrogen atom) known as an **exciton**. An exciton has a small binding energy (0.1 to 1 eV), and a very short lifetime on the order of nanoseconds. When an exciton "decays" (the negative electron and positive hole combine), a photon is emitted. These photons are the useful output.

The energy hf of the photon, and its frequency corresponding to the color, depends on the energy structure of the exciton. The energy gap, LUMO–HOMO, sets an upper limit on hf, but the vibrational energy levels of the molecules reduce that by varying amounts, as does the binding energy of the exciton. The spectrum has a peak, like those in Fig. 29–34a, but is wider, 100–200 nm at half maximum. The organic molecules are chosen so that the photons have frequencies in the color range desired, say for a display subpixel: bluish (B), greenish (G), or red (R).

The conductive layer is also called the **hole transport layer** (**HTL**), which name expresses its purpose. The emissive layer, on the other hand (Fig. 29–38), serves two purposes: (1) it serves to transport electrons toward the junction, and (2) it is in this layer (near the junction) that holes meet electrons to form excitons and then combine and emit light. These two functions can be divided in a more sophisticated OLED that has three layers: Adjacent to the cathode is the **electron transport layer** (**ETL**), plus there is an **emissive layer** (**EML**) sandwiched between the ETL and the HTL. The emissive layer can be complex, containing a **host** material plus a **guest** compound in small concentration—a kind of doping— to fine-tune energy levels and efficiency.

*29–10 Transistors: Bipolar and MOSFETs

The **bipolar junction transistor** was invented in 1948 by J. Bardeen, W. Shockley, and W. Brattain. It consists of a crystal of one type of doped semiconductor sandwiched between two of the opposite type. Both *npn* and *pnp* transistors can be made, and they are shown schematically in Fig. 29–40a. The three semiconductors are given the names **collector**, **base**, and **emitter**. The symbols for *npn* and *pnp* transistors are shown in Fig. 29–40b. The arrow is always placed on the emitter and indicates the direction of (conventional) current flow in normal operation.

The operation of an *npn* transistor as an **amplifier** is shown in Fig. 29–41. A dc voltage V_{CE} is maintained between the collector and emitter by battery $\mathscr{E}_C$. The voltage applied to the base is called the *base bias voltage*, V_{BE}. If V_{BE} is positive, conduction electrons in the emitter are attracted into the base. The base region is very thin, much less than 1 μm, so most of these electrons flow right across into the collector which is maintained at a positive voltage. A large current, I_C, flows between collector and emitter and a much smaller current, I_B, through the base. In the steady state, I_B and I_C can be considered dc. But a small variation (= ac) in the base voltage due to an input signal attracts (or repels)

FIGURE 29–40 (a) Schematic diagram of *npn* and *pnp* transistors. (b) Symbols for *npn* and *pnp* transistors.

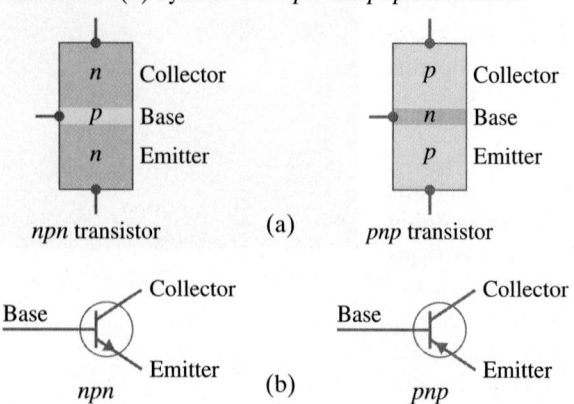

FIGURE 29–41 An *npn* transistor used as an amplifier. I_B is the current produced by $\mathscr{E}_B$ (in the absence of a signal), i_B is the ac signal current (= change in I_B).

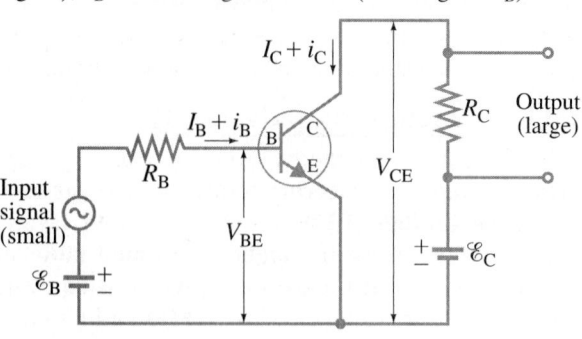

charge that passes through into the collector and thus can cause a large *change* in the collector current and a large change in the voltage drop across the output resistor R_C. Hence a transistor can *amplify* a small signal into a larger one.

Typically a small ac signal (call it i_B) is to be amplified, and when added to the base bias current (and voltage) causes the current and voltage at the collector to vary at the same rate but magnified. Thus, what is important for amplification is the *change* in collector current for a given input *change* in base current. We label these ac signal currents (= changes in I_C and I_B) as i_C and i_B. The **current gain** is defined as the ratio

$$\beta_I = \frac{\text{output (collector) ac current}}{\text{input (base) ac current}} = \frac{i_C}{i_B}.$$

β_I may be on the order of 10 to 100. Similarly, the **voltage gain** is

$$\beta_V = \frac{\text{output (collector) ac voltage}}{\text{input (base) ac voltage}}.$$

Transistors are the basic elements in modern electronic amplifiers of all sorts.

A *pnp* transistor operates like an *npn*, except that holes move instead of electrons. The collector voltage is negative, and so is the base voltage in normal operation.

Another kind of transistor, very important, is the **MOSFET** (metal-oxide semiconductor field-effect transistor) common in **digital circuits** as a type of switch. Its construction is shown in Fig. 29–42a, and its symbol in Fig. 29–42b. What is called the emitter in a bipolar transistor is called the **source** in a MOSFET, and the collector is called the **drain**. The base is called the **gate**. The gate acts to let a current flow, or not, from the source to the drain, depending on the electric field it (the gate) provides across an insulator that separates it from the *p*-type semiconductor below, Fig. 29–42a. Hence the name "field-effect transistor" (FET).[†] MOSFETs are often used like switches, on or off, which in digital circuits can allow the storage of a binary bit, a "1" or a "0". We discussed uses of MOSFETs relative to digital TV (Section 17–11) and computer memory storage (Section 21–8).

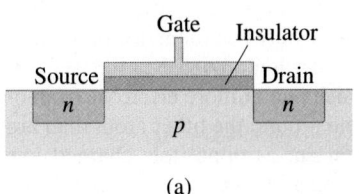

(a)

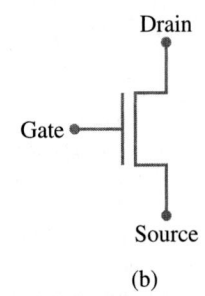

(b)

FIGURE 29–42 (a) Construction of a MOSFET of *n*- and *p*-type semiconductors and a gate of metal or heavily doped silicon (= a good conductor). (b) Symbol for a MOSFET which suggests its function.

*29–11 Integrated Circuits, 22-nm Technology

Although individual transistors are very small compared to the once-used vacuum tubes, they are huge compared to **integrated circuits** or **chips** (photo at start of this Chapter), invented in 1959 independently by Jack Kilby and Robert Noyce. Tiny amounts of impurities can be inserted or injected at particular locations within a single silicon crystal or wafer. These can be arranged to form diodes, transistors, resistors (undoped semiconductors), and very thin connecting "wires" (= conductors) which are heavily doped thin lines. Capacitors and inductors can also be formed, but also can be connected separately. Integrated circuits are the heart of computers, televisions, calculators, cameras, and the electronic instruments that control aircraft, space vehicles, and automobiles.

A tiny chip, a few millimeters on a side, may contain billions of transistors and other circuit elements. The number of elements/mm^2 has been doubling every 2 or 3 years. We often hear of the **technology generation**, which is a number that refers to the minimum width of a conducting line ("wire"). The gate of a MOSFET may be even smaller. Since 2003 we have passed from 90-nm technology to 65-nm, to 45-nm, to 32-nm, to 22-nm, every 2 to 3 years, and now 16-nm technology which—being only a few atoms wide—may involve new structures and quantum-mechanical effects. Smaller means more diodes and transistors per mm^2 and therefore greater speed (faster response time) because the distance signals have to travel is less. Smaller also means lower power consumption. Size, speed, and power have all been improved 10 to 100 million times in the last 40 years.

[†] The "MOS" comes from a version with a **M**etal gate, silicon di**O**xide insulator, and a **S**emiconductor (*p*-type shown in Fig. 29–42a). The gate can also be heavily doped silicon (= good conductor).

*Summary

Quantum mechanics explains the bonding together of atoms to form **molecules**. In a **covalent bond**, the atoms share electrons. The electron clouds of two or more atoms overlap because of constructive interference between the electron waves. The positive nuclei are attracted to this concentration of negative charge between them, forming the bond.

An **ionic bond** is an extreme case of a covalent bond in which one or more electrons from one atom spend much more time around the other atom than around their own. The atoms then act as oppositely charged ions that attract each other, forming the bond.

These **strong bonds** hold molecules together, and also hold atoms and molecules together in solids. Also important are **weak bonds** (or **van der Waals bonds**), which are generally dipole attractions between molecules.

When atoms combine to form molecules, the energy levels of the outer electrons are altered because they now interact with each other. Additional energy levels also become possible because the atoms can vibrate with respect to each other, and the molecule as a whole can rotate. The energy levels for both vibrational and rotational motion are quantized, and are very close together (typically, 10^{-1} eV to 10^{-3} eV apart). Each atomic energy level thus becomes a set of closely spaced levels corresponding to the vibrational and rotational motions. Transitions from one level to another appear as many very closely spaced lines. The resulting spectra are called **band spectra**.

The quantized rotational energy levels are given by

$$E_{\text{rot}} = \ell(\ell + 1)\frac{\hbar^2}{2I}, \quad \ell = 0, 1, 2, \cdots, \quad \textbf{(29–1)}$$

where I is the moment of inertia of the molecule.

The energy levels for vibrational motion are given by

$$E_{\text{vib}} = \left(\nu + \tfrac{1}{2}\right)hf, \quad \nu = 0, 1, 2, \cdots, \quad \textbf{(29–3)}$$

where f is the classical natural frequency of vibration for the molecule. Transitions between energy levels are subject to the selection rules $\Delta\ell = \pm 1$ and $\Delta\nu = \pm 1$.

Some **solids** are bound together by covalent and ionic bonds, just as molecules are. In metals, the electrostatic force between free electrons and positive ions helps form the **metallic bond**.

In the free-electron theory of metals, electrons occupy the possible energy states according to the exclusion principle. At $T = 0$ K, all possible states are filled up to a maximum energy level called the **Fermi energy**, E_F, the magnitude of which is typically a few eV. All states above E_F are vacant at $T = 0$ K.

In a crystalline solid, the possible energy states for electrons are arranged in **bands**. Within each band the levels are very close together, but between the bands there may be forbidden **energy gaps**. Good conductors are characterized by the highest occupied band (the **conduction band**) being only partially full, so lots of states are available to electrons to move about and accelerate when a voltage is applied. In a good insulator, the highest occupied energy band (the **valence band**) is completely full, and there is a large energy gap (5 to 10 eV) to the next highest band, the *conduction band*. At room temperature, molecular kinetic energy (thermal energy) available due to collisions is only about 0.04 eV, so almost no electrons can jump from the valence to the conduction band in an insulator. In a **semiconductor**, the gap between valence and conduction bands is much smaller, on the order of 1 eV, so a few electrons can make the transition from the essentially full valence band to the nearly empty conduction band, allowing a small amount of conductivity.

In a **doped** semiconductor, a small percentage of impurity atoms with five or three valence electrons replace a few of the normal silicon atoms with their four valence electrons. A five-electron impurity produces an **n-type** semiconductor with negative electrons as carriers of current. A three-electron impurity produces a **p-type** semiconductor in which positive **holes** carry the current. The energy level of impurity atoms lies slightly below the conduction band in an *n*-type semiconductor, and acts as a **donor** from which electrons readily pass into the conduction band. The energy level of impurity atoms in a *p*-type semiconductor lies slightly above the valence band and acts as an **acceptor** level, since electrons from the valence band easily reach it, leaving holes behind to act as charge carriers.

A semiconductor **diode** consists of a *pn* **junction** and allows current to flow in one direction only; *pn* junction diodes are used as **rectifiers** to change ac to dc, as photovoltaic cells to produce electricity from sunlight, and as lasers. **Light-emitting diodes** (**LED**) use compound semiconductors which can emit light when a forward-bias voltage is applied; uses include read-outs, infrared remote controls, visible lighting (flashlights, street lights), and very large TV screens. LEDs using organic molecules or polymers (**OLED**) are used as screens on cell phones and other displays. Common **transistors** consist of three semiconductor sections, either as *pnp* or *npn*. Transistors can amplify electrical signals and in computers serve as switches or **gates** for the 1s and 0s of digital bits. An integrated circuit consists of a tiny semiconductor crystal or **chip** on which many transistors, diodes, resistors, and other circuit elements are constructed by placement of impurities.

Questions

1. What type of bond would you expect for (*a*) the N_2 molecule, (*b*) the HCl molecule, (*c*) Fe atoms in a solid?

2. Describe how the molecule $CaCl_2$ could be formed.

3. Does the H_2 molecule have a permanent dipole moment? Does O_2? Does H_2O? Explain.

4. Although the molecule H_3 is not stable, the ion $H_3{}^+$ is. Explain, using the Pauli exclusion principle.

5. Would you expect the molecule $H_2{}^+$ to be stable? If so, where would the single electron spend most of its time?

6. Explain why the carbon atom ($Z = 6$) usually forms four bonds with hydrogen-like atoms.

7. The energy of a molecule can be divided into four categories. What are they?

8. If conduction electrons are free to roam about in a metal, why don't they leave the metal entirely?

9. Explain why the resistivity of metals increases with increasing temperature whereas the resistivity of semiconductors may decrease with increasing temperature.

10. Compare the resistance of a *pn* junction diode connected in forward bias to its resistance when connected in reverse bias.

11. Explain how a transistor can be used as a switch.

12. Figure 29–43 shows a "bridge-type" full-wave rectifier. Explain how the current is rectified and how current flows during each half cycle.

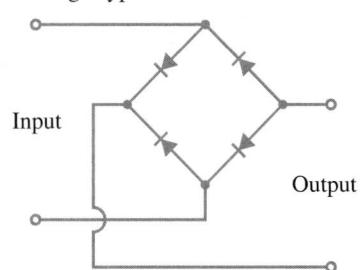

FIGURE 29–43
Question 12.

13. What is the main difference between n-type and p-type semiconductors?

14. Explain on the basis of energy bands why the sodium chloride crystal is a good insulator. [*Hint*: Consider the shells of Na^+ and Cl^- ions.]

15. In a transistor, the base–emitter junction and the base–collector junction are essentially diodes. Are these junctions reverse-biased or forward-biased in the application shown in Fig. 29–41?

16. A transistor can amplify an electronic signal, meaning it can increase the power of an input signal. Where does it get the energy to increase the power?

17. A silicon semiconductor is doped with phosphorus. Will these atoms be donors or acceptors? What type of semiconductor will this be?

18. Do diodes and transistors obey Ohm's law? Explain.

19. Can a diode be used to amplify a signal? Explain.

MisConceptual Questions

1. What holds molecules together?
(*a*) Gravitational forces.
(*b*) Magnetic forces.
(*c*) Electric forces.
(*d*) Glue.
(*e*) Nuclear forces.

2. Which of the following is true for covalently bound diatomic molecules such as H_2?
(*a*) All electrons in atoms have identical quantum numbers.
(*b*) The molecule has fewer electrons than the two separate atoms do.
(*c*) The molecule has less energy than two separate atoms.
(*d*) The energy of the molecule is greatest when the atoms are separated by one bond length.

3. A hydrogen atom ($Z = 1$) is bonded to a lithium atom ($Z = 3$) in lithium hydride, LiH. Which of the following are possible spin states of the two shared electrons?
(*a*) $+\frac{1}{2}, +\frac{1}{2}$.
(*b*) $-\frac{1}{2}, -\frac{1}{2}$.
(*c*) $+\frac{1}{2}, -\frac{1}{2}$.
(*d*) Both (*a*) and (*b*).
(*e*) Any of the above.

4. Ionic bonding is related to
(*a*) magnetic dipole interactions.
(*b*) the transfer of one or more electrons from one atom to another.
(*c*) the sharing of electrons between atoms.
(*d*) the transfer of electrons to the solid.
(*e*) oscillation dipoles.

5. Consider Fig. 29–10. As the last phosphate group approaches and then bonds to the ADP molecule, which of the following is true? Choose all that apply.
(*a*) The phosphate group is first repelled and then attracted to the ADP molecule.
(*b*) The phosphate group is always attracted to the ADP molecule.
(*c*) The phosphate group is always repelled by the ADP molecule.
(*d*) The system first loses and then stores potential energy.
(*e*) Both binding energy and activation energy are negative.
(*f*) Both binding energy and activation energy are positive.

6. Which type of bond holds the molecules of the DNA double helix together?
(*a*) Covalent bond.
(*b*) Ionic bond.
(*c*) Einstein bond.
(*d*) Van der Waals bond.

7. In a p-type semiconductor, a hole is
(*a*) a region in the molecular structure where an atom is missing.
(*b*) an extra electron from one of the donor atoms.
(*c*) an extra positively charged particle in the molecular structure.
(*d*) a region missing an electron relative to the rest of the molecular structure.

8. The electrical resistance of a semiconductor may decrease with increasing temperature because, at elevated temperature, more electrons
(*a*) collide with the crystal lattice.
(*b*) move faster.
(*c*) are able to jump across the energy gap.
(*d*) form weak van der Waals bonds.

9. Which of the following would *not* be used as an impurity in doping silicon?
(*a*) Germanium.
(*b*) Gallium.
(*c*) Boron.
(*d*) Phosphorus.
(*e*) Arsenic.

10. Why are metals good conductors?
(*a*) Gaining a tiny bit of energy allows their electrons to move.
(*b*) They have more electrons than protons, so some of the electrons are extra and free to move.
(*c*) They have more protons than electrons, so some of the protons are extra and free to move.
(*d*) Gaining a tiny bit of energy allows their protons to move.
(*e*) Electrons are tightly bound to their atoms.

Problems

*29–1 to 29–3 Molecular Bonds

1. (I) Estimate the binding energy of a KCl molecule by calculating the electrostatic potential energy when the K^+ and Cl^- ions are at their stable separation of 0.28 nm. Assume each has a charge of magnitude $1.0e$.

2. (II) The measured binding energy of KCl is 4.43 eV. From the result of Problem 1, estimate the contribution to the binding energy of the repelling electron clouds at the equilibrium distance $r_0 = 0.28$ nm.

3. (II) The equilibrium distance r_0 between two atoms in a molecule is called the **bond length**. Using the bond lengths of homogeneous molecules (like H_2, O_2, and N_2), one can estimate the bond length of heterogeneous molecules (like CO, CN, and NO). This is done by summing half of each bond length of the homogenous molecules to estimate that of the heterogeneous molecule. Given the following bond lengths: H_2 (= 74 pm), N_2 (= 145 pm), O_2 (= 121 pm), C_2 (= 154 pm), estimate the bond lengths for: HN, CN, and NO.

4. (II) Binding energies are often measured experimentally in kcal per mole, and then the binding energy in eV per molecule is calculated from that result. What is the conversion factor in going from kcal per mole to eV per molecule? What is the binding energy of KCl (= 4.43 eV) in kcal per mole?

5. (III) Estimate the binding energy of the H_2 molecule, assuming the two H nuclei are 0.074 nm apart and the two electrons spend 33% of their time midway between them.

6. (III) (a) Apply reasoning similar to that in the text for the $S = 0$ and $S = 1$ states in the formation of the H_2 molecule to show why the molecule He_2 is *not* formed. (b) Explain why the He_2^+ molecular ion *could* form. (Experiment shows it has a binding energy of 3.1 eV at $r_0 = 0.11$ nm.)

*29–4 Molecular Spectra

7. (I) Show that the quantity $\hbar^2/I$ has units of energy.

8. (II) (a) Calculate the "characteristic rotational energy," $\hbar^2/2I$, for the O_2 molecule whose bond length is 0.121 nm. (b) What are the energy and wavelength of photons emitted in an $\ell = 3$ to $\ell = 2$ transition?

9. (II) The "characteristic rotational energy," $\hbar^2/2I$, for N_2 is 2.48×10^{-4} eV. Calculate the N_2 bond length.

10. (II) The equilibrium separation of H atoms in the H_2 molecule is 0.074 nm (Fig. 29–8). Calculate the energies and wavelengths of photons for the rotational transitions (a) $\ell = 1$ to $\ell = 0$, (b) $\ell = 2$ to $\ell = 1$, and (c) $\ell = 3$ to $\ell = 2$.

11. (II) Determine the wavelength of the photon emitted when the CO molecule makes the rotational transition $\ell = 5$ to $\ell = 4$. [*Hint*: See Example 29–2.]

12. (II) Calculate the bond length for the NaCl molecule given that three successive wavelengths for rotational transitions are 23.1 mm, 11.6 mm, and 7.71 mm.

13. (II) (a) Use the curve of Fig. 29–17 to estimate the stiffness constant k for the H_2 molecule. (Recall that $PE = \frac{1}{2}kx^2$.) (b) Then estimate the fundamental wavelength for vibrational transitions using the classical formula (Chapter 11), but use only $\frac{1}{2}$ the mass of an H atom (because both H atoms move).

*29–5 Bonding in Solids

14. (II) Common salt, NaCl, has a density of 2.165 g/cm^3. The molecular weight of NaCl is 58.44. Estimate the distance between nearest neighbor Na and Cl ions. [*Hint: Each* ion can be considered to be at the corner of a cube.]

15. (II) Repeat Problem 14 for KCl whose density is 1.99 g/cm^3.

16. (II) The spacing between "nearest neighbor" Na and Cl ions in a NaCl crystal is 0.24 nm. What is the spacing between two nearest neighbor Na ions?

*29–7 Band Theory of Solids

17. (I) A semiconductor is struck by light of slowly increasing frequency and begins to conduct when the wavelength of the light is 620 nm. Estimate the energy gap E_g.

18. (I) Calculate the longest-wavelength photon that can cause an electron in silicon ($E_g = 1.12$ eV) to jump from the valence band to the conduction band.

19. (II) The energy gap between valence and conduction bands in germanium is 0.72 eV. What range of wavelengths can a photon have to excite an electron from the top of the valence band into the conduction band?

20. (II) The band gap of silicon is 1.12 eV. (a) For what range of wavelengths will silicon be transparent? (See Example 29–5.) In what region of the electromagnetic spectrum does this transparent range begin? (b) If window glass is transparent for all visible wavelengths, what is the minimum possible band gap value for glass (assume $\lambda = 400$ nm to 700 nm)? [*Hint*: If the photon has less energy than the band gap, the photon will pass through the solid without being absorbed.]

21. (II) The energy gap E_g in germanium is 0.72 eV. When used as a photon detector, roughly how many electrons can be made to jump from the valence to the conduction band by the passage of an 830-keV photon that loses all its energy in this fashion?

22. (III) We saw that there are $2N$ possible electron states in the 3s band of Na, where N is the total number of atoms. How many possible electron states are there in the (a) 2s band, (b) 2p band, and (c) 3p band? (d) State a general formula for the total number of possible states in any given electron band.

*29–8 Semiconductors and Doping

23. (III) Suppose that a silicon semiconductor is doped with phosphorus so that one silicon atom in 1.5×10^6 is replaced by a phosphorus atom. Assuming that the "extra" electron in every phosphorus atom is donated to the conduction band, by what factor is the density of conduction electrons increased? The density of silicon is 2330 kg/m^3, and the density of conduction electrons in pure silicon is about 10^{16} m^{-3} at room temperature.

24. (I) At what wavelength will an LED radiate if made from a material with an energy gap $E_g = 1.3\,\text{eV}$?

25. (I) If an LED emits light of wavelength $\lambda = 730\,\text{nm}$, what is the energy gap (in eV) between valence and conduction bands?

26. (I) A semiconductor diode laser emits 1.3-μm light. Assuming that the light comes from electrons and holes recombining, what is the band gap in this laser material?

27. (II) A silicon diode, whose current–voltage characteristics are given in Fig. 29–30, is connected in series with a battery and a 960-Ω resistor. What battery voltage is needed to produce a 14-mA current?

28. (II) An ac voltage of 120-V rms is to be rectified. Estimate very roughly the average current in the output resistor R ($= 31\,\text{k}\Omega$) for (a) a half-wave rectifier (Fig. 29–31), and (b) a full-wave rectifier (Fig. 29–32) without capacitor.

29. (III) Suppose that the diode of Fig. 29–30 is connected in series to a 180-Ω resistor and a 2.0-V battery. What current flows in the circuit? [Hint: Draw a line on Fig. 29–30 representing the current in the resistor as a function of the voltage across the diode; the intersection of this line with the characteristic curve will give the answer.]

30. (III) Sketch the resistance as a function of current, for $V > 0$, for the diode shown in Fig. 29–30.

31. (III) A 120-V rms 60-Hz voltage is to be rectified with a full-wave rectifier as in Fig. 29–32, where $R = 33\,\text{k}\Omega$, and $C = 28\,\mu\text{F}$. (a) Make a rough estimate of the average current. (b) What happens if $C = 0.10\,\mu\text{F}$? [Hint: See Section 19–6.]

32. (I) From Fig. 29–41, write an equation for the relationship between the base current (I_B), the collector current (I_C), and the emitter current (I_E, not labeled in Fig. 29–41). Assume $i_B = i_C = 0$.

33. (I) Draw a circuit diagram showing how a pnp transistor can operate as an amplifier, similar to Fig. 29–41 showing polarities, etc.

34. (II) If the current gain of the transistor amplifier in Fig. 29–41 is $\beta = i_C/i_B = 95$, what value must R_C have if a 1.0-μA ac base current is to produce an ac output voltage of 0.42 V?

35. (II) Suppose that the current gain of the transistor in Fig. 29–41 is $\beta = i_C/i_B = 85$. If $R_C = 3.8\,\text{k}\Omega$, calculate the ac output voltage for an ac input current of 2.0 μA.

36. (II) An amplifier has a voltage gain of 75 and a 25-kΩ load (output) resistance. What is the peak output current through the load resistor if the input voltage is an ac signal with a peak of 0.080 V?

37. (II) A transistor, whose current gain $\beta = i_C/i_B = 65$, is connected as in Fig. 29–41 with $R_B = 3.8\,\text{k}\Omega$ and $R_C = 7.8\,\text{k}\Omega$. Calculate (a) the voltage gain, and (b) the power amplification.

General Problems

38. Use the uncertainty principle to estimate the binding energy of the H_2 molecule by calculating the difference in kinetic energy of the electrons between (i) when they are in separate atoms and (ii) when they are in the molecule. Take Δx for the electrons in the separated atoms to be the radius of the first Bohr orbit, 0.053 nm, and for the molecule take Δx to be the separation of the nuclei, 0.074 nm. [Hint: Let $\Delta p \approx \Delta p_x$.]

39. The average translational kinetic energy of an atom or molecule is about $\text{KE} = \frac{3}{2}kT$ (see Section 13–9), where $k = 1.38 \times 10^{-23}\,\text{J/K}$ is Boltzmann's constant. At what temperature T will KE be on the order of the bond energy (and hence the bond easily broken by thermal motion) for (a) a covalent bond (say H_2) of binding energy 4.0 eV, and (b) a "weak" hydrogen bond of binding energy 0.12 eV?

40. A diatomic molecule is found to have an activation energy of 1.3 eV. When the molecule is disassociated, 1.6 eV of energy is released. Draw a potential energy curve for this molecule.

41. In the ionic salt KF, the separation distance between ions is about 0.27 nm. (a) Estimate the electrostatic potential energy between the ions assuming them to be point charges (magnitude $1e$). (b) When F "grabs" an electron, it releases 3.41 eV of energy, whereas 4.34 eV is required to ionize K. Find the binding energy of KF relative to free K and F atoms, neglecting the energy of repulsion.

42. The rotational absorption spectrum of a molecule displays peaks about 8.9×10^{11} Hz apart. Determine the moment of inertia of this molecule.

43. For O_2 with a bond length of 0.121 nm, what is the moment of inertia about the center of mass?

44. Must we consider quantum effects for everyday rotating objects? Estimate the differences between rotational energy levels for a spinning baton compared to the energy of the baton. Assume the baton consists of a uniform 32-cm-long bar with a mass of 230 g and two small end masses, each of mass 380 g, and it rotates at 1.8 rev/s about the bar's center.

45. For a certain semiconductor, the longest wavelength radiation that can be absorbed is 2.06 mm. What is the energy gap in this semiconductor?

46. When EM radiation is incident on diamond, it is found that light with wavelengths shorter than 226 nm will cause the diamond to conduct. What is the energy gap between the valence band and the conduction band for diamond?

47. The energy gap between valence and conduction bands in zinc sulfide is 3.6 eV. What range of wavelengths can a photon have to excite an electron from the top of the valence band into the conduction band?

48. Most of the Sun's radiation has wavelengths shorter than 1100 nm. For a solar cell to absorb all this, what energy gap ought the material have?

49. A TV remote control emits IR light. If the detector on the TV set is *not* to react to visible light, could it make use of silicon as a "window" with its energy gap $E_g = 1.12\,\text{eV}$? What is the shortest-wavelength light that can strike silicon without causing electrons to jump from the valence band to the conduction band?

50. Green and blue LEDs became available many years after red LEDs were first developed. Approximately what energy gaps would you expect to find in green (525 nm) and in blue (465 nm) LEDs?

51. Consider a monatomic solid with a weakly bound cubic lattice, with each atom connected to six neighbors, each bond having a binding energy of $3.4 \times 10^{-3}\,\text{eV}$. When this solid melts, its latent heat of fusion goes directly into breaking the bonds between the atoms. Estimate the latent heat of fusion for this solid, in J/mol. [*Hint:* Show that in a simple cubic lattice (Fig. 29–44), there are *three* times as many bonds as there are atoms, when the number of atoms is large.]

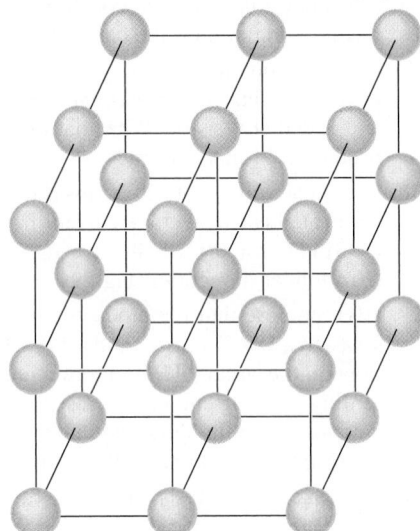

FIGURE 29–44
Problem 51.

Search and Learn

1. Explain why metals are shiny. (See Section 29–5.)

2. Compare the potential energy diagram for an H_2 molecule with the potential energy diagram for ATP formation from ADP and ℗. Explain the significance of the difference in shapes of the two diagrams. (See Section 29–2.)

3. (*a*) Why are weak bonds important in cells? (*b*) Explain why heating proteins too much may cause them to denature—that is, lose the specific shape they need to function. (See Section 29–3.) (*c*) What is the strongest weak bond, and why? (*d*) If this bond, and the other weak bonds, were stronger (that is, too strong), what would be the consequence for protein synthesis?

4. Assume conduction electrons in a semiconductor behave as an ideal gas. (This is not true for conduction electrons in a metal.) (*a*) Taking mass $m = 9 \times 10^{-31}\,\text{kg}$ and temperature $T = 300\,\text{K}$, determine the de Broglie wavelength of a semiconductor's conduction electrons. (*b*) Given that the spacing between atoms in a semiconductor's atomic lattice is on the order of 0.3 nm, would you expect room-temperature conduction electrons to travel in straight lines or diffract when traveling through this lattice? Explain.

5. A strip of silicon 1.6 cm wide and 1.0 mm thick is immersed in a magnetic field of strength 1.5 T perpendicular to the strip (Fig. 29–45). When a current of 0.28 mA is run through the strip, there is a resulting Hall effect voltage of 18 mV across the strip (Section 20–4). How many electrons per silicon atom are in the conduction band? The density of silicon is $2330\,\text{kg/m}^3$.

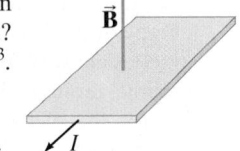

FIGURE 29–45
Search and Learn 5.

6. For an arsenic donor atom in a doped silicon semiconductor, assume that the "extra" electron moves in a Bohr orbit about the arsenic ion. For this electron in the ground state, take into account the dielectric constant $K = 12$ of the Si lattice (which represents the weakening of the Coulomb force due to all the other atoms or ions in the lattice), and estimate (*a*) the binding energy, and (*b*) the orbit radius for this extra electron. [*Hint:* Substitute $\epsilon = K\epsilon_0$ in Coulomb's law; see Section 17–8 and also 27–12.]

In this Chapter we begin our discussion of nuclear physics. We study the properties of nuclei, the various forms of radioactivity, and how radioactive decay can be used in a variety of fields to determine the age of old objects, from bones and trees to rocks and other mineral substances, and obtain information on the history of the Earth.

Shown is one version of a **Chart of the Nuclides**. Each horizontal row has a square for each known isotope (nuclide) of one element with a particular Z value (= number of electrons in the neutral atom = number of protons in the nucleus). At the far left is a white box with the average atomic weight (or a range if uncertain) of the naturally occurring isotopes of that element. Each vertical column contains nuclides with the same neutron number N. For $N = 1$ (to right of pencil), starting at the bottom, there is a lone neutron, then above it $^2_1$H, then $^3_2$He and $^4_3$Li. Each square is color coded: black means a stable nuclide. Radioactive nuclides are blue green for β^- decay, pink for β^+ decay or electron capture (ε) such as $^7_4$Be, yellow for α decay, and so on. Thus $^1_1$H and $^2_1$H are stable but $^3_1$H (tritium) undergoes β^- decay with half-life = 12.3 years ("a" is for Latin "anno" = year). The squares contain the atomic mass of that isotope, or half-life and energy released if radioactive. Other details may be alternate decay modes and certain cross sections (σ).

Nuclear Physics and Radioactivity

CHAPTER-OPENING QUESTION—Guess now!

If half of an 80-μg sample of $^{60}_{27}$Co decays in 5.3 years, how much $^{60}_{27}$Co is left in 10.6 years?

(a) 10 μg.

(b) 20 μg.

(c) 30 μg.

(d) 40 μg.

(e) 0 μg.

In the early part of the twentieth century, Rutherford's experiments (Section 27–10) led to the idea that at the center of an atom there is a tiny but massive nucleus with a positive charge. At the same time that the quantum theory was being developed and scientists were attempting to understand the structure of the atom and its electrons, investigations into the nucleus itself had also begun. In this Chapter and the next, we take a brief look at *nuclear physics*.

CONTENTS

30-1 Structure and Properties of the Nucleus

An important question for physicists was whether the nucleus had a structure, and what that structure might be. By the early 1930s, a model of the nucleus had been developed that is still useful. According to this model, a nucleus is made up of two types of particles: protons and neutrons. [These "particles" also have wave properties, but for ease of visualization and language, we usually refer to them simply as "particles."] A **proton** is the nucleus of the simplest atom, hydrogen. The proton has a positive charge ($= +e = +1.60 \times 10^{-19}$ C, the same magnitude as for the electron) and its mass is measured to be

$$m_p = 1.67262 \times 10^{-27} \text{ kg.}$$

The **neutron**, whose existence was ascertained in 1932 by the English physicist James Chadwick (1891–1974), is electrically neutral ($q = 0$), as its name implies. Its mass is very slightly larger than that of the proton:

$$m_n = 1.67493 \times 10^{-27} \text{ kg.}$$

These two constituents of a nucleus, neutrons and protons, are referred to collectively as **nucleons**.

Although a normal hydrogen nucleus consists of a single proton alone, the nuclei of all other elements consist of both neutrons and protons. The different nuclei are often referred to as **nuclides**. The number of protons in a nucleus (or nuclide) is called the **atomic number** and is designated by the symbol Z. The total number of nucleons, neutrons plus protons, is designated by the symbol A and is called the **atomic mass number**, or sometimes simply **mass number**. This name is used since the mass of a nucleus is very closely A times the mass of one nucleon. A nuclide with 7 protons and 8 neutrons thus has $Z = 7$ and $A = 15$. The **neutron number** N is $N = A - Z$.

To specify a given nuclide, we need give only A and Z. A special symbol is commonly used which takes the form

$$^A_Z X,$$

where X is the chemical symbol for the element (see Appendix B, and the Periodic Table inside the back cover), A is the atomic mass number, and Z is the atomic number. For example, $^{15}_7 N$ means a nitrogen nucleus containing 7 protons and 8 neutrons for a total of 15 nucleons. In a neutral atom, the number of electrons orbiting the nucleus is equal to the atomic number Z (since the charge on an electron has the same magnitude but opposite sign to that of a proton). The main properties of an atom, and how it interacts with other atoms, are largely determined by the number of electrons. Hence Z determines what kind of atom it is: carbon, oxygen, gold, or whatever. It is redundant to specify both the symbol of a nucleus and its atomic number Z as described above. If the nucleus is nitrogen, for example, we know immediately that $Z = 7$. The subscript Z is thus sometimes dropped and $^{15}_7 N$ is then written simply ^{15}N; in words we say "nitrogen fifteen."

For a particular type of atom (say, carbon), nuclei are found to contain different numbers of neutrons, although they all have the same number of protons. For example, carbon nuclei always have 6 protons, but they may have 5, 6, 7, 8, 9, or 10 neutrons. Nuclei that contain the same number of protons but different numbers of neutrons are called **isotopes**. Thus, $^{11}_6 C$, $^{12}_6 C$, $^{13}_6 C$, $^{14}_6 C$, $^{15}_6 C$, and $^{16}_6 C$ are all isotopes of carbon. The isotopes of a given element are not all equally common. For example, 98.9% of naturally occurring carbon (on Earth) is the isotope $^{12}_6 C$, and about 1.1% is $^{13}_6 C$. These percentages are referred to as the **natural abundances**.[†] Even hydrogen has isotopes: 99.99% of natural hydrogen is $^1_1 H$, a simple proton, as the nucleus; there are also $^2_1 H$, called **deuterium**, and $^3_1 H$, **tritium**, which besides the proton contain 1 or 2 neutrons. (The bare nucleus in each case is called the **deuteron** and **triton**.)

[†]The mass value for each element as given in the Periodic Table (inside back cover) is an average weighted according to the natural abundances of its isotopes.

Many isotopes that do not occur naturally can be produced in the laboratory by means of nuclear reactions (more on this later). Indeed, all elements beyond uranium ($Z > 92$) do not occur naturally on Earth and are only produced artificially (in the laboratory), as are many nuclides with $Z \leq 92$.

The approximate size of nuclei was determined originally by Rutherford from the scattering of charged particles by thin metal foils. We cannot speak about a definite size for nuclei because of the wave–particle duality (Section 27–7): their spatial extent must remain somewhat fuzzy. Nonetheless a rough "size" can be measured by scattering high-speed electrons off nuclei. It is found that nuclei have a roughly spherical shape with a radius that increases with A according to the approximate formula

$$r \approx (1.2 \times 10^{-15}\,\text{m})\left(A^{\frac{1}{3}}\right). \tag{30–1}$$

Since the volume of a sphere is $V = \frac{4}{3}\pi r^3$, we see that the volume of a nucleus is approximately proportional to the number of nucleons, $V \propto A$ (because $\left(A^{\frac{1}{3}}\right)^3 = A$). This is what we would expect if nucleons were like impenetrable billiard balls: if you double the number of balls, you double the total volume. Hence, all nuclei have nearly the same density, and it is enormous (see Example 30–2).

The metric abbreviation for $10^{-15}\,\text{m}$ is the fermi (after Enrico Fermi, Fig. 30–7) or the femtometer, fm (see Table 1–4 or inside the front cover). Thus $1.2 \times 10^{-15}\,\text{m} = 1.2\,\text{fm}$ or 1.2 fermis.

EXAMPLE 30–1 | **ESTIMATE** | **Nuclear sizes.** Estimate the diameter of the smallest and largest naturally occurring nuclei: (*a*) ^1_1H, (*b*) $^{238}_{92}\text{U}$.

APPROACH The radius r of a nucleus is related to its number of nucleons A by Eq. 30–1. The diameter $d = 2r$.

SOLUTION (*a*) For hydrogen, $A = 1$, Eq. 30–1 gives

$$d = \text{diameter} = 2r \approx 2(1.2 \times 10^{-15}\,\text{m})\left(A^{\frac{1}{3}}\right) = 2.4 \times 10^{-15}\,\text{m}$$

since $A^{\frac{1}{3}} = 1^{\frac{1}{3}} = 1$.

(*b*) For uranium $d \approx (2.4 \times 10^{-15}\,\text{m})(238)^{\frac{1}{3}} = 15 \times 10^{-15}\,\text{m}$.

The range of nuclear diameters is only from 2.4 fm to 15 fm.

NOTE Because nuclear radii vary as $A^{\frac{1}{3}}$, the largest nuclei (such as uranium with $A = 238$) have a radius only about $\sqrt[3]{238} \approx 6$ times that of the smallest, hydrogen ($A = 1$).

EXAMPLE 30–2 | **ESTIMATE** | **Nuclear and atomic densities.** Compare the density of nuclear matter to the density of normal solids.

APPROACH The density of normal liquids and solids is on the order of 10^3 to $10^4\,\text{kg/m}^3$ (see Table 10–1), and because the atoms are close packed, atoms have about this density too. We therefore compare the density (mass per volume) of a nucleus to that of its atom as a whole.

SOLUTION The mass of a proton is greater than the mass of an electron by a factor

$$\frac{1.67 \times 10^{-27}\,\text{kg}}{9.1 \times 10^{-31}\,\text{kg}} \approx 2000.$$

Thus, over 99.9% of the mass of an atom is in the nucleus, and for our estimate we can say the mass of the atom equals the mass of the nucleus, $m_{\text{nucl}}/m_{\text{atom}} = 1$. Atoms have a radius of about $10^{-10}\,\text{m}$ (Chapter 27) and nuclei on the order of $10^{-15}\,\text{m}$ (Eq. 30–1). Thus the ratio of nuclear density to atomic density is about

$$\frac{\rho_{\text{nucl}}}{\rho_{\text{atom}}} = \frac{(m_{\text{nucl}}/V_{\text{nucl}})}{(m_{\text{atom}}/V_{\text{atom}})} = \left(\frac{m_{\text{nucl}}}{m_{\text{atom}}}\right)\frac{\frac{4}{3}\pi r_{\text{atom}}^3}{\frac{4}{3}\pi r_{\text{nucl}}^3} \approx (1)\frac{(10^{-10})^3}{(10^{-15})^3} = 10^{15}.$$

The nucleus is 10^{15} times more dense than ordinary matter.

The masses of nuclei can be determined from the radius of curvature of fast-moving nuclei (as ions) in a known magnetic field using a mass spectrometer, as discussed in Section 20–11. Indeed the existence of different isotopes of the same element (different number of neutrons) was discovered using this device.

Nuclear masses can be specified in **unified atomic mass units** (u). On this scale, a neutral $^{12}_{6}C$ atom is given the exact value 12.000000 u. A neutron then has a measured mass of 1.008665 u, a proton 1.007276 u, and a neutral hydrogen atom $^{1}_{1}H$ (proton plus electron) 1.007825 u. The masses of many nuclides are given in Appendix B. It should be noted that the masses in this Table, as is customary, are for the *neutral atom* (including electrons), and not for a bare nucleus.

Masses may be specified using the electron-volt energy unit, $1\,eV = 1.6022 \times 10^{-19}\,J$ (Section 17–4). This can be done because mass and energy are related, and the precise relationship is given by Einstein's equation $E = mc^2$ (Chapter 26). Since the mass of a proton is $1.67262 \times 10^{-27}\,kg$, or 1.007276 u, then 1 u is equal to

$$1.0000\,u = (1.0000\,u)\left(\frac{1.67262 \times 10^{-27}\,kg}{1.007276\,u}\right) = 1.66054 \times 10^{-27}\,kg;$$

this is equivalent to an energy (see Table inside front cover) in MeV ($= 10^6\,eV$) of

$$E = mc^2 = \frac{(1.66054 \times 10^{-27}\,kg)(2.9979 \times 10^8\,m/s)^2}{(1.6022 \times 10^{-19}\,J/eV)} = 931.5\,MeV.$$

Thus,

$$1\,u = 1.6605 \times 10^{-27}\,kg = 931.5\,MeV/c^2.$$

The rest masses of some of the basic particles are given in Table 30–1. As a rule of thumb, to remember, the masses of neutron and proton are about $1\,GeV/c^2$ ($= 1000\,MeV/c^2$) which is about 2000 times the mass of an electron ($\approx \frac{1}{2}\,MeV/c^2$).

CAUTION
Masses are for neutral atom (nucleus plus electrons)

TABLE 30–1
Rest Masses in Kilograms, Unified Atomic Mass Units, and MeV/c^2

Object	Mass		
	kg	**u**	**MeV/c^2**
Electron	9.1094×10^{-31}	0.00054858	0.51100
Proton	1.67262×10^{-27}	1.007276	938.27
$^{1}_{1}H$ atom	1.67353×10^{-27}	1.007825	938.78
Neutron	1.67493×10^{-27}	1.008665	939.57

Just as an electron has intrinsic spin and angular momentum quantum numbers, so too do nuclei and their constituents, the proton and neutron. Both the proton and the neutron are spin $\frac{1}{2}$ particles, just like the electron. A nucleus, made up of protons and neutrons, has a **nuclear spin** quantum number, I, that can be either integer or half integer, depending on whether it is made up of an even or an odd number of nucleons.

30–2 Binding Energy and Nuclear Forces

Binding Energies

The total mass of a stable nucleus is always less than the sum of the masses of its separate protons and neutrons, as the following Example shows.

EXAMPLE 30–3 $^{4}_{2}He$ **mass compared to its constituents.** Compare the mass of a $^{4}_{2}He$ atom to the total mass of its constituent particles.

APPROACH The $^{4}_{2}He$ nucleus contains 2 protons and 2 neutrons. Tables normally give the masses of neutral atoms—that is, nucleus plus its Z electrons. We must therefore be sure to balance out the electrons when we compare masses. Thus we use the mass of $^{1}_{1}H$ rather than that of a proton alone. We look up the mass of the $^{4}_{2}He$ atom in Appendix B (it includes the mass of 2 electrons), as well as the mass for the 2 neutrons and 2 hydrogen atoms ($= 2$ protons $+ 2$ electrons).

PROBLEM SOLVING
Keep track of electron masses

SOLUTION The mass of a neutral $_2^4$He atom, from Appendix B, is 4.002603 u. The mass of two neutrons and two H atoms (2 protons including the 2 electrons) is

$$2m_n = 2(1.008665\ u) = 2.017330\ u$$
$$2m(_1^1H) = 2(1.007825\ u) = \underline{2.015650\ u}$$
$$sum = 4.032980\ u.$$

Thus the mass of $_2^4$He is measured to be less than the masses of its constituents by an amount 4.032980 u − 4.002603 u = 0.030377 u.

Where has this lost mass of 0.030377 u disappeared to? It must be $E = mc^2$.

If the four nucleons suddenly came together to form a $_2^4$He nucleus, the mass "loss" would appear as energy of another kind (such as radiation, or kinetic energy). The mass (or energy) difference in the case of $_2^4$He, given in energy units, is (0.030377 u)(931.5 MeV/u) = 28.30 MeV. This difference is referred to as the **total binding energy** of the nucleus. The total binding energy represents the amount of energy that must be put *into* a nucleus in order to break it apart into its constituents. If the mass of, say, a $_2^4$He nucleus were exactly equal to the mass of two neutrons plus two protons, the nucleus could fall apart without any input of energy. To be stable, the mass of a nucleus *must* be less than that of its constituent nucleons, so that energy input *is* needed to break it apart.

Binding energy is not something a nucleus has—it is energy it "lacks" relative to the total mass of its separate constituents.

[As a comparison, we saw in Chapter 27 that the binding energy of the one electron in the hydrogen atom is 13.6 eV; so the mass of a $_1^1$H atom is less than that of a single proton plus a single electron by 13.6 eV/c^2. The binding energies of nuclei are on the order of MeV, so the eV binding energies of electrons can be ignored. Nuclear binding energies, compared to nuclear masses, are on the order of (28 MeV/4000 MeV) ≈ 1 × 10^{-2}, where we used helium's binding energy of 28.3 MeV (see above) and mass ≈ 4 × 940 MeV ≈ 4000 MeV.]

EXERCISE A Determine how much less the mass of the $_3^7$Li nucleus is compared to that of its constituents. See Appendix B.

The **binding energy per nucleon** is defined as the total binding energy of a nucleus divided by A, the total number of nucleons. We calculated above that the binding energy of $_2^4$He is 28.3 MeV, so its binding energy per nucleon is 28.3 MeV/4 = 7.1 MeV. Figure 30–1 shows the measured binding energy per nucleon as a function of A for stable nuclei. The curve rises as A increases and reaches a plateau at about 8.7 MeV per nucleon above $A \approx 40$. Beyond $A \approx 80$, the curve decreases slowly, indicating that larger nuclei are held together less tightly than those in the middle of the Periodic Table. We will see later that these characteristics allow the release of nuclear energy in the processes of fission and fusion.

> ⚠ **CAUTION**
> *Mass of nucleus must be less than mass of constituents*

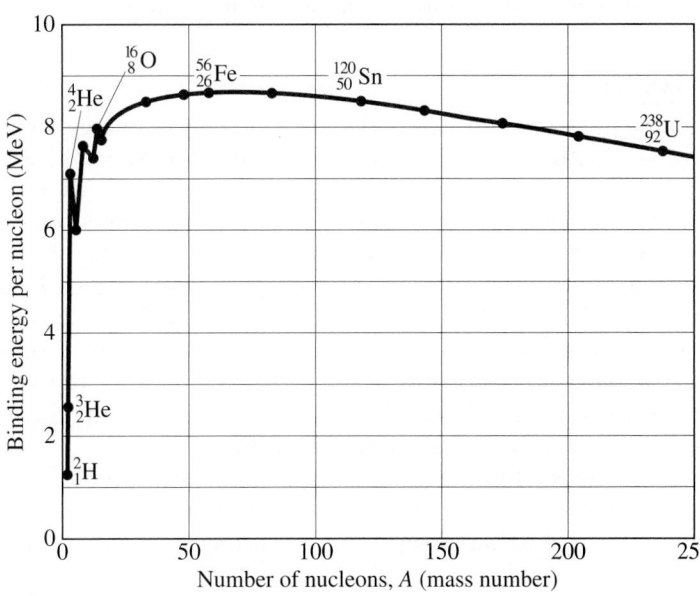

FIGURE 30–1 Binding energy per nucleon for the more stable nuclides as a function of mass number A.

EXAMPLE 30–4 **Binding energy for iron.** Calculate the total binding energy and the binding energy per nucleon for $^{56}_{26}$Fe, the most common stable isotope of iron.

APPROACH We subtract the mass of a $^{56}_{26}$Fe atom from the total mass of 26 hydrogen atoms and 30 neutrons, all found in Appendix B. Then we convert mass units to energy units; finally we divide by $A = 56$, the total number of nucleons.

SOLUTION $^{56}_{26}$Fe has 26 protons and 30 neutrons whose separate masses are

$$
\begin{aligned}
26m(^1_1\text{H}) &= (26)(1.007825\,\text{u}) = 26.20345\,\text{u} \quad (\text{includes 26 electrons}) \\
30m_\text{n} &= (30)(1.008665\,\text{u}) = \underline{30.25995\,\text{u}} \\
\text{sum} &= 56.46340\,\text{u.} \\
\text{Subtract mass of } ^{56}_{26}\text{Fe:} &= \underline{-55.93494\,\text{u}} \quad (\text{Appendix B}) \\
\Delta m &= 0.52846\,\text{u.}
\end{aligned}
$$

The total binding energy is thus

$$(0.52846\,\text{u})(931.5\,\text{MeV/u}) = 492.26\,\text{MeV}$$

and the binding energy per nucleon is

$$\frac{492.26\,\text{MeV}}{56\,\text{nucleons}} = 8.79\,\text{MeV.}$$

NOTE The binding energy per nucleon graph (Fig. 30–1) peaks about here, for iron. So the iron nucleus, and its neighbors, are the most stable of nuclei.

| **EXERCISE B** Determine the binding energy per nucleon for $^{16}_8$O.

EXAMPLE 30–5 **Binding energy of last neutron.** What is the binding energy of the last neutron in $^{13}_6$C?

APPROACH If $^{13}_6$C lost one neutron, it would be $^{12}_6$C. We subtract the mass of $^{13}_6$C from the masses of $^{12}_6$C and a free neutron.

SOLUTION Obtaining the masses from Appendix B, we have

$$
\begin{aligned}
\text{Mass } ^{12}_6\text{C} &= 12.000000\,\text{u} \\
\text{Mass } ^1_0\text{n} &= \underline{1.008665\,\text{u}} \\
\text{Total} &= 13.008665\,\text{u.} \\
\text{Subtract mass of } ^{13}_6\text{C:} &= \underline{-13.003355\,\text{u}} \\
\Delta m &= 0.005310\,\text{u.}
\end{aligned}
$$

which in energy is $(931.5\,\text{MeV/u})(0.005310\,\text{u}) = 4.95\,\text{MeV}$. That is, it would require 4.95 MeV input of energy to remove one neutron from $^{13}_6$C.

Nuclear Forces

We can analyze nuclei not only from the point of view of energy, but also from the point of view of the forces that hold them together. We might not expect a collection of protons and neutrons to come together spontaneously, since protons are all positively charged and thus exert repulsive electric forces on each other. Since stable nuclei *do* stay together, another force must be acting. This new force has to be stronger than the electric force in order to hold the nucleus together, and is called the **strong nuclear force**. The strong nuclear force acts as an attractive force between all nucleons, protons and neutrons alike. Thus protons attract each other via the strong nuclear force at the same time they repel each other via the electric force. Neutrons, because they are electrically neutral, only attract other neutrons or protons via the strong nuclear force.

The strong nuclear force turns out to be far more complicated than the gravitational and electromagnetic forces. One important aspect of the strong nuclear force is that it is a **short-range** force: it acts only over a very short distance.

It is very strong between two nucleons if they are less than about 10^{-15} m apart, but it is essentially zero if they are separated by a distance greater than this. Compare this to electric and gravitational forces, which decrease as $1/r^2$ but continue acting over any distances and are therefore called **long-range** forces.

The strong nuclear force has some strange features. For example, if a nuclide contains too many or too few neutrons relative to the number of protons, the binding of the nucleons is reduced; nuclides that are too unbalanced in this regard are unstable. As shown in Fig. 30–2, stable nuclei tend to have the same number of protons as neutrons $(N = Z)$ up to about $A = 30$. Beyond this, stable nuclei contain more neutrons than protons. This makes sense since, as Z increases, the electrical repulsion increases, so a greater number of neutrons—which exert only the attractive strong nuclear force—are required to maintain stability. For very large Z, no number of neutrons can overcome the greatly increased electric repulsion. Indeed, there are no completely stable nuclides above $Z = 82$.

What we mean by a *stable nucleus* is one that stays together indefinitely. What then is an *unstable nucleus*? It is one that comes apart; and this results in radioactive decay. Before we discuss the important subject of radioactivity (next Section), we note that there is a second type of nuclear force that is much weaker than the strong nuclear force. It is called the **weak nuclear force**, and we are aware of its existence only because it shows itself in certain types of radioactive decay. These two nuclear forces, the strong and the weak, together with the gravitational and electromagnetic forces, comprise the four fundamental types of force in nature.

FIGURE 30–2 Number of neutrons versus number of protons for stable nuclides, which are represented by dots. The straight line represents $N = Z$.

30–3 Radioactivity

Nuclear physics had its beginnings in 1896. In that year, Henri Becquerel (1852–1908) made an important discovery: in his studies of phosphorescence, he found that a certain mineral (which happened to contain uranium) would darken a photographic plate even when the plate was wrapped to exclude light. It was clear that the mineral emitted some new kind of radiation that, unlike X-rays (Section 25–11), occurred without any external stimulus. This new phenomenon eventually came to be called **radioactivity**.

Soon after Becquerel's discovery, Marie Curie (1867–1934) and her husband, Pierre Curie (1859–1906), isolated two previously unknown elements that were very highly radioactive (Fig. 30–3). These were named polonium and radium. Other radioactive elements were soon discovered as well. The radioactivity was found in every case to be unaffected by the strongest physical and chemical treatments, including strong heating or cooling or the action of strong chemicals. It was suspected that the source of radioactivity must be deep within the atom, coming from the nucleus. It became apparent that radioactivity is the result of the **disintegration** or **decay** of an unstable nucleus. Certain isotopes are not stable, and they decay with the emission of some type of radiation or "rays."

Many unstable isotopes occur in nature, and such radioactivity is called "natural radioactivity." Other unstable isotopes can be produced in the laboratory by nuclear reactions (Section 31–1); these are said to be produced "artificially" and to have "artificial radioactivity." Radioactive isotopes are sometimes referred to as **radioisotopes** or **radionuclides**.

Rutherford and others began studying the nature of the rays emitted in radioactivity about 1898. They classified the rays into three distinct types according to their penetrating power. One type of radiation could barely penetrate a piece of paper. The second type could pass through as much as 3 mm of aluminum. The third was extremely penetrating: it could pass through several centimeters of lead and still be detected on the other side. They named these three types of radiation alpha (α), beta (β), and gamma (γ), respectively, after the first three letters of the Greek alphabet.

FIGURE 30–3 Marie and Pierre Curie in their laboratory (about 1906) where radium was discovered.

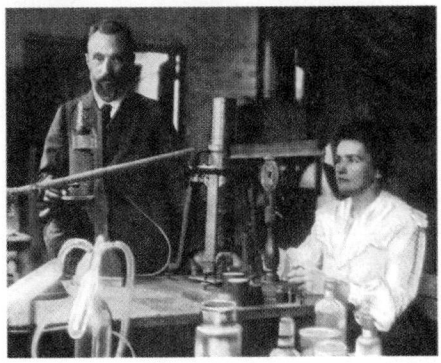

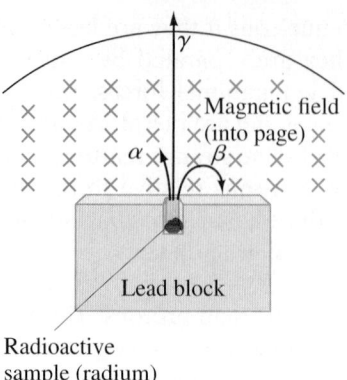

FIGURE 30–4 Alpha and beta rays are bent in opposite directions by a magnetic field, whereas gamma rays are not bent at all.

Each type of ray was found to have a different charge and hence is bent differently in a magnetic field, Fig. 30–4; α rays are positively charged, β rays are negatively charged, and γ rays are neutral. It was soon found that all three types of radiation consisted of familiar kinds of particles. Gamma rays are very high-energy *photons* whose energy is even higher than that of X-rays. Beta rays were found to be identical to *electrons* that orbit the nucleus, but they are created within the nucleus itself. Alpha rays (or α particles) are simply the nuclei of *helium* atoms, $^4_2\mathrm{He}$; that is, an α ray consists of two protons and two neutrons bound together.

We now discuss each of these three types of radioactivity, or decay, in more detail.

30–4 Alpha Decay

Experiments show that when nuclei decay, the number of nucleons (= mass number A) is conserved, as well as electric charge (= Ze). When a nucleus emits an α particle ($^4_2\mathrm{He}$), the remaining nucleus will be different from the original: it has lost two protons and two neutrons. Radium 226 ($^{226}_{88}\mathrm{Ra}$), for example, is an α emitter. It decays to a nucleus with $Z = 88 - 2 = 86$ and $A = 226 - 4 = 222$. The nucleus with $Z = 86$ is radon (Rn)—see Appendix B or the Periodic Table. Thus radium decays to radon with the emission of an α particle. This is written

$$^{226}_{88}\mathrm{Ra} \rightarrow {}^{222}_{86}\mathrm{Rn} + {}^4_2\mathrm{He}.$$

See Fig. 30–5.

FIGURE 30–5 Radioactive decay of radium to radon with emission of an alpha particle.

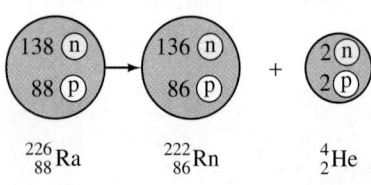

When α decay occurs, a different element is formed. The **daughter** nucleus ($^{222}_{86}\mathrm{Rn}$ in this case) is different from the **parent** nucleus ($^{226}_{88}\mathrm{Ra}$ in this case). This changing of one element into another is called **transmutation** of the elements.

Alpha decay can be written in general as

$$^A_Z\mathrm{N} \rightarrow {}^{A-4}_{Z-2}\mathrm{N}' + {}^4_2\mathrm{He} \qquad\qquad [\alpha \text{ decay}]$$

where N is the parent, N' the daughter, and Z and A are the atomic number and atomic mass number, respectively, of the parent.

> **EXERCISE C** $^{154}_{66}\mathrm{Dy}$ decays by α emission to what element? (*a*) Pb, (*b*) Gd, (*c*) Sm, (*d*) Er, (*e*) Yb.

Alpha decay occurs because the strong nuclear force is unable to hold very large nuclei together. The nuclear force is a short-range force: it acts only between neighboring nucleons. But the electric force acts all the way across a large nucleus. For very large nuclei, the large Z means the repulsive electric force becomes so large (Coulomb's law) that the strong nuclear force is unable to hold the nucleus together.

We can express the instability of the parent nucleus in terms of energy (or mass): the mass of the parent nucleus is greater than the mass of the daughter nucleus plus the mass of the α particle. The mass difference appears as kinetic energy, which is carried away by the α particle and the recoiling daughter nucleus. The total energy released is called the **disintegration energy**, Q, or the **Q-value** of the decay. From conservation of energy,

$$M_P c^2 = M_D c^2 + m_\alpha c^2 + Q,$$

where Q equals the kinetic energy of the daughter and α particle, and M_P, M_D, and m_α are the masses of the parent, daughter, and α particle, respectively. Thus

$$Q = M_P c^2 - (M_D + m_\alpha)c^2. \tag{30–2}$$

If the parent had *less* mass than the daughter plus the α particle (so $Q < 0$), the decay would violate conservation of energy. Such decays have never been observed, another confirmation of this great conservation law.

EXAMPLE 30–6 **Uranium decay energy release.** Calculate the disintegration energy when $^{232}_{92}\text{U}$ (mass = 232.037156 u) decays to $^{228}_{90}\text{Th}$ (228.028741 u) with the emission of an α particle. (As always, masses given are for neutral atoms.)

APPROACH We use conservation of energy as expressed in Eq. 30–2. $^{232}_{92}\text{U}$ is the parent, $^{228}_{90}\text{Th}$ is the daughter.

SOLUTION Since the mass of the ^4_2He is 4.002603 u (Appendix B), the total mass in the final state $(m_{Th} + m_{He})$ is

$$228.028741 \text{ u} + 4.002603 \text{ u} = 232.031344 \text{ u}.$$

The mass lost when the $^{232}_{92}\text{U}$ decays $(m_U - m_{Th} - m_{He})$ is

$$232.037156 \text{ u} - 232.031344 \text{ u} = 0.005812 \text{ u}.$$

Because 1 u = 931.5 MeV, the energy Q released is

$$Q = (0.005812 \text{ u})(931.5 \text{ MeV/u}) = 5.4 \text{ MeV}$$

and this energy appears as kinetic energy of the α particle and the daughter nucleus.

Additional Example

EXAMPLE 30–7 **Kinetic energy of the α in $^{232}_{92}\text{U}$ decay.** For the $^{232}_{92}\text{U}$ decay of Example 30–6, how much of the 5.4-MeV disintegration energy will be carried off by the α particle?

APPROACH In any reaction, momentum must be conserved as well as energy.

SOLUTION Before disintegration, the nucleus can be assumed to be at rest, so the total momentum was zero. After disintegration, the total vector momentum must still be zero so the magnitude of the α particle's momentum must equal the magnitude of the daughter's momentum (Fig. 30–6):

$$m_\alpha v_\alpha = m_D v_D.$$

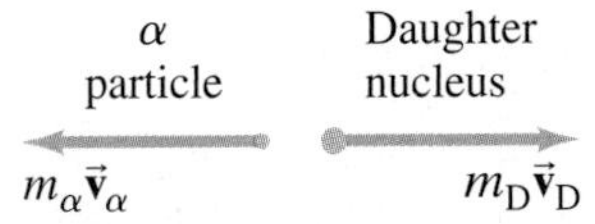

Thus $v_\alpha = m_D v_D / m_\alpha$ and the α's kinetic energy is

$$\text{KE}_\alpha = \tfrac{1}{2} m_\alpha v_\alpha^2 = \tfrac{1}{2} m_\alpha \left(\frac{m_D v_D}{m_\alpha}\right)^2 = \tfrac{1}{2} m_D v_D^2 \left(\frac{m_D}{m_\alpha}\right) = \left(\frac{m_D}{m_\alpha}\right)\text{KE}_D$$

$$= \left(\frac{228.028741 \text{ u}}{4.002603 \text{ u}}\right)\text{KE}_D = 57 \text{ KE}_D.$$

The total disintegration energy is $Q = \text{KE}_\alpha + \text{KE}_D = 57 \text{ KE}_D + \text{KE}_D = 58 \text{ KE}_D$. Hence

$$\text{KE}_\alpha = 57 \text{ KE}_D = \frac{57}{58}Q = 5.3 \text{ MeV}.$$

The lighter α particle carries off (57/58) or 98% of the total kinetic energy. The total energy released is 5.4 MeV, so the daughter nucleus, which recoils in the opposite direction, carries off only 0.1 MeV.

α particle   Daughter nucleus

$m_\alpha \vec{v}_\alpha$ $m_D \vec{v}_D$

FIGURE 30–6 Momentum conservation in Example 30–7.

Why α Particles?

Why, you may wonder, do nuclei emit this combination of four nucleons called an α particle? Why not just four separate nucleons, or even one? The answer is that the α particle is very strongly bound, so that its mass is significantly less than that of four separate nucleons. That helps the final state in α decay to have less total mass, thus allowing certain nuclides to decay which could not decay to, say, 2 protons plus 2 neutrons. For example, $^{232}_{92}U$ could not decay to $2p + 2n$ because the masses of the daughter $^{228}_{90}Th$ plus four separate nucleons is $228.028741 \, u + 2(1.007825 \, u) + 2(1.008665 \, u) = 232.061721 \, u$, which is greater than the mass of the $^{232}_{92}U$ parent ($232.037156 \, u$). Such a decay would violate the conservation of energy. Indeed, we have never seen $^{232}_{92}U \rightarrow {}^{228}_{90}Th + 2p + 2n$. Similarly, it is almost always true that the emission of a single nucleon is energetically not possible; see Example 30–5.

Smoke Detectors—An Application

One widespread application of nuclear physics is present in nearly every home in the form of an ordinary **smoke detector**. One type of smoke detector contains about 0.2 mg of the radioactive americium isotope, $^{241}_{95}Am$, in the form of AmO_2. The radiation continually ionizes the nitrogen and oxygen molecules in the air space between two oppositely charged plates. The resulting conductivity allows a small steady electric current. If smoke enters, the radiation is absorbed by the smoke particles rather than by the air molecules, thus reducing the current. The current drop is detected by the device's electronics and sets off the alarm. The radiation dose that escapes from an intact americium smoke detector is much less than the natural radioactive background, and so can be considered relatively harmless. There is no question that smoke detectors save lives and reduce property damage.

30–5 Beta Decay

β^- Decay

Transmutation of elements also occurs when a nucleus decays by β decay—that is, with the emission of an electron or β^- particle. The nucleus $^{14}_6C$, for example, emits an electron when it decays:

$$^{14}_6C \rightarrow {}^{14}_7N + e^- + \text{neutrino},$$

where e^- is the symbol for the electron. The particle known as the neutrino has charge $q = 0$ and a very small mass, long thought to be zero. It was not initially detected and was only later hypothesized to exist, as we shall discuss later in this Section. No nucleons are lost when an electron is emitted, and the total number of nucleons, A, is the same in the daughter nucleus as in the parent. But because an electron has been emitted from the nucleus itself, the charge on the daughter nucleus is $+1e$ greater than that on the parent. The parent nucleus in the decay written above had $Z = +6$, so from charge conservation the nucleus remaining behind must have a charge of $+7e$. So the daughter nucleus has $Z = 7$, which is nitrogen.

It must be carefully noted that the electron emitted in β decay is *not* an orbital electron. Instead, the electron is created *within the nucleus itself*. What happens is that one of the neutrons changes to a proton and in the process (to conserve charge) emits an electron. Indeed, free neutrons actually do decay in this fashion:

$$n \rightarrow p + e^- + \text{neutrino}.$$

To remind us of their origin in the nucleus, the electrons emitted in β decay are often referred to as "β particles." They are, nonetheless, indistinguishable from orbital electrons.

EXAMPLE 30–8 **Energy release in $^{14}_{6}$C decay.** How much energy is released when $^{14}_{6}$C decays to $^{14}_{7}$N by β emission?

APPROACH We find the mass difference before and after decay, Δm. The energy released is $E = (\Delta m)c^2$. The masses given in Appendix B are those of the neutral atom, and we have to keep track of the electrons involved. Assume the parent nucleus has six orbiting electrons so it is neutral; its mass is 14.003242 u. The daughter in this decay, $^{14}_{7}$N, is not neutral because it has the same six orbital electrons circling it but the nucleus has a charge of $+7e$. However, the mass of this daughter with its six electrons, plus the mass of the emitted electron (which makes a total of seven electrons), is just the mass of a neutral nitrogen atom.

SOLUTION The total mass in the final state is

$$\left(\text{mass of } {}^{14}_{7}\text{N nucleus} + 6 \text{ electrons}\right) + (\text{mass of 1 electron}),$$

and this is equal to

$$\text{mass of neutral } {}^{14}_{7}\text{N (includes 7 electrons)},$$

which from Appendix B is a mass of 14.003074 u. So the mass difference is 14.003242 u − 14.003074 u = 0.000168 u, which is equivalent to an energy change $\Delta m\, c^2 = (0.000168\ \text{u})(931.5\ \text{MeV/u}) = 0.156\ \text{MeV}$ or 156 keV.

NOTE The neutrino doesn't contribute to either the mass or charge balance because it has $q = 0$ and $m \approx 0$.

CAUTION

Be careful with atomic and electron masses in β decay

According to Example 30–8, we would expect the emitted electron to have a kinetic energy of 156 keV. (The daughter nucleus, because its mass is very much larger than that of the electron, recoils with very low velocity and hence gets very little of the kinetic energy—see Example 30–7.) Indeed, very careful measurements indicate that a few emitted β particles do have kinetic energy close to this calculated value. But the vast majority of emitted electrons have somewhat less energy. In fact, the energy of the emitted electron can be anywhere from zero up to the maximum value as calculated above. This range of electron kinetic energy was found for any β decay. It was as if the law of conservation of energy was being violated, and Bohr actually considered this possibility. Careful experiments indicated that linear momentum and angular momentum also did not seem to be conserved. Physicists were troubled at the prospect of giving up these laws, which had worked so well in all previous situations.

In 1930, Wolfgang Pauli proposed an alternate solution: perhaps a new particle that was very difficult to detect was emitted during β decay in addition to the electron. This hypothesized particle could be carrying off the energy, momentum, and angular momentum required to maintain the conservation laws. This new particle was named the **neutrino**—meaning "little neutral one"—by the great Italian physicist Enrico Fermi (1901–1954; Fig. 30–7), who in 1934 worked out a detailed theory of β decay. (It was Fermi who, in this theory, postulated the existence of the fourth force in nature which we call the *weak nuclear force*.) The neutrino has zero charge, spin of $\frac{1}{2}\hbar$, and was long thought to have zero mass, although today we are quite sure that it has a very tiny mass $\left(< 0.14\ \text{eV}/c^2\right)$. If its mass were zero, it would be much like a photon in that it is neutral and would travel at the speed of light. But the neutrino is very difficult to detect. In 1956, complex experiments produced further evidence for the existence of the neutrino; but by then, most physicists had already accepted its existence.

The symbol for the neutrino is the Greek letter nu (ν). The correct way of writing the decay of $^{14}_{6}$C is then

$$^{14}_{6}\text{C} \rightarrow {}^{14}_{7}\text{N} + \text{e}^- + \bar{\nu}.$$

The bar $(^-)$ over the neutrino symbol is to indicate that it is an "antineutrino." (Why this is called an antineutrino rather than simply a neutrino is discussed in Chapter 32.)

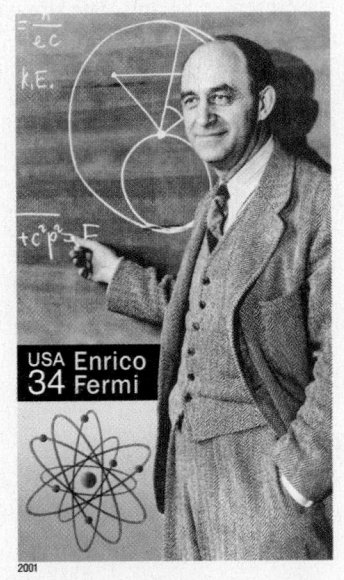

FIGURE 30–7 Enrico Fermi, as portrayed on a US postage stamp. Fermi contributed significantly to both theoretical and experimental physics, a feat almost unique in modern times: statistical theory of identical particles that obey the exclusion principle (= fermions); theory of the weak interaction and β decay; neutron physics; induced radioactivity and new elements; first nuclear reactor; first resonance of particle physics; led and inspired a vast amount of other nuclear research.

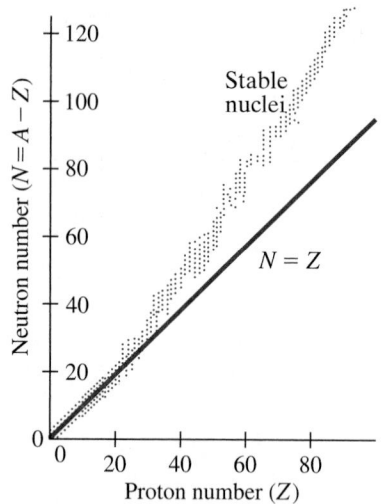

FIGURE 30–2 (Repeated.) Number of neutrons versus number of protons for stable nuclides, which are represented by dots. The straight line represents $N = Z$.

In β decay, it is the weak nuclear force that plays the crucial role. The neutrino is unique in that it interacts with matter only via the weak force, which is why it is so hard to detect.

β^+ Decay

Many isotopes decay by electron emission. They are always isotopes that have too many neutrons compared to the number of protons. That is, they are isotopes that lie above the stable isotopes plotted in Fig. 30–2. But what about unstable isotopes that have too few neutrons compared to their number of protons—those that fall below the stable isotopes of Fig. 30–2? These, it turns out, decay by emitting a **positron** instead of an electron. A positron (sometimes called an e^+ or β^+ particle) has the same mass as the electron, but it has a positive charge of $+1e$. Because it is so like an electron, except for its charge, the positron is called the **antiparticle**[†] to the electron. An example of a β^+ decay is that of $^{19}_{10}Ne$:

$$^{19}_{10}Ne \rightarrow {}^{19}_{9}F + e^+ + \nu,$$

where e^+ stands for a positron. Note that the ν emitted here is a neutrino, whereas that emitted in β^- decay is called an antineutrino. Thus an antielectron (= positron) is emitted with a neutrino, whereas an antineutrino is emitted with an electron; this gives a certain balance as discussed in Chapter 32.

We can write β^- and β^+ decay, in general, as follows:

$$^{A}_{Z}N \rightarrow {}^{A}_{Z+1}N' + e^- + \bar{\nu} \qquad [\beta^- \text{ decay}]$$

$$^{A}_{Z}N \rightarrow {}^{A}_{Z-1}N' + e^+ + \nu, \qquad [\beta^+ \text{ decay}]$$

where N is the parent nucleus and N' is the daughter.

Electron Capture

Besides β^- and β^+ emission, there is a third related process. This is **electron capture** (abbreviated EC in Appendix B) and occurs when a nucleus absorbs one of its orbiting electrons. An example is $^{7}_{4}Be$, which as a result becomes $^{7}_{3}Li$. The process is written

$$^{7}_{4}Be + e^- \rightarrow {}^{7}_{3}Li + \nu,$$

or, in general,

$$^{A}_{Z}N + e^- \rightarrow {}^{A}_{Z-1}N' + \nu. \qquad [\text{electron capture}]$$

Usually it is an electron in the innermost (K) shell that is captured, in which case the process is called **K-capture**. The electron disappears in the process, and a proton in the nucleus becomes a neutron; a neutrino is emitted as a result. This process is inferred experimentally by detection of emitted X-rays (due to other electrons jumping down to fill the state of the captured e^-).

30–6 Gamma Decay

Gamma rays are photons having very high energy. They have their origin in the decay of a nucleus, much like emission of photons by excited atoms. Like an atom, a nucleus itself can be in an excited state. When it jumps down to a lower energy state, or to the ground state, it emits a photon which we call a γ ray. The possible states of a nucleus are much farther apart in energy than those of an atom: on the order of keV or MeV, as compared to a few eV for electrons in an atom. Hence, the emitted photons have energies that can range from a few keV to several MeV. For a given decay, the γ ray always has the same energy. Since a γ ray carries no charge, there is no change in the element as a result of a γ decay.

How does a nucleus get into an excited state? It may occur because of a violent collision with another particle. More commonly, the nucleus remaining after a previous radioactive decay may be in an excited state. A typical example is shown

[†]Discussed in Chapter 32. Briefly, an antiparticle has the same mass as its corresponding particle, but opposite charge. A particle and its antiparticle can quickly annihilate each other, releasing energy in the form of two γ rays: $e^+ + e^- \rightarrow 2\gamma$.

in the energy-level diagram of Fig. 30–8. $^{12}_{5}B$ can decay by β decay directly to the ground state of $^{12}_{6}C$; or it can go by β decay to an excited state of $^{12}_{6}C$, written $^{12}_{6}C^*$, which itself decays by emission of a 4.4-MeV γ ray to the ground state of $^{12}_{6}C$.

We can write γ decay as

$$^{A}_{Z}N^* \rightarrow {^{A}_{Z}}N + \gamma, \qquad \text{[} \gamma \text{ decay]}$$

where the asterisk means "excited state" of that nucleus.

What, you may wonder, is the difference between a γ ray and an X-ray? They both are electromagnetic radiation (photons) and, though γ rays usually have higher energy than X-rays, their range of energies overlap to some extent. The difference is not intrinsic. We use the term X-ray if the photon is produced by an electron–atom interaction, and γ ray if the photon is produced in a nuclear process.

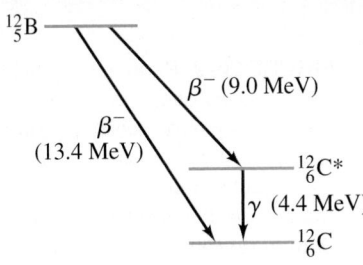

FIGURE 30–8 Energy-level diagram showing how $^{12}_{5}B$ can decay to the ground state of $^{12}_{6}C$ by β decay (total energy released = 13.4 MeV), or can instead β decay to an excited state of $^{12}_{6}C$ (indicated by *), which subsequently decays to its ground state by emitting a 4.4-MeV γ ray.

*Isomers; Internal Conversion

In some cases, a nucleus may remain in an excited state for some time before it emits a γ ray. The nucleus is then said to be in a **metastable state** and is called an **isomer**.

An excited nucleus can sometimes return to the ground state by another process known as **internal conversion** with no γ ray emitted. In this process, the excited nucleus interacts with one of the orbital electrons and ejects this electron from the atom with the same kinetic energy (minus the binding energy of the electron) that an emitted γ ray would have had.

30–7 Conservation of Nucleon Number and Other Conservation Laws

In all three types of radioactive decay, the classical conservation laws hold. Energy, linear momentum, angular momentum, and electric charge are all conserved. These quantities are the same before the decay as after. But a new conservation law is also revealed, the **law of conservation of nucleon number**. According to this law, the total number of nucleons (A) remains constant in any process, although one type can change into the other type (protons into neutrons or vice versa). This law holds in all three types of decay. [In Chapter 32 we will generalize this and call it conservation of baryon number.]

Table 30–2 gives a summary of α, β, and γ decay.

TABLE 30–2 The Three Types of Radioactive Decay

α decay:
$$^{A}_{Z}N \rightarrow {^{A-4}_{Z-2}}N' + {^{4}_{2}}He$$

β decay:
$$^{A}_{Z}N \rightarrow {^{A}_{Z+1}}N' + e^- + \bar{\nu}$$
$$^{A}_{Z}N \rightarrow {^{A}_{Z-1}}N' + e^+ + \nu$$
$$^{A}_{Z}N + e^- \rightarrow {^{A}_{Z-1}}N' + \nu \text{ [EC]}^{\dagger}$$

γ decay:
$$^{A}_{Z}N^* \rightarrow {^{A}_{Z}}N + \gamma$$

†Electron capture.
*Indicates the excited state of a nucleus.

30–8 Half-Life and Rate of Decay

A macroscopic sample of any radioactive isotope consists of a vast number of radioactive nuclei. These nuclei do not all decay at one time. Rather, they decay one by one over a period of time. This is a random process: we can not predict exactly when a given nucleus will decay. But we can determine, on a probabilistic basis, approximately how many nuclei in a sample will decay over a given time period, by assuming that each nucleus has the same probability of decaying in each second that it exists.

The number of decays ΔN that occur in a very short time interval Δt is then proportional to Δt and to the total number N of radioactive (parent) nuclei present:

$$\Delta N = -\lambda N \, \Delta t \qquad \textbf{(30–3a)}$$

where the minus sign means N is decreasing. We rewrite this to get the **rate of decay** (number of decays per second):

$$\frac{\Delta N}{\Delta t} = -\lambda N. \qquad \textbf{(30–3b)}$$

In these equations, λ is a measurable constant called the **decay constant**, which is different for different isotopes. The greater λ is, the greater the rate of decay, $\Delta N / \Delta t$, and the more "radioactive" that isotope is said to be.

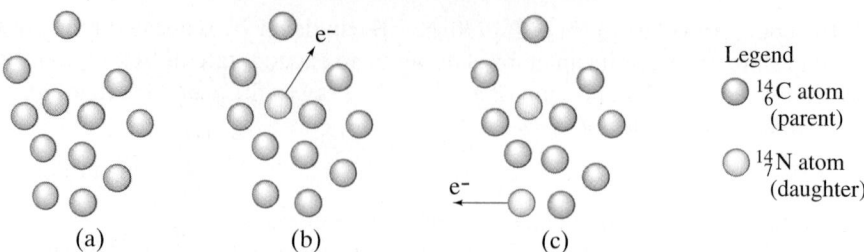

FIGURE 30–9 Radioactive nuclei decay one by one. Hence, the number of parent nuclei in a sample is continually decreasing. When a $^{14}_{6}C$ nucleus emits an electron (b), the nucleus becomes a $^{14}_{7}N$ nucleus. Another decays in (c).

Legend
● $^{14}_{6}C$ atom (parent)
○ $^{14}_{7}N$ atom (daughter)

(a)   (b)   (c)

The number of decays that occur in the short time interval Δt is designated ΔN because each decay that occurs corresponds to a decrease by one in the number N of parent nuclei present. That is, radioactive decay is a "one-shot" process, Fig. 30–9. Once a particular parent nucleus decays into its daughter, it cannot do it again.

Exponential Decay

Equation 30–3a or b can be solved for N (using calculus) and the result is

$$N = N_0 e^{-\lambda t}, \qquad (30\text{–}4)$$

where N_0 is the number of parent nuclei present at any chosen time $t = 0$, and N is the number remaining after a time t. The symbol e is the natural exponential (encountered earlier in Sections 19–6 and 21–12) whose value is $e = 2.718 \cdots$. Thus the number of parent nuclei in a sample decreases exponentially in time. This is shown in Fig. 30–10a for the decay of $^{14}_{6}C$. Equation 30–4 is called the **radioactive decay law**.

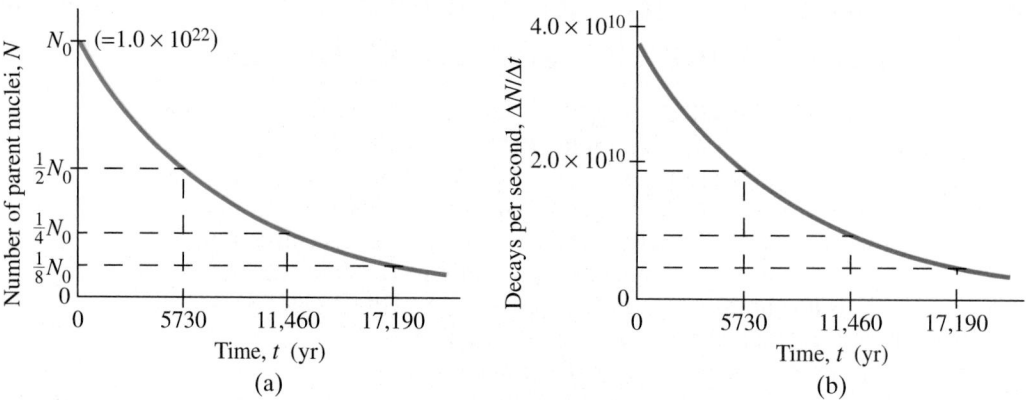

FIGURE 30–10 (a) The number N of parent nuclei in a given sample of $^{14}_{6}C$ decreases exponentially. We assume a sample that has $N_0 = 1.00 \times 10^{22}$ nuclei. (b) The number of decays per second also decreases exponentially. The half-life of $^{14}_{6}C$ is 5730 yr, which means that the number of parent nuclei, N, and the rate of decay, $\Delta N/\Delta t$, decrease by half every 5730 yr.

The number of decays per second, or decay rate R, is the magnitude of $\Delta N/\Delta t$, and is also called the **activity** of the sample. The magnitude (always positive) of a quantity is often indicated using vertical lines. The magnitude of $\Delta N/\Delta t$ is written $|\Delta N/\Delta t|$ and it is proportional to N (see Eq. 30–3b). So it too decreases exponentially in time at the same rate (Fig. 30–10b). The activity of a pure sample at time t is

$$R = \left| \frac{\Delta N}{\Delta t} \right| = R_0 e^{-\lambda t}, \qquad (30\text{–}5)$$

where $R_0 = |\Delta N/\Delta t|_0$ is the activity at $t = 0$.

Equation 30–5 is also referred to as the **radioactive decay law** (as is Eq. 30–4).

Half-Life

The rate of decay of any isotope is often specified by giving its "half-life" rather than the decay constant λ. The **half-life** of an isotope is defined as the time it takes for half the original amount of parent isotope in a given sample to decay.

For example, the half-life of $^{14}_{6}C$ is about 5730 years. If at some time a piece of petrified wood contains, say, 1.00×10^{22} nuclei of $^{14}_{6}C$, then 5730 years later it will contain half as many, 0.50×10^{22} nuclei. After another 5730 years it will contain 0.25×10^{22} nuclei, and so on. This is shown in Fig. 30–10a. Since the rate of decay $\Delta N/\Delta t$ is proportional to N, it, too, decreases by a factor of 2 every half-life (Fig. 30–10b).

The half-lives of known radioactive isotopes vary from very short ($\approx 10^{-22}$ s) to more than 10^{23} yr ($> 10^{30}$ s). The half-lives of many isotopes are given in Appendix B. It should be clear that the half-life (which we designate $T_{\frac{1}{2}}$) bears an inverse relationship to the decay constant. The longer the half-life of an isotope, the more slowly it decays, and hence λ is smaller. Conversely, very active isotopes (large λ) have very short half-lives. The precise relationship between half-life and decay constant is

$$T_{\frac{1}{2}} = \frac{\ln 2}{\lambda} = \frac{0.693}{\lambda}. \qquad (30\text{–}6)$$

We derive this in the next (optional) subsection.

EXERCISE D The half-life of $^{22}_{11}Na$ is 2.6 years. How much $^{22}_{11}Na$ will be left of a pure 1.0-μg sample after 7.8 yr? (a) None. (b) $\frac{1}{8}$ μg. (c) $\frac{1}{4}$ μg. (d) $\frac{1}{2}$ μg. (e) 0.693 μg.

EXERCISE E Return to the Chapter-Opening Question, page 857, and answer it again now. Try to explain why you may have answered differently the first time.

*Deriving the Half-Life Formula

We can derive Eq. 30–6 starting from Eq. 30–4 by setting $N = N_0/2$ at $t = T_{\frac{1}{2}}$:

$$\frac{N_0}{2} = N_0 e^{-\lambda T_{\frac{1}{2}}}$$

so

$$\frac{1}{2} = e^{-\lambda T_{\frac{1}{2}}}$$

and

$$e^{\lambda T_{\frac{1}{2}}} = 2.$$

We take natural logs of both sides ("ln" and "e" are inverse operations, meaning $\ln(e^x) = x$) and find

$$\ln\left(e^{\lambda T_{\frac{1}{2}}}\right) = \ln 2,$$

so

$$\lambda T_{\frac{1}{2}} = \ln 2 = 0.693$$

and

$$T_{\frac{1}{2}} = \frac{\ln 2}{\lambda} = \frac{0.693}{\lambda},$$

which is Eq. 30–6.

*Mean Life

Sometimes the **mean life** τ of an isotope is quoted, which is defined as $\tau = 1/\lambda$. Then Eq. 30–4 can be written $N = N_0 e^{-t/\tau}$, just as for RC and LR circuits (Chapters 19 and 21 where τ was called the time constant). The mean life of an isotope is then given by (see also Eq. 30–6)

$$\tau = \frac{1}{\lambda} = \frac{T_{\frac{1}{2}}}{0.693}. \qquad \text{[mean life]} \quad (30\text{–}7)$$

The mean life and half-life differ by a factor of 0.693, so confusing them can cause serious error (and has). The radioactive decay law, Eq. 30–5, can then be written as $R = R_0 e^{-t/\tau}$.

CAUTION

Do not confuse half-life and mean life

30–9 Calculations Involving Decay Rates and Half-Life

Let us now consider Examples of what we can determine about a sample of radioactive material if we know the half-life.

EXAMPLE 30–9 **Sample activity.** The isotope $^{14}_{6}C$ has a half-life of 5730 yr. If a sample contains 1.00×10^{22} carbon-14 nuclei, what is the activity of the sample?

APPROACH We first use the half-life to find the decay constant (Eq. 30–6), and use that to find the activity, Eq. 30–3b. The number of seconds in a year is $(60)(60)(24)(365\frac{1}{4}) = 3.156 \times 10^7\,\text{s}$.

SOLUTION The decay constant λ from Eq. 30–6 is

$$\lambda = \frac{0.693}{T_{\frac{1}{2}}} = \frac{0.693}{(5730\,\text{yr})(3.156 \times 10^7\,\text{s/yr})} = 3.83 \times 10^{-12}\,\text{s}^{-1}.$$

From Eqs. 30–3b and 30–5, the activity or rate of decay is

$$R = \left|\frac{\Delta N}{\Delta t}\right| = \lambda N = (3.83 \times 10^{-12}\,\text{s}^{-1})(1.00 \times 10^{22}) = 3.83 \times 10^{10}\,\text{decays/s}.$$

Notice that the graph of Fig. 30–10b starts at this value, corresponding to the original value of $N = 1.0 \times 10^{22}$ nuclei in Fig. 30–10a.

NOTE The unit "decays/s" is often written simply as s^{-1} since "decays" is not a unit but refers only to the number. This simple unit of activity is called the becquerel: $1\,\text{Bq} = 1\,\text{decay/s}$, as discussed in Chapter 31.

CONCEPTUAL EXAMPLE 30–10 **Safety: Activity versus half-life.** One might think that a short half-life material is safer than a long half-life material because it will not last as long. Is that true?

RESPONSE No. A shorter half-life means the activity is higher and thus more "radioactive" and can cause more biological damage. In contrast, a longer half-life for the same sample size N means a lower activity but we have to worry about it for longer and find safe storage until it reaches a safe (low) level of activity.

EXAMPLE 30–11 **A sample of radioactive $^{13}_{7}N$.** A laboratory has 1.49 μg of pure $^{13}_{7}N$, which has a half-life of 10.0 min (600 s). (a) How many nuclei are present initially? (b) What is the rate of decay (activity) initially? (c) What is the activity after 1.00 h? (d) After approximately how long will the activity drop to less than one per second $(= 1\,\text{s}^{-1})$?

APPROACH We use the definition of the mole and Avogadro's number (Sections 13–6 and 13–8) to find (a) the number of nuclei. For (b) we get λ from the given half-life and use Eq. 30–3b for the rate of decay. For (c) and (d) we use Eq. 30–5.

SOLUTION (a) The atomic mass is 13.0, so 13.0 g will contain 6.02×10^{23} nuclei (Avogadro's number). We have only 1.49×10^{-6} g, so the number of nuclei N_0 that we have initially is given by the ratio

$$\frac{N_0}{6.02 \times 10^{23}} = \frac{1.49 \times 10^{-6}\,\text{g}}{13.0\,\text{g}}.$$

Solving for N_0, we find $N_0 = 6.90 \times 10^{16}$ nuclei.

(b) From Eq. 30–6,

$$\lambda = 0.693/T_{\frac{1}{2}} = (0.693)/(600\,\text{s}) = 1.155 \times 10^{-3}\,\text{s}^{-1}.$$

Then, at $t = 0$ (see Eqs. 30–3b and 30–5)

$$R_0 = \left|\frac{\Delta N}{\Delta t}\right|_0 = \lambda N_0 = (1.155 \times 10^{-3}\,\text{s}^{-1})(6.90 \times 10^{16}) = 7.97 \times 10^{13}\,\text{decays/s}.$$

(c) After $1.00\,h = 3600\,s$, the magnitude of the activity will be (Eq. 30–5)
$$R \;=\; R_0\,e^{-\lambda t} \;=\; \left(7.97 \times 10^{13}\ s^{-1}\right)e^{-(1.155\times 10^{-3}\ s^{-1})(3600\,s)} \;=\; 1.25 \times 10^{12}\ s^{-1}.$$

(d) We want to determine the time t when $R = 1.00\,s^{-1}$. From Eq. 30–5, we have
$$e^{-\lambda t} \;=\; \frac{R}{R_0} \;=\; \frac{1.00\ s^{-1}}{7.97 \times 10^{13}\ s^{-1}} \;=\; 1.25 \times 10^{-14}.$$

We take the natural log (ln) of both sides $\left(\ln e^{-\lambda t} = -\lambda t\right)$ and divide by λ to find
$$t \;=\; -\,\frac{\ln\!\left(1.25 \times 10^{-14}\right)}{\lambda} \;=\; 2.77 \times 10^{4}\ s \;=\; 7.70\,h.$$

Easy Alternate Solution to (c) $1.00\,h = 60.0$ minutes is 6 half-lives, so the activity will decrease to $\left(\tfrac{1}{2}\right)\left(\tfrac{1}{2}\right)\left(\tfrac{1}{2}\right)\left(\tfrac{1}{2}\right)\left(\tfrac{1}{2}\right)\left(\tfrac{1}{2}\right) = \left(\tfrac{1}{2}\right)^{6} = \tfrac{1}{64}$ of its original value, or $(7.97 \times 10^{13})/(64) = 1.25 \times 10^{12}$ per second.

30–10 Decay Series

It is often the case that one radioactive isotope decays to another isotope that is also radioactive. Sometimes this daughter decays to yet a third isotope which also is radioactive. Such successive decays are said to form a **decay series**. An important example is illustrated in Fig. 30–11. As can be seen, $^{238}_{92}\text{U}$ decays by α emission to $^{234}_{90}\text{Th}$, which in turn decays by β decay to $^{234}_{91}\text{Pa}$. The series continues as shown, with several possible branches near the bottom, ending at the stable lead isotope, $^{206}_{82}\text{Pb}$. The two last decays can be

$$^{206}_{81}\text{Tl} \;\rightarrow\; ^{206}_{82}\text{Pb} + e^{-} + \bar{\nu}, \qquad \left(T_{\frac{1}{2}} = 4.2\,\text{min}\right)$$

or

$$^{210}_{84}\text{Po} \;\rightarrow\; ^{206}_{82}\text{Pb} + \alpha. \qquad \left(T_{\frac{1}{2}} = 138\,\text{days}\right)$$

Other radioactive series also exist.

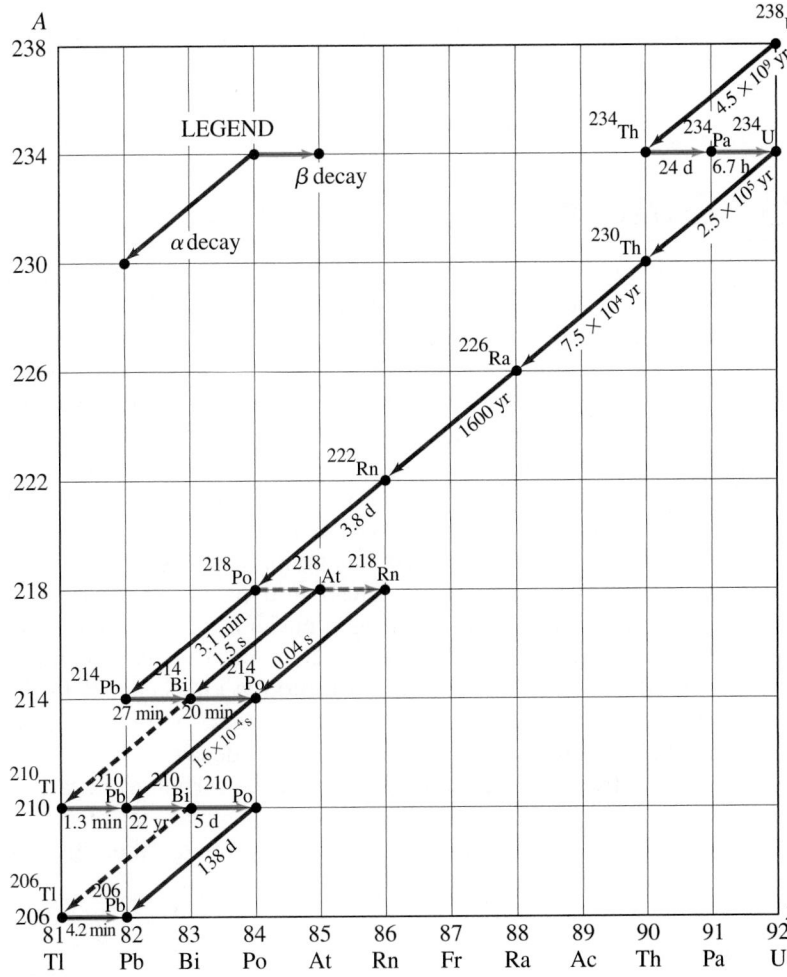

FIGURE 30–11 Decay series beginning with $^{238}_{92}\text{U}$. Nuclei in the series are specified by a dot representing A and Z values. Half-lives are given in seconds (s), minutes (min), hours (h), days (d), or years (yr). Note that a horizontal arrow represents β decay (A does not change), whereas a diagonal line represents α decay (A changes by 4, Z changes by 2). For the four nuclides shown that can decay by both α and β decay, the more prominent decay (in these four cases, >99.9%) is shown as a solid arrow and the less common decay (<0.1%) as a dashed arrow.

Because of such decay series, certain radioactive elements are found in nature that otherwise would not be. When the solar system (including Earth) was formed about 5 billion years ago, it is believed that nearly all nuclides were present, having been formed (by fusion and neutron capture, Sections 31–3 and 33–2) in a nearby supernova explosion (Section 33–2). Many isotopes with short half-lives decayed quickly and no longer are detected in nature today. But long-lived isotopes, such as $^{238}_{92}U$ with a half-life of 4.5×10^9 yr, still do exist in nature today. Indeed, about half of the original $^{238}_{92}U$ still remains. We might expect, however, that radium $\left(^{226}_{88}Ra\right)$, with a half-life of 1600 yr, would have disappeared from the Earth long ago. Indeed, the original $^{226}_{88}Ra$ nuclei must by now have all decayed. However, because $^{238}_{92}U$ decays (in several steps, Fig. 30–11) to $^{226}_{88}Ra$, the supply of $^{226}_{88}Ra$ is continually replenished, which is why it is still found on Earth today. The same can be said for many other radioactive nuclides.

CONCEPTUAL EXAMPLE 30–12 **Decay chain.** In the decay chain of Fig. 30–11, if we look at the decay of $^{234}_{92}U$, we see four successive nuclides with half-lives of 250,000 yr, 75,000 yr, 1600 yr, and a little under 4 days. Each decay in the chain has an alpha particle of a characteristic energy, and so we can monitor the radioactive decay rate of each nuclide. Given a sample that was pure $^{234}_{92}U$ a million years ago, which alpha decay would you expect to have the highest activity rate in the sample?

RESPONSE The first instinct is to say that the process with the shortest half-life would show the highest activity. Surprisingly, perhaps, the activities of the four nuclides in this sample are all the same. The reason is that in each case the decay of the parent acts as a bottleneck to the decay of the daughter. Compared to the 1600-yr half-life of $^{226}_{88}Ra$, for example, its daughter $^{222}_{86}Rn$ decays almost immediately, but it cannot decay until it is made. (This is like an automobile assembly line: if worker A takes 20 minutes to do a task and then worker B takes only 1 minute to do the next task, worker B still does only one car every 20 minutes.)

30–11 Radioactive Dating

Radioactive decay has many interesting applications. One is the technique of *radioactive dating* by which the age of ancient materials can be determined.

PHYSICS APPLIED

Carbon-14 dating

The age of any object made from once-living matter, such as wood, can be determined using the natural radioactivity of $^{14}_6C$. All living plants absorb carbon dioxide (CO_2) from the air and use it to synthesize organic molecules. The vast majority of these carbon atoms are $^{12}_6C$, but a small fraction, about 1.3×10^{-12}, is the radioactive isotope $^{14}_6C$. The ratio of $^{14}_6C$ to $^{12}_6C$ in the atmosphere has remained roughly constant over many thousands of years, in spite of the fact that $^{14}_6C$ decays with a half-life of about 5730 yr. This is because energetic nuclei in the cosmic radiation, which impinges on the Earth from outer space, strike nuclei of atoms in the atmosphere and break those nuclei into pieces, releasing free neutrons. Those neutrons can collide with nitrogen nuclei in the atmosphere to produce the nuclear transformation $n + {}^{14}_7N \rightarrow {}^{14}_6C + p$. That is, a neutron strikes and is absorbed by a $^{14}_7N$ nucleus, and a proton is knocked out in the process. The remaining nucleus is $^{14}_6C$. This continual production of $^{14}_6C$ in the atmosphere roughly balances the loss of $^{14}_6C$ by radioactive decay.

As long as a plant or tree is alive, it continually uses the carbon from carbon dioxide in the air to build new tissue and to replace old. Animals eat plants, so they too are continually receiving a fresh supply of carbon for their tissues.

Organisms cannot distinguish[†] $^{14}_{6}C$ from $^{12}_{6}C$, and because the ratio of $^{14}_{6}C$ to $^{12}_{6}C$ in the atmosphere remains nearly constant, the ratio of the two isotopes within the living organism remains nearly constant as well. When an organism dies, carbon dioxide is no longer taken in and utilized. Because the $^{14}_{6}C$ decays radioactively, the ratio of $^{14}_{6}C$ to $^{12}_{6}C$ in a dead organism decreases over time. The half-life of $^{14}_{6}C$ is about 5730 yr, so the $^{14}_{6}C/^{12}_{6}C$ ratio decreases by half every 5730 yr. If, for example, the $^{14}_{6}C/^{12}_{6}C$ ratio of an ancient wooden tool is half of what it is in living trees, then the object must have been made from a tree that was felled about 5730 years ago.

Actually, corrections must be made for the fact that the $^{14}_{6}C/^{12}_{6}C$ ratio in the atmosphere has not remained precisely constant over time. The determination of what this ratio has been over the centuries has required techniques such as comparing the expected ratio to the actual ratio for objects whose age is known, such as very old trees whose annual rings can be counted reasonably accurately.

EXAMPLE 30–13 **An ancient animal.** The mass of carbon in an animal bone fragment found in an archeological site is 200 g. If the bone registers an activity of 16 decays/s, what is its age?

APPROACH First we determine how many $^{14}_{6}C$ atoms there were in our 200-g sample when the animal was alive, given the known fraction of $^{14}_{6}C$ to $^{12}_{6}C$, 1.3×10^{-12}. Then we use Eq. 30–3b to find the activity back then, and Eq. 30–5 to find out how long ago that was by solving for the time t.

SOLUTION The 200 g of carbon is nearly all $^{12}_{6}C$; 12.0 g of $^{12}_{6}C$ contains 6.02×10^{23} atoms, so 200 g contains

$$\left(\frac{6.02 \times 10^{23} \text{ atoms/mol}}{12.0 \text{ g/mol}} \right)(200 \text{ g}) = 1.00 \times 10^{25} \text{ atoms.}$$

When the animal was alive, the ratio of $^{14}_{6}C$ to $^{12}_{6}C$ in the bone was 1.3×10^{-12}. The number of $^{14}_{6}C$ nuclei at that time was

$$N_0 = \left(1.00 \times 10^{25} \text{ atoms}\right)\left(1.3 \times 10^{-12}\right) = 1.3 \times 10^{13} \text{ atoms.}$$

From Eq. 30–3b with $\lambda = 3.83 \times 10^{-12} \text{ s}^{-1}$ (Example 30–9) the magnitude of the activity when the animal was still alive ($t = 0$) was

$$R_0 = \left| \frac{\Delta N}{\Delta t} \right|_0 = \lambda N_0 = \left(3.83 \times 10^{-12} \text{ s}^{-1}\right)\left(1.3 \times 10^{13}\right) = 50 \text{ s}^{-1}.$$

From Eq. 30–5

$$R = R_0 e^{-\lambda t}$$

where R, its activity now, is given as 16 s^{-1}. Then

$$16 \text{ s}^{-1} = \left(50 \text{ s}^{-1}\right)e^{-\lambda t}$$

or

$$e^{\lambda t} = \frac{50}{16}.$$

We take the natural logs of both sides (and divide by λ) to get

$$t = \frac{1}{\lambda} \ln\left(\frac{50}{16}\right) = \frac{1}{3.83 \times 10^{-12} \text{ s}^{-1}} \ln\left(\frac{50}{16}\right)$$

$$= 2.98 \times 10^{11} \text{ s} = 9400 \text{ yr,}$$

which is the time elapsed since the death of the animal.

PHYSICS APPLIED
Archeological dating

[†]Organisms operate almost exclusively via chemical reactions—which involve only the outer orbital electrons of the atom; extra neutrons in the nucleus have essentially no effect.

Geological Time Scale Dating

Carbon dating is useful only for determining the age of objects less than about 60,000 years old. The amount of $^{14}_{6}C$ remaining in objects older than that is usually too small to measure accurately, although new techniques are allowing detection of even smaller amounts of $^{14}_{6}C$, pushing the time frame further back. On the other hand, radioactive isotopes with longer half-lives can be used in certain circumstances to obtain the age of older objects. For example, the decay of $^{238}_{92}U$, because of its long half-life of 4.5×10^9 years, is useful in determining the ages of rocks on a geologic time scale. When molten material on Earth long ago solidified into rock as the temperature dropped, different compounds solidified according to the melting points, and thus different compounds separated to some extent. Uranium present in a material became fixed in position and the daughter nuclei that result from the decay of uranium were also fixed in that position. Thus, by measuring the amount of $^{238}_{92}U$ remaining in the material relative to the amount of daughter nuclei, the time when the rock solidified can be determined.

Radioactive dating methods using $^{238}_{92}U$ and other isotopes have shown the age of the oldest Earth rocks to be about 4×10^9 yr. The age of rocks in which the oldest fossilized organisms are embedded indicates that life appeared more than $3\frac{1}{2}$ billion years ago. The earliest fossilized remains of mammals are found in rocks 200 million years old, and humanlike creatures seem to have appeared more than 2 million years ago. Radioactive dating has been indispensable for the reconstruction of Earth's history.

*30–12 Stability and Tunneling

Radioactive decay occurs only if the mass of the parent nucleus is greater than the sum of the masses of the daughter nucleus and all particles emitted. For example, $^{238}_{92}U$ can decay to $^{234}_{90}Th$ because the mass of $^{238}_{92}U$ is greater than the mass of the $^{234}_{90}Th$ plus the mass of the α particle. Because systems tend to go in the direction that reduces their internal or potential energy (a ball rolls downhill, a positive charge moves toward a negative charge), you may wonder why an unstable nucleus doesn't fall apart immediately. In other words, why do $^{238}_{92}U$ nuclei $(T_{\frac{1}{2}} = 4.5 \times 10^9 \text{ yr})$ and other isotopes have such long half-lives? Why don't parent nuclei all decay at once?

The answer has to do with quantum theory and the nature of the forces involved. One way to view the situation is with the aid of a potential-energy diagram, as in Fig. 30–12. Let us consider the particular case of the decay $^{238}_{92}U \rightarrow ^{234}_{90}Th + ^4_2He$. The blue line represents the potential energy, including rest mass, where we imagine the α particle as a separate entity within the $^{238}_{92}U$ nucleus. The region labeled A in Fig. 30–12 represents the PE of the α particle when it is held within the uranium nucleus by the strong nuclear force (R_0 is the nuclear radius). Region C represents the PE when the α particle is free of the nucleus. The downward-curving PE (proportional to $1/r$) represents the electrical repulsion (Coulomb's law) between the positively charged α and the $^{234}_{90}Th$ nucleus. To get to region C, the α particle has to get through the "**Coulomb barrier**" shown. Since the PE just beyond $r = R_0$ (region B) is greater than the energy of the alpha particle (dashed line), the α particle could not escape the nucleus if it were governed by classical physics. It could escape only if there were an input of energy equal to the height of the barrier. Nuclei decay spontaneously, however, without any input of energy. How, then, does the α particle get from region A to region C? It actually passes through the barrier in a process known as quantum-mechanical **tunneling**. Classically, this could not happen, because an α particle in region B (within the barrier) would be violating the conservation-of-energy principle.[†]

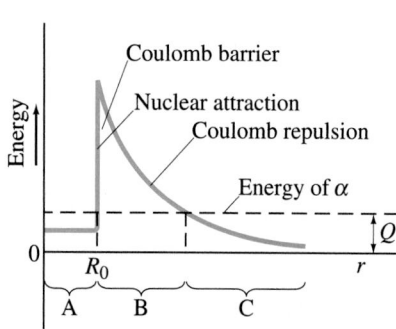

FIGURE 30–12 Potential energy for alpha particle and nucleus, showing the "Coulomb barrier" through which the α particle must tunnel to escape. The Q-value of the reaction is also indicated.

[†]The total energy E (dashed line in Fig. 30–12) would be less than the PE; because $\text{KE} = \frac{1}{2}mv^2 > 0$, then classically, $E = \text{KE} + \text{PE}$ could not be less than the PE.

The uncertainty principle, however, tells us that energy conservation can be violated by an amount ΔE for a length of time Δt given by

$$(\Delta E)(\Delta t) \approx \frac{h}{2\pi}.$$

We saw in Section 28–3 that this is a result of the wave–particle duality. Thus quantum mechanics allows conservation of energy to be violated for brief periods that may be long enough for an α particle to "tunnel" through the barrier. ΔE would represent the energy difference between the average barrier height and the particle's energy, and Δt the time to pass through the barrier. The higher and wider the barrier, the less time the α particle has to escape and the less likely it is to do so. It is therefore the height and width of this barrier that control the rate of decay and half-life of an isotope.

30–13 Detection of Particles

Individual particles such as electrons, protons, α particles, neutrons, and γ rays are not detected directly by our senses. Consequently, a variety of instruments have been developed to detect them.

Counters

One of the most common detectors is the **Geiger counter**. As shown in Fig. 30–13, it consists of a cylindrical metal tube filled with a certain type of gas. A long wire runs down the center and is kept at a high positive voltage ($\approx 10^3$ V) with respect to the outer cylinder. The voltage is just slightly less than that required to ionize the gas atoms. When a charged particle enters through the thin "window" at one end of the tube, it ionizes a few atoms of the gas. The freed electrons are attracted toward the positive wire, and as they are accelerated they strike and ionize additional atoms. An "avalanche" of electrons is quickly produced, and when it reaches the wire anode, it produces a voltage pulse. The pulse, after being amplified, can be sent to an electronic counter, which counts how many particles have been detected. Or the pulses can be sent to a loudspeaker and each detection of a particle is heard as a "click." Only a fraction of the radiation emitted by a sample is detected by any detector.

A **scintillation counter** makes use of a solid, liquid, or gas known as a **scintillator** or **phosphor**. The atoms of a scintillator are easily excited when struck by an incoming particle and emit visible light when they return to their ground states. Typical scintillators are crystals of NaI and certain plastics. One face of a solid scintillator is cemented to a photomultiplier tube, and the whole is wrapped with opaque material to keep it light-tight (in the dark) or is placed within a light-tight container. The **photomultiplier (PM) tube** converts the energy of the scintillator-emitted photon(s) into an electric signal. A PM tube is a vacuum tube containing several electrodes (typically 8 to 14), called *dynodes*, which are maintained at successively higher voltages as shown in Fig. 30–14. At its top surface is a photoelectric surface, called the *photocathode*, whose work function (Section 27–3) is low enough that an electron is easily released when struck by a photon from the scintillator. Such an electron is accelerated toward the positive voltage of the first dynode. When it strikes the first dynode, the electron has acquired sufficient kinetic energy so that it can eject two to five more electrons. These, in turn, are accelerated toward the higher voltage second dynode, and a multiplication process begins. The number of electrons striking the last dynode may be 10^6 or more. Thus the passage of a particle through the scintillator results in an electric signal at the output of the PM tube that can be sent to an electronic counter just as for a Geiger tube. Solid scintillators are much more dense than the gas of a Geiger counter, and so are much more efficient detectors—especially for γ rays, which interact less with matter than do α or β particles. Scintillators that can measure the total energy deposited are much used today and are called **calorimeters**.

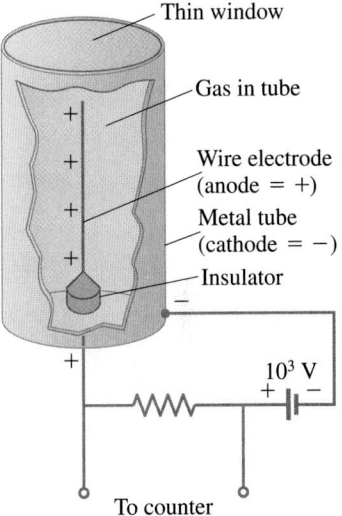

FIGURE 30–13 Diagram of a Geiger counter.

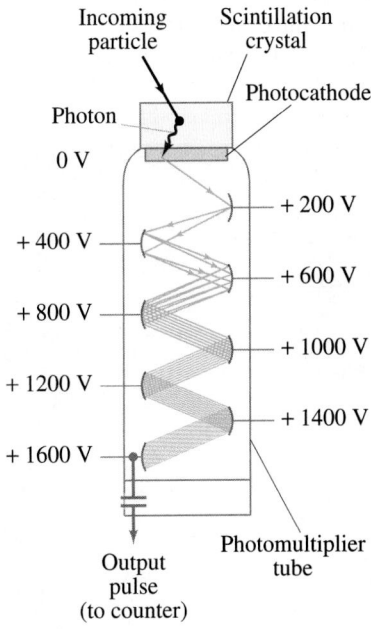

FIGURE 30–14 Scintillation counter with a photomultiplier tube.

In tracer work (Section 31–7), **liquid scintillators** are often used. Radioactive samples taken at different times or from different parts of an organism are placed directly in small bottles containing the liquid scintillator. This is particularly convenient for detection of β rays from ^3_1H and $^{14}_6C$, which have very low energies and have difficulty passing through the outer covering of a crystal scintillator or Geiger tube. A PM tube is still used to produce the electric signal from the liquid scintillator.

A **semiconductor detector** consists of a reverse-biased *pn* junction diode (Section 29–9). A charged particle passing through the junction can excite electrons into the conduction band, leaving holes in the valence band. The freed charges produce a short electric current pulse that can be counted as for Geiger and scintillation counters.

Hospital workers and others who work around radiation may carry *film badges* which detect the accumulation of radiation exposure. The film inside is periodically replaced and developed, the darkness of the developed film being related to total exposure (see Section 31–5).

Visualization

The devices discussed so far are used for counting the number of particles (or decays of a radioactive isotope). There are also devices that allow the track of charged particles to be *seen*. Very important are semiconductor detectors. **Silicon wafer semiconductors** have their surface etched into separate tiny pixels, each providing particle position information. They are much used in elementary particle physics (Chapter 32) to track the positions of particles produced and to determine their point of origin and/or their momentum (with the help of a magnetic field). The pixel arrangement can be CCD or CMOS (Section 25–1), the latter able to incorporate electronics inside, allowing fast readout.

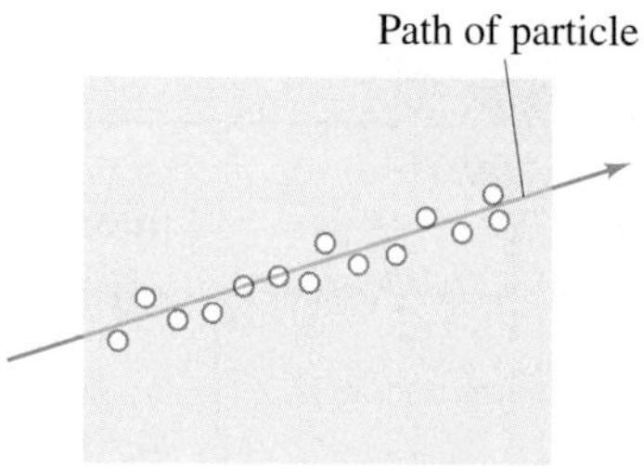

One of the oldest tracking devices is the **photographic emulsion**, which can be small and portable, used now particularly for cosmic-ray studies from balloons. A charged particle passing through an emulsion ionizes the atoms along its path. These points undergo a chemical change, and when the emulsion is developed (like film) the particle's path is revealed.

In a **cloud chamber**, used in the early days of nuclear physics, a gas is cooled to a temperature slightly below its usual condensation point ("supercooled"). Tiny droplets form around ions produced when a charged particle passes through (Fig. 30–15). Light scattering from these droplets reveals the track of the particle.

The **bubble chamber**, invented in 1952 by Donald A. Glaser (1926–2013), makes use of a superheated liquid kept close to its normal boiling point. Bubbles characteristic of boiling form around ions produced by the passage of a charged particle, revealing paths of particles that recently passed through. Because a bubble chamber uses a liquid, often liquid hydrogen, many more interactions can occur than in a cloud chamber. A magnetic field applied across the chamber makes charged particle paths curve (Chapter 20) and allows the momentum of charged particles to be determined from the radius of curvature of their paths.

A **multiwire**[†] **chamber** consists of a set of closely spaced fine wires immersed in a gas (Fig. 30–16). Many wires are grounded, and the others between are kept at very high voltage. A charged particle passing through produces ions in the gas. Freed electrons drift toward the nearest high voltage wire, creating an "avalanche" of many more ions, and producing an electric pulse or signal at that wire. The positions of the particles are determined electronically by the position of the wire and by the time it takes the pulses to reach "readout" electronics at the ends of the wires. The paths of the particles are reconstructed electronically by computers which can "draw" a picture of the tracks, as shown in Fig. 32–15, Chapter 32. An external magnetic field curves the paths, allowing the momentum of the particles to be measured.

Many detectors are also **calorimeters** which measure the energy of the particles.

Path of particle

FIGURE 30–15 In a cloud chamber or bubble chamber, droplets or bubbles are formed around ions produced by the passage of a charged particle.

FIGURE 30–16 Multiwire chamber inside the Collider Detector at Fermilab (CDF). Figure 32–15 in Chapter 32 was made with this detector.

[†]Also called *wire drift chamber* or *wire proportional chamber*.

Summary

Nuclear physics is the study of atomic nuclei. Nuclei contain **protons** and **neutrons**, which are collectively known as **nucleons**. The total number of nucleons, A, is the nucleus's **atomic mass number**. The number of protons, Z, is the **atomic number**. The number of neutrons equals $A - Z$. **Isotopes** are nuclei with the same Z, but with different numbers of neutrons. For an element X, an isotope of given Z and A is represented by

$$^A_Z X.$$

The nuclear radius is approximately proportional to $A^{\frac{1}{3}}$, indicating that all nuclei have about the same density. Nuclear masses are specified in **unified atomic mass units** (u), where the mass of $^{12}_6 C$ (including its 6 electrons) is defined as exactly 12.000000 u. In terms of the energy equivalent (because $E = mc^2$),

$$1\,u \;=\; 931.5\,MeV/c^2 \;=\; 1.66 \times 10^{-27}\,kg.$$

The mass of a stable nucleus is less than the sum of the masses of its constituent nucleons. The difference in mass (times c^2) is the **total binding energy**. It represents the energy needed to break the nucleus into its constituent nucleons. The **binding energy per nucleon** averages about 8 MeV per nucleon, and is lowest for low mass and high mass nuclei.

Unstable nuclei undergo **radioactive decay**; they change into other nuclei with the emission of an α, β, or γ particle. An α particle is a $^4_2 He$ nucleus; a β particle is an electron or positron; and a γ ray is a high-energy photon. In β decay, a **neutrino** is also emitted. The transformation of **parent** nuclei into **daughter** nuclei is called **transmutation** of the elements. Radioactive decay occurs spontaneously only when the mass of the products is less than the mass of the parent nucleus. The loss in mass appears as kinetic energy of the products.

Nuclei are held together by the **strong nuclear force**. The **weak nuclear force** makes itself apparent in β decay. These two forces, plus the gravitational and electromagnetic forces, are the four known types of force.

Electric charge, linear and angular momentum, mass–energy, and **nucleon number** are **conserved** in all decays.

Radioactive decay is a statistical process. For a given type of radioactive nucleus, the number of nuclei that decay (ΔN) in a time Δt is proportional to the number N of parent nuclei present:

$$\Delta N \;=\; -\lambda N\,\Delta t; \tag{30–3a}$$

the minus sign means N *decreases* in time.

The proportionality constant λ is called the **decay constant** and is characteristic of the given nucleus. The number N of nuclei remaining after a time t decreases exponentially,

$$N \;=\; N_0 e^{-\lambda t}, \tag{30–4}$$

as does the **activity**, $R = $ magnitude of $\Delta N/\Delta t$:

$$R \;=\; \left| \frac{\Delta N}{\Delta t} \right|_0 e^{-\lambda t}. \tag{30–5}$$

The **half-life**, $T_{\frac{1}{2}}$, is the time required for half the nuclei of a radioactive sample to decay. It is related to the decay constant by

$$T_{\frac{1}{2}} \;=\; \frac{0.693}{\lambda}. \tag{30–6}$$

Radioactive dating is the use of radioactive decay to determine the age of certain objects, such as carbon dating.

[*Alpha decay occurs via a purely quantum-mechanical process called **tunneling** through a barrier.]

Particle **detectors** include **Geiger counters**, **scintillators** with attached **photomultiplier tubes**, and **semiconductor detectors**. Detectors that can image particle tracks include **semiconductors**, photographic **emulsions**, **bubble chambers**, and **multiwire chambers**.

Questions

1. What do different isotopes of a given element have in common? How are they different?

2. What are the elements represented by the X in the following: (a) $^{232}_{92}X$; (b) $^{18}_7 X$; (c) $^1_1 X$; (d) $^{86}_{38}X$; (e) $^{252}_{100}X$?

3. How many protons and how many neutrons do each of the isotopes in Question 2 have?

4. Identify the element that has 87 nucleons and 50 neutrons.

5. Why are the atomic masses of many elements (see the Periodic Table) not close to whole numbers?

6. Why are atoms much more likely to emit an alpha particle than to emit separate neutrons and protons?

7. What are the similarities and the differences between the strong nuclear force and the electric force?

8. What is the experimental evidence in favor of radioactivity being a nuclear process?

9. The isotope $^{64}_{29}Cu$ is unusual in that it can decay by γ, β^-, and β^+ emission. What is the resulting nuclide for each case?

10. A $^{238}_{92}U$ nucleus decays via α decay to a nucleus containing how many neutrons?

11. Describe, in as many ways as you can, the difference between α, β, and γ rays.

12. Fill in the missing particle or nucleus:
    (a) $^{45}_{20}Ca \rightarrow ? + e^- + \bar{\nu}$
    (b) $^{58}_{29}Cu^* \rightarrow ? + \gamma$
    (c) $^{46}_{24}Cr \rightarrow ^{46}_{23}V + ?$
    (d) $^{234}_{94}Pu \rightarrow ? + \alpha$
    (e) $^{239}_{93}Np \rightarrow ^{239}_{94}Pu + ?$

13. Immediately after a $^{238}_{92}U$ nucleus decays to $^{234}_{90}Th + ^4_2 He$, the daughter thorium nucleus may still have 92 electrons circling it. Since thorium normally holds only 90 electrons, what do you suppose happens to the two extra ones?

14. When a nucleus undergoes either β^- or β^+ decay, what happens to the energy levels of the atomic electrons? What is likely to happen to these electrons following the decay?

15. The alpha particles from a given alpha-emitting nuclide are generally monoenergetic; that is, they all have the same kinetic energy. But the beta particles from a beta-emitting nuclide have a spectrum of energies. Explain the difference between these two cases.

16. Do isotopes that undergo electron capture generally lie above or below the stable nuclides in Fig. 30–2?

17. Can hydrogen or deuterium emit an α particle? Explain.

18. Why are many artificially produced radioactive isotopes rare in nature?

19. An isotope has a half-life of one month. After two months, will a given sample of this isotope have completely decayed? If not, how much remains?

20. Why are none of the elements with $Z > 92$ stable?

21. A proton strikes a $_3^6$Li nucleus. As a result, an α particle and another particle are released. What is the other particle?

22. Can $_6^{14}$C dating be used to measure the age of stone walls and tablets of ancient civilizations? Explain.

23. Explain the absence of β^+ emitters in the radioactive decay series of Fig. 30–11.

24. As $_{86}^{222}$Rn decays into $_{82}^{206}$Pb, how many alpha and beta particles are emitted? Does it matter which path in the decay series is chosen? Why or why not?

25. A ^{238}U nucleus (initially at rest) decays into a ^{234}Th nucleus and an alpha particle. Which has the greater (i) momentum, (ii) velocity, (iii) kinetic energy? Explain.
    (a) The ^{234}Th nucleus.
    (b) The alpha particle.
    (c) Both the same.

MisConceptual Questions

1. Elements of the Periodic Table are distinguished by
   (a) the number of protons in the nucleus.
   (b) the number of neutrons in the nucleus.
   (c) the number of electrons in the atom.
   (d) Both (a) and (b).
   (e) (a), (b), and (c).

2. A nucleus has
   (a) more energy than its component neutrons and protons have.
   (b) less energy than its component neutrons and protons have.
   (c) the same energy as its component neutrons and protons have.
   (d) more energy than its component neutrons and protons have when the nucleus is at rest but less energy than when it is moving.

3. Which of the following will generally create a more stable nucleus?
   (a) Having more nucleons.
   (b) Having more protons than neutrons.
   (c) Having a larger binding energy per nucleon.
   (d) Having the same number of electrons as protons.
   (e) Having a larger total binding energy.

4. There are 82 protons in a lead nucleus. Why doesn't the lead nucleus burst apart?
   (a) Coulomb repulsive force doesn't act inside the nucleus.
   (b) Gravity overpowers the Coulomb repulsive force inside the nucleus.
   (c) The negatively charged neutrons balance the positively charged protons.
   (d) Protons lose their positive charge inside the nucleus.
   (e) The strong nuclear force holds the nucleus together.

5. The half-life of a radioactive nucleus is
   (a) half the time it takes for the entire substance to decay.
   (b) the time it takes for half of the substance to decay.
   (c) the same as the decay constant.
   (d) Both (a) and (b) (they are the same).
   (e) All of the above.

6. As a radioactive sample decays,
   (a) the half-life increases.
   (b) the half-life decreases.
   (c) the activity remains the same.
   (d) the number of radioactive nuclei increases.
   (e) None of the above.

7. If the half-life of a radioactive sample is 10 years, then it should take _____ years for the sample to decay completely.
   (a) 10. (b) 20. (c) 40.
   (d) Cannot be determined.

8. A sample's half-life is 1 day. What fraction of the original sample will have decayed after 3 days?
   (a) $\frac{1}{8}$. (b) $\frac{1}{4}$. (c) $\frac{1}{2}$. (d) $\frac{3}{4}$. (e) $\frac{7}{8}$. (f) All of it.

9. After three half-lives, what fraction of the original radioactive material is left?
   (a) None. (b) $\frac{1}{16}$. (c) $\frac{1}{8}$. (d) $\frac{1}{4}$. (e) $\frac{3}{4}$. (f) $\frac{7}{8}$.

10. Technetium $_{43}^{98}$Tc has a half-life of 4.2×10^6 yr. Strontium $_{38}^{90}$Sr has a half-life of 28.79 yr. Which statements are true?
    (a) The decay constant of Sr is greater than the decay constant of Tc.
    (b) The activity of 100 g of Sr is less than the activity of 100 g of Tc.
    (c) The long half-life of Tc means that it decays by alpha decay.
    (d) A Tc atom has a higher probability of decaying in 1 yr than a Sr atom.
    (e) 28.79 g of Sr has the same activity as 4.2×10^6 g of Tc.

11. A material having which decay constant would have the shortest half-life?
    (a) 100/second.
    (b) 5/year.
    (c) 8/century.
    (d) 10^9/day.

12. Uranium-238 decays to lead-206 through a series of
    (a) alpha decays.
    (b) beta decays.
    (c) gamma decays.
    (d) some combination of alpha, beta, and gamma decays.

13. Carbon dating is useful only for determining the age of objects less than about _____ years old.
    (a) 4.5 million.
    (b) 1.2 million.
    (c) 600,000.
    (d) 60,000.
    (e) 6000.

14. Radon has a half-life of about 1600 years. The Earth is several billion years old, so why do we still find radon on this planet?
    (a) Ice-age temperatures preserved some of it.
    (b) Heavier unstable isotopes decay into it.
    (c) It is created in lightning strikes.
    (d) It is replenished by cosmic rays.
    (e) Its half-life has increased over time.
    (f) Its half-life has decreased over time.

15. How does an atom's nucleus stay together and remain stable?
    (a) The attractive gravitational force between the protons and neutrons overcomes the repulsive electrostatic force between the protons.
    (b) Having just the right number of neutrons overcomes the electrostatic force between the protons.
    (c) A strong covalent bond develops between the neutrons and protons, because they are so close to each other.
    (d) None of the above.

16. What has greater mass?
    (a) A neutron and a proton that are far from each other (unbound).
    (b) A neutron and a proton that are bound together in a hydrogen (deuterium) nucleus.
    (c) Both the same.

For assigned homework and other learning materials, go to the MasteringPhysics website.

Problems

[See Appendix B for masses]

30–1 Nuclear Properties

1. (I) A pi meson has a mass of $139 \, \text{MeV}/c^2$. What is this in atomic mass units?

2. (I) What is the approximate radius of an α particle (^4_2He)?

3. (I) By what % is the radius of $^{238}_{92}\text{U}$ greater than the radius of $^{232}_{92}\text{U}$?

4. (II) (a) What is the approximate radius of a $^{112}_{48}\text{Cd}$ nucleus? (b) Approximately what is the value of A for a nucleus whose radius is $3.7 \times 10^{-15} \, \text{m}$?

5. (II) What is the mass of a bare α particle (without electrons) in MeV/c^2?

6. (II) Suppose two alpha particles were held together so they were just "touching" (use Eq. 30–1). Estimate the electrostatic repulsive force each would exert on the other. What would be the acceleration of an alpha particle subjected to this force?

7. (II) (a) What would be the radius of the Earth if it had its actual mass but had the density of nuclei? (b) By what factor would the radius of a $^{238}_{92}\text{U}$ nucleus increase if it had the Earth's density?

8. (II) What stable nucleus has approximately half the radius of a uranium nucleus? [Hint: Find A and use Appendix B to get Z.]

9. (II) If an alpha particle were released from rest near the surface of a $^{257}_{100}\text{Fm}$ nucleus, what would its kinetic energy be when far away?

10. (II) (a) What is the fraction of the hydrogen atom's mass (^1_1H) that is in the nucleus? (b) What is the fraction of the hydrogen atom's volume that is occupied by the nucleus?

11. (II) Approximately how many nucleons are there in a 1.0-kg object? Does it matter what the object is made of? Why or why not?

12. (III) How much kinetic energy, in MeV, must an α particle have to just "touch" the surface of a $^{232}_{92}\text{U}$ nucleus?

30–2 Binding Energy

13. (I) Estimate the total binding energy for $^{63}_{29}\text{Cu}$, using Fig. 30–1.

14. (I) Use Fig. 30–1 to estimate the total binding energy of (a) $^{238}_{92}\text{U}$, and (b) $^{84}_{36}\text{Kr}$.

15. (II) Calculate the binding energy per nucleon for a $^{15}_{7}\text{N}$ nucleus, using Appendix B.

16. (II) Use Appendix B to calculate the binding energy of ^2_1H (deuterium).

17. (II) Determine the binding energy of the last neutron in a $^{23}_{11}\text{Na}$ nucleus.

18. (II) Calculate the total binding energy, and the binding energy per nucleon, for (a) ^7_3Li, (b) $^{195}_{78}\text{Pt}$. Use Appendix B.

19. (II) Compare the average binding energy of a nucleon in $^{23}_{11}\text{Na}$ to that in $^{24}_{11}\text{Na}$, using Appendix B.

20. (III) How much energy is required to remove (a) a proton, (b) a neutron, from $^{15}_{7}\text{N}$? Explain the difference in your answers.

21. (III) (a) Show that the nucleus ^8_4Be (mass = 8.005305 u) is unstable and will decay into two α particles. (b) Is $^{12}_{6}\text{C}$ stable against decay into three α particles? Show why or why not.

30–3 to 30–7 Radioactive Decay

22. (I) The ^7_3Li nucleus has an excited state 0.48 MeV above the ground state. What wavelength gamma photon is emitted when the nucleus decays from the excited state to the ground state?

23. (II) Show that the decay $^{11}_{6}\text{C} \rightarrow {}^{10}_{5}\text{B} + \text{p}$ is not possible because energy would not be conserved.

24. (II) Calculate the energy released when tritium, ^3_1H, decays by β^- emission.

25. (II) What is the maximum kinetic energy of an electron emitted in the β decay of a free neutron?

26. (II) Give the result of a calculation that shows whether or not the following decays are possible:
    (a) $^{233}_{92}\text{U} \rightarrow {}^{232}_{92}\text{U} + \text{n}$;
    (b) $^{14}_{7}\text{N} \rightarrow {}^{13}_{7}\text{N} + \text{n}$;
    (c) $^{40}_{19}\text{K} \rightarrow {}^{39}_{19}\text{K} + \text{n}$.

27. (II) $^{24}_{11}\text{Na}$ is radioactive. (a) Is it a β^- or β^+ emitter? (b) Write down the decay reaction, and estimate the maximum kinetic energy of the emitted β.

28. (II) A $^{238}_{92}$U nucleus emits an α particle with kinetic energy = 4.20 MeV. (a) What is the daughter nucleus, and (b) what is the approximate atomic mass (in u) of the daughter atom? Ignore recoil of the daughter nucleus.

29. (II) Calculate the maximum kinetic energy of the β particle emitted during the decay of $^{60}_{27}$Co.

30. (II) How much energy is released in electron capture by beryllium: $^{7}_{4}$Be + e$^-$ → $^{7}_{3}$Li + ν?

31. (II) The isotope $^{218}_{84}$Po can decay by either α or β^- emission. What is the energy release in each case? The mass of $^{218}_{84}$Po is 218.008973 u.

32. (II) The nuclide $^{32}_{15}$P decays by emitting an electron whose maximum kinetic energy can be 1.71 MeV. (a) What is the daughter nucleus? (b) Calculate the daughter's atomic mass (in u).

33. (II) A photon with a wavelength of 1.15×10^{-13} m is ejected from an atom. Calculate its energy and explain why it is a γ ray from the nucleus or a photon from the atom.

34. (II) How much recoil energy does a $^{40}_{19}$K nucleus get when it emits a 1.46-MeV gamma ray?

35. (II) Determine the maximum kinetic energy of β^+ particles released when $^{11}_{6}$C decays to $^{11}_{5}$B. What is the maximum energy the neutrino can have? What is the minimum energy of each?

36. (III) Show that when a nucleus decays by β^+ decay, the total energy released is equal to

$$(M_P - M_D - 2m_e)c^2,$$

where M_P and M_D are the masses of the parent and daughter atoms (neutral), and m_e is the mass of an electron or positron.

37. (III) When $^{238}_{92}$U decays, the α particle emitted has 4.20 MeV of kinetic energy. Calculate the recoil kinetic energy of the daughter nucleus and the Q-value of the decay.

30–8 to 30–11 Half-Life, Decay Rates, Decay Series, Dating

38. (I) (a) What is the decay constant of $^{238}_{92}$U whose half-life is 4.5×10^9 yr? (b) The decay constant of a given nucleus is 3.2×10^{-5} s^{-1}. What is its half-life?

39. (I) A radioactive material produces 1120 decays per minute at one time, and 3.6 h later produces 140 decays per minute. What is its half-life?

40. (I) What fraction of a sample of $^{68}_{32}$Ge, whose half-life is about 9 months, will remain after 2.5 yr?

41. (I) What is the activity of a sample of $^{14}_{6}$C that contains 6.5×10^{20} nuclei?

42. (I) What fraction of a radioactive sample is left after exactly 5 half-lives?

43. (II) The iodine isotope $^{131}_{53}$I is used in hospitals for diagnosis of thyroid function. If 782 μg are ingested by a patient, determine the activity (a) immediately, (b) 1.50 h later when the thyroid is being tested, and (c) 3.0 months later. Use Appendix B.

44. (II) How many nuclei of $^{238}_{92}$U remain in a rock if the activity registers 420 decays per second?

45. (II) In a series of decays, the nuclide $^{235}_{92}$U becomes $^{207}_{82}$Pb. How many α and β^- particles are emitted in this series?

46. (II) $^{124}_{55}$Cs has a half-life of 30.8 s. (a) If we have 8.7 μg initially, how many Cs nuclei are present? (b) How many are present 2.6 min later? (c) What is the activity at this time? (d) After how much time will the activity drop to less than about 1 per second?

47. (II) Calculate the mass of a sample of pure $^{40}_{19}$K with an initial decay rate of 2.4×10^5 s^{-1}. The half-life of $^{40}_{19}$K is 1.248×10^9 yr.

48. (II) Calculate the activity of a pure 6.7-μg sample of $^{32}_{15}$P ($T_{\frac{1}{2}} = 1.23 \times 10^6$ s).

49. (II) A sample of $^{233}_{92}$U ($T_{\frac{1}{2}} = 1.59 \times 10^5$ yr) contains 4.50×10^{18} nuclei. (a) What is the decay constant? (b) Approximately how many disintegrations will occur per minute?

50. (II) The activity of a sample drops by a factor of 6.0 in 9.4 minutes. What is its half-life?

51. (II) A 345-g sample of pure carbon contains 1.3 parts in 10^{12} (atoms) of $^{14}_{6}$C. How many disintegrations occur per second?

52. (II) A sample of $^{238}_{92}$U is decaying at a rate of 4.20×10^2 decays/s. What is the mass of the sample?

53. (II) **Rubidium–strontium dating**. The rubidium isotope $^{87}_{37}$Rb, a β emitter with a half-life of 4.75×10^{10} yr, is used to determine the age of rocks and fossils. Rocks containing fossils of ancient animals contain a ratio of $^{87}_{38}$Sr to $^{87}_{37}$Rb of 0.0260. Assuming that there was no $^{87}_{38}$Sr present when the rocks were formed, estimate the age of these fossils.

54. (II) Two of the naturally occurring radioactive decay sequences start with $^{232}_{90}$Th and with $^{235}_{92}$U. The first five decays of these two sequences are:

$$\alpha, \beta, \beta, \alpha, \alpha$$

and

$$\alpha, \beta, \alpha, \beta, \alpha.$$

Determine the resulting intermediate daughter nuclei in each case.

55. (II) An ancient wooden club is found that contains 73 g of carbon and has an activity of 7.0 decays per second. Determine its age assuming that in living trees the ratio of ^{14}C/^{12}C atoms is about 1.3×10^{-12}.

56. (II) Use Fig. 30–11 and calculate the relative decay rates for α decay of $^{218}_{84}$Po and $^{214}_{84}$Po.

57. (III) The activity of a radioactive source decreases by 5.5% in 31.0 hours. What is the half-life of this source?

58. (III) $^{7}_{4}$Be decays with a half-life of about 53 d. It is produced in the upper atmosphere, and filters down onto the Earth's surface. If a plant leaf is detected to have 350 decays/s of $^{7}_{4}$Be, (a) how long do we have to wait for the decay rate to drop to 25 per second? (b) Estimate the initial mass of $^{7}_{4}$Be on the leaf.

59. (III) At $t = 0$, a pure sample of radioactive nuclei contains N_0 nuclei whose decay constant is λ. Determine a formula for the number of daughter nuclei, N_D, as a function of time; assume the daughter is stable and that $N_D = 0$ at $t = 0$.

General Problems

60. Which radioactive isotope of lead is being produced if the measured activity of a sample drops to 1.050% of its original activity in 4.00 h?

61. An old wooden tool is found to contain only 4.5% of the $^{14}_{6}C$ that an equal mass of fresh wood would. How old is the tool?

62. A neutron star consists of neutrons at approximately nuclear density. Estimate, for a 10-km-diameter neutron star, (*a*) its mass number, (*b*) its mass (kg), and (*c*) the acceleration of gravity at its surface.

63. **Tritium dating**. The $^{3}_{1}H$ isotope of hydrogen, which is called *tritium* (because it contains three nucleons), has a half-life of 12.3 yr. It can be used to measure the age of objects up to about 100 yr. It is produced in the upper atmosphere by cosmic rays and brought to Earth by rain. As an application, determine approximately the age of a bottle of wine whose $^{3}_{1}H$ radiation is about $\frac{1}{10}$ that present in new wine.

64. Some elementary particle theories (Section 32–11) suggest that the proton may be unstable, with a half-life $\geq 10^{33}$ yr. (*a*) How long would you expect to wait for one proton in your body to decay (approximate your body as all water)? (*b*) Of the roughly 7 billion people on Earth, about how many would have a proton in their body decay in a 70 yr lifetime?

65. The original experiments which established that an atom has a heavy, positive nucleus were done by shooting alpha particles through gold foil. The alpha particles had a kinetic energy of 7.7 MeV. What is the closest they could get to the center of a gold nucleus? How does this compare with the size of the nucleus?

66. How long must you wait (in half-lives) for a radioactive sample to drop to 2.00% of its original activity?

67. If the potassium isotope $^{40}_{19}K$ gives 42 decays/s in a liter of milk, estimate how much $^{40}_{19}K$ and regular $^{39}_{19}K$ are in a liter of milk. Use Appendix B.

68. Strontium-90 is produced as a nuclear fission product of uranium in both reactors and atomic bombs. Look at its location in the Periodic Table to see what other elements it might be similar to chemically, and tell why you think it might be dangerous to ingest. It has too many neutrons to be stable, and it decays with a half-life of about 29 yr. How long will we have to wait for the amount of $^{90}_{38}Sr$ on the Earth's surface to reach 1% of its current level, assuming no new material is scattered about? Write down the decay reaction, including the daughter nucleus. The daughter is radioactive: write down its decay.

69. The activity of a sample of $^{35}_{16}S$ $(T_{\frac{1}{2}} = 87.37\text{ days})$ is 4.28×10^4 decays per second. What is the mass of the sample?

70. The nuclide $^{191}_{76}Os$ decays with β^- energy of 0.14 MeV accompanied by γ rays of energy 0.042 MeV and 0.129 MeV. (*a*) What is the daughter nucleus? (*b*) Draw an energy-level diagram showing the ground states of the parent and daughter and excited states of the daughter. (*c*) To which of the daughter states does β^- decay of $^{191}_{76}Os$ occur?

71. Determine the activities of (*a*) 1.0 g of $^{131}_{53}I$ $(T_{\frac{1}{2}} = 8.02\text{ days})$ and (*b*) 1.0 g of $^{238}_{92}U$ $(T_{\frac{1}{2}} = 4.47 \times 10^9\text{ yr})$.

72. Use Fig. 30–1 to estimate the total binding energy for copper and then estimate the energy, in joules, needed to break a 3.0-g copper penny into its constituent nucleons.

73. Instead of giving atomic masses for nuclides as in Appendix B, some Tables give the **mass excess**, Δ, defined as $\Delta = M - A$, where A is the atomic mass number and M is the mass in u. Determine the mass excess, in u and in MeV/c^2, for: (*a*) $^{4}_{2}He$; (*b*) $^{12}_{6}C$; (*c*) $^{86}_{38}Sr$; (*d*) $^{235}_{92}U$. (*e*) From a glance at Appendix B, can you make a generalization about the sign of Δ as a function of Z or A?

74. When water is placed near an intense neutron source, the neutrons can be slowed down to almost zero speed by collisions with the water molecules, and are eventually captured by a hydrogen nucleus to form the stable isotope called **deuterium**, $^{2}_{1}H$, giving off a gamma ray. What is the energy of the gamma ray?

75. The practical limit for carbon-14 dating is about 60,000 years. If a bone contains 1.0 kg of carbon, and the animal died 60,000 years ago, what is the activity today?

76. Using Section 30–2 and Appendix B, determine the energy required to remove one neutron from $^{4}_{2}He$. How many times greater is this energy than the binding energy of the last neutron in $^{13}_{6}C$?

77. (*a*) If all of the atoms of the Earth were to collapse and simply become nuclei, what would be the Earth's new radius? (*b*) If all of the atoms of the Sun were to collapse and simply become nuclei, what would be the Sun's new radius?

78. (*a*) A 72-gram sample of natural carbon contains the usual fraction of $^{14}_{6}C$. Estimate roughly how long it will take before there is only one $^{14}_{6}C$ nucleus left. (*b*) How does the answer in (*a*) change if the sample is 340 grams? What does this tell you about the limits of carbon dating?

79. If the mass of the proton were just a little closer to the mass of the neutron, the following reaction would be possible even at low collision energies:
$$e^- + p \rightarrow n + \nu.$$
(*a*) Why would this situation be catastrophic? (See last paragraph of Chapter 33.) (*b*) By what percentage would the proton's mass have to be increased to make this reaction possible?

80. What is the ratio of the kinetic energies for an alpha particle and a beta particle if both make tracks with the same radius of curvature in a magnetic field, oriented perpendicular to the paths of the particles?

81. A 1.00-g sample of natural samarium emits α particles at a rate of 120 s^{-1} due to the presence of $^{147}_{62}Sm$. The natural abundance of $^{147}_{62}Sm$ is 15%. Calculate the half-life for this decay process.

82. Almost all of naturally occurring uranium is $^{238}_{92}U$ with a half-life of 4.468×10^9 yr. Most of the rest of natural uranium is $^{235}_{92}U$ with a half-life of 7.04×10^8 yr. Today a sample contains 0.720% $^{235}_{92}U$. (*a*) What was this percentage 1.0 billion years ago? (*b*) What percentage of uranium will be $^{235}_{92}U$ 100 million years from now?

83. A banana contains about 420 mg of potassium, of which a small fraction is the radioactive isotope $^{40}_{19}K$ (Appendix B). Estimate the activity of an average banana due to $^{40}_{19}K$.

84. When $^{23}_{10}Ne$ (mass = 22.9947 u) decays to $^{23}_{11}Na$ (mass = 22.9898 u), what is the maximum kinetic energy of the emitted electron? What is its minimum energy? What is the energy of the neutrino in each case? Ignore recoil of the daughter nucleus.

85. (a) In α decay of, say, a $^{226}_{88}Ra$ nucleus, show that the nucleus carries away a fraction $1/(1 + \frac{1}{4} A_D)$ of the total energy available, where A_D is the mass number of the daughter nucleus. [Hint: Use conservation of momentum as well as conservation of energy.] (b) Approximately what percentage of the energy available is thus carried off by the α particle when $^{226}_{88}Ra$ decays?

86. Decay series, such as that shown in Fig. 30–11, can be classified into four families, depending on whether the mass numbers have the form $4n$, $4n + 1$, $4n + 2$, or $4n + 3$, where n is an integer. Justify this statement and show that for a nuclide in any family, all its daughters will be in the same family.

Search and Learn

1. Describe in detail why we think there is a strong nuclear force.

2. (a) Under what circumstances could a fermium nucleus decay into an einsteinium nucleus? (b) What about the reverse, an Es nucleus decaying into Fm?

3. Using the uncertainty principle and the radius of a nucleus, estimate the minimum possible kinetic energy of a nucleon in, say, iron. Ignore relativistic corrections. [Hint: A particle can have a momentum at least as large as its momentum uncertainty.]

4. In Fig. 30–17, a nucleus decays and emits a particle that enters a region with a uniform magnetic field of 0.012 T directed into the page. The path of the detected particle is shown. (a) What type of radioactive decay is this? (b) If the radius of the circular arc is 4.7 mm, what is the velocity of the particle?

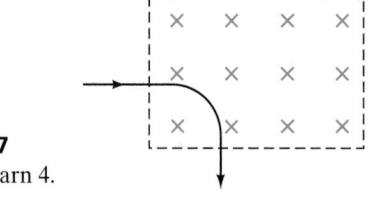

FIGURE 30–17
Search and Learn 4.

5. In both internal conversion and β decay, an electron is emitted. How could you determine which decay process occurred?

6. Suppose we discovered that several thousand years ago cosmic rays had bombarded the Earth's atmosphere a lot more than we had thought. Compared to previous calculations of the carbon-dated age of organic matter, we would now calculate it to be older, younger, or the same age as previously calculated? Explain.

7. In 1991, the frozen remains of a Neolithic-age man, nicknamed Otzi, were found in the Italian Alps by hikers. The body was well preserved, as were his bow, arrows, knife, axe, other tools, and clothing. The date of his death can be determined using carbon-14 dating. (a) What is the decay constant for $^{14}_{6}C$? (b) How many $^{14}_{6}C$ atoms per gram of $^{12}_{6}C$ are there in a living organism? (c) What is the activity per gram in naturally occurring carbon for a living organism? (d) For Otzi, the activity per gram of carbon was measured to be 0.121. How long ago did he live?

8. Some radioactive isotopes have half-lives that are greater than the age of the universe (like gadolinium or samarium). The only way to determine these half-lives is to monitor the decay rate of a sample that contains these isotopes. For example, suppose we find an asteroid that currently contains about 15,000 kg of $^{152}_{64}Gd$ (gadolinium) and we detect an activity of 1 decay/s. Estimate the half-life of gadolinium (in years).

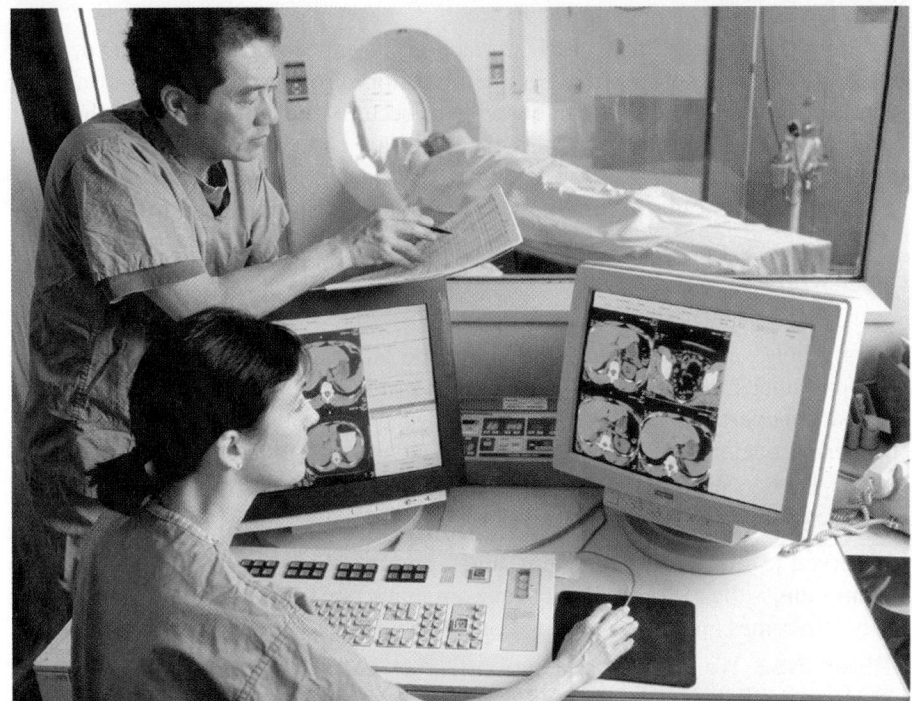

Technicians are looking at an MRI image of sections through a patient's body. MRI is one of several powerful types of medical imaging based on physics used by doctors to diagnose illnesses.

This Chapter opens with basic and important physics topics of nuclear reactions, nuclear fission, and nuclear fusion, and how we obtain nuclear energy. Then we examine the health aspects of radiation—dosimetry, therapy, and imaging: MRI, PET, and SPECT.

Nuclear Energy; Effects and Uses of Radiation

CONTENTS

CHAPTER-OPENING QUESTION—Guess now!

The Sun is powered by
- **(a)** nuclear alpha decay.
- **(b)** nuclear beta decay.
- **(c)** nuclear gamma decay.
- **(d)** nuclear fission.
- **(e)** nuclear fusion.

We continue our study of nuclear physics in this Chapter. We begin with a discussion of nuclear reactions, and then we examine the important huge energy-releasing processes of fission and fusion. We also deal with the effects of nuclear radiation passing through matter, particularly biological matter, and how radiation is used medically for therapy, diagnosis, and imaging techniques.

31–1 Nuclear Reactions and the Transmutation of Elements

When a nucleus undergoes α or β decay, the daughter nucleus is a different element from the parent. The transformation of one element into another, called **transmutation**, also occurs via nuclear reactions. A **nuclear reaction** is said to occur when a nucleus is struck by another nucleus, or by a simpler particle such as a γ ray, neutron, or proton, and an interaction takes place. Ernest Rutherford was the first to report seeing a nuclear reaction. In 1919 he observed that some of the α particles passing through nitrogen gas were absorbed and protons emitted. He concluded that nitrogen nuclei had been transformed into oxygen nuclei via the reaction

$$ {}^{4}_{2}\text{He} + {}^{14}_{7}\text{N} \rightarrow {}^{17}_{8}\text{O} + {}^{1}_{1}\text{H}, $$

where ${}^{4}_{2}\text{He}$ is an α particle, and ${}^{1}_{1}\text{H}$ is a proton.

Since then, a great many nuclear reactions have been observed. Indeed, many of the radioactive isotopes used in the laboratory are made by means of nuclear reactions. Nuclear reactions can be made to occur in the laboratory, but they also occur regularly in nature. In Chapter 30 we saw an example: $^{14}_{6}\text{C}$ is continually being made in the atmosphere via the reaction $n + ^{14}_{7}\text{N} \rightarrow ^{14}_{6}\text{C} + p$.

Nuclear reactions are sometimes written in a shortened form: for example,

$$n + ^{14}_{7}\text{N} \rightarrow ^{14}_{6}\text{C} + p$$

can be written

$$^{14}_{7}\text{N} \, (n, \, p) \, ^{14}_{6}\text{C}.$$

The symbols outside the parentheses on the left and right represent the initial and final nuclei, respectively. The symbols inside the parentheses represent the bombarding particle (first) and the emitted small particle (second).

In any nuclear reaction, both electric charge and nucleon number are conserved. These conservation laws are often useful, as the following Example shows.

CONCEPTUAL EXAMPLE 31–1 | **Deuterium reaction.** A neutron is observed to strike an $^{16}_{8}\text{O}$ nucleus, and a deuteron is given off. (A **deuteron**, or **deuterium**, is the isotope of hydrogen containing one proton and one neutron, $^{2}_{1}\text{H}$; it is sometimes given the symbol d or D.) What is the nucleus that results?

RESPONSE We have the reaction $n + ^{16}_{8}\text{O} \rightarrow ? + ^{2}_{1}\text{H}$. The total number of nucleons initially is $1 + 16 = 17$, and the total charge is $0 + 8 = 8$. The same totals apply after the reaction. Hence the product nucleus must have $Z = 7$ and $A = 15$. From the Periodic Table, we find that it is nitrogen that has $Z = 7$, so the nucleus produced is $^{15}_{7}\text{N}$.

| **EXERCISE A** Determine the resulting nucleus in the reaction $n + ^{137}_{56}\text{Ba} \rightarrow ? + \gamma$.

Energy and momentum are also conserved in nuclear reactions, and can be used to determine whether or not a given reaction can occur. For example, if the total mass of the final products is less than the total mass of the initial particles, this decrease in mass (recall $\Delta E = \Delta m \, c^2$) is converted to kinetic energy (KE) of the outgoing particles. But if the total mass of the products is greater than the total mass of the initial reactants, the reaction requires energy. The reaction will then not occur unless the bombarding particle has sufficient kinetic energy. Consider a nuclear reaction of the general form

$$a + X \rightarrow Y + b, \tag{31–1}$$

where particle a is a moving projectile particle (or small nucleus) that strikes nucleus X, producing nucleus Y and particle b (typically, p, n, α, γ). We define the **reaction energy**, or **Q-value**, in terms of the masses involved, as

$$Q = (M_a + M_X - M_b - M_Y)c^2. \tag{31–2a}$$

For a γ ray, $M = 0$. If energy is released by the reaction, $Q > 0$. If energy is required, $Q < 0$.

Because energy is conserved, Q has to be equal to the change in kinetic energy (final minus initial):

$$Q = \text{KE}_b + \text{KE}_Y - \text{KE}_a - \text{KE}_X. \tag{31–2b}$$

If X is a target nucleus at rest (or nearly so) struck by incoming particle a, then $\text{KE}_X = 0$. For $Q > 0$, the reaction is said to be *exothermic* or *exoergic*; energy is released in the reaction, so the total kinetic energy is greater after the reaction than before. If Q is negative, the reaction is said to be *endothermic* or *endoergic*: an energy input is required to make the reaction happen. The energy input comes from the kinetic energy of the initial colliding particles (a and X).

EXAMPLE 31–2 | **A slow-neutron reaction.** The nuclear reaction

$$n + ^{10}_{5}\text{B} \rightarrow ^{7}_{3}\text{Li} + ^{4}_{2}\text{He}$$

is observed to occur even when very slow-moving neutrons (mass $M_n = 1.0087 \text{ u}$) strike boron atoms at rest. For a particular reaction in which $\text{KE}_n \approx 0$, the outgoing helium ($M_{\text{He}} = 4.0026 \text{ u}$) is observed to have a speed of $9.30 \times 10^6 \text{ m/s}$. Determine (*a*) the kinetic energy of the lithium ($M_{\text{Li}} = 7.0160 \text{ u}$), and (*b*) the Q-value of the reaction.

APPROACH Since the neutron and boron are both essentially at rest, the total momentum before the reaction is zero; momentum is conserved and so must be zero afterward as well. Thus,

$$M_{Li} v_{Li} = M_{He} v_{He}.$$

We solve this for v_{Li} and substitute it into the equation for kinetic energy. In (b) we use Eq. 31–2b.

SOLUTION (a) We can use classical kinetic energy with little error, rather than relativistic formulas, because $v_{He} = 9.30 \times 10^6$ m/s is not close to the speed of light c. And v_{Li} will be even less because $M_{Li} > M_{He}$. Thus we can write the KE of the lithium, using the momentum equation just above, as

$$\text{KE}_{Li} = \frac{1}{2} M_{Li} v_{Li}^2 = \frac{1}{2} M_{Li} \left(\frac{M_{He} v_{He}}{M_{Li}} \right)^2 = \frac{M_{He}^2 v_{He}^2}{2 M_{Li}}.$$

We put in numbers, changing the mass in u to kg and recall that 1.60×10^{-13} J = 1 MeV:

$$\text{KE}_{Li} = \frac{(4.0026\,\text{u})^2 (1.66 \times 10^{-27}\,\text{kg/u})^2 (9.30 \times 10^6\,\text{m/s})^2}{2(7.0160\,\text{u})(1.66 \times 10^{-27}\,\text{kg/u})}$$

$$= 1.64 \times 10^{-13}\,\text{J} = 1.02\,\text{MeV}.$$

(b) We are given the data $\text{KE}_a = \text{KE}_X = 0$ in Eq. 31–2b, so $Q = \text{KE}_{Li} + \text{KE}_{He}$, where

$$\text{KE}_{He} = \frac{1}{2} M_{He} v_{He}^2 = \frac{1}{2}(4.0026\,\text{u})(1.66 \times 10^{-27}\,\text{kg/u})(9.30 \times 10^6\,\text{m/s})^2$$

$$= 2.87 \times 10^{-13}\,\text{J} = 1.80\,\text{MeV}.$$

Hence, $Q = 1.02\,\text{MeV} + 1.80\,\text{MeV} = 2.82\,\text{MeV}$.

EXAMPLE 31–3 **Will the reaction "go"?** Can the reaction

$$p + {}^{13}_{6}\text{C} \rightarrow {}^{13}_{7}\text{N} + n$$

occur when ${}^{13}_{6}\text{C}$ is bombarded by 2.0-MeV protons?

APPROACH The reaction will "go" if the reaction is exothermic ($Q > 0$) and even if $Q < 0$ if the input momentum and kinetic energy are sufficient. First we calculate Q from the difference between final and initial masses using Eq. 31–2a, and look up the masses in Appendix B.

SOLUTION The total masses before and after the reaction are:

Before		After	
$M({}^{13}_{6}\text{C}) =$	13.003355	$M({}^{13}_{7}\text{N}) =$	13.005739
$M({}^{1}_{1}\text{H}) =$	1.007825	$M(n) =$	1.008665
	14.011180		14.014404

(We must use the mass of the ${}^{1}_{1}\text{H}$ atom rather than that of the bare proton because the masses of ${}^{13}_{6}\text{C}$ and ${}^{13}_{7}\text{N}$ include the electrons, and we must include an equal number of electron masses on each side of the equation.) The products have an excess mass of

$$(14.014404 - 14.011180)\text{u} = 0.003224\,\text{u} \times 931.5\,\text{MeV/u} = 3.00\,\text{MeV}.$$

Thus $Q = -3.00$ MeV, and the reaction is endothermic. This reaction requires energy, and the 2.0-MeV protons do not have enough to make it go.

NOTE The incoming proton in this Example would need more than 3.00 MeV of kinetic energy to make this reaction go; 3.00 MeV would be enough to conserve energy, but a proton of this energy would produce the ${}^{13}_{7}\text{N}$ and n with no kinetic energy and hence no momentum. Since an incident 3.0-MeV proton has momentum, conservation of momentum would be violated. A calculation using conservation of energy *and* of momentum, as we did in Examples 30–7 and 31–2, shows that the minimum proton energy, called the **threshold energy**, is 3.23 MeV in this case.

(a)

Neutron captured by $^{238}_{92}$U.

(b)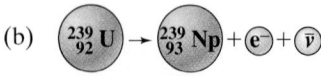

$^{239}_{92}$U decays by β decay to neptunium-239.

(c)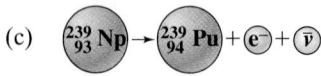

$^{239}_{93}$Np itself decays by β decay to produce plutonium-239.

FIGURE 31–1 Neptunium and plutonium are produced in this series of reactions, after bombardment of $^{238}_{92}$U by neutrons.

FIGURE 31–2 Projectile particles strike a target of area A and thickness ℓ made up of n nuclei per unit volume.

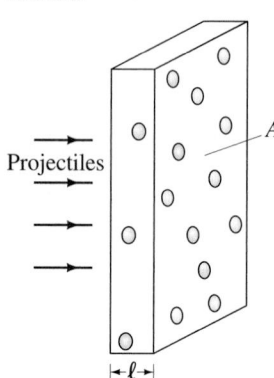

Neutron Physics

The artificial transmutation of elements took a great leap forward in the 1930s when Enrico Fermi realized that neutrons would be the most effective projectiles for causing nuclear reactions and in particular for producing new elements. Because neutrons have no net electric charge, they are not repelled by positively charged nuclei as are protons or alpha particles. Hence the probability of a neutron reaching the nucleus and causing a reaction is much greater than for charged projectiles,[†] particularly at low energies. Between 1934 and 1936, Fermi and his co-workers in Rome produced many previously unknown isotopes by bombarding different elements with neutrons. Fermi realized that if the heaviest known element, uranium, is bombarded with neutrons, it might be possible to produce new elements with atomic numbers greater than that of uranium. After several years of hard work, it was suspected that two new elements had been produced, neptunium ($Z = 93$) and plutonium ($Z = 94$). The full confirmation that such "transuranic" elements could be produced came several years later at the University of California, Berkeley. The reactions are shown in Fig. 31–1.

It was soon shown that what Fermi had actually observed when he bombarded uranium was an even stranger process—one that was destined to play an extraordinary role in the world at large. We discuss it in Section 31–2.

*Cross Section

Some reactions have a higher probability of occurring than others. The reaction probability is specified by a quantity called the collision **cross section**. Although the size of a nucleus, like that of an atom, is not a clearly defined quantity since the edges are not distinct like those of a tennis ball or baseball, we can nonetheless define a *cross section* for nuclei undergoing collisions by using an analogy. Suppose that projectile particles strike a stationary target of total area A and thickness ℓ, as shown in Fig. 31–2. Assume also that the target is made up of identical objects (such as marbles or nuclei), each of which has a cross-sectional area σ, and we assume the incoming projectiles are small by comparison. We assume that the target objects are fairly far apart and the thickness ℓ is so small that we don't have to worry about overlapping. This is often a reasonable assumption because nuclei have diameters on the order of 10^{-14} m but are at least 10^{-10} m (atomic size) apart even in solids. If there are n nuclei per unit volume, the total cross-sectional area of all these tiny targets is

$$A' = nA\ell\sigma$$

since $nA\ell = (n)(\text{volume})$ is the total number of targets and σ is the cross-sectional area of each. If $A' \ll A$, most of the incident projectile particles will pass through the target without colliding. If R_0 is the rate at which the projectile particles strike the target (number/second), the rate at which collisions occur, R, is

$$R = R_0 \frac{A'}{A} = R_0 \frac{nA\ell\sigma}{A}$$

so

$$R = R_0 n\ell\sigma.$$

Thus, by measuring the collision rate, R, we can determine σ:

$$\sigma = \frac{R}{R_0 n\ell}.$$

The cross section σ is an "effective" target area. It is a *measure of the probability of a collision or of a particular reaction occurring* per target nucleus, independent of the dimensions of the entire target. The concept of cross section is useful

---

[†]That is, positively charged particles. Electrons rarely cause nuclear reactions because they do not interact via the strong nuclear force.

because σ depends only on the properties of the interacting particles, whereas R depends on the thickness and area of the physical (macroscopic) target, on the number of particles in the incident beam, and so on.

31–2 Nuclear Fission; Nuclear Reactors

In 1938, the German scientists Otto Hahn and Fritz Strassmann made an amazing discovery. Following up on Fermi's work, they found that uranium bombarded by neutrons sometimes produced smaller nuclei that were roughly half the size of the original uranium nucleus. Lise Meitner and Otto Frisch quickly realized what had happened: the uranium nucleus, after absorbing a neutron, actually had split into two roughly equal pieces. This was startling, for until then the known nuclear reactions involved knocking out only a tiny fragment (for example, n, p, or α) from a nucleus.

Nuclear Fission and Chain Reactions

This new phenomenon was named **nuclear fission** because of its resemblance to biological fission (cell division). It occurs much more readily for $^{235}_{92}U$ than for the more common $^{238}_{92}U$. The process can be visualized by imagining the uranium nucleus to be like a liquid drop. According to this **liquid-drop model**, the neutron absorbed by the $^{235}_{92}U$ nucleus (Fig. 31–3a) gives the nucleus extra internal energy (like heating a drop of water). This intermediate state, or **compound nucleus**, is $^{236}_{92}U$ (because of the absorbed neutron), Fig. 31–3b. The extra energy of this nucleus—it is in an excited state—appears as increased motion of the individual nucleons inside, which causes the nucleus to take on abnormal elongated shapes. When the nucleus elongates (in this model) into the shape shown in Fig. 31–3c, the attraction of the two ends via the short-range nuclear force is greatly weakened by the increased separation distance. Then the electric repulsive force becomes dominant, and the nucleus splits in two (Fig. 31–3d). The two resulting nuclei, X_1 and X_2, are called **fission fragments**, and in the process a number of neutrons (typically two or three) are also given off. The reaction can be written

$$n + {}^{235}_{92}U \rightarrow {}^{236}_{92}U \rightarrow X_1 + X_2 + \text{neutrons.} \qquad \textbf{(31–3)}$$

The compound nucleus, $^{236}_{92}U$, exists for less than 10^{-12} s, so the process occurs very quickly. The two fission fragments, X_1 and X_2, rarely split the original uranium mass precisely half and half, but more often as about 40%–60%. A typical fission reaction is

$$n + {}^{235}_{92}U \rightarrow {}^{141}_{56}Ba + {}^{92}_{36}Kr + 3n, \qquad \textbf{(31–4)}$$

although many others also occur.

| CONCEPTUAL EXAMPLE 31–4 | **Counting nucleons.** Identify the element X in the fission reaction $n + {}^{235}_{92}U \rightarrow {}^{A}_{Z}X + {}^{93}_{38}Sr + 2n$.

RESPONSE The number of nucleons is conserved (Section 30–7). The uranium nucleus with 235 nucleons plus the incoming neutron make $235 + 1 = 236$ nucleons. So there must be 236 nucleons after the reaction. The Sr has 93 nucleons, and the two neutrons make 95 nucleons, so X has $A = 236 - 95 = 141$. Electric charge is also conserved: before the reaction, the total charge is $92e$. After the reaction the total charge is $(Z + 38)e$ and must equal $92e$. Thus $Z = 92 - 38 = 54$. The element with $Z = 54$ is xenon (see Appendix B or the Periodic Table inside the back cover), so the isotope is $^{141}_{54}Xe$.

| **EXERCISE B** In the fission reaction $n + {}^{235}_{92}U \rightarrow {}^{137}_{53}I + {}^{96}_{39}Y + \text{neutrons}$, how many neutrons are produced?

Figure 31–4 shows the measured distribution of $^{235}_{92}U$ fission fragments according to mass. Only rarely (about 1 in 10^4) does a fission result in equal mass fragments (arrow in Fig. 31–4).

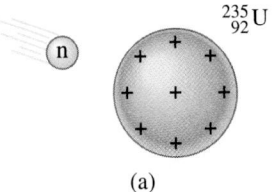

(a)

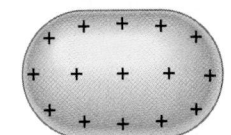

(b) $^{236}_{92}U$ (compound nucleus)

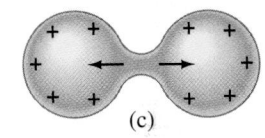

(c)

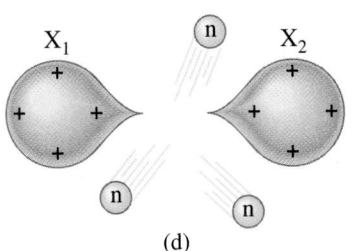

(d)

FIGURE 31–3 Fission of a $^{235}_{92}U$ nucleus after capture of a neutron, according to the liquid-drop model.

FIGURE 31–4 Mass distribution of fission fragments from $^{235}_{92}U$ + n. The small arrow indicates equal mass fragments ($\frac{1}{2} \times (236 - 2) = 117$, assuming 2 neutrons are liberated). Note that the vertical scale is logarithmic.

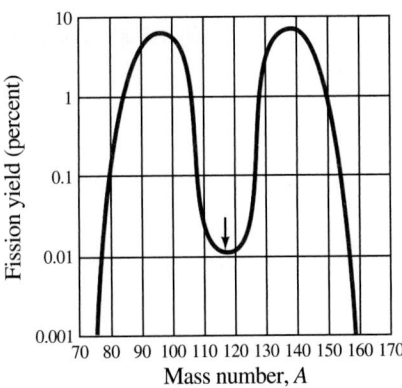

A tremendous amount of energy is released in a fission reaction because the mass of $^{235}_{92}U$ is considerably greater than the total mass of the fission fragments plus released neutrons. This can be seen from the binding-energy-per-nucleon curve of Fig. 30–1; the binding energy per nucleon for uranium is about 7.6 MeV/nucleon, but for fission fragments that have intermediate mass (in the center portion of the graph, $A \approx 100$), the average binding energy per nucleon is about 8.5 MeV/nucleon. Since the fission fragments are more tightly bound, the sum of their masses is less than the mass of the uranium. The difference in mass, or energy, between the original uranium nucleus and the fission fragments is about $8.5 - 7.6 = 0.9$ MeV per nucleon. Because there are 236 nucleons involved in each fission, the total energy released per fission is

$$(0.9 \,\text{MeV/nucleon})(236 \,\text{nucleons}) \approx 200 \,\text{MeV}. \qquad (31–5)$$

This is an enormous amount of energy for one single nuclear event. At a practical level, the energy from one fission is tiny. But if many such fissions could occur in a short time, an enormous amount of energy at the macroscopic level would be available. A number of physicists, including Fermi, recognized that the neutrons released in each fission (Eqs. 31–3 and 31–4) could be used to create a **chain reaction**. That is, one neutron initially causes one fission of a uranium nucleus; the two or three neutrons released can go on to cause additional fissions, so the process multiplies as shown schematically in Fig. 31–5.

If a **self-sustaining chain reaction** was actually possible in practice, the enormous energy available in fission could be released on a larger scale. Fermi and his co-workers (at the University of Chicago) showed it was possible by constructing the first **nuclear reactor** in 1942 (Fig. 31–6).

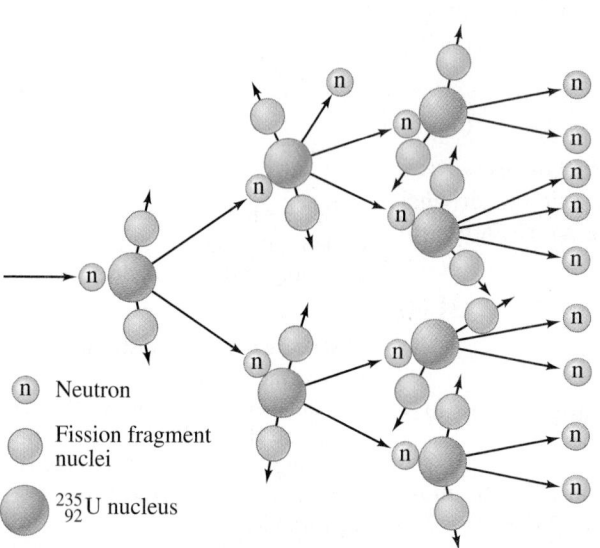

n Neutron

  Fission fragment
  nuclei

$^{235}_{92}U$ nucleus

FIGURE 31–5 Chain reaction.

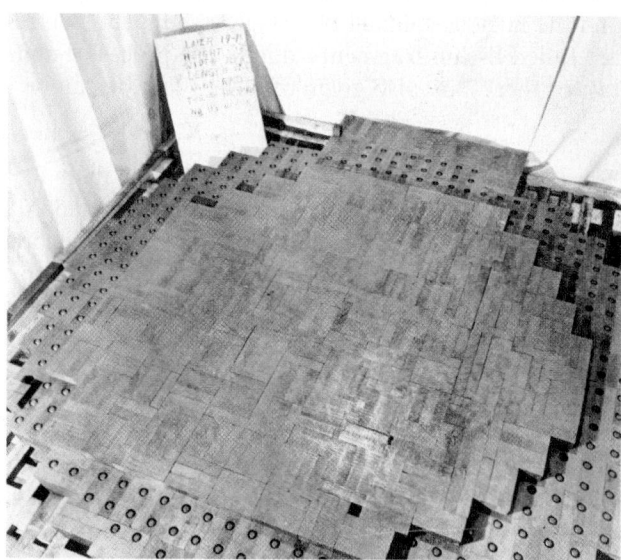

FIGURE 31–6 This is the only photograph of the first nuclear reactor, built by Fermi under the grandstand of Stagg Field at the University of Chicago. It is shown here under construction as a layer of graphite (used as moderator) was being placed over a layer of natural uranium. On December 2, 1942, Fermi slowly withdrew the cadmium control rods and the reactor went critical. This first self-sustaining chain reaction was announced to Washington, via telephone, by Arthur Compton who witnessed the event and reported: "The Italian navigator has just landed in the new world."

Nuclear Reactors

Several problems have to be overcome to make any nuclear reactor function. First, the probability that a $^{235}_{92}$U nucleus will absorb a neutron is large only for slow neutrons, but the neutrons emitted during a fission (which are needed to sustain a chain reaction) are moving very fast. A substance known as a **moderator** must be used to slow down the neutrons. The most effective moderator will consist of atoms whose mass is as close as possible to that of the neutrons. (To see why this is true, recall from Chapter 7 that a billiard ball striking an equal mass ball at rest can itself be stopped in one collision; but a billiard ball striking a heavy object bounces off with nearly unchanged speed.) The best moderator would thus contain $^1_1$H atoms. Unfortunately, $^1_1$H tends to absorb neutrons. But the isotope of hydrogen called *deuterium*, $^2_1$H, does not absorb many neutrons and is thus an almost ideal moderator. Either $^1_1$H or $^2_1$H can be used in the form of water. In the latter case, it is **heavy water**, in which the hydrogen atoms have been replaced by deuterium. Another common moderator is *graphite*, which consists of $^{12}_6$C atoms.

A second problem is that the neutrons produced in one fission may be absorbed and produce other nuclear reactions with other nuclei in the reactor, rather than produce further fissions. In a "light-water" reactor, the $^1_1$H nuclei absorb neutrons, as does $^{238}_{92}$U to form $^{239}_{92}$U in the reaction $n + {}^{238}_{92}U \rightarrow {}^{239}_{92}U + \gamma$. Naturally occurring uranium[†] contains 99.3% $^{238}_{92}$U and only 0.7% fissionable $^{235}_{92}$U. To increase the probability of fission of $^{235}_{92}$U nuclei, natural uranium can be **enriched** to increase the percentage of $^{235}_{92}$U by using processes such as diffusion or centrifugation. Enrichment is not usually necessary for reactors using heavy water as moderator because heavy water doesn't absorb neutrons.

The third problem is that some neutrons will escape through the surface of the reactor core before they can cause further fissions (Fig. 31–7). Thus the mass of fuel must be sufficiently large for a self-sustaining chain reaction to take place. The minimum mass of uranium needed is called the **critical mass**. The value of the critical mass depends on the moderator, the fuel ($^{239}_{94}$Pu may be used instead of $^{235}_{92}$U), and how much the fuel is enriched, if at all. Typical values are on the order of a few kilograms (that is, neither grams nor thousands of kilograms). Critical mass depends also on the average number of neutrons released per fission: 2.5 for $^{235}_{92}$U, 2.9 for $^{239}_{94}$Pu so the critical mass for $^{239}_{94}$Pu is smaller.

To have a self-sustaining chain reaction, on average at least one neutron produced in each fission must go on to produce another fission. The average number of neutrons per fission that do go on to produce further fissions is called the **neutron multiplication factor**, f. For a self-sustaining chain reaction, we must have $f \geq 1$. If $f < 1$, the reactor is "subcritical." If $f > 1$, it is "supercritical" (and could become dangerously explosive). Reactors are equipped with movable **control rods** (good neutron absorbers like cadmium or boron), whose function is to absorb neutrons and maintain the reactor at just barely "critical," $f = 1$.

The release of neutrons and subsequent fissions occur so quickly that manipulation of the control rods to maintain $f = 1$ would not be possible if it weren't for the small percentage ($\approx 1\%$) of so-called **delayed neutrons**. They come from the decay of neutron-rich fission fragments (or their daughters) having lifetimes on the order of seconds—sufficient to allow enough reaction time to operate the control rods and maintain $f = 1$.

Nuclear reactors have been built for use in research and to produce electric power. Fission produces many neutrons and a "research reactor" is basically an intense source of neutrons. These neutrons can be used as projectiles in nuclear reactions to produce nuclides not found in nature, including isotopes used as tracers and for therapy. A "power reactor" is used to produce electric power.

[†]$^{238}_{92}$U will fission, but only with fast neutrons ($^{238}_{92}$U is more stable than $^{235}_{92}$U). The probability of absorbing a fast neutron and producing a fission is too low to produce a self-sustaining chain reaction.

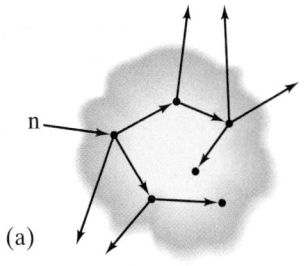

(a)

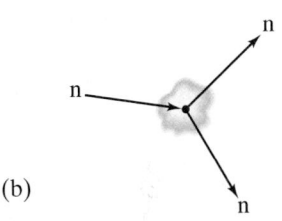

(b)

FIGURE 31–7 If the amount of uranium exceeds the critical mass, as in (a), a sustained chain reaction is possible. If the mass is less than critical, as in (b), too many neutrons escape before additional fissions occur, and the chain reaction is not sustained.

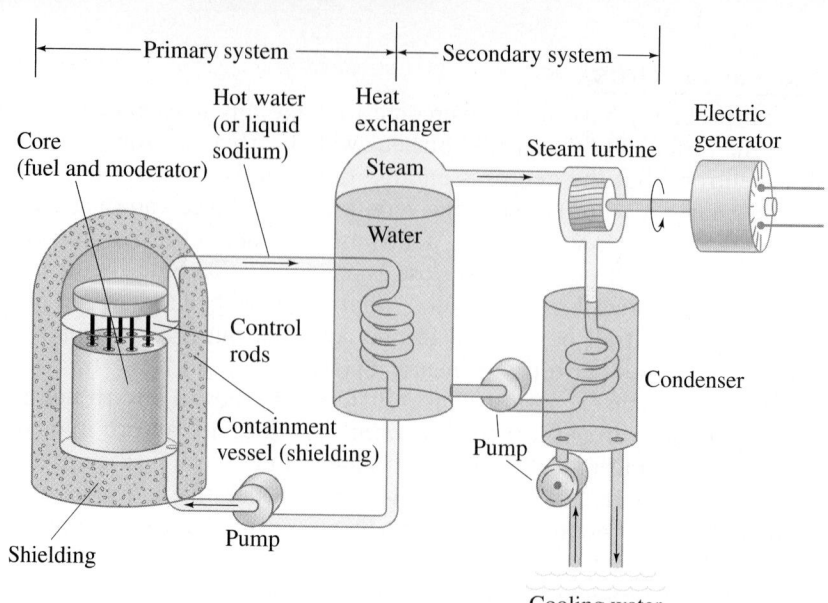

FIGURE 31–8 A nuclear reactor. The heat generated by the fission process in the fuel rods is carried off by hot water or liquid sodium and is used to boil water to steam in the heat exchanger. The steam drives a turbine to generate electricity and is then cooled in the condenser (to reduce pressure on the back side of the turbine blades).

The energy released in the fission process appears as heat, which is used to boil water and produce steam to drive a turbine connected to an electric generator (Fig. 31–8). The **core** of a nuclear reactor consists of the fuel and a moderator (water in most U.S. commercial reactors). The fuel is usually uranium enriched so that it contains 2 to 4 percent $^{235}_{92}\text{U}$. Water at high pressure or other liquid (such as liquid sodium) is allowed to flow through the core. The thermal energy it absorbs is used to produce steam in the heat exchanger, so the fissionable fuel acts as the heat input for a heat engine (Chapter 15).

There are problems associated with nuclear power plants. Besides the usual thermal pollution associated with any heat engine (Section 15–11), there is the serious problem of disposal of the radioactive fission fragments produced in the reactor, plus radioactive nuclides produced by neutrons interacting with the structural parts of the reactor. Fission fragments, like their uranium or plutonium parents, have about 50% more neutrons than protons. Nuclei with atomic number in the typical range for fission fragments ($Z \approx 30$ to 60) are stable only if they have more nearly equal numbers of protons and neutrons (see Fig. 30–2). Hence the highly neutron-rich fission fragments are very unstable and decay radioactively. The accidental release of highly radioactive fission fragments into the atmosphere poses a serious threat to human health (Section 31–4), as does possible leakage of the radioactive wastes when they are disposed of. The accidents at Three Mile Island, Pennsylvania (1979), at Chernobyl, Russia (1986), and at Fukushima, Japan (2011), have illustrated some of the dangers and have shown that nuclear plants must be located, constructed, maintained, and operated with great care and precision (Fig. 31–9).

Finally, the lifetime of nuclear power plants is limited to 30-some years, due to buildup of radioactivity and the fact that the structural materials themselves are weakened by the intense conditions inside. The cost of "decommissioning" a power plant is very great.

So-called **breeder reactors** were proposed as a solution to the problem of limited supplies of fissionable uranium, $^{235}_{92}\text{U}$. A breeder reactor is one in which some of the neutrons produced in the fission of $^{235}_{92}\text{U}$ are absorbed by $^{238}_{92}\text{U}$, and $^{239}_{94}\text{Pu}$ is produced via the set of reactions shown in Fig. 31–1. $^{239}_{94}\text{Pu}$ is fissionable with slow neutrons, so after separation it can be used as a fuel in a nuclear reactor. Thus a breeder reactor "breeds" new fuel[†] ($^{239}_{94}\text{Pu}$) from otherwise useless $^{238}_{92}\text{U}$. Natural uranium is 99.3 percent $^{238}_{92}\text{U}$, which in a breeder becomes useful fissionable $^{239}_{94}\text{Pu}$, thus increasing the supply of fissionable fuel by more than a factor of 100. But breeder reactors have the same problems as other reactors, plus other serious problems. Not only is plutonium a serious health hazard in itself (radioactive with a half-life of 24,000 years), but plutonium produced in a reactor can readily be used in a bomb, increasing the danger of nuclear proliferation and theft of fuel to produce a bomb.

FIGURE 31–9 Smoke rising from Fukushima, Japan, after the nuclear power plant meltdown in 2011.

[†]A breeder reactor does *not* produce more fuel than it uses.

Nuclear power presents risks. Other large-scale energy-conversion methods, such as conventional oil and coal-burning steam plants, also present health and environmental hazards; some of them were discussed in Section 15–11, and include air pollution, oil spills, and the release of CO_2 gas which can trap heat as in a greenhouse to raise the Earth's temperature. The solution to the world's needs for energy is not only technological, but also economic and political. A major factor surely is to "conserve"—to minimize our energy use. "Reduce, reuse, recycle."

EXAMPLE 31–5 | **Uranium fuel amount.** Estimate the minimum amount of $^{235}_{92}U$ that needs to undergo fission in order to run a 1000-MW power reactor per year of continuous operation. Assume an efficiency (Chapter 15) of about 33%.

APPROACH At 33% efficiency, we need 3×1000 MW $= 3000 \times 10^6$ J/s input. Each fission releases about 200 MeV (Eq. 31–5), so we divide the energy for a year by 200 MeV to get the number of fissions needed per year. Then we multiply by the mass of one uranium atom.

SOLUTION For 1000 MW output, the total power generation needs to be 3000 MW, of which 2000 MW is dumped as "waste" heat. Thus the total energy release in 1 yr $(3 \times 10^7 \text{ s})$ from fission needs to be about

$$(3 \times 10^9 \text{ J/s})(3 \times 10^7 \text{ s}) \approx 10^{17} \text{ J}.$$

If each fission releases 200 MeV of energy, the number of fissions required for a year is

$$\frac{(10^{17} \text{ J})}{(2 \times 10^8 \text{ eV/fission})(1.6 \times 10^{-19} \text{ J/eV})} \approx 3 \times 10^{27} \text{ fissions}.$$

The mass of a single uranium atom is about $(235 \text{ u})(1.66 \times 10^{-27} \text{ kg/u}) \approx 4 \times 10^{-25}$ kg, so the total uranium mass needed is

$$(4 \times 10^{-25} \text{ kg/fission})(3 \times 10^{27} \text{ fissions}) \approx 1000 \text{ kg},$$

or about a ton of $^{235}_{92}U$.

NOTE Because $^{235}_{92}U$ makes up only 0.7% of natural uranium, the yearly requirement for uranium is on the order of a hundred tons. This is orders of magnitude less than coal, both in mass and volume. Coal releases 2.8×10^7 J/kg, whereas $^{235}_{92}U$ can release 10^{17} J per ton, as we just calculated, or 10^{17} J/10^3 kg $= 10^{14}$ J/kg. For natural uranium, the figure is 100 times less, 10^{12} J/kg.

EXERCISE C A nuclear-powered submarine needs 6000-kW input power. How many $^{235}_{92}U$ fissions is this per second?

Atom Bomb

The first use of fission, however, was not to produce electric power. Instead, it was first used as a fission bomb (called the "atomic bomb"). In early 1940, with Europe already at war, Germany's leader, Adolf Hitler, banned the sale of uranium from the Czech mines he had recently taken over. Research into the fission process suddenly was enshrouded in secrecy. Physicists in the United States were alarmed. A group of them approached Einstein—a man whose name was a household word—to send a letter to President Franklin Roosevelt about the possibilities of using nuclear fission for a bomb far more powerful than any previously known, and inform him that Germany might already have begun development of such a bomb. Roosevelt responded by authorizing the program known as the Manhattan Project, to see if a bomb could be built. Work began in earnest after Fermi's demonstration in 1942 that a sustained chain reaction was possible. A new secret laboratory was developed on an isolated mesa in New Mexico known as Los Alamos. Under the direction of J. Robert Oppenheimer (1904–1967; Fig. 31–10), it became the home of famous scientists from all over Europe and the United States.

FIGURE 31–10 J. Robert Oppenheimer, on the left, with General Leslie Groves, who was the administrative head of Los Alamos during World War II. The photograph was taken at the Trinity site in the New Mexico desert, where the first atomic bomb was exploded.

FIGURE 31–11 Photo taken a month after the bomb was dropped on Nagasaki. The shacks were constructed afterwards from debris in the ruins. The bombs dropped on Hiroshima and Nagasaki were each equivalent to about 20,000 tons of the common explosive TNT ($\sim 10^{14}$ J).

To build a bomb that was subcritical during transport but that could be made supercritical (to produce a chain reaction) at just the right moment, two pieces of uranium were used, each less than the critical mass but together greater than the critical mass. The two masses, kept separate until the moment of detonation, were then forced together quickly by a kind of gun, and a chain reaction of explosive proportions occurred. An alternate bomb detonated conventional explosives (TNT) surrounding a plutonium sphere to compress it by implosion to double its density, making it more than critical and causing a nuclear explosion. The first fission bomb was tested in the New Mexico desert in July 1945. It was successful. In early August, a fission bomb using uranium was dropped on Hiroshima and a second, using plutonium, was dropped on Nagasaki (Fig. 31–11), both in Japan. World War II ended shortly thereafter.

Besides its destructive power, a fission bomb produces many highly radioactive fission fragments, as does a nuclear reactor. When a fission bomb explodes, these radioactive isotopes are released into the atmosphere as **radioactive fallout**.

Testing of nuclear bombs in the atmosphere after World War II was a cause of concern, because the movement of air masses spread the fallout all over the globe. Radioactive fallout eventually settles to the Earth, particularly in rainfall, and is absorbed by plants and grasses and enters the food chain. This is a far more serious problem than the same radioactivity on the exterior of our bodies, because α and β particles are largely absorbed by clothing and the outer (dead) layer of skin. But inside our bodies as food, the isotopes are in contact with living cells. One particularly dangerous radioactive isotope is $^{90}_{38}\text{Sr}$, which is chemically much like calcium and becomes concentrated in bone, where it causes bone cancer and destroys bone marrow. The 1963 treaty signed by over 100 nations that bans nuclear weapons testing in the atmosphere was motivated because of the hazards of fallout.

31–3 Nuclear Fusion

The mass of every stable nucleus is less than the sum of the masses of its constituent protons and neutrons. For example, the mass of the helium isotope ^4_2He is less than the mass of two protons plus two neutrons, Example 30–3. If two protons and two neutrons were to come together to form a helium nucleus, there would be a loss of mass. This mass loss is manifested in the release of energy.

Nuclear Fusion; Stars

The process of building up nuclei by bringing together individual protons and neutrons, or building larger nuclei by combining small nuclei, is called **nuclear fusion**. In Fig. 31–12 (same as Fig. 30–1), we can see why small nuclei can combine to form larger ones with the release of energy: it is because the binding energy per nucleon is less for light nuclei than it is for heavier nuclei (up to about $A \approx 60$).

FIGURE 31–12 Average binding energy per nucleon as a function of mass number A for stable nuclei. Same as Fig. 30–1.

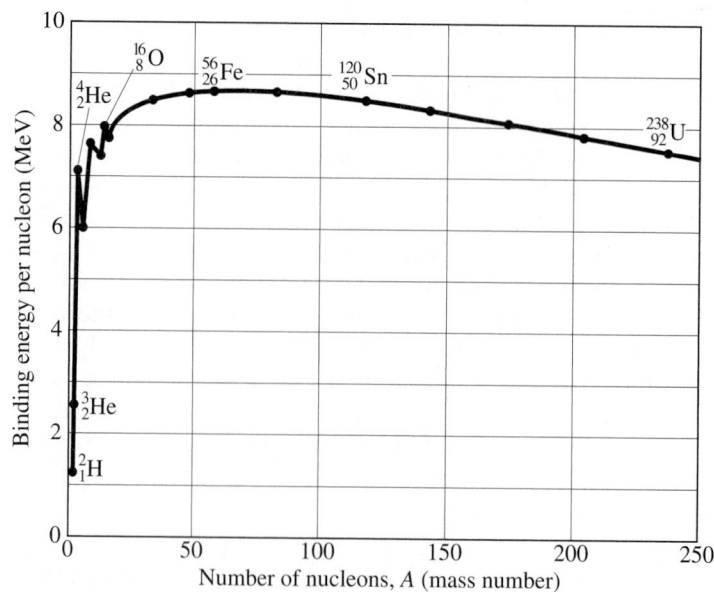

For two positively charged nuclei to get close enough to fuse, they must have very high kinetic energy to overcome the electric repulsion. It is believed that many of the elements in the universe were originally formed through the process of fusion in stars (see Chapter 33) where the temperature is extremely high, corresponding to high KE (Eq. 13–8). Today fusion is still producing the prodigious amounts of light energy (EM waves) stars emit, including our Sun.

EXAMPLE 31–6 **Fusion energy release.** One of the simplest fusion reactions involves the production of deuterium, ^2_1H, from a neutron and a proton: $^1_1H + n \rightarrow \, ^2_1H + \gamma$. How much energy is released in this reaction?

APPROACH The energy released equals the difference in mass (times c^2) between the initial and final masses.

SOLUTION From Appendix B, the initial mass is

$$1.007825 \, u \, + \, 1.008665 \, u \, = \, 2.016490 \, u,$$

and after the reaction the mass is that of the ^2_1H, namely 2.014102 u (the γ is massless). The mass difference is

$$2.016490 \, u \, - \, 2.014102 \, u \, = \, 0.002388 \, u,$$

so the energy released is

$$(\Delta m)c^2 \, = \, (0.002388 \, u)(931.5 \, \text{MeV}/u) \, = \, 2.22 \, \text{MeV},$$

and it is carried off by the ^2_1H nucleus and the γ ray.

The energy output of our Sun is believed to be due principally to the following sequence of fusion reactions:

$$^1_1H + \, ^1_1H \rightarrow \, ^2_1H + e^+ + \nu \qquad (0.42 \, \text{MeV}) \qquad \textbf{(31–6a)}$$

$$^1_1H + \, ^2_1H \rightarrow \, ^3_2He + \gamma \qquad (5.49 \, \text{MeV}) \qquad \textbf{(31–6b)}$$

$$^3_2He + \, ^3_2He \rightarrow \, ^4_2He + \, ^1_1H + \, ^1_1H. \qquad (12.86 \, \text{MeV}) \qquad \textbf{(31–6c)}$$

Proton– proton chain

where the energy released (Q-value) for each reaction is given in parentheses. These reactions are between nuclei (without electrons at these very high temperatures); the first reaction can be written as

$$p + p \rightarrow d + e^+ + \nu$$

where p = proton and d = deuteron. The net effect of this sequence, which is called the **proton–proton chain**, is for four protons to combine to form one ^4_2He nucleus plus two positrons, two neutrinos, and two gamma rays:

$$4 \, ^1_1H \rightarrow \, ^4_2He + 2e^+ + 2\nu + 2\gamma. \qquad \textbf{(31–7)}$$

Note that it takes two of each of the first two reactions (Eqs. 31–6a and b) to produce the two ^3_2He for the third reaction. So the total energy release for the net reaction, Eq. 31–7, is $(2 \times 0.42 \, \text{MeV} + 2 \times 5.49 \, \text{MeV} + 12.86 \, \text{MeV}) = 24.7 \, \text{MeV}$. In addition, each of the two e^+ (Eq. 31–6a) quickly annihilates with an electron to produce 2 γ rays (Section 27–6) with total energy $2m_e c^2 = 1.02 \, \text{MeV}$; so the total energy released is $(24.7 \, \text{MeV} + 2 \times 1.02 \, \text{MeV}) = 26.7 \, \text{MeV}$. The first reaction, the formation of deuterium from two protons (Eq. 31–6a), has a very low probability, and so limits the rate at which the Sun produces energy. (Thank goodness! This is why the Sun is still shining brightly.)

EXERCISE D Return to the Chapter-Opening Question, page 885, and answer it again now. Try to explain why you may have answered it differently the first time.

EXAMPLE 31–7 **ESTIMATE** **Estimating fusion energy.** Estimate the energy released if the following reaction occurred:

$$^2_1H + \, ^2_1H \rightarrow \, ^4_2He.$$

APPROACH We use Fig. 31–12 for a quick estimate.

SOLUTION We see in Fig. 31–12 that each ^2_1H has a binding energy of about $1\frac{1}{4} \, \text{MeV}/\text{nucleon}$, which for 2 nuclei of mass 2 is $4 \times \left(1\frac{1}{4}\right) \approx 5 \, \text{MeV}$. The ^4_2He has a binding energy per nucleon (Fig. 31–12) of about 7 MeV for a total of $4 \times 7 \, \text{MeV} \approx 28 \, \text{MeV}$. Hence the energy release is about $28 \, \text{MeV} - 5 \, \text{MeV} \approx 23 \, \text{MeV}$.

In stars hotter than the Sun, it is more likely that the energy output comes principally from the **carbon** (or **CNO**) **cycle**, which comprises the following sequence of reactions:

$$^{12}_{6}C + ^{1}_{1}H \rightarrow ^{13}_{7}N + \gamma$$

$$^{13}_{7}N \rightarrow ^{13}_{6}C + e^{+} + \nu$$

Carbon

$$^{13}_{6}C + ^{1}_{1}H \rightarrow ^{14}_{7}N + \gamma$$

cycle

$$^{14}_{7}N + ^{1}_{1}H \rightarrow ^{15}_{8}O + \gamma$$

$$^{15}_{8}O \rightarrow ^{15}_{7}N + e^{+} + \nu$$

$$^{15}_{7}N + ^{1}_{1}H \rightarrow ^{12}_{6}C + ^{4}_{2}He.$$

No net carbon is consumed in this cycle and the net effect is the same as the proton–proton chain, Eq. 31–7 (plus one extra γ). The theory of the proton–proton chain and of the carbon cycle as the source of energy for the Sun and stars was first worked out by Hans Bethe (1906–2005) in 1939.

CONCEPTUAL EXAMPLE 31–8 | **Stellar fusion.** What is the heaviest element likely to be produced in fusion processes in stars?

RESPONSE Fusion is possible if the final products have more binding energy (less mass) than the reactants, because then there is a net release of energy. Since the binding energy curve in Fig. 31–12 (or Fig. 30–1) peaks near $A \approx 56$ to 58 which corresponds to iron or nickel, it would not be energetically favorable to produce elements heavier than that. Nevertheless, in the center of massive stars or in supernova explosions, there is enough initial kinetic energy available to drive endothermic reactions that produce heavier elements as well.

EXERCISE E If the Sun is generating a constant amount of energy via fusion, the mass of the Sun must be (*a*) increasing, (*b*) decreasing, (*c*) constant, (*d*) irregular.

Possible Fusion Reactors

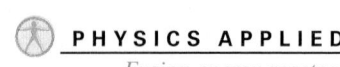

PHYSICS APPLIED

Fusion energy reactors

The possibility of utilizing the energy released in fusion to make a power reactor is very attractive. The fusion reactions most likely to succeed in a reactor involve the isotopes of hydrogen, $^{2}_{1}H$ (deuterium) and $^{3}_{1}H$ (tritium), and are as follows, with the energy released given in parentheses:

$$^{2}_{1}H + ^{2}_{1}H \rightarrow ^{3}_{1}H + ^{1}_{1}H \qquad (4.03 \text{ MeV}) \qquad \textbf{(31–8a)}$$

$$^{2}_{1}H + ^{2}_{1}H \rightarrow ^{3}_{2}He + n \qquad (3.27 \text{ MeV}) \qquad \textbf{(31–8b)}$$

$$^{2}_{1}H + ^{3}_{1}H \rightarrow ^{4}_{2}He + n. \qquad (17.59 \text{ MeV}) \qquad \textbf{(31–8c)}$$

Comparing these energy yields with that for the fission of $^{235}_{92}U$, we can see that the energy released in fusion reactions can be greater for a given mass of fuel than in fission. Furthermore, as fuel, a fusion reactor could use deuterium, which is very plentiful in the water of the oceans (the natural abundance of $^{2}_{1}H$ is 0.0115% on average, or about 1 g of deuterium per 80 L of water). The simple proton–proton reaction of Eq. 31–6a, which could use a much more plentiful source of fuel, $^{1}_{1}H$, has such a small probability of occurring that it cannot be considered a possibility on Earth.

Although a useful fusion reactor has not yet been achieved, considerable progress has been made in overcoming the inherent difficulties. The problems are associated with the fact that all nuclei have a positive charge and repel each other. However, if they can be brought close enough together so that the short-range attractive strong nuclear force can come into play, it can pull the nuclei together and fusion can occur. For the nuclei to get close enough together, they must have large kinetic energy to overcome the electric repulsion. High kinetic energies are readily attainable with particle accelerators (Chapter 32), but the number of particles involved is too small. To produce realistic amounts of energy, we must deal with matter in bulk, for which high kinetic energy means higher temperatures.

Indeed, very high temperatures are required for sustained fusion to occur, and fusion devices are often referred to as **thermonuclear devices**. The interiors of the Sun and other stars are very hot, many millions of degrees, so the nuclei are moving fast enough for fusion to take place, and the energy released keeps the temperature high so that further fusion reactions can occur. The Sun and the stars represent huge self-sustaining thermonuclear reactors that stay together because of their great gravitational mass. But on Earth, containment of the fast-moving nuclei at the high temperatures and densities required has proven difficult.

It was realized after World War II that the temperature produced within a fission (or "atomic") bomb was close to 10^8 K. This suggested that a fission bomb could be used to ignite a fusion bomb (popularly known as a thermonuclear or hydrogen bomb) to release the vast energy of fusion. The uncontrollable release of fusion energy in an H-bomb (in 1952) was relatively easy to obtain. But to realize usable energy from fusion at a slow and controlled rate has turned out to be a serious challenge.

EXAMPLE 31–9 | **ESTIMATE** | **Temperature needed for d–t fusion.** Estimate the temperature required for deuterium–tritium fusion (d–t) to occur.

APPROACH We assume the nuclei approach head-on, each with kinetic energy KE, and that the nuclear force comes into play when the distance between their centers equals the sum of their nuclear radii. The electrostatic potential energy (Chapter 17) of the two particles at this distance equals the minimum total kinetic energy of the two particles when far apart. The average kinetic energy is related to Kelvin temperature by Eq. 13–8.

SOLUTION The radii of the two nuclei ($A_d = 2$ and $A_t = 3$) are given by Eq. 30–1: $r_d \approx 1.5$ fm, $r_t \approx 1.7$ fm, so $r_d + r_t = 3.2 \times 10^{-15}$ m. We equate the kinetic energy of the two initial particles to the potential energy when at this distance:

$$2\text{KE} \approx \frac{1}{4\pi\epsilon_0} \frac{e^2}{(r_d + r_t)}$$

$$\approx \left(9.0 \times 10^9 \, \frac{\text{N} \cdot \text{m}^2}{\text{C}^2}\right) \frac{\left(1.6 \times 10^{-19} \, \text{C}\right)^2}{\left(3.2 \times 10^{-15} \, \text{m}\right)\left(1.6 \times 10^{-19} \, \text{J/eV}\right)} \approx 0.45 \, \text{MeV}.$$

Thus, KE ≈ 0.22 MeV, and if we ask that the average kinetic energy be this high, then from Eq. 13–8, $\frac{3}{2}kT = \overline{\text{KE}}$, we have a temperature of

$$T = \frac{2\overline{\text{KE}}}{3k} = \frac{2(0.22 \, \text{MeV})\left(1.6 \times 10^{-13} \, \text{J/MeV}\right)}{3\left(1.38 \times 10^{-23} \, \text{J/K}\right)} \approx 2 \times 10^9 \, \text{K}.$$

NOTE More careful calculations show that the temperature required for fusion is actually about an order of magnitude less than this rough estimate, partly because it is not necessary that the *average* kinetic energy be 0.22 MeV—a small percentage of nuclei with this much energy (in the high-energy tail of the Maxwell distribution, Fig. 13–20) would be sufficient. Reasonable estimates for a usable fusion reactor are in the range $T \gtrsim 1$ to 4×10^8 K.

A high temperature is required for a fusion reactor. But there must also be a high density of nuclei to ensure a sufficiently high collision rate. A real difficulty with controlled fusion is to contain nuclei long enough and at a high enough density for sufficient reactions to occur so that a usable amount of energy is obtained. At the temperatures needed for fusion, the atoms are ionized, and the resulting collection of nuclei and electrons is referred to as a **plasma**. Ordinary materials vaporize at a few thousand degrees at most, and hence cannot be used to contain a high-temperature plasma. Two major containment techniques are *magnetic confinement* and *inertial confinement*.

In **magnetic confinement**, magnetic fields are used to try to contain the hot plasma. A simple approach is the "magnetic bottle" shown in Fig. 31–13. The paths of the charged particles in the plasma are bent by the magnetic field; where magnetic field lines are close together, the force on the particles reflects them

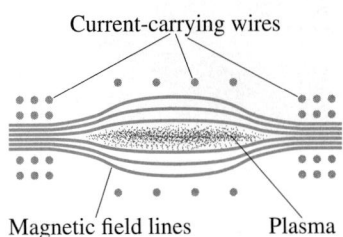

FIGURE 31–13 "Magnetic bottle" used to confine a plasma.

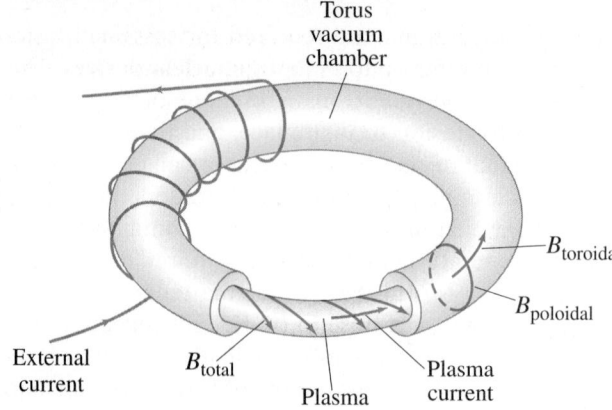

FIGURE 31–14 Tokamak configuration, showing the total $\vec{B}$ field due to external current plus current in the plasma itself.

Torus vacuum chamber

B_{toroidal}

B_{poloidal}

External current

B_{total}

Plasma

Plasma current

FIGURE 31–15 (a) Tokamak: split image view of the Joint European Torus (JET) located near Oxford, England. Interior, on the left, and an actual plasma in there (T $\approx 1 \times 10^8$ K) on the right. (b) A 2-mm-diameter round d–t (deuterium–tritium) inertial target, being filled through a thin glass tube from above, at the National Ignition Facility (NIF), Lawrence Livermore National Laboratory, California.

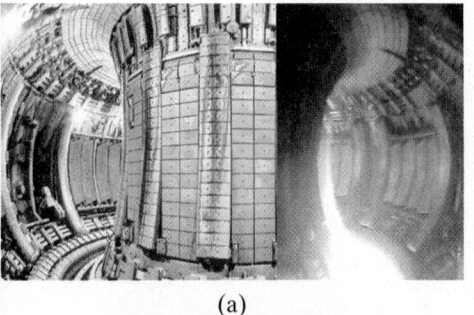

(a)

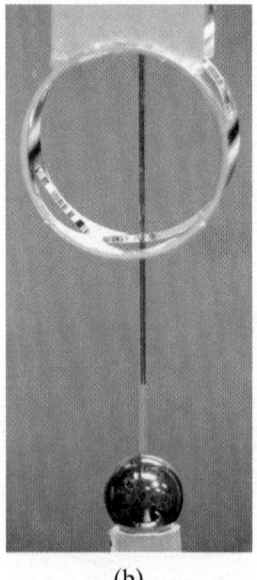

(b)

back toward the center. Unfortunately, magnetic bottles develop "leaks" and the charged particles slip out before sufficient fusion takes place. The most promising design today is the **tokamak**, first developed in Russia. A tokamak (Fig. 31–14) is toroid-shaped (a torus, which is like a donut) and involves complicated magnetic fields: current-carrying conductors produce a magnetic field directed along the axis of the toroid ("toroidal" field); an additional field is produced by currents within the plasma itself ("poloidal" field). The combination produces a helical field as shown in Fig. 31–14, confining the plasma, at least briefly, so it doesn't touch the vacuum chamber's metal walls (Fig. 31–15a).

In 1957, J. D. Lawson showed that the product of ion density n (= ions/m^3) and confinement time τ must exceed a minimum value of approximately

$$n\tau \ \gtrsim \ 3 \times 10^{20} \ \text{s/m}^3.$$

This **Lawson criterion** must be reached to produce **ignition**, meaning fusion that continues after all external heating is turned off. Practically, it is expected to be achieved with $n \approx 1$ to $3 \times 10^{20} \ \text{m}^{-3}$ and $\tau \approx 1$ to 3 s. To reach **break-even**, the point at which the energy output due to fusion is equal to the energy input to heat the plasma, requires an $n\tau$ about an order of magnitude less. The break-even point was very closely approached in the 1990s at the Tokamak Fusion Test Reactor (TFTR) at Princeton, and the very high temperature needed for ignition $(4 \times 10^8 \ \text{K})$ was exceeded—although not both of these at the same time.

Magnetic confinement fusion research continues throughout the world. This research will help us in developing the huge multinational test device (European Union, India, Japan, South Korea, Russia, China, and the U.S.), called ITER (International Thermonuclear Experimental Reactor). It is hoped that ITER will be finished and running by 2020, in France, with an expected power output of about 500 MW, 10 times the input energy. ITER is planned to be the final research step before building a working reactor.

The second method for containing the fuel for fusion is **inertial confinement fusion** (ICF): a small pellet or capsule of deuterium and tritium (Fig. 31–15b) is struck simultaneously from hundreds of directions by very intense laser beams. The intense influx of energy heats and ionizes the pellet into a plasma, compressing it and heating it to temperatures at which fusion can occur $(> 10^8 \ \text{K})$. The confinement time is on the order of 10^{-11} to 10^{-9} s, during which time the ions do not move appreciably because of their own inertia, and fusion can take place.

31–4 Passage of Radiation Through Matter; Biological Damage

When we speak of *radiation*, we include α, β, γ, and X-rays, as well as protons, neutrons, and other particles such as pions (see Chapter 32). Because charged particles can ionize the atoms or molecules of any material they pass through, they are referred to as **ionizing radiation**. And because radiation produces ionization, it can cause considerable damage to materials, particularly to biological tissue.

Charged particles, such as α and β rays and protons, cause ionization because of electric forces. That is, when they pass through a material, they can attract or repel electrons strongly enough to remove them from the atoms of the material. Since the α and β rays emitted by radioactive substances have energies on the order of 1 MeV (10^4 to 10^7 eV), whereas ionization of atoms and molecules requires on the order of 10 eV (Chapter 27), we see that a single α or β particle can cause thousands of ionizations.

Neutral particles also give rise to ionization when they pass through materials. For example, X-ray and γ-ray photons can ionize atoms by knocking out electrons by means of the photoelectric and Compton effects (Chapter 27). Furthermore, if a γ ray has sufficient energy (greater than 1.02 MeV), it can undergo pair production: an electron and a positron are produced (Section 27–6). The charged particles produced in all of these processes can themselves go on to produce further ionization. Neutrons, on the other hand, interact with matter mainly by collisions with nuclei, with which they interact strongly. Often the nucleus is broken apart by such a collision, altering the molecule of which it was a part. The fragments produced can in turn cause ionization.

Radiation passing through matter can do considerable damage. Metals and other structural materials become brittle and their strength can be weakened if the radiation is very intense, as in nuclear reactor power plants and for space vehicles that must pass through areas of intense cosmic radiation.

Biological Damage

The radiation damage produced in biological organisms is due primarily to ionization produced in cells. Several related processes can occur. Ions or radicals are produced that are highly reactive and take part in chemical reactions that interfere with the normal operation of the cell. All forms of radiation can ionize atoms by knocking out electrons. If these are bonding electrons, the molecule may break apart, or its structure may be altered so that it does not perform its normal function or may perform a harmful function. In the case of proteins, the loss of one molecule is not serious if there are other copies of the protein in the cell and additional copies can be made from the gene that codes for it. However, large doses of radiation may damage so many molecules that new copies cannot be made quickly enough, and the cell dies.

Damage to the DNA is more serious, since a cell may have only one copy. Each alteration in the DNA can affect a gene and alter the molecule that gene codes for (Section 29–3), so that needed proteins or other molecules may not be made at all. Again the cell may die. The death of a single cell is not normally a problem, since the body can replace it with a new one. (There are exceptions, such as neurons, which are mostly not replaceable, so their loss is serious.) But if many cells die, the organism may not be able to recover. On the other hand, a cell may survive but be defective. It may go on dividing and produce many more defective cells, to the detriment of the whole organism. Thus radiation can cause cancer—the rapid uncontrolled production of cells.

The possible damage done by the medical use of X-rays and other radiation must be balanced against the medical benefits and prolongation of life as a result of their diagnostic use.

31–5 Measurement of Radiation— Dosimetry

Although the passage of ionizing radiation through the human body can cause considerable damage, radiation can also be used to treat certain diseases, particularly cancer, often by using very narrow beams directed at a cancerous tumor in order to destroy it (Section 31–6). It is therefore important to be able to quantify the amount, or **dose**, of radiation. This is the subject of **dosimetry**.

The strength of a source can be specified at a given time by stating the **source activity**: how many nuclear decays (or disintegrations) occur per second. The traditional unit is the **curie** (Ci), defined as

$$1 \, \text{Ci} = 3.70 \times 10^{10} \text{ decays per second.}$$

(This number comes from the original definition as the activity of exactly one gram of radium.) Although the curie is still in common use, the SI unit for source activity is the **becquerel** (Bq), defined as

$$1 \, \text{Bq} = 1 \, \text{decay/s.}$$

Commercial suppliers of **radionuclides** (radioactive nuclides) used as tracers specify the activity at a given time. Because the activity decreases over time, more so for short-lived isotopes, it is important to take this decrease into account.

The magnitude of the source activity, $\Delta N / \Delta t$, is related to the number of radioactive nuclei present, N, and to the half-life, $T_{\frac{1}{2}}$, by (see Section 30–8):

$$\frac{\Delta N}{\Delta t} = \lambda N = \frac{0.693}{T_{\frac{1}{2}}} N.$$

EXAMPLE 31–10 **Radioactivity taken up by cells.** In a certain experiment, $0.016 \, \mu\text{Ci}$ of $^{32}_{15}\text{P}$ is injected into a medium containing a culture of bacteria. After 1.0 h the cells are washed and a 70% efficient detector (counts 70% of emitted β rays) records 720 counts per minute from the cells. What percentage of the original $^{32}_{15}\text{P}$ was taken up by the cells?

APPROACH The half-life of $^{32}_{15}\text{P}$ is about 14 days (Appendix B), so we can ignore any loss of activity over 1 hour. From the given activity, we find how many β rays are emitted. We can compare 70% of this to the $(720/\text{min})/(60 \, \text{s/min}) = 12$ per second detected.

SOLUTION The total number of decays per second originally was $(0.016 \times 10^{-6})(3.7 \times 10^{10}) = 590$. The counter could be expected to count 70% of this, or 410 per second. Since it counted $720/60 = 12$ per second, then $12/410 = 0.029$ or 2.9% was incorporated into the cells.

Another type of measurement is the exposure or **absorbed dose**—that is, the *effect* the radiation has on the absorbing material. The earliest unit of dosage was the **roentgen** (R), defined in terms of the amount of ionization produced by the radiation ($1 \, \text{R} = 1.6 \times 10^{12}$ ion pairs per gram of dry air at standard conditions). Today, 1 R is defined as the amount of X-ray or γ radiation that deposits $0.878 \times 10^{-2} \, \text{J}$ of energy per kilogram of air. The roentgen was largely superseded by another unit of absorbed dose applicable to any type of radiation, the **rad**: *1 rad is that amount of radiation which deposits energy per unit mass of $1.00 \times 10^{-2} \, \text{J/kg}$ in any absorbing material.* (This is quite close to the roentgen for X- and γ rays.) The proper SI unit for absorbed dose is the **gray** (Gy):

$$1 \, \text{Gy} = 1 \, \text{J/kg} = 100 \, \text{rad.} \tag{31–9}$$

The absorbed dose depends not only on the energy per particle and on the strength of a given source or of a radiation beam (number of particles per second), but also on the type of material that is absorbing the radiation. Bone, for example, absorbs more of X-ray or γ radiation normally used than does flesh, so the same beam passing through a human body deposits a greater dose (in rads or grays) in bone than in flesh.

The gray and the rad are physical units of dose—the energy deposited per unit mass of material. They are, however, not the most meaningful units for measuring the biological damage produced by radiation because equal doses of different types of radiation cause differing amounts of damage. For example, 1 rad of α radiation does 10 to 20 times the amount of damage as 1 rad of β or γ rays. This difference arises largely because α rays (and other heavy particles such as protons and neutrons) move much more slowly than β and γ rays of equal energy due to their greater mass. Hence, ionizing collisions occur closer together, so more irreparable damage can be done. The **relative biological effectiveness** (RBE) of a given type of radiation is defined as the number of rads of X-ray or γ radiation that produces the same biological damage as 1 rad of the given radiation. For example, 1 rad of slow neutrons does the same damage as 5 rads of X-rays. Table 31–1 gives the RBE for several types of radiation. The numbers are approximate because they depend somewhat on the energy of the particles and on the type of damage that is used as the criterion.

The **effective dose** can be given as the product of the dose in rads and the RBE, and this unit is known as the **rem** (which stands for *rad equivalent man*):

$$\text{effective dose (in rem)} = \text{dose (in rad)} \times \text{RBE.} \quad \textbf{(31–10a)}$$

This unit is being replaced by the SI unit for "effective dose," the **sievert** (Sv):

$$\text{effective dose (Sv)} = \text{dose (Gy)} \times \text{RBE} \quad \textbf{(31–10b)}$$

so

$$1 \text{ Sv} = 100 \text{ rem} \qquad \text{or} \qquad 1 \text{ rem} = 10 \text{ mSv.}$$

By these definitions, 1 rem (or 1 Sv) of any type of radiation does approximately the same amount of biological damage. For example, 50 rem of fast neutrons does the same damage as 50 rem of γ rays. But note that 50 rem of fast neutrons is only 5 rads, whereas 50 rem of γ rays is 50 rads.

Human Exposure to Radiation

We are constantly exposed to low-level radiation from natural sources: cosmic rays, natural radioactivity in rocks and soil, and naturally occurring radioactive isotopes in our food, such as $^{40}_{19}$K. **Radon**, $^{222}_{86}$Rn, is of considerable concern today. It is the product of radium decay and is an intermediate in the decay series from uranium (see Fig. 30–11). Most intermediates remain in the rocks where formed, but radon is a gas that can escape from rock (and from building material like concrete) to enter the air we breathe, and damage the interior of the lung.

The **natural radioactive background** averages about 0.30 rem (300 mrem) per year per person in the U.S., although there are large variations. From medical X-rays and scans, the average person receives about 50 to 60 mrem per year, giving an average total dose of about 360 mrem (3.6 mSv) per person. U.S. government regulators suggest an upper limit of allowed radiation for an individual in the general populace at about 100 mrem (1 mSv) per year in addition to natural background. It is believed that even low doses of radiation increase the chances of cancer or genetic defects; there is *no safe level* or threshold of radiation exposure.

The upper limit for people who work around radiation—in hospitals, in power plants, in research—has been set higher, a maximum of 20 mSv (2 rem) whole-body dose, averaged over some years (a maximum of 50 mSv (5 rem/yr) in any one year). To monitor exposure, those people who work around radiation generally carry some type of dosimeter, one common type being a **radiation film badge** which is a piece of film wrapped in light-tight material. The passage of ionizing radiation through the film changes it so that the film is darkened upon development, and thus indicates the received dose. Newer types include the *thermoluminescent dosimeter* (TLD). Dosimeters and badges do not protect the worker, but high levels detected suggest reassignment or modified work practices to reduce radiation exposure to acceptable levels.

Large doses of radiation can cause unpleasant symptoms such as nausea, fatigue, and loss of body hair, because of cellular damage. Such effects are sometimes referred to as **radiation sickness**. Very large doses can be fatal, although the time span of the dose is important. A brief dose of 10 Sv (1000 rem) is nearly always fatal. A 3-Sv (300-rem) dose in a short period of time is fatal in about 50% of patients within a month. However, the body possesses remarkable repair processes, so that a 3-Sv dose spread over several weeks is usually not fatal. It will, nonetheless, cause considerable damage to the body.

The effects of low doses over a long time are difficult to determine and are not well known as yet.

TABLE 31–1 Relative Biological Effectiveness (RBE)	
Type	**RBE**
X- and γ rays	1
β (electrons)	1
Protons	2
Slow neutrons	5
Fast neutrons	≈ 10
α particles and heavy ions	≈ 20

PHYSICS APPLIED
Radon

PHYSICS APPLIED
Human radiation exposure

PHYSICS APPLIED
Radiation worker exposure
Film badge

PHYSICS APPLIED
Radiation sickness

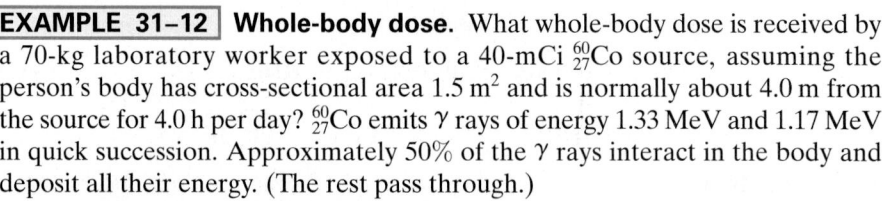

CONCEPTUAL EXAMPLE 31–11 | **Limiting the dose.** A worker in an environment with a radioactive source is warned that she is accumulating a dose too quickly and will have to lower her exposure by a factor of ten to continue working for the rest of the year. If the worker is able to work farther away from the source, how much farther away is necessary?

RESPONSE If the energy is radiated uniformly in all directions, then the intensity (dose/area) should decrease as the distance squared, just as it does for sound and light. If she can work four times farther away, the exposure lowers by a factor of sixteen, enough to make her safe.

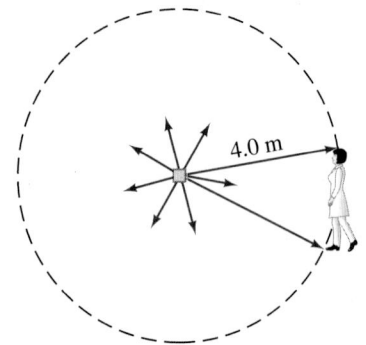

FIGURE 31–16 Radiation spreads out in all directions. A person 4.0 m away intercepts only a fraction: her cross-sectional area divided by the area of a sphere of radius 4.0 m. Example 31–12.

EXAMPLE 31–12 | **Whole-body dose.** What whole-body dose is received by a 70-kg laboratory worker exposed to a 40-mCi $^{60}_{27}$Co source, assuming the person's body has cross-sectional area 1.5 m^2 and is normally about 4.0 m from the source for 4.0 h per day? $^{60}_{27}$Co emits γ rays of energy 1.33 MeV and 1.17 MeV in quick succession. Approximately 50% of the γ rays interact in the body and deposit all their energy. (The rest pass through.)

APPROACH Of the given energy emitted, only a fraction passes through the worker, equal to *her* area divided by the total area (or $4\pi r^2$) over a full sphere of radius $r = 4.0$ m (Fig. 31–16).

SOLUTION The total γ-ray energy per decay is $(1.33 + 1.17)$ MeV $= 2.50$ MeV, so the total energy emitted by the source per second is

$$(0.040 \text{ Ci})(3.7 \times 10^{10} \text{ decays/Ci} \cdot \text{s})(2.50 \text{ MeV}) = 3.7 \times 10^9 \text{ MeV/s}.$$

The proportion of this energy intercepted by the body is its 1.5-m^2 area divided by the area of a sphere of radius 4.0 m (Fig. 31–16):

$$\frac{1.5 \text{ m}^2}{4\pi r^2} = \frac{1.5 \text{ m}^2}{4\pi(4.0 \text{ m})^2} = 7.5 \times 10^{-3}.$$

So the rate energy is deposited in the body (remembering that only 50% of the γ rays interact in the body) is

$$E = (\tfrac{1}{2})(7.5 \times 10^{-3})(3.7 \times 10^9 \text{ MeV/s})(1.6 \times 10^{-13} \text{ J/MeV}) = 2.2 \times 10^{-6} \text{ J/s}.$$

Since $1 \text{ Gy} = 1 \text{ J/kg}$, the whole-body dose rate for this 70-kg person is $(2.2 \times 10^{-6} \text{ J/s})/(70 \text{ kg}) = 3.1 \times 10^{-8}$ Gy/s. In 4.0 h, this amounts to a dose of

$$(4.0 \text{ h})(3600 \text{ s/h})(3.1 \times 10^{-8} \text{ Gy/s}) = 4.5 \times 10^{-4} \text{ Gy}.$$

RBE ≈ 1 for gammas, so the effective dose is 450 μSv (Eqs. 31–10b and 31–9) or:

$$(100 \text{ rad/Gy})(4.5 \times 10^{-4} \text{ Gy})(1 \text{ rem/rad}) = 45 \text{ mrem} = 0.45 \text{ mSv}.$$

NOTE This 45-mrem effective dose is almost 50% of the normal allowed dose for a whole year (100 mrem/yr), or 1% of the maximum one-year allowance for radiation workers. This worker should not receive such a large dose every day and should seek ways to reduce it (shield the source, vary the work, work farther from the source, work less time this close to source, etc.).

We have assumed that the intensity of radiation decreases as the square of the distance. It actually falls off faster than $1/r^2$ because of absorption in the air, so our answers are a slight overestimate of dose received.

PHYSICS APPLIED

Radon exposure

EXAMPLE 31–13 | **Radon exposure.** In the U.S., yearly deaths from radon exposure (the second leading cause of lung cancer) are estimated to exceed the yearly deaths from drunk driving. The Environmental Protection Agency recommends taking action to reduce the radon concentration in living areas if it exceeds 4 pCi/L of air. In some areas 50% of houses exceed this level from naturally occurring radon in the soil. Estimate (*a*) the number of decays/s in 1 m^3 of air and (*b*) the mass of radon that emits 4.0 pCi of $^{222}_{86}$Rn radiation.

APPROACH We can use the definition of the curie to determine how many decays per second correspond to 4 pCi, then Eq. 30–3b to determine how many nuclei of radon it takes to have this activity $\Delta N / \Delta t$.

SOLUTION (a) We saw at the start of this Section that $1\,\text{Ci} = 3.70 \times 10^{10}\,\text{decays/s}$. Thus

$$\frac{\Delta N}{\Delta t} = 4.0\,\text{pCi} = (4.0 \times 10^{-12}\,\text{Ci})(3.70 \times 10^{10}\,\text{decays/s/Ci})$$

$$= 0.148\,\text{s}^{-1}$$

per liter of air. In $1\,\text{m}^3$ of air $(1\,\text{m}^3 = 10^6\,\text{cm}^3 = 10^3\,\text{L})$ there would be $(0.148\,\text{s}^{-1})(1000) = 150\,\text{decays/s}$.

(b) From Eqs. 30–3b and 30–6

$$\frac{\Delta N}{\Delta t} = \lambda N = \frac{0.693}{T_{\frac{1}{2}}} N.$$

Appendix B tells us $T_{\frac{1}{2}} = 3.8235\,\text{days}$ for radon, so

$$N = \left(\frac{\Delta N}{\Delta t}\right) \frac{T_{\frac{1}{2}}}{0.693}$$

$$= (0.148\,\text{s}^{-1}) \frac{(3.8235\,\text{days})(8.64 \times 10^4\,\text{s/day})}{0.693}$$

$$= 7.06 \times 10^4\,\text{atoms of radon-222}.$$

The molar mass (222 u) and Avogadro's number are used to find the mass:

$$m = \frac{(7.06 \times 10^4\,\text{atoms})(222\,\text{g/mol})}{6.02 \times 10^{23}\,\text{atoms/mol}} = 2.6 \times 10^{-17}\,\text{g}$$

or 26 attograms in 1 L of air at the limit of 4 pCi/L. This $2.6 \times 10^{-17}\,\text{g/L}$ is 2.6×10^{-14} grams of radon per m^3 of air.

NOTE Each radon atom emits 4 α particles and 4 β particles, each one capable of causing many harmful ionizations, before the sequence of decays reaches a stable element.

*31–6 Radiation Therapy

The medical application of radioactivity and radiation to human beings involves two basic aspects: (1) **radiation therapy**—the treatment of disease (mainly cancer)—which we discuss in this Section; and (2) the *diagnosis* of disease, which we discuss in the following Sections of this Chapter.

Radiation can cause cancer. It can also be used to treat it. Rapidly growing cancer cells are especially susceptible to destruction by radiation. Nonetheless, large doses are needed to kill the cancer cells, and some of the surrounding normal cells are inevitably killed as well. It is for this reason that cancer patients receiving radiation therapy often suffer side effects characteristic of radiation sickness. To minimize the destruction of normal cells, a narrow beam of γ or X-rays is often used when a cancerous tumor is well localized. The beam is directed at the tumor, and the source (or body) is rotated so that the beam passes through various parts of the body to keep the dose at any one place as low as possible—except at the tumor and its immediate surroundings, where the beam passes at all times (Fig. 31–17). The radiation may be from a radioactive source such as $^{60}_{27}\text{Co}$, or it may be from an X-ray machine that produces photons in the range 200 keV to 5 MeV. Protons, neutrons, electrons, and pions, which are produced in particle accelerators (Section 32–1), are also being used in cancer therapy.

PHYSICS APPLIED
Radiation therapy

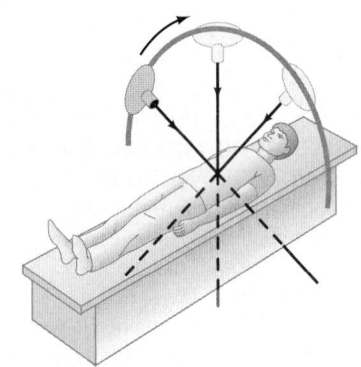

FIGURE 31–17 Radiation source rotates so that the beam always passes through the diseased tissue, but minimizes the dose in the rest of the body.

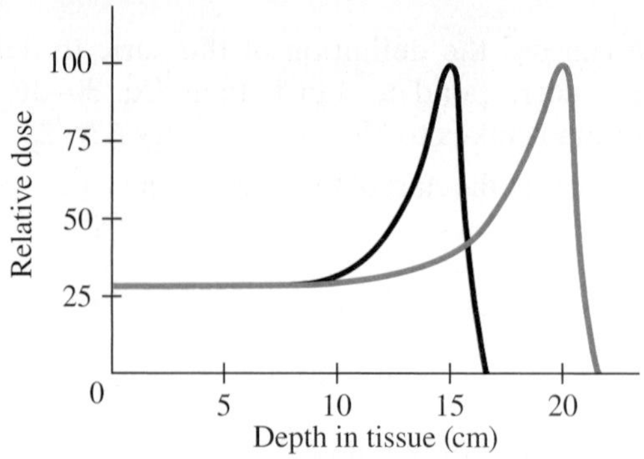

FIGURE 31–18 Energy deposited in tissue as a function of depth for 170-MeV protons (red curve) and 190-MeV protons (green). The peak of each curve is often called the Bragg peak.

PHYSICS APPLIED
Proton therapy

Protons used to kill tumors have a special property that makes them particularly useful. As shown in Fig. 31–18, when protons enter tissue, most of their energy is deposited at the end of their path. The protons' initial kinetic energy can be chosen so that most of the energy is deposited at the depth of the tumor itself, to destroy it. The incoming protons deposit only a small amount of energy in the tissue in front of the tumor, and none at all behind the tumor, thus having less negative effect on healthy tissue than X- or γ rays. Because tumors have physical size, even several centimeters in diameter, a range of proton energies is often used. Heavier ions, such as α particles or carbon ions, are similarly useful. This **proton therapy** technique is more than a half century old, but the necessity of having a large accelerator has meant that few hospitals have used the technique until now. Many such "proton centers" are now being built.

Another form of treatment is to insert a tiny radioactive source directly inside a tumor, which will eventually kill the majority of the cells. A similar technique is used to treat cancer of the thyroid with the radioactive isotope ${}^{131}_{53}\mathrm{I}$. The thyroid gland concentrates iodine present in the bloodstream, particularly in any area where abnormal growth is taking place. Its intense radioactivity can destroy the defective cells.

Another application of radiation is for sterilizing bandages, surgical equipment, and even packaged foods such as ground beef, chicken, and produce, because bacteria and viruses can be killed or deactivated by large doses of radiation.

*31–7 Tracers in Research and Medicine

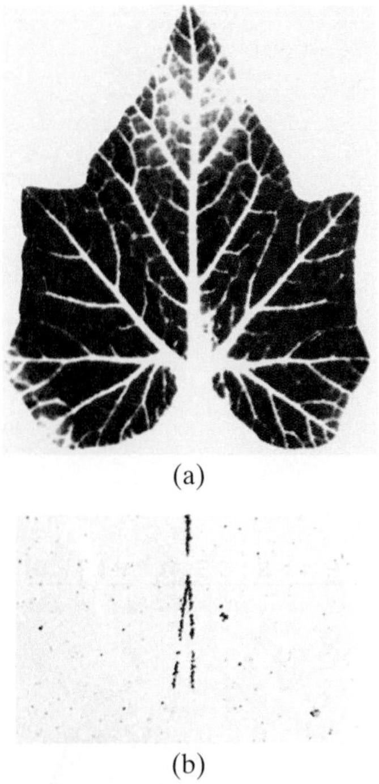

FIGURE 31–19 (a) Autoradiograph of a leaf exposed for 30 s to ${}^{14}CO_2$. Only the tissue where the CO_2 has been taken up, to be used in photosynthesis (Example 27–7), has become radioactive. The non-metabolizing tissue of the veins is free of ${}^{14}_{6}\mathrm{C}$ and does not blacken the X-ray sheet. (b) Autoradiograph of chromosomal DNA. The dashed arrays of film grains show the Y-shaped growing point of replicating DNA.

Radioactive isotopes are used in biological and medical research as **tracers**. A given compound is artificially synthesized incorporating a radioactive isotope such as ${}^{14}_{6}\mathrm{C}$ or ${}^{3}_{1}\mathrm{H}$. Such "tagged" molecules can then be traced as they move through an organism or as they undergo chemical reactions. The presence of these tagged molecules (or parts of them, if they undergo chemical change) can be detected by a Geiger or scintillation counter, which detects emitted radiation (see Section 30–13). How food molecules are digested, and to what parts of the body they are diverted, can be traced in this way.

Radioactive tracers have been used to determine how amino acids and other essential compounds are synthesized by organisms. The permeability of cell walls to various molecules and ions can be determined using radioactive tracers: the tagged molecule or ion is injected into the extracellular fluid, and the radioactivity present inside and outside the cells is measured as a function of time.

In a technique known as **autoradiography**, the position of the radioactive isotopes is detected on film. For example, the distribution of carbohydrates produced in the leaves of plants from absorbed CO_2 can be observed by keeping the plant in an atmosphere where the carbon atom in the CO_2 is ${}^{14}_{6}\mathrm{C}$. After a time, a leaf is placed firmly on a photographic plate and the emitted radiation darkens the film most strongly where the isotope is most strongly concentrated (Fig. 31–19a). Autoradiography using labeled nucleotides (components of DNA) has revealed much about the details of DNA replication (Fig. 31–19b). Today gamma cameras are used in a similar way—see next page.

For medical diagnosis, the radionuclide commonly used today is $^{99m}_{43}$Tc, a long-lived excited state of technetium-99 (the "m" in the symbol stands for "metastable" state). It is formed when $^{99}_{42}$Mo decays. The great usefulness of $^{99m}_{43}$Tc derives from its convenient half-life of 6 h (short, but not too short) and the fact that it can combine with a large variety of compounds. The compound to be labeled with the radionuclide is so chosen because it concentrates in the organ or region of the anatomy to be studied. Detectors outside the body then record, or image, the distribution of the radioactively labeled compound. The detection could be done by a single detector (Fig. 31–20a) which is moved across the body, measuring the intensity of radioactivity at a large number of points. The image represents the relative intensity of radioactivity at each point. The relative radioactivity is a diagnostic tool. For example, high or low radioactivity may represent overactivity or underactivity of an organ or part of an organ, or in another case may represent a lesion or tumor. More complex **gamma cameras** make use of many detectors which simultaneously record the radioactivity at many points. The measured intensities can be displayed on a TV or computer monitor. The image is sometimes called a scintigram (after scintillator), Fig. 31–20b. Gamma cameras are relatively inexpensive, but their resolution is limited—by non-perfect collimation[†]. Yet they allow "dynamic" studies: images that change in time, like a movie.

PHYSICS APPLIED
Medical diagnosis

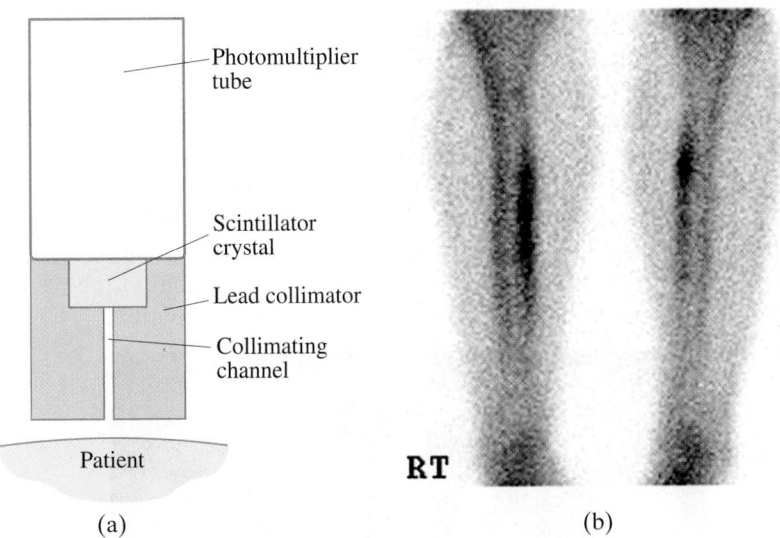

FIGURE 31–20 (a) Collimated gamma-ray detector for scanning (moving) over a patient. The collimator selects γ rays that come in a (nearly) straight line from the patient. Without the collimator, γ rays from all parts of the body could strike the scintillator, producing a poor image. Detectors today usually have many collimator tubes and are called *gamma cameras*. (b) Gamma camera image (scintigram), of both legs of a patient with shin splints, detecting γs from $^{99m}_{43}$Tc.

*31–8 Emission Tomography: PET and SPECT

The images formed using the standard techniques of nuclear medicine, as briefly discussed in the previous Section, are produced from radioactive tracer sources within the *volume* of the body. It is also possible to image the radioactive emissions from a single plane or slice through the body using the computed tomography techniques discussed in Section 25–12. A gamma camera measures the radioactive intensity from the tracer at many points and angles around the patient. The data are processed in much the same way as for X-ray CT scans (Section 25–12). This technique is referred to as **single photon emission computed tomography** (SPECT), or simply SPET (single photon emission tomography).

Another important technique is **positron emission tomography** (PET), which makes use of positron emitters such as $^{11}_{6}$C, $^{13}_{7}$N, $^{15}_{8}$O, and $^{18}_{9}$F whose half-lives are short. These isotopes are incorporated into molecules that, when inhaled or injected, accumulate in the organ or region of the body to be studied.

[†]To "collimate" means to "make parallel," usually by blocking non-parallel rays with a narrow tube inside lead, as in Fig. 31–20a.

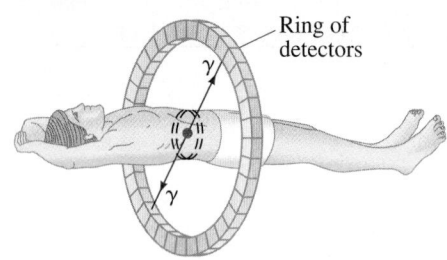

FIGURE 31–21 Positron emission tomography (PET) system showing a ring of detectors to detect the two annihilation γ rays $(e^+ + e^- \rightarrow 2\gamma)$ emitted at $180°$ to each other.

When such a nuclide undergoes β^+ decays, the emitted positron travels at most a few millimeters before it collides with a normal electron. In this collision, the positron and electron are annihilated, producing two γ rays $(e^+ + e^- \rightarrow 2\gamma)$, each having an energy of $511\,\mathrm{keV}\ (= m_e c^2)$. The two γ rays fly off in opposite directions $(180° \pm 0.25°)$ since they must have almost exactly equal and opposite momenta to conserve momentum (the momenta of the initial e^+ and e^- are essentially zero compared to the momenta afterward of the γ rays). Because the photons travel along the same line in opposite directions, their detection in coincidence by rings of detectors surrounding the patient (Fig. 31–21) readily establishes the line along which the emission took place. If the difference in time of arrival of the two photons could be determined accurately, the actual position of the emitting nuclide along that line could be calculated. Present-day electronics can measure times to at best $\pm 300\,\mathrm{ps}$, so at the γ ray's speed $(c = 3 \times 10^8\,\mathrm{m/s})$, the actual position could be determined to an accuracy on the order of about $d = vt \approx (3 \times 10^8\,\mathrm{m/s})(300 \times 10^{-12}\,\mathrm{s}) \approx 10\,\mathrm{cm}$, which is not very useful. Although there may be future potential for *time-of-flight* measurements to determine position, today computed tomography techniques are used instead, similar to those for X-ray CT, which can reconstruct PET images with a resolution on the order of 2–5 mm. One big advantage of PET is that no collimators are needed (as for detection of a single photon—see Fig. 31–20a). Thus, fewer photons are "wasted" and lower doses can be administered to the patient with PET.

Both PET and SPECT systems can give images that relate to biochemistry, metabolism, and function. This is to be compared to X-ray CT scans, whose images reflect shape and structure—that is, the anatomy of the imaged region.

Figure 31–22 shows PET scans of the same person's brain (a) when using a cell phone near the ear and (b) with the cell phone off. The bright red spots in (a) indicate a higher rate of glucose metabolism, suggesting excitability of brain tissue (the glucose was tagged with a radioactive tracer). Emfs from the cell phone antenna thus seem to affect metabolism and may be harmful to us!

The colors shown here are faked (only visible light has colors). The original images are various shades of gray, representing intensity (or counts).

FIGURE 31–22 False-color PET scans of a horizontal section through a brain showing glucose metabolism rates (red is high) by a person (a) using a cell phone near the ear, and (b) with the cell phone off.

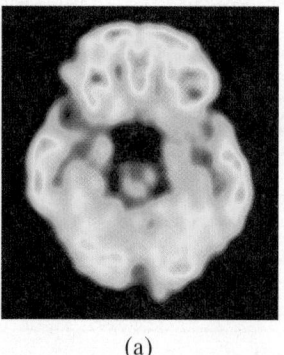

(a)

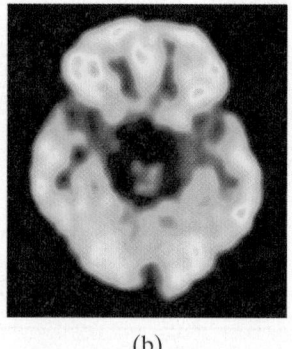

(b)

31–9 Nuclear Magnetic Resonance (NMR) and Magnetic Resonance Imaging (MRI)

Nuclear magnetic resonance (NMR) is a phenomenon which soon after its discovery in 1946 became a powerful research tool in a variety of fields from physics to chemistry and biochemistry. It is also an important medical imaging technique. We first briefly discuss the phenomenon, and then look at its applications.

*Nuclear Magnetic Resonance (NMR)

We saw in Chapter 28 (Section 28–6) that when atoms are placed in a magnetic field, atomic energy levels split into several closely spaced levels (see Fig. 28–8). Nuclei, too, exhibit these magnetic properties. We examine only the simplest, the hydrogen (H) nucleus, since it is the one most used, even for medical imaging.

The $^1_1$H nucleus consists of a single proton. Its spin angular momentum (and its magnetic moment), like that of the electron, can take on only two values when placed in a magnetic field: we call these "spin up" (parallel to the field) and "spin down" (antiparallel to the field), as suggested in Fig. 31–23. When a magnetic field is present, the energy of the nucleus splits into two levels as shown in Fig. 31–24, with the spin up (parallel to field) having the lower energy. (This is like the Zeeman effect for atomic levels, Fig. 28–8.) The difference in energy ΔE between these two levels is proportional to the total magnetic field B_T at the nucleus:

$$\Delta E = kB_T,$$

where k is a proportionality constant that is different for different nuclides.

In a standard **nuclear magnetic resonance** (NMR) setup, the sample to be examined is placed in a static magnetic field. A radiofrequency (RF) pulse of electromagnetic radiation (that is, photons) is applied to the sample. If the frequency, f, of this pulse corresponds precisely to the energy difference between the two energy levels (Fig. 31–24), so that

$$hf = \Delta E = kB_T, \tag{31–11}$$

then the photons of the RF beam will be absorbed, exciting many of the nuclei from the lower state to the upper state. This is a resonance phenomenon because there is significant absorption only if f is very near $f = kB_T/h$. Hence the name "nuclear magnetic resonance." For free $^1_1$H nuclei, the frequency is 42.58 MHz for a magnetic field $B_T = 1.0$ T. If the H atoms are bound in a molecule the total magnetic field B_T at the H nuclei will be the sum of the external applied field (B_{ext}) plus the local magnetic field (B_{local}) due to electrons and nuclei of neighboring atoms. Since f is proportional to B_T, the value of f for a given external field will be slightly different for bound H atoms than for free atoms:

$$hf = k(B_{ext} + B_{local}).$$

This small change in frequency can be measured, and is called the "chemical shift." A great deal has been learned about the structure of molecules and bonds using this NMR technique.

*Magnetic Resonance Imaging (MRI)

For producing medically useful NMR images—now commonly called MRI, or **magnetic resonance imaging**—the element most used is hydrogen since it is the commonest element in the human body and gives the strongest NMR signals. The experimental apparatus is shown in Fig. 31–25. The large coils set up the static magnetic field, and the RF coils produce the RF pulse of electromagnetic waves (photons) that cause the nuclei to jump from the lower state to the upper one (Fig. 31–24). These same coils (or another coil) can detect the absorption of energy or the emitted radiation (also of frequency $f = \Delta E/h$, Eq. 31–11) when the nuclei jump back down to the lower state.

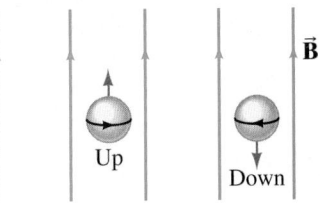

FIGURE 31–23 Schematic picture of a proton in a magnetic field $\vec{\mathbf{B}}$ (pointing upward) with the two possible states of proton spin, up and down.

FIGURE 31–24 Energy E_0 in the absence of a magnetic field splits into two levels in the presence of a magnetic field.

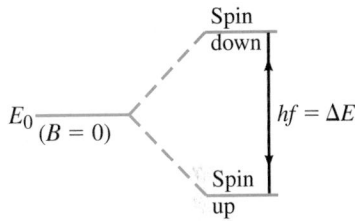

PHYSICS APPLIED

NMR imaging (MRI)

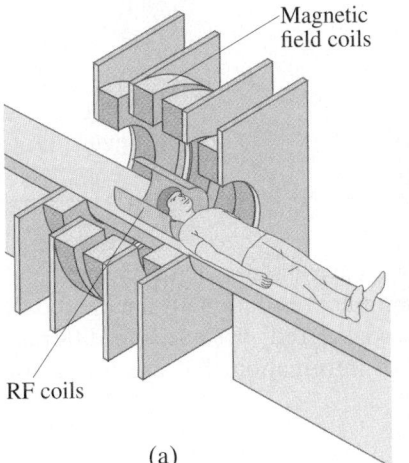

(a)

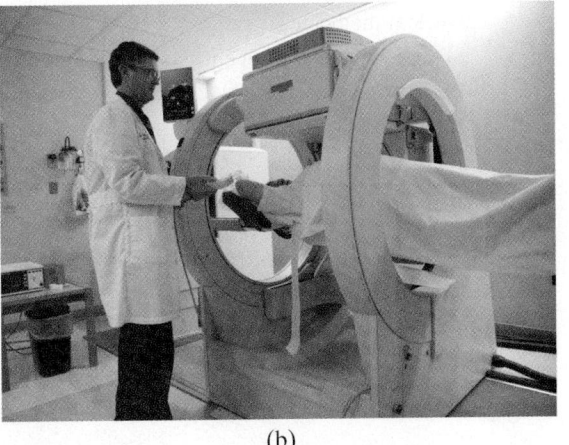

(b)

FIGURE 31–25 NMR imaging setup: (a) diagram; (b) photograph.

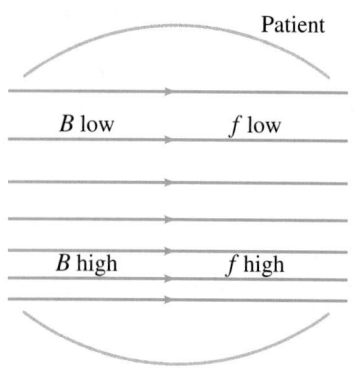

FIGURE 31–26 A static field that is stronger at the bottom than at the top. The frequency of absorbed or emitted radiation is proportional to B in NMR.

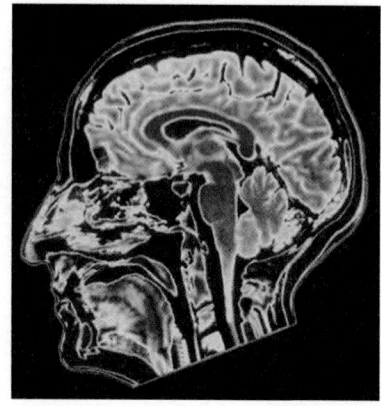

FIGURE 31–27 False-color NMR image (MRI) through the head showing structures in the brain.

The formation of a two-dimensional or three-dimensional image can be done using techniques similar to those for computed tomography (Section 25–12). The simplest thing to measure for creating an image is the intensity of absorbed and/or reemitted radiation from many different points of the body, and this would be a measure of the density of H atoms at each point. But how do we determine from what part of the body a given photon comes? One technique is to give the static magnetic field a gradient; that is, instead of applying a uniform magnetic field, B_T, the field is made to vary with position across the width of the sample (or patient), Fig. 31–26. Because the frequency absorbed by the H nuclei is proportional to B_T (Eq. 31–11), only one plane within the body will have the proper value of B_T to absorb photons of a particular frequency f. By varying f, absorption by different planes can be measured. Alternately, if the field gradient is applied *after* the RF pulse, the frequency of the emitted photons will be a measure of where they were emitted. If a magnetic field gradient in one direction is applied during excitation (absorption of photons) and photons of a single frequency are transmitted, only H nuclei in one thin slice will be excited. By applying a gradient during reemission in a direction perpendicular to the first, the frequency f of the reemitted radiation will represent depth in that slice. Other ways of varying the magnetic field throughout the volume of the body can be used in order to correlate NMR frequency with position.

A reconstructed image based on the density of H atoms (that is, the intensity of absorbed or emitted radiation) is not very interesting. More useful are images based on the rate at which the nuclei decay back to the ground state, and such images can produce resolution of 1 mm or better. This NMR technique (sometimes called **spin-echo**) produces images of great diagnostic value, both in the delineation of structure (anatomy) and in the study of metabolic processes. An NMR image is shown in Fig. 31–27.

NMR imaging is considered to be noninvasive. We can calculate the energy of the photons involved: as mentioned above, in a 1.0-T magnetic field, $f = 42.58\,\text{MHz}$ for ^1_1H. This corresponds to an energy of $hf = (6.6 \times 10^{-34}\,\text{J} \cdot \text{s})(43 \times 10^6\,\text{Hz}) \approx 3 \times 10^{-26}\,\text{J}$ or about $10^{-7}\,\text{eV}$. Since molecular bonds are on the order of 1 eV, the RF photons can cause little cellular disruption. This should be compared to X- or γ rays, whose energies are 10^4 to $10^6\,\text{eV}$ and thus can cause significant damage. The static magnetic fields, though often large (as high as 1.0 to 1.5 T), are believed to be harmless (except for people who wear heart pacemakers).

TABLE 31–2 Medical Imaging Techniques

Technique	Where Discussed in This Book	Optimal Resolution
Conventional X-ray	Section 25–12	$\frac{1}{2}$ mm
CT scan, X-ray	Section 25–12	$\frac{1}{2}$ mm
Nuclear medicine (tracers)	Section 31–7	1 cm
SPECT (single photon emission)	Section 31–8	1 cm
PET (positron emission)	Section 31–8	2–5 mm
MRI (NMR)	Section 31–9	$\frac{1}{2}$–1 mm
Ultrasound	Section 12–9	0.3–2 mm

Table 31–2 lists the major techniques we have discussed for imaging the interior of the human body, along with the optimum resolution attainable today. Resolution is only one factor that must be considered, because the different imaging techniques provide different types of information that are useful for different types of diagnosis.

Summary

A **nuclear reaction** occurs when two nuclei collide and two or more other nuclei (or particles) are produced. In this process, as in radioactivity, **transmutation** (change) of elements occurs.

The **reaction energy** or **Q-value** of a reaction $a + X \rightarrow Y + b$ is

$$Q = (M_a + M_X - M_b - M_Y)c^2 \qquad \textbf{(31-2a)}$$
$$= KE_b + KE_Y - KE_a - KE_X. \qquad \textbf{(31-2b)}$$

In **fission**, a heavy nucleus such as uranium splits into two intermediate-sized nuclei after being struck by a neutron. $^{235}_{92}U$ is fissionable by slow neutrons, whereas some fissionable nuclei require fast neutrons. Much energy is released in fission (≈ 200 MeV per fission) because the binding energy per nucleon is lower for heavy nuclei than it is for intermediate-sized nuclei, so the mass of a heavy nucleus is greater than the total mass of its fission products. The fission process releases neutrons, so that a **chain reaction** is possible. The **critical mass** is the minimum mass of fuel needed so that enough emitted neutrons go on to produce more fissions and sustain a chain reaction. In a **nuclear reactor** or nuclear weapon, a **moderator** is used to slow down the released neutrons.

The **fusion** process, in which small nuclei combine to form larger ones, also releases energy. The energy from our Sun originates in the fusion reactions known as the **proton–proton chain** in which four protons fuse to form a $^{4}_{2}He$ nucleus producing 25 MeV of energy. A useful fusion reactor for power generation has not yet proved possible because of the difficulty in containing the fuel (e.g., deuterium) long enough at the extremely high temperature required ($\approx 10^8$ K). Nonetheless, progress has been made in confining the collection of charged ions known as a **plasma**. The two main methods are **magnetic confinement**, using a magnetic field in a device such as the donut-shaped **tokamak**, and **inertial confinement** in which intense laser beams compress a fuel pellet of deuterium and tritium.

Radiation can cause damage to materials, including biological tissue. Quantifying amounts of radiation is the subject of **dosimetry**. The **curie** (Ci) and the **becquerel** (Bq) are units that measure the **source activity** or rate of decay of a sample: 1 Ci $= 3.70 \times 10^{10}$ decays per second, whereas 1 Bq $= 1$ decay/s. The **absorbed dose**, often specified in **rads**, measures the amount of energy deposited per unit mass of absorbing material: 1 rad is the amount of radiation that deposits energy at the rate of 10^{-2} J/kg of material. The SI unit of absorbed dose is the **gray**: 1 Gy $= 1$ J/kg $= 100$ rad. The **effective dose** is often specified by the **rem** $=$ rad $\times$ RBE, where RBE is the "relative biological effectiveness" of a given type of radiation; 1 rem of any type of radiation does approximately the same amount of biological damage. The average dose received per person per year in the United States is about 360 mrem. The SI unit for effective dose is the **sievert**: 1 Sv $= 100$ rem.

[*Nuclear radiation is used in medicine for cancer therapy, and for imaging of biological structure and processes. Tomographic imaging of the human body, which can provide 3-dimensional detail, includes several types: PET, SPET (= SPECT), MRI, and CT scans (discussed in Chapter 25). MRI makes use of **nuclear magnetic resonance** (NMR).]

Questions

1. Fill in the missing particles or nuclei:
   (a) $n + {}^{232}_{90}Th \rightarrow ? + \gamma$;
   (b) $n + {}^{137}_{56}Ba \rightarrow {}^{137}_{55}Cs + ?$;
   (c) $d + {}^{2}_{1}H \rightarrow {}^{4}_{2}He + ?$;
   (d) $\alpha + {}^{197}_{79}Au \rightarrow ? + d$
   where d stands for deuterium.

2. When $^{22}_{11}Na$ is bombarded by deuterons $\left({}^{2}_{1}H\right)$, an α particle is emitted. What is the resulting nuclide? Write down the reaction equation.

3. Why are neutrons such good projectiles for producing nuclear reactions?

4. What is the Q-value for radioactive decay reactions?
   (a) $Q < 0$.   (b) $Q > 0$.   (c) $Q = 0$.
   (d) The sign of Q depends on the nucleus.

5. The energy from nuclear fission appears in the form of thermal energy—but the thermal energy of what?

6. (a) If $^{235}_{92}U$ released only 1.5 neutrons per fission on average (instead of 2.5), would a chain reaction be possible? (b) If so, how would the chain reaction be different than if 3 neutrons were released per fission?

7. Why can't uranium be enriched by chemical means?

8. How can a neutron, with practically no kinetic energy, excite a nucleus to the extent shown in Fig. 31–3?

9. Why would a porous block of uranium be more likely to explode if kept under water rather than in air?

10. A reactor that uses highly enriched uranium can use ordinary water (instead of heavy water) as a moderator and still have a self-sustaining chain reaction. Explain.

11. Why must the fission process release neutrons if it is to be useful?

12. Why are neutrons released in a fission reaction?

13. What is the reason for the "secondary system" in a nuclear reactor, Fig. 31–8? That is, why is the water heated by the fuel in a nuclear reactor not used directly to drive the turbines?

14. What is the basic difference between fission and fusion?

15. Discuss the relative merits and disadvantages, including pollution and safety, of power generation by fossil fuels, nuclear fission, and nuclear fusion.

16. Why do gamma particles penetrate matter more easily than beta particles do?

17. Light energy emitted by the Sun and stars comes from the fusion process. What conditions in the interior of stars make this possible?

18. How do stars, and our Sun, maintain confinement of the plasma for fusion?

19. People who work around metals that emit alpha particles are trained that there is little danger from proximity or touching the material, but they must take extreme precautions against ingesting it. Why? (Eating and drinking while working are forbidden.)

20. What is the difference between absorbed dose and effective dose? What are the SI units for each?

21. Radiation is sometimes used to sterilize medical supplies and even food. Explain how it works.

*22. How might radioactive tracers be used to find a leak in a pipe?

MisConceptual Questions

1. In a nuclear reaction, which of the following is *not* conserved?
   (a) Energy.
   (b) Momentum.
   (c) Electric charge.
   (d) Nucleon number.
   (e) None of the above.

2. Fission fragments are typically
   (a) β^+ emitters.
   (b) β^- emitters.
   (c) Both.
   (d) Neither.

3. Which of the following properties would decrease the critical mass needed to sustain a nuclear chain reaction?
   (a) Low boiling point.
   (b) High melting point.
   (c) More neutrons released per fission.
   (d) Low nuclear density.
   (e) Filled valence shell.
   (f) All of the above.

4. Rather than having a maximum at about $A \approx 60$, as shown in Fig. 31–12, suppose the average binding energy per nucleon continually increased with increasing mass number. Then,
   (a) fission would still be possible, but not fusion.
   (b) fusion would still be possible, but not fission.
   (c) both fission and fusion would still be possible.
   (d) neither fission nor fusion would be possible.

5. Why is a moderator needed in a normal uranium fission reactor?
   (a) To increase the rate of neutron capture by uranium-235.
   (b) To increase the rate of neutron capture by uranium-238.
   (c) To increase the rate of production of plutonium-239.
   (d) To increase the critical mass of the fission fuel.
   (e) To provide more neutrons for the reaction.
   (f) All of the above.

6. What is the difference between nuclear fission and nuclear fusion?
   (a) Nuclear fission is used for bombs; nuclear fusion is used in power plants.
   (b) There is no difference. Fission and fusion are different names for the same physical phenomenon.
   (c) Nuclear fission refers to using deuterium to create a nuclear reaction.
   (d) Nuclear fusion occurs spontaneously, as happens to the C^{14} used in carbon dating.
   (e) In nuclear fission, a nucleus splits; in nuclear fusion, nucleons or nuclei and nucleons join to form a new nucleus.

7. A primary difficulty in energy production by fusion is
   (a) the scarcity of necessary fuel.
   (b) the disposal of radioactive by-products produced.
   (c) the high temperatures necessary to overcome the electrical repulsion of protons.
   (d) the fact that it is possible in volcanic regions only.

8. If two hydrogen nuclei, $^2_1$H, each of mass m_H, fuse together and form a helium nucleus of mass m_{He},
   (a) $m_{He} < 2m_H$.
   (b) $m_{He} = 2m_H$.
   (c) $m_{He} > 2m_H$.
   (d) All of the above are possible.

9. Which radiation induces the most biological damage for a given amount of energy deposited in tissue?
   (a) Alpha particles.
   (b) Gamma radiation.
   (c) Beta radiation.
   (d) All do the same damage for the same deposited energy.

10. Which would produce the most energy in a single reaction?
   (a) The fission reaction associated with uranium-235.
   (b) The fusion reaction of the Sun (two hydrogen nuclei fused to one helium nucleus).
   (c) Both (a) and (b) are about the same.
   (d) Need more information.

11. The fuel necessary for fusion-produced energy could be derived from
   (a) water.
   (b) superconductors.
   (c) uranium.
   (d) helium.
   (e) sunlight.

12. Which of the following is true?
   (a) Any amount of radiation is harmful to living tissue.
   (b) Radiation is a natural part of the environment.
   (c) All forms of radiation will penetrate deep into living tissue.
   (d) None of the above is true.

13. Which of the following would reduce the cell damage due to radiation for a lab technician who works with radioactive isotopes in a hospital or lab?
   (a) Increase the worker's distance from the radiation source.
   (b) Decrease the time the worker is exposed to the radiation.
   (c) Use shielding to reduce the amount of radiation that strikes the worker.
   (d) Have the worker wear a radiation badge when working with the radioactive isotopes.
   (e) All of the above.

14. If the same dose of each type of radiation was provided over the same amount of time, which type would be most harmful?
   (a) X-rays.
   (b) γ rays.
   (c) β rays.
   (d) α particles.

15. $^{235}_{92}$U releases an average of 2.5 neutrons per fission compared to 2.9 for $^{239}_{94}$Pu. Which has the smaller critical mass?
   (a) $^{235}_{92}$U.
   (b) $^{239}_{94}$Pu.
   (c) Both the same.

Problems

(NOTE: Masses are found in Appendix B.)

31–1 Nuclear Reactions, Transmutation

1. (I) Natural aluminum is all $^{27}_{13}Al$. If it absorbs a neutron, what does it become? Does it decay by β^+ or β^-? What will be the product nucleus?

2. (I) Determine whether the reaction $^2_1H + {}^1_1H \rightarrow {}^3_2He + n$ requires a threshold energy, and why.

3. (I) Is the reaction $n + {}^{238}_{92}U \rightarrow {}^{239}_{92}U + \gamma$ possible with slow neutrons? Explain.

4. (II) (a) Complete the following nuclear reaction, $p + ? \rightarrow {}^{32}_{16}S + \gamma$. (b) What is the Q-value?

5. (II) The reaction $p + {}^{18}_8O \rightarrow {}^{18}_9F + n$ requires an input of energy equal to 2.438 MeV. What is the mass of $^{18}_9F$?

6. (II) (a) Can the reaction $n + {}^{24}_{12}Mg \rightarrow {}^{23}_{11}Na + d$ occur if the bombarding particles have 18.00 MeV of kinetic energy? (d stands for deuterium, ^2_1H.) (b) If so, how much energy is released? If not, what kinetic energy is needed?

7. (II) (a) Can the reaction $p + {}^7_3Li \rightarrow {}^4_2He + \alpha$ occur if the incident proton has kinetic energy = 3100 keV? (b) If so, what is the total kinetic energy of the products? If not, what kinetic energy is needed?

8. (II) In the reaction $\alpha + {}^{14}_7N \rightarrow {}^{17}_8O + p$, the incident α particles have 9.85 MeV of kinetic energy. The mass of $^{17}_8O$ is 16.999132 u. (a) Can this reaction occur? (b) If so, what is the total kinetic energy of the products? If not, what kinetic energy is needed?

9. (II) Calculate the Q-value for the "capture" reaction $\alpha + {}^{16}_8O \rightarrow {}^{20}_{10}Ne + \gamma$.

10. (II) Calculate the total kinetic energy of the products of the reaction $d + {}^{13}_6C \rightarrow {}^{14}_7N + n$ if the incoming deuteron has kinetic energy KE = 41.4 MeV.

11. (II) Radioactive $^{14}_6C$ is produced in the atmosphere when a neutron is absorbed by $^{14}_7N$. Write the reaction and find its Q-value.

12. (II) An example of a **stripping** nuclear reaction is $d + {}^6_3Li \rightarrow X + p$. (a) What is X, the resulting nucleus? (b) Why is it called a "stripping" reaction? (c) What is the Q-value of this reaction? Is the reaction endothermic or exothermic?

13. (II) An example of a **pick-up** nuclear reaction is $^3_2He + {}^{12}_6C \rightarrow X + \alpha$. (a) Why is it called a "pick-up" reaction? (b) What is the resulting nucleus? (c) What is the Q-value of this reaction? Is the reaction endothermic or exothermic?

14. (II) Does the reaction $p + {}^7_3Li \rightarrow {}^4_2He + \alpha$ require energy, or does it release energy? How much energy?

15. (II) Calculate the energy released (or energy input required) for the reaction $\alpha + {}^9_4Be \rightarrow {}^{12}_6C + n$.

31–2 Nuclear Fission

16. (I) What is the energy released in the fission reaction of Eq. 31–4? (The masses of $^{141}_{56}Ba$ and $^{92}_{36}Kr$ are 140.914411 u and 91.926156 u, respectively.)

17. (I) Calculate the energy released in the fission reaction $n + {}^{235}_{92}U \rightarrow {}^{88}_{38}Sr + {}^{136}_{54}Xe + 12n$. Use Appendix B, and assume the initial kinetic energy of the neutron is very small.

18. (I) How many fissions take place per second in a 240-MW reactor? Assume 200 MeV is released per fission.

19. (I) The energy produced by a fission reactor is about 200 MeV per fission. What fraction of the mass of a $^{235}_{92}U$ nucleus is this?

20. (II) Suppose that the average electric power consumption, day and night, in a typical house is 960 W. What initial mass of $^{235}_{92}U$ would have to undergo fission to supply the electrical needs of such a house for a year? (Assume 200 MeV is released per fission, as well as 100% efficiency.)

21. (II) Consider the fission reaction

$$^{235}_{92}U + n \rightarrow {}^{133}_{51}Sb + {}^{98}_{41}Nb + ?n.$$

(a) How many neutrons are produced in this reaction? (b) Calculate the energy release. The atomic masses for Sb and Nb isotopes are 132.915250 u and 97.910328 u, respectively.

22. (II) How much mass of $^{235}_{92}U$ is required to produce the same amount of energy as burning 1.0 kg of coal (about 3×10^7 J)?

23. (II) What initial mass of $^{235}_{92}U$ is required to operate a 950-MW reactor for 1 yr? Assume 34% efficiency.

24. (II) If a 1.0-MeV neutron emitted in a fission reaction loses one-half of its kinetic energy in each collision with moderator nuclei, how many collisions must it make to reach thermal energy $(\frac{3}{2}kT = 0.040 \text{ eV})$?

25. (II) Assuming a fission of $^{236}_{92}U$ into two roughly equal fragments, estimate the electric potential energy just as the fragments separate from each other. Assume that the fragments are spherical (see Eq. 30–1) and compare your calculation to the nuclear fission energy released, about 200 MeV.

26. (III) Suppose that the neutron multiplication factor is 1.0004. If the average time between successive fissions in a chain of reactions is 1.0 ms, by what factor will the reaction rate increase in 1.0 s?

31–3 Nuclear Fusion

27. (I) What is the average kinetic energy of protons at the center of a star where the temperature is 2×10^7 K? [*Hint:* See Eq. 13–8.]

28. (II) Show that the energy released in the fusion reaction $^2_1H + {}^3_1H \rightarrow {}^4_2He + n$ is 17.59 MeV.

29. (II) Show that the energy released when two deuterium nuclei fuse to form ^3_2He with the release of a neutron is 3.27 MeV (Eq. 31–8b).

30. (II) Verify the Q-value stated for each of the reactions of Eqs. 31–6. [*Hint*: Use Appendix B; be careful with electrons (included in mass values except for p, d, t).]

31. (II) (*a*) Calculate the energy release per gram of fuel for the reactions of Eqs. 31–8a, b, and c. (*b*) Calculate the energy release per gram of uranium $^{235}_{92}\text{U}$ in fission, and give its ratio to each reaction in (*a*).

32. (II) How much energy is released when $^{238}_{92}\text{U}$ absorbs a slow neutron (kinetic energy ≈ 0) and becomes $^{239}_{92}\text{U}$?

33. (II) If a typical house requires 960 W of electric power on average, what minimum amount of deuterium fuel would have to be used in a year to supply these electrical needs? Assume the reaction of Eq. 31–8b.

34. (II) If ^6_3Li is struck by a slow neutron, it can form ^4_2He and another nucleus. (*a*) What is the second nucleus? (This is a method of generating this isotope.) (*b*) How much energy is released in the process?

35. (II) Suppose a fusion reactor ran on "d–d" reactions, Eqs. 31–8a and b in equal amounts. Estimate how much natural water, for fuel, would be needed per hour to run a 1150-MW reactor, assuming 33% efficiency.

36. (III) Show that the energies carried off by the ^4_2He nucleus and the neutron for the reaction of Eq. 31–8c are about 3.5 MeV and 14 MeV, respectively. Are these fixed values, independent of the plasma temperature?

37. (III) How much energy (J) is contained in 1.00 kg of water if its natural deuterium is used in the fusion reaction of Eq. 31–8a? Compare to the energy obtained from the burning of 1.0 kg of gasoline, about 5×10^7 J.

38. (III) (*a*) Give the ratio of the energy needed for the first reaction of the *carbon cycle* to the energy needed for a deuterium–tritium reaction (Example 31–9). (*b*) If a deuterium–tritium reaction actually requires a temperature $T \approx 3 \times 10^8$ K, estimate the temperature needed for the first carbon-cycle reaction.

31–5 Dosimetry

39. (I) 350 rads of α-particle radiation is equivalent to how many rads of X-rays in terms of biological damage?

40. (I) A dose of 4.0 Sv of γ rays in a short period would be lethal to about half the people subjected to it. How many grays is this?

41. (I) How many rads of slow neutrons will do as much biological damage as 72 rads of fast neutrons?

42. (II) How much energy is deposited in the body of a 65-kg adult exposed to a 2.5-Gy dose?

43. (II) A cancer patient is undergoing radiation therapy in which protons with an energy of 1.2 MeV are incident on a 0.20-kg tumor. (*a*) If the patient receives an effective dose of 1.0 rem, what is the absorbed dose? (*b*) How many protons are absorbed by the tumor? Assume RBE ≈ 1.

44. (II) A 0.035-μCi sample of $^{32}_{15}\text{P}$ is injected into an animal for tracer studies. If a Geiger counter intercepts 35% of the emitted β particles, what will be the counting rate, assumed 85% efficient?

45. (II) About 35 eV is required to produce one ion pair in air. Show that this is consistent with the two definitions of the roentgen given in the text.

46. (II) A 1.6-mCi source of $^{32}_{15}\text{P}$ (in NaHPO$_4$), a β emitter, is implanted in a tumor where it is to administer 32 Gy. The half-life of $^{32}_{15}\text{P}$ is 14.3 days, and 1.0 mCi delivers about 10 mGy/min. Approximately how long should the source remain implanted?

47. (II) What is the mass of a 2.50-μCi $^{14}_6\text{C}$ source?

48. (II) $^{57}_{27}\text{Co}$ emits 122-keV γ rays. If a 65-kg person swallowed 1.55 μCi of $^{57}_{27}\text{Co}$, what would be the dose rate (Gy/day) averaged over the whole body? Assume that 50% of the γ-ray energy is deposited in the body. [*Hint*: Determine the rate of energy deposited in the body and use the definition of the gray.]

49. (II) Ionizing radiation can be used on meat products to reduce the levels of microbial pathogens. Refrigerated meat is limited to 4.5 kGy. If 1.6-MeV electrons irradiate 5 kg of beef, how many electrons would it take to reach the allowable limit?

50. (III) Huge amounts of radioactive $^{131}_{53}\text{I}$ were released in the accident at Chernobyl in 1986. Chemically, iodine goes to the human thyroid. (It can be used for diagnosis and treatment of thyroid problems.) In a normal thyroid, $^{131}_{53}\text{I}$ absorption can cause damage to the thyroid. (*a*) Write down the reaction for the decay of $^{131}_{53}\text{I}$. (*b*) Its half-life is 8.0 d; how long would it take for ingested $^{131}_{53}\text{I}$ to become 5.0% of the initial value? (*c*) Absorbing 1 mCi of $^{131}_{53}\text{I}$ can be harmful; what mass of iodine is this?

51. (III) Assume a liter of milk typically has an activity of 2000 pCi due to $^{40}_{19}\text{K}$. If a person drinks two glasses (0.5 L) per day, estimate the total effective dose (in Sv and in rem) received in a year. As a crude model, assume the milk stays in the stomach 12 hr and is then released. Assume also that roughly 10% of the 1.5 MeV released per decay is absorbed by the body. Compare your result to the normal allowed dose of 100 mrem per year. Make your estimate for (*a*) a 60-kg adult, and (*b*) a 6-kg baby.

52. (III) Radon gas, $^{222}_{86}\text{Rn}$, is considered a serious health hazard (see discussion in text). It decays by α-emission. (*a*) What is the daughter nucleus? (*b*) Is the daughter nucleus stable or radioactive? If the latter, how does it decay, and what is its half-life? (See Fig. 30–11.) (*c*) Is the daughter nucleus also a noble gas, or is it chemically reactive? (*d*) Suppose 1.4 ng of $^{222}_{86}\text{Rn}$ seeps into a basement. What will be its activity? If the basement is then sealed, what will be the activity 1 month later?

31–9 NMR

53. (II) Calculate the wavelength of photons needed to produce NMR transitions in free protons in a 1.000-T field. In what region of the spectrum is this wavelength?

General Problems

54. Consider a system of nuclear power plants that produce 2100 MW. (a) What total mass of $^{235}_{92}U$ fuel would be required to operate these plants for 1 yr, assuming that 200 MeV is released per fission? (b) Typically 6% of the $^{235}_{92}U$ nuclei that fission produce strontium-90, $^{90}_{38}Sr$, a β^- emitter with a half-life of 29 yr. What is the total radioactivity of the $^{90}_{38}Sr$, in curies, produced in 1 yr? (Neglect the fact that some of it decays during the 1-yr period.)

55. J. Chadwick discovered the neutron by bombarding $^{9}_{4}Be$ with the popular projectile of the day, alpha particles. (a) If one of the reaction products was the then unknown neutron, what was the other product? (b) What is the Q-value of this reaction?

56. Fusion temperatures are often given in keV. Determine the conversion factor from kelvins to keV using, as is common in this field, $\overline{KE} = kT$ without the factor $\frac{3}{2}$.

57. One means of enriching uranium is by diffusion of the gas UF_6. Calculate the ratio of the speeds of molecules of this gas containing $^{235}_{92}U$ and $^{238}_{92}U$, on which this process depends.

58. (a) What mass of $^{235}_{92}U$ was actually fissioned in the first atomic bomb, whose energy was the equivalent of about 20 kilotons of TNT (1 kiloton of TNT releases 5×10^{12} J)? (b) What was the actual mass transformed to energy?

59. The average yearly background radiation in a certain town consists of 32 mrad of X-rays and γ rays plus 3.4 mrad of particles having a RBE of 10. How many rem will a person receive per year on average?

60. A shielded γ-ray source yields a dose rate of 0.048 rad/h at a distance of 1.0 m for an average-sized person. If workers are allowed a maximum dose of 5.0 rem in 1 year, how close to the source may they operate, assuming a 35-h work week? Assume that the intensity of radiation falls off as the square of the distance. (It actually falls off more rapidly than $1/r^2$ because of absorption in the air, so your answer will give a better-than-permissible value.)

61. Radon gas, $^{222}_{86}Rn$, is formed by α decay. (a) Write the decay equation. (b) Ignoring the kinetic energy of the daughter nucleus (it's so massive), estimate the kinetic energy of the α particle produced. (c) Estimate the momentum of the alpha and of the daughter nucleus. (d) Estimate the kinetic energy of the daughter, and show that your approximation in (b) was valid.

62. In the net reaction, Eq. 31–7, for the proton–proton chain in the Sun, the neutrinos escape from the Sun with energy of about 0.5 MeV. The remaining energy, 26.2 MeV, is available to heat the Sun. Use this value to calculate the "heat of combustion" per kilogram of hydrogen fuel and compare it to the heat of combustion of coal, about 3×10^7 J/kg.

63. Energy reaches Earth from the Sun at a rate of about 1300 W/m². Calculate (a) the total power output of the Sun, and (b) the number of protons consumed per second in the reaction of Eq. 31–7, assuming that this is the source of all the Sun's energy. (c) Assuming that the Sun's mass of 2.0×10^{30} kg was originally all protons and that all could be involved in nuclear reactions in the Sun's core, how long would you expect the Sun to "glow" at its present rate? See Problem 62. [Hint: Use $1/r^2$ law.]

64. Estimate how many solar neutrinos pass through a 180-m² ceiling of a room, at latitude 44°, for an hour around midnight on midsummer night. [Hint: See Problems 62 and 63.]

65. Estimate how much total energy would be released via fission if 2.0 kg of uranium were enriched to 5% of the isotope $^{235}_{92}U$.

66. Some stars, in a later stage of evolution, may begin to fuse two $^{12}_{6}C$ nuclei into one $^{24}_{12}Mg$ nucleus. (a) How much energy would be released in such a reaction? (b) What kinetic energy must two carbon nuclei each have when far apart, if they can then approach each other to within 6.0 fm, center-to-center? (c) Approximately what temperature would this require?

67. An average adult body contains about 0.10 μCi of $^{40}_{19}K$, which comes from food. (a) How many decays occur per second? (b) The potassium decay produces beta particles with energies of around 1.4 MeV. Estimate the dose per year in sieverts for a 65-kg adult. Is this a significant fraction of the 3.6-mSv/yr background rate?

68. When the nuclear reactor accident occurred at Chernobyl in 1986, 2.0×10^7 Ci were released into the atmosphere. Assuming that this radiation was distributed uniformly over the surface of the Earth, what was the activity per square meter? (The actual activity was not uniform; even within Europe wet areas received more radioactivity from rainfall.)

69. A star with a large helium abundance can burn helium in the reaction $^{4}_{2}He + ^{4}_{2}He + ^{4}_{2}He \rightarrow ^{12}_{6}C$. What is the Q-value for this reaction?

70. A 1.2-μCi $^{137}_{55}Cs$ source is used for 1.4 hours by a 62-kg worker. Radioactive $^{137}_{55}Cs$ decays by β^- decay with a half-life of 30 yr. The average energy of the emitted betas is about 190 keV per decay. The β decay is quickly followed by a γ with an energy of 660 keV. Assuming the person absorbs all emitted energy, what effective dose (in rem) is received?

71. Suppose a future fusion reactor would be able to put out 1000 MW of electrical power continuously. Assume the reactor will produce energy solely through the reaction given in Eq. 31–8a and will convert this energy to electrical energy with an efficiency of 33%. Estimate the minimum amount of deuterium needed to run this facility per year.

72. If a 65-kg power plant worker has been exposed to the maximum slow-neutron radiation for a given year, how much total energy (in J) has that worker absorbed? What if he were exposed to fast protons?

73. Consider the fission reaction

$$n + ^{235}_{92}U \rightarrow ^{92}_{38}Sr + X + 3n.$$

(a) What is X? (b) If this were part of a chain reaction in a fission power reactor running at "barely critical," what would happen on average to the three produced neutrons? (c) (optional) What is the Q-value of this reaction? [Hint: Mass values can be found at www.nist.gov/pml/data/comp.cfm.]

74. A large amount of $^{90}_{38}$Sr was released during the Chernobyl nuclear reactor accident in 1986. The $^{90}_{38}$Sr enters the body through the food chain. How long will it take for 85% of the $^{90}_{38}$Sr released during the accident to decay? See Appendix B.

75. Three radioactive sources have the same activity, 35 mCi. Source A emits 1.0-MeV γ rays, source B emits 2.0-MeV γ rays, and source C emits 2.0-MeV alphas. What is the relative danger of these sources?

76. A 55-kg patient is to be given a medical test involving the ingestion of $^{99m}_{43}$Tc (Section 31–7) which decays by emitting a 140-keV gamma. The half-life for this decay is 6 hours. Assuming that about half the gamma photons exit the body without interacting with anything, what must be the initial activity of the Tc sample if the whole-body dose cannot exceed 50 mrem? Make the rough approximation that biological elimination of Tc can be ignored.

Search and Learn

1. Referring to Section 31–2, (a) state three problems that must be overcome to make a functioning fission nuclear reactor; (b) state three environmental problems or dangers that do or could result from the operation of a nuclear fission reactor; (c) describe an additional problem or danger associated with a breeder reactor.

2. Referring to Section 31–3, (a) why can small nuclei combine to form larger ones, releasing energy in the process? (b) Why does the first reaction in the proton–proton chain limit the rate at which the Sun produces energy? (c) What are the heaviest elements for which energy is released if the elements are created by fusion of lighter elements? (d) What keeps the Sun and stars together, allowing them to sustain fusion? (e) What two methods are currently being investigated to contain high-temperature plasmas on the Earth to create fusion in the laboratory?

3. Deuterium makes up 0.0115% of natural hydrogen on average. Make a rough estimate of the total deuterium in the Earth's oceans and estimate the total energy released if all of it were used in fusion reactors.

4. The energy output of massive stars is believed to be due to the *carbon cycle* (see text). (a) Show that no carbon is consumed in this cycle and that the net effect is the same as for the proton–proton chain. (b) What is the total energy release? (c) Determine the energy output for each reaction and decay. (d) Why might the carbon cycle require a higher temperature $(\approx 2 \times 10^7 \text{ K})$ than the proton–proton chain $(\approx 1.5 \times 10^7 \text{ K})$?

5. Consider the effort by humans to harness nuclear fusion as a viable energy source. (a) What are some advantages of using nuclear fusion rather than nuclear fission? (b) What is the major technological problem with using controlled nuclear fusion as a source of energy? (c) Discuss two different approaches to solving this problem. (d) What fuel is necessary in a nuclear fusion reaction? (e) Write a nuclear reaction using two nuclei of the fuel in part (d) to create a third nucleus. (f) Calculate the Q-value of the reaction in part (e).

6. (a) Explain how each of the following can cause damage to materials: beta particles, alpha particles, energetic neutrons, and gamma rays. (b) How might metals be damaged? (c) How can the damage affect living cells?

ANSWERS TO EXERCISES

A: $^{138}_{56}$Ba.
B: 3 neutrons.
C: 2×10^{17}.

D: (e).
E: (b).

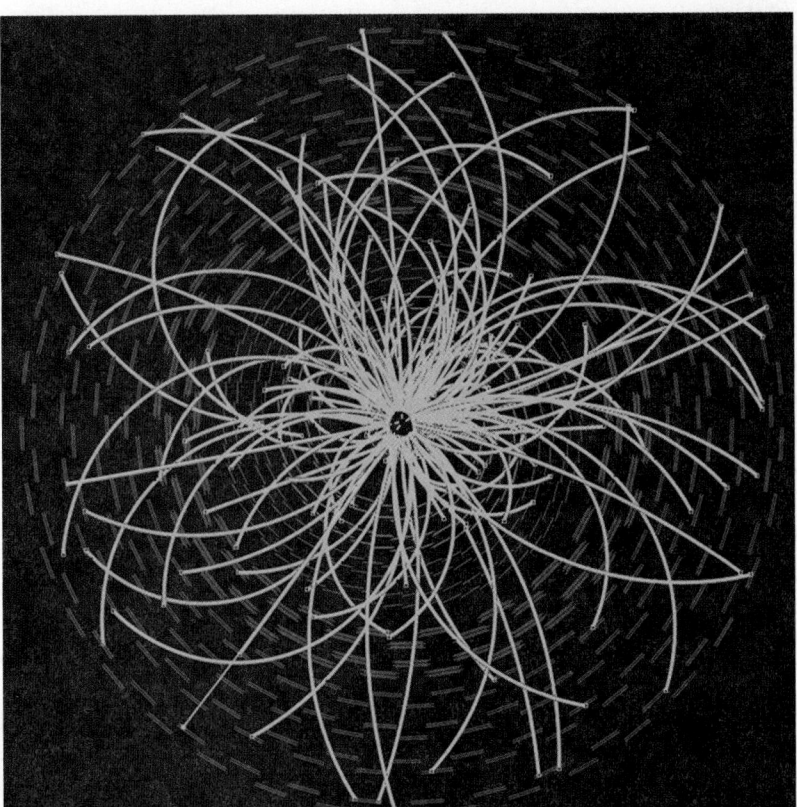

This photo is a computer reconstruction of particles produced due to a 7 TeV proton–proton collision at the Large Hadron Collider (LHC). It is a candidate for having produced the long-sought Higgs boson (plus other particles). The Higgs in this case could have decayed (very quickly $\sim 10^{-22}$ s) into two Z bosons (which are carriers of the weak force):

$$H^0 \rightarrow Z^0 + Z^0.$$

We don't see the tracks of the Z^0 particles because (1) they are neutral and (2) they decay too quickly ($\sim 10^{-24}$ s), in this case:

$$Z^0 \rightarrow e^- + e^+.$$

The tracks of the 2 electrons and 2 positrons are shown as green lines. The Higgs is thought to play a fundamental role in the Standard Model of particle physics, importantly providing mass to fundamental particles.

The CMS detector of this photo uses a combination of the detector types discussed in Section 30–13. A magnetic field causes particles to move in curved paths so the momentum of each can be measured (Section 20–4). Tracks of particles with very large momentum, such as our electrons here, are barely curved.

In this Chapter we will study elementary particle physics from its beginnings until today, including antiparticles, neutrinos, quarks, the Standard Model, and theories that go beyond. We start with the great machines that accelerate particles so they can collide at high energies.

Elementary Particles

CHAPTER-OPENING QUESTIONS—Guess now!

1. Physicists reserve the term "fundamental particle" for particles with a special property. What do you think that special property is?
   (a) Particles that are massless.
   (b) Particles that possess the minimum allowable electric charge.
   (c) Particles that have no internal structure.
   (d) Particles that produce no force on other objects.

2. The fundamental particles as we see them today, besides the long-sought-for Higgs boson, are
   (a) atoms and electrons.
   (b) protons, neutrons, and electrons.
   (c) protons, neutrons, electrons, and photons.
   (d) quarks, leptons, and gauge bosons (carriers of force).
   (e) hadrons, leptons, and gauge bosons.

In the final two Chapters of this book we discuss two of the most exciting areas of contemporary physics: elementary particles in this Chapter, and cosmology and astrophysics in Chapter 33. These are subjects at the forefront of knowledge—elementary particles treats the smallest objects in the universe; cosmology treats the largest (and oldest) aspects of the universe. The reader who wants an understanding of the great beauties of present-day science (and its limits) will want to read these Chapters. So will those who want to be good citizens, even if there is not time to cover them in a physics course.

CONTENTS

In this penultimate Chapter we discuss *elementary particle* physics, which represents the human endeavor to understand the basic building blocks of all matter, and the fundamental forces that govern their interactions.

Almost a century ago, by the 1930s, it was accepted that all atoms can be considered to be made up of neutrons, protons, and electrons. The basic constituents of the universe were no longer considered to be atoms (as they had been for 2000 years) but rather the proton, neutron, and electron. Besides these three "elementary particles," several others were also known: the positron (a positive electron), the neutrino, and the γ particle (or photon), for a total of six elementary particles.

By the 1950s and 1960s many new types of particles similar to the neutron and proton were discovered, as well as many "midsized" particles called *mesons* whose masses were mostly less than nucleon masses but more than the electron mass. (Other mesons, found later, have masses greater than nucleons.) Physicists felt that these particles could not all be fundamental, and must be made up of even smaller constituents (later confirmed by experiment), which were given the name *quarks*.

By the term **fundamental particle**, we mean a particle that is so simple, so basic, that it has no internal structure[†] (is not made up of smaller subunits)—see Chapter-Opening Question 1.

Today, the fundamental constituents of matter are considered to be **quarks** (they make up protons and neutrons as well as mesons) and **leptons** (a class that includes electrons, positrons, and neutrinos). There are also the "carriers of force" known as **gauge bosons**, including the photon, gluons, and W and Z bosons. In addition there is the elusive **Higgs** boson, predicted in the 1960s but whose first suggestions of experimental detection came only in 2011–2013. The theory that describes our present view is called the **Standard Model**. How we came to our present understanding of elementary particles is the subject of this Chapter.

One of the exciting developments of the last few years is an emerging synthesis between the study of elementary particles and astrophysics (Chapter 33). In fact, recent observations in astrophysics have led to the conclusion that the greater part of the mass–energy content of the universe is not ordinary matter but two mysterious and invisible forms known as "dark matter" and "dark energy" which cannot be explained by the Standard Model in its present form.

Indeed, we are now aware that the Standard Model is not sufficient. There are problems and important questions still unanswered, and we will mention some of them in this Chapter and how we hope to answer them.

32–1 High-Energy Particles and Accelerators

In the late 1940s, after World War II, it was found that if the incoming particle in a nuclear reaction (Section 31–1) has sufficient energy, new types of particles can be produced. The earliest experiments used **cosmic rays**—particles that impinge on the Earth from space. In the laboratory, various types of particle accelerators have been constructed to accelerate protons or electrons to high energies so they can collide with other particles—often protons (the hydrogen nucleus). Heavy ions, up to lead (Pb), have also been accelerated. These **high-energy accelerators** have been used to probe more deeply into matter, to produce and study new particles, and to give us information about the basic forces and constituents of nature. The particles produced in high-energy collisions can be detected by a variety of special detectors, discussed in Section 30–13, including scintillation counters, bubble chambers, multiwire chambers, and semiconductors. The rate of production of

---

[†]Recall from Section 13–1 that the word "atom" comes from the Greek meaning "indivisible." Atoms have a substructure (protons, neutrons) so are not fundamental. Yet an atom is still the smallest "piece" of an element that has the characteristics of that material.

any group of particles is quantified using the concept of *cross section*, Section 31–1. Because the projectile particles are at high energy, this field is sometimes called **high-energy physics**.

Wavelength and Resolution

Particles accelerated to high energy can probe the interior of nuclei and nucleons or other particles they strike. An important factor is that faster-moving projectiles can reveal more detail. The wavelength of projectile particles is given by de Broglie's wavelength formula (Eq. 27–8),

$$\lambda = \frac{h}{p}, \tag{32–1}$$

showing that the greater the momentum p of the bombarding particle, the shorter its wavelength. As discussed in Chapter 25 on optical instruments, resolution of details in images is limited by the wavelength: the shorter the wavelength, the finer the detail that can be obtained. This is one reason why particle accelerators of higher and higher energy have been built in recent years: to probe ever deeper into the structure of matter, to smaller and smaller size.

EXAMPLE 32–1 **High resolution with electrons.** What is the wavelength, and hence the expected resolution, for 1.3-GeV electrons?

APPROACH Because 1.3 GeV is much larger than the electron mass, we must be dealing with relativistic speeds. The momentum of the electrons is found from Eq. 26–9, and the wavelength is $\lambda = h/p$.

SOLUTION Each electron has $\text{KE} = 1.3\,\text{GeV} = 1300\,\text{MeV}$, which is about 2500 times the rest energy of the electron $(mc^2 = 0.51\,\text{MeV})$. Thus we can ignore the term $(mc^2)^2$ in Eq. 26–9, $E^2 = p^2c^2 + m^2c^4$, and we solve for p:

$$p = \sqrt{\frac{E^2 - m^2c^4}{c^2}} \approx \sqrt{\frac{E^2}{c^2}} = \frac{E}{c}.$$

Therefore the de Broglie wavelength is

$$\lambda = \frac{h}{p} = \frac{hc}{E},$$

where $E = 1.3\,\text{GeV}$. Hence

$$\lambda = \frac{(6.63 \times 10^{-34}\,\text{J·s})(3.0 \times 10^8\,\text{m/s})}{(1.3 \times 10^9\,\text{eV})(1.6 \times 10^{-19}\,\text{J/eV})} = 0.96 \times 10^{-15}\,\text{m},$$

or 0.96 fm. This resolution of about 1 fm is on the order of the size of nuclei (see Eq. 30–1).

NOTE The maximum possible resolution of this beam of electrons is far greater than for a light beam in a light microscope ($\lambda \approx 500\,\text{nm}$).

| **EXERCISE A** What is the wavelength of a proton with $\text{KE} = 1.00\,\text{TeV}$?

A major reason today for building high-energy accelerators is that new particles of greater mass can be produced at higher collision energies, transforming the kinetic energy of the colliding particles into massive particles by $E = mc^2$, as we will discuss shortly. Now we look at particle accelerators.

Cyclotron

The cyclotron was developed in 1930 by E. O. Lawrence (1901–1958; Fig. 32–1) at the University of California, Berkeley. It uses a magnetic field to maintain charged ions—usually protons—in nearly circular paths. Although particle physicists no longer use simple cyclotrons, they are used in medicine for treating cancer, and their operating principles are useful for understanding modern accelerators.

FIGURE 32–1 Ernest O. Lawrence, left, with Donald Cooksey and the "dees" of an early cyclotron.

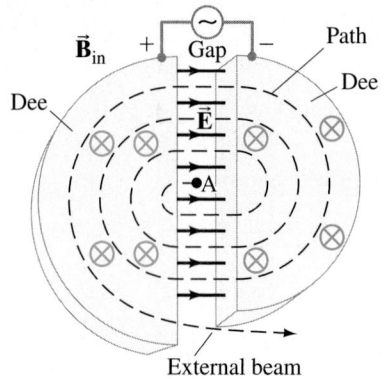

FIGURE 32–2 Diagram of a cyclotron. The magnetic field, applied by a large electromagnet, points into the page. The protons start at A, the ion source. The red electric field lines shown are for the alternating electric field in the gap at a certain moment.

The protons move in a vacuum inside two D-shaped cavities, as shown in Fig. 32–2. Each time they pass into the gap between the "dees," a voltage accelerates them (the electric force), increasing their speed and increasing the radius of curvature of their path in the magnetic field. After many revolutions, the protons acquire high kinetic energy and reach the outer edge of the cyclotron where they strike a target. The protons speed up only when they are in the gap *between* the dees, and the voltage must be alternating. When protons are moving to the right across the gap in Fig. 32–2, the right dee must be electrically negative and the left one positive. A half-cycle later, the protons are moving to the left, so the left dee must be negative in order to accelerate them.

The frequency, f, of the applied voltage must be equal to the frequency of the circulating protons. When ions of charge q are circulating *within* the hollow dees, the net force F on each ion is due to the magnetic field B, so $F = qvB$, where v is the speed of the ion at a given moment (Eq. 20–4). The magnetic force is perpendicular to both $\vec{v}$ and $\vec{B}$, and does not speed up the ions but causes them to move in circles; the acceleration within the dees is centripetal and equals v^2/r, where r is the radius of the ion's path at a given moment. We use Newton's second law, $F = ma$, and find that

$$F = ma$$
$$qvB = \frac{mv^2}{r}$$

when the protons are within the dees (not the gap), so their (constant) speed at radius r is

$$v = \frac{qBr}{m}.$$

The time required for a complete revolution is the period T and is equal to

$$T = \frac{\text{distance}}{\text{speed}} = \frac{2\pi r}{qBr/m} = \frac{2\pi m}{qB}.$$

Hence the frequency of revolution f is

$$f = \frac{1}{T} = \frac{qB}{2\pi m}. \tag{32–2}$$

This is known as the **cyclotron frequency**.

EXAMPLE 32–2 **Cyclotron.** A small cyclotron of maximum radius $R = 0.25$ m accelerates protons in a 1.7 T magnetic field. Calculate (*a*) the frequency needed for the applied alternating voltage, and (*b*) the kinetic energy of protons when they leave the cyclotron.

APPROACH The frequency of the protons revolving within the dees (Eq. 32–2) must equal the frequency of the voltage applied across the gap if the protons are going to increase in speed.

SOLUTION (*a*) From Eq. 32–2,

$$f = \frac{qB}{2\pi m} = \frac{(1.6 \times 10^{-19}\,\text{C})(1.7\,\text{T})}{(6.28)(1.67 \times 10^{-27}\,\text{kg})} = 2.6 \times 10^7\,\text{Hz} = 26\,\text{MHz},$$

which is in the radio-wave region of the EM spectrum (Fig. 22–8).
(*b*) The protons leave the cyclotron at $r = R = 0.25$ m. From $qvB = mv^2/r$ (see above), we have $v = qBr/m$, so their kinetic energy is

$$\text{KE} = \frac{1}{2}mv^2 = \frac{1}{2}m\frac{q^2B^2R^2}{m^2} = \frac{q^2B^2R^2}{2m}$$
$$= \frac{(1.6 \times 10^{-19}\,\text{C})^2(1.7\,\text{T})^2(0.25\,\text{m})^2}{(2)(1.67 \times 10^{-27}\,\text{kg})} = 1.4 \times 10^{-12}\,\text{J} = 8.7\,\text{MeV}.$$

NOTE The kinetic energy is much less than the rest energy of the proton (938 MeV), so relativity is not needed.

NOTE The magnitude of the voltage applied to the dees does not appear in the formula for KE, and so does not affect the final energy. But the higher this voltage, the fewer the revolutions required to bring the protons to full energy.

An important aspect of the cyclotron is that the frequency of the applied voltage, as given by Eq. 32–2, does not depend on the radius r of the particle's path. Thus the frequency does not have to be changed as the protons or ions start from the source and are accelerated to paths of larger and larger radii. But this is only true at nonrelativistic energies. At higher speeds, the momentum (Eq. 26–4) is $p = \gamma mv = mv/\sqrt{1 - v^2/c^2}$, so m in Eq. 32–2 has to be replaced by γm and the cyclotron frequency f (Eq. 32–2) depends on speed v because γ does. To keep the particles in sync, machines called **synchrocyclotrons** reduce the frequency in time to correspond to the increase of γm (in Eq. 32–2) as a packet of charged particles increases in speed more slowly at larger orbits.

Synchrotron

Another way to accelerate relativistic particles is to increase the magnetic field B in time so as to keep f (Eq. 32–2) constant as the particles speed up. Such devices are called **synchrotrons**; the particles move in a circle of fixed radius, which can be very large. The larger the radius, the greater the KE of the particles can be for a given magnetic field strength (see argument on previous page). The biggest synchrotron of all is at the European Center for Nuclear Research (CERN) in Geneva, Switzerland, the Large Hadron Collider (LHC). It is 4.3 km in radius, and 27 km in circumference, and accelerates protons to 4 TeV (soon to be 7 TeV).

The *Tevatron* accelerator at Fermilab (Fermi National Accelerator Laboratory, near Chicago, Illinois, has a radius of 1.0 km.[†] The Tevatron accelerated protons to about 1000 GeV = 1 TeV (hence its name, 1 TeV = 10^{12} eV). It was shut down in 2011.

These large synchrotrons use a narrow ring of magnets (see Fig. 32–3) with each magnet placed at the same radius from the center of the circle. The magnets are interrupted by gaps where high voltage accelerates the particles to higher speeds. Another way to describe the acceleration is to say the particles "surf" on a traveling electromagnetic wave within radiofrequency (RF) cavities. (The particles are first given considerable energy in smaller accelerators, "injectors," before being injected into the large ring of the large synchrotron.)

One problem of any accelerator is that accelerating electric charges radiate electromagnetic energy (see Chapter 22). Since ions or electrons are accelerated in an accelerator, we can expect considerable energy to be lost by radiation. The effect increases with energy and is especially important in circular machines where centripetal acceleration is present, such as synchrotrons, and hence is called **synchrotron radiation**. Synchrotron radiation can be useful, however. Intense beams of photons (γ rays) are sometimes needed, and they are often obtained from an electron synchrotron. Strong sources of such photons are referred to as **light sources**.

FIGURE 32–3 The interior of the tunnel of the main accelerator at Fermilab, showing (red) the ring of superconducting magnets used to keep particles moving in a circular path at the 1-TeV Tevatron.

[†]Robert Wilson, who helped design the Tevatron, and founded the field of proton therapy (Section 31–6), expressed his views on accelerators and national security in this exchange with Senator John Pastore during testimony before a Congressional Committee in 1969:

> Pastore: "Is there anything connected with the hopes of this accelerator [the Tevatron] that in any way involves the security of the country?"
>
> Robert Wilson: "No sir, I don't believe so."
>
> Pastore: "Nothing at all?"
>
> Wilson: "Nothing at all. ... "
>
> Pastore: "It has no value in that respect?"
>
> Wilson: "It has only to do with the respect with which we regard one another, the dignity of men, our love of culture. ... It has to do with are we good painters, good sculptors, great poets? I mean all the things we really venerate in our country and are patriotic about ... it has nothing to do directly with defending our country except to make it worth defending."

Linear Accelerators

In a **linear accelerator** (linac), electrons or ions are accelerated along a straight-line path, Fig. 32–4, passing through a series of tubular conductors. Voltage applied to the tubes is alternating so that when electrons (say) reach a gap, the tube in front of them is positive and the one they just left is negative. At low speeds, the particles cover less distance in the same amount of time, so the tubes are shorter at first. Electrons, with their small mass, get close to the speed of light quickly, $v \approx c$, and the tubes are nearly equal in length. Linear accelerators are particularly important for accelerating electrons to avoid loss of energy due to synchrotron radiation. The largest electron linear accelerator has been at Stanford University (Stanford Linear Accelerator Center, or SLAC), about 3 km (2 mi) long, accelerating electrons to 50 GeV. Linacs accelerating protons are used as injectors into circular machines to provide initial kinetic energy. Many hospitals have 10-MeV electron linacs that strike a metal foil to produce γ ray photons to irradiate tumors.

FIGURE 32–4 Diagram of a simple linear accelerator.

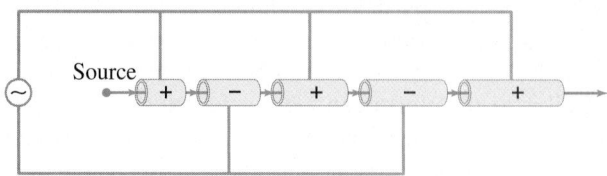

Colliding Beams

High-energy physics experiments were once done by aiming a beam of particles from an accelerator at a stationary target. But to obtain the maximum possible collision energy from a given accelerator, two beams of particles are now accelerated to very high energy and are steered so that they collide head-on. One way to accomplish such **colliding beams** with a single accelerator is through the use of **storage rings**, in which oppositely circulating beams can be repeatedly brought into collision with one another at particular points. For example, in the experiments that provided strong evidence for the top quark (Section 32–9 and Fig. 32–15), the Fermilab Tevatron accelerated protons and antiprotons each to 900 GeV, so that the combined energy of head-on collisions was 1.8 TeV.

The largest collider is the Large Hadron Collider (LHC) at CERN, with a circumference of 26.7 km (Fig. 32–5). The two colliding beams are designed to each carry 7-TeV protons for a total interaction energy of 14 TeV. For the experiments in 2011 and 2012 the total interaction energy was 7 TeV and 8 TeV. The protons for each of the beams, moving in opposite directions, are accelerated in several stages. The penultimate is SPS (Super Proton Synchrotron), seen in Fig. 32–5a which accelerates protons from 28 GeV to the 450 GeV at which they are injected into the LHC itself.

FIGURE 32–5 (a) The large circle represents the position of the tunnel, about 100 m below the ground at CERN (near Geneva) on the French–Swiss border, which houses the LHC. The smaller circle shows the position of the Super Proton Synchrotron used for accelerating protons prior to injection into the LHC. (b) Circulating proton beams, in opposite directions, inside the vacuum tube within the LHC tunnel.

(a)

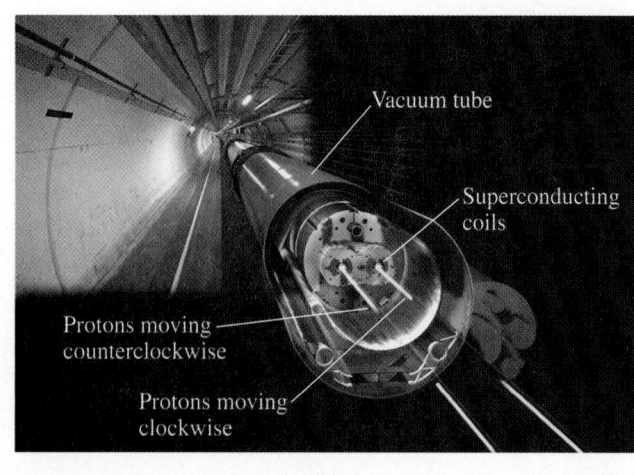

(b)

Figure 32–6 shows part of one of the detectors (ATLAS) as it was being constructed at the LHC. The detectors within ATLAS include silicon semiconductor detectors with huge numbers of pixels used to track particle paths and find their point of interaction, and to measure their radius of curvature in a magnetic field and thus determine their momentum (Section 20–4). Their energy is determined in "calorimeters" utilizing plastic, liquid, or dense metal compound crystal scintillators (Section 30–13).

In the planning stage is the International Linear Collider (ILC) which would have colliding beams of e^- and e^+ at around 0.3 to 1 TeV. It would utilize semiconductor detectors using CMOS (Section 25–1) with embedded transistors to allow fast readout.

FIGURE 32–6 ATLAS, one of the large complex detectors at the LHC, is shown here as it was being built. In 2012 it was used to provide evidence for the Higgs boson. Note the people near the bottom. From the outside, the CMS detector at the LHC looks similar.

| **EXAMPLE 32–3** | **Protons at relativistic speeds.** Determine the energy required to accelerate a proton in a high-energy accelerator (a) from rest to $v = 0.900c$, and (b) from $v = 0.900c$ to $v = 0.999c$. (c) What is the kinetic energy achieved by the proton in each case?

APPROACH We use the work-energy principle, which is still valid relativistically as mentioned in Section 26–9: $W = \Delta \mathrm{KE}$.

SOLUTION The kinetic energy of a proton of mass m is given by Eq. 26–5,

$$\mathrm{KE} = (\gamma - 1)mc^2,$$

where the relativistic factor γ is

$$\gamma = \frac{1}{\sqrt{1 - v^2/c^2}}.$$

The work-energy theorem becomes

$$W = \Delta \mathrm{KE} = (\gamma_2 - 1)mc^2 - (\gamma_1 - 1)mc^2 = (\gamma_2 - \gamma_1)mc^2$$

where γ_1 and γ_2 are for the initial and final speeds, $v_1 = 0$, $v_2 = 0.900c$.
(a) For $v = v_1 = 0$, $\gamma_1 = 1$; and for $v_2 = 0.900c$

$$\gamma_2 = \frac{1}{\sqrt{1 - (0.900)^2}} = 2.29.$$

For a proton, $mc^2 = 938\ \mathrm{MeV}$, so the work (or energy) needed to accelerate it from rest to $v_2 = 0.900c$ is

$$W = \Delta \mathrm{KE} = (\gamma_2 - \gamma_1)mc^2$$
$$= (2.29 - 1.00)(938\ \mathrm{MeV}) = 1.21\ \mathrm{GeV}.$$

(b) To go from $v_2 = 0.900c$ to $v_3 = 0.999c$, we need

$$\gamma_3 = \frac{1}{\sqrt{1 - (0.999)^2}} = 22.4.$$

So the work needed to accelerate a proton from $0.900c$ to $0.999c$ is

$$W = \Delta \mathrm{KE} = (\gamma_3 - \gamma_2)mc^2$$
$$= (22.4 - 2.29)(938\ \mathrm{MeV}) = 18.9\ \mathrm{GeV},$$

which is 15 times as much.

(c) The kinetic energy reached by the proton in (a) is just equal to the work done on it, $\mathrm{KE} = 1.21\ \mathrm{GeV}$. The final kinetic energy of the proton in (b), moving at $v_3 = 0.999c$, is

$$\mathrm{KE} = (\gamma_3 - 1)mc^2 = (21.4)(938\ \mathrm{MeV}) = 20.1\ \mathrm{GeV}.$$

NOTE This result makes sense because, starting from rest, we did work

$$W = 1.21\ \mathrm{GeV} + 18.9\ \mathrm{GeV} = 20.1\ \mathrm{GeV}$$

on it.

32–2 Beginnings of Elementary Particle Physics—Particle Exchange

The accepted model for elementary particles today views *quarks* and *leptons* as the fundamental constituents of ordinary matter. To understand our present-day view of elementary particles, it is necessary to understand the ideas leading up to its formulation.

Elementary particle physics might be said to have begun in 1935 when the Japanese physicist Hideki Yukawa (1907–1981) predicted the existence of a new particle that would in some way mediate the strong nuclear force. To understand Yukawa's idea, we first consider the electromagnetic force. When we first discussed electricity, we saw that the electric force acts over a distance, without contact. To better perceive how a force can act over a distance, we used the idea of a **field**. The force that one charged particle exerts on a second can be said to be due to the electric field set up by the first. Similarly, the magnetic field can be said to carry the magnetic force. Later (Chapter 22), we saw that electromagnetic (EM) fields can travel through space as waves. Finally, in Chapter 27, we saw that electromagnetic radiation (light) can be considered as either a wave or as a collection of particles called *photons*. Because of this wave–particle duality, it is possible to imagine that the electromagnetic force between charged particles is due to

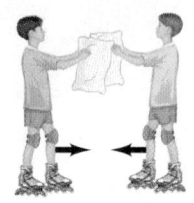

(a) Repulsive force (children throwing pillows)

(b) Attractive force (children grabbing pillows from each other's hands)

FIGURE 32–7 Forces equivalent to particle exchange. (a) Repulsive force (children on roller skates throwing pillows at each other). (b) Attractive force (children grabbing pillows from each other's hands).

(1) the EM field set up by one charged particle and felt by the other, or

(2) an exchange of photons (γ particles) between them.

It is (2) that we want to concentrate on here, and a crude analogy for how an exchange of particles could give rise to a force is suggested in Fig. 32–7. In part (a), two children start throwing heavy pillows at each other; each throw and each catch results in the child being pushed backward by the impulse. This is the equivalent of a repulsive force. On the other hand, if the two children exchange pillows by grabbing them out of the other person's hand, they will be pulled toward each other, as when an attractive force acts.

For the electromagnetic force, it is photons exchanged between two charged particles that give rise to the force between them. A simple diagram describing this photon exchange is shown in Fig. 32–8. Such a diagram, called a **Feynman diagram** after its inventor, the American physicist Richard Feynman (1918–1988), is based on the theory of **quantum electrodynamics** (QED).

FIGURE 32–8 Feynman diagram showing a photon acting as the carrier of the electromagnetic force between two electrons. This is sort of an *x* vs. *t* graph, with *t* increasing upward. Starting at the bottom, two electrons approach each other. As they get close, momentum and energy get transferred from one to the other, carried by a photon (or more than one), and the two electrons bounce apart.

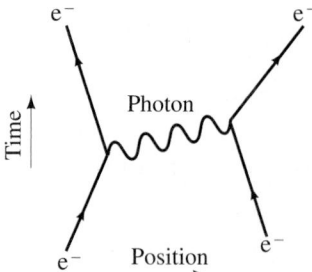

Figure 32–8 represents the simplest case in QED, in which a single photon is exchanged. One of the charged particles emits the photon and recoils somewhat as a result; and the second particle absorbs the photon. In such a collision or *interaction*, energy and momentum are transferred from one charged particle to the other, carried by the photon. The photon is absorbed by the second particle after it is emitted by the first particle and is not observable. Hence the photon is referred to as a **virtual** photon, in contrast to one that is free and can be detected by instruments. The photon is said to **mediate**, or **carry**, the electromagnetic force.

By analogy with photon exchange that mediates the electromagnetic force, Yukawa argued that there ought to be a particle that mediates the strong nuclear force—the force that holds nucleons together in the nucleus. Yukawa called this predicted particle a **meson** (meaning "medium mass"). Figure 32–9 is a Feynman diagram showing the original model of meson exchange: a meson carrying the strong force between a neutron and a proton.

A rough estimate of the mass of the meson can be made as follows. Suppose the proton on the left in Fig. 32–9 is at rest. For it to emit a meson would require energy (to make the meson's mass) which, coming from nowhere, would violate conservation of energy. But the uncertainty principle allows nonconservation of energy by an amount ΔE if it occurs only for a time Δt given by $(\Delta E)(\Delta t) \approx h/2\pi$. We set ΔE equal to the energy needed to create the mass m of the meson: $\Delta E = mc^2$. Conservation of energy is violated only as long as the meson exists, which is the time Δt required for the meson to pass from one nucleon to the other, where it is absorbed and disappears. If we assume the meson travels at relativistic speed, close to the speed of light c, then Δt need be at most about $\Delta t = d/c$, where d is the maximum distance that can separate the interacting nucleons. Thus we can write

$$\Delta E \, \Delta t \approx \frac{h}{2\pi}$$

$$mc^2 \left(\frac{d}{c}\right) \approx \frac{h}{2\pi}$$

or

$$mc^2 \approx \frac{hc}{2\pi d}. \tag{32-3}$$

The range of the strong nuclear force (the maximum distance away it can be felt) is small—not much more than the size of a nucleon or small nucleus (see Eq. 30–1)—so let us take $d \approx 1.5 \times 10^{-15}$ m. Then from Eq. 32–3,

$$mc^2 \approx \frac{hc}{2\pi d} = \frac{(6.6 \times 10^{-34}\,\text{J}\cdot\text{s})(3.0 \times 10^8\,\text{m/s})}{(6.28)(1.5 \times 10^{-15}\,\text{m})} \approx 2.1 \times 10^{-11}\,\text{J} = 130\,\text{MeV}.$$

The mass of the predicted meson, roughly $130\,\text{MeV}/c^2$, is about 250 times the electron mass of $0.51\,\text{MeV}/c^2$.

> EXERCISE B What effect does an increase in the mass of the virtual exchange particle have on the range of the force it mediates? (*a*) Decreases it; (*b*) increases it; (*c*) has no appreciable effect; (*d*) decreases the range for charged particles and increases the range for neutral particles.

Note that since the electromagnetic force has infinite range, Eq. 32–3 with $d = \infty$ tells us that the exchanged particle for the electromagnetic force, the photon, will have zero mass, which it does.

The particle predicted by Yukawa was discovered in cosmic rays by C. F. Powell and G. Occhialini in 1947, and is called the "π" or pi meson, or simply the **pion**. It comes in three charge states: $+e$, $-e$, or 0, where $e = 1.6 \times 10^{-19}$ C. The π^+ and π^- have mass of $139.6\,\text{MeV}/c^2$ and the π^0 a mass of $135.0\,\text{MeV}/c^2$, all close to Yukawa's prediction. All three interact strongly with matter. Reactions observed in the laboratory, using a particle accelerator, include

$$\text{p} + \text{p} \rightarrow \text{p} + \text{p} + \pi^0,$$
$$\text{p} + \text{p} \rightarrow \text{p} + \text{n} + \pi^+. \tag{32-4}$$

The incident proton from the accelerator must have sufficient energy to produce the additional mass of the free pion.

Yukawa's theory of pion exchange as carrier of the strong force has been superseded by *quantum chromodynamics* in which protons, neutrons, and other strongly interacting particles are made up of basic entities called *quarks*, and the basic carriers of the strong force are *gluons*, as we shall discuss shortly. But the basic idea of the earlier theory, that forces can be understood as the exchange of particles, remains valid.

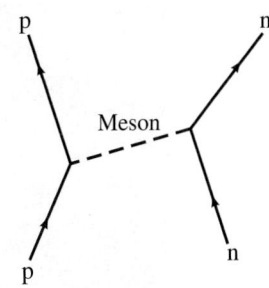

FIGURE 32–9 Early model showing meson exchange when a proton and neutron interact via the strong nuclear force. (Today, as we shall see shortly, we view the strong force as carried by gluons between quarks.)

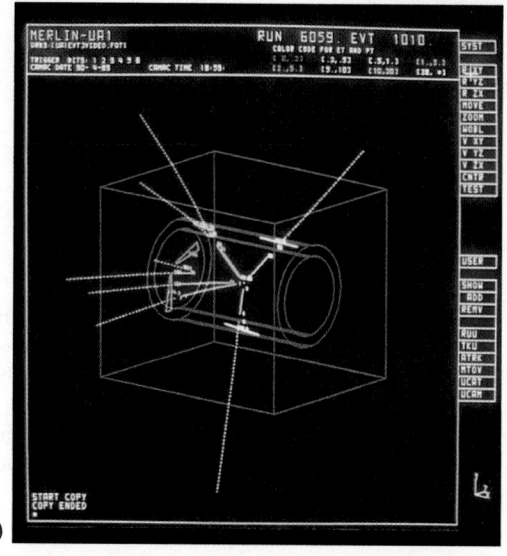

FIGURE 32–10 (a) Computer reconstruction of a Z-particle decay into an electron and a positron $(Z^0 \rightarrow e^+ + e^-)$ whose tracks are shown in white, which took place in the UA1 detector at CERN. (b) Photo of the UA1 detector at CERN as it was being built. 1980s.

There are four known types of force—or interactions—in nature. The electromagnetic force is carried by the photon, the strong force by gluons. What about the other two: the weak force and gravity? These too are believed to be mediated by particles. The particles that transmit the weak force are referred to as the W^+, W^-, and Z^0, and were detected in 1983 (Fig. 32–10). The quantum (or carrier) of the gravitational force has been named the **graviton**, but its existence has not been detected and it may not be detectable.

A comparison of the four forces is given in Table 32–1, where they are listed according to their (approximate) relative strengths. Although gravity may be the most obvious force in daily life (because of the huge mass of the Earth), on a nuclear scale gravity is by far the weakest of the four forces and its effect at the particle level can nearly always be ignored.

TABLE 32–1 The Four Forces in Nature

Type	Relative Strength (approx., for 2 protons in nucleus)	Field Particle
Strong	1	Gluons (= g)
Electromagnetic	10^{-2}	Photon (= γ)
Weak	10^{-6}	$W^\pm$ and Z^0
Gravitational	10^{-38}	Graviton (?)

32–3 Particles and Antiparticles

The positron, as we discussed in Sections 27–6 (pair production) and 30–5 (β^+ decay), is basically a positive electron. That is, many of its properties are the same as for the electron, such as mass, but it has the opposite electric charge $(+e)$. Other quantum numbers that we will discuss shortly are also reversed. The positron is said to be the **antiparticle** to the electron.

The positron was first detected as a curved path in a cloud chamber in a magnetic field by Carl Anderson in 1932. It was predicted that other particles also would have antiparticles. It was decades before another type was found.

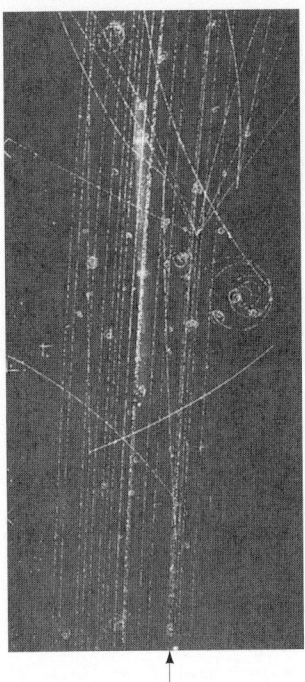
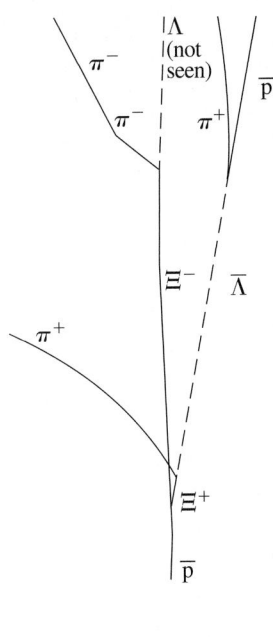

FIGURE 32–11 Liquid-hydrogen bubble-chamber photograph of an antiproton ($\bar{\text{p}}$) colliding with a proton at rest, producing a Xi—anti-Xi pair ($\bar{\text{p}} + \text{p} \rightarrow \Xi^- + \Xi^+$) that subsequently decay into other particles. The drawing indicates the assignment of particles to each track, which is based on how or if that particle decays, and on mass values estimated from measurement of momentum (curvature of track in magnetic field) and energy (thickness of track, for example). Neutral particle paths are shown by dashed lines since neutral particles rarely ionize atoms, around which bubbles form, and hence leave no tracks. 1950s.

Finally, in 1955 the antiparticle to the proton, the **antiproton** ($\bar{\text{p}}$), which carries a negative charge (Fig. 32–11), was discovered at the University of California, Berkeley, by Emilio Segrè (1905–1989, Fig. 32–12) and Owen Chamberlain (1920–2006). A bar, such as over the p, is used to indicate the antiparticle ($\bar{\text{p}}$). Soon after, the antineutron ($\bar{\text{n}}$) was found. All particles have antiparticles. But a few, like the photon, the π^0, and the Higgs, do not have distinct antiparticles—we say that they are their own antiparticles.

Antiparticles are produced in nuclear reactions when there is sufficient energy available to produce the required mass, and they do not live very long in the presence of matter. For example, a positron is stable when by itself but rarely survives for long; as soon as it encounters an electron, the two annihilate each other. The energy of their vanished mass, plus any kinetic energy they possessed, is usually converted into the energy of two γ rays. Annihilation also occurs for all other particle–antiparticle pairs.

Antimatter is a term referring to material that would be made up of "antiatoms" in which antiprotons and antineutrons would form the nucleus around which positrons (antielectrons) would move. The term is also used for antiparticles in general. If there were pockets of antimatter in the universe, a huge explosion would occur if it should encounter normal matter. It is believed that antimatter was prevalent in the very early universe (Section 33–7).

FIGURE 32–12 Emilio Segrè: he worked with Fermi in the 1930s, later discovered the first "man-made" element, technetium, and other elements, and then the antiproton. The inscription below the photo is from a book by Segrè given to this book's author.

*Negative Sea of Electrons; Vacuum State

The original idea for antiparticles came from a relativistic wave equation developed in 1928 by the Englishman P. A. M. Dirac (1902–1984). Recall that, as we saw in Chapter 26, the total energy E of a particle with mass m and momentum p and zero potential energy is given by Eq. 26–9, $E^2 = p^2c^2 + m^2c^4$. Thus

$$E = \pm\sqrt{p^2c^2 + m^2c^4}.$$

Dirac applied his new equation and found that it included solutions with both $+$ and $-$ signs. He could not ignore the solution with the negative sign, which we might have thought unphysical. If those negative energy states are real, then we would expect normal free electrons to drop down into those states, emitting

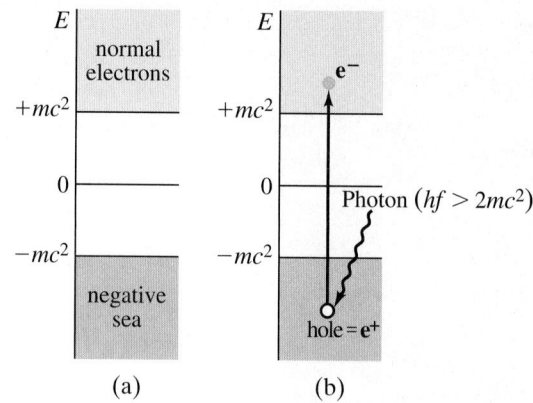

FIGURE 32–13 (a) Possible energy states for an electron. Note the vast sea of fully occupied electron states at $E < -mc^2$. (b) An electron in the negative sea is hit by a photon $(E > 2mc^2)$ and knocks it up to a normal positive energy state. The positive "hole" left behind acts like a positive electron—it is a positron.

photons—never experimentally seen. To deal with this difficulty, Dirac postulated that all those negative energy states are *normally occupied*. That is, what we thought was the **vacuum** is instead a vast **sea of electrons** in negative energy states (Fig. 32–13a). These electrons are not normally observable. But if a photon strikes one of these negative energy electrons, that electron can be knocked up to a normal $(E > mc^2)$ energy state as shown in Fig. 32–13b. Note in Fig. 32–13 that there are no energy states between $E = -mc^2$ and $E = +mc^2$ (because p^2 cannot be negative in the equation $E = \pm\sqrt{p^2c^2 + m^2c^4}$). The photon that knocks an e^- from the negative sea up to a normal state (Fig. 32–13b) must have an energy greater than $2mc^2$. What is left behind is a hole (as in semiconductors, Sections 29–7 and 29–8) with positive charge. We call that "hole" a **positron**, and it can move around as a free particle with positive energy. Thus Fig. 32–13b represents (Section 27–6) **pair production**: $\gamma \rightarrow e^-e^+$.

The vast sea of electrons with negative energy in Fig. 32–13 is the vacuum (or **vacuum state**). According to quantum mechanics, the vacuum is not empty, but contains electrons and other particles as well. The uncertainty principle allows a particle to jump briefly up to a normal energy, thus creating a **particle–antiparticle** pair. It is possible that they could be the source of the recently discovered *dark energy* that fills the universe (Chapter 33). We still have a lot to learn.

32–4 Particle Interactions and Conservation Laws

One of the important uses of high-energy accelerators and colliders is to study the interactions of elementary particles with each other. As a means of ordering this subnuclear world, the conservation laws are indispensable. The laws of conservation of energy, of momentum, of angular momentum, and of electric charge are found to hold precisely in all particle interactions.

A study of particle interactions has revealed a number of new conservation laws which (just like the old ones) are ordering principles: they help to explain why some reactions occur and others do not. For example, the following reaction has never been observed:

$$p + n \nrightarrow p + p + \bar{p}$$

even though charge, energy, and so on, are conserved ($\nrightarrow$ means the reaction does not occur). To understand why such a reaction does not occur, physicists hypothesized a new conservation law, the conservation of **baryon number**. (Baryon number is a generalization of nucleon number, which we saw earlier is conserved in nuclear reactions and decays.) All nucleons are defined to have baryon number $B = +1$, and all antinucleons (antiprotons, antineutrons) have $B = -1$. All other types of particles, such as photons, mesons, and electrons and other leptons, have $B = 0$. The reaction shown at the start of this paragraph does not conserve baryon number since the left side has $B = (+1) + (+1) = +2$, and the right has $B = (+1) + (+1) + (-1) = +1$. On the other hand, the following reaction does conserve B and *does* occur if the incoming proton has sufficient energy:

$$p + p \rightarrow p + p + \bar{p} + p,$$
$$B = +1 + 1 = +1 + 1 - 1 + 1.$$

As indicated, $B = +2$ on both sides of this equation. From these and other reactions, the **conservation of baryon number** has been established as a basic principle of physics.

Also useful are conservation laws for the three **lepton numbers**, associated with weak interactions including decays. In ordinary β decay, an electron or positron is emitted along with a neutrino or antineutrino. In another type of decay, a particle known as a "μ" or **muon**, can be emitted instead of an electron. The muon (discovered in 1937) seems to be much like an electron, except its mass is 207 times larger $(106 \, \text{MeV}/c^2)$. The neutrino (ν_e) that accompanies an emitted electron is found to be different from the neutrino (ν_μ) that accompanies an emitted muon. Each of these neutrinos has an antiparticle: $\bar{\nu}_e$ and $\bar{\nu}_\mu$. In ordinary β decay we have, for example,

$$n \rightarrow p + e^- + \bar{\nu}_e$$

CAUTION
The different types of neutrinos are not identical

but not $n \not\rightarrow p + e^- + \bar{\nu}_\mu$. To explain why these do not occur, the concept of **electron lepton number**, L_e, was invented. If the electron (e^-) and the electron neutrino (ν_e) are assigned $L_e = +1$, and e^+ and $\bar{\nu}_e$ are assigned $L_e = -1$, whereas all other particles have $L_e = 0$, then all observed decays conserve L_e. For example, in $n \rightarrow p + e^- + \bar{\nu}_e$, initially $L_e = 0$, and afterward $L_e = 0 + (+1) + (-1) = 0$. Decays that do not conserve L_e, even though they would obey the other conservation laws, are not observed to occur.

In a decay involving muons, such as

$$\pi^+ \rightarrow \mu^+ + \nu_\mu,$$

a second quantum number, **muon lepton number** (L_μ), is conserved. The μ^- and ν_μ are assigned $L_\mu = +1$, and their antiparticles μ^+ and $\bar{\nu}_\mu$ have $L_\mu = -1$, whereas all other particles have $L_\mu = 0$. L_μ too is conserved in interactions and decays. Similar assignments can be made for the **tau lepton number**, L_τ, associated with the τ lepton (discovered in 1976 with mass more than 3000 times the electron mass) and its neutrino, ν_τ.

Antiparticles have not only opposite electric charge from their particles, but also opposite B, L_e, L_μ, and L_τ. For example, a neutron has $B = +1$, an antineutron has $B = -1$ (and all the L's are zero).

CONCEPTUAL EXAMPLE 32–4 | **Lepton number in muon decay.** Which of the following decay schemes is possible for muon decay: (a) $\mu^- \rightarrow e^- + \bar{\nu}_e$; (b) $\mu^- \rightarrow e^- + \bar{\nu}_e + \nu_\mu$; (c) $\mu^- \rightarrow e^- + \nu_e$? All of these particles have $L_\tau = 0$.

RESPONSE A μ^- has $L_\mu = +1$ and $L_e = 0$. This is the initial state for all decays given, and the final state must also have $L_\mu = +1$, $L_e = 0$. In (a), the final state has $L_\mu = 0 + 0 = 0$, and $L_e = +1 - 1 = 0$; L_μ would not be conserved and indeed this decay is not observed to occur. The final state of (b) has $L_\mu = 0 + 0 + 1 = +1$ and $L_e = +1 - 1 + 0 = 0$, so both L_μ and L_e are conserved. This is in fact the most common decay mode of the μ^-. Lastly, (c) does not occur because L_e $(= +2$ in the final state) is not conserved, nor is L_μ.

EXAMPLE 32–5 **Energy and momentum are conserved.** In addition to the "number" conservation laws which help explain the decay schemes of particles, we can also apply the laws of conservation of energy and momentum. The decay of a Σ^+ particle at rest with mass $1189 \text{ MeV}/c^2$ (Table 32–2 in Section 32–6) can yield a proton ($m_\text{p} = 938 \text{ MeV}/c^2$) and a neutral pion, π^0 ($m_{\pi^0} = 135 \text{ MeV}/c^2$):

$$\Sigma^+ \rightarrow \text{p} + \pi^0.$$

Determine the kinetic energies of the proton and π^0.

APPROACH We find the energy release from the change in mass ($E = mc^2$) as we did for nuclear processes (Eq. 30–2 or 31–2a). Then we apply conservation of energy and momentum, using relativistic formulas as the energies are large.

SOLUTION The energy released, $Q = \text{KE}_\text{p} + \text{KE}_{\pi^0}$, is the change in mass $\times c^2$:

$$Q = \left[m_{\Sigma^+} - \left(m_\text{p} + m_{\pi^0}\right)\right]c^2 = \left[1189 - (938 + 135)\right] \text{MeV} = 116 \text{ MeV}.$$

Next we apply conservation of momentum: the initial particle Σ^+ is at rest, so the π^0 and p have opposite momentum but are equal in magnitude: $p_{\pi^0} = p_\text{p}$. We square this equation, $p_{\pi^0}^2 = p_\text{p}^2$, which becomes, using Eq. 26–9 ($p^2c^2 = E^2 - m^2c^4$),

$$E_{\pi^0}^2 - m_{\pi^0}^2 c^4 = E_\text{p}^2 - m_\text{p}^2 c^4.$$

Solving for $E_{\pi^0}^2$:

$$E_{\pi^0}^2 = E_\text{p}^2 - m_\text{p}^2 c^4 + m_{\pi^0}^2 c^4.$$

We substitute Eq. 26–6a, $E = \text{KE} + mc^2$, for both the π^0 and the p:

$$\left(\text{KE}_{\pi^0} + m_{\pi^0}c^2\right)^2 = \left(\text{KE}_\text{p} + m_\text{p}c^2\right)^2 - m_\text{p}^2 c^4 + m_{\pi^0}^2 c^4$$

$$\text{KE}_{\pi^0}^2 + 2\text{KE}_{\pi^0}m_{\pi^0}c^2 + \overline{m_{\pi^0}^2 c^4} = \text{KE}_\text{p}^2 + 2\text{KE}_\text{p}m_\text{p}c^2 + \overline{m_\text{p}^2 c^4} - \overline{m_\text{p}^2 c^4} + \overline{m_{\pi^0}^2 c^4}.$$

Next (after cancelling as shown) we substitute $\text{KE}_\text{p} = Q - \text{KE}_{\pi^0}$:

$$\cancel{\text{KE}_{\pi^0}^2} + 2\text{KE}_{\pi^0}m_{\pi^0}c^2 = Q^2 - 2Q\text{KE}_{\pi^0} + \cancel{\text{KE}_{\pi^0}^2} + 2Qm_\text{p}c^2 - 2\text{KE}_{\pi^0}m_\text{p}c^2.$$

After cancelling as shown, we solve for KE_{π^0}:

$$\text{KE}_{\pi^0} = \frac{Q^2 + 2Qm_\text{p}c^2}{2m_{\pi^0}c^2 + 2Q + 2m_\text{p}c^2} = \frac{(116 \text{ MeV})^2 + 2(116 \text{ MeV})(938 \text{ MeV})}{2(135 \text{ MeV}) + 2(116 \text{ MeV}) + 2(938 \text{ MeV})}$$

which gives $\text{KE}_{\pi^0} = 97 \text{ MeV}$. Then $\text{KE}_\text{p} = 116 \text{ MeV} - 97 \text{ MeV} = 19 \text{ MeV}$.

32–5 Neutrinos

We first met neutrinos with regard to β decay in Section 30–5. The study of neutrinos is a "hot" subject today. Experiments are being carried out in deep underground laboratories, sometimes in deep mine shafts. The thick layer of earth above is meant to filter out all other "background" particles, leaving mainly the very weakly interacting neutrinos to arrive at the detectors.

Some very important results have come to the fore in recent years. First there was the **solar neutrino problem**. The energy output of the Sun is believed to be due to the nuclear fusion reactions discussed in Chapter 31, Eqs. 31–6 and 31–7. The neutrinos emitted in these reactions are all ν_e (accompanied by e^+). But the rate at which ν_e arrive at Earth was measured starting in the late 1960s to be much less than expected based on the power output of the Sun. It was then proposed that, on the long trip between Sun and Earth, ν_e might turn into ν_μ or ν_τ. Subsequent experiments, definitive only in 2001, confirmed this hypothesis. Thus the three neutrinos, $\nu_\text{e}, \nu_\mu, \nu_\tau$, can change into one another in certain circumstances, a phenomenon called **neutrino flavor oscillation**[†]. (Each of the three neutrino types is called, whimsically, a different "flavor.") This result suggests that the lepton numbers L_e, L_μ, and L_τ are not perfectly conserved. But the sum, $L_\text{e} + L_\mu + L_\tau$, is believed to be always conserved.

[†]Neutrino oscillations had first been proposed in 1957 by Bruno Pontecorvo. He also proposed that the electron and muon neutrinos are different species; and he also suggested a way to confirm the existence of neutrinos by detecting $\bar{\nu}_\text{e}$ emitted in huge numbers by a nuclear reactor, an experiment carried out by Frederick Reines and Clyde Cowan in the 1950s. The experimentalists who confirmed these two predictions were awarded the Nobel Prize, but not the theorist who proposed them.

The second exceptional result has long been speculated on: are neutrinos massless as originally thought, or do they have a nonzero mass? Rough upper limits on the masses have been made. Today astrophysical experiments show that the sum of all three neutrino masses combined is less than about $0.14 \, \text{eV}/c^2$. But can all the masses be zero? Not if there are the flavor oscillations discussed above. It seems that at most, one type could have zero mass, and it is likely that at least one neutrino type has a mass of at least $0.04 \, \text{eV}/c^2$.

As a result of neutrino oscillations, the three types of neutrino may not be exactly what we thought they were (e, μ, τ). If not, the three basic neutrinos, called 1, 2, and 3, are combinations of ν_e, ν_μ, and ν_τ.

Another outstanding question is whether or not neutrinos are in the category called **Majorana particles**,[†] meaning they would be their own antiparticles, like γ, π^0, and Higgs. If so, a lot of other questions (and answers) would appear.

*Neutrino Mass Estimate from a Supernova

The explosion of a supernova in the outer parts of our Galaxy in 1987 (Section 33–2) released lots of neutrinos and offered an opportunity to estimate electron neutrino mass. If neutrinos do have mass, then their speed would be less than c, and neutrinos of different energy would take different times to travel the 170,000 light-years from the supernova to Earth. To get an idea of how such a measurement could be done, suppose two neutrinos from "SN1987A" were emitted at the same time and were actually detected on Earth (via the reaction $\bar{\nu}_e + p \rightarrow n + e^+$) 10 seconds apart, with measured kinetic energies of about 20 MeV and 10 MeV. From other laboratory measurements we expect the neutrino mass to be less than 100 eV; and since our neutrinos have kinetic energy of 20 MeV and 10 MeV, we can make the approximation $m_\nu c^2 \ll E$, so that E (the total energy) is essentially equal to the kinetic energy. We use Eq. 26–6b, which tells us

$$E = \frac{m_\nu c^2}{\sqrt{1 - v^2/c^2}}.$$

We solve this for v, the velocity of a neutrino with energy E:

$$v = c\left(1 - \frac{m_\nu^2 c^4}{E^2}\right)^{\frac{1}{2}} = c\left(1 - \frac{m_\nu^2 c^4}{2E^2} + \cdots\right),$$

where we have used the binomial expansion $(1 - x)^{\frac{1}{2}} = 1 - \frac{1}{2}x + \cdots$, and we ignore higher-order terms since $m_\nu^2 c^4 \ll E^2$. The time t for a neutrino to travel a distance d $(= 170{,}000 \, \text{ly})$ is

$$t = \frac{d}{v} = \frac{d}{c\left(1 - \frac{m_\nu^2 c^4}{2E^2}\right)} \approx \frac{d}{c}\left(1 + \frac{m_\nu^2 c^4}{2E^2}\right),$$

where again we used the binomial expansion $\left[(1 - x)^{-1} = 1 + x + \cdots\right]$. The difference in arrival times for our two neutrinos of energies $E_1 = 20 \, \text{MeV}$ and $E_2 = 10 \, \text{MeV}$ is

$$t_2 - t_1 = \frac{d}{c}\frac{m_\nu^2 c^4}{2}\left(\frac{1}{E_2^2} - \frac{1}{E_1^2}\right).$$

We solve this for $m_\nu c^2$ and set $t_2 - t_1 = 10 \, \text{s}$:

$$m_\nu c^2 = \left[\frac{2c(t_2 - t_1)}{d}\frac{E_1^2 E_2^2}{E_1^2 - E_2^2}\right]^{\frac{1}{2}} = 22 \times 10^{-6} \, \text{MeV} = 22 \, \text{eV}.$$

This calculation, with its optimistic assumptions, estimates the mass of the neutrino to be $22 \, \text{eV}/c^2$. But there would be experimental uncertainties, and even worse there is the unwarranted assumption that the two neutrinos were emitted at the same time.

[†]The brilliant young physicist Ettore Majorana (1906–1938) disappeared from a ship under mysterious circumstances in 1938 at the age of 31.

Theoretical models of supernova explosions suggest that the neutrinos are emitted in a burst that lasts from a second or two up to perhaps 10 s. If we assume the neutrinos are not emitted simultaneously but rather at any time over a 10-s interval, then that 10-s difference in arrival times could be due to a 10-s difference in their emission time. In this case the data would be consistent with zero neutrino mass, and it put an approximate *upper limit* of $22 \text{ eV}/c^2$.

The actual detection of these neutrinos was brilliant—it was a rare event that allowed us to detect something other than EM radiation from beyond the solar system, and was an exceptional confirmation of theory. In the experiments, the most sensitive detector consisted of several thousand tons of water in an underground chamber. It detected 11 events in 12 seconds, probably via the reaction $\bar{\nu}_e + p \rightarrow n + e^+$. There was not a clear correlation between energy and time of arrival. Nonetheless, a careful analysis of that experiment set a rough upper limit on the electron antineutrino mass of about $4 \text{ eV}/c^2$. The more recent results mentioned above are much more definitive—they provide evidence that mass is much smaller, and that it is *not zero*—but precise neutrino masses still elude us.

32–6 Particle Classification

In the decades after the discovery of the π meson in the late 1940s, hundreds of other subnuclear particles were discovered. One way to categorize the particles is according to their interactions, since not all particles interact via all four of the forces known in nature (though all interact via gravity). Table 32–2 (next page) lists some of the more common particles classified in this way along with many of their properties. At the top of Table 32–2 are the so-called "fundamental" particles which we believe have no internal structure. Below them are some of the "composite" particles which are made up of quarks, according to the Standard Model.

The **fundamental particles** include the **gauge bosons** (so-named after the theory that describes them, *gauge theory*), which include the gluons, the photon, and the W and Z particles; these are the particles that mediate (or "carry") the strong, electromagnetic, and weak interactions, respectively.

Also fundamental are the **leptons**, which are particles that do not interact via the strong force but do interact via the weak nuclear force. Leptons that carry electric charge also interact via the electromagnetic force. The leptons include the electron, the muon, and the tau, and three types of neutrino: the electron neutrino (ν_e), the muon neutrino (ν_μ), and the tau neutrino (ν_τ). Each lepton has an antiparticle. Finally, the recently detected Higgs boson is also considered to be fundamental, with no internal structure.

The second category of particle in Table 32–2 is the **hadrons**, which are **composite** particles (made up of quarks as we will discuss shortly). Hadrons are particles that interact via the strong nuclear force and are said to be **strongly interacting particles**. They also interact via the other forces, but the strong force predominates at short distances. The hadrons include the proton, neutron, pion, and many other particles. They are divided into two subgroups: **baryons**, which are particles that have baryon number $+1$ (or -1 in the case of their antiparticles) and, as we shall see, are each made up of three quarks; and **mesons**, which have baryon number $= 0$, and are made up of a quark and an antiquark.

Only a few of the hundreds of hadrons (a veritable "zoo") are included in Table 32–2. Notice that the baryons Λ, Σ, Ξ, and Ω all decay to lighter-mass baryons, and eventually to a proton or neutron. All these processes conserve baryon number. Since there is no particle lighter than the proton with $B = +1$, if baryon number is strictly conserved, the proton itself cannot decay and is stable. (But see Section 32–11.) Note that Table 32–2 gives the **mean life** (τ) of each particle (as is done in particle physics), not the half-life $(T_{\frac{1}{2}})$. Recall that they differ by a factor 0.693: $\tau = T_{\frac{1}{2}}/\ln 2 = T_{\frac{1}{2}}/0.693$, Eq. 30–7. The term **lifetime** in particle physics means the mean life τ (= mean lifetime).

The baryon and lepton numbers (B, L_e, L_μ, L_τ), as well as strangeness S (Section 32–8), as given in Table 32–2 are for particles; their antiparticles have opposite sign for these numbers.

TABLE 32–2 Particles (selected)[†]

Category	Forces involved	Particle name	Symbol	Anti-particle	Spin	Mass (MeV/c^2)	B	L_e	L_μ	L_τ	S	Mean life (s)	Principal Decay Modes
							B [antiparticles have opposite sign]						

Category	Forces involved	Particle name	Symbol	Anti-particle	Spin	Mass (MeV/c^2)	B [antiparticles	L_e have	L_μ opposite	L_τ sign]	S	Mean life (s)	Principal Decay Modes
Fundamental													
Gauge bosons (force carriers)	str	Gluons	g	Self	1	0	0	0	0	0	0	Stable	
	em	Photon	γ	Self	1	0	0	0	0	0	0	Stable	
	w, em	W	W^+	W^-	1	80.385×10^3	0	0	0	0	0	3×10^{-25}	$e\nu_e, \mu\nu_\mu, \tau\nu_\tau$, hadrons
	w	Z	Z^0	Self	1	91.19×10^3	0	0	0	0	0	3×10^{-25}	$e^+e^-, \mu^+\mu^-, \tau^+\tau^-$, hadrons
Higgs boson	w, str	Higgs	H^0	Self	0	125×10^3	0	0	0	0	0	1.6×10^{-22}	$b\bar{b}, Z^0Z^0, W^+W^-, g\bar{g}, \tau\bar{\tau}, \gamma\gamma$
Leptons	w, em[‡]	Electron	e^-	e^+	$\frac{1}{2}$	0.511	0	+1	0	0	0	Stable	
		Neutrino (e)	ν_e	$\bar{\nu}_e$	$\frac{1}{2}$	$0\ (<0.14\,\text{eV}/c^2)$[‡]	0	+1	0	0	0	Stable	
		Muon	μ^-	μ^+	$\frac{1}{2}$	105.7	0	0	+1	0	0	2.20×10^{-6}	$e^-\bar{\nu}_e\nu_\mu$
		Neutrino (μ)	ν_μ	$\bar{\nu}_\mu$	$\frac{1}{2}$	$0\ (<0.14\,\text{eV}/c^2)$[‡]	0	0	+1	0	0	Stable	
		Tau	τ^-	τ^+	$\frac{1}{2}$	1777	0	0	0	+1	0	2.91×10^{-13}	$\mu^-\bar{\nu}_\mu\nu_\tau, e^-\bar{\nu}_e\nu_\tau$, hadrons $+\nu_\tau$
		Neutrino (τ)	ν_τ	$\bar{\nu}_\tau$	$\frac{1}{2}$	$0\ (<0.14\,\text{eV}/c^2)$[‡]	0	0	0	+1	0	Stable	
Quarks	w, em, str	(see Table 32–3)											
Hadrons (composite), selected													
Mesons (quark–antiquark)	str, em, w	Pion	π^+	π^-	0	139.6	0	0	0	0	0	2.60×10^{-8}	$\mu^+\nu_\mu$
			π^0	Self	0	135.0	0	0	0	0	0	0.85×10^{-16}	2γ
		Kaon	K^+	K^-	0	493.7	0	0	0	0	+1	1.24×10^{-8}	$\mu^+\nu_\mu, \pi^+\pi^0$
			K^0_S	$\bar{K}^0_S$	0	497.6	0	0	0	0	+1	0.895×10^{-10}	$\pi^+\pi^-, 2\pi^0$
			K^0_L	$\bar{K}^0_L$	0	497.6	0	0	0	0	+1	5.12×10^{-8}	$\pi^\pm e^\mp \overset{(-)}{\nu}_e, \pi^\pm \mu^\mp \overset{(-)}{\nu}_\mu, 3\pi$
		Eta	η^0	Self	0	547.9	0	0	0	0	0	5.1×10^{-19}	$2\gamma, 3\pi^0, \pi^+\pi^-\pi^0$
		Rho	ρ^0	Self	1	775	0	0	0	0	0	4.4×10^{-24}	$\pi^+\pi^-, 2\pi^0$
			ρ^+	ρ^-	1	775	0	0	0	0	0	4.4×10^{-24}	$\pi^+\pi^0$
		and others											
Baryons (3 quarks)	str, em, w	Proton	p	$\bar{p}$	$\frac{1}{2}$	938.3	+1	0	0	0	0	Stable	
		Neutron	n	$\bar{n}$	$\frac{1}{2}$	939.6	+1	0	0	0	0	882	$pe^-\bar{\nu}_e$
		Lambda	Λ^0	$\bar{\Lambda}^0$	$\frac{1}{2}$	1115.7	+1	0	0	0	−1	2.63×10^{-10}	$p\pi^-, n\pi^0$
		Sigma	Σ^+	$\bar{\Sigma}^-$	$\frac{1}{2}$	1189.4	+1	0	0	0	−1	0.80×10^{-10}	$p\pi^0, n\pi^+$
			Σ^0	$\bar{\Sigma}^0$	$\frac{1}{2}$	1192.6	+1	0	0	0	−1	7.4×10^{-20}	$\Lambda^0\gamma$
			Σ^-	$\bar{\Sigma}^+$	$\frac{1}{2}$	1197.4	+1	0	0	0	−1	1.48×10^{-10}	$n\pi^-$
		Xi	Ξ^0	$\bar{\Xi}^0$	$\frac{1}{2}$	1314.9	+1	0	0	0	−2	2.90×10^{-10}	$\Lambda^0\pi^0$
			Ξ^-	Ξ^+	$\frac{1}{2}$	1321.7	+1	0	0	0	−2	1.64×10^{-10}	$\Lambda^0\pi^-$
		Omega	Ω^-	Ω^+	$\frac{3}{2}$	1672.5	+1	0	0	0	−3	0.82×10^{-10}	$\Xi^0\pi^-, \Lambda^0 K^-, \Xi^-\pi^0$
		and others											

[†]See also Table 32–4 for particles with charm and bottom. *S* in this Table stands for "strangeness" (see Section 32–8). More detail online at: pdg.lbl.gov.

[‡]Neutrinos partake only in the weak interaction. Experimental upper limits on neutrino masses are given in parentheses, as obtained mainly from the WMAP survey (Chapter 33). Detection of neutrino oscillations suggests that at least one type of neutrino has a nonzero mass greater than $0.04\,\text{eV}/c^2$.

EXAMPLE 32–6 **Baryon decay.** Show that the decay modes of the Σ^+ baryon given in Table 32–2 do not violate the conservation laws we have studied up to now: energy, charge, baryon number, lepton numbers.

APPROACH Table 32–2 shows two possible decay modes, (a) $\Sigma^+ \to p + \pi^0$, (b) $\Sigma^+ \to n + \pi^+$. All the particles have lepton numbers equal to zero.

SOLUTION (a) Energy: for $\Sigma^+ \to p + \pi^0$ the change in mass-energy is

$$\Delta(Mc^2) = m_\Sigma c^2 - m_p c^2 - m_{\pi^0} c^2$$
$$= 1189.4\,\text{MeV} - 938.3\,\text{MeV} - 135.0\,\text{MeV} = +116.1\,\text{MeV},$$

so energy can be conserved with the resulting particles having kinetic energy.

Charge: $+e = +e + 0$, so charge is conserved.

Baryon number: $+1 = +1 + 0$, so baryon number is conserved.

(b) Energy: for $\Sigma^+ \to n + \pi^+$, the mass-energy change is

$$\Delta(Mc^2) = m_\Sigma c^2 - m_n c^2 - m_{\pi^+} c^2$$
$$= 1189.4\,\text{MeV} - 939.6\,\text{MeV} - 139.6\,\text{MeV} = +110.2\,\text{MeV}.$$

This reaction releases 110.2 MeV of energy as kinetic energy of the products.

Charge: $+e = 0 + e$, so charge is conserved.

Baryon number: $+1 = +1 + 0$, so baryon number is conserved.

32–7 Particle Stability and Resonances

Many particles listed in Table 32–2 are unstable. The lifetime of an unstable particle depends on which force is most active in causing the decay. When a stronger force influences a decay, that decay occurs more quickly. Decays caused by the weak force typically have lifetimes of 10^{-13} s or longer (W and Z decay directly and more quickly). Decays via the electromagnetic force have much shorter lifetimes, typically about 10^{-16} to 10^{-19} s, and normally involve a γ (photon). Most of the unstable particles included in Table 32–2 decay either via the weak or the electromagnetic interaction.

Many particles have been found that decay via the strong interaction, with very short lifetimes, typically about 10^{-23} s. Their lifetimes are so short they do not travel far enough to be detected before decaying. The existence of such short-lived particles is inferred from their decay products. Consider the first such particle discovered (by Fermi), using a beam of π^+ particles with varying amounts of energy directed through a hydrogen target (protons). The number of interactions (π^+ scattered) plotted versus the pion's kinetic energy is shown in Fig. 32–14. The large number of interactions around 200 MeV led Fermi to conclude that the π^+ and proton combined momentarily to form a short-lived particle before coming apart again, or at least that they resonated together for a short time. Indeed, the large peak in Fig. 32–14 resembles a resonance curve (see Figs. 11–18 and 21–46), and this new "particle"—now called the Δ—is referred to as a **resonance**. Hundreds of other resonances have been found, and are regarded as excited states of lighter mass particles such as a nucleon.

FIGURE 32–14 Number of π^+ particles scattered elastically by a proton target as a function of the incident π^+ kinetic energy. The resonance shape represents the formation of a short-lived particle, the Δ, which has a charge in this case of $+2e\ (\Delta^{++})$.

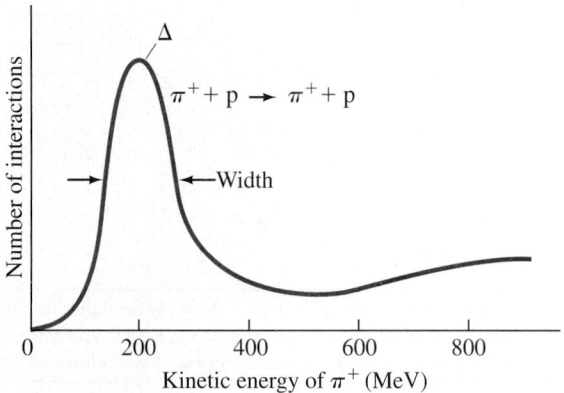

The **width** of a resonance—in Fig. 32–14 the full width of the Δ peak at half the maximum is on the order of 100 MeV—is an interesting application of the uncertainty principle. If a particle lives only 10^{-23} s, then its mass (i.e., its rest energy) will be uncertain by an amount

$$\Delta E \approx \frac{h}{2\pi\,\Delta t} \approx \frac{(6.6 \times 10^{-34}\,\text{J}\cdot\text{s})}{(6)(10^{-23}\,\text{s})} \approx 10^{-11}\,\text{J} \approx 100\,\text{MeV},$$

which is what is observed. Actually, the lifetimes of $\approx 10^{-23}$ s for such resonances are inferred by the reverse process: from the measured width being ≈ 100 MeV.

32–8 Strangeness? Charm? Towards a New Model

In the early 1950s, the newly found particles K, Λ, and Σ were found to behave rather strangely in two ways. First, they were always produced in pairs. For example, the reaction

$$\pi^- + p \rightarrow K^0 + \Lambda^0$$

occurred with high probability, but the similar reaction $\pi^- + p \nrightarrow K^0 + n$ was never observed to occur even though it did not violate any known conservation law. The second feature of these **strange particles**, as they came to be called, was that they were produced via the strong interaction (that is, at a high interaction rate), but did not decay at a fast rate characteristic of the strong interaction (even though they decayed into strongly interacting particles).

To explain these observations, a new quantum number, **strangeness**, and a new conservation law, **conservation of strangeness**, were introduced. By assigning the strangeness numbers (S) indicated in Table 32–2, the production of strange particles in pairs was explained. Antiparticles were assigned opposite strangeness from their particles. For example, in the reaction $\pi^- + p \rightarrow K^0 + \Lambda^0$, the initial state has strangeness $S = 0 + 0 = 0$, and the final state has $S = +1 - 1 = 0$, so strangeness is conserved. But for $\pi^- + p \nrightarrow K^0 + n$, the initial state has $S = 0$ and the final state has $S = +1 + 0 = +1$, so strangeness would not be conserved; and this reaction is not observed.

To explain the decay of strange particles, it is assumed that strangeness is conserved in the strong interaction but is *not conserved in the weak interaction*. Thus, strange particles were forbidden by strangeness conservation to decay to nonstrange particles of lower mass via the strong interaction, but could decay by means of the weak interaction at the observed longer lifetimes of 10^{-10} to 10^{-8} s.

The conservation of strangeness was the first example of a **partially conserved** quantity. In this case, the quantity strangeness is conserved by strong interactions but not by weak.

CAUTION

Partially conserved quantities

CONCEPTUAL EXAMPLE 32–7 | **Guess the missing particle.** Using the conservation laws for particle interactions, determine the possibilities for the missing particle in the reaction

$$\pi^- + p \rightarrow K^0 + ?$$

in addition to $K^0 + \Lambda^0$ mentioned above.

RESPONSE We write equations for the conserved numbers in this reaction, with B, L_e, S, and Q as unknowns whose determination will reveal what the possible particle might be:

Baryon number:	$0 + 1 = 0 + B$
Lepton number:	$0 + 0 = 0 + L_e$
Charge:	$-1 + 1 = 0 + Q$
Strangeness:	$0 + 0 = 1 + S.$

The unknown product particle would have to have these characteristics:

$$B = +1 \qquad L_e = 0 \qquad Q = 0 \qquad S = -1.$$

In addition to Λ^0, a neutral sigma particle, Σ^0, is also consistent with these numbers.

In the next Section we will discuss another partially conserved quantity which was given the name **charm**. The discovery in 1974 of a particle with charm helped solidify a new theory involving quarks, which we now discuss.

32–9 Quarks

One difference between leptons and hadrons is that the hadrons interact via the strong interaction, whereas the leptons do not. There is an even more fundamental difference. The six leptons $\left(e^-, \mu^-, \tau^-, \nu_e, \nu_\mu, \nu_\tau\right)$ are considered to be truly fundamental particles because they do not show any internal structure, and have no measurable size. (Attempts to determine the size of leptons have put an upper limit of about 10^{-18} m.) On the other hand, there are hundreds of hadrons, and experiments indicate they do have an internal structure. When an electron collides with another electron, it scatters off as per Coulomb's law. But electrons scattering off a proton reveal a more complex pattern, implying internal parts within the proton (= quarks).

FIGURE 32–15 This computer-generated reconstruction of a proton–antiproton collision at Fermilab occurred at an energy of nearly 2 TeV. It is one of the events that provided evidence for the top quark (1995). The multiwire chamber (Section 30–13) is in a magnetic field, and the radius of curvature of the charged particle tracks is a measure of each particle's momentum (Chapter 20). The white dots represent signals seen on the electric wires of the multiwire chamber. The colored lines are particle paths. The top quark (t) has too brief a lifetime ($\approx 10^{-23}$ s) to be detected itself, so we look for its possible decay products. Analysis indicates the following interaction and subsequent decays:

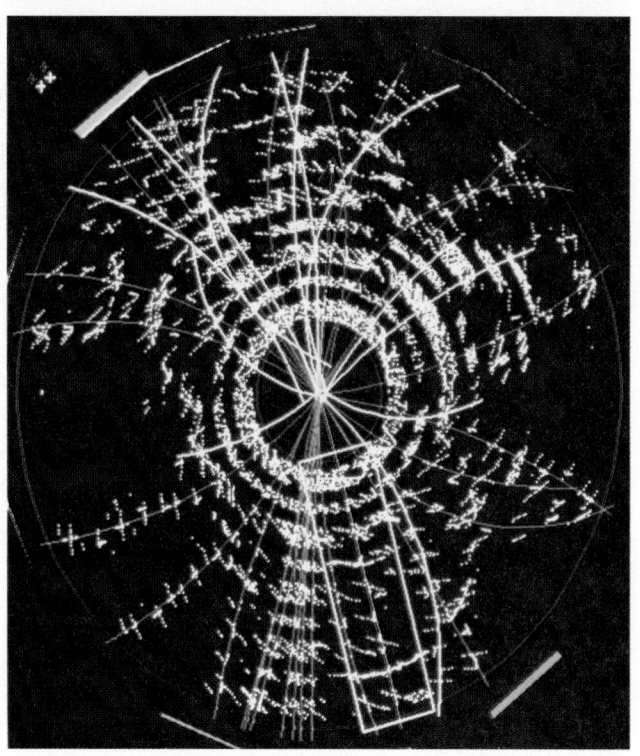

$$
\begin{array}{l}
p + \bar{p} \longrightarrow t + \bar{t} \\
\qquad\qquad\quad \hookrightarrow W^- + \bar{b} \\
\qquad\qquad\qquad\quad \hookrightarrow \text{jet} \\
\qquad\qquad\qquad\qquad\quad \longrightarrow \mu^- + \bar{\nu}_\mu \\
\qquad\qquad\quad \hookrightarrow W^+ + b \\
\qquad\qquad\qquad\quad \hookrightarrow \text{jet} \\
\qquad\qquad\qquad\quad \hookrightarrow u + \bar{d} \\
\qquad\qquad\qquad\qquad\quad \hookrightarrow \text{jet} \\
\qquad\qquad\qquad\qquad\quad \longrightarrow \text{jet}
\end{array}
$$

The tracks in the photo include jets (groups of particles moving in roughly the same direction), and a muon (μ^-) whose track is the pink one enclosed by a yellow rectangle to make it stand out.

In 1963, M. Gell-Mann and G. Zweig proposed that none of the hadrons, not even the proton and neutron, are truly fundamental, but instead are made up of combinations of three more fundamental pointlike entities called (somewhat whimsically) **quarks**.[†] Today, the quark theory is well-accepted, and quarks are considered truly fundamental particles, like leptons. The three quarks originally proposed were named **up**, **down**, and **strange**, with abbreviations u, d, s. The theory today has six quarks, just as there are six leptons—based on a presumed *symmetry* in nature. The other three quarks are called **charm**, **bottom**, and **top** (c, b, t). The names apply also to new properties of each quark (quantum numbers c, b, t) that distinguish these new quarks from the 3 original quarks (see Table 32–3). These properties (like strangeness) are conserved in strong, but not weak, interactions. Figure 32–15 shows one of the events that provided evidence for the top quark.

All quarks have spin $\frac{1}{2}$ and an electric charge of either $+\frac{2}{3}e$ or $-\frac{1}{3}e$ (that is, a fraction of the previously thought smallest charge e). Antiquarks have opposite sign of electric charge Q, baryon number B, strangeness S, charm c, bottom b, and top t. Other properties of quarks are shown in Table 32–3.

[†]Gell-Mann chose the word from a phrase in James Joyce's *Finnegans Wake*.

TABLE 32–3 Properties of Quarks (Antiquarks have opposite sign *Q, B, S, c, b, t*)

					Quarks			
Name	Symbol	Mass (MeV/c^2)	Charge Q	Baryon Number B	Strangeness S	Charm c	Bottom b	Top t
Up	u	2.3	$+\frac{2}{3}e$	$\frac{1}{3}$	0	0	0	0
Down	d	4.8	$-\frac{1}{3}e$	$\frac{1}{3}$	0	0	0	0
Strange	s	95	$-\frac{1}{3}e$	$\frac{1}{3}$	-1	0	0	0
Charm	c	1275	$+\frac{2}{3}e$	$\frac{1}{3}$	0	$+1$	0	0
Bottom	b	4180	$-\frac{1}{3}e$	$\frac{1}{3}$	0	0	-1	0
Top[†]	t	173,500	$+\frac{2}{3}e$	$\frac{1}{3}$	0	0	0	$+1$

[†]The top quark, with its extremely short lifetime of 5×10^{-25} s, does not live long enough to form hadrons.

TABLE 32–4 **Partial List of Heavy Hadrons, with Charm and Bottom** ($L_e = L_\mu = L_\tau = 0$)

Category	Particle	Anti-particle	Spin	Mass (MeV/c^2)	Baryon Number B	Strangeness S	Charm c	Bottom b	Mean life (s)	Principal Decay Modes
Mesons	D^+	D^-	0	1869.6	0	0	+1	0	10.4×10^{-13}	K + others, e + others
	D^0	$\overline{D}^0$	0	1864.9	0	0	+1	0	4.1×10^{-13}	K + others, μ or e + others
	D_S^+	D_S^-	0	1968.5	0	+1	+1	0	5.0×10^{-13}	K + others
	J/ψ (3097)	Self	1	3096.9	0	0	0	0	0.71×10^{-20}	Hadrons, e^+e^-, $\mu^+\mu^-$
	Υ (9460)	Self	1	9460.3	0	0	0	0	1.2×10^{-20}	Hadrons, $\mu^+\mu^-$, e^+e^-, $\tau^+\tau^-$
	B^-	B^+	0	5279.3	0	0	0	-1	1.6×10^{-12}	D^0 + others
	B^0	$\overline{B}^0$	0	5279.6	0	0	0	-1	1.5×10^{-12}	D^0 + others
Baryons	Λ_c^+	Λ_c^-	$\frac{1}{2}$	2286	+1	0	+1	0	2.0×10^{-13}	Hadrons (e.g., Λ + others)
	Σ_c^{++}	Σ_c^{--}	$\frac{1}{2}$	2454	+1	0	+1	0	2.9×10^{-22}	$\Lambda_c^+\pi^+$
	Σ_c^+	Σ_c^-	$\frac{1}{2}$	2453	+1	0	+1	0	$>1.4 \times 10^{-22}$	$\Lambda_c^+\pi^0$
	Σ_c^0	$\overline{\Sigma}_c^0$	$\frac{1}{2}$	2454	+1	0	+1	0	3.0×10^{-22}	$\Lambda_c^+\pi^-$
	Λ_b^0	$\overline{\Lambda}_b^0$	$\frac{1}{2}$	5619	+1	0	0	-1	1.4×10^{-12}	$J/\psi\Lambda^0$, $pD^0\pi^-$, $\Lambda_c^+\pi^+\pi^-\pi^-$

All hadrons are considered to be made up of combinations of quarks (plus the gluons that hold them together), and their properties are described by looking at their quark content. Mesons consist of a quark–antiquark pair. For example, a π^+ meson is a $u\overline{d}$ combination: note that for the $u\overline{d}$ pair (Table 32–3), $Q = \frac{2}{3}e + \frac{1}{3}e = +1e$, $B = \frac{1}{3} - \frac{1}{3} = 0$, $S = 0 + 0 = 0$, as they must for a π^+; and a $K^+ = u\overline{s}$, with $Q = +1$, $B = 0$, $S = +1$. A π^0 can be made of $u\overline{u}$ or $d\overline{d}$.

Baryons, on the other hand, consist of three quarks. For example, a neutron is n = ddu, whereas an antiproton is $\overline{p} = \overline{u}\,\overline{u}\,\overline{d}$. See Fig. 32–16. Strange particles all contain an s or $\overline{s}$ quark, whereas charm particles contain a c or $\overline{c}$ quark. A few of these hadrons are listed in Table 32–4.

Current models suggest that quarks may be so tightly bound together that they may not ever exist singly in the free state. But quarks can be detected indirectly when they turn into narrow **jets** of other particles, as in Fig. 32–15. Also, observations of very high energy electrons scattered off protons suggest that protons are indeed made up of constituents.

Today, the truly **fundamental particles** are considered to be the six quarks, the six leptons, the gauge bosons that carry the fundamental forces, and the Higgs (page 939). See Table 32–5, where the quarks and leptons are arranged in three "families" or "generations." Ordinary matter—atoms made of protons, neutrons, and electrons—is contained in the "first generation." The others are thought to have existed in the very early universe, but are seen by us today only at powerful colliders or in cosmic rays. All of the hundreds of hadrons can be accounted for by combinations of the six quarks and six antiquarks.

EXERCISE C Return to the Chapter-Opening Questions, page 915, and answer them again now. Try to explain why you may have answered differently the first time.

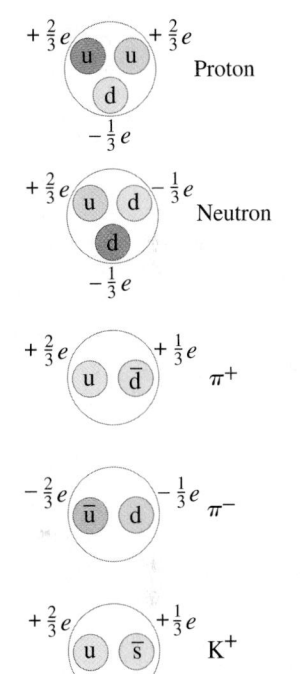

FIGURE 32–16 Quark compositions for several particles.

TABLE 32–5 The Fundamental Particles† of the Standard Model

Gauge bosons	γ	g		Z^0	$W^\pm$
Higgs boson	H^0				

	Quarks		and	Leptons	
First generation	u	d		e	ν_e
Second generation	s	c		μ	ν_μ
Third generation	b	t		τ	ν_τ

†The graviton (G^0) has been hypothesized but not detected and may not be detectable. It is not part of the Standard Model.

Quark combinations. Find the baryon number, charge, and strangeness for the following quark combinations, and identify the hadron particle that is made up of these quark combinations: (*a*) udd, (*b*) u$\bar{\text{u}}$, (*c*) uss, (*d*) sdd, and (*e*) b$\bar{\text{u}}$.

RESPONSE We use Table 32–3 to get the properties of the quarks, then Table 32–2 or 32–4 to find the particle that has these properties.

(*a*) udd has

$$Q = +\tfrac{2}{3}e - \tfrac{1}{3}e - \tfrac{1}{3}e = 0,$$
$$B = \tfrac{1}{3} + \tfrac{1}{3} + \tfrac{1}{3} = 1,$$
$$S = 0 + 0 + 0 = 0,$$

as well as $c = 0$, bottom $= 0$, top $= 0$. The only baryon ($B = +1$) that has $Q = 0$, $S = 0$, etc., is the neutron (Table 32–2).

(*b*) u$\bar{\text{u}}$ has $Q = \tfrac{2}{3}e - \tfrac{2}{3}e = 0$, $B = \tfrac{1}{3} - \tfrac{1}{3} = 0$, and all other quantum numbers $= 0$. Sounds like a π^0 (d$\bar{\text{d}}$ also gives a π^0).

(*c*) uss has $Q = 0$, $B = +1$, $S = -2$, others $= 0$. This is a Ξ^0.

(*d*) sdd has $Q = -1$, $B = +1$, $S = -1$, so must be a Σ^-.

(*e*) b$\bar{\text{u}}$ has $Q = -1$, $B = 0$, $S = 0$, $c = 0$, bottom $= -1$, top $= 0$. This must be a B$^-$ meson (Table 32–4).

| EXERCISE D What is the quark composition of a K$^-$ meson?

32–10 The Standard Model: QCD and Electroweak Theory

Not long after the quark theory was proposed, it was suggested that quarks have another property (or quality) called **color**, or "color charge" (analogous to electric charge). The distinction between the six types of quark (u, d, s, c, b, t) was referred to as **flavor**. According to theory, each flavor of quark can have one of three colors, usually designated red, green, and blue. (These are the three primary colors which, when added together in appropriate amounts, as on a TV screen, produce white.) Note that the names "color" and "flavor" have nothing to do with our senses, but are purely whimsical—as are other names, such as charm, in this new field. (We did, however, "color" the quarks in Fig. 32–16.) The antiquarks are colored antired, antigreen, and antiblue. Baryons are made up of three quarks, one of each color. Mesons consist of a quark–antiquark pair of a particular color and its anticolor. Both baryons and mesons are thus colorless or white.

Originally, the idea of quark color was proposed to preserve the Pauli exclusion principle (Section 28–7). Not all particles obey the exclusion principle. Those that do, such as electrons, protons, and neutrons, are called **fermions**. Those that don't are called **bosons**. These two categories are distinguished also in their spin: bosons have integer spin (0, 1, 2, etc.) whereas fermions have half-integer spin, usually $\tfrac{1}{2}$ as for electrons and nucleons, but other fermions have spin $\tfrac{3}{2}, \tfrac{5}{2}$, etc. Matter is made up mainly of fermions, but the carriers of the forces (γ, W, Z, and gluons) are all bosons. Quarks are fermions (they have spin $\tfrac{1}{2}$) and therefore should obey the exclusion principle. Yet for three particular baryons (uuu, ddd, and sss), all three quarks would have the same quantum numbers, and at least two quarks have their spin in the same direction (since there are only two choices, spin up $[m_s = +\tfrac{1}{2}]$ or spin down $[m_s = -\tfrac{1}{2}]$). This would seem to violate the exclusion principle; but if quarks have that additional quantum number *color*, which is different for each quark, it would serve to distinguish them and allow the exclusion principle to hold. Although quark color, and the resulting threefold increase in the number of quarks, was originally an *ad hoc* idea, it also served to bring the theory into better agreement with experiment, such as predicting the correct lifetime of the π^0 meson, and the measured rate of hadron production in observed e^+e^- collisions at accelerators. The idea of color soon became a central feature of the theory as determining the force binding quarks together in a hadron.

Each quark is assumed to carry a *color charge*, analogous to electric charge, and the strong force between quarks is referred to as the **color force**. This theory of the strong force is called **quantum chromodynamics** (*chroma* = color in Greek), or **QCD**, to indicate that the force acts between color charges (and not between, say, electric charges). The strong force between two hadrons is considered to be a force between the quarks that make them up, as suggested in Fig. 32–17.

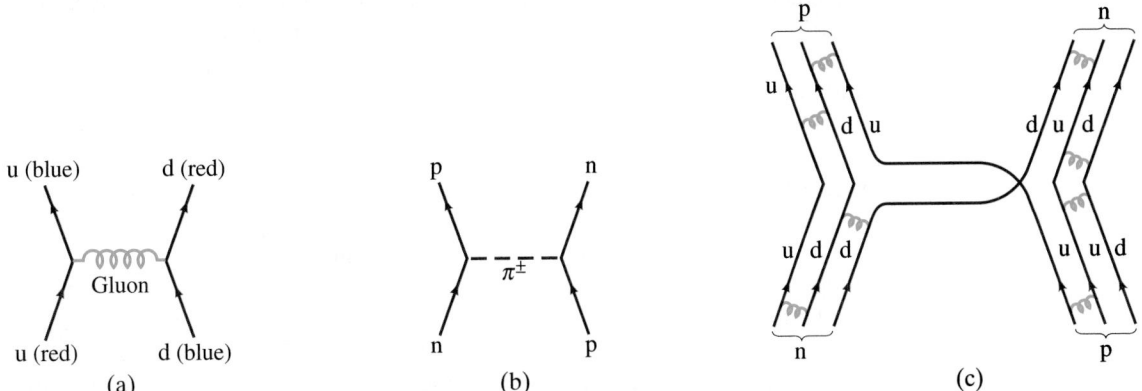

FIGURE 32–17 (a) The force between two quarks holding them together as part of a proton, for example, is carried by a gluon, which in this case involves a change in color. (b) Strong interaction $n + p \rightarrow n + p$ with the exchange of a charged π meson (+ or −, depending on whether it is considered moving to the left or to the right). (c) Quark representation of the same interaction $n + p \rightarrow n + p$. The blue coiled lines between quarks represent gluon exchanges holding the hadrons together. (The exchanged meson may be regarded as $\bar{u}d$ emitted by the n and absorbed by the p, or as $u\bar{d}$ emitted by p and absorbed by n, because a u (or d) quark going to the left in the diagram is equivalent to a $\bar{u}$ (or $\bar{d}$) going to the right.)

The particles that transmit the color force (analogous to photons for the EM force) are called **gluons** (a play on "glue"). They are included in Tables 32–2 and 32–5. There are eight gluons, according to the theory, all massless and all have color charge.[†]

You might ask what would happen if we try to see a single quark with color by reaching deep inside a hadron and extracting a single quark. Quarks are so tightly bound to other quarks that extracting one would require a tremendous amount of energy, so much that it would be sufficient to create more quarks $(E = mc^2)$. Indeed, such experiments are done at modern particle colliders and all we get is more hadrons (quark–antiquark pairs, or triplets, which we observe as mesons or baryons), never an isolated quark. This property of quarks, that they are always bound in groups that are colorless, is called **confinement**.

The color force has the interesting property that, as two quarks approach each other very closely (equivalently, have high energy), the force between them becomes small. This aspect is referred to as **asymptotic freedom**.

The weak force, as we have seen, is thought to be mediated by the W^+, W^-, and Z^0 particles. It acts between the "weak charges" that each particle has. Each fundamental particle can thus have electric charge, weak charge, color charge, and gravitational mass, although one or more of these could be zero. For example, all leptons have color charge of zero, so they do not interact via the strong force.

[†]Compare to the EM interaction, where the photon has no electric charge. Because gluons have color charge, they could attract each other and form composite particles (photons cannot). Such "glueballs" are being searched for.

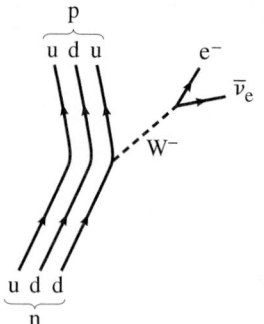

FIGURE 32–18 Quark representation of the Feynman diagram for β decay of a neutron into a proton. Example 23–9.

CONCEPTUAL EXAMPLE 32–9 **Beta decay.** Draw a Feynman diagram, showing what happens in beta decay using quarks.

RESPONSE Beta decay is a result of the weak interaction, and the mediator is either a $W^{\pm}$ or Z^0 particle. What happens, in part, is that a neutron (udd quarks) decays into a proton (uud). Apparently a d quark (charge $-\frac{1}{3}e$) has turned into a u quark (charge $+\frac{2}{3}e$). Charge conservation means that a negatively charged particle, namely a W^-, was emitted by the d quark. Since an electron and an antineutrino appear in the final state, they must have come from the decay of the virtual W^-, as shown in Fig. 32–18.

To summarize, the Standard Model says that the truly fundamental particles (Table 32–5) are the leptons, the quarks, the gauge bosons (photon, W and Z, and the gluons), and the Higgs boson. The photon, leptons, W^+, W^-, and Z^0 have all been observed in experiments. But only combinations of quarks (baryons and mesons) have been observed in the free state, and it seems likely that free quarks and gluons cannot be observed in isolation.

One important aspect of theoretical work is the attempt to find a **unified** basis for the different forces in nature. This was a long-held hope of Einstein, which he was never able to fulfill. A so-called **gauge theory** that unifies the weak and electromagnetic interactions was put forward in the 1960s by S. Weinberg, S. Glashow, and A. Salam. In this **electroweak theory**, the weak and electromagnetic forces are seen as two different manifestations of a single, more fundamental, *electroweak* interaction. The electroweak theory has had many successes, including the prediction of the $W^{\pm}$ particles as carriers of the weak force, with masses of $80.38 \pm 0.02 \text{ GeV}/c^2$ in excellent agreement with the measured values of $80.385 \pm 0.015 \text{ GeV}/c^2$ (and similar accuracy for the Z^0).

The combination of electroweak theory plus QCD for the strong interaction is referred to today as the **Standard Model (SM)** of fundamental particles.

EXAMPLE 32–10 ESTIMATE Range of weak force. The weak nuclear force is of very short range, meaning it acts over only a very short distance. Estimate its range using the masses (Table 32–2) of the $W^{\pm}$ and Z: $m \approx 80$ or $90 \text{ GeV}/c^2 \approx 10^2 \text{ GeV}/c^2$.

APPROACH We assume the $W^{\pm}$ or Z^0 exchange particles can exist for a time Δt given by the uncertainty principle (Section 28–3), $\Delta t \approx \hbar/\Delta E$, where $\Delta E \approx mc^2$ is the energy needed to create the virtual particle $(W^{\pm}, Z)$ that carries the weak force.

SOLUTION Let Δx be the distance the virtual W or Z can move before it must be reabsorbed within the time $\Delta t \approx \hbar/\Delta E$. To find an upper limit on Δx, and hence the maximum range of the weak force, we let the W or Z travel close to the speed of light, so $\Delta x \lesssim c \Delta t$. Recalling that $1 \text{ GeV} = 1.6 \times 10^{-10} \text{ J}$, then

$$\Delta x \lesssim c \Delta t \approx \frac{c\hbar}{\Delta E} \approx \frac{(3 \times 10^8 \text{ m/s})(10^{-34} \text{ J·s})}{(10^2 \text{ GeV})(1.6 \times 10^{-10} \text{ J/GeV})} \approx 10^{-18} \text{ m}.$$

This is indeed a very small range.

NOTE Compare this to the range of the electromagnetic force whose range is infinite ($1/r^2$ never becomes zero for any finite r), which makes sense because the mass of its virtual exchange particle, the photon, is zero (in the denominator of the above equation).

[We did a similar calculation for the strong force in Section 32–2, estimating the mass of the π meson as exchange particle. In our deeper view of the strong force, namely the color force between quarks within a nucleon, the gluons have zero mass, which implies infinite range (see formula in Example 32–10). We might have expected a range of about 10^{-15} m (nuclear size). But according to the Standard Model, the color force is weak at very close distances and increases greatly with distance (causing quark confinement). Thus its range could be infinite.]

Theoreticians have wondered why the W and Z have large masses rather than being massless like the photon. Peter Higgs and others in 1964 used electroweak theory to suggest an explanation by means of a **Higgs field** and its particle, the **Higgs boson**, which interact with the W and Z to "slow them down." In being forced to go slower than the speed of light, they would have to have mass ($m = 0$ only if $v = c$). Indeed, the Higgs field is thought to permeate the vacuum ("empty space") and to perhaps confer mass on particles that now have mass by slowing them down. In 2012 strong evidence was announced at CERN's Large Hadron Collider (Section 32–1) for a particle of mass $125\ \mathrm{GeV}/c^2$ that is thought to be the long-sought Higgs boson of the Standard Model. But intense research continues, not only to better understand this particle, but to search for additional Higgs-like particles suggested by theories that go beyond the Standard Model such as *supersymmetry* (Section 32–12).

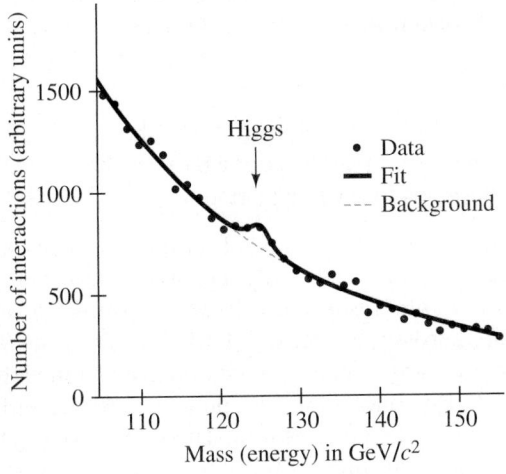

FIGURE 32–19 Evidence for the Higgs boson.

FIGURE 32–20 Fabiola Gianotti, leader of the ATLAS team (3000 physicists), at the LHC with theorist Peter Higgs, July 4, 2012, when the long hoped-for boson was announced.

Figure 32–19 shows the "resonance" bump (Section 32–7) that represents the Higgs boson as detected by the CMS team at the LHC. A second experiment, ATLAS, came up with the same mass. ATLAS is considered the largest scientific experiment ever (see Fig. 32–20).

There is no way to know if the Chapter-Opening Photo of a possible Higgs event, page 915, is actually a Higgs or is a background event. As can be seen in Fig. 32–19, there are many more background events around 125 GeV than there are in the resonance bump representing the Higgs boson.

32–11 Grand Unified Theories

The Standard Model, for all its success, cannot explain some important issues—such as why the charge on the electron has *exactly* the same magnitude as the charge on the proton. This is crucial, because if the charge magnitudes were even a little different, atoms would not be neutral and the resulting large electric forces would surely have made life impossible. Indeed, the Standard Model is now considered to be a low-energy approximation to a more complete theory.

With the success of unified electroweak theory, theorists are trying to incorporate it and QCD for the strong (color) force into a so-called **grand unified theory (GUT)**.

One type of such a grand unified theory of the electromagnetic, weak, and strong forces has been proposed in which there is only one class of particle—leptons and quarks belong to the same family and are able to change freely from one type to the other—and the three forces are different aspects of a single underlying force. The unity is predicted to occur, however, only on a scale of less than about 10^{-31} m, corresponding to a typical particle energy of about 10^{16} GeV.

If two elementary particles (leptons or quarks) approach each other to within this **unification scale**, the apparently fundamental distinction between them would not exist at this level, and a quark could readily change to a lepton, or vice versa. Baryon and lepton numbers would not be conserved. The weak, electromagnetic, and strong (color) force would blend to a force of a single strength.

What happens between the unification distance of 10^{-31} m and more normal (larger) distances is referred to as **symmetry breaking**. As an analogy, consider an atom in a crystal. Deep within the atom, there is much symmetry—in the innermost regions the electron cloud is spherically symmetric (Chapter 28). Farther out, this symmetry breaks down—the electron clouds are distributed preferentially along the lines (bonds) joining the atoms in the crystal. In a similar way, at 10^{-31} m the force between elementary particles is theorized to be a single force—it is symmetrical and does not single out one type of "charge" over another. But at larger distances, that symmetry is broken and we see three distinct forces. (In the "Standard Model" of electroweak interactions, Section 32–10, the symmetry breaking between the electromagnetic and the weak interactions occurs at about 10^{-18} m.)

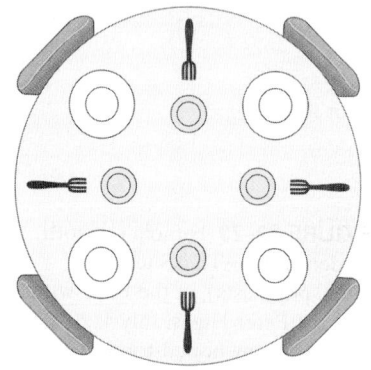

FIGURE 32–21 Symmetry around a table. Example 32–11.

CONCEPTUAL EXAMPLE 32–11 **Symmetry.** The table in Fig. 32–21 has four identical place settings. Four people sit down to eat. Describe the symmetry of this table and what happens to it when someone starts the meal.

RESPONSE The table has several kinds of symmetry. It is symmetric to rotations of 90°: that is, the table will look the same if everyone moved one chair to the left or to the right. It is also north–south symmetric and east–west symmetric, so that swaps across the table don't affect the way the table looks. It also doesn't matter whether any person picks up the fork to the left of the plate or the fork to the right. But once that first person picks up either fork, the choice is set for all the rest at the table as well. The symmetry has been *broken*. The underlying symmetry is still there—the blue glasses could still be chosen either way—but some choice must get made and at that moment the symmetry of the diners is broken.

Another example of symmetry breaking is a pencil standing on its point before falling. Standing, it looks the same from any horizontal direction. From above, it is a tiny circle. But when it falls to the table, it points in one particular direction—the symmetry is broken.

Proton Decay

Since unification is thought to occur at such tiny distances and huge energies, the theory is difficult to test experimentally. But it is not completely impossible. One testable prediction is the idea that the proton might decay (via, for example, $p \rightarrow \pi^0 + e^+$) and violate conservation of baryon number. This could happen if two quarks within a proton approached to within 10^{-31} m of each other. But it is very unlikely at normal temperature and energy, so the decay of a proton can only be an unlikely process. In the simplest form of GUT, the theoretical estimate of the proton mean life for the decay mode $p \rightarrow \pi^0 + e^+$ is about 10^{31} yr, and this is now within the realm of testability.[†] Proton decays have still not been seen, and experiments put the lower limit on the proton mean life for the above mode to be about 10^{33} yr, somewhat greater than this prediction. This may seem a disappointment, but on the other hand, it presents a challenge. Indeed more complex GUTs may resolve this conflict.

[†]This is much larger than the age of the universe ($\approx 14 \times 10^9$ yr). But we don't have to wait 10^{31} yr to see. Instead we can wait for one decay among 10^{31} protons over a year (see Eqs. 30–3a and 30–7, $\Delta N = \lambda N \, \Delta t = N \, \Delta t / \tau$, and Example 32–12).

EXAMPLE 32–12 ESTIMATE **Proton decay.** An experiment uses 3300 tons of water waiting to see a proton decay of the type $p \rightarrow \pi^0 + e^+$. If the experiment is run for 4 years without detecting a decay, estimate the lower limit on the proton mean life.

APPROACH As with radioactive decay, the number of decays is proportional to the number of parent species (N), the time interval (Δt), and the decay constant (λ) which is related to the mean life τ by (see Eqs. 30–3 and 30–7):

$$\Delta N = -\lambda N \Delta t = -\frac{N \Delta t}{\tau}.$$

SOLUTION Dealing only with magnitudes, we solve for τ:

$$\tau = \frac{N \Delta t}{\Delta N}.$$

Thus for $\Delta N < 1$ (we don't see even one decay) over the four-year trial,

$$\tau > N (4\,\text{yr}),$$

where N is the number of protons in 3300 tons of water. To determine N, we note that each molecule of H_2O contains $2 + 8 = 10$ protons. So one mole of water (18 g, 6×10^{23} molecules) contains $10 \times 6 \times 10^{23} = 6 \times 10^{24}$ protons in 18 g of water (= 18 g/1000 g = 1/56 of a kg), or about 3×10^{26} protons per kilogram. One ton is 10^3 kg, so the 3300 tons contains $(3.3 \times 10^6\,\text{kg})(3 \times 10^{26}\,\text{protons/kg}) \approx 1 \times 10^{33}$ protons. Then our very rough estimate for a lower limit on the proton mean life is $\tau > (10^{33})(4\,\text{yr}) \approx 4 \times 10^{33}\,\text{yr}$.

*GUT and Cosmology

An interesting prediction of unified theories relates to cosmology (Chapter 33). It was thought by many theorists that during the first 10^{-35} s after the theorized Big Bang that created the universe, the temperature was so extremely high that particles had energies corresponding to the unification scale. Baryon number would not have been conserved then, perhaps allowing an imbalance that might account for the observed predominance of matter ($B > 0$) over antimatter ($B < 0$) in the universe. The fact that we are surrounded by matter, with no significant antimatter in sight, is considered a problem in search of an explanation (not given by the Standard Model). We call this the **matter–antimatter problem**. To understand it may require still undiscovered phenomena—perhaps related to quarks or neutrinos, or the Higgs boson or supersymmetry (next Section).

Many theorists no longer think the Big Bang was sufficiently hot to create unification. Nonetheless we see that there is a deep connection between investigations at either end of the size scale: theories about the tiniest objects (elementary particles) have a strong bearing on the understanding of the universe on a large scale. We look at this more in the next Chapter.

Figure 32–22 is a rough diagram indicating how the four fundamental forces in nature might have "condensed out" (a symmetry was broken) as time went on after the Big Bang (Chapter 33), and as the mean temperature of the universe and the typical particle energy decreased.

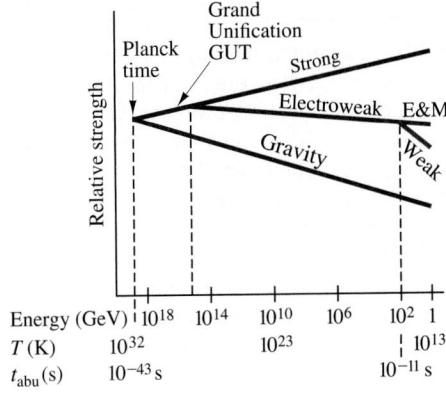

FIGURE 32–22 Time and energy plot of the four fundamental forces, perhaps unified at the "Planck time" (10^{-43} s after the birth of the universe), and how each condensed out, assuming a very hot Big Bang. The symbol t_{abu} means time after the birth of the universe. Note that the typical particle energy (and average temperature of the universe) decreases to the right, as time increases.

32–12 Strings and Supersymmetry

We have seen that the Standard Model is unable to address important experimental issues, and that theoreticians are attacking the problem as experimenters search for new data, new particles, new concepts.

Even more ambitious than grand unified theories are attempts to also incorporate gravity, and thus unify all four forces in nature into a single theory. (Such theories are sometimes referred to misleadingly as **theories of everything**.) A major attempt to unify all four forces is called **string theory**, introduced by Gabriele Veneziano in 1968: Each fundamental particle (Table 32–5) is imagined not as a point but as a one-dimensional **string**, perhaps 10^{-35} m long, which vibrates in a particular standing wave pattern. (You might say each particle is a different note on a tiny stretched string.) More sophisticated theories propose the fundamental entities as being multidimensional **branes** (after 2-D membranes).

A related idea that also goes way beyond the Standard Model is **supersymmetry**, which applied to strings is known as **superstring theory**. Supersymmetry, developed by Bruno Zumino (1923–) and Julius Wess (1934–2007), predicts that interactions exist that would change fermions into bosons and vice versa, and that each known fermion would have a supersymmetric boson partner. Thus, for each quark (a fermion), there would be a **squark** (a boson) or "supersymmetric" quark. For every lepton there would be a **slepton**. Likewise, for every known boson (photons and gluons, for example), there would be a supersymmetric fermion (**photinos** and **gluinos**). Supersymmetry predicts also that a *graviton*, which transmits the gravity force, has a partner, the **gravitino**. Supersymmetry (often abbreviated SUSY) offers solutions to a number of important theoretical problems. Supersymmetric particles are a candidate for the "dark matter" of the universe (discussed in Chapter 33). But why hasn't this "missing part" of the universe ever been detected? The best guess is that supersymmetric particles might be heavier than their conventional counterparts, perhaps too heavy to have been produced in today's accelerators. A search for supersymmetric particles is already being done at CERN's new Large Hadron Collider.

Versions of these theories predict other interesting properties, such as that space has 11 dimensions, but 7 of them are "coiled up" so we normally only notice the 4-D of space–time. We would like to know if and how many extra dimensions there are, and how and why they are hidden. We hope to have some answers from the new LHC (Section 32–1).

Some theorists think SUSY and other theories are approximations to a more fundamental, still undiscovered, **M-theory**. Edward Witten coined the term when proposing an 11 dimensional approximation, but never said what "M" stands for.

The world of elementary particles is opening new vistas. What happens in the future is bound to be exciting.

Summary

Particle accelerators are used to accelerate charged particles, such as electrons and protons, so they can have very high energy collisions with other particles. High-energy particles have short wavelength and so can be used to probe the structure of matter in great detail (very small distances). High kinetic energy also allows the creation of new particles through collisions (via $E = mc^2$).

Cyclotrons and **synchrotrons** use a magnetic field to keep the particles in a circular path and accelerate them at intervals by high voltage. **Linear accelerators** accelerate particles along a line. **Colliding beams** allow higher interaction energy.

An **antiparticle** has the same mass as a particle but opposite charge. Certain other properties may also be opposite: for example, the antiproton has **baryon number** (nucleon number) opposite ($B = -1$) to that for the proton ($B = +1$).

In all nuclear and particle reactions, the following conservation laws hold: momentum, angular momentum, mass–energy, electric charge, baryon number, and **lepton numbers**.

Certain particles have a property called **strangeness**, which is conserved by the strong force but not by the weak force. The properties **charm**, **bottom**, and **top** also are conserved by the strong force but not by the weak force.

Just as the electromagnetic force can be said to be due to an exchange of photons, the strong nuclear force is carried by massless **gluons**. The W and Z particles carry the weak force. These fundamental force carriers (photon, W and Z, gluons) are called **gauge bosons**.

Other particles can be classified as either *leptons* or *hadrons*. **Leptons** participate only in gravity, the weak, and the electromagnetic interactions. **Hadrons**, which today are considered **composite** particles, are made up of **quarks**, and participate in all four interactions, including the strong interaction. The hadrons can be classified as **mesons**, with baryon number zero, and **baryons**, with nonzero baryon number.

Most particles, except for the photon, electron, neutrinos, and proton, decay with measurable mean lives varying from 10^{-25} s to 10^3 s. The mean life depends on which force is predominant. Weak decays usually have mean lives greater than about 10^{-13} s. Electromagnetic decays typically have mean lives on the order of 10^{-16} to 10^{-19} s. The shortest lived particles, called **resonances**, decay via the strong interaction and live typically for only about 10^{-23} s.

Today's **Standard Model** of elementary particles considers **quarks** as the basic building blocks of the hadrons. The six quark "flavors" are called **up**, **down**, **strange**, **charm**, **bottom**, and **top**. It is expected that there are the same number of quarks as leptons (six of each), and that quarks and leptons are truly fundamental particles along with the gauge bosons (γ, W, Z, gluons) and the Higgs boson.

Quarks are said to have **color**, and, according to **quantum chromodynamics** (QCD), the strong color force acts between their color charges and is transmitted by **gluons**. **Electroweak theory** views the weak and electromagnetic forces as two aspects of a single underlying interaction. QCD plus the electroweak theory are referred to as the *Standard Model* of the fundamental particles.

Grand unified theories of forces suggest that at very short distance (10^{-31} m) and very high energy, the weak, electromagnetic, and strong forces would appear as a single force, and the fundamental difference between quarks and leptons would disappear.

According to **string theory**, the fundamental particles may be tiny strings, 10^{-35} m long, distinguished by their standing wave pattern. **Supersymmetry** predicts that each fermion (or boson) has a corresponding boson (or fermion) partner.

Questions

1. Give a reaction between two nucleons, similar to Eq. 32–4, that could produce a π^-.

2. If a proton is moving at very high speed, so that its kinetic energy is much greater than its rest energy (mc^2), can it then decay via $p \rightarrow n + \pi^+$?

3. What would an "antiatom," made up of the antiparticles to the constituents of normal atoms, consist of? What might happen if *antimatter*, made of such antiatoms, came in contact with our normal world of matter?

4. What particle in a decay signals the electromagnetic interaction?

5. (*a*) Does the presence of a neutrino among the decay products of a particle necessarily mean that the decay occurs via the weak interaction? (*b*) Do all decays via the weak interaction produce a neutrino? Explain.

6. Why is it that a neutron decays via the weak interaction even though the neutron and one of its decay products (proton) are strongly interacting?

7. Which of the four interactions (strong, electromagnetic, weak, gravitational) does an electron take part in? A neutrino? A proton?

8. Verify that charge and baryon number are conserved in each of the decays shown in Table 32–2.

9. Which of the particle decays listed in Table 32–2 occur via the electromagnetic interaction?

10. Which of the particle decays listed in Table 32–2 occur by the weak interaction?

11. The Δ baryon has spin $\frac{3}{2}$, baryon number 1, and charge $Q = +2, +1, 0,$ or -1. Why is there no charge state $Q = -2$?

12. Which of the particle decays in Table 32–4 occur via the electromagnetic interaction?

13. Which of the particle decays in Table 32–4 occur by the weak interaction?

14. Quarks have spin $\frac{1}{2}$. How do you account for the fact that baryons have spin $\frac{1}{2}$ or $\frac{3}{2}$, and mesons have spin 0 or 1?

15. Suppose there were a kind of "neutrinolet" that was massless, had no color charge or electrical charge, and did not feel the weak force. Could you say that this particle even exists?

16. Is it possible for a particle to be both (*a*) a lepton and a baryon? (*b*) a baryon and a hadron? (*c*) a meson and a quark? (*d*) a hadron and a lepton? Explain.

17. Using the ideas of quantum chromodynamics, would it be possible to find particles made up of two quarks and no antiquarks? What about two quarks and two antiquarks?

18. Why can neutrons decay when they are free, but not when they are inside a stable nucleus?

19. Is the reaction $e^- + p \rightarrow n + \bar{\nu}_e$ possible? Explain.

20. Occasionally, the Λ will decay by the following reaction: $\Lambda^0 \rightarrow p^+ + e^- + \bar{\nu}_e$. Which of the four forces in nature is responsible for this decay? How do you know?

MisConceptual Questions

1. There are six kinds (= flavors) of quarks: up, down, strange, charm, bottom, and top. Which flavors make up most of the known matter in the universe?
   (*a*) Up and down quarks.
   (*b*) Strange and charm quarks.
   (*c*) Bottom and top quarks.
   (*d*) All of the above.

2. Which of the following particles can not be composed of quarks?
   (*a*) Proton.
   (*b*) Electron.
   (*c*) π meson.
   (*d*) Neutron.
   (*e*) Higgs boson.

3. If gravity is the weakest force, why is it the one we notice most?
   (a) Our bodies are not sensitive to the other forces.
   (b) The other forces act only within atoms and therefore have no effect on us.
   (c) Gravity may be "very weak" but always attractive, and the Earth has enormous mass. The strong and weak nuclear forces have very short range. The electromagnetic force has a long range, but most matter is electrically neutral.
   (d) At long distances, the gravitational force is actually stronger than the other forces.
   (e) The other forces act only on elementary particles, not on objects our size.

4. Is it possible for a tau lepton (whose mass is almost twice that of a proton) to decay into only hadrons?
   (a) Yes, because it is so massive it could decay into a proton and pions.
   (b) Yes, it could decay into pions and nothing else.
   (c) No, such a decay would violate lepton number; all of its decay products must be leptons.
   (d) No, its decay products must include a tau neutrino but could include hadrons such as pions.
   (e) No, the tau lepton is too massive to decay.

5. Many particle accelerators are circular because:
   (a) particles accelerate faster around circles.
   (b) in order to move in a circle, acceleration is required.
   (c) a circular accelerator has a shorter length than a square one.
   (d) the particles can be accelerated through the same potential difference many times, making the accelerator more compact.
   (e) a particle moving in a circle needs more energy than a particle moving in a straight line.

6. Which of the following are today considered fundamental particles (that is, not composed of smaller components)? Choose as many as apply.
   (a) Atoms. (b) Electrons. (c) Protons. (d) Neutrons. (e) Quarks. (f) Photon. (g) Higgs boson.

7. The electron's antiparticle is called the positron. Which of the following properties, if any, are the same for electrons and positrons?
   (a) Mass.
   (b) Charge.
   (c) Lepton number.
   (d) None of the above.

8. The strong nuclear force between a neutron and a proton is due to
   (a) the exchange of π mesons between the neutron and the proton.
   (b) the conservation of baryon number.
   (c) the beta decay of the neutron into the proton.
   (d) the exchange of gluons between the quarks within the neutron and the proton.
   (e) Both (a) and (d) at different scales.

9. Electrons are still considered fundamental particles (in the group called leptons). But protons and neutrons are no longer considered fundamental; they have substructure and are made up of
   (a) pions. (b) leptons. (c) quarks. (d) bosons. (e) photons.

10. Which of the following will interact via the weak nuclear force *only*?
    (a) Quarks. (b) Gluons. (c) Neutrons. (d) Neutrinos. (e) Electrons. (f) Muons. (g) Higgs boson.

For assigned homework and other learning materials, go to the MasteringPhysics website.

Problems

32–1 Particles and Accelerators

1. (I) What is the total energy of a proton whose kinetic energy is 4.65 GeV?

2. (I) Calculate the wavelength of 28-GeV electrons.

3. (I) If α particles are accelerated by the cyclotron of Example 32–2, what must be the frequency of the voltage applied to the dees?

4. (I) What is the time for one complete revolution for a very high-energy proton in the 1.0-km-radius Fermilab accelerator?

5. (II) What strength of magnetic field is used in a cyclotron in which protons make 3.1×10^7 revolutions per second?

6. (II) (a) If the cyclotron of Example 32–2 accelerated α particles, what maximum energy could they attain? What would their speed be? (b) Repeat for deuterons ($^2_1$H). (c) In each case, what frequency of voltage is required?

7. (II) Which is better for resolving details of the nucleus: 25-MeV alpha particles or 25-MeV protons? Compare each of their wavelengths with the size of a nucleon in a nucleus.

8. (II) What is the wavelength (= minimum resolvable size) of 7.0-TeV protons at the LHC?

9. (II) The 1.0-km radius Fermilab Tevatron took about 20 seconds to bring the energies of the stored protons from 150 GeV to 1.0 TeV. The acceleration was done once per turn. Estimate the energy given to the protons on each turn. (You can assume that the speed of the protons is essentially c the whole time.)

10. (II) A cyclotron with a radius of 1.0 m is to accelerate deuterons ($^2_1$H) to an energy of 12 MeV. (a) What is the required magnetic field? (b) What frequency is needed for the voltage between the dees? (c) If the potential difference between the dees averages 22 kV, how many revolutions will the particles make before exiting? (d) How much time does it take for one deuteron to go from start to exit? (e) Estimate how far it travels during this time.

11. (III) Show that the energy of a particle (charge e) in a synchrotron, in the relativistic limit ($v \approx c$), is given by E (in eV) $= Brc$, where B is the magnetic field and r is the radius of the orbit (SI units).

32–2 to 32–6 Particle Interactions, Particle Exchange

12. (I) About how much energy is released when a Λ^0 decays to n + π^0? (See Table 32–2.)

13. (I) How much energy is released in the decay
    $$\pi^+ \rightarrow \mu^+ + \nu_\mu?$$
    See Table 32–2.

14. (I) Estimate the range of the strong force if the mediating particle were the kaon instead of a pion.

15. (I) How much energy is required to produce a neutron–antineutron pair?

16. (II) Determine the total energy released when Σ^0 decays to Λ^0 and then to a proton.

17. (II) Two protons are heading toward each other with equal speeds. What minimum kinetic energy must each have if a π^0 meson is to be created in the process? (See Table 32–2.)

18. (II) What minimum kinetic energy must a proton and an antiproton each have if they are traveling at the same speed toward each other, collide, and produce a K^+K^- pair in addition to themselves? (See Table 32–2.)

19. (II) What are the wavelengths of the two photons produced when a proton and antiproton at rest annihilate?

20. (II) The Λ^0 cannot decay by the following reactions. What conservation laws are violated in each of the reactions?
(a) $\Lambda^0 \nrightarrow n + \pi^-$
(b) $\Lambda^0 \nrightarrow p + K^-$
(c) $\Lambda^0 \nrightarrow \pi^+ + \pi^-$

21. (II) What would be the wavelengths of the two photons produced when an electron and a positron, each with 420 keV of kinetic energy, annihilate in a head-on collision?

22. (II) Which of the following reactions and decays are possible? For those forbidden, explain what laws are violated.
(a) $\pi^- + p \rightarrow n + \eta^0$
(b) $\pi^+ + p \rightarrow n + \pi^0$
(c) $\pi^+ + p \rightarrow p + e^+$
(d) $p \rightarrow e^+ + \nu_e$
(e) $\mu^+ \rightarrow e^+ + \bar{\nu}_\mu$
(f) $p \rightarrow n + e^+ + \nu_e$

23. (II) Antiprotons can be produced when a proton with sufficient energy hits a stationary proton. Even if there is enough energy, which of the following reactions will not happen?
$$p + p \rightarrow p + \bar{p}$$
$$p + p \rightarrow p + p + \bar{p}$$
$$p + p \rightarrow p + p + p + \bar{p}$$
$$p + p \rightarrow p + e^+ + e^+ + \bar{p}$$

24. (III) In the rare decay $\pi^+ \rightarrow e^+ + \nu_e$, what is the kinetic energy of the positron? Assume the π^+ decays from rest and $m_\nu = 0$.

25. (III) For the decay $\Lambda^0 \rightarrow p + \pi^-$, calculate (a) the Q-value (energy released), and (b) the kinetic energy of the p and π^-, assuming the Λ^0 decays from rest. (Use relativistic formulas.)

26. (III) Calculate the maximum kinetic energy of the electron when a muon decays from rest via $\mu^- \rightarrow e^- + \bar{\nu}_e + \nu_\mu$. [*Hint*: In what direction do the two neutrinos move relative to the electron in order to give the electron the maximum kinetic energy? Both energy and momentum are conserved; use relativistic formulas.]

32–7 to 32–11 Resonances, Standard Model, Quarks, QCD, GUT

27. (I) The mean life of the Σ^0 particle is 7×10^{-20} s. What is the uncertainty in its rest energy? Express your answer in MeV.

28. (I) The measured width of the ψ (3686) meson is about 300 keV. Estimate its mean life.

29. (I) The measured width of the J/ψ meson is 88 keV. Estimate its mean life.

30. (I) The B^- meson is a $b\bar{u}$ quark combination. (a) Show that this is consistent for all quantum numbers. (b) What are the quark combinations for B^+, B^0, $\overline{B}^0$?

31. (I) What is the energy width (or uncertainty) of (a) η^0, and (b) ρ^+? See Table 32–2.

32. (II) Which of the following decays are possible? For those that are forbidden, explain which laws are violated.
(a) $\Xi^0 \rightarrow \Sigma^+ + \pi^-$
(b) $\Omega^- \rightarrow \Sigma^0 + \pi^- + \nu$
(c) $\Sigma^0 \rightarrow \Lambda^0 + \gamma + \gamma$

33. (II) In ordinary radioactive decay, a W particle may be created even though the decaying particle has less mass than the W particle. If you assume $\Delta E \approx$ mass of the virtual W, what is the expected lifetime of the W?

34. (II) What quark combinations produce (a) a Ξ^0 baryon and (b) a Ξ^- baryon?

35. (II) What are the quark combinations that can form (a) a neutron, (b) an antineutron, (c) a Λ^0, (d) a $\overline{\Sigma}^0$?

36. (II) What particles do the following quark combinations produce: (a) uud, (b) $\bar{u}\bar{u}\bar{s}$, (c) $\bar{u}s$, (d) $d\bar{u}$, (e) $\bar{c}s$?

37. (II) What is the quark combination needed to produce a D^0 meson ($Q = B = S = 0$, $c = +1$)?

38. (II) The D_S^+ meson has $S = c = +1$, $B = 0$. What quark combination would produce it?

39. (II) Draw a possible Feynman diagram using quarks (as in Fig. 32–17c) for the reaction $\pi^- + p \rightarrow \pi^0 + n$.

40. (II) Draw a Feynman diagram for the reaction $n + \nu_\mu \rightarrow p + \mu^-$.

General Problems

41. What is the total energy of a proton whose kinetic energy is 15 GeV? What is its wavelength?

42. The mean lifetimes listed in Table 32–2 are in terms of *proper time*, measured in a reference frame where the particle is at rest. If a tau lepton is created with a kinetic energy of 950 MeV, how long would its track be as measured in the lab, on average, ignoring any collisions?

43. (a) How much energy is released when an electron and a positron annihilate each other? (b) How much energy is released when a proton and an antiproton annihilate each other? (All particles have KE ≈ 0.)

44. If 2×10^{14} protons moving at $v \approx c$, with KE = 4.0 TeV, are stored in the 4.3-km-radius ring of the LHC, (a) how much current (amperes) is carried by this beam? (b) How fast would a 1500-kg car have to move to carry the same kinetic energy as this beam?

45. Protons are injected into the 4.3-km-radius Large Hadron Collider with an energy of 450 GeV. If they are accelerated by 8.0 MV each revolution, how far do they travel and approximately how much time does it take for them to reach 4.0 TeV?

46. Which of the following reactions are possible, and by what interaction could they occur? For those forbidden, explain why.
    (a) $\pi^- + p \rightarrow K^0 + p + \pi^0$
    (b) $K^- + p \rightarrow \Lambda^0 + \pi^0$
    (c) $K^+ + n \rightarrow \Sigma^+ + \pi^0 + \gamma$
    (d) $K^+ \rightarrow \pi^0 + \pi^0 + \pi^+$
    (e) $\pi^+ \rightarrow e^+ + \nu_e$

47. Which of the following reactions are possible, and by what interaction could they occur? For those forbidden, explain why.
    (a) $\pi^- + p \rightarrow K^+ + \Sigma^-$
    (b) $\pi^+ + p \rightarrow K^+ + \Sigma^+$
    (c) $\pi^- + p \rightarrow \Lambda^0 + K^0 + \pi^0$
    (d) $\pi^+ + p \rightarrow \Sigma^0 + \pi^0$
    (e) $\pi^- + p \rightarrow p + e^- + \bar{\nu}_e$

48. One decay mode for a π^+ is $\pi^+ \rightarrow \mu^+ + \nu_\mu$. What would be the equivalent decay for a π^-? Check conservation laws.

49. Symmetry breaking occurs in the electroweak theory at about 10^{-18} m. Show that this corresponds to an energy that is on the order of the mass of the $W^\pm$.

50. Calculate the Q-value for each of the reactions, Eq. 32–4, for producing a pion.

51. How many fundamental fermions are there in a water molecule?

52. The mass of a π^0 can be measured by observing the reaction $\pi^- + p \rightarrow \pi^0 + n$ with initial kinetic energies near zero. The neutron is observed to be emitted with a kinetic energy of 0.60 MeV. Use conservation of energy and momentum to determine the π^0 mass.

53. (a) Show that the so-called unification distance of 10^{-31} m in grand unified theory is equivalent to an energy of about 10^{16} GeV. Use the uncertainty principle, and also de Broglie's wavelength formula, and explain how they apply. (b) Calculate the temperature corresponding to 10^{16} GeV.

54. Calculate the Q-value for the reaction $\pi^- + p \rightarrow \Lambda^0 + K^0$, when negative pions strike stationary protons. Estimate the minimum pion kinetic energy needed to produce this reaction. [Hint: Assume Λ^0 and K^0 move off with the same velocity.]

55. A proton and an antiproton annihilate each other at rest and produce two pions, π^- and π^+. What is the kinetic energy of each pion?

56. For the reaction $p + p \rightarrow 3p + \bar{p}$, where one of the initial protons is at rest, use relativistic formulas to show that the threshold energy is $6m_p c^2$, equal to three times the magnitude of the Q-value of the reaction, where m_p is the proton mass. [Hint: Assume all final particles have the same velocity.]

57. At about what kinetic energy (in eV) can the rest energy of a proton be ignored when calculating its wavelength, if the wavelength is to be within 1.0% of its true value? What are the corresponding wavelength and speed of the proton?

58. Use the quark model to describe the reaction
    $$\bar{p} + n \rightarrow \pi^- + \pi^0.$$

59. Identify the missing particle in the following reactions.
    (a) $p + p \rightarrow p + n + \pi^+ + ?$ (b) $p + ? \rightarrow n + \mu^+$

60. What fraction of the speed of light c is the speed of a 7.0-TeV proton?

61. Using the information in Section 32–1, show that the Large Hadron Collider's two colliding proton beams can resolve details that are less than 1/10,000 the size of a nucleus.

62. Searches are underway for a process called **neutrinoless double beta decay**, in which a nucleus decays by emitting two electrons. (a) If the parent nucleus is $^{96}_{40}$Zr, what would the daughter nucleus be? (b) What conservation laws would be violated during this decay? (c) How could $^{96}_{40}$Zr decay to the same daughter nucleus without violating any conservation laws?

63. Estimate the lifetime of the Higgs boson from the width of the "bump" in Fig. 32–19, using the uncertainty principle. [Note: This is not a realistic estimate because the underlying processes are very complicated.]

Search and Learn

1. (a) What are the two major classes of particles that make up the matter of the universe? (b) Name six types, or flavors, of each class of particles. (c) What are the four known fundamental forces in the universe? (d) Name the particles that carry the forces in part c. Which force is much weaker than the other three?

2. (a) What property characterizes all hadrons? (b) What property characterizes all baryons? (c) What property characterizes all mesons?

3. Show that all conservation laws hold for all the decays described in Fig. 32–15 for the decays of the top quark.

4. The Higgs boson, Section 32–10, has very probably been detected at the CERN LHC. (a) If a Higgs boson at rest decays into two tau leptons, what is the kinetic energy of each tau? Follow the analysis of Example 32–5. See Table 32–2. (b) What are the signs of the electric charges of the two tau leptons? (c) Could a Higgs boson decay into two Z bosons (Table 32–2)?

5. (a) Show, by conserving momentum and energy, that it is impossible for an isolated electron to radiate only a single photon. (b) With this result in mind, how can you defend the photon exchange diagram in Fig. 32–8?

6. What magnetic field is required for the 4.25-km-radius Large Hadron Collider (LHC) to accelerate protons to 7.0 TeV? [Hint: Use relativity, Chapter 26.]

ANSWERS TO EXERCISES

A: 1.24×10^{-18} m = 1.24 am.
B: (a).

C: (c); (d).
D: $s\bar{u}$.

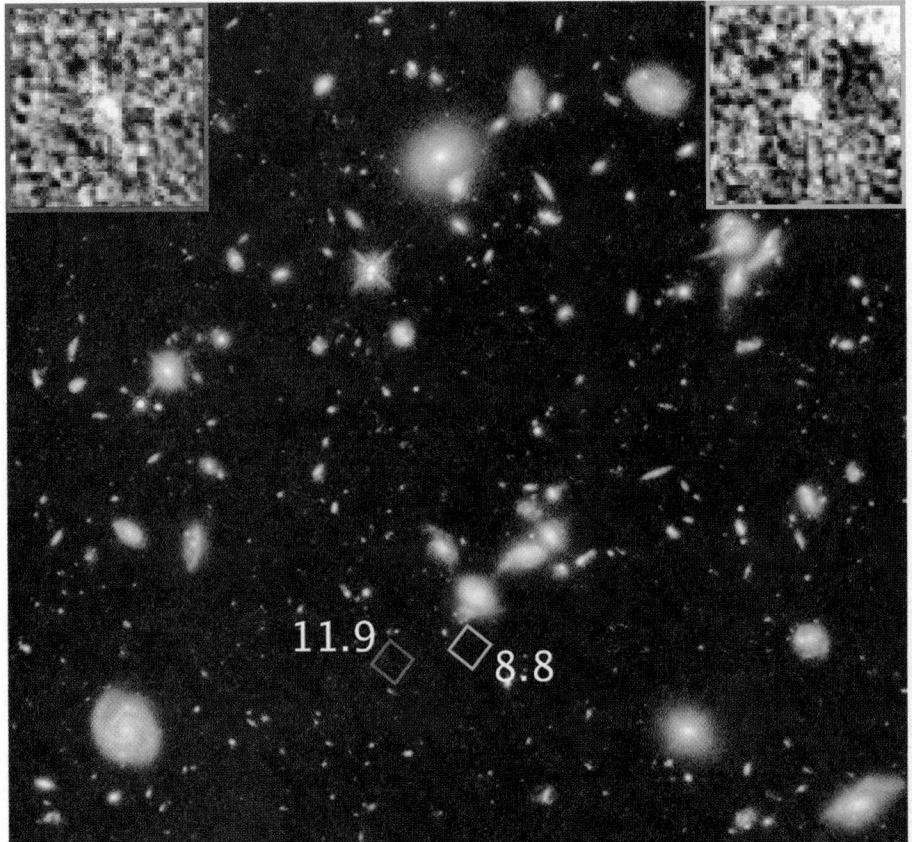

$z = 11.9$ $z = 8.8$

11.9 ◇   ◇ 8.8

This Hubble eXtreme Deep Field (XDF) photograph is of a very small part of the sky. It includes what may be the most distant galaxies observable by us (small red and green squares, and shown enlarged in the corners), with $z \approx 8.8$ and 11.9, that already existed when the universe was about 0.4 billion years old. We see these galaxies as they appeared then, 13.4 billion years ago, which is when they emitted this light. The most distant galaxies were young and small and grew to become large galaxies by colliding and merging with other small galaxies.

We examine the latest theories on how stars and galaxies form and evolve, including the role of nucleosynthesis, as well as Einstein's general theory of relativity which deals with gravity and curvature of space. We take a thorough look at the evidence for the expansion of the universe, and the Standard Model of the universe evolving from an initial Big Bang. We point out some unsolved problems, including the nature of dark matter and dark energy that make up most of our universe.

Astrophysics and Cosmology

CHAPTER-OPENING QUESTIONS—Guess now!

1. Until recently, astronomers expected the expansion rate of the universe would be decreasing. Why?
(a) Friction.
(b) The second law of thermodynamics.
(c) Gravity.
(d) The electromagnetic force.

2. The universe began expanding right at the beginning. How long will it continue to expand?
(a) Until it runs out of room.
(b) Until friction slows it down and brings it to a stop.
(c) Until all galaxies are moving at the speed of light relative to the center.
(d) Possibly forever.

I n the previous Chapter, we studied the tiniest objects in the universe—the elementary particles. Now we leap to the grandest objects in the universe— stars, galaxies, and clusters of galaxies—plus the history and structure of the universe itself. These two extreme realms, elementary particles and the cosmos, are among the most intriguing and exciting subjects in science. And, surprisingly, these two extreme realms are related in a fundamental way, as was already hinted in Chapter 32.

CONTENTS

Use of the techniques and ideas of physics to study the night sky is often referred to as **astrophysics**. Central to our present theoretical understanding of the universe (or cosmos) is Einstein's *general theory of relativity* which represents our most complete understanding of gravitation. Many other aspects of physics are involved, from electromagnetism and thermodynamics to atomic and nuclear physics as well as elementary particles. General Relativity serves also as the foundation for modern **cosmology**, which is the study of the universe as a whole. Cosmology deals especially with the search for a theoretical framework to understand the observed universe, its origin, and its future. The questions posed by cosmology are profound and difficult; the possible answers stretch the imagination. They are questions like "Has the universe always existed, or did it have a beginning in time?" Either alternative is difficult to imagine: time going back indefinitely into the past, or an actual moment when the universe began (but, then, what was there before?). And what about the size of the universe? Is it infinite in size? It is hard to imagine infinity. Or is it finite in size? This is also hard to imagine, for if the universe is finite, it does not make sense to ask what is beyond it, because the universe is all there is.

In the last 10 to 20 years, so much progress has occurred in astrophysics and cosmology that many scientists are calling recent work a "Golden Age" for cosmology. Our survey will be qualitative, but we will nonetheless touch on the major ideas. We begin with a look at what can be seen beyond the Earth.

33–1 Stars and Galaxies

According to the ancients, the stars, except for the few that seemed to move relative to the others (the planets), were fixed on a sphere beyond the last planet. The universe was neatly self-contained, and we on Earth were at or near its center. But in the centuries following Galileo's first telescopic observations of the night sky in 1609, our view of the universe has changed dramatically. We no longer place ourselves at the center, and we view the universe as vastly larger. The distances involved are so great that we specify them in terms of the time it takes light to travel the given distance: for example,

$$1 \text{ light-second} = (3.0 \times 10^8 \text{ m/s})(1.0 \text{ s}) = 3.0 \times 10^8 \text{ m} = 300{,}000 \text{ km};$$
$$1 \text{ light-minute} = (3.0 \times 10^8 \text{ m/s})(60 \text{ s}) = 18 \times 10^6 \text{ km}.$$

The most common unit is the **light-year** (ly):

$$1 \text{ ly} = (2.998 \times 10^8 \text{ m/s})(3.156 \times 10^7 \text{ s/yr})$$
$$= 9.46 \times 10^{15} \text{ m} \approx 10^{13} \text{ km} \approx 10^{16} \text{ m}.$$

For specifying distances to the Sun and Moon, we usually use meters or kilometers, but we could specify them in terms of light seconds or minutes. The Earth–Moon distance is 384,000 km, which is 1.28 light-seconds. The Earth–Sun distance is 1.50×10^{11} m, or 150,000,000 km; this is equal to 8.3 light-minutes (it takes 8.3 min for light emitted by the Sun to reach us). Far out in our solar system, Pluto is about 6×10^9 km from the Sun, or 6×10^{-4} ly.[†] The nearest star to us, other than the Sun, is Proxima Centauri, about 4.2 ly away.

On a clear moonless night, thousands of stars of varying degrees of brightness can be seen, as well as the long cloudy stripe known as the Milky Way (Fig. 33–1). Galileo first observed, with his telescope, that the Milky Way is comprised of countless individual stars. A century and a half later (about 1750), Thomas Wright suggested that the Milky Way was a flat disk of stars extending to great distances in a plane, which we call the **Galaxy** (Greek for "milky way").

[†]We can also say this is about 5 light-hours.

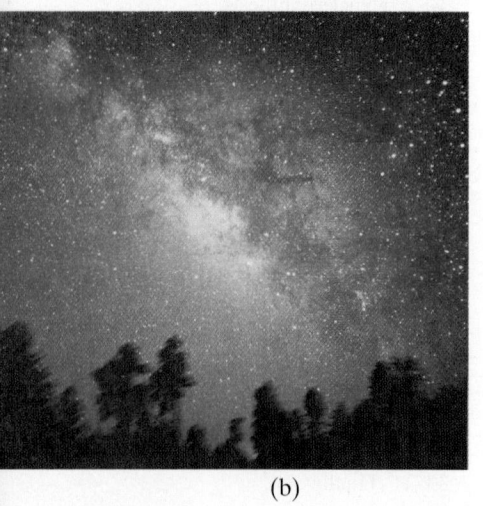

FIGURE 33–1 Sections of the Milky Way. In (a), the thin line is the trail of an artificial Earth satellite in this long time exposure. The dark diagonal area is due to dust absorption of visible light, blocking the view. In (b) the view is toward the center of the Galaxy (taken in summer from Arizona).

(a)

(b)

Our Galaxy has a diameter of almost 100,000 light-years and a thickness of roughly 2000 ly. It has a central bulge and spiral arms (Fig. 33–2). Our Sun, which is a star like many others, is located about halfway from the galactic center to the edge, some 26,000 ly from the center. Our Galaxy contains roughly 400 billion (4×10^{11}) stars. The Sun orbits the galactic center approximately once every 250 million years, so its speed is roughly 200 km/s relative to the center of the Galaxy. The total mass of all the stars in our Galaxy is estimated to be about 4×10^{41} kg of ordinary matter. There is also strong evidence that our Galaxy is permeated and surrounded by a massive invisible "halo" of "dark matter" (Section 33–9).

FIGURE 33–2 Our Galaxy, as it would appear from the outside: (a) "edge view," in the plane of the disk; (b) "top view," looking down on the disk. (If only we could see it like this—from the outside!) (c) Infrared photograph of the inner reaches of the Milky Way, showing the central bulge and disk of our Galaxy. This very wide angle photo taken from the COBE satellite (Section 33–6) extends over 360° of sky. The white dots are nearby stars.

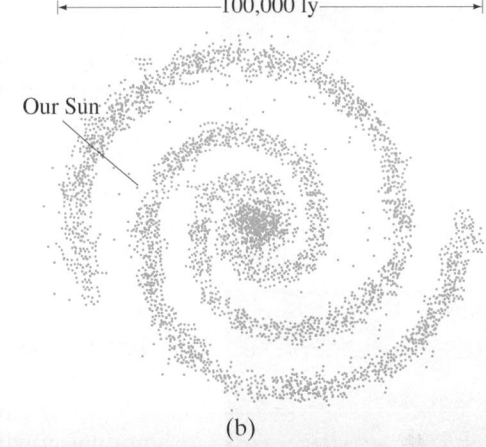

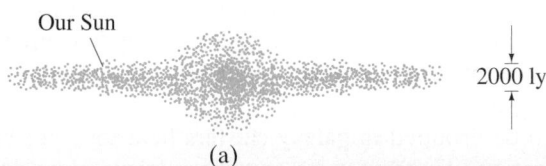

(a)

(b)

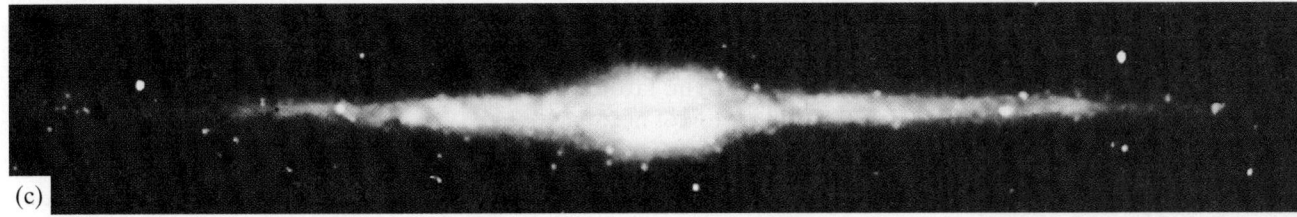

(c)

EXAMPLE 33–1 ESTIMATE **Our Galaxy's mass.** Estimate the total mass of our Galaxy using the orbital data above for the Sun about the center of the Galaxy. Assume the mass of the Galaxy is concentrated in the central bulge.

APPROACH We assume that the Sun (including our solar system) has total mass m and moves in a circular orbit about the center of the Galaxy (total mass M), and that the mass M can be considered as being located at the center of the Galaxy. We then apply Newton's second law, $F = ma$, with a being the centripetal acceleration, $a = v^2/r$, and for F we use the universal law of gravitation (Chapter 5).

SOLUTION Our Sun and solar system orbit the center of the Galaxy, according to the best measurements as mentioned above, with a speed of about $v = 200$ km/s at a distance from the Galaxy center of about $r = 26,000$ ly. We use Newton's second law:

$$F = ma$$
$$G\frac{Mm}{r^2} = m\frac{v^2}{r}$$

where M is the mass of the Galaxy and m is the mass of our Sun and solar system. Solving this, we find

$$M = \frac{rv^2}{G} \approx \frac{(26,000 \text{ ly})(10^{16} \text{ m/ly})(2 \times 10^5 \text{ m/s})^2}{6.67 \times 10^{-11} \text{ N} \cdot \text{m}^2/\text{kg}^2} \approx 2 \times 10^{41} \text{ kg.}$$

NOTE In terms of *numbers* of stars, if they are like our Sun $(m = 2.0 \times 10^{30} \text{ kg})$, there would be about $(2 \times 10^{41} \text{ kg})/(2 \times 10^{30} \text{ kg}) \approx 10^{11}$ or very roughly on the order of 100 billion stars.

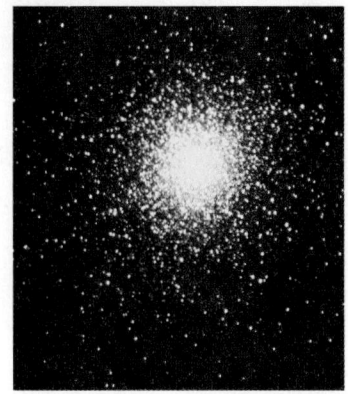

FIGURE 33–3 This globular star cluster is located in the constellation Hercules.

FIGURE 33–4 This gaseous nebula, found in the constellation Carina, is about 9000 light-years from us.

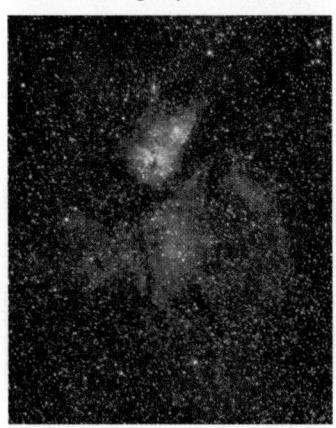

In addition to stars both within and outside the Milky Way, we can see by telescope many faint cloudy patches in the sky which were all referred to once as "nebulae" (Latin for "clouds"). A few of these, such as those in the constellations Andromeda and Orion, can actually be discerned with the naked eye on a clear night. Some are **star clusters** (Fig. 33–3), groups of stars that are so numerous they appear to be a cloud. Others are glowing clouds of gas or dust (Fig. 33–4), and it is for these that we now mainly reserve the word **nebula**.

Most fascinating are those that belong to a third category: they often have fairly regular elliptical shapes. Immanuel Kant (about 1755) guessed they are faint because they are a great distance beyond our Galaxy. At first it was not universally accepted that these objects were **extragalactic**—that is, outside our Galaxy. But the very large telescopes constructed in the twentieth century revealed that individual stars could be resolved within these extragalactic objects and that many contain spiral arms. Edwin Hubble (1889–1953) did much of this observational work in the 1920s using the 2.5-m (100-inch) telescope[†] on Mt. Wilson near Los Angeles, California, then the world's largest. Hubble demonstrated that these objects were indeed extragalactic because of their great distances. The distance to our nearest large galaxy,[‡] Andromeda, is over 2 million light-years, a distance 20 times greater than the diameter of our Galaxy. It seemed logical that these nebulae must be **galaxies** similar to ours. (Note that it is usual to capitalize the word "galaxy" only when it refers to our own.) Today it is thought there are roughly 10^{11} galaxies in the observable universe—that is, roughly as many galaxies as there are stars in a galaxy. See Fig. 33–5.

Many galaxies tend to be grouped in **galaxy clusters** held together by their mutual gravitational attraction. There may be anywhere from a few dozen to many thousands of galaxies in each cluster. Furthermore, clusters themselves seem to be organized into even larger aggregates: clusters of clusters of galaxies, or **superclusters**. The farthest detectable galaxies are more than 10^{10} ly distant. See Table 33–1 (top of next page).

[†]2.5 m (= 100 inches) refers to the diameter of the curved objective mirror. The bigger the mirror, the more light it collects (greater brightness) and the less diffraction there is (better resolution), so more and fainter stars can be seen. See Chapter 25. Until recently, photographic films or plates were used to take long time exposures. Now large solid-state CCD or CMOS sensors (Section 25–1) are available containing hundreds of millions of pixels (compared to 10 million pixels in a good-quality digital camera).

[‡]The *Magellanic clouds* are much closer than Andromeda, but are small and are usually considered small satellite galaxies of our own Galaxy.

FIGURE 33–5 Photographs of galaxies. (a) Spiral galaxy in the constellation Hydra. (b) Two galaxies: the larger and more dramatic one is known as the Whirlpool galaxy. (c) An infrared image (given "false" colors) of the same galaxies as in (b), here showing the arms of the spiral as having more substance than in the visible light photo (b); the different colors correspond to different light intensities. Visible light is scattered and absorbed by interstellar dust much more than infrared is, so infrared gives us a clearer image.

(a)   (b)   (c)

CONCEPTUAL EXAMPLE 33–2 | **Looking back in time.** Astronomers often think of their telescopes as time machines, looking back toward the origin of the universe. How far back do they look?

RESPONSE The distance in light-years measures how long in years the light has been traveling to reach us, so Table 33–1 tells us also how far back in time we are looking. For example, if we saw Proxima Centauri explode into a supernova today, then the event would have really occurred about 4.2 years ago. The most distant galaxies emitted the light we see now roughly 13×10^9 years ago. What we see was how they were then, 13×10^9 yr ago.

EXERCISE A Suppose we could place a huge mirror 1 light-year away from us. What would we see in this mirror if it is facing us on Earth? When did what we see in the mirror take place? (This might be called a "time machine.")

Besides the usual stars, clusters of stars, galaxies, and clusters and superclusters of galaxies, the universe contains many other interesting objects. Among these are stars known as *red giants, white dwarfs, neutron stars,* exploding stars called *novae* and *supernovae,* and *black holes* whose gravity is so strong that even light cannot escape them. In addition, there is electromagnetic radiation that reaches the Earth but does not come from the bright pointlike objects we call stars: particularly important is the microwave background radiation that arrives nearly uniformly from all directions in the universe.

Finally, there are **active galactic nuclei (AGN)**, which are very luminous pointlike sources of light in the centers of distant galaxies. The most dramatic examples of AGN are **quasars** ("quasistellar objects" or QSOs), which are so luminous that the surrounding starlight of the galaxy is drowned out. Their luminosity is thought to come from matter falling into a giant black hole at a galaxy's center.

33–2 Stellar Evolution: Birth and Death of Stars, Nucleosynthesis

The stars appear unchanging. Night after night the night sky reveals no significant variations. Indeed, on a human time scale, the vast majority of stars change very little (except for novae, supernovae, and certain variable stars). Although stars *seem* fixed in relation to each other, many move sufficiently for the motion to be detected. Speeds of stars relative to neighboring stars can be hundreds of km/s, but at their great distance from us, this motion is detectable only by careful measurement. There is also a great range of brightness among stars, due to differences in the rate stars emit energy and to their different distances from us.

Luminosity and Brightness of Stars

Any star or galaxy has an **intrinsic luminosity**, L (or simply **luminosity**), which is its total power radiated in watts. Also important is the **apparent brightness**, b, defined as the power crossing unit area at the Earth perpendicular to the path of the light. Given that energy is conserved, and ignoring any absorption in space, the total emitted power L when it reaches a distance d from the star will be spread over a sphere of surface area $4\pi d^2$. If d is the distance from the star to the Earth, then L must be equal to $4\pi d^2$ times b (power per unit area at Earth). That is,

$$b = \frac{L}{4\pi d^2}. \tag{33–1}$$

EXAMPLE 33–3 | **Apparent brightness.** Suppose a star has luminosity equal to that of our Sun. If it is 10 ly away from Earth, how much dimmer will it appear?

APPROACH We use the inverse square law in Eq. 33–1 to determine the relative brightness $(b \propto 1/d^2)$ since the luminosity L is the same for both stars.

SOLUTION Using the inverse square law, the star appears dimmer by a factor

$$\frac{b_{\text{star}}}{b_{\text{Sun}}} = \frac{d_{\text{Sun}}^2}{d_{\text{star}}^2} = \frac{(1.5 \times 10^8 \,\text{km})^2}{(10 \,\text{ly})^2 (10^{13} \,\text{km/ly})^2} \approx 2 \times 10^{-12}.$$

Careful study of nearby stars has shown that the luminosity for most stars depends on the mass: *the more massive the star, the greater its luminosity*[†]. Another important parameter of a star is its surface temperature, which can be determined from the spectrum of electromagnetic frequencies it emits. As we saw in Chapter 27, as the temperature of a body increases, the spectrum shifts from predominantly lower frequencies (and longer wavelengths, such as red) to higher frequencies (and shorter wavelengths such as blue). Quantitatively, the relation is given by Wien's law (Eq. 27–2): the wavelength λ_P at the peak of the spectrum of light emitted by a blackbody (we often approximate stars as blackbodies) is inversely proportional to its Kelvin temperature T; that is, $\lambda_P T = 2.90 \times 10^{-3}\,\text{m}\cdot\text{K}$. The surface temperatures of stars typically range from about 3000 K (reddish) to about 50,000 K (UV).

EXAMPLE 33–4 **Determining star temperature and star size.** Suppose that the distances from Earth to two nearby stars can be reasonably estimated, and that their measured apparent brightnesses suggest the two stars have about the same luminosity, L. The spectrum of one of the stars peaks at about 700 nm (so it is reddish). The spectrum of the other peaks at about 350 nm (bluish). Use Wien's law (Eq. 27–2) and the Stefan-Boltzmann equation (Section 14–8) to determine (*a*) the surface temperature of each star, and (*b*) how much larger one star is than the other.

APPROACH We determine the surface temperature T for each star using Wien's law and each star's peak wavelength. Then, using the Stefan-Boltzmann equation (power output or luminosity $\propto AT^4$ where A = surface area of emitter), we can find the surface area ratio and relative sizes of the two stars.

SOLUTION (*a*) Wien's law (Eq. 27–2) states that $\lambda_P T = 2.90 \times 10^{-3}\,\text{m}\cdot\text{K}$. So the temperature of the reddish star is

$$T_r = \frac{2.90 \times 10^{-3}\,\text{m}\cdot\text{K}}{\lambda_P} = \frac{2.90 \times 10^{-3}\,\text{m}\cdot\text{K}}{700 \times 10^{-9}\,\text{m}} = 4140\,\text{K}.$$

The temperature of the bluish star will be double this because its peak wavelength is half (350 nm vs. 700 nm):

$$T_b = 8280\,\text{K}.$$

(*b*) The Stefan-Boltzmann equation, Eq. 14–6, states that the power radiated *per unit area* of surface from a blackbody is proportional to the fourth power of the Kelvin temperature, T^4. The temperature of the bluish star is double that of the reddish star, so the bluish one must radiate $(T_b/T_r)^4 = 2^4 = 16$ times as much energy per unit area. But we are given that they have the same luminosity (the same total power output); so the surface area of the blue star must be $\frac{1}{16}$ that of the red one. The surface area of a sphere is $4\pi r^2$, so the radius of the reddish star is $\sqrt{16} = 4$ times larger than the radius of the bluish star (or $4^3 = 64$ times the volume).

H–R Diagram

An important astronomical discovery, made around 1900, was that for most stars, the color is related to the intrinsic luminosity and therefore to the mass. A useful way to present this relationship is by the so-called Hertzsprung–Russell (H–R) diagram. On the H–R diagram, the horizontal axis shows the surface temperature T and the vertical axis is the luminosity L; each star is represented by a point

[†]Applies to "main-sequence" stars (see next page). The mass of a star can be determined by observing its gravitational effects on other visible objects. Many stars are part of a cluster, the simplest being a binary star in which two stars orbit around each other, allowing their masses to be determined using rotational mechanics.

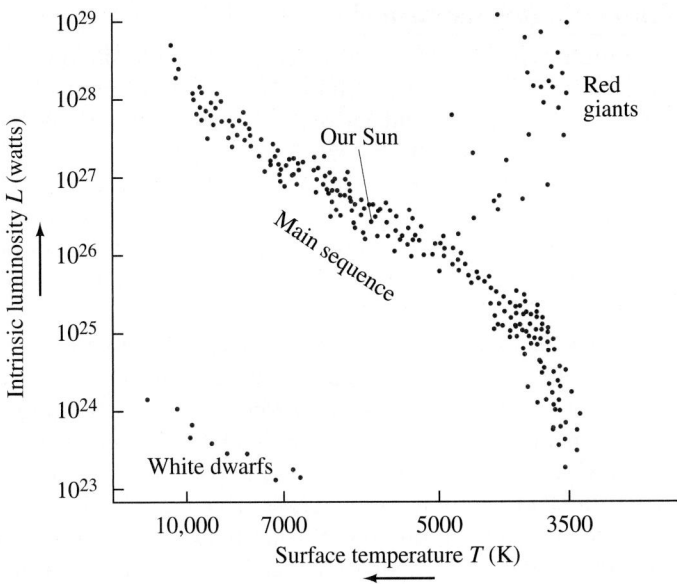

FIGURE 33–6 Hertzsprung–Russell (H–R) diagram is a logarithmic graph of luminosity vs. surface temperature T of stars (note that T increases to the left).

on the diagram, Fig. 33–6. Most stars fall along the diagonal band termed the **main sequence**. Starting at the lower right we find the coolest stars: by Wien's law, $\lambda_P T = $ constant, their light output peaks at long wavelengths, so they are reddish in color. They are also the least luminous and therefore of low mass. Farther up toward the left we find hotter and more luminous stars that are whitish, like our Sun. Still farther up we find even more luminous and more massive stars, bluish in color. Stars that fall on this diagonal band are called *main-sequence stars*. There are also stars that fall outside the main sequence. Above and to the right we find extremely large stars, with high luminosities but with low (reddish) color temperature: these are called **red giants**. At the lower left, there are a few stars of low luminosity but with high temperature: these are the **white dwarfs**.

EXAMPLE 33–5 ESTIMATE **Distance to a star using the H–R diagram and color.** Suppose that detailed study of a certain star suggests that it most likely fits on the main sequence of an H–R diagram. Its measured apparent brightness is $b = 1.0 \times 10^{-12}\,\mathrm{W/m^2}$, and the peak wavelength of its spectrum is $\lambda_P \approx 600\,\mathrm{nm}$. Estimate its distance from us.

APPROACH We find the temperature using Wien's law, Eq. 27–2. The luminosity is estimated for a main-sequence star on the H–R diagram of Fig. 33–6, and then the distance is found using the relation between brightness and luminosity, Eq. 33–1.

SOLUTION The star's temperature, from Wien's law (Eq. 27–2), is

$$T \approx \frac{2.90 \times 10^{-3}\,\mathrm{m \cdot K}}{600 \times 10^{-9}\,\mathrm{m}} \approx 4800\,\mathrm{K}.$$

A star on the main sequence of an H–R diagram at this temperature has luminosity of about $L \approx 1 \times 10^{26}\,\mathrm{W}$, read off of Fig. 33–6. Then, from Eq. 33–1,

$$d = \sqrt{\frac{L}{4\pi b}} \approx \sqrt{\frac{1 \times 10^{26}\,\mathrm{W}}{4(3.14)(1.0 \times 10^{-12}\,\mathrm{W/m^2})}} \approx 3 \times 10^{18}\,\mathrm{m}.$$

Its distance from us in light-years is

$$d = \frac{3 \times 10^{18}\,\mathrm{m}}{10^{16}\,\mathrm{m/ly}} \approx 300\,\mathrm{ly}.$$

EXERCISE B Estimate the distance to a 6000-K main-sequence star with an apparent brightness of $2.0 \times 10^{-12}\,\mathrm{W/m^2}$.

Stellar Evolution; Nucleosynthesis

Why are there different types of stars, such as red giants and white dwarfs, as well as main-sequence stars? Were they all born this way, in the beginning? Or might each different type represent a different age in the life cycle of a star? Astronomers and astrophysicists today believe the latter is the case. Note, however, that we cannot actually follow any but the tiniest part of the life cycle of any given star because they live for ages vastly greater than ours, on the order of millions or billions of years. Nonetheless, let us follow the process of **stellar evolution** from the birth to the death of a star, as astrophysicists have theoretically reconstructed it today.

Stars are born, it is believed, when gaseous clouds (mostly hydrogen) contract due to the pull of gravity. A huge gas cloud might fragment into numerous contracting masses, each mass centered in an area where the density is only slightly greater than that at nearby points. Once such "globules" form, gravity causes each to contract in toward its center of mass. As the particles of such a *protostar* accelerate inward, their kinetic energy increases. Eventually, when the kinetic energy is sufficiently high, the Coulomb repulsion between the positive charges is not strong enough to keep all the hydrogen nuclei apart, and nuclear fusion can take place.

In a star like our Sun, the fusion of hydrogen (sometimes referred to as "burning")[†] occurs via the *proton–proton chain* (Section 31–3, Eqs. 31–6), in which four protons fuse to form a $_2^4$He nucleus with the release of γ rays, positrons, and neutrinos: $4\,_1^1\text{H} \rightarrow\ _2^4\text{He} + 2\,e^+ + 2\nu_e + 2\gamma$. These reactions require a temperature of about $10^7\,\text{K}$, corresponding to an average kinetic energy ($\approx kT$) of about $1\,\text{keV}$ (Eq. 13–8). In more massive stars, the carbon cycle produces the same net effect: four $_1^1$H produce a $_2^4$He—see Section 31–3. The fusion reactions take place primarily in the core of a star, where T may be on the order of 10^7 to $10^8\,\text{K}$. (The surface temperature is much lower—on the order of a few thousand kelvins.) The tremendous release of energy in these fusion reactions produces an outward pressure sufficient to halt the inward gravitational contraction. Our protostar, now really a young *star*, stabilizes on the *main sequence*. Exactly where the star falls along the main sequence depends on its mass. The more massive the star, the farther up (and to the left) it falls on the H–R diagram of Fig. 33–6. Our Sun required perhaps 30 million years to reach the main sequence, and is expected to remain there about 10 billion years ($10^{10}\,\text{yr}$). Although most stars are billions of years old, evidence is strong that stars are actually being born at this moment. More massive stars have shorter lives, because they are hotter and the Coulomb repulsion is more easily overcome, so they use up their fuel faster. Our Sun may remain on the main sequence for 10^{10} years, but a star ten times more massive may reside there for only 10^7 years.

As hydrogen fuses to form helium, the helium that is formed is denser and tends to accumulate in the central core where it was formed. As the core of helium grows, hydrogen continues to fuse in a shell around it: see Fig. 33–7. When much of the hydrogen within the core has been consumed, the production of energy decreases at the center and is no longer sufficient to prevent the huge gravitational forces from once again causing the core to contract and heat up. The hydrogen in the shell around the core then fuses even more fiercely because of this rise in temperature, allowing the outer envelope of the star to expand and to cool. The surface temperature, thus reduced, produces a spectrum of light that peaks at longer wavelength (reddish).

This process marks a new step in the evolution of a star. The star has become redder, it has grown in size, and it has become more luminous, which means it has left the main sequence. It will have moved to the right and upward on the

FIGURE 33–7 A shell of "burning" hydrogen (fusing to become helium) surrounds the core where the newly formed helium gravitates.

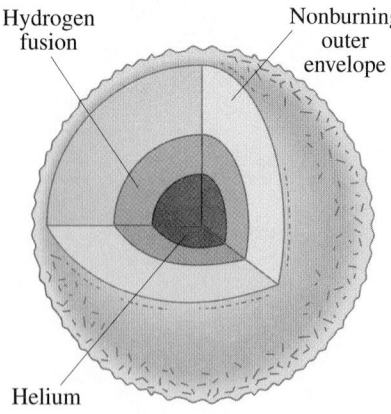

Hydrogen fusion

Nonburning outer envelope

Helium

[†]The word "burn," meaning fusion, is put in quotation marks because these high-temperature fusion reactions occur via a *nuclear* process, and must not be confused with ordinary burning (of, say, paper, wood, or coal) in air, which is a *chemical* reaction, occurring at the *atomic* level (and at a much lower temperature).

H–R diagram, as shown in Fig. 33–8. As it moves upward, it enters the **red giant** stage. Thus, theory explains the origin of red giants as a natural step in a star's evolution. Our Sun, for example, has been on the main sequence for about $4\frac{1}{2}$ billion years. It will probably remain there another 5 or 6 billion years. When our Sun leaves the main sequence, it is expected to grow in diameter (as it becomes a red giant) by a factor of 100 or more, possibly swallowing up inner planets such as Mercury and possibly Venus and even Earth.

If the star is like our Sun, or larger, further fusion can occur. As the star's outer envelope expands, its core continues to shrink and heat up. When the temperature reaches about 10^8 K, even helium nuclei, in spite of their greater charge and hence greater electrical repulsion, can come close enough to each other to undergo fusion. The reactions are

$$_2^4\text{He} + {}_2^4\text{He} \rightarrow {}_4^8\text{Be}$$
$$_2^4\text{He} + {}_4^8\text{Be} \rightarrow {}_6^{12}\text{C} \qquad \textbf{(33–2)}$$

with the emission of two γ rays. These two reactions must occur in quick succession (because ${}_4^8\text{Be}$ is very unstable), and the net effect is

$$3\,{}_2^4\text{He} \rightarrow {}_6^{12}\text{C} + 2\gamma. \qquad (Q = 7.3\,\text{MeV})$$

This fusion of helium causes a change in the star which moves rapidly to the "horizontal branch" on the H–R diagram (Fig. 33–8). Further fusion reactions are possible, with ${}_2^4\text{He}$ fusing with ${}_6^{12}\text{C}$ to form ${}_8^{16}\text{O}$. In more massive stars, higher Z elements like ${}_{10}^{20}\text{Ne}$ or ${}_{12}^{24}\text{Mg}$ can be made. This process of creating heavier nuclei from lighter ones (or by absorption of neutrons which tends to occur at higher Z) is called **nucleosynthesis**.

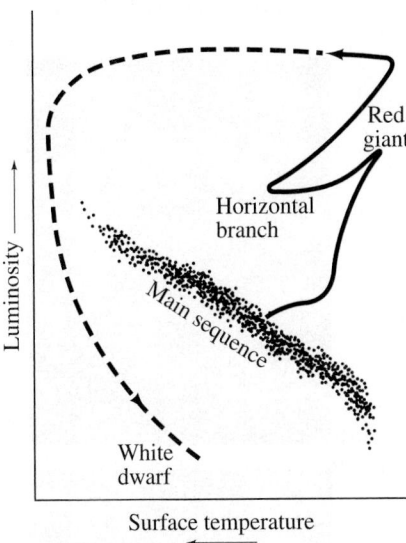

FIGURE 33–8 Evolutionary "track" of a star like our Sun represented on an H–R diagram.

Low Mass Stars—White Dwarfs

The final fate of a star depends on its mass. Stars can lose mass as parts of their outer envelope move off into space. Stars born with a mass less than about 8 solar masses ($8\times$ the mass of our Sun) eventually end up with a residual mass less than about 1.4 solar masses. A residual mass of 1.4 solar masses is known as the **Chandrasekhar limit**. For stars smaller than this, no further fusion energy can be obtained because of the large Coulomb repulsion between nuclei. The core of such a "low mass" star (original mass $\lesssim 8$ solar masses) contracts under gravity. The outer envelope expands again and the star becomes an even brighter and larger red giant, Fig. 33–8. Eventually the outer layers escape into space, and the newly revealed surface is hotter than before. So the star moves to the left in the H–R diagram (horizontal dashed line in Fig. 33–8). Then, as the core shrinks the star cools, and typically follows the downward dashed route shown on the left in Fig. 33–8, becoming a **white dwarf**. A white dwarf with a residual mass equal to that of the Sun would be about the size of the Earth. A white dwarf contracts to the point at which the electrons start to overlap, but no further because, by the Pauli exclusion principle, no two electrons can be in the same quantum state. At this point the star is supported against further collapse by this **electron degeneracy** pressure. A white dwarf continues to lose internal energy by radiation, decreasing in temperature and becoming dimmer until it glows no more. It has then become a cold dark chunk of extremely dense material.

High Mass Stars—Supernovae, Neutron Stars, Black Holes

Stars whose original mass is greater than about 8 solar masses are thought to follow a very different scenario. A star with this great a mass can contract under gravity and heat up even further. At temperatures $T \approx 3$ or 4×10^9 K, nuclei as heavy as ${}_{26}^{56}\text{Fe}$ and ${}_{28}^{56}\text{Ni}$ can be made. But here the formation of heavy nuclei from lighter ones, by fusion, ends. As we saw in Fig. 30–1, the average binding energy per nucleon begins to decrease for A greater than about 60. Further fusions would *require* energy, rather than release it.

(a)

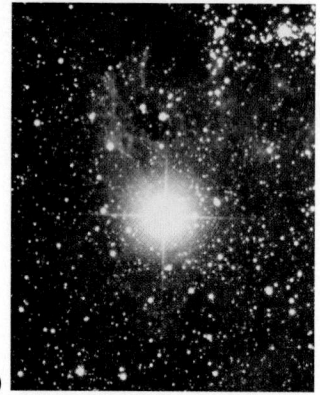

(b)

FIGURE 33–9 The star indicated by the arrow in (a) exploded in 1987 as a supernova (SN1987A), as shown in (b). The bright spot in (b) indicates a huge release of energy but does not represent the physical size.

FIGURE 33–10 Hypothetical model for novae and Type Ia supernovae, showing how a white dwarf could pull mass from its normal companion.

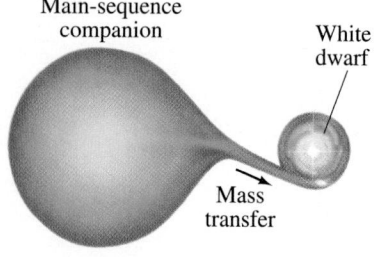

Main-sequence companion

White dwarf

Mass transfer

At these extremely high temperatures, well above 10^9 K, high-energy collisions can cause the breaking apart of iron and nickel nuclei into He nuclei, and eventually into protons and neutrons:

$$^{56}_{26}\text{Fe} \rightarrow 13\,^4_2\text{He} + 4\text{n}$$

$$^4_2\text{He} \rightarrow 2\text{p} + 2\text{n}.$$

These are energy-requiring (endothermic) reactions, which rob energy from the core, allowing gravitational contraction to begin. This then can force electrons and protons together to form neutrons in **inverse β decay**:

$$\text{e}^- + \text{p} \rightarrow \text{n} + \nu.$$

As a result of these reactions, the pressure in the core drops precipitously. As the core collapses under the huge gravitational forces, the tremendous mass becomes essentially an enormous nucleus made up almost exclusively of neutrons. The size of the star is no longer limited by the exclusion principle applied to electrons, but rather by **neutron degeneracy** pressure, and the star contracts rapidly to form an enormously dense **neutron star**. The core of a neutron star contracts to the point at which all neutrons are as close together as they are in an atomic nucleus. That is, the density of a neutron star is on the order of 10^{14} times greater than normal solids and liquids on Earth. A cupful of such dense matter would weigh billions of tons. A neutron star that has a mass 1.5 times that of our Sun would have a diameter of only about 20 km. (Compare this to a white dwarf with 1 solar mass whose diameter would be $\approx 10^4$ km, as mentioned on the previous page.)

The contraction of the core of a massive star would mean a great reduction in gravitational potential energy. Somehow this energy would have to be released. Indeed, it was suggested in the 1930s that the final core collapse to a neutron star could be accompanied by a catastrophic explosion known as a **supernova** (plural = supernovae). The tremendous energy release (Fig. 33–9) could form virtually all elements of the Periodic Table (see below) and blow away the entire outer envelope of the star, spreading its contents into interstellar space. The presence of heavy elements on Earth and in our solar system suggests that our solar system formed from the debris of many such supernova explosions.

The elements heavier than Ni are thought to form mainly by **neutron capture** in these exploding supernovae (rather than by fusion, as for elements up to Ni). Large numbers of free neutrons, resulting from nuclear reactions, are present inside those highly evolved stars and they can readily combine with, say, a $^{56}_{26}\text{Fe}$ nucleus to form (if three are captured) $^{59}_{26}\text{Fe}$, which decays to $^{59}_{27}\text{Co}$. The $^{59}_{27}\text{Co}$ can capture neutrons, also becoming neutron rich and decaying by β^- to the next higher Z element, and so on to the highest Z elements.

The final state of a neutron star depends on its mass. If the final mass is less than about three solar masses, the subsequent evolution of the neutron star is thought to resemble that of a white dwarf. If the mass is greater than this (original mass $\gtrsim 40$ solar masses), the neutron star collapses under gravity, overcoming even neutron degeneracy. Gravity would then be so strong that emitted light could not escape—it would be pulled back in by the force of gravity. Since no radiation could escape from such a "star," we could not see it— it would be black. An object may pass by it and be deflected by its gravitational field, but if the object came too close it would be swallowed up, never to escape. This is a **black hole**.

Novae and Supernovae

Novae (singular is *nova*, meaning "new" in Latin) are faint stars that have suddenly increased in brightness by as much as a factor of 10^6 and last for a month or two before fading. Novae are thought to be faint white dwarfs that have pulled mass from a nearby companion (they make up a *binary* system), as illustrated in Fig. 33–10. The captured mass of hydrogen suddenly fuses into helium at a high rate for a few weeks. Many novae (maybe all) are *recurrent*—they repeat their bright glow years later.

Supernovae are also brief explosive events, but release millions of times more energy than novae, up to 10^{10} times more luminous than our Sun. The peak of brightness may exceed that of the entire galaxy in which they are located, but lasts only a few days or weeks. They slowly fade over a few months. Many supernovae form by core collapse to a neutron star as described above. See Fig. 33–9.

Type Ia supernovae are different. They all seem to have very nearly the same luminosity. They are believed to be binary stars, one of which is a white dwarf that pulls mass from its companion, much like for a nova, Fig. 33–10. The mass is higher, and as mass is captured and the total mass approaches the Chandrasekhar limit of 1.4 solar masses, it explodes as a "white-dwarf" supernova by undergoing a "thermonuclear runaway"—an uncontrolled chain of nuclear reactions that entirely destroys the white dwarf. Type Ia supernovae are useful to us as "standard candles" in the night sky to help us determine distance—see next Section.

33–3 Distance Measurements

Parallax

We have talked about the vast distances of objects in the universe. But how do we measure these distances? One basic technique employs simple geometry to measure the **parallax** of a star. By parallax we mean the apparent motion of a star, against the background of much more distant stars, due to the Earth's motion around the Sun. As shown in Fig. 33–11, we can measure the angle 2ϕ that the star appears to shift, relative to very distant stars, when viewed 6 months apart. If we know the distance d from Earth to Sun, we can reconstruct the right triangles shown in Fig. 33–11 and can then determine the distance D to the star. This is essentially the way the heights of mountains are determined, by "triangulation": see Example 1–8.

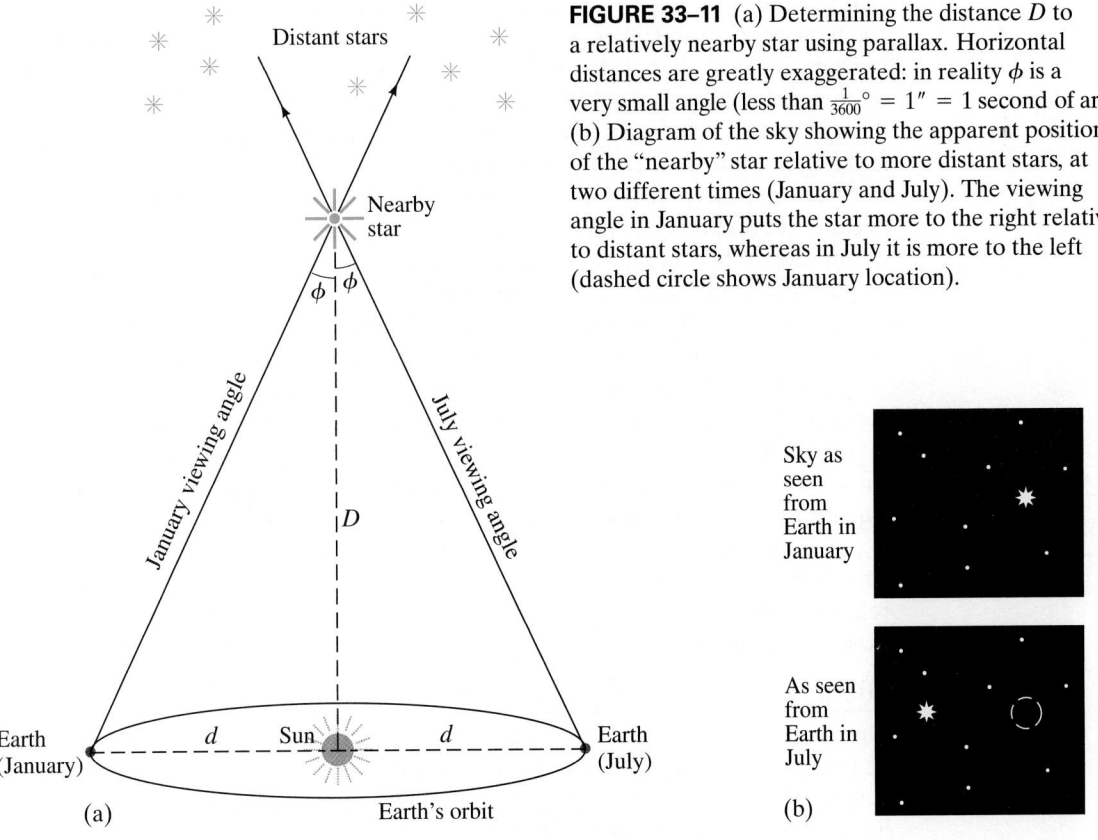

FIGURE 33–11 (a) Determining the distance D to a relatively nearby star using parallax. Horizontal distances are greatly exaggerated: in reality ϕ is a very small angle (less than $\frac{1}{3600}^{\circ} = 1'' = 1$ second of arc). (b) Diagram of the sky showing the apparent position of the "nearby" star relative to more distant stars, at two different times (January and July). The viewing angle in January puts the star more to the right relative to distant stars, whereas in July it is more to the left (dashed circle shows January location).

EXAMPLE 33–6 | **ESTIMATE** | **Distance to a star using parallax.** Estimate the distance D to a star if the angle 2ϕ in Fig. 33–11a is measured to be $2\phi = 0.00012°$.

APPROACH From trigonometry, $\tan\phi = d/D$ in Fig. 33–11a. The Sun–Earth distance is $d = 1.5 \times 10^8$ km (inside front cover).

SOLUTION The angle $\phi = 0.00006°$, or about $(0.00006°)(2\pi \text{ rad}/360°) = 1.0 \times 10^{-6}$ radians. We can use $\tan\phi \approx \phi$ because ϕ is very small. We solve for D in $\tan\phi = d/D$. The distance D to the star is

$$D = \frac{d}{\tan\phi} \approx \frac{d}{\phi} = \frac{1.5 \times 10^8 \text{ km}}{1.0 \times 10^{-6} \text{ rad}} = 1.5 \times 10^{14} \text{ km,}$$

or about 15 ly.

*Parsec

Distances to stars are often specified in terms of parallax angle (ϕ in Fig. 33–11a) given in seconds of arc: 1 second ($1''$) is $\frac{1}{60}$ of one minute ($1'$) of arc, which is $\frac{1}{60}$ of a degree, so $1'' = \frac{1}{3600}$ of a degree. The distance is then specified in **parsecs** (pc) (meaning *par*allax angle in *sec*onds of arc): $D = 1/\phi$ with ϕ in seconds of arc. In Example 33–6, $\phi = (6 \times 10^{-5})°(3600) = 0.22''$ of arc, so we would say the star is at a distance of $1/0.22'' = 4.5$ pc. One parsec is given by (recall $D = d/\phi$, and we set the Sun–Earth distance (Fig. 33–11a) as $d = 1.496 \times 10^{11}$ m):

$$1 \text{ pc} = \frac{d}{1''} = \frac{1.496 \times 10^{11} \text{ m}}{(1'')\left(\frac{1'}{60''}\right)\left(\frac{1°}{60'}\right)\left(\frac{2\pi \text{ rad}}{360°}\right)} = 3.086 \times 10^{16} \text{ m}$$

$$1 \text{ pc} = (3.086 \times 10^{16} \text{ m})\left(\frac{1 \text{ ly}}{9.46 \times 10^{15} \text{ m}}\right) = 3.26 \text{ ly.}$$

Distant Stars and Galaxies

Parallax can be used to determine the distance to stars as far away as about 100 light-years from Earth, and from an orbiting spacecraft perhaps 5 to 10 times farther. Beyond that distance, parallax angles are too small to measure. For greater distances, more subtle techniques must be employed. We might compare the apparent brightnesses of two stars, or two galaxies, and use the *inverse square law* (apparent brightness drops off as the square of the distance) to roughly estimate their relative distances. We can't expect this technique to be very precise because we don't expect any two stars, or two galaxies, to have the same intrinsic luminosity. When comparing galaxies, a perhaps better estimate assumes the brightest stars in all galaxies (or the brightest galaxies in galaxy clusters) are similar and have about the same intrinsic luminosity. Consequently, their *apparent brightness* would be a measure of how far away they were.

Another technique makes use of the H–R diagram. Measurement of a star's surface temperature (from its spectrum) places it at a certain point (within 20%) on the H–R diagram, assuming it is a main-sequence star, and then its luminosity can be estimated from the vertical axis (Fig. 33–6). Its apparent brightness and Eq. 33–1 give its approximate distance; see Example 33–5.

A better estimate comes from comparing *variable stars*, especially *Cepheid variables* whose luminosity varies over time with a period that is found to be related to their average luminosity. Thus, from their period and apparent brightness we get their distance.

Distance via SNIa, Redshift

The largest distances are estimated by comparing the apparent brightnesses of Type Ia supernovae ("SNIa"). Type Ia supernovae all have a similar origin (as described on the previous page and Fig. 33–10), and their brief explosive burst of light is expected to be of nearly the same luminosity. They are thus sometimes referred to as "standard candles."

Another important technique for estimating the distance of very distant galaxies is from the "redshift" in the line spectra of elements and compounds. The redshift is related to the expansion of the universe, as we shall discuss in Section 33–5. It is useful for objects farther than 10^7 to 10^8 ly away.

As we look farther and farther away, measurement techniques are less and less reliable, so there is more uncertainty in the measurements of large distances.

33–4 General Relativity: Gravity and the Curvature of Space

We have seen that the force of gravity plays an important role in the processes that occur in stars. Gravity too is important for the evolution of the universe as a whole. The reasons gravity plays a dominant role in the universe, and not one of the other of the four forces in nature, are (1) it is long-range and (2) it is always attractive. The strong and weak nuclear forces act over very short distances only, on the order of the size of a nucleus; hence they do not act over astronomical distances (they do act between nuclei and nucleons in stars to produce nuclear reactions). The electromagnetic force, like gravity, acts over great distances. But it can be either attractive or repulsive. And since the universe does not seem to contain large areas of net electric charge, a large net force does not occur. But gravity acts only as an *attractive* force between *all* masses, and there are large accumulations of mass in the universe. The force of gravity as Newton described it in his law of universal gravitation was modified by Einstein. In his general theory of relativity, Einstein developed a theory of gravity that now forms the basis of cosmological dynamics.

In the *special theory of relativity* (Chapter 26), Einstein concluded that there is no way for an observer to determine whether a given frame of reference is at rest or is moving at constant velocity in a straight line. Thus the laws of physics must be the same in different inertial reference frames. But what about the more general case of motion where reference frames can be *accelerating*?

Einstein tackled the problem of accelerating reference frames in his **general theory of relativity** and in it also developed a theory of gravity. The mathematics of General Relativity is complex, so our discussion will be mainly qualitative.

We begin with Einstein's **principle of equivalence**, which states that

> **no experiment can be performed that could distinguish between a uniform gravitational field and an equivalent uniform acceleration.**

If observers sensed that they were accelerating (as in a vehicle speeding around a sharp curve), they could not prove by any experiment that in fact they weren't simply experiencing the pull of a gravitational field. Conversely, we might think we are being pulled by gravity when in fact we are undergoing an acceleration having nothing to do with gravity.

As a thought experiment, consider a person in a freely falling elevator near the Earth's surface. If our observer held out a book and let go of it, what would happen? Gravity would pull it downward toward the Earth, but at the same rate $(g = 9.8 \text{ m/s}^2)$ at which the person and elevator were falling. So the book would hover right next to the person's hand (Fig. 33–12). The effect is exactly the same as if this reference frame was at rest and *no* forces were acting. On the other hand, if the elevator was out in space where the gravitational field is essentially zero, the released book would float, just as it does in Fig. 33–12. Next, if the elevator (out in space) is accelerated upward (using rockets) at an acceleration of 9.8 m/s^2, the book as seen by our observer would fall to the floor with an acceleration of 9.8 m/s^2, just as if it were falling due to gravity at the surface of the Earth. According to the principle of equivalence, the observer could not determine whether the book fell because the elevator was accelerating upward, or because a gravitational field was acting downward and the elevator was at rest. The two descriptions are equivalent.

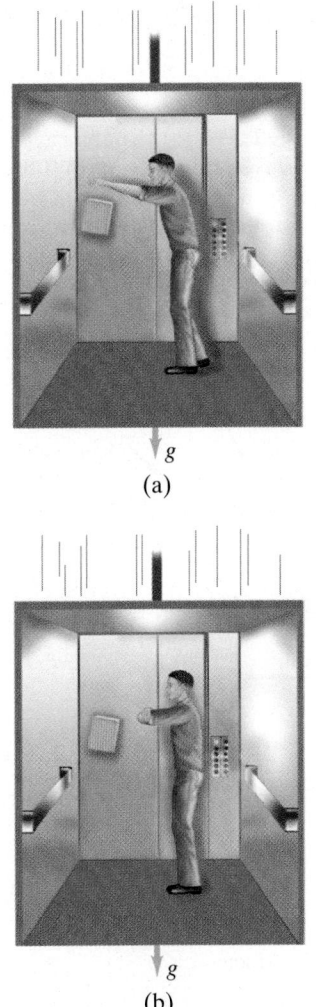

FIGURE 33–12 In an elevator falling freely under gravity, (a) a person releases a book; (b) the released book hovers next to the owner's hand; (b) is a few moments after (a).

(a)

(b)

The principle of equivalence is related to the concept that there are two types of mass. Newton's second law, $F = ma$, uses **inertial mass**. We might say that inertial mass represents "resistance" to any type of force. The second type of mass is **gravitational mass**. When one object attracts another by the gravitational force (Newton's law of universal gravitation, $F = Gm_1 m_2/r^2$, Chapter 5), the strength of the force is proportional to the product of the *gravitational masses* of the two objects. This is much like Coulomb's law for the electric force between two objects which is proportional to the product of their electric charges. The electric charge on an object is not related to its inertial mass; so why should we expect that an object's gravitational mass (call it gravitational charge if you like) be related to its inertial mass? All along we have assumed they were the same. Why? Because no experiment—not even of high precision—has been able to discern any measurable difference between inertial mass and gravitational mass. (For example, in the absence of air resistance, all objects fall at the same acceleration, g, on Earth.) This is another way to state the equivalence principle: *gravitational mass is equivalent to inertial mass.*

FIGURE 33–13 (a) Light beam goes straight across an elevator which is not accelerating. (b) The light beam bends (exaggerated) according to an observer in an accelerating elevator whose speed increases in the upward direction.

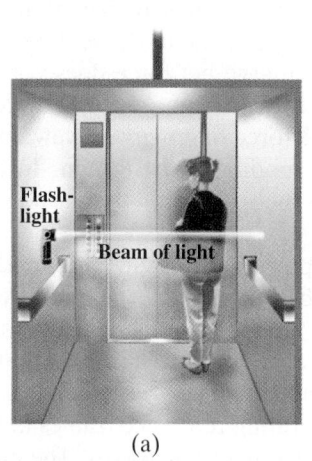
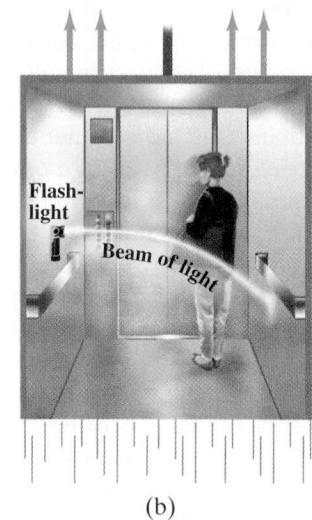

(a)   (b)

FIGURE 33–14 (a) Two stars in the sky observed from Earth. (b) If the light from one of these stars passes very near the Sun, whose gravity bends the rays, the star will appear higher than it actually is (follow the ray backwards). [Not to scale.]

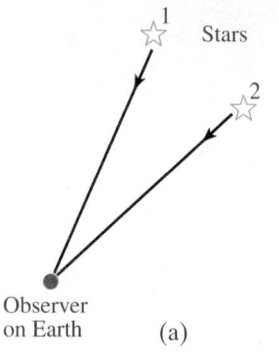

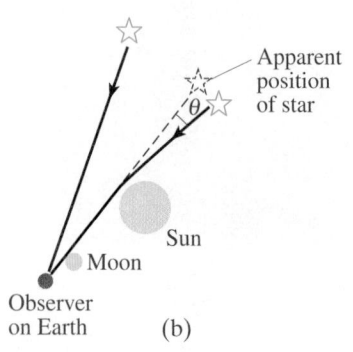

The principle of equivalence can be used to show that light ought to be deflected by the gravitational force due to a massive object. Consider another thought experiment, in which an elevator is in free space where virtually no gravity acts. If a light beam is emitted by a flashlight attached to the side of the elevator, the beam travels straight across the elevator and makes a spot on the opposite side if the elevator is at rest or moving at constant velocity (Fig. 33–13a). If instead the elevator is accelerating upward, as in Fig. 33–13b, the light beam still travels straight across in a reference frame at rest. In the upwardly accelerating elevator, however, the beam is observed to curve downward. Why? Because during the time the light travels from one side of the elevator to the other, the elevator is moving upward at a vertical speed that is increasing relative to the light. Next we note that according to the equivalence principle, an upwardly accelerating reference frame is equivalent to a downward gravitational field. Hence, we can picture the curved light path in Fig. 33–13b as being due to the effect of a gravitational field. Thus, from the principle of equivalence, we expect gravity to exert a force on a beam of light and to bend it out of a straight-line path!

That light is affected by gravity is an important prediction of Einstein's general theory of relativity. And it can be tested. The amount a light beam would be deflected from a straight-line path must be small even when passing a massive object. (For example, light near the Earth's surface after traveling 1 km is predicted to drop only about 10^{-10} m, which is equal to the diameter of a small atom and not detectable.) The most massive object near us is the Sun, and it was calculated that light from a distant star would be deflected by 1.75″ of arc (tiny but detectable) as it passed by the edge of the Sun (Fig. 33–14). However, such a measurement could be made only during a total eclipse of the Sun, so that the Sun's tremendous brightness would not obscure the starlight passing near its edge.

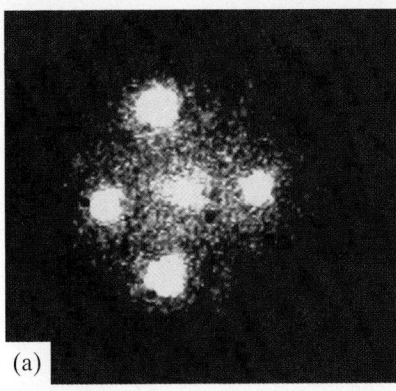

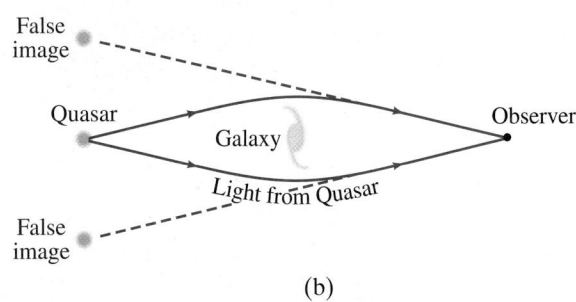

FIGURE 33–15 (a) Hubble Space Telescope photograph of the so-called "Einstein cross," thought to represent "gravitational lensing": the central spot is a relatively nearby galaxy, whereas the four other spots are thought to be images of a single quasar *behind* the galaxy. (b) Diagram showing how the galaxy could bend the light coming from the quasar behind it to produce the four images. See also Fig. 33–14. [If the shape of the nearby galaxy and distant quasar were perfect spheres and perfectly aligned, we would expect the "image" of the distant quasar to be a circular ring or halo instead of the four separate images seen here. Such a ring is called an "Einstein ring."]

An opportune eclipse occurred in 1919, and scientists journeyed to the South Atlantic to observe it. Their photos of stars just behind the Sun revealed shifts in accordance with Einstein's prediction. Another example of gravitational deflection of light is **gravitational lensing**, as described in Fig. 33–15. The very distant galaxies shown in the XDF photo at the start of this Chapter, page 947, are thought to be visible only because of gravitational lensing (and magnification of their emitted light) by nearer galaxies—as if the nearby galaxies acted as a magnifying glass.

The mathematician Fermat showed in the 1600s that optical phenomena, including reflection, refraction, and effects of lenses, can be derived from a simple principle: that light traveling between two points follows the shortest path in space. Thus if gravity curves the path of light, then gravity must be able to curve space itself. That is, *space itself can be curved*, and it is gravitational mass that causes the curvature. Indeed, the curvature of space—or rather, of four-dimensional space-time—is a basic aspect of Einstein's General Relativity.

What is meant by **curved space**? To understand, recall that our normal method of viewing the world is via Euclidean plane geometry. In Euclidean geometry, there are many axioms and theorems we take for granted, such as that the sum of the angles of any triangle is 180°. Non-Euclidean geometries, which involve curved space, have also been imagined by mathematicians. It is hard enough to imagine three-dimensional curved space, much less curved four-dimensional space-time. So let us try to understand the idea of curved space by using two-dimensional surfaces.

Consider, for example, the two-dimensional surface of a sphere. It is clearly curved, Fig. 33–16, at least to us who view it from the outside—from our three-dimensional world. But how would hypothetical two-dimensional creatures determine whether their two-dimensional space was flat (a plane) or curved? One way would be to measure the sum of the angles of a triangle. If the surface is a plane, the sum of the angles is 180°, as we learn in plane geometry. But if the space is curved, and a sufficiently large triangle is constructed, the sum of the angles will *not* be 180°. To construct a triangle on a curved surface, say the sphere of Fig. 33–16, we must use the equivalent of a straight line: that is, the shortest distance between two points, which is called a **geodesic**. On a sphere, a geodesic is an arc of a great circle (an arc in a plane passing through the center of the sphere) such as the Earth's equator and the Earth's longitude lines. Consider, for example, the large triangle of Fig. 33–16: its sides are two longitude lines passing from the north pole to the equator, and the third side is a section of the equator as shown. The two longitude lines make 90° angles with the equator (look at a world globe to see this more clearly). They make an angle with each other at the north pole, which could be, say, 90° as shown; the sum of these angles is 90° + 90° + 90° = 270°. This is clearly *not* a Euclidean space. Note, however, that if the triangle is small in comparison to the radius of the sphere, the angles will add up to nearly 180°, and the triangle (and space) will seem flat.

FIGURE 33–16 On a two-dimensional curved surface, the sum of the angles of a triangle may not be 180°.

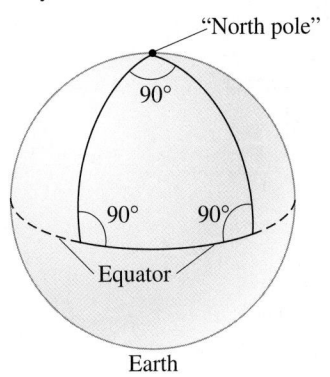

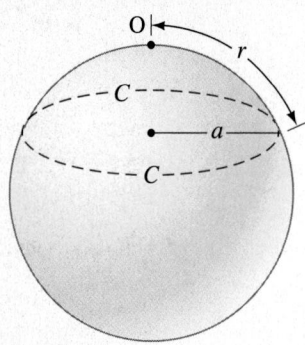

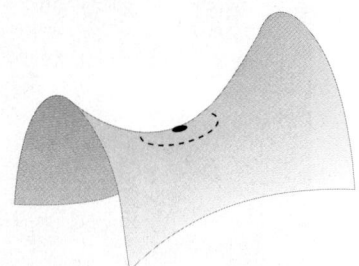

FIGURE 33–17 On a spherical surface (a two-dimensional world) a circle of circumference C is drawn (red) about point O as the center. The radius of the circle (not the sphere) is the distance r along the surface. (Note that in our three-dimensional view, we can tell that $C = 2\pi a$. Since $r > a$, then $C < 2\pi r$.)

FIGURE 33–18 Example of a two-dimensional surface with negative curvature.

Another way to test the curvature of space is to measure the radius r and circumference C of a large circle. On a plane surface, $C = 2\pi r$. But on a two-dimensional spherical surface, C is *less* than $2\pi r$, as can be seen in Fig. 33–17. The proportionality between C and r is *less* than 2π. Such a surface is said to have *positive curvature*. On the saddlelike surface of Fig. 33–18, the circumference of a circle is greater than $2\pi r$, and the sum of the angles of a triangle is less than 180°. Such a surface is said to have a *negative curvature*.

Curvature of the Universe

What about our universe? On a large scale (not just near a large mass), what is the overall curvature of the universe? Does it have positive curvature, negative curvature, or is it flat (zero curvature)? We perceive our world as Euclidean (flat), but we can not exclude the possibility that space could have a curvature so slight that we don't normally notice it. This is a crucial question in cosmology, and it can be answered only by precise experimentation.

If the universe had a positive curvature, the universe would be *closed*, or *finite* in volume. This would *not* mean that the stars and galaxies extended out to a certain boundary, beyond which there is empty space. There is no boundary or edge in such a universe. The universe is all there is. If a particle were to move in a straight line in a particular direction, it would eventually return to the starting point—perhaps eons of time later.

On the other hand, if the curvature of space was zero or negative, the universe would be *open*. It could just go on forever. An open universe could be *infinite*; but according to recent research, even that may not necessarily be so.

Today the evidence is very strong that the universe on a large scale is very close to being flat. Indeed, it is so close to being flat that we can't tell if it might have very slightly positive or very slightly negative curvature.

Black Holes

According to Einstein's theory of general relativity (sometimes abbreviated GR), space-time is curved near massive objects. We might think of space as being like a thin rubber sheet: if a heavy weight is placed on the sheet, it sags as shown in Fig. 33–19a (top of next page). The weight corresponds to a huge mass that causes space (space itself!) to curve. Thus, in the context of general relativity[†] we do not speak of the "force" of gravity acting on objects. Instead we say that objects and light rays move as they do because space-time is curved. An object starting at rest or moving slowly near the great mass of Fig. 33–19a would follow a geodesic (the equivalent of a straight line in plane geometry) toward that great mass.

The extreme curvature of space-time shown in Fig. 33–19b could be produced by a **black hole**. A black hole, as we mentioned in Section 33–2, has such strong gravity that even light cannot escape from it. To become a black hole, an object of mass M must undergo **gravitational collapse**, contracting by gravitational self-attraction to within a radius called the **Schwarzschild radius**,

$$R = \frac{2GM}{c^2},$$

where G is the gravitational constant and c the speed of light. If an object collapses to within this radius, it is predicted by general relativity to collapse to a point at $r = 0$, forming an infinitely dense singularity. This prediction is uncertain, however, because in this realm we need to combine quantum mechanics with gravity, a unification of theories not yet achieved (Section 32–12).

| **EXERCISE C** What is the Schwarzschild radius for an object with 10 solar masses?

The Schwarzschild radius also represents the event horizon of a black hole. By **event horizon** we mean the surface beyond which no emitted signals can ever reach us, and thus inform us of events that happen beyond that surface. As a star collapses toward a black hole, the light it emits is pulled harder and harder by gravity, but we can still see it. Once the matter passes within the event horizon, the emitted light cannot escape but is pulled back in by gravity (= curvature of space-time).

All we can know about a black hole is its mass, its angular momentum (rotating black holes), and its electric charge. No other information, no details of its structure or the kind of matter it was formed of, can be known because no information can escape.

How might we observe black holes? We cannot see them because no light can escape from them. They would be black objects against a black sky. But they do exert a gravitational force on nearby objects, and also on light rays (or photons) that pass nearby (just like in Fig. 33–15). The black hole believed to be at the center of our Galaxy ($M \approx 4 \times 10^6 \, M_{Sun}$) was discovered by examining the motion of matter in its vicinity. Another technique is to examine stars which appear to move as if they were one member of a *binary system* (two stars rotating about their common center of mass), but without a visible companion. If the unseen star is a black hole, it might be expected to pull off gaseous material from its visible companion (as in Fig. 33–10). As this matter approached the black hole, it would be highly accelerated and should emit X-rays of a characteristic type before plunging inside the event horizon. Such X-rays, plus a sufficiently high mass estimate from the rotational motion, can provide evidence for a black hole. One of the many candidates for a black hole is in the binary-star system Cygnus X-1. It is widely believed that the center of most galaxies is occupied by a black hole with a mass 10^6 to 10^9 times the mass of a typical star like our Sun.

| **EXERCISE D** A black hole has radius R. Its mass is proportional to (a) R, (b) R^2, (c) R^3. Justify your answer.

[†]Alexander Pope (1688–1744) wrote an epitaph for Newton:
   "Nature, and Nature's laws lay hid in night:
   God said, *Let Newton be!* and all was light."
Sir John Squire (1884–1958), perhaps uncomfortable with Einstein's profound thoughts, added:
   "It did not last: the Devil howling '*Ho!*
   Let Einstein be!' restored the status quo."

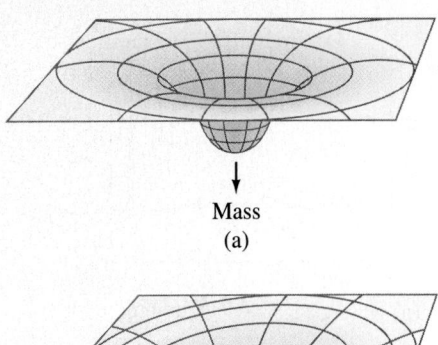

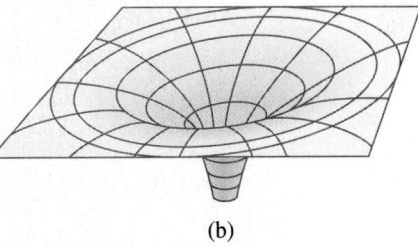

FIGURE 33–19 (a) Rubber-sheet analogy for space-time curved by matter. (b) Same analogy for a black hole, which can "swallow up" objects that pass near.

Mass
(a)

(b)

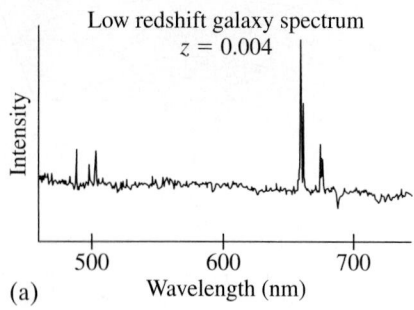

Low redshift galaxy spectrum
$z = 0.004$

(a)

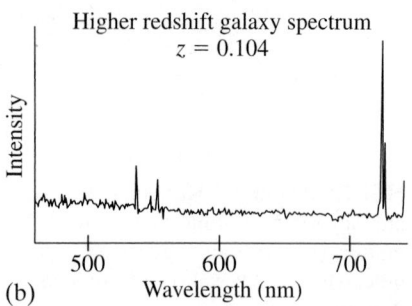

Higher redshift galaxy spectrum
$z = 0.104$

(b)

FIGURE 33–20 Atoms and molecules emit and absorb light of particular frequencies depending on the spacing of their energy levels, as we saw in Chapters 27 to 29. (a) The spectrum of light received from a relatively slow-moving galaxy. (b) Spectrum of a galaxy moving away from us at a much higher speed. Note how the peaks (or lines) in the spectrum have moved to longer wavelengths. The redshift is $z = (\lambda_{obs} - \lambda_{rest})/\lambda_{rest}$.

HUBBLE'S LAW

33–5 The Expanding Universe: Redshift and Hubble's Law

We discussed in Section 33–2 how individual stars evolve from their birth to their death as white dwarfs, neutron stars, or black holes. But what about the universe as a whole: is it static, or does it change? One of the most important scientific discoveries of the twentieth century was that distant galaxies are racing away from us, and that the farther they are from us at a given time, the faster they are moving away. How astronomers arrived at this astonishing idea, and what it means for the past history of the universe as well as its future, will occupy us for the remainder of the book.

Observational evidence that the universe is expanding was first put forth by Edwin Hubble in 1929. This idea was based on distance measurements of galaxies (Section 33–3), and determination of their velocities by the Doppler shift of spectral lines in the light received from them (Fig. 33–20). In Chapter 12 we saw how the frequency of sound is higher and the wavelength shorter if the source and observer move toward each other. If the source moves away from the observer, the frequency is lower and the wavelength longer. The **Doppler effect** occurs also for light, but the formula for light is slightly different than for sound and is given by[†]

$$\lambda_{obs} = \lambda_{rest}\sqrt{\frac{1 + v/c}{1 - v/c}}, \qquad \begin{bmatrix} \text{source and observer moving} \\ \text{away from each other} \end{bmatrix} \quad \textbf{(33–3)}$$

where λ_{rest} is the emitted wavelength as seen in a reference frame at rest with respect to the source, and λ_{obs} is the wavelength observed in a frame moving with velocity v away from the source along the line of sight. (For relative motion *toward* each other, $v < 0$ in this formula.) When a distant source emits light of a particular wavelength, and the source is moving away from us, the wavelength appears longer to us: the color of the light (if it is visible) is shifted toward the red end of the visible spectrum, an effect known as a **redshift**. (If the source moves toward us, the color shifts toward the blue or shorter wavelength.)

In the spectra of stars in other galaxies, lines are observed that correspond to lines in the known spectra of particular atoms (see Section 27–11 and Figs. 24–28 and 27–23). What Hubble found was that the lines seen in the spectra from distant galaxies were generally *redshifted*, and that the amount of shift seemed to be approximately proportional to the distance of the galaxy from us. That is, the velocity v of a galaxy moving away from us is proportional to its distance d from us:

$$v = H_0 d. \quad \textbf{(33–4)}$$

This is **Hubble's law**, one of the most fundamental astronomical ideas. It was first suggested, in 1927, by Georges Lemaître, a Belgian physics professor and priest, who also first proposed what later came to be called the Big Bang. The constant H_0 is called the **Hubble parameter**.

The value of H_0 until recently was uncertain by over 20%, and thought to be between 15 and 25 km/s/Mly. But recent measurements now put its value more precisely at

$$H_0 = 21 \text{ km/s/Mly}$$

(that is, 21 km/s per million light-years of distance). The current uncertainty is about 2%, or ± 0.5 km/s/Mly. [H_0 can be written in terms of parsecs (Section 33–3) as $H_0 = 67$ km/s/Mpc (that is, 67 km/s per megaparsec of distance) with an uncertainty of about ± 1.2 km/s/Mpc.]

[†]For light there is no medium and we can make no distinction between motion of the source and motion of the observer (special relativity), as we did for sound which travels in a medium.

Redshift Origins

Galaxies very near us seem to be moving randomly relative to us: some move towards us (blueshifted), others away from us (redshifted); their speeds are on the order of $0.001c$. But for more distant galaxies, the velocity of recession is much greater than the velocity of local random motion, and so is dominant and Hubble's law (Eq. 33–4) holds very well. More distant galaxies have higher recession velocity and a larger redshift, and we call their redshift a **cosmological redshift**. We interpret this redshift today as due to the *expansion of space* itself. We can think of the originally emitted wavelength λ_{rest} as being stretched out (becoming longer) along with the expanding space around it, as suggested in Fig. 33–21. Although Hubble thought of the redshift as a Doppler shift, now we prefer to understand it in this sense of expanding space. (But note that atoms in galaxies do not expand as space expands; they keep their regular size.)

There is a third way to produce a redshift, which we mention for completeness: a **gravitational redshift**. Light leaving a massive star is gaining in gravitational potential energy (just like a stone thrown upward from Earth). So the kinetic energy of each photon, hf, must be getting smaller (to conserve energy). A smaller frequency f means a larger (longer) wavelength λ $(= c/f)$, which is a redshift.

The amount of a redshift is specified by the **redshift parameter**, z, defined as

$$z = \frac{\lambda_{obs} - \lambda_{rest}}{\lambda_{rest}} = \frac{\Delta\lambda}{\lambda_{rest}}, \qquad \textbf{(33–5a)}$$

where λ_{rest} is a wavelength as seen by an observer at rest relative to the source, and λ_{obs} is the wavelength measured by a moving observer. Equation 33–5a can be written as

$$z = \frac{\lambda_{obs}}{\lambda_{rest}} - 1 \qquad \textbf{(33–5b)}$$

and

$$z + 1 = \frac{\lambda_{obs}}{\lambda_{rest}}. \qquad \textbf{(33–5c)}$$

For low speeds not close to the speed of light ($v \lesssim 0.1\,c$), the Doppler formula (Eq. 33–3) can be used to show (Problem 32) that z is proportional to the speed of the source toward or away from us:

$$z = \frac{\lambda_{obs} - \lambda_{rest}}{\lambda_{rest}} = \frac{\Delta\lambda}{\lambda_{rest}} \approx \frac{v}{c}. \qquad [v \ll c] \quad \textbf{(33–6)}$$

But redshifts are not always small, in which case the approximation of Eq. 33–6 is not valid. For high z galaxies, not even Eq. 33–3 applies because the redshift is due to the expansion of space (cosmological redshift), not the Doppler effect. Our Chapter-Opening Photograph, page 947, shows two very distant high z galaxies, $z = 8.8$ and 11.9, which are also shown enlarged.

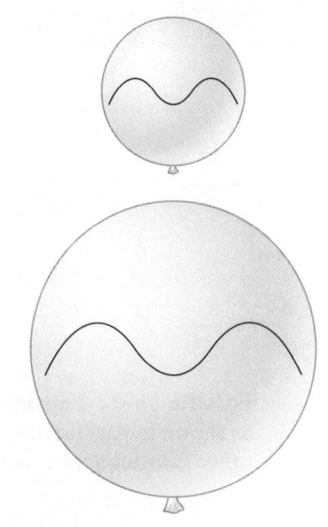

FIGURE 33–21 Simplified model of a 2-dimensional universe, imagined as a balloon. As you blow up the balloon (= expanding universe), the wavelength of a wave on its surface gets longer (redshifted).

*Scale Factor (advanced)

The expansion of space can be described as a scaling of the typical distance between two points or objects in the universe. If two distant galaxies are a distance d_0 apart at some initial time, then a time t later they will be separated by a greater distance $d(t)$. The **scale factor** is the same as for light, expressed in Eq. 33–5a:

$$\frac{d(t) - d_0}{d_0} = \frac{\Delta\lambda}{\lambda} = z$$

or

$$\frac{d(t)}{d_0} = 1 + z.$$

Thus, for example, if a galaxy has $z = 3$, then the scale factor is now $(1 + 3) = 4$ times larger than when the light was emitted from that galaxy. That is, the average distance between galaxies has become 4 times larger. Thus the factor by which the wavelength has increased since it was emitted tells us by what factor the universe (or the typical distance between objects) has increased.

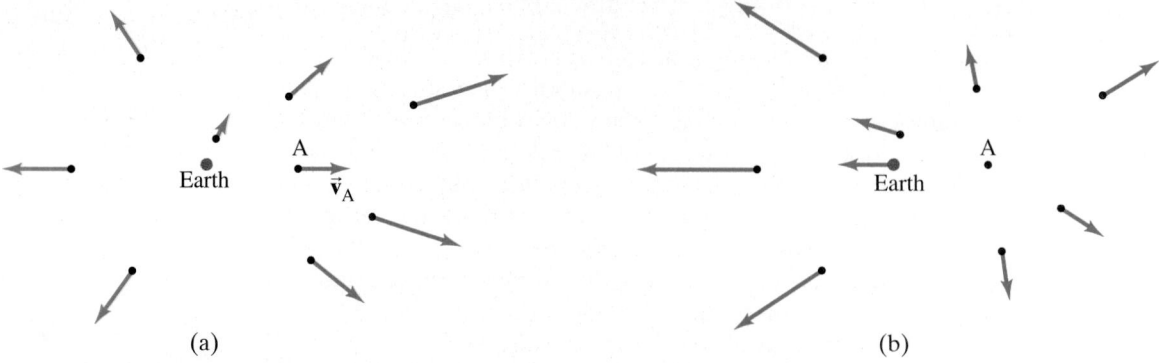

FIGURE 33–22 Expansion of the universe looks the same from any point in the universe. If you are on Earth as shown in part (a), or you are instead at galaxy A (which is at rest in the reference frame shown in (b)), all other galaxies appear to be racing away from you.

Expansion, and the Cosmological Principle

What does it mean that distant galaxies are all moving away from us, and with ever greater speed the farther they are from us? It seems to suggest some kind of explosive expansion that started at some very distant time in the past. And at first sight we seem to be in the middle of it all. But we aren't. The expansion appears the same from any other point in the universe. To understand why, see Fig. 33–22. In Fig. 33–22a we have the view from Earth (or from our Galaxy). The velocities of surrounding galaxies are indicated by arrows, pointing away from us, and the arrows are longer (faster speeds) for galaxies more distant from us. Now, what if we were on the galaxy labeled A in Fig. 33–22a? From Earth, galaxy A appears to be moving to the right at a velocity, call it $\vec{\mathbf{v}}_A$, represented by the arrow pointing to the right. If we were *on* galaxy A, Earth would appear to be moving to the left at velocity $-\vec{\mathbf{v}}_A$. To determine the velocities of other galaxies relative to A, we vectorially add the velocity vector, $-\vec{\mathbf{v}}_A$, to all the velocity arrows shown in Fig. 33–22a. This yields Fig. 33–22b, where we see that the universe is expanding away from galaxy A as well; and the velocities of galaxies receding from A are proportional to their current distance from A. *The universe looks pretty much the same from different points.*

Thus the expansion of the universe can be stated as follows: all galaxies are racing away from *each other* at an average rate of about 21 km/s per million light-years of distance between them. The ramifications of this idea are profound, and we discuss them in a moment.

A basic assumption in cosmology has been that on a large scale, the universe would look the same to observers at different places at the same time. In other words, the universe is both *isotropic* (looks the same in all directions) and *homogeneous* (would look the same if we were located elsewhere, say in another galaxy). This assumption is called the **cosmological principle**. On a local scale, say in our solar system or within our Galaxy, it clearly does not apply (the sky looks different in different directions). But it has long been thought to be valid if we look on a large enough scale, so that the average population density of galaxies and clusters of galaxies ought to be the same in different areas of the sky. This seems to be valid on distances greater than about 700 Mly. The expansion of the universe (Fig. 33–22) is consistent with the cosmological principle; and the near uniformity of the cosmic microwave background radiation (discussed in Section 33–6) supports it. Another way to state the cosmological principle is that *our place in the universe is not special.*

The expansion of the universe, as described by Hubble's law, strongly suggests that galaxies must have been closer together in the past than they are now. This is, in fact, the basis of the *Big Bang* theory of the origin of the universe, which pictures the universe as a relentless expansion starting from a very hot and compressed beginning. We discuss the Big Bang in detail shortly, but first let us see what can be said about the age of the universe.

One way to estimate the age of the universe uses the Hubble parameter. With $H_0 \approx 21$ km/s per 10^6 light-years, the time required for the galaxies to arrive at their present separations would be approximately (starting with $v = d/t$ and using Hubble's law, Eq. 33–4),

$$t = \frac{d}{v} = \frac{d}{H_0 d} = \frac{1}{H_0} \approx \frac{(10^6\,\text{ly})(0.95 \times 10^{13}\,\text{km/ly})}{(21\,\text{km/s})(3.16 \times 10^7\,\text{s/yr})} \approx 14 \times 10^9\,\text{yr},$$

or 14 billion years. The age of the universe calculated in this way is called the *characteristic expansion time* or "Hubble age." It is a very rough estimate and assumes the rate of expansion of the universe was constant (which today we are quite sure is not true). Today's best measurements give the age of the universe as about 13.8×10^9 yr, in remarkable agreement with the rough Hubble age estimate.

*Steady-State Model

Before discussing the Big Bang in detail, we mention one alternative to the Big Bang—the **steady-state model**—which assumed that the universe is infinitely old and on average looks the same now as it always has. (This assumed uniformity in time as well as space was called the *perfect cosmological principle*.) According to the steady-state model, no large-scale changes have taken place in the universe as a whole, particularly no Big Bang. To maintain this view in the face of the recession of galaxies away from each other, matter would need to be created continuously to maintain the assumption of uniformity. The rate of mass creation required is very small—about one nucleon per cubic meter every 10^9 years.

The steady-state model provided the Big Bang model with healthy competition in the mid-twentieth century. But the discovery of the cosmic microwave background radiation (next Section), as well as other observations of the universe, has made the Big Bang model universally accepted.

33–6 The Big Bang and the Cosmic Microwave Background

The expansion of the universe suggests that typical objects in the universe were once much closer together than they are now. This is the basis for the idea that the universe began about 14 billion years ago as an expansion from a state of very high density and temperature known affectionately as the **Big Bang**.

The birth of the universe was not an explosion, because an explosion blows pieces out into the surrounding space. Instead, the Big Bang was the start of an expansion of space itself. The observable universe was relatively very small at the start and has been expanding, getting ever larger, ever since. The initial tiny universe of extremely dense matter is not to be thought of as a concentrated mass in the midst of a much larger space around it. The initial tiny but dense universe was the *entire universe*. There wouldn't have been anything else. When we say that the universe was once smaller than it is now, we mean that the average separation between objects (such as electrons or galaxies) was less. The universe may have been infinite in extent even then, and it may still be now (only bigger). The **observable universe** (that which we have the possibility of observing because light has had time to reach us) is, however, finite.

A major piece of evidence supporting the Big Bang is the **cosmic microwave background** radiation (or CMB) whose discovery came about as follows.

In 1964, Arno Penzias and Robert Wilson pointed their horn antenna for detecting radio waves (Fig. 33–23) into the sky. With it they detected widespread emission, and became convinced that it was coming from outside our Galaxy. They made precise measurements at a wavelength $\lambda = 7.35$ cm, in the microwave region of the electromagnetic spectrum (Fig. 22–8). The intensity of this radiation was found initially not to vary by day or night or time of year, nor to depend on direction. It came from all directions in the universe with equal intensity, to a precision of better than 1%. It could only be concluded that this radiation came from the universe as a whole.

FIGURE 33–23 Photo of Arno Penzias (right, who signed it "Arno") and Robert Wilson. Behind them their "horn antenna."

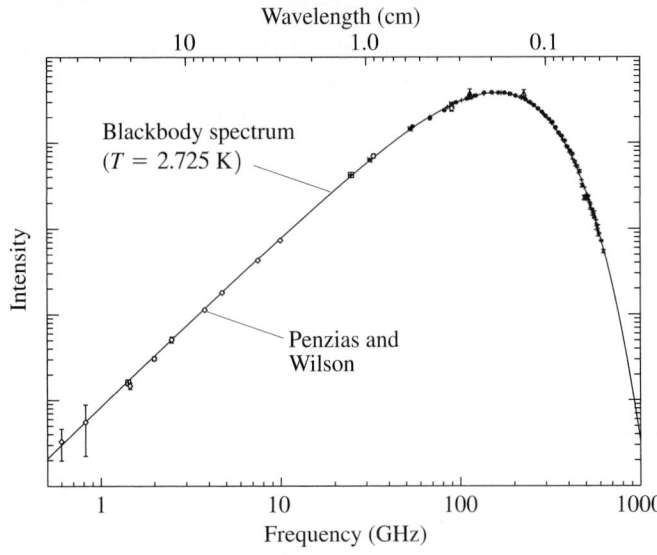

Wavelength (cm)

FIGURE 33–24 Spectrum of cosmic microwave background radiation, showing blackbody curve and experimental measurements including at the frequency detected by Penzias and Wilson. (Thanks to G. F. Smoot and D. Scott. The vertical bars represent the most recent experimental uncertainty in a measurement.)

FIGURE 33–25 COBE scientists John Mather (chief scientist and responsible for measuring the blackbody form of the spectrum) and George Smoot (chief investigator for anisotropy experiment) shown here during celebrations for their Dec. 2006 Nobel Prize, given for their discovery of the spectrum and anisotropy of the CMB using the COBE instrument.

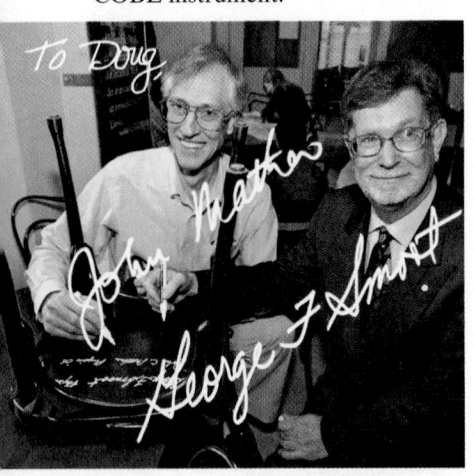

The intensity of this CMB measured at $\lambda = 7.35$ cm corresponds to blackbody radiation (see Section 27–2) at a temperature of about 3 K. When radiation at other wavelengths was measured by the COBE satellite (COsmic Background Explorer), the intensities were found to fall on a nearly perfect blackbody curve as shown in Fig. 33–24, corresponding to a temperature of 2.725 K (± 0.002 K).

The remarkable uniformity of the CMB was in accordance with the cosmological principle. But theorists felt that there needed to be some small inhomogeneities, or "anisotropies," in the CMB that would have provided "seeds" at which galaxy formation could have started. Small areas of slightly higher density, which could have contracted under gravity to form clusters of galaxies, were indeed found. These tiny inhomogeneities in density and temperature were detected first by the COBE satellite experiment in 1992, led by George Smoot and John Mather (Fig. 33–25).

This discovery of the **anisotropy** of the CMB ranks with the discovery of the CMB itself in the history of cosmology. The blackbody fit and the anisotropy were the culmination of decades of research by pioneers such as Richard Muller, Paul Richards, and David Wilkinson. Subsequent experiments gave us greater detail in 2003, 2006, and 2012 with the WMAP (Wilkinson Microwave Anisotropy Probe) results, Fig. 33–26, and even more recently with the European Planck satellite results in 2013.

The CMB provides strong evidence in support of the Big Bang, and gives us information about conditions in the very early universe. In fact, in the late 1940s, George Gamow and his collaborators calculated that a Big Bang origin of the universe should have generated just such a microwave background radiation.

To understand why, let us look at what a Big Bang might have been like. (Today we usually use the term "Big Bang" to refer to the *process*, starting from a moment after the birth of the universe through the subsequent expansion.) The temperature must have been extremely high at the start, so high that there could not have been any atoms in the very early stages of the universe (high energy collisions would have broken atoms apart into nuclei and free electrons). Instead, the universe would have consisted solely of radiation (photons) and a plasma of charged electrons and other elementary particles. The universe would have been

FIGURE 33–26 Measurements of the cosmic microwave background radiation over the entire sky, color-coded to represent differences in temperature from the average 2.725 K: the color scale ranges from $+200\,\mu$K (red) to $-200\,\mu$K (dark blue), representing slightly hotter and colder spots (associated with variations in density). Results are from the WMAP satellite in 2012: the angular resolution is 0.2°.

opaque—the photons in a sense "trapped," traveling very short distances before being scattered again, primarily by electrons. Indeed, the details of the microwave background radiation provide strong evidence that matter and radiation were once in equilibrium at a very high temperature. As the universe expanded, the energy spread out over an increasingly larger volume and the temperature dropped. Not long before the temperature had fallen to ~ 3000 K, some 380,000 years later, could nuclei and electrons combine together as stable atoms. With the disappearance of free electrons, as they combined with nuclei to form atoms, the radiation would have been freed—**decoupled** from matter, we say. The universe became *transparent* because photons were now free to travel nearly unimpeded straight through the universe.

It is this radiation, from 380,000 years after the birth of the universe, that we now see as the CMB. As the universe expanded, so too the wavelengths of the radiation lengthened, thus redshifting to longer wavelengths that correspond to lower temperature (recall Wien's law, $\lambda_P T$ = constant, Section 27–2), until they would have reached the 2.7-K background radiation we observe today.

Looking Back toward the Big Bang—Lookback Time

Figure 33–27 shows our Earth point of view, looking out in all directions back toward the Big Bang and the brief (380,000-year-long) period when radiation was trapped in the early plasma (yellow band). The time it takes light to reach us from an event is called its **lookback time**. The "close-up" insert in Fig. 33–27 shows a photon scattering repeatedly inside that early plasma and then exiting the plasma in a straight line. No matter what direction we look, our view of the very early universe is blocked by this wall of plasma. It is like trying to look into a very thick fog or into the surface of the Sun—we can see only as far as its surface, called the **surface of last scattering**, but not into it. Wavelengths from there are redshifted by $z \approx 1100$. Time $\Delta t'$ in Fig. 33–27 is the lookback time (not real time that goes forward).

Recall that when we view an object far away, we are seeing it as it was then, when the light was emitted, not as it would appear today.

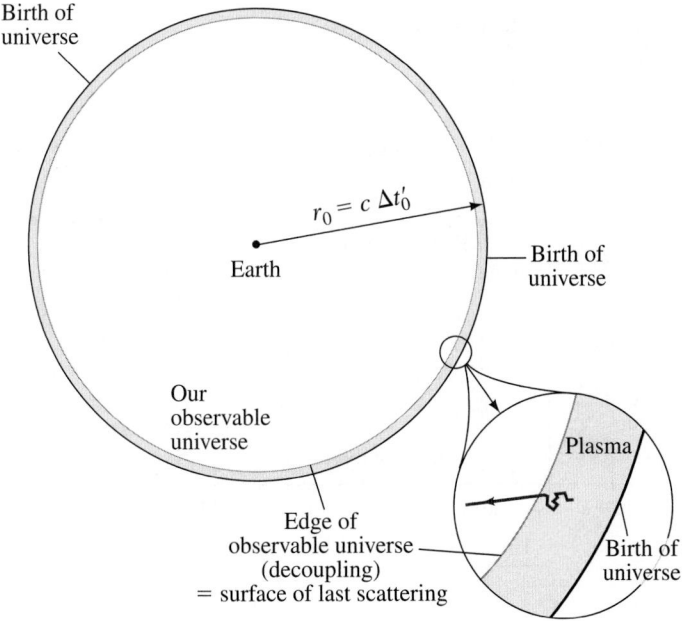

FIGURE 33–27 When we look out from the Earth, we look back in time. Any other observer in the universe would see more or less the same thing. The farther an object is from us, the longer ago the light we see had to have left it. We cannot see quite as far as the Big Bang; we can see only as far as the "surface of last scattering," which radiated the CMB. The insert on the lower right shows the earliest 380,000 years of the universe when it was opaque: a photon is shown scattering many times and then (at decoupling, 380,000 yr after the birth of the universe) becoming free to travel in a straight line. If this photon wasn't heading our way when "liberated," many others were. Galaxies are not shown, but would be concentrated close to Earth in this diagram because they were created relatively recently. *Note:* This diagram is not a normal map. Maps show a section of the world as might be seen all *at a given time*. This diagram shows space (like a map), but each point is *not* at the same time. The light coming from a point a distance r from Earth took a time $\Delta t' = r/c$ to reach Earth, and thus shows an event that took place long ago, a time $\Delta t' = r/c$ in the past, which we call its "lookback time." The universe began $\Delta t'_0 = 13.8\,\text{Gyr}$ ago.

The Observable Universe

Figure 33–27 can easily be misinterpreted: it is not a picture of the universe at a given instant, but is intended to suggest how we look out in all directions from our observation point (the Earth, or near it). Be careful not to think that the birth of the universe took place in a circle or a sphere surrounding us as if Fig. 33–27 were a photo taken at a given moment. What Fig. 33–27 does show is what we can see, the *observable universe*. Better yet, it shows the *most* we could see.

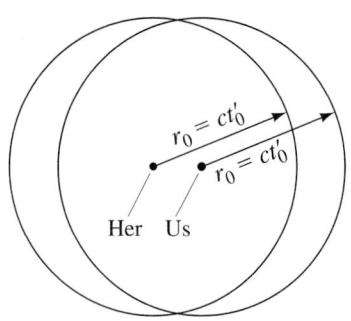

FIGURE 33–28 Two observers, on widely separated galaxies, have different horizons, different observable universes.

We would undoubtedly be arrogant to think that we could see the entire universe. Indeed, theories assume that we cannot see everything, that the **entire universe** is greater than the **observable universe**, which is a sphere of radius $r_0 = ct_0$ centered on the observer, with t_0 being the age of the universe. We can never see further back than the time it takes light to reach us.

Consider, for example, an observer in another galaxy, very far from us, located to the left of our observation point in Fig. 33–27. That observer would not yet have seen light coming from the far right of the large circle in Fig. 33–27 that we see—it will take some time for that light to reach her. But she will have already, some time ago, seen the light coming from the left that we are seeing now. In fact, her observable universe, superimposed on ours, is suggested by Fig. 33–28.

The edge of our observable universe is called the **horizon**. We could, in principle, see as far as the horizon, but not beyond it. An observer in another galaxy, far from us, will have a different horizon.

33–7 The Standard Cosmological Model: Early History of the Universe

In the last decade or two, a convincing theory of the origin and evolution of the universe has been developed, now called the **Standard Cosmological Model**. Part of this theory is based on recent theoretical and experimental advances in elementary particle physics, and part from observations of the universe including COBE, WMAP, and Planck. Indeed, cosmology and elementary particle physics have cross-fertilized to a surprising extent.

Let us go back to the earliest of times—as close as possible to the Big Bang—and follow a Standard Model theoretical scenario of events as the universe expanded and cooled after the Big Bang. Initially we talk of extremely small time intervals as well as extremely high temperatures, far higher than any temperature in the universe today. Figure 33–29 is a compressed graphical representation of the events, and it may be helpful to consult it as we go along.

FIGURE 33–29 Compressed graphical representation of the development of the universe after the Big Bang, according to modern cosmology. [The time scale is mostly logarithmic (each factor of 10 in time gets equal treatment), except at the start (there can be no $t = 0$ on a log scale), and during inflation (to save space).] The vertical height is a rough indication of the size of the universe, mainly to suggest expansion of the universe: Early on (after inflation) the universe is decelerating in its expansion (note slight downward curve); but for the last 7 Gyr (= thin strip on right) it has been accelerating, so the size line on the top curves upward at upper right.

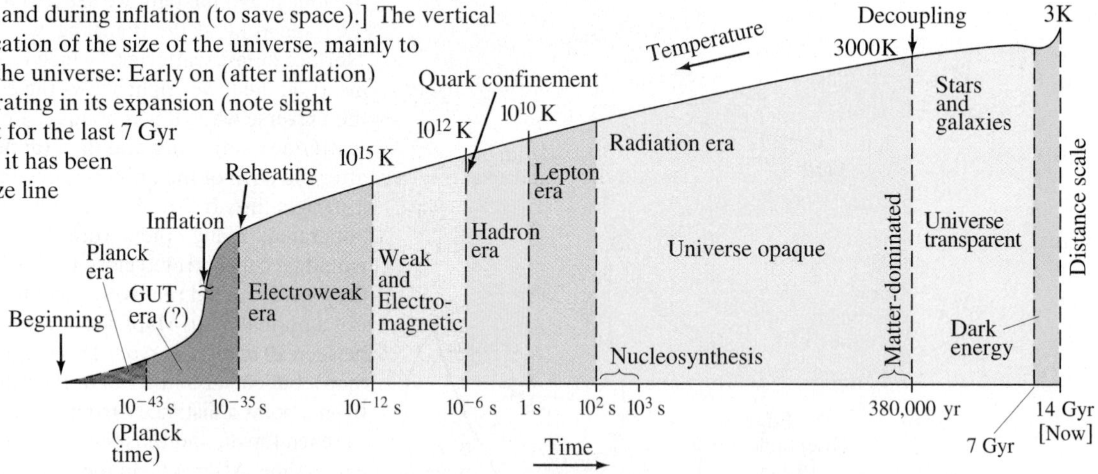

Quark confinement 10^{10} K
10^{12} K
10^{15} K
Reheating
Inflation
Temperature
Decoupling 3K
3000K
Radiation era
Lepton era
Hadron era
Stars and galaxies
Universe transparent
Matter-dominated
Planck era
GUT era (?)
Electroweak era
Weak and Electromagnetic
Universe opaque
Dark energy
Beginning
Nucleosynthesis
Distance scale

10^{-43} s (Planck time) 10^{-35} s 10^{-12} s 10^{-6} s 1 s 10^2 s 10^3 s 380,000 yr 14 Gyr [Now] 7 Gyr

Time

The History

We begin at a time only a minuscule fraction of a second after the "beginning" of the universe, 10^{-43} s. This time (sometimes referred to as the **Planck time**) is an unimaginably short time, and predictions can be only speculative. Earlier, we can say nothing because we do not have a theory of quantum gravity which would be needed for the incredibly high densities and temperatures during this "Planck era."

The first theories of the Big Bang assumed the universe was extremely hot in the beginning, maybe 10^{32} K, and then gradually cooled down while expanding. In those first moments after 10^{-43} s, the four forces of nature were thought to be united—there was only one force (Chapter 32, Fig. 32–22). Then a kind of "phase transition" would have occurred during which the gravitational force would have "condensed out" as a separate force. This and subsequent phase transitions, as shown in Fig. 32–22, are analogous to phase transitions water undergoes as it cools from a gas condensing into a liquid, and with further cooling freezes into ice.[†] The *symmetry* of the four forces would have been broken leaving the strong, weak, and electromagnetic forces still unified, and the universe would have entered the **grand unified era** (GUT—see Section 32–11).

This scenario of a *hot* Big Bang is now doubted by some important theorists, such as Andreí Linde, whose theories suggest the universe was much cooler at the Planck time. But what happened next to the universe, though very strange, is accepted by most cosmologists: a brilliant idea, suggested by Linde and Alan Guth in the early 1980s, proposed that the universe underwent an incredible exponential expansion, increasing in size by a factor of 10^{30} or maybe much more, in a tiny fraction of a second, perhaps 10^{-35} s or 10^{-32} s. The usefulness of this **inflationary scenario** is that it solved major problems with earlier Big Bang models, such as explaining why the universe is flat, as well as the thermal equilibrium to provide the nearly uniform CMB, as discussed below.

When inflation ended, whatever energy caused it then ended up being transformed into elementary particles with very high kinetic energy, corresponding to very high temperature (Eq. 13–8, $\overline{KE} = \frac{3}{2}kT$). That process is referred to as **reheating**, and the universe was now a "soup" of leptons, quarks, and other particles. We can think of this "soup" as a plasma of particles and antiparticles, as well as photons—all in roughly equal numbers—colliding with one another frequently and exchanging energy.

The temperature of the universe at the end of inflation was much lower than that expected by the hot Big Bang theory. But it would have been high enough so that the weak and electromagnetic forces were unified into a single force, and this stage of the universe is sometimes called the **electroweak era**. Approximately 10^{-12} s after the Big Bang, the temperature dropped to about 10^{15} K corresponding to randomly moving particles with an average kinetic energy KE of about 100 GeV (see Eq. 13–8):

$$\text{KE} \approx kT \approx \frac{(1.4 \times 10^{-23}\,\text{J/K})(10^{15}\,\text{K})}{1.6 \times 10^{-19}\,\text{J/eV}} \approx 10^{11}\,\text{eV} = 100\,\text{GeV}.$$

(As an estimate, we usually ignore the factor $\frac{3}{2}$ in Eq. 13–8.) At that time, symmetry between weak and electromagnetic forces would have broken down, and the weak force separated from the electromagnetic.

As the universe cooled down to about 10^{12} K (KE $\approx$ 100 MeV), approximately 10^{-6} s after the Big Bang, quarks stop moving freely and begin to "condense" into more normal particles: nucleons and the other hadrons and their antiparticles. With this **confinement of quarks**, the universe entered the **hadron era**. But it did not last long. Very soon the vast majority of hadrons disappeared. To see why, let us focus on the most familiar hadrons: nucleons and their antiparticles. When the average kinetic energy of particles was somewhat higher than 1 GeV, protons, neutrons, and their antiparticles were continually being created out of the energies of collisions involving photons and other particles, such as

$$\text{photons} \rightarrow p + \bar{p}$$
$$\rightarrow n + \bar{n}.$$

But just as quickly, particles and antiparticles would annihilate: for example

$$p + \bar{p} \rightarrow \text{photons or leptons.}$$

So the processes of creation and annihilation of nucleons were in equilibrium. The numbers of nucleons and antinucleons were high—roughly as many as there were electrons, positrons, or photons. But as the universe expanded and cooled, and the average kinetic energy of particles dropped below about 1 GeV, which is the minimum energy needed in a typical collision to create nucleons and anti-nucleons (about 940 MeV each), the process of nucleon creation could not continue.

[†]It may be interesting to point out that this story of origins here bears some resemblance to ancient accounts (nonscientific) that mention the "void," "formless wasteland" (or "darkness over the deep"), "abyss," "divide the waters" (= a phase transition?), not to mention the sudden appearance of light.

Annihilation could continue, however, with antinucleons annihilating nucleons, until almost no nucleons were left. But not quite zero. Somehow we need to explain our present world of matter (nucleons and electrons) with very little antimatter in sight.

To explain our world of matter, we might suppose that earlier in the universe, after the inflationary period, a slight excess of quarks over antiquarks was formed.[†] This would have resulted in a slight excess of nucleons over antinucleons. And it is these "leftover" nucleons that we are made of today. The excess of nucleons over antinucleons was probably about one part in 10^9. During the hadron era, there should have been about as many nucleons as photons. After it ended, the "leftover" nucleons thus numbered only about one nucleon per 10^9 photons, and this ratio has persisted to this day. Protons, neutrons, and all other heavier particles were thus tremendously reduced in number by about 10^{-6} s after the Big Bang. The lightest hadrons, the pions, soon disappeared, about 10^{-4} s after the Big Bang; because they are the lightest mass hadrons (140 MeV), pions were the last hadrons able to be created as the temperature (and average kinetic energy) dropped. Lighter particles, including electrons and neutrinos, were the dominant form of matter, and the universe entered the **lepton era**.

By the time the first full second had passed (clearly the most eventful second in history!), the universe had cooled to about 10 billion degrees, 10^{10} K. The average kinetic energy was about 1 MeV. This was still sufficient energy to create electrons and positrons and balance their annihilation reactions, since their masses correspond to about 0.5 MeV. So there were about as many e^+ and e^- as there were photons. But within a few more seconds, the temperature had dropped sufficiently so that e^+ and e^- could no longer be formed. Annihilation $(e^+ + e^- \rightarrow \text{photons})$ continued. And, like nucleons before them, electrons and positrons all but disappeared from the universe—except for a slight excess of electrons over positrons (later to join with nuclei to form atoms). Thus, about $t = 10$ s after the Big Bang, the universe entered the **radiation era** (Fig. 33–29). Its major constituents were photons and neutrinos. But the neutrinos, partaking only in the weak force, rarely interacted. So the universe, until then experiencing significant amounts of energy in matter and in radiation, now became **radiation-dominated**: much more energy was contained in radiation than in matter, a situation that would last more than 50,000 years.

FIGURE 33–29 (Repeated.) Compressed graphical representation of the development of the universe after the Big Bang, according to modern cosmology.

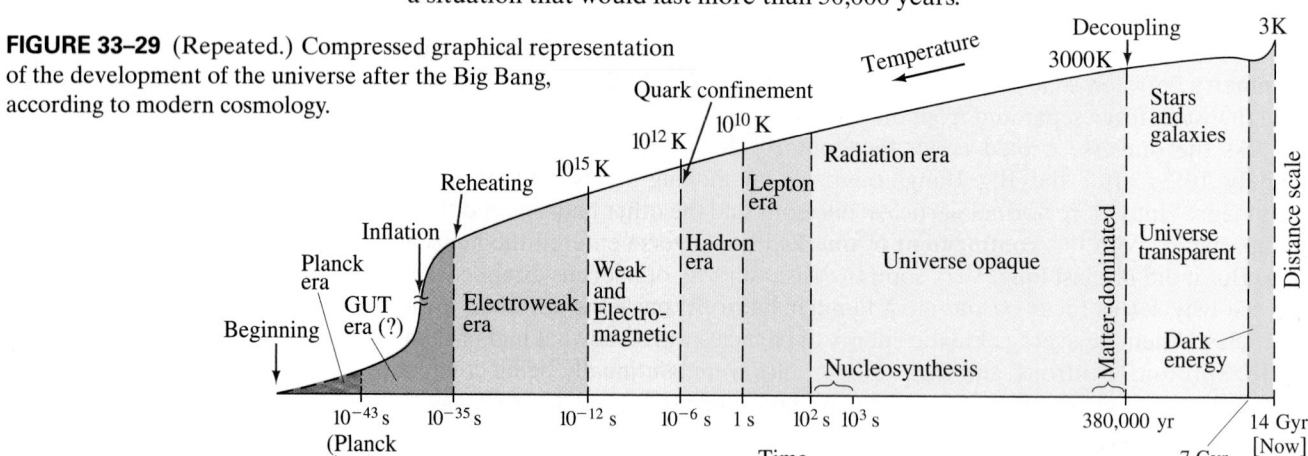

Meanwhile, during the next few minutes, crucial events were taking place. Beginning about 2 or 3 minutes after the Big Bang, nuclear fusion began to occur. The temperature had dropped to about 10^9 K, corresponding to an average kinetic energy $\overline{\text{KE}} \approx 100$ keV, where nucleons could strike each other and be able to fuse (Section 31–3), but now cool enough so newly formed nuclei would not be immediately broken apart by subsequent collisions. Deuterium, helium, and very tiny amounts of lithium nuclei were made. But the universe was cooling too quickly, and larger nuclei were not made. After only a few minutes, probably not even a quarter of an hour after the Big Bang, the temperature dropped far enough that nucleosynthesis stopped, not to start again for millions of years (in stars).

[†]Why this could have happened is a question for which we are seeking an answer today.

Thus, after the first quarter hour or so of the universe, matter consisted mainly of bare nuclei of hydrogen (about 75%) and helium (about 25%)[†] as well as electrons. But radiation (photons) continued to dominate.

Our story is almost complete. The next important event is thought to have occurred 380,000 years later. The universe had expanded to about $\frac{1}{1000}$ of its present scale, and the temperature had cooled to about 3000 K. The average kinetic energy of nuclei, electrons, and photons was less than an electron volt. Since ionization energies of atoms are on the order of eV, then as the temperature dropped below this point, electrons could orbit the bare nuclei and remain there (without being ejected by collisions), thus forming atoms. This period is often called the **recombination** epoch (a misnomer since electrons had never before been combined with nuclei to form atoms). With the disappearance of free electrons and the birth of atoms, the photons—which had been continually scattering from the free electrons—now became free to spread throughout the universe. As mentioned in the previous Section, we say that the photons became **decoupled** from matter. Thus *decoupling* occurred at *recombination*. The energy contained in radiation had been decreasing (lengthening in wavelength as the universe expanded); and at about $t = 56,000\,\text{yr}$ (even before decoupling) the energy contained in matter became dominant over radiation. The universe was said to have become **matter-dominated** (marked on Fig. 33–29). As the universe continued to expand, the electromagnetic radiation cooled further, to 2.7 K today, forming the cosmic microwave background radiation we detect from everywhere in the universe.

After the birth of atoms, then stars and galaxies could begin to form: by self-gravitation around mass concentrations (inhomogeneities). Stars began to form about 200 million years after the Big Bang, galaxies after almost 10^9 years. The universe continued to evolve until today, some 14 billion years after it started.

* * *

This scenario, like other scientific models, cannot be said to be "proven." Yet this model is remarkably effective in explaining the evolution of the universe we live in, and makes predictions which can be tested against the next generation of observations.

A major event, and something only discovered recently, is that when the universe was about half as old as it is now (about 7 Gyr ago), its expansion began to accelerate. This was a big surprise because it was assumed the expansion of the universe would slow down due to gravitational attraction of all objects toward each other. This acceleration in the expansion of the universe is said to be due to "dark energy," as we discuss in Section 33–9. On the right in Fig. 33–29 is a narrow vertical strip that represents the most recent 7 billion years of the universe, during which *dark energy* seems to have dominated.

33–8 Inflation: Explaining Flatness, Uniformity, and Structure

The idea that the universe underwent a period of exponential inflation early in its life, expanding by a factor of 10^{30} or more (previous Section), was first put forth by Alan Guth and Andreí Linde. Many sophisticated models based on this general idea have since been proposed. The energy required for this wild expansion may have been due to fields somewhat like the Higgs field (Section 32–10). So far, the evidence for inflation is indirect; yet it is a feature of most viable cosmological models because it alone is able to provide natural explanations for several remarkable features of our universe.

[†]This Standard Model prediction of a 25% primordial production of helium agrees with what we observe today—the universe *does* contain about 25% He—and it is strong evidence in support of the Standard Big Bang Model. Furthermore, the theory says that 25% He abundance is fully consistent with there being three neutrino types, which is the number we observe. And it sets an upper limit of four to the maximum number of possible neutrino types. This is a striking example of the powerful connection between particle physics and cosmology.

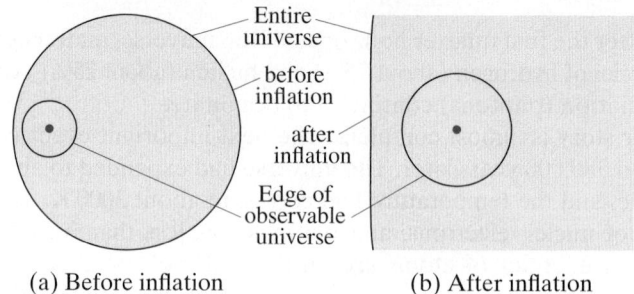

FIGURE 33–30 (a) Simple 2-D model of the entire universe; the observable universe is suggested by the small circle centered on us (blue dot). (b) Edge of entire universe is essentially flat after the 10^{30}-fold expansion during inflation.

(a) Before inflation   (b) After inflation

Flatness

First of all, our best measurements suggest that the universe is flat, that it has zero curvature. As scientists, we would like some reason for this remarkable result. To see how inflation explains flatness, consider a simple 2-dimensional model of the universe as we did earlier in Figs. 33–16 and 33–21. A circle in this 2-dimensional universe (= surface of a sphere, Fig. 33–30a) represents the *observable* universe as seen by an observer at the blue dot. A possible hypothesis is that inflation occurred over a time interval that very roughly doubled the age of the universe from, let us say, $t = 1 \times 10^{-35}$ s to $t = 2 \times 10^{-35}$ s. The size of the *observable* universe ($r = ct$) would have increased by a factor of two during inflation, while the radius of curvature of the *entire* universe increased by an enormous factor of 10^{30} or more. Thus the edge of our 2-D sphere representing the entire universe would have seemed flat to a high degree of precision, as shown in Fig. 33–30b. Even if the time of inflation was a factor of 10 or 100 (instead of 2), the expansion factor of 10^{30} or more would have blotted out any possibility of observing anything but a flat universe.

CMB Uniformity

Inflation also explains why the CMB is so uniform. Without inflation, the tiny universe at 10^{-35} s would not have been small enough for all parts of it to have been in contact and so reach the same temperature (information cannot travel faster than c). To see this, suppose that the currently observable universe came from a region of space about 1 cm in diameter at $t = 10^{-36}$ s, as per original Big Bang theory. In that 10^{-36} s, light could have traveled $d = ct = (3 \times 10^8 \, \text{m/s})(10^{-36} \, \text{s}) = 10^{-27}$ m, way too small for the opposite sides of a 1-cm-wide "universe" to have been in communication. But if that region had been 10^{30} times smaller $(= 10^{-32} \, \text{m})$, as proposed by the inflation model, there could have been contact and thermal equilibrium to produce the observed nearly uniform CMB. Inflation, by making the very early universe extremely small, assures that all parts of that region which is today's observable universe could have been in thermal equilibrium. And after inflation the universe could be large enough to give us today's observable universe.

Galaxy Seeds, Fluctuations, Magnetic Monopoles

Inflation also gives us a clue as to how the present structure of the universe (galaxies and clusters of galaxies) came about. We saw earlier that, according to the uncertainty principle, energy might be not conserved by an amount ΔE for a time $\Delta t \approx \hbar/\Delta E$. Forces, whether electromagnetic or other types, can undergo such tiny **quantum fluctuations** according to quantum theory, but they are so tiny they are not detectable unless magnified in some way. That is what inflation might have done: it could have magnified those fluctuations perhaps 10^{30} times in size, which would give us the density irregularities seen in the cosmic microwave background (WMAP, Fig. 33–26). That would be very nice, because the density variations we see in the CMB are what we believe were the seeds that later coalesced under gravity into galaxies and galaxy clusters, and our models fit the data extremely well.

Sometimes it is said that the quantum fluctuations occurred in the **vacuum state** or vacuum energy. This could be possible because the vacuum is no longer considered to be empty, as we discussed in Section 32–3 relative to positrons as holes in a negative energy sea of electrons. Indeed, the vacuum is thought to be filled with fields and particles occupying all the possible negative energy states.

Also, the virtual exchange particles that carry the forces, as discussed in Chapter 32, could leave their brief virtual states and actually become real as a result of the 10^{30} magnification of space (according to inflation) and the very short time over which it occurred ($\Delta t = \hbar/\Delta E$).

Inflation helps us too with the puzzle of why **magnetic monopoles** (Section 20–1) have never been observed, yet isolated magnetic poles may well have been copiously produced at the start. After inflation, they would have been so far apart that we have never stumbled on one.

Inflation may solve outstanding problems, but we may need new physics to understand how inflation occurred. Many predictions of inflationary theory have been confirmed by recent cosmological observations.

33–9 Dark Matter and Dark Energy

According to the Standard Big Bang Model, the universe is evolving and changing. Individual stars are being created, evolving, and then dying to become white dwarfs, neutron stars, or black holes. At the same time, the universe as a whole is expanding. One important question is whether the universe will continue to expand forever. Until the late 1990s, the universe was thought to be dominated by matter which interacts by gravity, and the fate of the universe was connected to the curvature of space-time (Section 33–4). If the universe had *negative* curvature, the expansion of the universe would never stop, although the rate of expansion would decrease due to the gravitational attraction of its parts. Such a universe would be *open* and infinite. If the universe is *flat* (no curvature), it would still be open and infinite but its expansion would slowly approach a zero rate. If the universe had *positive* curvature, it would be *closed* and finite; the effect of gravity would be strong enough that the expansion would eventually stop and the universe would begin to contract, collapsing back onto itself in a **big crunch**.

Critical Density

According to the above scenario (which does not include inflation or the recently discovered acceleration of the universe), the fate of the universe would depend on the average mass–energy density in the universe. For an average mass density greater than a critical value known as the **critical density**, estimated to be about

$$\rho_c \approx 10^{-26}\,\text{kg/m}^3$$

(i.e., a few nucleons/m³ on average throughout the universe), space-time would have a positive curvature and gravity would prevent expansion from continuing forever. Eventually (if $\rho > \rho_c$) gravity would pull the universe back into a big crunch. If instead the actual density was equal to the critical density, $\rho = \rho_c$, the universe would be flat and open, just barely expanding forever. If the actual density was less than the critical density, $\rho < \rho_c$, the universe would have negative curvature and would easily expand forever. See Fig. 33–31. Today we believe the universe is very close to flat. But recent evidence suggests the universe is expanding at an *accelerating* rate, as discussed below.

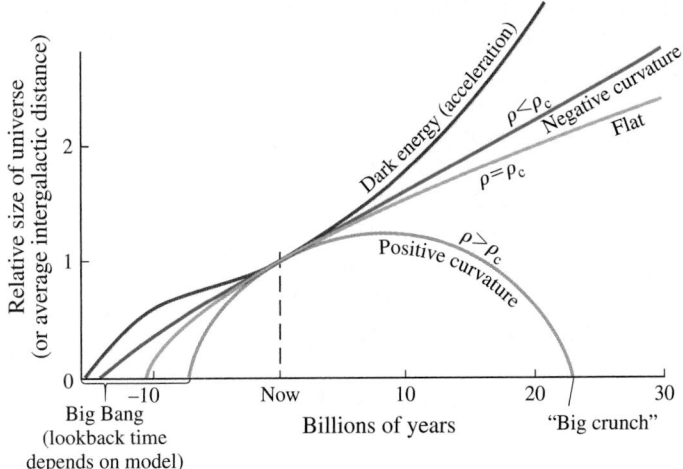

FIGURE 33–31 Three future possibilities for the universe, depending on the density ρ of ordinary matter, plus a fourth possibility that includes dark energy. Note that all curves have been chosen to have the same slope (= H_0, the Hubble parameter) right now. Looking back in time, the Big Bang occurs where each curve touches the horizontal (time) axis.

EXERCISE E Return to the Chapter-Opening Questions, page 947, and answer them again. Try to explain why you may have answered differently the first time.

Dark Matter

WMAP and other experiments have convinced scientists that the universe is flat and $\rho = \rho_c$. But this ρ cannot be only normal baryonic matter (atoms are 99.9% baryons—protons and neutrons—by weight). These recent experiments put the amount of normal baryonic matter in the universe at only about 5% of the critical density. What is the other 95%? There is strong evidence for a significant amount of nonluminous matter in the universe referred to as **dark matter**, which acts normally under gravity, but does not absorb or radiate light sufficiently to be visible. For example, observations of the rotation of galaxies suggest that they rotate as if they had considerably more mass than we can see. Recall from Chapter 5, Example 5–12, that for a satellite of mass m revolving around Earth (mass M)

$$m\frac{v^2}{r} = G\frac{mM}{r^2}$$

and hence $v = \sqrt{GM/r}$. If we apply this equation to stars in a galaxy, we see that their speed depends on galactic mass. Observations show that stars farther from the galactic center revolve much faster than expected if there is only the pull of visible matter, suggesting a great deal of invisible matter. Similarly, observations of the motion of galaxies within clusters also suggest that they have considerably more mass than can be seen. Furthermore, theory suggests that without dark matter, galaxies and stars probably would not have formed and would not exist. Dark matter seems to hold the universe together.

What might this nonluminous matter in the universe be? We don't know yet. But we hope to find out soon. It cannot be made of ordinary (baryonic) matter, so it must consist of some other sort of elementary particle, perhaps created at a very early time. Perhaps it is made up of previously undetected *weakly interacting massive particles* (**WIMPs**), possibly supersymmetric particles (Section 32–12) such as neutralinos. We are anxiously awaiting the results of intense searches for such particles, looking both at what arrives from far out in the cosmos with underground detectors[†], and by producing them in particle colliders (the LHC, Section 32–1).

Dark matter makes up roughly 25% of the mass–energy of the universe, according to the latest observations and models. Thus the total mass–energy is 25% dark matter plus 5% baryons for a total of about 30%, which does not bring ρ up to ρ_c. What is the other 70%? We are not sure about that either, but we have given it a name: "dark energy."

FIGURE 33–32 Saul Perlmutter, center, flanked by Adam G. Riess (left) and Brian P. Schmidt, at the Nobel Prize celebrations, December 2011.

Dark Energy—Cosmic Acceleration

In 1998, just before the turn of the millennium, two groups, one led by Saul Perlmutter and the other by Brian Schmidt and Adam Riess (Fig. 33–32), reported a huge surprise. Gravity was assumed to be the predominant force on a large scale in the universe, and it was thought that the expansion of the universe ought to be slowing down in time because gravity acts as an attractive force between objects. But measurements of Type Ia supernovae (our best standard candles—see Section 33–3) unexpectedly showed that very distant (high z) supernovae were dimmer than expected. That is, given their great distance d as determined from their low brightness, their speed v as determined from the measured z was less than expected according to Hubble's law. This result suggests that nearer galaxies are moving away from us relatively faster than those very distant ones, meaning the expansion of the universe in more recent epochs has sped up.

[†]In deep mines and under mountains to block out most other particles.

This **acceleration** in the expansion of the universe (in place of the expected deceleration due to gravitational attraction between masses) seems to have begun roughly 7 billion years ago (7 Gyr, which would be about halfway back to what we call the Big Bang).

What could be causing the universe to accelerate in its expansion, against the attractive force of gravity? Does our understanding of gravity need to be revised? We don't know the answers to these questions. Many scientists say dark energy is the biggest mystery facing physical science today. There are several speculations. But somehow it seems to have a long-range *repulsive* effect on space, like a negative gravity, causing objects to speed away from each other ever faster. Whatever it is, it has been given the name **dark energy**.

One idea is a sort of quantum field given the name **quintessence**. Another possibility suggests an energy latent in space itself (**vacuum energy**) and relates to an aspect of General Relativity known as the **cosmological constant** (symbol Λ). When Einstein developed his equations, he found that they offered no solutions for a static universe. In those days (1917) it was thought the universe was static—unchanging and everlasting. Einstein added an arbitrary constant (Λ) to his equations to provide solutions for a static universe.[†] A decade later, when Hubble showed us an expanding universe, Einstein discarded his cosmological constant as no longer needed ($\Lambda = 0$). But today, measurements are consistent with dark energy being due to a nonzero cosmological constant, although further measurements are needed to see subtle differences among theories.

There is increasing evidence that the effects of some form of dark energy are very real. Observations of the CMB, supernovae, and large-scale structure (Section 33–10) agree well with theories and computer models when they input dark energy as providing about 70% of the mass–energy in the universe, and when the total mass–energy density equals the critical density ρ_c.

Today's best estimate of how the mass–energy in the universe is distributed is approximately (see also Fig. 33–33):

> 70% dark energy
>
> 30% matter, subject to the known gravitational force.
>
>> Of this 30%, about
>>
>> 25% is dark matter
>>
>> 5% is baryons (what atoms are made of); of this 5% only $\frac{1}{10}$ is readily visible matter—stars and galaxies (that is, 0.5% of the total); the other $\frac{9}{10}$ of ordinary matter, which is not visible, is mainly gaseous plasma.

It is remarkable that only 0.5% of all the mass–energy in the universe is visible as stars and galaxies.

The idea that the universe is dominated by completely unknown forms of matter and energy seems bizarre. Nonetheless, the ability of our present model to precisely explain observations of the CMB anisotropy, cosmic expansion, and large-scale structure (next Section) presents a compelling case.

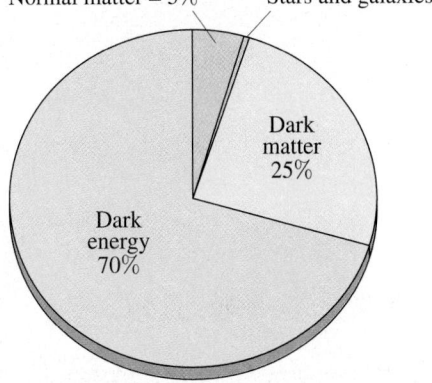

FIGURE 33–33 Portions of total mass–energy in the universe (approximate).

33–10 Large-Scale Structure of the Universe

The beautiful WMAP pictures of the sky (Fig. 33–26) show small but significant inhomogeneities in the temperature of the cosmic microwave background (CMB). These anisotropies reflect compressions and expansions in the primordial plasma just before decoupling (Fig. 33–29), from which galaxies and clusters of galaxies formed. Analyses of the irregularities in the CMB using mammoth computer

[†]It seems strange that Einstein and other scientists believed in a static universe. The ancients, including the Roman Lucretius argued against it: *If there was no birth-time of earth and heaven and they have been from everlasting, why before the Theban war and the destruction of Troy have not other poets as well sung other themes?* [The reference is to Homer being the oldest known writings.] The ancient Hebrews also argued for a beginning (like our Big Bang): see Genesis.

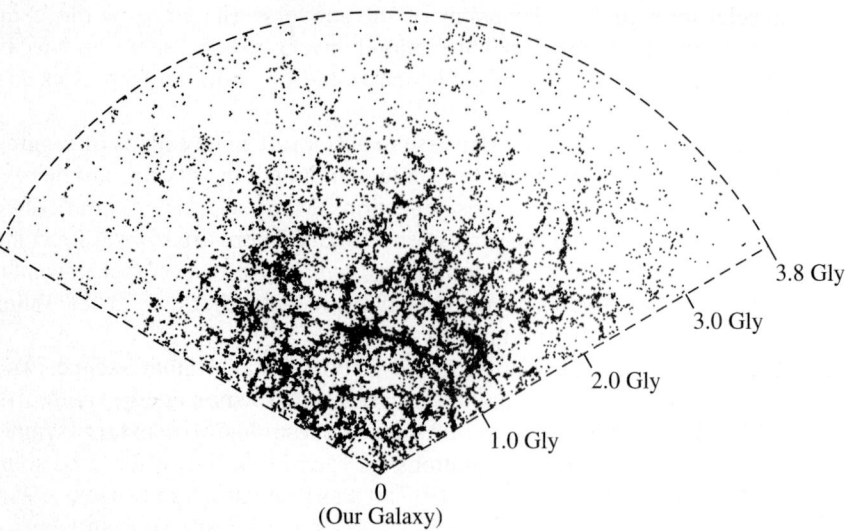

FIGURE 33–34 Distribution of some 50,000 galaxies in a 2.5° slice through almost half of the sky above the equator, as measured by the Sloan Digital Sky Survey (SDSS). Each dot represents a galaxy. The distance from us is obtained from the redshift and Hubble's law, and is given in units of 10^9 light-years (Gly). The point 0 represents us, our observation point. This diagram may seem to put us at the center, but remember that at greater distances, fewer galaxies are bright enough to be detected, thus resulting in an apparent thinning out of galaxies. Note the "walls" and "voids" of galaxies.

3.8 Gly

3.0 Gly

2.0 Gly

1.0 Gly

0
(Our Galaxy)

simulations predict a large-scale distribution of galaxies very similar to what is seen today (Fig. 33–34). These simulations are very successful if they contain dark energy and dark matter; and the dark matter needs to be *cold* (slow speed—think of Eq. 13–8, $\frac{1}{2}m\overline{v^2} = \frac{3}{2}kT$ where T is temperature), rather than "hot" dark matter such as neutrinos which move at or very near the speed of light. Indeed, the modern **cosmological model** is called the Λ**CDM** model, where lambda (Λ) stands for the cosmological constant, and CDM is **cold dark matter**.

Cosmologists have gained substantial confidence in this cosmological model from such a precise fit between observations and theory. They can also extract very precise values for cosmological parameters which previously were only known with low accuracy. The CMB is such an important cosmological observable that every effort is being made to extract all of the information it contains. A new generation of ground, balloon, and satellite experiments is observing the CMB with greater resolution and sensitivity. They may detect interaction of **gravity waves** (produced in the inflationary epoch) with the CMB and thereby provide direct evidence for cosmic inflation, and also provide information about elementary particle physics at energies far beyond the reach of man-made accelerators and colliders.

33–11 Finally . . .

When we look up into the night sky, we see stars; and with the best telescopes, we see galaxies and the exotic objects we discussed earlier, including rare supernovae. But even with our best instruments we do not see the processes going on inside stars and supernovae that we hypothesized (and believe). We are dependent on brilliant theorists who come up with viable theories and verifiable models. We depend on complicated computer models whose parameters are varied until the outputs compare favorably with our observations and analyses of WMAP and other experiments. And we now have a surprisingly precise idea about some aspects of our universe: it is flat, it is about 14 billion years old, it contains only 5% "normal" baryonic matter (for atoms), and so on.

The questions raised by cosmology are difficult and profound, and may seem removed from everyday "reality." We can always say, "the Sun is shining, it's going to shine on for an unimaginably long time, all is well." Nonetheless, the questions of cosmology are deep ones that fascinate the human intellect. One aspect that is especially intriguing is this: calculations on the formation and evolution of the universe have been performed that deliberately varied the values—just slightly—of certain fundamental physical constants. The result?

A universe in which life as we know it could not exist. [For example, if the difference in mass between a proton and a neutron were zero, or less than the mass of the electron, $0.511\ \text{MeV}/c^2$, there would be no atoms: electrons would be captured by protons to make neutrons.] Such results have contributed to a philosophical idea called the **anthropic principle**, which says that if the universe were even a little different than it is, we could not be here. We physicists are trying to find out if there are some undiscovered fundamental laws that determined those conditions that allowed us to exist. A poet might say that the universe is exquisitely tuned, almost as if to accommodate us.

Summary

The night sky contains myriads of stars including those in the Milky Way, which is a "side view" of our **Galaxy** looking along the plane of the disk. Our Galaxy includes over 10^{11} stars. Beyond our Galaxy are billions of other galaxies.

Astronomical distances are measured in **light-years** ($1\ \text{ly} \approx 10^{13}\ \text{km}$). The nearest star is about 4 ly away and the nearest large galaxy is 2 million ly away. Our Galactic disk has a diameter of about 100,000 ly. [Distances are sometimes specified in **parsecs**, where 1 parsec = 3.26 ly.]

Stars are believed to begin life as collapsing masses of gas (protostars), largely hydrogen. As they contract, they heat up (potential energy is transformed to kinetic energy). When the temperature reaches about 10 million degrees, nuclear fusion begins and forms heavier elements (**nucleosynthesis**), mainly helium at first. The energy released during these reactions heats the gas so its outward pressure balances the inward gravitational force, and the young star stabilizes as a **main-sequence** star. The tremendous luminosity of stars comes from the energy released during these thermonuclear reactions. After billions of years, as helium is collected in the core and hydrogen is used up, the core contracts and heats further. The outer envelope expands and cools, and the star becomes a **red giant** (larger diameter, redder color).

The next stage of stellar evolution depends on the mass of the star, which may have lost much of its original mass as its outer envelope escaped into space. Stars of residual mass less than about 1.4 solar masses cool further and become **white dwarfs**, eventually fading and going out altogether. Heavier stars contract further due to their greater gravity: the density approaches nuclear density, the huge pressure forces electrons to combine with protons to form neutrons, and the star becomes essentially a huge nucleus of neutrons. This is a **neutron star**, and the energy released during its final core collapse is believed to produce **supernova** explosions. If the star is very massive, it may contract even further and form a **black hole**, which is so dense that no matter or light can escape from it.

In the **general theory of relativity**, the **equivalence principle** states that an observer cannot distinguish acceleration from a gravitational field. Said another way, gravitational and inertial masses are the same. The theory predicts gravitational bending of light rays to a degree consistent with experiment. Gravity is treated as a curvature in space and time, the curvature being greater near massive objects. The universe as a whole may be curved. With sufficient mass, the curvature of the universe would be positive, and the universe is *closed* and *finite*; otherwise, it would be *open* and *infinite*. Today we believe the universe is **flat**.

Distant galaxies display a **redshift** in their spectral lines, originally interpreted as a Doppler shift. The universe is observed to be **expanding**, its galaxies racing away from each other at speeds (v) proportional to the distance (d) between them:

$$v = H_0 d, \qquad (33\text{--}4)$$

which is known as **Hubble's law** (H_0 is the **Hubble parameter**). This expansion of the universe suggests an explosive origin, the **Big Bang**, which occurred about 13.8 billion years ago. It is not like an ordinary explosion, but rather an expansion of space itself.

The **cosmological principle** assumes that the universe, on a large scale, is homogeneous and isotropic.

Important evidence for the Big Bang model of the universe was the discovery of the **cosmic microwave background** radiation (CMB), which conforms to a blackbody radiation curve at a temperature of 2.725 K.

The **Standard Model** of the Big Bang provides a possible scenario as to how the universe developed as it expanded and cooled after the Big Bang. Starting at 10^{-43} seconds after the Big Bang, according to this model, the universe underwent a brief but rapid exponential expansion, referred to as **inflation**. Shortly thereafter, quarks were **confined** into hadrons (the **hadron era**). About $10^{-4}\ \text{s}$ after the Big Bang, the majority of hadrons disappeared, having combined with anti-hadrons, producing photons, leptons, and energy, leaving mainly photons and leptons to freely move, thus introducing the **lepton era**. By the time the universe was about 10 s old, the electrons too had mostly disappeared, having combined with their antiparticles; the universe was **radiation-dominated**. A couple of minutes later, nucleosynthesis began, but lasted only a few minutes. It then took almost four hundred thousand years before the universe was cool enough for electrons to combine with nuclei to form atoms (**recombination**). Photons, up to then continually being scattered off of free electrons, could now move freely—they were **decoupled** from matter and the universe became transparent. The background radiation had expanded and cooled so much that its total energy became less than the energy in matter, and **matter dominated** increasingly over radiation. Then stars and galaxies formed, producing a universe not much different than it is today—some 14 billion years later.

Recent observations indicate that the universe is essentially flat, that it contains an as-yet unknown type of **dark matter**, and that it is dominated by a mysterious **dark energy** which exerts a sort of negative gravity causing the expansion of the universe to accelerate. The total contributions of baryonic (normal) matter, dark matter, and dark energy sum up to the **critical density**.

Questions

1. The Milky Way was once thought to be "murky" or "milky" but is now considered to be made up of point sources. Explain.

2. A star is in equilibrium when it radiates at its surface all the energy generated in its core. What happens when it begins to generate more energy than it radiates? Less energy? Explain.

3. Describe a red giant star. List some of its properties.

4. Does the H–R diagram directly reveal anything about the core of a star?

5. Why do some stars end up as white dwarfs, and others as neutron stars or black holes?

6. If you were measuring star parallaxes from the Moon instead of Earth, what corrections would you have to make? What changes would occur if you were measuring parallaxes from Mars?

7. *Cepheid variable* stars change in luminosity with a typical period of several days. The period has been found to have a definite relationship with the average intrinsic luminosity of the star. How could these stars be used to measure the distance to galaxies?

8. What is a geodesic? What is its role in General Relativity?

9. If it were discovered that the redshift of spectral lines of galaxies was due to something other than expansion, how might our view of the universe change? Would there be conflicting evidence? Discuss.

10. Almost all galaxies appear to be moving away from us. Are we therefore at the center of the universe? Explain.

11. If you were located in a galaxy near the boundary of our observable universe, would galaxies in the direction of the Milky Way appear to be approaching you or receding from you? Explain.

12. Compare an explosion on Earth to the Big Bang. Consider such questions as: Would the debris spread at a higher speed for more distant particles, as in the Big Bang? Would the debris come to rest? What type of universe would this correspond to, open or closed?

13. If nothing, not even light, escapes from a black hole, then how can we tell if one is there?

14. The Earth's age is often given as about 4.6 billion years. Find that time on Fig. 33–29. Modern humans have lived on Earth on the order of 200,000 years. Where is that on Fig. 33–29?

15. Why were atoms, as opposed to bare nuclei, unable to exist until hundreds of thousands of years after the Big Bang?

16. (*a*) Why are Type Ia supernovae so useful for determining the distances of galaxies? (*b*) How are their distances actually measured?

17. Under what circumstances would the universe eventually collapse in on itself?

18. (*a*) Why did astronomers expect that the expansion rate of the universe would be decreasing (decelerating) with time? (*b*) How, in principle, could astronomers hope to determine whether the universe used to expand faster than it does now?

MisConceptual Questions

1. Which one of the following is *not* expected to occur on an H–R diagram during the lifetime of a single star?
   (*a*) The star will move off the main sequence toward the upper right of the diagram.
   (*b*) Low-mass stars will become white dwarfs and end up toward the lower left of the diagram.
   (*c*) The star will move along the main sequence from one place to another.
   (*d*) All of the above.

2. When can parallax be used to determine the approximate distance from the Earth to a star?
   (*a*) Only during January and July.
   (*b*) Only when the star's distance is relatively small.
   (*c*) Only when the star's distance is relatively large.
   (*d*) Only when the star appears to move directly toward or away from the Earth.
   (*e*) Only when the star is the Sun.
   (*f*) Always.
   (*g*) Never.

3. Observations show that all galaxies tend to move away from Earth, and that more distant galaxies move away from Earth at faster velocities than do galaxies closer to the Earth. These observations imply that
   (*a*) the Earth is the center of the universe.
   (*b*) the universe is expanding.
   (*c*) the expansion of the universe will eventually stop.
   (*d*) All of the above.

4. Which process results in a tremendous amount of energy being emitted by the Sun?
   (*a*) Hydrogen atoms burn in the presence of oxygen— that is, hydrogen atoms oxidize.
   (*b*) The Sun contracts, decreasing its gravitational potential energy.
   (*c*) Protons in hydrogen atoms fuse, forming helium nuclei.
   (*d*) Radioactive atoms such as uranium, plutonium, and cesium emit gamma rays with high energy.
   (*e*) None of the above.

5. Which of the following methods can be used to find the distance from us to a star outside our galaxy? Choose all that apply.
   (*a*) Parallax.
   (*b*) Using luminosity and temperature from the H–R diagram and measuring the apparent brightness.
   (*c*) Using supernova explosions as a "standard candle."
   (*d*) Redshift in the line spectra of elements and compounds.

6. The history of the universe can be determined by observing astronomical objects at various (large) distances from the Earth. This method of discovery works because
   (*a*) time proceeds at different rates in different regions of the universe.
   (*b*) light travels at a finite speed.
   (*c*) matter warps space.
   (*d*) older galaxies are farther from the Earth than are younger galaxies.

7. Where did the Big Bang occur?
   (a) Near the Earth.
   (b) Near the center of the Milky Way Galaxy.
   (c) Several billion light-years away.
   (d) Throughout all space.
   (e) Near the Andromeda Galaxy.

8. When and how were virtually all of the elements of the Periodic Table formed?
   (a) In the very early universe a few seconds after the Big Bang.
   (b) At the centers of stars during their main-sequence phases.
   (c) At the centers of stars during novae.
   (d) At the centers of stars during supernovae.
   (e) On the surfaces of planets as they cooled and hardened.

9. We know that there must be dark matter in the universe because
   (a) we see dark dust clouds.
   (b) we see that the universe is expanding.
   (c) we see that stars far from the galactic center are moving faster than can be explained by visible matter.
   (d) we see that the expansion of the universe is accelerating.

10. Acceleration of the universe's expansion rate is due to
    (a) the repulsive effect of dark energy.
    (b) the attractive effect of dark matter.
    (c) the attractive effect of gravity.
    (d) the thermal expansion of stellar cores.

For assigned homework and other learning materials, go to the MasteringPhysics website.

Problems

33–1 to 33–3 Stars, Galaxies, Stellar Evolution, Distances

1. (I) The parallax angle of a star is 0.00029°. How far away is the star?

2. (I) A star exhibits a parallax of 0.27 seconds of arc. How far away is it?

3. (I) If one star is twice as far away from us as a second star, will the parallax angle of the farther star be greater or less than that of the nearer star? By what factor?

4. (II) What is the relative brightness of the Sun as seen from Jupiter, as compared to its brightness from Earth? (Jupiter is 5.2 times farther from the Sun than the Earth is.)

5. (II) When our Sun becomes a red giant, what will be its average density if it expands out to the orbit of Mercury $(6 \times 10^{10} \text{ m}$ from the Sun)?

6. (II) We saw earlier (Chapter 14) that the rate energy reaches the Earth from the Sun (the "solar constant") is about $1.3 \times 10^3 \text{ W/m}^2$. What is (a) the apparent brightness b of the Sun, and (b) the intrinsic luminosity L of the Sun?

7. (II) Estimate the angular width that our Galaxy would subtend if observed from the nearest galaxy to us (Table 33–1). Compare to the angular width of the Moon from Earth.

8. (II) Assuming our Galaxy represents a good average for all other galaxies, how many stars are in the observable universe?

9. (II) Calculate the density of a white dwarf whose mass is equal to the Sun's and whose radius is equal to the Earth's. How many times larger than Earth's density is this?

10. (II) A neutron star whose mass is 1.5 solar masses has a radius of about 11 km. Calculate its average density and compare to that for a white dwarf (Problem 9) and to that of nuclear matter.

*11. (II) A star is 56 pc away. What is its parallax angle? State (a) in seconds of arc, and (b) in degrees.

*12. (II) What is the parallax angle for a star that is 65 ly away? How many parsecs is this?

*13. (II) A star is 85 pc away. How long does it take for its light to reach us?

14. (III) Suppose two stars of the same apparent brightness b are also believed to be the same size. The spectrum of one star peaks at 750 nm whereas that of the other peaks at 450 nm. Use Wien's law and the Stefan-Boltzmann equation (Eq. 14–6) to estimate their relative distances from us. [*Hint:* See Examples 33–4 and 33–5.]

15. (III) Stars located in a certain cluster are assumed to be about the same distance from us. Two such stars have spectra that peak at $\lambda_1 = 470 \text{ nm}$ and $\lambda_2 = 720 \text{ nm}$, and the ratio of their apparent brightness is $b_1/b_2 = 0.091$. Estimate their relative sizes (give ratio of their diameters) using Wien's law and the Stefan-Boltzmann equation, Eq. 14–6.

33–4 General Relativity, Gravity and Curved Space

16. (I) Show that the Schwarzschild radius for Earth is 8.9 mm.

17. (II) What is the Schwarzschild radius for a typical galaxy (like ours)?

18. (II) What mass will give a Schwarzschild radius equal to that of the hydrogen atom in its ground state?

19. (II) What is the maximum sum-of-the-angles for a triangle on a sphere?

20. (II) Describe a triangle, drawn on the surface of a sphere, for which the sum of the angles is (a) 359°, and (b) 179°.

21. (III) What is the apparent deflection of a light beam in an elevator (Fig. 33–13) which is 2.4 m wide if the elevator is accelerating downward at 9.8 m/s^2?

33–5 Redshift, Hubble's Law

22. (I) The redshift of a galaxy indicates a recession velocity of 1850 km/s. How far away is it?

23. (I) If a galaxy is traveling away from us at 1.5% of the speed of light, roughly how far away is it?

24. (II) A galaxy is moving away from Earth. The "blue" hydrogen line at 434 nm emitted from the galaxy is measured on Earth to be 455 nm. (a) How fast is the galaxy moving? (b) How far is it from Earth based on Hubble's law?

25. (II) Estimate the wavelength shift for the 656.3-nm line in the Balmer series of hydrogen emitted from a galaxy whose distance from us is (a) $7.0 \times 10^6 \text{ ly}$, (b) $7.0 \times 10^7 \text{ ly}$.

26. (II) If an absorption line of calcium is normally found at a wavelength of 393.4 nm in a laboratory gas, and you measure it to be at 423.4 nm in the spectrum of a galaxy, what is the approximate distance to the galaxy?

27. (II) What is the speed of a galaxy with $z = 0.060$?

28. (II) What would be the redshift parameter z for a galaxy traveling away from us at $v = 0.075c$?

29. (II) Estimate the distance d from the Earth to a galaxy whose redshift parameter $z = 1$.

30. (II) Estimate the speed of a galaxy, and its distance from us, if the wavelength for the hydrogen line at 434 nm is measured on Earth as being 610 nm.

31. (II) Radiotelescopes are designed to observe 21-cm waves emitted by atomic hydrogen gas. A signal from a distant radio-emitting galaxy is found to have a wavelength that is 0.10 cm longer than the normal 21-cm wavelength. Estimate the distance to this galaxy.

32. (III) Starting from Eq. 33–3, show that the Doppler shift in wavelength is $\Delta\lambda/\lambda_{rest} \approx v/c$ (Eq. 33–6) for $v \ll c$. [Hint: Use the binomial expansion.]

33–6 to 33–8 The Big Bang, CMB, Universe Expansion

33. (I) Calculate the wavelength at the peak of the blackbody radiation distribution at 2.7 K using Wien's law.

34. (II) Calculate the peak wavelength of the CMB at 1.0 s after the birth of the universe. In what part of the EM spectrum is this radiation?

35. (II) The critical density for closure of the universe is $\rho_c \approx 10^{-26}$ kg/m^3. State ρ_c in terms of the average number of nucleons per cubic meter.

36. (II) The scale factor of the universe (average distance between galaxies) at any given time is believed to have been inversely proportional to the absolute temperature. Estimate the size of the universe, compared to today, at (a) $t = 10^6$ yr, (b) $t = 1$ s, (c) $t = 10^{-6}$ s, and (d) $t = 10^{-35}$ s.

37. (II) At approximately what time had the universe cooled below the threshold temperature for producing (a) kaons $(M \approx 500 \text{ MeV}/c^2)$, (b) $\Upsilon (M \approx 9500 \text{ MeV}/c^2)$, and (c) muons $(M \approx 100 \text{ MeV}/c^2)$?

33–9 Dark Matter, Dark Energy

38. (II) Only about 5% of the energy in the universe is composed of baryonic matter. (a) Estimate the average density of baryonic matter in the observable universe with a radius of 14 billion light-years that contains 10^{11} galaxies, each with about 10^{11} stars like our Sun. (b) Estimate the density of dark matter in the universe.

General Problems

39. Use conservation of angular momentum to estimate the angular velocity of a neutron star which has collapsed to a diameter of 16 km, from a star whose core radius was equal to that of Earth $(6 \times 10^6 \text{ m})$. Assume its mass is 1.5 times that of the Sun, and that it rotated (like our Sun) about once a month.

40. By what factor does the rotational kinetic energy change when the star in Problem 39 collapses to a neutron star?

41. Suppose that three main-sequence stars could undergo the three changes represented by the three arrows, A, B, and C, in the H–R diagram of Fig. 33–35. For each case, describe the changes in temperature, intrinsic luminosity, and size.

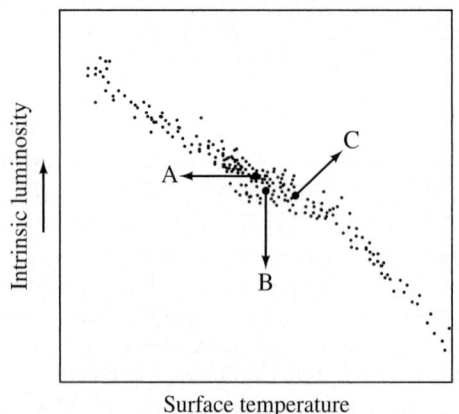

FIGURE 33–35 Problem 41.

42. Assume that the nearest stars to us have an intrinsic luminosity about the same as the Sun's. Their apparent brightness, however, is about 10^{11} times fainter than the Sun. From this, estimate the distance to the nearest stars.

43. A certain pulsar, believed to be a neutron star of mass 1.5 times that of the Sun, with diameter 16 km, is observed to have a rotation speed of 1.0 rev/s. If it loses rotational kinetic energy at the rate of 1 part in 10^9 per day, which is all transformed into radiation, what is the power output of the star?

44. The nearest large galaxy to our Galaxy is about 2×10^6 ly away. If both galaxies have a mass of 4×10^{41} kg, with what gravitational force does each galaxy attract the other? Ignore dark matter.

45. How large would the Sun be if its density equaled the critical density of the universe, $\rho_c \approx 10^{-26}$ kg/m^3? Express your answer in light-years and compare with the Earth–Sun distance and the diameter of our Galaxy.

46. Two stars, whose spectra peak at 660 nm and 480 nm, respectively, both lie on the main sequence. Use Wien's law, the Stefan-Boltzmann equation, and the H–R diagram (Fig. 33–6) to estimate the ratio of their diameters.

47. (a) In order to measure distances with parallax at 100 ly, what minimum angular resolution (in degrees) is needed? (b) What diameter mirror or lens would be needed?

48. (*a*) What temperature would correspond to 14-TeV collisions at the LHC? (*b*) To what era in cosmological history does this correspond? [*Hint*: See Fig. 33–29.]

49. In the later stages of stellar evolution, a star (if massive enough) will begin fusing carbon nuclei to form, for example, magnesium:

$$^{12}_{6}C + ^{12}_{6}C \rightarrow ^{24}_{12}Mg + \gamma.$$

(*a*) How much energy is released in this reaction (see Appendix B)? (*b*) How much kinetic energy must each carbon nucleus have (assume equal) in a head-on collision if they are just to "touch" (use Eq. 30–1) so that the strong force can come into play? (*c*) What temperature does this kinetic energy correspond to?

50. Consider the reaction

$$^{16}_{8}O + ^{16}_{8}O \rightarrow ^{28}_{14}Si + ^{4}_{2}He,$$

and answer the same questions as in Problem 49.

51. Use *dimensional analysis* with the fundamental constants c, G, and $\hbar$ to estimate the value of the so-called *Planck time*. It is thought that physics as we know it can say nothing about the universe before this time.

52. Estimate the mass of our observable universe using the following assumptions: Our universe is spherical in shape, it has been expanding at the speed of light since the Big Bang, and its density is the critical density.

Search and Learn

1. Estimate what neutrino mass (in eV/c^2) would provide the critical density to close the universe. Assume the neutrino density is, like photons, about 10^9 times that of nucleons, and that nucleons make up only (*a*) 2% of the mass needed, or (*b*) 5% of the mass needed.

2. Describe how we can estimate the distance from us to other stars. Which methods can we use for nearby stars, and which can we use for very distant stars? Which method gives the most accurate distance measurements for the most distant stars?

3. The evolution of stars, as discussed in Section 33–2, can lead to a white dwarf, a neutron star, or even a black hole, depending on the mass. (*a*) Referring to Sections 33–2 and 33–4, give the radius of (i) a white dwarf of 1 solar mass, (ii) a neutron star of 1.5 solar masses, and (iii) a black hole of 3 solar masses. (*b*) Express these three radii as ratios $(r_i : r_{ii} : r_{iii})$.

4. (*a*) Describe some of the evidence that the universe began with a "Big Bang." (*b*) How does the curvature of the universe affect its future destiny? (*c*) How does dark energy affect the possible future of the universe?

5. When stable nuclei first formed, about 3 minutes after the Big Bang, there were about 7 times more protons than neutrons. Explain how this leads to a ratio of the mass of hydrogen to the mass of helium of 3:1. This is about the actual ratio observed in the universe.

6. Explain what the 2.7-K cosmic microwave background radiation is. Where does it come from? Why is its temperature now so low?

7. We cannot use Hubble's law to measure the distances to nearby galaxies, because their random motions are larger than the overall expansion. Indeed, the closest galaxy to us, the Andromeda Galaxy, 2.5 million light-years away, is approaching us at a speed of about 130 km/s. (*a*) What is the shift in wavelength of the 656-nm line of hydrogen emitted from the Andromeda Galaxy, as seen by us? (*b*) Is this a redshift or a blueshift? (*c*) Ignoring the expansion, how soon will it and the Milky Way Galaxy collide?

ANSWERS TO EXERCISES

A: Our Earth and ourselves, 2 years ago.

B: 600 ly (estimating L from Fig. 33–6 as $L \approx 8 \times 10^{26}$ W; note that on a log scale, 6000 K is closer to 7000 K than it is to 5000 K).

C: 30 km.

D: (*a*); not the usual R^3, but R: see formula for the Schwarzschild radius.

E: (*c*); (*d*).

Mathematical Review

A–1 Relationships, Proportionality, and Equations

One of the important aspects of physics is the search for relationships between different quantities—that is, determining how one quantity affects another.

As a simple example, the ancients found that if one circle has twice the diameter of a second circle, the first also has twice the circumference. If the diameter is three times as large, the circumference is also three times as large. In other words, an increase in the diameter results in a proportional increase in the circumference. We say that the circumference is **directly proportional** to the diameter. This can be written in symbols as $C \propto D$, where "$\propto$" means "is proportional to," and C and D refer to the circumference and diameter of a circle, respectively. The next step is to change this proportionality to an equation, which will make it possible to link the two quantities numerically. This means inserting a proportionality constant, which in many cases is determined by measurement. The ancients found that the ratio of the circumference to the diameter of any circle was 3.1416 (to keep only the first few decimal places). This number is designated by the Greek letter π. It is the constant of proportionality for the relationship $C \propto D$. To obtain an equation, we insert π into the proportion and change the $\propto$ to $=$. Thus,

$$C = \pi D.$$

Other kinds of proportionality occur as well. For example, the area of a circle is proportional to the *square* of its radius. That is, if the radius is doubled, the area becomes four times as large; and so on. In this case we can write $A \propto r^2$, where A stands for the area and r for the radius of the circle. The constant of proportionality is found to be π again: $A = \pi r^2$.

Sometimes two quantities are related in such a way that an increase in one leads to a proportional *decrease* in the other. This is called **inverse proportion**. For example, the time required to travel a given distance is inversely proportional to the speed of travel. The greater the speed, the less time it takes. We can write this inverse proportion as

$$\text{time} \propto 1/\text{speed}.$$

The larger the denominator of a fraction, the lower the value of the fraction is as a whole. For example, $\frac{1}{4}$ is less than $\frac{1}{2}$. Thus, if the speed is doubled, the time is halved, which is what we want to express by this inverse proportionality relationship.

If you suspect that a relationship exists between two or more quantities, you can try to determine the precise nature of this relationship by varying one of the quantities and measuring how the other varies as a result. Sometimes a given quantity is affected by two or more quantities; for instance, the acceleration of an object is related to both its mass and the applied force. In such a case, only one quantity is varied at a time, while the others are held constant.

When one quantity affects another, we often use the expression **is a function of** to indicate this dependence; for example, we say that the pressure in a tire is a function of the temperature.

Whatever kind of proportion is found to hold, it can be changed to an equality by finding the proper proportionality constant. Quantitative statements or predictions about the physical world can then be made with the equation.

A–2 Exponents

When we write 10^4, we mean that you multiply 10 by itself four times: $10^4 = 10 \times 10 \times 10 \times 10 = 10{,}000$. The superscript 4 is called an **exponent**, and 10 is said to be raised to the fourth power. Any number or symbol can be raised to a power. Special names are used when the exponent is 2 (a^2 is "a squared") or 3 (a^3 is "a cubed"). For any other power, we say a^n is "a to the nth power." If the exponent is 1, it is usually dropped: $a^1 = a$, since no multiplication is involved.

The rules for multiplying numbers expressed as powers are as follows: first,

$$(a^n)(a^m) = a^{n+m}. \tag{A–1}$$

That is, the exponents are added. To see why, consider the result of the multiplication of 3^3 by 3^4:

$$(3^3)(3^4) = (3)(3)(3) \times (3)(3)(3)(3) = (3)^7.$$

Here the sum of the exponents is $3 + 4 = 7$, so rule A–1 works. Notice that this rule works only if the base numbers (a in Eq. A–1) are the same. Thus we *cannot* use the rule of summing exponents for $(6^3)(5^2)$; these numbers would have to be written out. However, if the base numbers are different but the exponents are the same, we can write a second rule:

$$(a^n)(b^n) = (ab)^n. \tag{A–2}$$

For example, $(5^3)(6^3) = (30)^3$ since

$$(5)(5)(5)(6)(6)(6) = (30)(30)(30).$$

The third rule involves a power raised to another power: $(a^3)^2$ means $(a^3)(a^3)$, which is equal to $a^{3+3} = a^6$. The general rule is then

$$(a^n)^m = a^{nm}. \tag{A–3}$$

In this case, the exponents are multiplied.

Negative exponents are used for reciprocals. Thus,

$$\frac{1}{a} = a^{-1}, \qquad \frac{1}{a^3} = a^{-3},$$

and so on. The reason for using negative exponents is to allow us to use the multiplication rules given above. For example, $(a^5)(a^{-3})$ means

$$\frac{(a)(a)(a)(a)(a)}{(a)(a)(a)} = a^2,$$

after canceling 3 of the a's. Rule A–1 gives us the same result:

$$(a^5)(a^{-3}) = a^{5-3} = a^2.$$

What does an exponent of zero mean? That is, what is a^0? Any number raised to the zeroth power is defined as being equal to 1:

$$a^0 = 1.$$

This definition is used because it follows from the rules for adding exponents. For example,

$$a^3 a^{-3} = a^{3-3} = a^0 = 1.$$

But *does* $a^3 a^{-3}$ actually equal 1? Yes, because

$$a^3 a^{-3} = \frac{a^3}{a^3} = 1.$$

Fractional exponents are used to represent *roots*. For example, $a^{\frac{1}{2}}$ means the square root of a; that is, $a^{\frac{1}{2}} = \sqrt{a}$. Similarly, $a^{\frac{1}{3}}$ means the cube root of a, and so on. The fourth root of a means that if you multiply the fourth root of a by itself four times, you again get a:

$$\left(a^{\frac{1}{4}}\right)^4 = a.$$

This is consistent with rule A–3 since $\left(a^{\frac{1}{4}}\right)^4 = a^{\frac{4}{4}} = a^1 = a$.

A–3 Powers of 10, or Exponential Notation

Writing out very large and very small numbers such as the distance of Neptune from the Sun, 4,500,000,000 km, or the diameter of a typical atom, 0.00000001 cm, is inconvenient and prone to error. It also leaves in question (see Section 1–4) the number of significant figures. (How many of the zeros are significant in the number 4,500,000,000 km?)

For these reasons we make use of the "powers of 10," or exponential notation. The distance from Neptune to the Sun is then expressed as 4.50×10^9 km (assuming that the value is significant to three digits), and the diameter of an atom 1.0×10^{-8} cm. This way of writing numbers is based on the use of exponents, where a^n signifies a multiplied by itself n times. For example, $10^4 = 10 \times 10 \times 10 \times 10 = 10,000$. Thus, $4.50 \times 10^9 = 4.50 \times 1,000,000,000 = 4,500,000,000$. Notice that the exponent (9 in this case) is just the number of places the decimal point is moved to the right to obtain the fully written-out number (4.500,000,000.)

When two numbers are multiplied (or divided), you first multiply (or divide) the simple parts and then the powers of 10. Thus, 2.0×10^3 multiplied by 5.5×10^4 equals $(2.0 \times 5.5) \times (10^3 \times 10^4) = 11 \times 10^7$, where we have used the rule for adding exponents (Appendix A–2). Similarly, 8.2×10^5 divided by 2.0×10^2 equals

$$\frac{8.2 \times 10^5}{2.0 \times 10^2} = \frac{8.2}{2.0} \times \frac{10^5}{10^2} = 4.1 \times 10^3.$$

For numbers less than 1, say 0.01, the exponent power of 10 is written with a negative sign (see previous page): $0.01 = 1/100 = 1/10^2 = 1 \times 10^{-2}$. Similarly, $0.002 = 2 \times 10^{-3}$. The decimal point has again been moved the number of places expressed in the exponent. For example, $0.020 \times 3600 = 72$, or in exponential notation $(2.0 \times 10^{-2}) \times (3.6 \times 10^3) = 7.2 \times 10^1 = 72$.

Notice also that $10^1 \times 10^{-1} = 10 \times 0.1 = 1$, and by the law of exponents, $10^1 \times 10^{-1} = 10^0$. Therefore, $10^0 = 1$.

When writing a number in exponential notation, it is usual to make the simple number be between 1 and 10. Thus it is conventional to write 4.5×10^9 rather than 45×10^8, although they are the same number.[†] This notation also allows the number of *significant figures* to be clearly expressed. We write 4.50×10^9 if this value is accurate to three significant figures, but 4.5×10^9 if it is accurate to only two.

A–4 Algebra

Physical relationships between quantities can be represented as equations involving symbols (usually letters of the Greek or Roman alphabet) that represent the quantities. The manipulation of such equations is the field of algebra, and it is used a great deal in physics. An equation involves an equals sign, which tells us that the quantities on either side of the equals sign have the same value. Examples of equations are

$$3 + 8 = 11$$
$$2x + 7 = 15$$
$$a^2b + c = 6.$$

The first equation involves only numbers, so is called an arithmetic equation. The other two equations are algebraic since they involve symbols. In the third equation, the quantity a^2b means the product of a times a times b: $a^2b = a \times a \times b$.

[†] Another convention used, particularly with computers, is that the simple number be between 0.1 and 1. Thus we could write 4,500,000,000 as 0.450×10^{10}. This is slightly less compact.

Solving for an Unknown

Often we wish to solve for one (or more) symbols, and we treat it as an *unknown*. For example, in the equation $2x + 7 = 15$, x is the unknown; this equation is true, however, only when $x = 4$. Determining what value (or values) the unknown can have to satisfy the equation is called *solving the equation*. To solve an equation, the following rule can be used:

> *An equation will remain true if any operation performed on one side is also performed on the other side.* For example: (*a*) addition or subtraction of a number or symbol; (*b*) multiplication or division by a number or symbol; (*c*) raising each side of the equation to the same power, or taking the same root (such as square root).

EXAMPLE A–1 Solve for x in the equation

$$2x + 7 = 15.$$

APPROACH We perform the same operations on both sides of the equation to isolate x as the only variable on the left side of the equals sign.

SOLUTION We first subtract 7 from both sides:

$$2x + 7 - 7 = 15 - 7$$

or

$$2x = 8.$$

Then we divide both sides by 2 to get

$$\frac{2x}{2} = \frac{8}{2},$$

or, carrying out the divisions,

$$x = 4,$$

and this solves the equation.

EXAMPLE A–2 (*a*) Solve the equation

$$a^2b + c = 24$$

for the unknown a in terms of b and c. (*b*) Solve for a assuming that $b = 2$ and $c = 6$.

APPROACH We perform operations to isolate a as the only variable on the left side of the equals sign.

SOLUTION (*a*) We are trying to solve for a, so we first subtract c from both sides:

$$a^2b = 24 - c,$$

then divide by b:

$$a^2 = \frac{24 - c}{b},$$

and finally take square roots:

$$a = \sqrt{\frac{24 - c}{b}}.$$

(*b*) If we are given that $b = 2$ and $c = 6$, then

$$a = \sqrt{\frac{24 - 6}{2}} = 3.$$

But this is not the only answer. Whenever we take a square root, the number can be either positive or negative. Thus $a = -3$ is also a solution. Why? Because $(-3)^2 = 9$, just as $(+3)^2 = 9$. So we actually get two solutions: $a = +3$ and $a = -3$.

NOTE When an unknown appears squared in an equation, there are generally two solutions for that unknown.

To check a solution, we put it back into the original equation (this is really a check that we did all the manipulations correctly). In the equation

$$a^2b + c = 24,$$

we put in $a = 3$, $b = 2$, $c = 6$ and find

$$(3)^2(2) + (6) \stackrel{?}{=} 24$$
$$24 = 24,$$

which checks.

> **EXERCISE A** Put $a = -3$ into the equation of Example A–2 and show that it works too.

Two or More Unknowns

If we have two or more unknowns, one equation is not sufficient to find them. In general, if there are n unknowns, n independent equations are needed. For example, if there are two unknowns, we need two equations. If the unknowns are called x and y, a typical procedure is to solve one equation for x in terms of y, and substitute this into the second equation.

> **EXAMPLE A–3** Solve the following pair of equations for x and y:
>
> $$3x - 2y = 19$$
> $$x + 4y = -3.$$
>
> **APPROACH** We have two unknowns and two equations; we can start by solving the second equation for x in terms of y. Then we substitute this result for x into the first equation.
>
> **SOLUTION** We subtract $4y$ from both sides of the second equation:
>
> $$x = -3 - 4y.$$
>
> We substitute this expression for x into the first equation, and simplify:
>
> $$3(-3 - 4y) - 2y = 19$$
> $$-9 - 12y - 2y = 19 \quad \text{(carried out the multiplication by 3)}$$
> $$-14y = 28 \quad \text{(added 9 to both sides)}$$
> $$y = -2. \quad \text{(divided both sides by } -14)$$
>
> Now that we know $y = -2$, we substitute this into the expression for x:
>
> $$x = -3 - 4y$$
> $$= -3 - 4(-2) = -3 + 8 = 5.$$
>
> Our solution is $x = 5$, $y = -2$. We check this solution by putting these values back into the original equations:
>
> $$3x - 2y \stackrel{?}{=} 19$$
> $$3(5) - 2(-2) \stackrel{?}{=} 19$$
> $$15 + 4 \stackrel{?}{=} 19$$
> $$19 = 19 \quad \text{(it checks)}$$
>
> and
>
> $$x + 4y \stackrel{?}{=} -3$$
> $$5 + 4(-2) \stackrel{?}{=} -3$$
> $$-3 = -3. \quad \text{(it checks)}$$

Other methods for solving two or more equations, such as the method of determinants, can be found in an algebra textbook.

The Quadratic Formula

We sometimes encounter equations that involve an unknown, say x, that appears not only to the first power, but squared as well. Such a **quadratic equation** can be written in the general form

$$ax^2 + bx + c = 0.$$

The quantities a, b, and c are typically numbers or constants that are given.[†] The general solutions to such an equation are given by the **quadratic formula**:

$$x = \frac{-b \pm \sqrt{b^2 - 4ac}}{2a}. \qquad \textbf{(A–4)}$$

The $\pm$ sign indicates that there are two solutions for x: one where the plus sign is used, the other where the minus sign is used.

EXAMPLE A–4 Find the solutions for x in the equation

$$3x^2 - 5x = 2.$$

APPROACH Here x appears both to the first power and squared, so we use the quadratic equation.

SOLUTION First we write this equation in the standard form

$$ax^2 + bx + c = 0$$

by subtracting 2 from both sides:

$$3x^2 - 5x - 2 = 0.$$

In this case, a, b, and c in the standard formula take the values $a = 3$, $b = -5$, and $c = -2$. The two solutions for x are, using Eq. A–4,

$$x = \frac{+5 + \sqrt{25 - (4)(3)(-2)}}{(2)(3)} = \frac{5 + 7}{6} = 2$$

and

$$x = \frac{+5 - \sqrt{25 - (4)(3)(-2)}}{(2)(3)} = \frac{5 - 7}{6} = -\frac{1}{3}.$$

In this Example, the two solutions are $x = 2$ and $x = -\frac{1}{3}$. In physics problems, it sometimes happens that only one of the solutions corresponds to a real-life situation; in this case, the other solution is discarded. In other cases, both solutions may correspond to physical reality.

Notice, incidentally, that b^2 must be greater than $4ac$, so that $\sqrt{b^2 - 4ac}$ yields a real number. If $(b^2 - 4ac)$ is less than zero (negative), there is no real solution. The square root of a negative number is called **imaginary**.

A second-order equation—one in which the highest power of x is 2—has two solutions; a third-order equation—involving x^3—has three solutions; and so on.

A–5 The Binomial Expansion

Sometimes we end up with a quantity of the form $(1 + x)^n$. That is, the quantity $(1 + x)$ is raised to the nth power. This can be written as an infinite sum of terms known as the **binomial expansion**:

$$(1 + x)^n = 1 + nx + \frac{n(n - 1)}{2}x^2 + \cdots. \qquad \textbf{(A–5)}$$

This formula is useful for us mainly when x is very small compared to one ($x \ll 1$). In this case, each successive term is much smaller than the preceding

[†]Or one or more of them could be variables, in which case additional equations are needed.

term. For example, let $x = 0.01$, and $n = 2$. Then whereas the first term equals 1, the second term is $nx = (2)(0.01) = 0.02$, and the third term is $[(2)(1)/2](0.01)^2 = 0.0001$, and so on. Thus, when x is small, we can ignore all but the first two (or three) terms and can write

$$(1 + x)^n \approx 1 + nx. \qquad \text{(A–6)}$$

This approximation often allows us to solve an equation easily that otherwise might be very difficult. Some examples of the binomial expansion are

$$(1 + x)^2 \approx 1 + 2x,$$
$$\frac{1}{1 + x} = (1 + x)^{-1} \approx 1 - x,$$
$$\sqrt{1 + x} = (1 + x)^{\frac{1}{2}} \approx 1 + \tfrac{1}{2}x,$$
$$\frac{1}{\sqrt{1 + x}} = (1 + x)^{-\frac{1}{2}} \approx 1 - \tfrac{1}{2}x,$$

where $x \ll 1$.

As a numerical example, let us evaluate $\sqrt{1.02}$ using the binomial expansion since $x = 0.02$ is much smaller than 1:

$$\sqrt{1.02} = (1.02)^{\frac{1}{2}} = (1 + 0.02)^{\frac{1}{2}} \approx 1 + \tfrac{1}{2}(0.02) = 1.01.$$

You can check with a calculator (and maybe not even more quickly) that $\sqrt{1.02} \approx 1.01$.

A–6 Plane Geometry

We review here a number of theorems involving angles and triangles that are useful in physics.

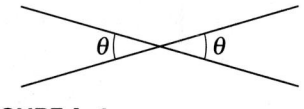

FIGURE A–1

FIGURE A–2

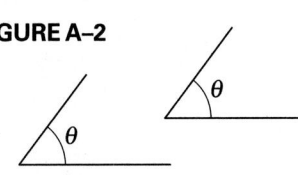

1. **Equal angles.** Two angles are equal if any of the following conditions are true:
   (a) They are vertical angles (Fig. A–1); *or*
   (b) the left side of one is parallel to the left side of the other, and the right side of one is parallel to the right side of the other (Fig. A–2; the left and right sides are as seen from the vertex, where the two sides meet); *or*
   (c) the left side of one is perpendicular to the left side of the other, and the right sides are likewise perpendicular (Fig. A–3).

FIGURE A–3

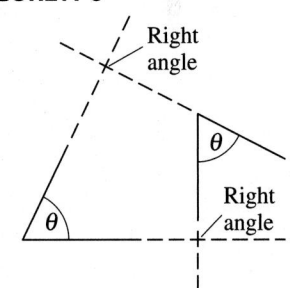

2. *The sum of the angles* in any plane triangle is 180°.

3. **Similar triangles.** Two triangles are said to be similar if all three of their angles are equal (in Fig. A–4, $\theta_1 = \phi_1$, $\theta_2 = \phi_2$, and $\theta_3 = \phi_3$). Similar triangles thus have the same basic shape but may be different sizes and have different orientations. Two useful theorems about similar triangles are:
   (a) Two triangles are similar if any two of their angles are equal. (This follows because the third angles must also be equal since the sum of the angles of a triangle is 180°.)
   (b) The ratios of corresponding sides of two similar triangles are equal. That is (Fig. A–4),

$$\frac{a_1}{b_1} = \frac{a_2}{b_2} = \frac{a_3}{b_3}.$$

FIGURE A–4

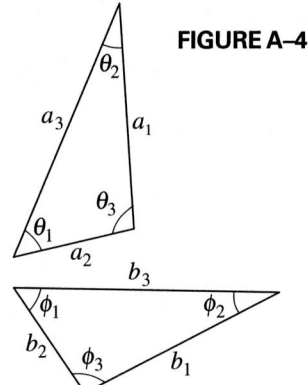

4. **Congruent triangles.** Two triangles are congruent if one can be placed precisely on top of the other. That is, they are similar triangles and they have the same size. Two triangles are congruent if any of the following holds:
   (a) The three corresponding sides are equal.
   (b) Two sides and the enclosed angle are equal ("side-angle-side").
   (c) Two angles and the enclosed side are equal ("angle-side-angle").

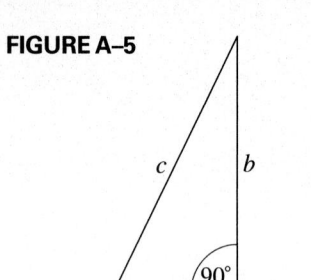

5. Right triangles. A right triangle has one angle that is 90° (a **right angle**); that is, the two sides that meet at the right angle are perpendicular (Fig. A–5). The two other (acute) angles in the right triangle add up to 90°.

6. Pythagorean theorem. In any right triangle, the square of the length of the hypotenuse (the side opposite the right angle) is equal to the sum of the squares of the lengths of the other two sides. In Fig. A–5,

$$c^2 = a^2 + b^2.$$

A–7 Trigonometric Functions and Identities

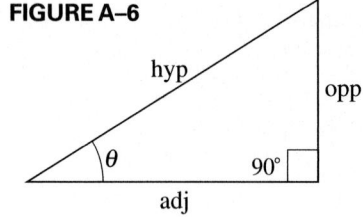

Trigonometric functions for any angle θ are defined by constructing a right triangle about that angle as shown in Fig. A–6; opp and adj are the lengths of the sides opposite and adjacent to the angle θ, and hyp is the length of the hypotenuse:

$$\sin \theta = \frac{\text{opp}}{\text{hyp}} \qquad\qquad \csc \theta = \frac{1}{\sin \theta} = \frac{\text{hyp}}{\text{opp}}$$

$$\cos \theta = \frac{\text{adj}}{\text{hyp}} \qquad\qquad \sec \theta = \frac{1}{\cos \theta} = \frac{\text{hyp}}{\text{adj}}$$

$$\tan \theta = \frac{\text{opp}}{\text{adj}} = \frac{\sin \theta}{\cos \theta} \qquad\qquad \cot \theta = \frac{1}{\tan \theta} = \frac{\text{adj}}{\text{opp}}$$

$$\text{adj}^2 + \text{opp}^2 = \text{hyp}^2 \qquad \text{(Pythagorean theorem)}.$$

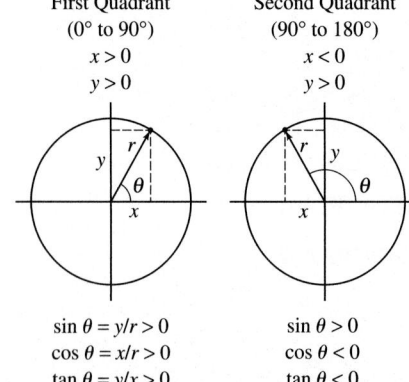

First Quadrant
(0° to 90°)
$x > 0$
$y > 0$

$\sin \theta = y/r > 0$
$\cos \theta = x/r > 0$
$\tan \theta = y/x > 0$

Second Quadrant
(90° to 180°)
$x < 0$
$y > 0$

$\sin \theta > 0$
$\cos \theta < 0$
$\tan \theta < 0$

Third Quadrant
(180° to 270°)
$x < 0$
$y < 0$

$\sin \theta < 0$
$\cos \theta < 0$
$\tan \theta > 0$

Fourth Quadrant
(270° to 360°)
$x > 0$
$y < 0$

$\sin \theta < 0$
$\cos \theta > 0$
$\tan \theta < 0$

Figure A–7 shows the signs (+ or −) that cosine, sine, and tangent take on for angles θ in the four quadrants (0° to 360°). Note that angles are measured counterclockwise from the x axis as shown; negative angles are measured from *below* the x axis, clockwise: for example, $-30° = +330°$, and so on.

The following are some useful identities among the trigonometric functions:

$$\sin^2 \theta + \cos^2 \theta = 1$$

$$\sin 2\theta = 2 \sin \theta \cos \theta$$

$$\cos 2\theta = \cos^2 \theta - \sin^2 \theta = 2 \cos^2 \theta - 1 = 1 - 2 \sin^2 \theta$$

$$\tan 2\theta = \frac{2 \tan \theta}{1 - \tan^2 \theta}$$

$$\sin(A \pm B) = \sin A \cos B \pm \cos A \sin B$$

$$\cos(A \pm B) = \cos A \cos B \mp \sin A \sin B$$

$$\tan(A \pm B) = \frac{\tan A \pm \tan B}{1 \mp \tan A \tan B}$$

$$\sin(180° - \theta) = \sin \theta$$

$$\cos(180° - \theta) = -\cos \theta$$

$$\sin(90° - \theta) = \cos \theta$$

$$\cos(90° - \theta) = \sin \theta$$

$$\sin \tfrac{1}{2}\theta = \sqrt{\frac{1 - \cos \theta}{2}}$$

$$\cos \tfrac{1}{2}\theta = \sqrt{\frac{1 + \cos \theta}{2}}$$

$$\tan \tfrac{1}{2}\theta = \sqrt{\frac{1 - \cos \theta}{1 + \cos \theta}}$$

$$\sin A \pm \sin B = 2 \sin\left(\frac{A \pm B}{2}\right) \cos\left(\frac{A \mp B}{2}\right).$$

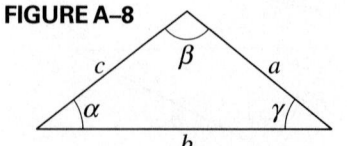

For any triangle (see Fig. A–8):

$$\frac{\sin \alpha}{a} = \frac{\sin \beta}{b} = \frac{\sin \gamma}{c} \qquad \text{(law of sines)}$$

$$c^2 = a^2 + b^2 - 2ab \cos \gamma. \qquad \text{(law of cosines)}$$

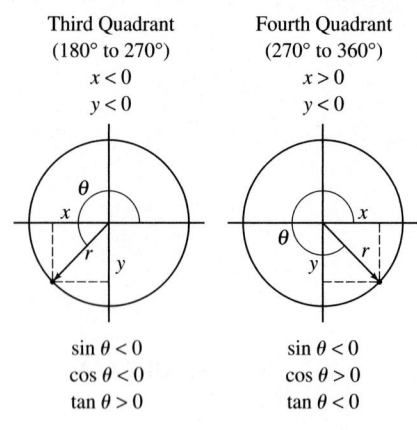

Trigonometric Table: Numerical Values of Sin, Cos, Tan

Angle in Degrees	Angle in Radians	Sine	Cosine	Tangent	Angle in Degrees	Angle in Radians	Sine	Cosine	Tangent
0°	0.000	0.000	1.000	0.000					
1°	0.017	0.017	1.000	0.017	46°	0.803	0.719	0.695	1.036
2°	0.035	0.035	0.999	0.035	47°	0.820	0.731	0.682	1.072
3°	0.052	0.052	0.999	0.052	48°	0.838	0.743	0.669	1.111
4°	0.070	0.070	0.998	0.070	49°	0.855	0.755	0.656	1.150
5°	0.087	0.087	0.996	0.087	50°	0.873	0.766	0.643	1.192
6°	0.105	0.105	0.995	0.105	51°	0.890	0.777	0.629	1.235
7°	0.122	0.122	0.993	0.123	52°	0.908	0.788	0.616	1.280
8°	0.140	0.139	0.990	0.141	53°	0.925	0.799	0.602	1.327
9°	0.157	0.156	0.988	0.158	54°	0.942	0.809	0.588	1.376
10°	0.175	0.174	0.985	0.176	55°	0.960	0.819	0.574	1.428
11°	0.192	0.191	0.982	0.194	56°	0.977	0.829	0.559	1.483
12°	0.209	0.208	0.978	0.213	57°	0.995	0.839	0.545	1.540
13°	0.227	0.225	0.974	0.231	58°	1.012	0.848	0.530	1.600
14°	0.244	0.242	0.970	0.249	59°	1.030	0.857	0.515	1.664
15°	0.262	0.259	0.966	0.268	60°	1.047	0.866	0.500	1.732
16°	0.279	0.276	0.961	0.287	61°	1.065	0.875	0.485	1.804
17°	0.297	0.292	0.956	0.306	62°	1.082	0.883	0.469	1.881
18°	0.314	0.309	0.951	0.325	63°	1.100	0.891	0.454	1.963
19°	0.332	0.326	0.946	0.344	64°	1.117	0.899	0.438	2.050
20°	0.349	0.342	0.940	0.364	65°	1.134	0.906	0.423	2.145
21°	0.367	0.358	0.934	0.384	66°	1.152	0.914	0.407	2.246
22°	0.384	0.375	0.927	0.404	67°	1.169	0.921	0.391	2.356
23°	0.401	0.391	0.921	0.424	68°	1.187	0.927	0.375	2.475
24°	0.419	0.407	0.914	0.445	69°	1.204	0.934	0.358	2.605
25°	0.436	0.423	0.906	0.466	70°	1.222	0.940	0.342	2.747
26°	0.454	0.438	0.899	0.488	71°	1.239	0.946	0.326	2.904
27°	0.471	0.454	0.891	0.510	72°	1.257	0.951	0.309	3.078
28°	0.489	0.469	0.883	0.532	73°	1.274	0.956	0.292	3.271
29°	0.506	0.485	0.875	0.554	74°	1.292	0.961	0.276	3.487
30°	0.524	0.500	0.866	0.577	75°	1.309	0.966	0.259	3.732
31°	0.541	0.515	0.857	0.601	76°	1.326	0.970	0.242	4.011
32°	0.559	0.530	0.848	0.625	77°	1.344	0.974	0.225	4.331
33°	0.576	0.545	0.839	0.649	78°	1.361	0.978	0.208	4.705
34°	0.593	0.559	0.829	0.675	79°	1.379	0.982	0.191	5.145
35°	0.611	0.574	0.819	0.700	80°	1.396	0.985	0.174	5.671
36°	0.628	0.588	0.809	0.727	81°	1.414	0.988	0.156	6.314
37°	0.646	0.602	0.799	0.754	82°	1.431	0.990	0.139	7.115
38°	0.663	0.616	0.788	0.781	83°	1.449	0.993	0.122	8.144
39°	0.681	0.629	0.777	0.810	84°	1.466	0.995	0.105	9.514
40°	0.698	0.643	0.766	0.839	85°	1.484	0.996	0.087	11.43
41°	0.716	0.656	0.755	0.869	86°	1.501	0.998	0.070	14.301
42°	0.733	0.669	0.743	0.900	87°	1.518	0.999	0.052	19.081
43°	0.750	0.682	0.731	0.933	88°	1.536	0.999	0.035	28.636
44°	0.768	0.695	0.719	0.966	89°	1.553	1.000	0.017	57.290
45°	0.785	0.707	0.707	1.000	90°	1.571	1.000	0.000	∞

A–8 Logarithms

Logarithms are defined in the following way:

$$\text{if } y = A^x, \quad \text{then } x = \log_A y.$$

That is, the logarithm of a number y to the base A is that number which, as the exponent of A, gives back the number y. For **common logarithms**, the base is 10, so

$$\text{if } y = 10^x, \quad \text{then } x = \log y.$$

The subscript 10 on $\log_{10}$ is usually omitted when dealing with common logs. Another base sometimes used is the exponential base $e = 2.718\cdots$, a natural number.[†] Such logarithms are called **natural logarithms** and are written "ln". Thus,

$$\text{if } y = e^x, \quad \text{then } x = \ln y.$$

For any number y, the two types of logarithm are related by

$$\ln y = 2.3026 \log y.$$

Some simple rules for logarithms include:

$$\log(ab) = \log a + \log b. \tag{A–7}$$

This is true because if $a = 10^n$ and $b = 10^m$, then $ab = 10^{n+m}$. From the definition of logarithm, $\log a = n$, $\log b = m$, and $\log(ab) = n + m$; hence, $\log(ab) = n + m = \log a + \log b$. In a similar way, we can show the rules

$$\log\left(\frac{a}{b}\right) = \log a - \log b \tag{A–8}$$

and

$$\log a^n = n \log a. \tag{A–9}$$

These three rules apply not only to common logs but to natural or any other kind of logarithm.

Logs were once used as a technique for simplifying certain types of calculation. Because of the advent of electronic calculators and computers, they are not often used any more for that purpose. However, logs do appear in certain physical equations, so it is helpful to know how to deal with them. If you do not have a calculator that calculates logs, you can use a **log table**, such as the small one shown here (Table A–1). The number N is given to two digits (some tables give N to three or more digits); the first digit is in the vertical column to the left, the second digit is in the horizontal row across the top. For example, the Table tells us that $\log 1.0 = 0.000$, $\log 1.1 = 0.041$, and $\log 4.1 = 0.613$. The Table gives logs for numbers between 1.0 and 9.9; for larger or smaller numbers, we use rule A–7:

$$\log(ab) = \log a + \log b.$$

For example,

$$\log(380) = \log(3.8 \times 10^2) = \log(3.8) + \log(10^2).$$

From the Table, $\log 3.8 = 0.580$; and from rule A–9,

$$\log(10^2) = 2 \log(10) = 2,$$

since $\log(10) = 1$. [This follows from the definition of the logarithm: if

---

[†]The exponential base e can be written as an infinite series:

$$e = 1 + \frac{1}{1} + \frac{1}{1 \cdot 2} + \frac{1}{1 \cdot 2 \cdot 3} + \frac{1}{1 \cdot 2 \cdot 3 \cdot 4} + \cdots.$$

TABLE A-1 Short Table of Common Logarithms

N	0.0	0.1	0.2	0.3	0.4	0.5	0.6	0.7	0.8	0.9
1	.000	.041	.079	.114	.146	.176	.204	.230	.255	.279
2	.301	.322	.342	.362	.380	.398	.415	.431	.447	.462
3	.477	.491	.505	.519	.531	.544	.556	.568	.580	.591
4	.602	.613	.623	.633	.643	.653	.663	.672	.681	.690
5	.699	.708	.716	.724	.732	.740	.748	.756	.763	.771
6	.778	.785	.792	.799	.806	.813	.820	.826	.833	.839
7	.845	.851	.857	.863	.869	.875	.881	.886	.892	.898
8	.903	.908	.914	.919	.924	.929	.935	.940	.944	.949
9	.954	.959	.964	.968	.973	.978	.982	.987	.991	.996

$10 = 10^1$, then $1 = \log(10)$.] Thus,

$$\log(380) = \log(3.8) + \log(10^2)$$
$$= 0.580 + 2$$
$$= 2.580.$$

Similarly,

$$\log(0.081) = \log(8.1) + \log(10^{-2})$$
$$= 0.908 - 2 = -1.092.$$

Sometimes we need to do the reverse process: find the number N whose log is, say, 2.670. This is called "taking the **antilogarithm**." To do so, we separate our number 2.670 into two parts, making the separation at the decimal point:

$$\log N = 2.670 = 2 + 0.670$$
$$= \log 10^2 + 0.670.$$

We now look at Table A–1 to see what number has its log equal to 0.670; none does, so we must **interpolate**: we see that $\log 4.6 = 0.663$ and $\log 4.7 = 0.672$. So the number we want is between 4.6 and 4.7, and closer to the latter by $\frac{7}{9}$. Approximately we can say that $\log 4.68 = 0.670$. Thus

$$\log N = 2 + 0.670$$
$$= \log(10^2) + \log(4.68) = \log(4.68 \times 10^2),$$

so $N = 4.68 \times 10^2 = 468$.

If the given logarithm is negative, say, -2.180, we proceed as follows:

$$\log N = -2.180 = -3 + 0.820$$
$$= \log 10^{-3} + \log 6.6 = \log 6.6 \times 10^{-3},$$

so $N = 6.6 \times 10^{-3}$. Notice that we added to our given logarithm the next largest integer (3 in this case) so that we have an integer, plus a decimal number between 0 and 1.0 whose antilogarithm can be looked up in the Table.

APPENDIX B

Selected Isotopes

(1) Atomic Number Z	(2) Element	(3) Symbol	(4) Mass Number A	(5) Atomic Mass†	(6) % Abundance (or Radioactive Decay‡ Mode)	(7) Half-life (if radioactive)
0	(Neutron)	n	1	1.008665	β^-	10.183 min
1	Hydrogen	H	1	1.007825	99.9885%	
	[proton	p	1	1.007276]		
	Deuterium	$^2_1$H	2	2.014102	0.0115%	
	[deuteron	d or D	2	2.013553]		
	Tritium	$^3_1$H	3	3.016049	β^-	12.32 yr
	[triton	t or T	3	3.015500]		
2	Helium	He	3	3.016029	0.000137%	
			4	4.002603	99.999863%	
3	Lithium	Li	6	6.015123	7.59%	
			7	7.016003	92.41%	
4	Beryllium	Be	7	7.016929	EC, γ	53.24 days
			9	9.012183	100%	
5	Boron	B	10	10.012937	19.9%	
			11	11.009305	80.1%	
6	Carbon	C	11	11.011434	β^+, EC	20.334 min
			12	12.000000	98.93%	
			13	13.003355	1.07%	
			14	14.003242	β^-	5730 yr
7	Nitrogen	N	13	13.005739	β^+, EC	9.965 min
			14	14.003074	99.632%	
			15	15.000109	0.368%	
8	Oxygen	O	15	15.003066	β^+, EC	122.24 s
			16	15.994915	99.757%	
			18	17.999160	0.205%	
9	Fluorine	F	19	18.998403	100%	
10	Neon	Ne	20	19.992440	90.48%	
			22	21.991385	9.25%	
11	Sodium	Na	22	21.994437	β^+, EC, γ	2.6027 yr
			23	22.989769	100%	
			24	23.990963	β^-, γ	14.997 h
12	Magnesium	Mg	24	23.985042	78.99%	
13	Aluminum	Al	27	26.981539	100%	
14	Silicon	Si	28	27.976927	92.223%	
			31	30.975363	β^-, γ	157.3 min
15	Phosphorus	P	31	30.973762	100%	
			32	31.973908	β^-	14.262 days

† The masses (atomic mass units) given in column (5) are those for the neutral atom, including the Z electrons (except for the proton, deuteron, triton).

‡ Chapter 30; EC = electron capture.

(1) Atomic Number Z	(2) Element	(3) Symbol	(4) Mass Number A	(5) Atomic Mass	(6) % Abundance (or Radioactive Decay Mode)	(7) Half-life (if radioactive)
16	Sulfur	S	32	31.972071	94.99%	
			35	34.969032	β^-	87.37 days
17	Chlorine	Cl	35	34.968853	75.76%	
			37	36.965903	24.24%	
18	Argon	Ar	40	39.962383	99.6035%	
19	Potassium	K	39	38.963706	93.2581%	
			40	39.963998	0.0117%	
					β^-, EC, γ, β^+	1.248×10^9 yr
20	Calcium	Ca	40	39.962591	96.94%	
21	Scandium	Sc	45	44.955908	100%	
22	Titanium	Ti	48	47.947942	73.72%	
23	Vanadium	V	51	50.943957	99.750%	
24	Chromium	Cr	52	51.940506	83.789%	
25	Manganese	Mn	55	54.938044	100%	
26	Iron	Fe	56	55.934936	91.754%	
27	Cobalt	Co	59	58.933194	100%	
			60	59.933816	β^-, γ	5.2713 yr
28	Nickel	Ni	58	57.935342	68.077%	
			60	59.930786	26.223%	
29	Copper	Cu	63	62.929598	69.15%	
			65	64.927790	30.85%	
30	Zinc	Zn	64	63.929142	49.17%	
			66	65.926034	27.73%	
31	Gallium	Ga	69	68.925574	60.108%	
32	Germanium	Ge	72	71.922076	27.45%	
			74	73.921178	36.50%	
33	Arsenic	As	75	74.921595	100%	
34	Selenium	Se	80	79.916522	49.61%	
35	Bromine	Br	79	78.918338	50.69%	
36	Krypton	Kr	84	83.911498	56.987%	
37	Rubidium	Rb	85	84.911790	72.17%	
38	Strontium	Sr	86	85.909261	9.86%	
			88	87.905612	82.58%	
			90	89.907730	β^-	28.90 yr
39	Yttrium	Y	89	88.905840	100%	
40	Zirconium	Zr	90	89.904698	51.45%	
41	Niobium	Nb	93	92.906373	100%	
42	Molybdenum	Mo	98	97.905405	24.39%	
43	Technetium	Tc	98	97.907212	β^-, γ	4.2×10^6 yr
44	Ruthenium	Ru	102	101.904344	31.55%	
45	Rhodium	Rh	103	102.905498	100%	
46	Palladium	Pd	106	105.903480	27.33%	
47	Silver	Ag	107	106.905092	51.839%	
			109	108.904755	48.161%	
48	Cadmium	Cd	114	113.903365	28.73%	
49	Indium	In	115	114.903879	95.71%; β^-	4.41×10^{14} yr
50	Tin	Sn	120	119.902202	32.58%	
51	Antimony	Sb	121	120.903812	57.21%	

(1) Atomic Number Z	(2) Element	(3) Symbol	(4) Mass Number A	(5) Atomic Mass	(6) % Abundance (or Radioactive Decay Mode)	(7) Half-life (if radioactive)
52	Tellurium	Te	130	129.906223	34.08%; $\beta^-\beta^-$	$>3.0 \times 10^{24}$ yr
53	Iodine	I	127	126.904472	100%	
			131	130.906126	β^-, γ	8.0252 days
54	Xenon	Xe	132	131.904155	26.9086%	
			136	135.907214	8.8573%; $\beta^-\beta^-$	$>2.4 \times 10^{21}$ yr
55	Cesium	Cs	133	132.905452	100%	
56	Barium	Ba	137	136.905827	11.232%	
			138	137.905247	71.698%	
57	Lanthanum	La	139	138.906356	99.9119%	
58	Cerium	Ce	140	139.905443	88.450%	
59	Praseodymium	Pr	141	140.907658	100%	
60	Neodymium	Nd	142	141.907729	27.152%	
61	Promethium	Pm	145	144.912756	EC, α	17.7 yr
62	Samarium	Sm	152	151.919740	26.75%	
63	Europium	Eu	153	152.921238	52.19%	
64	Gadolinium	Gd	158	157.924112	24.84%	
65	Terbium	Tb	159	158.925355	100%	
66	Dysprosium	Dy	164	163.929182	28.260%	
67	Holmium	Ho	165	164.930329	100%	
68	Erbium	Er	166	165.930300	33.503%	
69	Thulium	Tm	169	168.934218	100%	
70	Ytterbium	Yb	174	173.938866	31.026%	
71	Lutetium	Lu	175	174.940775	97.401%	
72	Hafnium	Hf	180	179.946557	35.08%	
73	Tantalum	Ta	181	180.947996	99.98799%	
74	Tungsten (wolfram)	W	184	183.950931	30.64%; α	$>8.9 \times 10^{21}$ yr
75	Rhenium	Re	187	186.955750	62.60%; β^-	4.33×10^{10} yr
76	Osmium	Os	191	190.960926	β^-, γ	15.4 days
			192	191.961477	40.78%	
77	Iridium	Ir	191	190.960589	37.3%	
			193	192.962922	62.7%	
78	Platinum	Pt	195	194.964792	33.78%	
79	Gold	Au	197	196.966569	100%	
80	Mercury	Hg	199	198.968281	16.87%	
			202	201.970643	29.86%	
81	Thallium	Tl	205	204.974428	70.48%	
82	Lead	Pb	206	205.974466	24.1%	
			207	206.975897	22.1%	
			208	207.976652	52.4%	
			210	209.984189	β^-, γ, α	22.20 yr
			211	210.988737	β^-, γ	36.1 min
			212	211.991898	β^-, γ	10.64 h
			214	213.999806	β^-, γ	26.8 min
83	Bismuth	Bi	209	208.980399	100%	
			211	210.987270	α, γ, β^-	2.14 min
84	Polonium	Po	210	209.982874	α, γ, EC	138.376 days
			214	213.995202	α, γ	164.3 μs
85	Astatine	At	218	218.008695	α, β^-	1.5 s

(1) Atomic Number Z	(2) Element	(3) Symbol	(4) Mass Number A	(5) Atomic Mass	(6) % Abundance (or Radioactive Decay Mode)	(7) Half-life (if radioactive)
86	Radon	Rn	222	222.017578	α, γ	3.8235 days
87	Francium	Fr	223	223.019736	β^-, γ, α	22.00 min
88	Radium	Ra	226	226.025410	α, γ	1600 yr
89	Actinium	Ac	227	227.027752	β^-, γ, α	21.772 yr
90	Thorium	Th	228	228.028741	α, γ	1.9116 yr
			232	232.038056	100%; α, γ	1.40×10^{10} yr
91	Protactinium	Pa	231	231.035884	α, γ	3.276×10^4 yr
92	Uranium	U	232	232.037156	α, γ	68.9 yr
			233	233.039636	α, γ	1.592×10^5 yr
			235	235.043930	0.7204%; α, γ	7.04×10^8 yr
			236	236.045568	α, γ	2.342×10^7 yr
			238	238.050788	99.2742%; α, γ	4.468×10^9 yr
			239	239.054294	β^-, γ	23.45 min
93	Neptunium	Np	237	237.048174	α, γ	2.144×10^6 yr
			239	239.052939	β^-, γ	2.356 days
94	Plutonium	Pu	239	239.052164	α, γ	24,110 yr
			244	244.064205	α	8.00×10^7 yr
95	Americium	Am	243	243.061381	α, γ	7370 yr
96	Curium	Cm	247	247.070354	α, γ	1.56×10^7 yr
97	Berkelium	Bk	247	247.070307	α, γ	1380 yr
98	Californium	Cf	251	251.079589	α, γ	898 yr
99	Einsteinium	Es	252	252.082980	α, EC, γ	471.7 days
100	Fermium	Fm	257	257.095106	α, γ	100.5 days
101	Mendelevium	Md	258	258.098431	α, γ	51.5 days
102	Nobelium	No	259	259.101030	α, EC	58 min
103	Lawrencium	Lr	262	262.109610	α, EC, fission	≈ 4 h
104	Rutherfordium	Rf	263	263.112500	fission	10 min
105	Dubnium	Db	268	268.125670	fission	32 h
106	Seaborgium	Sg	271	271.133930	α, fission	2.4 min
107	Bohrium	Bh	274	274.143550	α, fission	0.9 min
108	Hassium	Hs	270	270.134290	α	22 s
109	Meitnerium	Mt	278	278.156310	α, fission	8 s
110	Darmstadtium	Ds	281	281.164510	α, fission	20 s
111	Roentgenium	Rg	281	281.166360	α, fission	26 s
112	Copernicium	Cn	285	285.177120	α	30 s
113[†]			286	286.18210	α, fission	20 s
114	Flerovium	Fl	289	289.190420	α	2.7 s
115[†]			289	289.193630	α, fission	0.22 s
116	Livermorium	Lv	293	293.204490	α	53 ms
117[†]			294	294.210460	α	0.08 s
118[†]			294	294.213920	α, fission	0.9 ms

[†] Preliminary evidence (unconfirmed) has been reported for elements 113, 115, 117, and 118.

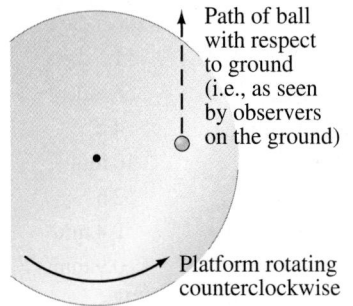

People on ground appear to move this way

Path of ball with respect to rotating platform (i.e., as seen by observer on platform)

(a) Rotating reference frame

Path of ball with respect to ground (i.e., as seen by observers on the ground)

Platform rotating counterclockwise

(b) Earth reference frame

FIGURE C–1 Path of a ball released on a rotating merry-go-round (a) in the reference frame of the merry-go-round, and (b) in a reference frame fixed on the ground.

Rotating Frames of Reference; Inertial Forces; Coriolis Effect

Inertial and Noninertial Reference Frames

In Chapters 5 and 8 we examined the motion of objects, including circular and rotational motion, from the outside, as observers fixed on the Earth. Sometimes it is convenient to place ourselves (in theory, if not physically) into a reference frame that is rotating. Let us examine the motion of objects from the point of view, or frame of reference, of persons seated on a rotating platform such as a merry-go-round. It looks to them as if the rest of the world is going around *them*. But let us focus on what they observe when they place a tennis ball on the floor of the rotating platform, which we assume is frictionless. If they put the ball down gently, without giving it any push, they will observe that it accelerates from rest and moves outward as shown in Fig. C–1a. According to Newton's first law, an object initially at rest should stay at rest if no net force acts on it. But, according to the observers on the rotating platform, the ball starts moving even though there is no net force acting on it. To observers on the ground, this is all very clear: the ball has an initial velocity when it is released (because the platform is moving), and it simply continues moving in a straight-line path as shown in Fig. C–1b, in accordance with Newton's first law.

But what shall we do about the frame of reference of the observers on the rotating platform? Newton's first law, the law of inertia, does not hold in this rotating frame of reference since the ball starts moving with no net force on it. For this reason, such a frame is called a **noninertial reference frame**. An **inertial reference frame** (as discussed in Chapter 4) is one in which the law of inertia—Newton's first law—does hold, and so do Newton's second and third laws. In a noninertial reference frame, such as our rotating platform, Newton's second law also does not hold. For instance in the situation described above, there is no net force on the ball; yet, with respect to the rotating platform, the ball accelerates.

Fictitious (Inertial) Forces

Because Newton's laws do not hold when observations are made with respect to a rotating frame of reference, calculation of motion can be complicated. However, we can still apply Newton's laws in such a reference frame if we make use of a trick. The ball on the rotating platform of Fig. C–1a flies outward when released (even though no force is actually acting on it). So the trick we use is to write down the equation $\Sigma F = ma$ as if a force equal to mv^2/r (or $m\omega^2 r$) were acting radially outward on the object in addition to any other forces that may be acting. This extra force, which might be designated as "centrifugal force" since it *seems* to act outward, is called a **fictitious force** or **pseudoforce**. It is a pseudoforce ("pseudo" means "false") because there is no object that exerts this force. Furthermore, when viewed from an inertial reference frame, the effect doesn't exist at all. We have made up this pseudoforce so that we can make calculations in a noninertial frame using Newton's second law, $\Sigma F = ma$. Thus the observer in the noninertial frame of Fig. C–1a uses Newton's second law for the ball's outward motion by assuming that a force equal to mv^2/r acts on it. Such pseudoforces are also called **inertial forces** since they arise only because the reference frame is not an inertial one.

In Section 5–3 we discussed the forces on a person in a car going around a curve (Fig. 5–11) from the point of view of an inertial frame. The car, on the other hand, is not an inertial frame. Passengers in such a car could interpret this being pressed outward as the effect of a "centrifugal" force. But they need to recognize that it is a pseudoforce because there is no identifiable object exerting it. It is an effect of being in a noninertial frame of reference.

The Earth itself is rotating on its axis. Thus, strictly speaking, Newton's laws are not valid on the Earth. However, the effect of the Earth's rotation is usually so small that it can be ignored, although it does influence the movement of large air masses and ocean currents. Because of the Earth's rotation, the material of the Earth is concentrated slightly more at the equator. The Earth is thus not a perfect sphere but is slightly fatter at the equator than at the poles.

Coriolis Effect

In a reference frame that rotates at a constant angular speed ω (relative to an inertial frame), there exists another pseudoforce known as the *Coriolis force*. It appears to act on an object in a rotating reference frame only if the object is moving relative to that rotating reference frame, and it acts to deflect the object sideways. It, too, is an effect of the rotating reference frame being noninertial and hence is referred to as an *inertial force*. It also affects the weather.

To see how the Coriolis force arises, consider two people, A and B, at rest on a platform rotating with angular speed ω, as shown in Fig. C–2a. They are situated at distances r_A and r_B from the axis of rotation (at O). The woman at A throws a ball with a horizontal velocity $\vec{v}$ (in her reference frame) radially outward toward the man at B on the outer edge of the platform. In Fig. C–2a, we view the situation from an inertial reference frame. The ball initially has not only the velocity $\vec{v}$ radially outward, but also a tangential velocity $\vec{v}_A$ due to the rotation of the platform. Now Eq. 8–4 tells us that $v_A = r_A \omega$, where r_A is the woman's radial distance from the axis of rotation at O. If the man at B had this same velocity v_A, the ball would reach him perfectly. But his speed is $v_B = r_B \omega$, which is greater than v_A because $r_B > r_A$. Thus, when the ball reaches the outer edge of the platform, it passes a point that the man at B has already gone by because his speed in that direction is greater than the ball's. So the ball passes behind him.

Figure C–2b shows the situation as seen from the rotating platform as frame of reference. Both A and B are at rest, and the ball is thrown with velocity $\vec{v}$ toward B, but the ball deflects to the right as shown and passes behind B as previously described. This is not a centrifugal-force effect, because that would act radially outward. Instead, this effect acts sideways, perpendicular to $\vec{v}$, and is called a **Coriolis acceleration**; it is said to be due to the Coriolis force, which is a fictitious inertial force. Its explanation as seen from an inertial system was given above: it is an effect of being in a rotating system, for which a point farther from the rotation axis has a higher linear speed. On the other hand, when viewed from the rotating system, the motion can be described using Newton's second law, $\Sigma \vec{F} = m\vec{a}$, if we add a "pseudoforce" term corresponding to this Coriolis effect.

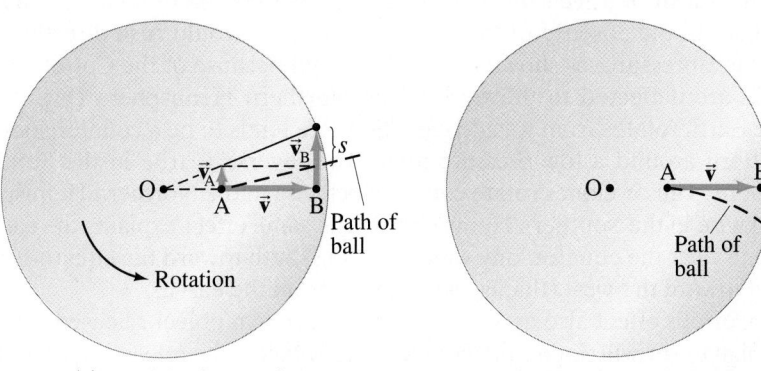

(a) Inertial reference frame   (b) Rotating reference frame

FIGURE C–2 The origin of the Coriolis effect. Looking down on a rotating platform, (a) as seen from a nonrotating inertial reference frame, and (b) as seen from the rotating platform as frame of reference.

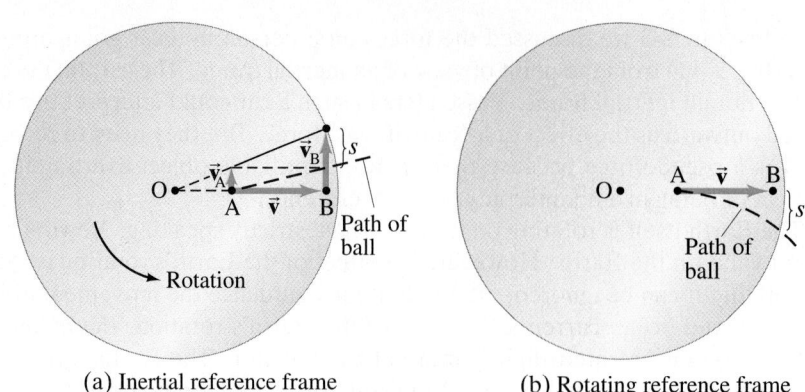

FIGURE C–2 (Repeated.) The origin of the Coriolis effect. Looking down on a rotating platform, (a) as seen from a nonrotating inertial reference frame, and (b) as seen from the rotating platform as frame of reference.

(a) Inertial reference frame

(b) Rotating reference frame

Let us determine the magnitude of the Coriolis acceleration for the simple case described above. (We assume v is large and distances short, so we can ignore gravity.) We do the calculation from the inertial reference frame (Fig. C–2a). The ball moves radially outward a distance $r_B - r_A$ at speed v in a short time t given by

$$r_B - r_A = vt.$$

During this time, the ball moves to the side a distance s_A given by

$$s_A = v_A t.$$

The man at B, in this time t, moves a distance

$$s_B = v_B t.$$

The ball therefore passes behind him a distance s (Fig. C–2a) given by

$$s = s_B - s_A = (v_B - v_A)t.$$

We saw earlier that $v_A = r_A \omega$ and $v_B = r_B \omega$, so

$$s = (r_B - r_A)\omega t.$$

We substitute $r_B - r_A = vt$ (see above) and get

$$s = \omega v t^2. \tag{C–1}$$

This same s equals the sideways displacement as seen from the noninertial rotating system (Fig. C–2b).

Equation C–1 corresponds to motion at constant acceleration, because as we saw in Chapter 2 (Eq. 2–11b), $y = \frac{1}{2}at^2$ for a constant acceleration (with zero initial velocity in the y direction). Thus, if we write Eq. C–1 in the form $s = \frac{1}{2}a_{Cor}t^2$, we see that the Coriolis acceleration a_{Cor} is

$$a_{Cor} = 2\omega v. \tag{C–2}$$

This relation is valid for any velocity in the plane of rotation perpendicular to the axis of rotation (in Fig. C–2, the axis through point O perpendicular to the page).

Because the Earth rotates, the Coriolis effect has some interesting manifestations on the Earth. It affects the movement of air masses and thus has an influence on weather. In the absence of the Coriolis effect, air would rush directly into a region of low pressure, as shown in Fig. C–3a. But because of the Coriolis effect, the winds are deflected to the right in the Northern Hemisphere (Fig. C–3b), since the Earth rotates from west to east. So there tends to be a counterclockwise wind pattern around a low-pressure area. The reverse is true in the Southern Hemisphere. Thus cyclones rotate counterclockwise in the Northern Hemisphere and clockwise in the Southern Hemisphere. The same effect explains the easterly trade winds near the equator: any winds heading south toward the equator will be deflected toward the west (that is, as if coming from the east).

The Coriolis effect also acts on a falling object. An object released from the top of a high tower will not hit the ground directly below the release point, but will be deflected slightly to the east. Viewed from an inertial frame, this happens because the top of the tower revolves with the Earth at a slightly higher speed than the bottom of the tower.

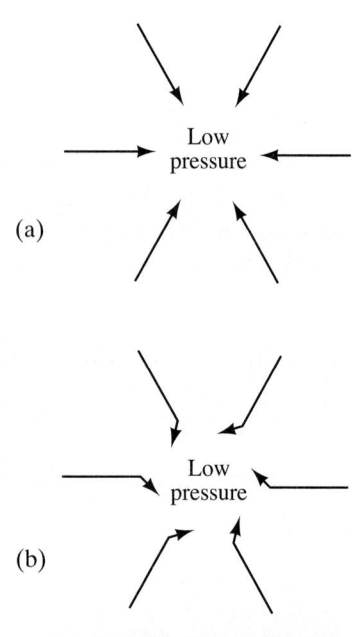

(a)

(b)

(c)

FIGURE C–3 (a) Winds (moving air masses) would flow directly toward a low-pressure area if the Earth did not rotate; (b) and (c): because of the Earth's rotation, the winds are deflected to the right in the Northern Hemisphere (as in Fig. C–2) as if a fictitious (Coriolis) force were acting.

Molar Specific Heats for Gases, and the Equipartition of Energy

Molar Specific Heats for Gases

The values of the specific heats for gases depend on how the thermodynamic process is carried out. Two important processes are those in which either the volume or the pressure is kept constant, and Table D–1 shows how different they can be.

The difference in specific heats for gases is nicely explained in terms of the first law of thermodynamics and kinetic theory. For gases we usually use **molar specific heats**, C_V and C_P, which are defined as the heat required to raise 1 mol of a gas by 1 K (or 1 C°) at constant volume and at constant pressure, respectively. In analogy to Eq. 14–2, the heat Q needed to raise the temperature of n moles of gas by ΔT is

$$Q = nC_V \Delta T \qquad \text{[volume constant]} \quad \textbf{(D–1a)}$$

$$Q = nC_P \Delta T. \qquad \text{[pressure constant]} \quad \textbf{(D–1b)}$$

We can see from the definition of molar specific heat (compare Eqs. 14–2 and D–1) that

$$C_V = Mc_V$$

$$C_P = Mc_P,$$

where M is the molecular mass of the gas ($M = m/n$ in grams/mol).[†] The values for molar specific heats are included in Table D–1. These values are nearly the same for different gases that have the same number of atoms per molecule.

Now we use kinetic theory and imagine that an ideal gas is slowly heated via two different processes—first at constant volume, and then at constant pressure. In both processes, we let the temperature increase by the same amount, ΔT.

[†]For example, $M = 2\,\text{g/mol}$ for He, and $M = 32\,\text{g/mol}$ for O_2.

TABLE D–1 Specific Heats of Gases at 15°C					
Gas	Specific Heats (kcal/kg · K)		Molar Specific Heats (cal/mol · K)		$C_P - C_V$ (cal/mol · K)
	c_V	c_P	C_V	C_P	
Monatomic					
He	0.75	1.15	2.98	4.97	1.99
Ne	0.148	0.246	2.98	4.97	1.99
Diatomic					
N_2	0.177	0.248	4.96	6.95	1.99
O_2	0.155	0.218	5.03	7.03	2.00
Triatomic					
CO_2	0.153	0.199	6.80	8.83	2.03
H_2O (100°C)	0.350	0.482	6.20	8.20	2.00
Polyatomic					
C_2H_6	0.343	0.412	10.30	12.35	2.05

In the constant-volume process, no work is done since $\Delta V = 0$. Thus, according to the first law of thermodynamics ($Q = \Delta U + W$, Section 15–1), the heat added (which we denote by Q_V) all goes into increasing the internal energy of the gas:

$$Q_V = \Delta U.$$

In the constant-pressure process, work *is* done. Hence the heat added, Q_P, must not only increase the internal energy but also is used to do work $W = P\,\Delta V$. Thus, to increase the temperature by the same ΔT, more heat must be added in the process at constant pressure than in the process at constant volume. For the process at constant pressure, the first law of thermodynamics gives

$$Q_P = \Delta U + P\,\Delta V.$$

Since ΔU is the same in the two processes (we chose ΔT to be the same), we can combine the two above equations:

$$Q_P - Q_V = P\,\Delta V.$$

From the ideal gas law, $V = nRT/P$, so for a process at constant pressure $\Delta V = nR\,\Delta T/P$. Putting this into the last equation and using Eqs. D–1, we find

$$nC_P\,\Delta T - nC_V\,\Delta T = P\!\left(\frac{nR\,\Delta T}{P}\right)$$

or, after cancellations,

$$C_P - C_V = R. \tag{D–2}$$

Since the gas constant $R = 8.314\,\text{J/mol}\cdot\text{K} = 1.99\,\text{cal/mol}\cdot\text{K}$, our prediction is that C_P will be larger than C_V by about $1.99\,\text{cal/mol}\cdot\text{K}$. Indeed, this is very close to what is obtained experimentally, as shown in the last column in Table D–1.

Now we calculate the molar specific heat of a monatomic gas using kinetic theory. For a process carried out at constant volume, no work is done, so the first law of thermodynamics tells us that

$$\Delta U = Q_V.$$

For an ideal monatomic gas, the internal energy, U, is the total kinetic energy of all the molecules,

$$U = N\!\left(\tfrac{1}{2}m\overline{v^2}\right) = \tfrac{3}{2}nRT$$

as we saw in Section 14–2. Then, using Eq. D–1a, we write $\Delta U = Q_V$ as

$$\Delta U = \tfrac{3}{2}nR\,\Delta T = nC_V\,\Delta T \tag{D–3}$$

or

$$C_V = \tfrac{3}{2}R. \tag{D–4}$$

Since $R = 8.314\,\text{J/mol}\cdot\text{K} = 1.99\,\text{cal/mol}\cdot\text{K}$, kinetic theory predicts that $C_V = 2.98\,\text{cal/mol}\cdot\text{K}$ for an ideal monatomic gas. This is very close to the experimental values for monatomic gases such as helium and neon (Table D–1). From Eq. D–2, C_P is predicted to be $R + C_V = (1.99 + 2.98)\,\text{cal/mol}\cdot\text{K} = 4.97\,\text{cal/mol}\cdot\text{K}$, also in agreement with experiment (Table D–1).

Equipartition of Energy

The measured molar specific heats for more complex gases (Table D–1), such as diatomic (two atoms) and triatomic (three atoms) gases, increase with the increased number of atoms per molecule. We can explain this by assuming that the internal energy includes not only translational kinetic energy but other forms of energy as well. For example, in a diatomic gas (Fig. D–1), the two atoms can rotate about two different axes (but rotation about a third axis passing through the two atoms would give rise to very little energy since the moment of inertia is so small). The molecules can have rotational as well as translational kinetic energy.

It is useful to introduce the idea of **degrees of freedom**, by which we mean the number of independent ways molecules can possess energy. For example, a monatomic gas has three degrees of freedom, because an atom can have velocity along the x, y, and z axes. These are considered to be three independent motions because a change in any one of the components would not affect the others. A diatomic molecule has the same three degrees of freedom associated with translational kinetic energy plus two more degrees of freedom associated with rotational kinetic energy (Fig. D–1), for a total of five degrees of freedom.

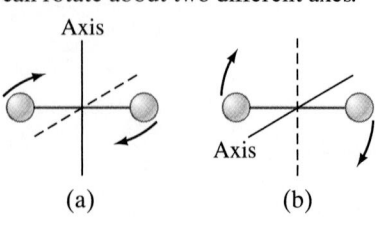

FIGURE D–1 A diatomic molecule can rotate about two different axes.

Axis

(a) (b)

Axis

Table D–1 indicates that the C_V for diatomic gases is about $\frac{5}{3}$ times as great as for a monatomic gas—that is, in the same ratio as their degrees of freedom. This result led nineteenth-century physicists to the **principle of equipartition of energy**. This principle states that energy is shared equally among the active degrees of freedom, and each active degree of freedom of a molecule has on average an energy equal to $\frac{1}{2}kT$. Thus, the average energy for a molecule of a monatomic gas would be $\frac{3}{2}kT$ (which we already knew) and of a diatomic gas $\frac{5}{2}kT$. Hence the internal energy of a diatomic gas would be $U = N\left(\frac{5}{2}kT\right) = \frac{5}{2}nRT$, where n is the number of moles. Using the same argument we did for monatomic gases, we see that for diatomic gases the molar specific heat at constant volume would be $\frac{5}{2}R = 4.97\ \text{cal/mol·K}$, close to measured values (Table D–1). More complex molecules have even more degrees of freedom and thus greater molar specific heats.

However, measurements showed that for diatomic gases at very low temperatures, C_V has a value of only $\frac{3}{2}R$, as if it had only three degrees of freedom. And at very high temperatures, C_V was about $\frac{7}{2}R$, as if there were seven degrees of freedom. The explanation is that at low temperatures, nearly all molecules have only translational kinetic energy; that is, no energy goes into rotational energy and only three degrees of freedom are "active." At very high temperatures, all five degrees of freedom are active plus two additional ones. We interpret the two new degrees of freedom as being associated with the two atoms vibrating, as if they were connected by a spring (Fig. D–2). One degree of freedom comes from the kinetic energy of the vibrational motion, and the second comes from the potential energy of vibrational motion $\left(\frac{1}{2}kx^2\right)$. At room temperature, these two degrees of freedom are apparently not active (Fig. D–3). Why fewer degrees of freedom are "active" at lower temperatures was eventually explained by Einstein using quantum theory.

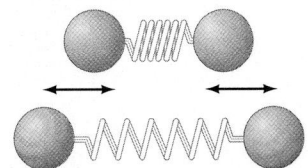

FIGURE D–2 A diatomic molecule can vibrate, as if the two atoms were connected by a spring. They aren't, but rather they exert forces on each other that are electrical in nature—of a form that resembles a spring force.

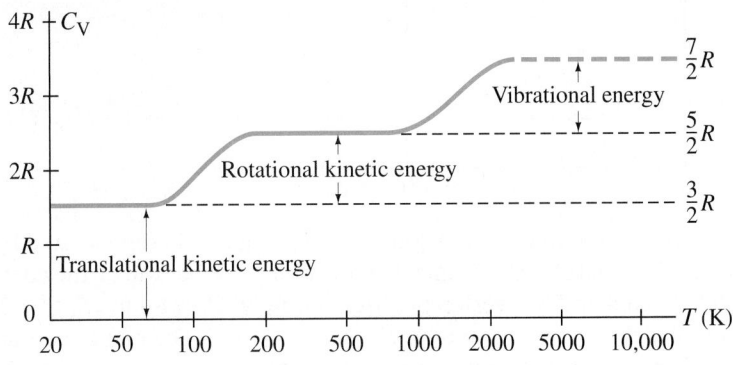

FIGURE D–3 Molar specific heat C_V as a function of temperature for hydrogen molecules (H_2). As the temperature is increased, some of the translational kinetic energy can be transferred in collisions into rotational kinetic energy and, at still higher temperature, into vibrational kinetic and potential energy. [Note: H_2 dissociates into two atoms at about 3200 K, so the last part of the curve is shown dashed.]

Solids

The principle of equipartition of energy can be applied to solids as well. The molar specific heat of any solid at high temperature is close to $3R$ ($6.0\ \text{cal/mol·K}$), Fig. D–4. This is called the *Dulong and Petit value* after the scientists who first measured it in 1819. (Note that Table 14–1 gave the specific heats per kilogram, not per mole.) At high temperatures, each atom apparently has six degrees of freedom, although some are not active at low temperatures. Each atom in a crystalline solid can vibrate about its equilibrium position as if it were connected by springs to each of its neighbors (Fig. D–5). Thus it can have three degrees of freedom for kinetic energy and three more associated with potential energy of vibration in each of the x, y, and z directions, which is in accord with measured values.

FIGURE D–4 Molar specific heats of solids as a function of temperature.

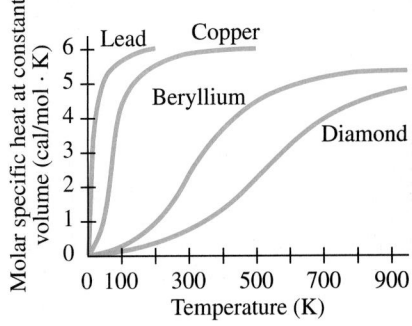

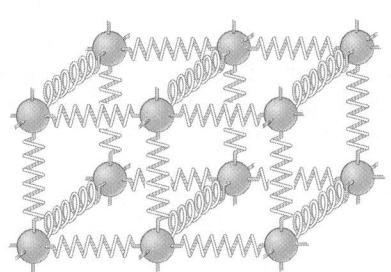

FIGURE D–5 The atoms in a crystalline solid can vibrate about their equilibrium positions as if they were connected to their neighbors by springs. (The forces between atoms are actually electrical in nature.)

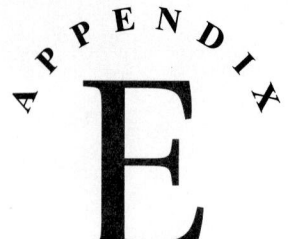

Galilean and Lorentz Transformations

We examine in detail the mathematics of relating quantities in one inertial reference frame to the equivalent quantities in another. In particular, we will see how positions and velocities *transform* (that is, change) from one frame of reference to the other.

We begin with the classical, or Galilean, viewpoint. Consider two inertial reference frames S and S′ which are each characterized by a set of coordinate axes, Fig. E–1. The axes x and y (z is not shown) refer to S, and x' and y' refer to S′. The x' and x axes overlap one another, and we assume that frame S′ moves to the right (in the x direction) at speed v with respect to S. For simplicity let us assume the origins 0 and 0′ of the two reference frames are superimposed at time $t = 0$.

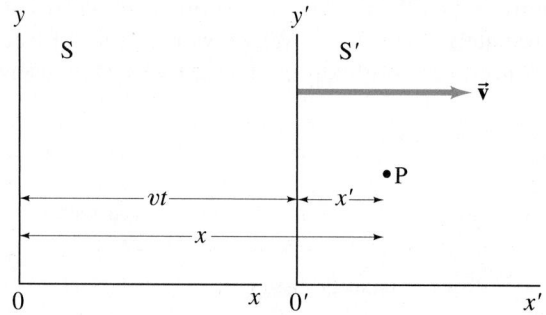

FIGURE E–1 Inertial reference frame S′ moves to the right at speed v with respect to inertial frame S.

Now consider an event that occurs at some point P (Fig. E–1) represented by the coordinates x', y', z' in reference frame S′ at the time t'. What will be the coordinates of P in S? Since S and S′ overlap precisely initially, after a time t', S′ will have moved a distance vt'. Therefore, at time t', $x = x' + vt'$. The y and z coordinates, on the other hand, are not altered by motion along the x axis; thus $y = y'$ and $z = z'$. Finally, since time is assumed to be absolute in Galilean–Newtonian physics, clocks in the two frames will agree with each other; so $t = t'$. We summarize these in the following **Galilean transformation equations**:

$$x = x' + vt'$$
$$y = y'$$
$$z = z' \qquad\qquad \text{[Galilean]} \quad \textbf{(E–1)}$$
$$t = t'.$$

These equations give the coordinates of an event in the S frame when those in the S′ frame are known. If those in the S frame are known, then the S′ coordinates are obtained from

$$x' = x - vt, \qquad y' = y, \qquad z' = z, \qquad t' = t. \qquad \text{[Galilean]}$$

These four equations are the "inverse" transformation, and are obtained from Eqs. E–1 by exchanging primed and unprimed quantities and replacing v by $-v$. This makes sense because, as seen from the S′ frame, S moves to the left (negative x direction) with speed v.

Now suppose that the point P in Fig. E–1 represents an object that is moving. Let the components of its velocity vector in S' be u'_x, u'_y, and u'_z. (We use u to distinguish it from the relative velocity of the two frames, v.) Now $u'_x = \Delta x'/\Delta t'$, $u'_y = \Delta y'/\Delta t'$, and $u'_z = \Delta z'/\Delta t'$, where all quantities are as measured in the S' frame. For example, if at time t'_1 the particle is at x'_1 and a short time later, t'_2, it is at x'_2, then

$$u'_x = \frac{x'_2 - x'_1}{t'_2 - t'_1} = \frac{\Delta x'}{\Delta t'}.$$

The velocity of P as seen from S will have components u_x, u_y, and u_z. We can show how these are related to the velocity components in S' by using Eqs. E–1. For example,

$$u_x = \frac{\Delta x}{\Delta t} = \frac{x_2 - x_1}{t_2 - t_1} = \frac{(x'_2 + vt'_2) - (x'_1 + vt'_1)}{t'_2 - t'_1}$$

$$= \frac{(x'_2 - x'_1) + v(t'_2 - t'_1)}{t'_2 - t'_1}$$

$$= \frac{\Delta x'}{\Delta t'} + v = u'_x + v.$$

For the other components, $u'_y = u_y$ and $u'_z = u_z$, so we have

$$u_x = u'_x + v$$
$$u_y = u'_y \qquad\qquad\qquad \text{[Galilean]} \quad \textbf{(E–2)}$$
$$u_z = u'_z.$$

These are known as the **Galilean velocity transformation equations**. We see that the y and z components of velocity are unchanged, but the x components differ by v. This is just what we have used before when dealing with relative velocity (Section 3–8). For example, if S' is a train and S the Earth, and the train moves with speed v with respect to Earth, a person walking toward the front of the train with speed u'_x will have a speed with respect to the Earth of $u_x = u'_x + v$.

The Galilean transformations, Eqs. E–1 and E–2, are accurate only when the velocities involved are not relativistic (Chapter 26)—that is, much less than the speed of light, c. We can see, for example, that the first of Eqs. E–2 will not work for the speed of light, c, which is the same in all inertial reference frames (a basic postulate in the theory of relativity). That is, light traveling in S' with speed $u'_x = c$ will have speed $c + v$ in S, according to Eq. E–2, whereas the theory of relativity insists it must be c in S. Clearly, then, a new set of transformation equations is needed to deal with relativistic velocities.

We will derive the required equations, again looking at Fig. E–1. We assume the transformation is linear and for x is of the form

$$x = \gamma(x' + vt'). \qquad\qquad\qquad\qquad\qquad \textbf{(i)}$$

That is, we modify the first of Eqs. E–1 by multiplying by a factor γ which is yet to be determined.[†] We assume the y and z equations are unchanged

$$y = y', \quad z = z'$$

because there is no length contraction in these directions. We will not assume a form for t, but will derive it. The inverse equations must have the same form with v replaced by $-v$. (The principle of relativity demands it, since S' moving to the right with respect to S is equivalent to S moving to the left with respect to S'.) Therefore

$$x' = \gamma(x - vt). \qquad\qquad\qquad\qquad\qquad \textbf{(ii)}$$

Suppose a light pulse leaves the common origin of S and S' at time $t = t' = 0$.

[†] We are NOT assuming γ is $1/\sqrt{1 - v^2/c^2}$, as in Chapter 26. Our minds are open. Let's see.

Then after a time t it will have traveled along the x axis a distance $x = ct$ (in S), or $x' = ct'$ (in S'). Therefore, from Eqs. (i) and (ii) above,

$$ct = x = \gamma(ct' + vt') = \gamma(c + v)t', \qquad \text{(iii)}$$
$$ct' = x' = \gamma(ct - vt) = \gamma(c - v)t. \qquad \text{(iv)}$$

From Eq. (iv), $t' = \gamma(c - v)(t/c)$, and we substitute this into Eq. (iii) and find $ct = \gamma(c + v)\gamma(c - v)(t/c) = \gamma^2(c^2 - v^2)t/c$. We cancel out the t on each side and solve for γ to find

$$\gamma = \frac{1}{\sqrt{1 - v^2/c^2}}.$$

We have found that γ is, in fact, the value for γ we used in Chapter 26, Eq. 26–2.

Now that we have found γ, we need only find the relation between t and t'. To do so, we combine $x' = \gamma(x - vt)$ with $x = \gamma(x' + vt')$:

$$x' = \gamma(x - vt) = \gamma\big[\gamma(x' + vt') - vt\big].$$

We solve for t, doing some algebra, and find $t = \gamma(t' + vx'/c^2)$. In summary,

LORENTZ TRANSFORMATIONS

$$
\begin{aligned}
x &= \frac{1}{\sqrt{1 - v^2/c^2}}(x' + vt') \\
y &= y' \\
z &= z' \\
t &= \frac{1}{\sqrt{1 - v^2/c^2}}\left(t' + \frac{vx'}{c^2}\right).
\end{aligned}
\qquad \text{(E–3)}
$$

These are called the **Lorentz transformation equations**. They were first proposed, in a slightly different form, by Lorentz in 1904 to explain the null result of the Michelson–Morley experiment and to make Maxwell's equations take the same form in all inertial reference frames. A year later, Einstein derived them independently based on his theory of relativity. Notice that not only is the x equation modified as compared to the Galilean transformation, but so is the t equation. Indeed, we see directly in this last equation how the space and time coordinates mix.

The relativistically correct velocity equations are obtained using Eqs. E–3 (we let $\gamma = 1/\sqrt{1 - v^2/c^2}$) and $u_x = \Delta x/\Delta t$, $u'_x = \Delta x'/\Delta t'$:

$$
\begin{aligned}
u_x &= \frac{\Delta x}{\Delta t} = \frac{\gamma(\Delta x' + v\,\Delta t')}{\gamma(\Delta t' + v\,\Delta x'/c^2)} = \frac{(\Delta x'/\Delta t') + v}{1 + (v/c^2)(\Delta x'/\Delta t')} \\
&= \frac{u'_x + v}{1 + vu'_x/c^2}.
\end{aligned}
$$

The others are obtained in the same way, and we collect them here:

RELATIVISTIC VELOCITY TRANSFORMATIONS

$$
\begin{aligned}
u_x &= \frac{u'_x + v}{1 + vu'_x/c^2} \\
u_y &= \frac{u'_y\sqrt{1 - v^2/c^2}}{1 + vu'_x/c^2} \\
u_z &= \frac{u'_z\sqrt{1 - v^2/c^2}}{1 + vu'_x/c^2}.
\end{aligned}
\qquad \text{(E–4)}
$$

The first of these equations is Eq. 26–11, which we used in Section 26–10 where we discussed how velocities do not add in our commonsense (Galilean) way, because of the denominator $(1 + vu'_x/c^2)$. We can now also see that the y and z components of velocity are also altered and that they depend on the x' component of velocity.

EXAMPLE E–1 **Length contraction.** Derive the length contraction formula, Eq. 26–3, from the Lorentz transformation equations.

APPROACH We consider measurements in two reference frames, S and S', that move with speed v relative to each other, as in Fig. E–1.

SOLUTION Let an object of length ℓ_0 be at rest on the x axis in S. The coordinates of its two end points are x_1 and x_2, so that $x_2 - x_1 = \ell_0$. At any instant in S', the end points will be at x_1' and x_2' as given by the Lorentz transformation equations. The length measured in S' is $\ell = x_2' - x_1'$. An observer in S' measures this length by measuring x_2' and x_1' at the same time (in the S' frame), so $t_2' = t_1'$. Then, from the first of Eqs. E–3,

$$\ell_0 = x_2 - x_1 = \frac{1}{\sqrt{1 - v^2/c^2}}\left(x_2' + vt_2' - x_1' - vt_1'\right).$$

Since $t_2' = t_1'$, we have

$$\ell_0 = \frac{1}{\sqrt{1 - v^2/c^2}}\left(x_2' - x_1'\right) = \frac{\ell}{\sqrt{1 - v^2/c^2}},$$

or

$$\ell = \ell_0\sqrt{1 - v^2/c^2},$$

which is Eq. 26–3a: the length contraction formula.

EXAMPLE E–2 **Time dilation.** Derive the time dilation formula, Eq. 26–1, from the Lorentz transformation equations.

APPROACH Again we compare measurements in two reference frames, S and S', that move with speed v relative to each other, Fig. E–1.

SOLUTION The time Δt_0 between two events that occur at the same place $\left(x_2' = x_1'\right)$ in S' is measured to be $\Delta t_0 = t_2' - t_1'$. Since $x_2' = x_1'$, then from the last of Eqs. E–3, the time Δt between the events as measured in S is

$$\Delta t = t_2 - t_1 = \frac{1}{\sqrt{1 - v^2/c^2}}\left(t_2' + \frac{vx_2'}{c^2} - t_1' - \frac{vx_1'}{c^2}\right)$$

$$= \frac{1}{\sqrt{1 - v^2/c^2}}\left(t_2' - t_1'\right)$$

$$= \frac{\Delta t_0}{\sqrt{1 - v^2/c^2}},$$

which is the time dilation formula, Eq. 26–1. Notice that we chose S' to be the frame in which the two events occur at the same place, so that $x_1' = x_2'$, and then the terms containing x_1' and x_2' cancel out.

Answers to Odd-Numbered Problems

Chapter 16

1. 2.7×10^{-3} N.
3. 2.2×10^{4} N.
5. (1.9×10^{-13})%.
7. 3.76 cm.
9. -4.6×10^{8} C, 0.
11. $F_{\text{left}} = 120$ N, to the left;
    $F_{\text{center}} = 560$ N, to the right;
    $F_{\text{right}} = 450$ N, to the left.
13. 2.1×10^{12} electrons.
15. $10.1 \dfrac{kQ^2}{\ell^2}$, at $61°$.
17. (a) 88.8×10^{-6} C, 1.2×10^{-6} C;
    (b) 91.1×10^{-6} C, -1.1×10^{-6} C.
19. 3.94×10^{-16} N, west.
21. 6.30×10^{6} N/C, upward.
23. 1.33×10^{14} m/s^2, opposite to the field.
25.

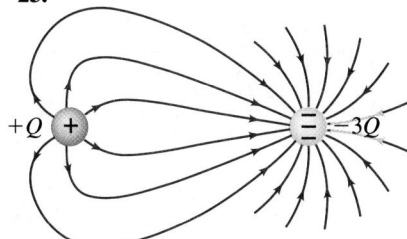

27. 5.97×10^{-10} N/C, south.
29. Upper right corner,
    $E = 3.76 \times 10^{4}$ N/C, at $45.0°$.
31. $\dfrac{4kQxa}{(x^2 - a^2)^2}$, to the left.
33. 3.7×10^{7} N/C, $330°$.
35. $E_A = 3.0 \times 10^{6}$ N/C, at $90°$;
    $E_B = 7.8 \times 10^{7}$ N/C, at $56°$; yes.
37. (a) 5×10^{-10} N;
    (b) 7×10^{-10} N;
    (c) 6×10^{-5} N.
39. (a) -1.1×10^{5} N·m^2/C;
    (b) 0.
41. 8.3×10^{-10} C.
43. (a) $k\dfrac{Q}{r^2}$;
    (b) 0;
    (c) $k\dfrac{Q}{r^2}$;
    (d) The shell causes the field to be 0 in the shell material. The charge polarizes the shell.

45. 4.0×10^{9} C.
47. 6.8×10^{5} C, negative.
49. 1.0×10^{7} electron charges.
51. 5.2×10^{-11} m.
53. 4.3 m.
55. 0.14 N, rightward.
57. 8.2×10^{-7} C, positive.
59. (a) 4×10^{10} particles,
    (b) 4×10^{-5} kg.
61. 9.90×10^{6} N/C, downward.
63. $x = d\left(\sqrt{2} + 1\right) \approx 2.41d$.
65. $QE\ell$, counterclockwise.
67. 8.94×10^{-19}.

Chapter 17

1. 5.0×10^{-4} J.
3. -1.0 V.
5. 4030 V, plate B.
7. 5.78 V.
9. -4.25×10^{4} V.
11. -157 V.
13. 9.0×10^{5} m/s.
15. 3000 V; only a small amount of charge was transferred.
17.

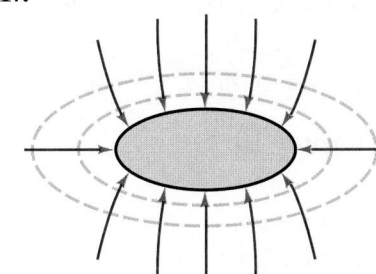

19. 2.8×10^{-9} C.
21. (a) 5.8×10^{5} V;
    (b) 9.2×10^{-14} J.
23. 9.15×10^{6} m/s.
25. (a) 18 cm from $-$ charge, on opposite side from $+$ charge;
    (b) 1.6 cm from $-$ charge, toward $+$ charge, and 8.0 cm from $-$ charge, away from $+$ charge.
27. (a) 1.6×10^{4} V;
    (b) 9.9×10^{4} V/m, $64°$.
29. 4.2×10^{6} V.
31. (a) 27 V;
    (b) 2.2×10^{-18} J, or 14 eV;

(c) -2.2×10^{-18} J, or -14 eV;
(d) 2.2×10^{-18} J, or 14 eV.
33. (a) 6.6×10^{-3} V;
    (b) 4.6×10^{-3} V;
    (c) -4.6×10^{-3} V.
35. 2.6×10^{-6} F.
37. 6.00×10^{-5} C.
39. 6.3×10^{-7} F.
41. 0.24 m^2.
43. 9×10^{-16} m, no.
45. $V_{2.50\,\mu F} = V_{6.80\,\mu F} = 611$ V,
    $Q_{2.50\,\mu F} = 1.53 \times 10^{-3}$ C,
    $Q_{6.80\,\mu F} = 4.16 \times 10^{-3}$ C.
47. 4.7×10^{-11} F.
49. 9.5 V.
51. 4.20×10^{-9} F, 0.247 m^2.
53. 9.6×10^{-5} F.
55. (a) 9×10^{-12} F;
    (b) 8×10^{-11} C;
    (c) 200 V/m;
    (d) 4×10^{-10} J;
    (e) capacitance, charge, work done.
57. 1.0×10^{-7} J/m^3.
59. 1110100.
61. 43,690.
63. (a) 65,536;
    (b) 16,777,216;
    (c) 16,777,216.
65. (b) 56 Hz.
67. $+2.0 \times 10^{5}$ V/m to -2.0×10^{5} V/m.
69. Yes, 1.3×10^{-12} V.
71. (a) Multiplied by 2;
    (b) multiplied by 2.
73. Alpha particle, 2.
75. Left: $-6.85kQ/\ell$, top: $-3.46kQ/\ell$, right: $-5.15kQ/\ell$.
77. (a) 17 cm from $-$ charge, on opposite side from $+$ charge;
    (b) 1.1 cm from $-$ charge, toward $+$ charge, and 8.1 cm from $-$ charge, away from $+$ charge.
79. (a) 31 J;
    (b) 5.9×10^{5} W.
81. 1.8 J.
83. 3.7×10^{-10} C.

85. (a) 6.4×10^{-11} C;
  (b) 6.4×10^{-11} C;
  (c) 18 V;
  (d) 2×10^{-10} J.
87. (a) 3.6×10^3 m/s;
  (b) 2.8×10^3 m/s.
89. 1.7×10^6 V.
91. 1.3×10^{-6} C.
93. 16°.
95. (a) $0.32\ \mu m^2$;
  (b) 59 megabytes.

Chapter 18

1. 1.00×10^{19} electrons/s.
3. 6.2×10^{-11} A.
5. 1200 V.
7. (a) 28 A;
  (b) 8.4×10^4 C.
9. (a) 8.9 Ω;
  (b) 1.2×10^4 C.
11. (a) 4.8 A;
  (b) 6.6 A.
13. 5.1×10^{-2} Ω.
15. Yes, for length 4.0 mm.
17. 2.0 V.
19. (a) 3.8×10^{-4} Ω;
  (b) 1.5×10^{-3} Ω;
  (c) 6.0×10^{-3} Ω.
21. 18C°.
23. 2400°C.
25. $R_{\text{carbon}} = 1.42$ kΩ,
  $R_{\text{Nichrome}} = 1.78$ kΩ.
27. 0.72 W.
29. 31 V.
31. 1.7×10^5 C.
33. (a) 950 W;
  (b) 15 Ω;
  (c) 9.9 Ω.
35. (a) 1.1 A;
  (b) 110 Ω.
37. 0.046 kWh; 6.6 cents per month.
39. 2.8×10^6 J.
41. 24 bulbs.
43. 1.5 m; power increases 36× and could start a fire.
45. (a) 7.2 A;
  (b) 1.7 Ω.
47. 0.12 A.
49. (a) Infinite resistance;
  (b) 96 Ω.
51. (a) 930 V;
  (b) 3.9 A.
53. (a) 3300 W;
  (b) 9.7 A.

55. 6.0×10^{-10} m/s.
57. $2.2\ \text{A/m}^2$, north.
59. 32 m/s (possible delay between nerve stimulation and generation of action potential).
61. 9.8 h.
63. 6.22 A.
65. 2.4×10^{-4} m.
67. $3200 per hour per meter.
69. 4.2×10^{-3} m.
71. (a) 33 Hz;
  (b) 0.990 A;
  (c) $V = (33.6 \sin 210t)$ V.
73. 2.25 Ω.
75. (b) As large as possible.
77. (a) 7.4 hp;
  (b) 220 km.
79. 1.7×10^{-4} m.
81. 32% increase.
83. (a) $I_A = 0.33$ A, $I_B = 3.3$ A;
  (b) $R_A = 360$ Ω, $R_B = 3.6$ Ω;
  (c) $Q_A = 1.2 \times 10^3$ C,
    $Q_B = 1.2 \times 10^4$ C;
  (d) $E_A = E_B = 1.4 \times 10^5$ J;
  (e) Bulb B.
85. (a) 4×10^6 J;
  (b) 2×10^4 m.
87. (a) 12 W;
  (b) 4.6 W.
89. 1.34×10^{-4} Ω.
91. $f = 1 - \dfrac{V}{V_0}$.

Chapter 19

1. (a) 5.92 V;
  (b) 5.99 V.
3. 0.034 Ω; 0.093 Ω.
5. (a) 330 Ω;
  (b) 8.9 Ω.
7. 2.
9. Connect 18 resistors in series; then measure voltage across 7 consecutive series resistors.
11. 0.3 Ω.
13. 560 Ω, 0.020.
15. 32 Ω.
17. 140 Ω.
19. $\frac{13}{8} R$.
21. 4.8 kΩ.
23. 55 V.
25. 0.35 A.
27. 0.

29. (a) 34 V;
  (b) 85-V battery: 82 V;
    45-V battery: 43 V.
31. $I_1 = 0.68$ A, left; $I_2 = 0.33$ A, left.
33. (a) $\mathscr{E}/R$;
  (b) R.
35. 0.71 A.
37. 3 parallel sets, each with 100 cells in series.
39. 3.71×10^{-6} F.
41. 2.0×10^{-9} F, yes.
43. 1.90×10^{-8} F in parallel,
  1.7×10^{-9} F in series.
45. 2:1.
47. In parallel, 750 pF.
49. 29.3 μF, 5.7 μF.
51. (a) $\frac{3}{5}C$;
  (b) $Q_1 = Q_2 = \frac{1}{5}CV$, $Q_3 = \frac{2}{5}CV$,
    $Q_4 = \frac{3}{5}CV$; $V_1 = V_2 = \frac{1}{5}V$,
    $V_3 = \frac{2}{5}V$, $V_4 = \frac{3}{5}V$.
53. 1.0×10^6 Ω.
55. 7.4×10^{-3} s.
57. (a) $I_1 = \dfrac{2\mathscr{E}}{3R}$, $I_2 = I_3 = \dfrac{\mathscr{E}}{3R}$

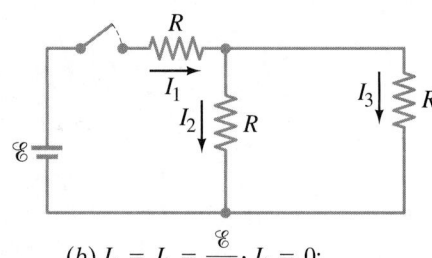

  (b) $I_1 = I_2 = \dfrac{\mathscr{E}}{2R}$, $I_3 = 0$;

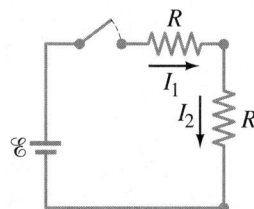

  (c) $\frac{1}{2}\mathscr{E}$.
59. (a) 2.9×10^{-5} A;
  (b) 8.8×10^6 Ω.
61. Add 710 Ω in series with ammeter, 29 Ω/V.
63. 9.60×10^{-4} A, 4.8 V;
  current: +20%, voltage: −20%.
65. 9.8 V.
67. Put 9.0 kΩ in series with the body.
69. $\frac{1}{4}C$, $\frac{2}{5}C$, $\frac{3}{5}C$, $\frac{3}{4}C$, C, $\frac{4}{3}C$, $\frac{5}{3}C$, $\frac{5}{2}C$, $4C$.
71. 9.2×10^4 Ω.
73. (a) 3.6 Ω;
  (b) 14 W.

77. 600 cells; 0.54 m², 4 banks in parallel, each containing 150 cells in series.

79. (a) 6.0 Ω;
(b) 2.2 V.

81. 11 V.

83. 100 Ω.

87. 9.0 Ω.

89. $Q_{12\,\mu F} = 1.0 \times 10^{-4}$ C,
$Q_{48\,\mu F} = 4.1 \times 10^{-4}$ C.

91. (a) 1.9×10^{-4} J;
(b) 4.0×10^{-5} J;
(c) $Q_a = 16\,\mu$C; $Q_b = 3.3\,\mu$C.

93. $Q_1 = \dfrac{C_1 C_2}{C_2 + C_1} V_0$,

$Q_2 = \dfrac{C_2^2}{C_2 + C_1} V$.

95. (a) In parallel;
(b) 7.7 pF to 35 pF.

97. $Q_1 = 11\,\mu$C, $V_1 = 11$ V;
$Q_2 = 13\,\mu$C, $V_2 = 6.3$ V;
$Q_3 = 13\,\mu$C, $V_3 = 5.2$ V.

Chapter 20

1. (a) 5.8 N/m;
(b) 3.3 N/m.

3. 1.3 N.

5. 27°.

7. (a) South pole;
(b) 3.86 A;
(c) 8.50×10^{-2} N.

9. 5.6×10^{-14} N, north.

11. 0.24 T.

13. (a) To the right;
(b) downward;
(c) into the page.

15. (a) 6.0×10^5 m/s;
(b) 3.6×10^{-2} m;
(c) 3.8×10^{-7} s.

17. 0.59 m.

19. $r_{\text{proton}}/r_{\text{electron}} = 42.8$.

21. 1.97×10^{-6} m.

23. (a) Sign determines polarity but not magnitude of Hall emf.
(b) 0.56 m/s.

25. 2.9×10^{-4} T, about 5.8 times larger.

27. 7.8×10^{-2} N, toward other wire.

29.

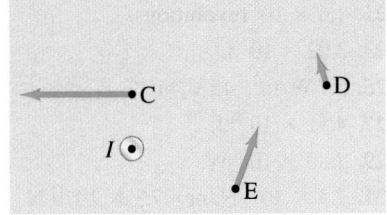

31. 5.1×10^{-6} N, toward wire.

33. 3.8×10^{-5} T, 17° below horizontal.

35. (a) $(2.0 \times 10^{-5}$ T/A$)(I - 25$ A$)$;
(b) $(2.0 \times 10^{-5}$ T/A$)(I + 25$ A$)$.

37.

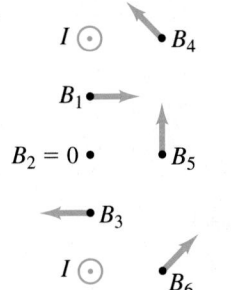

39. 15 A, downward.

41. Closer wire: 4.4×10^{-2} N/m, attract; farther wire: 2.2×10^{-2} N/m, repel.

43. 4.66×10^{-5} T.

45. 1.19 A.

47. 0.12 N, south.

49. (c) No; inversely as distance from center of toroid: $B \propto 1/R$.

51. 1.18 T.

53. 69.7 μA.

55. 1.87×10^6 V/m; perpendicular to velocity and magnetic field, and in opposite direction to magnetic force on protons.

57. 1.3×10^{-3} m; 6.5×10^{-4} m.

59. $^2_1$H nucleus or $^4_2$He nucleus.

61. 0.5 T.

63.

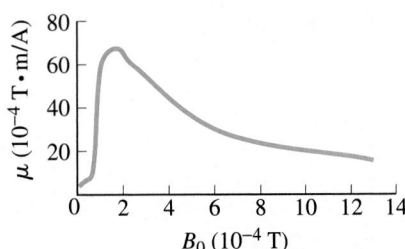

65. 2.7×10^{-2} T, upward.

67. 0.30 N, northerly, 68° above horizontal.

69. 7.7×10^{-6} N.

71. 1.1×10^{-6} m/s, west.

73. $\dfrac{2\,\mu_0 I}{\pi \ell}$, to the left.

75. They will exit above or below second tube; 52°.

77. -2.1×10^{-20} J.

79. $r = 5.3 \times 10^{-5}$ m,
$p = 3.3 \times 10^{-4}$ m.

81. 1.9×10^{-3} T.

83. $\dfrac{5.0mg}{\ell B}$, to the left.

85. (a) Negative;
(b) $\dfrac{qB_0 \left(d^2 + \ell^2\right)}{2d}$.

87. 1.2 A; downward.

89. (a) M: 5.8×10^{-4} N/m, upward;
N: 3.4×10^{-4} N/m, at 300°;
P: 3.4×10^{-4} N/m, at 240°;
(b) 1.75×10^{-4} T, at $-14°$.

91. 2.9×10^{-6} T, $B_{\text{wire}} \approx 0.06 B_{\text{Earth}}$.

Chapter 21

1. 560 V.

3. Counterclockwise.

5. 0.20 V.

7. (a) 1.0×10^{-2} Wb;
(b) 48°;
(c) 6.7×10^{-3} Wb.

9. (a) 1.73×10^{-2} V;
(b) 0.114 V/m, downward.

11. (a) 0;
(b) clockwise;
(c) counterclockwise;
(d) clockwise.

13. 0.65 mV, east or west.

15. (a) Magnetic force on current in moving bar, $B^2\ell^2 v/R$;
(b) $B^2\ell^2 v^2/R$.

17. (a) 0.17 V;
(b) 7.1×10^{-3} A;
(c) 7.5×10^{-4} N, to the right.

19. 5.23 C.

21. (a) 810 V;
(b) double the rotation frequency.

23. 17 rotations per second.

25. 92 V.

27. 1.71×10^4 turns.

29. $I_S = 0.21 I_P$.

31. (a) 6.2 V;
(b) step-down.

33. 450 V, 56 A.

35. 6×10^9 m.

37. 55 MW.

39. 6.9 V.

41. 0.10 H.

43. (a) 1.5×10^{-2} H;
  (b) 75 turns.

45. 46 m, 21 km; 0.70 kΩ.

47. 23 J.

49. 5×10^{15} J.

51. 3.7.

53. (a) 2.3;
  (b) 4.6;
  (c) 6.9.

55. 3300 Hz.

57. 1.6×10^4 Ω, 1.47×10^{-2} A.

59. (a) 7400 Ω;
  (b) 0.38 A.

61. (a) 3.6×10^4 Ω;
  (b) 3.7×10^4 Ω.

63. 205 Ω.

65. 270 Hz; the voltages are out of phase.

67. (a) 1.77×10^{-2} A;
  (b) $-12.6°$;
  (c) $V_R = 117$ V, $V_C = 26.1$ V.

69. 3.6×10^5 Hz.

71. (a) 1.3×10^{-7} F;
  (b) 37 A.

73. (a) 0.032 H;
  (b) 0.032 A;
  (c) 16 μJ.

75. 6.01×10^{-3} J.

77. Coil radius = 1.5 cm, 10,000 turns.

79. 200 kV.

81. (a) 41 kV;
  (b) 31 MW;
  (c) 1.0 MW;
  (d) 30 MW.

85. Put a 98-mH inductor in series with it.

87. 82 V.

89. 93 mH.

91. 2.

Chapter 22

1. 1.1×10^5 V/m/s.

3. 1.7×10^{15} V/m/s.

5. 2.4×10^{-13} T.

7. 90.0 kHz, 2.33 V/m, along the horizontal north–south line.

9. 1.25 s.

11. 4.20×10^{-7} m, violet visible light.

13. 2.00×10^{10} Hz.

15. (a) 1.319×10^{-2} m;
  (b) 2.5×10^{18} Hz.

17. 4.0×10^{16} m.

19. 9600 wavelengths; 3.54×10^{-15} s.

21. 1.6×10^6 revolutions/s.

23. 3.02×10^7 s.

25. 5.7 W/m², 46 V/m.

27. 4.51×10^{-6} J.

29. 3.80×10^{26} W.

31. 7.3×10^{-7} N/m²; 7.3×10^{-11} N away from bulb.

33. 400 m².

35. 0.16 m.

37. Channel 2: 5.56 m, Channel 51: 0.434 m.

39. 1.5×10^{-12} F.

41. (a) 1.3×10^{-6} H;
  (b) 9.9×10^{-11} F.

43. 6.25×10^{-4} V/m; 1.04×10^{-9} W/m².

45. 10 ns.

47. 1.36 s; inside.

49. 5.00×10^2 s.

51. 13 V/m, 4.4×10^{-8} T.

53. (a) 1.2×10^{-10} J;
  (b) 8.7×10^{-6} V/m, 2.9×10^{-14} T.

55. 61 km.

57. (a) 2.8×10^{-3} J/s;
  (b) 1.0 V/m;
  (c) 1.0 V;
  (d) 2.0×10^{-2} V.

59. 6×10^{10} W.

61. 35 kW.

Chapter 23

1.

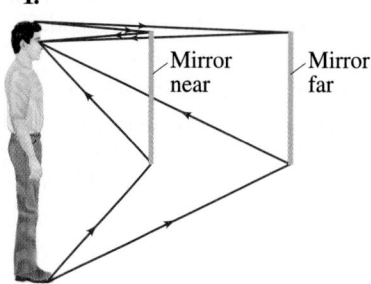

3. 11°.

5.

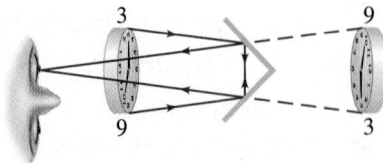

7. 4.0×10^{-6} m².

9. 10.5 cm.

11. (a) $d_i \approx -5$ cm;

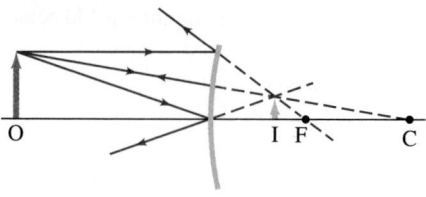

  (b) $d_i = -5.3$ cm;
  (c) 1.0 mm.

13. -6.8 m.

15. 5.7 m.

17. 2.0 cm behind ball's front surface; virtual; upright.

19. 1.0 m.

21. (a) Concave;
  (b) upright, virtual, and magnified;
  (c) 1.40 m.

23.

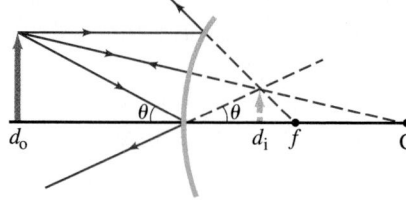

25. 1.31.

27. 1.62.

29. 50.1°.

31. 38.6°.

33. 81.9°.

35. 61.0°, crown glass.

37. At least 93.5 cm away.

39. (a) 1.4;
  (b) no;
  (c) 1.9.

41. (a) ~500 mm;

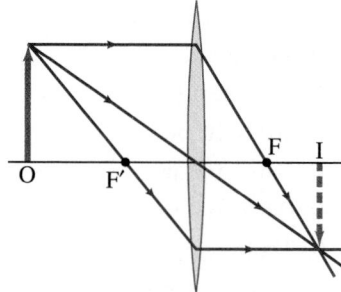

  (b) 478 mm.

43. (a) 3.08 D, converging;
  (b) -0.148 m, diverging.

45. (a) 106 mm;
  (b) 109 mm;
  (c) 117 mm;
  (d) 513 mm.

47. (a) 37 cm behind lens;

(b) +2.3×.

49. $d_i = -6.67$ cm behind lens, virtual and upright, $h_i = 0.534$ mm.

51. (a) 70.0 mm;

(b) 30.0 mm.

53. 64 cm.

55. 21.3 cm or 64.7 cm from object.

57. 0.105 m; 5.8 m.

59. 18.5 cm beyond second lens; −0.651× (inverted).

61. (a) 10 cm beyond second lens;

(b) −1.0×;

(c)

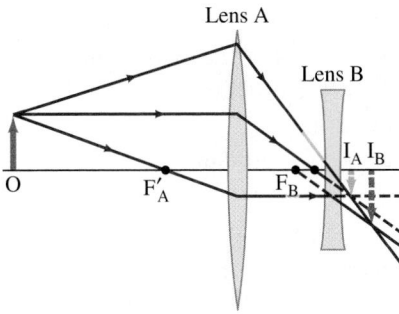

63. −29.7 cm.

65. 9.0 cm.

67. 0.34 m.

69. 1.25 s.

71. 6.04 m.

73. 1.58, light flint glass.

75. (a) Convex;

(b) 25 cm behind mirror;

(c) −110 cm;

(d) −220 cm.

77. 67°.

79. 9 cm, 12 cm.

81. $n \geq 1.60$.

83. (a) −0.33 mm;

(b) −0.47 mm;

(c) −0.98 mm.

85. Left: converging; right: diverging.

89. 5.7 cm from object, between it and lens.

Chapter 24

1. 5.4×10^{-7} m.

3. 6.3×10^{-7} m, 4.8×10^{14} Hz.

5. 17° and 64°.

7. 1.5×10^{-4} m.

9. 3.1 cm.

11.

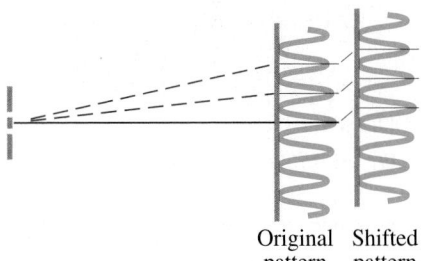

Original pattern   Shifted pattern

13. 2.2×10^{-3} m.

15. 14 mm.

17. 570 nm.

19. 0.23°.

21. 1.8°.

23. 1.0×10^{-6} m.

25. 7.38×10^{-2} m.

27. (a) λ;

(b) 400 nm.

29. 3.4×10^{-6} m.

31. Entire pattern is shifted, with central maximum at 28.0° to the normal.

33. 330 nm.

35. 556 nm.

37. 490 nm, 610 nm, 640 nm, 670 nm.

39. $(3.0 \times 10^1)°$.

41. Second order.

43. 7140 slits/cm.

45. 0.878 cm.

47. 230 nm.

49. 9 lenses.

51. 33 dark bands.

53. 110 nm; 230 nm.

55. 482 nm.

57. 691 nm.

59. 1.004328.

61. 57.3°.

63. 48.1%.

65. No; for diamond, $\theta_p = 59.4°$.

67. 61.2°.

69. 36.9°; 48.8°; 53.1°.

71. 28° relative to first polarizer.

73. (a) 1.3×10^{-4} m;

(b) 3.9×10^{-7} m.

75. H: 13.7°; Ne: 13.5°; Ar: 14.5°.

77. 480 nm.

79. $\lambda_2 > 600$ nm overlaps with $\lambda_3 < 467$ nm.

81. (a) 82 nm;

(b) 130 nm.

83. 4.8×10^4 m.

85. 3.19×10^{-5} m.

87. 580 nm.

89. 0.6 m.

91. 658 nm; 782 slits/cm.

93. 400 nm, 600 nm.

95. (a) 0;

(b) 0.11;

(c) 0.

Chapter 25

1. $f/8$.

3. 3.00 mm to 46 mm.

5. 110 mm; 220 mm.

7. 5.1 m to infinity.

9. 16 mm.

11. $f/2.5$.

13. 1.1 D.

15. (a) −1.2 D;

(b) 35 cm.

17. −8.3 D; −7.1 D.

19. (a) 2.0 cm;

(b) 1.9 cm.

21. 18.4 cm, 1.00 m.

23. 1.6×.

25. 6.2 cm from lens, 4.0×.

27. (a) 3.6×;

(b) 14 mm;

(c) 6.9 cm.

29. 4 cm toward contract.

31. −29×; 85 cm.

33. 25 cm.

35. −22×.

37. −110×.

39. Objective: 1.09 m; eyepiece: 9.09 cm.

41. 3.0 m.

43. $(4.0 \times 10^2)×$.

45. 520×.

47. (a) $(9.00 \times 10^2)×$;

(b) eyepiece: 1.8 cm; objective, 0.300 cm;

(c) 0.306 cm from objective.

49. (a) 14 cm;

(b) 130×.

51. (a) Converging;

(b) 281 cm.

53. $(1.54 \times 10^{-5})°$.

55. 1.7×10^{11} m.

57. 1.0×10^4 m.

59. 8.5 cm, 6.2×10^{-6} rad (distance of 2.4 km).

61. (a) 53.8°;

(b) 0.19 nm.

63. (a) $1\times$;
    (b) $1\times$ (at back of body) to
       $2.7\times$ (at front of body).

65. $\frac{1}{6}$ s.

67. (a) 1.8×10^4 m;
    (b) $23''$; atmospheric effects and
       aberrations in the eye.

69. 0.82 mm.

71. 8.2.

73. -0.75 D (upper part),
    $+2.0$ D (lower part).

75. (a) $-2.8\times$;
    (b) $+5.5$-D lens.

77. $+3.9$ D.

79. 4.8×10^3 C°/min.

81. $-15\times$.

83. 0.4 m.

85. 110 m.

87. 2 m.

89. -8.4 cm.

Chapter 26

1. 83.9 m.

3. 35 ly.

5. $0.70c$.

7. (a) $0.141c$;
    (b) $0.140c$.

9. (a) 6.92 m, 1.35 m;
    (b) 13.9 s;
    (c) $0.720c$;
    (d) 13.9 s.

11. (a) 3.6 yr;
    (b) 7.0 yr.

13. $(6.97 \times 10^{-8})\%$.

15. $0.9716c$.

17. (a) -1.1%;
    (b) -34%.

19. $0.95c$.

21. 8.209×10^{-14} J, 0.512 MeV.

23. 1000 kg.

25. 5.36×10^{-13} kg.

27. 1.6 GeV/c.

29. (a) 4.5 GeV;
    (b) 5.4 GeV/c.

31. $0.866c$, 0.886 MeV/c.

33. 9.0×10^{13} J; 9.2×10^9 kg.

35. 1.30 MeV/c^2.

37. $0.333c$.

39. (a) $0.866c$;
    (b) $0.745c$.

41. (a) 1.8×10^{19} J;
    (b) -1.7%.

43. $\dfrac{2m}{\sqrt{1 - v^2/c^2}}$; 0;
    $\left(\dfrac{1}{\sqrt{1 - v^2/c^2}} - 1\right)2mc^2$.

47. $0.69c$.

49. (a) $0.959c$;
    (b) $0.36c$.

51. $0.962c$.

53. $0.3c$.

55. (a) $0.66c$;
    (b) 6.5 yr.

57. $0.91c$.

59. 1.022 MeV.

61. 0.79 MeV.

63. (a) 4×10^9 kg/s;
    (b) 4×10^7 yr;
    (c) 1×10^{13} yr.

67. 5.38×10^{-12} kg; $(1.5 \times 10^{-8})\%$.

69. (a)

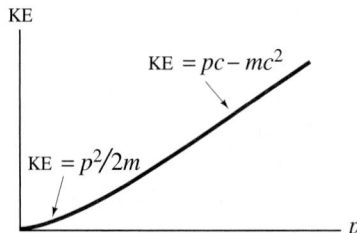

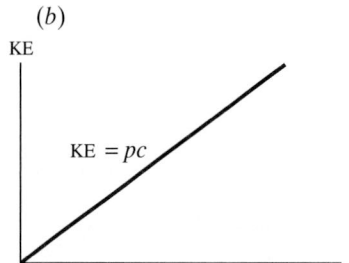

(b)

71. $0.981c$.

73. 1.7×10^{21} J; $\sim 20\times$ greater.

75. (a) 15 m;
    (b) 42 min.

77. (a) $0.986c$;
    (b) $(1 - 5 \times 10^{-7})c$.

79. Yes, in barn's reference frame, if
    his speed is $\geq 0.639c$; no, in boy's
    reference frame.

81. 8.0×10^{-8} s.

Chapter 27

1. 6.2×10^4 C/kg.

3. 5 electrons.

5. (a) 10.6 μm, far infrared;
    (b) 940 nm, near infrared;
    (c) 0.7 mm, microwave.

7. 5.4×10^{-20} J, 0.34 eV.

9. 9.35×10^{-6} m.

11. 2.7×10^{-19} J to 5.0×10^{-19} J,
    1.7 eV to 3.1 eV.

13. 1.14×10^{-27} kg·m/s.

17. 7.2×10^{14} Hz.

19. 429 nm.

21. (a) 2.3 eV;
    (b) 0.85 V.

23. 0.92 eV; 5.7×10^5 m/s.

25. (a) 0.78 eV;
    (b) no ejected electrons.

27. 3.32 eV.

31. (a) 2.43×10^{-12} m;
    (b) 1.32×10^{-15} m.

33. 2.62 MeV.

35. 212 MeV; 5.86×10^{-15} m.

37. 1.592 MeV, 7.81×10^{-13} m.

39. 4.7×10^{-12} m.

41. 1840.

43. (a) 4×10^{-10} m;
    (b) 1×10^{-10} m;
    (c) 3.9×10^{-11} m.

45. 6.4×10^{-12} m; yes;
    much less than 5 cm; no.

47. 22 V.

49. (a) Absorption; final state;
       largest energy photon;
    (b) emission, initial state;
    (c) absorption, final state.

51. $n = 5$ to $n' = 3$.

53. (a) 486 nm;
    (b) 103 nm;
    (c) 434 nm.

55. 91.2 nm.

57. Yes; from $n = 1$ to $n = 3$.

59.

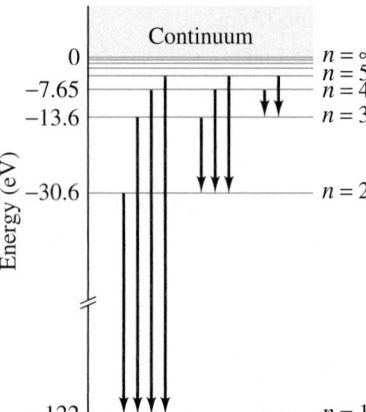

61. -0.544 eV.

63. Yes; $(7.30 \times 10^{-3})c$,
    $1/\gamma = 0.99997$.

65. 1.20×10^{29} m, -4.22×10^{-97} J.

67. 3.28×10^{15} Hz.

69. 4.4×10^{26} photons/s.

71. 5.5×10^{18} photons/s.

73. (a) 1.7 eV;

    (b) 3.0 eV.

75. 0.34 MeV for both.

77. 4.0×10^{-14} m.

79. 4.40×10^{-40}; yes.

81. 0.32 V.

83. 4.0×10^{-7} W/m²;

    1.7×10^{-2} V/m.

85. 8.2×10^{-10} m.

87. 3.5×10^{-12} m.

89. 1200 m/s.

91. Paschen series, level 4.

93. 1.8×10^{11} C/kg.

95. $E_n = -\dfrac{2.84 \times 10^{165}\,\text{J}}{n^2}$,

    $r_n = n^2(5.16 \times 10^{-129}\,\text{m})$;
    not apparent.

Chapter 28

1. 4.5×10^{-7} m.

3. 5.3×10^{-11} m.

5. 7×10^{-8} eV.

7. 3.00×10^{-10} eV/c².

9. Electron: 1.5×10^{-3} m;
    baseball: 9.7×10^{-33} m;
    $\Delta x_{\text{electron}}/\Delta x_{\text{baseball}} = 1.5 \times 10^{29}$.

11. 7700 m/s; 1.7×10^{-4} eV.

13. 5.53×10^{-10} m.

15. $\ell = 0,\ 1,\ 2,\ 3,\ 4,\ 5.$

17. 14 electrons.

19. (a) $\left(1, 0, 0, -\frac{1}{2}\right)$, $\left(1, 0, 0, +\frac{1}{2}\right)$,
    $\left(2, 0, 0, -\frac{1}{2}\right)$, $\left(2, 0, 0, +\frac{1}{2}\right)$,
    $\left(2, 1, -1, -\frac{1}{2}\right)$, $\left(2, 1, -1, +\frac{1}{2}\right)$;

    (b) $\left(1, 0, 0, -\frac{1}{2}\right)$, $\left(1, 0, 0, +\frac{1}{2}\right)$,
    $\left(2, 0, 0, -\frac{1}{2}\right)$, $\left(2, 0, 0, +\frac{1}{2}\right)$,
    $\left(2, 1, -1, -\frac{1}{2}\right)$, $\left(2, 1, -1, +\frac{1}{2}\right)$,
    $\left(2, 1, 0, -\frac{1}{2}\right)$, $\left(2, 1, 0, +\frac{1}{2}\right)$,
    $\left(2, 1, 1, -\frac{1}{2}\right)$, $\left(2, 1, 1, +\frac{1}{2}\right)$,
    $\left(3, 0, 0, -\frac{1}{2}\right)$, $\left(3, 0, 0, +\frac{1}{2}\right)$,
    $\left(3, 1, -1, -\frac{1}{2}\right)$.

21. $12\hbar$, or 3.65×10^{-34} J·s.

23. $n \geq 4$; $3 \leq \ell \leq n - 1$;
    $m_s = -\frac{1}{2}, +\frac{1}{2}.$

25. (a) $1s^2 2s^2 2p^6 3s^2 3p^6 3d^6 4s^2$;

    (b) $1s^2 2s^2 2p^6 3s^2 3p^6 3d^{10} 4s^2 4p^4$;

    (c) $1s^2 2s^2 2p^6 3s^2 3p^6 3d^{10} 4s^2 4p^6 5s^2$.

27. (a) 5;

    (b) -0.544 eV;

    (c) $\sqrt{6}\,\hbar$, 2;

    (d) $-2,\ -1,\ 0,\ 1,\ 2.$

31. $n = 3$, $\ell = 2$.

33. 4.36×10^{-11} m; 1×10^{-9} m.

37. 1.798×10^{-10} m.

39. 6.12×10^{-11} m; partial shielding
    by $n = 2$ shell.

41. 0.017 J, 5.5×10^{16} photons.

43. 634 nm.

45. $n \geq 7$, $\ell = 6$, $m_\ell = 2$.

47. $L_{\text{max}} = \sqrt{30}\,\hbar$, $L_{\text{min}} = 0$.

49. 5.3×10^{-27} kg·m/s,
    1.3×10^{-7} m.

51. 5.75×10^{-13} m, 115 keV.

53. 22 electrons.

57. $L_{\text{Bohr}} = 2\hbar$; $L_{\text{qm}} = 0$ or $\sqrt{2}\,\hbar$.

59. 3.7×10^{-37} m.

61. $\Delta p_{\text{electron}}/\Delta p_{\text{proton}} = 0.0234$.

63. 2.0×10^{-35} m/s; yes; 10^{34} s.

65. Copper.

Chapter 29

1. 5.1 eV.

3. HN: 110 pm, CN: 150 pm,
    NO: 133 pm.

5. 4.6 eV.

9. 1.10×10^{-10} m.

11. 5.22×10^{-4} m.

13. (a) 680 N/m;

    (b) 2.1×10^{-6} m.

15. 0.315 nm.

17. 2.0 eV.

19. $\lambda \leq 1.7\ \mu\text{m}$.

21. 1.2×10^6 electrons.

23. 3×10^6.

25. 1.7 eV.

27. 14 V.

29. 7.3 mA.

31. (a) 5.1 mA;

    (b) 3.6 mA.

33.

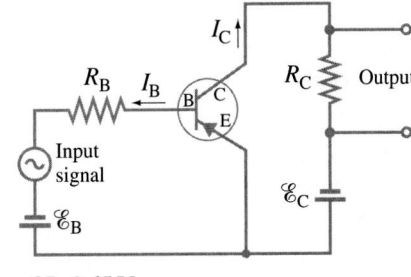

35. 0.65 V.

37. (a) 130;

    (b) 8700.

39. (a) 3.1×10^4 K;

    (b) 930 K.

41. (a) -5.3 eV;

    (b) 4.4 eV.

43. 1.94×10^{-46} kg·m².

45. 6.03×10^{-4} eV.

47. $\lambda \leq 3.5 \times 10^{-7}$ m.

49. Yes; 1.11×10^{-6} m.

51. 980 J/mol.

Chapter 30

1. 0.149 u.

3. 0.855%.

5. 3727 MeV/c².

7. (a) 180 m;

    (b) 3.5×10^4.

9. 30 MeV.

11. 6.0×10^{26} nucleons; no;
    all nucleons have about the
    same mass.

13. 550 MeV.

15. 7.699 MeV/nucleon.

17. 12.42 MeV.

19. $^{23}_{11}$Na: 8.113 MeV/nucleon;
    $^{24}_{11}$Na: 8.063 MeV/nucleon.

21. (b) Yes, binding energy is positive.

25. 0.782 MeV.

27. (a) β^- emitter;

    (b) $^{24}_{11}$Na $\rightarrow$ $^{24}_{12}$Mg $+ \beta^- + \bar{\nu}$,
    5.515 MeV.

29. 2.822 MeV.

31. α: 6.114 MeV; β^-: 0.259 MeV.

33. 10.8 MeV.

35. For both: $\text{KE}_{\text{max}} = 0.9612$ MeV,
    $\text{KE}_{\text{min}} = 0$.

37. $\text{KE}_{\text{recoil}} = 0.0718$ MeV,
    $Q = 4.27$ MeV.

39. 1.2 h.

41. 2.5×10^9 decays/s.

43. (a) 3.60×10^{12} decays/s;

    (b) 3.58×10^{12} decays/s;

    (c) 1.34×10^9 decays/s.

45. 7 α particles; 4 β^- particles.

47. 0.91 g.

49. (a) 1.38×10^{-13} s⁻¹;

    (b) 3.73×10^7 decays/min.

51. 86 decays/s.

53. 1.78×10^9 yr.

55. 7900 yr.

57. 15.8 d.

59. $N_D = N_0\left(1 - e^{-\lambda t}\right)$.

61. 2.6×10^4 yr.

63. 41 yr.

65. 3.0×10^{-14} m,
    $4.2\times$ nuclear radius.

67. $^{40}_{19}$K: 0.16 mg; $^{39}_{19}$K: 1.2 g.

69. 2.71×10^{-11} g.

71. (a) 4.6×10^{15} decays/s;
   (b) 1.2×10^4 decays/s.

73. (a) 0.002603 u, 2.425 MeV/c^2;
   (b) 0, 0;
   (c) -0.090739 u, -84.52 MeV/c^2;
   (d) 0.043930 u, 40.92 MeV/c^2;
   (e) $\Delta \geq 0$ for $0 \leq Z \leq 8$
   and $Z \geq 85$;
   $\Delta < 0$ for $9 \leq Z \leq 84$;
   $\Delta \geq 0$ for $0 \leq A \leq 15$
   and $A \geq 218$;
   $\Delta < 0$ for $16 \leq A < 218$.

75. 0.2 decays/s.

77. (a) 180 m;
   (b) 13 km.

79. (a) There would be no atoms—just neutrons;
   (b) 0.083%.

81. 1.1×10^{11} yr.

83. 13 decays/s.

85. (b) 98.2%.

Chapter 31

1. $^{28}_{13}$Al; β^-; $^{28}_{14}$Si.

3. Yes, $Q > 0$.

5. 18.000937 u.

7. (a) Yes;
   (b) 20.4 MeV.

9. 4.730 MeV.

11. n + $^{14}_{7}$N $\rightarrow$ $^{14}_{6}$C + p, 0.626 MeV.

13. (a) He picks up a neutron from C;
   (b) $^{11}_{6}$C;
   (c) 1.856 MeV; exothermic.

15. 5.702 MeV released.

17. 126.5 MeV.

19. 1/1100.

21. (a) 5 neutrons;
   (b) 171.1 MeV.

23. 1100 kg.

25. 260 MeV; about 30% > fission
   energy released.

27. 3000 eV.

31. (a) a: 6.03×10^{23} MeV/g;
   b: 4.89×10^{23} MeV/g;
   c: 2.11×10^{24} MeV/g;
   (b) 5.13×10^{23} MeV/g;
   a: 0.851; b: 1.05; c: 0.243.

33. 0.39 g.

35. 5.6×10^3 kg/h.

37. 2.46×10^9 J; $50\times$ > gasoline.

39. 7000 rads.

41. 144 rads.

43. (a) 1.0 rad, or 0.010 Gy;
   (b) 1.0×10^{10} protons.

47. 5.61×10^{-10} kg.

49. 9×10^{16} e$^-$.

51. (a) 2×10^{-7} Sv/yr,
   2×10^{-5} rem/yr;
   $(2 \times 10^{-4}) \times$ allowed dose;
   (b) 2×10^{-6} Sv/yr,
   2×10^{-4} rem/yr;
   $(2 \times 10^{-3}) \times$ allowed dose.

53. 7.041 m; radio wave.

55. (a) $^{12}_{6}$C;
   (b) 5.702 MeV.

57. 1.0043 : 1.

59. 6.6×10^{-2} rem/yr.

61. (a) $^{226}_{88}$Ra $\rightarrow$ $^{4}_{2}$He + $^{222}_{86}$Rn;
   (b) 4.871 MeV;
   (c) 190.6 MeV/c for both;
   (d) 8.78×10^{-2} MeV.

63. (a) 3.7×10^{26} W;
   (b) 3.5×10^{38} protons/s;
   (c) 1.1×10^{11} yr.

65. 8×10^{12} J.

67. (a) 3700 decays/s;
   (b) 4.0×10^{-4} Sv/yr;
   11% of background.

69. 7.274 MeV.

71. 990 kg.

73. (a) $^{141}_{54}$Xe;
   (b) 2 neutrons escape or are
   absorbed, 1 causes another
   fission;
   (c) 176.0 MeV.

75. Most to least dangerous:
   C > B > A.

Chapter 32

1. 5.59 GeV.

3. 1.3×10^7 Hz.

5. 2.0 T.

7. Alpha particles;
   $\lambda_\alpha \approx d_{\text{nucleon}}$, $\lambda_p \approx 2d_{\text{nucleon}}$.

9. 0.9 MeV/rev.

13. 33.9 MeV.

15. 1879.2 MeV.

17. 67.5 MeV.

19. 1.32×10^{-15} m.

21. 1.3×10^{-12} m.

23. First, second, and fourth will not
   happen.

25. (a) 37.8 MeV;
   (b) $\text{KE}_p = 5.4$ MeV,
   $\text{KE}_{\pi^-} = 32.4$ MeV.

27. 9×10^{-3} MeV.

29. 7.5×10^{-21} s.

31. (a) 1300 eV;
   (b) 150 MeV.

33. 8×10^{-27} s.

35. (a) udd;
   (b) $\bar{u}\bar{d}\bar{d}$;
   (c) uds;
   (d) $\bar{u}\bar{d}\bar{s}$.

37. c$\bar{u}$.

39.

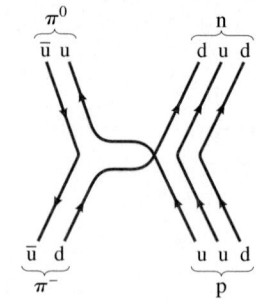

41. 16 GeV; 7.8×10^{-17} m.

43. (a) 1.022 MeV;
   (b) 1876.6 MeV.

45. 1.2×10^{10} m, 4.0×10^1 s.

47. (a) Possible, strong interaction;
   (b) possible, strong interaction;
   (c) possible, strong interaction;
   (d) not possible; charge is not
   conserved;
   (e) possible, weak interaction.

49. 10^{-18} m corresponds to 200 GeV.

51. 64 fundamental fermions.

53. (b) 10^{29} K.

55. 798.7 MeV.

57. 9.3×10^{10} eV;
   1.3×10^{-17} m, $0.99995c$.

59. (a) π^0;
   (b) $\bar{\nu}_\mu$.

63. 10^{-25} s.

Chapter 33

1. 3.1 ly.

3. Less than; by a factor of 2.

5. 2×10^{-3} kg/m^3.

7. 4.2×10^{-2} rad, or 2.4°;
   about $4.5\times$ Moon's width.

9. 1.83×10^9 kg/m^3;
   3.33×10^5 times larger.

11. (a) 0.018";
   (b) $(5.0 \times 10^{-6})°$.

13. 280 yr.

15. $D_1/D_2 = 0.13$.

17. 3×10^{14} m.

19. 540°.

21. 3.1×10^{-16} m.

23. 2.1×10^8 ly.

25. (a) 0.3 nm;

(b) 3.2 nm.

27. $0.058c$.

29. 9×10^9 ly.

31. 6.8×10^7 ly.

33. 1.1×10^{-3} m.

35. 6 nucleons/m^3.

37. (a) 10^{-5} s;

(b) 10^{-7} s;

(c) 10^{-4} s.

39. 0.2 rev/s.

41. A: T increases, L doesn't change, size decreases.

B: T unchanged, L decreases, size decreases.

C: T decreases, L increases, size increases.

43. 1.7×10^{25} W.

45. 400 ly; $r_{Sun}/d_{Earth-Sun} = 2 \times 10^7$, $r_{Sun}/d_{Galaxy} = 4 \times 10^{-3}$.

47. (a) $(9 \times 10^{-6})°$;

(b) 4 m.

49. (a) 13.93 MeV;

(b) 4.71 MeV;

(c) 5.46×10^{10} K.

51. $t_P = 5.38 \times 10^{-44}$ s.

Index

Note: The abbreviation *defn* means the page cited gives the definition of the term; *fn* means the reference is in a footnote; *pr* means it is found in a Problem or Question; *ff* means "also the following pages."

A (atomic mass number), 858
Aberration:
    chromatic, 725 *fn*, 728, 732
    of lenses, 727–28, 729, 731
    spherical, 650, 727, 728, 732
Absolute pressure, 264
Absolute space, 746, 748
Absolute temperature scale, 362, 368
Absolute time, 746
Absolute zero, 368, 424
    kinetic energy near, 376
Absorbed dose, 900
Absorption lines, 692–93, 787, 838
Absorption spectrum, 692–93, 787, 838
Absorption wavelength, 793
Abundances, natural, 858
Ac (*defn*), 514
Ac circuits, 514–15, 526 *fn*, 611–16
Ac generator, 597, 599
Ac motor, 577
Accademia del Cimento, 361
Accelerating reference frames, 77, 80, A-16–A-18
Acceleration, 26–38, 40, 58–63
    angular, 201–4, 208–12
    average, 26–27, 40
    centripetal, 110 *ff*
    constant, 28–38
    constant angular, 203–4
    Coriolis, A-17–A-18
    cosmic, 976–77
    of expansion of the universe, 976–77
    as a function of time (SHM), 301
    in *g*'s, 37
    due to gravity, 33–38, 58–63, 79 *fn*, 84, 121–22
    instantaneous (*defn*), 26, 40
    of the Moon, 112, 119
    motion at constant, 28–38, 58–64
    radial, 110 *ff*, 118
    related to force, 78–80
    of simple harmonic oscillator, 301
    and slope, 40
    tangential, 118, 201–3
    uniform, 28–38, 58–64
    velocity vs., 27

Accelerators, particle, 916–21
Accelerometer, 92
Acceptor level, 845
Accommodation of eye, 719
Accuracy, 8
    precision vs., 8
Achromatic doublet, 728
Achromatic lens, 728
Actinides, 817
Action at a distance, 119
Action potential, 518–19
Action–reaction (Newton's third law), 81–83
Activation, on an LCD screen, 491
Activation energy, 377, 833, 834
Active galactic nuclei (AGN), 951
Active matrix, 492
Active solar heating, 435
Activity, 870
    and half-life, 872
    source, 900
ADC, 488–89
Addition of vectors, 50–57, 87, 450
Addition of velocities:
    classical, 65–66
    relativistic, 764
Addressing pixels, 491–92
Adenine, 460
Adhesion (*defn*), 281–82
ADP, 833
AF signal, 637
AFM, 786
AGN, 951
Air circulation, underground, 278
Air columns, vibrations of, 337–40
Air conditioners, 425–27
Air gap, 694
Air pollution, 434–35
Air resistance, 33
Airplane:
    noise, 333
    wing, 277
Airy disk, 729
Algebra, review of, A-3–A-6
Alkali metals, 817
Allowed transitions, 814, 838–39

Alpenhorn, 358 *pr*
Alpha decay, 864–66, 869
    and tunneling, 876
Alpha particle (or ray), 786–87, 864–66
Alternating current (ac), 514–15, 526 *fn*, 611–16
Alternators, 598
AM, 737
AM radio, 637
Amino acids, 836–37
Ammeter, 546–48, 576
    digital, 546, 548
    connecting, 547
    resistance, effect of, 547–488
AMOLED, 849
Amorphous solids, 840
Ampère, André, 504, 573
Ampere (A) (unit), 504, 572
    operational definition of, 572
Ampere-hour (A·h) (unit), 505
Ampère's law, 573–75, 626–27
Amplifiers, 850–51
Amplitude, 294, 306, 319
    intensity related to, 333
    of vibration, 294
    of wave, 294, 306, 310, 319, 333, 804–6
Amplitude modulation (AM), 637
Analog meters, 546–48, 576
Analog signals, 488–89, 604
Analog-to-digital converter (ADC), 488–89
Analyzer (of polarized light), 700
Anderson, Carl, 924
Andromeda, 950, 983 *pr*
Aneroid barometer, 266
Aneroid gauge, 266
Angle, 11 *fn*, 199
    attack, 277
    Brewster's, 702, 710 *pr*
    critical, 659
    of dip, 562
    of incidence (*defn*), 313, 317, 645, 657
    phase, 615
    polarizing, 702
    radian measure of, 199–200
    of reflection (*defn*), 313, 645
    of refraction, 317, 657

solid, 11 *fn*
Angstrom (Å) (unit), 20 *pr*, 685 *fn*
Angular acceleration, 201–4, 208–12
    average, 201
    constant, 203–4
    instantaneous, 201
Angular displacement, 200, 302
Angular magnification, 722
Angular momentum, 215–18, 789, 795
    in atoms, 789, 812–14
    conservation, law of, 215–17, 869
    quantized in atoms, 812–13
    quantized in molecules, 837–38
    vector, 218
Angular position, 199
Angular quantities, 199 *ff*
    vector nature, 217–18
Angular velocity, 200–3
    average, 200
    instantaneous, 200
Animals, and sound waves, 309
Anisotropy of CMB, 968, 977
Annihilation of particles, 781, 925, 971–72
Anode, 490
Antenna, 627–28, 631, 638
Anthropic principle, 979
Antiatoms, 925
Anticodon, 836
Antilock brakes, 116
Antilogarithm, A-3, A-11
Antimatter, 925, 941, 943 *pr* (*see also* Antiparticle)
Antineutrino, 867–68, 930
Antineutron, 925
Antinodes, 315, 337, 338, 339
Antiparticle, 868, 924–26, 930–31 (*see also* Antimatter)
Antiproton, 924–25, 934
Antiquark, 930–31, 934–35, 936
Apparent brightness, 951–52, 958
Apparent weight, 124–25, 270
Apparent weightlessness, 124–25
Approximations, 8, 13–15
Arago, F., 687
Archeological dating, 875

Wave–particle duality
(*continued*)
of matter, 782–84, 795–96, 804–9
Waveform, 340–41
Wavelength (*defn*), 306, 314 *fn*
absorption, 793
Compton, 780
cutoff, 818–19
de Broglie, 782–83, 795–96, 805, 917
depending on index of refraction, 681, 686
as limit to resolution, 732, 917
of material particles, 782–83, 795–96
of spectral lines, 792–93
Weak bonds, 460, 461, 834–37, 840
Weak charge, 937
Weak nuclear force, 129, 863, 867, 924–42, 959
range of, 938
Weakly interacting massive particles (WIMPS), 976
Weather, 381
and Coriolis effect, A-18
forecasting, and Doppler effect, 348
Weber (Wb) (unit), 592
Weight, 76, 78, 84–86, 121–22
apparent, 124–25, 270
atomic, 360
as a force, 78, 84
force of gravity, 76, 84–86, 121–22
mass compared to, 78, 84
molecular, 360
Weightlessness, 124–25
Weinberg, S., 938
Wess, J., 942
Whales, echolocation in, 309

Wheatstone bridge, 556 *pr*
Whirlpool galaxy, 950
White dwarfs, 951, 953, 955–57
White light, 686
White-light holograms, 824
White-light LED, 848
Whole-body dose, 902
Wide-angle lens, 718, 728
Width, of resonance, 932
Wien's (displacement) law, 774, 952, 953
Wilkinson, D., 968
Wilkinson Microwave Anisotropy Probe (WMAP), 931 *fn*, 968
Wilson, Robert, 919 *fn*, 967–68
WIMPS, 976
Wind:
as convection, 402
and Coriolis effect, A-18
noise, 340
power, 435
Wind instruments, 317, 337–40
Windings, 577
Windows:
heat loss through, 401
thermal, 401
Windshield wipers, intermittent, 543
Wing of an airplane, lift on, 277
Wire, ground, 544–45
Wire drift chamber, 878 *fn*
Wire proportional chamber, 878 *fn*
Wire-wound resistor, 506
Wireless communication, 625, 636–39
Wireless transmission of power, 604
Wiring, electrical, 545
Witten, Edward, 942
WMAP, 931 *fn*, 968
Word-line, 605

Work, 138–45, 155, 391, 412–19
to bring positive charges together, 480
compared to heat, 412
defined, 139, 412 *ff*
done by a constant force (*defn*), 139–42
done by an electric field, 474
done by a gas, 414 *ff*
done by torque, 214
done by a varying force, 142
done in volume changes, 415–17
in first law of thermodynamics, 413–19
graphical analysis for, 142
from heat engines, 420 *ff*
on the Moon, 142
negative, 140
and power, 159–61
relation to energy, 142–47, 155, 157–61
units of, 139
Work function, 776–77
Work-energy principle, 142–45, 150, 760, 921
energy conservation vs., 157
as reformulation of Newton's laws, 144
Working off calories, 392
Working substance (*defn*), 421
Wrench, 223 *pr*
Wright, Thomas, 948
Writing data, 605

XDF (Hubble eXtreme Deep Field), 947, 961
Xerox (*see* Photocopier)
Xi (particle), 931
Xi—anti-Xi pair, 925

X-rays, 630, 733–36, 817–19, 869
and atomic number, 817–19
characteristic, 818
in electromagnetic spectrum, 630
spectra, 817–19
X-ray crystallography, 734
X-ray diffraction, 733–35
X-ray image, normal, 735
X-ray scattering, 780

YBCO superconductor, 517
Yerkes Observatory, 724
Yosemite Falls, 155
Young, Thomas, 682, 685
Young's double-slit experiment, 682–85, 690, 805–6
Young's modulus, 241–42
Yo-yo, 227 *pr*
Yttrium, barium, copper, oxygen superconductor (YBCO), 517
Yukawa, Hideki, 922–23
Yukawa particle, 922–23

Z (atomic number), 815, 817–19, 858
Z^0 particle, 826 *pr*, 915, 924, 930–32, 937
Z-particle decay, 924
Zeeman effect, 812
Zener diode, 846
Zero, absolute, temperature of, 368, 424
Zero-point energy, 839
Zeroth law of thermodynamics, 363
Zoom, digital, 718
Zoom lens, 718
Zumino, B., 942
Zweig, G., 934

Photo Credits

Front cover D. Giancoli **Cover inset** Science Photo Library/Alamy **Back cover** D. Giancoli

p. iii Reuters/NASA **p. iv** Scott Boehm/BCI **p. v** Pixland/age fotostock **p. vi** AFP/Getty Images/Newscom **p. vii** Giuseppe Molesini, Istituto Nazionale di Ottica, Florence **p. viii** Professors Pietro M. Motta & Silvia Correr/Photo Researchers, Inc. **p. ix** NASA, ESA, R. Ellis (Caltech), and the UDF 2012 Team **p. xvii** D. Giancoli

CO–16 Mike Dunning/Dorling Kindersley **16–37** Peter Menzel/Photo Researchers, Inc. **16–38** Dr. Gopal Murti/Science Source/Photo Researchers, Inc. **16–48** American Association of Physics Teachers/Matthew Claspill

CO–17 Emily Michot/Miami Herald/MCT/Newscom **17–8** D. Giancoli **17–13c, 17–18, 17–20** Eric Schrader/Pearson Education **17–21** tunart/iStockphoto **17–31 left** Eric Schrader/Pearson Education **17–31 right** Robnil **17–36** beerkoff/Shutterstock **17–44** Andrea Sordini

CO–18 left Mahaux Photography **CO–18 right** Eric Schrader/Pearson Education **18–1** Jean-Loup Charmet/Science Photo Library/Photo Researchers, Inc. **18–2** The Burndy Library Collection/Huntington Library **18–6a** Richard Megna/Fundamental Photographs **18–11** T. J. Florian/Rainbow Image Library **18–15** Richard Megna/Fundamental Photographs **18–16** Tony Freeman/PhotoEdit **18–18** Clint Spencer/iStockphoto **18–32** Alexandra Truitt & Jerry Marshall **18–34** Scott T. Smith/Corbis **18–36** Jim Wehtje/Getty Images

CO–19 Patrik Stoffarz/AFP/Getty Images/Newscom **19–15** David R. Frazier/Photolibrary, Inc./Alamy **19–24** Apogee/Photo Researchers, Inc. **19–27a** Photodisc/Getty Images **19–27b, 19–28, 19–30a** Eric Schrader/Pearson Education **19–30b** Olaf Doring/Imagebroker/AGE Fotostock **19–67** Raymond Forbes/AGE fotostock **19–74** Eric Schrader/Pearson Education

CO–20 Richard Megna/Fundamental Photographs **20–1** Dorling Kindersley **20–4a** Stephen Oliver/Dorling Kindersley **20–6** Mary Teresa Giancoli **20–8a/b, 20–18** Richard Megna/Fundamental Photographs **20–20b** Jack Finch/Science Photo Library/Photo Researchers, Inc. **20–28b, 20–43** Richard Megna/Fundamental Photographs **20–50** Clive Streeter/Dorling Kindersley, Courtesy of The Science Museum, London

CO–21 Richard Megna/Fundamental Photographs **21–7** Photo Courtesy of Diva de Provence, Toronto, ON, Canada **21–12** Jeff Hunter/Getty Images **21–21** Associated Press Photo/Robert F. Bukaty **21–22** Photograph by Robert Fenton Houser **21–27a** Terence Kearey **21–34a** 4kodiak/iStockphoto **21–34b** Eric Schrader/Pearson Education

CO–22 NASA **22–1** Original photograph in the possession of Sir Henry Roscoe, courtesy AIP Emilio Segrè Visual Archives **22–9** The Image Works Archives **22–12** Time Life Pictures/Getty Images **22–19** David J. Green/Alamy Images **22–20** Don Baida **22–22** NASA

CO–23, 23–6 D. Giancoli **23–11a** Mary Teresa Giancoli and Suzanne Saylor **23–11b** Paul Silverman/Fundamental Photographs **23–20** John Lawrence/Travel Pix Ltd. **23–22a** Shannon Fagan/age fotostock **23–23** Giuseppe Molesini, Istituto Nazionale di Ottica, Florence **23–30b** Garo/Phanie/Photo Researchers, Inc. **23–31c/d** D. Giancoli **23–32** D. Giancoli and Howard Shugat **23–34** Kari Erik Marttila Photography **23–38a/b, 23–47** D. Giancoli **23–50** Mary Teresa Giancoli **23–65** American Association of Physics Teachers/Annacy Wilson **23–66** American Association of Physics Teachers/Matt Buck **23–69** American Association of Physics Teachers/Sarah Lampen **23–70a/b** Scott Dudley

CO–24 Giuseppe Molesini, Istituto Nazionale di Ottica, Florence **24–4a** Kent Wood/Photo Researchers, Inc. **24–9a** Bausch & Lomb Incorporated **24–13** David Parker/Science Photo Library/Photo Researchers, Inc. **24–16b** Lewis Kemper Photography **24–17** George Diebold **24–19a** P. M. Rinard/American Journal of Physics **24–19b** Richard Megna/Fundamental Photographs **24–19c** Ken Kay/Fundamental Photographs **24–28** Wabash Instrument Corp./Fundamental Photographs **24–29a** Giuseppe Molesini, Istituto Nazionale di Ottica, Florence **24–29b** Richard Megna/Fundamental Photographs **24–29c** Paul Silverman/Fundamental Photographs **24–31b** Ken Kay/Fundamental Photographs **24–33b/c** Bausch & Lomb Incorporated **24–35** D. Hurst/Alamy **24–44** Diane Schiumo/Fundamental Photographs **24–47a/b** JiarenLau Photography, http://creativecommons.org/licenses/by/2.0/deed.en **24–52** Suunto **24–53** Daniel Rutter/Dan's Data

CO–25, 25–5, 25–6a/b Mary Teresa Giancoli **25–08** Leonard Lessin/Photolibrary **25–19a/b** Museo Galileo - Istituto e Museo di Storia della Scienza **25–21** Yerkes Observatory, University of Chicago **25–22c** Sandy Huffaker/Stringer/Getty Images **25–22d** Inter-University Centre for Astronomy and Astrophysics/Laurie Hatch **25–24b** Leica Microsystems **25–29a/b** Reproduced by permission from M. Cagnet, M. Francon, and J. Thrier, The Atlas of Optical Phenomena. Berlin:

Useful Geometry Formulas — Areas, Volumes

Circumference of circle $\quad C = \pi d = 2\pi r$

Area of circle $\qquad\qquad A = \pi r^2 = \dfrac{\pi d^2}{4}$

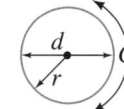

Area of rectangle $\qquad A = \ell w$

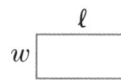

Area of parallelogram $\quad A = bh$

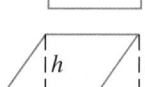

Area of triangle $\qquad A = \frac{1}{2} hb$

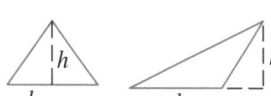

Right triangle
  (Pythagoras) $\quad c^2 = a^2 + b^2$

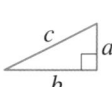

Sphere: surface area $\quad A = 4\pi r^2$
  volume $\qquad\qquad V = \frac{4}{3} \pi r^3$

Rectangular solid:
  volume $\qquad\qquad V = \ell w h$

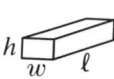

Cylinder (right):
  surface area $\qquad A = 2\pi r\ell + 2\pi r^2$
  volume $\qquad\qquad V = \pi r^2 \ell$

Right circular cone:
  surface area $\qquad A = \pi r^2 + \pi r \sqrt{r^2 + h^2}$
  volume $\qquad\qquad V = \frac{1}{3} \pi r^2 h$

Exponents [See Appendix A–2 for details]

$(a^n)(a^m) = a^{n+m}$ $\quad$ [Example: $(a^3)(a^2) = a^5$]
$(a^n)(b^n) = (ab)^n$ $\quad$ [Example: $(a^3)(b^3) = (ab)^3$]
$(a^n)^m = a^{nm}$ $\qquad \begin{bmatrix}\text{Example: } (a^3)^2 = a^6 \\ \text{Example: } (a^{\frac{1}{4}})^4 = a\end{bmatrix}$

$a^{-1} = \dfrac{1}{a} \qquad a^{-n} = \dfrac{1}{a^n} \qquad a^0 = 1$

$a^{\frac{1}{2}} = \sqrt{a} \qquad a^{\frac{1}{4}} = \sqrt{\sqrt{a}}$

$(a^n)(a^{-m}) = \dfrac{a^n}{a^m} = a^{n-m}$ $\quad$ [Ex.: $(a^5)(a^{-2}) = a^3$]

$\dfrac{a^n}{b^n} = \left(\dfrac{a}{b}\right)^n$

Quadratic Formula [Appendix A–4]

Equation with unknown x, in the form
$$ax^2 + bx + c = 0,$$
has solutions
$$x = \frac{-b \pm \sqrt{b^2 - 4ac}}{2a}.$$

Logarithms [Appendix A–8; Table p. A–11]

If $y = 10^x$, $\quad$ then $\quad x = \log_{10} y = \log y$.
If $y = e^x$, $\quad$ then $\quad x = \log_e y = \ln y$.

$\log(ab) = \log a + \log b$

$\log\left(\dfrac{a}{b}\right) = \log a - \log b$

$\log a^n = n \log a$

Binomial Expansion [Appendix A–5]

$(1 + x)^n = 1 + nx + \dfrac{n(n-1)}{2 \cdot 1} x^2 + \dfrac{n(n-1)(n-2)}{3 \cdot 2 \cdot 1} x^3 + \cdots$ $\quad$ [for $x^2 < 1$]

$\qquad\qquad \approx 1 + nx \quad$ if $x \ll 1$

$\qquad\qquad$ [Example: $(1 + 0.01)^3 \approx 1.03$]

$\qquad\qquad$ [Example: $\dfrac{1}{\sqrt{0.99}} = \dfrac{1}{\sqrt{1 - 0.01}} = (1 - 0.01)^{-\frac{1}{2}} \approx 1 - (-\frac{1}{2})(0.01) \approx 1.005$]

Fractions

$\dfrac{a}{b} = \dfrac{c}{d}$ is the same as $ad = bc$

$\dfrac{\left(\dfrac{a}{b}\right)}{\left(\dfrac{c}{d}\right)} = \dfrac{ad}{bc}$

Trigonometric Formulas [Appendix A–7]

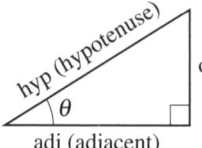

$\sin\theta = \dfrac{\text{opp}}{\text{hyp}}$

$\cos\theta = \dfrac{\text{adj}}{\text{hyp}}$

$\tan\theta = \dfrac{\text{opp}}{\text{adj}}$

$\text{adj}^2 + \text{opp}^2 = \text{hyp}^2 \quad$ (Pythagorean theorem)

$\tan\theta = \dfrac{\sin\theta}{\cos\theta}$

$\sin^2\theta + \cos^2\theta = 1$

$\sin 2\theta = 2\sin\theta\cos\theta$

$\cos 2\theta = (\cos^2\theta - \sin^2\theta) = (1 - 2\sin^2\theta) = (2\cos^2\theta - 1)$

$\sin(180° - \theta) = \sin\theta \qquad\qquad \cos(180° - \theta) = -\cos\theta$
$\sin(90° - \theta) = \cos\theta$
$\cos(90° - \theta) = \sin\theta$
$\sin\frac{1}{2}\theta = \sqrt{(1 - \cos\theta)/2} \qquad \cos\frac{1}{2}\theta = \sqrt{(1 + \cos\theta)/2}$
$\sin\theta \approx \theta \quad$ [for small $\theta \lesssim 0.2$ rad]
$\cos\theta \approx 1 - \dfrac{\theta^2}{2} \quad$ [for small $\theta \lesssim 0.2$ rad]
$\sin(A \pm B) = \sin A \cos B \pm \cos A \sin B$
$\cos(A \pm B) = \cos A \cos B \mp \sin A \sin B$

For any triangle:
$c^2 = a^2 + b^2 - 2ab\cos\gamma \quad$ (law of cosines)
$\dfrac{\sin\alpha}{a} = \dfrac{\sin\beta}{b} = \dfrac{\sin\gamma}{c} \quad$ (law of sines)

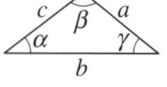